Springer Series in Translational Stroke Research

Series Editor
John Zhang
Loma Linda, USA

More information about this series at http://www.springer.com/series/10064

Paul A. Lapchak • John H. Zhang
Editors

Neuroprotective Therapy for Stroke and Ischemic Disease

Springer

Editors
Paul A. Lapchak
Department of Neurology and Neurosurgery
Cedars-Sinai Medical Center
Los Angeles, CA, USA

John H. Zhang
Loma Linda University School of Medicine
Loma Linda, CA, USA

ISSN 2363-958X ISSN 2363-9598 (electronic)
Springer Series in Translational Stroke Research
ISBN 978-3-319-45344-6 ISBN 978-3-319-45345-3 (eBook)
DOI 10.1007/978-3-319-45345-3

Library of Congress Control Number: 2016955837

Printed on acid-free paper

This Springer imprint is published by Springer Nature
The registered company is Springer International Publishing AG
The registered company address is: Gewerbestrasse 11, 6330 Cham, Switzerland

Preface

This is a very timely compilation of cutting-edge aspects of neuroprotective therapy for ischaemic stroke in a myriad of clinical and experimental situations. It was assembled by basic and clinical scientists at a time well after the dust had settled on the subject, following the difficulties of translating neuroprotection from bench to bedside. However, we are still mystified as to why this translational gap persists although numerous possible explanations have been given, in what now must be thousands of review articles. The book is logically and easily divided into four sections, ranging from the historical aspects of neuroprotection right through to the very latest aspects of clinical trial design.

Several positives have come from the difficulties experienced in translation of neuroprotective therapies, and these are all brought out nicely in this volume. First, there is now a vastly increased rigour in preclinical trial design which led to the STAIR (Stroke Treatment Academic Industry Roundtable) series of recommendations. These have provided valuable benchmarks for all those now conducting research in this area. Second, the idea of aggregating preclinical animal model data using meta-analysis techniques as used in the clinical sphere was pioneered by the Melbourne and Edinburgh groups led by Malcolm McLeod and David Howells, which later exposed problems with sample sizes, publication bias and lack of blinding as only some of the issues that need to be addressed to improve the quality of the preclinical research data. These findings have increased the likelihood of translation into the clinic. The Melbourne and Edinburgh groups have formalised their efforts in this area with the CAMARADES (Collaborative Approach to Meta-Analysis and Review of Animal Data from Experimental Studies) collaboration which acts as a preclinical form of the Cochrane Collaboration. The groups have also published guidelines for experimental design in a number of journals; indeed the movement has become so influential that other research disciplines are rapidly following suit. Third, translational failures are stimulating the search for new therapeutic targets such as autophagy as well as existing pathways and processes such as uric acid metabolism and the immune system.

An important component of the book is the recognition that modern endovascular treatment and its ischaemic consequences play an important part in the neuroprotection story. Indeed, the entirety of Sect. 4 is dedicated to combination therapies. Clearly, the outstanding success of recanalisation approaches with devices has re-energised the debate about reperfusion injury. There already existed an extensive experimental background well before these interventions were shown to be so clinically successful. On balance, the likelihood that an effective form of neuroprotection of clinical importance would arise from this type of research would seem to be quite high. Time will tell.

The compilation of such a complete range of aspects of neuroprotection raises the question as to whom the readership should be directed and who would most benefit. It certainly would be an ideal reference for any current researcher in any aspect of neuroprotection, either experienced or just beginning. It would also be a useful compendium for interested students, clinicians or scientists from other disciplines who'd like to dip into its rich matrix.

The quality of the information assembled shines through the experiences of the authors who are clearly leaders in their field. This alone makes this volume one that anyone would be proud to have on their bookshelf or, as is often the case today, readily accessible on their laptop.

Melbourne, VIC, Australia
2016

Geoffrey A. Donnan

Contents

Contributors

Todd A. Abruzzo, M.D. Department of Neurosurgery, College of Medicine, University of Cincinnati (UC), Cincinnati, OH, USA

Comprehensive Stroke Center at UC Neuroscience Institute, Cincinnati, OH, USA

Mayfield Clinic, Cincinnati, OH, USA

Gregory W. Albers Department of Neurology, Stanford University, Stanford, CA, USA

Alberto Alcázar Department of Investigation, Hospital Ramón y Cajal, IRYCIS, Madrid, Spain

Andrei V. Alexandrov, M.D. Department of Neurology, University of Tennessee Health Science Center, Memphis, TN, USA

Stroke Unit, Metropolitan Hospital, Piraeus, Greece

Carine Ali Normandie Université, UNICAEN, INSERM UMR-S U919 "Serine Proteases and Pathophysiology of the Neurovascular Unit", Caen, France

Raeed Altaee Department of Neuroscience, Psychology and Behaviour, University of Leicester, Leicester, UK

Diana Amantea, Ph.D. Section of Preclinical and Translational Pharmacology, Department of Pharmacy, Health and Nutritional Sciences, University of Calabria, Rende (CS), Calabria, Italy

Sergio Amaro, M.D., Ph.D. Hospital Clinic of Barcelona Neurosciences Institute, Functional Unit of Cerebrovascular Diseases, Hospital Clínic of Barcelona, Barcelona, Spain

Institut d'Investigacions Biomediques August Pi i Sunyer, Barcelona, Spain

Fika Tri Anggraini Department of Physiology, Wayne State University, Detroit, MI, USA

Anders Bach, Department of Drug Design and Pharmacology, Faculty of Health and Medical Sciences, University of Copenhagen, Copenhagen, Denmark

Giacinto Bagetta, Section of Preclinical and Translational Pharmacology, Department of Pharmacy, Health and Nutritional Sciences, University of Calabria, Rende (CS), Calabria, Italy

Taura L. Barr, Ph.D., R.N., F.A.H.A. CereDx Inc., Eight Medical Center Drive, Morgantown, WV, USA

Philip M.W. Bath, Stroke, Division of Clinical Neuroscience, University of Nottingham, Nottingham, UK

Paul D. Boitano, B.Sc. Department of Neurology and Neurosurgery, Cedars-Sinai Medical Center, Los Angeles, CA, USA

Hormozd Bozorgchami, Oregon Health & Science University, Portland, OR, USA

Alastair M. Buchan, D.Sc. (Oxon), L.L.D. (Hon), F.R.C.P., F.R.C.P.Ed. Acute Stroke Programme, Radcliffe Department of Medicine, University of Oxford, Oxford, UK

Medical Sciences Division, University of Oxford, Oxford, Oxfordshire, UK

Acute Vascular Imaging Centre, University of Oxford, Oxford University Hospitals, Oxford, UK

E. Candeias, CNC—Center for Neuroscience and Cell Biology, Rua Larga, Faculty of Medicine (Pólo 1, 1st Floor), University of Coimbra, Coimbra, Portugal

Institute for Interdisciplinary Research (IIIUC), University of Coimbra, Coimbra, Portugal

Stephen P. Cary, Omniox Inc., San Carlos, CA, USA

Ángel Chamorro, M.D., Ph.D. Functional Unit of Cerebrovascular Diseases, Clínic Institute of Neurosciences (ICN), Hospital Clínic de Barcelona, Barcelona, Spain

Institut d'Investigacions Biomediques August Pi i Sunyer, Barcelona, Spain

Medical Department, School of Medicine, University of Barcelona, Barcelona, Spain

Anjali Chauhan, Ph.D. McGovern Medical School—The University of Texas Health Science Center at Houston, Houston, TX, USA

Mourad Chioua, Laboratory of Medicinal Chemistry, Institute of General Organic Chemistry (CSIC), Madrid, Spain

Daisy Chou, M.D. Division of Cardiothoracic Surgery, Cedars-Sinai Medical Center, Los Angeles, CA, USA

Steven C. Cramer, M.D., M.M.Sc. Department of Neurology, University of California, Irvine, Orange, CA, USA

Gillian L. Currie, Centre for Clinical Brain Sciences, University of Edinburgh, Edinburgh, UK

Donald J. DeGracia, Department of Physiology, Wayne State University, Detroit, MI, USA

Center for Molecular Medicine and Genetics, Wayne State University, Detroit, MI, USA

Ana I. Duarte, CNC—Center for Neuroscience and Cell Biology, Rua Larga, Faculty of Medicine (Pólo 1, 1st Floor), University of Coimbra, Coimbra, Portugal

Institute for Interdisciplinary Research (IIIUC), University of Coimbra, Coimbra, Portugal

Alejandro Escobar-Peso, Laboratory of Medicinal Chemistry, Institute of General Organic Chemistry (CSIC), Madrid, Spain

Chizoba J. Ezepue, M.D. Department of Neurology, Medical College of Georgia, Augusta University, Augusta, GA, USA

Zhong-Ping Feng, Department of Physiology, Faculty of Medicine, University of Toronto, Toronto, Ontario, Canada

Eduardo F.A. Fernandes, Department of Drug Design and Pharmacology, University of Copenhagen, Copenhagen, Denmark

Federica Ferrari, B.Sc., M.Sc., Ph.D., A.C.C.P. Laboratory of Pharmacology and Molecular Medicine of Central Nervous System, Department of Biology and Biotechnology, University of Pavia, Pavia, Italy

Mark Fisher, M.D. Departments of Neurology, Anatomy & Neurobiology, and Pathology & Laboratory Medicine, University of California Irvine, Irvine, CA, USA

Department of Neurology, UC Irvine Medical Center, Orange, CA, USA

Valerio Frezza, Department of Investigation, Hospital Ramón y Cajal, IRYCIS, Madrid, Spain

Claire L. Gibson, Department of Neuroscience, Psychology and Behaviour, University of Leicester, Leicester, UK

R. Giersch, CereDx Inc., Eight Medical Center Drive, Morgantown, WV, USA

V. Gionis, CereDx Inc., Eight Medical Center Drive

Steven L. Gogela, M.D. Department of Neurosurgery, College of Medicine, University of Cincinnati (UC), Cincinnati, OH, USA

Rosaria Greco, Laboratory of Neurophysiology of Integrative Autonomic Systems, Headache Science Centre, "C. Mondino" National Neurological Institute, Pavia, Italy

Linda M. Haugaard-Kedström, Department of Drug Design and Pharmacology, University of Copenhagen, Copenhagen, Denmark

David C. Hess, M.D. Department of Neurology, Medical College of Georgia, Augusta University, Augusta, GA, USA

David W. Howells, Florey Institute of Neuroscience and Mental Health, Heidelberg, VIC, Australia

Faculty of Health, School of Medicine, University of Tasmania, Medical Science Precinct, Hobart, TAS, Australia

Zhi-Feng Huang, Department of Physics and Astronomy, Wayne State University, Detroit, MI, USA

Jennifer L.H. Johnson, Ph.D. Thayer Medical Corporation, Tucson, AZ, USA

Ali Khoynezhad, M.D., Ph.D., F.A.C.S. Division of Cardiothoracic Surgery, Cedars-Sinai Medical Center, Los Angeles, CA, USA

Ana Krtolica, Omniox Inc., San Carlos, CA, USA

Paul A. Lapchak, Ph.D., F.A.H.A. Department of Neurology and Neurosurgery, Cedars-Sinai Medical Center, Los Angeles, CA, USA

JungSeok Lee, M.D., Ph.D. Department of Neurology, JeJu National University School of Medicine, Jeju, South Korea

Department of Neurology, University of California Irvine, Irvine, CA, USA

Eloise Lemarchand, Normandie Université, UNICAEN, INSERM UMR-S U919 "Serine Proteases and Pathophysiology of the Neurovascular Unit", Caen, France

Philberta Y. Leung, Omniox Inc., San Carlos, CA, USA

Morgane Louessard, Normandie Université, UNICAEN, INSERM UMR-S U919 "Serine Proteases and Pathophysiology of the Neurovascular Unit", Caen, France

Helmi L. Lutsep, Oregon Health & Science University, Portland, OR, USA

Malcolm R. Macleod, Centre for Clinical Brain Sciences, University of Edinburgh, Edinburgh, UK

José Marco-Contelles, Laboratory of Medicinal Chemistry, Institute of General Organic Chemistry (CSIC), Madrid, Spain

Emma Martínez-Alonso, Department of Investigation, Hospital Ramón y Cajal, IRYCIS, Madrid, Spain

Sarah K. McCann, Centre for Clinical Brain Sciences, University of Edinburgh, Edinburgh, UK

Louise D. McCullough, M.D., Ph.D. McGovern Medical School—The University of Texas Health Science Center at Houston, Houston, TX, USA

Memorial Hermann Hospital—Texas Medical Center, Houston, TX, USA

Natacha Le Moan, Ph.D. Omniox Inc., San Carlos, CA, USA

Paula I. Moreira, CNC—Center for Neuroscience and Cell Biology, Rua Larga, Faculty of Medicine (Pólo 1, 1st Floor), University of Coimbra, Coimbra, Portugal

Institute of Physiology, Faculty of Medicine, University of Coimbra, Coimbra, Portugal

Antonio Moretti, D.Pharm., Ph.D. Laboratory of Pharmacology and Molecular Medicine of Central Nervous System, Department of Biology and Biotechnology, University of Pavia, Pavia, Italy

Hope Moser, R.N., D.N.P. Memorial Hermann Hospital—Texas Medical Center, Houston, TX, USA

Anja Muehle, M.D. Division of Cardiothoracic Surgery, Cedars-Sinai Medical Center, Los Angeles, CA, USA

Ain A. Neuhaus, M.A. Acute Stroke Programme, Radcliffe Department of Medicine, University of Oxford, Oxford, UK

Victoria E. O'Collins, Florey Neuroscience Institutes, Parkville, VIC, Australia

C.R. Oliveira, CNC—Center for Neuroscience and Cell Biology, Rua Larga, Faculty of Medicine (Pólo 1, 1st Floor), University of Coimbra, Coimbra, Portugal

Institute of Biochemistry, Faculty of Medicine, University of Coimbra, Coimbra, Portugal

Roy Poblete, M.D. Department of Neurology, Keck Hospital of the University of Southern California, Los Angeles, CA, USA

Ali Razmara, M.D., Ph.D. Department of Neurology, University of California, Irvine, Orange, CA, USA

Sarah Rewell, Florey Institute of Neuroscience and Mental Health, Heidelberg, VIC, Australia

Natalia Rost, J. Philip Kistler Stroke Research Center, Department of Neurology, Massachusetts General Hospital, Boston, MA, USA

Benoit Denis Roussel, Normandie Université, UNICAEN, INSERM UMR-S U919 "Serine Proteases and Pathophysiology of the Neurovascular Unit", Caen, France

Apostolos Safouris, M.D. Second Department of Neurology, National and Kapodistrian University of Athens, School of Medicine, Athens, Greece

Stroke Unit, Metropolitan Hospital, Piraeus, Greece

M.S. Santos, CNC—Center for Neuroscience and Cell Biology, Rua Larga, Faculty of Medicine (Pólo 1, 1st Floor), University of Coimbra, Coimbra, Portugal

Life Sciences Department, University of Coimbra, Coimbra, Portugal

Wolf-Rüdiger Schäbitz, Department of Neurology, Bethel-EvKB, University of Münster, Bielefeld, Germany

I. Sebastião, CNC—Center for Neuroscience and Cell Biology, Rua Larga, Faculty of Medicine (Pólo 1, 1st Floor), University of Coimbra, Coimbra, Portugal

Emily S. Sena, Centre for Clinical Brain Sciences, University of Edinburgh, Edinburgh, UK

Kristian Strømgaard, Ph.D. in Medicinal Chemistry Department of Drug Design and Pharmacology, University of Copenhagen, Copenhagen, Denmark

Hong-Shuo Sun, M.D., M.Sc., Ph.D. Department of Surgery, Faculty of Medicine, University of Toronto, Toronto, Ontario, Canada

Department of Physiology, Faculty of Medicine, University of Toronto, Toronto, ON, Canada

Department of Pharmacology, Faculty of Medicine, University of Toronto, Toronto, ON, Canada

Institute of Medical Science, Faculty of Medicine, University of Toronto, Toronto, ON, Canada

Gene Sung, M.D., M.P.H. Department of Neurology, University of Southern California, Los Angeles, CA, USA

Brad A. Sutherland, Ph.D. Acute Stroke Programme, Radcliffe Department of Medicine, University of Oxford, Oxford, UK

School of Medicine, University of Tasmania, Hobart, Australia

Doaa Taha, Department of Physics and Astronomy, Wayne State University, Detroit, MI, USA

Cristina Tassorelli, Laboratory of Neurophysiology of Integrative Autonomic Systems, Headache Science Centre, "C. Mondino" National Neurological Institute, Pavia, Italy

Department of Brain and Behavioural Sciences, University of Pavia, Pavia, Italy

George Trendelenburg, Department of Neurology, University Medical Center Göttingen, Göttingen, Germany

Kreiskrankenhaus Freiberg gGmbH, Freiberg, Germany

Jenny P. Tsai, Department of Neurology, Stanford University, Stanford, CA, USA

Georgios Tsivgoulis, M.D. Department of Neurology, University of Tennessee Health Science Center, Memphis, TN, USA

Second Department of Neurology, National and Kapodistrian University of Athens, School of medicine, Athens, Greece

Department of Neurology, Attikon Hospital, Athens, Greece

Xabier Urra, M.D., Ph.D. Hospital Clinic of Barcelona Neurosciences Institute, Functional Unit of Cerebrovascular Diseases, Hospital Clínic of Barcelona, Barcelona, Spain

Institut d'Investigacions Biomediques August Pi i Sunyer, Barcelona, Spain

Victor V. Uteshev, Ph.D. Institute for Healthy Aging, Center for Neuroscience Discovery, University of North Texas Health Science Center, Fort Worth, TX, USA

Roberto Federico Villa, D.Sc., M.D., Ph.D., A.C.C.P. Laboratory of Pharmacology and Molecular Medicine of Central Nervous System, Department of Biology and Biotechnology, University of Pavia, Pavia, Italy

Denis Vivien, Normandie Université, UNICAEN, INSERM UMR-S U919 "Serine Proteases and Pathophysiology of the Neurovascular Unit", Caen, France

Jonathan A. Winger, Omniox Inc., San Carlos, CA, USA

Burak Yulug, Department of Neurology and Regenerative and Restorative Medical Research Center, Experimental Neurology Laboratory, University of Istanbul-Medipol, Istanbul, Turkey

Mario Zuccarello, M.D. Department of Neurosurgery, College of Medicine, University of Cincinnati (UC), Cincinnati, OH, USA

Comprehensive Stroke Center at UC Neuroscience Institute, Cincinnati, OH, USA

Mayfield Clinic, Cincinnati, OH, USA

Part I
Stroke Neuroprotection: The History and Learning Experience

Chapter 1
Reflections on Neuroprotection Research and the Path Toward Clinical Success

Paul A. Lapchak and Paul D. Boitano

"Hubris and science are incompatible"
Douglas J. Preston

(American Author, novelist, journalist, 1956-)

Abstract Translational neuroprotection research is currently undergoing a rebirth, a much needed revival, in part due to the efficacy of both thrombolytic and endovascular procedures in subpopulations of ischemic stroke patients. Stroke is currently treated with the Food and Drug administration (FDA)-approved thrombolytic, tissue plasminogen activator (rt-PA), and can be treated with endovascular approaches using the MERCI stent retriever or the Solitaire FR stent retriever, with the application of thrombolytic (i.e., rt-PA or urokinase) prior to embolectomy for rt-PA eligible patients. Moreover, from retrospective analysis in rt-PA ineligible stroke patients, embolectomy alone has proven safe and beneficial if completed within 6 h.

Despite many decades of research into the identification and translational development of neuroprotective compounds, only few strategies have progressed into appropriately designed unbiased, randomized, placebo-controlled clinical trials. The FDA has still not been able to afford approval to a neuroprotectant to treat ischemic disease, primarily because of exaggerated overestimation of efficacy in rodent models that did not translate into efficacy in humans. During the process of

P.A. Lapchak, Ph.D., F.A.H.A. (✉) • P.D. Boitano, B.Sc.
Department of Neurology and Neurosurgery, Cedars-Sinai Medical Center, Advanced Health Sciences Pavilion, Suite 8305, 127S. San Vicente Blvd,, Los Angeles, CA 90048, USA
e-mail: Paul.Lapchak@cshs.org

P.A. Lapchak, J.H. Zhang (eds.), *Neuroprotective Therapy for Stroke and Ischemic Disease*, Springer Series in Translational Stroke Research,
DOI 10.1007/978-3-319-45345-3_1

developing neuroprotective compounds to treat ischemic diseases, stroke in particular, numerous problems have emerged including the absolute failure to translate preclinical animal efficacy into efficacy in stroke victims, and in some cases, both significant adverse events and unforeseen toxicities have hindered drug development and approval. This chapter describes successes and failures in the stroke neuroprotection research, provides a comprehensive tabulated assessment of select neuroprotectants that have been tested in clinical trials, and proposes recommendations and essential checklists to both guide and improve the quality of science being conducted in preclinical and translational laboratories worldwide. The ultimate goal is to reap the benefits of a worldwide concerted neuroprotection research effort to provide superior care to stroke victims.

Keywords Translational • Neuroprotection • Neuroprotective • Cytoprotective • Brain • Stroke • Hemorrhage • SAH • ICH • Clinical trial • NIHSS • STAIR • RIGOR • Transparency • Rodin • Penumbra

1 Introduction: A Brief Chronological History of Stroke

Stroke, or a condition referred to as apoplexy, or the sudden onset of paralysis was first "reported" by Hippocrates between 460 and 370BC and can be found documented in Hippocratic transcripts [1]. In 1658, Wepfer redescribed apoplexy or apoplectic seizure which formed the basis for stroke classification into cerebral infarction and hemorrhagic stroke [2]. The interesting account by Wepfer, discussed in detail by Gurdjian and Gurdjian [3] communicates the idea that body-derived "natural spirits" also known as "vital spirits" are transported into brain via the carotid and vertebral arteries and a network of arteries at the base of the brain. Wepfer hypothesized that oxygenated blood was transported into the brain as a vital factor, as a source of nutrition and thereafter Willis suggested that the "cerebrovascular system" included a network of arteries, including the "circle of Willis, which he described in 1664 [4]. In autopsy records from the 1700s there was confirmation of two types of apoplexy; the first suggested to be apoplexia serosa defined as serous apoplexy and the second apoplexia sanguinea defined as sanguineous apoplexy (i.e., ischemic stroke and hemorrhagic stroke, respectively). In 1856, Virchow [5] was the first physician to recognize that an "embolus" could result in a thromboembolism, and coined both terms related to the pathogenesis of ischemic stroke. The 1866 *Dictionnaire Encyclopédique des Sciences Médicales of Dechambre* summarized over 150 apoplexy references [6] documented from the 1600 to 1800s. More recently, but still of historical significance, Bramwell and Symonds published scientific articles in 1886 [7] and 1924 [8] that described "spontaneous" meningeal and subarachnoid hemorrhage (SAH). It is currently estimated that brain hemorrhage categorized into SAH, intracerebral hemorrhage

(ICH), and intraventricular hemorrhage (IVH) occurs in approximately 17–20 % of all stroke victims [9–11]; hemorrhage is usually associated with a higher mortality rate than ischemic stroke and a rapid decline after vessel rupture. The 30-day mortality rate for ischemic stroke is estimated to be 8–12 %, whereas hemorrhagic stroke is estimated to be 50 % [12–15].

1.1 *Of Neurons and Time*

The term neuroprotection has been in use for more than 50 years [see [16–19]]. The first pharmacological "modern" approach to therapeutic neuroprotection, or method to prevent neuronal death was the application of barbiturate drugs, which were neuroprotective and thought to target and reduce oedema (edema), free radicals, altered fatty acid metabolism, and even stabilize membranes [18]. Currently, "neuroprotection" is used interchangeably with "cytoprotection," especially when applied to the treatment of stroke where there has been the realization that the neurovascular unit (neurons, glial cells, and vascular connectivity) requires protection after an ischemic event [20–23].

1.1.1 Extrapolated Stroke and Cerebrovascular Disease Incidence

We all agree that there is a critical need for new neuroprotective or cytoprotective strategies to reduce the morbidity and mortality incidence related to ischemic stroke, and to ultimately improve stroke victim quality of life, not just select clinical measures routinely used for 30- and 90-day evaluation, but every day "quality" for all victims. Until recently, stroke has been described as the fifth leading cause of mortality and leading cause of adult morbidity in the United States and it is estimated that 0.8 million people suffer a stroke in the USA [24], and 15 million people worldwide [9]. In the United Kingdom, stroke is the 4th largest cause of death with an annual incidence of 152,000 and 12.5 % of stroke victims die within 30 days.

The societal impact of stroke becomes even more devastating and overwhelming if we consider the updated definition of stroke from the American Heart Association (AHA)/American Stroke Association (ASA) [25], which now includes "central nervous system infarction of brain, spinal cord or retinal cell death attributable to ischemia." The AHA study authors also suggest that nonsymptomatic silent infarcts should be included in the statistical analysis of cerebrovascular disease. With silent infarcts included, 15–20 % of the US population would have some form of cerebrovascular disease that eventually becomes apparent as cognitive impairment, dementia, and Alzheimer's disease [25]. With a current worldwide population of 7.4 billion people [26], the estimated population with cerebrovascular disease escalates to 1.4 billon.

Table 1.1 Recent successes and failure in stroke treatment

Clinical Trial	Time to Treatment (min)	Outcome	Estimated # Neurons Lost (million)	Reference
NINDS rt-PA	180 hours	Success	270	(30)
ECASS III	270 hours	Success	405	(29, 46, 47)
Endovascular Procedures (with thrombolysis)	rt-PA endovascular			
ESCAPE	Endo GP 185 (116-315) rt-PA 110 (80-142)	↑		
EXTEND	Endo GP 210 (166-251) rt-PA 127 (93-162)		Endo DTGPT (range 232-680)	
MR CLEAN	Endo GP 260 (210-313) rt-PA 85 (67-110)	Success		(31-35)
REVASCAT	Endo GP 269 (201-340) rt-PA 117.5 (90-150)		rt-PA (range 134-324)	
SWIFT PRIME	Endo GP 224 (165-275) rt-PA 110.5 (85-156)	↓		
Endovascular Procedures (without thrombolysis)	Endo GP (inclusive of 360)	Success	720	(36) but see also (37)
FAST-MAG	0.75 hours	Failure	90	(38)
SAINT-II	3.76 hours	Failure	338.4	(39, 40)
NEST-3	16 hours	Failure	1440	(41-44)

DTGPT-door to groin puncture time (median); rt-PA-initiation of thrombolytic administration

1.1.2 Time–brain matters!

Time is of the essence according to Saver [27], Holscher et al. [28], and Lapchak [29], among others, all researchers who have emphasized that an ischemic event is devastating to the brain, and that rapid treatment is essential. It is estimated that two million neurons die in the human brain per minute after hypoxia, the majority of which are located in the "penumbra" the primary and possibly only target of neuroprotectants, and 14 billion synapses are lost every minute following an ischemic event. As shown in Table 1.1, there is no correlation between time to treatment efficacious outcome, and estimated cellular loss in a select few recently conducted clinical trials. Within the current clinically recommended door-to-needle-time (DTNT) for rt-PA, it can be estimated that a stroke patient loses 120×10^6 neurons if treated with 60 min. The losses escalate to fraction of billions, to billions of lost neurons in many other recent clinical trials, when the initiation of treatment is delayed to hours. This is particularly important when one considers that recent reperfusion therapies including endovascular procedures demonstrated considerable efficacy up to 6 h following a stroke, but even a classical neuroprotection trial attempt (i.e., FAST-MAG) was ineffective even when drug administration starting as soon as 45 min following a stroke.

- Is this in any way related to choice of drug/device, target, or patient population selection?

All of these questions remain to be answered and can only be answered when there is conclusive evidence that a neuroprotective therapy is efficacious in stroke victims.

2 Limited Benefit Treatment Options for Acute Ischemic Stroke Victims: Successes!

This section will provide an overview of currently accepted and utilized treatments for acute ischemic stroke. They can both be categorized as reperfusion therapies: (1) thrombolysis with rt-PA or other proteins with similar activity [30, 45, 46], and (2) endovascular procedures (thrombectomy or embolectomy) in combination with thrombolysis [31–35] or without rt-PA administration [36].

2.1 Thrombolysis: Thrombolytic Therapy

The thrombolytic, rt-PA (Alteplase™) was first approved by the FDA in 1996 and is now widely accepted as a standard of care, but drastically underutilized in almost all communities worldwide. Alteplase has been shown to be effective up to 4.5 h after a stroke [45, 46], but it is currently FDA approved for use within a 3-h therapeutic window. It has been difficult to estimate the actual use and application of rt-PA in eligible stroke victims, but it has been suggested that less than 7–10 % of stroke patients are being treated with rt-PA in the United States [47–49] despite the fact that rt-PA may be beneficial in up to 50 % of patients provided the drug as a treatment option [30].

Clua-Espuny and colleagues have recently reported an important gender difference in survival response after a stroke that was correlated with benefit when rt-PA was provided after a first stroke [50]. Based upon a 1272 patient cohort, the authors analyzed survival outcome in male and females with first strokes in Spain and found that thrombolysis increased survival in both groups, but there was a pronounced survival effect in women that was not observed in men over an 8-year period. Clearly, rt-PA should be provided to both genders if they are eligible at time of admission.

Cost analysis of rt-PA utilization within 3–4.5 h after stroke onset clearly shows incremental benefit in patients with National Institutes of Health Stroke Scale (NIHSS) scores of 0–19, compared to no treatment [51]. This translates into quality-adjusted life-years (QALY) benefit for the stroke victim [51]. While rt-PA is beneficial, it does have a few shortcomings, primarily a significant risk of hemorrhagic transformation (HT) or intracerebral hemorrhage (ICH) in up to 6 % of patients treated within 3–4.5 h of a stroke [52], an increase in the odds ratio for mortality rate after 4 h [52], and minimal [53, 54] or lack of neuroprotective activities, or possibly detrimental biological properties and adverse effects under some circumstances [55].

Table 1.2 Thrombolysis trials—efficacy analysis: mRS 90 day outcome

mRS/OHS Score		No Symptoms--► Death						
Study	**Treatment**	**0**	**1**	**2**	**3**	**4**	**5**	**6**
NINDS rt-PA	**Control (312)**	**26**		**25**		27		21
	Intervention (312)	**39**		**21**		23		17
ECASS	**Control (403)**	**21.8**	**23.3**	**16.4**	11.4	13.7	5.2	8.2
	Intervention (418)	**27.5**	**24.9**	**14.1**	9.3	9.3	8.1	6.7
IST-3 (OHS scale)	**Control (1520)**	**8**	**13**	**14**	13	9	17	27
	Intervention (1515)	**9**	**15**	**13**	16	8	13	27

mRS: modified Rankin scale (%); OHS: Oxford Handicap Scale
Highlighted Boxes indicate mRS/OHS 0-2 functional independence.

Of importance to the topic of this Springer volume of Neuroprotection is the measure commonly referred to as DTNT. In the literature, the recommended DTNT for standard rt-PA thrombolytic therapy is less than 60 min [56–58]. However, according to a recent Cochrane review article, rt-PA is being administered in a time frame in great excess of that recommendation [59]. In the original National Institute of Neurological Disorders and Stroke (NINDS) rt-PA clinical trial [30], the administration time was stratified between 0–90 and 91–180 min. Subsequent clinical trials [European Cooperative Acute Stroke Study (ECASS trials)] determined whether rt-PA would be efficacious with an expanded therapeutic window [45, 60], rather than reduce time to treatment, and rt-PA was shown to retain efficacy in specific patient populations.

Table 1.2 summarizes the historical efficacy data from the original NINDS rt-PA clinical trial [30], and ECASS III [45, 61, 62], directly comparing 90-day outcome on the modified Rankin scale (mRS). In ECASS III [45, 61, 62], the safety and efficacy of rt-PA was studied in patients when administered up to 4.5 h following an ischemic stroke. The trial showed that there was a significant shift in the mRS score 0–3 in 66.5 % of rt-PA-treated patients compared to 61.5 % in the control group. This represented an absolute change of 5 %. As recently described by the Stroke Thrombolysis Trialists Group [63], administration of a thrombolytic is most efficacious when provided within 3 h of a stroke. An additional eight intravenous rt-PA clinical trials enrolling 6729 patients also provided data that thrombolysis can promote significant improvement in patients when administered up to 6 h following a stroke. Since time is brain, all efforts should be made to administer therapy as soon as possible.

It has now been a milestone 20 years since the FDA approval of rt-PA; somewhat of a platinum anniversary, and the treatment is still underutilized worldwide [47, 49], but there has been some improvement in DTNT [64, 65]. A leader in the field of stroke clinical trials, Dr. Saver (UCLA, Los Angeles) has published on an extensive data analysis set collected using more than 58,000 stroke patients receiving rt-PA within 4.5 h of symptom onset. The analysis of onset to time of treatment provided some evidence for a direct correlation between the rapidity of "thrombolytic" treatment and benefit on measures important to the patient that included increased functional independence and increase time of discharge from the hospital, in addition to reduced hemorrhage incidence and mortality. The argument can then be made that Time is Brain and the faster reperfusion therapy is administered, the greater benefit to the patient [27, 29, 66–69].

Nothing is a waste of time if you use the experience wisely
Francois-Auguste-Rene Rodin
(French born sculptor and progenitor of modern sculpture. 1840–1917)

2.2 *Endovascular Procedures and Thrombolysis*

The great ESCAPE, EXTEND-IA, MR CLEAN, REVASCAT, and SWIFT PRIME endovascular trials demonstrated that a specific well defined, but heterogeneous population of acute ischemic stroke patient with a large vessel occlusion (LVO) can now be offered additional therapy to successfully improve functional independence. Endovascular procedures with clot retrievers can promote functional independence at 90 days as indicated by a significant shift in modified Rankin Scale score (mRS) 0–2 (common odds ratio range of 1.7–3.1) in 13.5–31 % of patients undergoing the endovascular procedure. Moreover, the procedure has now been shown to be safe in patients with LVOS, salvageable brain tissue (i.e., large penumbra) with small infarct areas ASPECTS score 7–10, and median NINDS score of 16–17; therapy was neither age, nor gender specific. This section will review salient aspects of the efficacy and safety results from the trials, discuss enrollment criteria, and propose future stroke development strategies incorporating neuroprotection.

Until recently, the use of mechanical embolectomy in patients with documented LVO was complicated by lack of efficacy [70–75], either due to clinical trial design flaws, patient selection, inadequate device design, or possibly due to methodological problems arising from interpretation of recanalization success rates [76]. In previous embolectomy trials, the procedure was performed more than 6 h after stroke onset and did not appear to provide benefit to patients [77–79]. However, with the introduction and use of new devices including the Mechanical Embolus Removal in Cerebral Ischemia (MERCI retriever; Concentric Medical Inc., Mountain View, CA, USA) and Solitaire FR Revascularization Device (Ev3/Covidien, Paris France) retrievable stent, patients had substantial benefit when recanalization was initiated within 6 h of stroke onset. Now, use of mechanical embolectomy, with or without thrombolysis, is considered the newest form of standard of care therapy in patients with a documented LVO [31–36].

Table 1.3 ESCAPE-embolectomy–thrombolysis enrollment—end point analysis

Trial	Embolectomy + Thrombolysis	Thrombolysis
ESCAPE(32) **N=316 (238 received intravenous rt-PA)**	120 patients	118 patients
Patient population selection: Small infarct core (ASPECTS 6-10), an occluded proximal artery in the anterior circulation (middle cerebral artery trunk and immediate branches with or without internal carotid artery occlusion), and moderate-to-good collateral circulation (filling of 50% or more of the middle cerebral artery pial artery circulation).		
% Achieving Reperfusion **% Achieving Recanalization @ 24hours**	72.4 NA	31.2 NA
% Achieving mRS 0-2 (90 days)	53	29.3 [p<0.001]
Age range (yrs): **NIHSS range interquartile range (median)]**	60-81 (71) 13-20 (16)	60-81 (70) 12-20 (17)
Time-to-Treat [range (median)] min	**Endo GP 185 (116-315)** rt-PA 110 (80-142)	rt-PA 125 (89-183)
Median Time to Reperfusion following stroke (min)	**241 (176-359)**	
Symptomatic Intracerebral Hemorrhage %	3.6	2.7
Serious Adverse Event Excluding death % **(SAE definiton for ESCAPE- resulting in a prolonged hospital stay, re-admission, were severe or life threatening (large or malignant MCAO stroke, hematoma at access site, perforation of the MCA).**	7.2	10.7
Mortality Rate % (90 days)	10.4	19.0

2.2.1 ESCAPE [32]: Table 1.3

The Canadian-based ESCAPE trial supported by Covidien enrolled a total of 316 patients with a proximal intracranial occlusion in the anterior circulation (ICA and M1 middle cerebral artery, or M1 or M2 middle cerebral artery segments, and moderate-to-good collateral circulation. Importantly, as a measure of a small ischemic core and large penumbra, median ASPECTS on CT was 9, interquartile range of 8–10 in the embolectomy arm and 8–10 in the rt-PA arm. Patients were enrolled up to 12 h after symptom onset; 238 patients received rt-PA (120 in the embolectomy plus IV rt-PA arm and 118 in the rt-PA control group) within a median time of 110 min for the embolectomy arm and 125 min for the rt-PA control arm. The median time from CT head to the first noted reperfusion was 84 min for embolectomy (65–115 interquartile range), defined as the first visualization of reflow in the middle cerebral artery, which in most patients was coincident with the deployment of a retrievable stent. For embolectomy, the median time from symptom onset to groin puncture was 51 min (39–68 interquartile range) in the intervention group, and rt-PA was initiated within 110 min (80–142 interquartile range). In the control group, IV rt-PA was administered within 125 min (89–183 interquartile range) from stroke onset. The median time of stroke onset to first reperfusion was 241 min (176–359 range) in the embolectomy arm. The trial used retrievable stents or balloon catheters for suction clot removal. Notably, both groups included 87.3 % of white race patients, and 52.1–52.7 % of females. In both groups, the majority of patients had hypertension (63.6–72 %), a minority were diabetic (20–26 %) and had atrial fibrillation (37–40 %).

In the combination treatment group, 53 % of patients achieved mRS of 0–2 (90 days) compared to only 29.3 % in the rt-PA group, showing greater benefit of embolectomy plus rt-PA compared to rt-PA alone at 90 days. This benefit could be associated with a greater population of patients achieving reperfusion in the embolectomy/rt-PA group 72.4 % (Thrombolysis in Cerebral Infarction [TICI] scale score of 2b or 3) compared to 31.2 % (modified arterial occlusive lesion [mAOL] score of 2 or 3) in the control group. Both genders, male and female responded equally to treatment with a common odds ratio (95 % Cl) of 2.5 (1.4–4.5 male), to 2.6 (1.5–4.4 female) favoring embolectomy/rt-PA. While symptomatic intracranial hemorrhage (sICH) rates specifically defined as a new intracranial hemorrhage associated with evidence of clinical worsening, in which the hemorrhage is judged to be the most important cause of clinical worsening, were not significantly different between groups ($P = 0.75$), and were lower (by 50 %) than reported for the NINDS-rt-PA trial [30], mortality was greatly reduced by embolectomy/rt-PA compared to rt-PA 10.4 % vs. 19 % control ($P = 0.04$). The rate ratio for sICH and mortality was 1.4; 95 % CI, 0.4–4.7 and 0.5; 95 % CI, 0.3–1.0; $P = 0.04$, respectively.

2.2.2 EXTEND-IA [33]: Table 1.4

The Australian-based EXTEND-IA clinical trial was a small 70 patient trial, with patients receiving rt-PA within 4.5 h of stroke symptom onset. Inclusion criteria included occlusion of the internal carotid or middle cerebral artery and evidence of salvageable brain tissue and ischemic core of less than 70 mL on CT perfusion imaging. Endovascular therapy had to be initiated via groin puncture within 6 h of stroke onset and completed within 8 h of onset. The trial demonstrated significant efficacy after enrolling 35 patients in each of the two treatment groups. The trial enrolled 49 % male/51 % female in both groups, and there was also a history of hypertension (rt-PA 66; embolectomy 60 %), diabetes (23 % rt-PA; 6 % embolectomy), and atrial fibrillation (31 % rt-PA; 34 % embolectomy). Thus, the trial was stopped at interim analysis because of significant benefit in the endovascular therapy arm and publication of MR CLEAN trial with similar efficacy (see Sect. 2.2.3). Patients were treated with the Solitaire FR retrievable stent within a median time of 248 min from stroke onset to modified TICI2b or 3 or completion of the procedure (interquartile range 204–277) for endovascular therapy in combination with rt-PA (within 127 min median) compared to a median time of 145 min for rt-PA. The median time from stroke to groin puncture was 210 min (166–251 min) and 74 min from initiation of rt-PA to groin puncture.

Table 1.4 EXTEND-IA embolectomy–thrombolysis enrollment—end point analysis

Trial	Embolectomy + Thrombolysis	Thrombolysis
EXTEND-IA (33) **N=70**	35 patients	35 patients
Patient population selection: Occlusion in the proximal anterior circulation (middle cerebral artery or internal carotid artery) with evidence of salvageable brain tissue and small ischemic core (<70ml on CT).		
% Achieving Reperfusion **% Achieving Recanalization @24hrs**	89 (reperfusion >90% without SICH) 94	34 (reperfusion >90% without SICH) [p< 0.001] 42
% Achieving mRS 0-2 (90 days)	71	40 [p=0.01]
Age range (yrs): **NIHSS range interquartile range (median)]**	68.6 ±12.3 13-20 (17)	70.2 ±11.8 9-19 (13)
Time-to-Treat range (median) min	**Endo GP 210 (166-251)** rt-PA 127 (93-162)	rt-PA 145 (105-180)
Median Time to Reperfusion following stroke (min)	**248 (204-277)**	
Symptomatic Intracerebral Hemorrhage %	0	6
Serious Adverse Event % **(SAE definiton for EXTEND-IA-parenchymal hematoma)**	11	9
Mortality Rate %	9	20

In the endovascular therapy arm, 94 % of patients achieved recanalization at 24 h, 89 % had >90 % reperfusion at 24 h and were without symptomatic intracerebral hemorrhage, and 71 % of patients achieved mRS of 0–2 at 90 days compared to 40 % in control arm. This is compared to 34 % of rt-PA patients having extensive reperfusion at 24 h and a lower recanalization rate at the same time point (43 % vs. 94 % in the endovascular group). As shown in Table 1.4, there were no significant differences in sICH rate (sICH defined as a large parenchymal hematoma where blood occupied >30 % of the infarct volume with mass effect, and an increase of 4 points or more in the NIHSS score) or mortality between the groups. The patients included in the study were randomized 70 min sooner (30 vs. 100 min) and time from stroke onset to groin puncture was faster compared to MR CLEAN (210 vs. 260 min. see later).

2.2.3 MR CLEAN [31]: Table 1.5

The Netherlands-based MR CLEAN trial enrolled 500 patients with a proximal artery occlusion in the anterior cerebral circulation within 6 h after symptom onset: 233 enrolled in the intra-arterial treatment arm (mechanical embolectomy and thrombolytic) and 267 receiving thrombolytic treatment (either rt-PA maximum dose of 90 mg or 1.2 million IU of urokinase with a median time to treatment of 85 min for the embolectomy arm and 87 min for the rt-PA/urokinase arm.

Table 1.5 MR CLEAN-embolectomy–thrombolysis enrollment—end point analysis

Trial	Embolectomy + Thrombolysis	Thrombolysis
MR CLEAN(31) N=500	233 patients	267 patients
Patient population: Occlusion of the distal intracranial carotid artery, middle cerebral artery (M1 or M2) or anterior cerebral artery (A1 or A2) and small ischemic cores ASPECTS (7-10).		
% Achieving Reperfusion **% Achieving Recanalization @ 24hours**	58.7 (TICI score of 2b or 3) 75.4	57.5 (mAOL score of 2 or 3) 32.9
% Achieving mRS 0-2 (90 days)	32.6	19.1
Age range (yrs) **NIHSS range interquartile range (median)]**	54.5-76 (65.8) 14-21 (17)	55.5-76.4 (65.7) 14-22 (18)
Time-to-Treat [range (median)] min	**Endo GP 260 (210-313)** rt-PA 85 (67-110)	rt-PA 87 (65-116)
Median Time to groin puncture following stroke (min)	**260 (210-313)**	
Symptomatic Intracerebral Hemorrhage %	7.7	6.4
Serious Adverse Event Excluding Death %	25.3	18.0
Mortality Rate % (30 days)	18.9	18.4

Importantly, as a measure of a small ischemic core and large penumbra, median ASPECTS on CT was 9, interquartile range of 7–10 in the embolectomy arm and 8–10 in the rt-PA arm. Like the ESCAPE and EXTEND-IA trials, the patient population was diverse with certain common comorbidities including diabetes mellitus (rt-PA 12.7; embolectomy 14.6 %), hypertension (42.1–48.3 %), and atrial fibrillation (rt-PA 28.3: embolectomy 25.8). The trial was funded and supported in part by Covidien/ev3, Medac/Lamepro, and Penumbra.

This comparison study used mechanical treatment in 83.7 % of patients: retrievable stents were used in 81.5 % of patients and other devices were used in 2.1 % patients assigned to the intra-arterial treatment. There was no difference in the proportion of patients achieving reperfusion between the two groups: 58.7 % (TICI score of 2b or 3) vs. 57.5 % (mAOL score of 2 or 3), but there was a statistically significant difference in the rate of functional independence (mRS 0–2) in favor of combined intervention (32.6 % vs. 19.1 %), an absolute difference of 13.5 %. Interestingly, in the intervention group, 75.4 % of patients showed an absence of residual occlusion of the target site compared to only 32.9 % in rt-PA patients.

Treatment effect favored endovascular intervention for both age groups >80 and <80, and surprisingly, patients in the >80 age group did much better on the endovascular intervention adjusted conditional odds ratio (acOR) 95 % CI 3.24 vs. 1.6). The number of serious adverse events in this trial was high for both groups (42.3–47.2 %), but sICH defined as type 1—one or more blood clots in 30 % or less of the infarcted area with a mild space-occupying effect, or type-2 blood clots in more than 30 % of the infarcted area with a clinically significant space-occupying effect, were in line with the 6.4 % observed in the NINDS rt-PA trial (7.7 % in the embolectomy arm and 6.4 % in the rt-PA arm), and mortality rate (30 days) was similar in both groups (18.9 embolectomy vs. 18.4 rt-PA). However, thrombolytic agents and embolectomy procedure were not limited to tPA and stent retrievers only and minimum NIHSS inclusion criterion was 2, which is relatively low compared to other trials.

2.2.4 REVASCAT [34]: Table 1.6

The Spanish REVASCAT trial enrolled 206 patients with a proximal anterior circulation occlusion were randomized to receive embolectomy plus IV rt-PA when eligible and control rt-PA within a median time for groin puncture 269 min for the embolectomy arm plus 117.5 min for rt-PA compared to 105 min for the rt-PA arm. Importantly, as a measure of a small ischemic core and large penumbra, median ASPECTS on CT was 7, interquartile range of 6–9 in the embolectomy arm and 8 with an interquartile range of 6–9 in the rt-PA arm. The study allowed for inclusion of patients within 8 h of stroke symptom onset. The Solitaire stent retriever device was used in this study, which was funded by Covidien. The study enrolled in the embolectomy and control groups, respectively, 53.4 and 52.4 % male participant, including 21.4 and 18.4 % diabetic patients, 60.2 and 69.9 % having hypertension, and 34.0 and 35.9 % having atrial fibrillation.

Table 1.6 REVASCAT-embolectomy–thrombolysis enrollment—end point analysis

Trial	Embolectomy + Thrombolysis	Thrombolysis
REVASCAT(34) N=206	103 patients	103 patients
Patient population: Occlusion in the proximal anterior circulation (middle cerebral artery or internal carotid artery) and small ischemic cores ASPECTS (7-8).		
% Achieving Reperfusion **% Achieving Recanalization @ 24 hours**	66 (core assessment) vs. 80% (interventionalist assessment) NA	NA NA
% Achieving mRS 0-2 (90 days)	43.7	28.1
Age range (yrs) **NIHSS range interquartile range (median)]**	65.7 ±11.3 14-20 (17)	67.2 ±9.5 12-19 (17)
Time-to-Treat range (median) min	**Endo GP 269 (201-340)** rt-PA 117.5 (90-150)	rt-PA 105 (86-137.5)
Median Time to Reperfusion following stroke (min)	**355 (269-430)**	
Symptomatic Intracerebral Hemorrhage %	1.9	1.9
Serious Adverse Event % **(SAE definition for REVASCAT-neurologic worsening, malignant cerebral edema, and recurrent stroke)**	30.1	25.2
Mortality Rate (90 days)	18.4	15.5

The rates of revascularization in the embolectomy group were reported to be 66 %, but there was no value provided for the rt-PA group. The reperfusion rate was lower when compared to other trials which could be accounted due to delay in treatment (30 min.) for improvement by IV-rt-PA therapy before randomization. As a measure of functional independence (mRS 0–2) at 90 days, 43.7 % of patients in the embolectomy combined arm and 28.1 % in the standard care arms achieved a "cure." Statistical analysis pointed to an adjusted odds ratio for 1-point improvement of 1.7 (1.05–2.8) in favor of embolectomy. The sICH rate of 1.9 % was low in both groups when using Safe Implementation of Thrombolysis in Stroke: A Multinational Multicentre Monitoring Study of Safety and Efficacy of Thrombolysis (SITS-MOST) criteria, while serious adverse events (neurologic worsening, edema, and recurrent stroke) were 30.1 % in the embolectomy group and 25.2 % in the control group. The mortality rate at 90 days was 18.4 % in the embolectomy and 15.5 % in the control group (risk ratio 95 % CI 1.2 (0.6–2.2).

2.2.5 SWIFT-PRIME [35]: Table 1.7

The US and Europe-based SWIFT PRIME enrolled 98 patients in the embolectomy plus IV rt-PA arm within a median time of 252 min (190–300 interquartile range) from stroke onset to first deployment of the stent, and 98 patients in the IV rt-PA

Table 1.7 SWIFT PRIME—embolectomy–thrombolysis enrollment—end point analysis

Trial	Embolectomy + Thrombolysis	Thrombolysis
SWIFT PRIME (35) **N=196**	98 patients	98 patients
Patient population: Proximal anterior intracranial circulation occlusion in the absence of large ischemic core lesions Small ischemic cores ASPECTS (7-10).		
% Achieving Reperfusion **% Achieving Recanalization @ 24 hours**	82.8 (reperfusion ≥90%) 88	40.4 (reperfusion ≥ 90%) [p< 0.0001] NA
% Achieving mRS 0-2 (90 days)	60.2	35.5 [p = 0.0008]
Age range (yrs): **NIHSS range (median)**	65.0 ± 12.5 13-20 (17)	66.3 ± 11.3 13-19 (17)
Time-to-Treat range (median) min	**Endo GP 224 (165-275)** rt-PA 110.5 (85-156)	rt-PA 117 (80-155)
Median Time to stent deployment following stroke (min)	**252 (190-300)**	
Symptomatic Intracerebral Hemorrhage %	0	3
Serious Adverse Event % **(SAE definition for SWIFT PRIME-an event that led to death, life-threatening illness or injury, permanent impairment, prolonged hospitalization).**	36	31
Mortality Rate %	9	12

arm. Patients were enrolled with a proximal anterior intracranial circulation occlusion in the absence of large ischemic core lesions and were randomized within 6 h after stroke symptom onset. Importantly, as a measure of a small ischemic core and large penumbra, median ASPECTS on CT was 9, interquartile range of 7–10 in the embolectomy arm and 8–10 in the rt-PA arm. In the embolectomy group, the median time from stroke onset to groin punctures was 224 min. The Solitaire revascularization device (self-expanding stent retriever) was used for embolectomy. Like all other trials, the majority of patients had one or more comorbidities: hypertension (rt-PA 58; embolectomy 67 %), diabetes (rt-PA 15; embolectomy 12 %), and atrial fibrillation (rt-PA 39; embolectomy 36 %). The study demonstrated a significant improvement in functional independence using the mRS scale (range 0–2) at 90 days, and greatly enhanced reperfusion in the majority of patients in the combined treatment group (83 % at 27 h) with embolectomy, and 40 % (at 27 h) in the rt-PA group. Moreover, 60 % of patients treated with embolectomy and intravenous thrombolysis had a mRS score of 0–2 compared with standard rt-PA care alone (35 %).

The rate of sICH assessed radiologically at 27 h was 0 % in the intervention group and 3 % in the rt-PA group (Risk ratio 0.00) and mortality rate at 90 days was 9 % and 12 % in the intervention and rt-PA group, respectively ($P = 0.50$). Serious adverse events defined as an event that led to death, life-threatening illness or injury, permanent impairment, prolonged hospitalization was 36 % in the embolectomy group and 31 % in the rt-PA group [Risk ratio 1.15 (0.78–1.720 $P = 0.54$)].

2.2.6 Endovascular Procedure Benefit

As shown in Tables 1.8 and 1.9, there are substantial differences in the rate of functional independence (achieving mRS 0–2) in the five embolectomy trials. The range of absolute improvement is 13.5–31 % points, when a direct comparison is made between the thrombolytic "control" group and the embolectomy/thrombolytic "intervention groups."

In Table 1.8, odds ratio and significance levels for each of the embolectomy trials are presented. Comparison of the odds ratio mean of 2.89 for embolectomy trials points to significant benefit over previous thrombolysis trials (mean odds ratio = 1.46). Thus, embolectomy in combination with IA or IV thrombolysis (rt-PA or urokinase) is beneficial in a patients presenting with a large vessel stroke within approximately 6 h.

Retrospective meta-analysis (Table 1.9) published by the HERMES collaboration [80] demonstrate that the best embolectomy outcome is achieved when ASPECTS was 6–8 or 9–10 indicating a significant amount of penumbra, when the clot location was either the ICA or M1 segment middle cerebral artery, and when intervention was initiated ≤300 min. There were no significant gender differences, but age-dependent improvement was observed. There was benefit in patients 50–80 years of age, but less benefit between 18 and 49 years of age.

Table 1.8 Embolectomy–thrombolysis trial—efficacy analysis: mRS outcome (90 day)

mRS Score		No Symptoms--► Death						
Study	**Treatment**	**0**	**1**	**2**	**3**	**4**	**5**	**6**
ESCAPE	**Control (150)**	**7**	**10**	**12**	15	24	12	19
	Intervention (165)	**15**	**21**	**18**	16	13	7	10
EXTEND-IA	**Control (35)**	**17**	**11**	**11**	11	17	11	20
	Intervention (35)	**26**	**26**	**20**	17	3	0	9
MR CLEAN	**Control (267)**	**0**	**6**	**13**	16	30	12	22
	Intervention (233)	**3**	**9**	**21**	18	22	6	21
REVASCAT	**Control (103)**	**5.8**	**6.8**	**15.5**	19.4	16.5	20.4	15.5
	Intervention (103)	**6.8**	**17.5**	**19.4**	18.4	7.8	11.7	18.4
SWIFT PRIME	**Control (98)**	**9**	**11**	**16**	17	22	26	
	Intervention (98)	**17**	**26**	**17**	12	15	12	

mRS: modified Rankin scale (%)
Highlighted Boxes indicate mRS 0-2 functional independence.

Table 1.9 Embolectomy–thrombolysis trial—hermes efficacy analysis [80]: mRS outcome (90 day)

mRS Score		No Symptoms--► Death						
Study	**Treatment**	**0**	**1**	**2**	**3**	**4**	**5**	**6**
OVERALL	**Control (645)**	**5.0**	**7.9**	**13.6**	16.4	24.7	13.5	18.9
	Intervention (633)	**10.0**	**16.9**	**19.1**	16.9	15.6	6.2	15.3
tPA eligible	**Control (565)**	**5.1**	**8.1**	**13.8**	17.5	23.7	13.3	18.4
	Intervention (520)	**9.9**	**17.1**	**19.4**	16.6	17.3	5.9	13.7
rt-PA ineligible	**Control (80)**	**3.6**	**6.2**	**12.5**	**8.7**	31.2	15.0	22.5
	Intervention (108)	**10.2**	**15.7**	**17.6**	**18.5**	7.4	7.4	23.1

2.3 *Embolectomy Concerns*

Clinical trial data presented in Sects. 2.2.1–2.2.7 point to the only major efficacious achievement in the treatment of embolic stroke since the approval of rt-PA. While the use of embolectomy has somewhat expanded the window for treatment opportunities for patients, there are still challenges that must be overcome to further improve long-term clinical outcome. The following sections address patient population selection and limitations of the recent clinical trials.

2.3.1 Patient Population Selection for Embolectomy and Thrombolysis

Inclusion and exclusion criteria for embolectomy as an adjunct to thrombolysis are routinely used in clinical trials and the 2015 AHA embolectomy guidelines have been published [81]. Specific criteria for thrombolysis have also been documented [81–84].

The landmark NINDS rt-PA trial established that rt-PA improved neurological function (NIHSS and mRS) when administered within 3 h of stroke onset, and additional trials have established efficacy within 6 h of stroke onset [45, 85–93]. In addition, the recent embolectomy trials were conducted with patients enrolled to receive the procedure within 6 h. Retrospective data analysis of rt-PA trials has shown reduced benefit in stroke patients with an NIHSS score >19, and there was decreased or no benefit in diabetic patients or patients with atrial fibrillation [51]. However, based upon five successful embolectomy clinical trials, all of which included diabetic patients and patients with atrial fibrillation, it appears that embolectomy/thrombolysis should not exclude either patient population. There may be less value of endovascular procedures in patients >80 years of age, and patients with NIHSS >17, but neither is significant reason to withhold treatment.

2.3.2 Limitations of the Trials

Before deeming the recent embolectomy trials a complete success, it is necessary to briefly further examine two main limitations of the trials, specifically the methodology used to assess patients for inclusion as well as time to treatment, although other limitations and discrepancies have been identified by investigators [94]. At first glance, it is apparent that the five studies cannot easily be generalized to all stroke patients. The reasons are as follows in the next sections.

Patient Selection Bias

As shown in Tables 1.3–1.7, patients were included in the embolectomy trials if the infarct core was small, and in some cases documentation to demonstrate that there was brain tissue to salvage. Notably, patients with large core ischemic strokes were excluded from the trials. It is interesting to note that the majority of patients screened for EXTEND IA and ESCAPE were excluded from the trials because of adherence to strict inclusion criteria [32, 33, 94]. In the open population of stroke patient in the United States, approximately 30–35 % have large vessels occlusions [30, 94, 95], making the patient selection limitation of the studies troublesome!

Second, each trial had varied methodology used to define the stroke regions and "penumbra." The lack of standardization across the studies can be seen as problematic or even beneficial. For example, in the REVASCAT trial, computed tomography (CT) was utilized with ASPECTS; patients with a score less than 7 were excluded, or a score of 6 with diffusion-weighted magnetic resonance imaging (MRI) [34], and in the EXTEND trial ASPECTS was also used (inclusion criteria score of 6–10). In the EXTEND-IA trial, CT angiography was used, and automated (RAPID) CT perfusion imaging was used to identify salvageable brain tissue [33, 94]. The MR CLEAN trial used various methods for inclusion of patients including CT, angiography (CTA), MRI, or digital-subtraction angiography (DSA) [12]. Despite the use of various techniques, the defined and very limited patient population improved significantly following the endovascular procedure in combination with thrombolysis.

Saving Penumbra

The integrity of the ischemic penumbra appears to be key in the success of the endovascular trials and will no doubt be critical in advancing many new therapeutic approaches to treat ischemic stroke victims in the near future. In 4 of the 5 trials, ASPECTS scoring [96–99] was utilized to define the extent of an ischemic core with scoring ranges of 0–5 a large ischemic core, 6–7 a modest sized core, 8–10 a small or minimal ischemic score. In the endovascular trials, the median score of 9 was indicative of small core ischemic strokes with ample penumbra at the time of randomization into the two treatment groups. This is a very important factor not only for effective thrombolysis, but also for combination therapy where substantial additional stent retriever efficacy was demonstrated.

While not openly discussed to any extent in the endovascular trial papers [31–35], there are other important considerations when offering the stroke victim reperfusion therapy, or even enrolling the patient in a randomized clinical trial. It is likely that patients with eloquent brain structure damage, either due to the main ischemic event or worsening of the ischemic core may specifically affect sensorimotor, language, and cognitive functions due to damage in cortical or subcortical brain regions. Due to the rapidity now required for endovascular and thrombolytic procedures (DTNT or DTGPT of <60 min), patients with the damage described earlier are typically enrolled in clinical trials without the benefit of extensive brain mapping using functional MRI (fMRI) and diffusion tensor imaging (DTI) [100]. If the damage is severe, then the response of the patient will be less than optimal, thereby skewing clinical outcome results and the procedures may offer little long-term benefit to the patient. Nevertheless, the patient should be provided the best possible care available because it would be unethical not to do so.

Time to Treatment

As provided in Tables 1.3–1.7, the success of the trials is attributed to rapid treatment of stroke patients identified within approximately 6 h, and the imaging of the infarct and salvageable area accomplished within 60 min or less. The process required a well-disciplined multidisciplinary team to efficiently move the patient from enrollment to treatment. The concern now is that the clinical trials were "artificial" and the rapidity of the process is not the norm in most worldwide facilities. Thus, reproducibility of safe and effective endovascular procedures will be a challenge. As described in a few recent publications, there are a limited number of sites worldwide that can accomplish the treatment procedures [29, 94, 101–103]. The safe use of rt-PA must be expanded in the USA and worldwide to promote recanalization prior to endovascular procedures or other novel procedures targeting cytoprotection.

Coadministration of Thrombolytics

In the five trials, rt-PA or urokinase was administered IV well in advance of embolectomy. In the five trials, the median time of thrombolytic administration was in the range of 85–127 min in the embolectomy arm and within 87–145 min in the thrombolysis arm, both well within current FDA guidelines. Moreover, in the embolectomy arm, the initiation of "thrombolysis" occurred well before the procedure.

There is now some reasonable evidence resulting from retrospective analysis of the embolectomy trial database [31–35], demonstrating that embolectomy alone in patients ineligible for rt-PA is beneficial [36] based upon mRS scores, and reperfusion measures (Table 1.10). Notably, patients "selected" for the procedures have ASPECTS scores of 8–9 [36] indicative of large penumbral areas as a substrate for therapy. However, there was a higher mortality rate (61 %) in the embolectomy group alone compared to rt-PA + embolectomy.

- Can another adjunct "neuroprotective" therapy complement thrombolysis in the clinical setting for the select population selected to receive embolectomy? These pertinent questions remain to be addressed and answered.

2.3.3 Future of Embolectomy

Five embolectomy trials conducted worldwide showed efficacy (Australia, Canada, Europe, United States) in large vessel stroke patients enrolled within 6 h. This is important to note and suggests reproducibility of the results and that the effect is not patient population specific. As discussed by Mocco et al. [104], the efficacy may be related to the enhanced ability of stroke centers to perform both thrombolysis and embolectomy in stroke patients. Currently, this includes an interventionalist, a clinician or nurse trained in the administration of IV rt-PA. The future may see additional benefit from neuroprotection administered IV, IA, transcranially or via

Table 1.10 Efficacy and safety of embolectomy in rt-PA ineligible patients [36]

Treatment Time-Range	Embolectomy	Embolectomy with rt-PA
0-3 hours	**N=40**	**N=60**
MCA M1/M2/ICA-T (%)	**74**	**91**
Baseline NIHSS	**19 (15-21)**	**21 (17-24)**
Mean Time IV-rt-PA (min)	**NA**	**92 (69-105)**
Last normal to GP (min)	**135 (101-167)**	**150 (125-165)**
mRS 0-2 (%)	**40**	**50**
Mortality 90 days (%)	**37**	**15**
0-4 hours + 5 hours	**N=102**	**N=177**
MCA M1/M2/ICA-T (%)	**85**	**92**
Baseline NIHSS	**19 (15-24)**	**20 (16-23)**
Mean Time IV-rt-PA (min)	**NA**	**108 (81-135)**
Last normal to GP (min)	**202 (160-246)**	**207 (165-240)**
mRS 0-2 (%)	**35**	**45**
Mortality 90 days (%)	**36**	**45**
0-6 hours	**N=169**	**N=253**
MCA M1/M2/ICA-T (%)	**86**	**90**
Baseline NIHSS	**19 (15-24)**	**20 (15-23)**
Mean Time IV-rt-PA (min)	**NA**	**119 (90-160)**
Last normal to GP (min)	**250 (181-304)**	**235 (181-289)**
mRS 0-2 (%)	**38**	**45**
Mortality 90 days (%)	**34**	**21**

another method yet to be described. To obtain efficacy, some centers may have to adopt drip and ship procedures where thrombolysis is provided rapidly at one site and the patient is then transported to another site for endovascular procedures. This method would result in longer treatment interval times and reduced efficacy compared to sites offering drip and keep procedures where both thrombolysis and embolectomy can be performed rapidly. Clearly, the future of stroke treatment will utilize a multidisciplinary team approach to provide the best care to patients, and centers will have to be expert with rt-PA and retrievers.

As informed steering committees develop new stroke trials based upon high-quality translational studies that incorporate STAIR [105] and RIGOR [106–108] guidelines, perhaps the limited "therapeutic window" for embolectomy and thrombolytic efficacy described herein should be recognized as critical. Previous arguments have been made for rapid administration of therapy [27, 29, 68] and the recommended DTNT for thrombolytic therapy administration is less than 1 h [56–58, 68]. Nevertheless, all embolectomy trials described in this chapter used a therapeutic window 5 or more fold in excess of that recommendation, and efficacy was still demonstrated, either with or without rt-PA. The fact that enhanced efficacy and safety were measured when a thrombolytic was preadministered to patients may be an important factor, but it is not a critical component to demonstrate significant clinical efficacy as reviewed earlier and in HERMES [36].

Endovascular Procedure Summary

Rapid advances are being made in the field of endovascular procedures with new devices currently under development. The five embolectomy trials discussed earlier will lead the way for new therapeutic options to be provided to stroke patients. The second phase of endovascular procedure trials with expanded therapeutic windows (6–8 h) is already enrolling patients. Nevertheless, with current clinical embolectomy and thrombolysis efficacy data within 6 h, only 13.5–31 % of patients achieve an mRS score of 0–2. 70–85 % of patients are not adequately treated and do not return to normal. It is doubtful that expanding the therapeutic window for embolectomy/thrombolysis will have a significant effect on the proportion of patients improving to the point of functional independence. Brain tissue still needs to be preserved, and the literature by Saver clearly argues that time is brain and time is extremely important [27]. The future of translational and clinical stroke research will be to provide an adjuvant therapeutic to patients either in combination with endovascular procedures, in the absence or presence of thrombolysis to further improve clinical function, and increase the mRS 0-2 population.

If at first, the idea is not absurd, then there is no hope for it
Albert Einstein
(German born American Physicist. 1879–1955)

3 Neuroprotection

Because effective neuroprotection or adjuvant neuroprotection has still not reached fruition in stroke victims, this section will only very briefly summarize in the form of an extensive, but not all inclusive table, the diverse types of neuroprotective compounds studied in clinical trials, and their outcomes, most quite disappointing except for select thrombolytics and a few neuroprotectives. The compilation of this encyclopedia of neuroprotective drug candidates was inspired, in part, by the seminal paper entitled "*1026 experimental treatments in acute stroke*" in 2006 [109], which was published 10 years ago.

It has now been 10 years since the exhaustive review on the art of neuroprotection in animal models, and the list continues to grow almost exponentially. We have taken a closer look at a selection of drugs, some with demonstrated efficacy in animals [109], and have compiled a Table of drugs listed with proposed mechanism for each drug (Table 1.11). Where possible, the Table includes a link to the www.clinicaltrials.gov database or another online database where trial information can be found, so that the reader can mine data, and review patient selection criteria and other study design information. Since many of the drugs and their targets are

Table 1.11 Neuroprotection clinical trials

Drug name	Proposed mechanism	Clinical Trial Information	Date	Design (if published)	Sample size (if known)	Result (Safety and/or Efficacy)	References
Excitotoxicity (Pathways)							
ACEA 1021 (licostinel)	NMDA glycine site antagonist	—	1999	Licostinel *vs* Placebo	64	Confirmation of safety; similar improvement of NIHSS scores versus placebo	(110)
AR-R15896AR (AKA-ARL 15896, or AR-A15896AR)	NMDA antagonist	—	2002	AR-R15896AR *vs* Placebo	103	No significant difference in neurological recovery; More side effects than placebo.	(111)
		—	2001	AR-R15896AR *vs* Placebo	175	No significant difference in neurological recovery; dose ranging study.	(112)
Baclofen	GABA-B agonist	—	2001	Baclofen *vs* Placebo	21	Maintains the reduction of the spastic hypertonia	(113)
		Study of the Effects on Motor Recovery of Early Post-stroke Spasticity Treatment (BacloTox)	2015	botulinum toxin and placebo baclofen *vs* oral baclofen and placebo botulinum toxin *vs* placebo baclofen and placebo botulinum toxin	300	Recruiting	(114)
		SISTERS: Spasticity In Stroke Study - Randomized Study (SISTERS)	2016	Intrathecal Baclofen *vs* Best Medical Treatment	88	Ongoing but not recruiting	(115)
BIII-890-CL	Sodium channel blocker	—	2002	Unknown			(116)
		Safety and Tolerability of BIII-890 in Patients With Acute Ischemic Stroke	2014	BIII-890 *Vs* Placebo	97	No designated safety issues reported	(117)
BMS-204352	Potassium channel opener	POST	2002	BMS-204352 *vs* Placebo	1,978	No significant difference in neurological recovery	(118)
CGS 19755 (Selfotel)	NMDA antagonist	—	2000	Selfotel *vs* Placebo	567	No significant difference in neurological recovery; might have neurotoxic effect	(119)
Clomethiazole (CMZ, Zendra)	GABA agonist	CLASS-I	2002	Clomethiazole *vs* Placebo	1,198	Does not improve outcome in neurological recovery.	(120)
CNS1102 (Cerestat, Aptiganel)	NMDA ion channel blocker	—	2001	Aptiganel *vs* Placebo	628	No significant difference in neurological recovery, and may be harmful.	(121)
CP101, 606 (Traxiprodil)	NMDA ion channel blocker	A Study to Evaluate the Efficacy and Safety of CP-101,606 in Subjects With an Acute Stroke	2004	CP101, 606 *vs* Placebo	300	Study Terminated, unreported results	(122)
Dextrorphan	NMDA ion channel blocker	—	1995	Dextrorphan *vs* Placebo	37	No difference in neurological outcome; dose ranging study	(123)
Dextremethorphan	NMDA ion channel blocker/ Neuroprotection	Evaluation of the neuroprotective effect of dextromethorphan in the acute phase of ischaemic stroke	2011	Dextremethorphan *vs* Placebo	40	Not neuroprotective, but does not worsen condition or neurological outcome; reduction in seizures, and increase of MI and renal failure versus placebo.	(124)
Diazepam (valium)	Benzodiazepine	—	2006	Diazepam *vs* Placebo	880	Safe in acute ischemic stroke, but may better be avoided in intracerebral hemorrhage.	(125)
Eliprodil (SL 82.0715)	NMDA polyamine antagonist Sigma ligand	Eliprodil trial	1996	Eliprodil *vs* Placebo	483	No significant difference in neurological recovery	(126)
Fosphenytoin	Sodium channel blocker, Glutamate release inhibitor.	—	2008	Fosphenytoin *vs* Placebo	462	No significant difference in neurological recovery	(127)

(continued)

Table 1.11 (continued)

Gavestinel (GV150526A)	NMDA glycine antagonist	GAIN International Trial	2005	Gavestinel *vs* Placebo	3450	No significant difference in neurological recovery	(128)
Glycine	NMDA antagonist	—	2000	Glycine *vs* Placebo	200	Confirmation of safety; Improvement of neurological recovery	(129)
GSK249320 (mAb)	Antagonises or neutralises myelin-associated glycoprotein (MAG) - mediated inhibition	MAG111539	2013	GSK249320 *vs* Placebo	42	Confirmation of safety	(130)
Lifarizine (RS-87476)	Sodium/calcium channel blocker	—	1995	Lifarizine *vs* Placebo	147	Confirmation of safety; Mortality and functional data suggest favorable trends	(131, 132)
Lubeluzole	Sodium/calcium channel blocker NOS inhibitor	—	2001	Lubeluzole + alteplase *vs* Placebo + alteplase	89	No significant difference in neurological recovery; confirmation of safety	(133)
		—	2000	Lubeluzole *vs* Placebo	1,786	No significant difference in neurological recovery	(134)
		—	1998	Lubeluzole *vs* Placebo	725	No significant difference in neurological recovery; did not increase morbidity.	(135)
Magnesium	NMDA ion channel blocker. Calcium antagonist	—	2013	Magnesium sulphate *vs* Placebo	107	Improvement of neurological recovery; confirmation of safety.	(136)
		Field Administration of Stroke Therapy - Magnesium (FAST-MAG) Trial (FAST-MAG)	2015	Magnesium sulphate *vs* Saline	1700	No significant difference in neurological recovery	(38)
NA-1 (Tat-NR2B9c)	Postsynaptic density-95 protein inhibitor	ENACT	2012	NA-1 *vs* Placebo	197	Fewer ischemic infarcts vs placebo.	(137)
		FRONTIER	2016	NA-1 *vs* Placebo	558	Recruiting	(138)
Nalmefene	Opioid antagonist	—	2000	Nalmefene *vs* Placebo	368	No significant difference in neurological recovery	(139)
Naloxone	Opioid antagonist	—	1991	Naloxone *vs* Placebo	24	No significant difference in neurological recovery	(140)
Nicergoline	α2 adrenoceptor agonist Enhances glutamate uptake	—	1997	Nicergoline *vs* Placebo	136	Improvement of neurological recovery	(141)
			2002	Nicergoline *vs* Placebo	25	Improvement of neurological recovery	(142)
NPS 1506	NMDA ion channel blocker	—	2000	NPS 1506 *vs* Placebo	36	Confirmation of safety	(143, 144)
Remacemide	NMDA ion channel blocker Sodium channel blocker	—	1999	Remacemide *vs* Placebo	61	Confirmation of safety	(145)
Sipatrigine (BW619C89)	Sodium channel antagonist Glutamate release inhibitor	—	2000	Sipatrigine *vs* Placebo	27	Neuropsychic events higher versus placebo; No significant difference in neurological recovery	(146)
YM872	AMPA antagonist	ARTIST	2006	YM872 + Alteplse *vs* Placebo + Alteplase	NA	This study was abandoned after failing an interim futility analysis.	(147)
		ARTIST MRI	2006	YM872 *vs* Placebo	NA	This study was abandoned after failing an interim futility analysis.	(147)
ZK200775 (MPQX)	AMPA antagonist	—	2002	ZK200775 *vs* Placebo	61	Transiently worsened the neurological condition	(148)

(continued)

Table 1.11 (continued)

Anti-inflammatory/Vascular Adhesion/Anti-bacterial/Immune suppression							
Cyclosporin A	Immunosuppres sant	CsAStroke	2014	Cyclosporin A *vs* Saline	126	Not effective in reducing infarct size. However, a smaller infarct size was observed in patients with proximal cerebral artery occlusion and efficient recanalization.	(149, 150)
Dapsone (diamino-diphenyl sulfone, DDS)	Antibacterial	DAISY	2010	Dapsone *vs* Placebo	300	Unknown	(151)
		Pilot	2007	Dapsone *vs* Placebo	30	Confirmation of safety; Improvement of neurological recovery	(152)
Dexamethasone	Glucocorticoid, antiinflammatory	—	2011	Dexamethasone *vs* Placebo	60	Improvement of level of consciousness was statistically significant in Dexamethasone treated group, but did not reduce volume of hypodense area.	(153)
		—	2001	Dexamethasone *vs* Placebo	40	Study failed to demonstrate any benefit of a short-course of high dose steroid in improving the mortality of stroke patients.	(154)
Enlimomab (anti–ICAM-1 antibody)	Leukocyte migration and adhesion inhibitor	EAST	2001	Enlimomab *vs* Placebo	625	No significant difference inn eurological recovery; may significantly worsen stroke outcome	(155)
FK506 (Tacrolimus)	Immunosuppres sant	—	2012	Unknown		Stopped in Phase II-adverse side effects	See review (156)
Fludrocortisone	Mineralocorticoid	—	1989	Fludrocortisone *vs* None	91	Although fludrocortisone treatment tended to diminish the decrease in plasma volume, the difference was not significant.	(157, 158)
Ganglioside GM1	Metabolism, growth	SASS	1994	Ganglioside GM1 *vs* Placebo	287	Confirmation of safety; only certain post hoc tests showed statistically significant differences or trends favoring GM1.	(159)
		IASS-H	1989	Ganglioside GM1 *vs* Placebo	502	No significant difference in neurological recovery at 120 days.	(160)
Hu23F2G (LeukArrest)	Leukocyte adhesion inhibitor	—	2002	LeukArrest + Alteplase *vs* Placebo + Alteplase	Unknown	The trial was terminated after interim analysis	(161)
Interleukin-1 receptor antagonist(IL-1 RA)	IL-1 inhibitor	—	2005	IL-1 RA *vs* Placebo	34	Confirmation of safety	(162)
Minocycline	Broad-spectrum tetracycline antibiotic	—	2007	Minocycline *vs* Placebo	152	Improvement of neurological recovery	(163)
		NeuMAST	2012	Minocycline *vs* Placebo	139	Study Terminated; interim analysis shows futility	(164)
		—	2015	Minocycline *vs* Control	53	Improvement of neurological recovery; no significance between females and males.	(165)
		—	2013	Minocycline *vs* Control	95	Intravenous minocycline was safe but not efficacious	(166)
		—	2010	Minocycline *Vs* Enoxaparin *vs* Minocycline + Enoxaparin *Vs* Placebo	6	Terminated; too few acute stroke patients available to meet enrollment requirements	(167)
		MINOS	2010	Open label dose-escalation.	60	Dose Finding study	(168)

(continued)

Table 1.11 (continued)

Neutrophil inhibitory factor (rNIF, UK-279.276)	Neutrophil inhibitor	ASTIN	2003	UK-279.276 *vs* Placebo	966	No significant difference in neurological recovery; stopped early for futility.	(169)
Niacin	Vitamin B(3): metabolic enhancer	—	1980	Xanthinol nicotinate *vs* Pentoxifylline	58	Xanthinol nicotinate was inferior to pentoxifylline in neurological recovery	(170)
Paracetemol (Acetaminophen)	Analgesic/antipyretic COX inhibitor	PAIS	2009	Paracetemol *vs* Placebo	1,400	Paracetamol might have a beneficial effect on functional outcome in patients admitted with a body temperature 37-39 degrees C, but results do not support the use of high dose paracetamol for acute stroke.	(171)
		PAIS 2	2015	Paracetemol *vs* Placebo	1,500	Unknown	(172)
Antioxidant-free radicals scavengers							
Albumin	Antioxidant Improvement of microcirculation	ALIAS	2011	Albumin *vs* Placebo	434	Preliminary results from the ALIAS Part 1 suggest a trend toward a favorable primary outcome in subjects treated with ALB and support the validity of the statistical assumptions that underlie the ALIAS Part 2 Trial.	(173)
		ALIAS-2	2013	Albumin *vs* Placebo	843	No Clinical benefit; trial halted due to futility	(174)
Deferoxamine	Iron chelator; bacterial siderophore	TANDEM-1	2012	Deferoxamine + Alteplase *vs* Placebo + Alteplase	62	Unknown	(175, 176)
Ebselen	Free radical scavenger; synthetic organo-selenium anti-inflammatory, anti-oxidant and cytoprotective activity; mimic glutathione peroxidase	—	2009	Ebselen *vs* Placebo	394	Unknown	(177)
		—	1998	Ebselen *vs* Placebo	302	Early treatment improved the neurological recovery	(178)
Edaravone	Free radical scavenger; nootropic and neuroprotective agent	–	On going	Compound Edaravone *Vs* Control	1200	Unknown	(179)
		–	2015	Edaravone Low/Medium/High Dose *Vs* Control	400	Unknown	(180)
		CER	2014	Edaravone *vs* Butylphthalide *vs* Cerebrolysin *vs* Citicoline *vs* Piracetam	20,000	Unknown	(181)
		YAMATO	On going	Edaravone (pre-administaration) + Alteplase *vs* Edaravone (post-administration) + Alteplase	200	Unknown	(182)
		PROTECT4.5	On going	Edaravone (pre-administration) + Alteplase *vs* Edaravone (post-administaration) + Alteplase *vs* Edaarvone + Argatroban *vs* Edaarvone + Sodium ozagrel	10,000	Unknown	(183)

(continued)

Table 1.11 (continued)

		–	2012	Edaravone (pre-administration) + Alteplase *vs* Edaravone (post-administaration) + Alteplase	40	Improvement of neurological recovery	(184)
			2011	Edaravone *Vs* Placebo	50	Improvement of neurological recovery	(185)
		–	2010	Edaravone (long-term-administration) *vs* Edaravone (short-term-administaration)	47	Suppresses the progression of disuse muscle atrophy and improves leg locomotor function to a greater extent than shorter-term treatment in acute stroke patients.	(186)
		–	2010	Edaravone (long-term-administration) *vs* Edaravone (short-term-administaration)	72	Retrospective study shows that Edaravone dose-dependently increases rehabilitation gain	(187)
		EDO	2009	Edaravone *vs* Sodium ozagrel	401	Verified that edaravone was not inferior to ozagrel. Edaravone was at least as effective as ozagrel for the treatment of acute noncardioembolic ischemic stroke.	(188)
		EAST	2008	Edaravone + Argatroban *Vs* Argatroban	814	No favorable effects of edaravone when added to baseline treatment with argatroban.	(189)
		EAIS	2003	Edaravone *vs* Placebo	250	Improvement of neurological recovery	(190)
		–	2003	Edaravone + Sodium ozagrel *vs* Sodium ozagrel	138	Improvement of neurological recovery in combination vs monotherapy with ozagrel.	(191)
N-acetyl-cysteine (NAC)	Free radical scavenger	—	2015	N-acetylcysteine *vs* Acetaminophen *vs* N-acetylcysteine + Acetaminophen *vs* Placebo	120	Unknown	(192)
Nicaraven (AVS)	Free radical scavenger; hydroxyl radical scavenger	—	1996	AVS *vs* Placebo	162	Marked improvement in the Glasgow Outcome Scalescores at 1 month, as well as a reduction in the cumulative incidence of death by 3 months.	(193)
NXY-059	Nitrone free radical scavenger	SAINT II	2008	NXY-059 *vs* Placebo	3,306	Ineffective within 6 hours of symptom onset	(40, 195)
		SAINT I	2007	NXY-059 *vs* Placebo	1,722	NXY-059 significantly improved the overall distribution of scores on the modified Rankin scale, as compared with placebo (P=0.038).	(39, 194)
PG2 (Polysaccharides of Astragalus membranaceus)	Chinese Herb Antiinflammatory	Pass	2015	PG2 *vs* Placebo	300	Unknown	(196)
Tirilazad (U74006F)	Free radical scavenger; nonglucocorticoid, 21-aminosteroid that inhibits lipid peroxidation	RANTTAS and RANTAS II	1996	Tirilazad *vs* Placebo (Randomized, double-blinded, vehicle-controlled trial)	556 and 126	No significant difference in neurological recovery (worsens stroke outcome)	(197)

(continued)

Table 1.11 (continued)

Anti-apoptotic/regeneration							
Basic fibroblast growth factor (Trafermin, Fiblast)	Growth factor	—	2002	Trafermin *vs* Placebo	286	Confirmation of safety.	(198)
Erythropoietin (EPO)	Controls erythropoiesis, or red blood cell production		2016 on going	EPO + G-CSF *vs* Placebo	60	Recruiting	(199)
		—	2009	EPO *vs* Placebo	522	No significant difference in neurological recovery. Significantly increased mortality rate and safety concerns.	(200)
		BETAS	2008	Beta-hCG+ erythropoietin	15	Confirmation of safety	(201, 202)
GM602	Anti-inflammatory	GMAIS	2016 (ongoing)	GM602 *vs* Placebo	36	Unknown	(203)
Granulocyte-colony stimulating factor (G-CSF)	Regeneration; stimulates the bone marrow to produce granulocytes and stem cells	AXIS-2	2013	AX200 *vs* Placebo	328	No significant difference in neurological recovery	(204, 205)
Lu AA24493 (carbamylated erythropoietinC EPO)	Controls erythropoiesis, or red blood cell production	CEPO	2011	Lu AA24493 *vs* Placeb	16/24	Unknown-Toxicity claims halted development	(206, 207)
MP-124	PARP-1 inhibitor	—	2011	Unknown	72	Unknown	(208)
NTx®-265	Regeneration; Human Chorionic Gonadotropin (hCG) and Epoetin Alfa (EPO)	REGENESIS	2009	NTx®-265 *vs* Placebo	134	Unknown	(209)
		REGENESIS-LED	2014	NTx®-265+ EPO *vs* Placebo	96	Confirmation of safety; study halted; No significant difference in neurological recovery	(210, 211)
		REGENESIS	2010	NTx®-265+rHCG+ EPO *vs* Placebo	30	Study withdrawn prior to enrollment.	(212)
PS519/MLN519	Proteasome inhibitor	—	2002	PS519	39	Confirmation of safety	(213)
Calcium channel/adrenergic modulators/antihypertensives							
Atenolol (Tenormin)	Selective β1 receptor antagonist	—	1993	Atenol + aspirin *vs* Placebo + aspirin	1,473	No significant difference in the occurrence of stroke events	(214)
Candesartan cilexetil (TCV-116, Blopress, CV-11974)	AT1 receptor antagonist Antihypertensive	SCAST	2012	Candesartan cilexetil *vs* Placebo	2,029	No significant difference in neurological recovery; harmful effect suggested.	(215, 216)
Cyclandelate	Vasodilator (calcium modulator)	—	1987	Cyclandelate	303	Improvement of neurological recovery; uncontrolled study design.	(217)
DP-b99 (DP-BAPA)	Calcium chelator	MACSI	2012	DP-b99 *vs* Placebo	446	Interim futility analysis; no evidence of efficacy in treating human ischemic stroke	(218-220)
		—	2008	DP-b99 *vs* Placebo	150	Improvement of neurological recovery at 90 days.	(221)
Flunarizine	Calcium channel blocker-non-selective and histamine H1 blocker	FIST	1996	Flunarizine *vs* Placebo	331	Flunarizine did not improve neurologic and functional outcome in patients with acute ischemic stroke	(222)
ILS-920	Calcium channel blocker (Rapamycin Analog)	—	2009	ILS-920	16	Unknown	(223)
Irbesartan	AT1 receptor antagonist Antihypertensive	—	2012	Irbesartan *vs* Placebo	43	Agent did not appear to substantially modify infarct growth.	(224)
Nicardipine	Calcium antagonist (dihydropyridine class of calcium channel blockers)	—	2001	Nicardipine + Aspirin *vs* Aspirin	264	At six months, the cumulative incidence of recurrent ischemic cerebrovascular events was significantly lower in the aspirin-plus-nicardipine group than in the aspirin-only group. This difference was no longer significant at 12 months.	(225)

(continued)

Table 1.11 (continued)

		—	2004	Double concentrated *Vs* Triple concentrated IV Nicardipine	50	Unknown	(226)
Nimodipine	Calcium channel blocker (dihydropyridine calcium channel)	—	2016 on going	Nimodipine *vs* Saline + citicoline	72	Recruiting	(227)
		NICE	2012	Nimodipine *vs* Placebo	665	Unknown	(228)
		INWEST	2003	Nimodipine *vs* Placebo	295	No significant difference in neurological recovery	(229, 230)
		VENUS	2001	Nimodipine *vs* Placebo	454	No significant difference in neurological recovery	(231)
		NIMPAS	1999	Nimodipine *vs* Placebo	50	No significant difference in neurological recovery	(232)
		—	1994	Nimodipine *vs* Placebo	350	No significant difference in neurological recovery	(233, 234)
		ANS	1992	Nimodipine *vs* Placebo	1,064	No significant difference in neurological recovery	(235, 236)
		TRUST	1990	Nimodipine *vs* Placebo	1,215	No significant difference in neurological recovery	(237, 238)
		—	1990	Nimodipine *vs* Placebo	164	No significant difference in neurological recovery	(239, 240)
		—	1988	Nimodipine *vs* Placebo	186	Improvement of neurological recovery	(241, 242)
Papaverine	Opium alkaloid antispasmodic; selective phosphodiesterase inhibitor for the PDE10A)	—	1979	Papaverine *vs* Vincamine	263	Papaverine was inferior to vincamine in neurological recovery	(243)
Propranolol	β-adrenergic blockade, Membrane stabilization	BEST	1988	Propranolol *vs* Atenol *vs* Placebo	302	No significant difference in neurological recovery	(244)
		BIAS	2013	Propranolol *vs* Placebo	20	Unknown	(245)
PY 108-068	Calcium antagonist; dihydro-pyridine derivative	—	1989	PY 108-068 *vs* Placebo	19	Safe, but no significant difference in neurological recovery	(246)
SUN-N8075	Anti-oxidant	—	2006	Unknown			(247)
S-0139 (SB-737004)	selective ET-A endothelin receptor antagonist	—	2002	Unknown			(248)
TS-011	20-Hydroxyeicosatetraenoic acid synthesis inhibitor	Effect of a New Inhibitor of the Synthesis of 20-HETE on Cerebral Ischemia Reperfusion Injury	2006	Phase I development abandoned			(249)
Verapamil	Calcium channel blocker (Phenylalkylamine calcium channel)	SAVER-I	2016 on going	Verapamil	30	Recruiting	(250)
Vinpocetine (ethyl apovincaminate)	Calcium inhibitor, Vasodilator, Sodium blocker; semisynthetic derivative of the vinca alkaloid vincamine, an extract from the lesser periwinkle plant.	CAVIN	2013	Vinpocetine + Citicoline + Aspirin or Clopidogrel *vs* Citicoline + Aspirin or Clopidogrel	610	Unknown	(251)
		—	2001	Vinpocetine + Dextran *vs* Dextran	30	Improvement of neurological recovery	(252)

(continued)

Table 1.11 (continued)

Coagulation Cascade/Clot Stability and Lysis							
Abciximab (reopro, c7E 3Fab)	Antiplatelet: glycoprotein inhibitor	AbESTT-II	2008	Abciximab *vs* Placebo	808	Trial terminated prematurely, trial did not demonstrate either safety or efficacy of intravenous administration of abciximab for the treatment of patients with acute ischemic stroke	(253)
Ancrod	Defibrinogenating agent derived from the venom of the Malayan pit viper	ESTAT	2006	Ancrod *vs* Placebo	1,222	No significant difference in neurological recovery	(254)
		STAT	2000	Ancrod *vs* Placebo	500	Improvement of neurological recovery	(255, 256)
		ASPI/ASPII	2010	Ancrod *vs* Placebo	277/311	No significant difference in neurological recovery	(257, 258)
Argatroban	Anticoagulant-small molecule direct thrombin inhibitor	ARTSS-2	2015	Argatroban + Alteplase *vs* Alteplase	90	Unknown; terminated to begin a new study to test the safety of combining IA with tPA and Argatroban	(259)
		ARTSS	2012	Argatroban + Alteplase	65	Confirmation of safety	(260)
		ARTSS-IA	2015, ongoing	Argatroban + Alteplase	10	Unknown	(259)
		ARGIS-1	2007	Argatroban *vs* Heparin *vs* No anti-coagulation treatment	2529	Argatroban was inferior to heparin in neurological recovery	(261, 262)
		ARGIS-1	2004	Argatroban *vs* Placebo	171	Confirmation of safety	(262)
Aspirin (acetylsalicylic acid)	Antiplatelet	CAST-IST	2000	Aspirin *vs* Placebo (CAST), Open (IST)	40,000	Secondary prevention of stroke	(263)
Batroxobin (Defibrase, DF-521)	Snake venom produced by Bothrops atrox and Bothrops moojeni, venomous species of pit viper. It is a hemotoxin which acts as a serine protease closely related to thrombin		2005	Defibrase *vs* Placebo	1,053	Improvement of neurological recovery	(264)
Certoparin	A low molecular weight heparin, primarily active against factor Xa	PROTECT	2006	Certoparin *vs* Unfractionated heparin	545	Confirmation of safety	(265)
Clopidogrel	Antiplatelet; thienopyridine-class antiplatelet agent	SPS3	2012	Clopidogrel + Aspirin *vs* Aspirin	3,020	No significant difference in secondary prevention of stroke	(266)
Dalteparin	Anticoagulant; low molecular weight heparin	HAEST	2000	Dalteparin *vs* Aspirin	449	No significant difference in neurological recovery	(267)
Eptifibatide (cromafiban; Integrilin)	Antiplatelet: glycoprotein IIb/IIIa inhibitor class; a cyclic heptapeptide derived from a protein found in the venom of the southeastern pygmy rattlesnake	CLEAR-ER	2015	Eptifibatide + Alteplase *vs* Alteplase	126	Confirmation of safety	(268, 269)
		CLEAR-FDR	2015	Eptifibatide + Alteplase *vs* Alteplase	27	Confirmation of safety	(270)
		CLEAR	2008	Eptifibatide + Alteplase *vs* Alteplase	94	Confirmation of safety	(271)
Heparin	Anticoagulant	IST	1997	Heparin *vs* No heparin	19,435	No significant difference in neurological recovery	(272)
MRX-801	Microbubbles w/ultrasound	TUCSON	2009	MRX-801 + Alteplase *vs* Placebo + Alteplase	72	The confirmation of the safety; study terminated by sponsor	(273)
Pentoxifylline	Improve capillary flow; Blood Viscosity Reducer	—	1988	Pentoxifylline *vs* Placebo	297	No significant difference in neurological recovery	(274)
			1985	Acetylsalicylic acid and dipyridamole *vs* Pentoxifylline	138	23 ASAD patients and 9 PTX patients suffered a recurrence. This difference was significant (p=0.019) in favor of the PTX group.	(275)

(continued)

Table 1.11 (continued)

Prosatacyclin	Antiplatelet: Eicosanoid vasodilator	—	1987	Prosatacyclin *vs* Placebo	80	No significant difference in neurological recovery	(276)
Tinzaparin	Anticoagulant; an anti-thrombotic; a low molecular weight heparin	TAIST	2001	Tinzaparin *vs* Aspirin	1,486	No significant difference in neurological recovery	(277)
		CATALIST	2009	Tinzaparin + Aspirin + tPA + eptifibatide	18	Safety and dose finding	(278)
Tirofiban (MK-383, aggrastat)	Antiplatelet: glycoprotein inhibitor; glycoprotein IIb/IIIa inhibitor.	SaTIS	2011	Tirofiban *vs* Placebo	260	No significant difference in neurological/functional outcome	(279)
Triflusal (2-acetoxy-4-trifluoromethylb enzonic acid)	Arachidonic acid metabolism inhibitor (antiplatelet; platelet aggregation inhibitor)	MAESTRO	2016	Triflusal *vs* Clopidgrel	795	Unknown	(280, 281)
		TAPIRSS	2004	Triflusal *vs* Aspirin	431	No significant difference in prevention of secondary stroke	(282)
		TACIP	2003	Triflusal *vs* Aspirin	2113	No significant difference in prevention of secondary stroke	(283)
Fluid regulation							
Glycerol	Hyperosmolar agent	—	1993	Glycerol *vs* Placebo	113	No significant difference in neurological recovery	(284)
Dextran	Hemodilution	—	1992	Hemodilution with dextran *vs* No treatment	102	Improvement of neurological recovery	(285)
Hydroxyethyl starch pentastarch	Hemodilution	—	2002	Hemodilution with hydroxyethyl starch pentastarch *vs* Hemodilution with saline	106	Confirmation of safety; No significant difference in neurological recovery	(286)
Mannitol	Hyperosmotic agent. Reduces edema and ICP	—	2003	Mannitol *vs* No mannitol	805	No significant difference in neurological recovery	(287)
Oxygen Carriers and delivery							
Diaspirin cross-linked hemoglobin	Oxygen delivery Free radical scavenger	—	1999	Diaspirin cross-linked hemoglobin *vs* Placebo	85	Significantly worse in neurological recovery	(288)
Hyperbaric oxygen treatment	Oxygen delivery	—	2003	Hyperbaric oxygen treatment *vs* Sham treatment	33	No significant difference in neurological recovery, may be harmful in patients with acute ischemic stroke	(289)
Normobaric oxygen treatment	Oxygen delivery	—	2010	Normobaric oxygen treatment *vs* Control (room air or oxygen 2L/min)	40	No significant difference in neurological recovery	(290)
			2009	Normobaric oxygen treatment *vs* Control (room air or oxygen 2L/min)	85	Terminated; Imbalance in deaths favoring control arm; deaths not attributed to treatment by the blinded external medical monitor;	(291)
Oxygengated flurocarbon nutrient emulsion	Oxygen delivery	—	2006	Oxygengated flurocarbon nutrient emulsion	4	Confirmation of safety	(292)
Miscellaneous Mechanisms							
3K3A-APC	a protease with anticoagulant and cytoprotective activities	RHAPSOD Y	2014	3K3A-APC *Vs* Placebo	100	Ongoing	(293, 294)
Atorvastatin	HMGCoA reductase inhibitor Antioxidant	—	2010	Atorvastatin *vs* Simvastatin	371	Improvement of neurological recovery	(295)
			2014	Atorvastatin *vs* no treatment	50	Unknown	(296)
		ASSORT	2015	Atorvastatin *Vs* Pitavastain *Vs* Rosuvastain *Vs* Non treated	300	Unknown	(297)
		SEATIS	2016 On going	Atorvastatin High dose (80mg) *Vs* Atorvastatin low dose (20mg)	256	Unknown	(298)

(continued)

Table 1.11 (continued)

BAY-387271	Cannabinoid agonist; analgesic and neuroprotective. High affinity for CB1 and CB2 receptors.		Bayer Phase 1 trial- abandoned.				(299)
Caffeinol	Stimulant, depressant, diuretic adenosine receptor modulator	—	2009	Caffeinol + Alteplase + Hypothemia	20	Confirmation of safety	(300)
		—	2011	Caffeinol + Hypothemia	30	Confirmation of safety	(301)
Cerebrolysin	A mixture of peptides purified from pig brains, including brain-derived neurotrophic factor, glial cell line-derived neurotrophic factor, nerve growth factor, and ciliary neurotrophic factor	CASTA	2012	Cerebrolysin *vs* Placebo	1,070	No significant difference in neurological recovery; favorable outcome trend seen in severely affected patients	(302)
Cilostazol	A quinolinone-derivative; Phosphodiestera se 3 Inhibitor	—	2012	Cilostazol + Aspirin *vs* Aspirin	76	Improvement of neurological recovery	(303)
Citalopram	Selective serotonin reuptake inhibitor	TALOS	2016 On going	Citalopram *vs* Placebo	600	Recruiting	(304)
Citicoline (CDP choline)	Membrane precursor, antioxidant Vascular insufficiency Immunostimulat ory Nootropic	ICTUS	2012	Citicoline *vs* Placebo	2,078	No significant difference in neurological recovery	(305)
Dexamphetamine	Stimulant and amphetamine enantiomer	—	2003	Dexamphetamine *vs* Placebo	45	Confirmation of safety; increase in blood pressure and heart rate versus placebo	(306)
EGB-761 (Gingko biloba extract)	MAO inhibitor Antiplatelet. Antioxidant Reduces leukocyte activation Increases cerebral blood flow	—	1995	Gingko biloba extract *vs* Placebo	55	No significant difference in neurological recovery	(307)
Estradiol	Antioxidant Antiinflammatory Antiapoptosis	WEST	2006	Estradiol *vs* Placebo	664	Slightly worse in neurological recovery	(308, 309)
Glyburide (RP-1127)	Ca-ATP channel blocker; Glibenclamide is a sulfonylurea antidiabetic drug	GAMES-PILOT	2014	RP-1127	10	Confirmation of safety	(310, 311)
		GAMES RP	On going	RP-1127 *Vs* Placebo	83	Unknown	(312, 313)
Glyceryl trinitrate (Nitroglycerin, GTN)	NO donor	ENOS	On going	Glyceryl trinitrate *vs* No glyceryl trinitrate	3,500	Safe to administer and associated with improved functional outcome and fewer deaths when administered within 6 hours of stroke onset.	(314-316)
Hydergine	Nootropic- a mixture of the methanesulfonat e salts of three dihydrogenated ergot alkaloids	—	1978	Hydergine *vs* Dexamethasone *vs* Mannitol *vs* Placebo	300	No significant difference in neurological recovery	(317)
Hypothermia	Reduce reducing cerebral oxygen demand (CMRO2), Metabolic and synaptic transmission inhibitor.	ICTuS-L	2010	Hypothemia *vs* Normothemia	59	No significant difference in neurological recovery. Initial studies not powered for significance.	(318)

(continued)

Table 1.11 (continued)

Lithium	Stimulus of the secretion of BDNF	—	On going	Lithium	35	Unknown	(319, 320)
Nafronyl oxalate (Naftidrofuryl)	Enhance cellular oxidative capacity, and may also be a 5-HT2 receptor antagonist.	—	1990	Naftidrofuryl *vs* Placebo	100	No significant difference in neurological recovery	(321)
Neuroaid (MLC601, Danqi Piantang Jiaonang)	Stimulus of the secretion of BDNF	CHIMES	2012	Neuroaid *vs* Placebo	1,100	Unknown	(322, 323)
		—	2011	MLC601 *vs* Placebo	150	Improvement of neurological recovery; confirmation of safety	(324)
		—	2009	Danqi Piantang Jiaonang *vs* Buchang Naoxintong Jiaonang.	605	Improvement of neurological recovery	(325)
ONO-2506	Astrocyte modulating agent Anenuates extracellular Monamine; homolog of valproic acid	—	2012	ONO-2506 *vs* Placebo	1,320	Unknown	(326)
			2006	ONO-2506 *vs* Placebo	92	Confirmation of safety	(327)
			2008	ONO-2506 *vs* Placebo	757	Unknown	(328)
Pioglitazone	PPAR- γ agonist	—	2016	Pioglitazone *vs* Placebo	152	Recruiting	(329)
Piracetam	AMPA (NA+) modulator; a cyclic derivative of GABA	PASS	1997	Piracetam *vs* Placebo	927	Post hoc analyses suggest that piracetam may confer benefit when given within 7 hours of onset, particularly in patients with stroke of moderate	(330, 331)
		PASS II	Unknown			Active not recruiting	(331)
Propentofylline (HWA 285)	A xanthine derivative with neuroprotective effects; a phosphodiesterase inhibitor; an adenosine reuptake inhibitor	—	1993	Propentofylline *vs* Placebo	30	No significant difference in neurological recovery	(332)
Repinotan (BAY × 3072)	Serotonin agonist; selective 5-HT1A receptor full agonist	mRECT	2009	Repinotan *vs* Placebo	681	No significant difference in neurological recovery	(333)
Rosuvastatin	HMGCoA reductase inhibitor	JUPITER	2010	Rosuvastatin *vs* Placebo	17802	Reduces the incidence of ischemic stroke	(334)
		EUREKA	2013	Rosuvastatin *vs* Placebo	318	Efficacy in reducing recurrence in acute stroke was inconclusive	(335, 336)
		—	Ongoing	Rosuvastatin *vs* Placebo	100	Recruiting	(337)
		REPAIRS	Ongoing	Rosuvastatin *vs* Non-treated	456	Recruiting	(338)
		STAMINA-MRI	Ongoing	Rosuvastatin *Vs* Atorvastatin	80	Recruiting	(339)
		—	2012	Rosuvastatin	75	Statin exposure following ischemic stroke was not associated with ICH.	(340, 341)
		HOPE-3	2016	Rosuvastatin + Candesartan/HCT *vs* Placebo	12705	Significantly lower risk of cardiovascular events than placebo	(342)
Semax	Derivative of ACTH-4-10; nootropic, neuroprotective, and neurogenic/neurorestorative properties	—	1997	Semax *vs* Conventional therapy	110	Confirmation of safety.	(343)

(continued)

Table 1.11 (continued)

Simvastatin	HMGCoA reductase inhibitor Antioxidant	STARS07	On going	Simvastatin *vs* Placebo	340	Unknown	(344)
		—	2014	Simvastatin *vs* Placebo	104	Unknown	(345)
		MISTICS	2008	Simvastatin *vs* Placebo	60	Confirmation of safety	(346)
Spheno-Palatine Ganglion (SPG) stimulation	Induction of cerebral vasodilatation	ImpACT-24	On going	The Ischemic Stroke System *vs* Sham control	800	Unknown	(347)
		ImpACT-24Bt	On going	The Ischemic Stroke System + rtPA *vs* Sham control + rtPA	150	Unknown	(348)
SUN N4057 (Piclozotan)	A partial serotonin 1A receptor agonist	—	2007	Piclozotan high dose/ low dose *vs* Placebo	43	Unknown	(349)
Transcranial laser therapy (TLT)	Mitochondrial stimulation, enhanced blood flow, neuron repair	NEST-3	2012-Ended	NeuroThera® Laser System *Vs* Sham treatment	1,000	Futile	(44)
		NEST-2	2009	NeuroThera® Laser System *vs* Sham treatment	660	No overall difference, but benefit noted when NIHSS stratified < 16	(42)
		NEST-1	2007	NeuroThera® Laser System *vs* Sham treatment	120	Benefit improved neurological function	(43)
Trazodone (Desyrel)	Serotonin reuptake inhibitor	—	1986	Trazodone *vs* Placebo	49	No significant difference in neurological recovery	(350)
Thrombolytic							
Drug name	**Drug proposed mechanism**	**Trial name**	**Date**	**Design**	**Sample size**	**Result**	**References**
Alfimeprase	A fibrolase derivative with thrombolytic activity	—	2008	Alfimeprase	7	Phase 2 data did not show efficacy.	(351)
Alteplase (rt-PA)	Thrombolytic; a serine protease	EXTEND	On going	Alteplase *vs* Placebo	400	Unknown	(352)
		NINDS rt-PA	1995	Alteplase vs placebo	624	Improvement of neurological recovery	(30)
		IST-3	2012	Alteplase *vs* Placebo	3,035	Improvement of neurological recovery	(353)
		ECASS III	2008	Alteplase *vs* Placebo	821	Improvement of neurological recovery	(45)
		SITS-MOST	2007	Alteplase *vs* Placebo	6,483	Improvement of neurological recovery	(354)
Desmoteplase	A highly fibrin-specific thrombolytic agent from the common vampire bat Desmodus rotundus	DIAS-4	2014	Desmoteplase *vs* Placebo	270	Study halted; due to DIAS-3 indicated primary endpoint unlikely to be reached	(356)
		DIAS-3	2014	Desmoteplase *vs* Placebo	492	No significant difference in neurological recovery	(355)
		DIAS-2	2009	Desmoteplase *vs* Placebo	193	No significant difference in neurological recovery	(357)
Enoxaparin	Heparin-based anticoagulant	PREVAIL	2009	Enoxaparin *vs* Unfractionated heparin	1762	Clinical benefits enoxaparin are not associated with poorer long-term neurological outcomes or increased rates of symptomatic intracranial hemorrhage	(358, 359)
Microplasmin	Thrombolytic; a molecule created from plasminogen	—	2009	Microplasmin *vs* Placebo	40	Confirmation of safety	(360)
Nadroparin	Heparin-based anticoagulant	FISS-tris	2007	Nadroparin *vs* Aspirin	353	No significant difference in neurological recovery	(361, 362)
		Heparinas	2013	Nadroparin *vs* Heparin *vs* placebo	150	Unknown	(363)
		FISS	2000	High Dose Nadroparin *Vs* Low Dose Nadroparin *Vs* Placebo	312	Significant dose dependent effect	(364, 365)

(continued)

Table 1.11 (continued)

Org 10172 (Danaparoid, Orgaran)	Heparin-based anticoagulant	TOAST	1998	Org 10172 *vs* Placebo	1,281	No significant difference in neurological recovery	(366)
Prourokinase	Thrombolytic; a serine protease	PROACT II	1999	Prourokinase + Heparin *vs* Heparin	180	Improvement of neurological recovery	(367)
Reteplase	Thrombolytic; non-glycosylated form of human tPA	—	2011	Reteplase + Alteplase *vs* Urokinase + Alteplase	197	Confirmation of safety	(368)
			2002	ReoPro +Reteplase *Vs* ReoPro	72	Unknown	(369)
SAR104772	Thrombolytic; a thrombin-activable fibrinolysis inhibitor	—	2009	Unknown			(370)
Solulin	Thrombomodulin analog	—	2011	Solulin *vs* Placebo	42	Confirmation of safety	(371)
Streptokinase	Thrombolytic; bind and activate human plasminogen	ASK	1996	Streptokinase *vs* Placebo	340	No significant difference in neurological recovery	(372)
		MAST-E	1996	Streptokinase *vs* Placebo	310	No significant difference in neurological recovery; Increased mortality	(373)
		MAST-I	1995	Streptokinase *vs* Aspirin *vs* Streptokinase + Asprin *vs* Neither	622	No significant difference in neurological recovery	(374)
TAL-05-00018 (human plasma-derived fibrinolysin)	Thrombolytic; human plasma-derived plasmin	—	2014	TAL-05-00018	61	Unknown	(375)
		—	2014	20 mg of Plasmin, 40 mg of Plasmin, 80 mg of Plasmin	40	Safety and dose finding for ICH	(376)
THR-18	plasminogen activator inhibitor-1 peptide	—	2010	THR-18 + Alteplase *vs* Placebo + Alteplase	22	Unknown	(377)
			2013	THR + tPA *Vs* tPA + Placebo	30	Unknown	(378)
Tenecteplase	Thrombolytic; a recombinant fibrin-specific plasminogen activator		2012	Tenecteteplase *vs* Alteplase	75	Improvement of neurological recovery	(379)
		NOR-TEST	2015	Tenecteteplase *vs* Alteplase	954	Recruiting	(380)
		ATTEST	2015	Tenecteteplase *vs* Alteplase	104	Neurological and radiological outcomes did not differ between the tenecteplase and alteplase groups	(381, 382)
		DIVA	2015	Rt-PA + Tenecteteplase *Vs* Rt-PA	60	Not open	(383)
		TNK-S2B	2005	Tenecteteplase +tPA *Vs* tPA	112	Prematurely terminated; Neither promise nor futility could be established.	(384)
		TEMPO-1	2014	Tenecteteplase + tPA	50	Confirmation of safety	(385, 386)
		TEMPO-2	2015	TNK-tPA *Vs* Control (antiplatelets)	1274	Recruiting	(387)
		EXTEND-IA TNK	2015	TNK *vs* tPA	64	Recruiting	(388)
Urokinase (Abbokinase)	Thrombolytic; a serine protease	SWISS	2015	Urokinase (Intravenous) *vs* Urokinase (Intraarterial)	1368	Unknown	(389)
		—	2012	Urokinase *vs* Antiplatelet drugs	143	Improvement of neurological recovery	(390)
		—	2002	Urokinase *vs* Placebo	511	Improvement of neurological recovery	(391)
		AUST-PILOT	2009	Urokinase+heparin *vs* Heparin	16	A good outcome was observed in 4 of 8 patients who received intra-arterial urokinase compared with 1 of 8 patients in the control group	(392)
		AUST	1998	Urokinase+heparin *vs* Heparin	200	Unknown/ suspended	(393)

discussed in detail by other contributors to this landmark neuroprotection volume, there will be limited discussion of the Table contents.

Upon review of the published literature [109], and noting some indication of the estimation of the quality of STAIR criteria used in preclinical and translational studies, we propose that a short list of previously clinically tested compounds be resurrected and further studied in preclinical stroke models, and then tested for efficacy in combination with modern endovascular reperfusion therapy (See *Evolving Opportunities* later). Further study is necessary in stroke victim pertinent animal models because many of the therapies never did undergo systematic testing in multiple species, mixed gender populations, or animals with predisposing conditions (see Sect. 3.3). Since most studies were not conducted using the now standard rigorous criteria adopted by the majority of the scientific community, funding agencies should not only support research into new avenues, but also support research to further develop therapies for which there is already a substantial knowledge base.

3.1 Evolving Opportunities

Over the last decade, there have been many promising neuroprotective treatments developed through one or more animal species, all of which have failed in clinical trials for one or more reasons [29, 109, 394–397]. The positive endovascular/thrombolytic trials represent an opportunity to revisit, even revive some of the therapeutics that were promising in early clinical trials, but eventually failed in larger clinical trials.

A few of the most promising approaches that require new investigations are as follows

1. NXY-059 (SAINT I–II trials) [39, 40, 398–401]
2. Radicut (Edaravone; no organized randomized double-blind trial), which is already approved in Japan and other parts of Asia [402, 403]
3. Transcranial laser therapy (NEST 1–3 trials) [43, 395, 404]
4. Albumin (ALIAS trials) [174, 405, 406]
5. Magnesium (FAST-mag trials) [38, 102, 103, 407]

The FAST-mag trial by Saver and colleagues established a network of capable paramedics to be utilized in the field for rapid administration of proposed neuroprotective compounds; a component that may be key in obtaining efficacy for a neuroprotective [27, 29, 68, 408], but the FAST-mag treatment strategy failed.

Currently, there are a few promising target-selective and pleiotropic "neuroprotective agents" advancing through the drug development pipeline that could potentially be utilized in conjunction with endovascular procedures, with or without thrombolysis. For instance, because current embolectomy procedures target patients with large penumbral areas (ASPECTS >8), it would be most important to add on a neuroprotective that could enhance *brain tissue oxygenation* [409, 410]; and/or *attenuate reperfusion-injury and free radical production* [39, 40,

398, 399, 402, 403]. Since current embolectomy procedures are most effective in the presence of thrombolytic therapy, *adjuvant thrombolytics or fibrinolytics*, which may also promote clot destabilization and further promote recanalization [379, 411] could eventually be used in combination with endovascular procedures.

Lastly, one must be open to novel therapeutic and physical approaches to promote cytoprotection including compounds that have shown efficacy in multiple species such as compounds that uncouple postsynaptic density protein PSD-95 from neurotoxic signaling pathways (PSD-95 inhibitors) [412], promote synaptic plasticity [413–415], or possibly improve the safety of current thrombolytic standard-of-care therapy [416–418]. Alternative approaches including hypothermia, which has inherent delivery and recovery problems has not been validated for stroke. Physically induced hypothermia has not been shown to be beneficial as a method of cerebral protection during surgery [419], has limited temporary efficacy in aged rodents [420], and even promotes infarct growth [420] and cell death [421] under certain conditions. However, new pharmacological-induced hypothermia strategies are promising and safer and are now being pursued using cannabinoid and neurotensin-receptor ligands [422–427]. Some of these topics and drug development opportunities for neuroprotection are discussed by experts contributing chapters to this volume.

3.2 The RIGORs of Drug Development: Guideline Adherence

There have been numerous iterations of a set of recommendations proposed by stroke therapy academic industry roundtable (STAIR), a collaborative effort between academics and industry in an effort to provide guidance to researchers with the intent to develop effective stroke treatments [428]. STAIR recommends that preclinical studies be randomized and blinded, efficacy should be established in two or more laboratories, and efficacy should be replicated in a second species. STAIR suggested the possible use of small species (rodents), and large species (rabbits and nonhuman primates) to test treatment strategies measuring two outcomes: (1) functional response and, (2) histological outcome in both the acute stroke phase (1–3 days) and long term (7–30 days). The use of multiple species allows investigators to bridge the gap between animals and humans and demonstrate irrefutable efficacy and reproducibility of therapy effect in multiple species. Sharp and colleagues [429] emphasized the utility of the rabbit (*Oryctolagus cuniculus*), a large animal nonrodent species, to test stroke therapies because the model was used effectively for the preclinical development of rt-PA [429–431]. The proposal suggests that the rabbit small clot embolic stroke model (RSCEM) be used for primary screening of therapies alone or in combination with tPA, which is a bona fide "positive control" for stroke. For additional drug development, nonhuman primates have also been used as bridge of the translational gap between animals and humans [432]. While this has not been validated for any therapy, the process is underway for a PSD95 inhibitor [38, 412].

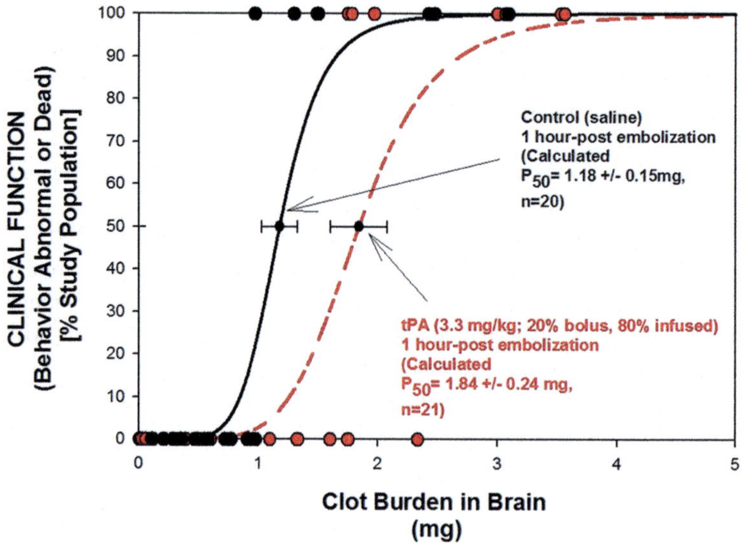

Fig. 1.1 Effect of standard dose tPA on behavioral outcome in embolized rabbits. For the superimposed graphs, behaviorally normal animals are plotted on the *y*-axis at 0 and abnormal animals are plotted at 100. The figure shows that there is positive correlation between the raw data and the statistically fit sigmoidal quantal curves. (**a**) *Black circles*-control; *red circles* tPA. The slope of each quantal curve is defined by the statistical quantal analysis curve-fitting program [430, 431, 433, 436, 437] and is a direct representation of the behavioral response of the population around the P_{50} point

An example of the high quality of translational research obtained using the RSCEM is demonstrated in Fig. 1.1. Figure 1.1 shows a pair of heterogeneous population response curves from vehicle and tPA-treated rabbits. tPA results clot lysis, and significant behavioral improvement when administered 60 min following embolization.

The use of a clinical rating score is a desirable primary end point to use when a novel therapeutic is being tested for further development and to support a clinical trial [430, 431]. Clinical scores in combination with quantal analysis is a sophisticated statistical analysis method to determine how a large population of stroke "patients," in this case, rabbits, will respond to a treatment. To evaluate the quantitative relationship between clot dose lodged in the brain and behavioral deficits or clinical scores, logistic sigmoidal (S-shaped) quantal analysis curves were fit to dose–response data as originally described by Waud [433]. To construct a quantal analysis curve, a wide range of clot doses were injected via the indwelling carotid catheter in order to produce a spectrum of behaviorally normal to abnormal animals, which included death on the continuum of embolization-induced effects. In the absence of a neuroprotective treatment regimen, small numbers of microclots lodged in the brain vasculature cause no grossly apparent neurologic dysfunction. However, when large numbers of microclots become lodged in the vasculature, they

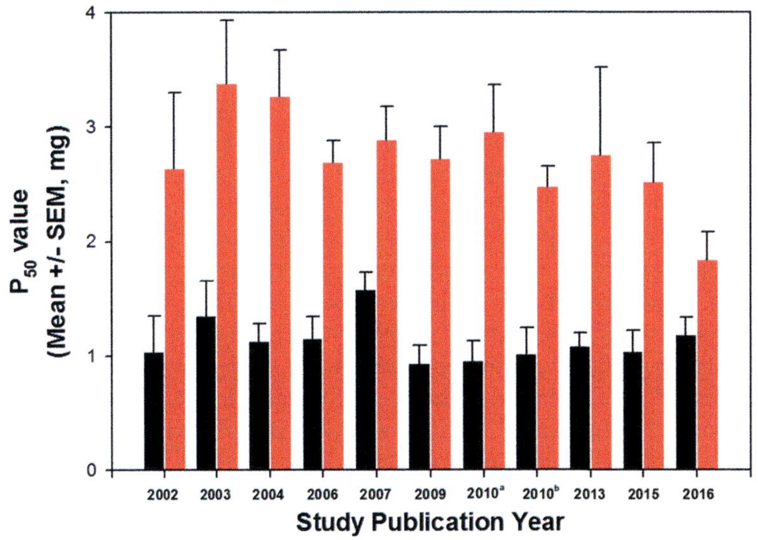

Fig. 1.2 Effect of standard dose tPA on behavioral outcome in embolized rabbits: a historical perspective. The graph documents the P_{50} value of vehicle-treated rabbits compared to standard dose tPA-treated rabbits for neuroprotection studies published in the peer-reviewed literature between 2002 and 2016. References: 2002 [399]; 2003 [438]; 2004 [411]; 2006 [439]; 2007 [440]; 2009 [403]; 2010a [441]; 2010b [442]; 2013 [443]; 2015 [444]; 2016 [435]

invariably caused encephalopathy due to ischemia, neuronal degeneration, depletion of ATP, and cell death [434, 435]. A three-tiered clinical scoring scale is used for analysis of embolized rabbits: normal, abnormal, or dead. Embolized rabbits are scored as abnormal if they have one or more of the following symptoms: ataxia, leaning, circling, lethargy, nystagmus, loss of balance, loss of limb/facial sensation and occasionally, hind-limb paraplegia. Using a simple dichotomous rating system, with a reproducible composite result and low interrater variability (<5 %), each surviving animal is rated as either behaviorally normal or abnormal by an observer naïve to the study design and/or treatment groups. For construction of quantal analysis curves, *clot burden* is plotted against *clinical function* (behavior) in order to define the P_{50} value (in mg). A separate quantal curve was generated for each treatment condition and a statistically significant increase in the P_{50} value (or the amount of blood clots in brain that produce neurologic dysfunction in 50 % of a group of animals) compared to a control curve is indicative of neuroprotection or a behavioral improvement in the study population.

Figure 1.2 is a graphic compilation of historical data demonstrating the reproducibility of tPA response in the RSECM. Data was mined from Lapchak et al. articles from 2000 to 2016. In the original landmark Science article from Zivin and colleagues, there was significant behavioral improvement when tPA was administered 45 min after an embolic stroke; behavioral improvement correlated with a 67.1 % increase in ED_{50}, now commonly referred to as a P_{50} value in all subsequent articles using the quantal analysis technique. However, in the original study, 100's of mg of small-sized blood clots were injected to produce an embolic stroke in catheterized rabbits. Today, the model has been refined, and only mg levels of blood clots produce the full spectrum of behavioral deficit in embolized rabbits.

As noted in Fig. 1.2, there is some level of inherent variability in both the vehicle baseline values as well as the tPA-response values, and this is common in the model, as it is in all stroke clinical trials. The RSCEM is a model built on assessing the response of a population of rabbits with embolism-induced stroke, and unlike most animal models being used for research purposes, all embolized animals are included in final quantal analysis, without exclusions to select specific stroke deficits or magnitude of stroke deficits or to tighten data sets. Thus, data resulting from blinded-randomized testing in the RSCEM is a true reflection of population response to a novel therapeutic, a critical step in the drug development process.

There is still no consensus regarding the optimal path to achieve success with a neuroprotective agent, which is evidenced by the path of recent clinical trials that have been initiated, which have foregone adequate testing in species other than small rodents [293, 294], and did not fulfill minimum STAIR criteria. Moreover, with a high level of uncertainty regarding the development of a neuroprotectant, even funding organizations have faltered in direction, and have proposed the use of few animals to develop a neuroprotective strategy. The divergence from a standardized systematic approach to develop an effective stroke therapy is quite disconcerting!

To aid the stroke researcher with development challenges, this section of the chapter will deal with possible guidelines that should be considered when designing translational studies. Table 1.12 is a simplified checklist and development plan based upon recommendations that have been published in the literature [106, 108, 445–448].

3.3 Meta-analysis of Translational Studies

While this topic will be covered in another chapter in this volume by Howells and colleagues, it is important to briefly survey some of the historical findings of the Collaborative Approach to Meta-Analysis and Review of Animal Data from Experimental Studies (CAMARADES) project [449, 450]. The authors have delved into possible reasons for the failure of neuroprotectants in clinical trials by investigating the quality of the preclinical studies that were conducted to "support" initiation of a clinical trial.

- Numerous meta-analysis publications indicate that investigators report larger measured efficacy effects or improvement on behavioral or histological end points when the study was not randomized or when study end points were assessed by nonnaïve raters. For example, between 2004 and 2011, STAIR continued to report on recommendations for developing neuroprotective therapies, expanding treatment options, utilizing combination therapies, and designing

Table 1.12 Translational study design and requirement characteristics

A) Species Us ed in Investigation	
1) Small animals	
Mice (common: C57BL/6J)	
Gerbil	
Rat (Sprague-Dawley; Wistar; Fisher 344)	
1.1) Large animals	
Dog	
Cat	
Rabbit (common: New Zealand white)	
Swine (common: white Yorkshire swine)	
Non-human primate lissencephalic and gyrencephalic (common: Common marmoset; cynomologous; rhesus macaque)	
1.2) Gender Analysis	
Male	
Female	
1.3) Co-morbidities	
Age (3 ages required for an aging study; i.e. 3 month, 12 month and 24 month)	
Diabetes (diet and/or streptozotocin-induced Type 2 Type 1	
Hypertension (genetic or induced)	
B) Experimental quality	
Study Randomization	
Surgeon naïve to study details	
Assessment of outcome by naïve raters	
Temperature controlled (if required to maintain homeostasis)	
Conflict of interest statement	
Funding statement (Government, industry, donation, private)	
Ethics approval (Formal review and approval of animal protocol)	
Power Analysis: Sample-size calculation	
Pre-defined exclusion criteria	
Identified Statistical Analysis (ANOVA, Bonferonni correction)	
D) Co-administration with Standard-of-Care Therapy	
Thrombolysis (rt-PA, urokinase, streptokinase)	
Anti-hypertensive i.e.: Angiotensin converting enzyme (ACE) inhibitors, Angiotensin II Receptor Blockers (ARBs), calcium channel antagonist, β-adrenergic receptor [BAR] antagonist (blocker)	
Anti-diabetic i.e.: Metformin, insulin	

clinical trials [445–448]. The recommendations attempted to address the problems resulting from the devastating failure of NXY-059 [40]. While the development of NXY-059 was deemed as an acceptable, but not optimal plan [394, 396, 451–453] according to original STAIR criteria, retrospective analysis showed that the efficacy of was greatly overestimated in preclinical studies [394]. Thus, even though these new criteria were put into place, subsequent clinical trials, relying on this more stringent criteria, failed to produce any benefit?

- Specific examples that can be applied to translational stroke research demonstrate that only 36 % of published studies reported randomization and only 29 % of published studies were blinded.
- Moreover, prior to the institution of NINDS transparency guidelines [106–108], there was the absence of power analysis for studies, and there was inadequate statistical analyses for many published studies. The guidelines are now being considered by most authors, and in most articles there is now a comprehensive statistical analysis section.
- Despite guidelines to publish "negative" data [454], there also appears to be significant amounts of unpublished negative or neutral data in the hands of researchers, while only positive data is published. One of the many reasons for "negative" data is the observation that many researches do not conduct dose–response analysis in their model of choice. In many published studies, only a single dose is tested; this is insufficient analysis for translational purposes! Suppression of negative data causes an overestimation of drug efficacy and provides erroneous support for expensive clinical trials. There is still some apprehension by Editors when negative study data is presented without adequate control (negative and positive).
- Importantly, there was no significant association ($r^2 = 0.06$) between the quality of science being published in journals and the impact factor of the journal.
- There was also no significant association ($r^2 = 0.004$) between the quality of science and the number of citations of a particular study article.

Based upon the findings cited earlier, recommendations for effective translational research, which should be routinely incorporated into translational research programs are proposed herein. The following recommendations should ensure reproducible, rigorous research findings (Table 1.13).

- The investigator should provide sufficient rationale for the model selection and end points being measured.
- The studies should incorporate justification of sample size, including complete power analysis calculations for primary end points. Power analysis for secondary end points may be of little benefit.
- An adequate number of control groups using an appropriate route of administration and timing of intervention delivery should be incorporated into the study to reflect eventual use in the patient population (i.e., oral, intravenous, acute, chronic).

Table 1.13 Drug development

Drug Development Profiles	Selected References
Chemical Profile for CNS active drugs Blood Brain Barrier Penetrating Drugs	(471-473)
BBB Penetration Profile (MDCK analysis in vitro)	(474)
BBB Penetration Profile (Multiple species in normal animals and animals with stroke)	(470)
Dose-Response Analysis in Multiple Species (Required for estimating dosing in clinical trials-human equivalent dosing) Dose-extrapolation: Allometric-Body Surface Area.	(475)
Therapeutic window Analysis (Clinically relevant in humans) (See NINDS rt-PA and endovascular procedure trials)	(30-35)
Standard ADME (minimal half-life, distribution, secretion)	(476)
hERG inhibitory activity	(477-481)
CP450 inhibition analysis	(482-484)
Toxicity screen (in vitro CeeTox analysis)	(414, 415)
Toxicity screen (in vivo-2 species)	(485, 486)

- The studies should be fully randomized and blinded with respect to surgical interventions, drug treatments, in vivo analysis, and postmortem biochemical and histochemical analysis.
- The statistical analysis for results interpretation should be consistent with study design. For example, if multiple groups are being compared, ANOVA with a post hoc test incorporating the Bonferroni correction may be required.

4 Comorbidities and Models

There have been STAIR recommendations to address comorbidities that are common in stroke patients, and suggestions that some comorbidities be included in clinically relevant translational stroke studies. The long-standing recommendations are population based and are centered around the idea that studies in animal models with comorbidities such as hypertension, diabetes, and hypercholesterolemia would better reproduce the "pathophysiological" state of some portion of patients presenting with strokes [see, for example, patient enrollment in [31–36]].

Using the NINDS rt-PA, ECASSIII and recent endovascular trials as examples, the stroke patient population had the characteristics described in Table 1.14. The Table was populated with data when the investigators reported the data either in the primary article or supplemental article information. The most pertinent aspects are that strokes occur in a mixed gender aged population, particularly those with a history of hypertension or diabetes, which is usually controlled by one or more pharmaceuticals. Moreover, the majority of stroke patients are not antihypertensive naive [455, 456], they may receive antihypertensive agents, anticoagulants, and statins [31–36, 455], but the studies do not explicitly include the mode of regulation of hypertension or diabetes? These points are important when considerations are made regarding the use of an appropriate animal model for drug development.

Table 1.14 Clinical trial patient characteristics

Characteristic	NINDS rt-PA(30)	ECASS III (45)	Endovascular Trials (31-36)				
			MR CLEAN	SWIFT PRIME	ESCAPE	EXTEND-IA	REVASCAT
Age (years)	>66	>64.9	54.5-76	65 ± 12.5	60-81	68.6 ± 12.3	65.7 ± 11.3
Gender (male)	57-60%	57.3-63.2%	58-67%	52.4-53.4%	47.3-47.9%	49%	52.4-53.4%
Mean NIHSS	14-15 (median)	10.7-11.6	14-21	13-20	13-20	13-20	14-20
Hypertension	64-66%	62.4-62.8%	42.1-48.3%	58-67%	63.6-72%	60-66%	60.2-69.9%
Diabetes	20-24%	14.8-16.6%	12.7-14.6%	12-15%	20-26%	6-23%	18.4-21.4%
Prior use of drugs:							
1) aspirin/antiplatelet	26-41%	31.1-32.5%	27.5-30.0%	–	–	–	25.2-30.1%
2) anticoagulant	–	–	7.7-7.9%	–	–	–	15.5-22.3%
3) tobacco (current)	37-43%	28.8-30.6%	28.9-31.0%	–	48.8-49.3%	34-43%	22.5-25.2%
4) tobacco (Ex-smoker)	–	20.6-24.6%	–	42-43%	–	–	–
5) anti-hypertensives	–	–	–	–	–	–	–
6) anti-diabetics	–	–	–	–	–	–	
7) statin	–	–	27.9-29.2%	–	–	–	40.8-46.6%
8) hyperlipidemia	–	–	24.9-26.6%	22.7-24.5%	35.2-44.5%	–	52.4-60.2%

4.1 Hypertension Models

The exact hypertension model (s) to be used to screen and develop new therapeutic strategies have become quite controversial, in part due to the "limited value" of conducting studies in models that are not representative of hypertension in humans, and the exorbitant cost of conducting comorbidity analysis studies [457]. It is becoming apparent that commonly used inbred spontaneously hypertensive rats (SHR) and spontaneously hypertensive stroke prone (SHRSP) are inferior choices for a translational stroke model due to the unfortunate pathophysiological characteristic that the animals have little to no penumbra following an ischemic event [458, 459], and infarcts are larger. Thus, if the penumbra is indeed the target for neuroprotection, there is little substrate available. This is exemplified by numerous reports in the literature, including the study by McCabe and colleagues [460, 461] and Letourneur et al. [458]. In the studies, the authors showed that rodents with either genetic [i.e., spontaneously hypertensive rats (SHR), spontaneously hypertensive stroke-prone (SHRSP)] have smaller salvageable penumbra and larger ischemic cores compared to control Wistar–Kyoto (WKY) rats. One might suggest an induced hypertension model to be more representative of slowly developing hypertension observed in humans, but a recent article [458] suggests that renal-clamp-induced hypertensive rats [i.e., renovascular chronic arterial hypertension (CAH)] will also lead to pathophysiology which limits penumbra (see also [459]).

- Is there any evidence that hypertensive patients have altered pathophysiological consequences of a stroke due to increased stroke volume and low APECTS compared to nonhypertensive stroke patients?

4.2 Diabetic Models

There is also significant information available in the literature regarding the group of refractory diabetic patients enrolled in the NINDS rt-PA trial [30] and the ECASS III trial [45], which includes a small percentage of patients with diabetes (Table 1.14). It is extremely difficult to treat diabetic stroke patients since they do not respond to rt-PA [51] or have an attenuated response to standard dose thrombolytic therapy [462–464]. Patients with diabetes have been shown to be independently associated with poor neurological outcome and higher mortality in the absence of thrombolytic treatment [462–464] and among patients treated with intravenous tPA [30], the presence of diabetes significantly reduces the odds of favorable outcome at 3 months. It is important to note that glucose levels in stroke patients are important to regulate since recent studies have shown that elevated glucose levels differentially affects outcome in diabetic and nondiabetic stroke patients [465]. Importantly, the authors demonstrated that nondiabetic patients with hyperglycemia had significantly increased infarct growth compared to nonhyperglycemic patients; importantly in diabetics, there was no significant difference between infarct size in the absence or presence of hyperglycemia. Thus, the glucose state of all patients must be closely monitored during the therapy development process. These points are critical to the development of a translational research program, when an investigator must utilize the animal model most representative of the disease condition for drug development.

4.2.1 Coadministration of Drugs in Stroke Models

Based upon the recent endovascular trials, the stroke patient population included in the trials were a mixed gender aged population, had a history of hypertension or diabetes, which may have been controlled by one or more prescription drug or other pharmaceutical. Basic researchers have now begun to address this new avenue of drug coadministration. For example, the Fagan group has discussed the treatment of type 2 diabetic Goto–Kakizaki (GK) rats [378, 466–468] to restore "normal" cellular signaling mechanisms [466, 469]. Considering the information provided earlier and the refractory phenomenon in diabetic patients, will using a standard naive hypertensive rodent be appropriate or sufficient to predict drug efficacy in a heterogeneous population of stroke patients?

- Should translational studies attempt to address the diabetic population presenting with a stroke, or should "proof of concept" efficacy first be obtained for the larger mixed gender aged population, leaving the quest for efficacy in diabetics for post drug approval?

Considering the extensive population information discussed earlier, the stroke research community will need to address animal models and determine which animal model (s) best reproduce the embolic stroke target population. Will using a standard "drug" naive hypertensive rodent be sufficient to predict drug efficacy in a heterogeneous population of stroke patients? There is no clear answer other than translational studies should incorporate aged mixed gender species for preliminary therapy investigation studies.

5 Drug Development Considerations

In addition to current translational research guidelines (i.e., STAIR and RIGOR), standard industry drug development guidelines should be considered by researchers interested in applying their research to developing a drug through to fruition, a clinical trial end point. The development of CNS-active drugs to treat stroke require special attention since they must be able to cross the blood–brain barrier (BBB) to penetrate into the penumbra. This can be taken into consideration when developing molecules using the Lipinski rules as well as utilizing BBB penetration assays, first in vitro for candidate selection and then in vivo [470–472]. Table 1.13 provides references for many useful drug development profiling tools.

6 Conclusion

In conclusion, while many neuroprotective strategies have been assessed in animals and humans, to date, none have been FDA approved. With significant efficacy of embolectomy alone and in combination with tPA, a select few previously tested strategies should be reconsidered for trials as well as novel new strategies that are currently being developed. To promote good laboratory practices, the RIGOR guidelines discussed earlier, most importantly, method of blinding, study group randomization, complete Power analysis, and statistical analysis should be incorporated into translational research programs. Science should be conducted at the highest possible level with full confidence that the path forward will eventually allow for the effective treatment of stroke victims. Randomized, blinded, controlled clinical trials should not be initiated for any compound or device until the new neuroprotective strategies are thoroughly investigated in multiple species, including a rodent, and one or more large animal species models representative of the target stroke population. This strategy will allow investigators to derisk the drug development process, reduce the continuing trend for failure in stroke victims, and provide some level of patient satisfaction and benefit so they are not blindsided by the culmination of poorly conducted or inadequately conducted science when they consent to be the final test subject for a new therapy. *Ethics in translational science and clinical trials must guide the process, not Hubris.*

Acknowledgments and Funding This article was written without financial support from government sources.

Conflict of Interest Statements There are no conflicts of interest to disclose.

References

1. Poirier J, Derouesné D (1993) VIII.9: apoplexy and stroke. The Cambridge World History of Human Disease. 10.1017/CHOL9780521332866.071
2. Wepfer J (1658) Observationes anatomicae ex cadaveribus eorum quos sustulit apoplexia (cum exercitatione de eious loco affecto). J C Suteri, Schaffussii, p. 670
3. Gurdjian ES, Gurdjian ES (1979) History of occlusive cerebrovascular disease I. From Wepfer to Moniz. Arch Neurol 36(6):340–343
4. Willis T (1664) Cerebri Anatome: cui accessit nervorum descriptio et usus. London
5. Virchow R (1856) Thrombose und Embolie: Gefässenzýndung und Septische Infektion, in Gesammelte Abhandlungen zur wissen schaftlichen Medicin: Frankfurt, a. M., Meidinger, Sohn und Co
6. Dechambre A (1866) Dictionnaire Encyclopédique des Sciences Médicales. Paris, France
7. Bramwell B (1886) Spontaneous meningeal haemorrhage. Edinb Med J 32:101
8. Symonds CP (1924) Spontaneous sub-arachnoid haemorrhage. Proc R Soc Med 17(Neurol Sect):39–52
9. Lloyd-Jones D, Adams R, Carnethon M, De Simone G, Ferguson TB, Flegal K et al (2009) Heart disease and stroke statistics—2009 update: a report from the American Heart Association Statistics Committee and Stroke Statistics Subcommittee. Circulation 119(3):480–486
10. Lloyd-Jones D, Adams RJ, Brown TM, Carnethon M, Dai S, De Simone G et al (2010) Executive summary: heart disease and stroke statistics—2010 update: a report from the American Heart Association. Circulation 121(7):948–954
11. Lloyd-Jones D, Adams RJ, Brown TM, Carnethon M, Dai S, De Simone G et al (2010) Heart disease and stroke statistics—2010 update: a report from the American Heart Association. Circulation 121(7):e46–e215
12. Lapchak PA (2002) Hemorrhagic transformation following ischemic stroke: significance, causes, and relationship to therapy and treatment. Curr Neurol Neurosci Rep 2(1):38–43
13. Lyden PD, Zivin JA (1993) Hemorrhagic transformation after cerebral ischemia: mechanisms and incidence. Cerebrovasc Brain Metab Rev 5(1):1–16
14. Bernstein RA, Del-Signore M (2005) Recent advances in the management of acute intracerebral hemorrhage. Curr Neurol Neurosci Rep 5(6):483–487
15. van Gijn J, Kerr RS, Rinkel GJ (2007) Subarachnoid haemorrhage. Lancet 369(9558): 306–318
16. Freeman WD, Karnatovskaia LV, Vadeboncoeur TF (2014) Therapeutic hypothermia and targeted temperature management after cardiac arrest. JAMA Neurol 71(12):1577–1578
17. Karnatovskaia LV, Wartenberg KE, Freeman WD (2014) Therapeutic hypothermia for neuroprotection: history, mechanisms, risks, and clinical applications. Neurohospitalist 4(3):153–163
18. Hall R, Murdoch J (1990) Brain protection: physiological and pharmacological considerations. Part II: the pharmacology of brain protection. Can J Anaesth 37(7):762–777
19. Murdoch J, Hall R (1990) Brain protection: physiological and pharmacological considerations. Part I: the physiology of brain injury. Can J Anaesth 37(6):663–671
20. Lok J, Wang XS, Xing CH, Maki TK, Wu LM, Guo SZ et al (2015) Targeting the neurovascular unit in brain trauma. CNS Neurosci Ther 21(4):304–308
21. Maki T, Hayakawa K, Pham LD, Xing C, Lo EH, Arai K (2013) Biphasic mechanisms of neurovascular unit injury and protection in CNS diseases. CNS Neurol Disord Drug Targets 12(3):302–315
22. Lo EH, Rosenberg GA (2009) The neurovascular unit in health and disease: introduction. Stroke 40(3 suppl):S2–S3
23. Lok J, Gupta P, Guo S, Kim WJ, Whalen MJ, van Leyen K et al (2007) Cell-cell signaling in the neurovascular unit. Neurochem Res 32(12):2032–2045

24. Mozaffarian D, Benjamin EJ, Go AS, Arnett DK, Blaha MJ, Cushman M et al (2015) Heart disease and stroke statistics—2015 update: a report from the American Heart Association. Circulation 131(4):e29–e322
25. Sacco RL, Kasner SE, Broderick JP, Caplan LR, Connors JJ, Culebras A et al (2013) An updated definition of stroke for the 21st century: a statement for healthcare professionals from the American Heart Association/American Stroke Association. Stroke 44(7):2064–2089
26. Worldometer (2016) http://www.worldometers.info/world-population/
27. Saver JL (2006) Time is brain—quantified. Stroke 37(1):263–266
28. Holscher T, Dunford JV, Schlachetzki F, Boy S, Hemmen T, Meyer BC et al (2013) Prehospital stroke diagnosis and treatment in ambulances and helicopters-a concept paper. Am J Emerg Med 31(4):743–747
29. Lapchak PA (2013) Fast neuroprotection (fast-NPRX) for acute ischemic stroke victims: the time for treatment is now. Transl Stroke Res 4(6):704–709
30. NINDS (1995) Tissue plasminogen activator for acute ischemic stroke. The National Institute of Neurological Disorders and Stroke rt-PA Stroke Study Group. N Engl J Med 333(24):1581–1587
31. Berkhemer OAFP, Beumer D et al (2015) A randomized trial of intraarterial treatment for acute ischemic stroke. N Engl J Med 372(1):11–20
32. Goyal M, Demchuk AM, Menon BK, Eesa M, Rempel JL, Thornton J et al (2015) Randomized assessment of rapid endovascular treatment of ischemic stroke. N Engl J Med 372(11):1019–1030
33. Campbell BC, Mitchell PJ, Kleinig TJ, Dewey HM, Churilov L, Yassi N et al (2015) Endovascular therapy for ischemic stroke with perfusion-imaging selection. N Engl J Med 372(11):1009–1018
34. Jovin TG, Chamorro A, Cobo E, de Miquel MA, Molina CA, Rovira A et al (2015) Thrombectomy within 8 hours after symptom onset in ischemic stroke. N Engl J Med 372(24):2296–2306
35. Saver JL, Goyal M, Bonafe A, Diener HC, Levy EI, Pereira VM et al (2015) Stent-retriever thrombectomy after intravenous t-PA vs. t-PA alone in stroke. N Engl J Med 372(24):2285–2295
36. Rebello LC, Haussen DC, Grossberg JA, Belagaje S, Lima A, Anderson A et al (2016) Early endovascular treatment in intravenous tissue plasminogen activator-ineligible patients. Stroke 47(4):1131–1134
37. Lahiri S, Nuno M, Lyden PD (2015) Letter by Lahiri et al Regarding Article, "Endovascular thrombectomy for anterior circulation stroke: systematic review and meta-analysis". Stroke 46(12), e258
38. Saver JL, Starkman S, Eckstein M, Stratton SJ, Pratt FD, Hamilton S et al (2015) Prehospital use of magnesium sulfate as neuroprotection in acute stroke. N Engl J Med 372(6):528–536
39. Lees KR, Zivin JA, Ashwood T, Davalos A, Davis SM, Diener HC et al (2006) NXY-059 for acute ischemic stroke. N Engl J Med 354(6):588–600
40. Shuaib A, Lees KR, Lyden P, Grotta J, Davalos A, Davis SM et al (2007) NXY-059 for the treatment of acute ischemic stroke. N Engl J Med 357(6):562–571
41. Kasner SE, Rose DZ, Skokan A, Walker MG, Shi J, Streeter J et al (2013) Transcranial laser therapy and infarct volume. Stroke 44(7):2025–2027
42. Zivin JA, Albers GW, Bornstein N, Chippendale T, Dahlof B, Devlin T et al (2009) Effectiveness and safety of transcranial laser therapy for acute ischemic stroke. Stroke 40(4):1359–1364
43. Lampl Y, Zivin JA, Fisher M, Lew R, Welin L, Dahlof B et al (2007) Infrared laser therapy for ischemic stroke: a new treatment strategy: results of the NeuroThera Effectiveness and Safety Trial-1 (NEST-1). Stroke 38(6):1843–1849
44. Hacke W, Schellinger PD, Albers GW, Bornstein NM, Dahlof BL, Fulton R et al (2014) Transcranial laser therapy in acute stroke treatment: results of neurothera effectiveness and safety trial 3, a phase III clinical end point device trial. Stroke 45(11):3187–3193

45. Hacke W, Kaste M, Bluhmki E, Brozman M, Davalos A, Guidetti D et al (2008) Thrombolysis with alteplase 3 to 4.5 hours after acute ischemic stroke. N Engl J Med 359(13):1317–1329
46. Lansberg MG, Bluhmki E, Thijs VN (2009) Efficacy and safety of tissue plasminogen activator 3 to 4.5 hours after acute ischemic stroke: a metaanalysis. Stroke 40(7):2438–2441
47. Fang MC, Cutler DM, Rosen AB (2010) Trends in thrombolytic use for ischemic stroke in the United States. J Hosp Med 5(7):406–409
48. Reeves MJ, Arora S, Broderick JP, Frankel M, Heinrich JP, Hickenbottom S et al (2005) Acute stroke care in the US: results from 4 pilot prototypes of the Paul Coverdell National Acute Stroke Registry. Stroke 36(6):1232–1240
49. Schwamm LH, Ali SF, Reeves MJ, Smith EE, Saver JL, Messe S et al (2013) Temporal trends in patient characteristics and treatment with intravenous thrombolysis among acute ischemic stroke patients at get with the guidelines-stroke hospitals. Circ Cardiovasc Qual Outcomes 6(5):543–549
50. Clua-Espuny JL, Ripolles-Vicente R, Forcadell-Arenas T, Gil-Guillen VF, Queralt-Tomas ML, Gonzalez-Henares MA et al (2015) Sex differences in long-term survival after a first stroke with intravenous thrombolysis: ebrictus study. Cerebrovasc Dis Extra 5(3):95–102
51. Boudreau DM, Guzauskas G, Villa KF, Fagan SC, Veenstra DL (2012) A model of cost-effectiveness of tissue plasminogen activator in patient subgroups 3 to 4.5 hours after onset of acute ischemic stroke. Ann Emerg Med 61:46–55
52. Lees KR, Bluhmki E, von Kummer R, Brott TG, Toni D, Grotta JC et al (2010) Time to treatment with intravenous alteplase and outcome in stroke: an updated pooled analysis of ECASS, ATLANTIS, NINDS, and EPITHET trials. Lancet 375(9727):1695–1703
53. Kim YH, Park JH, Hong SH, Koh JY (1999) Nonproteolytic neuroprotection by human recombinant tissue plasminogen activator. Science 284(5414):647–650
54. Wu F, Torre E, Cuellar-Giraldo D, Cheng L, Yi H, Bichler EK et al (2015) Tissue-type plasminogen activator triggers the synaptic vesicle cycle in cerebral cortical neurons. J Cereb Blood Flow Metab 35(12):1966–1976
55. Lopez-Atalaya JP, Roussel BD, Levrat D, Parcq J, Nicole O, Hommet Y et al (2008) Toward safer thrombolytic agents in stroke: molecular requirements for NMDA receptor-mediated neurotoxicity. J Cereb Blood Flow Metab 28(6):1212–1221
56. Desai JA, Smith EE (2013) Prenotification and other factors involved in rapid tPA administration. Curr Atheroscler Rep 15(7):337
57. Olson DM, Constable M, Britz GW, Lin CB, Zimmer LO, Schwamm LH et al (2011) A qualitative assessment of practices associated with shorter door-to-needle time for thrombolytic therapy in acute ischemic stroke. J Neurosci Nurs 43(6):329–336
58. Fonarow GC, Smith EE, Saver JL, Reeves MJ, Hernandez AF, Peterson ED et al (2011) Improving door-to-needle times in acute ischemic stroke: the design and rationale for the American Heart Association/American Stroke Association's Target: stroke initiative. Stroke 42(10):2983–2989
59. Wardlaw JM, Koumellis P, Liu M (2013) Thrombolysis (different doses, routes of administration and agents) for acute ischaemic stroke. Cochrane Database Syst Rev (5):CD000514
60. Bluhmki E, Chamorro A, Davalos A, Machnig T, Sauce C, Wahlgren N et al (2009) Stroke treatment with alteplase given 3.0-4.5 h after onset of acute ischaemic stroke (ECASS III): additional outcomes and subgroup analysis of a randomised controlled trial. Lancet Neurol 8(12):1095–1102
61. Cronin CA, Smith EE (2015) Response to letter regarding article, "Adherence to third European cooperative acute stroke study 3- to 4.5-hour exclusions and association with outcome: data from get with the guidelines-stroke". Stroke 46(1), e16
62. Cronin CA, Sheth KN, Zhao X, Messe SR, Olson DM, Hernandez AF et al (2014) Adherence to Third European Cooperative Acute Stroke Study 3- to 4.5-hour exclusions and association with outcome: data from Get with the Guidelines-Stroke. Stroke 45(9):2745–2749
63. Emberson J, Lees KR, Lyden P, Blackwell L, Albers G, Bluhmki E et al (2014) Effect of treatment delay, age, and stroke severity on the effects of intravenous thrombolysis with

alteplase for acute ischaemic stroke: a meta-analysis of individual patient data from randomised trials. Lancet 384(9958):1929–1935
64. Ricci S, Cenciarelli S, Mazzoli T (2013) Italian guidelines on thrombolysis indications in ischaemic stroke have been revised after IST 3 trial and Cochrane Review: PROS. Intern Emerg Med 8(8):653–654
65. Costantino G, Podda GM, Bonzi M, Sbrojavacca R, Gruppo di Autoformazione Metodologica (GrAM) (2013) Italian guidelines on thrombolysis indications in ischemic stroke have been revised after IST-3 trial and Cochrane revision: cons. Intern Emerg Med 8(8):651–652
66. Saver JL (2011) Improving reperfusion therapy for acute ischaemic stroke. J Thromb Haemost 9(suppl 1):333–343
67. Saver JL, Fonarow GC, Smith EE, Reeves MJ, Grau-Sepulveda MV, Pan W et al (2013) Time to treatment with intravenous tissue plasminogen activator and outcome from acute ischemic stroke. JAMA 309(23):2480–2488
68. Saver JL, Smith EE, Fonarow GC, Reeves MJ, Zhao X, Olson DM et al (2010) The "golden hour" and acute brain ischemia: presenting features and lytic therapy in >30,000 patients arriving within 60 minutes of stroke onset. Stroke 41(7):1431–1439
69. Lapchak PA (2015) Critical early thrombolytic & endovascular reperfusion therapy for acute ischemic stroke victims: a call for adjunct neuroprotection. Transl Stroke Res 6:345–354. doi:10.1007/s12975-015-0419-5
70. Brinjikji W, Rabinstein AA, Kallmes DF, Cloft HJ (2011) Patient outcomes with endovascular embolectomy therapy for acute ischemic stroke: a study of the national inpatient sample: 2006 to 2008. Stroke 42(6):1648–1652
71. Singh B, Parsaik AK, Prokop LJ, Mittal MK (2013) Endovascular therapy for acute ischemic stroke: a systematic review and meta-analysis. Mayo Clin Proc 88(10):1056–1065
72. Nogueira RG, Yoo AJ, Buonanno FS, Hirsch JA (2009) Endovascular approaches to acute stroke, part 2: a comprehensive review of studies and trials. AJNR Am J Neuroradiol 30(5):859–875
73. Osanai T, Pasupuleti V, Deshpande A, Thota P, Roman Y, Hernandez AV et al (2015) Acute endovascular reperfusion therapy in ischemic stroke: a systematic review and meta-analysis of randomized controlled trials. PLoS One 10(4), e0122806
74. Punal-Rioboo J, Atienza G, Blanco M (2015) Safety and efficacy of mechanical thrombectomy using stent retrievers in the endovascular treatment of acute ischaemic stroke: a systematic review. Interv Neurol 3(3–4):149–164
75. Rouchaud A, Mazighi M, Labreuche J, Meseguer E, Serfaty JM, Laissy JP et al (2011) Outcomes of mechanical endovascular therapy for acute ischemic stroke: a clinical registry study and systematic review. Stroke 42(5):1289–1294
76. Tomsick T (2007) TIMI, TIBI, TICI: I came, I saw, I got confused. AJNR Am J Neuroradiol 28(2):382–384
77. Kidwell CSJR, Alger JR (2014) Design and rationale of the mechanical retrieval and recanalization of stroke clots using embolectomy (MR RESCUE) trial. Int J Stroke 9(1):110–116
78. Ciccone AVL, Nichelatti M (2013) Endovascular treatment for acute ischemic stroke. N Engl J Med 368(10):904–913
79. Broderick JPPY, Demchuk AM (2013) Endovascular therapy after intravenous t-PA versus t-PA alone for stroke. N Engl J Med 368(10):893–903
80. Goyal M, Menon BK, van Zwam WH, Dippel DW, Mitchell PJ, Demchuk AM et al (2016) Endovascular thrombectomy after large-vessel ischaemic stroke: a meta-analysis of individual patient data from five randomised trials. Lancet 387(10029):1723–1731
81. Powers WJ, Derdeyn CP, Biller J, Coffey CS, Hoh BL, Jauch EC et al (2015) 2015 American Heart Association/American Stroke Association focused update of the 2013 guidelines for the early management of patients with acute ischemic stroke regarding endovascular treatment: a guideline for healthcare professionals from the American Heart Association/American Stroke Association. Stroke 46(10):3020–3035
82. Paciaroni M (2014) Italian guidelines on thrombolysis indications in ischemic stroke have been revised after the IST-3 trial and Cochrane review. Intern Emerg Med 9(7):823–824

83. Xu J, Zhang Y, Wei H, Xu Y, Wang M, Cai Z et al (2015) A comparison of rt-PA thrombolysis guidelines between China and the USA: are changes needed? Neurol Res 37(1):57–63
84. Jauch EC, Saver JL, Adams HP Jr, Bruno A, Connors JJ, Demaerschalk BM et al (2013) Guidelines for the early management of patients with acute ischemic stroke: a guideline for healthcare professionals from the American Heart Association/American Stroke Association. Stroke 44(3):870–947
85. Albers GW, Bates VE, Clark WM, Bell R, Verro P, Hamilton SA (2000) Intravenous tissue-type plasminogen activator for treatment of acute stroke: the Standard Treatment with Alteplase to Reverse Stroke (STARS) study. JAMA 283(9):1145–1150
86. Alberts MJ (1998) tPA in acute ischemic stroke: United States experience and issues for the future. Neurology 51(3 suppl 3):S53–S55
87. Christou I, Alexandrov AV, Burgin WS, Wojner AW, Felberg RA, Malkoff M et al (2000) Timing of recanalization after tissue plasminogen activator therapy determined by transcranial doppler correlates with clinical recovery from ischemic stroke. Stroke 31(8): 1812–1816
88. Clark WM, Albers GW, Madden KP, Hamilton S (2000) The rtPA (alteplase) 0- to 6-hour acute stroke trial, part A (A0276g): results of a double-blind, placebo-controlled, multicenter study. Thrombolytic therapy in acute ischemic stroke study investigators. Stroke 31(4):811–816
89. Grotta JC, Alexandrov AV (1998) tPA-associated reperfusion after acute stroke demonstrated by SPECT. Stroke 29(2):429–432
90. Grotta JC, Burgin WS, El-Mitwalli A, Long M, Campbell M, Morgenstern LB et al (2001) Intravenous tissue-type plasminogen activator therapy for ischemic stroke: Houston experience 1996 to 2000. Arch Neurol 58(12):2009–2013
91. Hacke W, Brott T, Caplan L, Meier D, Fieschi C, von Kummer R et al (1999) Thrombolysis in acute ischemic stroke: controlled trials and clinical experience. Neurology 53(7):S3–S14
92. Hacke W, Kaste M, Fieschi C, Toni D, Lesaffre E, von Kummer R et al (1995) Intravenous thrombolysis with recombinant tissue plasminogen activator for acute hemispheric stroke. The European Cooperative Acute Stroke Study (ECASS). JAMA 274(13):1017–1025
93. Lapchak PA (2002) Development of thrombolytic therapy for stroke: a perspective. Expert Opin Investig Drugs 11(11):1623–1632
94. Campbell BC, Donnan GA, Lees KR, Hacke W, Khatri P, Hill MD et al (2015) Endovascular stent thrombectomy: the new standard of care for large vessel ischaemic stroke. Lancet Neurol 14(8):846–854
95. Smith WS, Lev MH, English JD, Camargo EC, Chou M, Johnston SC et al (2009) Significance of large vessel intracranial occlusion causing acute ischemic stroke and TIA. Stroke 40(12):3834–3840
96. Weir NU, Pexman JH, Hill MD, Buchan AM, CASES Investigators (2006) How well does ASPECTS predict the outcome of acute stroke treated with IV tPA? Neurology 67(3):516–518
97. Coutts SB, Lev MH, Eliasziw M, Roccatagliata L, Hill MD, Schwamm LH et al (2004) ASPECTS on CTA source images versus unenhanced CT: added value in predicting final infarct extent and clinical outcome. Stroke 35(11):2472–2476
98. Coutts SB, Hill MD, Demchuk AM, Barber PA, Pexman JH, Buchan AM (2003) ASPECTS reading requires training and experience. Stroke 34(10), e179, author reply e179
99. Pexman JH, Barber PA, Hill MD, Sevick RJ, Demchuk AM, Hudon ME et al (2001) Use of the Alberta Stroke Program Early CT Score (ASPECTS) for assessing CT scans in patients with acute stroke. AJNR Am J Neuroradiol 22(8):1534–1542
100. Klein E (2015) Eloquent brain, ethical challenges: functional brain mapping in neurosurgery. Semin Ultrasound CT MR 36(3):291–295
101. Schlachetzki F, Hoelscher T, Lange M, Kasprzak P (2000) Ultrasound finding of a mobile atheroma in the common carotid artery. Circulation 102(15), E105
102. Saver JL, Starkman S, Eckstein M, Stratton S, Pratt F, Hamilton S et al (2014) Methodology of the Field Administration of Stroke Therapy—Magnesium (FAST-MAG) phase 3 trial: Part 2—prehospital study methods. Int J Stroke 9(2):220–225

103. Saver JL, Kidwell C, Eckstein M, Starkman S, F-MPT Investigators (2004) Prehospital neuroprotective therapy for acute stroke: results of the Field Administration of Stroke Therapy-Magnesium (FAST-MAG) pilot trial. Stroke 35(5):e106–e108
104. Mocco J, Fargen KM, Goyal M, Levy EI, Mitchell PJ, Campbell BC et al (2015) Neurothrombectomy trial results: stroke systems, not just devices, make the difference. Int J Stroke 10(7):990–993
105. STAIR (1999) Recommendations for standards regarding preclinical neuroprotective and restorative drug development. Stroke 30(12):2752–2758
106. Landis SC, Amara SG, Asadullah K, Austin CP, Blumenstein R, Bradley EW et al (2012) A call for transparent reporting to optimize the predictive value of preclinical research. Nature 490(7419):187–191
107. Lapchak PA (2012) Scientific rigor recommendations for optimizing the clinical applicability of translational research. J Neurol Neurophysiol 3, e111
108. Lapchak PA (2013) Recommendations and practices to optimize stroke therapy: developing effective translational research programs. Stroke 44(3):841–843
109. O'Collins VE, Macleod MR, Donnan GA, Horky LL, van der Worp BH, Howells DW (2006) 1,026 experimental treatments in acute stroke. Ann Neurol 59(3):467–477
110. Albers GW, Clark WM, Atkinson RP, Madden K, Data JL, Whitehouse MJ (1999) Dose escalation study of the NMDA glycine-site antagonist licostinel in acute ischemic stroke. Stroke 30(3):508–513
111. Diener HC, AlKhedr A, Busse O, Hacke W, Zingmark PH, Jonsson N et al (2002) Treatment of acute ischaemic stroke with the low-affinity, use-dependent NMDA antagonist AR-R15896AR. A safety and tolerability study. J Neurol 249(5):561–568
112. Lees KR, Dyker AG, Sharma A, Ford GA, Ardron ME, Grosset DG (2001) Tolerability of the low-affinity, use-dependent NMDA antagonist AR-R15896AR in stroke patients: a dose-ranging study. Stroke 32(2):466–472
113. Meythaler JM, Guin-Renfroe S, Brunner RC, Hadley MN (2001) Intrathecal baclofen for spastic hypertonia from stroke. Stroke 32(9):2099–2109
114. Marque P (2015) Study of the effects on motor recovery of early post-stroke spasticity treatment (BacloTox). Available from https://clinicaltrials.gov/ct2/show?term=baclofen+stroke&rank=1. Last accessed 26 May 2016
115. Salturari L (2016) SISTERS: spasticity in stroke study—randomized study (SISTERS) 2016 [ClinicalTrials.gov Identifier: NCT01032239]. Available from https://clinicaltrials.gov/ct2/show/NCT01032239?term=baclofen+stroke&rank=2. Accessed 26 May 2016
116. Meythaler J (2002) BIII-890-CL. Boehringer Ingelheim. Curr Opin Investig Drugs (London, England: 2000) 3(12):1733–1735
117. Ingelheim B (2014) Safety and tolerability of BIII 890 in patients with acute ischemic stroke. Available from https://clinicaltrials.gov/ct2/show/NCT02251197?term=BIII-890-CL&rank=5. Accessed 26 May 2016
118. Jensen BS (2002) BMS-204352: a potassium channel opener developed for the treatment of stroke. CNS Drug Rev 8(4):353–360
119. Davis SM, Lees KR, Albers GW, Diener HC, Markabi S, Karlsson G et al (2000) Selfotel in acute ischemic stroke: possible neurotoxic effects of an NMDA antagonist. Stroke 31(2):347–354
120. Lyden P, Shuaib A, Ng K, Levin K, Atkinson RP, Rajput A et al (2002) Clomethiazole Acute Stroke Study in ischemic stroke (CLASS-I): final results. Stroke 33(1):122–128
121. Albers GW, Goldstein LB, Hall D, Lesko LM (2001) Aptiganel hydrochloride in acute ischemic stroke: a randomized controlled trial. JAMA 286(21):2673–2682
122. Saltarelli MD, Weaver JJ, Hsu C, Bednar MM (2004) Randomized double-blind, placebo-controlled study to evaluate the safety and efficacy of CP-101,606 (traxoprodil), an NR2B-selective N-methyl D-aspartate receptor antagonist, in subjects with acute ischemic stroke. Stroke 35(1):241
123. Albers GW, Atkinson RP, Kelley RE, Rosenbaum DM (1995) Safety, tolerability, and pharmacokinetics of the N-methyl-D-aspartate antagonist dextrorphan in patients with acute stroke. Dextrorphan Study Group. Stroke 26(2):254–258

124. Mousavi SA, Saadatnia M, Khorvash F, Hoseini T, Sariaslani P (2011) Evaluation of the neuroprotective effect of dextromethorphan in the acute phase of ischaemic stroke. Arch Med Sci 7(3):465–469
125. Lodder J, van Raak L, Hilton A, Hardy E, Kessels A, Group ES (2006) Diazepam to improve acute stroke outcome: results of the early GABA-Ergic activation study in stroke trial. a randomized double-blind placebo-controlled trial. Cerebrovasc Dis 21(1–2):120–127
126. Synthélabo R (1996) Eliprodil trial "eliprodil phase 3". Available from http://www.strokecenter.org/trials/clinicalstudies/eliprodil-trial. Last accessed 26 May 2016
127. Pulsinelli WA (2008) Multicenter study to evaluate the safety and efficacy of intravenous fosphenytoin in patients with acute ischemic stroke, "Fosphenytoin Phase 3." Available from http://www.strokecenter.org/trials/clinicalstudies/multicenter-study-to-evaluate-the-safety-and-efficacy-of-intravenous-fosphenytoin-in-patients-with-acute-ischemic-stroke/description. Accessed 26 May 2016
128. Haley EC Jr, Thompson JL, Levin B, Davis S, Lees KR, Pittman JG et al (2005) Gavestinel does not improve outcome after acute intracerebral hemorrhage: an analysis from the GAIN International and GAIN Americas studies. Stroke 36(5):1006–1010
129. Gusev EI, Skvortsova VI, Dambinova SA, Raevskiy KS, Alekseev AA, Bashkatova VG et al (2000) Neuroprotective effects of glycine for therapy of acute ischaemic stroke. Cerebrovasc Dis 10(1):49–60
130. Cramer SC, Abila B, Scott NE, Simeoni M, Enney LA, MAGS Investigators (2013) Safety, pharmacokinetics, and pharmacodynamics of escalating repeat doses of GSK249320 in patients with stroke. Stroke 44(5):1337–1342
131. Squire IB, Lees KR, Pryse-Phillips W, Kertesz A, Bamford J (1995) Efficacy and tolerability of lifarizine in acute ischemic stroke. A pilot study. Lifarizine Study Group. Ann N Y Acad Sci 765:317–318
132. Squire IB, Lees KR, PrysePhillips W, Kertesz A, Bamford J, Reid JL et al (1996) The effects of lifarizine in acute cerebral infarction: a pilot safety study. Cerebrovasc Dis 6(3):156–160
133. Grotta J (2001) Combination Therapy Stroke Trial: recombinant tissue-type plasminogen activator with/without lubeluzole. Cerebrovasc Dis 12(3):258–263
134. Diener HC, Cortens M, Ford G, Grotta J, Hacke W, Kaste M et al (2000) Lubeluzole in acute ischemic stroke treatment: a double-blind study with an 8-hour inclusion window comparing a 10-mg daily dose of lubeluzole with placebo. Stroke 31(11):2543–2551
135. Diener HC (1998) Multinational randomised controlled trial of lubeluzole in acute ischaemic stroke. European and Australian Lubeluzole Ischaemic Stroke Study Group. Cerebrovasc Dis 8(3):172–181
136. Afshari D, Moradian N, Rezaei M (2013) Evaluation of the intravenous magnesium sulfate effect in clinical improvement of patients with acute ischemic stroke. Clin Neurol Neurosurg 115:400–404
137. Hill MD, Martin RH, Mikulis D, Wong JH, Silver FL, Terbrugge KG et al (2012) Safety and efficacy of NA-1 in patients with iatrogenic stroke after endovascular aneurysm repair (ENACT): a phase 2, randomised, double-blind, placebo-controlled trial. Lancet Neurol 11(11):942–950
138. Christenson J (2015) Field randomization of NA-1 therapy in early responders (FRONTIER). Available from https://clinicaltrials.gov/ct2/show/study/NCT02315443. Last accessed 26 May 2016
139. Clark WM, Raps EC, Tong DC, Kelly RE (2000) Cervene (Nalmefene) in acute ischemic stroke : final results of a phase III efficacy study. The Cervene Stroke Study Investigators. Stroke 31(6):1234–1239
140. Federico F, Lucivero V, Lamberti P, Fiore A, Conte C (1991) A double blind randomized pilot trial of naloxone in the treatment of acute ischemic stroke. Ital J Neurol Sci 12(6):557–563
141. Herrmann WM, Stephan K, Gaede K, Apeceche M (1997) A multicenter randomized double-blind study on the efficacy and safety of nicergoline in patients with multi-infarct dementia. Dement Geriatr Cogn Disord 8(1):9–17
142. Pilkowska E, Jakubowska T, Witkowska K, Kulczycki J (2002) Nicergoline in the treatment of patients after a mild ischemic stroke. Neurol Neurochir Pol 36(6):1075–1085

143. Mueller AL, Artman LD, Balandrin MF, Brady E, Chien Y, DelMar EG et al (2000) NPS 1506, a moderate affinity uncompetitive NMDA receptor antagonist: preclinical summary and clinical experience. Amino Acids 19(1):177–179
144. Mueller AL, Artman LD, Balandrin MF, Brady E, Chien Y, Delmar EG et al (1999) NPS 1506, a novel NMDA receptor antagonist and neuroprotectant. Review of preclinical and clinical studies. Ann N Y Acad Sci 890:450–457
145. Dyker AG, Lees KR (1999) Remacemide hydrochloride: a double-blind, placebo-controlled, safety and tolerability study in patients with acute ischemic stroke. Stroke 30(9):1796–1801
146. Muir KW, Holzapfel L, Lees KR (2000) Phase II clinical trial of sipatrigine (619C89) by continuous infusion in acute stroke. Cerebrovasc Dis 10(6):431–436
147. Inc Astellas Phanma (2006) A study to evaluate the effects of YM872 on Brain function and disability when administered in combination with alteplase (tissue plasminogen activator). Available from https://clinicaltrials.gov/ct2/show/NCT00044057. Last accessed 26 May 2016
148. Elting JW, Sulter GA, Kaste M, Lees KR, Diener HC, Hommel M et al (2002) AMPA antagonist ZK200775 in patients with acute ischemic stroke: possible glial cell toxicity detected by monitoring of S-100B serum levels. Stroke 33(12):2813–2818
149. Nighoghossian N (2014) Neuroprotection impact of cyclosporin A in cerebral infarction (CsAStroke). Available from https://clinicaltrials.gov/show/NCT01527240. Last accessed 26 May 2016
150. Nighoghossian N, Berthezene Y, Mechtouff L, Derex L, Cho TH, Ritzenthaler T et al (2015) Cyclosporine in acute ischemic stroke. Neurology 84(22):2216–2223
151. Nader JA (2010) Dapsone for Acute Ischemia Stroke Study (DAISY). Available from https://clinicaltrials.gov/ct2/show/NCT01144650. Last accessed 26 May 2016
152. Nader-Kawachi J, Gongora-Rivera F, Santos-Zambrano J, Calzada P, Rios C (2007) Neuroprotective effect of dapsone in patients with acute ischemic stroke: a pilot study. Neurol Res 29(3):331–334
153. Shaikh AK, Mohammad QD, Ullah MA, Ahsan MM, Rahman A, Shakoor MA (2011) Effect of dexamethasone on brain oedema following acute ischemic stroke. Mymensingh Med J 20(3):450–458
154. Ogun SA, Odusote KA (2001) Effectiveness of high dose dexamethasone in the treatment of acute stroke. West Afr J Med 20(1):1–6
155. ICAM-1 (2001) Use of anti-ICAM-1 therapy in ischemic stroke: results of the Enlimomab Acute Stroke Trial. Neurology 57(8):1428–1434
156. Katsura K, Suda S, Abe A, Kanamaru T, Toda Y, Katayama Y (2012) Brain protection therapy in acute cerebral infarction. J Nippon Med Sch 79(2):104–110
157. Hasan D, Lindsay KW, Wijdicks EF, Murray GD, Brouwers PJ, Bakker WH et al (1989) Effect of fludrocortisone acetate in patients with subarachnoid hemorrhage. Stroke 20(9):1156–1161
158. Hasan D (2004) Effect of fludrocortisone acetate in patients with subarachnoid hemorrhage. Available from http://www.strokecenter.org/trials/clinicalstudies/effect-of-fludrocortisone-acetate-in-patients-with-subarachnoid-hemorrhage. Last accessed 26 May 2016
159. TS Trial (1994) Ganglioside GM1 in acute ischemic stroke. The SASS Trial. Stroke 25(6):1141–1148
160. Argentino C, Sacchetti ML, Toni D, Savoini G, D'Arcangelo E, Erminio F et al (1989) GM1 ganglioside therapy in acute ischemic stroke. Italian Acute Stroke Study—Hemodilution+Drug. Stroke 20(9):1143–1149
161. Becker KJ (2002) Anti-leukocyte antibodies: LeukArrest (Hu23F2G) and Enlimomab (R6.5) in acute stroke. Curr Med Res Opin 18(suppl 2):s18–s22
162. Emsley HC, Smith CJ, Georgiou RF, Vail A, Hopkins SJ, Rothwell NJ et al (2005) A randomised phase II study of interleukin-1 receptor antagonist in acute stroke patients. J Neurol Neurosurg Psychiatry 76(10):1366–1372
163. Lampl Y, Boaz M, Gilad R, Lorberboym M, Dabby R, Rapoport A et al (2007) Minocycline treatment in acute stroke: an open-label, evaluator-blinded study. Neurology 69(14):1404–1410

164. Singh R (2012) Neuroprotection with minocycline therapy for acute stroke recovery trial (NeuMAST). https://clinicaltrials.gov/ct2/show/NCT00930020?term=NeuMAST&rank=1
165. Amiri-Nikpour MR, Nazarbaghi S, Hamdi-Holasou M, Rezaei Y (2015) An open-label evaluator-blinded clinical study of minocycline neuroprotection in ischemic stroke: gender-dependent effect. Acta Neurol Scand 131(1):45–50
166. Kohler E, Prentice DA, Bates TR, Hankey GJ, Claxton A, van Heerden J et al (2013) Intravenous minocycline in acute stroke: a randomized, controlled pilot study and meta-analysis. Stroke 44(9):2493–2499
167. Jonas S (2010) Pilot study of treatment with intravenous enoxaparin and/or oral minocycline to limit infarct size after ischemic stroke. Available from https://clinicaltrials.gov/ct2/show/NCT00836355. Last accessed 26 May 2016
168. Fagan SC, Waller JL, Nichols FT, Edwards DJ, Pettigrew LC, Clark WM et al (2010) Minocycline to improve neurologic outcome in stroke (MINOS): a dose-finding study. Stroke 41(10):2283–2287
169. Krams M, Lees KR, Hacke W, Grieve AP, Orgogozo JM, Ford GA (2003) Acute stroke therapy by inhibition of neutrophils (ASTIN): an adaptive dose-response study of UK-279,276 in acute ischemic stroke. Stroke 34(11):2543–2548
170. Janaki S (1980) Pentoxifylline in strokes: a clinical study. J Int Med Res 8(1):56–62
171. den Hertog HM, van der Worp HB, van Gemert HM, Algra A, Kappelle LJ, van Gijn J et al (2009) The paracetamol (acetaminophen) in stroke (PAIS) trial: a multicentre, randomised, placebo-controlled, phase III trial. Lancet Neurol 8(5):434–440
172. de Ridder IR, de Jong FJ, den Hertog HM, Lingsma HF, van Gemert HM, Schreuder AH et al (2015) Paracetamol (Acetaminophen) in stroke 2 (PAIS 2): protocol for a randomized, placebo-controlled, double-blind clinical trial to assess the effect of high-dose paracetamol on functional outcome in patients with acute stroke and a body temperature of 36.5 degrees C or above. Int J Stroke 10(3):457–462
173. Hill MD, Martin RH, Palesch YY, Tamariz D, Waldman BD, Ryckborst KJ et al (2011) The Albumin in Acute Stroke Part 1 Trial: an exploratory efficacy analysis. Stroke 42(6):1621–1625
174. Ginsberg MD, Palesch YY, Hill MD, Martin RH, Moy CS, Barsan WG et al (2013) High-dose albumin treatment for acute ischaemic stroke (ALIAS) Part 2: a randomised, double-blind, phase 3, placebo-controlled trial. Lancet Neurol 12(11):1049–1058
175. Selim M (2010) Treatment with the iron chelator, deferoxamine mesylate, alters serum markers of oxidative stress in stroke patients. Transl Stroke Res 1(1):35–39
176. Torné MM (2012) Thrombolysis and deferoxamine in middle cerebral artery occlusion (TANDEM-1). Available from https://clinicaltrials.gov/ct2/show/study/NCT00777140. Last accessed 26 May 2016
177. Yamaguchi T (2009) Ebselen trial—phase III. Available from http://www.strokecenter.org/trials/clinicalstudies/298. Last accessed 26 May 2016
178. Yamaguchi T, Sano K, Takakura K, Saito I, Shinohara Y, Asano T et al (1998) Ebselen in acute ischemic stroke: a placebo-controlled, double-blind clinical trial. Ebselen Study Group. Stroke 29(1):12–17
179. Wang Y (2016) Study of compound edaravone injection for treatment of acute ischemic stroke. Available from https://clinicaltrials.gov/ct2/show/NCT02430350. Last accessed 26 May 2016
180. Wang Y (2015) Compound edaravone injection for acute ischemic stroke. Available from https://clinicaltrials.gov/ct2/show/NCT01929096. Last accessed 26 May 2016
181. Wang Y (2014) Comparative effectiveness of neuroprotectants on acute ischemic stroke (CER). Available from https://clinicaltrials.gov/ct2/show/NCT01556854. Last accessed 26 May 2016
182. Yamato S (2016) YAMATO 2016. Available from http://www.kawasaki-m.ac.jp/yamato/. Last accessed 26 May 2016
183. Protect S (2016) PROTECT 4.5 2016. Available from http://medical.mt-pharma.co.jp/di/radicut/rct-protect.pdf. Accessed 26 May 2016

184. Kimura K, Aoki J, Sakamoto Y, Kobayashi K, Sakai K, Inoue T et al (2012) Administration of edaravone, a free radical scavenger, during t-PA infusion can enhance early recanalization in acute stroke patients—a preliminary study. J Neurol Sci 313(1–2):132–136
185. Sharma P, Sinha M, Shukla R, Garg RK, Verma R, Singh MK (2011) A randomized controlled clinical trial to compare the safety and efficacy of edaravone in acute ischemic stroke. Ann Indian Acad Neurol 14(2):103–106
186. Naritomi H, Moriwaki H, Metoki N, Nishimura H, Higashi Y, Yamamoto Y et al (2010) Effects of edaravone on muscle atrophy and locomotor function in patients with ischemic stroke: a randomized controlled pilot study. Drugs R&D 10(3):155–163
187. Unno Y, Katayama M, Shimizu H (2010) Does functional outcome in acute ischaemic stroke patients correlate with the amount of free-radical scavenger treatment? A retrospective study of edaravone therapy. Clin Drug Investig 30(3):143–155
188. Shinohara Y, Saito I, Kobayashi S, Uchiyama S (2009) Edaravone (radical scavenger) versus sodium ozagrel (antiplatelet agent) in acute noncardioembolic ischemic stroke (EDO trial). Cerebrovasc Dis 27(5):485–492
189. Yamaguchi T (2008) Edaravone and Argatroban Stroke Therapy Study for acute ischemic stroke "EAST". Available from http://www.strokecenter.org/trials/clinicalstudies/edaravone-and-argatroban-stroke-therapy-study-for-acute-ischemic-stroke/results. Last accessed 26 May 2016
190. Otomo E, Tohgi H, Kogure K, Hirai S, Takakura K, Terashi A et al (2003) Effect of a novel free radical scavenger, edaravone (MCI-186), on acute brain infarction—randomized, placebo-controlled, double-blind study at multicenters. Cerebrovasc Dis 15(3):222–229
191. Takabatake Y, Uno E, Wakamatsu K, Yamazaki N, Hashimoto N, Tsuchiya Y (2003) [The clinical effect of combination therapy with edaravone and sodium ozagrel for acute cerebral infarction]. Brain Nerve 55(7):589–593
192. Oates JA (2015) Acetaminophen in aSAH to inhibit lipid peroxidation and cerebral vasospasm. Available from //www.strokecenter.org/trials/clinicalstudies/acetaminophen-in-asah-to-inhibit-lipid-peroxidation-and-cerebral-vasospasm. Last accessed 26 May 2016
193. Asano T, Takakura K, Sano K, Kikuchi H, Nagai H, Saito I et al (1996) Effects of a hydroxyl radical scavenger on delayed ischemic neurological deficits following aneurysmal subarachnoid hemorrhage: results of a multicenter, placebo-controlled double-blind trial. J Neurosurg 84(5):792–803
194. SAINT (2007) Safety and effectiveness of NXY-059 for the treatment of patients who have suffered from a stroke. Available from https://clinicaltrials.gov/ct2/show/NCT00061022. Last accessed 26 May 2016
195. Diener HC, Lees KR, Lyden P, Grotta J, Davalos A, Davis SM et al (2008) NXY-059 for the treatment of acute stroke: pooled analysis of the SAINT I and II Trials. Stroke 39(6):1751–1758
196. Hsu CY (2015) PG2 injection 500 mg in acute stroke study (pass) (pass). Available from https://clinicaltrials.gov/ct2/show/NCT01603667. Last accessed 26 May 2016
197. RANTTAS Investigators (1996) A randomized trial of tirilazad mesylate in patients with acute stroke (RANTTAS). The RANTTAS Investigators. Stroke 27(9):1453–1458
198. Bogousslavsky J, Victor SJ, Salinas EO, Pallay A, Donnan GA, Fieschi C et al (2002) Fiblast (trafermin) in acute stroke: results of the European-Australian phase II/III safety and efficacy trial. Cerebrovasc Dis 14(3–4):239–251
199. Cho S-R (2016) Establishment of clinical basis for hematopoietic growth factors therapy in brain injury. Available from https://clinicaltrials.gov/ct2/show/NCT02018406. Last accessed 26 May 2016
200. Ehrenreich H, Weissenborn K, Prange H, Schneider D, Weimar C, Wartenberg K et al (2009) Recombinant human erythropoietin in the treatment of acute ischemic stroke. Stroke 40(12):e647–e656
201. Cramer SC, Fitzpatrick C, Warren M, Hill MD, Brown D, Whitaker L, Ryckborst KJ, Pion L. The beta-hCG+erythropoietin in acute stroke (BETAS) study: a 3- center, single-dose, open-label, noncontrolled, phase lla safety trial. Stroke. 2010;41(5):927-31. doi: 10.1161/ STROKEAHA.109 .574343 . Epub 2010 Mar 4

202. Cramer SC (2008) Beta-hCG + Erythropoietin in acute stroke (BETAS). Available from https://clinicaltrials.gov/ct2/show/NCT00362414. Last accessed 26 May 2016
203. Ohanian AG (2016) Efficacy and safety study of GM602 in patients with acute middle cerebral artery ischemic stroke within 18 hours (GMAIS). Available from https://clinicaltrials.gov/ct2/show/NCT01221246. Last accessed 26 May 2016
204. Ringelstein EB, Thijs V, Norrving B, Chamorro A, Aichner F, Grond M et al (2013) Granulocyte colony-stimulating factor in patients with acute ischemic stroke: results of the AX200 for Ischemic Stroke trial. Stroke 44(10):2681–2687
205. Laage R (2012) AXIS 2: AX200 for the treatment of ischemic stroke (AXIS-2). Available from https://clinicaltrials.gov/show/NCT00927836. Last accessed 26 May 2016
206. Lundbeck (2010) Safety study of carbamylated erythropoietin (CEPO) to treat patients with acute ischemic stroke. Available from http://www.strokecenter.org/trials/clinicalstudies/safety-study-of-carbamylated-erythropoietin-cepo-to-treat-patients-with-acute-ischemic-stroke. Last accessed 26 May 2016
207. Lundbeck H (2011) Safety and pharmacokinetic study of carbamylated erythropoietin (CEPO) to treat patients with acute ischemic stroke. Available from https://clinicaltrials.gov/ct2/show/NCT00870844. Last accessed 26 May 2016
208. Wilson SE (2011) Clinical Study MP-124-A01 2011 [Clinical Study MP-124-A01. This Phase I, first-in-human study will assess the safety, tolerability, and pharmacokinetics of ascending 24- and 72-hour, continuous infusions of various doses of the neuroprotectant MP-in acute, ischemic stroke patients. 72 patients will be enrolled across twenty-three centers]. Available from http://stroke.ucla.edu/body.cfm?id=36. https://uncch.pure.elsevier.com/en/projects/mp-124-a01-a-phase-i-double-blind-placebo-controlled-ascending-do. Last accessed 26 May 2016
209. Hill MD (2009) REGENESIS (CA): a study of NTx™-265: human chorionic gonadotropin (hCG) and epoetin alfa (EPO) in acute ischemic stroke patients (REGENESIS). Available from https://clinicaltrials.gov/ct2/show/NCT00663416. Last accessed 26 May 2016
210. Cramer SC, Hill MD, Investigators R-L (2014) Human choriogonadotropin and epoetin alfa in acute ischemic stroke patients (REGENESIS-LED trial). Int J Stroke 9(3):321–327
211. Cramer SC (2011) Study of NTx®-265: human chorionic gonadotropin (hCG) and epoetin alfa (EPO) in acute ischemic stroke patients (REGENESIS-LED). Available from https://clinicaltrials.gov/ct2/show/NCT00938314. Last accessed 26 May 2016
212. Cramer SC (2010) REGENESIS (US): a phase IIb prospective, randomized, double-blind, placebo controlled study of NTx™-265: human chorionic gonadotropin (hCG) and epoetin alfa (EPO) in acute ischemic stroke patients (REGENESIS). Available from https://clinicaltrials.gov/ct2/show/NCT00715364. Last accessed 26 May 2016
213. Shah IM, Lees KR, Pien CP, Elliott PJ (2002) Early clinical experience with the novel proteasome inhibitor PS-519. Br J Clin Pharmacol 54(3):269–276
214. DTTS Group (1993) Trial of secondary prevention with atenolol after transient ischemic attack or nondisabling ischemic stroke. The Dutch TIA Trial Study Group. Stroke 24(4):543–548
215. Sandset EC, Bath PM, Boysen G, Jatuzis D, Korv J, Luders S et al (2011) The angiotensin-receptor blocker candesartan for treatment of acute stroke (SCAST): a randomised, placebo-controlled, double-blind trial. Lancet 377(9767):741–750
216. Berge E (2012) Scandinavian Candesartan Acute Stroke Trial (SCAST). Available from https://clinicaltrials.gov/ct2/show/NCT00120003. Last accessed 26 May 2016
217. Blakemore CB (1987) Cyclandelate in the treatment of multi-infarct dementia. Interim findings from a multicentre study in general practice. Drugs 33(suppl 2):110–113
218. Rosenberg G, Bornstein N, Diener HC, Gorelick PB, Shuaib A, Lees K (2011) The Membrane-Activated Chelator Stroke Intervention (MACSI) Trial of DP-b99 in acute ischemic stroke: a randomized, double-blind, placebo-controlled, multinational pivotal phase III study. Int J Stroke 6(4):362–367
219. Lees KR, Bornstein N, Diener HC, Gorelick PB, Rosenberg G, Shuaib A et al (2013) Results of Membrane-Activated Chelator Stroke Intervention randomized trial of DP-b99 in acute ischemic stroke. Stroke 44(3):580–584

220. Shuaib A (2012) Efficacy and safety study of DP-b99 in treating acute ischemic stroke (MACSI). Available from https://clinicaltrials.gov/ct2/show/NCT00893867. Last accessed 26 May 2016
221. Diener HC, Schneider D, Lampl Y, Bornstein NM, Kozak A, Rosenberg G (2008) DP-b99, a membrane-activated metal ion chelator, as neuroprotective therapy in ischemic stroke. Stroke 39(6):1774–1778
222. Franke CL, Palm R, Dalby M, Schoonderwaldt HC, Hantson L, Eriksson B et al (1996) Flunarizine in stroke treatment (FIST): a double-blind, placebo-controlled trial in Scandinavia and the Netherlands. Acta Neurol Scand 93(1):56–60
223. Study evaluating single ascending doses of ILS-920 2009. Available from https://clinicaltrials.gov/show/NCT00827190. Last accessed 26 May 2016
224. Beer C, Blacker D, Bynevelt M, Hankey GJ, Puddey IB (2012) A randomized placebo controlled trial of early treatment of acute ischemic stroke with atorvastatin and irbesartan. Int J Stroke 7(2):104–111
225. Masso JFM (2001) Nicardipine in the prevention of cerebral infarction. Available from http://www.strokecenter.org/trials/clinicalstudies/nicardipine-in-the-prevention-of-cerebral-infarction. Last accessed 26 May 2016
226. Wang D (2004) IV double and triple concentrated nicardipine for stroke and ICH. Available from https://clinicaltrials.gov/ct2/show/NCT00325793. Last accessed 26 May 2016
227. Li R (2016) Nimodipine for treating acute massive cerebral infarction. Available from https://clinicaltrials.gov/ct2/show/NCT02248233. Last accessed 26 May 2016
228. Wang P, Wang Y, Feng T, Zhao X, Zhou Y, Wang Y et al (2012) Rationale and design of a double-blind, placebo-controlled, randomized trial to evaluate the safety and efficacy of nimodipine in preventing cognitive impairment in ischemic cerebrovascular events (NICE). BMC Neurol 12(1):88
229. Ahmed N, Wahlgren NG (2003) Effects of blood pressure lowering in the acute phase of total anterior circulation infarcts and other stroke subtypes. Cerebrovasc Dis 15(4):235–243
230. Wahlgren N (2008) Intravenous nimodipine West European Stroke Trial "INWEST". Available from http://www.strokecenter.org/trials/clinicalstudies/intravenous-nimodipine-west-european-stroke-trial. Last accessed 26 May 2016
231. Horn J, de Haan RJ, Vermeulen M, Limburg M (2001) Very Early Nimodipine Use in Stroke (VENUS): a randomized, double-blind, placebo-controlled trial. Stroke 32(2):461–465
232. NIMPAS (1999) Nimodipine and perfusion changes after stroke "NIMPAS". Available from http://www.strokecenter.org/trials/clinicalstudies/nimodipine-and-perfusion-changes-after-stroke. Last accessed 26 May 2016
233. Kaste M, Fogelholm R, Erila T, Palomaki H, Murros K, Rissanen A et al (1994) A randomized, double-blind, placebo-controlled trial of nimodipine in acute ischemic hemispheric stroke. Stroke 25(7):1348–1353
234. Kaste M (1994) Randomized, double-blind, placebo-controlled trial of nimodipine in acute ischemic hemispheric stroke. Available from http://www.strokecenter.org/trials/clinicalstudies/randomized-double-blind-placebo-controlled-trial-of-nimodipine-in-acute-ischemic-hemispheric-stroke. Last accessed 26 May 2016
235. TANS Group (1992) Clinical trial of nimodipine in acute ischemic stroke. The American Nimodipine Study Group. Stroke 23(1):3–8
236. AN Study (1992) American Nimodipine Study "ANS". Available from http://www.strokecenter.org/trials/clinicalstudies/american-nimodipine-study. Last accessed 26 May 2016
237. TS GROUP (1990) Randomised, double-blind, placebo-controlled trial of nimodipine in acute stroke. Trust Study Group. Lancet 336(8725):1205–1209
238. Murphy JJ (1990) Randomized, double-blind, placebo-controlled trial of nimodipine in acute stroke "TRUST". Available from http://www.strokecenter.org/trials/clinicalstudies/randomized-double-blind-placebo-controlled-trial-of-nimodipine-in-acute-stroke. Last accessed 26 May 2016
239. Martinez-Vila E, Guillen F, Villanueva JA, Matias-Guiu J, Bigorra J, Gil P et al (1990) Placebo-controlled trial of nimodipine in the treatment of acute ischemic cerebral infarction. Stroke 21(7):1023–1028

240. Martinez-Vila E (1990) Placebo-controlled trial of Nimodipine in the treatment of acute ischemic cerebral infarction. Available from http://www.strokecenter.org/trials/clinicalstudies/placebo-controlled-trial-of-nimodipine-in-the-treatment-of-acute-ischemic-cerebral-infarction. Last accessed 26 May 2016
241. Gelmers HJ, Gorter K, de Weerdt CJ, Wiezer HJ (1988) A controlled trial of nimodipine in acute ischemic stroke. N Engl J Med 318(4):203–207
242. Gelmers HJ (1998) Controlled trial of nimodipine in acute ischemic stroke. Available from http://www.strokecenter.org/trials/clinicalstudies/controlled-trial-of-nimodipine-in-acute-ischemic-stroke. Last accessed 26 May 2016
243. Thiery E, Otte G, Vander EH (1979) Comparative study of the clinical effect of vincamine versus papaverine given parenterally in the acute phase of stroke. Arzneimittelforschung 29(3):671–674
244. Barer DH, Cruickshank JM, Ebrahim SB, Mitchell JR (1988) Low dose beta blockade in acute stroke ("BEST" trial): an evaluation. Br Med J (Clin Res Ed) 296(6624):737–741
245. Haverkamp W (2013) Beta blockers in acute ischemic stroke (BIAS). Available from https://clinicaltrials.gov/show/NCT01061190. Last accessed 26 May 2016
246. Oczkowski WJ, Hachinski VC, Bogousslavsky J, Barnett HJ, Carruthers SG (1989) A double-blind, randomized trial of PY108-068 in acute ischemic cerebral infarction. Stroke 20(5):604–608
247. Daiichi Asubio Pharmaceuticals Inc. (2006) New data confirm PK profile of SUN N8075, a multi-functional compound in development for acute ischemic stroke. Available from http://www.prnewswire.com/news-releases/new-data-confirm-pk-profile-of-sun-n8075-a-multi-functional-compound-in-development-for-acute-ischemic-stroke-56205602.html. Last accessed 26 May 2016
248. Wu-Wong JR (2002) S-0139 (Shionogi). Curr Opin Investig Drugs (London, England : 2000) 3(7):1051–1056
249. Omura T, Tanaka Y, Miyata N, Koizumi C, Sakurai T, Fukasawa M et al (2006) Effect of a new inhibitor of the synthesis of 20-HETE on cerebral ischemia reperfusion injury. Stroke 37(5):1307–1313
250. Fraser JF (2016) Superselective administration of verapamil during recanalization in acute ischemic stroke (SAVER-I) 2016. Available from https://clinicaltrials.gov/ct2/show/NCT02235558. Last accessed 26 May 2016
251. Zhang W (2013) Chinese assessment for vinpocetine in neurology. Available from https://clinicaltrials.gov/ct2/show/NCT01400035. Last accessed 26 May 2016
252. Feigin VL, Doronin BM, Popova TF, Gribatcheva EV, Tchervov DV (2001) Vinpocetine treatment in acute ischaemic stroke: a pilot single-blind randomized clinical trial. Eur J Neurol 8(1):81–85
253. Adams HP Jr, Effron MB, Torner J, Davalos A, Frayne J, Teal P et al (2008) Emergency administration of abciximab for treatment of patients with acute ischemic stroke: results of an international phase III trial: Abciximab in Emergency Treatment of Stroke Trial (AbESTT-II). Stroke 39(1):87–99
254. Hennerici MG, Kay R, Bogousslavsky J, Lenzi GL, Verstraete M, Orgogozo JM (2006) Intravenous ancrod for acute ischaemic stroke in the European Stroke Treatment with Ancrod Trial: a randomised controlled trial. Lancet 368(9550):1871–1878
255. Sherman DG, Atkinson RP, Chippendale T, Levin KA, Ng K, Futrell N et al (2000) Intravenous ancrod for treatment of acute ischemic stroke: the STAT study: a randomized controlled trial. Stroke Treatment with Ancrod Trial. JAMA 283(18):2395–2403
256. Levy DE, del Zoppo GJ, Demaerschalk BM, Demchuk AM, Diener HC, Howard G et al (2009) Ancrod in acute ischemic stroke results of 500 subjects beginning treatment within 6 hours of stroke onset in the Ancrod Stroke Program. Stroke 40(12):3796–3803
257. Wasiewski W (2010) Ancrod (Viprinex™) for the treatment of acute, ischemic stroke. Available from https://clinicaltrials.gov/ct2/show/NCT00141011. Last accessed 26 May 2016

258. Wasiewski W (2010) ASP-II: Ancrod Stroke Program: Ancrod (Viprinex™) for the treatment of acute, ischemic stroke. Available from https://clinicaltrials.gov/ct2/show/NCT00300196. Last accessed 26 May 2016
259. Barreto AD (2015) Safety and feasibility of argatroban, tissue plasminogen activator and intra-arterial therapy in stroke (ARTSS-IA). Available from https://clinicaltrials.gov/ct2/show/NCT02448069. Last accessed 26 May 2016
260. Barreto AD, Alexandrov AV, Lyden P, Lee J, Martin-Schild S, Shen L et al (2012) The argatroban and tissue-type plasminogen activator stroke study: final results of a pilot safety study. Stroke 43(3):770–775
261. Hosomi N, Naya T, Kohno M, Kobayashi S, Koziol JA (2007) Efficacy of anti-coagulant treatment with argatroban on cardioembolic stroke. J Neurol 254(5):605–612
262. LaMonte MP, Nash ML, Wang DZ, Woolfenden AR, Schultz J, Hursting MJ et al (2004) Argatroban anticoagulation in patients with acute ischemic stroke (ARGIS-1): a randomized, placebo-controlled safety study. Stroke 35(7):1677–1682
263. Chen ZM, Sandercock P, Pan HC, Counsell C, Collins R, Liu LS et al (2000) Indications for early aspirin use in acute ischemic stroke: a combined analysis of 40 000 randomized patients from the chinese acute stroke trial and the international stroke trial. On behalf of the CAST and IST collaborative groups. Stroke 31(6):1240–1249
264. Cooperative Group for Reassessment of Defibrase (2005) Reassessment of defibrase in treatment of acute cerebral infarction: a multicenter, randomized, double-blind, placebo-controlled trial. Chin Med Sci J 20(3):151–158
265. Diener HC, Ringelstein EB, von Kummer R, Landgraf H, Koppenhagen K, Harenberg J et al (2006) Prophylaxis of thrombotic and embolic events in acute ischemic stroke with the low-molecular-weight heparin certoparin: results of the PROTECT Trial. Stroke 37(1):139–144
266. Benavente OR, Hart RG, McClure LA, Szychowski JM, Coffey CS, Pearce LA (2012) Effects of clopidogrel added to aspirin in patients with recent lacunar stroke. N Engl J Med 367(9):817–825
267. Berge E, Abdelnoor M, Nakstad PH, Sandset PM (2000) Low molecular-weight heparin versus aspirin in patients with acute ischaemic stroke and atrial fibrillation: a double-blind randomised study. HAEST Study Group. Heparin in Acute Embolic Stroke Trial. Lancet 355(9211):1205–1210
268. Pancioli AM, Adeoye O, Schmit PA, Khoury J, Levine SR, Tomsick TA et al (2013) Combined approach to lysis utilizing eptifibatide and recombinant tissue plasminogen activator in acute ischemic stroke-enhanced regimen stroke trial. Stroke 44(9):2381–2387
269. Adeoye O, Sucharew H, Khoury J, Vagal A, Schmit PA, Ewing I et al (2015) Combined approach to lysis utilizing eptifibatide and recombinant tissue-type plasminogen activator in acute ischemic stroke-full dose regimen stroke trial. Stroke 46(9):2529–2533
270. Pancioli AM (2015) Study of the combination therapy of Rt-PA and eptifibatide to treat acute ischemic stroke (CLEAR-ER). https://clinicaltrials.gov/ct2/show/NCT00894803?term=CLEAR-ER&rank=1
271. Pancioli AM, Broderick J, Brott T, Tomsick T, Khoury J, Bean J et al (2008) The combined approach to lysis utilizing eptifibatide and rt-PA in acute ischemic stroke: the CLEAR stroke trial. Stroke 39(12):3268–3276
272. (IST) TIST (1997) The International Stroke Trial (IST): a randomised trial of aspirin, subcutaneous heparin, both, or neither among 19435 patients with acute ischaemic stroke. International Stroke Trial Collaborative Group. Lancet 349(9065):1569–1581
273. Barreto AD, Sharma VK, Lao AY, Schellinger PD, Amarenco P, Sierzenski P et al (2009) Safety and dose-escalation study design of Transcranial Ultrasound in Clinical SONolysis for acute ischemic stroke: the TUCSON Trial. Int J Stroke 4(1):42–48
274. Hsu CY, Norris JW, Hogan EL, Bladin P, Dinsdale HB, Yatsu FM et al (1988) Pentoxifylline in acute nonhemorrhagic stroke. A randomized, placebo-controlled double-blind trial. Stroke 19(6):716–722
275. Herskovits E (1985) Pentoxifylline versus conventional antiaggregants in secondary stroke prevention. Available from http://www.strokecenter.org/trials/clinicalstudies/pentoxifylline-

versus-conventional-antiaggregants-in-secondary-stroke-prevention. Last accessed 26 May 2016
276. Hsu CY, Faught RE Jr, Furlan AJ, Coull BM, Huang DC, Hogan EL et al (1987) Intravenous prostacyclin in acute nonhemorrhagic stroke: a placebo-controlled double-blind trial. Stroke 18(2):352–358
277. Bath PM, Lindenstrom E, Boysen G, De Deyn P, Friis P, Leys D et al (2001) Tinzaparin in acute ischaemic stroke (TAIST): a randomised aspirin-controlled trial. Lancet 358(9283):702–710
278. Warach S (2009) Combination anti-platelet and anti-coagulation treatment after lysis of ischemic stroke trial (CATALIST) (CATALIST). Available from https://clinicaltrials.gov/ct2/show/study/NCT00061373. Last accessed 26 May 2016
279. Siebler M, Hennerici MG, Schneider D, von Reutern GM, Seitz RJ, Rother J et al (2011) Safety of Tirofiban in acute Ischemic Stroke: the SaTIS trial. Stroke 42(9):2388–2392
280. Han SW, Kim YJ, Ahn SH, Seo WK, Yu S, Oh SH et al (2016) Protocol for the comparison of triflusal and clopidogrel in secondary prevention of stroke based on cytochrome P450 2C19 genotyping (MASETRO study): a multicenter, randomized, open-label, parallel-group trial. Int J Stroke 11:485–491
281. Lee KY (2015) Comparison of triflusal and clopidogrel in secondary prevention of stroke based on the genotyping (MAESTRO). Available from https://clinicaltrials.gov/ct2/show/NCT01174693. Last accessed 26 May 2016
282. Culebras A, Rotta-Escalante R, Vila J, Dominguez R, Abiusi G, Famulari A et al (2004) Triflusal vs aspirin for prevention of cerebral infarction: a randomized stroke study. Neurology 62(7):1073–1080
283. Matias-Guiu J, Ferro JM, Alvarez-Sabin J, Torres F, Jimenez MD, Lago A et al (2003) Comparison of triflusal and aspirin for prevention of vascular events in patients after cerebral infarction: the TACIP Study: a randomized, double-blind, multicenter trial. Stroke 34(4):840–848
284. Yu YL, Kumana CR, Lauder IJ, Cheung YK, Chan FL, Kou M et al (1993) Treatment of acute cortical infarct with intravenous glycerol. A double-blind, placebo-controlled randomized trial. Stroke 24(8):1119–1124
285. Strand T (1992) Evaluation of long-term outcome and safety after hemodilution therapy in acute ischemic stroke. Stroke 23(5):657–662
286. Rudolf J (2002) Hydroxyethyl starch for hypervolemic hemodilution in patients with acute ischemic stroke: a randomized, placebo-controlled phase II safety study. Cerebrovasc Dis 14(1):33–41
287. Bereczki D, Mihalka L, Szatmari S, Fekete K, Di Cesar D, Fulesdi B et al (2003) Mannitol use in acute stroke: case fatality at 30 days and 1 year. Stroke 34(7):1730–1735
288. Saxena R, Wijnhoud AD, Carton H, Hacke W, Kaste M, Przybelski RJ et al (1999) Controlled safety study of a hemoglobin-based oxygen carrier, DCLHb, in acute ischemic stroke. Stroke 30(5):993–996
289. Rusyniak DE, Kirk MA, May JD, Kao LW, Brizendine EJ, Welch JL et al (2003) Hyperbaric oxygen therapy in acute ischemic stroke: results of the Hyperbaric Oxygen in Acute Ischemic Stroke Trial Pilot Study. Stroke 34(2):571–574
290. Padma MV, Bhasin A, Bhatia R, Garg A, Singh MB, Tripathi M et al (2010) Normobaric oxygen therapy in acute ischemic stroke: a pilot study in Indian patients. Ann Indian Acad Neurol 13(4):284–288
291. Singhal AB (2009) Normobaric oxygen therapy in acute ischemic stroke trial. Available from https://clinicaltrials.gov/ct2/show/study/NCT00414726. Last accessed 26 May 2016
292. Bell RD, Powers BL, Brock D, Provencio JJ, Flanders A, Benetiz R et al (2006) Ventriculo-lumbar perfusion in acute ischemic stroke. Neurocrit Care 5(1):21–29
293. Williams PD, Zlokovic BV, Griffin JH, Pryor KE, Davis TP (2012) Preclinical safety and pharmacokinetic profile of 3K3A-APC, a novel, modified activated protein C for ischemic stroke. Curr Pharm Des 18(27):4215–4222
294. Lyden P (2014) Safety evaluation of 3K3A-APC in ischemic stroke (RHAPSODY). Available from https://clinicaltrials.gov/ct2/show/NCT02222714. Last accessed May 2016

295. Lampl Y, Lorberboym M, Gilad R, Vysberg I, Tikozky A, Sadeh M et al (2010) Early outcome of acute ischemic stroke in hyperlipidemic patients under atorvastatin versus simvastatin. Clin Neuropharmacol 33(3):129–134
296. Tuttolomondo A (2014) Atorvastatin in acute stroke treatment. Available from https://clinicaltrials.gov/ct2/show/NCT02225834. Last accessed 26 May 2016
297. Yoshimura S (2015) AdminiStration of Statin On Acute Ischemic stRoke patienT Trial (ASSORT). Available from https://clinicaltrials.gov/ct2/show/NCT02549846. Last accessed 26 May 2016
298. Lou M (2016) The safety and efficacy study of high dose atorvastatin after thrombolytic treatment in acute ischemic stroke (SEATIS). Available from https://clinicaltrials.gov/ct2/show/NCT02452502. Last accessed 26 May 2016
299. Makriyannis A, Mechoulam R, Piomelli D (2005) Therapeutic opportunities through modulation of the endocannabinoid system. Neuropharmacology 48(8):1068–1071
300. Martin-Schild S, Hallevi H, Shaltoni H, Barreto AD, Gonzales NR, Aronowski J et al (2009) Combined neuroprotective modalities coupled with thrombolysis in acute ischemic stroke: a pilot study of caffeinol and mild hypothermia. J Stroke Cerebrovasc Dis 18(2):86–96
301. Grotta J (2011) Caffeinol hypothermia protocol. Available from https://clinicaltrials.gov/ct2/show/study/NCT00299416. Last accessed 26 May 2016
302. Heiss WD, Brainin M, Bornstein NM, Tuomilehto J, Hong Z (2012) Cerebrolysin in patients with acute ischemic stroke in Asia: results of a double-blind, placebo-controlled randomized trial. Stroke 43(3):630–636
303. Nakamura T, Tsuruta S, Uchiyama S (2012) Cilostazol combined with aspirin prevents early neurological deterioration in patients with acute ischemic stroke: a pilot study. J Neurol Sci 313(1–2):22–26
304. Kraglund KL (2016) The efficacy of citalopram treatment in acute stroke (TALOS). Available from https://clinicaltrials.gov/ct2/show/NCT01937182. Last accessed 26 May 2016
305. Davalos A, Alvarez-Sabin J, Castillo J, Diez-Tejedor E, Ferro J, Martinez-Vila E et al (2012) Citicoline in the treatment of acute ischaemic stroke: an international, randomised, multicentre, placebo-controlled study (ICTUS trial). Lancet 380(9839):349–357
306. Martinsson L, Wahlgren NG (2003) Safety of dexamphetamine in acute ischemic stroke: a randomized, double-blind, controlled dose-escalation trial. Stroke 34(2):475–481
307. Garg RK, Nag D, Agrawal A (1995) A double blind placebo controlled trial of ginkgo biloba extract in acute cerebral ischaemia. J Assoc Physicians India 43(11):760–763
308. Viscoli CM, Brass LM, Kernan WN, Sarrel PM, Suissa S, Horwitz RI (2005) Estrogen therapy and risk of cognitive decline: results from the Women's Estrogen for Stroke Trial (WEST). Am J Obstet Gynecol 192(2):387–393
309. Horwitz RI (2006) Women's Estrogen for Stroke Trial (WEST). Available from https://clinicaltrials.gov/ct2/show/NCT00026039. Last accessed 26 May 2016
310. Sheth KN, Kimberly WT, Elm JJ, Kent TA, Mandava P, Yoo AJ et al (2014) Pilot study of intravenous glyburide in patients with a large ischemic stroke. Stroke 45(1):281–283
311. Sheth KN (2013) Glyburide advantage in malignant edema and stroke pilot (GAMES-PILOT). Available from https://clinicaltrials.gov/ct2/show/NCT01268683. Last accessed 26 May 2016
312. Sheth KN, Elm JJ, Beslow LA, Sze GK, Kimberly WT (2016) Glyburide Advantage in Malignant Edema and Stroke (GAMES-RP) Trial: rationale and design. Neurocrit Care 24(1):132–139
313. Sheth KN (2015) Glyburide advantage in malignant edema and stroke—remedy pharmaceuticals (GAMES-RP). Available from https://clinicaltrials.gov/ct2/show/NCT01794182. Last accessed 26 May 2016
314. Woodhouse L, Scutt P, Krishnan K, Berge E, Gommans J, Ntaios G et al (2015) Effect of hyperacute administration (within 6 hours) of transdermal glyceryl trinitrate, a nitric oxide donor, on outcome after stroke: subgroup analysis of the Efficacy of Nitric Oxide in Stroke (ENOS) Trial. Stroke 46(11):3194–3201

315. Krishnan K, Scutt P, Woodhouse L, Adami A, Becker JL, Berge E et al (2016) Glyceryl trinitrate for acute intracerebral hemorrhage: results from the Efficacy of Nitric Oxide in Stroke (ENOS) Trial, a subgroup analysis. Stroke 47(1):44–52
316. Bath P (2012) The efficacy of nitric oxide in stroke (ENOS) Trial (ENOS). Available from https://clinicaltrials.gov/ct2/show/NCT00989716. Last accessed 26 May 2016
317. Santambrogio S, Martinotti R, Sardella F, Porro F, Randazzo A (1978) Is there a real treatment for stroke? Clinical and statistical comparison of different treatments in 300 patients. Stroke 9(2):130–132
318. Hemmen TM, Raman R, Guluma KZ, Meyer BC, Gomes JA, Cruz-Flores S et al (2010) Intravenous thrombolysis plus hypothermia for acute treatment of ischemic stroke (ICTuS-L): final results. Stroke 41(10):2265–2270
319. Gold AB, Hermann L, Lanctôt KL (2011) Lithium and its neuroprotective and neurotrophic effects: potential treatment for post-ischemic stroke sequelae. Curr Drug Targets 12(2):243–255
320. Herrmann N (2016) The neurotrophic effects of lithium carbonate following stroke: a feasibility study. Available from https://clinicaltrials.gov/ct2/show/NCT01112813. Last accessed 26 May 2016
321. Gray CS, French JM, Venables GS, Cartlidge NE, James OF, Bates D (1990) A randomized double-blind controlled trial of naftidrofuryl in acute stroke. Age Ageing 19(6):356–363
322. Siddiqui FJ, Venketasubramanian N, Chan ESY, Chen C (2013) Efficacy and safety of MLC601 (NeuroAiD (R)), a traditional Chinese medicine, in poststroke recovery: a systematic review. Cerebrovasc Dis 35:8–17
323. Bousser MG (2012) CHInese Medicine NeuroAid Efficacy on Stroke Recovery (CHIMES). Available from https://clinicaltrials.gov/ct2/show/NCT00554723. Last accessed 26 May 2016
324. Harandi AA, Abolfazli R, Hatemian A, Ghragozlee K, Ghaffar-Pour M, Karimi M et al (2011) Safety and efficacy of MLC601 in Iranian patients after stroke: a double-blind. Placebo-Controlled Clinical Trial. Stroke Res Treat 2011:721613
325. Chen C, Venketasubramanian N, Gan RN, Lambert C, Picard D, Chan BP et al (2009) Danqi Piantang Jiaonang (DJ), a traditional Chinese medicine, in poststroke recovery. Stroke 40(3):859–863
326. Due B (2012) A study to evaluate the effects of ONO-2506 intravenous infusion in patients with acute ischemic stroke. Available from https://clinicaltrials.gov/ct2/show/NCT00046761. Last accessed 26 May 2016
327. Pettigrew LC, Kasner SE, Albers GW, Gorman M, Grotta JC, Sherman DG et al (2006) Safety and tolerability of arundic acid in acute ischemic stroke. J Neurol Sci 251(1–2):50–56
328. ONO (2008) Study of ONO-2506 in patients with acute ischemic stroke. Available from https://clinicaltrials.gov/ct2/show/NCT00229177. Last accessed 26 May 2016
329. Chi NF (2016) Neuroprotection of pioglitazone in acute ischemic stroke. Available from https://clinicaltrials.gov/ct2/show/NCT02195791. Last accessed 26 May 2016
330. De Deyn PP, Reuck JD, Deberdt W, Vlietinck R, Orgogozo JM (1997) Treatment of acute ischemic stroke with piracetam. Members of the Piracetam in Acute Stroke Study (PASS) Group. Stroke 28(12):2347–2352
331. Orgogozo JM (2009) Piracetam acute stroke study II "PASS II". Available from http://www.strokecenter.org/trials/clinicalstudies/piracetam-acute-stroke-study-ii. Last accessed 26 May 2016
332. Huber M, Kittner B, Hojer C, Fink GR, Neveling M, Heiss WD (1993) Effect of propentofylline on regional cerebral glucose metabolism in acute ischemic stroke. J Cereb Blood Flow Metab 13(3):526–530
333. Teal P, Davis S, Hacke W, Kaste M, Lyden PD, Fierus M (2009) A randomized, double-blind, placebo-controlled trial to evaluate the efficacy, safety, tolerability, and pharmacokinetic/pharmacodynamic effects of a targeted exposure of intravenous repinotan in patients with acute ischemic stroke: modified Randomized Exposure Controlled Trial (mRECT). Stroke 40(11):3518–3525

334. Everett BM, Glynn RJ, MacFadyen JG, Ridker PM (2010) Rosuvastatin in the prevention of stroke among men and women with elevated levels of C-reactive protein justification for the use of statins in prevention: an intervention trial evaluating rosuvastatin (JUPITER). Circulation 121(1):143–150
335. Heo JH, Song D, Nam HS, Kim EY, Kim YD, Lee KY et al (2016) Effect and safety of rosuvastatin in acute ischemic stroke. J Stroke 18(1):87–95
336. Heo JH (2013) The effects of very early use of rosuvastatin in preventing recurrence of ischemic stroke (EUREKA). Available from https://clinicaltrials.gov/ct2/show/NCT01364220. Last accessed 26 May 2016
337. Hospital R (2015) Exploratory study on the effects of early rosuvastatin treatment in patients with acute ischemic stroke (AIS). Available from https://clinicaltrials.gov/ct2/show/NCT02643784. Last accessed 26 May 2016
338. Liu CF (2015) Effect of rosuvastatin on prognosis of clinical response in acute ischemic stroke patients (REPAIRS). Available from https://clinicaltrials.gov/ct2/show/NCT02484027. Last accessed 26 May 2016
339. Bang OY (2016) Intensive statin treatment in acute ischemic stroke patients with intracranial atherosclerosis (STAMINA-MRI). Available from https://clinicaltrials.gov/ct2/show/NCT02458755. Last accessed 26 May 2016
340. Hackam DG, Austin PC, Huang A, Juurlink DN, Mamdani MM, Paterson JM et al (2012) Statins and intracerebral hemorrhage: a retrospective cohort study. Arch Neurol 69(1):39–45
341. Hackam DG, Woodward M, Newby LK, Bhatt DL, Shao M, Smith EE et al (2011) Statins and intracerebral hemorrhage: collaborative systematic review and meta-analysis. Circulation 124(20):2233–2242
342. Yusuf S, Bosch J, Dagenais G, Zhu J, Xavier D, Liu L et al (2016) Cholesterol lowering in intermediate-risk persons without cardiovascular disease. N Engl J Med 374:2021–2031
343. Gusev EI, Skvortsova VI, Miasoedov NF, Nezavibat'ko VN, Zhuravleva E, Vanichkin AV (1997) [Effectiveness of semax in acute period of hemispheric ischemic stroke (a clinical and electrophysiological study)]. Zhurnal nevrologii i psikhiatrii imeni SS Korsakova/Ministerstvo zdravookhraneniia i meditsinskoi promyshlennosti Rossiiskoi Federatsii, Vserossiiskoe obshchestvo nevrologov [i] Vserossiiskoe obshchestvo psikhiat 97(6):26–34
344. Ribó M (2010) Stroke treatment with acute reperfusion and simvastatin "STARS07". Available from http://www.strokecenter.org/trials/clinicalstudies/stroke-treatment-with-acute-reperfusion-and-simvastatin. Last accessed 26 May 2016
345. Montaner J (2014) Stroke treatment with acute reperfusion and simvastatin. Available from https://clinicaltrials.gov/ct2/show/NCT01073007. Last accessed 26 May 2016
346. Montaner J, Chacon P, Krupinski J, Rubio F, Millan M, Molina CA et al (2008) Simvastatin in the acute phase of ischemic stroke: a safety and efficacy pilot trial. Eur J Neurol 15(1):82–90
347. Corey R (2015) Implant for augmentation of cerebral blood flow trial, effectiveness and safety in a 24 hour window (ImpACT-24). Available from https://clinicaltrials.gov/ct2/show/NCT00272909. Last accessed 26 May 2016
348. Shai E (2015) The IMPACT-24Bt Post rtPA TRIAL (implant augmenting cerebral blood flow trial 24 hours from stroke onset, post IV-rtPA) (IMPACT-24Bt). Available from https://clinicaltrials.gov/ct2/show/NCT01874093. Last accessed 26 May 2016
349. Corey R (2007) Efficacy of SUN N4057 in subjects with acute ischemic stroke and measurable penumbra on magnetic resonance imaging (MRI). Available from https://clinicaltrials.gov/ct2/show/NCT00272909. Last access 26 May 2016
350. Ramirez-Lessepas M, Patrick BK, Snyder BD, Lakatua DJ (1986) Failure of central nervous system serotonin blockage to influence outcome in acute cerebral infarction. A double blind randomized trial. Stroke 17(5):953–956
351. Nuvelo (2008) Alfimeprase. Drugs R&D 9(3):185–190
352. Donnan G (2015) EXTEND (international): extending the time for thrombolysis in emergency neurological deficits. Available from https://clinicaltrials.gov/ct2/show/NCT01580839. Last accessed 26 May 2016

353. IST-3 Collaborative Group, Sandercock P, Wardlaw JM, Lindley RI, Dennis M, Cohen G et al (2012) The benefits and harms of intravenous thrombolysis with recombinant tissue plasminogen activator within 6 h of acute ischaemic stroke (the third international stroke trial [IST-3]): a randomised controlled trial. Lancet 379(9834):2352–2363
354. Wahlgren N, Ahmed N, Davalos A, Ford GA, Grond M, Hacke W et al (2007) Thrombolysis with alteplase for acute ischaemic stroke in the Safe Implementation of Thrombolysis in Stroke-Monitoring Study (SITS-MOST): an observational study. Lancet 369(9558): 275–282
355. Albers GW, von Kummer R, Truelsen T, Jensen JK, Ravn GM, Gremning BN, Chabriat H, Chang KC, Davalos AP, Ford GN , Grotta J, Kaste M, Schwamm LH, Shuaib A; DIAS-3 Investigators. Safety and efficacy of desmoteplase given 3-9h after ischaemic stroke in patients with occlusion or high-grade stenosis in major cerebral arteries (DIAS-3): a double-blind, randomised, placebo controlled phase 3 trial. Lancet Neural. 2015;14(6):575-84. doi:10.1016/$1474-4422(15)00047-2. Epub 2015 Apr 30.
356. Lundbeck H (2014) Efficacy and safety study of desmoteplase to treat acute ischemic stroke (DIAS-4). Available from https://clinicaltrials.gov/ct2/show/NCT00856661. Last accessed 26 May 2016
357. Hacke W, Furlan AJ, Al-Rawi Y, Davalos A, Fiebach JB, Gruber F et al (2009) Intravenous desmoteplase in patients with acute ischaemic stroke selected by MRI perfusion-diffusion weighted imaging or perfusion CT (DIAS-2): a prospective, randomised, double-blind, placebo-controlled study. Lancet Neurol 8(2):141–150
358. Kase CS, Albers GW, Bladin C, Fieschi C, Gabbai AA, O'Riordan W et al (2009) Neurological outcomes in patients with ischemic stroke receiving enoxaparin or heparin for venous thromboembolism prophylaxis: subanalysis of the Prevention of VTE after Acute Ischemic Stroke with LMWH (PREVAIL) study. Stroke 40(11):3532–3540
359. PREVAIL (2009) PREVAIL: prevention of VTE after acute ischemic stroke with LMWH enoxaparin (VTE: venous thromboembolism—LMWH: low molecular weight heparin). Available from https://clinicaltrials.gov/ct2/show/NCT00077805. Last accessed 26 May 2016
360. Thijs VN, Peeters A, Vosko M, Aichner F, Schellinger PD, Schneider D et al (2009) Randomized, placebo-controlled, dose-ranging clinical trial of intravenous microplasmin in patients with acute ischemic stroke. Stroke 40(12):3789–3795
361. Wong KS, Chen C, Ng PW, Tsoi TH, Li HL, Fong WC et al (2007) Low-molecular-weight heparin compared with aspirin for the treatment of acute ischaemic stroke in Asian patients with large artery occlusive disease: a randomised study. Lancet Neurol 6(5):407–413
362. Wang QS, Chen C, Chen XY, Han JH, Soo Y, Leung TW et al (2012) Low-molecular-weight heparin versus aspirin for acute ischemic stroke with large artery occlusive disease: subgroup analyses from the Fraxiparin in Stroke Study for the treatment of ischemic stroke (FISS-tris) study. Stroke 43(2):346–349
363. Nosal V (2013) Safety and efficacy of heparin and nadroparin in the acute phase of ischemic stroke (Heparinas). Available from https://clinicaltrials.gov/ct2/show/NCT01862978. Last accessed 26 May 2016
364. Kay R, Wong KS, Yu YL, Chan YW, Tsoi TH, Ahuja AT et al (1995) Low-molecular-weight heparin for the treatment of acute ischemic stroke. N Engl J Med 333(24):1588–1593
365. FISS (2000) Fraxiparin in Stroke Study "FISS". Available from http://www.strokecenter.org/trials/clinicalstudies/fraxiparin-in-stroke-study. Last accessed 26 May 2016
366. TOAST (1998) Low molecular weight heparinoid, ORG 10172 (danaparoid), and outcome after acute ischemic stroke: a randomized controlled trial. The Publications Committee for the Trial of ORG 10172 in Acute Stroke Treatment (TOAST) Investigators. JAMA 279(16):1265–1272
367. Furlan A, Higashida R, Wechsler L, Gent M, Rowley H, Kase C et al (1999) Intra-arterial prourokinase for acute ischemic stroke. The PROACT II study: a randomized controlled trial. Prolyse in acute cerebral thromboembolism. JAMA 282(21):2003–2011

368. Misra V, El Khoury R, Arora R, Chen PR, Suzuki S, Harun N et al (2011) Safety of high doses of urokinase and reteplase for acute ischemic stroke. AJNR Am J Neuroradiol 32(6):998–1001
369. Warach S (2002) ReoPro and retavase to treat acute stroke. Available from https://clinicaltrials.gov/show/NCT00046293. Last accessed 26 May 2016
370. Willemse JL, Heylen E, Nesheim ME, Hendriks DF (2009) Carboxypeptidase U (TAFIa): a new drug target for fibrinolytic therapy? J Thromb Haemost 7(12):1962–1971
371. van Iersel T, Stroissnig H, Giesen P, Wemer J, Wilhelm-Ogunbiyi K (2011) Phase I study of Solulin, a novel recombinant soluble human thrombomodulin analogue. Thromb Haemost 105(2):302–312
372. Donnan GA, Davis SM, Chambers BR, Gates PC, Hankey GJ, McNeil JJ et al (1996) Streptokinase for acute ischemic stroke with relationship to time of administration: Australian Streptokinase (ASK) Trial Study Group. JAMA 276(12):961–966
373. MAST-E (1996) Thrombolytic therapy with streptokinase in acute ischemic stroke. The Multicenter Acute Stroke Trial—Europe Study Group. N Engl J Med 335(3):145–150
374. MAST-I (1995) Randomised controlled trial of streptokinase, aspirin, and combination of both in treatment of acute ischaemic stroke. Multicentre Acute Stroke Trial—Italy (MAST-I) Group. Lancet 346(8989):1509–1514
375. Courtney K (2014) A phase 1/2a, open label, dose escalation, safety study of intra-thrombus plasmin (human) administration in acute, middle cerebral artery, ischemic stroke. Available from https://www.clinicaltrialsregister.eu/ctr-search/trial/2009-012086-66/DE#C. Last accessed 26 May 2016
376. Saver J (2014) A safety and dose finding study of plasmin (human) administered into the middle cerebral artery of stroke patients. Available from https://clinicaltrials.gov/show/NCT01014975. Last accessed 26 May 2016
377. Leker RR (2010) Study to assess the safety and efficacy of thr-18 when administered to patients suffering acute ischemic stroke and treated with tissue plasminogen activator (tPA). Available from https://clinicaltrials.gov/ct2/show/NCT01253512. Last accessed 26 May 2016
378. Kelly-Cobbs AI, Prakash R, Li W, Pillai B, Hafez S, Coucha M et al (2013) Targets of vascular protection in acute ischemic stroke differ in type 2 diabetes. Am J Physiol Heart Circ Physiol 304(6):H806–H815
379. Parsons M, Spratt N, Bivard A, Campbell B, Chung K, Miteff F et al (2012) A randomized trial of tenecteplase versus alteplase for acute ischemic stroke. N Engl J Med 366(12):1099–1107
380. Thomassen L (2015) Study of tenecteplase versus alteplase for thrombolysis (clot dissolving) in acute ischemic stroke (NOR-TEST). Available from https://clinicaltrials.gov/ct2/show/NCT01949948?term=Tenecteplase&recr=Open&rank=3. Last accessed 26 May 2016
381. Huang X, Cheripelli BK, Lloyd SM, Kalladka D, Moreton FC, Siddiqui A et al (2015) Alteplase versus tenecteplase for thrombolysis after ischaemic stroke (ATTEST): a phase 2, randomised, open-label, blinded endpoint study. Lancet Neurol 14(4):368–376
382. Muir KW (2011) Alteplase-tenecteplase trial evaluation for stroke thrombolysis- (ATTEST) (ATTEST) 2011. Available from https://clinicaltrials.gov/ct2/show/NCT01472926. Last accessed 26 May 2016
383. Joux J (2015) Superiority of Rt-PA + tenecteplase in comparison with Rt-PA only in proximal middle cerebral artery occlusion (DIVA). Available from https://clinicaltrials.gov/ct2/show/NCT02338466. Last accessed 26 May 2016
384. Haley EC (2005) Study of tenecteplase (TNK) in acute ischemic stroke (TNK-S2B). Available from https://clinicaltrials.gov/ct2/show/results/NCT00252239. Last accessed 26 May 2016
385. Coutts SB, Dubuc V, Mandzia J, Kenney C, Demchuk AM, Smith EE et al (2015) Tenecteplase-tissue-type plasminogen activator evaluation for minor ischemic stroke with proven occlusion. Stroke 46(3):769–774
386. Hill MD (2014) NK-tPA evaluation for minor ischemic stroke with proven occlusion (TEMPO-1). Available from https://clinicaltrials.gov/ct2/show/NCT01654445. Last accessed 26 May 2016

387. Hill MD (2015) A randomized controlled trial of TNK-tPA versus standard of care for minor ischemic stroke with proven occlusion (TEMPO-2). Available from https://clinicaltrials.gov/ct2/show/NCT02398656. Last accessed 26 May 2016
388. Campbell B (2015) Tenecteplase versus alteplase before endovascular therapy for ischemic stroke (EXTEND-IA TNK). Available from https://clinicaltrials.gov/ct2/show/NCT02388061. Last accessed 26 May 2016
389. Arnold M (2015) Swiss Intravenous and Intra-arterial Thrombolysis for Treatment of Acute Ischemic Stroke Registry (SWISS). Available from https://clinicaltrials.gov/ct2/show/NCT00811538. Last accessed 26 May 2016
390. Lee RS, Ok YC, Lim JS, Lim BC, Cho KY, Lee MC (2012) Outcome evaluation of intravenous infusion of urokinase for acute ischemic stroke. Chonnam Med J 48(1):52–56
391. (2002) Intravenous thrombolysis with urokinase for acute cerebral infarctions: the study group of a 5-year National Project of the People's Republic of China. Available from http://www.strokecenter.org/trials/clinicalstudies/intravenous-thrombolysis-with-urokinase-for-acute-cerebral-infarctions-the-study-group-of-a-5-year-national-project-of-the-peoples-republic-of-china. Last accessed 26 May 2016
392. Donnan G (2009) Australian urokinase stroke trial "AUST Pilot". Available from http://www.strokecenter.org/trials/clinicalstudies/australian-urokinase-stroke-trial--2. Last accessed 26 May 2016
393. AUST (1998) Australian Urokinase Stroke Trial "AUST". Available from http://www.strokecenter.org/trials/clinicalstudies/australian-urokinase-stroke-trial. Last accessed 26 May 2016
394. Bath PM, Gray LJ, Bath AJ, Buchan A, Miyata T, Green AR (2009) Effects of NXY-059 in experimental stroke: an individual animal meta-analysis. Br J Pharmacol 157(7):1157–1171
395. Lapchak PA (2012) Transcranial near-infrared laser therapy applied to promote clinical recovery in acute and chronic neurodegenerative diseases. Expert Rev Med Devices 9(1):71–83
396. Feuerstein GZ, Zaleska MM, Krams M, Wang X, Day M, Rutkowski JL et al (2008) Missing steps in the STAIR case: a translational medicine perspective on the development of NXY-059 for treatment of acute ischemic stroke. J Cereb Blood Flow Metab 28(1):217–219
397. Ginsberg MD (2009) Current status of neuroprotection for cerebral ischemia: synoptic overview. Stroke 40(3 suppl):S111–S114
398. Proctor PH, Tamborello LP (2007) SAINT-I worked, but the neuroprotectant is not NXY-059. Stroke 38(10), e109, author reply e110
399. Lapchak PA, Araujo DM, Song D, Wei J, Purdy R, Zivin JA (2002) Effects of the spin trap agent disodium- [tert-butylimino)methyl]benzene-1,3-disulfonate N-oxide (generic NXY-059) on intracerebral hemorrhage in a rabbit Large clot embolic stroke model: combination studies with tissue plasminogen activator. Stroke 33(6):1665–1670
400. Lapchak PA, Song D, Wei J, Zivin JA (2004) Coadministration of NXY-059 and tenecteplase six hours following embolic strokes in rabbits improves clinical rating scores. Exp Neurol 188(2):279–285
401. Lapchak PA, Araujo DM (2003) Development of the nitrone-based spin trap agent NXY-059 to treat acute ischemic stroke. CNS Drug Rev 9(3):253–262
402. Lapchak PA (2010) A critical assessment of edaravone acute ischemic stroke efficacy trials: is edaravone an effective neuroprotective therapy? Expert Opin Pharmacother 11(10):1753–1763
403. Lapchak PA, Zivin JA (2009) The lipophilic multifunctional antioxidant edaravone (radicut) improves behavior following embolic strokes in rabbits: a combination therapy study with tissue plasminogen activator. Exp Neurol 215(1):95–100
404. Lapchak PA, Wei J, Zivin JA (2004) Transcranial infrared laser therapy improves clinical rating scores after embolic strokes in rabbits. Stroke 35(8):1985–1988
405. Hill MD, Martin RH, Palesch YY, Moy CS, Tamariz D, Ryckborst KJ et al (2015) Albumin administration in acute ischemic stroke: safety analysis of the ALIAS part 2 multicenter trial. PLoS One 10(9), e0131390

406. Ginsberg MD, Palesch YY, Martin RH, Hill MD, Moy CS, Waldman BD et al (2011) The albumin in acute stroke (ALIAS) multicenter clinical trial: safety analysis of part 1 and rationale and design of part 2. Stroke 42(1):119–127
407. Kidwell CS, Lees KR, Muir KW, Chen C, Davis SM, De Silva DA et al (2009) Results of the MRI substudy of the intravenous magnesium efficacy in stroke trial. Stroke 40(5):1704–1709
408. Saver JL (2010) Targeting the brain: neuroprotection and neurorestoration in ischemic stroke. Pharmacotherapy 30(7 pt 2):62S–69S
409. Sanchez EC (2013) Mechanisms of action of hyperbaric oxygenation in stroke: a review. Crit Care Nurs Q 36(3):290–298
410. Michalski D, Pelz J, Weise C, Kacza J, Boltze J, Grosche J et al (2011) Early outcome and blood-brain barrier integrity after co-administered thrombolysis and hyperbaric oxygenation in experimental stroke. Exp Transl Stroke Med 3(1):5
411. Lapchak PA, Araujo DM, Zivin JA (2004) Comparison of Tenecteplase with Alteplase on clinical rating scores following small clot embolic strokes in rabbits. Exp Neurol 185(1):154–159
412. Cook DJ, Teves L, Tymianski M (2012) Treatment of stroke with a PSD-95 inhibitor in the gyrencephalic primate brain. Nature 483(7388):213–217
413. Yoshii A, Constantine-Paton M (2010) Postsynaptic BDNF-TrkB signaling in synapse maturation, plasticity, and disease. Dev Neurobiol 70(5):304–322
414. Lapchak PA, Schubert DR, Maher PA (2011) Delayed treatment with a novel neurotrophic compound reduces behavioral deficits in rabbit ischemic stroke. J Neurochem 116(1):122–131
415. Lapchak PA, Bombien R, Rajput PS (2013) J-147 a novel hydrazide lead compound to treat neurodegeneration: ceetox safety and genotoxicity analysis. J Neurol Neurophysiol. 4(3). pii: 158
416. Asahi M, Asahi K, Jung JC, del Zoppo GJ, Fini ME, Lo EH (2000) Role for matrix metalloproteinase 9 after focal cerebral ischemia: effects of gene knockout and enzyme inhibition with BB-94. J Cereb Blood Flow Metab 20(12):1681–1689
417. Lapchak PA, Chapman DF, Zivin JA (2000) Metalloproteinase inhibition reduces thrombolytic (tissue plasminogen activator)-induced hemorrhage after thromboembolic stroke. Stroke 31(12):3034–3040
418. Lapchak PA, Araujo DM (2001) Reducing bleeding complications after thrombolytic therapy for stroke: clinical potential of metalloproteinase inhibitors and spin trap agents. CNS Drugs 15(11):819–829
419. Galvin IM, Levy R, Boyd JG, Day AG, Wallace MC (2015) Cooling for cerebral protection during brain surgery. Cochrane Database Syst Rev (1):CD006638
420. Sandu RE, Buga AM, Balseanu AT, Moldovan M, Popa-Wagner A (2016) Twenty-four hours hypothermia has temporary efficacy in reducing brain infarction and inflammation in aged rats. Neurobiol Aging 38:127–140
421. Alonso-Alconada D, Broad KD, Bainbridge A, Chandrasekaran M, Faulkner SD, Kerenyi A et al (2015) Brain cell death is reduced with cooling by 3.5 degrees C to 5 degrees C but increased with cooling by 8.5 degrees C in a piglet asphyxia model. Stroke 46(1):275–278
422. Cao Z, Balasubramanian A, Marrelli SP (2014) Pharmacologically induced hypothermia via TRPV1 channel agonism provides neuroprotection following ischemic stroke when initiated 90 min after reperfusion. Am J Physiol Regul Integr Comp Physiol 306(2):R149–R156
423. Sun J, Fang YQ, Ren H, Chen T, Guo JJ, Yan J et al (2013) WIN55,212-2 protects oligodendrocyte precursor cells in stroke penumbra following permanent focal cerebral ischemia in rats. Acta Pharmacol Sin 34(1):119–128
424. Sun J, Fang Y, Chen T, Guo J, Yan J, Song S et al (2013) WIN55, 212-2 promotes differentiation of oligodendrocyte precursor cells and improve remyelination through regulation of the phosphorylation level of the ERK 1/2 via cannabinoid receptor 1 after stroke-induced demyelination. Brain Res 1491:225–235
425. Katz LM, Young A, Frank JE, Wang Y, Park K (2004) Neurotensin-induced hypothermia improves neurologic outcome after hypoxic-ischemia. Crit Care Med 32(3):806–810

426. Katz LM, Wang Y, McMahon B, Richelson E (2001) Neurotensin analog NT69L induces rapid and prolonged hypothermia after hypoxic ischemia. Acad Emerg Med 8(12): 1115–1121
427. Han HS, Park J, Kim JH, Suk K (2012) Molecular and cellular pathways as a target of therapeutic hypothermia: pharmacological aspect. Curr Neuropharmacol 10(1):80–87
428. STAIR (1999) Recommendations for standards regarding preclinical neuroprotective and restorative drug development. Stroke 30(12):2752–2758
429. Turner R, Jickling G, Sharp F (2011) Are underlying assumptions of current animal models of human stroke correct: from STAIRS to high hurdles? Transl Stroke Res 2(2):138–143
430. Lapchak PA (2010) Translational stroke research using a rabbit embolic stroke model: a correlative analysis hypothesis for novel therapy development. Transl Stroke Res 1(2):96–107
431. Lapchak PA (2012) A clinically relevant rabbit embolic stroke model for acute ischemic stroke therapy development: mechanisms & targets. In: Lapchak PA, Zhang JH (eds) Translational stroke research: from target selection to clinical trials. Springer, New York, pp 541–584
432. Cook DJ, Tymianski M (2012) Nonhuman primate models of stroke for translational neuroprotection research. Neurotherapeutics 9(2):371–379
433. Waud DR (1972) On biological assays involving quantal responses. J Pharmacol Exp Ther 183(3):577–607
434. Lapchak PA, De Taboada L (2010) Transcranial near infrared laser treatment (NILT) increases cortical adenosine-5′-triphosphate (ATP) content following embolic strokes in rabbits. Brain Res 1306:100–105
435. Lapchak PA, Boitano PD (2016) A novel method to promote behavioral improvement and enhance mitochondrial function following an embolic stroke. Brain Res. 1646:125–131. doi:10.1016/j.brainres.2016.04.039
436. Zivin JA, Waud DR (1992) Quantal bioassay and stroke. Stroke 23(5):767–773
437. Lapchak PA (2015) A cost-effective rabbit embolic stroke bioassay: insight into the development of acute ischemic stroke therapy. Transl Stroke Res 6(2):99–103
438. Lapchak PA, Zivin JA (2003) Ebselen, a seleno-organic antioxidant, is neuroprotective after embolic strokes in rabbits: synergism with low-dose tissue plasminogen activator. Stroke 34(8):2013–2018
439. Lapchak PA (2006) 3alpha-OL-5-beta-pregnan-20-one hemisuccinate, a steroidal low-affinity NMDA receptor antagonist improves clinical rating scores in a rabbit multiple infarct ischemia model: synergism with tissue plasminogen activator. Exp Neurol 197(2):531–537
440. Lapchak PA (2007) The phenylpropanoid micronutrient chlorogenic acid improves clinical rating scores in rabbits following multiple infarct ischemic strokes: synergism with tissue plasminogen activator. Exp Neurol 205(2):407–413
441. Lapchak PA, Han MK (2010) Simvastatin improves clinical scores in a rabbit multiple infarct ischemic stroke model: synergism with a ROCK inhibitor but not the thrombolytic tissue plasminogen activator. Brain Res 1344:217–225
442. Lapchak PA (2010) Efficacy and safety profile of the carotenoid trans sodium crocetinate administered to rabbits following multiple infarct ischemic strokes: a combination therapy study with tissue plasminogen activator. Brain Res 1309:136–145
443. Lapchak PA, Doyan S, Fan X, Woods CM (2013) Synergistic effect of AJW200, a von Willebrand factor neutralizing antibody with low dose (0.9 mg/mg) thrombolytic therapy following embolic stroke in rabbits. J Neurol Neurophysiol. 4(2). doi:10.4172/2155-9562.1000146
444. Lapchak PA, Daley JT, Boitano PD (2015) A blinded, randomized study of L-arginine in small clot embolized rabbits. Exp Neurol 266:143–146
445. Fisher M (2003) Recommendations for advancing development of acute stroke therapies: Stroke Therapy Academic Industry Roundtable 3. Stroke 34(6):1539–1546
446. Fisher M, Hanley DF, Howard G, Jauch EC, Warach S (2007) Recommendations from the STAIR V meeting on acute stroke trials, technology and outcomes. Stroke 38(2):245–248

447. Saver JL, Albers GW, Dunn B, Johnston KC, Fisher M, STAIR IV Consortium (2009) Stroke Therapy Academic Industry Roundtable (STAIR) recommendations for extended window acute stroke therapy trials. Stroke 40(7):2594–2600
448. Albers GW, Goldstein LB, Hess DC, Wechsler LR, Furie KL, Gorelick PB et al (2011) Stroke Treatment Academic Industry Roundtable (STAIR) recommendations for maximizing the use of intravenous thrombolytics and expanding treatment options with intra-arterial and neuroprotective therapies. Stroke 42(9):2645–2650
449. Kilkenny C, Browne W, Cuthill IC, Emerson M, Altman DG, National Centre for the Replacement, Refinement and Reduction of Amimals in Research (2011) Animal research: reporting in vivo experiments—the ARRIVE guidelines. J Cereb Blood Flow Metab 31(4):991–993
450. Kilkenny C, Parsons N, Kadyszewski E, Festing MF, Cuthill IC, Fry D et al (2009) Survey of the quality of experimental design, statistical analysis and reporting of research using animals. PLoS One 4(11), e7824
451. Ginsberg MD (2007) Life after cerovive: a personal perspective on ischemic neuroprotection in the post-NXY-059 era. Stroke 38(6):1967–1972
452. Savitz SI (2007) A critical appraisal of the NXY-059 neuroprotection studies for acute stroke: a need for more rigorous testing of neuroprotective agents in animal models of stroke. Exp Neurol 205(1):20–25
453. Savitz SI (2009) Cosmic implications of NXY-059. Stroke 40(3 suppl):S115–S118
454. Lapchak PA, Zhang JH (2011) Resolving the negative data publication dilemma in translational stroke research. Transl Stroke Res 2(1):1–6
455. Pardo Cabello AJ, Bermudo Conde S, Manzano Gamero V, Gomez Jimenez FJ, de la Higuera Torres-Puchol J (2013) Implementation of clinical practice guidelines for acute ischaemic stroke in specialist care centres. Neurologia 28:137–144
456. Sipahi I, Swaminathan A, Natesan V, Debanne SM, Simon DI, Fang JC (2012) Effect of antihypertensive therapy on incident stroke in cohorts with prehypertensive blood pressure levels: a meta-analysis of randomized controlled trials. Stroke 43(2):432–440
457. O'Collins VE, Donnan GA, Macleod MR, Howells DW (2013) Hypertension and experimental stroke therapies. J Cereb Blood Flow Metab 33(8):1141–1147
458. Letourneur A, Roussel S, Toutain J, Bernaudin M, Touzani O (2011) Impact of genetic and renovascular chronic arterial hypertension on the acute spatiotemporal evolution of the ischemic penumbra: a sequential study with MRI in the rat. J Cereb Blood Flow Metab 31(2):504–513
459. Yao H, Nabika T (2012) Standards and pitfalls of focal ischemia models in spontaneously hypertensive rats: with a systematic review of recent articles. J Transl Med 10:139
460. Reid E, Graham D, Lopez-Gonzalez MR, Holmes WM, Macrae IM, McCabe C (2012) Penumbra detection using PWI/DWI mismatch MRI in a rat stroke model with and without comorbidity: comparison of methods. J Cereb Blood Flow Metab 32(9):1765–1777
461. McCabe C, Gallagher L, Gsell W, Graham D, Dominiczak AF, Macrae IM (2009) Differences in the evolution of the ischemic penumbra in stroke-prone spontaneously hypertensive and Wistar-Kyoto rats. Stroke 40(12):3864–3868
462. Kiers L, Davis SM, Larkins R, Hopper J, Tress B, Rossiter SC et al (1992) Stroke topography and outcome in relation to hyperglycaemia and diabetes. J Neurol Neurosurg Psychiatry 55(4):263–270
463. Baird TA, Parsons MW, Barber PA, Butcher KS, Desmond PM, Tress BM et al (2002) The influence of diabetes mellitus and hyperglycaemia on stroke incidence and outcome. J Clin Neurosci 9(6):618–626
464. Lees KR (2005) Walters MR acute stroke and diabetes. Cerebrovasc Dis 20(suppl 1):9–14
465. Shimoyama T, Kimura K, Uemura J, Saji N, Shibazaki K (2014) Elevated glucose level adversely affects infarct volume growth and neurological deterioration in non-diabetic stroke patients, but not diabetic stroke patients. Eur J Neurol 21(3):402–410

466. Abdelsaid M, Prakash R, Li W, Coucha M, Hafez S, Johnson MH et al (2015) Metformin treatment in the period after stroke prevents nitrative stress and restores angiogenic signaling in the brain in diabetes. Diabetes 64(5):1804–1817
467. Ergul A, Abdelsaid M, Fouda AY, Fagan SC (2014) Cerebral neovascularization in diabetes: implications for stroke recovery and beyond. J Cereb Blood Flow Metab 34(4):553–563
468. Ergul A, Kelly-Cobbs A, Abdalla M, Fagan SC (2012) Cerebrovascular complications of diabetes: focus on stroke. Endocr Metab Immune Disord Drug Targets 12(2):148–158
469. Prakash R, Li W, Qu Z, Johnson MA, Fagan SC, Ergul A (2013) Vascularization pattern after ischemic stroke is different in control versus diabetic rats: relevance to stroke recovery. Stroke 44(10):2875–2882
470. Lapchak PA (2013) Drug-like property profiling of novel neuroprotective compounds to treat acute ischemic stroke: guidelines to develop pleiotropic molecules. Transl Stroke Res 4(3):328–342
471. Lipinski CA (2000) Drug-like properties and the causes of poor solubility and poor permeability. J Pharmacol Toxicol Methods 44(1):235–249
472. Lipinski CA (2004) Lead- and drug-like compounds: the rule-of-five revolution. Drug Discov Today Technol 1(4):337–341
473. Lipinski CA, Lombardo F, Dominy BW, Feeney PJ (2001) Experimental and computational approaches to estimate solubility and permeability in drug discovery and development settings. Adv Drug Deliv Rev 46(1–3):3–26
474. Hellinger E, Veszelka S, Toth AE, Walter F, Kittel A, Bakk ML et al (2012) Comparison of brain capillary endothelial cell-based and epithelial (MDCK-MDR1, Caco-2, and VB-Caco-2) cell-based surrogate blood-brain barrier penetration models. Eur J Pharm Biopharm 82(2):340–351
475. Sharma V, McNeill JH (2009) To scale or not to scale: the principles of dose extrapolation. Br J Pharmacol 157(6):907–921
476. Pellegatti M (2012) Preclinical in vivo ADME studies in drug development: a critical review. Expert Opin Drug Metab Toxicol 8(2):161–172
477. Brimecombe JC, Kirsch GE, Brown AM (2009) Test article concentrations in the hERG assay: losses through the perfusion, solubility and stability. J Pharmacol Toxicol Methods 59(1):29–34
478. Cheng CS, Alderman D, Kwash J, Dessaint J, Patel R, Lescoe MK et al (2002) A high-throughput HERG potassium channel function assay: an old assay with a new look. Drug Dev Ind Pharm 28(2):177–191
479. Gintant G (2011) An evaluation of hERG current assay performance: translating preclinical safety studies to clinical QT prolongation. Pharmacol Ther 129(2):109–119
480. Goineau S, Legrand C, Froget G (2012) Whole-cell configuration of the patch-clamp technique in the hERG channel assay to predict the ability of a compound to prolong QT interval. Curr Protoc Pharmacol. Chapter 10:Unit 10.5
481. Yu HB, Zou BY, Wang XL, Li M (2016) Investigation of miscellaneous hERG inhibition in large diverse compound collection using automated patch-clamp assay. Acta Pharmacol Sin 37(1):111–123
482. McDonnell AM, Dang CH (2013) Basic review of the cytochrome p450 system. J Adv Pract Oncol 4(4):263–268
483. Hedlund E, Gustafsson JA, Warner M (2001) Cytochrome P450 in the brain; a review. Curr Drug Metab 2(3):245–263
484. Evers R, Dallas S, Dickmann LJ, Fahmi OA, Kenny JR, Kraynov E et al (2013) Critical review of preclinical approaches to investigate cytochrome p450-mediated therapeutic protein drug-drug interactions and recommendations for best practices: a white paper. Drug Metab Dispos 41(9):1598–1609
485. Meador V, Jordan W, Zimmermann J (2002) Increasing throughput in lead optimization in vivo toxicity screens. Curr Opin Drug Discov Devel 5(1):72–78
486. Yoon M, Campbell JL, Andersen ME, Clewell HJ (2012) Quantitative in vitro to in vivo extrapolation of cell-based toxicity assay results. Crit Rev Toxicol 42(8):633–652

Chapter 2
Systematic Review and Meta-analysis: Important Tools in Understanding Drug Development for Stroke

Sarah K. McCann, Emily S. Sena, Gillian L. Currie, Malcolm R. Macleod, and David W. Howells

Abstract Animal models of ischaemic stroke have become an integral part of the preclinical pipeline for identifying novel neuroprotective drug targets and drugs. As the process serves as a filter, researchers do not expect complete concordance between the experimental animal and human clinical trial data. However, the paucity of clear examples of translation of promising animal results into drugs that work in a clinical setting has raised concerns about the utility of this translational paradigm. Preclinical systematic reviews have been used in response to these concerns to identify weaknesses in animal studies and provide empirical evidence supporting improvements to the design and conduct of preclinical animal experiments. We propose that further strategic development and application of data analysis methods can help continue this process of improvement and help identify the most promising therapeutic targets and drugs. These next steps in systematic review aim to tighten the focus of preclinical research, streamline the drug development process, and minimise research waste.

Keywords Stroke • Preclinical systematic review • Meta-analysis • Translational research • Bias

S.K. McCann • E.S. Sena • G.L. Currie • M.R. Macleod
Centre for Clinical Brain Sciences, University of Edinburgh, Edinburgh, UK

D.W. Howells (✉)
Florey Institute of Neuroscience and Mental Health, Heidelberg, VIC, Australia

Faculty of Health, School of Medicine, University of Tasmania, Medical Science Precinct, 17 Liverpool Street, Hobart, TAS, Australia
e-mail: david.howells@utas.edu.au

P.A. Lapchak, J.H. Zhang (eds.), *Neuroprotective Therapy for Stroke and Ischemic Disease*, Springer Series in Translational Stroke Research,
DOI 10.1007/978-3-319-45345-3_2

Abbreviations

ARRIVE	Animal Research: Reporting of In Vivo Experiments
CAMARADES	Collaborative Approach to Meta-analysis and Review of Animal Data from Experimental Studies
CONSORT	Consolidated Standards of Reporting Trials
EBPM	Evidence-based Preclinical Medicine
EDA	Experimental Design Assistant
EuroHYP-1	European multicentre randomised, phase III clinical trial of hypothermia plus best medical treatment versus best medical treatment alone for acute ischaemic stroke
FOCUS	Fluoxetine or Control Under Supervision
IL-1 RA	Interleukin-1 receptor antagonist
MeSH	Medical Subject Headings
NC3Rs	National Centre for the Replacement Refinement and Reduction of Animals in Research
NIH	National Institutes of Health
NINDS	National Institutes of Neurological Disorders and Stroke
NMDA	*N*-methyl-D-aspartate
PRISMA	Preferred Reporting Items for Systematic Reviews and Meta-Analyses
QUOROM	Quality of Reporting of Meta-analyses
SAINT	Stroke -Acute Ischemic NXY Treatment
STAIR	Stroke Treatment academic industry roundtable
SYRCLE	Systematic Review Centre for Laboratory Animal Experimentation
three Rs	Replacement reduction and refinement
tPA	Tissue plasminogen activator
UK	United Kingdom
US	United States

1 Introduction

Preclinical experiments involving animal models are used to investigate the pathophysiology of stroke, identify new therapeutic targets, and evaluate new treatments for clinical application. In the first and second of these research efforts, the contribution of animal experiments has been critical for our understanding of many aspects of human stroke [1]. However, when used to select candidate drugs for clinical trial, apparently successful animal experiments have generally not yielded clinically useful stroke drugs.

Reasons for the disparity between preclinical and clinical treatment efficacy are hypothesised to include that the animal experiments are falsely positive; clinical trials are falsely negative; or we cannot model stroke in animals with sufficient fidelity to provide a useful tool for translation. The hypothesis that animal experiments may report overly or falsely positive results has been the focus of intense scrutiny in recent years, fuelled largely by the results of preclinical systematic reviews. Systematic review and meta-analysis are summary techniques developed to combine the results of clinical trials and inform evidence-based health care policy and practice. Their application to preclinical research is relatively recent, but increasingly used not only to summarise available evidence, but to understand poor translation and identify approaches to improve the robustness of animal experiments.

One of the earliest systematic reviews and meta-analyses of preclinical stroke studies illustrates the high attrition rate when moving from the bench to the clinic. O'Collins and colleagues identified over 500 experimental stroke treatments effective in animals, of which only one, tissue plasminogen activator (tPA), proved successful in the clinic [2]. This exception highlights the important principal that animal experiments can be highly predictive of human outcomes. However, it should be noted that while the animal and human data are highly concordant, commencement of the clinical trial preceded the majority of the preclinical data. For tPA, the human and animal outcome data are almost identical with respect to the critical issues of timing of therapy and contraindicated comorbidities [3]. Similar parallels have been drawn by Dirnagl and Endres for the predictive value of preclinical studies of statins and antiplatelet agents and for the harmful interaction between tPA and erythropoietin [1]. However, systematic review and meta-analysis with follow-up experimentation also suggest that the presence of unrecognised biases within the preclinical datasets probably account for the failure of clinical trials of the free radical scavenger NXY-059 [4, 5] and of magnesium as an inhibitor of the NMDA receptor [6, 7]. Over-optimistic reportings of benefit when researchers fail to prevent bias in experimental design, or use cohort sizes too small to provide adequate statistical power for their animal experiments, are common themes in many analyses [8, 9]. Performing only those experiments most likely to provide a positive outcome and ignoring more difficult experiments in animals with comorbidities prevalent in the clinic (e.g. old age, hypertension, diabetes) and known to yield smaller treatment effects [8] are also likely to contribute to these difficulties. Overall, these observations suggest a failure of the drug evaluation process rather than failure of the animal models themselves.

The consequences of low-quality preclinical evidence are broader than failure to translate efficacy in clinical trials, extending to waning investment [10] and wasting of scarce research funds [11]. Strategies are needed to improve the quality of preclinical data collection and to allow more effective selection of therapeutic targets and drugs for stroke. Increased use of systematic review and meta-analysis should be part of this strategy.

2 Evidence-Based Approaches to Understanding Translational Stroke Research

The process and value of systematic review and meta-analysis is exemplified by Cochrane [12]. Cochrane, coupled with impetus from the CONSORT statement for better performance and reporting of clinical trials [13], has been highly successful in synthesising clinical trial data and providing outcomes in an easily assimilated, widely recognised format readily useable for healthcare policy and day-to-day clinical practice decisions. The systematic review process has been invaluable in identifying and understanding deficiencies and areas for improvement in clinical trial design, conduct, and reporting; this has driven unprecedented improvements in clinical research quality [14].

Systematic review aims to capture all evidence relevant to a pre-specified research question and the statistical pooling of outcome data (meta-analysis) provides a summary estimate of effect. These are transparent, reproducible methods to objectively synthesise and interpret scientific evidence. The processes and conclusions of a systematic review have a reduced risk of bias when compared to traditional, narrative reviews. In contrast to clinical systematic reviews, our primary interest is generally not in the 'headline' summary effect of an intervention, but investigating differences in efficacy between groups of studies, known as heterogeneity. Using these methods, we can assess the quality and range of available evidence; assess for publication bias; identify gaps in a field and inform the design of future experiments; try to explain discrepancies between preclinical and clinical trial results; and inform clinical trial design.

Prior to 1996, only 13 systematic reviews and/or meta-analyses of animal studies were identified through systematic searching. By 2006, this number had risen to 103 [15] and there are now over 500 reviews of in vivo animal experiments [16], a high proportion of these examining focal ischaemic stroke studies. Increasing the value of existing research, informing future research, and reducing research waste are just a few of the benefits derived through conducting these reviews.

Over the past 10 years, we have witnessed the evolution of methodologies [17] and applications for preclinical systematic review, which is now emerging as a field in its own right. Ongoing advances signal the promotion of these methods to prerequisites in the process of experimental design.

To date, preclinical systematic review and meta-analysis in stroke have helped identify animal experiments that are at high risk of experimental bias (leading to overstatement of efficacy) [5, 18–21] and publication bias [22]. They have identified that many researchers use cohort sizes too small to provide adequate statistical power [8, 9]. These analyses also suggest that modelling variables such as the choice of anaesthetic used have an impact on outcome [3, 5, 19, 23] and imply that we do not fully understand the animal models. Moreover, the observation by O'Collins that the effect sizes for those drugs taken forward to clinical trial were not substantially different from the average of all drugs tested in animals suggests that we may not even be prioritising the most promising treatments for progress to clinical trial [2].

2.1 *Assessing the Quality and Range of Available Evidence*

Quality and range are central to evaluating the evidence supporting a new pathophysiological mechanism, therapeutic target, or therapy for stroke. Quality is related to internal validity and is assessed at the level of individual studies or outcomes—how well were they conducted and how reliable are their results? Range of evidence refers to the body of evidence as a whole and speaks to its external validity and generalisability—has the target or therapy been tested over a number of experimental models and animal species, and in a clinically relevant context?

2.1.1 Quality of Evidence

The usefulness of animal data is often impaired by limited methodological quality. Problems with the design and execution of experiments, including risks of bias, can lead to exaggeration of the effect under investigation. Bias can introduce systematic error in the results of a study and can occur due to inadequacies in design, conduct, analysis, or reporting. In clinical research, assessing study quality helps to identify what evidence can be used to reliably inform healthcare decisions.

Assessing the quality of preclinical experiments is a critical component of systematic review, which allows us to evaluate the reliability of evidence and the level of confidence in our conclusions. This has proven itself a useful tool to highlight and investigate the impact of deficiencies in study quality. Quality assessment involves critically appraising studies based on the implementation of experimental design measures to reduce biases affecting selection, performance, measurement, attrition, and outcome reporting (Table 2.1). Random allocation of animals to experimental groups helps to ensure that groups are balanced for baseline characteristics. Measured effects are therefore more likely due to the intervention than pre-existing group characteristics. Blinding experimenters to group allocation ensures that biases in the way that animals are handled, stroke is induced, and outcomes are measured are reduced. To prevent attrition and reporting biases, exclusion criteria and primary

Table 2.1 Examples of types of bias that can threaten internal validity and measures to reduce their impact

Type of bias	Description	Study design and reporting measures to mitigate bias
Selection bias	Systematic differences between study groups at the start of an experiment	Randomise animals to groups
Performance bias	Systematic differences between how groups are handled during an experiment	Blind induction of ischaemia
		Blind conduct of experiment
Measurement or detection bias	Systematic differences between groups in how outcomes are measured	Blind assessment of outcome
Attrition bias	Systematic differences between groups in the number of animals excluded	Pre-specify exclusion criteria
		Report all exclusions
Outcome reporting bias	Systematic differences in the nature or direction of reported vs. unreported outcomes	Pre-specify outcomes
		Report all outcomes

and secondary outcomes should be pre-specified. All animals involved in an experiment should be accounted for and all outcomes reported, not only those that are significant or interesting.

Sample size calculation is also a critical element of study design; the number of animals included in an experiment affects both its validity and external factors. Too few animals and the result will be imprecise and lack the statistical power to detect meaningful differences between groups; too many and resources—animals, money, time—are wasted. While bias introduces systematic error, imprecision refers to random error and, despite smaller studies being less precise, they are not necessarily more biased [24]. To date, the impact of carrying out a sample size calculation has been difficult to measure in stroke studies due to low prevalence of reporting (Table 2.3, Fig. 2.1). Performing a sample size calculation and taking measures to reduce the biases in Table 2.1 are widely accepted to form a core set of standards that should be addressed in all experimental stroke studies [25, 26].

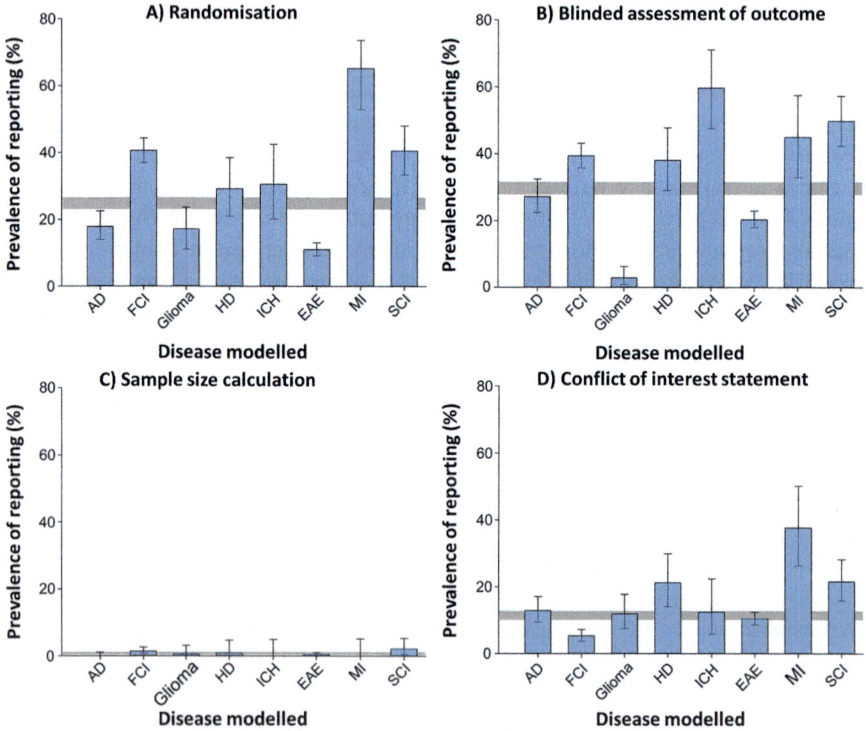

Fig. 2.1 Prevalence of reporting selected risks of bias in 2671 publications describing interventions in animal models of disease [28]. (**a**) Randomisation, (**b**) blinded assessment of outcome, (**c**) sample size calculations, and (**d**) conflict of interest; Alzheimer's disease (AD, $n = 324$ publications), focal cerebral ischaemia (FCI, 704), glioma (175), Huntington's disease (HD, 113), intracerebral haemorrhage (ICH, 72), experimental autoimmune encephalomyelitis (EAE, 1029), myocardial infarction (MI, 69), and spinal cord injury (SCI, 185). *Vertical error bars* represent 95% confidence intervals, and the *horizontal grey bar* represents the 95% confidence interval of the overall estimate

Table 2.2 Quality checklist items relating to stroke study conduct, reporting, and publication

Quality item	Description
Study conduct—stroke induction	
Control of temperature	To ensure that outcome effects are dependent on the intervention and not hyper- or hypothermic disturbances
Monitoring of blood pressure and blood gases	To ensure that physiological changes are not contributing to the treatment effect
Confirmation of cerebral blood flow reduction and (for transient ischaemia) restoration	Via laser Doppler or perfusion imaging, to establish that baseline ischaemic insult is similar in all animals
Avoidance of anaesthetics with marked intrinsic neuroprotective properties	To ensure that any intervention effect is not confounded by the action of anaesthesia
Study reporting and publication	
Publication in peer-reviewed journal	Indicating that the methods, results, and interpretation have undergone independent third-party appraisal
Statement of compliance with regulatory requirements	Indicating experiments have been conducted in accordance with ethical standards
Statement regarding possible conflict of interest	To help identify possible bias introduced by vested interests of the researchers involved

Further quality characteristics of stroke study conduct, reporting, and publication are described in Table 2.2. These items cover control of variables that have been shown to influence stroke outcome including temperature, physiological parameters, and cerebral blood flow. While anaesthetics have also been shown to impact stroke outcomes, the use of this item as a measure of quality can be contentious given the reported neuroprotective properties of many anaesthetics [27].

A systematic review by Krauth and colleagues identified 30 distinct instruments developed to assess quality and risk of bias in published animal studies [24]. Nine stroke-specific instruments were identified, the highest representation for disease-specific assessment tools; however, the validity and reliability of most of these tools has yet to be tested. In Table 2.3, we show the prevalence of reporting quality items that fulfil the Collaborative Approach to Meta-Analysis and Review of Animal Data from Experimental Studies (CAMARADES) checklist from systematic reviews of experimental ischaemic stroke studies. Within this cohort of studies, randomisation, blinding, allocation concealment, and a statement of potential conflict of interest are all reported in less than 50 % of publications, and a sample size calculation reported in only 3 %.

Within publications from different biomedical fields, identified in the context of systematic review, the prevalence of reporting measures to reduce selected biases varies considerably (Fig. 2.1) [28]. In this cohort of studies, focal cerebral ischaemia studies perform close to the average with slightly higher rates of randomisation (40.1 %) and blinding (38.5 %) than other fields and lower reporting of potential conflicts of interest (5.2 %). Across the board, reporting a sample size calculation is almost non-existent in preclinical studies (overall 0.8 %). There is room for substantial improvement in reporting these measures in the range of disease models investigated and likely across the biomedical sciences.

Table 2.3 Number (n) and proportion (%) of publications reporting individual components of the CAMARADES study quality checklist in preclinical systematic reviews of focal ischaemic stroke

	No.	(1)		(2)		(3)		(4)		(5)		(6)		(7)		(8)		(9)	
	pubs	*n*	%	*n*	%	*n*	%	*n*	%	*n*	%	*n*	%	*n*	%	*n*	%	*n*	%
Nicotinamide 2004	14	11	79	9	64	3	21	2	14	3	21	10	71	0	0	10	71	0	0
Melatonin 2005	13	12	92	11	85	4	31	2	15	4	31	11	85	0	0	10	77	0	0
FK506 2005	29	24	83	22	76	6	21	1	3	2	7	24	83	0	0	15	52	0	0
Tirilazad 2007	18	17	94	17	94	12	67	1	6	13	72	15	83	0	0	11	61	0	0
Hypothermia 2007	101	86	85	101	100	36	36	4	4	38	38	84	83	0	0	48	48	65	64
Piracetam 2008	6	6	100	3	50	2	33	0	0	1	17	6	100	0	0	5	83	0	0
NXY-059 2008	9	9	100	5	56	3	33	5	56	4	44	7	78	2	22	7	78	0	0
tPA 2010	104	95	91	63	61	45	43	16	15	27	26	72	69	7	7	72	69	0	0
Stem cells 2012	117	117	100	76	65	52	44	22	19	57	49	79	68	1	1	84	72	13	11
Antidepressants 2014	44	44	100	35	80	24	55	15	34	27	61	38	86	0	0	35	80	4	9
IL-1 RA 2016	25	24	96	22	88	8	32	6	24	12	48	23	92	3	12	17	68	7	28
Total	480	445	93	364	76	195	41	74	15	188	39	369	77	13	3	314	65	89	19

(1) Publication in a peer-reviewed journal, (2) Statement of control of temperature, (3) Randomisation to intervention or control, (4) Allocation concealment (blinded induction of ischaemia), (5) Blinded assessment of outcome, (6) Avoidance of anaesthetics with marked intrinsic neuroprotective properties (ketamine), (7) Sample size calculation, (8) Statement of compliance with ethical regulatory requirements, and (9) Statement regarding possible conflict of interest

2.1.2 Range of Evidence

Systematic reviews summarise comprehensively the range of evidence supporting a specific research question. No animal model of stroke can encapsulate the heterogeneous human disease in its entirety. Evidence that mechanisms of disease or treatments work under a range of conditions in multiple models is therefore necessary. Consistent results across diverse experimental designs inspire confidence that humans might respond in a similar manner. This ability to generalise findings across different measures, settings, and times is known as external validity and is assessed over a body of evidence. Construct validity refers to how well the experiment models human stroke and can be threatened when only specific characteristics of the disease are represented. Recommendations regarding the range of evidence that should be reported before research into a particular therapeutic target or drug advances through the translational pipeline have been discussed in detail by the STAIR collaboration and others and are listed in Table 2.4 [25, 29].

Some of the headline recommendations include evidence of an effect using clinically relevant endpoints, including sensorimotor and cognitive outcomes, at time points delayed from stroke onset to ensure sustained effects of treatment. In addition, efficacy should be shown in aged and comorbid animals of both sexes to avoid the notable discordance that currently exists between subjects of in vivo stroke studies and the human stroke population. Healthy, young, male animals are not representative of the often elderly stroke patient who may have co-existing condi-

Table 2.4 Checklist of the range of circumstances neuroprotective mechanisms or drugs should be tested under

Range of evidence
Assessment of histological outcome
Assessment of behavioural outcomes
Assessment of outcomes at acute and delayed time points
Replication in two or more laboratories
Tested in models of temporary and permanent ischaemia
Tested in males and females
Tested in animals with comorbidities including hypertension and diabetes
Tested in multiple species
Clinically appropriate route of administration tested
Dose–response relationship tested
Optimal time window of treatment investigated
Interaction studies with medications commonly used in stroke patients
Relevant biomarker endpoints examined

tions such as hypertension, diabetes, or hypercholesterolemia. In animals with comorbid conditions including hypertension and diabetes, a number of treatments show reduced or no effect [5, 19–21, 30, 31]. While it is recommended to show evidence of efficacy in multiple species to illustrate conservation of effect, the benefit of testing in higher order species is yet to be proven and increases markedly the cost and ethical considerations of experiments. The usefulness of testing in gyrencephalic species may be dependent on intervention mechanism of action. Evidence of efficacy using clinically relevant times and routes of treatment administration and investigation of the dose–response relationship are all recommended. Lastly, it is recommended that any effects should be replicated in at least two independent laboratories to avoid basing decisions on irreproducible results, a widespread problem in biomedicine [32]. By investigating the impact of different experimental design parameters and modelling characteristics, we can identify the range of circumstances where an intervention is effective and the limits to this efficacy.

2.2 *Assessing for Publication Bias*

Where data are collected but remain unpublished, they cannot contribute to our working knowledge of disease mechanisms and the efficacy of new treatments. Publication bias occurs when the results of published and unpublished studies differ systematically; typically the selective publication of significant, positive, or hypothesis-supporting results. This distorts scientific evidence and contributes to exaggerated intervention efficacy in research summaries. Research waste also occurs when investigators duplicate experiments with negative or neutral results that are not freely available in the public domain. An analysis of publication bias in systematic reviews of experimental ischaemic stroke by Sena and colleagues found that, of 525 unique publications, only 2 % reported no significant results. Further, they estimated that, in this dataset, published studies overestimated efficacy by approximately 30 % [22].

Assessment of publication bias is increasingly used in systematic reviews to detect and quantify the impact of selective publication [16]. This commonly includes assessing funnel plot asymmetry and estimating the number and effect magnitude of theoretically missing studies (Fig. 2.2). There are important limitations to these tests, including low power, factors other than publication bias causing funnel plot asymmetry, and both known and unknown additional factors confounding effect size estimates; they do, however, provide useful tools [22]. For ischaemic stroke reviews carried out using the CAMARADES database from 2010, publication bias was detected in 5 out of 9 datasets assessed. Bias resulted in relative overestimations of treatment efficacy of up to 65.3 % for structural outcomes [33] and 52.1 % for functional outcomes [23].

Publication bias is now recognised as one of a group of publication-related biases that can be introduced during the dissemination of preclinical evidence. These include biases caused by time, type and language of publication, multiple

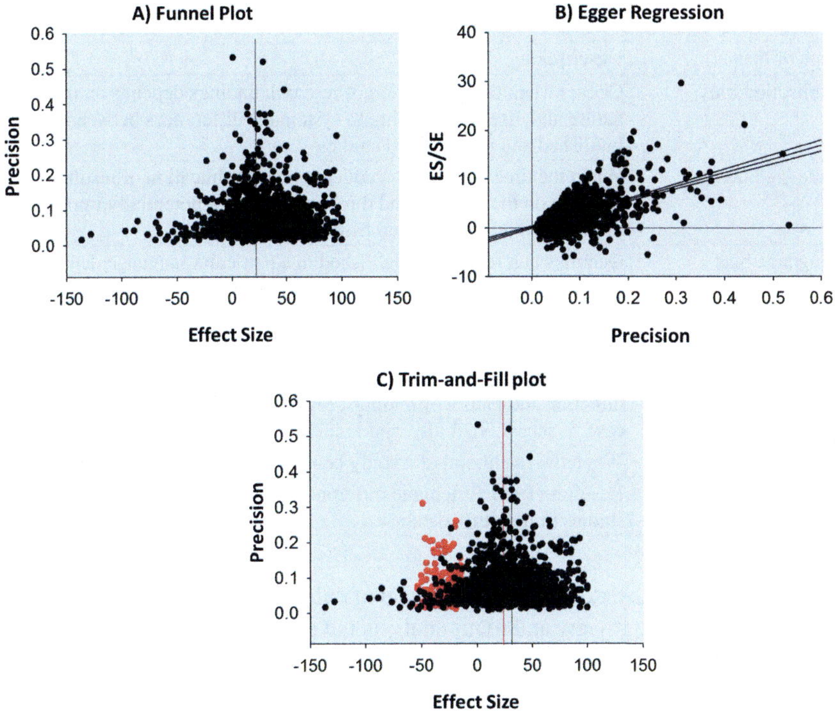

Fig. 2.2 Examples of plots used to assess publication bias. (**a**) Asymmetry in a *funnel plot* can be caused by non-publication of smaller studies that lack significant effects. The *horizontal line* represents the global estimate of efficacy. (**b**) This asymmetry can be tested using *Egger's test*; a regression line that does not pass through the origin suggests the presence of publication bias. (**c**) The number and effect sizes of theoretically 'missing' studies (*red points*) are estimated using *trim and fill*. An adjusted global estimate can be calculated, taking into account these missing studies (*red line*). Data are from a range of preclinical systematic reviews of interventions for focal ischaemic stroke [22]

publication, selective citation, and database indexing (Table 2.5) [34, 35]. Selective outcome reporting bias is also often considered under the umbrella of dissemination bias, in addition to threatening internal validity. Some of these biases can be mitigated during systematic review by ensuring methods are robust: searching multiple literature databases, not specifying language restrictions, and checking for duplicate publications. However, detection of many biases is difficult and no formal tests or research exist in the preclinical literature on, for example, rates of multiple publication, selective citation, or database indexing biases.

An alternative approach to detecting publication-related biases in large datasets is to test for excess significance by examining whether too many studies report significant results compared to what is expected. Using this method, substantial excess significance was found in focal ischaemic stroke meta-analyses: 626 results were observed to be positive, where only 370 would be expected a priori [9].

Table 2.5 Types of publication-related or dissemination biases

Type of bias	Description
Publication bias	Occurs when the publication of research findings depends on their nature and direction, leading to systematic differences between published and unpublished studies
Time lag bias	When the time between the collection and publication of results depends on their nature and direction, meaning potentially important information remains hidden from public scrutiny
Language bias	Occurs when results are published in a particular language depending on their direction and strength
Multiple (duplicate) publication bias	Resulting from significant results generating duplicate or multiple publications
Location bias	Occurs when the location (journal) of publication depends on the direction and nature of findings, resulting in systematic differences in ease of access, visibility, and indexing
Citation bias	When the likelihood of a study being cited is influenced by its results
Indexing bias	Resulting from systematic differences in the indexing of published studies in literature databases

By quantifying the extent and likely impact of publication biases in the preclinical literature, systematic review and meta-analysis have served as a catalyst for debate and changes in publication policy and practice. There are a growing number of journals and journal sections devoted to negative or null results, including the BioMed Central Journal of Negative Results in Biomedicine, PLOS ONE's Missing Pieces Collection, and the Journal of Cerebral Blood Flow and Metabolism's Negative Results Section. Additional measures include new publication models focussing on transparency, data sharing, and the pre-registration of preclinical studies with guaranteed publication of final results.

When trying to mitigate these biases, it is also important to acknowledge that much of science is built upon perverse incentives that reward scientists for 'impact' and 'productivity' rather than for the quality of their research or its reproducibility. Non-publication is a prevailing issue in the biomedical sciences, which can just as likely result from researchers not submitting neutral studies for publication as editors not deeming them worthy. Only half of completed preclinical and clinical studies are reported at all with selective publication of positive or significant results further biasing the literature [36]. A shift in attitudes and reward mechanisms in science is needed to address this problem. Standards for reporting study protocols, all study outcomes, and underlying data should be endorsed and enforced by all stakeholders including funding bodies, animal ethics committees, academic institutions, journals, and publishers [36].

2.3 *Impacts of Systematic Reviews*

Systematic reviews have proven useful to identify where data are lacking in terms of both the quality and range of evidence and to inform the design of future primary studies that fill these gaps. In 2006, Banwell and colleagues conducted a systematic review of IL-1 RA in models of stroke [37]. The review suggested a number of shortcomings in the range of conditions under which efficacy was tested, specifically in terms of comorbidities and time to treatment, study quality was modest, and there was evidence consistent with substantial publication bias. Leaders in the field have suggested that this initial review had a meaningful impact on their research focus and the manner in which they reported future in vivo studies. A recent update, 10 years later, suggests that many of the evidence gaps have indeed been filled [33]. The quality of the primary studies appears to have improved substantially; evidence of efficacy of IL-1 RA in stroke has now been demonstrated in comorbid animals, at much later times of administration and where delivery is via a clinically feasible route.

Establishing the prevalence and impact of risks of bias in the animal literature modelling stroke has had broader implications than the design of primary stroke studies and practice within specific research programmes. In response to the accumulating evidence from systematic reviews, Macleod and colleagues, in 2009, developed Good Laboratory Practice Guidelines in an attempt to address the limitations in the design, conduct, and reporting of animal experiments in stroke that were published in three major stroke journals [38–40]. The following year, the ARRIVE guidelines were developed by the NC3Rs to address these issues more broadly across the use of animals in experiments of the life sciences [41]. The necessity to improve the rigour with which we undertake and disseminate preclinical studies is now appreciated more widely. The ARRIVE guidelines have been endorsed in the UK by all major scientific journals, funding bodies, universities, and learned societies. In 2013, the Nature Publishing Group changed its editorial policy to try and address the replication crisis by requesting much more robust reporting by contributors to their journals [42].

The plethora of evidence has made substantial strides in highlighting the potential impact of bias on research findings, but informal discussions suggested that practical assistance to overcome these limitations was missing. In late 2015, the NC3Rs launched the Experimental Design Assistant (EDA) to engage researchers to think more critically about experimental design and provide a platform that assists in this process.

Granting bodies have also taken heed of the evidence from systematic review and concerns of replication crisis. The US National Institutes of Neurological Disorders and Stroke (NINDS) at the National Institutes of Health (NIH) and Research Councils UK now require substantially more detail regarding the rigour of experimental design including, for instance, randomisation, blinding, and meaningful power calculations.

Some potential impacts are yet to be realised. An important consideration in the use of animals in biomedical research is the application of the three Rs. Systematic review has potential to lead to Refinement and Reduction in the modelling of stroke. On Reduction, very few (less than 1 in 50) publications report a sample size calculation; systematic reviews can provide reliable data to support sample size calculations for various outcome measures. On Refinement, systematic reviews can allow comparison of the statistical performance of different outcome measures. All other things being equal, investigators should choose the outcome measure with the greatest precision. Further, different outcome measures do measure different aspects of biology; but since more than one outcome is reported for many experiments, and because a diverse range of biological activity is usually represented in the candidate interventions, we can use meta-regression to understand the relationship between different outcomes at different times. We can therefore start to characterise the extent to which subjecting the animal to an additional test results in additional knowledge. One benefit of this approach is that it uses existing data, rather than requiring additional animal experiments to develop and validate the approach.

2.4 Providing Potential Explanations for Discrepancies Between Preclinical and Clinical Trial Results

When a clinical trial does not show the efficacy which was predicted from animal models, it may be useful to identify possible reasons for this. These fall into three categories; that the clinical trial was falsely negative; or the animal studies were falsely positive; or that the animal model used did not recapitulate key features of human disease with sufficient fidelity to be useful in predicting efficacy in humans. By identifying limits to the efficacy of treatments in animal models, for example relating to dose or time to treatment, we can investigate whether clinical trials operated outside of these limits, as a possible reason for translational failure. To provide contrasting examples, NXY-059 was tested in humans on the basis of promising animal data, but—in addition to problems with the rigour of many of the animal experiments—only 7 % of the animals studied had high blood pressure [5]. This compares with two thirds of participants in the SAINT 2 clinical trial who had a history of hypertension at study entry [4]. As a second example, there are large datasets (involving more than 3000 subjects each) from animal and human studies of tPA. In thrombotic occlusion models in animals, the decline in efficacy in 90 min epochs from the time of occlusion is very similar to that seen in individual patient meta-analysis of clinical trials, suggesting that the impact of "duration of occlusion" on the development of brain injury is broadly similar between human and non-human animals, thus providing validation for the use of animal models [3].

2.5 Informing Clinical Trial Design

Of course, it makes more sense to set up our clinical trials to match—as closely as practicable—the circumstances under which efficacy was observed in animals when the clinical trial is designed, than to use such differences, identified after the event, to explain why clinical trials failed. There are several examples of where this has been done, including in the design of the EuroHYP-1 trial of brain cooling for stroke [43], where several key trial parameters (including the target temperature and the allowable delay before the initiation of treatment) were determined after taking into account findings from a large systematic review of animal data [44]. Additionally, the use of pethidine to manage shivering was allowed following subsequent targeted animal experiments [45].

Other examples in stroke include antidepressants [46], tested in the FOCUS trial [47]; and this approach has also been used in trials in spinal cord injury [48, 49] and multiple sclerosis ([28], https://clinicaltrials.gov/ct2/show/NCT01910259?term=MS-SMART&rank=1).

While the exclusive use of this approach limits the clinical investigator to considering those data which already exist, it might effectively be combined with phase III preclinical studies, in a development path which sees the supporting evidence for candidate drugs summarised using systematic review and meta-analysis; before key findings are confirmed and areas of uncertainty resolved in high quality large multicentre animal studies. While the costs of this approach have to date, at least in stroke, deterred either industry or academic funders from adopting this approach, it is likely that it would represent substantial value [50].

3 Limitations of Systematic Review and Meta-analysis

While preclinical systematic review and meta-analysis can provide valuable information, they are not without limitations. A frequent criticism is the grouping together of studies with sometimes vastly different characteristics. While this certainly limits the usefulness of an overall estimate of effect, by explaining this heterogeneity we can draw conclusions about differences in efficacy between groups. There is also an argument for including only high quality studies in a systematic review to establish conclusions with the highest degree of certainty and reliability. While this is a common strategy in clinical systematic review, where often inclusion is restricted to randomised controlled trials, we are not at a point in preclinical research where this is a feasible option. We consider exploring the impact of quality items and establishing guidance to improve study quality more valuable than excluding studies based on quality, at this time. An important caveat occurs where it is unethical or unfeasible to carry out research on humans and preclinical data solely inform clinical guidelines, for example, neonatal exposure to anaesthetics. In this context, it is paramount to include only high quality preclinical studies in decision-making.

As with any research domain, systematic reviews and meta-analyses are susceptible to bias and the quality of a review itself relies on robust methodology and reporting. While there are groups that have assessed the quality of systematic reviews [15, 16], this is based on expert opinion rather than an evidence-based approach. Mueller and colleagues reported poor methodology in preclinical systematic reviews, especially assessment of dissemination bias. Review quality was observed to be improving, however, with increased implementation of Preferred Reporting Items for Systematic Reviews and Meta-Analyses (PRISMA) and Quality of Reporting of Meta-analyses (QUOROM) checklists [16]. As yet, there are no evidence-based quality assessment tools to assess risks of bias in preclinical systematic reviews.

A major limitation of systematic review methodology as it currently stands is the sheer number of person-hours required, especially screening publications for inclusion and extracting data. When delays in publication are added to this equation, a search may be many years out of date before the results of a review are published. Clinical systematic reviews are generally not considered out of date as swiftly as preclinical reviews due to the fact that clinical trials are larger, take longer to complete than preclinical experiments, and are fewer in number.

Systematic review can only provide evaluation of reported study characteristics. These items may be confounded in some instances by reporting criteria set by some journals, but not others, and article word count restrictions. It is argued, for example, that failing to report blinding does not necessarily mean a study was conducted unblinded. Evidence suggests, however, that poor quality reporting is correlated with exaggerated results. Despite best practice, it can also be difficult to locate and procure all relevant articles, leading to publication bias in review results and conclusions. Finally, although we can detect and quantify heterogeneity, it is not always possible to determine its exact sources, and in some reviews, much heterogeneity remains unexplained. The methods for carrying out a preclinical systematic review and meta-analysis are evolving to address many of these limitations, discussed in detail in Sect. 4.

4 Future of Preclinical Systematic Review and Meta-analysis

Improvements to the range and sophistication of methodologies and applications for preclinical systematic review will help to guide the next generation of neuroprotection research. As more authors begin to report the facets of experimental design essential for its rigorous evaluation, the development of an evidence-based risk of bias tool will become possible. We will be able to determine which aspects of experimental design are most important by untangling collinearity between risks of bias. Moreover, it will become possible to properly examine the underlying biology of a problem, adjust for the quality of evidence from different studies and determine the range of circumstances under which a therapy has the potential to be effective in the clinic. Identifying well-designed and performed experiments will become easier and the value of the published literature will increase. This is already being recognised by publishers by establishment of journals such as Evidence-based Preclinical Medicine (EBPM) and the introduction of a specific Meta-Research section to PLOS Biology. No systematic or evidence-based criteria exist to inform the decision to venture from preclinical testing into clinical development. Recent initiatives have hypothesised that multicentre preclinical trials, which are of high internal and external validity, may be an appropriate and robust framework in which to test candidate interventions prior to clinical trial. Systematic reviews have been proposed as an important factor in this process to ensure that candidate interventions have been tested under a range of conditions and experiments are of sufficient quality prior to initiating high resource multicentre preclinical studies.

If we accept that this process of improvement will continue, what are the next big steps for advancing the technology itself and the value obtained from preclinical systematic review and meta-analysis?

Perhaps the most important issue is data volume. There is now more research published than ever before. The primary bibliographic database for biomedical research, PubMed, contains more than 24 million citations, and since 2005 has added 2000–4000 new references every working day [1]. A random sample of 2000 publications in PubMed suggests that ~31 % describe primary laboratory research. Eighteen percent of these describe in vivo research involving animals, of which 4 % are in pharmacology and 3 % are in the neurosciences. In the 750,000 PubMed citations indexed in 2014, approximately 30,000 will be neuroscience papers. No one individual can read, let alone appraise critically or use, even a small fraction of this new information, information which is the product of years of investigator effort and substantial investment of research funds [51].

This mismatch, between the amount of research produced and the amount that can be effectively used, is a major challenge to biomedical research. The sheer volume and publication rate of preclinical data predicate that methodology innovations are required beyond the largely manual processes that are currently adopted for most clinical systematic reviews. For example, in our systematic reviews of neuropathic pain, data from 229 clinical trials required extraction [52], whereas for the corresponding on-going preclinical systematic review, over 40,000 publications had to be screened and data are being extracted from ~4000 of these.

Automation of these processes by machine learning and data mining to identify relevant publications is essential. However, this is hindered by the fact that no single database covers all information needs [53–55]. Moreover, the controlled vocabularies of MeSH and BIOSIS and the Natural Language searching of Emtree are all subtly different and can change without backward compatibility, rendering reproducible searching over time a challenge. Libraries of verified search strategies, as exemplified by the SYstematic Review Centre for Laboratory animal Experimentation (SYRCLE) filter used to find data for animal experiments [56], that can be redeployed in different combinations will help. Better indexing and search engines for citation indices will also provide a part of the solution and this will be augmented by machine learning algorithms adept at finding critical words and phrases within that proportion of papers accessible to electronic searching. However, once relevant papers have been found, the daunting task of data extraction still remains. For newer electronic publication formats, identification and extraction of this information is becoming a possibility. However, a more logical long-term approach is linking of publications to the raw data, held in accessible public data repositories open to secondary data analysis.

Another often held, but difficult to accomplish, desire for systematic reviewers is the wish to rank the effectiveness of different treatment strategies when the background to each experiment is slightly different and the outcome measures used do not naturally align. This will be aided by the incorporation of a subset of standardised outcome assessments in experiments to facilitate development of better "normalisation" tools. Inclusion of common positive and negative controls within experiments will permit the use of network meta-analysis and thus the ability to perform secondary comparisons never envisaged by the original experimenters. Both will combine to improve our ability to select candidate drugs for future experiments, and in particular, inform the process of collaborative multicentre phase III preclinical drug trialling.

5 Conclusions

Preclinical systematic review and meta-analysis are proving to be valuable tools for understanding how our animal models work and the strengths and weaknesses of the data we use to make decisions to move forward. This includes selecting drugs for more advanced preclinical testing and for moving from the laboratory to the clinic, where these tools also permit evidence-based design of clinical trials. Importantly, by aggregating and summarising large volumes of data in a rigorous way, they have also provided the ammunition to motivate structural changes in the way preclinical research is funded, performed, and reported. However, much work is still needed to refine the statistical methodologies of preclinical systematic review and meta-analysis and to address the complexities of data access in a world where how we store data is rapidly changing and the volume of those data is expanding at an unprecedented rate.

6 Resources

- To our knowledge, there are two main groups who work to assist and educate preclinical systematic review authors and produce methodological tools and guidelines to generate high quality reviews: CAMARADES (http://www.dcn.ed.ac.uk/camarades/) and SYRCLE (https://www.radboudumc.nl/Research/Organisationofresearch/Departments/cdl/SYRCLE/Pages/default.aspx).
- The CAMARADES-NC3Rs SYstematic Review Facility (SYRF), a fully integrated online platform for performing systematic reviews of preclinical studies, has recently launched: http://syrf.org.uk/.
- Evidence-based Preclinical Medicine publishes preclinical systematic reviews and meta-analyses (http://onlinelibrary.wiley.com/journal/10.1002/%28ISSN%292054-703X) and have a helpdesk to answer review author questions: ebpm.helpdesk@wiley.com.

References

1. Dirnagl U, Endres M (2014) Found in translation: preclinical stroke research predicts human pathophysiology, clinical phenotypes, and therapeutic outcomes. Stroke 45(5):1510–1518
2. O'Collins VE, Macleod MR, Donnan GA, Horky LL, van der Worp BH, Howells DW (2006) 1,026 experimental treatments in acute stroke. Ann Neurol 59(3):467–477
3. Sena ES, Briscoe CL, Howells DW, Donnan GA, Sandercock PA, Macleod MR (2010) Factors affecting the apparent efficacy and safety of tissue plasminogen activator in thrombotic occlusion models of stroke: systematic review and meta-analysis. J Cereb Blood Flow Metab 30(12):1905–1913
4. Shuaib A, Lees KR, Lyden P, Grotta J, Davalos A, Davis SM et al (2007) NXY-059 for the treatment of acute ischemic stroke. N Engl J Med 357(6):562–571
5. Macleod MR, van der Worp HB, Sena ES, Howells DW, Dirnagl U, Donnan GA (2008) Evidence for the efficacy of NXY-059 in experimental focal cerebral ischaemia is confounded by study quality. Stroke 39(10):2824–2829

6. Greb E (2014) Magnesium treatment may not improve stroke outcome. Neurol Rev 22(3):5
7. O'Collins VE, Macleod MR, Cox SF, Van Raay L, Aleksoska E, Donnan GA et al (2011) Preclinical drug evaluation for combination therapy in acute stroke using systematic review, meta-analysis, and subsequent experimental testing. J Cereb Blood Flow Metab 31(3):962–975
8. Howells DW, Sena ES, Macleod MR (2014) Bringing rigour to translational medicine. Nat Rev Neurol 10(1):37–43
9. Tsilidis KK, Panagiotou OA, Sena ES, Aretouli E, Evangelou E, Howells DW et al (2013) Evaluation of excess significance bias in animal studies of neurological diseases. PLoS Biol 11(7), e1001609
10. Wegener G, Rujescu D (2013) The current development of CNS drug research. Int J Neuropsychopharmacol 16(7):1687–1693
11. Chalmers I, Bracken MB, Djulbegovic B, Garattini S, Grant J, Gulmezoglu AM et al (2014) How to increase value and reduce waste when research priorities are set. Lancet 383(9912):156–165
12. Cochrane|Trusted evidence (c2004–2015) Informed decisions. Better health. [Internet]. The Cochrane Collaboration, Oxford. Available from: http://www.cochrane.org/
13. Schulz KF, Altman DG, Moher D, for the CG (2010) CONSORT 2010 statement: updated guidelines for reporting parallel group randomised trials. PLoS Med 7(3):e1000251
14. Bunn F, Trivedi D, Alderson P, Hamilton L, Martin A, Pinkney E et al (2015) The impact of Cochrane Reviews: a mixed-methods evaluation of outputs from Cochrane Review Groups supported by the National Institute for Health Research. Health Technol Assess 19(28):1–100
15. Peters JL, Sutton AJ, Jones DR, Rushton L, Abrams KR (2006) A systematic review of systematic reviews and meta-analyses of animal experiments with guidelines for reporting. J Environ Sci Health B 41(7):1245–1258
16. Mueller KF, Briel M, Strech D, Meerpohl JJ, Lang B, Motschall E et al (2014) Dissemination bias in systematic reviews of animal research: a systematic review. PLoS One 9(12), e116016
17. Vesterinen HM, Sena ES, Egan KJ, Hirst TC, Churolov L, Currie GL et al (2014) Meta-analysis of data from animal studies: a practical guide. J Neurosci Methods 221:92–102
18. Jerndal M, Forsberg K, Sena ES, Macleod MR, O'Collins VE, Linden T et al (2010) A systematic review and meta-analysis of erythropoietin in experimental stroke. J Cereb Blood Flow Metab 30(5):961–968
19. Macleod MR, O'Collins T, Horky LL, Howells DW, Donnan GA (2005) Systematic review and metaanalysis of the efficacy of FK506 in experimental stroke. J Cereb Blood Flow Metab 25(6):713–721
20. Macleod MR, O'Collins T, Howells DW, Donnan GA (2004) Pooling of animal experimental data reveals influence of study design and publication bias. Stroke 35(5):1203–1208
21. Sena E, Wheble P, Sandercock P, Macleod M (2007) Systematic review and meta-analysis of the efficacy of tirilazad in experimental stroke. Stroke 38(2):388–394
22. Sena ES, van der Worp HB, Bath PM, Howells DW, Macleod MR (2010) Publication bias in reports of animal stroke studies leads to major overstatement of efficacy. PLoS Biol 8(3), e1000344
23. Lees JS, Sena ES, Egan KJ, Antonic A, Koblar SA, Howells DW et al (2012) Stem cell-based therapy for experimental stroke: a systematic review and meta-analysis. Int J Stroke 7(7):582–588
24. Krauth D, Woodruff TJ, Bero L (2013) Instruments for assessing risk of bias and other methodological criteria of published animal studies: a systematic review. Environ Health Perspect 121(9):985–992
25. Fisher M, Feuerstein G, Howells DW, Hurn PD, Kent TA, Savitz SI et al (2009) Update of the stroke therapy academic industry roundtable preclinical recommendations. Stroke 40(6):2244–2250
26. Landis SC, Amara SG, Asadullah K, Austin CP, Blumenstein R, Bradley EW et al (2012) A call for transparent reporting to optimize the predictive value of preclinical research. Nature 490(7419):187–191
27. Schifilliti D, Grasso G, Conti A, Fodale V (2010) Anaesthetic-related neuroprotection: intravenous or inhalational agents? CNS Drugs 24(11):893–907

28. Vesterinen HM, Connick P, Irvine CM, Sena ES, Egan KJ, Carmichael GG et al (2015) Drug repurposing: a systematic approach to evaluate candidate oral neuroprotective interventions for secondary progressive multiple sclerosis. PLoS One 10(4), e0117705
29. Sena E, van der Worp HB, Howells D, Macleod M (2007) How can we improve the pre-clinical development of drugs for stroke? Trends Neurosci 30(9):433–439
30. Egan KJ, Janssen H, Sena ES, Longley L, Speare S, Howells DW et al (2014) Exercise reduces infarct volume and facilitates neurobehavioral recovery: results from a systematic review and meta-analysis of exercise in experimental models of focal ischemia. Neurorehabil Neural Repair 28(8):800–812
31. Wheble PC, Sena ES, Macleod MR (2008) A systematic review and meta-analysis of the efficacy of piracetam and piracetam-like compounds in experimental stroke. Cerebrovasc Dis 25(1–2):5–11
32. Baker M (2016) 1,500 scientists lift the lid on reproducibility. Nature 533(7604):452–454
33. McCann SK, Cramond F, Macleod MR, Sena ES (2016) Systematic review and meta-analysis of the efficacy of interleukin-1 receptor antagonist in animal models of stroke: an update. Transl Stroke Res doi:10.1007/s12975-016-0489-z
34. Higgins JPT, Green S (2011) Cochrane handbook for systematic reviews of interventions version 5.1.0 [updated March 2011]: The Cochrane Collaboration. Available from: www.cochrane-handbook.org
35. Song F, Parekh S, Hooper L, Loke YK, Ryder J, Sutton AJ et al (2010) Dissemination and publication of research findings: an updated review of related biases. Health Technol Assess 14(8):Iii, ix–xi, 1–193
36. Chan AW, Song F, Vickers A, Jefferson T, Dickersin K, Gotzsche PC et al (2014) Increasing value and reducing waste: addressing inaccessible research. Lancet 383(9913):257–266
37. Banwell V, Sena ES, Macleod MR (2009) Systematic review and stratified meta-analysis of the efficacy of interleukin-1 receptor antagonist in animal models of stroke. J Stroke Cerebrovasc Dis 18(4):269–276
38. Macleod MR, Fisher M, O'Collins V, Sena ES, Dirnagl U, Bath PM et al (2009) Reprint: good laboratory practice: preventing introduction of bias at the bench. Int J Stroke 4(1):3–5
39. Macleod MR, Fisher M, O'Collins V, Sena ES, Dirnagl U, Bath PM et al (2009) Reprint: good laboratory practice: preventing introduction of bias at the bench. J Cereb Blood Flow Metab 29(2):221–223
40. Macleod MR, Fisher M, O'Collins V, Sena ES, Dirnagl U, Bath PM et al (2009) Good laboratory practice: preventing introduction of bias at the bench. Stroke 40(3):e50–e52
41. Kilkenny C, Browne WJ, Cuthill IC, Emerson M, Altman DG (2010) Improving bioscience research reporting: the ARRIVE guidelines for reporting animal research. PLoS Biol 8(6), e1000412
42. Editorial (2013) Announcement: reducing our irreproducibility. Nature 496(7446):398.
43. van der Worp HB, Macleod MR, Bath PM, Demotes J, Durand-Zaleski I, Gebhardt B et al (2014) EuroHYP-1: European multicenter, randomized, phase III clinical trial of therapeutic hypothermia plus best medical treatment vs. best medical treatment alone for acute ischemic stroke. Int J Stroke 9:642–645
44. van der Worp HB, Sena ES, Donnan GA, Howells DW, Macleod MR (2007) Hypothermia in animal models of acute ischaemic stroke: a systematic review and meta-analysis. Brain 130(pt 12):3063–3074
45. Sena ES, Jeffreys AL, Cox SF, Sastra SA, Churilov L, Rewell S et al (2013) The benefit of hypothermia in experimental ischemic stroke is not affected by pethidine. Int J Stroke 8(3):180–185
46. McCann SK, Irvine C, Mead GE, Sena ES, Currie GL, Egan KE et al (2014) Efficacy of antidepressants in animal models of ischemic stroke: a systematic review and meta-analysis. Stroke 45(10):3055–3063
47. Mead G, Hackett ML, Lundstrom E, Murray V, Hankey GJ, Dennis M (2015) The FOCUS, AFFINITY and EFFECTS trials studying the effect(s) of fluoxetine in patients with a recent stroke: a study protocol for three multicentre randomised controlled trials. Trials 16:369

48. Batchelor PE, Skeers P, Antonic A, Wills TE, Howells DW, Macleod MR et al (2013) Systematic review and meta-analysis of therapeutic hypothermia in animal models of spinal cord injury. PLoS One 8(8), e71317
49. Batchelor PE, Wills TE, Skeers P, Battistuzzo CR, Macleod MR, Howells DW et al (2013) Meta-analysis of pre-clinical studies of early decompression in acute spinal cord injury: a battle of time and pressure. PLoS One 8(8), e72659
50. Howells DW, Sena ES, O'Collins V, Macleod MR (2012) Improving the efficiency of the development of drugs for stroke. Int J Stroke 7(5):371–377
51. Rottingen JA, Regmi S, Eide M, Young AJ, Viergever RF, Ardal C et al (2013) Mapping of available health research and development data: what's there, what's missing, and what role is there for a global observatory? Lancet 382(9900):1286–1307
52. Finnerup NB, Attal N, Haroutounian S, McNicol E, Baron R, Dworkin RH et al (2015) Pharmacotherapy for neuropathic pain in adults: a systematic review and meta-analysis. Lancet Neurol 14(2):162–173
53. Betran AP, Say L, Gulmezoglu AM, Allen T, Hampson L (2005) Effectiveness of different databases in identifying studies for systematic reviews: experience from the WHO systematic review of maternal morbidity and mortality. BMC Med Res Methodol 5(1):6
54. Kwon Y, Powelson SE, Wong H, Ghali WA, Conly JM (2014) An assessment of the efficacy of searching in biomedical databases beyond MEDLINE in identifying studies for a systematic review on ward closures as an infection control intervention to control outbreaks. Syst Rev 3:135
55. Beynon R, Leeflang MM, McDonald S, Eisinga A, Mitchell RL, Whiting P et al (2013) Search strategies to identify diagnostic accuracy studies in MEDLINE and EMBASE. Cochrane Database Syst Rev (9):MR000022
56. Hooijmans CR, Tillema A, Leenaars M, Ritskes-Hoitinga M (2010) Enhancing search efficiency by means of a search filter for finding all studies on animal experimentation in PubMed. Lab Anim 44(3):170–175

Chapter 3
Neuroprotection Is Technology, Not Science

Donald J. DeGracia, Doaa Taha, Fika Tri Anggraini, and Zhifeng Huang

Abstract All human clinical trials of neuroprotection after brain ischemia and reperfusion injury have failed. Brain ischemia is currently conceptualized as an "ischemic cascade" and therapy is directed to treating one or another element of this cascade. This approach conflates the science of cell injury with the development of neuroprotective technologies. Here we review a theory that describes the generic nonlinear dynamics of acute cell injury. This approach clearly demarcates the science of cell injury from any possible downstream technological applications. We begin with a discussion that contrasts the qualitative, descriptive approach of biology to the quantitative, mathematical approach used in physics. Next we discuss ideas from quantitative biology that underlie the theory. After briefly reviewing the autonomous theory, we present, for the first time, a non-autonomous theory that describes multiple injuries over time and can simulate pre- or post-conditioning or post-injury pharmacologics. The non-autonomous theory provides a foundation for three-dimensional spatial models that can simulate complex tissue injuries such as stroke. The cumulative theoretical formulations suggest new technologies. We outline possible prognosticative and neuroprotective technologies that would operate with engineering precision and function on a patient-by-patient basis, hence personalized medicine. Thus, we contend that a generic, mathematical approach to acute cell injury will accomplish what highly detailed descriptive biology has so far failed to accomplish: successful neuroprotective technology.

Keywords Brain ischemia • Ischemic cascade • Autonomous theory • Lac operon • Boolean network • Acute cell injury • Bistable bifurcation

D.J. DeGracia (✉)
Department of Physiology, Wayne State University, Detroit, MI 48201, USA

Center for Molecular Medicine and Genetics, Wayne State University, Detroit, MI 48201, USA
e-mail: ddegraci@med.wayne.edu

D. Taha • Z.-F. Huang
Department of Physics and Astronomy, Wayne State University, Detroit, MI 48201, USA

F.T. Anggraini
Department of Physiology, Wayne State University, Detroit, MI 48201, USA

P.A. Lapchak, J.H. Zhang (eds.), *Neuroprotective Therapy for Stroke and Ischemic Disease*, Springer Series in Translational Stroke Research,
DOI 10.1007/978-3-319-45345-3_3

1 Introduction

In the setting of brain ischemia, neuroprotection can be defined as taking a post-ischemic brain region we know will die and performing some intervention to prevent it from dying. While many neuroprotective interventions are described in the preclinical literature, none have successfully "translated" to clinical stroke neuroprotection in humans [1–4]. Analogous failures have plagued other biomedical fields, such as cardiac and renal ischemia [5, 6]. These are not isolated cases, but part of a larger pattern of deficiencies in biomedical research. The "lack of reproducibility" of biomedical results has garnered national media attention and serious reform efforts from journal editorial boards and the NIH [4, 7–11]. All of this, we suggest, is part of the same picture. Attempts to remedy these deficiencies have focused almost exclusively on technical details of experimental execution [12–14]. That empirical ambiguity needs to be minimized should go without saying. We have argued that an equal, or even greater, contributing factor is the lack of theoretical foundation in biomedical research [15–19].

The purpose of this chapter is to consider what neuroprotection might look like in a world that possessed a working theoretical biomedicine. We have offered such a theory and summarize it below. The theory has not yet been empirically validated. Nonetheless, the theoretical construct clarifies a number of critical issues. Perhaps most importantly, it makes clear the distinction between the science of cell injury and therapeutic technology. Developing and confirming the theory constitute the science. Any application that stems from the scientific results constitutes technology. Hence, neuroprotection becomes a technological goal that comes *after* the science is completed. Efforts that conflate science and technology in biomedical research have only served to confound both.

This chapter has two main parts. In the first part, we discuss the broader views that underlie our approach. We consider how physics and biology differ and why it matters to the idea of neuroprotection. We begin at the root of the problem and briefly compare the scientific cultures of physics and biology. We then briefly discuss the foundations of our approach which are grounded in network theory. In the second part, we illustrate an approach to acute cell injury that utilizes the principles and notions described in the first part. The solutions of this theory offer paradigm-transforming insight into the nature of acute cell injury. The theoretical findings provide clear directions for technological application, of which we consider three facets: (1) therapy as sublethal injury, (2) the technology of prognostication, and (3) the technology of neuroprotection.

2 Part 1: Physics and Biology

2.1 A Tale of Two Cultures

Since their respective inceptions in the modern era, biology and physics have developed along separate tracts. Biology was burdened with the task of describing the myriad biological organisms and the almost infinite variations of their structures, functions, niches, and so forth [20]. Physics, on the other hand, since the time of

Galileo, sought to find mathematical patterns that described some general aspect of entire classes of phenomena. In Galileo's own words [quoted in [21]]:

> Philosophy [nature] is written in that great book whichever is before our eyes—I mean the universe—but we cannot understand it if we do not first learn the language and grasp the symbols in which it is written. The book is written in mathematical language...without whose help it is impossible to comprehend a single word of it; without which one wanders in vain through a dark labyrinth.

The essence of the method of physics was captured succinctly by the mathematician Morris Kline [21]:

> The bold new plan, proposed by Galileo and pursued by his successors, is that of obtaining *quantitative descriptions* of scientific phenomena *independently of any physical explanation.*

The distinction between the qualitative classifications of biology and the quantitative descriptions of physics illuminates the crisis in biomedical research as explained by entrepreneur Bill Frezza [22]:

> [We must]...shift the life sciences over to practices that have been advancing the physical sciences for years. In order to do this the culture must change. Mathematics is the language of engineering and life scientists can no longer take a pass on it. A system that cannot be modeled cannot be understood, and hence cannot be controlled. Statistical modeling is not enough, for the simple reason that correlation is not causation. Life science engineers need to catch up with their peers in the physical sciences when it comes to developing abstract mathematical representations of the systems they are studying. Progress comes from constantly refining these models through ever more detailed measurements.

2.2 What Is Measured?

Since the time of Newton, physicists have struggled with the link between the physical objects we perceive and their description by mathematical patterns [21]. The dominance of mathematical abstraction over physical intuition was definitively established in the 1920s. The founders of quantum mechanics were forced by the empirical behavior of light and atoms to abandon appeals to everyday physical intuition. Quantum mechanics made clear that physics provides mathematical descriptions of phenomena whether or not they make sense to everyday intuition. For example, a physical intuition of the superposition of quantum states is not possible [23]. This approach has been justified by the overwhelming scientific and technological success of quantum mechanics.

Physics of course did not come to this realization overnight. The transition occurred over centuries. However, from the time of Galileo, physical objects were idealized into mathematical quantities. Physical objects were imagined to be points (center of mass) moving through frictionless media, whose motion traced out idealized geometric curves. Idealizing perceptions of the real world in this fashion allowed the recognition that the same mathematical pattern could be applied, for example, to projectiles and planetary motion, as Newton famously showed [24].

Thus, from its inception, what was to be measured in physics was intimately bound to mathematical description. It did not matter if it was an apple, cannon ball, or the Moon. The qualitative differences were ignored and each was to be treated as a point-like object distinguished only by its quantitative mass. Mass is a nonsensory abstraction [25]. We perceive weight, which is a function of position in a gravity field. A cannon ball becomes weightless in outer space, but the mass remains constant. The perception of weight was the intuitive forerunner of mass. Mass is a mathematical quantity in the equations of physics. Thus, what is measured in physics are *operational quantities that are defined exclusively in mathematical terms.*

In contrast, in modern biomedicine, organisms are dissected/homogenized into relatively stable pieces that are measured by a variety of means (centrifugation, Western blot, etc.), and intuitively thought of as classical objects with classical functions. There is no underlying mathematical theory to guide the definition of such objects. The nature of the objects is wholly dependent on the methods of isolation. Different methods can result in different manifestations of supposedly the same object [26]. Some subcellular components are not yet amenable to biochemical isolation [27]. The employment of the arsenal of molecular biological tools without an underlying mathematical conception has led to empirical chaos where definitions rest not on sound theory, but on the methods used to generate the objects of study. This approach is not systematic, and it is no wonder that control of such systems has been elusive.

We present below an approach to acute cell injury based on the method of mathematical abstraction and idealization used in physics, where the main concepts and objects of measurement are defined by the theory. Before describing the theory, we discuss the foundations on which it rests.

2.3 The Mathematizing of Biology

It is a common assertion that biology is too complex to treat mathematically the way physics treats physical phenomena. The truth of this assertion has continuously eroded over time. This section briefly reviews one line of development that has successfully mathematized biology. Some of this material was discussed in greater detail elsewhere [19], so only salient aspects are presented here.

The first major step was the discovery of graph theory by the famous eighteenth century mathematician Euler, which initiated the study of networks as mathematical entities [28]. However, the relevance of network mathematics to biology did not become explicit until the 1960s.

After the discovery of the structure of DNA, and the cracking of the genetic code in the 1960s, the physical and informational structure of chromosomes was at least partially revealed. Even in the context of the now discredited "one gene one protein" model, a central paradox became apparent in the construction of organisms. Using current numbers, there are on order 20,000–30,000 genes in multicellular organisms. Yet an organism is made of a much smaller number of cell types, on the order of several dozens. The question thus arose [29]: how can such a large number of genes lead to a much smaller number of cell phenotypes? This question has a paradoxical character because there are an astronomical number of combinations of gene expression patterns. Yet, the limited number of cell phenotypes indicates that most of these possibilities play no role in organismic biology. Somehow, only an extremely small percentage of possible gene expression patterns actually mattered. Could there be a basic theoretical principle behind this observation?

The discovery leading to the resolution of this seeming paradox was the Nobel Prize winning work of Jacob and Monad. As is well-known, Jacob and Monad discovered the Lac operon [30], providing the first clear example of gene regulation. The key finding was that the product of the *lacI* gene, the lac repressor protein, could control the transcription of the lac operon. Binding of lac repressor protein inhibited transcription of the genes contained in the lac operon. But lactose binding to the lac repressor protein dissociated it from the DNA, thereby inducing lac operon transcription. This discovery revealed that gene products could regulate the expression of other genes and showed that genes were functionally interlinked to form a self-regulating network of mutual influences.

Shortly after discovery of the Lac operon, in 1969 Stuart Kauffman demonstrated that Boolean networks, in which each node is either "on" or "off", could model gene networks [29]. Kauffman did not use Boolean networks to model any specific gene network. Instead, he studied the generic mathematical properties of random Boolean networks. A random Boolean network of N nodes has 2^N possible states. For example, a network of $N=25$ nodes will have $2^{25}=33{,}554{,}432$ possible states.

Kaufmann's main finding was that, of all the possible states, only a small number of them were stable. A stable network state, also called an *attractor state*, is one that, once obtained, no longer changes to another state [31]. He found that the number of attractor states was on the order of $\sqrt{N}$ [29, 32]. Thus, for $N=25$ nodes, there would be ~5 attractor states. This represents ~150 parts per billion of the possible network states, a vanishingly small fraction.

Kauffman's theoretical finding gave insight into how gene networks could operate. If each gene was taken as a node, and the network was approximated as Boolean (i.e. each gene was simply on or off), there would be $2^{20,000}$ possible states. However, there would only be ~ $\sqrt{20,000}$ or 141 stable states, a not unrealistic number of cell types in an organism. Thus, Kauffman's work led to a critical new insight: *the stable states of a network, the attractor states, could be associated with the phenotypes of cells*. If validated, this would stand as a basic principle in theoretical biology based on the mathematical properties of networks.

The empirical demonstration of Kauffmann's theory had to await the advent of omic technology where thousands of genes could be measured simultaneously. In 2007, Huang and colleagues provided compelling evidence that changes in gene expression could be modeled as changes in the gene network from one stable attractor state to another [33]. The initial and final gene expression patterns, each associated with a distinct cell type, were stable, but the intermediate gene changes between the two phenotypes were dynamic and followed a precise mathematical pattern of change. Below, we utilize the same mathematical pattern to theoretically model acute cell injury.

These ideas were a critical advance towards quantitatively abstracting biological systems. Cell phenotypes correspond to gene network attractor states. This thinking accounts for two levels of biological action simultaneously: (1) the level of the gene network and its potentially very complex molecular interactions, and (2) the level of the phenotype which represents the *net action* of the underlying molecular network. In physics, this is sometimes called a "dual" representation, where a problem that may be overwhelmingly complex in one representation is considerably simplified in the other representation [34]. Phenotypes and genotypes have been linked since Mendel. The view above cracks the barrier to precisely quantifying what has until now been treated in qualitative terms. Our theory of acute cell injury is grounded in this quantitative network view that links the gene network and cell phenotype.

3 Part 2: A Theory of Acute Cell Injury

3.1 Introduction to the Theory

We have presented our theory in detail elsewhere [35] and so here summarize salient points. We begin with a qualitative heuristic, and then present the autonomous form of the theory, where "autonomous" is a technical mathematical term that means time, t, is *absent* from the right hand side of a differential equation [31]. We briefly review the solutions of the autonomous theory. We then discuss three possible technological directions suggested by the theory:

1. A non-autonomous version of the theory allows for sequential injuries. This simulates preconditioning and other clinically relevant conditions. However, this result has much broader significance by indicating that *therapy in general is synonymous with sublethal injury.*
2. To develop a quantitative approach to prognostication of acute cell injury, such as stroke, we outline an externally perturbed spatial model on a 3D connected lattice.
3. Based on the previous two discussions, we describe a possible quantitative approach that would use precisely targeted radiation to affect neuroprotection.

The "take home message" of our presentation is that the main advantage of a theory-driven biomedicine is that *the theory provides a clear, stepwise roadmap from the science of acute cell injury to therapeutic technology.*

3.2 Qualitative Description of the Theory

The theory is an idealization of what happens when a single cell is acutely injured. Three features of cell injury are abstracted as continuous mathematical quantities: (1) the *intensity of the injury*, I, (2) the *total amount of cell damage*, D, and (3) the *total amount of all stress responses induced by the injured cell*, S. The theory addresses how D and S change over time, t, as a function of I. How to characterize the specific injury and specific cell type naturally emerges as we proceed.

Imagine a generic cell acutely injured by a generic injury mechanism with intensity, *I*. This will activate many simultaneous damage and stress response pathways. The sum of all damage at any instant is *D*. The sum of all activated stress responses at any instant is *S*. By definition, *D* and *S* are mutually antagonistic. The function of stress responses is to combat damage, but the damage products can inhibit or destroy the stress responses. Therefore, *D* and *S* "battle", and the level of each changes over time. The "battle" concludes with only one of two possible outcomes: $D>S$ or $S>D$ (the special case of $D=S$ is discussed below). If $D>S$, damage wins out over the stress responses and, unable to overcome the damage, the cell dies. If $S>D$, then stress responses win, the cell repairs itself and survives.

We can summarize the qualitative idea with a circuit diagram (Fig. 3.1). The core of the diagram is the mutual antagonism of *D* and *S*. *D* is positively driven, and *S* is negatively driven, by *I*. This is intuitive: the stronger an injury (the higher the value of *I*), the more damage it will produce, and the less the cell will be able to respond effectively to the injury.

Tying into the general ideas discussed above, our theory recognizes an uninjured cell as a stable phenotype generated by a stable pattern of gene expression. Application of an acute injury is an extrinsic perturbation (of intensity *I*) to the system. The genetic changes associated with cell injury are explained as a deviation from the stable gene network state into a series of unstable states. The instability of the gene network resolves itself either by returning to its original stable state, or by becoming so unstable that the system can no longer maintain integrity and so disintegrates, i.e. dies.

In addition to the gene network, we posit that the damage pathways generated by acute injury also act as a network [19]. The various damage pathways do not act independent of each other. Instead, each damage pathway has a point or points of contact with others that link them into a unified network. Hence, *D*, the damage network that seeks to destroy the cell, interacts with *S*, the gene network that seeks to maintain homeostasis. Thus, *we need only consider the net action of the two competing networks and not concern ourselves with the details of the specific pathways that instantiate the networks.*

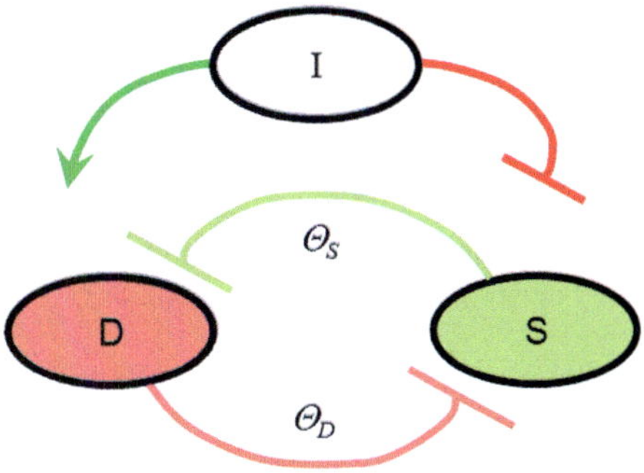

Fig. 3.1 Circuit diagram of the theory of acute cell injury

3.3 The Autonomous Theory

The above qualitative picture becomes a mathematical theory via the following four postulates:

1. D and S exist.
2. The mutual antagonism of D and S follows S-shaped curves (Hill functions).
3. D and S are exponentially driven by I and $-I$, respectively.
4. Decay from the attractor state, (D^*, S^*), is a function of $|D^*-S^*|$.

Postulate 1 was justified in the previous section. Postulate 2 is expressed by Eq. (3.1), a system of autonomous ordinary differential equations that is well-known [36] to model two mutually antagonistic factors.

$$\frac{dD}{dt}=v_{\mathrm{D}}\frac{\Theta_{\mathrm{D}}^{n}}{\Theta_{\mathrm{D}}^{n}+S^{n}}-k_{\mathrm{D}}D$$
$$\frac{dS}{dt}=v_{\mathrm{S}}\frac{\Theta_{\mathrm{S}}^{n}}{\Theta_{\mathrm{S}}^{n}+D^{n}}-k_{\mathrm{S}}S \tag{3.1}$$
$$\text{net rate}=\text{formation rate}-\text{decay rate}$$

Equation (3.1) specifies that the net rate of change of D or S equals the rate of formation minus the rate of decay. The rate of formation is given by Hill functions with a Hill coefficient n, threshold Θ, and velocity v. Θ_{D} is the value of D at a 50 % decrease in S. Θ_{S} is the value of S at a 50 % decrease in D (Fig. 3.2). The mutual inhibition is captured by the inverse relationships $dD/dt \propto 1/S^n$ and $dS/dt \propto 1/D^n$. Eq. (3.1) treats the decay rate as first order, with decay constants k_{D} and k_{S}.

Postulate 3 is expressed by having *the thresholds change as a function of injury intensity, I*. The minimal assumption is an exponential relationship. The threshold of D, Θ_{D}, is proportional to Ie^{I}. The threshold of S goes as Ie^{-I}. To convert to equality, multiplier (c) and exponential (λ) proportionality constants are required:

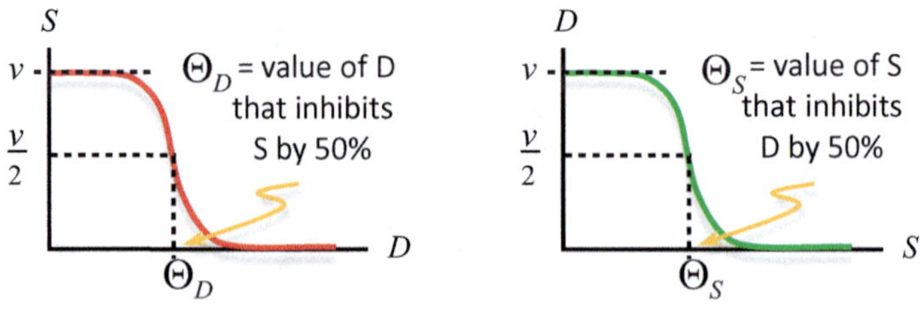

Fig. 3.2 The threshold of D, Θ_D, represents the strength of D to inhibit S by 50 %. The threshold of S, Θ_S, represents the strength of S to inhibit 50 % of D

$$\Theta_{\mathrm{D}} = c_{\mathrm{D}} I \mathrm{e}^{I\lambda_{\mathrm{D}}}$$
$$\Theta_{\mathrm{S}} = c_{\mathrm{S}} I \mathrm{e}^{-I\lambda_{\mathrm{S}}} \tag{3.2}$$

The specific injury and the specific cell type naturally emerge from Eq. (3.2) via the proportionality constants [15]. (c_{D}, λ_{D}) quantify the *lethality* of the injury mechanism, analogous to how LD_{50} quantifies the lethality of a substance. Larger values of (c_{D}, λ_{D}) mean an injury mechanism is more lethal than one with lower values. (c_{S}, λ_{S}) quantify the *strength* of a cell's intrinsic stress responses. Larger c_{S} and smaller λ_{S} correspond to stronger stress responses. Thus, every possible acute injury mechanism and every specific cell type can, in principle, be given numerical values for (c_{D}, λ_{D}) and (c_{S}, λ_{S}), respectively. This step is analogous to how an apple, cannon ball, or planet is abstracted to its numerical mass, which eliminates the qualitative distinctions. Similarly, specific qualitative injures such as ischemia, head trauma, or poisoning will each have distinct values of (c_{D}, λ_{D}). Specific cell types, such as neurons, myocytes, or endothelial cells will each have distinct values of (c_{S}, λ_{S}).

Substituting Eq. (3.2) into Eq. (3.1) gives the autonomous version of the nonlinear dynamical theory of acute cell injury:

$$\frac{dD}{dt} = v_{\mathrm{D}} \frac{\left(c_{\mathrm{D}} I \mathrm{e}^{I\lambda_{\mathrm{D}}}\right)^n}{\left(c_{\mathrm{D}} I \mathrm{e}^{I\lambda_{\mathrm{D}}}\right)^n + S^n} - k_{\mathrm{D}} D$$
$$\frac{dS}{dt} = v_{\mathrm{S}} \frac{\left(c_{\mathrm{S}} I \mathrm{e}^{-I\lambda_{\mathrm{S}}}\right)^n}{\left(c_{\mathrm{S}} I \mathrm{e}^{-I\lambda_{\mathrm{S}}}\right)^n + D^n} - k_{\mathrm{S}} S \tag{3.3}$$

The autonomous theory is based on the first three postulates. Postulate 4 is introduced after studying the solutions to Eq. (3.3). By studying the solutions to Eq. (3.3) and giving them biological interpretations, the theory provides a universal understanding of acute cell injury dynamics.

3.4 Solutions of the Autonomous Theory

Equation (3.3) is solved by Runge-Kutta numerical methods [32] that apply the input parameters to Eq. (3.3) and output the solution as a *phase plane* containing all possible *trajectories*. By varying I as the *control parameter*, the theory can model an injury system over a range of injury intensities. A single trajectory starts at an *initial condition* (D_0, S_0). Each trajectory converts to a pair of covarying D and S *time courses*. The time course pairs converge to some steady state. This steady state is called a *fixed point*, which, by definition, is where the *rates of all variables simultaneously equal zero*. There are two types of fixed points, notated (D^*, S^*), relevant to our theory. Trajectories converge to *attractors* and diverge from *repellers* [31].

Plotting fixed points vs. the control parameter is called a *bifurcation diagram* and shows quantitative and qualitative changes in fixed points. A *qualitative* change in fixed points is called a *bifurcation*.

Above we said if $D>S$, the cell dies, but if $S>D$ the cell recovers. In the solutions to Eq. (3.3), *outcome is determined by the D and S values of the attractors*. If $S^*>D^*$, the cell recovers. If $D^*>S^*$, the cell dies. We demonstrated for all numerical combinations of (c_D, λ_D, c_S, λ_S, I), that Eq. (3.3) outputs only *four types* of bifurcation diagrams [35]. Here we discuss two of the bifurcation diagram types to illustrate our main theoretical findings.

3.5 Monostable Outcome

We now explain how the theory mathematically defines cell death, illustrated by the bifurcation diagrams. Figure 3.3, panel 1, shows a bifurcation diagram plotting D^* vs. I (red) and S^* vs. I (green). As I increases continuously, at each I the phase plane is *monostable*, containing only one attractor. Four phase planes are shown (at values of I indicated by orange dashed lines on bifurcation diagrams). Trajectories from $(D_0, S_0)=(0, 0)$ either recover (green) or die (red). Initial conditions (0, 0) correspond to the uninjured state, or "the control condition".

The special case of $D=S$ occurs when the bifurcation curves cross at $D^*=S^*$ (Fig. 3.3, panel 1, arrow). This crossing occurs at a specific value of I termed I_X, *the tipping-point value of I*. I_X is intuitively understood as the "cell death threshold". Technically, I_X is *not* a threshold (thresholds are illustrated in Fig. 3.2). I_X is the tipping point value of injury intensity, I, defined as that value of I where $D^*=S^*$. For $I<I_X$, $S^*>D^*$ and the cell recovers. For $I>I_X$, $D^*>S^*$, and the cell dies.

At $D^*=S^*$, $\Theta_D=\Theta_S$, and Eq. (3.2) can be solved to calculate I_X:

$$I_X = \frac{\ln(c_S) - \ln(c_D)}{\lambda_D + \lambda_S} \tag{3.4}$$

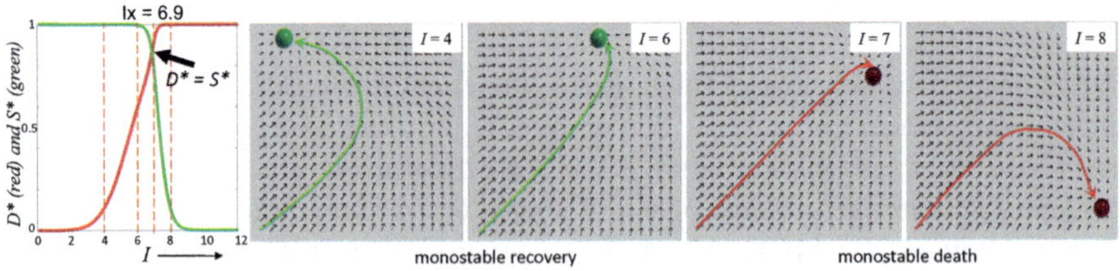

Fig. 3.3 A monostable bifurcation diagram describes the case where there is only one outcome at each value of injury intensity, I. The phase planes in panels 2–5 are marked by *orange dashed lines* on the bifurcation diagram in *panel 1*. The attractor states are indicated by *green* ($S^*>D^*$) or *red* ($D^*>S^*$) *circles* on each phase plane

This is a rather amazing result in the context of descriptive biomedicine. Given the parameters for any injury mechanism (c_D, λ_D) and any cell type (c_S, λ_S), we can, in principle, *calculate beforehand* the "cell death threshold" or tipping point intensity, I_X, for that combination. Additionally, Eq. (3.4) makes explicit that the "cell death threshold", I_X, is a function of *both* the injury mechanism and cell type. That means different cell types in a tissue (e.g. cortical vs. hippocampal neurons in brain) will die at different values of injury intensity. This is, at best, only intuitively understood in descriptive biomedicine. Our theory derives it as a mathematical fact.

The monostable case in Fig. 3.3 matches the general intuition that a cell will survive injury intensities less than the "cell death threshold" (I_X), but die if injury intensity is greater than I_X.

3.6 Bistable Outcome

We now state the most important feature of the autonomous theory: For some parameter sets, *both a survival and a death attractor are present on the same phase plane*. The scenario of two attractors on a phase plane is called *bistability* [31, 32]. This result is *counterintuitive*. The monostable case corresponds to our intuition that a single injury intensity leads to *either* a survival *or* a death outcome. Our theoretical investigation *demonstrates* that some injury magnitudes are bistable and have *both* a survival *and* a death outcome.

Bistability is illustrated in Fig. 3.4, panel 1. As I increases, there is a value of I at which the phase planes transform from having one attractor to having two attractors and one repeller. The system is said to *bifurcate* at this value of I. The system is bistable for a range of I (yellow region) and then bifurcates again to monostable. The value of I_X falls exactly in the middle of the bistable range of I values.

The monostable bifurcation diagram (Fig. 3.3) displays only two dynamical states: (1) recovery for $I<I_X$, and (2) death for $I>I_X$. The bistable bifurcation diagram shows *four* dynamical states: (1) monostable recovery, (2) bistable recovery, (3) bistable death, and (4) monostable death.

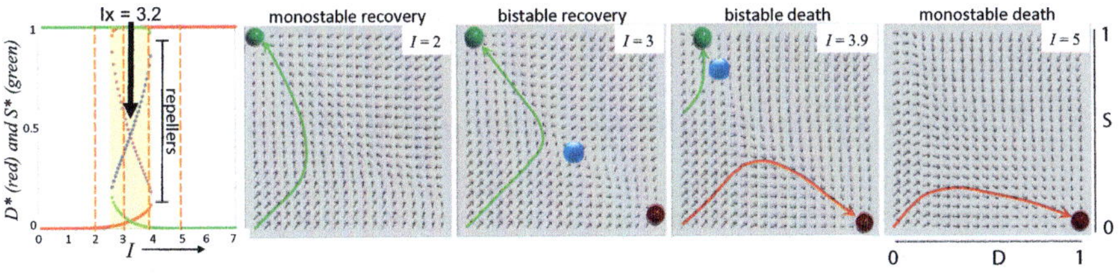

Fig. 3.4 A bistable bifurcation diagram describes the case where, for some range of injury intensities, both death and survival outcomes are possible for each injury intensity in that range

3.7 Pre-treatment Therapies

The mathematical results described in the previous section provide a new, paradigm-transforming definition of therapy. In the monostable case, there is only one possible outcome at each value of injury intensity, either recovery for $I<I_X$, or death if $I>I_X$. *No therapy is possible* for injured cells with monostable dynamics because there is no attractor with $S^*>D^*$ when $I>I_X$. For the bistable bifurcation diagram, *a range of injury intensities contain both the capacity to recover and to die*. In other words, our theory demonstrates that *bistability is required* if we wish to take a system on a pro-death trajectory and convert it to a pro-survival trajectory.

For the autonomous theory, whether a trajectory leads to recovery or death depends on *initial conditions* (D_0, S_0). Altering initial conditions corresponds to a *pre-treatment therapy*. The effect of altering initial conditions was extensively studied [15], so only salient points are outlined here. When $D_0<0$ (D_0 is a negative number), this represents pre-administering an agent (drug) that inhibits damage. Starting at $S_0>0$ (S_0 is a positive number) corresponds to a manipulation that activates stress responses prior to inducing injury, for example, by transfecting a protective gene (e.g. heat shock 70).

For the bistable death range, all trajectories from $(D_0, S_0)=(0, 0)$ will die. But if initial conditions $D_0<0$ or $S_0>0$, then the system may follow a trajectory to the recovery attractor (Fig. 3.4, panel 4, green trajectory). Thus, the theory clearly indicates that *the ability to convert a cell on a pro-death trajectory to a pro-survival trajectory is due to the bistable dynamics of the system*.

This is truly a paradigm-transforming insight. It mathematically expresses the intuition that some injury magnitudes that are lethal if untreated can be reversed by therapy. Such insight is, at best, only intuitively apprehended using the prevalent qualitative molecular pathways approach and certainly cannot be calculated. Our theory defines therapeutics mathematically and calculates the range of lethal injury intensities that can be recovered by *any* possible therapy.

However, a limitation of the autonomous version is that it cannot model a post-injury treatment. The need to model a post-treatment provided major motivation to develop the non-autonomous theory. We described this new theory for the first time below. However, before giving the new theory, we turn attention to the main limitation of the autonomous version.

3.8 Closed Trajectories and the Autonomous Theory

The solutions to Eq. (3.3) capture only *half* of what happens to an acutely injured cell. Acute injury displaces the cell from its stable phenotype through a continuous series of unstable states. The attractor solutions to Eq. (3.3) represent the *maximum deviation from the stable state* and specify whether outcome will be recovery ($S^*>D^*$) or death ($D^*>S^*$). However, the cell must "decay" from maximum instability either to its initial phenotype (recover) or cease to exist (death). We developed [35] a stop-gap solution to overcome this limitation which is expressed by Postulate 4: decay from the attractor state, (D^*, S^*), is a function of $|D^*-S^*|$.

We noted that, from the perspective of the theory, *both* recovery and death are represented when $D=S=0$. Recovery is the state where there is no longer any damage ($D=0$) and stress responses are no longer expressed ($S=0$). Death is the complete disappearance of the cell, and hence all of its variables, including D and S, will equal zero.

Thus, we asked: what determines *how long* it takes (the *decay time*, τ_D) for the cell to go from the attractor (D^*, S^*) back to (D, S) = (0, 0)? If D^* is much greater than S^* ($D^* \gg S^*$), damage is so great the cell will die quickly (τ_D is short), which is a condition widely recognized as *necrosis*. If $S^* \gg D^*$, stress responses overwhelm the small amount of damage, and the cell recovers quickly (τ_D is short). If D^* is only *slightly larger* than S^*, it will take a longer time for damage to overcome the stress responses (τ_D is long), and vice versa if S^* is only slightly larger than D^*. We thus noticed the importance of the magnitude $|D^*-S^*|$. The decay time, τ_D, from the attractor state (D^*, S^*) is an inverse function of $|D^*-S^*|$.

The simplest assumption was an exponential decay of the cell from the attractor to (D, S) = (0, 0):

$$\begin{aligned} D &= D^* e^{-|D^*-S^*|t} \\ S &= S^* e^{-|D^*-S^*|t} \end{aligned} \tag{3.5}$$

We then concatenated solutions from Eq. (3.5) to Eq. (3.3) to obtain closed trajectories (Fig. 3.5). In Fig. 3.5, the trajectory solution to Eq. (3.5) is overlaid on the phase plane solution to Eq. (3.3). In actual fact, Eqs. (3.3) and (3.5) would have different phase planes, but depicting it as we have makes clear the need to have a closed loop trajectory on the phase plane to fully model the temporal progression of acute cell injury and recovery or death.

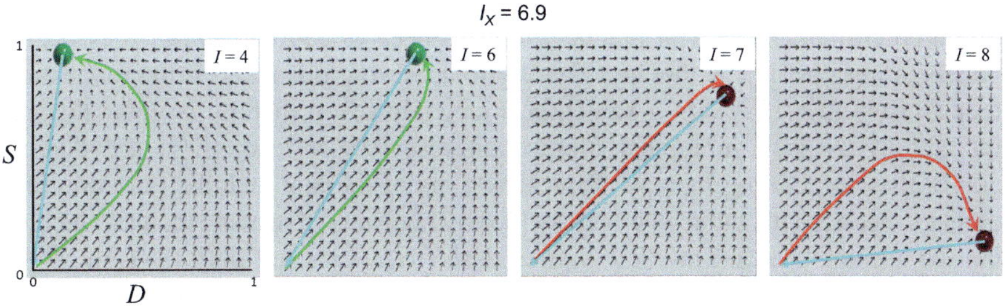

Fig. 3.5 Closed trajectories obtained by applying postulate 4 where the decay rate to recovery or death goes as $e^{-|D^*-S^*|}$

This line of questioning showed that the empirically observed facts of rapid (necrotic) and *delayed death* after injury emerged naturally from our theory. Delayed neuronal death (DND) occurs in hippocampal CA1 after global ischemia [37] and in penumbra after stroke [38]. Many molecular pathways have been advanced to explain DND such as apoptosis, aponecrosis, protein aggregation, and so on [39–42]. However, the *main feature* of DND is *time*, the quintessential dynamical quantity. The theory indicates that if $|D^* - S^*|$ is small, the cell will take a long time to die. Thus, *delayed death after injury is a consequence of the injury dynamics*. Biological factors are *not* causative, but instead merely mediate *what is intrinsically a dynamical effect*.

However, even though the theory could address rapid and delayed death after injury, the need to concatenate equations was artificial. Thus, an important motivation behind the non-autonomous theory was to develop a mathematical framework that could automatically calculate closed loop trajectories that model the full sequence of injury and recovery or death. Solving this problem automatically solved the post-injury treatment issue, as we now discuss.

4 Technological Applications

We now describe possible technological applications of the dynamical theory of acute cell injury. These require modifications of the autonomous formulation. We discuss: (1) multiple sequential injuries, (2) a three-dimensional version of the theory to model tissue injury, and (3) a possible neuroprotective technology based on the former two results.

4.1 Approaches to Therapy

The autonomous theory, Eq. (3.3), can only deal with pre-treatment by altering the initial conditions. In clinical practice, injuries need to be treated *after* they occur, necessitating post-injury therapy. The key to achieving this result is the realization that *therapy is necessarily a sublethal injury*. However, therapy is generally not thought of as such in typical biomedical research. Instead, there are diverse approaches to defining therapy. Typical notions of therapy include: (1) inhibiting damage mechanisms induced by injury [43], (2) bolstering stress responses (including the immune system) to allow injured tissue to repair itself [44], (3) some combination of 1 and 2 (e.g. multiple drug treatments) [45, 46], and (4) treatments designed to decrease the intensity or duration of the injury [47]. Our theory clearly demarcates control of injury intensity, I, from control of D and S.

4.2 Ascertaining Injury Intensity

In clinical practice, controlling injury intensity, *I*, is often not an option. For a given injury mechanism (ischemia, trauma, poisoning), a large range of continuous values of injury intensity, *I*, can present clinically. The clinical problem is therefore to estimate the magnitude of *I* (e.g. duration of ischemia, force of trauma, concentration of poison) and to determine if an intervention is possible, e.g. tPA or surgery for stroke [48]. It is important to know the duration from time of injury to clinical presentation (e.g. 2 h time window for stroke).

Bifurcation diagrams (Figs. 3.3 and 3.4) provide a systematic framework for estimating injury intensity, *I*. A given injury will correspond to specific value of *I* and its accompanying phase plane. The phase plane provides time courses from different initial conditions, one of which will describe the injury evolution over time. Thus, in principle, it is possible to achieve engineering levels of prognosticative precision by using our theory.

4.3 Protective Therapeutics

The most common view of post-injury therapy is that there exists a "silver bullet" treatment (pharmacologic or otherwise) to stop cell death [49]. We note this notion fails completely to account for the range of injury intensities that are possible for a given acute injury. In the context of stroke, such a treatment to stop neuron death is called neuroprotection. It is almost universally agreed that a drug will specifically target *the* molecular mechanism that causes cell death. This definition of therapy presumes a detailed understanding of the biological specifics of the injury *and* the drug action. We assert that this view is wholly incorrect.

Instead, our theory clearly indicates that a drug is a nonspecific form of injury. If given at high enough concentration ($\gg LD_{50}$), a drug is lethal. This is well-known. Ideas of "on-target" and "off-target" effects are arbitrary distinctions that reflect only our ignorance of the total biological action of a drug [17]. For example, how many neuroprotectants were thought to have "specific" actions inhibiting a damage pathway but later discovered to simply lower temperature? In the most general terms, any drug will interfere with normal cell function and act as an injury mechanism. At low concentrations (low values of *I*), it is sublethal. At high concentrations (high values of *I*), it is lethal.

We also note that the logic of "specific targets" has proven elusive with the two most neuroprotective stroke therapies: hypothermia and pre-conditioning. Preconditioning is, by definition, sub-lethal injury. Hypothermia too is an injury mechanism. If the intensity of hypothermia is too large (e.g. temperature is reduced too much), the system will die. Therapeutic hypothermia is the application of a sublethal dose (intensity) of hypothermia.

Therefore, through the lens of our theory, all possible therapies fall into only two categories: (1) *efforts to reduce injury intensity*, which have limited practicality when an injury has already occurred prior to clinical presentation, and (2) *sublethal injuries* whose main effect is to alter the trajectories of D and S. This understanding provides very strong motivation to reformulate our theory so that we can apply sequential injuries over time. The solution to this problem is the non-autonomous version of the theory.

4.4 The Non-autonomous Dynamical Theory of Acute Cell Injury

The non-autonomous theory builds stepwise on the autonomous theory. The autonomous theory treats the threshold parameters, Θ, in Eq. (3.2) as functions of injury intensity, I, but treats the v and k parameters as constants. The non-autonomous theory assigns functions to the v and k parameters. These are captured by modifying postulate 4 and adding a new postulate 5:

4. The decay parameter, k, is a function of the instantaneous value of $|D-S|$.
5. The velocity parameter, v, decays exponentially with time: $v \propto e^{-t}$.

With regard to new postulate 4, the importance of the term $|D^*-S^*|$ for realistic injury dynamics (e.g. necrosis vs. DND) was discussed above. Instead of taking $|D-S|$ only at the attractor, we set the decay parameter k for both D and S equal to the *instantaneous value* of $|D-S|$ times a constant of proportionality c_2. The effect of $|D-S|$ is augmented by multiplying it by time, t, causing the injury to decay faster than if k was taken only as a function of $|D-S|$:

$$k_D = k_S = c_2 t|D-S| \tag{3.6}$$

For postulate 5, recall that the velocity parameter, v, gives the rate of D and S *formation in* Eq. (3.1). Setting $v=1$ (or any number) means D and S will continue to increase at a constant rate over time. This is physically unrealistic. We expect the rate of formation of both D and S to *slow down* with time after the injury. Of the possible functional forms, the simplest is an exponential decay:

$$v_D = v_S = v_0 e^{-c_1 t} \tag{3.7}$$

In Eq. (3.7), v_0 is the initial velocity of formation of both D and S, and c_1 is a time constant that specifies the rate of decay of velocity. The presence of time, t, in Eqs. (3.6) and (3.7) makes the theory non-autonomous.

Substituting Eqs. (3.6) and (3.7) into Eq. (3.3) gives the non-autonomous version of the dynamical theory of acute cell injury:

$$\frac{dD}{dt} = v_0 e^{-c_1 t} \frac{\left(c_D I e^{I\lambda_D}\right)^n}{\left(c_D I e^{I\lambda_D}\right)^n + S^n} - c_2 t\left|D - S\right| \cdot D$$
$$\frac{dS}{dt} = v_0 e^{-c_1 t} \frac{\left(c_S I e^{-I\lambda_S}\right)^n}{\left(c_S I e^{-I\lambda_S}\right)^n + D^n} - c_2 t\left|D - S\right| \cdot S \tag{3.8}$$

4.5 Sequential Injuries

Equation (3.8) solves both limitations of the autonomous theory: (1) it generates closed trajectories that automatically return to $(D, S) = (0, 0)$ and (2) it allows simulation of multiple injuries over time as, for example, with ischemic preconditioning. Injury 1 (at intensity, I_1) is applied at time zero, and injury 2 (at intensity, I_2) can be applied at any time thereafter. In general, the sequential injuries either interact or they do not interact. The non-interacting case is not realistic and so discarded on the grounds that the second injury will necessarily interact with the first.

There are multiple ways to model how injuries 1 and 2 interact. We consider only one form of interaction here where D and S from the second injury (D_2, S_2) antagonize each other as well as D and S from the first injury (D_1, S_1). This formulation is based on the approach of Zhou and colleagues [50], is represented by the circuit diagram in Fig. 3.6, and gives rise to a system of four coupled nonlinear, non-autonomous differential equations:

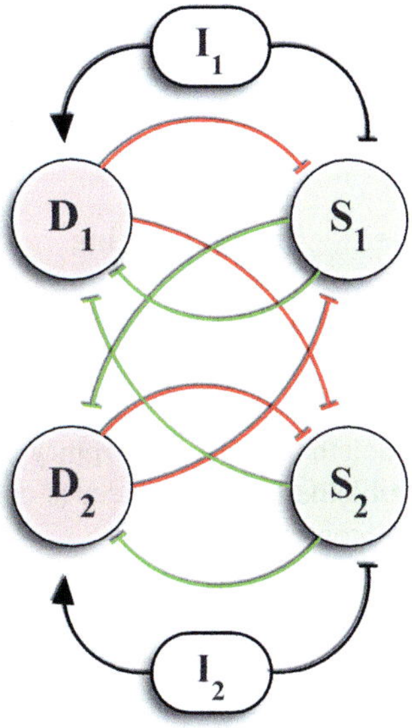

Fig. 3.6 Circuit diagram for two interacting injuries

$$\begin{aligned}\frac{dD_1}{dt} &= v\frac{\Theta_D^n}{\Theta_D^n + S_1^{\,n} + S_2^{\,n}} - kD_1\\ \frac{dS_1}{dt} &= v\frac{\Theta_S^n}{\Theta_S^n + D_1^{\,n} + D_2^{\,n}} - kS_1\\ \frac{dD_2}{dt} &= v\frac{\Theta_D^n}{\Theta_D^n + S_1^{\,n} + S_2^{\,n}} - kD_2\\ \frac{dS_2}{dt} &= v\frac{\Theta_S^n}{\Theta_S^n + D_1^{\,n} + D_2^{\,n}} - kS_2\end{aligned} \quad (3.9)$$

In Eq. (3.9), v is as given in Eq. (3.7). The k parameter is modified to allow D and S to interact:

$$k = c_2 t\left|\left(D_1 + D_2\right) - \left(S_1 + S_2\right)\right| \quad (3.10)$$

For Eqs. (3.9) and (3.10), the interaction of injury 1 and injury 2 is by addition of the values of D and S from each injury. Whether this is true or not must be empirically tested. For our purposes, it stands as an assumption. Our goal here is to study solutions of Eq. (3.9) and determine if they do or do not conform to what is already empirically established.

4.6 *Solutions of the Multiple Injury Model*

We present three examples of solutions to Eq. (3.9). Example 1 considers the case of preconditioning. Examples 2 and 3 simulate a post-injury drug treatment, where example 2 considers the effect of time of administration, and example 3 illustrates the effect of dose.

4.6.1 Preconditioning

Preconditioning is simulated by setting the parameters of injury 1 equal to those of injury 2, except for injury intensity, I. Equal values of (c_D, λ_D) and (c_S, λ_S) for both injuries mean applying the same injury mechanism to the same cell type. Injury 1 is sublethal ($I_1 < I_X$) and injury 2 is lethal ($I_2 > I_X$), thereby simulating the case where a sublethal injury precedes a lethal injury, which is the definition of ischemic preconditioning.

Injuries 1 and 2 *increasingly interact* as time between them (Δt) decreases. When $\Delta t = 250$ h, injury 1 mostly runs its course and has little effect on lethal injury 2, and the system dies (Fig. 3.7a). When Δt is 72 h, the injuries interact, but not enough for injury 1 to salvage injury 2 and thus D edges out S and the system dies (Fig. 3.7b).

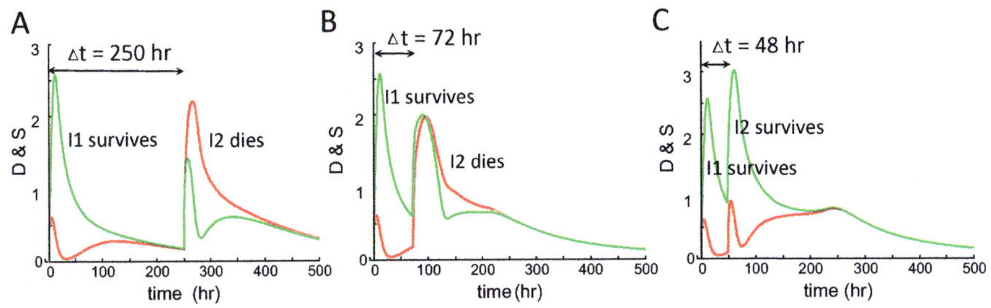

Fig. 3.7 Simulation of preconditioning where a sublethal insult is given before a lethal insult for three different time differences between injuries 1 and 2

However, at $\Delta t = 48$ h, the excess total stress responses of injury 1 adds, in a nonlinear fashion given by Eq. (3.9), to those of injury 2, allowing it to overcome total damage and survive (Fig. 3.7c). Therefore, Eq. (3.9) effectively simulates preconditioning.

4.6.2 Post-injury Drug Treatment

By modifying the input parameters to Eq. (3.9), it can be used to model a post-injury therapy. In this case, injury 1 is the lethal injury, and injury 2 is the therapy, treated as a sublethal injury. The parameter sets used in the example are:

INJURY 1 parameters	INJURY 2 parameters
$c_{S1}=0.25$; $\lambda_{S1}=0.9$; $n_1=4$	$c_{S2}=0.25$; $\lambda_{S2}=0.9$; $n_2=4$
$c_{D1}=0.1$; $\lambda_{D1}=0.1$	$c_{D2}=0.001$; $\lambda_{D2}=0.01$
$I_X=0.92$	$I_X=6.1$
$v_{0,1}=0.1$; $c_{1,1}=0.1$	$v_{0,2}=0.5$; $c_{1,2}=1$
$c_{2,1}=0.2$	$c_{2,2}=1$

To model applying the injuries to the same cell type, the parameters (c_S, λ_S) are set equal for the two injuries. The Hill coefficients are, arbitrarily, also kept equal. We assume that a drug, as a form of injury, will be considerably weaker than the main injury (which might be ischemia, or trauma, or etc.). Therefore, for injury 2, c_D is 1/100th and λ_D is 1/10th that of injury 1. Eq. (3.4) calculates the tipping point injury intensities for injury 1, $I_X=0.92$, and for injury 2, $I_X=6.1$. Thus, the main injury is lethal if $I_1>0.92$, and the therapy is lethal if $I_2>6.1$. The velocity parameters, v_0 and c_1, and decay parameter, c_2, control the overall rate of the injury. It is assumed that a drug will act quicker and so these parameters are 5×, 10×, and 5× that of injury 1. There are two additional parameters typically associated with a post-injury therapy. The time of administration of the therapy is set by the time after injury 1 when injury 2 is initiated. The dose of the therapy (e.g. drug dose) is set by the parameter, I, the injury intensity.

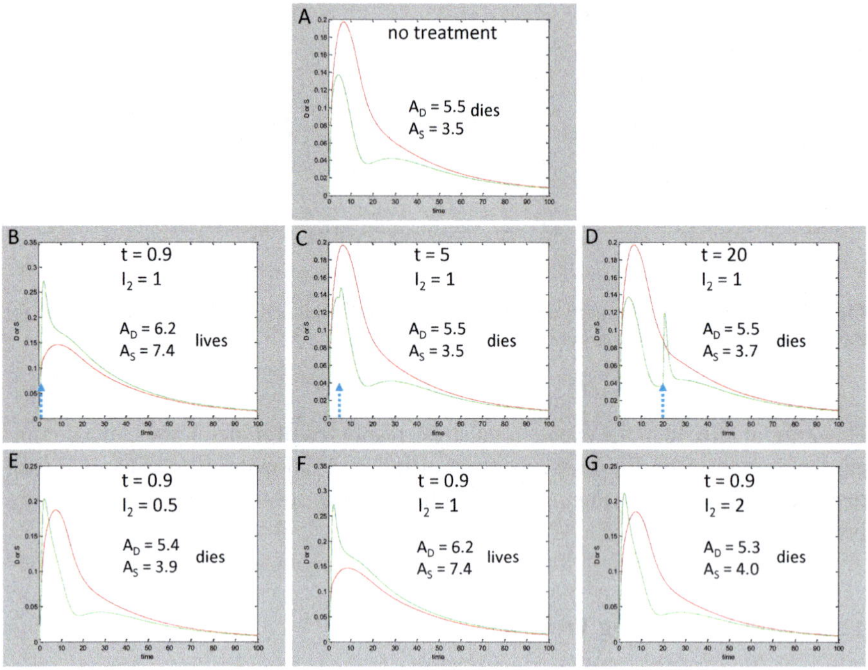

Fig. 3.8 Post-injury therapy of intensity I_2, applied to the main, lethal injury of lethal intensity, $I_1 = 1$. *Blue arrows* in (**b–d**) indicate time of application of second, sublethal injury

The results of running Eq. (3.9) with the above parameters are shown in Fig. 3.8. Figure 3.8a shows the main injury, injury 1, with lethal $I = 1$, and no treatment. The injury runs >90 % of its course over 100 time units. To precisely quantify whether the system survives or dies, we calculated the area under the D time course (A_D), the *cumulative total damage*, and the S time course (A_S), the *cumulative total stress responses*. We assert but do not attempt to justify here that $A_S > A_D$ is the condition of survival, and $A_D > A_S$ is the condition for death. When injury 1 with $I = 1$ is untreated, $A_D = 5.5 > A_S = 3.5$ and it dies.

Panels B–D of Fig. 3.8 show the effect of altering the time of administration. When the drug is given very early in the time course ($t = 0.9$), $A_S > A_D$, and the system survives. However, if the sublethal injury (i.e., drug) is given at $t = 5$ or $t = 20$, the system dies. This result is consistent with the common experience that a drug must be administered within a specific time window to be effective at halting cell death.

Panels E–G show the effect of concentration of the sublethal therapy. Panel F reproduces panel B showing that sublethal therapy of $I_2 = 1$ at $t = 0.9$ causes survival of lethal injury 1. However, if the dose of therapy is either halved ($I_2 = 0.5$, panel E) or doubled ($I_2 = 2$, panel G), damage dominates and the system dies. This result reproduces the notion of an optimal dose of therapy. If the dose is too small, it will be ineffective. If the dose is too large, it contributes to lethality.

These examples show proof of principle that the non-autonomous theory can simulate multiple sequential injuries. We are systematically studying the non-autonomous model and will present a full exposition at a later date. The above examples illustrate how the theory generalizes the notion of therapy as sublethal injury. As such, the theory provides a novel framework for pharmacodynamics.

The therapy (sublethal injury) may be applied before the lethal injury (preconditioning) or after (post-injury therapy), capturing mathematically several important empirical results. The theory provides a systematic framework to calculate beforehand all doses and times of administration, allowing optimization of outcome. Further, the theory is not confined only to two sequential injuries, but can model any number of injuries over time.

4.7 Spatial Applications

In this section, we briefly outline a spatial application of the theory that can be used to model three-dimensional (3D) tissue injury and thereby, for example, prognosticate stroke. We do not give a mathematical exposition, but instead merely outline the construction of this application.

Spatial applications involve two main components: (1) a 3D lattice with (2) a spatially distributed injury superimposed over the lattice. Each vertex of the lattice is taken as a biological cell, and multiple instances of the theory [e.g. Eq. (3.9)] are run in parallel at each lattice point. Figure 3.9 illustrates a simple 3D grid lattice. The injury mechanism is depicted as a colored solid in which the lattice is embedded. The legend maps the colors to a continuous distribution of injury intensity, I, which runs from sublethal to lethal and depicts a bistable system. The value of I at the location of a vertex serves as the I input parameter for that vertex.

The new feature of a spatial model is a coupling function between the cells/vertices, indicated by μ in Fig. 3.9. In general, the coupling function, μ, represents interaction between neighboring cells. Examples of what could be modeled by μ include an incremental addition of I to neighboring cells if the central cell dies, paracrine or autocrine influences between cells, or both.

Every aspect of a spatial application is constructible. The 3D lattice and superimposed injury mechanism can take on any geometry. The injury mechanism may be static or dynamic. The coupling functions can represent any possible interactions among cells. The common features of any spatial applications are: (1) a (x, y, z) dependence on both cell locations and the distribution of the injury mechanism (2), the coupling among cells, and (3) the need to run multiple instances of the theory in parallel. In general, spatial applications will be massively parallel computing problems.

We can envision a spatial application to invent technology to prognosticate stroke outcome. The cerebral blood flow (CBF) distribution, ascertained by PET or fMRI, could be used to generate the spatially distributed injury. The blood flow distribution could be mapped across an entire bifurcation diagram (Fig. 3.10a). The 3D cell

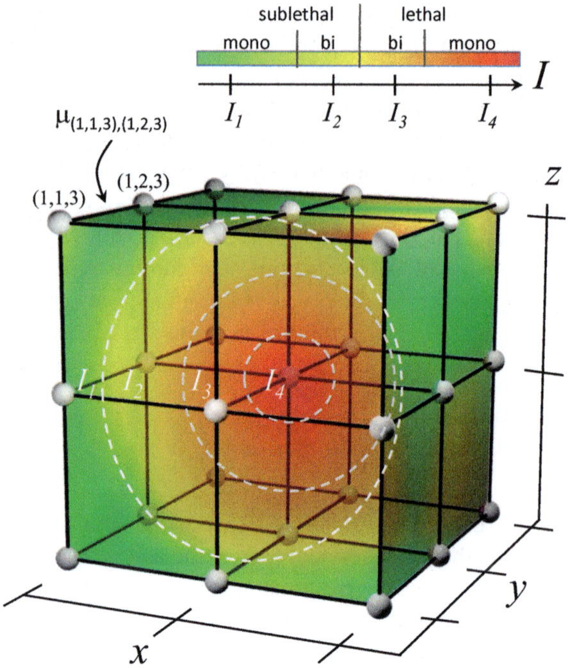

Fig. 3.9 A simple spatial model of acute cell injury. "Mono" indicates monostable attractors. "Bi" indicates bistability

lattice can approximate the distribution of cells in the affected brain tissue. The lattice is then superimposed with the patient's 3D CBF distribution (Fig. 3.10b, C). Blood or cerebrospinal fluid could also provide biomarkers to further parameterize the theory, e.g. by estimating time after injury. Such a simulation could, in principle, be generated and solved in near real time on a patient-by-patient basis directly at the patient bedside. Figure 3.10b–d show an initial time (say, patient presentation) used to parameterize the spatial application. Solving *D* & *S* time courses at each vertex, based on the patient's CBF gradient, could, in principle, *fully prognosticate outcome*. The blow-up in Fig. 3.10c shows calculated initial and final lesions. Here green points survive, yellow points are bistable and live or die depending on *I* (penumbra), and red is necrotic core. Figure 3.10d shows the final 3D reconstructions that could be generated in the clinic for physician use at the bedside.

Many neuroimaging efforts are directed at prognosticating stroke outcome [51, 52]. To date there has been no unequivocal success in this endeavor [53]. Our theoretical approach provides a missing link between the raw biological data of neuroimaging and biomarkers, on one hand, and prognostication on the other hand. The raw data needs to serve as input into a theory capable of prognosticating outcome, which is precisely the purpose of the dynamical theory of acute cell injury.

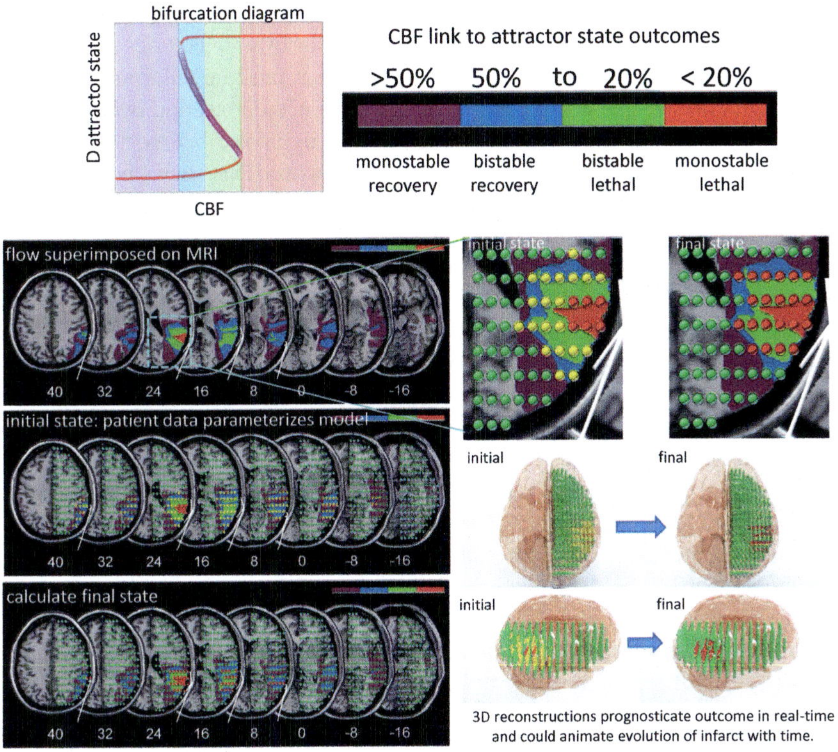

Fig. 3.10 The spatial model applied to stroke prognosis. (**a**) Mapping a bistable bifurcation diagram to cerebral blood flow (CBF) rates. (**b**) Superimposing a lattice of cells over real time CBF measurements, where CBF serves as the spatially distributed I parameter for the simulation. (**c**) Zoomed view from panels in (**b**). (**d**) Final 3D reconstruction calculating outcome would serve to prognosticate changes in lesion volume over time

4.8 A Possible Neuroprotective Technology

From the multiple injury models, we conclude that, theoretically, any form of protective therapy is necessarily a form of sublethal injury. Spatial applications hold out the promise to model, simulate, and prognosticate outcome in 3D tissue. We combine these with other insights offered by the theory and present a possible technology for stroke neuroprotection.

From studying bifurcation diagrams with I as the control parameter, it is necessarily the case that the behavior of an acutely injured system varies with injury intensity, I. This means a patient who experiences 40 % CBF is a completely different case than one with 20 % CBF or 0 % CBF. Simply stated, there will never be a "one size fits all single" or a "silver bullet" treatment for stroke. Each case will be different in the particulars of time after injury, degree and distribution of CBF reductions, comorbidities, etc. Furthermore, the theory unambiguously specifies that therapy is only possible when the injury dynamics are bistable. Prognostic efforts must be able to identify the bistable volumes of tissue. Then, the theory can calculate the conditions required to transform bistable lethal trajectories to survival trajectories.

Therefore, if clinical neuroprotection is possible, it must be tailor-made for each patient, on a patient-by-patient basis. That is, stroke neuroprotection is inherently an example of so-called "personalized medicine", not by choice but by necessity. The prognosticative technology considered in the previous sections shows how a patient-specific prognosis is possible in principle. It is only a small conceptual step from the prognosticative technology to neuroprotective technology that can be applied on a patient-by-patient basis. Our example will describe a technology that does not yet exist. It speaks to the power of our theory that it allows us to imagine new technology.

The example of a post-injury treatment given above (Fig. 3.8) considered the post-injury therapy to be a drug. But once we recognize that therapy is, in the context of our theory, always a form of sublethal injury, we are free to consider other ways to induce sublethal injury. We require a form of sublethal injury that can be precisely targeted and whose intensity can be precisely controlled. We also require something that can penetrate the skin and skull and be controlled to penetrate to any required depth in the brain tissue. The obvious possibility is some form of radiation that can be administered on an intensity continuum from nonlethal to lethal. We thus envision a machine that targets a brain-penetrant radiation, of variable intensity, to specific voxels in the brain. This is not an unheard of possibility. Recent work suggests that infrared or other radiation may be used as a neuroprotectant [54–57]. Our theory suggests that the "mechanism" of the neuroprotective action of radiation may precisely be its ability to impact acute injury dynamics.

Location, intensity, and duration of radiation are determined by the calculated prognosis (Fig. 3.11). If left untreated, the theory can calculate which bistable regions (e.g. penumbra) will survive and which will die, and therefore predict the maximum final lesion. Because the prognostic calculations can distinguish bistable from monostable lethal cases, the regions susceptible to neuroprotective therapy can be determined. Further, on a voxel by voxel basis, the intensity of the sublethal injury required to convert death trajectories to survival trajectories can be calculated (e.g. as illustrated in Fig. 3.8). Therapy would then be administered in the form of precision targeted sublethal injury (Fig. 3.11).

Ideally, all bistable regions susceptible to therapy would be shifted to pro-survival dynamics, resulting in the minimal lesion volume where only monostable lethal volumes die. It is well-recognized that core is not subject to salvage. The goal of neuroprotective therapy is to minimize the extent to which penumbra converts to additional core lesion. Our theory offers the possibility of engineering-level precision over these factors.

This section provides merely a rough sketch of a possible neuroprotective technology in the context of the dynamical theory of acute cell injury. Many scientific and technological hurdles exist before such technology can become reality. That the theory allows imagining such possibilities speaks of the strength of a mathematical and theoretical view of cell injury. Such technology cannot be envisioned from the qualitative, descriptive, pathway-centric view of cell injury that currently dominates biomedical research.

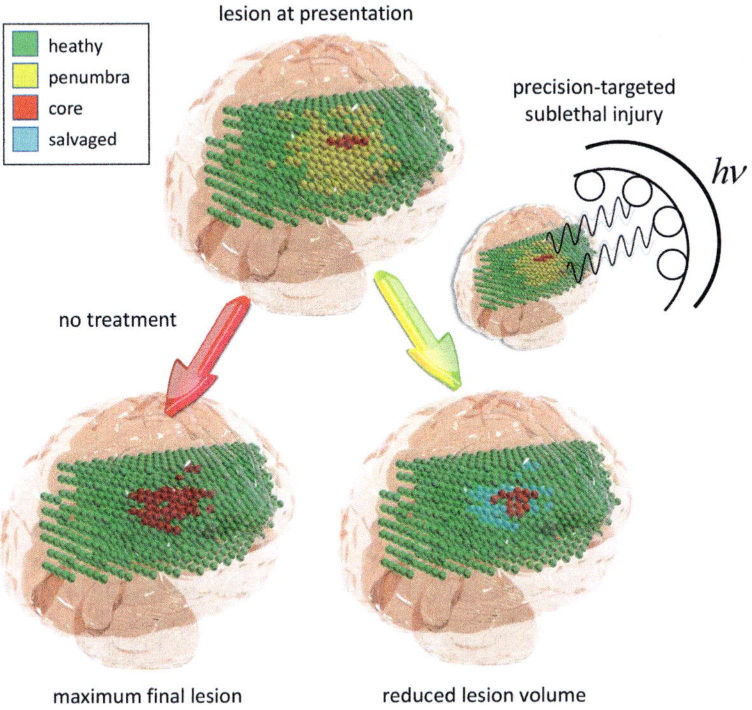

Fig. 3.11 A possible neuroprotective technology based on the nonlinear dynamical theory of acute cell injury. Penumbra is envisioned to consist of voxel dynamics that are either sublethal or lethal bistable, where lethal bistable voxels have the potential to be salvaged via a sublethal injury therapy

5 The Mathematical Road to Neuroprotection

This chapter has outlined the path from the science of acute cell injury to the technology of neuroprotection. A succinct summary of the steps is:

1. The correct mathematical form of the theory must be empirically validated for single injuries.
2. The correct mathematical form of the theory applied to multiple sequential injuries must be empirically validated.
3. The links between autonomous and non-autonomous versions must be systematically studied.
4. The theory provides the scientific basis to determine the parameters for combinations of injury mechanisms (c_D, λ_D) and cell types (c_S, λ_S) in the laboratory.
5. As injury dynamics become well-understood empirically, the doors to technological applications open. Possible technologies to prognosticate stroke outcome and to administer stroke neuroprotection were given as examples.

There is thus a stepwise progression from science to technology. The scientific step is to fully parameterize and validate the theory. This in turn provides the information *to calculate* outcomes at all injuries I (e.g. Fig. 3.11a). Technology then uses this information to design applications. The theory thus provides a quantitative and systematic platform to study therapeutics for all possible combinations of acute injury mechanisms and cell types.

By applying the method that Galileo advocated some 350 years ago to the study of cell injury, we have constructed a mathematical theory that has the potential to radically alter the understanding and treatment of acute cell injury. In short, the theory can usher in a new paradigm of acute cell injury that is firmly grounded in a mathematical paradigm of biology. We feel this direction will go a long way to overcome the weaknesses that are evident in the qualitative, pathways-based approach that currently dominates biomedical research.

References

1. Cheng YD, Al-Khoury L, Zivin JA (2004) Neuroprotection for ischemic stroke: two decades of success and failure. NeuroRx 1(1):36–45
2. O'Collins VE, Macleod MR, Donnan GA, Horky LL, van der Worp BH, Howells DW (2006) 1,026 experimental treatments in acute stroke. Ann Neurol 59(3):467–477
3. Turner RC, Dodson SC, Rosen CL, Huber JD (2013) The science of cerebral ischemia and the quest for neuroprotection: navigating past failure to future success. J Neurosurg 118(5):1072–1085
4. Xu SY, Pan SY (2013) The failure of animal models of neuroprotection in acute ischemic stroke to translate to clinical efficacy. Med Sci Monit Basic Res 19:37–45
5. Jo SK, Rosner MH, Okusa MD (2007) Pharmacologic treatment of acute kidney injury: why drugs haven't worked and what is on the horizon. Clin J Am Soc Nephrol 2(2):356–365
6. Kloner RA (2013) Current state of clinical translation of cardioprotective agents for acute myocardial infarction. Circ Res 113(4):451–463
7. Freedman LP, Gibson MC, Ethier SP, Soule HR, Neve RM, Reid YA (2015) Reproducibility: changing the policies and culture of cell line authentication. Nat Methods 12(6):493–497
8. Halsey LG, Curran-Everett D, Vowler SL, Drummond GB (2015) The fickle P value generates irreproducible results. Nat Methods 12(3):179–185
9. Landis SC, Amara SG, Asadullah K, Austin CP, Blumenstein R, Bradley EW et al (2012) A call for transparent reporting to optimize the predictive value of preclinical research. Nature 490(7419):187–191
10. Plant AL, Locascio LE, May WE, Gallagher PD (2014) Improved reproducibility by assuring confidence in measurements in biomedical research. Nat Methods 11(9):895–898
11. Vasilevsky NA, Brush MH, Paddock H, Ponting L, Tripathy SJ, Larocca GM et al (2013) On the reproducibility of science: unique identification of research resources in the biomedical literature. PeerJ 1, e148
12. Drummond GB, Paterson DJ, McGrath JC (2010) ARRIVE: new guidelines for reporting animal research. J Physiol 588(pt 14):2517
13. Kilkenny C, Browne W, Cuthill IC, Emerson M, Altman DG, National Centre for the Replacement, Refinement and Reduction of Animals in Research et al (2011) Animal research: reporting in vivo experiments—the ARRIVE guidelines. J Cereb Blood Flow Metab 31(4):991–993
14. Kilkenny C, Browne WJ, Cuthill IC, Emerson M, Altman DG (2010) Improving bioscience research reporting: the ARRIVE guidelines for reporting animal research. J Pharmacol Pharmacother 1(2):94–99

15. DeGracia DJ (2013) A program for solving the brain ischemia problem. Brain Sci 3(2):460–503
16. DeGracia DJ (2010) Towards a dynamical network view of brain ischemia and reperfusion. Part II: a post-ischemic neuronal state space. J Exp Stroke Transl Med 3(1):72–89
17. DeGracia DJ (2010) Towards a dynamical network view of brain ischemia and reperfusion. Part III: therapeutic implications. J Exp Stroke Transl Med 3(1):90–103
18. DeGracia DJ (2010) Towards a dynamical network view of brain ischemia and reperfusion. Part IV: additional considerations. J Exp Stroke Transl Med 3(1):104–114
19. DeGracia DJ (2010) Towards a dynamical network view of brain ischemia and reperfusion. Part I: background and preliminaries. J Exp Stroke Transl Med 3(1):59–71
20. Nordenskiöld E, Eyre LB (1935) The history of biology, a survey. Tudor, New York, pp 3–629
21. Kline M (1985) Mathematics and the search for knowledge. Oxford University Press, New York, p 257
22. Frezza W (2012) The skeptical outsider: how to rescue the life sciences from technological torpor. Updated 13 July 2012. Available from: http://www.bio-itworld.com/BioIT_Article.aspx?id=117147, http://www.bio-itworld.com2013
23. Bao L, Radish EF (2002) Understanding probabilistic interpretations of physical systems: a prerequisite to learning quantum physics. Am J Phys 70(3):210–217
24. Newton I, Cohen IB, Whitman AM (1999) The Principia: mathematical principles of natural philosophy. University of California Press, Berkeley, p 966
25. Weyl H (1934) Mind and nature. University of Pennsylvania Press, Philadelphia, p 100
26. Hillman H (1972) Certainty and uncertainty in biochemical techniques. Ann Arbor Science, Ann Arbor, p 126
27. Kedersha N, Anderson P (2007) Mammalian stress granules and processing bodies. Methods Enzymol 431:61–81
28. Biggs N, Lloyd EK, Wilson RJ (1986) Graph theory, 1736-1936. Oxfordshire/Clarendon Press, New York, p 239
29. Kauffman SA (1969) Metabolic stability and epigenesis in randomly constructed genetic nets. J Theor Biol 22(3):437–467
30. Jacob F, Perrin D, Sanchez C, Monod J (1960) Operon: a group of genes with the expression coordinated by an operator. C R Hebd Seances Acad Sci 250:1727–1729
31. Strogatz SH (1994) Nonlinear dynamics and Chaos: with applications to physics, biology, chemistry, and engineering. Addison-Wesley, Reading, p 498
32. Kaplan D, Glass L (1995) Understanding nonlinear dynamics. Springer, New York, p 420
33. Huang S, Guo YP, May G, Enver T (2007) Bifurcation dynamics in lineage-commitment in bipotent progenitor cells. Dev Biol 305(2):695–713
34. Vaughn MT (2007) Introduction to mathematical physics. Wiley-VCH, Weinheim, p 527
35. DeGracia DJ, Huang ZF, Huang S (2012) A nonlinear dynamical theory of cell injury. J Cereb Blood Flow Metab 32(6):1000–1013
36. Alon U (2007) An introduction to systems biology: design principles of biological circuits. Chapman & Hall/CRC, Boca Raton, p 301
37. Kirino T (2000) Delayed neuronal death. Neuropathology 20(suppl):S95–S97
38. Hossmann KA (1994) Viability thresholds and the penumbra of focal ischemia. Ann Neurol 36(4):557–565
39. Balduini W, Carloni S, Buonocore G (2012) Autophagy in hypoxia-ischemia induced brain injury. J Matern Fetal Neonatal Med 25(suppl 1):30–34
40. Chalmers-Redman RM, Fraser AD, Ju WY, Wadia J, Tatton NA, Tatton WG (1997) Mechanisms of nerve cell death: apoptosis or necrosis after cerebral ischaemia. Int Rev Neurobiol 40:1–25
41. Ge P, Zhang F, Zhao J, Liu C, Sun L, Hu B (2012) Protein degradation pathways after brain ischemia. Curr Drug Targets 13(2):159–165
42. Jouan-Lanhouet S, Riquet F, Duprez L, Vanden Berghe T, Takahashi N, Vandenabeele P (2014) Necroptosis, in vivo detection in experimental disease models. Semin Cell Dev Biol 35:2–13

43. Ginsberg MD (2008) Neuroprotection for ischemic stroke: past, present and future. Neuropharmacology 55(3):363–389
44. Alonso de Lecinana M, Diez-Tejedor E, Gutierrez M, Guerrero S, Carceller F, Roda JM (2005) New goals in ischemic stroke therapy: the experimental approach—harmonizing science with practice. Cerebrovasc Dis 20(suppl 2):159–168
45. Hermann DM, Bassetti CL (2007) Neuroprotection in the SAINT-II aftermath. Ann Neurol 62(6):677–678, author reply 8
46. Savitz SI, Fisher M (2007) Future of neuroprotection for acute stroke: in the aftermath of the SAINT trials. Ann Neurol 61(5):396–402
47. Molina CA, Saver JL (2005) Extending reperfusion therapy for acute ischemic stroke: emerging pharmacological, mechanical, and imaging strategies. Stroke 36(10):2311–2320
48. Molina CA, Alvarez-Sabin J (2009) Recanalization and reperfusion therapies for acute ischemic stroke. Cerebrovasc Dis 27(suppl 1):162–167
49. Rother J (2008) Neuroprotection does not work! Stroke 39(2):523–524
50. Zhou JX, Brusch L, Huang S (2011) Predicting pancreas cell fate decisions and reprogramming with a hierarchical multi-attractor model. PLoS One 6(3), e14752
51. Harston GW, Rane N, Shaya G, Thandeswaran S, Cellerini M, Sheerin F et al (2015) Imaging biomarkers in acute ischemic stroke trials: a systematic review. AJNR Am J Neuroradiol 36(5):839–843
52. Ward NS (2015) Does neuroimaging help to deliver better recovery of movement after stroke? Curr Opin Neurol 28(4):323–329
53. Hirano T (2014) Searching for salvageable brain: the detection of ischemic penumbra using various imaging modalities? J Stroke Cerebrovasc Dis 23(5):795–798
54. Brennan KM, Roos MS, Budinger TF, Higgins RJ, Wong ST, Bristol KS (1993) A study of radiation necrosis and edema in the canine brain using positron emission tomography and magnetic resonance imaging. Radiat Res 134(1):43–53
55. Lapchak PA (2010) Taking a light approach to treating acute ischemic stroke patients: transcranial near-infrared laser therapy translational science. Ann Med 42(8):576–586
56. Sanderson TH, Reynolds CA, Kumar R, Przyklenk K, Huttemann M (2013) Molecular mechanisms of ischemia-reperfusion injury in brain: pivotal role of the mitochondrial membrane potential in reactive oxygen species generation. Mol Neurobiol 47(1):9–23
57. Wong CS, Van der Kogel AJ (2004) Mechanisms of radiation injury to the central nervous system: implications for neuroprotection. Mol Interv 4(5):273–284

Chapter 4
History of Experimental Stroke Research

Victoria E. O'Collins

Abstract For a student investigating the history of experimental stroke research, the first challenge is developing an awareness of the evolving definition of *stroke* and of the multiple meanings of the word *history* as used in medical science papers. *Stroke* is a cerebrovascular disease involving the occlusion or rupture of a blood vessel supplying part of the brain accompanied by a decline in cerebral function (Stroke 44(7):2064–2089, 2013). Formal scientific classifications of stroke do not use the terms *part* or *whole,* but instead favor *focal* or *global,* respectively (Stroke 44(7):2064–2089, 2013). If wide regions of the brain are affected, it may be due to compromised cardiac or respiratory function and it is typically not designated a stroke. For the past few decades, much of stroke research has ignored physiology below the neck. With the rise of systems biology, this blinkered perspective is starting to decline.

Keywords Animal experiments • Cerebral ischemia • History • Stroke models • Experimental stroke • History • Stroke

Abbreviations

FDA	Food and Drug Administration
tPA or rtPA	Tissue plasminogen activator

1 Introduction

Early work by Johann Jakob Wepfer (1620–1695) and Giovanni Battista Morgagni (1682–1771) demonstrated that stroke can be caused by the blockage or rupture of vessels, a distinction which persists to this day [1]. The blockage (*ischemic stroke*) or rupture (*hemorrhagic stroke*) may occur in the arteries or veins supplying or

V.E. O'Collins (✉)
Florey Neuroscience Institutes, Level 2, Alan Gilbert Building, Parkville, VIC, Australia
e-mail: ocollins@unimelb.edu.au

P.A. Lapchak, J.H. Zhang (eds.), *Neuroprotective Therapy for Stroke and Ischemic Disease*, Springer Series in Translational Stroke Research,
DOI 10.1007/978-3-319-45345-3_4

draining blood from the brain, where the blockage or rupture is not caused by trauma and does not rapidly or automatically reverse. A third category is a *lacunar* stroke, characterized by fluid-filled cavities deep in the brain's white matter [2, 3].

2 Etymology, Origins, and Evolution of the Term Stroke

While the term *stroke* is commonly used now, many other terms have been used to describe the disease. These include *cerebrovascular accident, cerebrovascular insult, cerebral infarct, silent infarct, brain ischemia,* and more recently, *brain attack*—a term used to connote the urgency of stroke, by analogy with heart attack. Early uses of the word *stroke* date back to an 1599 recommendation for "an excellent Cinnamome water for the stroke of Gods hande" [4] and a medical essay by William Cole written in 1689 [5].

Historically, the term *apoplexy* was favored. Recorded as far back as Hippocrates in 400 BC [5], it continued to find use—through to the nineteenth century. In his 1658 essay, *Historiae Apoplecticorum*, Johann Jakob Wepfer [6] addresses the symptoms of apoplexy in Marcello Malpighi, the Pope's physician and professor of anatomy at Bologna. Without the window into the skull provided by modern brain imaging, it is difficult to determine the root cause of all historical references to strokes of apoplexy, but it is usually thought they are linked to hemorrhagic stroke (Fig. 4.1). The modern definition and classification of stroke has evolved greatly since the advent of brain imaging in the twentieth century [5].

A second trap for the new student is the use of the term *history*. History may refer to the case history of a patient, the information gained during a clinical examination. History may also indicate the "natural history" of stroke, the progression of stroke symptoms as the disease runs its course. Furthermore, history describes the names and dates of the men—and occasional woman—who developed the models of stroke, together with the culture, context, and meaning of those scientific advances [1, 7–10]. It is in this sense that the term history is used in this paper.

Histories of medicine often begin with Hippocrates (460–370 BC), while histories of experimental medical science often begin with Greek physician Claudius Galenus, also known as Galen of Pergamon (129 AD—c. 200/216) [11]. This is not necessarily because of the accuracy or primacy of Classical medicine, because many of the theories from other cultures may either be impenetrable—or incomparable thus far to those schooled in Western science. Our understanding of the history of stroke medicine is the poorer for it and waiting to be corrected.

Galen's ideas on vascular anatomy were developed through studies in Barbary macaques and oxen [12] as contemporary Roman law prohibited the use of human cadavers. His mechanistic theory explains blood flow as tidal ebb and flow from the body's core to the periphery and back. Medical writer for the New Yorker, Dr. Atal Gawande, argues that science is "beautifully self-correcting"[13]. Galen's theory took well over millennia to be displaced.

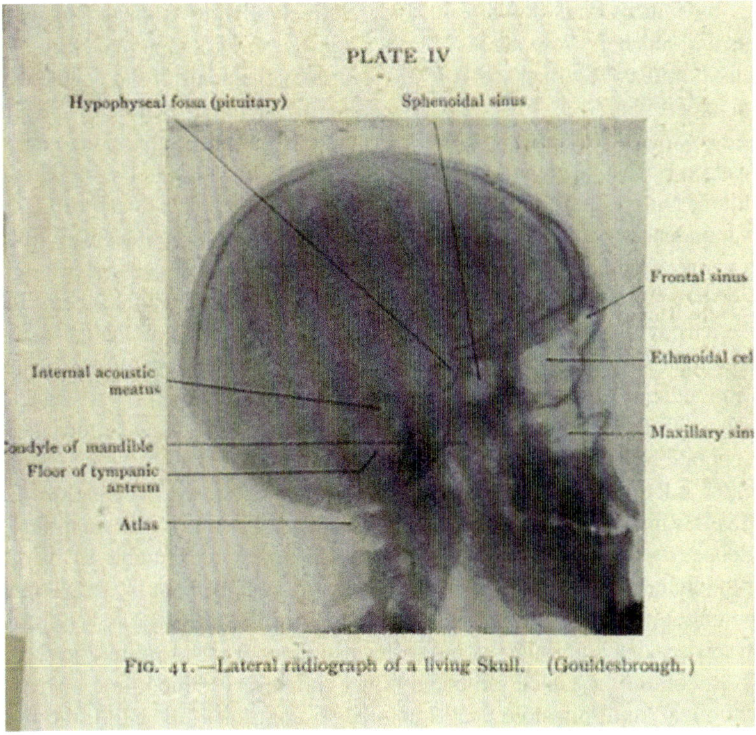

Fig. 4.1 The origins of modern brain imaging (1920)

Writing in 1940s, A.C. Ivy asserted that: "This evidence of the Renaissance from the barbarism and paganism of the Dark Ages was climaxed in 1628 by Harvey's 'discovery of the circulation' which involved controlled observations on animals and man" [11]. Such views on medieval Europe now seem rather quaint. While scientific advancement in the West might have been slowed by the religious mores of the ruling powers, it certainly had not stopped. William Harvey's work *On the Motion of the Heart and Blood in Animals* (*De Motu Cordis*) *1628* is a masterpiece and a revelation for the contemporary science student accustomed to the sterile and manufactured language of modern experimental science reports. However, his discovery built the knowledge of those before him. "Few working scientists can give a ground-up explanation of the phenomenon they study; they rely on information and techniques borrowed from other scientists. Knowledge and the virtues of the scientific orientation live far more in the community than the individual" [13].

Descriptions of the circulation of blood had already been given by Ibn an Nafis of Cairo (1210–1280), Michael Servetus (1509/1511–1553), who was burned as a heretic by Calvin in 1553 and the more fortunate Andrea Cesalpino (1919–1603), physician to Pope Clement VIII [14]. Yet the most profound influence on Harvey must have been the community of scientists he found in Padua. Founded in 1222, the University of Padua was established near the colossal Byzantine trading port of

Venice. Its fame, cosmopolitanism, and relatively liberal environment attracted students and teachers from across Europe. A professor at Padua, the Brussels born anatomist Andreas Vesalius (1514–1563) sparred with authorities to enable learning through experimentation (Fig. 4.2), the results of which may be read in *De Humani Coris Fabrica* (1543). Being a skilled teacher and communicator, his greatness can be measured by the success of those that came after him such as Harvey (Fig. 4.3).

William Harvey studied at Padua from 1599 to 1602 under the tutelage of Fabricius of Aquapondente, the surgeon and anatomist credited with the discovery of valves within veins. Also teaching at the University during Harvey's time was Galileo (1592–1610). While there is no evidence that Harvey knew Galileo, Fermi writes: "The nucleus of his audience was the medical student who took mathematics in order to understand cosmography and astrology; they needed astrology, for despite the new trends in medicine all good physicians were expected to draw on horoscopes" p 32 [15].[1] Harvey's teacher, Fabricius, was also Galileo's physician and colleague.

The parallels between Harvey's and Galileo's ideas are striking. Living near Venice, the tidal flows of the Mediterranean were of no small consequence. Galileo used the tides in his Dialogues between Salviati and Simplicio as an important piece of evidence to establish mathematically that flows were attributable to the circulation of the planets around the sun [16]. For tidal flow to be kept in circulation, they must be unbounded and infinite. Harvey recognised that the venous valves discovered by his teacher Fabricius would prevent unbounded and infinite flow in both directions, contrary to Galen's ebb and flow hypothesis of blood flow.

Anatomical and physiological knowledge continued to build through the seventieth to twentieth century. As Harvey did not have access to the compound microscope, he was not able to complete the circuit from arteries to veins via the capillaries. This discovery was made by Marcello Malpighi (1628–1694) [14]. Advances were then made in the understanding of blood pressure, in the coagulation process by Andrew Buchanan (1845), Alexander Schidt (1831–1894), and Olof Hammarsten (1875) and in identification of embolic causes of infarction by Rudolph Virchow (1821–1902) [1]. Up to the mid-twentieth century, scientific reports were predominantly physician-driven studies of blood flow in dogs and cats and clinical syndromes from patients. For instance, the towering figure of C. Miller Fisher was legendary for his clinical symptomatology of stroke [17].

New scientific techniques developed during the twentieth century enabled spatial or temporal information to be collected about the brain physiology. This led to a significant departure from the purely vascular-driven approach to stroke. Now, both vessels and the brain tissue they supplied featured in the theories of stroke. The most gripping of the new theories of stroke were the ischemic penumbra [18] and the ischemic cascade, initially reported in the kidney literature [19] and quickly adapted to the brain [20, 21]. The ischemic penumbra theory postulated that there were zones in the brain, a penumbra or shadow, where cells were electrically or

[1] Galileo was teaching not 20 years after the bubonic plagues claimed at least 16% of Venice's population and decimated the region, including Padua. "Universal" and "distant" causes in the motion of the planets were frequently applied to such calamities.

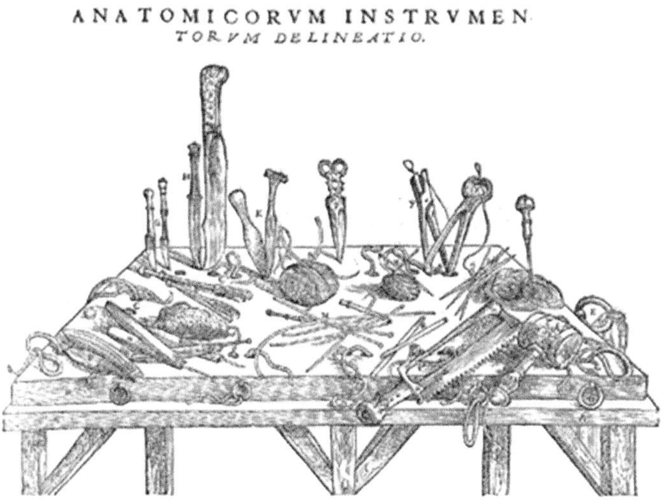

Fig. 4.2 The scientific tools of Vesalius from the "Studio of Titian" featured in De Humani Corporis Fabrica. Reproduced with permission of the National Library of Medicine (NML) (*source*: https://www.nlm.nih.gov/exhibition/historicalanatomies/vesalius_home.html)

Fig. 4.3 Portrait of Vesalius from De Humani Corporis Fabrica. Reproduced with permission of the National Library of Medicine (NML) (*source*: https://www.nlm.nih.gov/exhibition/historicalanatomies/vesalius_home.html)

functionally inactive but not yet dead [18]. "Possible reversibility of the functional inactivation in ischemia is the key problem of interest here" (p. 56), said the architects of this concept [18]. The theory of the ischemic cascade gave a timeline by which different, destructive physiological processes were set in action.

The development of modern therapies for stroke has been driven by physicians and scientists seeking to target blood flow in the ischemic penumbra or stymie destructive processes within the ischemic cascade. Many of the approaches have been developed from existing ideas adapted from other fields and applied in the context of stroke. Aspirin has been used to prevent the recurrence of stroke through the prevention of clots, yet was isolated in 1763 and the clinical use of the willow tree bark from which it was derived dates back to antiquity.[2] Clot busting therapy using tissue plasminogen activator (tPA or rtPA) received approval from the Food and Drug Administration (FDA) in 1996 for the treatment of ischemic stroke following the positive NINDS tPA Stroke Study [22]. It had been approved by the FDA for use in acute myocardial infarction (heart attack) 9 years earlier than stroke.

The possibility of treating stroke created a demand for developing potential stroke therapeutics and a demand for animal models of stroke in which to test them [23]. After the failure of a large number of high profile clinical trials including candidate NXY-059 [24], there was a retreat from the optimistic outlook and a re-evaluation of the animal models of stroke. This led to the further tightening of ethical and experimental guidelines and the greater alignment of standards for pre-clinical studies and clinical trials [25–27].

What is striking about reports of modern stroke experiments compared to the experiments reported by William Harvey is how narrow, formulaic, and confirmatory their methodology is.[3] For example, to explore the motion of blood, Harvey used the entire animal kingdom from flies, hornets, wasps, bees, hornets, crayfish, sponges, oysters, crabs, shrimps, frogs, partridge dogs, hogs, sheep, and ox [28]. In contrast, in a review of animal models of stroke, 80 % of animal experiments were conducted in the rat, and of these, 41 % used a single rat strain—the Sprague-Dawley [29]. The remainder were largely undertaken in mice [29]. We have not yet fully grasped how the generalizability of experimental results to humans is affected by the use of a quadrupedal species with a smaller brain, different allergens, circadian rhythms, and heat regulation systems. Such differences may or may not have profound effects.

The recent gap in translation of findings from animals to humans would suggest an apparent failure of positivist approach to stroke science where the determination of outcomes can be based purely on empirical evidence interpreted by reason and logic.

---

[2] The fate of Russian turned at least in part on the use of aspirin. Administered to the Russian Tsar's hemophiliac son Alexis by doctors, Rasputin reputedly halted the treatment, the bleeding subsided and his influence grew.

[3] A Chinese speaking colleague once commended me on ability to translate a Chinese animal stroke paper. I had to concede that I had no knowledge whatsoever, but on reading close to 10,000 or so of the papers, the reporting of methods and results in stroke papers was so predictable it was easy make a guess at extracting the information I was asking her to translate and confirm.

However, the tight formulation of introduction, methods, results, and discussion, which are governed by statistical testing, left little room for publication of information on outliers, anomalies and failed experiments. The importance and richness of such information cannot be underestimated in the drive for scientific advancement, yet it is typically only shared within the laboratory or informally at conferences.

To remedy this, scientists such as Ioannidis [30] have called for a greater rigor in experimental methodologies, including: better powered evidence through large studies and meta-analysis; enhanced research standards and diminution of bias; upfront registration of studies; hypothesis generation; networking of data collections; and use of a Bayesian approach to experimentation whereby prior information is included as part of the experimental data stream and inferences are updated regularly. Steps are being made to implement many of these changes, though the Bayesian approach is yet to find routine use in stroke. Systemic constraints on experimentation such as the use of standard operating procedures, the grant-funding process, and ethics committee approvals tend towards a slow and discrete approach.

While the dashed optimism is frequently attributed to the shortcomings in experimental models and procedures, there is room to question the way in which the targets for therapy are selected. The dominant theoretical frameworks in stoke—the ischemic penumbra and ischemic cascade—have great explanatory power. However, like the enduring hold of Galen's ebb and flow hypothesis, these frameworks with their readily graspable poetic references to clouds and waterfalls have power to blind scientists to contrary evidence. How many potential treatments may have been rejected because they did not fit neatly within these theories? How much contrary evidence was dismissed in the evaluation of parts of the cascade such as excitotoxicity [31], when much of the early evidence was drawn from small studies in the less relevant global model of ischemia? As noted by Kuhn (p. 122): "Given a paradigm, interpretation of data is central to the enterprise that explores it. But that interpretive enterprise...can only articulate a paradigm, not correct it" [32].

Almost 1000 years ago, Roger Bacon (c. 1219/1220–c. 1292) advised that: "He therefore who wishes to rejoice without doubt in regard to the truths underlying phenomena must know how devote himself to experiment" p. 584 [33]. In contrast, the ability of modern computing to capture large, structured, and unstructured data has led to pronouncements of the death of the experiment. The big data debate highlights the tension between expert knowledge and unfiltered data, between supervised, driven approaches and unsupervised, accumulated knowledge. Yet expert knowledge can be quantitatively distorted by bias, as shown by work of the international collaboration CAMARADES (Collaborative Approach to Meta-Analysis and Review of Animal Data from Experimental Studies) [34, 35]. Researchers need to arrive at a consensus on which experimental data is necessary to capture to parallel the personalized medicine approach facilitated by big data. Experimental controls will need to be scrutinized with reference to their validity and generalizability to the human condition [29]. Researchers will also need new ways of representing and processing large sets of data, while recognizing that patterns do not necessarily imply causation.

Recent years have also seen major breakthroughs in gene targeting technologies, which allow human genomes to be altered permanently [36, 37]. These tools also enable human tissue to be developed within animal hosts, blurring the line between animal model and the clinical condition it is mimicking. Such techniques raise profound ethical and safety concerns; gene editing and gene drive systems which alter whole populations are currently outpacing regulatory systems for analysis and oversight [38]. Given that stroke is a genetically heterogeneous disease primarily affecting the older population, it is not clear how such advances will play a role in stroke management, beyond the identification of genetic risk factors and targeting of individual genes in therapy.

3 Conclusion

Historical narrative is shaped by the present and future outlook and much has changed in scientific research over the past 10 years. There is now greater involvement of the public through open science and citizen science movements. There are challenges to the validity and ethics of animal studies. There is pressure from pharmaceutical companies to get a larger return on investment following a number of clinical trials failing to demonstrate a positive therapeutic effect. There are profound advances in technologies that decode and manipulate genes and data. Moving forward, different methods will be needed to measure progress in experimental stroke research in this changing environment. Keeping half an eye on the past can only help.

References

1. Caplan LR (2004) Cerebrovascular disease: historical background, with an eye to the future. Cleve Clin J Med 71(suppl 1):S22–S24
2. Fisher CM (1965) Lacunes: small, deep cerebral infarcts. Neurology 15:774–784
3. Fisher CM (1982) Lacunar strokes and infarcts: a review. Neurology 32(8):871–876
4. Pound P, Bury M, Ebrahim S (1997) From apoplexy to stroke. Age Ageing 26(5):331–337
5. Sacco RL, Kasner SE, Broderick JP, Caplan LR, Connors JJ, Culebras A et al (2013) An updated definition of stroke for the 21st century: a statement for healthcare professionals from the American Heart Association/American Stroke Association. Stroke 44(7):2064–2089
6. Pearce JM (1997) Wepfer's description of the apoplexy of Malpighi. J Neurol Neurosurg Psychiatry 62(4):394
7. Garrison FH (1929) History of medicine. W.B. Saunders, Philadelphia
8. Cunningham A (2002) The pen and the sword: recovering the disciplinary identity of physiology and anatomy before 1800 I: old physiology—the pen. Stud Hist Philos Biol Biomed Sci 33:631–665
9. Gerber CS (2003) Stroke: historical perspectives. Crit Care Nurs Q 26(4):268–275
10. O'Collins VE, Donnan GA, Howells DW (2011) History of animal models of stroke. Int J Stroke 6(1):77–78
11. Ivy AC (1948) The history and ethics of the use of human subjects in medical experiments. Science 108:1–5
12. Rocca J (1998) Galen and Greek neuroscience. Early Sci Med 3(3):216–240
13. Gawande A (2016) The mistrust of science. New Yorker

14. Ackerknecht EH (1968) A short history of medicine (Rev. print. edn). Ronald Press, New York, xviii, 275p
15. Fermi L, Bernardini G (1961) Galileo and the scientific revolution, 1st edn. Basic Books, New York, p 150
16. Galilei G, Crew H, de Salvio A (1914) Dialogues concerning two new sciences. Macmillan, New York, xxi, 300p
17. Greenberg SM (2013) Stroke history. C. Miller Fisher: an appreciation. Stroke 44:e171–e172
18. Astrup J, Symon L, Branston NM, Lassen NA (1977) Cortical evoked potential and extracellular K+ and H+ at critical levels of brain ischemia. Stroke 8(1):51–57
19. Fisher JW, Gross DM, Foley JE, Nelson PK, Rodgers GM, George WJ et al (1978) A concept for the control of kidney production of erythropoietin involving prostaglandins and cyclic nucleotides. Contrib Nephrol 13:37–59
20. Siesjo BK, Rehncrona S, Smith D (1980) Neuronal cell damage in the brain: possible involvement of oxidative mechanisms. Acta Physiol Scand Suppl 492:121–128
21. Hass WK (1983) The cerebral ischemic cascade. Neurol Clin 1(1):345–353
22. The National Institute of Neurological Disorders and Stroke rt-PA Stroke Study Group (1995) Tissue plasminogen activator for acute ischemic stroke. N Engl J Med 333(24):1581–1587
23. O'Collins VE, Macleod MR, Donnan GA, Horky LL, van der Worp BH, Howells DW (2006) 1,026 experimental treatments in acute stroke. Ann Neurol 59(3):467–477
24. Shuaib A, Lees KR, Lyden P, Grotta J, Davalos A, Davis SM et al (2007) NXY-059 for the treatment of acute ischemic stroke. N Engl J Med 357(6):562–571
25. Stroke Therapy Academic Industry Roundtable (1999) Recommendations for standards regarding preclinical neuroprotective and restorative drug development. Stroke 30(12):2752–2758
26. Macleod MM, Fisher M, O'Collins V, Sena ES, Dirnagl U, Bath PM et al (2009) Good laboratory practice. Preventing introduction of bias at the bench. Stroke 40(3):e50–e52
27. Lapchak PA, Zhang JH, Noble-Haeusslein LJ (2013) RIGOR guidelines: escalating STAIR and STEPS for effective translational research. Transl Stroke Res 4(3):279–285
28. Harvey W (1628) On the motion of the heart and blood in animals. Kessinger, Whitefish
29. O'Collins VE, Donnan GA, Macleod MR, Howells DW (2013) Animal models of stroke versus clinical stroke; comparison of infarct size, cause, location, study design and efficacy of experimental therapies. In: Conn M (ed) Animal models for the study of human disease, vol 1. Elsevier, London, pp 532–594
30. Ioannidis JP (2005) Why most published research findings are false. PLoS Med 2(8), e124
31. Olney JW (1969) Brain lesions, obesity, and other disturbances in mice treated with monosodium glutamate. Science 164(3880):719–721
32. Kuhn TS (1970) The structure of scientific revolutions. 2nd ed. University of Chicago Press, Chicago, xii, 210p
33. Bacon R (2002) Opus majus of Roger Bacon Part 2. In: Burke RB (ed). Kessinger Publishing, Whitefish
34. Macleod MR, O'Collins T, Howells DW, Donnan GA (2004) Pooling of animal experimental data reveals influence of study design and publication bias. Stroke 35(5):1203–1208
35. Sena ES, van der Worp HB, Bath PM, Howells DW, Macleod MR (2010) Publication bias in reports of animal stroke studies leads to major overstatement of efficacy. PLoS Biol 8(3), e1000344
36. Jinek M, Chylinski K, Fonfara I, Hauer M, Doudna JA, Charpentier E (2012) A programmable dual-RNA-guided DNA endonuclease in adaptive bacterial immunity. Science 337(6096): 816–821
37. Abudayyeh OO, Gootenberg JS, Konermann S, Joung J, Slaymaker IM, Cox DB, et al (2016) C2c2 is a single-component programmable RNA-guided RNA-targeting CRISPR effector. Science
38. Kuzma J (2016) New genetic engineering is slipping past old regulations. Aeon, 6 May 2016

Chapter 5
History of Neuroprotection: Trials and Tribulations

Ali Razmara and Steven C. Cramer

Abstract Neuroprotection is a strategy of interference, antagonism, and slowing down the sequence of molecular pathophysiological processes eventually resulting in irreversible cerebral ischemia. Over the past two decades, neuroprotection in ischemic stroke has emerged as a central topic of intense experimental animal studies and clinical trials in humans. Although rigorous animal studies have provided the proof of principle that neuroprotection is achievable, the novel agents and mechanisms investigated in human clinical trials have consistently failed to demonstrate a significant beneficial effect. Here we survey key neuroprotective trials and consider the strengths and shortcomings of these studies. Agents and mechanisms considered include calcium channel blockers, glutamate antagonists, GABA agonists, antioxidants and free radical scavengers, nitric oxide signal-transduction, modulation of inflammation, hemodilution, hypothermia, albumin therapy, and magnesium therapy. These human trials of neuroprotection therapies have been disappointing, unlike successful acute stroke approaches using reperfusion therapies such as thrombolytics or clot-retrieving devices. We highlight how improved clinical trial design and translational strategies and lessons learned from these negative trials will guide future directions including better clinical trial design and patient selection, multiple agent-combination therapies, and pre-hospital intervention.

Keywords Neuroprotection • Cerebral ischemia • Clinical trials • Lessons learned • Pre-hospital intervention

Abbreviations

ALIAS	Albumin in Acute Stroke
AMPA	Alpha-amino-3-hydroxy-5-methyl-4-isoxazole propionate
ARTIST	Alpha-amino-3-hydroxy-5-methyl-4-isoxazole propionate Receptor Antagonist Treatment in Ischemic Stroke Trial

A. Razmara, M.D., Ph.D. (✉) • S.C. Cramer, M.D., M.M.Sc.
Department of Neurology, University of California, Irvine,
200 S. Manchester Ave., Suite 206, Orange, CA 92868, USA
e-mail: arazmara@uci.edu; scramer@uci.edu

P.A. Lapchak, J.H. Zhang (eds.), *Neuroprotective Therapy for Stroke and Ischemic Disease*, Springer Series in Translational Stroke Research,
DOI 10.1007/978-3-319-45345-3_5

ASTIN	Acute Stroke Therapy by Inhibition of Neutrophils
CDP	Choline-cytidine 5-diphosphocholine
CLASS	Clomethiazole Acute Stroke Study
DNA	Deoxyribonucleic acid
Epo	Erythropoietin
FAST-MAG	Field Administration of Stroke Therapy-Magnesium
GABA	Gamma-aminobutyric acid
GAIN	Glycine Antagonist in Neuroprotection
ICTUS	International Citicoline Trial on Acute Stroke
ICTuS	Intravascular Cooling in the Treatment of Stroke
IMAGES	Intravenous Magnesium Efficacy in Stroke trial
MCA	Middle cerebral artery
MRI	Magnetic resonance imaging
mRS	modified Rankin scale
NIHSS	National Institutes of Health stroke scale
NMDA	*N*-methyl-D-aspartate
PBN	a-phenyl-*N*-tert-butyl nitrone
RANTTAS	Randomized Trial of Tirilazad mesylate in patients with Acute Stroke
SAINT	Stroke-Acute Ischemic NXY Treatment
STAIR	Stroke Therapy Academic Industry Roundtable
tPA	Tissue plasminogen activator
VENUS	Very Early Nimodipine Use in Stroke

1 Introduction

Over the last two decades, animal models of ischemic brain injury have elucidated the basis for the understanding of the pathophysiological mechanisms underlying ischemic stroke [1]. The concept of neuroprotection was an extension of the knowledge and strategic approach to interfere and hinder the underlying molecular cascade of events leading to irreversible cerebral ischemia [2]. The basic goal of neuroprotection is to protect brain tissue by impeding molecular targets in the ischemic cascade or by enhancing inherent protective mechanisms, until such time that cerebral perfusion is restored by spontaneous, pharmacological, or interventional procedures.

The definition of neuroprotection excludes reperfusion strategies including intravenous thrombolytics, antiplatelet and antithrombotic medications, and intra-arterial endovascular mechanical thrombectomy. The first and only pharmacological compound shown to be effective in acute stroke therapy is intravenous tissue plasminogen activator (tPA) within 3 h [3] and up to 4.5 h after the onset of stroke symptoms [4]. In addition, several recent positive trials [5–9] signal a new era of mechanical thrombectomy, with newer generation stent retrievers showing not only favorable radiological results, but now also improved functional outcomes. Endovascular thrombectomy alone was not beneficial, but only with pre-treatment with intravenous thrombolysis. Given these recent developments, the American

Heart Association/American Stroke Association recently updated their acute ischemic stroke treatment guidelines [10]. A critical distinction to be appreciated is that these reperfusion-based strategies target clots and arteries, while neuroprotection strategies target neurons and glia.

Neuroprotection was initially thought to hold the promise of assisting clinicians to reduce stroke morbidity and mortality in order to improve the quality of life of stroke survivors. Targeting molecular pathways appears to be relatively simple in vitro and in preclinical studies; however, data from numerous human trials have not been promising. Although a significant number of agents have been investigated, none has shown definite benefit in clinical trials of humans. Key agents tested in clinical trials in this context are described below.

2 Cerebrovascular Ischemia Cascade

A fundamental understanding of the ischemic cascade is important in order to formulate and design neuroprotection strategies. Cerebral ischemia is typically caused by an arterial occlusion leading to diminished cerebral blood flow, oxygen deprivation, and activation of a cascade of molecular events leading to tissue damage and ultimately neuronal death through necrosis and apoptosis. The goal of neuroprotection is to interfere in the molecular pathway that ischemic neurons undergo, leading to cell death and to salvage the ischemic penumbra [11]. The ischemic cascade has been extensively studied and a brief overview will aid in understanding the reasoning underlying clinical trial design of neuroprotection.

When there is a decrease in cerebral perfusion and oxygenation, neurons are unable to continue aerobic respiration in the mitochondria. Subsequently, there is a change to predominantly anaerobic metabolism and excess lactic acid production, leading to a decrease in pH due to unbuffered hydrogen ions [12]. This leads to decreasing efficiency of the electron transport chain, decreasing adenosine triphosphate production, and failure of the sodium-potassium pump. This leads to deterioration of membrane ion gradients, opening of various selective and unselective ion channels, and anoxic depolarization. Potassium leaves the cell, while sodium, chloride, and calcium enter, following which excitatory neurotransmitters such as glutamate and aspartate are released from cells in toxic concentrations [13]. Ultimately, failure of these energy-dependent mechanisms leads to cellular swelling with infusion of water and sodium flowing down osmotic as well as electrolyte gradients, leading to cytotoxic edema, which can be measured indirectly by changes in water on diffusion-weighted magnetic resonance imaging as a very early change following an acute ischemic stroke [14, 15].

Excitatory neurotransmitters, oxidative stress, and inflammation play important roles in the early stages of the ischemic pathway. Presynaptic glutamate is released early on in high concentrations [16, 17], binding to ligand-gated ion channels, including *N*-methyl-D-aspartate (NMDA), alpha-amino-3-hydroxy-5-methyl-4-isoxazole propionate (AMPA), and kainate receptors. Binding of glutamate to NMDA receptors leads to ion channel activation, large calcium influx, and eventual cell death as the culmination of these molecular cascades [18, 19]. Calcium-dependent processes include activation of proteases, lipid peroxidases, and eventual breakdown of DNA through activated endonucleases [20]. Reactive oxygen species are present during normal cellular respiration, but when free radical formation exceeds cellular antioxidant mechanisms, deleterious effects occur on DNA, lipids, and proteins. Inflammation is also an underlying mechanism associated with the ischemic cascade. Increased pro-inflammatory cytokines, such as tumor necrosis factor and interleukin-1, may be a late response in ischemic areas and the surrounding penumbra with damaging consequences, leading to disruption of the blood-brain barrier through upregulation of adhesion molecules [21].

Neuroprotection as a concept is not novel. The first recognized form of neuroprotection was hypothermia during the 1940s–1950s when a neurosurgeon, Dr. Temple Fay, observed improved outcomes in severe head trauma treated with hypothermia [22]. Since then, numerous agents have been shown to have efficacy in animal models of stroke, with ten classes of pharmacological agents reaching class III testing [23]. The focus of this chapter will not be on thrombolytics, antiplatelets, or anticoagulation therapies, but on agents acting on the different phases of the ischemic cascade that has been developed and tested in the laboratory and in clinical trials. The purpose of this chapter will be to highlight important agents and clinical trials as well as lessons learned and future directions for further advancing acute stroke therapies.

3 Drug Targets

3.1 Calcium Channel Blocker: Nimodipine

Calcium is a key player in the pathophysiologic mechanism underlying cardiac and cerebral ischemic injury [20, 24–26]. Nimodipine is a 1,4-dihydropyridine calcium channel antagonist that preferentially causes cerebral vasodilation with less systemic effects [27]. Nimodipine has been shown to reduce the severity of neurologic deficits due to cerebral vasospasm in subarachnoid hemorrhage [28]. Over 250 animal studies have been published examining nimodipine in cerebral ischemia, with only 20 deemed to be of adequate quality based on factors such as administering nimodipine after ischemia induction or inclusion of a control group [29]. Only half of these 20 studies favored nimodipine, and overall changes from treatment after 1 h stroke induction were not significant [29]. Five randomized trials of nimodipine in patients with acute ischemic stroke have been published [30–34]. A meta-analysis of these and other smaller trials found no overall benefit of nimodipine, however, with a subgroup of patients having beneficial effect on neurological and functional effects with 12 h of stroke onset [35]. However, a subsequent meta-analysis of over 7500 patients did not confirm these positive results and indeed showed that intravenous administration of calcium antagonists was associated with worse outcomes, and that oral nimodipine started within 12 h did not have any benefit [36]. The VENUS Trial aimed to test efficacy of early nimodipine administration with randomization of 454 patients to oral nimodipine or placebo with treatment within 6 h and continued for 10 days [34]; however, the trial was ended due to sample size concerns with primary end point of poor outcome not differing between treatment group and placebo [34]. There were several reasons for the failure of these trials including flawed trial design, timing of therapeutic efficacy, small sample size, critique of inclusion criteria, and outcome measures used. Prophylactic therapy with nimodipine is now the standard treatment in aneurysmal subarachnoid hemorrhage with the usual dosing of 60 mg orally every 4 h for 21 days. There is still uncertainty whether nimodipine has its primary effect in this clinical setting as a neuroprotective agent or as a vasodilator.

3.2 *Glutamate Antagonists*

As an excitatory neurotransmitter, glutamate can induce excitotoxic injury via the ischemic cascade [18, 37, 38] with relevant receptors including *N*-methyl-D-aspartate (NMDA) and the 3-hydroxy-5-methyl-4-isoxazole proprionic acid (AMPA). Dizocilpine (MK-801) is a non-competitive NMDA receptor antagonist that has been extensively studied in animal models of focal ischemia [39, 40], with evidence of ischemic volume reduction most effective prior to ischemia or within the first 1–2 h afterwards; however, it caused dose-related neuropsychological adverse events [41]. Dextromethorphan and its metabolite dextrorphan also have high affinity as non-competitive NMDA antagonists and have been shown to be neuroprotective in animal models [39, 40], but a variety of dose-related adverse effects occurred including nausea, vomiting, somnolence, hallucinations, agitation, and symptomatic hypotension.

Aptiganel (CNS-1102, Cerestat) is the only non-competitive NMDA antagonist that has been studied in clinical trials. In pre-clinical studies, aptiganel treatment 15 min after middle cerebral artery occlusion in rats showed significant decrease in infarction volumes by magnetic resonance imaging and post-mortem analysis [42]. The Aptiganel Acute Stroke Trial randomized to treatment within 6 h with high or low doses of aptiganel versus placebo [43]; however, there was no improvement in the modified Rankin score at 3 months and mortality rate at 4 months was higher in the high-dose aptiganel group (26 %) versus placebo (19 %) [43] and subsequently the trial was aborted.

One competitive NMDA antagonist, selfotel (CGS 19755), was investigated in an MCA occlusion model in rats with administration within 5 min of insult reducing infarct size and post-ischemia glucose hypermetabolism [44]. A safety and tolerability trial showed significant adverse effects including delirium, confusion, agitation, paranoia, and hallucinations [45] and a clinical trial showed higher mortality in the selfotel treatment group [46, 47].

Glycine is also an agonist at the NMDA receptor complex. Gavestinel (GV150526), a highly selective potent antagonist of this glycine site, has been investigated. In pre-clinical studies of MCA occlusion in rats, gavestinel administration up to 6 h after occlusion resulted in significant infarct volume reduction and protected somatosensory-evoked potential responses [48]. Two important clinical trials were performed: the GAIN International Trial [49] and the GAINS Americas Trial [50]. The two trials were similarly designed with age and stroke severity stratification and treatment within 6 h with either intravenous gavestinel versus placebo; however, both trials showed no difference for the primary outcome of Barthel index at 3 months or mortality [49, 50].

Several factors contributed to the discrepancies between the pre-clinical and clinical studies including the timing of intervention, with very early administration in pre-clinical studies, dose-limiting toxicity in humans, and also the possibility of potential detrimental effects of interference of NMDA receptor antagonism in survival mechanism of neurons [51]. In addition, pre-clinical studies were judged to be suboptimal [52] with difficult to replicate results [53].

AMPA receptor antagonists have shown evidence of neuroprotection in pre-clinical studies [54–56]. The AMPA antagonist ZK200755 was administered in a phase II trial [57] that was suspended due to safety concerns with worsening neurological exam. The AMPA antagonist YM872 was administered within 6 h in the AMPA Receptor Antagonist Treatment in Ischemic Stroke (ARTIST MRI Trial) and within 3 h in a related trial, ARTIST+; however, both trials were terminated after failed interim analysis, and the results have not been published.

3.3 Gamma-Aminobutyric acid (GABA) Agonists

Clomethiazole is a GABA(A)-agonist potentiating the activity of GABA, the main inhibitory neurotransmitter in the brain [58], and indeed this drug is used as a sedative. In pre-clinical evaluation models of focal ischemia in rats with 1-h MCA occlusion followed by 24 h reperfusion, clomethiazole was given either 1 h prior to occlusion, at 10 min after reperfusion, or at 70 min after ischemia onset [59] and tissue preservation was seen. In an MCA occlusion model, clomethiazole significantly reduced ischemic stroke volume at 1 h after occlusion, but was ineffective at 3 h [60]. In the CLASS Trial, clomethiazole was administered within 12 h to acute hemispheric stroke patients with no difference found in achieving functional independence at 90 days; however, in the subgroup of anterior circulation strokes, clomethiazole showed improved functional outcome, with sedation as the most common adverse effect [61]. Then in CLASS-I [62], patients with large ischemic strokes with limb weakness, cortical signs, and visual field deficits were randomized to clomethiazole or placebo with a 12-h time window, with final results showing no difference in Barthel scores at 90 days and no treatment effect with early intervention less than 6 h after stroke onset. Pre-clinical success appears to be dependent on very early administration, while the clinical trials allowed timing of treatment to be at 12 h. Diazepam, which is a GABA-ergic agent, has also been studied in stroke patients within 12 h of onset and showed to have no significant difference on modified Rankin scale at 3 months [63].

3.4 Antioxidants

Several antioxidant molecules have been evaluated clinically based on the assumption that impeding oxidative damage from free radicals will lead to decreased tissue damage in the ischemic cascade. Several free radical scavenging agents have undergone clinical trials that deserve to be reviewed. The first is NXY-059 (disodium 4-[tert-butylimino)methyl] benzene-1,3-disulfonate *N*-oxide), which is a proprietary nitrone spin-trap agent and a bis-sulfonated derivative of PBN [a-phenyl-*N*-tert-butyl nitrone] that has been shown to have neuroprotective roles in animal models. In pre-clinical MCA transient occlusion models in rats, NXY-059 administration 1 h after reperfusion had a dose-dependent effect on reducing infarct volume [64] that has also been confirmed in other similar models [65]. There is also evidence of reduction in infarct size after permanent MCA occlusions models [65, 66]. The pre-clinical studies were promising enough to justify clinical trials. Two large randomized, double blind trials studied NXY-059, SAINT I and SAINT II, in which both trials randomized patients to receive a 72-h infusion of intravenous NXY-059 or placebo within 6 h of stroke onset. In SAINT I [67], NXY-059 improved modified Rankin scores but did not improve NIHSS or Barthel index, while SAINT II [68] had entirely negative results [69].

The pre-clinical studies investigating NXY-059 were more extensive than for other candidate neuroprotective agents. First, the drug was studied in both transient and permanent focal ischemia models; in both rodents and non-human primates; a range of outcomes was measured including histological, physiological, and behavioral; and results were replicable; however, various deficiencies in preclinical investigations were also identified [70]. There are inherent shortcomings in the agent itself in terms of being polar, highly water-soluble, and with low blood-brain barrier permeability [64] and low cellular penetration [71], and being less potent as an antioxidant agent compared to other physiological compounds such as vitamin E, glutathione, and beta-carotene [72–74]. The discrepancy between the two clinical trials investigating NXY-059 is also interesting. The results of SAINT I showed quite modest clinical effect and others have criticized the trials design and statistical weakness of SAINT I [75, 76]. Likely, factors contributing to the failed clinical trials include the suboptimal biochemical profile of the agent as described above, as well as the likely use of longer time window for treatment than that used in pre-clinical studies.

Other antioxidant agents have also been investigated in larger (several hundred patients) trials. One example is tirilazad mesylate (U-74006F), which is a non-glucocorticoid 21-aminosteroid inhibitor of lipid peroxidation. A systemic review and meta-analysis of animal models of focal ischemia observed maximal efficacy with pre-treatment before onset of ischemia, although with potential sources of bias [77]. The RANTTAS trial was conducted to evaluate treatment with tirilazad within 6 h of onset of acute stroke [78]; however, the trial was terminated prematurely based on interim analysis with no effect on infarct volume or functional outcome. A subsequent trial was planned, but also not completed due to questions regarding safety. A Cochrane meta-analysis of four published and two unpublished trials [79] showed that tirilazad increased the odds of death or disability with borderline significance. Potential reasons for failure of these clinical trials include use of a wider dose range and a longer time to treatment compared to animal studies that administered the agent with a median time of only 10 min.

Another promising antioxidant agent was ebselen (2-phenyl-1,2-benzisoselenazol-3(2H)-one), which is a selenium compound with glutathione peroxidase-like activity that reacts with peroxynitrite [80]. In animal models, positive results were seen in terms of reducing ischemic damage when administered pre-ischemia [81], given within 30 min after ischemia [82], and 2 h after MCA occlusion in rats, where improvement of neurological deficits was seen at 24 h [83]. However, a clinical trial of acute stroke patients randomized to oral ebselen or placebo within 48 h and continued for 2 weeks did not show sustained effects at 3 months [84].

Edaravone (MCI-186; 3-methyl-1-phenyl-2-pyrazolin-5-one) is another oxygen-free radical scavenger and blocker of lipid peroxidation that has been investigated, with pre-clinical studies reporting positive results in focal and global animal models of ischemia [85]. However, these beneficial effects were related to timing of administration, with favorable effects apparent when drug was given prior to stroke onset, while no protection was observed when drug was initiated a few hours after ischemia onset [86]. A phase II trial showed administration of edaravone within a 72-h window resulted in improved modified Rankin scores at 3 months [87]. Another trial of this agent in lacunar stroke patients has reported decreased infarct volume and neurological improvement [88]. However, larger trials showing clinical efficacy and improving outcomes remain lacking [89].

3.5 A Phospholipid Precursor: Citicoline

Citicoline is an exogenous form of cytidine 5-diphosphocholine (CDP-choline), a precursor to choline. Citicoline is rapidly absorbed, crosses the blood-brain barrier, and is incorporated into the phospholipid membrane of neurons [90, 91]. Several studies of animal models of focal ischemia have shown promising results when the agent is administered early after stroke onset [92, 93]. Individual clinical trials have been inconclusive; however, a pooled analysis from the 4 randomized trials of oral citicoline given within the first 24 h and using a global 3-month outcome measure incorporating NIHSS and modified Rankin scale showed statistically significant effect on global recovery (25.2% versus 20.2%, OR 1.33, 95% CI 1.10–1.62, $p=0.0034$) [94]. Subsequently, in the International Citicoline Trial on Acute Stroke (ICTUS), patients with moderate to severe acute stroke were randomly assigned to either citicoline or placebo within the first 24 h and continued for 6 weeks. While citicoline had a good safety profile, disappointingly, at 90 days there were no significant differences between treatment arms in the modified Rankin scale scores or global test of combining Barthel Index, mRS, and NIHSS [95].

3.6 A Nitric Oxide Signal Transduction Modulator: Lubeluzole

Lubeluzole, the S-isomer of 3,4-difluoro benzothiazole, is a modulator of glutamate-activated nitric oxide synthase [96] and has been shown to protect hippocampal neurons against toxicity due to nitric oxide [97]. In pre-clinical studies, lubeluzole has been shown to reduce infarct size when administered early, i.e., within 15 or 30 min following MCA and carotid artery occlusions [98]; in addition, hippocampal CA1 neurons were protected in the setting of global cerebral ischemia in rats treated with lubeluzole within 5 min of ischemia onset [99]. In one clinical trial, patients with acute stroke who were randomized to lubeluzole or placebo within 6 h of symptom onset did not have a significant difference in mortality at 12 weeks, although there was improvement in NIHSS and Barthel index [100]. In two other clinical trials [101, 102], there was similarly no difference in mortality. A subsequent Cochrane Database meta-analysis reviewing five trials showed no effect on mortality or other key outcomes, but there was a significant increase in cardiac conduction disorders such as Q-T prolongation in the lubeluzole-treated patients [103]. There were several factors that likely contributed to the failed clinical trials of lubeluzole including the choice of mortality as the primary outcome measure, plus the use of a 6-h treatment window given that this was not strongly supported by pre-clinical trials.

3.7 Albumin

Albumin has been shown to have multiple potential neuroprotective mechanisms such as an antioxidant, hemodiluting agent and maintenance of colloid oncotic pressure, maintenance of vascular endothelial permeability, reducing cerebral swelling, improving perfusion, and decreasing erythrocyte sedimentation rate [104]. In a meta-analysis of focal ischemia animal studies, rats treated with albumin showed approximately 80 % decrease in infarct volume [105]. The Albumin in Acute Stroke (ALIAS) clinical trial [106] evaluated the safety of albumin administration within 16 h of stroke onset. Albumin therapy was safe and well-tolerated, there was a possible positive synergistic effect with intravenous thrombolysis in the half of patients who received intravenous thrombolysis, and a trend towards improved mRS score was also noted. In the ALIAS Part 2 trial [107], patients with acute ischemic stroke were randomized within 5 h of symptom onset to either intravenous administration of 25 % human albumin or placebo, with mRS and NIHSS scores used as primary outcome measures. The albumin-treated patients had more unfavorable complications including pulmonary edema, symptomatic intracerebral hemorrhage within 24 h, and there was no difference in the primary outcome measure.

3.8 Hyperbaric Oxygen

Oxygen has been shown to have various potential neuroprotective effects including blood-brain barrier stabilization, inhibition of inflammation, improved penumbra oxygenation, and decreasing excitotoxicity, while also having associated harmful effects including generation of reactive oxygen species and vasoconstriction. Hyperbaric oxygen therapy has shown contradictory results, although the sample size in such studies has been small [108, 109]. Introduction of oxygen after stroke has also been shown to reduce hemorrhage and edema associated with intravenous thrombolysis in a MCA occlusion experimental stroke model [110]. In a recent Cochrane review [111], 11 randomized controlled trials of hyperbaric oxygen for acute ischemic stroke were reviewed with no evidence of significant improvement in clinical outcomes.

3.9 Neurotrophic Factors

Various neurotrophic and growth factors have been investigated in experimental and clinical studies. Basic fibroblast growth factor has been shown to protect against excitotoxicity in vitro and in animal model of permanent focal ischemia by reducing ischemic volume [112]. Clinical safety and efficacy trials did not show any serious adverse effects [113], but no significant clinical improvement was shown thus the trials were halted [114], although patients enrolled in the latter hours of the time window appeared to benefit more than patients enrolled in the very earliest hours possibly suggesting a restorative mechanism. Recombinant erythropoietin (Epo) has also been shown to be safe and beneficial in experimental studies [115]. A clinical trial subsequently was negative with higher mortality and complications in patients receiving Epo likely due to interaction with intravenous thrombolysis, causing increased mortality [116]. Lastly, granulocyte colony-stimulating factor has also been studied in a dosing study with subjects receiving the agent within 12 h with reports of safe tolerability [117]; however, AX200, a larger follow-up study, did not find any benefit [118].

3.10 Modulation of Inflammation: Inhibition of Leukocytes

Modulation of inflammation represents an additional potential mechanism for achieving neuroprotection early after stroke onset. Enlimomab is an intercellular adhesion molecule-1 antibody that modulates leukocyte adhesion. In animal studies, rats undergoing transient MCA occlusion treated with this compound 1 h post-reperfusion showed a decrease in ischemic lesion size; however, no effect was seen in a permanent occlusion model [119]. However, in a multi-center clinical trial [120], acute stroke patients randomized to enlimomab within 6 h of stroke onset had a significantly worse outcome in terms of mRS scores and higher mortality. Reasons speculated for this negative trial include potential activation of innate neutrophils and complement in an acute ischemia [121]. In the ASTIN Trial [122], UK 279,276, a recombinant protein inhibitor of neutrophil activation on the CD11b/CD 18 receptors, was administered within 6 h of acute stroke; however, the trial was terminated early due to futility and lack of efficacy.

3.11 Hemodilution

Hemodilution, by decreasing blood viscosity, has been considered a theoretical means to increase oxygen delivery and cerebral perfusion; however, the concept has proven to be difficult to implement, with various hemodilution agents having wide-ranging results. Agents studied include dextran 40, 10 % hydroxyethyl starch, 20 % albumin with crystalloids, and venesection. Several small hemodilution trials for stroke have been studied in a Cochrane review [123]. The Scandinavian Stroke Study administered both dextran 40 and venesection to acute stroke patients within 48 h of symptom onset; however, there was no improvement in survival or outcome [124]. Similarly, in another large trial, both ischemic and hemorrhagic stroke patients were administered dextran 40 and venesection within 12 h with hematocrits 35 % or greater with no major differences in mortality or outcome improvement [125]. Hypervolemic hemodilution has also been studied in clinical trial without any benefit with 10 % hydroxyethyl starch [126].

3.12 Hypothermia

Fever has been shown in experimental and clinical studies to worsen outcome in stroke and other neurological injuries [127–130]. Hyperthermia has also been shown to be a predictor of poor neurological outcomes after acute ischemic stroke [131]. Moderate therapeutic hypothermia has been extensively studied over the last two decades with evidence suggesting neuroprotection in both focal and global ischemia models [132–138]. Two seminal clinical trials have shown that mild therapeutic hypothermia to 32–34 °C for 12 or 24 h after cardiac arrest significantly decreased mortality and improves neurological function [139, 140], and therapeutic hypothermia continues to be recommended for patients with an initial non-shockable arrest rhythm [141]. Also, hypothermia in cardiac arrest at a targeted temperature of 33 °C did not provide a benefit compared to 36 °C in a subsequent trial [142]. These data pertain to cardiac arrest; the application of therapeutic hypothermia in the treatment of patients with acute stroke has been difficult and complicated. Factors contributing to this include complex care of the patient often in an intensive care unit with management of intubation, sedation, shivering, cooling device management as well as potential adverse effects of hypothermia including cardiac arrhythmias, coagulopathies, and infection risk. A Cochrane review analyzed 8 trials of hypothermia in acute ischemic stroke and found no clinical efficacy and a non-significant risk of infection complications [143].

Advances in the management of shivering and cooling technology, such as endovascular cooling catheters, have recently advanced the feasibility of therapeutic hypothermia in acute stroke patients. In the Intravascular Cooling in the Treatment of Stroke (ICTuS) trial, stroke patients were successfully cooled using a novel endovascular cooling catheter and antishivering regimen [144]. In the ICTuS-L study, endovascular hypothermia was combined with intravenous thrombolysis and proved to be feasible, and hypothermia was associated with increased risk of pneumonia but no increased risk of bleeding complications [145]. The ICTuS2/3 trial still ongoing will assess whether the combination of endovascular hypothermia and thrombolysis is superior to thrombolysis alone with assessment of 90-day mRS score [146].

3.13 Magnesium

As an endogenous calcium antagonist, magnesium could theoretically provide neuroprotection through a number of divergent mechanisms, including blocking NMDA receptors, inhibiting excitatory neurotransmitter release, inhibiting calcium channels, and promoting vascular smooth muscle relaxation [147]. Clinically, magnesium has been used for the treatment of pre-eclampsia by reducing the risk of seizures, and thus, eclampsia by more than 50 % [148]. Although in several pre-clinical studies magnesium has been shown to reduce infarct volume with administration after the stroke insult [149–151], a subsequent review of these studies [152] showed mixed results and potential confounding factors including hypothermia. In clinical trials, the Intravenous Magnesium Efficacy in Stroke Trial (IMAGES) investigated intravenous treatment with magnesium sulfate within a 12-h window in patients with acute stroke [153], resulting in magnesium failing to improve disability at 90 days associated with a slight increase in mortality. Recently, the FAST-MAG trial showed pre-hospital administration of intravenous magnesium sulfate was safe and allowed initiation of therapy within 2 h of stroke symptom onset; however, there was no improvement in mortality or functional outcome at 90 days [154]. Although the results were disappointing in terms of magnesium activity as a neuroprotectant, FAST-MAG showed that it was possible to treat patients within the "golden hour" [155], that is, the first 60 min after the onset of stroke symptoms.

4 Reasons for Failure: Pre-clinical Experimental Studies

Many critics have asserted that the failure of neuroprotection trials is attributable to inadequate animal studies. Many variables influence the design and implementation of experimental investigations in animals that ultimately impact the quality, reliability, and results of preclinical experimental studies. First, there are animal-related factors including the species, strain, age, and sex. Although the vascular anatomy of rodents is similar to humans, it has been argued that there is significant difference in proportion of white matter between rodent and human brains [156], and note that the resting pulse of a rodent is five times that of a human. Then there are factors related to the model of ischemia: focal versus global ischemia, method of vessel occlusion, transient versus permanent occlusion, duration of ischemia and reperfusion, duration of survival, selection of anesthetic agent, and presence of multiple diseased arteries concomitantly. Almost all animal models are MCA occlusion-related, while stroke in humans is extremely heterogeneous. Factors influencing outcome assessment include behavioral testing methods, quantification of infarction, and neuronal survival and death. The quality of the study design is important with key factors including randomization, blinded versus unblinded assessment, biological variability of models, presence of multiple risk factors, age of subjects, statistical analysis, appropriate testing and evaluation, and correlation to outcome measures.

Furthermore, neuroprotective agents are required to have appropriate bioavailability, stability, and solubility for absorption across the blood-brain barrier. Also, the route of administration as well as pharmacokinetics and pharmacodynamics relating to plasma drug levels, timing of administration of agent with respect to ischemic insult, as well as activity and consequences of metabolites are all essential factors. Agents with poor blood-brain barrier permeability would require higher doses, likely increasing the incidence of systemic side effects, leading to lower therapeutic indices and any clinical gains being outweighed by adverse effects. Finally, approaching a single molecular target, as compared to multiple targets, may have less impact on complex ischemic cascade. As such, agents with multiple target or actions would likely have a more significant impact on ischemic damage prevention and recovery.

On the basis of these observations, the Stroke Therapy Academic Industry Roundtable (STAIR) recommendations were developed to provide guidance for potential therapeutic agents [157]. Even though these recommendations were published more than a decade ago, few drugs have met all the criteria, and more work on animal models is essential to improve the chances of successful translational studies for acute stroke patients.

5 Reasons for Failure: Clinical Trials

Over the last decade, clinical trials evaluating neuroprotective agents have progressive improvements in trial design, patient selection, and assessment of clinically meaningful outcome measures. The most recent clinical trials have incorporated the use of neurological outcomes and functional recovery as more appropriate outcome measures. However, earlier phase III trials had significant shortcomings, and it is important to review these. First, the therapeutic index of many agents were not clearly defined and thus doses tested in human clinical trials often differed and in many cases were much lower than doses that showed clinical efficacy in animal studies as evidenced with the adverse effects limiting dosing of the NMDA antagonists [158]. Moreover, while in animal models almost all neuroprotective agents showed efficacy only within an early limited time window, in human trials there was often extension of the treatment time window, e.g., with extension to many hours after acute stroke onset [59–61]. Also, since clinical trials often include patients with heterogeneous stroke types and severities, there may be a diluting effect of neuroprotective agents especially with agents with only marginal efficacy. Also, previous trials were not able to identify patients with salvageable penumbra that were unlikely to benefit from any intervention, thus further decreasing efficacy effect of the agent. Many previous trials also often had small sample size of patients that were not adequately powered in order to measure outcomes sufficiently.

Although previous clinical trials have been disappointing, the concept of neuroprotection has been emphatically proven in experimental models of ischemic stroke. Lessons learned include further understanding of enhanced drug development and improved clinical trial designs with well-selected acute stroke patients who are treated in a timely manner within the first "golden hour" [155]. Agents that can be administered in this hyperacute phase may be more useful in extending penumbra survival and increasing the applicability of reperfusion strategies. Most importantly, any potential neuroprotectant agent requires rigorous preclinical evaluation prior to entry into the clinical arena of human studies.

6 Future Directions and Lessons Learned

Insight gained from numerous experimental studies and clinical trials of neuroprotective agents in acute stroke will provide guidance for future endeavors. Basic scientists must develop potent agents or combinations of agents with different mechanisms of action with multiple targets in the complex pathophysiological mechanism underlying the ischemic cascade. Improved blood-brain barrier permeability and showing molecules reaching their intended target as well as innovative drug delivery strategies may provide improved therapeutic efficacy. Using the STAIR recommendations as a guide for the stringent translational preclinical testing will hopefully provide sound clinical trials. Additionally, it is important to use advanced imaging in both animal models to identify targeted delivery of molecules and in clinical trials to better identify patients with a salvageable penumbra in order to provide the best chance of clinical efficacy of putative neuroprotective agents.

Moreover, targeting mechanisms beyond the ischemic cascade, such as restorative therapies [159], will likely provide synergistic effects. Combinations of molecules and reperfusion strategies with better patient selection, improved clinical outcome measures, and more sound clinical trial design will further increase the chances of potentially identifying the next breakthrough in acute stroke therapies. It is also important that our expectations should be recalibrated in the setting of well-established data supporting simple control of various established vascular risk factors and thus these neuroprotective agents will hope to further augment and enhance our capability to improve patient outcome.

References

1. Dirnagl U, Iadecola C, Moskowitz MA (1999) Pathobiology of ischaemic stroke: an integrated view. Trends Neurosci 22(9):391–397
2. Green AR, Shuaib A (2006) Therapeutic strategies for the treatment of stroke. Drug Discov Today 11(15–16):681–693
3. (1995) Tissue plasminogen activator for acute ischemic stroke. The National Institute of Neurological Disorders and Stroke rt-PA Stroke Study Group. N Engl J Med 333(24):1581–1587
4. Hacke W, Kaste M, Bluhmki E, Brozman M, Davalos A, Guidetti D et al (2008) Thrombolysis with alteplase 3 to 4.5 hours after acute ischemic stroke. N Engl J Med 359(13):1317–1329
5. Berkhemer OA, Fransen PS, Beumer D, van den Berg LA, Lingsma HF, Yoo AJ et al (2015) A randomized trial of intraarterial treatment for acute ischemic stroke. N Engl J Med 372(1):11–20
6. Goyal M, Demchuk AM, Menon BK, Eesa M, Rempel JL, Thornton J et al (2015) Randomized assessment of rapid endovascular treatment of ischemic stroke. N Engl J Med 372(11):1019–1030
7. Campbell BC, Mitchell PJ, Kleinig TJ, Dewey HM, Churilov L, Yassi N et al (2015) Endovascular therapy for ischemic stroke with perfusion-imaging selection. N Engl J Med 372(11):1009–1018
8. Saver JL, Goyal M, Bonafe A, Diener HC, Levy EI, Pereira VM et al (2015) Stent-retriever thrombectomy after intravenous t-PA vs. t-PA alone in stroke. N Engl J Med 372(24):2285–2295
9. Jovin TG, Chamorro A, Cobo E, de Miquel MA, Molina CA, Rovira A et al (2015) Thrombectomy within 8 hours after symptom onset in ischemic stroke. N Engl J Med 372(24):2296–2306
10. Powers WJ, Derdeyn CP, Biller J, Coffey CS, Hoh BL, Jauch EC et al (2015) 2015 American Heart Association/American Stroke Association Focused Update of the 2013 Guidelines for the Early Management of Patients With Acute Ischemic Stroke Regarding Endovascular Treatment: A Guideline for Healthcare Professionals From the American Heart Association/American Stroke Association. Stroke 46(10):3020–3035
11. Felberg RA, Burgin WS, Grotta JC (2000) Neuroprotection and the ischemic cascade. CNS Spectr 5(3):52–58
12. Astrup J, Siesjo BK, Symon L (1981) Thresholds in cerebral ischemia—the ischemic penumbra. Stroke 12(6):723–725
13. Obrenovitch T, Sarna G, Symon L (1990) Ionic homeostasis and neurotransmitter changes in ischemic. In: Krieglstein J, Oberpichler H (eds) Pharmacology of cerebral ischemic. WVS, Stuttgart, Germany, pp 97–112
14. Kohno K, Hoehn-Berlage M, Mies G, Back T, Hossmann KA (1995) Relationship between diffusion-weighted MR images, cerebral blood flow, and energy state in experimental brain infarction. Magn Reson Imaging 13(1):73–80
15. von Kummer R (1999) Ischemic stroke and tissue hypodensity on computed tomography. Stroke 30(9):1974–1981
16. Butcher SP, Bullock R, Graham DI, McCulloch J (1990) Correlation between amino acid release and neuropathologic outcome in rat brain following middle cerebral artery occlusion. Stroke 21(12):1727–1733
17. Benveniste H, Drejer J, Schousboe A, Diemer NH (1984) Elevation of the extracellular concentrations of glutamate and aspartate in rat hippocampus during transient cerebral ischemia monitored by intracerebral microdialysis. J Neurochem 43(5):1369–1374
18. Rothman SM, Olney JW (1986) Glutamate and the pathophysiology of hypoxic—ischemic brain damage. Ann Neurol 19(2):105–111
19. Rothman SM (1992) Excitotoxins: possible mechanisms of action. Ann N Y Acad Sci 648:132–139

20. Siesjo BK, Bengtsson F (1989) Calcium fluxes, calcium antagonists, and calcium-related pathology in brain ischemia, hypoglycemia, and spreading depression: a unifying hypothesis. J Cereb Blood Flow Metab 9(2):127–140
21. Feuerstein GZ, Wang X, Barone FC (1998) Inflammatory mediators and brain injury: the role of cytokines and chemokines in stroke and CNS diseases. In: Ginsburg MD, Bogousslavsky J (eds) Cerebrovascular disease: pathophysiology, diagnosis, and management. Blackwell Science, Malden, pp 507–531
22. Fay T (1959) Early experiences with local and generalized refrigeration of the human brain. J Neurosurg 16(3):239–259; discussion 59–60
23. Grotta JC, Chiu D (1998) Pharmacologic modification of acute cerebral ischemia. In: Barnett HJM, Mohr JP, Stein BM, Yatsu FM (eds) Stroke: pathophysiology, diagnosis, and management, 3rd edn. Churchill Livingstone, Philadelphia, pp 1079–1088
24. Choi DW (1995) Calcium: still center-stage in hypoxic-ischemic neuronal death. Trends Neurosci 18(2):58–60
25. Tymianski M, Tator CH (1996) Normal and abnormal calcium homeostasis in neurons: a basis for the pathophysiology of traumatic and ischemic central nervous system injury. Neurosurgery 38(6):1176–1195
26. Erecinska M, Silver IA (1996) Calcium handling by hippocampal neurons under physiologic and pathologic conditions. Adv Neurol 71:119–136
27. Kazda S, Towart R (1982) Nimodipine: a new calcium antagonistic drug with a preferential cerebrovascular action. Acta Neurochir 63(1–4):259–265
28. Allen GS, Ahn HS, Preziosi TJ, Battye R, Boone SC, Boone SC et al (1983) Cerebral arterial spasm—a controlled trial of nimodipine in patients with subarachnoid hemorrhage. N Engl J Med 308(11):619–624
29. Horn J, de Haan RJ, Vermeulen M, Luiten PG, Limburg M (2001) Nimodipine in animal model experiments of focal cerebral ischemia: a systematic review. Stroke 32(10):2433–2438
30. Group TANS (1992) Clinical trial of nimodipine in acute ischemic stroke. The American Nimodipine Study Group. Stroke 23(1):3–8
31. Wahlgren NG, MacMahon DG, DeKeyser J, Indredavik B, Ryman T (1994) Intravenous Nimodipine West European Stroke Trial (INWEST) of nimodipine in the treatment of acute ischaemic stroke. Cerebrovasc Dis 4:204–210
32. Kaste M, Fogelholm R, Erila T, Palomaki H, Murros K, Rissanen A et al (1994) A randomized, double-blind, placebo-controlled trial of nimodipine in acute ischemic hemispheric stroke. Stroke 25(7):1348–1353
33. (1990) Randomised, double-blind, placebo-controlled trial of nimodipine in acute stroke. Trust Study Group. Lancet 336(8725):1205–1209
34. Horn J, de Haan RJ, Vermeulen M, Limburg M (2001) Very Early Nimodipine Use in Stroke (VENUS): a randomized, double-blind, placebo-controlled trial. Stroke 32(2):461–465
35. Mohr JP, Orgogozo JM, Harrison MJG, Hennerici M, Wahlgren NG, Gelmers HJ et al (1994) Meta-analysis of oral nimodipine trials in acute ischemic stroke. Cerebrovas Dis 4:197–203
36. Zhang J, Yang J, Zhang C, Jiang X, Zhou H, Liu M (2012) Calcium antagonists for acute ischemic stroke. Cochrane Database Syst Rev 5:Cd001928
37. Bruno V, Battaglia G, Copani A, D'Onofrio M, Di Iorio P, De Blasi A et al (2001) Metabotropic glutamate receptor subtypes as targets for neuroprotective drugs. J Cereb Blood Flow Metab 21(9):1013–1033
38. Arundine M, Tymianski M (2004) Molecular mechanisms of glutamate-dependent neurodegeneration in ischemia and traumatic brain injury. Cell Mol Life Sci 61(6):657–668
39. Ginsberg MD (1995) Neuroprotection in brain ischemia: an update. Part I. Neuroscientist 1:95–103
40. Ginsberg MD (1995) Neuroprotection in brain ischemia: an update. Part II. Neuroscientist 1:164–175
41. Olney JW (1994) Neurotoxicity of NMDA receptor antagonists: an overview. Psychopharmacol Bull 30(4):533–540

42. Minematsu K, Fisher M, Li L, Sotak CH (1993) Diffusion and perfusion magnetic resonance imaging studies to evaluate a noncompetitive N-methyl-D-aspartate antagonist and reperfusion in experimental stroke in rats. Stroke 24(12):2074–2081
43. Albers GW, Goldstein LB, Hall D, Lesko LM (2001) Aptiganel hydrochloride in acute ischemic stroke: a randomized controlled trial. JAMA 286(21):2673–2682
44. Simon R, Shiraishi K (1990) N-methyl-D-aspartate antagonist reduces stroke size and regional glucose metabolism. Ann Neurol 27(6):606–611
45. Grotta J, Clark W, Coull B, Pettigrew LC, Mackay B, Goldstein LB et al (1995) Safety and tolerability of the glutamate antagonist CGS 19755 (selfotel) in patients with acute ischemic stroke. Results of a phase IIa randomized trial. Stroke 26(4):602–605
46. Davis SM, Albers GW, Diener HC, Lees KR, Norris J (1997) Termination of acute stroke studies involving Selfotel treatment. ASSIST Steering Committed. Lancet 349(9044):32
47. Davis SM, Lees KR, Albers GW, Diener HC, Markabi S, Karlsson G et al (2000) Selfotel in acute ischemic stroke: possible neurotoxic effects of an NMDA antagonist. Stroke 31(2):347–354
48. Bordi F, Pietra C, Ziviani L, Reggiani A (1997) The glycine antagonist GV150526 protects somatosensory evoked potentials and reduces the infarct area in the MCAo model of focal ischemia in the rat. Exp Neurol 145(2 Pt 1):425–433
49. Lees KR, Asplund K, Carolei A, Davis SM, Diener HC, Kaste M et al (2000) Glycine antagonist (gavestinel) in neuroprotection (GAIN International) in patients with acute stroke: a randomised controlled trial. GAIN International Investigators. Lancet 355(9219):1949–1954
50. Sacco RL, DeRosa JT, Haley EC Jr, Levin B, Ordronneau P, Phillips SJ et al (2001) Glycine antagonist in neuroprotection for patients with acute stroke: GAIN Americas: a randomized controlled trial. JAMA 285(13):1719–1728
51. Hoyte L, Barber PA, Buchan AM, Hill MD (2004) The rise and fall of NMDA antagonists for ischemic stroke. Curr Mol Med 4(2):131–136
52. van der Worp HB, de Haan P, Morrema E, Kalkman CJ (2005) Methodological quality of animal studies on neuroprotection in focal cerebral ischaemia. J Neurol 252(9):1108–1114
53. Gorelick PB (2000) Neuroprotection in acute ischaemic stroke: a tale of for whom the bell tolls? Lancet 355(9219):1925–1926
54. Li H, Buchan AM (1993) Treatment with an AMPA antagonist 12 hours following severe normothermic forebrain ischemia prevents CA1 neuronal injury. J Cereb Blood Flow Metab 13(6):933–939
55. Gill R (1994) The pharmacology of alpha-amino-3-hydroxy-5-methyl-4-isoxazole propionate (AMPA)/kainate antagonists and their role in cerebral ischaemia. Cerebrovasc Brain Metab Rev 6(3):225–256
56. Takahashi M, Kohara A, Shishikura J, Kawasaki-Yatsugi S, Ni JW, Yatsugi S et al (2002) YM872: a selective, potent and highly water-soluble alpha-amino-3-hydroxy-5--methylisoxazole-4-propionic acid receptor antagonist. CNS Drug Rev 8(4):337–352
57. Elting JW, Sulter GA, Kaste M, Lees KR, Diener HC, Hommel M et al (2002) AMPA antagonist ZK200775 in patients with acute ischemic stroke: possible glial cell toxicity detected by monitoring of S-100B serum levels. Stroke 33(12):2813–2818
58. Wilby MJ, Hutchinson PJ (2004) The pharmacology of chlormethiazole: a potential neuroprotective agent? CNS Drug Rev 10(4):281–294
59. Sydserff SG, Cross AJ, West KJ, Green AR (1995) The effect of chlormethiazole on neuronal damage in a model of transient focal ischaemia. Br J Pharmacol 114(8):1631–1635
60. Sydserff SG, Cross AJ, Green AR (1995) The neuroprotective effect of chlormethiazole on ischaemic neuronal damage following permanent middle cerebral artery ischaemia in the rat. Neurodegeneration 4(3):323–328
61. Wahlgren NG, Ranasinha KW, Rosolacci T, Franke CL, van Erven PM, Ashwood T et al (1999) Clomethiazole acute stroke study (CLASS): results of a randomized, controlled trial of clomethiazole versus placebo in 1360 acute stroke patients. Stroke 30(1):21–28
62. Lyden P, Shuaib A, Ng K, Levin K, Atkinson RP, Rajput A et al (2002) Clomethiazole Acute Stroke Study in ischemic stroke (CLASS-I): final results. Stroke 33(1):122–128

63. Lodder J, van Raak L, Hilton A, Hardy E, Kessels A (2006) Diazepam to improve acute stroke outcome: results of the early GABA-Ergic activation study in stroke trial. A randomized double-blind placebo-controlled trial. Cerebrovasc Dis 21(1–2):120–127
64. Kuroda S, Tsuchidate R, Smith ML, Maples KR, Siesjo BK (1999) Neuroprotective effects of a novel nitrone, NXY-059, after transient focal cerebral ischemia in the rat. J Cereb Blood Flow Metab 19(7):778–787
65. Sydserff SG, Borelli AR, Green AR, Cross AJ (2002) Effect of NXY-059 on infarct volume after transient or permanent middle cerebral artery occlusion in the rat; studies on dose, plasma concentration and therapeutic time window. Br J Pharmacol 135(1):103–112
66. Zhao Z, Cheng M, Maples KR, Ma JY, Buchan AM (2001) NXY-059, a novel free radical trapping compound, reduces cortical infarction after permanent focal cerebral ischemia in the rat. Brain Res 909(1–2):46–50
67. Lees KR, Zivin JA, Ashwood T, Davalos A, Davis SM, Diener HC et al (2006) NXY-059 for acute ischemic stroke. N Engl J Med 354(6):588–600
68. Shuaib A, Lees KR, Lyden P, Grotta J, Davalos A, Davis SM et al (2007) NXY-059 for the treatment of acute ischemic stroke. N Engl J Med 357(6):562–571
69. Serebruany VL (2006) Hypokalemia, cardiac failure, and reporting NXY-059 safety for acute stroke. J Cardiovasc Pharmacol Ther 11(4):229–231
70. Savitz SI (2007) A critical appraisal of the NXY-059 neuroprotection studies for acute stroke: a need for more rigorous testing of neuroprotective agents in animal models of stroke. Exp Neurol 205(1):20–25
71. Dehouck MP, Cecchelli R, Richard Green A, Renftel M, Lundquist S (2002) In vitro blood-brain barrier permeability and cerebral endothelial cell uptake of the neuroprotective nitrone compound NXY-059 in normoxic, hypoxic and ischemic conditions. Brain Res 955(1–2):229–235
72. Buettner GR (1993) The pecking order of free radicals and antioxidants: lipid peroxidation, alpha-tocopherol, and ascorbate. Arch Biochem Biophys 300(2):535–543
73. Becker DA, Natero R, Echegoyen L, Lawson RC (1998) Redox behaviour of azulenyl nitrones: fully reversible one electron oxidation by cyclic voltammetry at potentials within the range of biological antioxidants. JCS Perkin Trans 2:1289–1291
74. Ginsberg MD (2007) Life after cerovive: a personal perspective on ischemic neuroprotection in the post-NXY-059 era. Stroke 38(6):1967–1972
75. Koziol JA, Feng AC (2006) On the analysis and interpretation of outcome measures in stroke clinical trials: lessons from the SAINT I study of NXY-059 for acute ischemic stroke. Stroke 37(10):2644–2647
76. Saver JL (2007) Clinical impact of NXY-059 demonstrated in the SAINT I trial: derivation of number needed to treat for benefit over entire range of functional disability. Stroke 38(5):1515–1518
77. Sena E, Wheble P, Sandercock P, Macleod M (2007) Systematic review and meta-analysis of the efficacy of tirilazad in experimental stroke. Stroke 38(2):388–394
78. Investigators R (1996) A randomized trial of tirilazad mesylate in patients with acute stroke (RANTTAS). The RANTTAS Investigators. Stroke 27(9):1453–1458
79. Committee TIS (2000) Tirilazad mesylate in acute ischemic stroke: a systematic review. Tirilazad International Steering Committee. Stroke 31(9):2257–2265
80. Parnham M, Sies H (2000) Ebselen: prospective therapy for cerebral ischaemia. Expert Opin Investig Drugs 9(3):607–619
81. Namura S, Nagata I, Takami S, Masayasu H, Kikuchi H (2001) Ebselen reduces cytochrome c release from mitochondria and subsequent DNA fragmentation after transient focal cerebral ischemia in mice. Stroke 32(8):1906–1911
82. Takasago T, Peters EE, Graham DI, Masayasu H, Macrae IM (1997) Neuroprotective efficacy of ebselen, an anti-oxidant with anti-inflammatory actions, in a rodent model of permanent middle cerebral artery occlusion. Br J Pharmacol 122(6):1251–1256

83. Imai H, Masayasu H, Dewar D, Graham DI, Macrae IM (2001) Ebselen protects both gray and white matter in a rodent model of focal cerebral ischemia. Stroke 32(9):2149–2154
84. Yamaguchi T, Sano K, Takakura K, Saito I, Shinohara Y, Asano T et al (1998) Ebselen in acute ischemic stroke: a placebo-controlled, double-blind clinical trial. Ebselen Study Group. Stroke 29(1):12–17
85. Watanabe T, Yuki S, Egawa M, Nishi H (1994) Protective effects of MCI-186 on cerebral ischemia: possible involvement of free radical scavenging and antioxidant actions. J Pharmacol Exp Ther 268(3):1597–1604
86. Nakajima H, Kakui N, Ohkuma K, Ishikawa M, Hasegawa T (2005) A newly synthesized poly(ADP-ribose) polymerase inhibitor, DR2313 [2-methyl-3,5,7,8-tetrahydrothiopyrano[4,3-d]-pyrimidine-4-one]: pharmacological profiles, neuroprotective effects, and therapeutic time window in cerebral ischemia in rats. J Pharmacol Exp Ther 312(2):472–481
87. Group EAIS (2003) Effect of a novel free radical scavenger, edaravone (MCI-186), on acute brain infarction. Randomized, placebo-controlled, double-blind study at multicenters. Cerebrovasc Dis 15(3):222–229
88. Nakase T, Yoshioka S, Suzuki A (2011) Free radical scavenger, edaravone, reduces the lesion size of lacunar infarction in human brain ischemic stroke. BMC Neurol 11:39
89. Lapchak PA (2010) A critical assessment of edaravone acute ischemic stroke efficacy trials: is edaravone an effective neuroprotective therapy? Expert Opin Pharmacother 11(10):1753–1763
90. Weiss GB (1995) Metabolism and actions of CDP-choline as an endogenous compound and administered exogenously as citicoline. Life Sci 56(9):637–660
91. (1995) Secades JJ, Frontera G. CDP-choline: pharmacological and clinical review. Meth Find Exp Clin Pharmacol 17 Suppl B:1–54
92. Schabitz WR, Weber J, Takano K, Sandage BW, Locke KW, Fisher M (1996) The effects of prolonged treatment with citicoline in temporary experimental focal ischemia. J Neurol Sci 138(1–2):21–25
93. Aronowski J, Strong R, Grotta JC (1996) Citicoline for treatment of experimental focal ischemia: histologic and behavioral outcome. Neurol Res 18(6):570–574
94. Davalos A, Castillo J, Alvarez-Sabin J, Secades JJ, Mercadal J, Lopez S et al (2002) Oral citicoline in acute ischemic stroke: an individual patient data pooling analysis of clinical trials. Stroke 33(12):2850–2857
95. Davalos A, Alvarez-Sabin J, Castillo J, Diez-Tejedor E, Ferro J, Martinez-Vila E et al (2012) Citicoline in the treatment of acute ischaemic stroke: an international, randomised, multicentre, placebo-controlled study (ICTUS trial). Lancet 380(9839):349–357
96. Lesage AS, Peeters L, Leysen JE (1996) Lubeluzole, a novel long-term neuroprotectant, inhibits the glutamate-activated nitric oxide synthase pathway. J Pharmacol Exp Ther 279(2):759–766
97. Maiese K, TenBroeke M, Kue I (1997) Neuroprotection of lubeluzole is mediated through the signal transduction pathways of nitric oxide. J Neurochem 68(2):710–714
98. Aronowski J, Strong R, Grotta JC (1996) Treatment of experimental focal ischemia in rats with lubeluzole. Neuropharmacology 35(6):689–693
99. Haseldonckx M, Van Reempts J, Van de Ven M, Wouters L, Borgers M (1997) Protection with lubeluzole against delayed ischemic brain damage in rats. A quantitative histopathologic study. Stroke 28(2):428–432
100. Grotta J (1997) Lubeluzole treatment of acute ischemic stroke. The US and Canadian Lubeluzole Ischemic Stroke Study Group. Stroke 28(12):2338–2346
101. Diener HC (1998) Multinational randomised controlled trial of lubeluzole in acute ischaemic stroke. European and Australian Lubeluzole Ischaemic Stroke Study Group. Cerebrovasc Dis 8(3):172–181
102. Diener HC, Cortens M, Ford G, Grotta J, Hacke W, Kaste M et al (2000) Lubeluzole in acute ischemic stroke treatment: a double-blind study with an 8-hour inclusion window comparing a 10-mg daily dose of lubeluzole with placebo. Stroke 31(11):2543–2551

103. Gandolfo C, Sandercock P, Conti M (2002) Lubeluzole for acute ischaemic stroke. Cochrane Database Syst Rev 1:Cd001924
104. Halliwell B (1988) Albumin—an important extracellular antioxidant? Biochem Pharmacol 37(4):569–571
105. Ginsberg MD, Belayev L, Bazan NG, Marcheselli VL, Hill MD, Palesch YY et al (2004) Albumin-based neurotherapeutics for acute ischemic stroke: from bench to bedside. In: Krieglstein J, Klumpp S (eds) Pharmacology of cerebral ischemia. Medpharm Scientific Publishers, Stuttgart, Germany, pp 421–433
106. Ginsberg MD, Palesch YY, Martin RH, Hill MD, Moy CS, Waldman BD et al (2011) The albumin in acute stroke (ALIAS) multicenter clinical trial: safety analysis of part 1 and rationale and design of part 2. Stroke 42(1):119–127
107. Ginsberg MD, Palesch YY, Hill MD, Martin RH, Moy CS, Barsan WG et al (2013) High-dose albumin treatment for acute ischaemic stroke (ALIAS) Part 2: a randomised, double-blind, phase 3, placebo-controlled trial. Lancet Neurol 12(11):1049–1058
108. Nighoghossian N, Trouillas P, Adeleine P, Salord F (1995) Hyperbaric oxygen in the treatment of acute ischemic stroke. A double-blind pilot study. Stroke 26(8):1369–1372
109. Rusyniak DE, Kirk MA, May JD, Kao LW, Brizendine EJ, Welch JL et al (2003) Hyperbaric oxygen therapy in acute ischemic stroke: results of the Hyperbaric Oxygen in Acute Ischemic Stroke Trial Pilot Study. Stroke 34(2):571–574
110. Qin Z, Karabiyikoglu M, Hua Y, Silbergleit R, He Y, Keep RF et al (2007) Hyperbaric oxygen-induced attenuation of hemorrhagic transformation after experimental focal transient cerebral ischemia. Stroke 38(4):1362–1367
111. Bennett MH, Weibel S, Wasiak J, Schnabel A, French C, Kranke P (2014) Hyperbaric oxygen therapy for acute ischaemic stroke. Cochrane Database Syst Rev 11:Cd004954
112. Ay H, Ay I, Koroshetz WJ, Finklestein SP (1999) Potential usefulness of basic fibroblast growth factor as a treatment for stroke. Cerebrovasc Dis 9(3):131–135
113. The FIBLAST Safety Study Group (1998) Clinical safety of intravenous bFGF in acute stroke. Stroke 29:287
114. Bogousslavsky J, Victor SJ, Salinas EO, Pallay A, Donnan GA, Fieschi C et al (2002) Fiblast (trafermin) in acute stroke: results of the European-Australian phase II/III safety and efficacy trial. Cerebrovasc Dis 14(3–4):239–251
115. Ehrenreich H, Hasselblatt M, Dembowski C, Cepek L, Lewczuk P, Stiefel M et al (2002) Erythropoietin therapy for acute stroke is both safe and beneficial. Mol Med 8(8):495–505
116. Ehrenreich H, Weissenborn K, Prange H, Schneider D, Weimar C, Wartenberg K et al (2009) Recombinant human erythropoietin in the treatment of acute ischemic stroke. Stroke 40(12):e647–e656
117. Schabitz WR, Laage R, Vogt G, Koch W, Kollmar R, Schwab S et al (2010) AXIS: a trial of intravenous granulocyte colony-stimulating factor in acute ischemic stroke. Stroke 41(11):2545–2551
118. Ringelstein EB, Thijs V, Norrving B, Chamorro A, Aichner F, Grond M et al (2013) Granulocyte colony-stimulating factor in patients with acute ischemic stroke: results of the AX200 for Ischemic Stroke trial. Stroke 44(10):2681–2687
119. Zhang RL, Chopp M, Jiang N, Tang WX, Prostak J, Manning AM et al (1995) Anti-intercellular adhesion molecule-1 antibody reduces ischemic cell damage after transient but not permanent middle cerebral artery occlusion in the Wistar rat. Stroke 26(8):1438–1442; discussion 43
120. Enlimomab Acute Stroke Trial Investigators (2001) Use of anti-ICAM-1 therapy in ischemic stroke: results of the Enlimomab Acute Stroke Trial. Neurology 57(8):1428–1434
121. Furuya K, Takeda H, Azhar S, McCarron RM, Chen Y, Ruetzler CA et al (2001) Examination of several potential mechanisms for the negative outcome in a clinical stroke trial of enlimomab, a murine anti-human intercellular adhesion molecule-1 antibody: a bedside-to-bench study. Stroke 32(11):2665–2674

122. Krams M, Lees KR, Hacke W, Grieve AP, Orgogozo JM, Ford GA (2003) Acute Stroke Therapy by Inhibition of Neutrophils (ASTIN): an adaptive dose-response study of UK-279,276 in acute ischemic stroke. Stroke 34(11):2543–2548
123. Chang TS, Jensen MB (2014) Haemodilution for acute ischaemic stroke. Cochrane Database Syst Rev 8:Cd000103
124. (1987) Multicenter trial of hemodilution in acute ischemic stroke. I. Results in the total patient population. Scandinavian Stroke Study Group. Stroke 18(4):691–699
125. (1988) Haemodilution in acute stroke: results of the Italian haemodilution trial. Italian Acute Stroke Study Group. Lancet 1(8581):318–321
126. Aichner FT, Fazekas F, Brainin M, Polz W, Mamoli B, Zeiler K (1998) Hypervolemic hemodilution in acute ischemic stroke: the Multicenter Austrian Hemodilution Stroke Trial (MAHST). Stroke 29(4):743–749
127. Reith J, Jorgensen HS, Pedersen PM, Nakayama H, Raaschou HO, Jeppesen LL et al (1996) Body temperature in acute stroke: relation to stroke severity, infarct size, mortality, and outcome. Lancet 347(8999):422–425
128. Castillo J, Davalos A, Marrugat J, Noya M (1998) Timing for fever-related brain damage in acute ischemic stroke. Stroke 29(12):2455–2460
129. Ginsberg MD, Busto R (1998) Combating hyperthermia in acute stroke: a significant clinical concern. Stroke 29(2):529–534
130. Wang Y, Lim LL, Levi C, Heller RF, Fisher J (2000) Influence of admission body temperature on stroke mortality. Stroke 31(2):404–409
131. Saini M, Saqqur M, Kamruzzaman A, Lees KR, Shuaib A (2009) Effect of hyperthermia on prognosis after acute ischemic stroke. Stroke 40(9):3051–3059
132. Ginsberg MD, Sternau LL, Globus MY, Dietrich WD, Busto R (1992) Therapeutic modulation of brain temperature: relevance to ischemic brain injury. Cerebrovasc Brain Metab Rev 4(3):189–225
133. Ginsberg MD, Belayev L (2005) Biological and molecular mechanisms of hypothermic neuroprotection. In: Mayer SA, Sessler DI (eds) Therapeutic hypothermia. Marcel Dekker, New York, pp 85–140
134. Colbourne F, Sutherland G, Corbett D (1997) Postischemic hypothermia. A critical appraisal with implications for clinical treatment. Mol Neurobiol 14(3):171–201
135. Barone FC, Feuerstein GZ, White RF (1997) Brain cooling during transient focal ischemia provides complete neuroprotection. Neurosci Biobehav Rev 21(1):31–44
136. Corbett D, Thornhill J (2000) Temperature modulation (hypothermic and hyperthermic conditions) and its influence on histological and behavioral outcomes following cerebral ischemia. Brain Pathol 10(1):145–152
137. Huh PW, Belayev L, Zhao W, Koch S, Busto R, Ginsberg MD (2000) Comparative neuroprotective efficacy of prolonged moderate intraischemic and postischemic hypothermia in focal cerebral ischemia. J Neurosurg 92(1):91–99
138. Gunn AJ, Battin M, Gluckman PD, Gunn TR, Bennet L (2005) Therapeutic hypothermia: from lab to NICU. J Perinat Med 33(4):340–346
139. Bernard SA, Gray TW, Buist MD, Jones BM, Silvester W, Gutteridge G et al (2002) Treatment of comatose survivors of out-of-hospital cardiac arrest with induced hypothermia. N Engl J Med 346(8):557–563
140. Hypothermia after Cardiac Arrest Study Group (2002) Mild therapeutic hypothermia to improve the neurologic outcome after cardiac arrest. N Engl J Med 346(8):549–556
141. Perman SM, Grossestreuer AV, Wiebe DJ, Carr BG, Abella BS, Gaieski DF (2015) The utility of therapeutic hypothermia for post-cardiac arrest syndrome patients with an initial nonshockable rhythm. Circulation 132(22):2146–2151
142. Nielsen N, Wetterslev J, Cronberg T, Erlinge D, Gasche Y, Hassager C et al (2013) Targeted temperature management at 33 degrees C versus 36 degrees C after cardiac arrest. N Engl J Med 369(23):2197–2206

143. Den Hertog HM, van der Worp HB, Tseng MC, Dippel DW (2009) Cooling therapy for acute stroke. Cochrane Database Syst Rev 1:Cd001247
144. Lyden PD, Allgren RL, Ng K, Akins P, Meyer B, Al-Sanani F et al (2005) Intravascular Cooling in the Treatment of Stroke (ICTuS): early clinical experience. J Stroke Cerebrovasc Dis 14(3):107–114
145. Hemmen TM, Raman R, Guluma KZ, Meyer BC, Gomes JA, Cruz-Flores S et al (2010) Intravenous thrombolysis plus hypothermia for acute treatment of ischemic stroke (ICTuS-L): final results. Stroke 41(10):2265–2270
146. Lyden PD, Hemmen TM, Grotta J, Rapp K, Raman R (2014) Endovascular therapeutic hypothermia for acute ischemic stroke: ICTuS 2/3 protocol. Int J Stroke 9(1):117–125
147. Ovbiagele B, Kidwell CS, Starkman S, Saver JL (2003) Neuroprotective agents for the treatment of acute ischemic stroke. Curr Neurol Neurosci Rep 3(1):9–20
148. Duley L, Gulmezoglu AM, Henderson-Smart DJ, Chou D (2010) Magnesium sulphate and other anticonvulsants for women with pre-eclampsia. Cochrane Database Syst Rev 11:Cd000025
149. Izumi Y, Roussel S, Pinard E, Seylaz J (1991) Reduction of infarct volume by magnesium after middle cerebral artery occlusion in rats. J Cereb Blood Flow Metab 11(6):1025–1030
150. Westermaier T, Zausinger S, Baethmann A, Schmid-Elsaesser R (2005) Dose finding study of intravenous magnesium sulphate in transient focal cerebral ischemia in rats. Acta Neurochir 147(5):525–532; discussion 32
151. Marinov MB, Harbaugh KS, Hoopes PJ, Pikus HJ, Harbaugh RE (1996) Neuroprotective effects of preischemia intraarterial magnesium sulfate in reversible focal cerebral ischemia. J Neurosurg 85(1):117–124
152. Meloni BP, Zhu H, Knuckey NW (2006) Is magnesium neuroprotective following global and focal cerebral ischaemia? A review of published studies. Magnes Res 19(2):123–137
153. Muir KW, Lees KR, Ford I, Davis S (2004) Magnesium for acute stroke (Intravenous Magnesium Efficacy in Stroke trial): randomised controlled trial. Lancet 363(9407): 439–445
154. Saver JL, Starkman S, Eckstein M, Stratton SJ, Pratt FD, Hamilton S et al (2015) Prehospital use of magnesium sulfate as neuroprotection in acute stroke. N Engl J Med 372(6):528–536
155. Saver JL, Smith EE, Fonarow GC, Reeves MJ, Zhao X, Olson DM et al (2010) The "golden hour" and acute brain ischemia: presenting features and lytic therapy in >30,000 patients arriving within 60 minutes of stroke onset. Stroke 41(7):1431–1439
156. Ho PW, Reutens DC, Phan TG, Wright PM, Markus R, Indra I et al (2005) Is white matter involved in patients entered into typical trials of neuroprotection? Stroke 36(12):2742–2744
157. Fisher M, Feuerstein G, Howells DW, Hurn PD, Kent TA, Savitz SI et al (2009) Update of the stroke therapy academic industry roundtable preclinical recommendations. Stroke 40(6):2244–2250
158. Albers GW, Atkinson RP, Kelley RE, Rosenbaum DM (1995) Safety, tolerability, and pharmacokinetics of the N-methyl-D-aspartate antagonist dextrorphan in patients with acute stroke. Dextrorphan Study Group. Stroke 26(2):254–258
159. Cramer SC, Sur M, Dobkin BH, O'Brien C, Sanger TD, Trojanowski JQ et al (2011) Harnessing neuroplasticity for clinical applications. Brain 134(Pt 6):1591–1609

Part II
Novel Neuroprotection Mechanisms and Targets

Chapter 6
Targeting PSD-95 as a Novel Approach in the Treatment of Stroke

Linda M. Haugaard-Kedström, Eduardo F.A. Fernandes, and Kristian Strømgaard

Abstract During ischemia and stroke, a large excess of glutamate is released from the presynaptic terminal into the synaptic cleft. The NMDA receptor is activated by glutamate in the presence of glycine and triggers an influx of Ca^{2+} and hence induces excitotoxicity. PSD-95 is a scaffolding protein comprising three PDZ domains, one SH3 and one GK domain, which modulates numerous physiological relevant protein–protein interactions by interacting with the C-terminus or an internal binding motif of protein partners. The GluN2B subunit of the NMDA receptor and nNOS forms a ternary complex together with PSD-95 by interacting with PDZ1 and PDZ2 of PSD-95, respectively. Upon Ca^{2+} influx, this ternary complex induces the production of the toxic substance nitric oxide and further on ischemic brain damage. Numerous attempts and strategies have been applied to inhibit PSD-95/nNOS/NMDA receptor-induced excitotoxicity. Here we will present previous and ongoing efforts to develop and evaluate peptide, peptidomimetics and small molecule inhibitors targeting PSD-95.

Keywords PSD-95 • PDZ • nNOS • NMDA receptor • Ischemic stroke • Inhibitors

Abbreviations

3-APS	3-Amino-1-propane sulfonic acid
3PVO	Three pial vessel occlusion
7-NI	7-Nitroindazole
AD	Alzheimer's disease
AMPA	α-2-Amino-3-(3-hydroxy-5-methylisoxazol-4-yl) propionate
BCEC	Brain capillary endothelial cells

L.M. Haugaard-Kedström • E.F.A. Fernandes
K. Strømgaard, Ph.D. in Medicinal Chemistry (✉)
Department of Drug Design and Pharmacology, University of Copenhagen, Universitetsparken 2, Copenhagen 2100, Denmark
e-mail: kristian.stromgaard@sund.ku.dk

P.A. Lapchak, J.H. Zhang (eds.), *Neuroprotective Therapy for Stroke and Ischemic Disease*, Springer Series in Translational Stroke Research,
DOI 10.1007/978-3-319-45345-3_6

CaMKII	Calcium calmodulin-dependent kinase II
cAMP	Cyclic adenosine monophosphate
CCI	Chronic construction of the sciatic nerve
CREB	cAMP response element-binding protein
CRIPT	Cysteine-rich interactor of PDZ 3
DAPK1	Death-associated protein kinase 1
ELISA	Enzyme-linked immunosorbent assay
ENACT	Evaluating neuroprotection in aneurysm coiling therapy
FAD	Flavin adenine dinucleotide
FMN	Flavin mononucleotide
FP	Fluorescence polarization
GK	Guanylate kinase
GLP-1	Glucagon-like peptide-1
HIV-1	Human immunodeficiency virus type 1
HSQC	Heteronuclear single quantum coherence
i.p.	Intraperitoneal
i.t.	Intrathecal
i.v.	Intravenous
iGluR	Ionotropic glutamate receptor
ITC	Isothermal titration calorimetry
KA	Kainate
KO	Knockout
MAGUK	Membrane-associated guanylate kinase
MCA	Middle cerebral artery
MCAO	Middle cerebral artery occlusion
mGluR	Metabotropic glutamate receptors
MRI	Magnetic resonance imaging
NADPH	Reduced nicotinamide adenine dinucleotide phosphate
NMDA	*N*-methyl-D-aspartate
NMDAR	*N*-methyl-D-aspartate receptor
NMR	Nuclear magnetic resonance
nNOS	Neuronal nitric oxidase synthase
NO	Nitric oxide
OGD	Oxygen-glucose-deprived
PDZ	PSD-95/discs large/ZO-1
PEG	Polyethylene glycol
pMCAO	Permanent middle cerebral artery occlusion
PPBP	4-Phenyl-1-(4-phenylbutyl)-piperidine
PPIs	Protein–protein interactions
PSD	Postsynaptic density
PSD-95	Postsynaptic density protein 95
s.c.	Subcutaneous
SAP	Synaptic-associated protein
SAR	Structure–activity relationship
SH3	Src homology 3

SynGAP	Synaptic Ras GTPase-activating protein
Tat	Nuclear transcription activation protein
TIP-1	Tax-interacting protein 1
tMCAO	Transient middle cerebral artery occlusion
ZO-1	Zonula occludens-1

1 Introduction

During cerebral ischemia, a complex system of biochemical and molecular mechanisms is triggered, such as excitotoxic glutamate signaling, ionic imbalance, free-radical reactions, and inflammation, which all results in impaired neurological function and ultimately the death of brain tissue [1, 2].

Glutamate is the major excitatory neurotransmitter in the brain and hence a key mediator of intracellular communication, plasticity, growth, and differentiation [3]. During normal physiological conditions, the glutamate concentration in the synaptic cleft is in the low micromolar range and is responsible for the synaptic signaling via activation of metabotropic (mGluR) and ionotropic glutamate receptors (iGluRs) [4]. The mGluRs are G protein-coupled receptors, whereas the iGluRs are cationic selective ligand-gated ion channels. Glutamate binds to three subtypes of iGluRs, namely *N*-methyl-D-aspartate (NMDA) receptor, α-2-amino-3-(3-hydroxy-5-methylisoxazol-4-yl) propionate (AMPA) receptor, and kainate (KA) receptor [3, 5, 6]. During ischemia, the concentration of glutamate in synaptic cleft is amplified due to the increased glutamate efflux and reduced uptake [7]. This build-up of extracellular glutamate triggers an influx of Ca^{2+} and further on intracellular secondary Ca^{2+}-mediated enzymatic activation reactions and subsequent cell damage [2]. Nearly all glutamate receptors have been implicated to be involved in neurotoxicity. However, it is generally accepted that the NMDA receptor is the key mediator of glutamate neurotoxicity, due to its high Ca^{2+} permeability, Fig. 6.1 [1, 8]. As the critical role of NMDA receptor-mediated calcium influx in stroke pathogenesis has been well-described, numerous attempts have been made to develop NMDA receptor antagonists as a direct protective treatment strategy for neuronal hypoxia. Different strategies have been used, such as developing competitive antagonists (e.g. selfotel [9] and midafotel [10]) and non-competitive antagonists (e.g. aptiganel/CNS 1102 [11]), uncompetitive channel blockers (e.g. dextrorphan [12] and remacemide [13]) and inhibitors of the glycine-binding site (e.g. gavelstinel [14] and licostinel [15]). However, none of the drugs, which target the NMDA receptor directly, has been approved for clinical use for the treatment of ischemic stroke. The reason for this is multifactorial, but mainly falls within two categories; (1) NMDA receptor blockers generate side effects related to the induced impairment of key brain functions, such as sedation and psychotomimetic side effects and (2) NMDA receptor blockers have a short therapeutic window for drug administration, as they are effective only when administrated before or shortly after a stroke. Other studies have also shown that the optimal plasma concentration cannot be achieved within a timely manner when

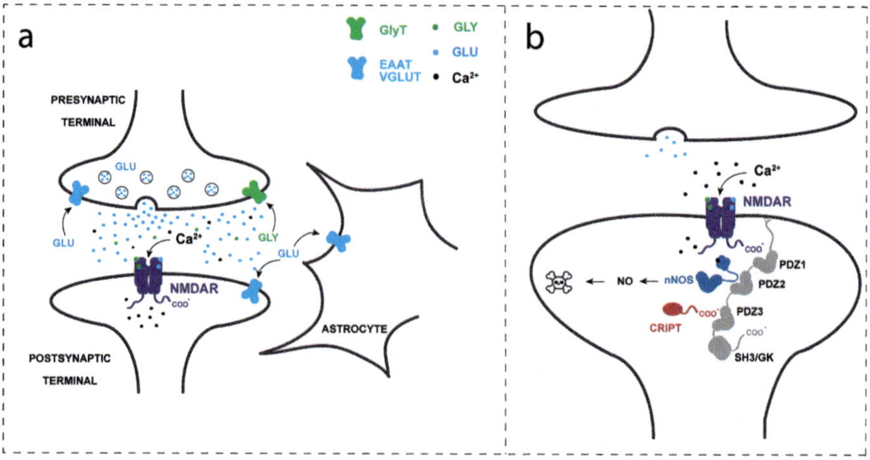

Fig. 6.1 Schematic representation of excitotoxicity induced by NMDA receptor activation. (**a**) During ischemia and stroke, a large excess of glutamate is released from the presynaptic terminal into the synaptic cleft. The NMDA receptor is activated by glutamate in the presence of glycine and induces an influx of Ca^{2+}. Glutamate and glycine are under normal physiological conditions rapidly removed from the synapse by uptake via glutamate (EAAT and VGLUT) and glycine transporters (GlyT), respectively. (**b**) Up on Ca^{2+} influx, the extreme C-terminal of the NMDA receptor subunit GluN2B and nNOS bind to PSD-95 PDZ1 and PDZ2, respectively, and induce the production of toxic nitric oxide (NO) and further cell death

administrated at the max tolerated dose and that the optimal neuroprotective concentration in animal models could not be translated to humans and further on failed to show any clinical efficacy [13, 15–19]. In an effort to circumvent the side effects related to the direct targeting of NMDA receptor, recent studies have targeted the downstream signaling cascade rather than the receptor itself. This includes targeting the transcription factors responsible for neuronal death upon NMDA receptor activation and direct activation of Akt and cyclic adenosine monophosphate (cAMP) response element-binding protein (CREB) [20–24]. In addition, targeting the interactions between the NMDA receptor C-terminal tail and its associated signaling proteins such as death-associated protein kinase 1 (DAPK1), calpain family members, and postsynaptic density protein 95 (PSD-95) has been investigated with promising results [25–31].

1.1 PSD-95/nNOS/NMDA Receptor Signaling Complex

The postsynaptic density (PSD) is a membrane-associated highly protein dense region mainly located at the head of the dendritic spines. The PSD contains receptors, adhesion molecules, signaling enzymes, and cytoskeletal elements, which are held together and organized by a set of abundant scaffolding proteins. The functional role of the PSD is to mediate the apposition of pre- and postsynaptic membranes, cluster postsynaptic receptors, and to facilitate the activation of postsynaptic receptors to biochemical signaling events in the postsynaptic neuron [32]. Numerous attempts have been made to try to analyze and identify the components of the PSD. The total numbers of putative proteins in the PSD range from a few hundred to thousands depending on analysis [33–35]. However, the most abounded proteins are members of the Ca^{2+}/calmodulin-dependent kinase II (CaMKII) family, the scaffolding protein PSD-95, and the synaptic Ras GTPase-activating protein (SynGAP) [32, 36]. In addition, a core set of 288 and 77 proteins, which are associated with PSD-95 and NMDA receptors, and largely overlapping, has been identified and is considered to include the key proteins in the PSD [37, 38]. It should, however, be noted that the composition of the PSD varies between brain regions and different cell types [39, 40].

1.1.1 NMDA Receptors

The NMDA receptors are typically found at postsynaptic and extrasynaptic sites and are essential mediators of synaptic transmission, synaptic plasticity, learning, and memory formation [41, 42]. There are seven different receptor subunits in three different subfamilies, namely GluN1, GluN2A-D, and GluN3A-B. The NMDA receptor assembles into a heterotetrameric ion channel complex by associating two GluN1 and two GluN2 subunits or two GluN2 and one GluN1 and one GluN3 [42, 43]. The composition of the subunits is the major determinant of the NMDA receptor functional heterogeneity, as the subunits display distinct agonist sensitivity, deactivation kinetic, gating, and permeation properties [44–46]. The NMDA receptors share an overall structure, comprising two large extracellular domains (the N-terminal domain and the agonist-binding domain), a three transmembrane spanning segment, a re-entrant loop which forms the ion channel, followed by a C-terminal tail involved in trafficking and binding to intracellular interaction partners [46, 47]. The opening and closing of the ion channels are primarily regulated by ligand binding, where glycine binds to the GluN1 or GluN3 subunits and glutamate to GluN2 subunit [48, 49], and the voltage-dependent Mg^{2+} blockade [50]. Upon channel activation, an influx of Na^{+} and Ca^{2+} and an efflux of K^{+} occurs [46].

1.1.2 PSD-95

PSD-95 is, together with PSD-93, synaptic-associated protein 97 (SAP-97) and SAP-102, members of the membrane-associated guanylate kinase (MAGUK) superfamily. Structurally, the MAGUK members contain three PSD-95/discs large/zonula occludens-1 (ZO-1) (PDZ) domains followed by one Src homology 3 (SH3) and one guanylate kinase (GK) domain, Fig. 6.2 [51–54]. The sequence identity among the MAGUK members are >70 % in the domain regions, thus being highly conserved, but <20 % in the linker regions [54].

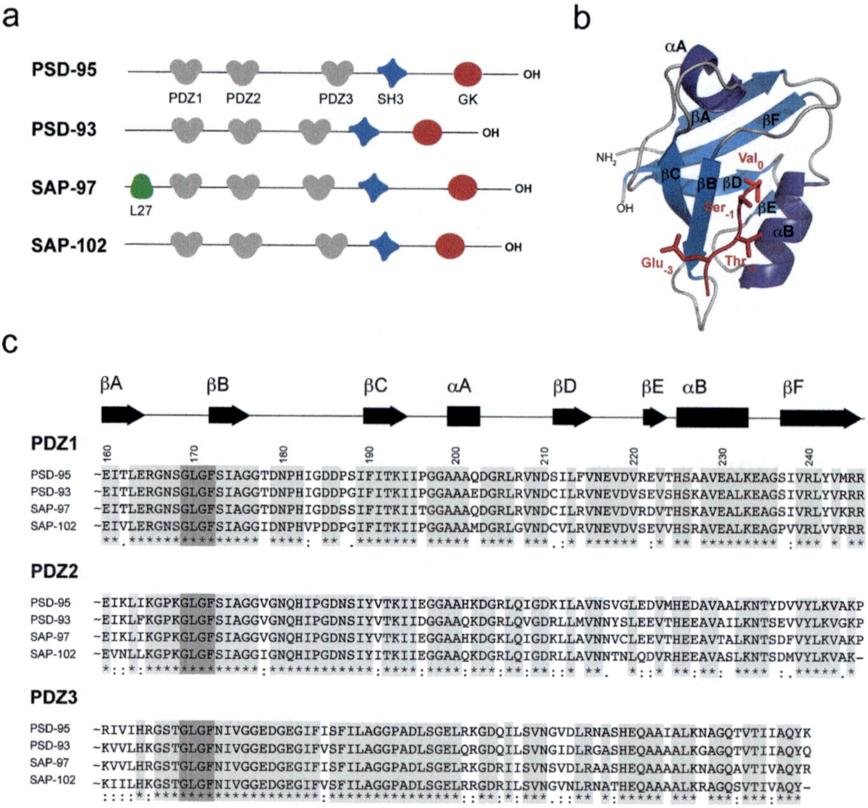

Fig. 6.2 Schematic representation of the domain arrangement of the MAGUK family members, tertiary structure of PSD-95 PDZ3, and sequence alignment of the MAGUK PDZ domains. (**a**) PSD-95, PSD-93, SAP-97, and SAP-102 contain three PDZ domains (*grey*) followed by one SH3 (*blue*) and one GK (*red*) domain. SAP-97 has an additional L27 (*green*) domain located in the N-terminal. (**b**) X-ray crystal structure of CRIPT binding to PSD-95 PDZ3. PDZ domains generally contain six β-sheets (A–F) and two α-helices (A–B) arranged in a partially open β-barrel capped by two α-helices. The ligand-binding site is located between the βB and the αB, indicated by the CRIPT ligand (*red*). PDB ID: 1be9. (**c**) Amino acid sequence alignment for the highly conserved PDZ domains for PSD-95, PSD-93, SAP-97, and SAP-102. The secondary structure is indicated by *arrows* and *boxes* above the sequence and based on PSD-95 PDZ1. The amino acid numbering reflects the numbering of PSD-95 PDZ1. The conserved GLGF loop is highlighted in *dark grey* and fully conserved residues among the MAGUK members are highlighted in *light grey*. The sequence alignment was generated using Clustal Omega. *Star* indicates a fully conserved residue; *colon* and *period* indicate conservation between group of strongly similar properties and conservation between groups of weakly similar properties, respectively

The main function of PSD-95 is to bind to and stabilize various membrane and signaling proteins in the PSD. Subsequently, PSD-95 is a key component and centrally involved in synaptic function and has been shown to be essential for the three-dimensional molecular organization of the PSD [55]. As such, 85

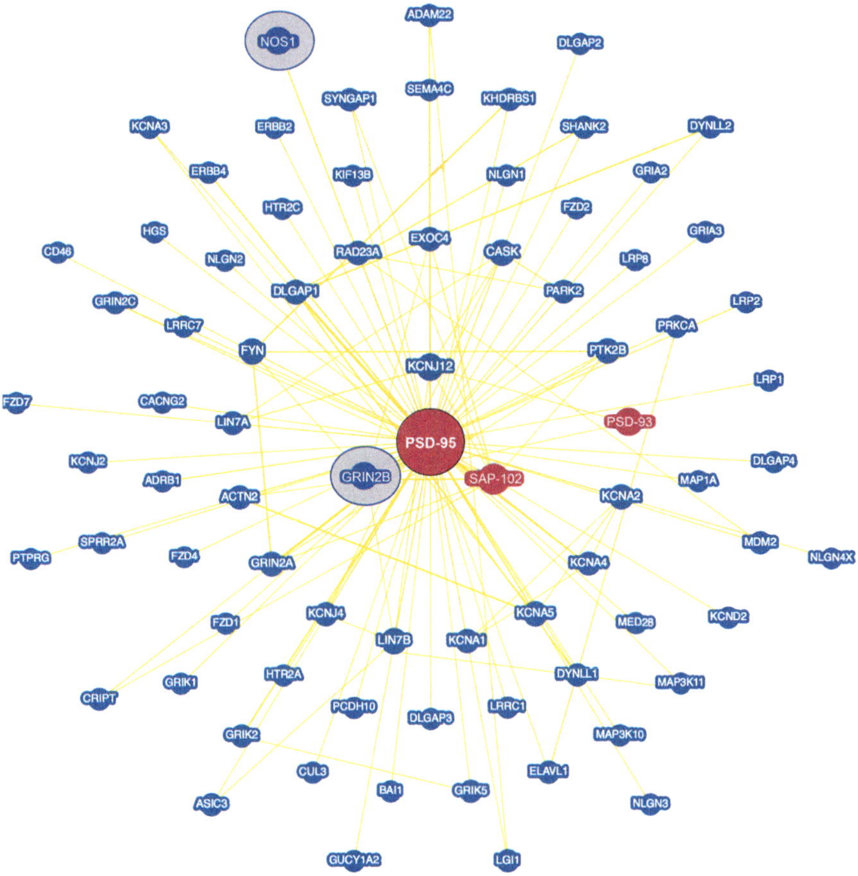

Fig. 6.3 The human PSD-95 interactome. Seventy-eight human proteins have been identified to interact with PSD-95 according to the database BioGRID[3.4]. The abbreviations for the proteins are based on the gene name for the corresponding protein, except for PSD-95, PSD-93, and SAP-102, which are denoted according to the protein abbreviation and displayed in *red*. NOS1 (nNOS) and GRIN2B (GluN2B) are highlighted in *grey*

different proteins have been identified to interact with PSD-95, Fig. 6.3 [56]. In particular, the clustering ability of PSD-95 to assemble NMDA receptor-associated protein complexes, which facilitates and regulates the functional coupling with downstream signaling molecules in the PSD to, e.g., neuronal nitric oxidase synthase (nNOS), is pivotal [57]. PSD-95 interacts with the extreme C-terminal of primarily GluN2B and GluN2A subunits of the NMDA receptors via the PDZ1 and PDZ2 domains [58, 59]. The PDZ2 domain also interacts with nNOS and hence brings the NMDA receptor, PSD-95 and nNOS in close proximity, Fig. 6.1. This complex assembly ensures an efficient activation of nNOS by the influx of calcium ions [57, 60].

1.1.3 nNOS and NO Production

nNOS is a member of the NOS enzyme family, which catalyzes the conversion of L-arginine to nitric oxide (NO) and citrulline, in the presence of O_2 and reduced nicotinamide adenine dinucleotide phosphate (NADPH) [61]. nNOS is highly important for the excitotoxicity and ischemic brain damage, as it has been shown that primary brain cultures treated with NOS inhibitors and nNOS knockout mice are resistant to NMDA-induced excitotoxicity [62, 63]. Multiple splice variants of nNOS, which are functionally distinct, have been reported. The main differences are found in the N-terminal domain, where the nNOSα protein isoform contains a PDZ domain, which is highly important for protein–protein interactions (PPIs) [64, 65]. In addition, nNOS consists of a heme, calmodulin, flavin mononucleotide (FMN), flavin adenine dinucleotide (FAD), and NADPH binding domains [66, 67]. nNOS is constitutively expressed and requires the formation of the Ca^{2+}/calmodulin complex for its activation [68]. nNOS is recruited to the PSD by PSD-95, where PDZ2 recognize the internal β-hairpin finger adjacent to the nNOS PDZ domain [60], Fig. 6.4. The production of NO under normal physiological conditions plays an important role in synaptogenesis, long-term potentiation, and long-term depression, hence modulating synaptic plasticity and memory function [68]. However, during ischemic stroke, when excessive levels of NO are produced, NO is neurotoxic. The neurotoxic effect is mediated via numerous signaling pathways, such as NO-induced potentiation of damage to the mitochondrial respiration system, glycolysis and DNA replication by inhibiting enzymes involved in these processes, and formation of peroxynitrite and free radicals leading to lipid peroxidation [69–71].

2 PSD-95 as Drug Target

PSD-95 has emerged as a new and exciting target for the treatment of ischemic brain damage after stroke [72], with promising results in both animal [73] and human [74] disease models. Inhibiting the PPIs in the ternary PSD-95/nNOS/NMDA receptor complex is believed to be the key mechanism underlying neuroprotection after glutamate excitotoxicity in stroke [75]. However, targeting protein–protein interfaces pose some challenges compared to conventional small molecule binding sites. PPIs have usually larger (1500–3000 $Å^2$) contact areas than protein–small molecules interactions (300–1000 $Å^2$) and these surfaces are generally shallow and lack deep grooves and pockets [76]. Due to their size and flexibility, peptide ligands appear to be more suited to inhibit PDZ domain-mediated PPIs and show higher binding affinities compared with small molecules [77, 78].

The PDZ domains of PSD-95 consist of approximately 90 amino acids arranged in a stable tertiary fold consisting of six anti-parallel β-strands (βA-βF) and two α-helixes (αA and αB), Fig. 6.2. The interaction partners of PSD-95 bind to the PDZ domains via two different binding modes: a so-called canonical and a non-canonical binding mode, respectively. The binding of free carboxyl-terminal peptides to PDZ domains is referred to as the canonical binding mode and binding of an internal protein motif is referred as the non-canonical binding mode. The canonical mode was elucidated by both X-ray crystallography [79] and solution nuclear magnetic resonance (NMR) [80] using PDZ3 of PSD-95 and the cysteine-rich interactor of PDZ three (CRIPT) peptide. C-terminal peptide ligands

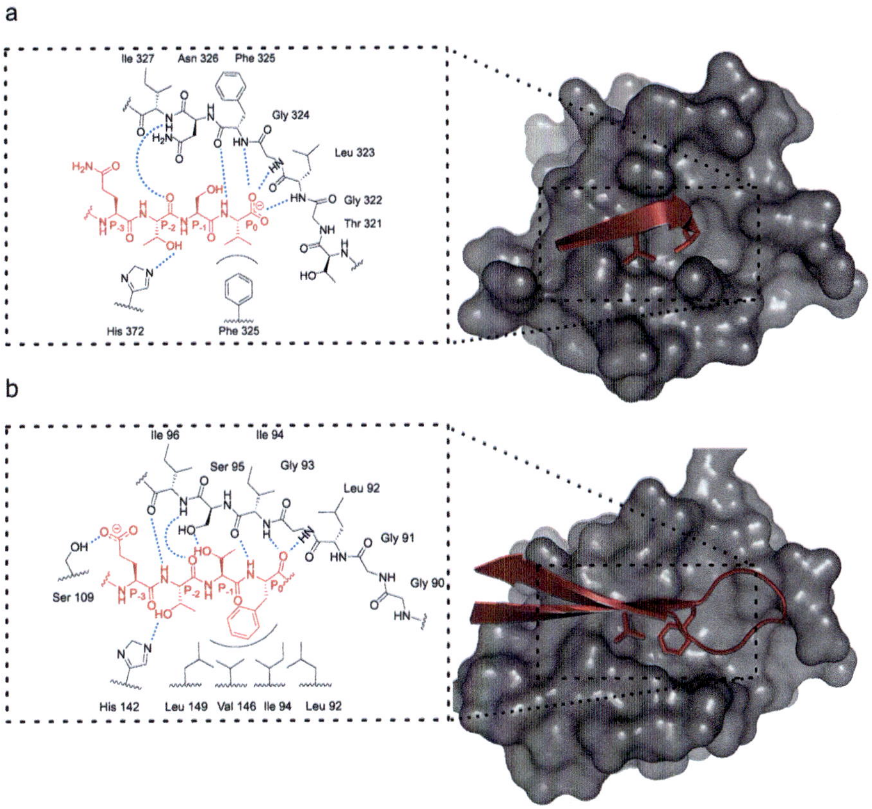

Fig. 6.4 Canonical and non-canonical binding modes of PDZ domains share similar interaction patterns. (**a**) Canonical binding mode represented by CRIPT peptide bound to PSD-95 PDZ3. Val (P_0) backbone and its carboxyl-terminal interact by hydrogen bonds with residues within the GLGF loop (Gly322-Phe325) and its side-chain interacts with Phe325 by hydrophobic interactions. Thr (P_{-2}) backbone forms a hydrogen bond interaction with the backbone of Ile327 and its side-chain forms a hydrogen bond with His372, located in αB. (**b**) X-ray structure of α-syntrophin PDZ domain bound to the nNOS β-finger by a non-canonical binding mode. The backbone carbonyl and amide of Phe (P_0) interacts with the backbone of Gly93 and Ile94 by hydrogen bonds and mimics Val (P_0) of CRIPT. In both binding modes, the side-chain of the residue at position P_0 inserts into a hydrophobic pocked and the hydroxyl of Thr side-chain at position P_{-2} makes a hydrogen bond with a His residue. Hydrogen bonds are shown in *blue* and hydrophobic interactions are shown by *black curves*. PDB ID: 1be9 (**a**) and 1qav (**b**)

bind as a β-strand between the βB and αA in the PDZ domain and the C-terminal carboxylate binds to a highly conserved loop (GLGF) between βA and βB, Fig. 6.4. This loop stabilizes the binding of the peptide carboxyl-terminal via backbone hydrogen bonds. Recent studies have demonstrated the importance of these backbone interactions by introduction of amide-to-ester mutations in the GLGF loop and in the βB strand. This study demonstrated a pivotal role for the GLGF backbone hydrogen bond interaction to the free carboxylate, whereas the non-canonical interaction was fundamentally different for the corresponding amide-to-ester mutations [81].

Although classification of PDZ domains is still a matter of debate, PDZ1 and PDZ2 of PSD-95 have been grouped into the class I PDZ domain [82]. In this class, PDZ domains recognize an S/T-X-ϕ peptide motif, with S/T representing a Ser or a Thr residue; X any amino acid and ϕ representing a hydrophobic carboxyl-terminal residue, preferentially Val or Leu [83]. Amino acids are numbered starting with the C-terminal amino acid being P_0 and subsequent amino acids P_{-1}, P_{-2}, P_{-3}, etc. It is generally agreed that amino acids in positions 0 and −2 are primarily responsible for the canonical interaction to PDZ domains. Truncations of the endogenous peptide ligand, the carboxyl-terminal GluN2B subunit of the NMDA receptor (YEKLSSIESDV), have been evaluated by yeast two-hybrid [58] and fluorescence polarization (FP) assays [84], revealing that only the last 5–7 amino acids are required for binding to PSD-95 with retained binding affinities of 18 and 4 μM for PDZ1 and PDZ2 of PSD-95, respectively. Interestingly, the tripeptide TAV still binds to PSD-95 PDZ2 with a K_i value of approximately 40 μM, but shorter peptides result in a pronounced affinity loss [84]. Within the five last amino acids, mutations on the carboxyl-terminal Val (P_0) and in the Ser/Thr residues (P_{-2}) generally abolish binding, while the presence of Gln or Glu at P_{-3} is shown to slightly increase binding affinity. Amino acids in position P_{-4} and further upstream are more promiscuous and a variety of different amino acid residues can be accommodated [85]. As in the case for the C-terminal carboxylate interaction with the GLGF loop, backbone hydrogen bonds from the peptide ligand to PDZ domains also play an important role for binding of peptides to the PDZ domains of PSD-95 [86].

A subclass of PDZ domains also recognizes internal peptide motifs, denoted the non-canonical binding mode, such as the nNOS PDZ β-hairpin structure (β-finger) binding to PSD-95 PDZ2 [87], where it is known that the entire folded nNOS β-finger is required for binding [66]. The β-finger is comprised of two β-strands connected by a sharp β-turn and occupies more than twice the surface area on the protein compared to peptides binding in the canonical binding mode. Both the β-finger and carboxyl-terminal peptides have key interactions with the GLGF binding loop in the PDZ domains, Fig. 6.4. Furthermore, the Phe residue in the nNOS β-finger has an analogous role to Val in the carboxyl-terminal position P_0. Interestingly, positions P_{-2} and P_{-3} in the β-finger also comprise the same residues as the carboxyl-terminal peptide ligands, Thr and Glu. Ala mutations in these three positions disrupted binding between nNOS and PSD-95 PDZ2, with Phe at position P_0 being the most sensitive position [60].

2.1 Neuroprotective PSD-95 Inhibitors: NA-1

The NA-1 peptide was the first peptide developed to prevent excitotoxicity in ischemic stroke by targeting PSD-95. NA-1 was developed in parallel with the elucidation of the role of the PSD-95/nNOS/NMDA receptor complex in excitotoxicity. Initially, it was observed that suppression of PSD-95 by antisense DNA in cultured neurons increased cell viability and reduced NO levels after NMDA induced toxicity without changing neuronal currents and calcium influx [57]. Later, NA-1 was designed by fusing the last nine amino acids from the C-terminal NMDA receptor subunit GluN2B (KLSSIESDV) with a cell-penetrating peptide sequence (YGRKKRRQRRR) derived from the nuclear transcription activation protein (Tat) from human immunodeficiency virus type 1 (HIV-1). Treatment of cortical neuron cultures with NA-1 reduced cell death induced by treatment with NMDA [29], and importantly, reduced brain infarct volumes in vivo in the transient middle cerebral artery occlusion (tMCAO) model in rats [29]. Proteomic analysis of potential binding partners of NA-1 among 145 expressed human PDZ domains revealed that NA-1 binds to seven PDZ domains, comprising the four MAGUKs (SAP-102, SAP-97, PSD-95, and PSD-93) and the human Tax-interacting protein 1 (TIP-1) at 7 nM. However, using an increased NA-1 concentration of 500 nM, the number of interacting PDZ domain proteins increased to 44. Suppression of MAGUKs, TIP-1, and nNOS proteins using RNAi did not affect the neuronal morphology and integrity. Furthermore, inhibiting MAGUKs and TIP-1 had no effect on neurons' viability after being challenged with NMDA. In contrast, inhibition of PSD-95 and nNOS expression decreased the NMDA-induced cytotoxic effect after the same treatment [75]. These results stimulated further studies of NA-1 using different dose regimens and different in vivo stroke models (Table **6.1**). These studies were pioneered by Tymianski and colleagues, who are currently also supporting NA-1 phase IIb/III clinical trials. The wide range of outcome measurements for the same compound reflects how complex the mechanisms of brain injury in stroke are. Ideally, the choice of an animal model to study neuroprotection must account for this complexity and poorly designed pre-clinical trials are one of the reasons why so many compounds that appeared effective in animal studies failed to translate its effectiveness into humans [88]. For example, the antioxidant NXY-059 showed 36–44 % infarct volume reduction in the rat permanent MCAO (pMCAO), even in treatments starting 4 h after occlusion [89, 90]. However, after having enrolled more than 3000 patients, a clinical trial showed no significant effects in functional outcome measurements [91].

In most pre-clinical studies, NA-1 was tested in rodent models using pMCAO, tMCAO, or the three-pial vessel occlusion (3PVO) lesion models. In the pMCAO model, the occlusion is maintained throughout the study and should cause a more severe brain damage than the other models [92]. In the tMCAO model, occlusion is kept for a fixed period with an intraluminal suture, which is removed and allows reperfusion. This should result in a smaller lesion that decreases progressively with time [93]. In the third model, three cortical branches of the middle cerebral artery (MCA) are permanently electrocauterized, and since some collateral branches of the MCA are not blocked, reperfusion is higher and the resulted lesions are generally

Table 6.1 Summary of pre-clinical assessments of NA-1 in vivo efficacy as a neuroprotectant for ischemic stroke

Reference	Occlusion method[a]	Animal model[b]	Dose and dose regimen[c]	Outcome measurements[d]		
				Infarct reduction (% of control)		Functional assays
Aarts et al. [29]	tMCAO (90 min)	SD rat	3 nmol/g (1 h AO)	67[e]	$n=8$	Postural reflex[e]
			3 nmol/g (45 min BO)	55[e]	$n=8$	Postural reflex[e]
Soriano et al. [96]	MCAO+	SD rat (12 d)	3 nmol/g (30 min BO) i.p.	15[NS]	$n=7$–20	NA
	3PVO	SD rat	3 nmol/g (15 min BO)	80[e]	$n=8$	NA
Sun et al. [94]	pMCAO	SD rat	0.3 nmol/g (1 h AO)	39[e]	$n=7$	NA
			3 nmol/g (1 h AO)	34[e]	$n=8$	NA
	tMCAO (90 min)	SD rat	0.03 nmol/g (3 h AO)	67[e]	$n=6$	NA
			0.3 nmol/g (3 h AO)	75[e]	$n=6$	NA
			1 nmol/g (3 h AO)	63[e]	$n=6$	NA
			3 nmol/g (3 h AO)	53[e]	$n=8$	Barnes maze[e]
	3PVO	SD rat	3 nmol/g (1 h AO)	60[e]	$n=8$	NA
			0.3 nmol/g (1 h AO)	5[NS]	$n=8$	NA
Bråtane et al. [95]	pMCAO	Wistar rat	3 nmol/g (1 h AO)	30[e]	$n=8$	Menzies index[e]
Bach et al. [31]	pMCAO	C57BL/6 mouse (7–8 w)	3 nmol/g (30 min AO)	20[NS]	6 h, $n=17$	NA
	pMCAO		3 nmol/g (30 min AO)	6[NS]	48 h, $n=17$	Grip STRENGTH[NS], rotarod[e]
Teves et al. [102]	tMCAO (30 min)	C57BL/6 mouse (8–10 w)	3 nmol/g (30 min AO)	0[NS]	$n=10$	Bederson scale[NS]
			10 nmol/g (30 min AO)	24[e]	$n=10$	
	tMCAO (60 min)		3 nmol/g (1 h AO)	0[NS]	$n=20$	
			10 nmol/g (1 h AO)	26[e]	$n=20$	
Bell et al. [136]	3VPO	SD rat	3 nmol/g (15 min BO)	80[e]	$n=8$, IA	NA

(continued)

Table 6.1 (continued)

Reference	Occlusion method[a]	Animal model[b]	Dose and dose regimen[c]	Outcome measurements[d]		
				Infarct reduction (% of control)		Functional assays
Cook et al. [73]	tMCAO(−p) (4.5 h)	Cynomologous macaques	1 nmol/g (1 h AO)	28[e]	7d, $n=6$, MRI	NHPSS[e]
	tMCAO(+p) (3.5 h)		1 nmol/g (3 h AO)	27[e]	14d, $n=12$, MRI	NHPSS[e]
	tMCAO(−p) (90 min)		1 nmol/g (1 h AO)	60[e]	30d, $n=10$	NHPSS[e]
Cook et al. [98]	Embolic	Cynomologous macaques	1 nmol/g (1 h AO)	60[e]	$n=5$, MRI	NHPSS[NS]

[a]*pMCAO* permanent middle cerebral artery occlusion, *3PVO* three pial vessel occlusion, *tMCAO* temporary middle cerebral artery occlusion, *MCAO+* severe middle artery occlusion, *tMCAO(+p)* transient middle cerebral artery occlusion with preserved penumbra, *tMCAO(−p)* transient middle cerebral artery occlusion with reduced penumbra, *Embolic* occlusion was archived by intracarotid injection of twenty 100 μm polystyrene spheres, *(time)* indicates occlusion period in transient models

[b]All animal models used were male adult individuals except when indicated; *SD* Sprague–Dawley rat strain, *d,w* animal age in days or weeks, respectively

[c]All drugs were administrated intravenously, except indicated; *(time) AO* drug administrated after a specific time after occlusion, *(time) BO* administration before a specific time before occlusion, (time), *i.p.* intraperitonial administration

[d]Infarct volumes were measured by histology after 24 h of occlusion, except indicated

[e]Significantly different from control; [NS] non-significantly different from control, *IA %* of infarct area reduction compared with control, *MRI* magnetic resonance imaging, *NA* not available, *NHPSS* non-human primate stroke scale

restricted to the cortex. In general, it is seen that NA-1 causes a higher reduction in infarct volumes when transient, rather than permanent, models are employed (Table **6.1**). Sun et al. [94] and Bråtane et al. [95] described reduction of infarct volumes of up to 40 % in the pMCAO model, whereas Aarts et al. [29] and Sun et al. [94] reported reductions of more than 70 % using the tMCAO model, while the highest reduction (80 %) was observed in the 3PVO model [96]. This indicates that increasing reperfusion could be correlated with increased neuroprotection by PSD-95 inhibitors. The in vivo neuroprotective efficacy of NA-1 was also assessed in gyrencephalic non-human primate models, developed by Tymianski and co-workers. This included two new tMCAO lesion models; a more severe tMCAO(−p) that results in a rapidly shrinking penumbra and second tMCAO(+p) model that allows for collateral circulation and increasing the ischemic penumbra area [73]. NA-1 was shown to be effective in both models, decreasing infarct volumes and improving functional measurements [97]. Interestingly, NA-1 reduced infarct sizes to a larger extent in the more severe model, which however could be related to the earlier administration of NA-1 in this model, namely 1 h after occlusion versus 3 h after in the tMCAO(+p) model. NA-1 was also evaluated in non-human primate model, specifically designed as a proof-of-concept study for the subsequent phase II clinical trial, the ENACT

(Evaluating Neuroprotection in Aneurysm Coiling Therapy) study [98]. In this model, it was investigated if NA-1 could reduce the number and volume of small embolic strokes lesions generated after intravascular manipulation. Stroke was induced in animals with intracarotid injection of small polystyrene spheres, generating multiple small ischemic areas that could be visualized by MRI and histology. NA-1 treatment reduced the number and volume (56 %) of stroke lesions, mainly in the cortex region. However, neurological scores showed no differences between drug and placebo treatment [98]. Finally, NA-1 safety and efficacy have been assessed in a human phase II clinical trial, the ENACT trial [74]. One hundred and eighty-five patients who underwent aneurism endovascular repair were treated with NA-1 (1 nmol/g) immediately after the procedure and infarct volumes and functional outcomes were compared after 30 days. No serious adverse effect could be attributed to NA-1 treatment, nor did the drug affect biochemical and hematological markers. NA-1 reduced the number of ischemic infarcts, but not the volume of the lesions or the values of the neurocognitive scores. Currently, NA-1 is being evaluated in a phase IIb/III clinical trial (The Frontier Trial), in which NA-1 is administrated i.v. by paramedics in the ambulance, i.e. prehospital treatment. If NA-1 is proven efficient, it has the potential of changing the paradigm in stroke therapy and represents a unique successful case after more than hundreds of failed clinical trials.

2.2 Multimeric Ligands Targeting PSD-95

The tandem arrangement of the first two PDZ domains of PSD-95, which is also found in the other MAGUKs [36], is suggested to form a supramodular structure with distinct binding proprieties from the simple sum of the two individual domains [53, 99]. The PSD-95 PDZ12 tandem construct showed limited rotational mobility in NMR studies and their binding pockets appear to face the same direction [53]. Furthermore, binding affinities between wild-type peptide ligands and PSD-95 PDZ12 were higher than between the same ligands and the individual PDZ1 and PDZ2 domains [53, 100]. This arrangement of PDZ domains was the basis of the design of dimeric inhibitors of PSD-95 [84]. These peptides would not only enhance binding affinity, but also increase selectivity, as a dimeric interaction with a tandem PDZ domain would be more unique than the multitude of possible interactions mediated by the S/T-X-ϕ motif alone [75, 101]. Currently, the only dimeric peptide ligand that has shown effect in stroke models is AVLX-144, while peptides described in this section all showed reduced in vitro affinity to PSD 95 compared to AVLX-144 and have not been examined as neuroprotective agents.

The design of AVLX-144 was based on the knowledge that a pentameric peptide showed wild-type affinity and that PDZ1 and PDZ2 bind to similar peptide sequences [84]. Thus, by cross-linking two pentameric peptides with a linker of appropriate length, it should be possible to generate dimeric peptides with increased affinity. Bach et al. [30] explored this principle, by connecting the IESDV pentapeptide with a polyethylene glycol (PEG) linker, which immediately led to a 100-fold affinity increase. Subsequent optimization of the linker length and peptide sequence (IETAV) led to a dimeric peptide, AVLX-125 (UCCB01-125), with a K_d of 10 nM, thus being by far the most potent inhibitor of PSD-95 and of any PDZ domain in general [30]. Later studies showed that AVLX-125 did not permeate the blood–brain barrier to a large extent, which led to the design and synthesis of the second-generation dimeric inhibitors. Key to this was the design of a custom-made PEG linker including a nitrogen as a chemical handle (*N*PEG-linker), which allowed the introduction of a cell-penetrating Tat peptide into the dimeric structure, Fig. 6.5 [31]. Gratifyingly, these modifications increased affinity relative to AVLX-125, and the resulting

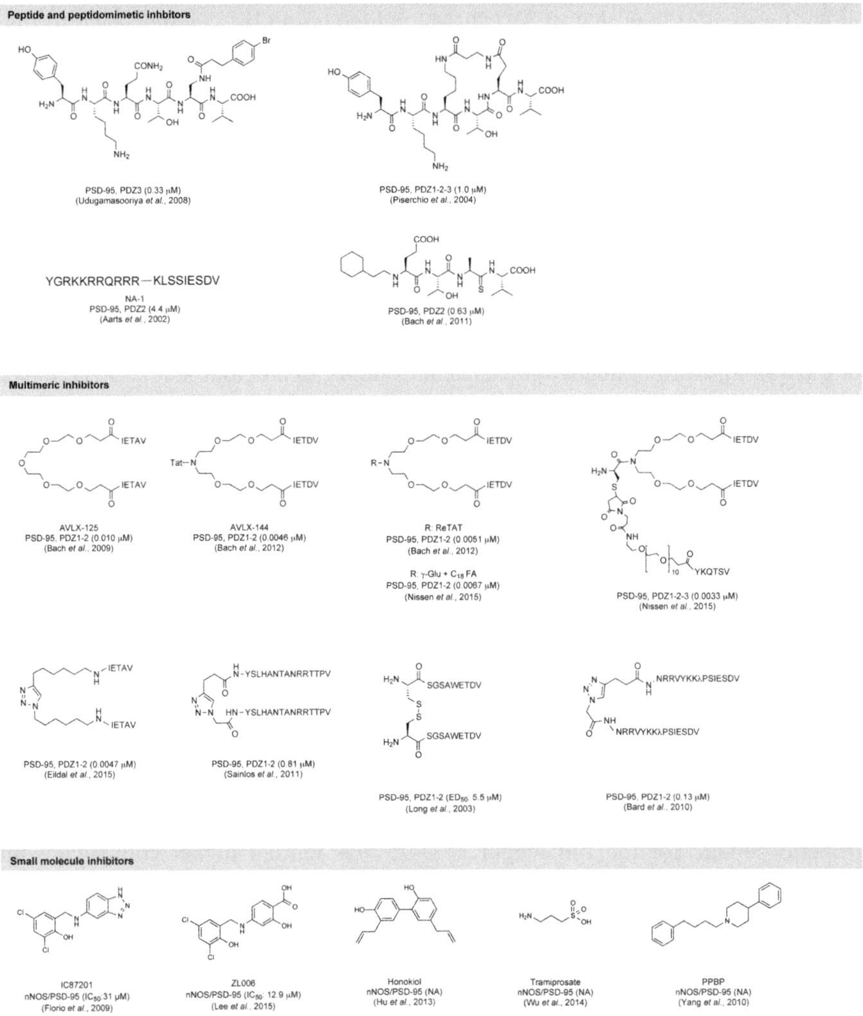

Fig. 6.5 Structures and binding affinities for PSD-95 and nNOS inhibitors. K_d and K_i values are given if otherwise not stated. Affinities for honokiol, tramiprosate, and PPBP have not been reported. *NA* not available

compound, AVLX-144, is currently the most potent PSD-95 inhibitor with a K_d of 4.6 nM, thus being 1000-fold more potent than NA-1 [31]. The increased in vitro affinity was also shown to translate into a more effective neuroprotection in a pMCAO mouse stroke model compared to NA-1. The mice were treated with either AVLX-144 or NA-1 at 3 nmol/g 30 min after occlusion. AVLX-144 resulted in reduced infarct volumes (40 %) and correspondingly improved neuromuscular functions, whereas NA-1 did not show significant effects [31]. Recently, Teves et al. [102] examined NA-1 in a tMCAO mouse model and confirmed the lack of efficacy of NA-1 at 3 nmol/g, but found a significant infarct reduction at 10 nmol/g. Thus, it seems that the therapeutic window of NA-1 might be smaller than that of AVLX-144, which could be of importance for the further development of these compounds; however, future studies of safety and efficacy should hopefully resolve this.

A number of other dimeric ligands targeting PDZ1 and PDZ2 of PSD-95 have been developed, Fig 6.5. The supramodular nature of PSD-95 PDZ12 domains was investigated by linking a PSD-95 PDZ12 ligand analog through Cys disulphide bonds, Fig. 6.5. A decapeptide comprising the six last C-terminal amino acids of the $K_v1.4$ K^+ channel, with a Val to Trp mutation in the P_{-4} position, was connected by a peptide linker (Ser-Gly-Ser) and a N-terminal Cys. The dimeric peptide showed ~20-fold higher affinity for PSD95 PDZ12 relative to the monomeric ligand [53]. Rigidified peptide dimeric ligands were prepared and the influence of linker length in PSD-95 PDZ12 binding was evaluated. Contrary to previous studies [31], linker length caused minor changes in binding affinities, and in all cases, the rigidified peptides retained nanomolar affinities to PSD-95 [103].

The role of PSD-95 in trafficking of membrane receptors in neurons was studied using dimeric ligands. The NMDAR subunit GluN2A was specifically disrupted from PSD-95 seemingly without affecting subunit GluN2B by a NRRVYKKλPSIESDV (λ = norleucine) sequence dimerized, Fig. 6.5 [100]. The dynamics of AMPA receptors in neurons were studied by inhibiting its interactions with PSD-95 and other MAGUK proteins with no influence on Shaker-type K^+ channels. In contrast, monomeric peptides failed to influence AMPA receptor stabilization and showed reduced binding affinity to PSD-95 PDZ12 [104].

The multimerization strategy was further explored by synthesis of trimeric ligands that simultaneously bind to all three PDZ domains of PSD-95 [105]. Starting from a dimeric ligand [30], a third binding moiety was added comprising a linker and a truncated C-terminal peptide derived from CRIPT. The binding affinity of this trimeric peptide for the full length PSD-95 was tenfold higher compared with dimeric ligands [105].

In order to enhance the in vivo stability of the dimeric peptide ligand AVLX-125, the secondary amine in the *N*PEG-linker of the dimeric ligand was acylated by different fatty acids, Fig. 6.5 [106]. Similar modifications have previously been shown to enhance the half-life of the glucagon-like peptide-1 (GLP-1) in pigs [107]. Addition of a fatty acid did not compromise binding to PSD-95 PDZ12, but the modification not only dramatically enhanced the in vitro plasma stability, but also prolonged the absorption and increased the elimination half-life in rats after subcutaneous (s.c.) administration [106].

2.3 Peptidomimetics Targeting PSD-95

Peptidomimetics are compounds that show structural similarity with peptides and can mimic the biological effect of a peptide [108], but offer advantages as drug candidates, such as increased stability against proteolysis, higher permeation through biological membranes, and the possibility to stabilize a favorable structure conformation [109]. In regard to PSD-95, extensive biophysical studies on its binding site identified key amino acid residues required for binding [53, 79, 80]. These residues, also called 'hot-spots' [110], were specifically targeted by peptidomimetic compounds in an attempt to design small, stable, and cell permeable compounds. While some peptidomimetics achieved binding affinities to PSD-95 comparable to wild-type ligands, none of them were as active as the dimeric peptide inhibitors [78].

The GluN2B C-terminal sequence was the starting point for the design of PSD-95 peptidomimetics inhibitors by Bach and co-workers [84]. First, sequential truncations reduced the natural ligand to the smaller IESAV pentapeptide, which retained wild-type affinity to PSD-95 PDZ1 and PDZ2. Then, the ESAV peptide was explored in a structure-activity relationship (SAR) study, demonstrating that D-amino acids and several non-proteinogenic amino acid substitutions reduced binding while N-terminal alkylations increased affinity. Further incorporation of bulky groups in N-terminal, such as the cyclohexylethyl group, led to a tenfold affinity increase ($K_i = 1$ μM) compared to the wild-type peptide. Finally, combining N-terminal alkylations with introduction of a thioamide between P_0 and P_{-1} in the ETAV peptide, Fig. 6.5, increased membrane permeability and plasma stability while retaining affinity towards PSD-95 PDZ2 [111].

A range of peptidomimetics was synthesized in parallel by acylating the amino side-chain at P_{-1} of the CRIPT analogue (YKQTKV) with different organic acids. In total, 272 compounds were prepared and screened in a high-throughput ELISA-type assay. The top-scoring hits were prepared and a derivative containing a bromobenzylethyl amide showed the highest affinity to PSD-95 PDZ3 (K_d=0.33 μM) compared to the CRIPT peptide (K_d=6.3 μM). Interestingly, the peptidomimetic compound shifted NMR signals from PSD-95 PDZ3 residues that are situated beyond the canonical binding pocket [112], which might indicate that these derivatives bind to additional regions not explored by the CRIPT peptide. In a similar study, halogenated phenylalanine derivatives were replaced at the P_{-1} position in the peptide KKETFV. 3,4-Difluoro analogues showed the highest affinity (K_d=0.28 μM) among the different peptidomimetics, but comparable to the previously reported bromobenzylethyl amide [113].

The CRIPT peptide was also used as a template for the synthesis of cyclic peptides that inhibited PSD-95 PDZ1 interactions. Positions P_{-1} (Ser) and P_{-3} (Gln) were replaced by Glu and Lys, respectively, and cyclized through amide bonds using a βAla as a linker, generating Tyr-Lys-c[-Lys-Thr-Glu(βAla)]-Val, Fig. 6.5. NMR-based titrations showed that this compound bound to PSD-95 PDZ1 with a K_d of 1 μM. Additionally, the cyclic peptide inhibited the interaction between GluK2 subunit of the kainate receptor and PSD-95 PDZ1 with a higher affinity. This inhibition was sustained for a longer period, compared with its linear counterpart, which could indicate increased peptide stability with no changes in the binding mode [114].

2.4 *Small Molecules Targeting the NMDA Receptor/nNOS/PSD-95 Complex*

In addition to the design and development of peptide and peptidomimetic-based inhibitors of the nNOS/PSD-95/NMDA receptor complex, a number of small molecule inhibitors have been investigated, although with significant lower affinity to PSD-95 compared to their peptidic counterparts.

2.4.1 IC87201

The first small molecule inhibitor targeting the nNOS/PSD-95 complex, IC87201 (2-((1*H*-benzo[*d*][1-3]triazol-5-ylamino)methyl)-4,6-dichlorophenol), Fig. 6.5, was identified by ICOS corporation through a high throughput screen of their library, consisting of 150,000 small molecules [115]. The screening assay was based on a 96-well format where the wells were coated with recombinant nNOS (1–299) followed by addition of biotinylated PSD-95. The small molecule inhibitors were then added in increasing concentrations to the nNOS/PSD-95 complex and the affinity was detected using fluorescence. IC87201 was determined to block the interaction between nNOS and PSD-95 with an IC_{50} of 31 μM. The in vitro screen was followed up by determining the rate of NMDA-induced increase in cGMP production, which is an indirect measurement of NO production, in neuronal cell culture and in primary rat hippocampal neurons. IC87201 was shown to dose-dependently inhibit the cGMP production with an IC_{50} of 2.7 μM. IC87201 was further characterized in vivo using several hyperalgesia models, such as, mouse thermal hyperalgesia model using intrathecal (i.t.) administration, tail flicking analgesia model, NMDA-induced nociceptive behavioral responses, and mechanical allodynia in rats with chronic construction of the sciatic nerve (CCI), which concluded that the IC87201-induced inhibition of thermal hyperalgesia could not be explained by inhibition of normal nociception. Pharmacokinetic studies in mice showed that peak plasma level was reached after 15 min following intraperitoneal (i.p.) administration, whereas the behavioral responses peaked after 45 min, suggested being due to the time needed for IC87201 to reach its target [115]. Lee et al. later verified that IC87201 disrupts the interaction between PSD-95 (1–392) and nNOS (1–199), with an EC_{50} of 23.9 μM, by using the fluorescent-based in vitro assay AlphaScreen [116].

2.4.2 ZL006

ZL006 is the result of further optimization of IC87201, Fig. 6.5, and structurally similar compared to IC87201 [117]. ZL006 has been described to be an efficient inhibitor of the PSD-95/nNOS complex and has shown promising results in vivo using ischemic stroke (tMCAO), depression (forced swimming test and tail suspension test), and regenerative repair after stroke (tMCAO) models [117–119].

In addition, ZL006 has also been shown to induce neurite outgrowth in vitro and inhibit oxidative stress and promote neuroprotection in Parkinson's disease (PD) models [120, 121]. Recently, Lee et al. followed up with an in vitro AlphaScreen binding study by using PSD-95 (1–392) and nNOS (1–299). The IC_{50} of ZL006 was determined to 12.9 μM [116]. ZL006 has, however, showed low cellular uptake [122], and to further enhance the therapeutic effect of ZL006, due to the poor permeability across the blood-brain barrier, Wang et al. used T7-conjugated PEGylated liposomes loaded with ZL006 (T7-P-LPs/ZL006) to increase the transferrin receptor-mediated endocytosis. The ZL006 loaded T7-P-LP liposomes showed a time and concentration-dependent increase in cellular uptake using brain capillary endothelial cells (BCEC), two to threefold increase in the in vivo biodistribution in the brain after i.v. administration, and significant reduced infarct volume in rats after tMCAO, compared to free ZL006, thus improving the therapeutic effect [122]. However, the mechanism of action of ZL006 has not been well-studied and the proposed inhibition mechanism is suggested to be a direct binding of ZL006 to the extended nNOS PDZ β-finger, which binds PSD-95, and thus uncouples the interaction between nNOS and PSD-95 [116]. It has previously been shown that a salt bridge interaction between Arg121 (in the nNOS β-finger) and Asp62 (in the nNOS PDZ domain) is crucial for the interaction between nNOS and PSD-95, as it has a stabilizing effect on the secondary structure of the β-finger [123]. As such, Zhou et al. have proposed that the carboxylate of ZL006 interrupts the salt bridge interaction between Arg121 and Asp62 [117]. To gain further insights into the ZL006 binding mechanism, Bach et al. utilized several biochemical and biophysical methods, but could not show that ZL006 directly interacts with the PDZ domains of nNOS (12–130) or PSD-95, nor inhibits the interaction between nNOS/PSD-95 [124]. This indicates that the mechanism of action of ZL006 might be through an indirect nNOS effect, rather a direct interaction between ZL006, nNOS, and PSD-95.

2.4.3 Tramiprosate

Tramiprosate (homotaurine, 3-amino-1-propane sulfonic acid (3-APS)) is a small aminosulfonate compound derived from red marine alga [125], Fig. 6.5. It has previously been investigated as a potential drug candidate for the treatment of Alzheimer's disease (AD), but failed in Phase III clinical trials [126]. Wu et al. further characterized the adverse effects of tramiprosate by investigating the neuroprotective effect on ischemic-induced brain damage and its modulatory effect on the PSD-95/nNOS/NMDA receptor complex. Tramiprosate was shown to dose-dependently reduce infarct volume in tMCAO-induced rats with a protective effective dose at 6 mg/kg and with a therapeutic time window (50 mg/kg) up to 6 h for cerebral ischemia. To investigate the potential effect on the PSD-95/nNOS/NMDA receptor complex by activating NMDA receptor, the non-selective NMDA receptor subunit antagonist MK801 (1 mg/kg) was administrated 4 h post MCAO. Tramiprosate alone reduced the infarct volume, whereas the co-administration of MK801 and tramiprosate abolished the effect, indicating that effect of tramiprosate is mediated by the NMDA receptor pathway. To further investigate the importance of PSD-95/nNOS/NMDA receptor complex, Wu et al. performed co-immunoprecipitation on lysate from rat brain tissue from the ipsilateral ischemic area 24 h post MCAO, primary neurons, and PC12 cells. Tramiprosate treatment showed that the translocation of nNOS to the membrane from the cytosol was significantly inhibited without affecting the total nNOS expression level in vivo and in vitro, indicating that neuroprotective effect of tramiprosate is potentially driven by the modulation of the PSD-95/nNOS interaction and nNOS translocation [127].

2.4.4 4-Phenyl-1-(4-Phenylbutyl)-Piperidine

The sigma receptor agonist 4-phenyl-1-(4-phenylbutyl)-piperidine (PPBP), Fig. 6.5, has also been shown to decrease the membrane recruitment of nNOS via modulating the PSD-95/nNOS interaction without affecting the NMDA receptor/PSD-95 interaction. Goyagu et al. found that continuous i.v. administration of PPBP after tMCAO attenuates infarct volume in rat. In addition, the [^{14}C]-citrulline recovery in the striatum was markedly reduced, which is an indirect measurement of reduced NO production, as NO is the side product of NOS-catalyzed conversion of arginine to citrulline. The decrease of infarct volume following PPBP treatment reflected the effect of 7-nitroindazole (7-NI), an inhibitor of nNOS. However, combining PPBP and 7-NI did not further alter the infarct volume. To get insights into the mechanism, Goyagu et al. used genetically deficient nNOS mice (nNOS knockout (KO)). Upon PPBP treatment, the nNOS KO mice showed a decrease in infarct volume similar to the decrease for WT animals treated with PPBP, indicating that PPBP likely acts upstream of neuronal NO production [128]. Yang et al. has further investigated the effect of PPBP by showing that a low continuously i.v. dose of PPBP has a neuroprotective effect in newborn piglets after hypoxia–ischemia brain injury, to the same effect as seen for MK801. However, a high dose of PPBP had only marginally improved effects. The authors speculate that this could be due to the side effects generated by a complete block of the NMDA receptor. In addition, a high dose of PPBP could potentially also affect the sigma-1 receptor and its interaction partners. Yang et al. also showed that the nNOS protein expression in the membrane fraction of putamen was significantly induced 3 h after induced hypoxia–ischemia in newborn piglets (H–I) and that this increase could be reduced by treatment of PPBP. However, the expression of nNOS in the cytosolic fraction was not affected 3 h after H–I or by PPBP treatment. To further investigate if the altered nNOS expression levels after H–I were related to the PSD-95/nNOS/NMDA receptor interaction, Yang et al. could show by co-immunoprecipitation that the level of PSD-95 coupled to nNOS, but not the amount PSD-95 coupled to GluN2A/2B, was significantly increased after H–I and further on that this increase could be reduced by PPBP treatment. However, the interactions between nNOS/PSD-95 and PSD-95/GluN2A/B were not affected in prefrontal cortex after H–I or PPBP treatment. This tissue-specific mechanism of anchoring nNOS to PSD-95 was shown to be regulated by phosphorylation of nNOS at Ser847. Interestingly, PPBP treatment reduced the phosphorylation level of Ser847. Thus, it is still up for debate if PPBP directly regulates the PSD-95/nNOS interaction or if the neuroprotective effect is mediated via a yet undefined indirect mechanism [129].

2.4.5 Honokiol

Honokiol is a natural biphenol compound, Fig. 6.5, which can be extracted and isolated from bark from several species of the magnolia family [130]. Honokiol has previously been shown to have anxiolytic, analgesic, anti-depressant, anti-inflammatory, anti-thrombic, anti-microbial, anti-tumorigenic, and neuroprotective properties [130]. Furthermore, multiple putative pathways for modulation of the neuroprotective effect have been investigated and relate to its described modulatory effect on $GABA_A$ receptor, NFkB, and NMDA receptor [131–134]. In 2013, Hu et al. investigated the effect of honokiol on the NMDA receptor-signaling pathway and showed that honokiol significantly protected rat cortical neurons from glutamate and oxygen-glucose-deprived (OGD)-induced injury. Hence, Hu et al. further investigated the effect of honokiol on the PSD-95/nNOS/NMDA receptor complex and found that honokiol disrupts the nNOS/PSD-95 interaction but not the PSD-95/NMDA receptor and inhibits the translocation of nNOS from the cytosol to the membrane, without affecting the total expression of nNOS in cultured cortical neurons. Furthermore, neurons pretreated with 10 μM honokiol for 1 h significantly inhibited the toxic NO production in neurons challenged with glutamate, which further supports a modulating effect of honokiol [135]. Nonetheless, there is no detailed information regarding the binding site for honokiol, nor any data showing a direct interaction between nNOS and/or PSD-95 and honokiol.

3 Conclusion and Outlook

Ischemic stroke is certainly among the largest and most challenging unmet medical needs in the Western world today. Historically, developing safe and efficient drugs to treat ischemic stroke has for many reasons been exceedingly difficult. In recent years, PSD-95 has emerged as one of the most promising strategies for neuroprotection and potential treatment of ischemic stroke and currently two compounds, NA-1 and AVLX-144, have shown particular promise. Thus, targeting PSD-95, combined with exciting developments in clinical trials of ischemic stroke, seems to hold great promise.

References

1. Sattler R, Tymianski M (2001) Molecular mechanisms of glutamate receptor-mediated excitotoxic neuronal cell death. Mol Neurobiol 24(1–3):107–129
2. Mehta SL, Manhas N, Raghubir R (2007) Molecular targets in cerebral ischemia for developing novel therapeutics. Brain Res Rev 54(1):34–66
3. Meldrum BS (2000) Glutamate as a neurotransmitter in the brain: review of physiology and pathology. J Nutr 130(4):1007S–1015S
4. Moussawi K, Riegel A, Nair S, Kalivas PW (2011) Extracellular glutamate: functional compartments operate in different concentration ranges. Front Syst Neurosci 5:94
5. Koh JY, Goldberg MP, Hartley DM, Choi DW (1990) Non-NMDA receptor-mediated neurotoxicity in cortical culture. J Neurosci 10(2):693–705
6. Monaghan DT, Holets VR, Toy DW, Cotman CW (1983) Anatomical distributions of four pharmacologically distinct ^{3}H-L-glutamate binding sites. Nature 306(5939):176–179
7. Faden AI, Demediuk P, Panter SS, Vink R (1989) The role of excitatory amino acids and NMDA receptors in traumatic brain injury. Science 244(4906):798–800
8. Choi DW (1988) Glutamate neurotoxicity and diseases of the nervous system. Neuron 1(8):623–634
9. Grotta J, Clark W, Coull B, Pettigrew LC, Mackay B, Goldstein LB et al (1995) Safety and tolerability of the glutamate antagonist CGS 19755 (Selfotel) in patients with acute ischemic stroke. Results of a phase IIa randomized trial. Stroke 26(4):602–605
10. Sveinbjornsdottir S, Sander JW, Upton D, Thompson PJ, Patsalos PN, Hirt D et al (1993) The excitatory amino acid antagonist D-CPP-ene (SDZ EAA-494) in patients with epilepsy. Epilepsy Res 16(2):165–174
11. Albers GW, Goldstein LB, Hall D, Lesko LM, Aptiganel Acute Stroke I (2001) Aptiganel hydrochloride in acute ischemic stroke: a randomized controlled trial. JAMA 286(21): 2673–2682
12. Albers GW, Atkinson RP, Kelley RE, Rosenbaum DM (1995) Safety, tolerability, and pharmacokinetics of the *N*-methyl-D-aspartate antagonist dextrorphan in patients with acute stroke. Dextrorphan Study Group. Stroke 26(2):254–258
13. Dyker AG, Lees KR (1999) Remacemide hydrochloride: a double-blind, placebo-controlled, safety and tolerability study in patients with acute ischemic stroke. Stroke 30(9):1796–1801
14. Sacco RL, DeRosa JT, Haley EC Jr, Levin B, Ordronneau P, Phillips SJ et al (2001) Glycine antagonist in neuroprotection for patients with acute stroke: GAIN Americas: a randomized controlled trial. JAMA 285(13):1719–1728
15. Albers GW, Clark WM, Atkinson RP, Madden K, Data JL, Whitehouse MJ (1999) Dose escalation study of the NMDA glycine-site antagonist licostinel in acute ischemic stroke. Stroke 30(3):508–513
16. De Keyser J, Sulter G, Luiten PG (1999) Clinical trials with neuroprotective drugs in acute ischaemic stroke: are we doing the right thing? Trends Neurosci 22(12):535–540
17. Lai TW, Wang YT (2010) Fashioning drugs for stroke. Nat Med 16(12):1376–1378
18. Diener HC, AlKhedr A, Busse O, Hacke W, Zingmark PH, Jonsson N et al (2002) Treatment of acute ischaemic stroke with the low-affinity, use-dependent NMDA antagonist AR-R15896AR. A safety and tolerability study. J Neurol 249(5):561–568
19. Lees KR, Dyker AG, Sharma A, Ford GA, Ardron ME, Grosset DG (2001) Tolerability of the low-affinity, use-dependent NMDA antagonist AR-R15896AR in stroke patients: a dose-ranging study. Stroke 32(2):466–472
20. Lai TW, Zhang S, Wang YT (2014) Excitotoxicity and stroke: identifying novel targets for neuroprotection. Prog Neurobiol 115:157–188
21. Dick O, Bading H (2010) Synaptic activity and nuclear calcium signaling protect hippocampal neurons from death signal-associated nuclear translocation of FoxO3a induced by extrasynaptic *N*-methyl-D-aspartate receptors. J Biol Chem 285(25):19354–19361

22. Zhang SJ, Steijaert MN, Lau D, Schutz G, Delucinge-Vivier C, Descombes P et al (2007) Decoding NMDA receptor signaling: identification of genomic programs specifying neuronal survival and death. Neuron 53(4):549–562
23. Jo H, Mondal S, Tan D, Nagata E, Takizawa S, Sharma AK et al (2012) Small molecule-induced cytosolic activation of protein kinase Akt rescues ischemia-elicited neuronal death. Proc Natl Acad Sci U S A 109(26):10581–10586
24. Hardingham GE, Fukunaga Y, Bading H (2002) Extrasynaptic NMDARs oppose synaptic NMDARs by triggering CREB shut-off and cell death pathways. Nat Neurosci 5(5):405–414
25. Tu W, Xu X, Peng L, Zhong X, Zhang W, Soundarapandian MM et al (2010) DAPK1 interaction with NMDA receptor NR2B subunits mediates brain damage in stroke. Cell 140(2):222–234
26. Wilbek TS, Skovgaard T, Sorrell FJ, Knapp S, Berthelsen J, Strømgaard K (2015) Identification and characterization of a small-molecule inhibitor of death-associated protein kinase 1. Chembiochem 16(1):59–63
27. Brorson JR, Marcuccilli CJ, Miller RJ (1995) Delayed antagonism of calpain reduces excitotoxicity in cultured neurons. Stroke 26(7):1259–1266
28. Nimmrich V, Szabo R, Nyakas C, Granic I, Reymann KG, Schroder UH et al (2008) Inhibition of calpain prevents *N*-methyl-D-aspartate-induced degeneration of the nucleus basalis and associated behavioral dysfunction. J Pharmacol Exp Ther 327(2):343–352
29. Aarts M, Liu Y, Liu L, Besshoh S, Arundine M, Gurd JW et al (2002) Treatment of ischemic brain damage by perturbing NMDA receptor-PSD-95 protein interactions. Science 298 (5594):846–850
30. Bach A, Chi CN, Pang GF, Olsen L, Kristensen AS, Jemth P et al (2009) Design and synthesis of highly potent and plasma-stable dimeric inhibitors of the PSD-95-NMDA receptor interaction. Angew Chem Int Ed 121(51):9865–9869
31. Bach A, Clausen BH, Møller M, Vestergaard B, Chi CN, Round A et al (2012) A high-affinity, dimeric inhibitor of PSD-95 bivalently interacts with PDZ1-2 and protects against ischemic brain damage. Proc Natl Acad Sci U S A 109(9):3317–3322
32. Sheng M, Kim E (2011) The postsynaptic organization of synapses. Cold Spring Harb Perspect Biol 3(12):1–20
33. Trinidad JC, Thalhammer A, Specht CG, Lynn AJ, Baker PR, Schoepfer R et al (2008) Quantitative analysis of synaptic phosphorylation and protein expression. Mol Cell Proteomics 7(4):684–696
34. Bayes A, van de Lagemaat LN, Collins MO, Croning MD, Whittle IR, Choudhary JS et al (2011) Characterization of the proteome, diseases and evolution of the human postsynaptic density. Nat Neurosci 14(1):19–21
35. Jordan BA, Fernholz BD, Boussac M, Xu C, Grigorean G, Ziff EB et al (2004) Identification and verification of novel rodent postsynaptic density proteins. Mol Cell Proteomics 3(9):857–871
36. Kim E, Sheng M (2004) PDZ domain proteins of synapses. Nat Rev Neurosci 5(10):771–781
37. Husi H, Ward MA, Choudhary JS, Blackstock WP, Grant SG (2000) Proteomic analysis of NMDA receptor-adhesion protein signaling complexes. Nat Neurosci 3(7):661–669
38. Dosemeci A, Makusky AJ, Jankowska-Stephens E, Yang X, Slotta DJ, Markey SP (2007) Composition of the synaptic PSD-95 complex. Mol Cell Proteomics 6(10):1749–1760
39. Zhang W, Vazquez L, Apperson M, Kennedy MB (1999) Citron binds to PSD-95 at glutamatergic synapses on inhibitory neurons in the hippocampus. J Neurosci 19(1):96–108
40. Cheng D, Hoogenraad CC, Rush J, Ramm E, Schlager MA, Duong DM et al (2006) Relative and absolute quantification of postsynaptic density proteome isolated from rat forebrain and cerebellum. Mol Cell Proteomics 5(6):1158–1170
41. Bliss TV, Collingridge GL (1993) A synaptic model of memory: long-term potentiation in the hippocampus. Nature 361(6407):31–39
42. Traynelis SF, Wollmuth LP, McBain CJ, Menniti FS, Vance KM, Ogden KK et al (2010) Glutamate receptor ion channels: structure, regulation, and function. Pharmacol Rev 62(3): 405–496

43. Cull-Candy SG, Leszkiewicz DN (2004) Role of distinct NMDA receptor subtypes at central synapses. Sci Signal 2004(255):1–9
44. Sheng M, Cummings J, Roldan LA, Jan YN, Jan LY (1994) Changing subunit composition of heteromeric NMDA receptors during development of rat cortex. Nature 368(6467): 144–147
45. Monyer H, Burnashev N, Laurie DJ, Sakmann B, Seeburg PH (1994) Developmental and regional expression in the rat brain and functional properties of four NMDA receptors. Neuron 12(3):529–540
46. Paoletti P, Bellone C, Zhou Q (2013) NMDA receptor subunit diversity: impact on receptor properties, synaptic plasticity and disease. Nat Rev Neurosci 14(6):383–400
47. Mony L, Kew JN, Gunthorpe MJ, Paoletti P (2009) Allosteric modulators of NR2B-containing NMDA receptors: molecular mechanisms and therapeutic potential. Br J Pharmacol 157(8):1301–1317
48. Kleckner NW, Dingledine R (1988) Requirement for glycine in activation of NMDA-receptors expressed in Xenopus oocytes. Science 241(4867):835–837
49. Furukawa H, Singh SK, Mancusso R, Gouaux E (2005) Subunit arrangement and function in NMDA receptors. Nature 438(7065):185–192
50. Burnashev N, Schoepfer R, Monyer H, Ruppersberg JP, Gunther W, Seeburg PH et al (1992) Control by asparagine residues of calcium permeability and magnesium blockade in the NMDA receptor. Science 257(5075):1415–1419
51. Zhang J, Lewis SM, Kuhlman B, Lee AL (2013) Supertertiary structure of the MAGUK core from PSD-95. Structure 21(3):402–413
52. Pan L, Chen J, Yu J, Yu H, Zhang M (2011) The structure of the PDZ3-SH3-GuK tandem of ZO-1 protein suggests a supramodular organization of the membrane-associated guanylate kinase (MAGUK) family scaffold protein core. J Biol Chem 286(46):40069–40074
53. Long JF, Tochio H, Wang P, Fan JS, Sala C, Niethammer M et al (2003) Supramodular structure and synergistic target binding of the N-terminal tandem PDZ domains of PSD-95. J Mol Biol 327(1):203–214
54. McCann JJ, Zheng L, Rohrbeck D, Felekyan S, Kuhnemuth R, Sutton RB et al (2012) Supertertiary structure of the synaptic MAGuK scaffold proteins is conserved. Proc Natl Acad Sci U S A 109(39):15775–15780
55. Chen X, Nelson CD, Li X, Winters CA, Azzam R, Sousa AA et al (2011) PSD-95 is required to sustain the molecular organization of the postsynaptic density. J Neurosci 31(17): 6329–6338
56. Chatr-Aryamontri A, Breitkreutz BJ, Oughtred R, Boucher L, Heinicke S, Chen D et al (2015) The BioGRID interaction database: 2015 update. Nucleic Acids Res 43:D470–D478
57. Sattler R, Xiong Z, Lu WY, Hafner M, MacDonald JF, Tymianski M (1999) Specific coupling of NMDA receptor activation to nitric oxide neurotoxicity by PSD-95 protein. Science 284 (5421):1845–1848
58. Kornau HC, Schenker LT, Kennedy MB, Seeburg PH (1995) Domain interaction between NMDA receptor subunits and the postsynaptic density protein PSD-95. Science 269(5231):1737–1740
59. Niethammer M, Kim E, Sheng M (1996) Interaction between the C terminus of NMDA receptor subunits and multiple members of the PSD-95 family of membrane-associated guanylate kinases. J Neurosci 16(7):2157–2163
60. Christopherson KS, Hillier BJ, Lim WA, Bredt DS (1999) PSD-95 assembles a ternary complex with the *N*-methyl-D-aspartic acid receptor and a bivalent neuronal NO synthase PDZ domain. J Biol Chem 274(39):27467–27473
61. Dawson TM, Dawson VL (1996) Nitric oxide synthase: role as a transmitter/mediator in the brain and endocrine system. Annu Rev Med 47:219–227
62. Dawson VL, Dawson TM, London ED, Bredt DS, Snyder SH (1991) Nitric oxide mediates glutamate neurotoxicity in primary cortical cultures. Proc Natl Acad Sci U S A 88(14):6368–6371
63. Samdani AF, Dawson TM, Dawson VL (1997) Nitric oxide synthase in models of focal ischemia. Stroke 28(6):1283–1288

64. Eliasson MJ, Blackshaw S, Schell MJ, Snyder SH (1997) Neuronal nitric oxide synthase alternatively spliced forms: prominent functional localizations in the brain. Proc Natl Acad Sci U S A 94(7):3396–3401
65. Putzke J, Seidel B, Huang PL, Wolf G (2000) Differential expression of alternatively spliced isoforms of neuronal nitric oxide synthase (nNOS) and *N*-methyl-D-aspartate receptors (NMDAR) in knockout mice deficient in nNOSα (nNOSα$^{\Delta/\Delta}$ mice). Brain Res Mol Brain Res 85(1–2):13–23
66. Hillier BJ, Christopherson KS, Prehoda KE, Bredt DS, Lim WA (1999) Unexpected modes of PDZ domain scaffolding revealed by structure of nNOS-syntrophin complex. Science 284(5415):812–815
67. Tochio H, Zhang Q, Mandal P, Li M, Zhang M (1999) Solution structure of the extended neuronal nitric oxide synthase PDZ domain complexed with an associated peptide. Nat Struct Biol 6(5):417–421
68. Calabrese V, Mancuso C, Calvani M, Rizzarelli E, Butterfield DA, Stella AM (2007) Nitric oxide in the central nervous system: neuroprotection versus neurotoxicity. Nat Rev Neurosci 8(10):766–775
69. Boissel JP, Bros M, Schrock A, Godtel-Armbrust U, Forstermann U (2004) Cyclic AMP-mediated upregulation of the expression of neuronal NO synthase in human A673 neuroepithelioma cells results in a decrease in the level of bioactive NO production: analysis of the signaling mechanisms that are involved. Biochemistry 43(22):7197–7206
70. Schulz JB, Matthews RT, Klockgether T, Dichgans J, Beal MF (1997) The role of mitochondrial dysfunction and neuronal nitric oxide in animal models of neurodegenerative diseases. Mol Cell Biochem 174(1–2):193–197
71. Pou S, Keaton L, Surichamorn W, Rosen GM (1999) Mechanism of superoxide generation by neuronal nitric-oxide synthase. J Biol Chem 274(14):9573–9580
72. Montaner J, Campos M, Cristobo I, Giralt D, Diaz-Guerra M (2013) Role of PSD-95 inhibitors in stroke and neuroprotection: a systematic view on NA-1 (Tat-NR2B9c). Drugs Future 38(7):485–497
73. Cook DJ, Teves L, Tymianski M (2012) Treatment of stroke with a PSD-95 inhibitor in the gyrencephalic primate brain. Nature 483(7388):213–217
74. Hill MD, Martin RH, Mikulis D, Wong JH, Silver FL, Terbrugge KG et al (2012) Safety and efficacy of NA-1 in patients with iatrogenic stroke after endovascular aneurysm repair (ENACT): a phase 2, randomised, double-blind, placebo-controlled trial. Lancet Neurol 11(11):942–950
75. Cui H, Hayashi A, Sun H-S, Belmares MP, Cobey C, Phan T et al (2007) PDZ protein interactions underlying NMDA receptor-mediated excitotoxicity and neuroprotection by PSD-95 inhibitors. J Neurosci 27(37):9901–9915
76. Wells JA, McClendon CL (2007) Reaching for high-hanging fruit in drug discovery at protein-protein interfaces. Nature 450(7172):1001–1009
77. Nevola L, Giralt E (2015) Modulating protein-protein interactions: the potential of peptides. Chem Commun 51(16):3302–3315
78. Chi CN, Bach A, Strømgaard K, Gianni S, Jemth P (2012) Ligand binding by PDZ domains. Biofactors 38(5):338–348
79. Doyle DA, Lee A, Lewis J, Kim E, Sheng M, MacKinnon R (1996) Crystal structures of a complexed and peptide-free membrane protein-binding domain: molecular basis of peptide recognition by PDZ. Cell 85(7):1067–1076
80. Tochio H, Hung F, Li M, Bredt DS, Zhang M (2000) Solution structure and backbone dynamics of the second PDZ domain of postsynaptic density-95. J Mol Biol 295(2):225–237
81. Pedersen SW, Pedersen SB, Anker L, Hultqvist G, Kristensen AS, Jemth P et al (2014) Probing backbone hydrogen bonding in PDZ/ligand interactions by protein amide-to-ester mutations. Nat Commun 5:3215
82. Stiffler MA, Chen JR, Grantcharova VP, Lei Y, Fuchs D, Allen JE et al (2007) PDZ domain binding selectivity is optimized across the mouse proteome. Science 317(5836):364–369
83. Songyang Z, Fanning AS, Fu C, Xu J, Marfatia SM, Chishti AH et al (1997) Recognition of unique carboxyl-terminal motifs by distinct PDZ domains. Science 275(5296):73–77

84. Bach A, Chi CN, Olsen TB, Pedersen SW, Røder MU, Pang GF et al (2008) Modified peptides as potent inhibitors of the postsynaptic density-95/*N*-methyl-D-aspartate receptor interaction. J Med Chem 51(20):6450–6459
85. Lim IA, Hall DD, Hell JW (2002) Selectivity and promiscuity of the first and second PDZ domains of PSD-95 and synapse-associated protein 102. J Biol Chem 277(24):21697–21711
86. Eildal JN, Hultqvist G, Balle T, Stuhr-Hansen N, Padrah S, Gianni S et al (2013) Probing the role of backbone hydrogen bonds in protein-peptide interactions by amide-to-ester mutations. J Am Chem Soc 135(35):12998–13007
87. Brenman JE, Chao DS, Gee SH, McGee AW, Craven SE, Santillano DR et al (1996) Interaction of nitric oxide synthase with the postsynaptic density protein PSD-95 and α1-syntrophin mediated by PDZ domains. Cell 84(5):757–767
88. Ginsberg MD (2008) Neuroprotection for ischemic stroke: past, present and future. Neurophar macology 55(3):363–389
89. Sydserff SG, Borelli AR, Green AR, Cross AJ (2002) Effect of NXY-059 on infarct volume after transient or permanent middle cerebral artery occlusion in the rat; studies on dose, plasma concentration and therapeutic time window. Br J Pharmacol 135(1):103–112
90. Zhao Z, Cheng M, Maples KR, Ma JY, Buchan AM (2001) NXY-059, a novel free radical trapping compound, reduces cortical infarction after permanent focal cerebral ischemia in the rat. Brain Res 909(1–2):46–50
91. Shuaib A, Lees KR, Lyden P, Grotta J, Davalos A, Davis SM et al (2007) NXY-059 for the treatment of acute ischemic stroke. N Engl J Med 357(6):562–571
92. Carmichael ST (2005) Rodent models of focal stroke: size, mechanism, and purpose. NeuroRX 2(3):396–409
93. Kastrup A, Engelhorn T, Beaulieu C, de Crespigny A, Moseley ME (1999) Dynamics of cerebral injury, perfusion, and blood-brain barrier changes after temporary and permanent middle cerebral artery occlusion in the rat. J Neurol Sci 166(2):91–99
94. Sun H-S, Doucette TA, Liu Y, Fang Y, Teves L, Aarts M et al (2008) Effectiveness of PSD95 inhibitors in permanent and transient focal ischemia in the rat. Stroke 39(9):2544–2553
95. Bråtane BT, Cui H, Cook DJ, Bouley J, Tymianski M, Fisher M (2011) Neuroprotection by freezing ischemic penumbra evolution without cerebral blood flow augmentation with a postsynaptic density-95 protein inhibitor. Stroke 42(11):3265–3270
96. Soriano FX, Martel M-A, Papadia S, Vaslin A, Baxter P, Rickman C et al (2008) Specific targeting of pro-death NMDA receptor signals with differing reliance on the NR2B PDZ ligand. J Neurosci 28(42):10696–10710
97. Roitberg B, Khan N, Tuccar E, Kompoliti K, Chu Y, Alperin N et al (2003) Chronic ischemic stroke model in cynomolgus monkeys: behavioral, neuroimaging and anatomical study. Neurol Res 25(1):68–78
98. Cook DJ, Teves L, Tymianski M (2012) A translational paradigm for the preclinical evaluation of the stroke neuroprotectant Tat-NR2B9c in gyrencephalic nonhuman primates. Sci Transl Med 4(154):1–8
99. Wang W, Weng J, Zhang X, Liu M, Zhang M (2009) Creating conformational entropy by increasing interdomain mobility in ligand binding regulation: a revisit to N-terminal tandem PDZ domains of PSD-95. J Am Chem Soc 131(2):787–796
100. Bard L, Sainlos M, Bouchet D, Cousins S, Mikasova L, Breillat C et al (2010) Dynamic and specific interaction between synaptic NR2-NMDA receptor and PDZ proteins. Proc Natl Acad Sci U S A 107(45):19561–19566
101. Wen W, Wang W, Zhang M (2006) Targeting PDZ domain proteins for treating NMDA receptor-mediated excitotoxicity. Curr Top Med Chem 6(7):711–721
102. Teves LM, Cui H, Tymianski M (2015) Efficacy of the PSD95 inhibitor Tat-NR2B9c in mice requires dose translation between species. J Cereb Blood Flow Metab 36(3):555–561
103. Eildal JNN, Bach A, Dogan J, Ye F, Zhang M, Jemth P et al (2015) Rigidified clicked dimeric ligands for studying the dynamics of the PDZ1-2 supramodule of PSD-95. Chembiochem 16(1):64–69

104. Sainlos M, Tigaret C, Poujol C, Olivier NB, Bard L, Breillat C et al (2011) Biomimetic divalent ligands for the acute disruption of synaptic AMPAR stabilization. Nat Chem Biol 7(2):81–91
105. Nissen KB, Haugaard-Kedström LM, Wilbek TS, Nielsen LS, Åberg E, Kristensen AS et al (2015) Targeting protein-protein interactions with trimeric ligands: high affinity inhibitors of the MAGUK protein family. PLoS One 10(2):e0117668
106. Nissen KB, Andersen JJ, Haugaard-Kedström LM, Bach A, Strømgaard K (2015) Design, synthesis, and characterization of fatty acid derivatives of a dimeric peptide-based postsynaptic density-95 (PSD-95) inhibitor. J Med Chem 58(3):1575–1580
107. Knudsen LB, Nielsen PF, Huusfeldt PO, Johansen NL, Madsen K, Pedersen FZ et al (2000) Potent derivatives of glucagon-like peptide-1 with pharmacokinetic properties suitable for once daily administration. J Med Chem 43(9):1664–1669
108. Giannis A, Rübsam F (1997) Peptidomimetics in drug design. In: Bernard T, Urs AM (eds) Advances in drug research, vol 29. Academic Press, New York, pp 1–78
109. Craik DJ, Fairlie DP, Liras S, Price D (2013) The future of peptide-based drugs. Chem Biol Drug Des 81(1):136–147
110. Arkin MR, Wells JA (2004) Small-molecule inhibitors of protein-protein interactions: progressing towards the dream. Nat Rev Drug Discov 3(4):301–317
111. Bach A, Eildal JNN, Stuhr-Hansen N, Deeskamp R, Gottschalk M, Pedersen SW et al (2011) Cell-permeable and plasma-stable peptidomimetic inhibitors of the postsynaptic density-95/*N*-methyl-D-aspartate receptor interaction. J Med Chem 54(5):1333–1346
112. Udugamasooriya DG, Sharma SC, Spaller MR (2008) A chemical library approach to organic-modified peptide ligands for PDZ domain proteins: a synthetic, thermodynamic and structural investigation. Chembiochem 9(10):1587–1589
113. Memic A, Spaller MR (2008) How do halogen substituents contribute to protein-binding interactions? A thermodynamic study of peptide ligands with diverse aryl halides. Chembiochem 9(17):2793–2795
114. Piserchio A, Salinas GD, Li T, Marshall J, Spaller MR, Mierke DF (2004) Targeting specific PDZ domains of PSD-95: structural basis for enhanced affinity and enzymatic stability of a cyclic peptide. Chem Biol 11(4):469–473
115. Florio SK, Loh C, Huang SM, Iwamaye AE, Kitto KF, Fowler KW et al (2009) Disruption of nNOS-PSD95 protein-protein interaction inhibits acute thermal hyperalgesia and chronic mechanical allodynia in rodents. Br J Pharmacol 158(2):494–506
116. Lee WH, Xu Z, Ashpole NM, Hudmon A, Kulkarni PM, Thakur GA et al (2015) Small molecule inhibitors of PSD95-nNOS protein-protein interactions as novel analgesics. Neuropharmacology 97:464–475
117. Zhou L, Li F, Xu HB, Luo CX, Wu HY, Zhu MM et al (2010) Treatment of cerebral ischemia by disrupting ischemia-induced interaction of nNOS with PSD-95. Nat Med 16(12):1439–1443
118. Doucet MV, Levine H, Dev KK, Harkin A (2013) Small-molecule inhibitors at the PSD-95/nNOS interface have antidepressant-like properties in mice. Neuropsychopharmacology 38(8):1575–1584
119. Luo CX, Lin YH, Qian XD, Tang Y, Zhou HH, Jin X et al (2014) Interaction of nNOS with PSD-95 negatively controls regenerative repair after stroke. J Neurosci 34(40):13535–13548
120. Doucet MV, O'Toole E, Connor T, Harkin A (2015) Small-molecule inhibitors at the PSD-95/nNOS interface protect against glutamate-induced neuronal atrophy in primary cortical neurons. Neuroscience 301:421–438
121. Hu W, Guan LS, Dang XB, Ren PY, Zhang YL (2014) Small-molecule inhibitors at the PSD-95/nNOS interface attenuate MPP^+-induced neuronal injury through Sirt3 mediated inhibition of mitochondrial dysfunction. Neurochem Int 79:57–64
122. Wang Z, Zhao Y, Jiang Y, Lv W, Wu L, Wang B et al (2015) Enhanced anti-ischemic stroke of ZL006 by T7-conjugated PEGylated liposomes drug delivery system. Sci Rep 5:12651–12666
123. Tochio H, Mok YK, Zhang Q, Kan HM, Bredt DS, Zhang M (2000) Formation of nNOS/PSD-95 PDZ dimer requires a preformed β-finger structure from the nNOS PDZ domain. J Mol Biol 303(3):359–370

124. Bach A, Pedersen SW, Dorr LA, Vallon G, Ripoche I, Ducki S et al (2015) Biochemical investigations of the mechanism of action of small molecules ZL006 and IC87201 as potential inhibitors of the nNOS-PDZ/PSD-95-PDZ interactions. Sci Rep 5:12157
125. Miyazawa K, Matsumoto F (1970) Occurrence of D-2-hydroxy-3-aminopropane sulfonic acid and 3-aminopropane sulfonic acid in a red alga, *Grateloupia livida*. Nippon Suisan Gakkaishi 36(1):109–114
126. Aisen PS, Gauthier S, Ferris SH, Saumier D, Haine D, Garceau D et al (2011) Tramiprosate in mild-to-moderate Alzheimer's disease—a randomized, double-blind, placebo-controlled, multi-centre study (the Alphase Study). Arch Med Sci 7(1):102–111
127. Wu S, Yue Y, Tian H, Tao L, Wang Y, Xiang J et al (2014) Tramiprosate protects neurons against ischemic stroke by disrupting the interaction between PSD95 and nNOS. Neuropharmacology 83:107–117
128. Goyagi T, Goto S, Bhardwaj A, Dawson VL, Hurn PD, Kirsch JR (2001) Neuroprotective effect of σ1-receptor ligand 4-phenyl-1-(4-phenylbutyl) piperidine (PPBP) is linked to reduced neuronal nitric oxide production. Stroke 32(7):1613–1620
129. Yang ZJ, Carter EL, Torbey MT, Martin LJ, Koehler RC (2010) Sigma receptor ligand 4-phe nyl-1-(4-phenylbutyl)-piperidine modulates neuronal nitric oxide synthase/postsynaptic density-95 coupling mechanisms and protects against neonatal ischemic degeneration of striatal neurons. Exp Neurol 221(1):166–174
130. Woodbury A, Yu SP, Wei L, Garcia P (2013) Neuro-modulating effects of honokiol: a review. Front Neurol 4:130–136
131. Zhang P, Liu X, Zhu Y, Chen S, Zhou D, Wang Y (2013) Honokiol inhibits the inflammatory reaction during cerebral ischemia reperfusion by suppressing NF-kB activation and cytokine production of glial cells. Neurosci Lett 534:123–127
132. Liou KT, Shen YC, Chen CF, Tsao CM, Tsai SK (2003) Honokiol protects rat brain from focal cerebral ischemia-reperfusion injury by inhibiting neutrophil infiltration and reactive oxygen species production. Brain Res 992(2):159–166
133. Harada S, Kishimoto M, Kobayashi M, Nakamoto K, Fujita-Hamabe W, Chen HH et al (2012) Honokiol suppresses the development of post-ischemic glucose intolerance and neuronal damage in mice. J Nat Med 66(4):591–599
134. Taferner B, Schuehly W, Huefner A, Baburin I, Wiesner K, Ecker GF et al (2011) Modulation of $GABA_A$-receptors by honokiol and derivatives: subtype selectivity and structure-activity relationship. J Med Chem 54(15):5349–5361
135. Hu Z, Bian X, Liu X, Zhu Y, Zhang X, Chen S et al (2013) Honokiol protects brain against ischemia-reperfusion injury in rats through disrupting PSD95-nNOS interaction. Brain Res 1491:204–212
136. Bell KFS, Bent RJ, Meese-Tamuri S, Ali A, Forder JP, Aarts MM (2013) Calmodulin kinase IV-dependent CREB activation is required for neuroprotection via NMDA receptor-PSD95 disruption. J Neurochem 126(2):274–287

Chapter 7
ATP-Sensitive Potassium Channels (K_{ATP}) Play a Role in Hypoxic Preconditioning Against Neonatal Hypoxic-Ischemic Brain Injury

Zhong-Ping Feng and Hong-Shuo Sun

Abstract ATP-sensitive potassium (K_{ATP}) channels are known as the potassium-conducting channels coupling cellular metabolic status to membrane electrical activity. Either an increase in ADP or decrease in ATP levels opens K_{ATP} channels and hyperpolarizes the membrane potential. Knocking out the inward rectifier K^+ channel (Kir6.2) subunit of the K_{ATP} channels or pharmacologically blocking K_{ATP} channels increases brain injury. Overexpression of the Kir6.2 subunit or pharmacologically opening K_{ATP} channel reduces neuronal injury from ischemic insults. Hypoxic preconditioning (HPC) provides neuroprotection against subsequent ischemic brain injury. Similar to its effects in heart, K_{ATP} channels contribute to the hypoxic preconditioning-induced neuroprotection. K_{ATP} channels may therefore serve as therapeutic targets in ischemic or hypoxic-ischemic brain injury.

Keywords Ion channels • K_{ATP} channels • Neuroprotection • Hypoxic preconditioning • Hypoxic-ischemic brain injury • In-vivo

Z.-P. Feng (✉)
Department of Physiology, Faculty of Medicine, University of Toronto,
1 King's College Circle, Toronto, Ontario, Canada, M5S 1A8
e-mail: zp.feng@utoronto.ca

H.-S. Sun, M.D., M.Sc., Ph.D.
Department of Physiology, Faculty of Medicine, University of Toronto,
1 King's College Circle, Toronto, Ontario, Canada, M5S 1A8

Department of Surgery, Faculty of Medicine, University of Toronto,
1 King's College Circle, Toronto, Ontario, Canada, M5S 1A8

Department of Pharmacology, Faculty of Medicine, University of Toronto,
1 King's College Circle, Toronto, Ontario, Canada, M5S 1A8

Institute of Medical Science, Faculty of Medicine, University of Toronto,
1 King's College Circle, Toronto, Ontario, Canada, M5S 1A8
e-mail: hss.sun@utoronto.ca

P.A. Lapchak, J.H. Zhang (eds.), *Neuroprotective Therapy for Stroke and Ischemic Disease*, Springer Series in Translational Stroke Research,
DOI 10.1007/978-3-319-45345-3_7

Abbreviations

ABC	ATP-binding cassette protein
ABCCx	ATP-binding cassette Cx gene
ADP	Adenosine diphosphate
ATP	Adenosine triphosphate
HPC	Hypoxic preconditioning
IPC	Ischemic preconditioning
K_{ATP}	ATP-sensitive potassium channels
KCN	Gene name for potassium channel
KCNJ11	Potassium channel J11 gene
Kir	Inward rectifier potassium channels
MCAO	Middle cerebral artery occlusion
NBD	Nucleotide binding domain
PIP2	Phosphatidylinositol 4,5-biphosphate
PKA	Protein kinase A
PKC	Protein kinase C
RT-PCR	Reverse transcription polymerase chain reaction
SNr	Substantia nigra
SUR	Sulfonylurea receptor
TM	Transmembrane region
TMD	Transmembrane domain

1 Introduction

Adenosine triphosphate (ATP)-sensitive potassium (K_{ATP}) channels conduct weak inward-rectifier potassium currents that are inhibited by intracellular ATP and linked the cellular metabolic status to the cell membrane electrical activity, thus regulating the cell function [1–4]. K_{ATP} channels belong to the Kir superfamily and were originally identified in the heart [1]. These channels were later reported in other tissues, including the skeletal muscle [5], smooth muscle [6], brain [7–10], pituitary gland [11, 12], kidney [13], and in pancreatic beta cells [2, 14–16]. Opening of K^+ channels hyperpolarizes the cell membrane potential toward the K^+ equilibrium potential (E_K), and thus, stabilizes the membrane [17]. A decrease in the ratio of ATP/ADP (adenosine diphosphate) opens K_{ATP} channels (leading to hyperpolarization), whereas an increase in the ATP/ADP ratio closes K_{ATP} channels (leading to depolarization). Thus, K_{ATP} channels are considered as a cellular metabolic sensor-regulating electrical activity of cell membranes.

K_{ATP} channels play a critical neuroprotective role in stroke or ischemic brain injury [18] as one of the non-glutamate mechanisms for stroke [19, 20]. Ischemic preconditioning (IPC) of the heart [21, 22] and ischemic tolerance of the brain [23, 24] are natural adaptive processes induced by a range of sublethal insults (such as brief ischemia or transient hypoxia) preceding sustained ischemia and can decrease infarct size [22, 25]. This chapter focuses on neuronal K_{ATP} channels and the neuroprotective role of K_{ATP} channels in pre-conditioning in cerebral ischemia and hypoxic-ischemic brain injury.

2 Classification and Structures of K_{ATP} Channels

K_{ATP} channel is a member of the inward rectifier K^+ (Kir) channel superfamily [1]. The functional K_{ATP} channels are hetero-octamers constituted of four pore-forming Kir6 subunits [26, 27] and four modulatory sulfonylurea receptor (SUR) subunits [3, 28, 29]. These subunits are organized in a one-to-one stoichiometry ratio.

2.1 *Kir6.1 and Kir 6.2 Subunits*

The Kir6.x includes two members, Kir6.1 and Kir6.2, encoded by *KCNJ8* gene and *KCNJ11 gene*, respectively. Kir6.1 gene (*KCNJ8*) is mainly found in the mitochondria [10, 30, 31] and Kir6.2 gene (*KCNJ11*) is located on chromosome 11. The human Kir6.2 submit shares 96 % identity with the sequences of that in mouse and rat [26]. Kir6.2 subunits are found in cell membranes (cell surface membranes, plasmaleminal or sarcolemmal membrane) [3, 10, 28, 29, 31]. Both sarcolemmal and mitochondrial K_{ATP} channels have been reported [3, 10, 28, 29, 31].

Kir6.2 subunit consists of two transmembrane domains (TM1, and TM2), which are linked by an extracellular loop (H5) to constitute the pore [17, 32] (Fig. 7.1). The K_{ATP} channel pore is formed within the four TM2 domains of the Kir6.2 subunit. The Kir6.2 channel selective filter is located in the H5 region and encoded the signature sequence GFG, which differs to the conserved GYG sequence in the other K^+ channels [17]. The cytoplasmic amino (N-) and carboxyl (C-) terminals of the Kir6.2 subunits link together as the cytoplasmic domain and are responsible for channel gating and ATP binding [17, 32]. Similar to other Kir subunits, Kir6 subunit does not contain the S4 voltage sensor segment that is critical for gating in all voltage-dependent calcium, sodium, and potassium channels. K_{ATP} channels are constitutively active when ATP are low or absent.

2.2 *SUR 1 and SUR 2 Subunits*

The SUR regulatory subunits of K_{ATP} channels belong to the ATP-binding cassette (ABC) transporter family [33, 34]. Two regulatory SUR subunits of K_{ATP} channels have been identified, SUR1 and SUR2 [34, 35]. SUR1 subunit is encoded by the ABC C8 (*ABCC8*) gene located on chromosome 11p15.1, and SUR2 encoded by the *ABCC9* gene located on chromosome 12p12.1.

The SUR regulatory subunits bind to Mg-ADP [17]. Each SUR subunit has 17 TM sections that are grouped into three transmembrane domains (TMDs) (Fig. 7.1). In contrast to TMD1 (TM6-11) and TMD2 (TM12-17), which are conserved among other members of the ABC family, TMD0 (TM1-5) of SUR1 is structurally unique to SUR. TMD0 is responsible for trafficking the Kir6.2 subunits to the membrane surface [36]. SUR subunits also contain two nucleotide-binding domains (NBDs), similar to all other ABC transporters. The dimer of NBDs creates a single nucleotide-binding site and a catalytic site each. Mg-dependent hydrolysis of the NBD dimer disinhibits ATP blockage on Kir6.2 [32]. The SUR regulatory subunits do not conduct any current, in contrast to the other traditional ABC transporters [34].

SUR1 subunit is mainly found in pancreatic beta cells and neurons [34, 35, 37]. The SUR2 subunit is mainly found in the heart and skeletal muscles, also in some neurons [17, 32]. SUR2A and SUR2B are the two alternative splice variants of SUR2 subunit. Both are structurally similar except the last 42 C-terminal residues (C42) [38]. However, the C42 of SUR2B are highly homologous to the C42 of SUR1 [38]. Since most K_{ATP} channels reported contain the Kir6.2 pore-forming subunit, the heterogeneity observed between the channels is mainly due to the differential expression of the regulatory SUR subunits. SUR1 subunit binds to sulfonylurea with a high affinity [34, 35, 37], whereas SUR2 has lower affinity for sulfonylureas binding [17, 32].

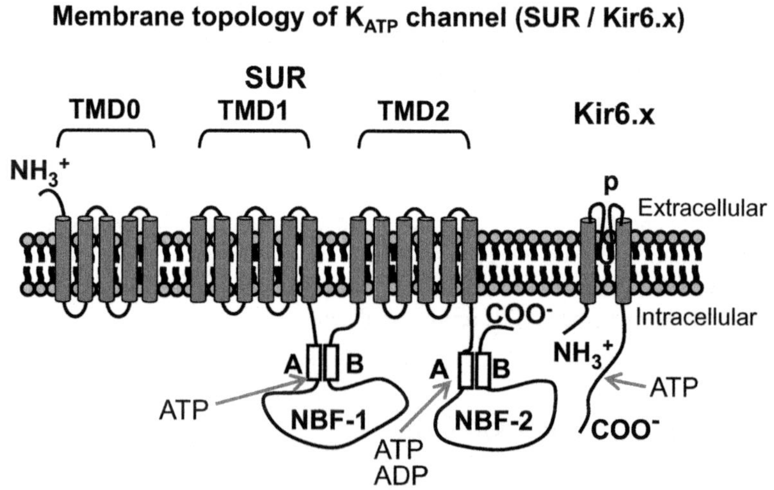

Fig. 7.1 Schematic display of membrane topology of K_{ATP} channel. Sulfonylurea receptor (SUR) has three transmembrane domains, TMD0, TMD1, and TMD3, with each domain with five, six, and six transmembrane segments. Nucleotide binding folds, NBF-1 and NBF-2, are in the intracellular loop between TMD1 and TMD2 domains and in the intracellular C-terminal region. Walker A motif (A) and Walker B motif (B) in the nucleotide-binding folds

3 Gating, Regulation, and Pharmacology of K_{ATP} Channels

3.1 Gating

A functional K_{ATP} channel requires $(SUR/Kir)_4$ [3, 28, 29], consisting of four pore-forming subunits (Kir6.1/Kir6.2) and four modulatory sulfonylurea receptor (SUR1/SUR2) subunits [2, 14]. The K_{ATP} channels consisting of Kir6.1 subunit exhibit a smaller unitary conductance than that of Kir6.2 subunit [28], and the most functional K_{ATP} channels reported are Kir6.2-based. The K_{ATP} channels conduct a weak inwardly rectifying K^+ current to hyperpolarize the membrane following energy depletion, and thus regulate the electrical activity of cell membranes as a cellular metabolic sensor [3, 28, 29, 39] (Fig. 7.2).

K_{ATP} channels are activated by Mg-ADP via binding to SUR regulatory subunit [3, 28, 29, 40]. K_{ATP} channels exhibit fast gating with long inter-burst closing kinetics [41–43]. The pore-loop near the selectivity filter of the Kir6.2 subunit determines the fast gating properties [41, 44], while ATP binding to the TM2 helices regulates the long last inter-burst closing kinetics [42, 45]. K_{ATP} channels exhibit one open and multiple closed states shown by the conformational modeling [32, 43]. All four Kir6.2 subunits in the open conformation are required

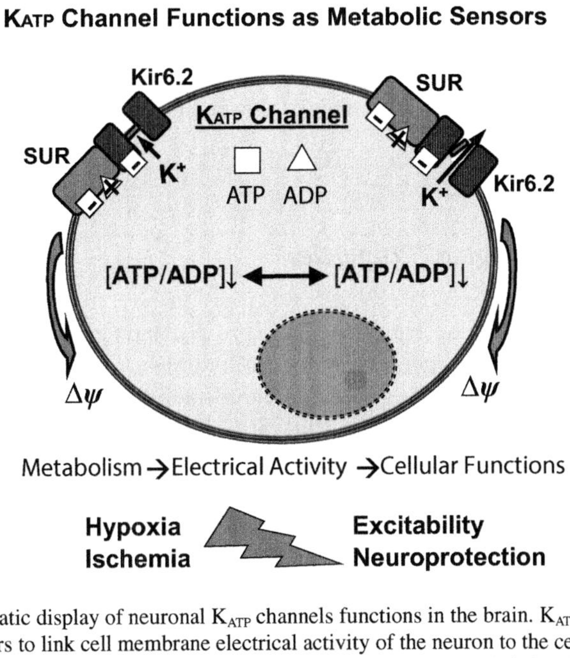

Fig. 7.2 Schematic display of neuronal K_{ATP} channels functions in the brain. K_{ATP} channels act as metabolic sensors to link cell membrane electrical activity of the neuron to the cell metabolic status. Under normal metabolic conditions, K_{ATP} channels close due to the high ATP/ADP ratio inside the cell. When the ATP/ADP ratio drops by either decreasing ATP levels or enhancing ADP levels inside the cell, K_{ATP} channels open and allows potassium (K+) ions to flow out of the cells, thus hyperpolarizing the neuronal cells (*SUR* sulfonylurea receptor)

for the open state of the channel. Once any one of the four subunits binds to ATP, the channel will be shifted from the open state of a closed state. SUR subunits couple their N-terminal bundle of five TM helices (TMD0-L0) to the outer helix and N-terminus of Kir6.2, to keep the gating of the ATP-inhibited Kir6.2 pores [37, 46]. This coupling bidirectionally modulates channel gating [37, 46].

3.2 *Regulation by Nucleotides, ATP and Mg-ADP*

Sensitivities of the K_{ATP} channels to the nucleotides ATP and ADP are dependent on the channel subtypes [28]. Kir6.2 subunit binding to ATP inhibits the K_{ATP} channels [47]. Each Kir6.2 subunit forms one ATP-binding pocket located at the cytoplasmic domain [36, 44, 48]. The ATP-binding pocket is composed of residues R50, I182, K185, R201, and G334 [17, 32]. Interruption of ATP binding releases channel inhibition and permits channel activation [17, 32].

The SUR subunits bind to Mg-ADP [3, 28, 29, 40] which is the endogenous activator of K_{ATP} channels [17]. The NBDs, Walker A, and Walker B motifs are located at the two large cytosolic loops following the TMD1 and TMD2 each (Fig. 7.1). Dimerization of the two NBDs creates the catalytic sites for ATP hydrolysis and is essential for transducing the ADP effect on Kir6.2 [49–51]. Mutations, including G1479R [40, 52] in the NBD2 of SUR1 and V187D [53] in the TMD0 of SUR1, disrupt NBD dimerization and reduce the ADP-mediated activation of K_{ATP} channels [47]. Under normal physiological ATP levels, the probability of opening K_{ATP} channels is extremely low in the absence of SUR regulatory subunits [17, 32].

3.3 *Pharmacological Properties*

K_{ATP} channels have distinctive pharmacology [28], and the major pharmacological targets are the SUR subunits. The sulfonylureas, including glibenclamide, acetohexamide, tolbutamide, glipizide, and glimepiride, bind to SUR subunits to block the channels with different affinity [28, 54]. Kir6.2/SUR1 K_{ATP} channels [55] bind to tolbutamide and glibenclamide with a high affinity ($K_i = 2$ μM), which is determined largely by the serine residue (S1237) at the C-terminus of SUR1. The SUR1 subunit binding to glibenclamide also requires a bivalent structure in the cytoplasmic loop 3 between TM5 and TM6 and the cytoplasmic loop 8 between TM15-TM16 [56]. Kir6.2 subunit binds to tolbutamide with a low affinity ($K_i = 1.8$ mM) [57].

K_{ATP} channel openers increase the channel conductance and hyperpolarize cell membrane. The typical channel openers include diazoxide, cromakalim, pinacidil, nicorandil, minoxidil sulfate, and iptakalim. Diazoxide [3] has been suggested to bind to SUR1 subunits to increase K^+ efflux through the Kir6.2/SUR1 channels

[58], while the site of action for diazoxide on K_{ATP} channels remains controversial. Cromakalim, pinacidil, and nicorandil bind to the TM2 domain of SUR2; specific residues that are essential for the drug binding in SUR2A include L1249 and T1253, and in SUR2B include T1286 and M1290 [59–61]. Iptakalim interacts with the SUR subunits [62] of Kir6.2-K_{ATP} and/or Kir6.1-mitoK_{ATP} channels [63], leading to cytoprotection [64]. In general, SUR1- and SUR2-based channels have different binding affinities to sulfonylureas drugs, whereas SUR2A- and SUR2B-based channels have different affinity to diazoxide [28].

3.4 Non-nucleotide Cellular Regulation

K_{ATP} channel activity is regulated via other cellular mechanisms, in addition to ATP and ADP. For instance, phosphatidylinositol 4,5-biphosphate (PIP_2) binds to channels via the cytoplasmic domain, decreases the ATP sensitivity of the channels, and prevents the channel from closing, thus stabilizing the channel open state [39, 65, 66]. K_{ATP} channel activity is regulated by protein kinase A (PKA) in smooth muscle cells [67, 68]. In β-pancreatic cells, the channels are regulated by leptin via both PKA [69, 70] and AMP-activated protein kinase [69]. In β-pancreatic cells, syntaxin 1A, a SNARE protein, binds to the SUR1 subunits, suppresses K_{ATP} channel activities [71–75], reduces the membrane expression level of K_{ATP} channels [76], and attenuates effects of the channel openers, including diazoxide, P1075, and cromakalim [77, 78].

4 Distribution of Neuronal K_{ATP} Channels

The neuronal K_{ATP} channels are mainly Kir6.2/SUR1 [3, 7, 10]. In the mammalian brain, Kir6.2-based K_{ATP} channels are found in various regions [7, 9, 79], including the substantia nigra (SNr) [80, 81], neocortex [82], hippocampus [83], and hypothalamus [84, 85], and in multiple type of brain cells, including glial cells [83, 86] and neurons in the hippocampus [87], dorsal vagus [88], hypothalamus [84, 85, 89], and SNr [90]. Kir6.2 mRNA, detected by single cell RT-PCR analysis, is predominantly expressed in interneurons, pyramidal, granule, and neuroglial cells in the hippocampus of young rats (10-13 days) [83]. Kir6.2 subunits are mainly located in the somata and dendrites of the central neurons [91–93].

The mitochondrial protein level of Kir6.1 subunit is 6–7 fold higher in the brain than that in the heart and liver in rat [10]. Radioligand binding assays showed the heterogeneous regional distribution of glibenclamide-sensitive K_{ATP} channels, with the highest densities in the substantia nigra and the globus pallidus regions, and the low level in hypothalamic nuclei and the main medullar oblongata areas [94].

5 The Roles of Neuronal K_{ATP} Channel Under Physiological Conditions

K_{ATP} channels exhibit two major roles under physiological condition, glucosensing and non-glucosensing. K_{ATP} channel is involved in glucose homeostasis in the hypothalamus by regulating the hormone secretion via the autonomic nervous system [95]. The non-glucosensing role of K_{ATP} channels [83, 89, 96] includes regulating cell excitability [97, 98]. Opening K_{ATP} channels by the ketone body R-beta-hydroxybutyrate increases neuronal electrical activity [99]. In brainstem inspiratory neurons, burst openings of single K_{ATP} channel synchronize with the rhythm of firing [100]. Blocking the Na–K ATPase suppresses the single channel conductance of K_{ATP} channels in the dentate granule (by strophanthidin) [99] and reduces the P_{open} of K_{ATP} channels [100]. It is likely that Na influx during action potentials activates Na–K ATPase, increases ATP consumption, and thus regulates the opening of K_{ATP} channels under physiological conditions.

6 Neuroprotective Role of K_{ATP} Channels in Cerebral Ischemia

K_{ATP} channels in the central neurons remain closed due to high intracellular ATP levels under physiological condition. Under conditions of severe metabolic deprivation, such as anoxia, ischemia, or hypoxia, energy failure reduces the ATP/ADP ratio and activates neuronal K_{ATP} channels (Fig. 7.2).

Consistent with other studies [87, 101–106], we reported that the K_{ATP} channel-mediated neuroprotection occurs during the initial phase of ischemia/hypoxia in focal and global ischemia models [92, 93]. During the first 5 min of ischemia, membrane hyperpolarization was observed in both hippocampal and cortical neurons in the wild-type animal [93]. The hypoxic hyperpolarization was absent in Kir6.2 knockout mice [92] or following pretreatment with the K_{ATP} channel blocker tolbutamide [93]. In addition, rapid depolarization and increased neuronal activity were observed in hippocampal CA1 neurons in Kir6.2 knockout mice [92]. In vivo study showed that cortical neurons in the Kir6.2 knockout mice [92] are more vulnerable to ischemic insults by middle cerebral artery occlusion than that in the wild-type [93]. In an independent study, LePoith group showed that overexpression of Kir6.2 in knock-in (KI) mice reduced the spontaneous electrical activity of the neurons from hippocampal and cortical regions and decreased the infarction area as comparing to that of the wild-type animals [107]. Opening of K_{ATP} channels causes cell membrane hyperpolarization and in turn suppresses neuronal activity and excitability [87, 92, 93, 103]. This hypoxia-ischemia-induced membrane hyperpolarization is considered as one of the cellular mechanisms underlying neuroprotection against stroke [92, 93]. Taken together, these findings using the pharmacological and genetic approaches strongly support the notion that Kir6.2-containing K_{ATP} channels are neuroprotective against ischemia.

Ischemic insults and stress cause anoxic membrane depolarization and excitotoxicity and resulted in neuronal damage and neurodegeneration [23, 104]. Low energy levels activate the cell membrane K_{ATP} channels to hyperpolarize neurons (Fig. 7.2) during ischemic insults and stress [108], thus protecting neurons from death. K_{ATP} channel proteins have been detected in different neurons in the hippocampus of adult rats [86]. The interneurons, CA3 neurons, and granule cells are relatively resistant to ischemic injury, especially in global ischemia [109, 110], which may be, at least in part, related to their higher expression level of K_{ATP} channels. In addition, K_{ATP} channels have been found in astrocytes [106] and reactive microglial cells [105]. Upregulation of SUR1 [105, 106], SUR2B [106], and Kir6.2 [106] subunits in the non-neuronal cells was also observed following ischemic insult. Taken together, neuroprotective role of K_{ATP} channels may be mediated via multiple mechanisms following ischemic brain injury and these mechanisms require further investigation.

7 K_{ATP} Channels in Preconditioning in Cerebral Ischemia or Hypoxic-Ischemic Brain Injury

7.1 Ischemic or Hypoxic Preconditioning

One of the fundamental properties of living cells is their adaptive capability for cytoprotection following recurrent sublethal stressors. Ischemic preconditioning (IPC) or hypoxic preconditioning (HPC) is one of these capacities that allow cells to survive exposure to recurrent brief ischemic or hypoxic stressors. Preconditioning induces ischemic tolerance that includes a rapid phase that lasts for an hour followed by a delayed sustained phase [111]. The delayed ischemic tolerance phase is normally detectable after 1 day and reaches to the peak at 3 days before fading out after 7 days [111], and thus, provides the time window for preconditioning-induced neuroprotection.

IPC was first reported in the heart [21, 22] as a natural adaptive cytoprotective process induced by a range of transient ischemic/hypoxic insult preceding sustained ischemia [22, 25]. Such inducible ischemic tolerance, HPC, was also found in the brain [23, 24]. Ischemic tolerance has been successfully induced in experimental models of cerebral ischemia in both adult [112, 113], and neonatal animals [114–117], and is considered as a promising approach for neuroprotection.

7.2 K_{ATP} Channels in Hypoxic Preconditioning-Induced Neuroprotection

We recently reported that K_{ATP} channel activity increased together with an upregulation of the protein level of Kir6.2 subunits in hippocampal neurons after HPC [118]. HPC 2 days prior to HI insults reduced subsequent HI-induced infarct volume and

TUNEL positive cell counts in the neonatal brains. The expression and activities of K_{ATP} channel reach a stable high level 2 days following the HPC [118], consistent with the time course for the delayed phase of ischemic tolerance [111]. A PKC inhibitor, chelerythrine, when applied 2 days after the preconditioning and before the HI, blocked the HPC-mediated protective effect in the hypoxic-ischemic brain injury [118]. HPC enhances the pro-survival and suppresses pro-apoptotic signaling pathways [119–121]. In the neonatal HI model, HPC 2 days prior to HI restored the Akt pro-survival signaling and inhibited the caspase-3 pro-apoptotic signaling and improved functional and behavioral recovery [115]. The ischemic tolerance induced by HPC was blocked by the K_{ATP} channel blocker tolbutamide, further supporting the notion that HPC in neonatal HI brain injury is mediated by neuronal K_{ATP} channels. The cell protective effects of preconditioning can be chemically mimicked by the K_{ATP} channel opener diazoxide, both in heart [122] and in brain in adult [123] and neonates [118].

Activation of mitochondrial K_{ATP} channels initiates IPC in heart and reduces the Ca^{2+}-dependent mitochondrial dysfunction during ischemic reperfusion [124–126]. Similarly, mitochondrial K_{ATP} openers, diazoxide or BMS-191095, reduce neuronal death [mice: [127]; rats: [124, 128–130]]. A selective mitochondrial K_{ATP} channel blocker, 5-hydroxydecanoate, blocks HPC-induced neuronal protection in MCAO focal cerebral ischemia [131]. In contrast, xenon-induced preconditioning is only mediated by cell membrane K_{ATP} channels [132]. The mechanisms underlying the K_{ATP} channel-related neuroprotective effects during HPC remain unclear. However, SUR1 subunits may not be directly involved because SUR1 knockout mice in adult exhibit preconditional ischemic tolerance [132].

8 Conclusion

Preconditioning increases the resistance of tissue against subsequent and potentially lethal ischemic attack and is a natural adaptive process against cell death preceding sustained ischemia. Strong evidence indicates both plasma membrane and mitochondrial K_{ATP} channels are involved in preconditioning, thus providing potential drug targets for preventing damage and enhancing functional recovery for neuroprotection against stroke-related brain injury. It is critical to understand the fundamental molecular and cellular mechanisms contributing K_{ATP} channel-mediated preconditioning, so that novel therapeutic methods can be established in future.

References

1. Noma A (1983) ATP-regulated K+ channels in cardiac muscle. Nature 305(5930):147–148
2. Ashcroft FM, Harrison DE, Ashcroft SJ (1984) Glucose induces closure of single potassium channels in isolated rat pancreatic beta-cells. Nature 312(5993):446–448

3. Seino S (1999) ATP-sensitive potassium channels: a model of heteromultimeric potassium channel/receptor assemblies. Annu Rev Physiol 61:337–362
4. Seino S, Miki T (2003) Physiological and pathophysiological roles of ATP-sensitive K+ channels. Prog Biophys Mol Biol 81(2):133–176
5. Spruce AE, Standen NB, Stanfield PR (1985) Voltage-dependent ATP-sensitive potassium channels of skeletal muscle membrane. Nature 316(6030):736–738
6. Standen NB, Quayle JM, Davies NW, Brayden JE, Huang Y, Nelson MT (1989) Hyperpolarizing vasodilators activate ATP-sensitive K+ channels in arterial smooth muscle. Science 245(4914):177–180
7. Ashford ML, Sturgess NC, Trout NJ, Gardner NJ, Hales CN (1988) Adenosine-5′-triphosphate-sensitive ion channels in neonatal rat cultured central neurones. Pflugers Arch 412(3):297–304
8. Bernardi H, Fosset M, Lazdunski M (1988) Characterization, purification, and affinity labeling of the brain (3H)glibenclamide-binding protein, a putative neuronal ATP-regulated K+ channel. Proc Natl Acad Sci U S A 85(24):9816–9820
9. Amoroso S, Schmid-Antomarchi H, Fosset M, Lazdunski M (1990) Glucose, sulfonylureas, and neurotransmitter release: role of ATP-sensitive K+ channels. Science 247(4944):852–854
10. Bajgar R, Seetharaman S, Kowaltowski AJ, Garlid KD, Paucek P (2001) Identification and properties of a novel intracellular (mitochondrial) ATP-sensitive potassium channel in brain. J Biol Chem 276(36):33369–33374
11. Bernardi H, De W Jr, Epelbaum J, Mourre C, Amoroso S, Slama A, Fosset M, Lazdunski M (1993) ATP-modulated K+ channels sensitive to antidiabetic sulfonylureas are present in adenohypophysis and are involved in growth hormone release. Proc Natl Acad Sci U S A 90(4):1340–1344
12. Inagaki N, Tsuura Y, Namba N, Masuda K, Gonoi T, Horie M, Seino Y, Mizuta M, Seino S (1995) Cloning and functional characterization of a novel ATP-sensitive potassium channel ubiquitously expressed in rat tissues, including pancreatic islets, pituitary, skeletal muscle, and heart. J Biol Chem 270(11):5691–5694
13. Hunter M, Giebisch G (1988) Calcium-activated K-channels of Amphiuma early distal tubule: inhibition by ATP. Pflugers Arch 412(3):331–333
14. Cook DL, Hales CN (1984) Intracellular ATP directly blocks K+ channels in pancreatic B-cells. Nature 311(5983):271–273
15. Rorsman P, Trube G (1985) Glucose dependent K+-channels in pancreatic beta-cells are regulated by intracellular ATP. Pflugers Arch 405(4):305–309
16. Kakei M, Kelly RP, Ashcroft SJ, Ashcroft FM (1986) The ATP-sensitivity of K+ channels in rat pancreatic B-cells is modulated by ADP. FEBS Lett 208(1):63–66
17. Hibino H, Inanobe A, Furutani K, Murakami S, Findlay I, Kurachi Y (2010) Inwardly rectifying potassium channels: their structure, function, and physiological roles. Physiol Rev 90(1):291–366
18. Sun XL, Hu G (2010) ATP-sensitive potassium channels: a promising target for protecting neurovascular unit function in stroke. Clin Exp Pharmacol Physiol 37(2):243–252
19. Besancon E, Guo S, Lok J, Tymianski M, Lo EH (2008) Beyond NMDA and AMPA glutamate receptors: emerging mechanisms for ionic imbalance and cell death in stroke. Trends Pharmacol Sci 29(5):268–275
20. Tymianski M (2011) Emerging mechanisms of disrupted cellular signaling in brain ischemia. Nat Neurosci 14(11):1369–1373
21. Lee SR, Tsuji K, Lee SR, Lo EH (2004) Role of matrix metalloproteinases in delayed neuronal damage after transient global cerebral ischemia. J Neurosci 24(3):671–678
22. Terzic A (1999) New frontiers of cardioprotection. Clin Pharmacol Ther 66(2):105–109
23. Blondeau N, Plamondon H, Richelme C, Heurteaux C, Lazdunski M (2000) K(ATP) channel openers, adenosine agonists and epileptic preconditioning are stress signals inducing hippocampal neuroprotection. Neuroscience 100(3):465–474
24. Dirnagl U, Iadecola C, Moskowitz MA (1999) Pathobiology of ischaemic stroke: an integrated view. Trends Neurosci 22(9):391–397

25. Murry CE, Jennings RB, Reimer KA (1986) Preconditioning with ischemia: a delay of lethal cell injury in ischemic myocardium. Circulation 74(5):1124–1136
26. Inagaki N, Gonoi T, Clement JP, Namba N, Inazawa J, Gonzalez G, Aguilar-Bryan L, Seino S, Bryan J (1995) Reconstitution of IKATP: an inward rectifier subunit plus the sulfonylurea receptor. Science 270(5239):1166–1170
27. Ammala C, Moorhouse A, Gribble F, Ashfield R, Proks P, Smith PA, Sakura H, Coles B, Ashcroft SJ, Ashcroft FM (1996) Promiscuous coupling between the sulphonylurea receptor and inwardly rectifying potassium channels. Nature 379(6565):545–548
28. Babenko AP, Guilar-Bryan L, Bryan J (1998) A view of sur/KIR6.X, KATP channels. Annu Rev Physiol 60:667–687
29. Miki T, Nagashima K, Seino S (1999) The structure and function of the ATP-sensitive K+ channel in insulin-secreting pancreatic beta-cells. J Mol Endocrinol 22(2):113–123
30. Inoue I, Nagase H, Kishi K, Higuti T (1991) ATP-sensitive K+ channel in the mitochondrial inner membrane. Nature 352(6332):244–247
31. Garlid KD (2000) Opening mitochondrial K(ATP) in the heart—what happens, and what does not happen. Basic Res Cardiol 95(4):275–279
32. Nichols CG (2006) KATP channels as molecular sensors of cellular metabolism. Nature 440(7083):470–476
33. Higgins CF (1992) ABC transporters: from microorganisms to man. Annu Rev Cell Biol 8:67–113
34. Aguilar-Bryan L, Nichols CG, Wechsler SW, Clement JP, Boyd AE III, Gonzalez G, Herrera-Sosa H, Nguy K, Bryan J, Nelson DA (1995) Cloning of the beta cell high-affinity sulfonylurea receptor: a regulator of insulin secretion. Science 268(5209):423–426
35. Inagaki N, Gonoi T, Clement JP, Wang CZ, Aguilar-Bryan L, Bryan J, Seino S (1996) A family of sulfonylurea receptors determines the pharmacological properties of ATP-sensitive K+ channels. Neuron 16(5):1011–1017
36. Antcliff JF, Haider S, Proks P, Sansom MS, Ashcroft FM (2005) Functional analysis of a structural model of the ATP-binding site of the KATP channel Kir6.2 subunit. EMBO J 24(2):229–239
37. Babenko AP, Bryan J (2003) Sur domains that associate with and gate KATP pores define a novel gatekeeper. J Biol Chem 278(43):41577–41580
38. Isomoto S, Kondo C, Yamada M, Matsumoto S, Higashiguchi O, Horio Y, Matsuzawa Y, Kurachi Y (1996) A novel sulfonylurea receptor forms with BIR (Kir6.2) a smooth muscle type ATP-sensitive K+ channel. J Biol Chem 271(40):24321–24324
39. Baukrowitz T, Schulte U, Oliver D, Herlitze S, Krauter T, Tucker SJ, Ruppersberg JP, Fakler B (1998) PIP2 and PIP as determinants for ATP inhibition of KATP channels. Science 282(5391):1141–1144
40. Nichols CG, Shyng SL, Nestorowicz A, Glaser B, Clement JP, Gonzalez G, Aguilar-Bryan L, Permutt MA, Bryan J (1996) Adenosine diphosphate as an intracellular regulator of insulin secretion. Science 272(5269):1785–1787
41. Proks P, Capener CE, Jones P, Ashcroft FM (2001) Mutations within the P-loop of Kir6.2 modulate the intraburst kinetics of the ATP-sensitive potassium channel. J Gen Physiol 118(4):341–353
42. Jiang Y, Lee A, Chen J, Cadene M, Chait BT, MacKinnon R (2002) The open pore conformation of potassium channels. Nature 417(6888):523–526
43. Enkvetchakul D, Nichols CG (2003) Gating mechanism of KATP channels: function fits form. J Gen Physiol 122(5):471–480
44. Drain P, Li L, Wang J (1998) KATP channel inhibition by ATP requires distinct functional domains of the cytoplasmic C terminus of the pore-forming subunit. Proc Natl Acad Sci U S A 95(23):13953–13958
45. Drain P, Geng X, Li L (2004) Concerted gating mechanism underlying KATP channel inhibition by ATP. Biophys J 86(4):2101–2112
46. Babenko AP, Bryan J (2002) SUR-dependent modulation of KATP channels by an N-terminal KIR6.2 peptide. Defining intersubunit gating interactions. J Biol Chem 277(46):43997–44004

47. Yamada M, Kurachi Y (2004) The nucleotide-binding domains of sulfonylurea receptor 2A and 2B play different functional roles in nicorandil-induced activation of ATP-sensitive K+ channels. Mol Pharmacol 65(5):1198–1207
48. Dong K, Tang LQ, MacGregor GG, Leng Q, Hebert SC (2005) Novel nucleotide-binding sites in ATP-sensitive potassium channels formed at gating interfaces. EMBO J 24(7):1318–1329
49. Campbell JD, Proks P, Lippiat JD, Sansom MS, Ashcroft FM (2004) Identification of a functionally important negatively charged residue within the second catalytic site of the SUR1 nucleotide-binding domains. Diabetes 53(suppl 3):S123–S127
50. Masia R, Nichols CG (2008) Functional clustering of mutations in the dimer interface of the nucleotide binding folds of the sulfonylurea receptor. J Biol Chem 283(44):30322–30329
51. Vergani P, Lockless SW, Nairn AC, Gadsby DC (2005) CFTR channel opening by ATP-driven tight dimerization of its nucleotide-binding domains. Nature 433(7028):876–880
52. Sharma N, Crane A, Gonzalez G, Bryan J, Aguilar-Bryan L (2000) Familial hyperinsulinism and pancreatic beta-cell ATP-sensitive potassium channels. Kidney Int 57(3):803–808
53. Otonkoski T, Ammala C, Huopio H, Cote GJ, Chapman J, Cosgrove K, Ashfield R, Huang E, Komulainen J, Ashcroft FM, Dunne MJ, Kere J, Thomas PM (1999) A point mutation inactivating the sulfonylurea receptor causes the severe form of persistent hyperinsulinemic hypoglycemia of infancy in Finland. Diabetes 48(2):408–415
54. Misler S, Giebisch G (1992) ATP-sensitive potassium channels in physiology, pathophysiology, and pharmacology. Curr Opin Nephrol Hypertens 1(1):21–33
55. Ashfield R, Gribble FM, Ashcroft SJ, Ashcroft FM (1999) Identification of the high-affinity tolbutamide site on the SUR1 subunit of the K(ATP) channel. Diabetes 48(6):1341–1347
56. Mikhailov MV, Mikhailova EA, Ashcroft SJ (2001) Molecular structure of the glibenclamide binding site of the beta-cell K(ATP) channel. FEBS Lett 499(1–2):154–160
57. Gribble FM, Tucker SJ, Ashcroft FM (1997) The interaction of nucleotides with the tolbutamide block of cloned ATP-sensitive K+ channel currents expressed in Xenopus oocytes: a reinterpretation. J Physiol 504(pt 1):35–45
58. Ashcroft FM, Gribble FM (2000) New windows on the mechanism of action of K(ATP) channel openers. Trends Pharmacol Sci 21(11):439–445
59. Moreau C, Jacquet H, Prost AL, D'hahan N, Vivaudou M (2000) The molecular basis of the specificity of action of K(ATP) channel openers. EMBO J 19(24):6644–6651
60. Moreau C, Gally F, Jacquet-Bouix H, Vivaudou M (2005) The size of a single residue of the sulfonylurea receptor dictates the effectiveness of K ATP channel openers. Mol Pharmacol 67(4):1026–1033
61. Moreau C, Prost AL, Derand R, Vivaudou M (2005) SUR, ABC proteins targeted by KATP channel openers. J Mol Cell Cardiol 38(6):951–963
62. Hu LF, Wang S, Shi XR, Yao HH, Sun YH, Ding JH, Liu SY, Hu G (2005) ATP-sensitive potassium channel opener iptakalim protected against the cytotoxicity of MPP+ on SH-SY5Y cells by decreasing extracellular glutamate level. J Neurochem 94(6):1570–1579
63. Costa AD (2009) Iptakalim: a new or just another KCO? Cardiovasc Res 83(3):417–418
64. Wang H, Zhang YL, Tang XC, Feng HS, Hu G (2004) Targeting ischemic stroke with a novel opener of ATP-sensitive potassium channels in the brain. Mol Pharmacol 66(5):1160–1168
65. Shyng SL, Cukras CA, Harwood J, Nichols CG (2000) Structural determinants of PIP(2) regulation of inward rectifier K(ATP) channels. J Gen Physiol 116(5):599–608
66. Shyng SL, Barbieri A, Gumusboga A, Cukras C, Pike L, Davis JN, Stahl PD, Nichols CG (2000) Modulation of nucleotide sensitivity of ATP-sensitive potassium channels by phosphatidylinositol-4-phosphate 5-kinase. Proc Natl Acad Sci U S A 97(2):937–941
67. Quinn KV, Giblin JP, Tinker A (2004) Multisite phosphorylation mechanism for protein kinase A activation of the smooth muscle ATP-sensitive K+ channel. Circ Res 94(10):1359–1366
68. Shi Y, Wu Z, Cui N, Shi W, Yang Y, Zhang X, Rojas A, Ha BT, Jiang C (2007) PKA phosphorylation of SUR2B subunit underscores vascular KATP channel activation by beta-adrenergic receptors. Am J Physiol Regul Integr Comp Physiol 293(3):R1205–R1214

69. Chen PC, Kryukova YN, Shyng SL (2013) Leptin regulates KATP channel trafficking in pancreatic beta-cells by a signaling mechanism involving AMP-activated protein kinase (AMPK) and cAMP-dependent protein kinase (PKA). J Biol Chem 288(47):34098–34109
70. Wu Y, Shyng SL, Chen PC (2015) Concerted trafficking regulation of Kv2.1 and KATP Channels by leptin in pancreatic beta-cells. J Biol Chem 290(50):29676–29690
71. Cui N, Kang Y, He Y, Leung YM, Xie H, Pasyk EA, Gao X, Sheu L, Hansen JB, Wahl P, Tsushima RG, Gaisano HY (2004) H3 domain of syntaxin 1A inhibits KATP channels by its actions on the sulfonylurea receptor 1 nucleotide-binding folds-1 and -2. J Biol Chem 279(51):53259–53265
72. Kang Y, Leung YM, Manning-Fox JE, Xia F, Xie H, Sheu L, Tsushima RG, Light PE, Gaisano HY (2004) Syntaxin-1A inhibits cardiac KATP channels by its actions on nucleotide binding folds 1 and 2 of sulfonylurea receptor 2A. J Biol Chem 279(45):47125–47131
73. Pasyk EA, Kang Y, Huang X, Cui N, Sheu L, Gaisano HY (2004) Syntaxin-1A binds the nucleotide-binding folds of sulphonylurea receptor 1 to regulate the KATP channel. J Biol Chem 279(6):4234–4240
74. Chao C, Liang T, Kang Y, Lin X, Xie H, Feng ZP, Gaisano HY (2011) Syntaxin-1A inhibits KATP channels by interacting with specific conserved motifs within sulfonylurea receptor 2A. J Mol Cell Cardiol 51(5):790–802
75. Chang N, Liang T, Lin X, Kang Y, Xie H, Feng ZP, Gaisano HY (2011) Syntaxin-1A interacts with distinct domains within nucleotide-binding folds of sulfonylurea receptor 1 to inhibit beta-cell ATP-sensitive potassium channels. J Biol Chem 286(26):23308–23318
76. Chen PC, Bruederle CE, Gaisano HY, Shyng SL (2011) Syntaxin 1A regulates surface expression of {beta}-cell ATP-sensitive potassium channels. Am J Physiol Cell Physiol 300(3):C506–C516
77. Ng B, Kang Y, Elias CL, He Y, Xie H, Hansen JB, Wahl P, Gaisano HY (2007) The actions of a novel potent islet beta-cell specific ATP-sensitive K+ channel opener can be modulated by syntaxin-1A acting on sulfonylurea receptor 1. Diabetes 56(8):2124–2134
78. Ng B, Kang Y, Xie H, Sun H, Gaisano HY (2008) Syntaxin-1A inhibition of P-1075, cromakalim, and diazoxide actions on mouse cardiac ATP-sensitive potassium channel. Cardiovasc Res 80(3):365–374
79. Bernardi H, Fosset M, Lazdunski M (1992) ATP/ADP binding sites are present in the sulfonylurea binding protein associated with brain ATP-sensitive K+ channels. Biochemistry 31(27):6328–6332
80. Roper J, Ashcroft FM (1995) Metabolic inhibition and low internal ATP activate K-ATP channels in rat dopaminergic substantia nigra neurones. Pflugers Arch 430(1):44–54
81. Stanford IM, Lacey MG (1996) Electrophysiological investigation of adenosine trisphosphate-sensitive potassium channels in the rat substantia nigra pars reticulata. Neuroscience 74(2):499–509
82. Ohno-Shosaku T, Yamamoto C (1992) Identification of an ATP-sensitive K+ channel in rat cultured cortical neurons. Pflugers Arch 422(3):260–266
83. Zawar C, Plant TD, Schirra C, Konnerth A, Neumcke B (1999) Cell-type specific expression of ATP-sensitive potassium channels in the rat hippocampus. J Physiol 514(Pt 2):327–341
84. Ashford ML, Boden PR, Treherne JM (1990) Tolbutamide excites rat glucoreceptive ventromedial hypothalamic neurones by indirect inhibition of ATP-K+ channels. Br J Pharmacol 101(3):531–540
85. Ashford ML, Boden PR, Treherne JM (1990) Glucose-induced excitation of hypothalamic neurones is mediated by ATP-sensitive K+ channels. Pflugers Arch 415(4):479–483
86. Zhou M, Tanaka O, Suzuki M, Sekiguchi M, Takata K, Kawahara K, Abe H (2002) Localization of pore-forming subunit of the ATP-sensitive K(+)-channel, Kir6.2, in rat brain neurons and glial cells. Brain Res Mol Brain Res 101(1–2):23–32
87. Fujimura N, Tanaka E, Yamamoto S, Shigemori M, Higashi H (1997) Contribution of ATP-sensitive potassium channels to hypoxic hyperpolarization in rat hippocampal CA1 neurons in vitro. J Neurophysiol 77(1):378–385
88. Trapp S, Ballanyi K (1995) KATP channel mediation of anoxia-induced outward current in rat dorsal vagal neurons in vitro. J Physiol 487(pt 1):37–50

89. Dunn-Meynell AA, Rawson NE, Levin BE (1998) Distribution and phenotype of neurons containing the ATP-sensitive K+ channel in rat brain. Brain Res 814(1–2):41–54
90. Liss B, Roeper J (2001) A role for neuronal K(ATP) channels in metabolic control of the seizure gate. Trends Pharmacol Sci 22(12):599–601
91. Thomzig A, Laube G, Pruss H, Veh RW (2005) Pore-forming subunits of K-ATP channels, Kir6.1 and Kir6.2, display prominent differences in regional and cellular distribution in the rat brain. J Comp Neurol 484(3):313–330
92. Sun HS, Feng ZP, Miki T, Seino S, French RJ (2006) Enhanced neuronal damage after ischemic insults in mice lacking Kir6.2-containing ATP-sensitive K+ channels. J Neurophysiol 95(4):2590–2601
93. Sun HS, Feng ZP, Barber PA, Buchan AM, French RJ (2007) Kir6.2-containing ATP-sensitive potassium channels protect cortical neurons from ischemic/anoxic injury in vitro and in vivo. Neuroscience 144(4):1509–1515
94. Mourre C, Ben AY, Bernardi H, Fosset M, Lazdunski M (1989) Antidiabetic sulfonylureas: localization of binding sites in the brain and effects on the hyperpolarization induced by anoxia in hippocampal slices. Brain Res 486(1):159–164
95. Taborsky GJ Jr, Ahren B, Havel PJ (1998) Autonomic mediation of glucagon secretion during hypoglycemia: implications for impaired alpha-cell responses in type 1 diabetes. Diabetes 47(7):995–1005
96. Karschin C, Ecke C, Ashcroft FM, Karschin A (1997) Overlapping distribution of K(ATP) channel-forming Kir6.2 subunit and the sulfonylurea receptor SUR1 in rodent brain. FEBS Lett 401(1):59–64
97. Pierrefiche O, Bischoff AM, Richter DW (1996) ATP-sensitive K+ channels are functional in expiratory neurones of normoxic cats. J Physiol 494(pt 2):399–409
98. Allen TG, Brown DA (2004) Modulation of the excitability of cholinergic basal forebrain neurones by KATP channels. J Physiol 554(pt 2):353–370
99. Tanner GR, Lutas A, Martinez-Francois JR, Yellen G (2011) Single K ATP channel opening in response to action potential firing in mouse dentate granule neurons. J Neurosci 31(23):8689–8696
100. Haller M, Mironov SL, Karschin A, Richter DW (2001) Dynamic activation of K(ATP) channels in rhythmically active neurons. J Physiol 537(pt 1):69–81
101. Yamamoto S, Tanaka E, Higashi H (1997) Mediation by intracellular calcium-dependent signals of hypoxic hyperpolarization in rat hippocampal CA1 neurons in vitro. J Neurophysiol 77(1):386–392
102. de Garcia AS, Franke H, Pissarek M, Nieber K, Illes P (1999) Neuroprotection by ATP-dependent potassium channels in rat neocortical brain slices during hypoxia. Neurosci Lett 273(1):13–16
103. Yamada K, Ji JJ, Yuan H, Miki T, Sato S, Horimoto N, Shimizu T, Seino S, Inagaki N (2001) Protective role of ATP-sensitive potassium channels in hypoxia-induced generalized seizure. Science 292(5521):1543–1546
104. Heurteaux C, Bertaina V, Widmann C, Lazdunski M (1993) K+ channel openers prevent global ischemia-induced expression of c-fos, c-jun, heat shock protein, and amyloid beta-protein precursor genes and neuronal death in rat hippocampus. Proc Natl Acad Sci U S A 90(20):9431–9435
105. Ortega FJ, Gimeno-Bayon J, Espinosa-Parrilla JF, Carrasco JL, Batlle M, Pugliese M, Mahy N, Rodriguez MJ (2012) ATP-dependent potassium channel blockade strengthens microglial neuroprotection after hypoxia-ischemia in rats. Exp Neurol 235(1):282–296
106. Simard JM, Chen M, Tarasov KV, Bhatta S, Ivanova S, Melnitchenko L, Tsymbalyuk N, West GA, Gerzanich V (2006) Newly expressed SUR1-regulated NC(Ca-ATP) channel mediates cerebral edema after ischemic stroke. Nat Med 12(4):433–440
107. Heron-Milhavet L, Xue-Jun Y, Vannucci SJ, Wood TL, Willing LB, Stannard B, Hernandez-Sanchez C, Mobbs C, Virsolvy A, LeRoith D (2004) Protection against hypoxic-ischemic injury in transgenic mice overexpressing Kir6.2 channel pore in forebrain. Mol Cell Neurosci 25(4):585–593

108. Gross GJ, Fryer RM (1999) Sarcolemmal versus mitochondrial ATP-sensitive K+ channels and myocardial preconditioning. Circ Res 84(9):973–979
109. Lipton P (1999) Ischemic cell death in brain neurons. Physiol Rev 79(4):1431–1568
110. Johansen FF (1993) Interneurons in rat hippocampus after cerebral ischemia. Morphometric, functional, and therapeutic investigations. Acta Neurol Scand Suppl 150:1–32
111. Obrenovitch TP (2008) Molecular physiology of preconditioning-induced brain tolerance to ischemia. Physiol Rev 88(1):211–247
112. Autheman D, Sheldon RA, Chaudhuri N, von Arx S, Siegenthaler C, Ferriero DM, Christen S (2012) Glutathione peroxidase overexpression causes aberrant ERK activation in neonatal mouse cortex after hypoxic preconditioning. Pediatr Res 72(6):568–575
113. Prass K, Scharff A, Ruscher K, Lowl D, Muselmann C, Victorov I, Kapinya K, Dirnagl U, Meisel A (2003) Hypoxia-induced stroke tolerance in the mouse is mediated by erythropoietin. Stroke 34(8):1981–1986
114. Sheldon RA, Aminoff A, Lee CL, Christen S, Ferriero DM (2007) Hypoxic preconditioning reverses protection after neonatal hypoxia-ischemia in glutathione peroxidase transgenic murine brain. Pediatr Res 61(6):666–670
115. Xu B, Xiao AJ, Chen W, Turlova E, Liu R, Barszczyk A, Sun CL, Liu L, Tymianski M, Feng ZP, Sun HS (2015) Neuroprotective effects of a PSD-95 inhibitor in neonatal hypoxic-ischemic brain injury. Mol Neurobiol. [Epub ahead of print]
116. Chen W, Xu B, Xiao A, Liu L, Fang X, Liu R, Turlova E, Barszczyk A, Zhong X, Sun CL, Britto LR, Feng ZP, Sun HS (2015) TRPM7 inhibitor carvacrol protects brain from neonatal hypoxic-ischemic injury. Mol Brain 8:11
117. Xiao AJ, Chen W, Xu B, Liu R, Turlova E, Barszczyk A, Sun CL, Liu L, Deurloo M, Wang GL, Feng ZP, Sun HS (2015) Marine compound xyloketal B reduces neonatal hypoxic-ischemic brain injury. Mar Drugs 13(1):29–47
118. Sun HS, Xu B, Chen W, Xiao A, Turlova E, Alibraham A, Barszczyk A, Bae CY, Quan Y, Liu B, Pei L, Sun CL, Deurloo M, Feng ZP (2015) Neuronal K(ATP) channels mediate hypoxic preconditioning and reduce subsequent neonatal hypoxic-ischemic brain injury. Exp Neurol 263:161–171
119. Brunet A, Bonni A, Zigmond MJ, Lin MZ, Juo P, Hu LS, Anderson MJ, Arden KC, Blenis J, Greenberg ME (1999) Akt promotes cell survival by phosphorylating and inhibiting a Forkhead transcription factor. Cell 96(6):857–868
120. Datta SR, Dudek H, Tao X, Masters S, Fu H, Gotoh Y, Greenberg ME (1997) Akt phosphorylation of BAD couples survival signals to the cell-intrinsic death machinery. Cell 91(2):231–241
121. Downward J (2004) PI 3-kinase, Akt and cell survival. Semin Cell Dev Biol 15(2):177–182
122. Horimoto H, Gaudette GR, Saltman AE, Krukenkamp IB (2000) The role of nitric oxide, K(+)(ATP) channels, and cGMP in the preconditioning response of the rabbit. J Surg Res 92(1):56–63
123. Munoz A, Nakazaki M, Goodman JC, Barrios R, Onetti CG, Bryan J, Guilar-Bryan L (2003) Ischemic preconditioning in the hippocampus of a knockout mouse lacking SUR1-based K(ATP) channels. Stroke 34(1):164–170
124. Asemu G, Papousek F, Ostadal B, Kolar F (1999) Adaptation to high altitude hypoxia protects the rat heart against ischemia-induced arrhythmias. Involvement of mitochondrial K(ATP) channel. J Mol Cell Cardiol 31(10):1821–1831
125. Testai L, Rapposelli S, Calderone V (2007) Cardiac ATP-sensitive potassium channels: a potential target for an anti-ischaemic pharmacological strategy. Cardiovasc Hematol Agents Med Chem 5(1):79–90
126. Kim KO, Choe G, Chung SH, Kim CS (2008) Delayed pharmacological pre-conditioning effect of mitochondrial ATP-sensitive potassium channel opener on neurologic injury in a rabbit model of spinal cord ischemia. Acta Anaesthesiol Scand 52(2):236–242
127. Liu D, Lu C, Wan R, Auyeung WW, Mattson MP (2002) Activation of mitochondrial ATP-dependent potassium channels protects neurons against ischemia-induced death by a mechanism involving suppression of Bax translocation and cytochrome c release. J Cereb Blood Flow Metab 22(4):431–443

128. Mayanagi K, Gaspar T, Katakam PV, Busija DW (2007) Systemic administration of diazoxide induces delayed preconditioning against transient focal cerebral ischemia in rats. Brain Res 1168:106–111
129. Shimizu K, Lacza Z, Rajapakse N, Horiguchi T, Snipes J, Busija DW (2002) MitoK(ATP) opener, diazoxide, reduces neuronal damage after middle cerebral artery occlusion in the rat. Am J Physiol Heart Circ Physiol 283(3):H1005–H1011
130. Mayanagi K, Gaspar T, Katakam PV, Kis B, Busija DW (2007) The mitochondrial K(ATP) channel opener BMS-191095 reduces neuronal damage after transient focal cerebral ischemia in rats. J Cereb Blood Flow Metab 27(2):348–355
131. Liu D, Slevin JR, Lu C, Chan SL, Hansson M, Elmer E, Mattson MP (2003) Involvement of mitochondrial K+ release and cellular efflux in ischemic and apoptotic neuronal death. J Neurochem 86(4):966–979
132. Bantel C, Maze M, Trapp S (2009) Neuronal preconditioning by inhalational anesthetics: evidence for the role of plasmalemmal adenosine triphosphate-sensitive potassium channels. Anesthesiology 110(5):986–995

Chapter 8
Targeting Oxidative Stress in Stroke

Anders Bach

Abstract Reactive oxygen species (ROS) and reactive nitrogen species (RNS) are produced under physiological conditions as innocent by-products and serve as useful redox signalling molecules. An advanced set of endogenous antioxidants safely handle these reactive species, which allows for them to be utilized by our cells and tissues. In many diseases, including ischaemic stroke, this fine balance is shifted so that our antioxidants are outnumbered by ROS and RNS; this results in oxidative stress, which can lead to toxicity and cell death with severe and potentially lethal consequences for the individual. Oxidative stress plays a crucial and central role in the pathophysiological cascade responsible for brain damage following a stroke by mediating severe toxicity in the acute phase and initiating and contributing to late-stage apoptosis and inflammation. This review will describe the chemistry, biochemical sources, and harmful effects of ROS/RNS in order to provide insight into the diverse and complex elements of oxidative stress. Then, oxidative stress' relationship to ischaemic stroke will be addressed, with a focus on the biochemical mechanisms that lead to neurotoxicity. Pharmacological strategies to inhibit oxidative stress in cerebral ischaemia are discussed, including examples that cover past and present efforts. Three potential drug targets against oxidative stress and ischaemic stroke are highlighted, namely PSD-95, NADPH oxidase, and Keap1, and the potential of multi-target drug discovery in relation to ischaemic stroke is outlined.

Keywords Oxidative stress • Ischaemic stroke • Reactive oxygen species • Reactive nitrogen species • Antioxidants • PSD-95 • NADPH oxidase • Keap1 • Drug discovery • Multi-target inhibitors

A. Bach (✉)
Department of Drug Design and Pharmacology, Faculty of Health and Medical Sciences, University of Copenhagen, Universitetsparken 2, Copenhagen 2100, Denmark
e-mail: anders.bach@sund.ku.dk

P.A. Lapchak, J.H. Zhang (eds.), *Neuroprotective Therapy for Stroke and Ischemic Disease*, Springer Series in Translational Stroke Research,
DOI 10.1007/978-3-319-45345-3_8

Abbreviations

AIF	Apoptosis-inducing factor
AKBA	Acetyl-11-keto-β-boswellic acid
AMPA	α-Amino-3-hydroxyl-5-methyl-4-isoxazolepropionic acid
Apaf-1	Apoptotic protease-activating factor-1
ARE	Antioxidant response element
BBB	Blood-brain barrier
cGMP	Cyclic guanosine 3′,5′-monophosphate
COPD	Chronic obstructive pulmonary disease
COX-2	Cyclooxygenase-2
CPP	Cell-penetrating peptide
CREB	cAMP response element-binding protein
CSD	Cortical spreading depression
DC	Direct current
DMF	Dimethyl fumarate
ENACT	Evaluating neuroprotection in aneurysm coiling therapy
eNOS	Endothelial nitric oxide synthase
FAD	Flavin adenine dinucleotide
$FADH_2$	Flavin adenine dinucleotide (reduced)
FBDD	Fragment-based drug discovery
FMN	Flavin mononucleotide
GPx	Glutathione peroxidases
GST	Glutathione *S*-transferase
HCAR2	Hydroxycarboxylic acid receptor 2
HNE	4-Hydroxy-2,3-nonenal
HO-1	Heme oxygenase 1
HTS	High-throughput screening
ICH	Intracerebral haemorrhage
iNOS	Inductive nitric oxide synthase
Keap1	Kelch-like ECH-associated protein 1
L-NAME	L-N^G-nitroarginine methyl ester
L-NNA	L-N^G-nitroarginine
LTP	Long-term potentiation
MMF	Monomethyl fumarate
MOMP	Mitochondrial outer membrane permeabilization
MPT	Mitochondrial permeability transition
MPTP	Mitochondrial permeability transition pore
NADH	Nicotine adenine dinucleotide (reduced)
NADPH	Nicotinamide adenine dinucleotide phosphate (reduced)
NMDA	*N*-methyl-D-aspartate
nNOS	Neuronal nitric oxide synthase
NOX	NADPH oxidase (catalytical subunit)
NQO1	NAD(P)H dehydrogenase (quinone) 1

Nrf2	Nuclear erythroid-related factor 2
8-OH-dG	8-Hydroxy-deoxy-guanidine
8-oxodG	8-Oxo-2′-deoxyguanosine
PARP	Poly (ADP-ribose) polymerase
PBN	α-Phenyl-*N*-tert-butyl nitrone
PDZ	PSD-95/Discs-large/ZO-1
PI3K	Phosphatidylinositol-3-kinase
PKC	Protein kinase C
PLA_2	Phospholipase A2
PLC	Phospholipase C
pMCAO	Permanent middle cerebral artery occlusion
PSD-95	Postsynaptic density protein-95
RET	Reverse electron transport
RNS	Reactive nitrogen species
ROS	Reactive oxygen species
RSCEM	Rabbit small clot embolic stroke model
rtPA	Recombinant tissue-plasminogen activator
RyR1	Ryanodine receptor Ca^{2+} channel
SDH	Succinate dehydrogenase
SOD	Superoxide dismutase
STAIR	Stroke therapy academic industry roundtable
STAZN	Stilbazulenyl nitrone
tBHQ	*tert*-butylhydroquinone
TCA	Tricarboxylic acid
tMCAO	Transient middle cerebral artery occlusion
TRPM7	Transient receptor potential cation channel M7
XO	Xanthine oxidase

1 Oxidative Stress

1.1 Introduction

Oxygen (O_2) is a vital molecule for plant and animal life. In aerobic organisms, oxygen is needed to drive the mitochondrial respiration process, which is essential for cellular energy metabolism. As a side effect, this process generates reactive oxygen species (ROS) that can react with and oxidize bioorganic molecules, such as DNA, protein, lipids, and enzymatic co-factors [1]. If it was not for a sophisticated myriad of endogenous defence molecules that neutralize ROS, and the related reactive nitrogen species (RNS), our cells would be damaged by normal energy metabolism. Interestingly, eukaryotic cells have not only adapted to tolerate oxygen metabolism, but also developed the ability to synthesize and exploit ROS for various beneficial purposes. For example, ROS are used as anti-bacterial weapons by immune cells, in the biosynthesis of hormones, and as signalling molecules (i.e., redox signalling) between and within cells [2–6]. Hydrogen peroxide, in particular, has been recognized as a signalling molecule by modifying proteins via oxidation of active site cysteines, leading to reversible activation or deactivation, similar to phosphorylation. Redox signalling functionally links cellular compartments and is required for important processes like cell proliferation, differentiation, and migration as well as wound healing and healthy immune responses [2, 4–7]. Obviously, this dichotomy of useful and harmful properties of ROS requires tight regulation of cellular oxidant and antioxidant levels; however, disease and exogenous sources (pollutants, drugs, smoke, ionizing radiation, UV light, heat, and diet) may lead to enhanced ROS/RNS production that strains the capacity of endogenous antioxidant systems. This situation is termed oxidative stress and, when prolonged or severe, leads to toxicity. The oxidant/antioxidant balance is actually very delicate, and certain tissues and organs like the brain [8, 9] are not particularly suited to handle dramatic ROS increases. Ultimately, increased ROS/RNS can irreversibly damage cells and tissues and are key players in the pathophysiology of diseases like atherosclerosis, cancer, diabetes, rheumatoid arthritis, chronic inflammation, multiple sclerosis, stroke, and CNS degenerative diseases [1, 3, 9, 10].

1.2 Chemistry of Oxidative Stress

ROS and RNS include a range of molecular species that can react with biomolecules or each other to form new reactive species. The addition of one electron to oxygen results in the charged radical superoxide ($^{\bullet}O_2^-$), and further reductions lead to the neutral but reactive molecule hydrogen peroxide (H_2O_2) and the very reactive hydroxyl radical ($^{\bullet}OH$) [3–5, 11, 12]. Superoxide can form hydrogen peroxide (H_2O_2) either spontaneously or when catalyzed by the enzyme superoxide dismutase (SOD) [3, 4, 13]. Hydrogen peroxide can convert to the highly reactive hydroxyl radical ($^{\bullet}OH$) when reacting with superoxide (the Haber–Weiss reaction) or ferrous (Fe^{2+}) and copper (Cu^+) ions (the Fenton reaction) [3, 4, 14, 15]. Nitric oxide ($^{\bullet}NO$) is produced by endothelial, inductive, or neuronal nitric oxide synthase (eNOS, iNOS, and nNOS, respectively) in a process that converts arginine to citrulline. This neutral RNS radical serves as a secondary messenger as well as a paracrine hormone during homeostasis [3, 10], but in the presence of superoxide reacts to give the negatively charged and very reactive peroxynitrite ($ONOO^-$) [3, 5, 10].

An important aspect of ROS and RNS biochemistry is their diffusion capability and membrane permeability. Hydrogen peroxide and nitric oxide are neutral and sufficiently stable to traverse lipid membranes and serve as signalling molecules between cells and cell compartments. Peroxynitrite is more reactive than nitric oxide and superoxide, but its half-life still allows its diffusion over 1–2 cell diameters. Interestingly, peroxynitrite can permeate membranes via anion channels and/or passively as peroxynitrous acid (*wide infra*). Superoxide is stable enough to allow for some diffusion inside cells; its charge generally disfavours membrane permeation, but in certain cases transport is facilitated by anion channels (*wide infra*). Hydroxyl radicals are not charged, but extremely reactive and are immediately quenched by surrounding macromolecules [3, 5, 10, 16].

1.3 ROS Sources

ROS can be generated from numerous cellular sources under both physiological and pathophysiological conditions, but three main categories exist (Fig. 8.1) [1–4, 11, 17]. *First*, a major source of ROS is the respiratory electron transport chain of mitochondria, which generates superoxide and hydrogen peroxide as by-products of normal respiration [1]. Electrons are donated by the reduced forms of nicotine adenine dinucleotide (NADH) and flavin adenine dinucleotide ($FADH_2$) and transported stepwise through four major transmembrane enzyme complexes situated in the inner membrane of mitochondria. Via this process, oxygen is reduced to water at complex IV, and a proton gradient is created that drives the production of ATP by complex V (ATP synthase). However, of the electrons that enter the respiratory electron transport chain, a small proportion (0.2–2 %) leaks away, primarily from complex I and III, and leads to the release of partially reduced oxygen in the form of superoxide in both the mitochondrial intermembrane space and matrix [1, 17]. Under normal conditions, superoxide is quickly dismutated to hydrogen peroxide (and oxygen) by SOD, and the resulting hydrogen peroxide is then reduced to water by glutathione peroxidases (GPx) [1, 3, 13]. However, if the superoxide level in the intermembrane space is extensive, some ROS may be transported to the cytosol via membrane pores like

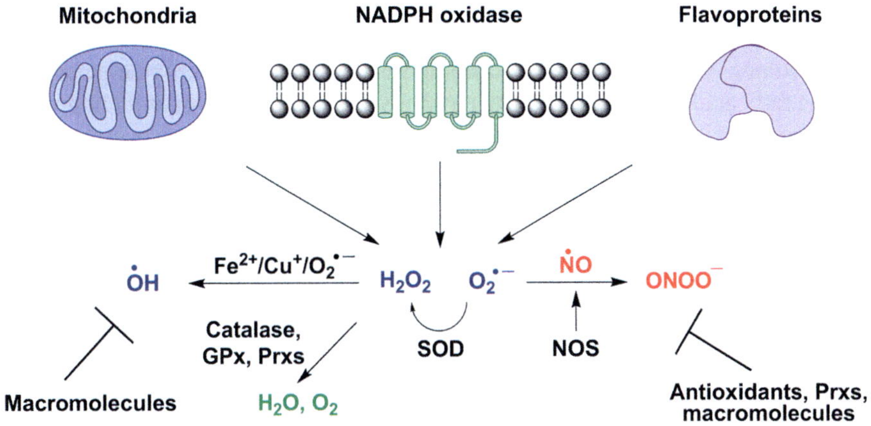

Fig. 8.1 The three main cellular producers of ROS (*blue*) and RNS (*red*) are mitochondria, NADPH oxidase, and the group of flavoproteins. Superoxide ($^{\bullet}O_2^-$) reacts with nitric oxide ($^{\bullet}NO$) produced by NOS (a flavoprotein) to give peroxynitrite ($ONOO^-$). Hydrogen peroxide (H_2O_2) can be converted to hydroxyl radicals ($^{\bullet}OH$) in reactions facilitated by metal ions or superoxide. ROS and RNS are quenched by endogenous antioxidants or macromolecules

voltage-dependent anion channels [18]. The amount of mitochondrial ROS is increased as a result of changes in the proton gradient-induced mitochondrial membrane potential, impaired metabolism, and changes in oxygen levels [1, 17]. *Secondly*, superoxide and hydrogen peroxide are generated by the enzyme nicotinamide adenine dinucleotide phosphate (NADPH) oxidase [19]. Seven types of NAPDH oxidase enzymes have been found (NOX1-5, DUOX1-2), which are expressed in tissues and cell types throughout the human body and are responsible for ROS production during a wide range of physiological processes as well as disease [19, 20]. *Thirdly*, ROS are generated by flavoproteins that use flavin-based co-enzymes, such as flavin adenine dinucleotide (FAD) and flavin mononucleotide (FMN), and are engaged in metabolic oxidation-reduction reactions [4, 11]. The human genome encodes for 90 flavoproteins; some important flavoproteins are succinate dehydrogenase (SDH) (complex II in the respiratory electron transport chain), monoamine oxidase A/B, xanthine dehydrogenase/oxidase, glutathione reductase, NOS, and NADPH oxidases [21]. Flavoproteins can also produce ROS (hydrogen peroxide and superoxide) by autoxidation when oxygen accidently reacts with dihydroflavin, i.e., the reduced form of flavin [11, 22, 23].

1.4 The Antioxidant System

ROS and RNS levels are regulated by a series of enzymatic and non-enzymatic antioxidants (Fig. 8.1). Antioxidant enzymes mainly include SODs, catalase, and peroxidases [1, 3, 24]. SOD neutralizes superoxide to hydrogen peroxide, and three isoforms exist: SOD1, found throughout the cell (nucleus, cytosol, and mitochondrial intermembrane space); SOD2, located in the mitochondrial matrix; and the extracellular SOD3 [1, 3]. Catalase is a heme protein mainly located in peroxisomes, where it efficiently decomposes the hydrogen peroxide generated during fatty acid degradation to water and oxygen and thus serves as an important regulator of cellular redox homeostasis [1, 3, 17]. Catalase has also been found in the mitochondria of the heart and skeletal muscle [25, 26]. Peroxidases comprise a family of enzymes that reduces hydrogen peroxide to water or organic peroxides (e.g., lipid peroxides) to corresponding alcohols by the aid of a cofactor that is being oxidized. For example, several GPx isoforms exist, with varying cellular locations and substrate specificity (hydrogen peroxide or lipid peroxides) [24]. GPx uses glutathione (GSH) as a cofactor, which is oxidized to its dimeric glutathione disulphide form (GSSG) during the catalytic cycle. The resulting GSSG is converted back to GSH by glutathione reductase in a NADPH-dependent reaction, and thus a sound ROS homeostasis also depends on the availability of glutathione reductase and NADPH [1, 3]. In addition to the above-mentioned enzymes, thioredoxin, with two active site cysteines, serves as an electron donor and antioxidant [1, 3, 24]. Thioredoxin facilitates the reduction of disulphide bridges of substrate proteins, which may have been formed as a result of enhanced

ROS levels. Several substrate proteins have been found for mammalian thioredoxin, e.g., ribonucleotide reductase, methionine sulfoxide reductases, and peroxiredoxins [27], which by themselves are involved in redox signalling or antioxidant processes. For example, peroxiredoxins directly scavenge hydrogen peroxide, organic peroxides, and peroxynitrite [5, 28, 29]. Similarly, glutaredoxins use GSH to reduce protein disulphide bonds and, in addition to thioredoxins, are able to reduce protein-SG adducts [28]. Another class of antioxidant proteins, though with a more indirect effect, is metal-ion storage and transporting proteins such as ferritin, ceruloplasmin, and lactoferrin, which reduce the levels of free metal ions and thus prevent unintended metal-catalyzed ROS production. For instance, ferritin stores iron in the ferric (Fe^{3+}) state and thus prevents excessive ROS production by Fenton chemistry, while still making iron ions available for important cellular processes [3].

In addition to the specific and efficient enzymatic anti-ROS defence system, our cells, blood, and tissues are equipped with non-enzymatic and highly abundant antioxidants such as vitamins (A, C, and E), GSH, and uric acid. These molecules directly scavenge ROS or serve as electron-donating co-enzymes and account for a great part of our total antioxidant capacity [3–5, 9, 10, 30, 31].

2 Harmful Effects of Oxidative Stress

2.1 ROS/RNS Reactivity

ROS and RNS can harm cells in several ways if they are not controlled. Hydrogen peroxide is a potent oxidizing agent of protein amino acids like cysteine and methionine, iron-sulphur clusters, and active site metals, although the reactivity greatly depends on the protein environment. The majority of hydrogen peroxide's toxicity, however, must be ascribed to its conversion to the highly reactive hydroxyl radical, which readily and unselectively reacts with surrounding macromolecules (lipid, DNA, and proteins), causing great damage [2–5]. Superoxide is not particularly reactive and its toxicity is mainly explained by its conversion to the highly reactive peroxynitrite and the resulting depletion of nitric oxide, which affects vascular functions. Also, superoxide conversion to hydrogen peroxide (and then hydroxyl radical) and its specific affinity for iron-sulphur clusters lead to loss of function in certain enzymes (e.g., mitochondrial complex II and aconitase) and iron release, which are important toxic mechanisms [2–5, 10, 32]. Likewise, the toxicity of nitric oxide, which is also not particularly reactive, can be understood by its conversion to peroxynitrite and also by its specific inhibition of enzymes where metal centres and/or radicals take part of the catalytic cycle. For example, nitric oxide can diffuse into mitochondria, where it specifically and reversibly inhibits the metal centres of complex IV in the electron transport chain, leading to enhanced ROS production due to slower respiration. Also, nitric oxide may react with mitochondrial superoxide to generate peroxynitrite, which will inhibit complexes I-III and V by oxidizing sulfhydryl groups, nitrate tyrosines, and damage iron-sulphur clusters, thereby inducing further superoxide production and impaired ATP production [1, 5, 10, 17, 33]. In contrast to nitric oxide, peroxynitrite is highly reactive and directly oxidizes amino acid side chains and enzyme metal centres, facilitates the depletion and inactivation of antioxidants like GSH, GPx, and SOD [34], and induces lipid peroxidation and DNA strand-breakage [10]. Also, peroxynitrite decomposes into other very reactive species, leading to tyrosine nitration and thus impaired enzymes, changes in protein structure, and inhibition of signal transduction, all of which can ultimately induce necrosis or apoptosis [3, 5, 10, 35].

2.2 Susceptibility of Macromolecules

Due to the diversity of ROS/RNS reactivity, basically all biomolecules can be modified by these species. *Proteins* are most susceptible to modifications at cysteine, methionine, tyrosine, tryptophan, and histidine residues, which are oxidized by hydrogen peroxide, superoxide, and hydroxyl radicals. Cysteine side chains may form new disulphide bridges, either within the protein itself or to other proteins, forming homo/hetero protein dimers. Also, proteins can be cross-linked via tyrosines (dityrosines). If hydroxyl radicals are created near proteins, these can be severely damaged, leading to ubiquitination and proteosomal degradation. Therefore, overall, these modifications can induce the inactivation, fragmentation, and aggregation of important proteins and form new antigens that are recognized as non-self by the immune system [2, 3, 36]. *DNA* can be oxidized at all four bases, but guanidine-adducts, such as 8-oxo-2′-deoxyguanosine (8-oxodG) and 8-hydroxy-deoxy-guanidine (8-OH-dG), serve as quantitative DNA markers for oxidative stress [37]. Often, it is either peroxynitrite or hydroxyl radicals (created via Fenton chemistry in the nucleus) that are responsible for DNA damage [3, 10, 22]. The ROS/RNS damage can cause mutations by base pair substitutions, strand breakage, and chromosomal rearrangements, which can lead to both carcinogenesis and blockage of DNA replication and apoptosis [38]. Unsaturated *lipids* in membranes are readily oxidized to lipid peroxides in a chain reaction called lipid peroxidation, where radicals like hydroxyl radicals, peroxynitrite, and protonated superoxide are quenched by the lipid, which then itself becomes a radical and thus reactive. The lipid radical is readily oxidized by oxygen to form lipid peroxyl radicals. This species absorbs hydrogen from surrounding lipids and is thereby converted to lipid hydroperoxide while creating a new lipid radical, which can continue this damaging process and influence membrane permeability. Also, lipid hydroperoxide can break down and form new reactive radicals and aldehyde-alkene end-products like 4-hydroxy-2,3-nonenal (HNE), leading to disturbed signal transduction, impaired macromolecules, and the triggering of apoptosis or necrosis [3, 39].

2.3 Cell Death Mechanisms

Prolonged exposure to ROS/RNS can initiate critical cellular mechanisms like apoptosis, autophagy, necrosis, and inflammation. The mitochondria are essential in these processes [1, 38]. ROS can lead to changes in mitochondrial inner and outer membrane permeability, whereby cytochrome c can leak out into the cytosol, where it binds to apoptotic protease-activating factor-1 (Apaf-1) to form apoptosomes that activate caspase-9 and thereby initiate apoptosis [1, 10, 17, 38, 40, 41]. Specifically, ROS like peroxynitrite can oxidize key sulfhydryl groups of the proteins that form the inner membrane mitochondrial permeability transition pore (MPTP), which together with Ca^{2+} overload leads to increased permeability of the inner membrane, a process known as the mitochondrial permeability transition (MPT). This causes a reduction of the mitochondrial membrane potential, electron transport, and ATP generation and increased ROS production. Also, the increased inner membrane permeability is followed by swelling of the matrix, which can induce membrane rupture and thus extensive release of cytochrome c and activation of apoptosis [1, 10, 17, 41, 42]. Furthermore, MPT and/or other apoptotic activators trigger the formation of outer membrane pores by proapoptotic proteins, such as Bid, Bax, and Bak, leading to mitochondrial outer membrane permeabilization (MOMP) and intensification of cytochrome c release [10, 17, 38, 40, 41]. In another route, ROS-mediated DNA damage is detected by p53, leading to activation of caspase-2 and enhancement of apoptosis [38]. Instead of apoptosis, the cell may rescue itself by encapsulating and degrading defective organelles like mitochondria by autophagy; but, if the oxidative stress is too severe and this process eliminates too many organelles, the cell may become non-sustainable and die anyway. Thus, apoptosis and autophagy are interlinked and the balance between the two is important for the final fate of the cell [38, 43]. Alternatively, high amounts of ROS can induce necrosis, which is generally considered more uncontrolled than apoptosis and often leads to increased cell membrane permeability and thus the leakage of cell components into the surrounding tissue and hence triggering of the immune system, inflammation, and worsening of the tissue damage [10, 36, 44–46]. Necrosis may be mediated by the enzyme poly (ADP-ribose) polymerase (PARP), which binds to DNA strand breaks and synthesizes ADP-ribose polymers from nearby proteins, such as histones, leading to the recruitment of DNA repair proteins. However, the polymerization reaction uses NAD+ as substrate, and if DNA damage is extensive, PARP will deplete the NAD+ stores, leading to ATP starvation. Hence, because apoptosis is ATP-dependent, extensive PARP activity leads to necrosis instead [10, 47, 48]. PARP may also induce apoptosis by triggering the mitochondrial release of apoptosis-inducing factor (AIF), which translocates to the nucleus and induces DNA fragmentation [10, 42, 49]. In addition, ROS can also activate the immune system more directly. Mitochondrial redox signalling is central in the activation of both adaptive and innate immune cells in a process facilitated by NADPH oxidase, and ROS may lead to the release of pro-inflammatory cytokines and chemokines and thus over-activation of the immune system and further tissue injury [1, 2, 10, 36, 50].

3 Ischaemic Stroke and Oxidative Stress

Stroke is the second leading cause of death and the leading cause of adult disability worldwide, affecting around 17 million people per year and imposing major costs on society [51–53]. Stroke can lead to loss of motor function, visual impairment, aphasia, headache, and cognitive dysfunctions and induces death in about 35 % of the cases. Two main types of stroke exist: *Ischaemic stroke*, responsible for about 87 % of stroke incidences, is caused by a blood clot (thrombus or embolus) that blocks the blood supply to the brain; and *Haemorrhagic stroke*, which accounts for 13 % of strokes and is caused by a rupture of brain blood vessels, leading to the dangerous leakage of blood into the brain [46, 54, 55].

3.1 The Ischaemic Cascade

Ischaemic stroke is a complex disease, where several mechanisms, together called the ischaemic cascade, are responsible for the neurological damage (Fig. 8.2a) [46, 56–59]. These devastating changes in the brain's biochemistry kick in within minutes of the blood flow reduction. The core of the affected brain area will be irreversibly damaged as the absence of blood leads to impaired energy metabolism and necrotic cell death. The less affected surrounding area, called the penumbra, is still partly perfused via collateral blood vessels and therefore viable (Fig. 8.2b). This area of the brain can be rescued from cell death by proper treatment, for example, by restoring the blood supply within 2–4 h after stroke onset [46, 56, 58, 60–62]; without treatment, however, the infarct core will expand into the penumbra and thus exacerbate the injury.

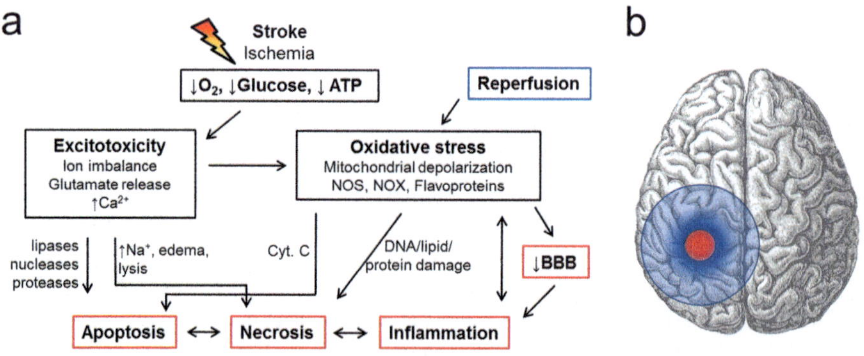

Fig. 8.2 (**a**) A complex and intertwined series of events, together called the ischaemic cascade, are responsible for the neurological damage during stroke. Excitotoxicity links to oxidative stress, and these biochemical changes lead to a mix of apoptosis, necrosis, and inflammation. Also, the BBB is affected. Oxidative stress is augmented by reperfusion. (**b**) Schematic illustration of the ischemic core (*red*) and the penumbra (*blue*)

Excitotoxicity, i.e., a pathological excess of excitatory amino acids, and oxidative stress are central factors of the ischaemic cascade (Fig. 8.2a) and related diseases, such as brain trauma, epilepsy, and chronic neurodegenerative disorders [10, 49, 63]. In ischaemic stroke, excitotoxicity is initiated by the lack of blood supply to the brain and thus depletion of oxygen and glucose levels. Thereby, energy levels are dramatically reduced with the immediate consequence that the ATP-dependent ion-channel Na^+/K^+-ATPase can no longer uphold the plasma membrane potential. As a result, neurons and glial cells depolarize and neuronal voltage-dependent Ca^{2+} channels are activated, leading to synaptic glutamate release. Also, reuptake mechanisms fail due to the impaired membrane potential, which contributes further to the accumulation of glutamate in the extracellular space [49, 56, 64, 65]. This overload of glutamate leads to excessive activation of ionotropic *N*-methyl-D-aspartate (NMDA) and α-amino-3-hydroxyl-5-methyl-4-isoxazolepropionic acid (AMPA) receptors and thus a massive influx of Ca^{2+} and Na^+ ions into neurons. The extensive influx of ions is followed by water and can give rise to cell lysis and cerebral oedema, which can be life-threatening [49, 56, 59]. The Ca^{2+} ions entering neurons via the NMDA receptor and other impaired ion channels situated in neurons and glial cells [46, 66, 67] are toxic to the cells when Ca^{2+} exceeds the capacity of intracellular storage organelles (e.g., endoplasmic reticulum and mitochondria) because Ca^{2+} activates a series of lipases, nucleases, and proteases that degrade cellular components and lead to irreversible damage (Fig. 8.2a) [56]. For example, phospholipase A2 (PLA_2) and C (PLC) are activated by Ca^{2+} and degrade membranes. Subsequently, the arachidonic acid released by PLA_2 is metabolized by cyclooxygenase-2 (COX-2) and lipoxygenases in a process that produces ROS, lipid peroxidation, toxic lipid-aldehyde species, and prostaglandins. This leads to further membrane damage and the process is linked to apoptosis and inflammation [10, 68–70]. Also, Ca^{2+} activates calpain that cleaves key structural proteins, such as cytoskeletal and microtubule subunits, and activates pro-death signalling pathways [49, 71].

3.2 *Ischaemia and ROS*

Cerebral ischaemia induces oxidative stress via several mechanisms, but generally, the three major routes behind physiological ROS production (Fig. 8.1) are also the most significant ones during stroke [10, 46, 69]. In *mitochondria*, NADH and $FADH_2$ levels are reduced due to the impaired blood supply. Normally, these reduced co-enzymes are generated by oxidative metabolism of glucose and the tricarboxylic acid (TCA) cycle and further processed by the electron transport chain to create ATP. Together with low levels of oxygen, this metabolic situation slows the electron transport chain, whereby superoxide generation from complexes I and III is enhanced [41, 42]. Additionally, the Ca^{2+} ions entering neurons are absorbed by mitochondria, where they cause membrane depolarization and further impairment of the electron transport chain, leading to more ROS and less ATP [41, 42, 72, 73]. The Ca^{2+}, together with ROS, facilitates the opening of MPTPs, and the resulting membrane leakage leads to complete mitochondrial depolarization and energy deprivation [10, 41, 42, 64]. Importantly, studies have shown that MPTPs are closed during ischaemic conditions, but open as reperfusion induces a burst of ROS, Ca^{2+}, and importantly, a rise and normalization of pH [74, 75]. Also, MPTPs and/or MOMP may induce mitochondrial rupture and thereby the release of cytochrome c, Ca^{2+}, and ROS into cytosol, causing cell damage and apoptosis (Fig. 8.2a) [10, 40–42]. Noticeably, the extent of mitochondrial injury and potential reversibility is directly proportional to the duration of ischaemia [73, 74]. If ATP impairment becomes too severe, the neurons will undergo abrupt necrosis [42], but in less affected areas, a more prolonged and ATP-dependent cell-death occurs [73]. Unfortunately, keeping up sufficient ATP levels is particularly challenging for mitochondria during neuronal ischaemia, because an extensive amount of ATP is used by cell membrane ion pumps to counteract NMDA receptor-mediated Ca^{2+} and Na^{+} influx. Also, ROS oxidize key thiol groups of the adenine nucleotide transporter, which is exacerbated by ROS-mediated GSH depletion and impairs ATP delivery from mitochondria to cytosol. Thus, the increased demand for ATP and the reduced capacity to deliver and produce it create a vicious cycle, which can ultimately lead to the collapse of energy resources and membrane ion flux mechanisms and cell death [64, 73].

In addition to mitochondria, a major source of superoxide and/or hydrogen peroxide during cerebral ischaemia is *NADPH oxidases*, primarily NOX2, which is situated in neuronal cell membranes and produces superoxide. NOX4 releases hydrogen peroxide and may also contribute to oxidative stress during ischaemia [76–78]. NOX4 has been detected in various subcellular compartments, such as mitochondria (although this has been questioned [79]), and the nucleus, endoplasmic reticulum, and plasma membrane [6, 80]. NOX2 is activated by NMDA receptor-mediated Ca^{2+} influx via protein kinase C (PKC) [81], and the resulting superoxide can cause membrane damage, convert to hydrogen peroxide (spontaneously or by SOD), or fuse with nitric oxide to produce the very harmful peroxynitrite species and cause cell death (Figs. 8.1 and 8.2a). NOX2 is located in neurons containing NMDA receptors and nNOS [82]. Interestingly, mitochondria were also found in close proximity to NOX2 [82], whereby ROS downstream of NOX2 may impair mitochondrial respiration and facilitate mitochondrial ROS generation [79, 82]. Also, mitochondrial ROS can activate NADPH oxidase, providing the basis for a malicious feed-forward ROS-producing cycle, although the mechanism behind this has not been fully elucidated [79].

Finally, several *flavoproteins* are activated during ischaemic conditions (Fig. 8.2a). For example, xanthine dehydrogenase is converted to the oxidoreductase xanthine oxidase (XO) by either oxidation of key sulfhydryl groups, mediated by the general pro-oxidant environment during ischaemia, or proteolytic cleavage [83, 84]. XO converts oxygen to superoxide, hydrogen peroxide, and the highly reactive carbonate radical anion using xanthine and hypoxanthine as substrates that accumulate due to ATP catabolism during ischaemia [72, 85, 86]. Also, the uncoupling of nNOS and eNOS converts these into superoxide-producing enzymes, which will contribute to vascular oxidative stress and endothelial dysfunction for eNOS and neuronal toxicity for nNOS. This uncoupling is mediated by peroxynitrite, which oxidizes the NOS cofactor tetrahydrobiopterin, and lack of the substrate L-arginine [46, 87–90].

Mitochondria have been considered the primary source of superoxide during NMDA-mediated neurotoxicity [41], but other studies of excitotoxicity considered NOS [89] or arachidonic acid [91] to be the main superoxide generators. More recent studies suggest that NADPH oxidase is the major supplier of NMDA-induced superoxide production in neurons [81]. In oxygen-glucose-deprived neurons, mitochondrial ROS dominate in the early phase, followed by a XO-mediated ROS burst, and then, upon reoxygenation, an NADPH oxidase-mediated ROS production. In this study, neuroprotection was achieved by inhibiting NADPH oxidase and XO, but not by inhibiting mitochondrial ROS [72].

3.3 Ischaemia and RNS

The main sources of RNS during cerebral ischaemia are the three nitric oxide-producing enzymes, eNOS, iNOS, and nNOS. In general, eNOS mediates vasodilation upon activation and plays a neuroprotective role by enhancing blood flow levels. However, in stroke, as in cardiovascular disease [87], eNOS uncoupling and superoxide-mediated nitric oxide depletion counteract this beneficial effect [88]. Induction of the immune system during stroke results in expression and activation of iNOS in brain-infiltrating macrophages and glial cells, and thus a damaging, late-stage nitric oxide production [10, 56, 89, 90]. Ca^{2+} ions that enter neurons via the NMDA receptor directly activate nNOS [92–94]. The importance of nitric oxide in excitotoxicity was elucidated by demonstrating that inhibition of nitric oxide synthesis impaired NMDA-mediated toxicity in neuronal cultures and reduced brain damage in mice following focal cerebral ischaemia [95, 96]. Also, nNOS knock-out mice demonstrated marked reduction in infarct size upon middle cerebral artery occlusion (MCAO) [97]. During ischaemic stroke, nitric oxide reacts with specific enzymes, such as caspases and metalloproteases, by *S*-nitrosylation of cysteines, leading to an overall inhibition of pro-survival pathways and activation of pro-death signalling [46, 49, 62]. Also, nitric oxide disturbs mitochondrial electron transport chain complexes, either directly or via its conversion to peroxynitrite [10], as described above. Peroxynitrite is central in mediating RNS neurotoxicity [10, 49, 90], as thoroughly demonstrated by preventing either nitric oxide or superoxide formation in various in vitro and in vivo models of stroke [81, 89, 98–102]. Peroxynitrite directly affects mitochondria and macromolecules, causing extensive cell damage and ultimately apoptosis or necrosis [10, 49]. Also, peroxynitrite activates the transient receptor potential cation channel M7 (TRPM7), leading to Ca^{2+} influx and thus contributing further to the ischaemia-induced Ca^{2+} overload [67, 103]. In particular, PARP activation triggered by peroxynitrite-mediated DNA damage is considered a major reason for neuronal death following excitotoxicity [10, 49, 71, 104–106].

In order for superoxide and nitric oxide to collide and generate peroxynitrite, they have to be present in the same cellular compartments. Nitric oxide is potentially produced directly by mitochondrial NOS [10, 33, 74, 107–109], or alternatively, it diffuses into mitochondria from nNOS-containing neurons, where it can react with superoxide [98, 110]. Extracellular or intraorganellar superoxide can be generated by NADPH oxidases and react with freely diffusible nitric oxide [19]. Cytosolic superoxide can originate from several sources, e.g., arachidonic acid metabolism, flavoproteins like XO, uncoupled NOS, and leaking mitochondria. Even superoxide produced by NADPH oxidase (e.g., NOX2), which is delivered extracellularly, can enter the cytosol via anion channels [19, 102, 111, 112]. Also, peroxynitrite generated outside cells can traverse cell membranes, partly via anion channels and partly by passive membrane diffusion as peroxynitrous acid [10, 102, 113–117].

3.4 Brain Damage

The brain contains low levels of antioxidants, but high levels of polyunsaturated fatty acids and transition metal ions and consumes a lot of oxygen relative to other organs. Therefore, the brain is particularly sensitive to ischaemia and the resulting oxidative stress [8, 9, 62, 69, 118, 119]. The situation is exacerbated by the acidic environment created as a result of ATP depletion and lactic acid accumulation [42], which leads to release of Fe^{3+} from ferritin inside neurons. Fe^{3+} is reduced to Fe^{2+} by superoxide, facilitating the conversion of the less severe hydrogen peroxide to the highly toxic hydroxyl radical by Fenton chemistry [68, 69]. Also, reperfusion of damaged brain tissue is, on the one hand, necessary for recovery, but on the other hand, leads to a secondary surge of ROS/RNS production as dysfunctional mitochondria and enzymes are provided with oxygen (Fig. 8.2a) [10, 42, 54, 58, 62, 118]. Mitochondria are particularly responsible for generating post-ischaemic ROS [6, 74, 86], but XO is also considered central during reperfusion [6, 84, 120]. However, the studies by Abramov et al. [72] (*wide supra*) clearly suggested that NADPH oxidase (NOX2), and not mitochondria or XO, is the main ROS producer upon reoxygenation [72].

ROS and RNS modify macromolecules, inactivate important enzymatic cofactors, and disturb mitochondrial respiration and beneficial vasoregulatory effects and thereby contribute to neuronal damage and cell death, as described. Whether the cells will recover or undergo apoptosis or necrosis depends on the severity and duration of the ischaemia and its location relative to the ischaemic core [46, 56]. In addition, blood-brain barrier (BBB) integrity is affected (Fig. 8.2a). The lower nitric oxide levels in endothelia, due to eNOS uncoupling, facilitate platelet aggregation and leukocyte adhesion and thereby vascular inflammation [46, 54, 118]. Also, ROS lead to activation of cytokines (e.g., via NFkB that is activated by hydrogen peroxide [118, 121]) and metalloproteases, leading to changes in the vascular structure, a leaking BBB, and activation of microglia. These conditions facilitate the invasion of immune system cells into the brain parenchyma and result in brain inflammation that persists for days [10, 46, 54, 56, 62, 118]. Inflammation is further substantiated by the release of cellular components after necrosis and oxidative stress, which are recognized by receptors on innate immune cells, and the inflammation contributes to further BBB degradation, cytokine production, cerebral oedema, and thus exacerbation of the brain damage (Fig. 8.2a). However, notably, some aspects of the inflammatory process are likely important for cellular repair and the recovery phase following a stroke, which poses a challenge for therapeutic interventions targeting inflammation [46, 56, 57, 62, 118].

4 Stroke Treatment and Drug Discovery

4.1 Reperfusion

The development of therapeutics against acute ischaemic stroke has been very unproductive so far; only recombinant tissue-plasminogen activator (rtPA) is approved in the EU and USA. rtPA converts plasminogen to plasmin, which promotes thrombolysis of blood clots and thus re-establishes brain blood supply, but this comes with a 2 % increased risk of fatal intracranial haemorrhage (total risk of intracranial haemorrhage is about 4–6 %, relative to ~1 % in non-treated groups [122, 123]), and therefore rtPA treatment of elderly and hypertensive patients should be done with care [124]. For the drug to be effective, it must be given within 4.5 h from stroke-onset, but preferably within 3 h [124, 125], and even then, a sufficient reperfusion is only obtained in less than 50 % of the cases [61, 126]. Due to safety concerns and its narrow therapeutic time window, less than 5 % of all stroke patients are treated with rtPA [126, 127], and a recent review of the clinical benefits of rtPA revealed a modest effect: 33 % of the treated patients showed good outcomes relative to 23 % of the non-treated patients [124]. Alternatively, five clinical trials have recently shown positive outcomes of thrombectomy as a treatment for acute ischaemic stroke. The mechanical removal of the blood clot was more efficient than rtPA treatment and allowed for a larger therapeutic window; thus, this endovascular procedure provides a complementary reperfusion strategy to rtPA [60, 61, 128]. Although rtPA and endovascular procedures benefit several individuals, acute stroke clearly remains a major challenge in clinical medicine [46, 126]. Therapeutic time windows, contraindications, clot sizes and locations, recanalization efficiency, and the risk of bleeding are still major complications, and importantly, the oxidative stress and excitotoxic events may still evolve even after reperfusion [54, 126]. Also, for haemorrhagic strokes and global ischaemia due to cardiac arrest, reperfusion therapy is obviously not relevant. Therefore, a majority of stroke patients are either not treated or do not benefit from currently available therapy. A drug that protects neurons from the devastating effects of ischaemia in an efficient and safe manner would be a huge contribution to stroke treatment. Such a neuroprotective agent could be used as a stand-alone drug for non-reperfused patients or in combination with rtPA. Combining reperfusion with protection against ischaemic conditions and reperfusion damage could provide an efficient combinatorial therapy in terms of effects and extended therapeutic window [58, 59, 129–134]. Particularly, if drug safety would allow early administration based on a symptom-based diagnosis [135, 136], in contrast to reperfusion therapy, where haemorrhagic stroke must be ruled out by neuroimaging, the clinical utility of neuroprotectants is obvious. Early administration could be facilitated by enhanced medical care logistics and ambulance treatments, which several cases showed was feasible [133, 137–139].

4.2 Past Failures

The complexity of ischaemic stroke poses a huge challenge for drug development. In 2006, an evaluation revealed that 114 drug candidates targeting the ischaemic cascade (e.g., 29 anti-excitotoxic and five radical scavengers) failed clinical trials of acute ischaemic stroke [140]; however, thorough analyses of these failures point out several flaws and shortcomings of the trial designs and the drug candidates themselves [56, 133, 140–145]. Often, the trials were initiated based on insufficient preclinical validation, the dose was too low because of side effects, patients were treated later than supported by preclinical data, or patient groups were too small or diverse in their aetiology and severity for significant improvements to be measurable. Also, some of the compounds were inferior with respect to target activity and BBB permeability in humans, and the quality and translational value of the animal models have been questioned too [131, 140, 141]. NMDA receptor antagonists are highly abundant on the list of past failures due to a combination of severe psychomimetic and cardiovascular side effects, which halted the trials and prevented effective dosing in humans, and the mentioned flaws in clinical trials [49, 56, 133, 142, 145, 146]. Additionally, the underlying mechanistic hypothesis was likely too simplified, as NMDA receptor antagonists inhibit neuroprotective and Ca^{2+}-mediated pro-survival pathways downstream of NMDA receptors and thereby contribute to neuronal cell death. The magnitude of activation, extrasynaptic versus synaptic location, and subunit composition of NMDA receptors are important for determining their role as either pro-death or pro-survival mediators [46, 49, 66, 133, 146–149].

4.3 New Development

The many failures have made drug companies and investors reluctant to pursue stroke therapies, despite their obvious commercial potential. However, based on our learning outcome from past failures, enhanced biological understanding, and identification of new intriguing targets [46, 49, 58, 67, 150, 151], there should be ground for optimism for future stroke drug discovery. Also, translational stroke research has advanced with detailed recommendations for preclinical validation serving as useful guidelines [131, 134, 141, 152–156], and new advancements in primate, rodent, and rabbit models may potentially result in better prediction value [131, 157–160]. Additionally, new clinical paradigms, such as iatrogenic stroke [136, 161] and early ambulance treatment [133, 137–139], allow for more controlled studies or hyper-acute treatment, enhancing study reliability and the chances for success.

In the following section, I analyse the option of targeting oxidative stress in future stroke therapy, discuss various strategies, and present some promising targets in this area.

5 Targeting Oxidative Stress in Stroke: Strategies and Examples

It is well elucidated that oxidative stress is critical for mediating neuronal cell death following brain ischaemia [46, 56, 59]. More than 22 knock-out and transgenic mouse studies clearly demonstrate that deleting SOD1/2/3 from the genome enhances brain damage, while overexpressing SOD1/2/3 leads to protection following transient or permanent focal cerebral ischaemia [118, 121]. Similarly, the neurotoxic effects of nNOS and iNOS and the neuroprotective role of eNOS have been

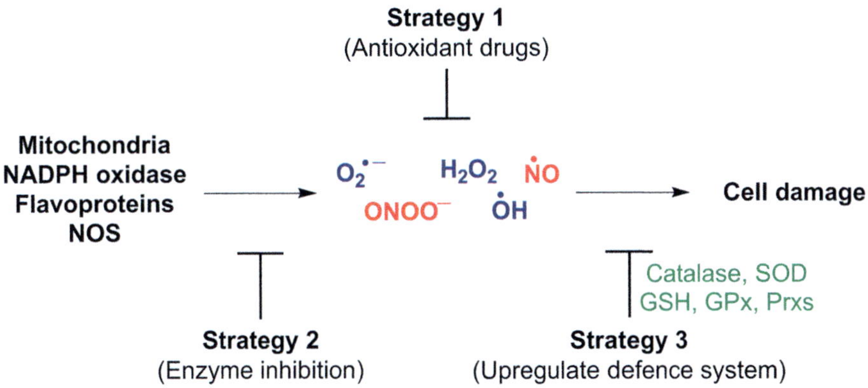

Fig. 8.3 Three overall strategies to inhibit oxidative stress

convincingly demonstrated in 13 studies using genetically manipulated mice; and knocking out NADPH oxidases or PLA_2 or overexpressing GPx, heme oxygenase 1 (HO-1), or thioredoxin reduces infarct volumes [10, 118, 121]. Inhibiting oxidative stress is a direct way of addressing acute neurotoxicity and reperfusion damage, and additionally, oxidative stress is linked to delayed processes, such as vascular damage, apoptosis, and inflammation (Fig. 8.2a). Thus, based on this, oxidative stress is likely an effective target for intervention against ischaemic stroke. So, how do we most efficiently inhibit oxidative stress in stroke? Three overall strategies are outlined here (Fig. 8.3): (1) scavenge ROS/RNS with antioxidants; (2) inhibit the enzymes that produce ROS/RNS; and (3) upregulate endogenous and protective antioxidant enzymes.

5.1 Strategy 1

Following this strategy, the administration of exogenous antioxidants will lead to reactions between the reactive species and the antioxidant, whereby the ROS/RNS are scavenged or converted into harmless molecules. This approach has been applied in many preclinical and experimental studies [6, 8, 162] and even pursued in large clinical stroke trials that tested tirilazad mesylate (U-74006F), ebselen, NXY-059, and edaravone (MCI-186) [140, 142, 163]. *Tirilazad* is an inhibitor of lipid peroxidation and was found to be neuroprotective in several models of focal cerebral ischaemia, especially transient occlusion models [142, 163, 164]. Despite living up to most of the STAIR (stroke therapy academic industry roundtable) criteria [140], the compound failed to show reductions in disability or infarct volume in a clinical study that included 556 ischaemic stroke patients [142, 163]. This could be because patients were treated later than suggested by the preclinical studies, and/or because tirilazad had a narrow effective dose range and was metabolized differently in women [142, 163, 164]. *Ebselen* is a selenium-containing compound that can convert hydrogen peroxide to water and scavenge peroxynitrite. In addition, ebselen enhances the expression of endogenous antioxidant enzymes by covalent modification of cysteines on kelch-like ECH-associated protein 1 (Keap1) and activation of the nuclear erythroid-related factor 2 (Nrf2) (*wide infra*) [165, 166]. Ebselen is neuroprotective in the transient MCAO (tMCAO) model when given before or at reperfusion and also shows modest effects when given 30 min post-occlusion in a permanent model [140, 142, 163]. Interestingly, ebselen showed a synergistic effect with rtPA in the rabbit small clot embolic stroke model (RSCEM) when co-administered 1 h after ischaemia, but alone it was only effective when administered 5 min post-stroke [167]. Despite the seemingly narrow time window in animals, humans were administered ebselen within 48 h post-stroke in a clinical trial that included 302 patients. After three months, there was no positive outcome among the treated patients compared to the placebo group, but the subgroup of patients treated within 24 h actually did show improved outcomes after one month [140, 142, 163]. Therefore, ebselen seems to provide some validation of the antioxidant strategy in humans, although more studies are needed to confirm that; follow-up studies were announced but have not been carried out. *NXY-059* represents one of the most devastating failures in translational stroke research. NXY-059 is a nitrone-based spin-trap molecule that can scavenge free radicals, and it showed remarkable neuroprotective properties in preclinical transient and permanent stroke models, including rats, rabbits (RSCEM model), and monkeys (marmosets) [130, 168–170]. Despite infarct reductions of >40 % in most of the animal studies, a 4–5 h therapeutic time window in rats and monkeys (but only 1 h in rabbits [159]), and a seemingly perfect adherence to the STAIR criteria [140, 142], NXY-059 failed after two large phase 3 clinical trials. The first study (SAINT I) [171, 172] indicated a positive effect relative to placebo, but the larger follow-up study (SAINT II) [173], which included 3195 patients, could not validate this. Subsequent analyses revealed many problems with the compound itself and its development, which retrospectively makes the failure less surprising. Concerns about the extent of blinding, randomization, inclusion/exclusion criteria, and confirmation of actual arterial occlusion during the preclinical assessment of NXY-059 have been expressed [174–176] and have led to more elaborate STAIR criteria in order to avoid bias in future studies [153]. Also, the SAINT trials have been criticized for using inappropriate outcome measures and statistics, applying an extensive time window (<6 h) and including very diverse stroke patients [132, 142, 159, 174]. Mechanistically, it is also important to note that NXY-059 does not permeate the BBB in rats due to its hydrophilicity, and its antioxidant and neuroprotective effects in animals were therefore believed to be exerted at the blood-endothelial cell interface and by interacting with inflammatory cells [168, 169]. Hence, NXY-059 has not tested the clinical relevance of reducing oxidative stress in the brain parenchyma. Also, the spin-trapping potency of NXY-059 is lower than vitamin E and the more hydrophobic and brain-permeable spin-trap compounds stilbazulenyl nitrone (STAZN) and α-phenyl-*N*-tert-butyl nitrone (PBN) [177–179]. Thus, the non-optimal potency and pharmacokinetic properties of NXY-059 could be the main reasons for its clinical failure [132, 142, 159, 174].

The final important compound in the category of clinically tested antioxidants is *edaravone*, which is able to scavenge oxygen radicals, including lipid peroxyl radicals. Edaravone improved neurobehavioral outcomes and reduced infarcts in several preclinical stroke studies [140, 142, 163]. There is some discrepancy in the literature with respect to its time window, as one study of mice exposed to 1 h tMCAO showed neuroprotective effect when edaravone was given 6 h after ischaemia [180], while a similar study in rats did not show effect of edaravone when given 1–2 h post-stroke [181]. In the RSCEM model, edaravone had a long therapeutic time window; it was effective when given 3 h post-embolization and earlier (but not at 6 h), which is about twice as long as that of rtPA and NXY-059 [182]. In an acute ischaemic stroke study with 250 patients, edaravone showed clear positive effects, even though a very broad 72-h treatment window was applied [183]. Also, edaravone resulted in smaller infarct volumes in another clinical study of acute ischaemic stroke when measured within 1 year, although the neurological score was only improved within 1–2 months [184]. A third study of cardioembolic stroke indicated modest improvement in mildly affected patients [185]. Additionally, a range of smaller clinical studies have been performed with positive outcomes [186], and therefore edaravone clearly supports the hypothesis that targeting oxidative stress is a viable strategy for treating acute ischaemic stroke patients [159, 186, 187]. Interestingly, edaravone is approved and widely used in Japan (where most of the clinical trials were conducted) and China as a stroke drug. However, additional large, randomized, and double-blinded clinical studies are warranted to establish a clearer relationship between dose, time window, and positive outcomes [159, 163, 186, 187]. Of interest, edaravone's mode of action is likely to be multifactorial. In addition to its direct antioxidant properties, such as inhibiting lipid peroxidation and protein oxidation, edaravone inhibits apoptosis, inflammation, and oedema in animals. Also, metalloproteinase-9 activity is reduced by edaravone, which potentially explains the inhibition of vascular damage and haemorrhage [126, 186]. Whether the many effects of edaravone are due to pleiotropy (i.e., that inhibition of ROS leads to several seemingly unrelated down-stream effects) or multi-target mechanisms (i.e., that edaravone directly affects several targets, e.g., ROS and specific proteins) have not been elucidated in detail. However, it is notable that the reported beneficial effects are mechanistically linked to oxidative stress, and thus the diverse biological impact caused by edaravone may illustrate the efficiency of targeting oxidative stress.

5.2 *Strategy 2*

The second strategy is to inhibit ROS/RNS-producing enzymes. Rather than trying to eliminate the short-lived, widespread, abundant, and highly reactive species that are already present, as in strategy 1, strategy 2 targets the evil at its root (Fig. 8.3) [76, 188]. By inhibiting pathologically related ROS/RNS production while leaving beneficial redox signalling intact, this could provide a specific and efficient approach [6, 46]. NOS inhibitors serve as illustrative examples, although they were never developed into clinical stroke trials [189, 190]. Initial studies of NOS inhibitors gave conflicting results due to lack of selectivity. Later, selective nNOS and iNOS inhibitors were shown to be neuroprotective in permanent and transient animal stroke models, while non-selective NOS inhibitors gave mixed results, likely due to eNOS activity and effects on cerebral blood flow [10, 90, 191]. The therapeutic time window was generally promising for NOS inhibitors (>1 h) [191] and especially extensive for certain iNOS inhibitors (6–18 h) [90, 192] and nNOS inhibitors, which also had antioxidant properties (8 h) [193]. Based on this, selective nNOS and iNOS inhibitors could have some potential in the treatment for acute ischaemic stroke; however, at the time they were developed, they gave mixed results and showed non-optimal selectivity and drug-like profiles [190, 194], which together with the complex biology of nitric oxide [10, 73, 90] and the many stroke failures probably hindered further development.

XO inhibitors constitute another example of the second strategy. Allopurinol and oxypurinol are close analogues of the natural XO substrates, xanthine and hypoxanthine; they serve as useful experimental XO inhibitors and allopurinol is even used in the treatment of gout. Both compounds have shown promise in transient and permanent stroke models and can reduce infarcts and neurological damage in several species [84]. Importantly, oxypurinol was protective, not only when given as a pretreatment but also when administered 1 h after ischaemia in a permanent MCAO (pMCAO) rat model [195]; however, oxypurinol failed to show neuroprotective effects in a gerbil global ischaemia model when administered 30 min after reperfusion [196], or when given after 2 h of MCAO occlusion followed by 2 h of reperfusion in rats [197]. In humans, allopurinol has shown beneficial effects in asphyxiated infants [198], and a phase IV study (XILO-FIST) has been scheduled to study the long-term effects of allopurinol on stroke recovery and recurrent stroke frequency [199]. Together, these studies indicate an important role for XO during cerebral ischaemia; however, it should be kept in mind that allopurinol and oxypurinol may also act via mechanisms other than XO inhibition, e.g., radical scavenging [84], although a study did show specificity for oxypurinol among the various ROS-producing phases in neurons [72]. Also, allopurinol and oxypurinol are not particularly potent XO inhibitors, and despite the fact that more potent inhibitors exist (e.g., BOF-4272 and febuxostat), they have not been tested in stroke models [84]. Thus, to further explore the potential of XO inhibition in relation to stroke treatment, new compounds need to be identified that are potent, selective, drug-like and sufficiently brain-permeable.

A final important target in this category is SDH, which participates in both the citric acid cycle by converting succinate to fumarate and the mitochondrial electron transport chain as complex II. During ischaemia, succinate accumulates due to the reverse activity of SDH, mediated by fumarate overproduction from the catabolism of purines, and partial reversal of the malate/aspartate shuttle. Upon reperfusion, the accumulated succinate is oxidized to fumarate by SDH, which initiates superoxide production at complex I by reverse electron transport (RET) [86]. Recently, it was convincingly demonstrated that dimethylmalonate inhibits SDH and thereby reduces the accumulation of succinate during ischaemia and the subsequent oxidation of succinate at reperfusion [86]. This strategy was also shown to reduce ischaemia/reperfusion-mediated ROS production via complex I. Dimethylmalonate administered right before and during ischaemia reduced brain infarct by ~50 % and improved neurological scores in rats exposed to 45 min of MCAO and 3 days of reperfusion. Similarly, dimethylmalonate reduced tissue damage in the heart following ischaemia/reperfusion damage [86]. Future studies will hopefully clarify if these promising results can be translated into safe and efficient drug-like molecules with a reasonable therapeutic time window. Regardless, this illustrates how ROS production can be directly inhibited by targeting mitochondrial enzymes.

5.3 *Strategy 3*

The third strategy is to upregulate endogenous and protective antioxidant enzymes instead of direct scavenging or preventing the formation of ROS/RNS (Fig. 8.3). The hypothesis is that the diverse set of ROS/RNS species are best tackled by a battery of antioxidants, and that the catalytic and consistent properties of our endogenous antioxidant enzymes make them exceptionally efficient, more so than drugs designed to scavenge or inhibit ROS/RNS formation. Also, this could offer a more generally applicable antioxidant strategy because knowing the exact source of the ROS/RNS is less crucial for a positive outcome.

In the following, three protein targets will be described in detail, representing the second (postsynaptic density protein-95 (PSD-95) and NADPH oxidase) and third (Keap1) strategies. These are promising targets against ischaemic stroke due to their central role in stroke pathophysiology and proven relevance in disease models of stroke.

5.4 *PSD-95*

PSD-95 is a scaffolding protein found in neuronal synapses, where it interacts with the NMDA receptor and nNOS through its PSD-95/Discs-large/ZO-1 (PDZ) domains. This ternary nNOS/PSD-95/NMDA receptor complex facilitates efficient and localized nitric oxide production as a result of glutamate-mediated Ca^{2+} influx via the NMDA receptor (Fig. 8.4a) [92, 200]. During cerebral ischaemia, excessive release of glutamate leads to extensive NMDA receptor activation and thereby harmful levels of Ca^{2+} and nitric oxide, which ultimately induce neuronal death and brain damage [92,

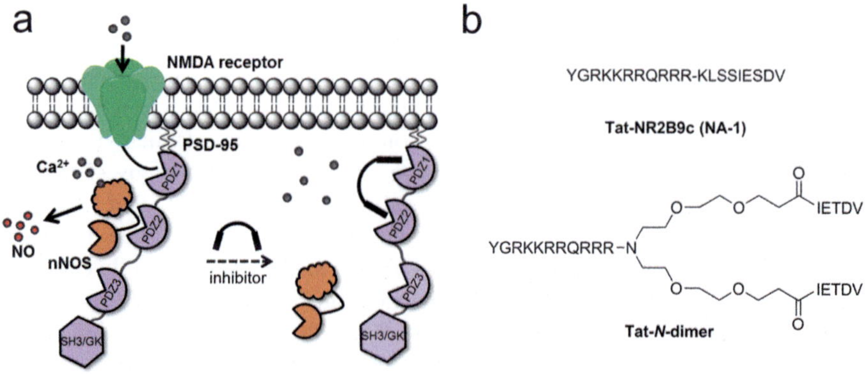

Fig. 8.4 (**a**) PSD-95 links glutamate-mediated Ca^{2+} influx via the NMDA receptor with nitric oxide (NO) production. PSD-95 inhibitors (here a dimeric one) bind PDZ1-2 and block the formation of the ternary nNOS/PSD-95/NMDA receptor complex and thereby reduce NO generation. (**b**) Tat-NR2B9c and Tat-*N*-dimer are peptide-based PSD-95 inhibitors under development

93, 95, 97]. Antisense suppression of PSD-95 in cultured cortical neurons inhibited NMDA receptor-induced neurotoxicity and also nitric oxide generation, as shown by using guanosine 3′,5′-monophosphate (cGMP) as a surrogate measure [92]. Later, similar results were obtained by pharmacological means using a 20-mer peptide, Tat-NR2B9c (Fig. 8.4b), corresponding to the 9 C-terminal amino acids of GluN2B fused to the HIV-1 Tat peptide to facilitate permeability across cell membranes and the BBB [93]. Tat-NR2B9c binds to PDZ1 and PDZ2 of PSD-95 and thereby blocks the formation of the nNOS/PSD-95/NMDA receptor complex and uncouples NMDA receptor activity from nitric oxide production. Tat-NR2B9c improved neurological function and reduced infarct volumes by 55 % when given before reperfusion and by 67 % when administered 1 h after reperfusion following a 90-min period of MCAO in rats [93]. Impressively, similar levels of neuroprotection were achieved when administering Tat-NR2B9c 3 h post-stroke and over a wide range of doses (0.03-3 nmol/g) in the rat tMCAO model [201]. Also, Tat-NR2B9c reduced infarcts and improved neurological scores in various permanent rodent stroke models [201–204], in a mouse tMCAO model [205], and in three versions of a newly developed non-human primate model with a 1–3 h therapeutic window [157, 206]. Based on this strong preclinical foundation, Tat-NR2B9c was tested in clinical trials under the name NA-1. In the ENACT (Evaluating Neuroprotection in Aneurysm Coiling Therapy) phase 2 trial, NA-1 met its primary safety endpoints and reduced the number of small infarcts in patients undergoing endovascular aneurysm repair. There were no observed effects on total infarct volume or clinical outcome, perhaps due to limited group size (185 patients in two groups) [161]. Further studies will hopefully be conducted to evaluate the potential of NA-1 as a neuroprotective drug.

The affinity of Tat-NR2B9c towards its target, PSD-95, is quite low for a drug candidate. Affinity (K_i) values of 5–10 μM were found for Tat-NR2B9c towards PDZ1 and PDZ2 of PSD-95 [207, 208], and an ELISA assay suggested inhibitory IC_{50} values of 0.2–8 μM against PDZ2 interactions with various NMDA receptor tails and nNOS [209]. This seemingly low affinity of Tat-NR2B9c prompted the design of a dimeric compound, Tat-*N*-dimer (Fig. 8.4b), which binds PDZ1 and PDZ2 of PSD-95 simultaneously, leading to 1000-fold higher affinity compared to the monomeric inhibitor Tat-NR2B9c [208]. Tat-*N*-dimer was neuroprotective in a pMCAO mouse model, where it reduced infarct volume by 40 % and improved motor function when administered 30 min post-ischaemia. Under the same conditions and dose (3 nmol/g), Tat-NR2B9c did not show significant protection [208]. Interestingly, both Tat-NR2B9c and Tat-*N*-dimer contain the cell-penetrating peptide (CPP) Tat to improve transport into cells and brain. This is a common trick used in experimental tool compounds, but no marketed drugs with a Tat moiety yet exist. Whether this is due to toxicity—e.g., related to histamine-release, which limited the human maximum tolerated dose of Tat-NR2B9c/NA-1 to 2.6 mg/kg [161, 210]—or the rather unspecific and limited membrane permeability [211] is not known. Currently, no small-molecule PSD-95 inhibitors with reasonable potency exist [212], and in general PDZ domains are difficult to target with conventional drug-like molecules; instead, their binding pockets are more prone to bind peptide-based ligands [213]. Hopefully, the promising results of Tat-NR2B9c/NA-1 in advanced primate models and humans indicate that Tat- and peptide-based PSD-95 inhibitors are sufficiently safe and efficient in the setting of acute ischaemic stroke.

The main protective effect of PSD-95 inhibition is believed to be reduction of nitric oxide and subsequent impairment of RNS-related toxicity. NMDA receptor ion-flux and nNOS expression were not affected by PSD-95 inhibition [92, 93], and Tat-NR2B9c was found to inhibit pro-death signalling pathways mediated by NMDA receptor and nNOS activation of p38, while not impairing pro-survival signalling via cAMP response element-binding protein (CREB) and Akt [202, 214]. Likewise, electrophysiological measurements in neurons treated with peptide-based PSD-95 inhibitors revealed no effects on basal synaptic transmission or induction of long-term potentiation (LTP) [202, 215]; therefore, based on this, PSD-95 inhibition is considered a safe and efficient alternative to NMDA receptor antagonists and nNOS inhibitors as drug candidates for ischaemic brain damage. However, mechanisms other than nitric oxide reduction may contribute to the neuroprotective effects of PSD-95 inhibition. PSD-95 interacts with a range of proteins in the brain, and homologues of PSD-95, such as PSD-93, SAP-97, and SAP-102, are likely also affected by PSD-95 inhibitors [209, 216–218]. Accordingly, cell-permeable peptide-based PSD-95 inhibitors reduced NMDA receptor surface expression by 20 % [219], reduced PSD-95/NMDA receptor colocalization [215], and affected the subunit composition of synaptic NMDA receptors in neurons [220]. The question is if PSD-95 inhibitors affect these systems to an extent that influence the neuroprotective properties—either positively or negatively. A few studies indicate that more than nitric oxide reduction might be going on; for example, Tat-NR2B9c protected striatal medium-sized spiny neurons from NMDA-mediated cell death without the concomitant impairment of nitric oxide levels (i.e., cGMP levels). Of note, this was only seen in neurons transfected with huntingtin to mimic Huntington's disease, which sensitizes neurons to excitotoxicity, and not in the corresponding wild-type neurons [219]. Also, Tat-NR2B9c enhanced CREB-dependent gene expression, which was important for Tat-NRB9c neuroprotection in vivo [204] and known to mediate synaptic NMDA receptor-dependent neuroprotection [221]. One mechanism behind this could be the uncoupling of PSD-95 (and associated proteins) from NMDA receptors, thereby preventing the activation of negative regulators of Ca^{2+}/calmodulin signalling and CREB activation [204].

Recently, it was demonstrated that the Tat-*N*-dimer can ameliorate key aspects of cortical spreading depression (CSD) [222], which is believed to contribute to infarct expansion during cerebral ischaemia [46, 62]. CSD is induced by high levels of extracellular glutamate and potassium ions, leading to a slowly propagating wave of depolarized neurons and glia. Tat-*N*-dimer reduced the drop in direct current (DC) potential by 33 % and partially preserved spontaneous neuronal activity following CSD, without affecting cerebral blood flow or oxygen consumption [222]. The molecular mechanisms behind these effects are not known, but could be due to changes in the dynamics and sorting of transmembrane ion channels. It is not clear if this amelioration of CSD contributes to the neuroprotective properties of Tat-*N*-dimer, but the experiments indicate that PSD-95 inhibition may induce other effects besides those initially presumed [222]. Finally, Tat-NR2B9c was shown to protect against NMDA-induced neurotoxicity by inhibiting p47^{phox} phosphorylation and thereby preventing NOX2 activation and superoxide production [223]. Importantly, the nNOS inhibitor L-N^G-nitroarginine methyl ester (L-NAME) did not suppress superoxide levels, but did protect the neurons. Therefore, these studies indicate that Tat-NR2B9c inhibits NOX2 activation, independent from its effect on nitric oxide production, and that both superoxide and nitric oxide are important for NMDA receptor-mediated toxicity, likely via peroxynitrite formation [223].

Overall, PSD-95 inhibitors are promising drug candidates against acute ischaemic stroke. Their main mode of action is likely via reduction of nitric oxide and RNS toxicity, but additional mechanisms may also contribute.

5.5 *NADPH Oxidase*

NADPH oxidases produce superoxide (NOX1-3, 5) or hydrogen peroxide (NOX4, DUOX1-2) for a range of physiological processes, such as migration, cell survival, and differentiation [19, 20, 76]. In the CNS, NOX2 is believed to play a modulatory role in LTP and memory and participate in intercellular signalling [19]. During ischaemia, the production of ROS is too extensive and causes toxicity. A direct way of attenuating oxidative stress is therefore to inhibit the activity of NADPH oxidase enzymes. All seven isoforms of NADPH oxidase comprise a transmembrane catalytic core subunit (NOX1-5, DUOX1-2) that can transfer an electron from cytosolic NADPH via a FAD coenzyme and two heme groups across the membrane, where oxygen acts as the final electron acceptor. Various membrane and cytosolic proteins are required to get fully active NADPH oxidase enzymes. For example, a maturation and stabilization partner is necessary for NOX1-4 (p22^{phox}) and DUOX1-2 (DUOXA1/2) function, and NOX1 and NOX2 depend on cytosolic organiser (NOXO1 and p47^{phox}, respectively) and activator (NOXA1 and p67^{phox}, respectively) subunits [19, 76]. NOX2 was initially found in phagocytes, where it produces ROS in relation to respiration bursts as a defence mechanism against bacterial infections. NOX2 is activated by phosphorylation of the organizer subunit p47^{phox}, leading to translocation of p47^{phox} and the activator subunit p67^{phox} to the membrane and thus assembling and activation of the entire multi-subunit NOX2 complex (Fig. 8.5a) [19, 76]. Later, NOX2 was identified in brain neurons, microglia, and astrocytes [19, 78, 82], and similarly to phagocytes, neuronal NOX2 is activated by NMDA receptor-mediated Ca^{2+}-influx, presumably via phosphatidylinositol-3-kinase (PI3K) activation of PKCζ, which then phosphorylates p47^{phox} and induces the assembly and activation of NOX2 [81, 224]. To support this, an adaptor protein named APPL1 was shown to interact with the NMDA receptor tail and PI3K simultaneously [225], providing a structural link for mediating efficient PI3K activation upon Ca^{2+} influx. One study showed that the NOS inhibitor L-N^G-nitroarginine (L-NNA) reduced NMDA-induced superoxide production in mouse neocortex and cultured neurons, indicating that nitric oxide also contributes to the activation of NOX2 after NMDA receptor stimulation [82]. However, other studies [81, 91, 223] showed

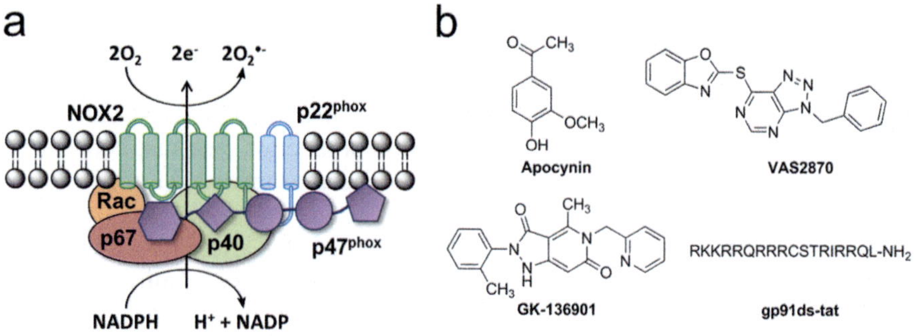

Fig. 8.5 (**a**) The activated and assembled NOX2 multi-subunit enzyme complex consists of membrane (NOX2, p22[phox]) and cytosolic (p47[phox], p67[phox], p40[phox], and Rac) proteins. (**b**) Key NADPH oxidase inhibitors

that NOS inhibitors did not reduce NMDA-induced superoxide production (but NOX2 inhibition did [81, 223]), thereby questioning the role of nitric oxide in NMDA receptor-mediated activation of NOX2.

NOX1, 2, and 4 are the most studied NOX proteins in relation to brain diseases and are all expressed in the cerebrovascular system and brain cells, such as neurons, microglia, and astrocytes [19, 78, 226]. Genetic knock-out studies have been instrumental in elucidating the importance of individual NOX isoforms [76, 78, 227]. NOX1 knock-out gave mixed results in four studies, indicating a minor role of this isoform [78, 227]. For NOX2, nine independent studies have shown that genetic knock-out protects against ischaemic brain damage in rodents, as infarct volumes were reduced by ~50 % on average [228–236]. Interestingly, infarct size was similar to that of wild-type mice when reintroducing NOX2-containing neutrophils to the NOX2 knock-out mice by bone marrow transplantation, indicating that migrating neutrophils are responsible for NOX2-mediated infarct development during ischaemia [228]. However, transplanting NOX2-deficient bone marrow into wild-type mice did not reduce infarct volume [228]. Thus, neutrophil NOX2-mediated ROS production is seemingly important, but not determinant for infarct development, and neuronal NOX2 is likely important too. Similar findings and conclusions have been reported in another study [235]. Knocking out NOX4 has been shown to be neuroprotective (75 % infarct reduction), and interestingly, this study also demonstrated a lack of neuroprotection by knocking out NOX2 [77]. However, the NOX4 knock-out results could not be repeated in another mouse strain [78]. The reasons for the contradicting results could be related to technical issues specific to the various knock-out models, expression levels of NOX2/4, and the differences in transient versus permanent models [78, 227, 237].

In oxygen-glucose-deprived neurons, NOX2 is clearly the primary source of ROS following reoxygenation, as both NOX inhibitors and knocking out the catalytic core subunit of NOX2 reduced ROS production and cell death [72]. Likewise, NOX2 (and not NOS, XO, or mitochondria) was the main producer of ROS and responsible for neuronal death in NMDA-treated cultured neurons and in situ mouse hippocampus, as demonstrated pharmacologically and by gene-deletion of $p47^{phox}$ [81]. Similarly, $p47^{phox}$ knock-out or pharmacological NOX inhibition in wild-type mice reduced superoxide production and neuronal death in vivo following transient cerebral ischaemia [238]. Overall, these studies strongly suggest that NOX2 is a crucial enzyme for producing ROS during cerebral ischaemia and reperfusion and is a potential future drug target against ischaemic stroke.

Current NADPH oxidase inhibitors suffer from many limitations, such as off-target binding, lack of isoform selectivity, and intrinsic anti/pro-oxidative effects, which obviously complicates the interpretation of biological results. Often, they have been identified by functional screening assays, wherefore their exact molecular mechanisms are not known [239–242]. Still, several pharmacological in vivo studies support the idea that NADPH oxidases play a key role in cerebral ischaemia [78, 227]. Apocynin is a naturally occurring organic molecule reported to inhibit the membrane translocation of $p47^{phox}$ and thus activation of the NOX2 complex (Fig. 8.5b) [242, 243]. It has been widely studied in MCAO models under various conditions and generally shows neuroprotective effects when given before or during reperfusion, even after a 2-h ischaemia period, but not when given >0.5 h after reperfusion [78, 227]. However, apocynin has several off-target effects, such as inhibition of rho kinases and both pro-oxidant and antioxidant effects (dependent on dose and cellular conditions), and its activity also depends on its conversion to a dimeric form by myeloperoxidase, which is not present in all cells [78, 239, 242, 243]. Therefore, results based on apocynin should be interpreted with caution. Nevertheless, apocynin reduced infarct volume by ~50 % and inhibited superoxide production in normal mice, but not in NOX2 knock-out mice, when exposed to 0.5-h ischaemia and reperfusion, indicating that NOX2 is the relevant protein for apocynin-mediated neuroprotection [232].

The small molecule triazolopyrimidine VAS2870 (Fig. 8.5b) inhibits NOX1/2/4/5 selectively over eNOS and XO and shows no ROS scavenging effects [242, 244]. The exact mechanism of action is not known, but VAS2870 has been suggested to inhibit NOX assembly or conformational changes important for NOX activity [245]. VAS2870 reduced infarct volume by 75 % in mice exposed to 1 h of tMCAO when administered intrathecally 2 h after occlusion and again at 12 h, and neurological function and motor coordination were also improved [77]. It was the same study that showed reduced infarct volumes in NOX4, but not in NOX2, knock-out mice. Accordingly, administering VAS2870 to NOX4 knock-out mice did not improve infarct volume further, which was interpreted as proof of VAS2870-mediated neuroprotection via NOX4 [77]. An alternative explanation is that maximal neuroprotection was already achieved by knocking out NOX4. Unfortunately, VAS2870 was not tested in their NOX2 knock-out model to explore if VAS2870 acted via NOX2 inhibition. Later, VAS2870 was found to thioalkylate cysteine residues in the ryanodine receptor Ca^{2+} channel (RyR1) and GSH, and it was illustrated that NOX4 activity was sensitive to cysteine alkylation [246]. However, it is not settled if this mechanism can explain the observed neuroprotective effects of VAS2870 [242]. Recently, questions were raised about the ability of this compound to inhibit NOX4 [78].

The pyrazolopyridine compounds GKT136901 (Fig. 8.5b) and GKT137831 potently inhibit NADPH oxidases with a ~10-fold selectivity towards NOX1, 4, and 5 over NOX2, and they do not inhibit XO or scavenge ROS/RNS (except for GKT136901, which reacts with peroxynitrite [247]) [239, 242, 248]. GKT137831 was tested in clinical trials of diabetic nephropathy and is currently explored in fibrotic indications. The neuroprotective properties of these compounds have not been elucidated.

Isoform-specific NOX inhibitors are highly valuable tools for unravelling the cellular roles of individual NADPH oxidases and could facilitate further drug development. NOX2 inhibitors can be achieved with compounds that can prevent the assembly of the NADPH oxidase complex by targeting the protein–protein interactions mediated by the $p47^{phox}$ and/or $p67^{phox}$ subunits [76]. Peptides derived from NOX2 (aka gp91) [249], $p22^{phox}$ [250], or $p47^{phox}$ [251] target the corresponding binding pockets on $p47^{phox}$ and $p67^{phox}$ and inhibit the assembly and thus activation of NOX2. The NOX2-derived gp91d-tat peptide (Fig. 8.5b) reduced superoxide levels in aorta and blood pressure in angiotensin II-induced hypertensive mice [249], inhibited neuronal death in hippocampus in mice exposed to global cerebral ischaemia [252, 253], and reduced oxidative stress and neuronal death in NMDA-treated cultured neurons [82, 223, 224]. It has been suggested that inhibition of the NOX2 subunits $p47^{phox}$/$p67^{phox}$ would also affect NOX1 due to homology and cross-reactivity among their respective activator and organizer subunits [76, 242, 254]; however, gp91ds-tat was shown to inhibit only NOX2 and not NOX1 (or NOX4) in cell-free reconstitution assays [255, 256]. Thus, $p47^{phox}$/$p67^{phox}$ peptide inhibitors demonstrate the feasibility of specifically affecting NOX2 by preventing the assembly of the multi-subunit complex. Interestingly, this strategy has been adapted to small-molecules targeting the $p67^{phox}$-Rac GTPase interaction [257], and in another study ebselen and analogues were able to inhibit the $p22^{phox}$/$p47^{phox}$ interaction likely by binding to the bis-SH3 domain of $p47^{phox}$ [258].

In conclusion, NADPH oxidases play a central role in mediating oxidative stress during cerebral ischaemia. The majority of in vivo data indicate that NOX2 is the relevant isoform, which is corroborated by a wealth of mechanistic studies in neurons; however, NOX4 could also be relevant. To advance the field, isoform-specific and drug-like NOX inhibitors are highly warranted. This could perhaps be achieved with small-molecule inhibitors of $p47^{phox}$ and $p67^{phox}$.

5.6 *Keap1*

Keap1 is a substrate adaptor protein, which under normal conditions binds to Nrf2 in cytosol and targets it for cullin 3-mediated ubiquitination and proteasomal degradation. Keap1 also serves as a redox sensor, as ROS modify sulfhydryl groups on Keap1 and induce a conformational change that prevents ubiquitination and degradation of Nrf2 [4, 259, 260]. Nrf2 then accumulates in the cytosol and translocates to the nucleus, where it forms a transcription factor complex that binds to the

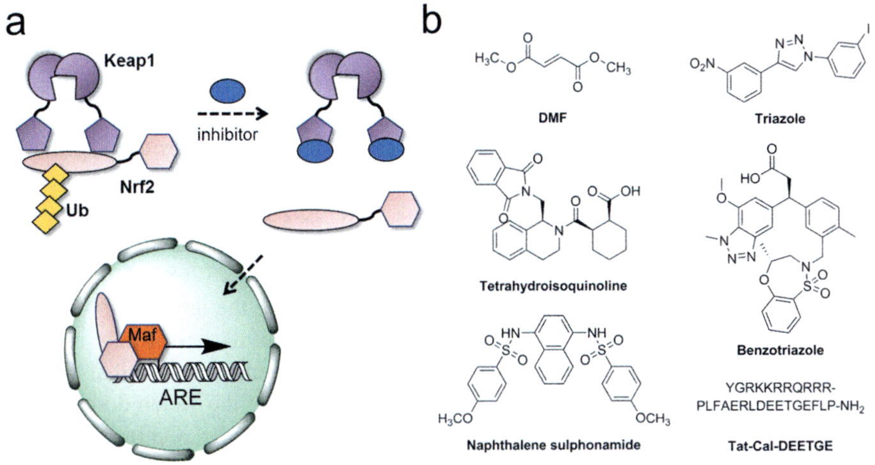

Fig. 8.6 (**a**) Keap1 targets Nrf2 for ubiquitination (Ub) and degradation. Compounds targeting the Kelch domain of Keap1 displace Nrf2 and allow Nrf2 to translocate to the nucleus, where it forms a heterodimer with Maf and initiates transcription of antioxidant enzymes. Covalent Keap1 modifiers and ROS/RNS are also able to affect the Keap1/Nrf2 interaction and thereby reduce Nrf2 degradation and augment Nrf2 translocation to the nucleus. (**b**) Key Keap1 inhibitors

antioxidant response element (ARE) promoter region and induces gene expression of detoxifying antioxidant enzymes, such as catalase, SOD, GPx, thioredoxin, HO-1, ferritin, glutathione reductase, NAD(P)H dehydrogenase (quinone) 1 (NQO1), and glutathione *S*-transferase (GST) (Fig. 8.6a). Amplifying the Nrf2 pathway provides a promising strategy to augment endogenous antioxidant enzymes, whereby oxidative stress and pro-inflammatory redox signalling are reduced (Strategy 3, Fig. 8.3) [54, 69, 259, 261].

Enhanced Nrf2 translocation and gene activation can be obtained with covalent Keap1 modifiers as well as reversible peptide or small-molecule Keap1-Nrf2 inhibitors (Fig. 8.6a) [262]. A plethora of electrophilic compounds, many of which are natural product compounds, covalently react with the key sulfhydryl groups of Keap1 and thus enhance Nrf2 transport to the nucleus [259, 263]. Dimethyl fumarate (DMF, Tecfidera®, Fig. 8.6b) is an important example and approved and marketed drug for multiple sclerosis [264]. DMF is taken orally and converts into monomethyl fumarate (MMF) during absorption, which can permeate blood cells and cross the BBB. MMF covalently affects Cys151 of Keap1, leading to Nrf2-mediated gene transactivation in brain neurons and glial cells, and attenuates oxidative stress and inflammation, explaining its neuroprotective effects and mode of action in relation to multiple sclerosis [265]. In general, the adduct-forming mechanism of covalent modifiers is a cause for concern due to potential unspecific reactions and subsequent toxicity or harmful immune response. Indeed, DMF leads to GSH depletion and affects hydroxycarboxylic acid receptor 2 (HCAR2)—side reactions that may contribute to the mode of action in relation to multiple sclerosis, but also explain some of the side effects [264]. Instead of covalent inhibitors, noncovalent Keap1-Nrf2 inhibitors represent an attractive strategy for future drug development. Several peptide-based inhibitors have been identified, which constitute useful tool compounds and starting points for further peptidomimetic drug discovery [262, 266, 267]. Also, a range of interesting small-molecule Keap1 inhibitors have been presented in the recent literature [268–279]; four series are especially promising (Fig. 8.6b). Tetrahydroisoquinoline-based compounds inhibit the interaction between the Keap1 Kelch domain and Nrf2-derived peptides in inhibition assays with good affinities (K_d/IC_{50} ~1 μM), and they potently induce Nrf2 nuclear translocation and ARE gene expression in cell-based assays [268, 269]. Several X-ray crystal structures displaying the molecular details of the interaction with the Kelch domain were presented [269], which provide useful data for future rational drug design. Compounds based on a naphthalene sulphonamide-scaffold also demonstrate low micromolar affinities in competition assays of the Kelch domain and Nrf2 peptides and are active in cellular ARE reporter assays [270]. Optimization has led to very potent and well-characterized analogues with low nanomolar affinities [271–274] and the ability to reduce inflammation in mice challenged with lipopolysaccharide [272, 274]. The triazole-based compounds (Fig. 8.6b) inhibit Keap1-Kelch/Nrf2 in vitro at low micromolar affinity and in live cells and induced Nrf2-mediated gene expression [275]. Finally, fragment-based drug discovery (FBDD) (*wide infra*) provided novel and very potent benzotriazole-containing Keap1 inhibitors ($K_d \geq 1.3$ nM) (Fig. 8.6b). Several useful X-ray crystal structures were generated through this work, and the optimized analogue induced Nrf2-dependent gene expression in cells and rats and reduced inflammation and restored GSH levels in a rat model of chronic obstructive pulmonary disease (COPD) [276]. The abilities of the above-mentioned compounds to cross the BBB or affect CNS disease have not been reported. A common caveat is that they all have carboxylic acid groups, which, due to its charge at physiological pH, is likely to prevent or decrease brain permeability.

Activation of Nrf2 protects neurons against cerebral ischaemia, haemorrhagic stroke, traumatic brain injury, and neurodegenerative disorders in animals [54, 57, 69, 259]. This has been elucidated by genetic knock-out studies and pharmacological inhibition using covalent modifiers and peptide inhibitors of Keap1. In a tMCAO model, infarct volumes were about 1.8-fold larger in Nrf2-deficient mice than in wild-type mice [280]. In another study, pre-treatment with *tert*-butylhydroquinone (tBHQ), a covalent Keap1 modifier, gave considerable smaller cortical infarcts in rats subjected to 1.5 h of MCAO followed by 24 h of reperfusion [281]. Accordingly, sensorimotor deficiency was reduced and cortical GSH levels were increased. Also, pMCAO induced larger infarcts in Nrf2 knock-out mice than in wild-type animals after 7 days, but not after 1 day, indicating that Nrf2 activity affects delayed processes, such as inflammation or apoptosis. Importantly, tBHQ did not reduce infarct volumes in Nrf2-deficient mice following endothelin-1-induced ischaemia [281], confirming that neuroprotection by tBHQ was obtained via Nrf2 activation. Similar results were obtained with acetyl-11-keto-β-boswellic acid (AKBA), which reduced infarct volume by 34 % when administered at reperfusion after 2 h of MCAO in rats.

Neurological scores were likewise improved and Nrf2-controlled genes upregulated by AKBA. AKBA also protected oxygen-glucose-deprived neurons, but not if the Nrf2 or HO-1 genes were deleted [282]. Further validation of the importance of Keap1 in ischaemia was obtained by in vitro studies. DMF and MMF protected hippocampal slices and neuronal cell lines from cell death following oxygen-glucose deprivation. The Nrf2-pathway was activated by DMF (but not MMF) as shown by increased mRNA and protein levels, and when the Nrf2 gene was silenced with RNA interference, DMF-mediated neuroprotection was abrogated [283].

In a mouse model of intracerebral haemorrhage (ICH), Nrf2-deficient animals experienced a 1.6-fold larger brain injury volume, which correlated with neurological deficits. Also, leukocyte infiltration and ROS production were increased in mice without Nrf2 [284]. In line with these findings, the covalent isothiocyanate Keap1 modifier sulforaphane, injected 30 min after ICH in mice, activated the Nrf2 pathway, reduced oxidative stress and neutrophils in the brain, and improved neurobehavioral scores [285]. Deletion of Nrf2 resulted in behavioural scores worse than wild-type mice after ICH, and noticeably, sulforaphane had no effect on these animals [285]. In a similar ICH study, DMF activated Nrf2 gene expression and reduced ICH-induced brain damage and neurological deficits when given 2 h after ICH in rats. Impressively, DMF improved neurological scores in mice when administered 24 h after ICH and had no effect in Nrf2-deficient animals [286].

The above validation studies were conducted with covalent modifiers. Importantly, similar findings were obtained with the reversible peptide inhibitor Tat-Cal-DEETGE (Fig. 8.6b). This peptide was designed by fusing an Nrf2-derived amino acid sequence (LDEETGEFLP) with Tat and a calpain (Cal) cleavage sequence [287]. The peptide induced expression of Nrf2-controlled genes in brain-injured mice (controlled cortical impact model), but only if the Cal-site was present, and not in uninjured animals. Tat-Cal-DEETGE reduced traumatic brain injury-associated disruption of the BBB following intracerebroventricular injection 2 h before injury and when given directly into the cortex 10 min after injury [287]. Later, this peptide was challenged in a global cerebral ischaemia rat model, where it reduced oxidative stress and neuronal cell death in the hippocampus and improved cognitive function when administered 30 min before ischaemia into the brain ventricles [288]. Mechanistically, Tat-Cal-DEETGE was shown to inhibit the interaction between Keap1 and Nrf2 in the cytosols of hippocampal neurons by using the proximity ligation assay, and it enhanced Nrf2 nuclear location, DNA binding, and gene expression. Tat-Cal-DEETGE also improved neuronal survival and cognition when administered subcutaneously with a mini-pump, starting 1 day after reperfusion and lasting for 9 days [288].

Overall, there is an extensive amount of data demonstrating that Keap1 inhibition leads to the amplification of Nrf2 pathways and neuroprotection in a range of conditions, including haemorrhagic, ischaemic, and traumatic brain injury. These studies were diligently conducted, as they combined pharmacological inhibition with genetic knock-out and biochemical characterization. However, the data were obtained with either covalent Keap1 modifiers or peptides. To advance drug discovery, future efforts should aim at developing drug-like, specific, and reversible Keap1-Nrf2 inhibitors that enter the brain following peripheral administration. Such compounds would be ideal tools to investigate if Keap1 inhibitors could become future drug candidates for treating ischaemic stroke and related brain injuries. Fortunately, a range of compounds already exist showing that the binding pocket of Keap1 is amenable to small-molecule drug discovery.

6 Multi-target Drug Discovery

Because ischaemic stroke is a multifactorial disease, it could be necessary to inhibit several pathways to obtain effects in ischaemic stroke patients [46, 134]. This can be done by polypharmacy, where more than one drug is used to achieve additive or synergistic therapeutic effects [131, 132], as illustrated to be efficient in several animal stroke studies using various compound combinations—e.g., PSD-95/JNK inhibition [202] and rtPA in combination with NXY-059 [130], ebselen [167], edaravone [182], or others [62]. While polypharmacy is promising in the cases where pharmacokinetic and dynamic profiles allow co-treatment, one drawback is the high costs and risks of developing two drugs rather than one (unless one of the compounds is already marketed) [59, 134, 289, 290]. Instead, finding a single compound that modulates several disease-relevant pathways is an attractive strategy [126, 134]. Interestingly, the neuroprotective properties of some of the most effective compounds discussed above (e.g., ebselen, edaravone, and PSD-95 inhibitors) and other treatments (e.g., statins [291] and hypothermia [59]) are believed to be mediated via multiple beneficial mechanisms [126]. A compound can have pleiotropic effects by inhibiting a target that connects to several unrelated down-stream pathways. For example, PSD-95 inhibitors link to nitric oxide and pro-survival pathways. Alternatively, multi-target compounds bind to several targets belonging to separate disease-relevant pathways. In both cases, the compound modulates several processes and thereby represents a potential powerful strategy towards ischaemic stroke.

Many well-established and clinically used drugs interact with more than one target [289]. The multi-target profiles of these drugs were not due to deliberate design strategies, but rather discovered retrospectively. Rational and intentional development of multi-target compounds is a relatively new approach. The main challenge is to tailor the selectivity profile of the compound so that desired targets are affected, without the compounds becoming promiscuous and binding irrelevant or even harmful targets. Also, finding multi-target starting points is not straightforward. One approach is high-throughput screening (HTS) against prespecified targets, followed by optimization to obtain the desired selectivity profile [289]. Functional screening assays followed by delineation of the mechanism of action are also an option [292]. Alternatively, two existing compounds binding to separate targets can be linked or merged together to produce one compound with

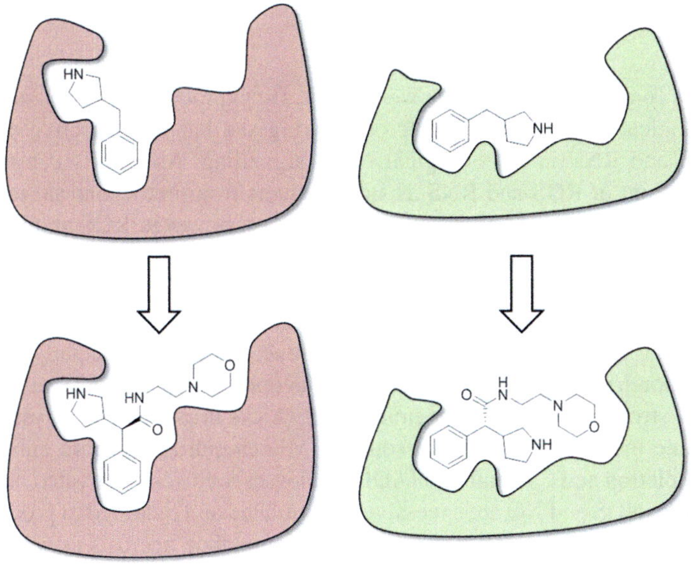

Fig. 8.7 The principle of using fragment-based drug discovery for finding multi-target inhibitors

dual activity [289]; an example of this is BN 80933, where a NOS inhibitor is linked to an antioxidant moiety [193]. The challenge in linking is that the molecules can become too big to cross the BBB or lose other drug-like properties. Merging molecules while still preserving activity toward the desired targets is a great medicinal chemistry challenge.

FBDD provides another interesting and promising approach for developing multi-target inhibitors (Fig. 8.6). FBDD has, in recent years, grown into a powerful strategy for identifying potent and drug-like small-molecule protein inhibitors and has provided several clinical candidates [293–295]. The principle of FBDD is to screen for small substructures (fragments) of drug-like compounds with molecular weights of 100–300 Da and few functional and hydrogen-bonding groups [296]. This increases the chances for finding hits, as long as sensitive detection methods are applied, and the fragment-hits often bind more closely and efficiently to the target protein due to fewer clashes and functionalities. The fragment-hits can then be converted to highly optimized and drug-like structures. For finding multi-target starting points, FBDD is particularly useful. By screening fragments towards the desired targets, fragments that recognize several targets can be found (Fig. 8.7). Importantly, due to the lower complexity of fragments, the chances of finding multi-target hits are higher than when screening normal-sized compounds [297–299]. Although the subsequent optimization is likely challenging, this approach provides a unique and direct search for multi-target inhibitors. Promisingly, examples of this approach have been presented in recent literature [300–302].

7 Conclusion and Perspectives

Oxygen is fundamental for life as we know it. By exploiting the chemistry of oxygen, biological systems are capable of utilizing and handling reactive molecules like ROS and RNS in physiological redox signalling. Also, cells can tolerate a certain amount of ROS and RNS as side products from mitochondrial respiration and other biochemical reactions. In many diseases, however, ROS and RNS overwhelm our antioxidant defence system, and their diverse and toxic reactivity can lead to cell death and physiological deterioration. Interestingly, it is often the same sources that produce ROS and RNS during physiological conditions, which are also responsible for oxidative stress in disease. Three groups are especially important: the mitochondria, NADPH oxidases, and flavoproteins (e.g., NOS and XO). In ischaemic stroke, oxidative stress immediately kicks in as a result of dysregulation of the three mentioned ROS/RNS sources. Mitochondrial ROS are enhanced by energy depletion and Ca^{2+} influx, NADPH oxidases are overly activated, and flavoproteins like NOS and XO are excessively stimulated and modified to produce ROS and RNS. The oxidative stress contributes to immediate necrosis as well as more delayed processes, like apoptosis and inflammation, and thereby plays a central function in the development of infarction and concomitant brain damage seen in stroke. A key question is if targeting oxidative stress can provide new treatment strategies against ischaemic stroke. Ischaemic stroke is a major cause of death and disability and imposes a huge burden on affected individuals and society. Unfortunately, there have been many failures in the ischaemic stroke drug discovery field, especially in the late 1990s, which have caused resignation and reluctance to invest resources into the area. However, detailed analyses of past clinical failures have identified the many shortcomings of these trials. New animal models have appeared and extensive guidelines for improving the quality and translational value of preclinical models now exist. Also, new treatment paradigms and logistics facilitate hyper-acute treatment. Therefore, there are reasons to be optimistic about future stroke drug discovery efforts. A vast amount of preclinical experiments and, importantly, even evidence from clinical studies suggest that oxidative stress plays a key role in ischaemic stroke. PSD-95, NADPH oxidase, and Keap1 represent specific protein targets with a clear mechanistic link to oxidative stress and cerebral ischaemia, and their inhibition leads to significant neuroprotection in animals. For PSD-95, there are indications from a clinical trial that corroborate its potential as a target for treating stroke. Finally, the multifactorial character of ischaemic stroke may necessitate the development of multi-target or pleiotropic drugs, and the development of such compounds should be a future direction for ischaemic stroke drug discovery.

Conflict of Interest The author is the cofounder of Avilex Pharma, which develops peptide-based PSD-95 inhibitors for stroke treatment.

References

1. Li XY, Fang P, Mai JT, Choi ET, Wang H, Yang XF (2013) Targeting mitochondrial reactive oxygen species as novel therapy for inflammatory diseases and cancers. J Hematol Oncol 6
2. Schieber M, Chandel NS (2014) ROS function in redox signaling and oxidative stress. Curr Biol 24(10):R453–R462
3. Novo E, Parola M (2008) Redox mechanisms in hepatic chronic wound healing and fibrogenesis. Fibrogenesis Tissue Repair 1(1):5
4. D'Autreaux B, Toledano MB (2007) ROS as signalling molecules: mechanisms that generate specificity in ROS homeostasis. Nat Rev Mol Cell Biol 8(10):813–824
5. Winterbourn CC (2008) Reconciling the chemistry and biology of reactive oxygen species. Nat Chem Biol 4(5):278–286
6. Kalogeris T, Bao Y, Korthuis RJ (2014) Mitochondrial reactive oxygen species: a double edged sword in ischemia/reperfusion vs preconditioning. Redox Biol 2:702–714
7. Rhee SG (2006) Cell signaling. H2O2, a necessary evil for cell signaling. Science 312(5782): 1882–1883
8. Rodrigo R, Fernandez-Gajardo R, Gutierrez R, Matamala JM, Carrasco R, Miranda-Merchak A et al (2013) Oxidative stress and pathophysiology of ischemic stroke: novel therapeutic opportunities. CNS Neurol Disord Drug Targets 12(5):698–714
9. Uttara B, Singh AV, Zamboni P, Mahajan RT (2009) Oxidative stress and neurodegenerative diseases: a review of upstream and downstream antioxidant therapeutic options. Curr Neuropharmacol 7(1):65–74
10. Pacher P, Beckman JS, Liaudet L (2007) Nitric oxide and peroxynitrite in health and disease. Physiol Rev 87(1):315–424
11. Imlay JA (2013) The molecular mechanisms and physiological consequences of oxidative stress: lessons from a model bacterium. Nat Rev Microbiol 11(7):443–454
12. Naqui A, Chance B, Cadenas E (1986) Reactive oxygen intermediates in biochemistry. Annu Rev Biochem 55:137–166
13. Cadenas E, Davies KJ (2000) Mitochondrial free radical generation, oxidative stress, and aging. Free Radic Biol Med 29(3-4):222–230
14. Walling C (1975) Fentons reagent revisited. Acc Chem Res 8(4):125–131
15. Thomas C, Mackey MM, Diaz AA, Cox DP (2009) Hydroxyl radical is produced via the Fenton reaction in submitochondrial particles under oxidative stress: implications for diseases associated with iron accumulation. Redox Rep 14(3):102–108
16. Bidmon HJ, Emde B, Kowalski T, Schmitt M, Mayer B, Kato K et al (2001) Nitric oxide synthase-I containing cortical interneurons co-express antioxidative enzymes and anti-apoptotic Bcl-2 following focal ischemia: evidence for direct and indirect mechanisms towards their resistance to neuropathology. J Chem Neuroanat 22(3):167–184
17. Turrens JF (2003) Mitochondrial formation of reactive oxygen species. J Physiol 552(Pt 2):335–344
18. Han D, Antunes F, Canali R, Rettori D, Cadenas E (2003) Voltage-dependent anion channels control the release of the superoxide anion from mitochondria to cytosol. J Biol Chem 278(8):5557–5563
19. Bedard K, Krause KH (2007) The NOX family of ROS-generating NADPH oxidases: physiology and pathophysiology. Phys Rev 87(1):245–313
20. Brown DI, Griendling KK (2009) Nox proteins in signal transduction. Free Radic Biol Med 47(9):1239–1253
21. Lienhart WD, Gudipati V, Macheroux P (2013) The human flavoproteome. Arch Biochem Biophys 535(2):150–162
22. Imlay JA (2003) Pathways of oxidative damage. Annu Rev Microbiol 57:395–418

23. Messner KR, Imlay JA (2002) Mechanism of superoxide and hydrogen peroxide formation by fumarate reductase, succinate dehydrogenase, and aspartate oxidase. J Biol Chem 277(45): 42563–42571
24. Muller FL, Lustgarten MS, Jang Y, Richardson A, Van Remmen H (2007) Trends in oxidative aging theories. Free Radic Biol Med 43(4):477–503
25. Phung CD, Ezieme JA, Turrens JF (1994) Hydrogen-peroxide metabolism in skeletal-muscle mitochondria. Arch Biochem Biophys 315(2):479–482
26. Radi R, Turrens JF, Chang LY, Bush KM, Crapo JD, Freeman BA (1991) Detection of catalase in rat heart mitochondria. J Biol Chem 266(32):22028–22034
27. Berndt C, Lillig CH, Holmgren A (2007) Thiol-based mechanisms of the thioredoxin and glutaredoxin systems: implications for diseases in the cardiovascular system. Am J Physiol Heart Circ Physiol 292(3):H1227–H1236
28. Arnér ES, Holmgren A (2000) Physiological functions of thioredoxin and thioredoxin reductase. Eur J Biochem 267(20):6102–6109
29. Wood ZA, Schroder E, Robin Harris J, Poole LB (2003) Structure, mechanism and regulation of peroxiredoxins. Trends Biochem Sci 28(1):32–40
30. Maxwell SRJ, Thomason H, Sandler D, Leguen C, Baxter MA, Thorpe GHG et al (1997) Antioxidant status in patients with uncomplicated insulin-dependent and non-insulin-dependent diabetes mellitus. Eur J Clin Invest 27(6):484–490
31. Gilgun-Sherki Y, Melamed E, Offen D (2004) The role of oxidative stress in the pathogenesis of multiple sclerosis: the need for effective antioxidant therapy. J Neurol 251(3):261–268
32. Guzik TJ, West NE, Pillai R, Taggart DP, Channon KM (2002) Nitric oxide modulates superoxide release and peroxynitrite formation in human blood vessels. Hypertension 39(6):1088–1094
33. Madamanchi NR, Runge MS (2007) Mitochondrial dysfunction in atherosclerosis. Circ Res 100(4):460–473
34. Grzelak A, Soszynski M, Bartosz G (2000) Inactivation of antioxidant enzymes by peroxynitrite. Scand J Clin Lab Invest 60(4):253–258
35. Peluffo G, Radi R (2007) Biochemistry of protein tyrosine nitration in cardiovascular pathology. Cardiovasc Res 75(2):291–302
36. West AP, Shadel GS, Ghosh S (2011) Mitochondria in innate immune responses. Nat Rev Immunol 11(6):389–402
37. Valavanidis A, Vlachogianni T, Fiotakis K, Loridas S (2013) Pulmonary oxidative stress, inflammation and cancer: respirable particulate matter, fibrous dusts and ozone as major causes of lung carcinogenesis through reactive oxygen species mechanisms. Int J Environ Res Public Health 10(9):3886–3907
38. Maiuri MC, Zalckvar E, Kimchi A, Kroemer G (2007) Self-eating and self-killing: crosstalk between autophagy and apoptosis. Nat Rev Mol Cell Biol 8(9):741–752
39. Dalleau S, Baradat M, Gueraud F, Huc L (2013) Cell death and diseases related to oxidative stress: 4-hydroxynonenal (HNE) in the balance. Cell Death Differ 20(12):1615–1630
40. Bernardi P, Petronilli V, Di Lisa F, Forte M (2001) A mitochondrial perspective on cell death. Trends Biochem Sci 26(2):112–117
41. Murphy AN, Fiskum G, Beal MF (1999) Mitochondria in neurodegeneration: bioenergetic function in cell life and death. J Cereb Blood Flow Metab 19(3):231–245
42. Sims NR, Muyderman H (2010) Mitochondria, oxidative metabolism and cell death in stroke. Biochem Biophys Acta 1802(1):80–91
43. Scherz-Shouval R, Elazar Z (2007) ROS, mitochondria and the regulation of autophagy. Trends Cell Biol 17(9):422–427
44. Hampton MB, Orrenius S (1997) Dual regulation of caspase activity by hydrogen peroxide: implications for apoptosis. FEBS Lett 414(3):552–556
45. Elmore S (2007) Apoptosis: a review of programmed cell death. Toxicol Pathol 35(4):495–516
46. Moskowitz MA, Lo EH, Iadecola C (2010) The science of stroke: mechanisms in search of treatments. Neuron 67(2):181–198
47. Virag L, Robaszkiewicz A, Rodriguez-Vargas JM, Oliver FJ (2013) Poly(ADP-ribose) signaling in cell death. Mol Aspects Med 34(6):1153–1167

48. Ekblad T, Camaioni E, Schuler H, Macchiarulo A (2013) PARP inhibitors: polypharmacology versus selective inhibition. FEBS J 280(15):3563–3575
49. Lai TW, Zhang S, Wang YT (2014) Excitotoxicity and stroke: identifying novel targets for neuroprotection. Prog Neurobiol 115:157–188
50. Kaminski MM, Roth D, Krammer PH, Gulow K (2013) Mitochondria as oxidative signaling organelles in T-cell activation: physiological role and pathological implications. Arch Immunol Ther Exp (Warsz) 61(5):367–384
51. Feigin VL, Forouzanfar MH, Krishnamurthi R, Mensah GA, Connor M, Bennett DA et al (2014) Global and regional burden of stroke during 1990-2010: findings from the Global Burden of Disease Study 2010. Lancet 383(9913):245–254
52. Lozano R, Naghavi M, Foreman K, Lim S, Shibuya K, Aboyans V et al (2012) Global and regional mortality from 235 causes of death for 20 age groups in 1990 and 2010: a systematic analysis for the Global Burden of Disease Study 2010. Lancet 380(9859):2095–2128
53. World Health Organization (WHO). The top ten causes of death. http://www.who.int/mediacentre/factsheets/fs310/en/ (Updated May 2014).
54. Lakhan SE, Kirchgessner A, Hofer M (2009) Inflammatory mechanisms in ischemic stroke: therapeutic approaches. J Transl Med 7:97
55. Mozaffarian D, Benjamin EJ, Go AS, Arnett DK, Blaha MJ, Cushman M et al (2015) Heart disease and stroke statistics—2015 update, a report from the American Heart Association. Circulation 131(4):E29–E322
56. Dirnagl U, Iadecola C, Moskowitz MA (1999) Pathobiology of ischaemic stroke: an integrated view. Trends Neurosci 22(9):391–397
57. Iadecola C, Anrather J (2011) Stroke research at a crossroad: asking the brain for directions. Nat Neurosci 14(11):1363–1368
58. Stankowski JN, Gupta R (2011) Therapeutic targets for neuroprotection in acute ischemic stroke: lost in translation? Antioxid Redox Signal 14(10):1841–1851
59. Sutherland BA, Minnerup J, Balami JS, Arba F, Buchan AM, Kleinschnitz C (2012) Neuroprotection for ischaemic stroke: translation from the bench to the bedside. Int J Stroke 7(5):407–418
60. Furlan AJ (2015) Endovascular therapy for stroke—it's about time. N Engl J Med 372(24):2347–2349
61. Saver JL, Goyal M, Bonafe A, Diener HC, Levy EI, Pereira VM et al (2015) Stent-retriever thrombectomy after intravenous t-PA vs. t-PA alone in stroke. N Engl J Med 372(24):2285–2295
62. Lo EH, Dalkara T, Moskowitz MA (2003) Mechanisms, challenges and opportunities in stroke. Nat Rev Neurosci 4(5):399–415
63. Aarts MM, Tymianski M (2003) Novel treatment of excitotoxicity: targeted disruption of intracellular signalling from glutamate receptors. Biochem Pharmacol 66(6):877–886
64. Nicholls DG (2008) Oxidative stress and energy crises in neuronal dysfunction. Ann N Y Acad Sci 1147:53–60
65. Choi DW, Rothman SM (1990) The role of glutamate neurotoxicity in hypoxic-ischemic neuronal death. Annu Rev Neurosci 13:171–182
66. Besancon E, Guo SZ, Lok J, Tymianski M, Lo EH (2008) Beyond NMDA and AMPA glutamate receptors: emerging mechanisms for ionic imbalance and cell death in stroke. Trends Pharmacol Sci 29(5):268–275
67. Tymianski M (2011) Emerging mechanisms of disrupted cellular signaling in brain ischemia. Nat Neurosci 14(11):1369–1373
68. Adibhatla RM, Hatcher JF, Dempsey RJ (2003) Phospholipase A2, hydroxyl radicals, and lipid peroxidation in transient cerebral ischemia. Antioxid Redox Signal 5(5):647–654
69. Adibhatla RM, Hatcher JF (2010) Lipid oxidation and peroxidation in CNS health and disease: from molecular mechanisms to therapeutic opportunities. Antioxid Redox Signal 12(1):125–169
70. Li WJ, Wu SS, Ahmad M, Jiang JF, Liu H, Nagayama T et al (2010) The cyclooxygenase site, but not the peroxidase site of cyclooxygenase-2 is required for neurotoxicity in hypoxic and ischemic injury. J Neurochem 113(4):965–977

71. Yuan J (2009) Neuroprotective strategies targeting apoptotic and necrotic cell death for stroke. Apoptosis 14(4):469–477
72. Abramov AY, Scorziello A, Duchen MR (2007) Three distinct mechanisms generate oxygen free radicals in neurons and contribute to cell death during anoxia and reoxygenation. J Neurosci 27(5):1129–1138
73. Nicholls DG (2004) Mitochondrial dysfunction and glutamate excitotoxicity studied in primary neuronal cultures. Curr Mol Med 4(2):149–177
74. Fiskum G, Murphy AN, Beal MF (1999) Mitochondria in neurodegeneration: acute ischemia and chronic neurodegenerative diseases. J Cereb Blood Flow Metab 19(4):351–369
75. Leung AWC, Halestrap AP (2008) Recent progress in elucidating the molecular mechanism of the mitochondrial permeability transition pore. Biochim Biophys Acta 1777(7-8):946–952
76. Drummond GR, Selemidis S, Griendling KK, Sobey CG (2011) Combating oxidative stress in vascular disease: NADPH oxidases as therapeutic targets. Nat Rev Drug Discov 10(6):453–471
77. Kleinschnitz C, Grund H, Wingler K, Armitage ME, Jones E, Mittal M et al. Post-stroke inhibition of induced NADPH oxidase type 4 prevents oxidative stress and neurodegeneration. PLoS Biol. 2010;8(9).
78. Kahles T, Brandes RP (2013) Which NADPH oxidase isoform is relevant for ischemic stroke? The case for Nox 2. Antioxid Redox Signal 18(12):1400–1417
79. Dikalov S (2011) Cross talk between mitochondria and NADPH oxidases. Free Radic Biol Med 51(7):1289–1301
80. Block K, Gorin Y, Abboud HE (2009) Subcellular localization of Nox4 and regulation in diabetes. Proc Natl Acad Sci U S A 106(34):14385–14390
81. Brennan AM, Suh SW, Won SJ, Narasimhan P, Kauppinen TM, Lee H et al (2009) NADPH oxidase is the primary source of superoxide induced by NMDA receptor activation. Nat Neurosci 12(7):857–863
82. Girouard H, Wang G, Gallo EF, Anrather J, Zhou P, Pickel VM et al (2009) NMDA receptor activation increases free radical production through nitric oxide and NOX2. J Neurosci 29(8):2545–2552
83. Nishino T, Okamoto K, Kawaguchi Y, Hori H, Matsumura T, Eger BT et al (2005) Mechanism of the conversion of xanthine dehydrogenase to xanthine oxidase: identification of the two cysteine disulfide bonds and crystal structure of a non-convertible rat liver xanthine dehydrogenase mutant. J Biol Chem 280(26):24888–24894
84. Pacher P, Nivorozhkin A, Szabo C (2006) Therapeutic effects of xanthine oxidase inhibitors: renaissance half a century after the discovery of allopurinol. Pharmacol Rev 58(1):87–114
85. Bonini MG, Miyamoto S, Di Mascio P, Augusto O (2004) Production of the carbonate radical anion during xanthine oxidase turnover in the presence of bicarbonate. J Biol Chem 279(50): 51836–51843
86. Chouchani ET, Pell VR, Gaude E, Aksentijevic D, Sundier SY, Robb EL et al (2014) Ischaemic accumulation of succinate controls reperfusion injury through mitochondrial ROS. Nature 515(7527):431–435
87. Forstermann U (2010) Nitric oxide and oxidative stress in vascular disease. Pflugers Arch 459(6):923–939
88. Endres M, Laufs U, Liao JK, Moskowitz MA (2004) Targeting eNOS for stroke protection. Trends Neurosci 27(5):283–289
89. Parathath SR, Gravanis I, Tsirka SE (2007) Nitric oxide synthase isoforms undertake unique roles during excitotoxicity. Stroke 38(6):1938–1945
90. Iadecola C (1997) Bright and dark sides of nitric oxide in ischemic brain injury. Trends Neurosci 20(3):132–139
91. Lafon-Cazal M, Pietri S, Culcasi M, Bockaert J (1993) NMDA-dependent superoxide production and neurotoxicity. Nature 364(6437):535–537
92. Sattler R, Xiong Z, Lu WY, Hafner M, MacDonald JF, Tymianski M (1999) Specific coupling of NMDA receptor activation to nitric oxide neurotoxicity by PSD-95 protein. Science 284(5421):1845–1848

93. Aarts M, Liu Y, Liu L, Besshoh S, Arundine M, Gurd JW et al (2002) Treatment of ischemic brain damage by perturbing NMDA receptor—PSD-95 protein interactions. Science 298(5594):846–850
94. Aarts MM, Tymianski M (2004) Molecular mechanisms underlying specificity of excitotoxic signaling in neurons. Curr Mol Med 4(2):137–147
95. Dawson VL, Dawson TM, London ED, Bredt DS, Snyder SH (1991) Nitric oxide mediates glutamate neurotoxicity in primary cortical cultures. Proc Natl Acad Sci U S A 88(14):6368–6371
96. Nowicki JP, Duval D, Poignet H, Scatton B (1991) Nitric oxide mediates neuronal death after focal cerebral ischemia in the mouse. Eur J Pharmacol 204(3):339–340
97. Huang Z, Huang PL, Panahian N, Dalkara T, Fishman MC, Moskowitz MA (1994) Effects of cerebral ischemia in mice deficient in neuronal nitric oxide synthase. Science 265(5180): 1883–1885
98. Dawson VL, Dawson TM, Bartley DA, Uhl GR, Snyder SH (1993) Mechanisms of nitric oxide-mediated neurotoxicity in primary brain cultures. J Neurosci 13(6):2651–2661
99. Dawson VL, Kizushi VM, Huang PL, Snyder SH, Dawson TM (1996) Resistance to neurotoxicity in cortical cultures from neuronal nitric oxide synthase-deficient mice. J Neurosci 16(8):2479–2487
100. Xia Y, Dawson VL, Dawson TM, Snyder SH, Zweier JL (1996) Nitric oxide synthase generates superoxide and nitric oxide in arginine-depleted cells leading to peroxynitrite-mediated cellular injury. Proc Natl Acad Sci U S A 93(13):6770–6774
101. Lipton SA, Choi YB, Pan ZH, Lei SZ, Chen HS, Sucher NJ et al (1993) A redox-based mechanism for the neuroprotective and neurodestructive effects of nitric oxide and related nitroso-compounds. Nature 364(6438):626–632
102. Beckman JS, Beckman TW, Chen J, Marshall PA, Freeman BA (1990) Apparent hydroxyl radical production by peroxynitrite: implications for endothelial injury from nitric oxide and superoxide. Proc Natl Acad Sci U S A 87(4):1620–1624
103. Aarts M, Iihara K, Wei WL, Xiong ZG, Arundine M, Cerwinski W et al (2003) A key role for TRPM7 channels in anoxic neuronal death. Cell 115(7):863–877
104. Szabo C (1996) DNA strand breakage and activation of poly-ADP ribosyltransferase: a cytotoxic pathway triggered by peroxynitrite. Free Radic Biol Med 21(6):855–869
105. Chiarugi A (2005) Poly (ADP-ribosyl)ation and stroke. Pharmacol Res 52(1):15–24
106. Mandir AS, Poitras MF, Berliner AR, Herring WJ, Guastella DB, Feldman A et al (2000) NMDA but not non-NMDA excitotoxicity is mediated by poly(ADP-ribose) polymerase. J Neurosci 20(21):8005–8011
107. Zaobornyj T, Ghafourifar P (2012) Strategic localization of heart mitochondrial NOS: a review of the evidence. Am J Physiol Heart Circ Physiol 303(11):H1283–H1293
108. Ghafourifar P, Cadenas E (2005) Mitochondrial nitric oxide synthase. Trends Pharmacol Sci 26(4):190–195
109. Brookes PS (2004) Mitochondrial nitric oxide synthase. Mitochondrion 3(4):187–204
110. Dalkara T, Endres M, Moskowitz MA (1998) Mechanisms of NO neurotoxicity. Prog Brain Res 118:231–239
111. Lynch RE, Fridovich I (1978) Permeation of the erythrocyte stroma by superoxide radical. J Biol Chem 253(13):4697–4699
112. Fisher AB (2009) Redox signaling across cell membranes. Antioxid Redox Signal 11(6): 1349–1356
113. Macfadyen AJ, Reiter C, Zhuang YX, Beckman JS (1999) A novel superoxide dismutase-based trap for peroxynitrite used to detect entry of peroxynitrite into erythrocyte ghosts. Chem Res Toxicol 12(3):223–229
114. Denicola A, Souza JM, Radi R (1998) Diffusion of peroxynitrite across erythrocyte membranes. Proc Natl Acad Sci U S A 95(7):3566–3571
115. Marla SS, Lee J, Groves JT (1997) Peroxynitrite rapidly permeates phospholipid membranes. Proc Natl Acad Sci U S A 94(26):14243–14248
116. Turko IV, Murad F (2002) Protein nitration in cardiovascular diseases. Pharmacol Rev 54(4):619–634

117. Lipton P (1999) Ischemic cell death in brain neurons. Physiol Rev 79(4):1431–1568
118. Wong CHY, Crack PJ (2008) Modulation of neuro-inflammation and vascular response by oxidative stress following cerebral ischemia-reperfusion injury. Curr Med Chem 15(1):1–14
119. Dringen R (2000) Metabolism and functions of glutathione in brain. Prog Neurobiol 62(6):649–671
120. Granger DN (1988) Role of xanthine oxidase and granulocytes in ischemia-reperfusion injury. Am J Physiol 255(6 Pt 2):H1269–H1275
121. Chan PH (2001) Reactive oxygen radicals in signaling and damage in the ischemic brain. J Cereb Blood Flow Metab 21(1):2–14
122. The National Institute of Neurological Disorders and Stroke rt-PA Stroke Study Group (1995) Tissue plasminogen activator for acute ischemic stroke. N Engl J Med 333(24):1581–1587
123. Lees KR, Bluhmki E, von Kummer R, Brott TG, Toni D, Grotta JC et al (2010) Time to treatment with intravenous alteplase and outcome in stroke: an updated pooled analysis of ECASS, ATLANTIS, NINDS, and EPITHET trials. Lancet 375(9727):1695–1703
124. Emberson J, Lees KR, Lyden P, Blackwell L, Albers G, Bluhmki E et al (2014) Effect of treatment delay, age, and stroke severity on the effects of intravenous thrombolysis with alteplase for acute ischaemic stroke: a meta-analysis of individual patient data from randomised trials. Lancet 384(9958):1929–1935
125. Hacke W, Kaste M, Bluhmki E, Brozman M, Davalos A, Guidetti D et al (2008) Thrombolysis with alteplase 3 to 4.5 hours after acute ischemic stroke. N Engl J Med 359(13):1317–1329
126. Lapchak PA (2011) Emerging therapies: pleiotropic multi-target drugs to treat stroke victims. Transl Stroke Res 2(2):129–135
127. Fonarow GC, Smith EE, Saver JL, Reeves MJ, Bhatt DL, Grau-Sepulveda MV et al (2011) Timeliness of tissue-type plasminogen activator therapy in acute ischemic stroke: patient characteristics, hospital factors, and outcomes associated with door-to-needle times within 60 minutes. Circulation 123(7):750–758
128. Berkhemer OA, Fransen PS, Beumer D, van den Berg LA, Lingsma HF, Yoo AJ et al (2015) A randomized trial of intraarterial treatment for acute ischemic stroke. N Engl J Med 372(1):11–20
129. Cheng YD, Al-Khoury L, Zivin JA (2004) Neuroprotection for ischemic stroke: two decades of success and failure. NeuroRx 1(1):36–45
130. Lapchak PA, Araujo DM, Song D, Wei J, Zivin JA (2002) Neuroprotective effects of the spin trap agent disodium-[(tert-butylimino)methyl]benzene-1,3-disulfonate *N*-oxide (generic NXY-059) in a rabbit small clot embolic stroke model: combination studies with the thrombolytic tissue plasminogen activator. Stroke 33(5):1411–1415
131. Kahle MP, Bix GJ (2012) Successfully climbing the "STAIRs": surmounting failed translation of experimental ischemic stroke treatments. Stroke Res Treat 2012:374098
132. Lapchak PA, Araujo DM (2007) Advances in ischemic stroke treatment: neuroprotective and combination therapies. Expert Opin Emerg Drugs 12(1):97–112
133. Garber K (2007) Stroke treatment-light at the end of the tunnel? Nat Biotechnol 25(8):838–840
134. Savitz SI, Fisher M (2007) Future of neuroprotection for acute stroke: in the aftermath of the SAINT trials. Ann Neurol 61(5):396–402
135. Tymianski M (2013) Novel approaches to neuroprotection trials in acute ischemic stroke. Stroke 44(10):2942–2950
136. Tymianski M (2010) Can molecular and cellular neuroprotection be translated into therapies for patients? Yes, but not the way we tried it before. Stroke 41(10):S87–S90
137. Saver JL, Starkman S, Eckstein M, Stratton SJ, Pratt FD, Hamilton S et al (2015) Prehospital use of magnesium sulfate as neuroprotection in acute stroke. N Engl J Med 372(6):528–536
138. Itrat A, Taqui A, Cerejo R, Briggs F, Cho SM, Organek N et al (2016) Telemedicine in prehospital stroke evaluation and thrombolysis: taking stroke treatment to the doorstep. JAMA Neurol 73(2):162–168
139. Ebinger M, Winter B, Wendt M, Weber JE, Waldschmidt C, Rozanski M et al (2014) Effect of the use of ambulance-based thrombolysis on time to thrombolysis in acute ischemic stroke: a randomized clinical trial. JAMA 311(16):1622–1631

140. O'Collins VE, Macleod MR, Donnan GA, Horky LL, van der Worp BH, Howells DW (2006) 1,026 experimental treatments in acute stroke. Ann Neurol 59(3):467–477
141. Dirnagl U (2006) Bench to bedside: the quest for quality in experimental stroke research. J Cereb Blood Flow Metab 26(12):1465–1478
142. Ginsberg MD (2008) Neuroprotection for ischemic stroke: past, present and future. Neuropharmacology 55(3):363–389
143. Ginsberg MD (2009) Current status of neuroprotection for cerebral ischemia: synoptic overview. Stroke 40(3 Suppl):S111–S114
144. Moskowitz MA (2010) Brain protection maybe yes, maybe no. Stroke 41(10):S85–S86
145. Muir KW (2006) Glutamate-based therapeutic approaches: clinical trials with NMDA antagonists. Curr Opin Pharmacol 6(1):53–60
146. Hoyte L, Barber PA, Buchan AM, Hill MD (2004) The rise and fall of NMDA antagonists for ischemic stroke. Curr Mol Med 4(2):131–136
147. Ikonomidou C, Turski L (2002) Why did NMDA receptor antagonists fail clinical trials for stroke and traumatic brain injury? Lancet Neurol 1(6):383–386
148. Hardingham GE, Bading H (2010) Synaptic versus extrasynaptic NMDA receptor signalling: implications for neurodegenerative disorders. Nat Rev Neurosci 11(10):682–696
149. Papadia S, Hardingham GE (2007) The dichotomy of NMDA receptor signaling. Neuroscientist 13(6):572–579
150. Mehta SL, Manhas N, Raghubir R (2007) Molecular targets in cerebral ischemia for developing novel therapeutics. Brain Res Rev 54(1):34–66
151. Watts LT, Lloyd R, Garling RJ, Duong T (2013) Stroke neuroprotection: targeting mitochondria. Brain Sci 3(2):540–560
152. STAIR (1999) Recommendations for standards regarding preclinical neuroprotective and restorative drug development. Stroke 30(12):2752–2758
153. Fisher M, Feuerstein G, Howells DW, Hurn PD, Kent TA, Savitz SI et al (2009) Update of the stroke therapy academic industry roundtable preclinical recommendations. Stroke 40(6):2244–2250
154. Turner RJ, Jickling GC, Sharp FR (2011) Are underlying assumptions of current animal models of human stroke correct: from STAIRs to high hurdles? Transl Stroke Res 2(2):138–143
155. Tymianski M. Neuroprotective therapies: preclinical reproducibility is only part of the problem. Sci Transl Med. 2015;7(299).
156. Llovera G, Hofmann K, Roth S, Salas-Perdomo A, Ferrer-Ferrer M, Perego C et al. Results of a preclinical randomized controlled multicenter trial (pRCT): Anti-CD49d treatment for acute brain ischemia. Sci Transl Med. 2015;7(299):299ra121.
157. Cook DJ, Teves L, Tymianski M (2012) Treatment of stroke with a PSD-95 inhibitor in the gyrencephalic primate brain. Nature 483(7388):213–217
158. Clarkson AN, Huang BS, Macisaac SE, Mody I, Carmichael ST (2010) Reducing excessive GABA-mediated tonic inhibition promotes functional recovery after stroke. Nature 468(7321):305–309
159. Lapchak PA (2010) Translational stroke research using a rabbit embolic stroke model: a correlative analysis hypothesis for novel therapy development. Transl Stroke Res 1(2):96–107
160. Lapchak PA (2015) A cost-effective rabbit embolic stroke bioassay: insight into the development of acute ischemic stroke therapy. Transl Stroke Res 6(2):99–103
161. Hill MD, Martin RH, Mikulis D, Wong JH, Silver FL, Terbrugge KG et al (2012) Safety and efficacy of NA-1 in patients with iatrogenic stroke after endovascular aneurysm repair (ENACT): a phase 2, randomised, double-blind, placebo-controlled trial. Lancet Neurol 11(11):942–950
162. Gilgun-Sherki Y, Rosenbaum Z, Melamed E, Offen D (2002) Antioxidant therapy in acute central nervous system injury: current state. Pharmacol Rev 54(2):271–284
163. Shirley R, Ord EN, Work LM (2014) Oxidative stress and the use of antioxidants in stroke. Antioxidants (Basel) 3(3):472–501
164. Sena E, Wheble P, Sandercock P, Macleod M (2007) Systematic review and meta-analysis of the efficacy of tirilazad in experimental stroke. Stroke 38(2):388–394

165. Parnham M, Sies H (2000) Ebselen: prospective therapy for cerebral ischaemia. Expert Opin Investig Drugs 9(3):607–619
166. Sakurai T, Kanayama M, Shibata T, Itoh K, Kobayashi A, Yamamoto M et al (2006) Ebselen, a seleno-organic antioxidant, as an electrophile. Chem Res Toxicol 19(9):1196–1204
167. Lapchak PA, Zivin JA (2003) Ebselen, a seleno-organic antioxidant, is neuroprotective after embolic strokes in rabbits: synergism with low-dose tissue plasminogen activator. Stroke 34(8):2013–2018
168. Kuroda S, Tsuchidate R, Smith ML, Maples KR, Siesjo BK (1999) Neuroprotective effects of a novel nitrone, NXY-059, after transient focal cerebral ischemia in the rat. J Cereb Blood Flow Metab 19(7):778–787
169. Marshall JWB, Cummings RM, Bowes LJ, Ridley RM, Green AR (2003) Functional and histological evidence for the protective effect of NXY-059 in a primate model of stroke when given 4 hours after occlusion. Stroke 34(9):2228–2233
170. Marshall JW, Duffin KJ, Green AR, Ridley RM (2001) NXY-059, a free radical-trapping agent, substantially lessens the functional disability resulting from cerebral ischemia in a primate species. Stroke 32(1):190–198
171. Lees KR, Zivin JA, Ashwood T, Davalos A, Davis SM, Diener H et al (2006) NXY-059 for acute ischemic stroke. N Engl J Med 354(6):588–600
172. Lees KR, Davalos A, Davis SM, Diener HC, Grotta J, Lyden P et al (2006) Additional outcomes and subgroup analyses of NXY-059 for acute ischemic stroke in the SAINT I trial. Stroke 37(12):2970–2978
173. Shuaib A, Lees KR, Lyden P, Grotta J, Davalos A, Davis SM et al (2007) NXY-059 for the treatment of acute ischemic stroke. N Engl J Med 357(6):562–571
174. Savitz SI (2007) A critical appraisal of the NXY-059 neuroprotection studies for acute stroke: a need for more rigorous testing of neuroprotective agents in animal models of stroke. Exp Neurol 205(1):20–25
175. Macleod MR, van der Worp HB, Sena ES, Howells DW, Dirnagl U, Donnan GA (2008) Evidence for the efficacy of NXY-059 in experimental focal cerebral ischaemia is confounded by study quality. Stroke 39(10):2824–2829
176. Bath PMW, Gray LJ, Bath AJG, Buchan A, Miyata T, Green AR et al (2009) Effects of NXY-059 in experimental stroke: an individual animal meta-analysis. Br J Pharmacol 157(7):1157–1171
177. Becker DA, Ley JJ, Echegoyen L, Alvarado R (2002) Stilbazulenyl nitrone (STAZN): a nitronyl-substituted hydrocarbon with the potency of classical phenolic chain-breaking antioxidants. J Am Chem Soc 124(17):4678–4684
178. Lapchak PA, Schubert DR, Maher PA (2011) De-risking of stilbazulenyl nitrone (STAZN), a lipophilic nitrone to treat stroke using a unique panel of in vitro assays. Transl Stroke Res 2(2):209–217
179. Maples KR, Ma F, Zhang YK (2001) Comparison of the radical trapping ability of PBN, S-PPBN and NXY-059. Free Radic Res 34(4):417–426
180. Zhang N, Komine-Kobayashi M, Tanaka R, Liu MZ, Mizuno Y, Urabe T (2005) Edaravone reduces early accumulation of oxidative products and sequential inflammatory responses after transient focal ischemia in mice brain. Stroke 36(10):2220–2225
181. Nakajima H, Kakui N, Ohkuma K, Ishikawa M, Hasegawa T (2005) A newly synthesized poly(ADP-ribose)polymerase inhibitor, DR2313 [2-methyl-3,5,7,8-tetrahydrothiopyrano[4,3-d]pyrimidine-4-one]: pharmacological profiles, neuroprotective effects, and therapeutic time window in cerebral ischemia in rats. J Pharmacol Exp Ther 312(2):472–481
182. Lapchak PA, Zivin JA (2009) The lipophilic multifunctional antioxidant edaravone (radicut) improves behavior following embolic strokes in rabbits: a combination therapy study with tissue plasminogen activator. Exp Neurol 215(1):95–100
183. Otomo E, Tohgi H, Kogure K, Hirai S, Takakura K, Terashi A et al (2003) Effect of a novel free radical scavenger, edaravone (MCI-186), on acute brain infarction—randomized, placebo-controlled, double-blind study at multicenters. Cerebrovasc Dis 15(3):222–229

184. Nakase T, Yoshioka S, Suzuki A (2011) Free radical scavenger, edaravone, reduces the lesion size of lacunar infarction in human brain ischemic stroke. BMC Neurol 11:39
185. Inatomi Y, Takita T, Yonehara T, Fujioka S, Hashimoto Y, Hirano T et al (2006) Efficacy of edaravone in cardioembolic stroke. Intern Med 45(5):253–257
186. Lapchak PA (2010) A critical assessment of edaravone acute ischemic stroke efficacy trials: is edaravone an effective neuroprotective therapy? Expert Opin Pharmacother 11(10):1753–1763
187. Feng SJ, Yang QW, Liu M, Li WZ, Yuan WM, Zhang SH et al (2011) Edaravone for acute ischaemic stroke. Cochrane Database Syst Rev 12, CD007230
188. Wingler K, Hermans JJ, Schiffers P, Moens A, Paul M, Schmidt HH (2011) NOX1, 2, 4, 5: counting out oxidative stress. Br J Pharmacol 164(3):866–883
189. Bath PM, Willmot M, Leonardi-Bee J, Bath FJ (2002) Nitric oxide donors (nitrates), L-arginine, or nitric oxide synthase inhibitors for acute stroke. Cochrane Database Syst Rev 4, CD000398
190. Terpolilli NA, Moskowitz MA, Plesnila N (2012) Nitric oxide: considerations for the treatment of ischemic stroke. J Cereb Blood Flow Metab 32(7):1332–1346
191. Willmot M, Gibson C, Gray L, Murphy S, Bath P (2005) Nitric oxide synthase inhibitors in experimental ischemic stroke and their effects on infarct size and cerebral blood flow: a systematic review. Free Radic Biol Med 39(3):412–425
192. Parmentier S, Bohme GA, Lerouet D, Damour D, Stutzmann JM, Margaill I et al (1999) Selective inhibition of inducible nitric oxide synthase prevents ischaemic brain injury. Br J Pharmacol 127(2):546–552
193. Chabrier PE, Auguet M, Spinnewyn B, Auvin S, Cornet S, Demerle-Pallardy C et al (1999) BN 80933, a dual inhibitor of neuronal nitric oxide synthase and lipid peroxidation: A promising neuroprotective strategy. Proc Natl Acad Sci U S A 96(19):10824–10829
194. Alderton WK, Cooper CE, Knowles RG (2001) Nitric oxide synthases: structure, function and inhibition. Biochem J 357:593–615
195. Lin Y, Phillis JW (1992) Deoxycoformycin and oxypurinol—protection against focal ischemic brain injury in the rat. Brain Res 571(2):272–280
196. Arai T, Mori H, Ishii H, Adachi T, Endo N, Makino K et al (1998) Oxypurinol, a xanthine oxidase inhibitor and a superoxide scavenger, did not attenuate ischemic neuronal damage in gerbils. Life Sci 63(7):1107–1112
197. Nakashima M, Niwa M, Iwai T, Uematsu T (1999) Involvement of free radicals in cerebral vascular reperfusion injury evaluated in a transient focal cerebral ischemia model of rat. Free Radic Biol Med 26(5-6):722–729
198. Van Bel F, Shadid M, Moison RMW, Dorrepaal CA, Fontijn J, Monteiro L et al (1998) Effect of allopurinol on postasphyxial free radical formation, cerebral hemodynamics, and electrical brain activity. Pediatrics 101(2):185–193
199. Higgins P, Ferguson LD, Walters MR (2011) Xanthine oxidase inhibition for the treatment of stroke disease: a novel therapeutic approach. Expert Rev Cardiovasc Ther 9(4):399–401
200. Christopherson KS, Hillier BJ, Lim WA, Bredt DS (1999) PSD-95 assembles a ternary complex with the *N*-methyl-D-aspartic acid receptor and a bivalent neuronal NO synthase PDZ domain. J Biol Chem 274(39):27467–27473
201. Sun HS, Doucette TA, Liu Y, Fang Y, Teves L, Aarts M et al (2008) Effectiveness of PSD95 inhibitors in permanent and transient focal ischemia in the rat. Stroke 39(9):2544–2553
202. Soriano FX, Martel MA, Papadia S, Vaslin A, Baxter P, Rickman C et al (2008) Specific targeting of pro-death NMDA receptor signals with differing reliance on the NR2B PDZ ligand. J Neurosci 28(42):10696–10710
203. Bratane BT, Cui H, Cook DJ, Bouley J, Tymianski M, Fisher M (2011) Neuroprotection by freezing ischemic penumbra evolution without cerebral blood flow augmentation with a postsynaptic density-95 protein inhibitor. Stroke 42(11):3265–3270
204. Bell KF, Bent RJ, Meese-Tamuri S, Ali A, Forder JP, Aarts MM (2013) Calmodulin kinase IV-dependent CREB activation is required for neuroprotection via NMDA receptor-PSD95 disruption. J Neurochem 126(2):274–287

205. Teves LM, Cui H, Tymianski M (2016) Efficacy of the PSD95 inhibitor Tat-NR2B9c in mice requires dose translation between species. J Cereb Blood Flow Metab 36(3):555–561
206. Cook DJ, Teves L, Tymianski M. A translational paradigm for the preclinical evaluation of the stroke neuroprotectant Tat-NR2B9c in gyrencephalic nonhuman primates. Sci Transl Med. 2012;4(154):154ra33.
207. Bach A, Chi CN, Olsen TB, Pedersen SW, Røder MU, Pang GF et al (2008) Modified peptides as potent inhibitors of the postsynaptic density-95/*N*-methyl-D-aspartate receptor interaction. J Med Chem 51(20):6450–6459
208. Bach A, Clausen BH, Møller M, Vestergaard B, Chi CN, Round A et al (2012) A high-affinity, dimeric inhibitor of PSD-95 bivalently interacts with PDZ1-2 and protects against ischemic brain damage. Proc Natl Acad Sci U S A 109(9):3317–3322
209. Cui H, Hayashi A, Sun HS, Belmares MP, Cobey C, Phan T et al (2007) PDZ protein interactions underlying NMDA receptor-mediated excitotoxicity and neuroprotection by PSD-95 inhibitors. J Neurosci 27(37):9901–9915
210. Tymianski M, Garman JD. Co-administration of an agent linked to an internalization peptide with an anti-inflammatory. 2011; Patent US 8,080,518 B2.
211. Sarko D, Beijer B, Garcia Boy R, Nothelfer EM, Leotta K, Eisenhut M et al (2010) The pharmacokinetics of cell-penetrating peptides. Mol Pharm 7(6):2224–2231
212. Bach A, Pedersen TB, Strømgaard K (2016) Design and synthesis of triazole-based peptidomimetics of a PSD-95 PDZ domain inhibitor. Med Chem Commun 7(3):531–536
213. Chi CN, Bach A, Strømgaard K, Gianni S, Jemth P (2012) Ligand binding by PDZ domains. Biofactors 38(5):338–348
214. Martel MA, Soriano FX, Baxter P, Rickman C, Duncan R, Wyllie DJA et al (2009) Inhibiting pro-death NMDA receptor signaling dependent on the NR2 PDZ ligand may not affect synaptic function or synaptic NMDA receptor signaling to gene expression. Channels 3(1):12–15
215. Lim IA, Merrill MA, Chen Y, Hell JW (2003) Disruption of the NMDA receptor-PSD-95 interaction in hippocampal neurons with no obvious physiological short-term effect. Neuropharmacology 45(6):738–754
216. Lim IA, Hall DD, Hell JW (2002) Selectivity and promiscuity of the first and second PDZ domains of PSD-95 and synapse-associated protein 102. J Biol Chem 277(24):21697–21711
217. Stiffler MA, Chen JR, Grantcharova VP, Lei Y, Fuchs D, Allen JE et al (2007) PDZ domain binding selectivity is optimized across the mouse proteome. Science 317(5836):364–369
218. Kim E, Sheng M (2004) PDZ domain proteins of synapses. Nat Rev Neurosci 5(10):771–781
219. Fan J, Cowan CM, Zhang LY, Hayden MR, Raymond LA (2009) Interaction of postsynaptic density protein-95 with NMDA receptors influences excitotoxicity in the yeast artificial chromosome mouse model of Huntington's disease. J Neurosci 29(35):10928–10938
220. Bard L, Sainlos M, Bouchet D, Cousins S, Mikasova L, Breillat C et al (2010) Dynamic and specific interaction between synaptic NR2-NMDA receptor and PDZ proteins. Proc Natl Acad Sci U S A 107(45):19561–19566
221. Hardingham GE (2006) Pro-survival signalling from the NMDA receptor. Biochem Soc Trans 34(Pt 5):936–938
222. Kucharz K, Rasmussen IS, Bach A, Strømgaard K, Lauritzen M (2016) PSD-95 uncoupling from NMDA receptors by Tat-*N*-dimer ameliorates neuronal depolarization in cortical spreading depression. J Cereb Blood Flow Metab. doi:10.1177/0271678X16645595
223. Chen Y, Brennan-Minnella AM, Sheth S, El-Benna J, Swanson RA (2015) Tat-NR2B9c prevents excitotoxic neuronal superoxide production. J Cereb Blood Flow Metab 35(5):739–742
224. Brennan-Minnella AM, Shen Y, Swanson RA (2013) Phosphoinositide 3-kinase couples NMDA receptors to superoxide release in excitotoxic neuronal death. Cell Death Dis 4, e580
225. Wang YB, Wang JJ, Wang SH, Liu SS, Cao JY, Li XM et al (2012) Adaptor protein APPL1 couples synaptic NMDA receptor with neuronal prosurvival phosphatidylinositol 3-kinase/Akt pathway. J Neurosci 32(35):11919–11929

226. Zhang L, Wu J, Duan X, Tian X, Shen H, Sun Q et al (2016) NADPH oxidase: a potential target for treatment of stroke. Oxid Med Cell Longev 2016:5026984
227. Radermacher KA, Wingler K, Kleikers P, Altenhofer S, Jr Hermans J, Kleinschnitz C et al (2012) The 1027th target candidate in stroke: will NADPH oxidase hold up? Exp Transl Stroke Med 4(1):11
228. Walder CE, Green SP, Darbonne WC, Mathias J, Rae J, Dinauer MC et al (1997) Ischemic stroke injury is reduced in mice lacking a functional NADPH oxidase. Stroke 28(11):2252–2258
229. Kunz A, Anrather J, Zhou P, Orio M, Iadecola C (2007) Cyclooxygenase-2 does not contribute to postischemic production of reactive oxygen species. J Cereb Blood Flow Metab 27(3):545–551
230. Kahles T, Luedike P, Endres M, Galla HJ, Steinmetz H, Busse R et al (2007) NADPH oxidase plays a central role in blood-brain barrier damage in experimental stroke. Stroke 38(11):3000–3006
231. Chen H, Song YS, Chan PH (2009) Inhibition of NADPH oxidase is neuroprotective after ischemia-reperfusion. J Cereb Blood Flow Metab 29(7):1262–1272
232. Jackman KA, Miller AA, De Silva TM, Crack PJ, Drummond GR, Sobey CG (2009) Reduction of cerebral infarct volume by apocynin requires pretreatment and is absent in Nox2-deficient mice. Br J Pharmacol 156(4):680–688
233. Brait VH, Jackman KA, Walduck AK, Selemidis S, Diep H, Mast AE et al (2010) Mechanisms contributing to cerebral infarct size after stroke: gender, reperfusion, T lymphocytes, and Nox2-derived superoxide. J Cereb Blood Flow Metab 30(7):1306–1317
234. Chen H, Kim GS, Okami N, Narasimhan P, Chan PH (2011) NADPH oxidase is involved in post-ischemic brain inflammation. Neurobiol Dis 42(3):341–348
235. Tang XN, Zheng Z, Giffard RG, Yenari MA (2011) Significance of marrow-derived nicotinamide adenine dinucleotide phosphate oxidase in experimental ischemic stroke. Ann Neurol 70(4):606–615
236. De Silva TM, Brait VH, Drummond GR, Sobey CG, Miller AA (2011) Nox2 oxidase activity accounts for the oxidative stress and vasomotor dysfunction in mouse cerebral arteries following ischemic stroke. PLoS One 6(12), e28393
237. Radermacher KA, Wingler K, Langhauser F, Altenhofer S, Kleikers P, Hermans JJR et al (2013) Neuroprotection after stroke by targeting NOX4 as a source of oxidative stress. Antioxid Redox Signal 18(12):1418–1427
238. Suh SW, Shin BS, Ma H, Van Hoecke M, Brennan AM, Yenari MA et al (2008) Glucose and NADPH oxidase drive neuronal superoxide formation in stroke. Ann Neurol 64(6):654–663
239. Cifuentes-Pagano E, Meijles DN, Pagano PJ (2014) The quest for selective Nox inhibitors and therapeutics: challenges, triumphs and pitfalls. Antioxid Redox Signal 20(17): 2741–2754
240. Wind S, Beuerlein K, Eucker T, Muller H, Scheurer P, Armitage ME et al (2010) Comparative pharmacology of chemically distinct NADPH oxidase inhibitors. Br J Pharmacol 161(4):885–898
241. Jaquet V, Scapozza L, Clark RA, Krause KH, Lambeth JD (2009) Small-molecule NOX Inhibitors: ROS-generating NADPH oxidases as therapeutic targets. Antioxid Redox Signal 11(10):2535–2552
242. Altenhofer S, Radermacher KA, Kleikers PWM, Wingler K, Schmidt HHHW (2015) Evolution of NADPH oxidase inhibitors: selectivity and mechanisms for target engagement. Antioxid Redox Signal 23(5):406–427
243. Aldieri E, Riganti C, Polimeni M, Gazzano E, Lussiana C, Campia I et al (2008) Classical inhibitors of NOX NAD(P)H oxidases are not specific. Curr Drug Metab 9(8):686–696
244. Wingler K, Altenhoefer SA, Kleikers PWM, Radermacher KA, Kleinschnitz C, Schmidt HHHW (2012) VAS2870 is a pan-NADPH oxidase inhibitor. Cell Mol Life Sci 69(18):3159–3160
245. Altenhofer S, Kleikers PW, Radermacher KA, Scheurer P, Rob Hermans JJ, Schiffers P et al (2012) The NOX toolbox: validating the role of NADPH oxidases in physiology and disease. Cell Mol Life Sci 69(14):2327–2343

246. Sun QA, Hess DT, Wang BL, Miyagi M, Stamler JS (2012) Off-target thiol alkylation by the NADPH oxidase inhibitor 3-benzyl-7-(2-benzoxazolyl)thio-1,2,3-triazolo[4,5-d]pyrimidine (VAS2870). Free Radic Biol Med 52(9):1897–1902
247. Schildknecht S, Weber A, Gerding HR, Pape R, Robotta M, Drescher M et al (2014) The NOX1/4 inhibitor GKT136901 as selective and direct scavenger of peroxynitrite. Curr Med Chem 21(3):365–376
248. Laleu B, Gaggini F, Orchard M, Fioraso-Cartier L, Cagnon L, Houngninou-Molango S et al (2010) First in class, potent, and orally bioavailable NADPH oxidase isoform 4 (Nox4) inhibitors for the treatment of idiopathic pulmonary fibrosis. J Med Chem 53(21):7715–7730
249. Rey FE, Cifuentes ME, Kiarash A, Quinn MT, Pagano PJ (2001) Novel competitive inhibitor of NAD(P)H oxidase assembly attenuates vascular O(2)(-) and systolic blood pressure in mice. Circ Res 89(5):408–414
250. Leto TL, Adams AG, de Mendez I (1994) Assembly of the phagocyte NADPH oxidase: binding of Src homology 3 domains to proline-rich targets. Proc Natl Acad Sci U S A 91(22):10650–10654
251. Finan P, Shimizu Y, Gout I, Hsuan J, Truong O, Butcher C et al (1994) An SH3 domain and proline-rich sequence mediate an interaction between two components of the phagocyte NADPH oxidase complex. J Biol Chem 269(19):13752–13755
252. Raz L, Zhang QG, Zhou CF, Han D, Gulati P, Yang LC et al (2010) Role of Rac1 GTPase in NADPH oxidase activation and cognitive impairment following cerebral ischemia in the rat. PLoS One 5(9), e12606
253. Zhang QG, Raz L, Wang RM, Han D, De Sevilla L, Yang F et al (2009) Estrogen attenuates ischemic oxidative damage via an estrogen receptor alpha-mediated inhibition of NADPH oxidase activation. J Neurosci 29(44):13823–13836
254. Selemidis S, Sobey CG, Wingler K, Schmidt HHHW, Drummond GR (2008) NADPH oxidases in the vasculature: Molecular features, roles in disease and pharmacological inhibition. Pharmacol Ther 120(3):254–291
255. Csanyi G, Cifuentes-Pagano E, Al Ghouleh I, Ranayhossaini DJ, Egana L, Lopes LR et al (2011) Nox2 B-loop peptide, Nox2ds, specifically inhibits the NADPH oxidase Nox2. Free Radic Biol Med 51(6):1116–1125
256. Cifuentes-Pagano E, Csanyi G, Pagano PJ (2012) NADPH oxidase inhibitors: a decade of discovery from Nox2ds to HTS. Cell Mol Life Sci 69(14):2315–2325
257. Bosco EE, Kumar S, Marchioni F, Biesiada J, Kordos M, Szczur K et al (2012) Rational design of small molecule inhibitors targeting the Rac GTPase-p67(phox) signaling axis in inflammation. Chem Biol 19(2):228–242
258. Smith SM, Min J, Ganesh T, Diebold B, Kawahara T, Zhu Y et al (2012) Ebselen and congeners inhibit NADPH oxidase 2-dependent superoxide generation by interrupting the binding of regulatory subunits. Chem Biol 19(6):752–763
259. Magesh S, Chen Y, Hu L (2012) Small molecule modulators of Keap1-Nrf2-ARE pathway as potential preventive and therapeutic agents. Med Res Rev 32(4):687–726
260. Baird L, Lleres D, Swift S, Dinkova-Kostova AT (2013) Regulatory flexibility in the Nrf2-mediated stress response is conferred by conformational cycling of the Keap1-Nrf2 protein complex. Proc Natl Acad Sci U S A 110(38):15259–15264
261. Johnson JA, Johnson DA, Kraft AD, Calkins MJ, Jakel RJ, Vargas MR et al (2008) The Nrf2-ARE pathway: an indicator and modulator of oxidative stress in neurodegeneration. Ann N Y Acad Sci 1147:61–69
262. Wells G (2015) Peptide and small molecule inhibitors of the Keap1-Nrf2 protein-protein interaction. Biochem Soc Trans 43:674–679
263. Surh YJ, Kundu JK, Na HK (2008) Nrf2 as a master redox switch in turning on the cellular signaling involved in the induction of cytoprotective genes by some chemopreventive phytochemicals. Planta Med 74(13):1526–1539
264. Prosperini L, Pontecorvo S (2016) Dimethyl fumarate in the management of multiple sclerosis: appropriate patient selection and special considerations. Ther Clin Risk Manag 12:339–350

265. Linker RA, Lee DH, Ryan S, van Dam AM, Conrad R, Bista P et al (2011) Fumaric acid esters exert neuroprotective effects in neuroinflammation via activation of the Nrf2 antioxidant pathway. Brain 134:678–692
266. Hancock R, Bertrand HC, Tsujita T, Naz S, El-Bakry A, Laoruchupong J et al (2012) Peptide inhibitors of the Keap1-Nrf2 protein-protein interaction. Free Radic Biol Med 52(2):444–451
267. Hancock R, Schaap M, Pfister H, Wells G (2013) Peptide inhibitors of the Keap1-Nrf2 protein-protein interaction with improved binding and cellular activity. Org Biomol Chem 11(21):3553–3557
268. Hu L, Magesh S, Chen L, Wang L, Lewis TA, Chen Y et al (2013) Discovery of a small-molecule inhibitor and cellular probe of Keap1-Nrf2 protein-protein interaction. Bioorg Med Chem Lett 23(10):3039–3043
269. Jnoff E, Albrecht C, Barker JJ, Barker O, Beaumont E, Bromidge S et al (2014) Binding mode and structure-activity relationships around direct inhibitors of the Nrf2-Keap1 complex. ChemMedChem 9(4):699–705
270. Marcotte D, Zeng W, Hus JC, McKenzie A, Hession C, Jin P et al (2013) Small molecules inhibit the interaction of Nrf2 and the Keap1 Kelch domain through a non-covalent mechanism. Bioorg Med Chem 21(14):4011–4019
271. Jiang ZY, Lu MC, Xu LL, Yang TT, Xi MY, Xu XL et al (2014) Discovery of potent Keap1-Nrf2 protein-protein interaction inhibitor based on molecular binding determinants analysis. J Med Chem 57(6):2736–2745
272. Jiang ZY, Xu LL, Lu MC, Chen ZY, Yuan ZW, Xu XL et al (2015) Structure-activity and structure-property relationship and exploratory in vivo evaluation of the nanomolar Keap1-Nrf2 protein-protein interaction inhibitor. J Med Chem 58(16):6410–6421
273. Jain AD, Potteti H, Richardson BG, Kingsley L, Luciano JP, Ryuzoji AF et al (2015) Probing the structural requirements of non-electrophilic naphthalene-based Nrf2 activators. Eur J Med Chem 103:252–268
274. Winkel AF, Engel CK, Margerie D, Kannt A, Szillat H, Glombik H et al (2015) Characterization of RA839, a noncovalent small molecule binder to Keapl and selective activator of Nrf2 signaling. J Biol Chem 290(47):28446–28455
275. Bertrand HC, Schaap M, Baird L, Georgakopoulos ND, Fowkes A, Thiollier C et al (2015) Design, synthesis, and evaluation of triazole derivatives that induce Nrf2 dependent gene products and inhibit the Keap1-Nrf2 protein-protein interaction. J Med Chem 58(18):7186–7194
276. Davies TG, Wixted WE, Coyle JE, Griffiths-Jones C, Hearn K, McMenamin R et al (2016) Monoacidic inhibitors of the Kelch-like ECH-associated protein 1: nuclear factor erythroid 2-related factor 2 (KEAP1:NRF2) protein-protein interaction with high cell potency identified by fragment-based discovery. J Med Chem 59(8):3991–4006
277. Shimozono R, Asaoka Y, Yoshizawa Y, Aoki T, Noda H, Yamada M et al (2013) Nrf2 activators attenuate the progression of nonalcoholic steatohepatitis-related fibrosis in a dietary rat model. Mol Pharmacol 84(1):62–70
278. Sun HP, Jiang ZY, Zhang MY, Lu MC, Yang TT, Pan Y et al (2014) Novel protein-protein interaction inhibitor of Nrf2-Keapl discovered by structure-based virtual screening. Med Chem Commun 5(1):93–98
279. Zhuang C, Narayanapillai S, Zhang W, Sham YY, Xing C (2014) Rapid identification of Keap1-Nrf2 small-molecule inhibitors through structure-based virtual screening and hit-based substructure search. J Med Chem 57(3):1121–1126
280. Shah ZA, Li RC, Thimmulappa RK, Kensler TW, Yamamoto M, Biswal S et al (2007) Role of reactive oxygen species in modulation of Nrf2 following ischemic reperfusion injury. Neuroscience 147(1):53–59
281. Shih AY, Li P, Murphy TH (2005) A small-molecule-inducible Nrf2-mediated antioxidant response provides effective prophylaxis against cerebral ischemia in vivo. J Neurosci 25(44):10321–10335

282. Ding Y, Chen MC, Wang M, Wang MM, Zhang TJ, Park JS et al (2014) Neuroprotection by acetyl-11-keto-beta-boswellic acid, in ischemic brain injury involves the Nrf2/HO-1 defense pathway. Sci Rep 4:7002
283. Lin-Holderer J, Li L, Gruneberg D, Marti HH, Kunze R (2016) Fumaric acid esters promote neuronal survival upon ischemic stress through activation of the Nrf2 but not the HIF-1 signaling pathway. Neuropharmacology 105:228–240
284. Wang JA, Fields J, Zhao CY, Langer J, Thimmulappa RK, Kensler TW et al (2007) Role of Nrf2 in protection against intracerebral hemorrhage injury in mice. Free Radic Biol Med 43(3):408–414
285. Zhao XR, Sun GH, Zhang J, Strong R, Dash PK, Kan YW et al (2007) Transcription factor Nrf2 protects the brain from damage produced by intracerebral hemorrhage. Stroke 38(12):3280–3286
286. Zhao XR, Sun GH, Zhang J, Ting SM, Gonzales N, Aronowski J (2015) Dimethyl fumarate protects brain from damage produced by intracerebral hemorrhage by mechanism involving Nrf2. Stroke 46(7):1923–1928
287. Zhao J, Redell JB, Moore AN, Dash PK (2011) A novel strategy to activate cytoprotective genes in the injured brain. Biochem Biophys Res Commun 407(3):501–506
288. Tu JY, Zhang X, Zhu Y, Dai YX, Li N, Yang F et al (2015) Cell-permeable peptide targeting the Nrf2-Keap1 interaction: a potential novel therapy for global cerebral ischemia. J Neurosci 35(44):14727–14739
289. Morphy R, Rankovic Z (2009) Designing multiple ligands—medicinal chemistry strategies and challenges. Curr Pharm Des 15(6):587–600
290. Cavalli A, Bolognesi ML, Minarini A, Rosini M, Tumiatti V, Recanatini M et al (2008) Multi-target-directed ligands to combat neurodegenerative diseases. J Med Chem 51(3):347–372
291. Cimino M, Gelosa P, Gianella A, Nobili E, Tremoli E, Sironi L (2007) Statins: multiple mechanisms of action in the ischemic brain. Neuroscientist 13(3):208–213
292. Lapchak PA (2013) Drug-like property profiling of novel neuroprotective compounds to treat acute ischemic stroke: guidelines to develop pleiotropic molecules. Transl Stroke Res 4(3):328–342
293. Congreve M, Chessari G, Tisi D, Woodhead AJ (2008) Recent developments in fragment-based drug discovery. J Med Chem 51(13):3661–3680
294. Hajduk PJ, Greer J (2007) A decade of fragment-based drug design: strategic advances and lessons learned. Nat Rev Drug Discov 6(3):211–219
295. Murray CW, Verdonk ML, Rees DC (2012) Experiences in fragment-based drug discovery. Trends Pharmacol Sci 33(5):224–232
296. Rees DC, Congreve M, Murray CW, Carr R (2004) Fragment-based lead discovery. Nat Rev Drug Discov 3(8):660–672
297. Bottegoni G, Favia AD, Recanatini M, Cavalli A (2012) The role of fragment-based and computational methods in polypharmacology. Drug Discov Today 17(1-2):23–34
298. Morphy R, Rankovic Z (2007) Fragments, network biology and designing multiple ligands. Drug Discov Today 12(3-4):156–160
299. Anighoro A, Bajorath J, Rastelli G (2014) Polypharmacology: challenges and opportunities in drug discovery. J Med Chem 57(19):7874–7887
300. Achenbach J, Klingler FM, Blocher R, Moser D, Hafner AK, Rodl CB et al (2013) Exploring the chemical space of multitarget ligands using aligned self-organizing maps. ACS Med Chem Lett 4(12):1169–1172
301. Prati F, De Simone A, Bisignano P, Armirotti A, Summa M, Pizzirani D et al (2015) Multitarget drug discovery for Alzheimer's disease: triazinones as BACE-1 and GSK-3 beta inhibitors. Angew Chem Int Ed 54(5):1578–1582
302. Shang EC, Yuan YX, Chen XY, Liu Y, Pei JF, Lai LH (2014) De novo design of multitarget ligands with an iterative fragment-growing strategy. J Chem Inf Model 54(4):1235–1241

Chapter 9
Nitrones, Old Fellows for New Therapies in Ischemic Stroke

Alejandro Escobar-Peso, Mourad Chioua, Valerio Frezza, Emma Martínez-Alonso, José Marco-Contelles, and Alberto Alcázar

Abstract Ischemic stroke is suffered by millions of people worldwide, being the second cause of death in 2012. To date, only recombinant tissue plasminogen activator and thrombectomy, as first-line recanalization approaches, are the only treatments approved for ischemic stroke therapy. Nevertheless, the low number of patients who can benefit from this treatment, as well as the limited beneficial outcome, even if proper recanalization rates are achieved, make evident the need for complementary therapeutic approaches. Among them, the neuroprotection strategy appeared as a promising approach which led to the development of drugs targeting distinct steps of the biochemical pathways that take place during and after the ischemic insult. However, no effective translation from preclinical studies to clinical use has been achieved until now, rising thus doubts about the suitability of the neuroprotection therapeutic strategy. Regarding the intrinsic complexity of ischemic stroke, pleiotropy has been proposed as a key issue, and nitrones, known to act as radical traps, have arisen as interesting drug candidates. From its widely known antioxidant behavior, new mechanisms of action have been proposed based on reported evidence. In this chapter, nitrones pleiotropy is reviewed. Specific results of nitrones developed by our group are reported and discussed.

Keywords Neuroprotection • Stroke • Oxidative stress • Nitrones • Brain ischemia • Therapy

---

A. Escobar-Peso • M. Chioua • J. Marco-Contelles
Laboratory of Medicinal Chemistry, Institute of General Organic Chemistry (CSIC), Juan de la Cierva 3, Madrid 28006, Spain
e-mail: a.escobar@iqog.csic.es; mchioua@gmail.com; iqoc21@iqog.csic.es

V. Frezza • E. Martínez-Alonso • A. Alcázar (✉)
Department of Investigation, Hospital Ramón y Cajal, IRYCIS, Madrid 28034, Spain
e-mail: valeriofrezza85@hotmail.it; emma.martinez@hrc.es; alberto.alcazar@hrc.es

P.A. Lapchak, J.H. Zhang (eds.), *Neuroprotective Therapy for Stroke and Ischemic Disease*, Springer Series in Translational Stroke Research,
DOI 10.1007/978-3-319-45345-3_9

1 Introduction

Stroke can be defined as a pathology in which a cerebral vessel affection produces an impairment in cerebral blood flow (CBF), altering the normal function of the whole brain or a specific part of it, either permanently or temporarily [1]. In the period 2010–2012, stroke was the second most common cause of death, accounting in 2012 for 11.9 % of deaths produced worldwide, just after ischaemic heart disease [2]. In the same period, stroke was the third largest cause of disability [3].

Among the two main types of stroke, hemorrhagic and ischemic, the latter accounts for more than 80 % of all the cases registered worldwide. Caused by an occluding agent formed within the brain (thrombus), in the heart (embolus), or by a systemic hypoperfusion [4], a variable impairment of blood flow is produced to the entire brain, in what is named a global ischemia, or to a specific part of it, focal ischemia. Usually, spontaneous reperfusion occurs quickly after the occlusion and no irreversible damage is produced, being called a transient ischemic attack (TIA) [5]. In contrast, a genuine acute ischemic stroke (AIS) is characterized by the presence of irreversibly damaged tissue named *ischemic core*, close to the vessel blockage [6]. The *penumbra*, or the *perinfarct* zone, surrounds the ischemic core and presents remaining CBF, allowing this area to maintain ATP stores, oxygen metabolism, and ionic gradient homeostasis even though electrical activity is depleted and the protein synthesis inhibited [6]. Therefore, within the penumbra, the cellular integrity is kept, but at high risk. Restoration of the normal circulatory conditions may reverse the damage produced within this area, but if ischemic insult is maintained over a longer period, it may also turn into infarcted tissue (reperfusion window) [7].

Under this perspective, achieving a proper reperfusion remains as the first goal for stroke treatment, but complementary therapies are needed. Despite the latest failures, neuroprotection remains as a promising approach. In this chapter, we will pay special attention to nitrones as well-known radical scavenging agents and pleiotropic drugs for ischemic stroke.

2 Ischemic and Reperfusion Damage

As for other diseases affecting the central nervous system, many important aspects of ischemic stroke have remained unknown for the scientific community for long time. It has been only recently, when advanced imaging techniques developed, that more about the physiopathology of the brain during an ischemic insult was really known. Even nowadays, while some aspects are being unfold, many others arise, creating an exhausting but challenging disease that is worth fighting against.

What is clear is that ischemic stroke is a very complex and multifactorial disease. In this sense, not only the etiology of the occlusion, but also its severity, its duration, and the vessel occluded play key roles. From that point, the whole brain can be affected (global ischemia), or just some part of it (focal ischemia). If the latter, specific affected areas and potential residual blood flow must be considered. A gradient of blood flow may take place and, therefore, different biochemical pathways can be activated depending on the CBF impairment. Additionally, cerebral tissue is formed by several cellular types, which react differently to the insult, and reperfusion, if happens, may influence the final outcome in one way or another, depending on the methodology used, its efficacy, the formation of secondary clots, possible hemorrhage, and so on. Of course, time windows are important variables and, with them, countless factors and parameters could be added.

Having presented its inherent complexity, in the following section we will discuss the most important biochemical pathways that take place after the ischemic stroke (Fig. 9.1) in order to properly understand neuroprotection and nitrone-based therapies that will be mentioned later. In particular, short-term and long-term damage contributions are explained for both ischemic core and penumbra in Sect. 2.1. Specific contribution of the following reperfusion to damage is considered in Sect. 2.2.

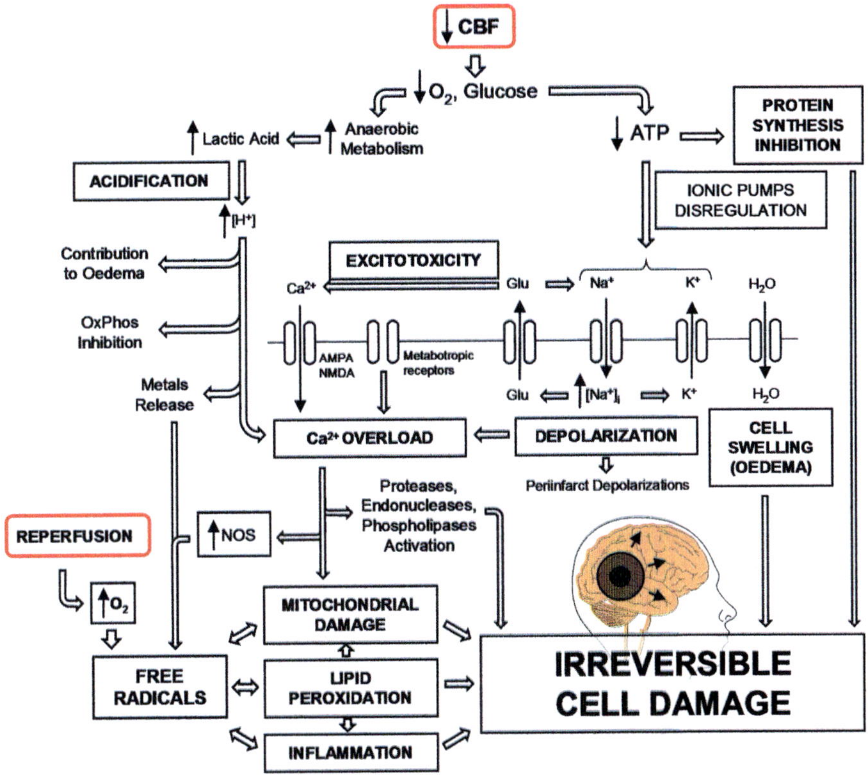

Fig. 9.1 Cascade of ischemia-reperfusion damage

2.1 Damage During Ischemia

During vessel occlusion, decreased nutrients supply, along with toxic metabolites elimination failure, causes a metabolic disruption within the affected tissue in what is called *ischemic damage*. As was previously mentioned, the specific brain regions affected as well as the severity and duration of the ischemic insult, among others, determine progression of ischemic injury, which can continue for minutes, hours, or even days [8].

Being conscious of this complexity, elucidation of metabolic consequences of the different grades of CBF impairment led to experimental determination in animal models of limits or thresholds. In this regard, in a first approach to this issue, Simon et al. reported the existence of two basic limits: a first 'functional threshold' (20–18 mL/100 g/min), in which brain functional activity is impaired due to the energy-consuming processes inhibition; and a second 'structural threshold', in which structural integrity is lost (<15 mL/100 g/min) [9]. According to this basic division, Astrup et al. identified the tissue that usually surrounds the ischemic core and places between these two thresholds as *penumbra*, meaning a tissue that is functionally compromised, but structurally intact [6]. Further studies of the penumbra raised the clinical relevance of this area, as blood supply restoration at any time was expected to reactivate its normal function [10]. Unfortunately, a correlation between higher blood flow and better functional outcome cannot be totally established, as we will see in the next sections.

Further investigation in the last 30 years reported a complete set of intermediate thresholds that trigger different and complex biochemical processes [10] and are referred as *ischemic cascade* (see Fig. 9.1). In rodents, from a normal CBF of 70–80 mL/100 g/min, the first consequence of O_2 and nutrients decrease is the inhibition of protein synthesis, the most sensitive parameter, which reaches its 50 % value at 55 mL/100 g/min [11]. Glucose consumption is enhanced below 35 mL/100 g/min, when anaerobic metabolism is activated. Nevertheless, progressive accumulation of lactate as an anaerobic metabolism by-product prompts glucose consumption inhibition below 25 mL/100 g/min [12]. At around 20–25 mL/100 g/min, tissue acidification inhibits oxidative phosphorylation (OxPhos) and causes severe energy depletion, along with protein uncoupling and iron release from intracellular stores [13].

Energy depletion causes loss of membrane potential in what is called *anoxic depolarization* of neurons and glia below 10–15 mL/100 g/min, limit of infarction. Consequently, intracellular sodium/potassium ratio increases and glutamate and other neurotransmitters are released into the extracellular space. Overstimulation of voltage-dependent calcium channels, as much as glutamate-regulated channels (AMPA, NMDA, and metabotropic receptors), causes a massive calcium, sodium, and chloride influx into the neurons (*excitotoxicity*) [14], a concomitant entrance of water that contributes to cellular edema [15] and depletion of calcium from the endoplasmic reticulum (ER) [16]. Cells within the ischemic core are not able to repolarize, whereas in the penumbra its higher CBF allows recovery of functional potential, intermittently lost due to extracellular glutamate in what are called *perinfarct depolarizations*.

Calcium ion is one of the major players in stroke damage. Massive calcium influx is produced by the previously mentioned voltage- and glutamate-dependent calcium channels, depletion from intracellular stores such as ER or mitochondria, and activation of acid-sensing ion channels [17]. This second-messenger overload further contributes to degradation by several pathways: (1) activation of proteolytic enzymes such as calpain, known to degrade cytoskeleton [18] and extracellular matrix-proteins [19]; (2) activation of endonucleases and phospholipases; (3) activation of reactive species-producing enzymes such as nitric oxide synthase (NOS) [20]; and (4) mitochondria degradation and release of cell-damaging product by mitochondrial permeability transition (MPT) pore opening and other means [21].

Nitric oxide (NO) role in ischemic/reperfusion injury has been subject of controversy since its proposal as a key mediator in stroke by Marshall and Kontos in 1990 [22]. Existence of different NOSs with variable roles depending on the time window and tissues present may have contributed. First described in neurons and endothelial tissue, nNOS and eNOS, respectively, are Ca^{2+}-activated enzymes [20]. In contrast, the inducible form of NOS (iNOS) was first discovered in macrophages [23]. Thus, ischemic activation of both nNOS and eNOS is expected within the first hours after stroke, when calcium accumulates in the cytoplasm, whereas delayed iNOS expression has been reported to occur in rodents 12 h after permanent medial cerebral artery occlusion [24], contributing to a long-term damage in inflammation injury. Apart from its combination with other free radicals (Fig. 9.2), NO production leads to cellular damage by oxygen exchange from the heme group of mitochondrial enzymes, ensuing mitochondrial transport chain inhibition and enhanced ATP depletion [25].

Increased amounts of other reactive species are also present within the penumbra under hypoxic conditions. Mitochondria dysfunction promotes electron leakage from electron transport chain, working as a major source for superoxide anion [26], as well as non-radical metabolites autoxidation catalyzed by released metals [27]. Calcium-mediated protease activation catalyzes xanthine dehydrogenase cleavage to xanthine oxidase, promoting xanthine conversion to uric acid, superoxide anion, and hydrogen peroxide [28]. In addition, the aforementioned enzyme NOS may produce superoxide [29], and stimulation of NADPH oxidase during inflammatory response is also an important source of the same radical [30].

These reactive oxygen or nitrogen species (ROS, RNS) play, therefore, important roles in cellular damage during the ischemic insult. Nevertheless, this deleterious role would be more pronounced during reperfusion, when massive oxygen supply facilitates even more its generation (see Sect. 2.2).

Additionally, during ischemia, resident microglia, astrocytes, and neurons in regions with residual blood supply upregulate their production of cytokines and chemokines due to increased intracellular calcium, ROS, RNS as well as hypoxia itself [31]. Among these inflammatory mediators, pro-inflammatory IL-1, IL-6, and TNF-α activate signaling pathways leading to adhesion molecules production

Fig. 9.2 ROS and RNS possible reactions

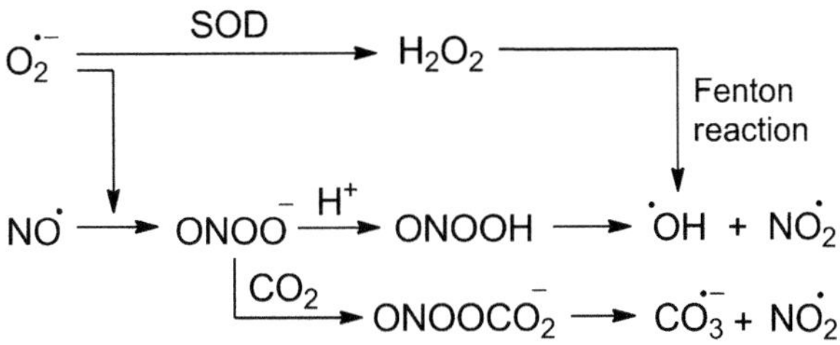

(selectins, integrins, immunoglobulins) by the endothelial surface that promotes neutrophil accumulation in the vessel lumen [32]. This recruitment can obstruct the microcirculation and may be responsible for a CBF decrease in the penumbra, as well as may prevent proper reperfusion when recanalization is produced [33].

Eventually, adhesion molecules expression, along with calcium-mediated blood-brain barrier (BBB) disruption [34], results in leukocyte infiltration into the damaged tissue, neutrophils being the first to cross the damaged vascular endothelium [32]. In the brain, they are known to enhance production of cytokines and release reactive species such as $NO^{\bullet-}$ by $iNOS^{-}$ and proteolytic enzymes such as protein kinase C [35] as well as further stimulate inflammatory mediators production.

Thus, inflammation contributes to exacerbate brain injury during ischemic insult. However, increasing evidence suggests that inflammation has both beneficial and deleterious effects, contributing to brain recovery [31]. In any way, inflammatory response is enhanced after reperfusion, being an important contribution to final patient outcome. That is why it will be also considered as part of the reperfusion-mediated damage (see below).

2.2 *Reperfusion Damage and Other Long-Term Contributions: Oxidative Stress*

During ischemia, previously mentioned phenomena take place, leading to cell damage. Necrosis in the ischemic core may occur within minutes, whereas in the penumbra, less severe ischemic insult promotes a delayed cell damage by longer-term processes such as apoptosis (Sect. 2.2.2) or inflammation (Sect. 2.2.3). However, if the ischemic insult remains, expansion of the ischemic core into the penumbra, increasing the infarct size, is only a matter of time [7]. That is why allowing reperfusion as soon as possible will avoid the ischemic core progression. Nevertheless, reperfusion does not necessarily improve patient outcome. As it was previously mentioned, complex signaling pathways take place in the penumbra, and longer-term processes triggered during the ischemic time (apoptosis, inflammation) may make difficult its complete recovery. Besides, ROS and RNS production is enhanced after reperfusion is produced, leading to *reperfusion damage* that must also be considered.

Regarding oxidative stress, sources of ROS and RNS activated during the ischemic insult have their activity enhanced because of the massive oxygen supply produced after spontaneous reperfusion or recanalization. This overwhelming concentration of oxidative species surpasses antioxidant capacity of endogenous systems, some of them shown in Table 9.1.

As a consequence, different reactive species are formed, as shown in Fig. 9.2.

Peroxynitrite forms by the combination of superoxide and nitric oxide, mainly after reperfusion [36]. Since its first proposal as a reactive species implicated in oxidative stress, peroxynitrite has been reported to directly oxidize and nitrate proteins, lipids, and DNA, as well as work as a source of hydroxyl radical ($^{\bullet}OH$), nitrogen dioxide ($NO_2^{\bullet}$), and carbonate radical ($CO_3^{\bullet-}$) [36, 37]. Free metals released during ischemia, such as iron or copper, are a source of hydroxyl radical (·OH) by the Fenton-type reaction (Fig. 9.3) [38].

Table 9.1 Main antioxidant systems present in brain

Enzymes	Cofactor	MW (Da)	Subcellular location	Reaction
Superoxide dismutase (SOD)				
Cu, Zn-SOD (SOD1)	Cu^{2+}, Zn^{2+}	15936	Cytosol, lysosomes, mitochondrial intermembrane space	$2O_2^{\bullet-} + 2H^+ \rightarrow H_2O_2 + O_2$
Mn-SOD (SOD2)	Mn^{2+}	24722	Mitochondrial matrix	
Extracellular SOD (SOD3)	Cu^{2+}, Zn^{2+}	25851	Extracellular	
Catalase (CAT)	Heme, $NADP^+$	59756	Peroxisomes	$2H_2O_2 \rightarrow O_2 + 2H_2O$
Glutathione peroxidase (GSHPx)				
GSHPx1-2		~22000	Cytoplasm	$2GSH + H_2O_2 \rightarrow GS–GS + 2H_2O$
GSHPx3,5-7		~21–25,000	Extracellular	
GSHPx4		22175 19525	Mitochondrion, cytoplasm	$2GSH + Lipid–OOH \rightarrow GS\text{-}GS + Lipid–OH + 2H_2O$
Peroxiredoxin				
Peroxiredoxin-1		22110	Cytoplasm, melanosome	$2\ R'–SH + ROOH \rightarrow R'–S–S–R' + H_2O + ROH$
Peroxiredoxin-2		21892	Cytoplasm	
Peroxiredoxin-3		27693	Mitochondrion	
Peroxiredoxin-4		30540	Cytoplasm, extracellular	
Peroxiredoxin-5		22086 17031	Mitochondrion, Cytoplasm, Peroxisome	
Peroxiredoxin-6		25035	Cytoplasm, Lysosome	
Antioxidant Molecules		MW (Da)		Structure
α-Tocopherol (vitamin E)		430		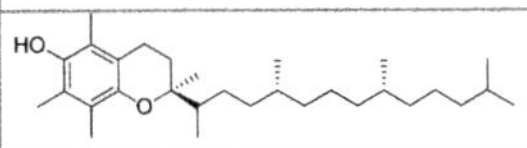
Ascorbic Acid (vitamin C)		176		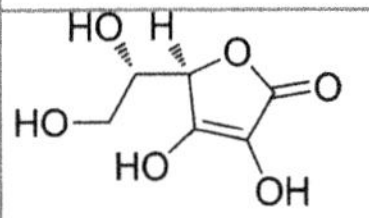
Glutathione (GSH)		307		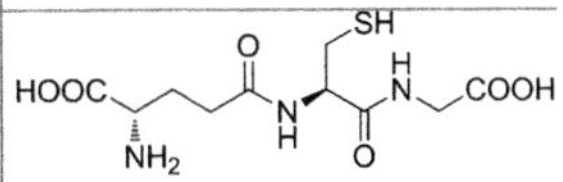
Melatonin		232		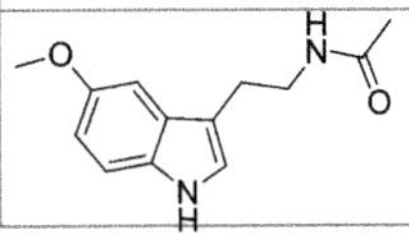

Fig. 9.3 Fenton type reaction catalyzed by iron salts

$$Fe^{2+} + H_2O_2 \longrightarrow {}^{\bullet}OH + {}^{-}OH + Fe^{3+}$$
$$Fe^{3+} + H_2O_2 \longrightarrow {}^{\bullet}OOH + H^{+} + Fe^{2+}$$
$$2\,H_2O_2 \xrightarrow{Fe^{2+}} {}^{\bullet}OH + {}^{\bullet}OOH + H_2O$$

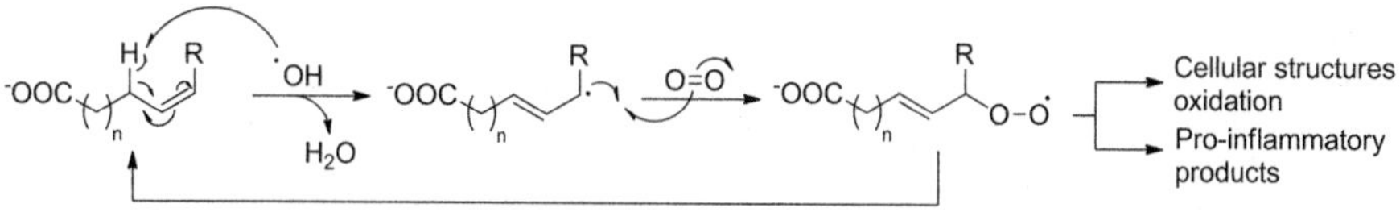

Fig. 9.4 Lipid peroxidation reaction

In general, these and other reactive species can directly oxidize cellular structures such as lipids, proteins, or nucleic acids, or they can interfere in diverse metabolic pathways acting as signaling molecules, as in apoptosis and inflammation.

2.2.1 Lipid Peroxidation

As it was previously mentioned, calcium-activated phospholipase A_2 [39] and depolarization-activated phospholipase C [40] release fatty acids, mainly arachidonic acid, by membrane phospholipid hydrolysis [41]. This not only produces a concomitant cellular damage or inhibits oxidative phosphorylation [42], but also serves as a source of substrates for the lipid peroxidation radical reaction. This effect is of great importance in brain since, compared with other organs, it has a high concentration of polyunsaturated fatty acids [43].

In particular, OH is able to abstract $H^{\bullet}$ from the polyunsaturated hydrocarbon chain of a membrane fatty acid, forming carbon-centered radicals that undergo oxygen addition to form lipid peroxides. These peroxides can attack double bonds in their own chain (forming cyclic peroxides), or in adjacent fatty acids and so on, propagating the free radical chain reaction [44], as shown in Fig. 9.4.

As a consequence of lipid peroxidation reaction, leakage of cellular bilayers and damage to membrane proteins are produced, altering normal membrane permeability. Additionally, cytotoxic by-products, such as 4-hydroxi-2-trans-nonenal, [45] are generated, along with more free radicals and pro-inflammatory products [46].

2.2.2 Apoptosis

As signaling molecules, ROS and RNS are known to promote a longer-term damage by apoptosis activation and neuronal death.

In contrast to the necrosis that usually takes place within the ischemic core, where CBF reaches the lowest thresholds for energetic maintenance, apoptosis appears in the penumbra, where some energetic metabolism is kept, as a result of varied death signals that trigger apoptotic cell death. One of these signals is the one regarding ROS and RNS. Apart from the damage produced, they are known to directly activate mitochondrial-mediated apoptosis (also referred as non-receptor-mediated apoptosis) by mitochondrial membrane degradation. Release of pro-apoptotic molecules such as cytochrome C [47], Smac/DIABLO factor [48], procaspases, and apoptosis-inducing factor (AIF) is produced consequently. This is translated by complex biochemical pathways that imply caspase route and poly(ADP-ribose) polymerase (PARP) cleavage [49], to DNA damage and, eventually, cell death [50].

2.2.3 Reperfusion Contribution to Inflammation

ROS and RNS are also known to play important roles in further contribution to inflammation.

As it was previously mentioned, leukocytes accumulate into the vascular endothelial surface as a consequence of adhesion molecules overexpression and BBB disruption. When the clot is dissolved and recanalization is allowed, this leukocyte plugging may make proper reperfusion to the brain difficult in what is known as the ischemic no-reflow phenomenon [32].

If leukocytes cross BBB, they release neurotoxic substances and pro-inflammatory mediators, such as cytokines, chemokines, and radical species, further contributing to inflammatory response. In these cells, as well as in activated microglia and ischemic neurons, radical producing enzymes are activated, and therefore, restoration of oxygen levels will enhance radical production. Among these, NO production is strongly induced after ischemic insult both by calcium-modulated isoforms of NOS or by iNOS synthesis de novo, in later stages of the inflammatory response [51]. Other enzymes, such as cyclooxygenase 2 (COX-2), are also highly expressed in neutrophils, vascular cells, and neurons, and it is known to produce superoxide anion and prostanoids, typical pro-inflammatory mediators [52]. In addition to all these factors, recanalization may increase flow of immune cells to the ischemic region, increasing cellular inflammatory response.

In conclusion, reperfusion-mediated contributions amplify inflammatory response and enhance deleterious effects. These may eventually lead to BBB breakage, cytotoxic edema, cell death, and hemorrhagic transformation, drastically worsening patients outcome [53].

2.3 Induced Reperfusion as Therapy for Stroke Treatment

The induction of reperfusion by endovascular mechanical procedures (thrombectomy) or systemic thrombolytic drugs is, to date, the only approved therapy for stroke treatment.

The majority of ischemic strokes result from thrombus occlusion. The thrombus is formed by blood cells in a fibrin matrix and its dispersion (fibrinolysis) is enzymatically driven by plasmin, a serine protease, which is activated from its zymogen plasminogen by naturally present tissue plasminogen activator (TPA). Genetic modifying techniques over these endogenous compounds led to the development of Alteplase®, a recombinant tissue plasminogen activator (rTPA), which is, to date, the only drug approved for acute stroke treatment by the Food and Drug Administration (FDA) and European Medical Agency [54].

In spite of being approved, rTPA present several deficiencies, such as short treatment time window of only 4.5 h after the symptoms onset [55] and neurotoxic side effects resulting from NMDA channel stimulation that contributes to excitotoxicity [56]. Additionally, rTPA use is associated with higher risk of subsequent hemorrhage (hemorrhagic transformation) and reocclusion, due to ineffective thrombolysis [57]. These factors, along with a low number of patients who can benefit from this therapy (approx. 5 %) and low recanalization rates (<50 %) [58], enhance the need for a proper reperfusion treatment.

As an alternative to systemic thrombolysis, besides being reported ineffective for large artery occlusions, thrombectomy methods try to remove the obstruction and eliminate the thrombus with the use of an intra-arterial mechanical device operated by high-skilled clinicians. Several devices have been approved by FDA: MERCI [59], PENUMBRA [60], SOLITAIRE FR [61], and TREVO [62]. Among these, most modern stent-retrievers (SOLITAIRE, TREVO) are preferred over the older generation (MERCI, PENUMBRA) and over systemic thrombolytics, since they show higher and faster recanalization rates and lower procedure complications, although phase III studies to prove better clinical outcome compared with standard 4.5 h rTPA-induced thrombolysis (Table 9.2) are still missing [63].

Apart from the previously mentioned methods, numerous researches have been done in the development of new recanalization approaches. Most promising approaches are summarized in Table 9.2 [82, 83].

As this summary tries to show, great advances have been made in the last fifteen years in order to find an effective therapy for reperfusion. Last thrombectomy techniques have arisen as promising alternatives to systemic thrombolysis because of higher recanalization efficacy and less procedure complications, which have led to an increasing number of patients who can benefit from this treatment. Nevertheless, there is a high percentage among them (>50 %) whose reperfusion does not translate into improved outcome, which is leading to an increasing number of patients with "*futile recanalization*" [84].

Complementary strategies guaranteeing improvement for those patients are, therefore, more required than ever due to the expansion of recanalization approaches. Neuroprotection strategies, which originally focused on recuperation of the ischemic penumbra, must now be aimed also to prevent damage produced during reperfusion and promote patient recovery. Pleiotropic drugs, as an interesting approach, could achieve these goals, benefiting not only patients with futile recanalization, but also spontaneously reperfused and even those directly benefited by clot dissolution.

Table 9.2 Up-to date summary of recanalization methodologies

Methodology	Current status	Study/clinical trial
Systemic thrombolysis		
Recombinant tissue plasminogen activator (rTPA)	First-line drug of choice to use within 4.5 h after the stroke onset [54]. To date, improvement between 4.5-6 h is not conclusive	*Phase III*
		NINDS [64] for 3 h
		ECASS-3 [55] for 4.5 h
		IST-3 [65] and SITS-ISTR [66] for 6 h
Tenecteplase	Phase III clinical trials ongoing [67]	*Phase II*
		ACTRN12608000466347 [68]
		ATTTEST [69]
		TEMPO-1 [70]
		Phase III
		NOR-TEST (NCT01949948)
		TASTE (ACTRN12613000243718)
		TEMPO-2 (NCT02398656)
Desmoteplase	Phase III studies showed good pharmacological profile but no significant improvement when compared with placebo. Testing in earlier time-windows was proposed [71]	*Phase II*
		DIAS [72]
		DEDAS [73]
		Phase III
		DIAS-2 [74]
		DIAS-3 [75]
		DIAS-4 (NCT00856661)
Argatroban	Argatroban and rTPA combination Phase II studies reported safety and higher canalization rates than rTPA alone. Further Phase II studies ongoing	*Phase II*
		ARTSS-1 [76]
		ARTSS-2 (NCT01464788)
		ARTSS-IA (NCT02448069)
Ultrasound + rTPA (Sonothrombolysis)	Ultrasound in combination with rTPA showed safety and good recanalization rates in Phase II studies. Phase III results pending	*Phase II*
		CLOTBUST-HF [77]
		Phase III
		CLOTBUST-ER (NCT01098981)
Endovascular reperfusion		
Thrombectomy	With higher and faster recanalization rates, several mechanical devices have been approved by the FDA, although Phase III studies proving safety and efficacy compared with 4.5 h rTPA are still ongoing	*Phase II*
		MERCI, Multi-MERCI [78]
		PENUMBRA [60]
		SWIFT [79]
		TREVO 2 [62]
		Phase III
		SWIFT PRIME [80]
		REVASCAT [81]

3 Neuroprotection: Nitrones as Promising Drug Candidates

Based on the current understanding of the biochemical processes taking place after the stroke onset and during reperfusion, many compounds have been developed targeting different ischemic/reperfusion cascade steps in order to lessen the neurological damage. In general, compounds aimed to avoid this and any other kind of neurodegeneration processes have been referred as *neuroprotective*.

Despite promising results in preclinical phases, neuroprotection drugs for stroke have failed in their attempt to reach the clinic. In 1999, this situation prompted the formation of the Stroke Therapy Academic Industry Roundtable (STAIR), which proposed a series of guidelines aimed to improve quality of preclinical and clinical studies [85]. Anyhow, since its formation, no neuroprotective drug has been granted approval by the FDA for stroke treatment. Numerous reviews, manuscripts, and meta-analysis have been published about this issue, leading to interesting conclusions questioning preclinical and clinical procedures.

3.1 *Nitrones as Antioxidant Agents and Spin Traps Deserve Special Attention*

Despite this disappointing situation, neuroprotection in stroke still remains an attractive and promising therapeutic field. In this context, different strategies have been developed aimed to target different components of the ischemic cascade (glutamate receptors, apoptosis inhibitors, inflammation modulators, etc.) [86]. However, among them, oxidative stress reduction or prevention has become one of the most interesting approaches because of its key relevance in ischemic/reperfusion injury (see Sects. 2.1 and 2.2). Additionally, the special sensitivity of the brain to this issue is due to its: (1) high oxygen consumption; (2) great cellular accumulation of iron, producer of radical species [87]; (3) high concentration of unsaturated lipids known to be targets for oxidation [43]; and (4) its relatively low amount of antioxidant mechanisms [88]. All these factors make oxidative stress an appealing target from where the development of stroke's pleiotropic drugs, meaning molecules able to show diverse polypharmacology, can begin.

In this regard, nitrones-derivatives, as well organic compounds known for their radical trapping capacity, have attracted the interest of a number of research groups and projects. These compounds were first developed by analytical chemists as radical detection tools when stable nitrone-radical adducts formation were observed [89]. Detection and characterization of such short-lived species became therefore possible by electron spin resonance methods applied to the more stable and paramagnetic "spin adducts" formed (Fig. 9.5) [90].

Shortly after this discovery, subcellular radicals detection using this technique [91–93] has led to an extensive knowledge of the potential of nitrones as therapeutic candidates for cancer [94–98], neurodegenerative disorders [99], hearing loss [98], visual impairment [100], and stroke [98, 101, 102], among others. In particular, the

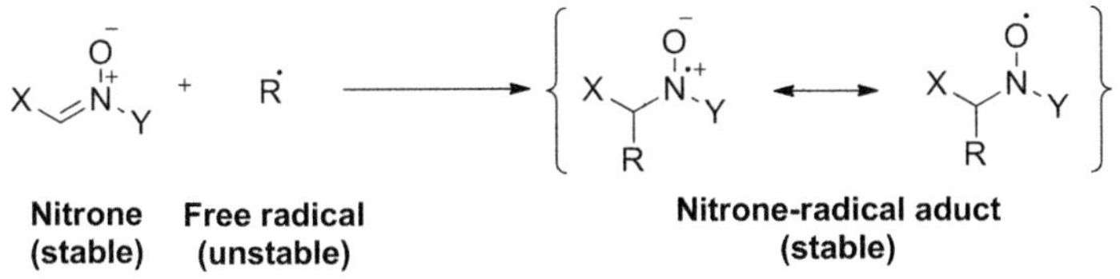

Fig. 9.5 Reaction of nitrones as free-radical trapping agents

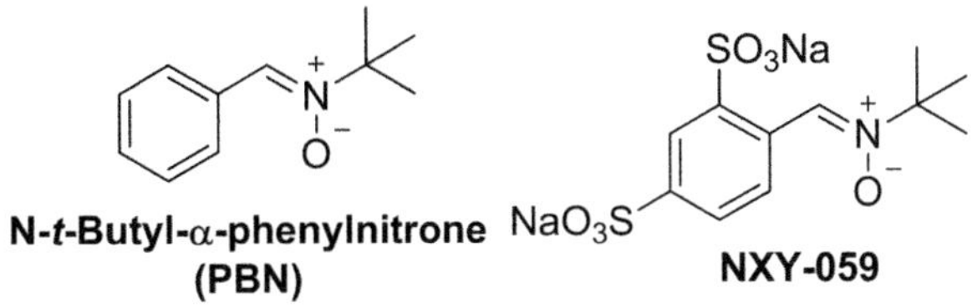

Fig. 9.6 PBN and its derivative, NXY-059, are the best known nitrones for stroke treatment

first proposal of nitrones as potential therapeutic agents for circulation issues was produced in 1986, when Novelli and coworkers found that phenyl-*t*-butylnitrone (PBN) (Fig. 9.6), a previously and widely used spin-trapping nitrone, prevented and reversed traumatic shock injury in rats [103].

3.2 *NXY-059 as the Furthest-Reaching Nitrone-Derived Drug Candidate to Date*

A PBN derivative, NXY-059 (Fig. 9.6) (Cerovive®, developed by Centaur Pharmaceuticals, later acquired by Renovis, which licensed it to AstraZeneca), was the first nitrone to reach clinical trials and, therefore, the first nitrone tested in humans.

Phase I and II studies reported no toxicity and a good pharmacokinetic profile, with low binding to plasma proteins (40 % approx.), short plasma elimination half-life (2–4 h), and low metabolic rate (about 80–90 % was excreted unchanged). Renal clearance was determined to be the most important parameter in NXY-059 elimination, which was not affected by urine pH or urine flow rate [104]. Regarding its tissue penetration, lower brain penetration than PBN was reported in several in vitro BBB penetration studies, but significantly increased during hypoxia or ischemic episodes [105]. From these results, no safety concern could be raised against this agent, as NXY-059 was demonstrated to be as safe as any other drug-candidate, which supposed great advances in nitrone-development field and promoted Phase III studies.

First Phase III studies, SAINT-1, began in USA in 2003 [106], with positive but not conclusive results. A second Phase III trial, SAINT-2, with larger number of patients was completed in 2006. Pooled analysis of both SAINT-1 and SAINT-2 trials concluded that NXY-059 was not effective as a treatment for ischemic stroke, despite all the previous results in preclinical and clinical stages [107]. This failure meant a disappointment in nitrone development, which casted doubts on neuroprotection as a strategy for stroke treatment [108, 109].

At this point, controversy was raised, as several studies published thereafter highlighted methodological defects, such as short delays between onset of ischemia and treatment [110, 111] and lack of rigor in individual studies of preclinical phases of NXY-059 [109], even though they were supposed to be developed according to the previously mentioned STAIR criteria [112]. Further meta-analysis pointed out that even NXY-059 showed neuroprotection in preclinical phases, questionable permeability, inadequate concentrations, and small sample size issues during these stages could have led to an overestimation of preclinical efficacy [113], in addition to inadequate design of clinical trials. Further, some authors raised the need for a proper characterization of the NXY-059 mechanism of action [114].

3.3 Current Development of Nitrones

Despite its concerns, NXY-059 set up the basis for the pharmacological development of new nitrones, which are known to possess several advantages with respect to other antioxidant candidates. First, nitrones present high versatility. They are easily afforded than different starting materials, and their reactivity and radical trapping properties can be easily modulated by simple chemical changes, leading to innovative designs hardly explored. Second, although nitrones are known for their radical trapping properties, its biological activity might go beyond this characteristic, as we will discuss in the next section, and therefore, multipotency may be affordable when combining with proper chemical motifs. Third, nitrones can be easily incorporated also in complex biochemical molecules, such as lipids, carbohydrates, proteins, etc., giving rise to interesting research lines regarding potential compartmentalization and specific biological targets. Fourth, latest advances in bio-organic chemistry have reported the use of nitrones as biorthogonal reagents, allowing the selective labeling of peptides and proteins [115, 116], creating interesting and challenging future perspectives. Finally, we must point out that, as it was previously mentioned, nitrone study started with analytical purposes and, therefore, they are well-documented probes for radical visualization. The combination of their diagnostic potential with their therapeutic activity could lead to interesting theranostics, not only for stroke, but also for many other diseases. In fact, and not surprisingly, numerous stroke research groups are investigating on new nitrones, such as TBN [128, 129] and STAZN [130, 131] (Fig. 9.7).

4 New Nitrones

After NXY-059 failure, but aware of the therapeutic potential of nitrones, our group decided to start in 2009 a medicinal chemistry program aimed to develop new potential compounds for ischemic stroke.

Although different scaffolds were tested, the most interesting results were achieved with compounds RP17, RP26 (*Phenylnitrones*), RP18, RP19

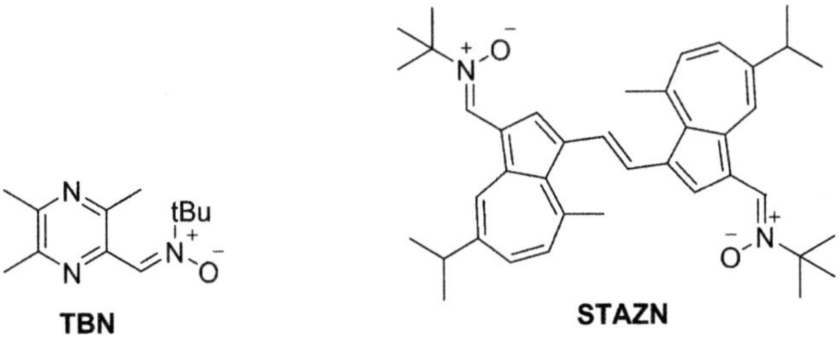

Fig. 9.7 Nitrones developed to date for the treatment of ischemic stroke

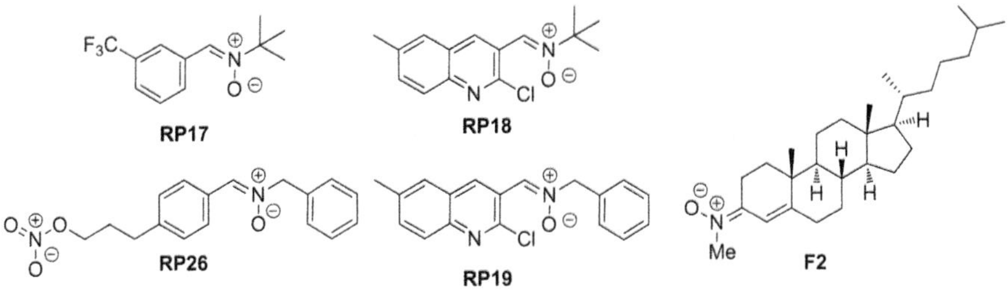

Fig. 9.8 Examples of nitrone-derived compounds assayed in our laboratory

(*Quinolylnitrones*), and F2 (*Cholesteronitrone*) [117], shown in Fig. 9.8. Previously reported [102, 118] and unpublished preclinical results obtained for these compounds are summarized and discussed in the next section.

4.1 In Vitro Tests for Antioxidant Effects

As radical-trapping agents, nitrones are expected to delay or prevent oxidation of easily oxidizable substrates, therefore being considered antioxidants. In vitro radical trapping and antioxidant activity were studied for nitrones RP17, RP26, RP18, RP19, F2, and PBN as reference compound, using the DPPH quenching, $^{\bullet}$OH and $O_2^{\bullet-}$ scavenging, and inhibition of lipid peroxidation by AAPH tests. Results are shown in Table 9.3 [102, 118]. Within the groups of nitrones tested, RP26 and RP18 are the best inhibitors of lipid peroxidation, according to AAPH test (81 and 85 %, respectively), in contrast to PBN, which showed no effect. DPPH and $O_2^{\bullet-}$ scavenging activities were low in general, with moderate values for RP17 and RP19 (0.9 % and 29 %, 42.3 % and 23 %, respectively). Scavenging of hydroxyl radical, as one of the most toxic radicals generated during ischemic stress, was also determined, showing that, except for RP17, higher trapping activities were achieved than reference compound PBN.

Table 9.3 In vitro antioxidant activity

Nitrone	AAPH (%)[a] (min)[b]	DPPH (%)[c]	$^{\bullet}OH$ (%)[d]	$O_2^{\bullet-}$ (%)[e]
RP17	no (0)	0.9	60	29
RP26	81 (36.6)	nd[f]	98	no
RP18	85 (56.7)	1.7	100	12
RP19	37 (78)	42.3	95	23
F2	55 (nd[f])	4	83	nd[f]
PBN	no (0)	nd[f]	90	15

[a]Determined at 0.1 mM
[b]Induction time (t_{inh}) produced by the tested nitrones
[c]Determined at 0.5 mM except for F2, that was determined at 0.1 mM
[d]Determined at 0.1 mM
[e]Determined at 0.1 mM
[f]*nd* not determined

Neuroprotective effect in stroke, however, may go beyond antioxidant capacity just reported, as was previously mentioned. Testing in living systems obviously casts a more precise image of real neuroprotective effect of the developed compounds and, for this reason, neuroprotective effect was also evaluated in neuronal cultures and in in vivo experiments.

4.2 *Neuroprotective Properties of Nitrones in Neuronal Cultures and in Animal Model of Transient Global Ischemia*

Neuroprotective effect of nitrones RP17, RP26, RP18, RP19, F2, and PBN and NXY-059 as reference compounds was studied in primary neuronal cultures from cerebral cortex, which were subjected to oxygen-glucose deprivation (OGD) as an in vitro model of ischemia. Cell viability was evaluated by quantification of living, metabolically active cells, as determined by the MTT assay. Neuroprotection is expressed as the percentage to reach the control value (100 %), from the untreated ischemic neurons value (0 %) (Table 9.4).

From these results, interesting neuroprotective effects in neuronal cultures were observed for all the nitrones tested, showing values in all cases higher than the one determined for PBN (−13.4 ± 1.9 % at 5 mM) and NXY-059 (56.8 ± 2.5 % at 250 μM). However, remarkable results, especially, were analyzed for RP19 (87.5 ± 3.2 % at 50 μM) and F2 (80.7 ± 2.7 % at 5 μM), a fact that prompted us to undertake further in vivo studies.

Transient global brain ischemia was performed on adult rats by the standard four-vessel occlusion model (4VO), in which carotid arteries are occluded during 15 min 24 h after the irreversible occlusion of vertebral arteries by electrocoagulation [119, 120]. Ischemic animals were treated with RP19, F2, and NXY-059 diluted

Table 9.4 Neuroprotection in neuronal cultures and in vivo model of ischemia

	Neuronal cultures		In vivo experiments					
Nitrone	Cc	Neuroprotection (%)	Species	Time after reperfusion	Cc (mg/kg)	Cell death reduction (%)	Apoptosis reduction (%)	Ref.
PBN	5 mM	−13.4 ± 1.9						[102]
NXY-059	250 μM	56.8 ± 2.5	*Rattus norvegicus*	5d	40	17 (CA1) 70 (C)	21 (CA1) 55 (C)	[118]
RP17	10 μM	63.1 ± 2.2						[122]
	100 μM	73.5 ± 2.2						
RP26	10 μM	77.6 ± 3.1						[102]
	100 μM	83.0 ± 2.7						
RP18	1 μM	71.2 ± 4.0						[102]
RP19	10 μM	70.9 ± 2.2	*Rattus norvegicus*	5d	0.5	35 (CA1)*** 63 (C)*	38 (CA1)** 79 (C)*	[122]
	50 μM	87.5 ± 3.2						
F2	1 μM	54.3 ± 1.3	*Rattus norvegicus*	5d	0.05	20 (CA1) 66 (C)	30 (CA1)* 89 (C)*	[118]
	5 μM	80.7 ± 2.7			0.1	27 (CA1)** 83 (C)*	35 (CA1)** 91 (C)*	

*$p < 0.05$, **$p < 0.01$ and ***$p < 0.001$, compared with vehicle by Dunnett's post-test after ANOVA

in 10 % ethanol in saline as vehicle by intraperitoneal injection when carotid arteries were un-clamped for reperfusion. Animals were studied after 5 d of reperfusion (R5d) after killing by transcardiac perfusion performed under deep anesthesia. Treatments were blindly and randomly performed and body temperature of 37 °C was maintained (Table 9.4).

Cell death and apoptosis were assessed in the hippocampal *cornu ammonis* 1 (CA1) region and cerebral cortex (C). Nitrones RP19 and F2 showed higher inhibition of cell death than for NXY-059. In particular, best results were obtained with F2 at 0.1 mg/kg concentration, and RP19 at 0.5 mg/kg concentration. Apoptosis reduction followed similar trend: F2 (35 % in hippocampal CA1, 91 % in cortex, at 0.1 mg/kg concentration) and RP19 (38 % in hippocampal CA1,79 % in cortex, at 0.5 mg/kg concentration) showing the best results, both higher than the values observed for nitrone NXY-059 (21 % in hippocampal CA1, 55 % in cortex) at the same concentration.

The potential therapeutic value of these results has been recently confirmed in ischemic mice using a model of transient focal cerebral ischemia by distal middle cerebral artery occlusion. The study was carried out as randomized and double-blind with compound F2 [121] and RP19 [122], opening the way to start with the pre-clinical phase in the future.

5 Nitrones Pleiotropy: New Perspectives

As it was previously mentioned, ischemic stroke is a multifactorial disease in which different metabolic pathways, as well as many other diverse aspects, determine patient's final outcome. However, candidate drugs for stroke treatment have been usually aimed to target only one of these disease aspects and many authors have pointed out that this monofunctionality, combined with a questionable design of preclinical and clinical studies, could be the cause of the complete lack of effective neuroprotection therapies nowadays [98, 123].

As a consequence, multiple efforts are being done in the last years in order to design pleiotropic drugs not only for stroke, but also for many other complex diseases without a current proper treatment. Additionally, new mechanisms of action are also being discovered for old-drugs, whose therapeutic effect is then being revealed much more complex than thought in its origins. In this sense, a stroke-directed drug only approved in Japan, edaravone (Radicut®), has received special attention regarding a possible multifunctionality as the cause of its therapeutic effects. In particular, apart from its antioxidant capacity derived from its radical trapping properties, edaravone has been suggested to suppress delayed neuronal death, reduce long-term inflammation, prevent edema growing, or protect brain vessel from vascular damage, among others. Taken together, these factors would undoubtedly make edaravone a perfect candidate for stroke treatment. Nevertheless, further clinical trials are needed to show its efficacy and be granted approval by the FDA [124].

As for edaravone, nitrones may also enter this multi-targeted drugs group. PBN and related compounds, despite its well-documented radical trapping activity, are not expected to act as radical traps in biological systems, since low reported reaction rates would mean high concentrations of nitrone, which hardly occurs in the lower neuroprotective concentrations used in biological tests [98]. For this reason, additional mechanisms have been added in order to account for the activity observed in biological systems. Among them, influence in membrane enzymes is one of the reasons proposed, due to high PBN lipophilicity, even though no clear evidences have been reported regarding ischemic stroke. Additionally, anti-inflammatory activity has been also pointed out, as PBN has been reported to inhibit iNOS and COX-2 in several biological models and systems [98].

But then, a question may arise: what makes a drug *pleiotropic*? Pleiotropic drugs are, by definition, compounds able to show diverse pharmacological effect. These functionalities depend on the physicochemical properties and the biological system tested and, therefore, any drug target profile could be modulated by varying any of these conditions. In the particular case of nitrones, pleiotropy has appeared as a necessity to justify a therapeutic effect that was unjustifiable based on previous knowledge, i.e., radical trapping activity. However, there are still great advances to be made in this aspect. Not only the exact mechanism of action of nitrones according to their chemical composition must be determined, but also the exact mechanism of cell death during ischemic stroke (Fig. 9.9). In doing so, more efficient and interesting drug developments could be performed, combining needed functionalities (Fig. 9.9) and making great advances that are required for the treatment of diseases such as stroke.

5.1 *In Silico Target Identification for Nitrones*

Regarding this high concern, one of the goals of our group is to elucidate the mechanism of action of the different nitrones developed. As case examples, we report herein interesting results obtained for previous nitrones RP17, RP26, RP18, RP19, and F2.

Computer-assisted methodologies are seen as promising predictive methodologies in drug discovery processes. Taken with precaution, results obtained, if

Fig. 9.9 Scheme of potential functionalities that could be added to nitrone-derived drugs

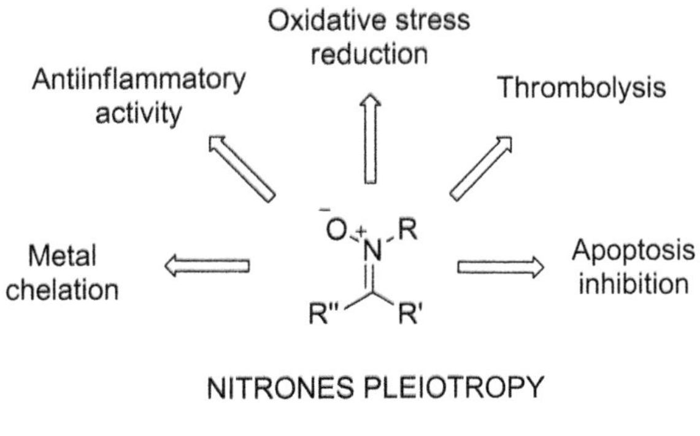

confirmed experimentally, can direct further drug developments and increase its efficacy. Pythia software, developed by Intelligent Pharma [125], helps in elucidating potential targets by comparison of a test compound chemical structure with existing molecules and drugs. Then, a set of targets is proposed for the tested compound, which are sorted according to a *MoA score*, parameter that describes the quality of the prediction taking into account both similarity and reported activity of the drug (Table 9.5, Predicted targets).

After particular set of targets is proposed for each molecule tested, Pythia searches similar target profiles that correspond to known drugs, and its similarity is defined by the *IT score* (from 0 to 100 % of correspondence). Based on the specific drug found, a therapeutic group is assigned to the tested compound, according to the Anatomical Therapeutic Chemical (ATC) classification system, defined by the World Health Organization Collaborating Center (WHOCC) for Drug Statistics Methodology (http://www.whocc.no) (Table 9.5, Target profile comparison).

Compounds RP17 and RP26 (phenyl), RP18 and RP19, (quinolyl), F2 (cholesterol), and reference compounds PBN and NXY-059 were subjected to Pythia algorithm prediction. In Table 9.5, *Predicted Targets*, those repeated in the different structural groups (phenyl, quinolyl, and cholesterol), as well as those present in their inactive precursors, were not shown, as an attempt to identify the targets responsible for their selective biological activity found. Highest MoA score for each target is reported. In Table 9.5, *Target Profile Comparison*, drugs showing target profiles with TI score >40 compared to the tested compound were shown. The drug found is reported according to its ChEMBL code (https://www.ebi.ac.uk/chembldb), along with its correspondent ATC code and therapeutic indication group according to ATC classification system (http://www.whocc.no/atc_ddd_index).

Regarding predicted targets, different types are proposed for each compound, which may account for a pleiotropic effect. Many of them are not known to play important roles in stroke disease, but some are interesting targets regarding an effective multifunctionality, as we will discuss later on. In addition, there is a huge variability among different structural groups, when the main structural scaffold is changed and also variations in substituents promote complete different target profiles.

Among all the targets proposed, three of them (metabotropic glutamate receptor 1, vascular endothelial grow factor (VEGF) receptor, and mitochondrial glutaminase) appear in different compounds corresponding to the same structural group and could account for interesting neuroprotective behavior. In particular, RP18 and RP19 are suggested to modulate metabotropic glutamate receptor, which is an obvious target for excitotoxicity reduction (see Sect. 2.1). Mitochondrial glutaminase, present in PBN and RP17, is known to play crucial role in glutamate metabolism in neurons, as it catalyzes glutamine conversion to glutamate in axonal terminals [126]. Vascular endothelial growth factor receptor (VEGFR), present in RP18 and RP19, is implied in vasculogenesis and angiogenesis [127], and therefore its stimulation could favor recovery of proper circulatory conditions after the ischemic insult.

From those targets only present in one molecule tested, special relevance has plasminogen activator inhibitor-1, proposed for RP26, whose modulation has obvious implications. Monoamine oxidase, suggested for RP17, is also of great importance.

Table 9.5 In silico target identification for nitrones

Predicted targets			Target profile comparison			
Target name[a]	Accession number[b]	MoA score[c]	TI score[c]	CHEMBL[d]	ATC code[e]	Chemical/therapeutic/pharmacological subgroup[f]
PBN						
ATP-dependent DNA helicase Q1	P46063	0.21	98.44	517	C01BA	Antiarrhythmics, class Ia
Glucose-6-phosphate 1-dehydrogenase	P11413	0.22	87.84	704	A07EC	Aminosalicylic acid and similar agents
Glutaminase kidney isoform, mitochondrial	O94925	0.21	87.84	1042	A11CC	Vitamin D and analogues
Thyroid hormone receptor beta-1	P10828	0.34	87.50	177756	S01JA	Colouring agents
Werner syndrome ATP-dependent helicase	Q14191	0.20	87.50	1200604	S01FA	Anticholinergics
			87.50	93268	V03AB	Antidotes
			87.50	76	P01BA	Aminoquinolines
			87.50	686	M01AG	Fenamates
			87.50	1314	R05CB	Mucolytics
			87.50	1200692	C09CA	Angiotensin II antagonists
NXY-059						
Cystic fibrosis transmembrane conductance regulator	P13569	0.18	88.87	1577	C03AA	Thiazides, plain

(continued)

Table 9.5 (continued)

Predicted targets			Target profile comparison			
Target name[a]	Accession number[b]	MoA score[c]	TI score[c]	CHEMBL[d]	ATC code[e]	Chemical/therapeutic/pharmacological subgroup[f]
Nuclear factor NF-kappa-B p105 subunit	P19838	0.16	87.88	1200617	S01BA	Corticosteroids, plain
			87.88	1200617	H02AB	Glucocorticoids
			87.76	1683	D07AB	Corticosteroids, moderately potent (group II)
			87.63	635	A07EA	Corticosteroids acting locally
			87.51	1239	P03AX	Other ectoparasiticides, incl. scabicides
			87.50	1670	L01XX	Other antineoplastic agents
			66.99	945	C03DB	Other potassium-sparing agents
			49.22	1532	C03BA	Sulfonamides, plain
			49.22	249837	J01AA	Tetracyclines
RP17						
Anandamide amidohydrolase	O00519	0.46	43.75	1525	P03AC	Pyrethrines, incl. synthetic compounds
Aryl hydrocarbon receptor	P35869	0.70	43.75	672	C10AB	Fibrates
ATP-dependent DNA helicase Q1	P46063	0.36				
Glutaminase kidney isoform, mitochondrial	O94925	0.43				
Monoamine oxidase (MAO)	P27338, P21397	0.34				
Neuropeptide Y receptor type 5	Q15761	0.37				
Serotonin 2b (5-HT2b) receptor	P41595	0.38				
Serotonin 2c (5-HT2c) receptor	P28335	0.37				

Target name[a]	Accession number[b]	MoA score[c]	TI score[c]	CHEMBL[d]	ATC code[e]	Chemical/therapeutic/pharmacological subgroup[f]
RP26						
Alpha-1d adrenergic receptor	P25100	0.23	92.97	750	N03AX	Other antiepileptics
Free fatty acid receptor 1	O14842	0.21	44.09	419	D06BA	Sulfonamides
Lanosterol synthase	P48449	0.21				
Peroxisome proliferator-activated receptor alpha	Q07869	0.20				
Plasminogen activator inhibitor-1	P05121	0.20				
Sphingosine 1-phosphate receptor Edg-1	P21453	0.28				
Sphingosine 1-phosphate receptor Edg-3	Q99500	0.21				
RP18						
Calpain 2	P17655	0.55	90.26	1200471	D11AX	Other dermatologicals
DNA polymerase eta	Q9Y253	0.49	87.50	178	L01DB	Anthracyclines and related substances
Metabotropic glutamate receptor 1	Q13255	0.74	87.50	1103	G01AX	Other antiinfectives and antiseptics
Vascular endothelial growth factor receptor 2	P35968	0.74	87.50	898	N02BA	Salicylic acid and derivatives
			87.50	1200468	P03AX	Other ectoparasiticides, incl. scabicides
			49.22	562	D01BA	Antifungals for systemic use
			49.22	562	D01AA	Antibiotics
			49.22	325041	L01XX	Other antineoplastic agents
			45.12	447	N05CA	Barbiturates
			45.12	1228	S01BC	Antiinflammatory agents, non-steroids

(continued)

Table 9.5 (continued)

Predicted targets			Target profile comparison			
Target name[a]	Accession number[b]	MoA score[c]	TI score[c]	CHEMBL[d]	ATC code[e]	Chemical/therapeutic/pharmacological subgroup[f]
RP19						
Bloom syndrome protein	P54132	0.51	90.23	964	P03AA	Sulfur containing products
DNA polymerase eta	Q9Y253	0.54	90.23	964	N07BB	Drugs used in alcohol dependence
Histone acetyltransferase GCN5	Q92830	0.53	88.87	1200751	L01BB	Purine analogues
Huntingtin	P42858	0.54	87.84	562	D01BA	Antifungals for systemic use
MAP kinase ERK2	P28482	0.50	87.84	562	D01AA	Antibiotics
Metabotropic glutamate receptor 1	Q13255	1.02	87.84	40	N03AA	Barbiturates and derivatives
Vascular endothelial growth factor receptor 2	P35968	0.63	87.84	1201129	L01BC	Pyrimidine analogues
			87.84	325041	L01XX	Other antineoplastic agents
			87.59	1228	S01BC	Antiinflammatory agents, non-steroids
			87.59	1228	M02AA	Antiinflammatory preparations, non-steroids
F2						
Cytochrome P450 2C19	P33261	0.67				
Estrogen receptor beta	Q92731	0.65				
Protein farnesyltransferase	P49354, P49356	0,66				

Notes: [a]Targets are ordered following alphabetical criteria within the same compound
[b]Accession number from UniProt Database (www.uniprot.org)
[c]MoA and TI scores from Pythia algorithm, Intelligent Pharma [125]
[d]ChEMBL code from ChEMBL Database (https://www.ebi.ac.uk/chembldb)
[e]ATC code from Anatomical Therapeutic Chemical (ATC) classification system (www.whocc.no/atc_ddd_index)
[f]Fourth level subgroups in the ATC classification system

As it is widely known, this enzyme catalyzes oxidation of monoamines, which are usually neurotransmitters such as dopamine, serotonin, or adrenaline and, since oxygen is used as a substrate, this enzyme could be implied in oxidative stress contribution to cell damage. Calpain, present for RP18, was considered as one of the enzymes activated by calcium and its inhibition could avoid cellular structures degradation (see Sect. 2.1). Finally, cytochrome C, which appears for F2, was mentioned in several stages of ischemic-reperfusion damage. It is not only implied in oxidative stress, as an important partner of the respiratory chain, but also is responsible for triggering apoptosis in later stages of cellular damage.

A remark must be made, however, on NXY-059, as it was not suggested to modulate any interesting target regarding stroke disease, despite being the furthest reaching nitrone drug candidate to date.

Target profile comparison also gave rise to interesting concerns, being the most noted the high similarity found for RP18, RP19, and NXY-059 with anti-inflammatory agents, both non-steroids anti-inflammatory agents and corticoids (see Target profile comparison, Table 9.5). In this regard, PBN target profile could also be identified with anti-inflammatory activity since, as it is well-known, salicylic acid and derivatives have this effect. Again, considering crucial relevance of inflammatory response in stroke outcome, it would be of great interest to confirm this anti-inflammatory activity, as it could partly justify those in vivo results found experimentally.

To summary these interesting results, some promising targets have arisen by using computer-assisted methodologies that could account for the neuroprotective behavior reported. Nevertheless, as an in silico prediction, it must be taken with precaution until confirmed experimentally.

5.2 *Exploring Nitrone-Based Multidrug Therapy*

In order to explore that nitrones tested may have different targets as deduced by in silico analysis, multidrug studies were performed. Variable concentrations of F2 were combined with the other nitrones at sub-optimal concentrations (Table 9.4) and evaluated in primary neuronal cultures subjected to OGD as in [102]. Results obtained are shown in Fig. 9.10.

From neuroprotection values of F2 (black line), a significant higher neuroprotection was achieved when combined with low concentrations of RP17 and RP19 (blue line and red line, respectively) at smallest concentrations of F2 (0.5 and 1 μM). Since these neuroprotection values were higher than highest neuroprotective concentrations for F2 (5 μM), different target modulation with complementary effect can be inferred. However, this effect seems to be dependent on concentration, as can be seen with RP19 when combined with F2 5 μM. On the other hand, opposite effect is achieved when combining F2 with RP26 or RP18, which means this combined modulation of targets was, in this case, antagonist.

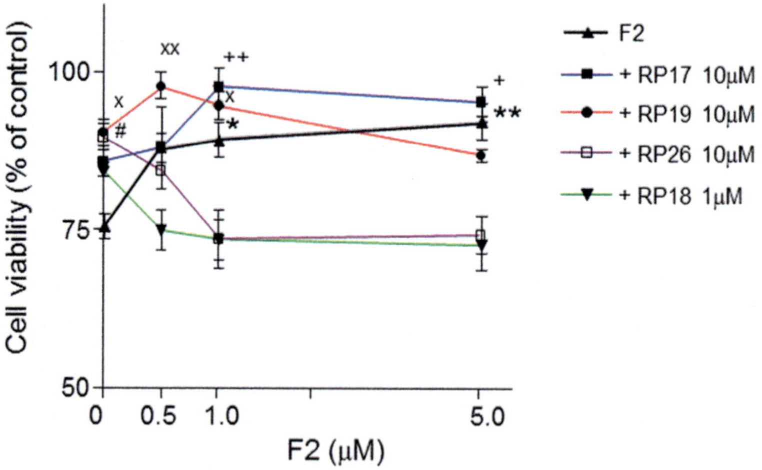

Fig. 9.10 Combination studies. Cell viability in OGD-exposed neuronal cultures treated with different concentrations of F2 in absence (*black line*) or combined with the indicated nitrones. *,$^{+}$, ^{x}, $^{\#}p<0.05$, and **, $^{++}$, $^{xx}p<0.01$ compared with the value without additions, by one-way analysis of variance ($p<0.001$) following Dunnett's post-test

From these results, it is feasible that nitrones, due to their potential multitargeting, can be combined properly to achieve better effects than performed alone. However, at the same time, if the multitargeting has opposite effects, these compounds in combination could yield pernicious effects.

6 Conclusions

In this chapter, we have updated and summarized the current status of the neuroprotection strategy centered on the oxidative stress, for the development of new drugs for stroke therapy. Particular attention has been devoted to old and currently neglected nitrones as potential drugs for the treatment of this pathology. Our most recent results confirm that the neuroprotective strategy and nitrones as therapeutic agents still offer strong and unexpected opportunities for improvement and therapeutic efficacy. The new and for the first time investigated *quinolylnitrones*, as well as the *cholesteronitrones*, have created great expectations that we must still reconfirm. Neuroprotection causes go beyond their radical trapping activity, but the exact target(s) modulated are still unknown. In silico studies show that pleiotropy—as multitargeting—may be a feature among the nitrones tested. Thus, computational methodologies can be useful in determining the target profile, although additional experimental studies must be performed.

These nitrones, and others that may come, appear as promising compounds due to robust antioxidant properties and potential pleiotropic activity, capable to target distinct steps of the biochemical pathways during and after the ischemic insult.

Acknowledgments We are indebted to Ph.D. D.J. Hadjipavlou-Litina for the in vitro antioxidants experiments and Ms. M. Gómez-Calcerrada for her assistance. This work was supported by the Institute of Health Carlos III from the Spanish Ministry of Economy and Competitiveness (MINECO) with the grant number PI14/00705 (State Plan), and the funding from FEDER grants. JMC and MC thank MINECO grants SAF2012-33304 and SAF2015-65586-R, European Union (COST Action 1103) and Universidad Camilo José Cela, for support.

References

1. World Health Organization. Stroke, Cerebrovascular accident. http://www.who.int/topics/cerebrovascular_accident/en/
2. World Health Organization (2014) Global health estimates 2014 summary tables: deaths by cause, age and sex, 2000-2012
3. World Health Organization (2014) Global health estimates 2014 summary tables: DALY by cause, age and sex, 2000-2012
4. Adams HP, Bendixen BH, Kappelle LJ, Biller J, Love BB, Gordon DL et al (1993) Classification of subtype of acute ischemic stroke. Definitions for use in a multicenter clinical trial. TOAST. Trial of Org 10172 in acute stroke treatment. Stroke 24(1):35–41
5. Easton JD, Saver JL, Albers GW, Alberts MJ, Chaturvedi S, Feldmann E et al (2009) Definition and evaluation of transient ischemic attack: a scientific statement for healthcare professionals from the American Heart Association/American Stroke Association Stroke Council; Council on Cardiovascular Surgery and Anesthesia; Council on Cardiovascular Radiology and Intervention; Council on Cardiovascular Nursing; and the Interdisciplinary Council on Peripheral Vascular Disease: The American Academy of Neurology affirms the value of this statement as an educational tool for neurologists. Stroke 40(6):2276–2293
6. Astrup J, Siesjö BK, Symon L (1981) Thresholds in cerebral ischemia—the ischemic penumbra. Stroke 12(6):723–725
7. Lassen NA, Fieschi C, Lenzi GL (1991) Ischemic penumbra and neuronal death: comments on the therapeutic window in acute stroke with particular reference to thrombolytic therapy. Cerebrovasc Dis 1(suppl 1):32–35
8. Moskowitz MA, Lo EH, Iadecola C (2010) The science of stroke: mechanisms in search of treatments. Neuron 67(2):181–198
9. Symon L, Branston NM, Strong AJ, Hope TD (1977) The concepts of thresholds of ischaemia in relation to brain structure and function. J Clin Pathol 11(1):s149–s154
10. Hossmann KA (1994) Viability thresholds and the penumbra of focal ischemia. Ann Neurol 36(4):557–565
11. Mies G, Ishimaru S, Xie Y, Seo K, Hossmann K-A (1991) Ischemic thresholds of cerebral protein synthesis and energy state following middle cerebral artery occlusion in rat. J Cereb Blood Flow Metab 11(5):753–761
12. Paschen W, Mies G, Hossmann KA (1992) Threshold relationship between cerebral blood flow, glucose utilization, and energy metabolites during development of stroke in gerbils. Exp Neurol 117(3):325–333
13. Siesjö BK (1988) Acidosis and ischemic brain damage. Neurochem Pathol 9(1):31–88
14. Katsura K, Kristián T, Siesjö BK (1994) Energy metabolism, ion homeostasis, and cell damage in the brain. Biochem Soc Trans 22(4):991–996

15. Schuier FJ, Hossmann KA (1980) Experimental brain infarcts in cats. II. Ischemic brain edema. Stroke 11(6):593–601
16. Paschen W, Mengesdorf T (2005) Endoplasmic reticulum stress response and neurodegeneration. Cell Calcium 38(3-4):409–415
17. Simon RP (2006) Acidotoxicity trumps excitotoxicity in ischemic brain. Arch Neurol 63(10):1368–1371
18. Chen ZL, Strickland S (1997) Neuronal death in the hippocampus is promoted by plasmin-catalyzed degradation of laminin. Cell 91(7):917–925
19. Choi DW (1988) Calcium-mediated neurotoxicity: relationship to specific channel types and role in ischemic damage. Trends Neurosci 11(10):465–469
20. Bredt DS, Snyder SH (1990) Isolation of nitric oxide synthetase, a calmodulin-requiring enzyme. Proc Natl Acad Sci U S A 87(2):682–685
21. Norenberg MD, Rao KVR (2007) The mitochondrial permeability transition in neurologic disease. Neurochem Int 50(7–8):983–997
22. Marshall JJ, Kontos HA (1990) Endothelium-derived relaxing factors. A perspective from in vivo data. Hypertension 16(4):371–386
23. Knowles RG, Moncada S (1994) Nitric oxide synthases in mammals. Biochem J 298(Pt 2):249–258
24. Iadecola C, Zhang F, Xu S, Casey R, Ross ME (1995) Inducible nitric oxide synthase gene expression in brain following cerebral ischemia. J Cereb Blood Flow Metab 15(3):378–384
25. Brown GC, Borutaite V (1999) Nitric oxide, cytochrome c and mitochondria. Biochem Soc Symp 66:17–25
26. Kudin AP, Debska-Vielhaber G, Kunz WS (2005) Characterization of superoxide production sites in isolated rat brain and skeletal muscle mitochondria. Biomed Pharmacother 59(4):163–168
27. Bolton JL, Trush MA, Penning TM, Dryhurst G, Monks TJ (2000) Role of quinones in toxicology. Chem Res Toxicol 13(3):135–160
28. Harrison R (2002) Structure and function of xanthine oxidoreductase: where are we now? Free Radic Biol Med 33(6):774–797
29. Pou S, Pou WS, Bredt DS, Snyder SH, Rosen GM (1992) Generation of superoxide by purified brain nitric oxide synthase. J Biol Chem 267(34):24173–24176
30. Lambeth JD (2004) NOX enzymes and the biology of reactive oxygen. Nat Rev Immunol 4(3):181–189
31. Jin R, Liu L, Zhang S, Nanda A, Li G (2013) Role of inflammation and its mediators in acute ischemic stroke. J Cardiovasc Transl Res 6(5):834–851
32. Huang J, Upadhyay UM, Tamargo RJ (2006) Inflammation in stroke and focal cerebral ischemia. Surg Neurol 66(3):232–245
33. del Zoppo GJ, Schmid-Schonbein GW, Mori E, Copeland BR, Chang CM (1991) Polymorphonuclear leukocytes occlude capillaries following middle cerebral artery occlusion and reperfusion in baboons. Stroke 22(10):1276–1283
34. Fukuda S, Fini CA, Mabuchi T, Koziol JA, Eggleston LL Jr, del Zoppo GJ (2004) Focal cerebral ischemia induces active proteases that degrade microvascular matrix. Stroke 35(4):998–1004
35. Chou WH, Choi DS, Zhang H, Mu D, McMahon T, Kharazia VN et al (2004) Neutrophil protein kinase Cdelta as a mediator of stroke-reperfusion injury. J Clin Invest 114(1):49–56
36. Beckman JS, Beckman TW, Chen J, Marshall PA, Freeman BA (1990) Apparent hydroxyl radical production by peroxynitrite: implications for endothelial injury from nitric oxide and superoxide. Proc Natl Acad Sci U S A 87(4):1620–1624
37. Alvarez B, Radi R (2003) Peroxynitrite reactivity with amino acids and proteins. Amino Acids 25(3-4):295–311
38. Halliwell B, Gutteridge JM (1992) Biologically relevant metal ion-dependent hydroxyl radical generation. An update. FEBS Lett 307(1):108–112

39. Moskowitz N, Schook W, Puszkin S (1984) Regulation of endogenous calcium-dependent synaptic membrane phospholipase A2. Brain Res 290(2):273–279
40. Kim H-H, Lee K-H, Lee D, Han Y-E, Lee S-H, Sohn J-W et al (2015) Costimulation of AMPA and metabotropic glutamate receptors underlies phospholipase C activation by glutamate in hippocampus. J Neurosci 35(16):6401–6412
41. Bazan NG (1976) Free arachidonic acid and other lipids in the nervous system during early ischemia and after electroshock. Adv Exp Med Biol 72:317–335
42. Wojtczak L, Schonfeld P (1993) Effect of fatty acids on energy coupling processes in mitochondria. Biochim Biophys Acta 1183(1):41–57
43. Anthonymuthu TS, Kenny EM, Bayir H (2016) Therapies targeting lipid peroxidation in traumatic brain injury. Brain Res 1640(Pt A):57–76
44. Fam SS, Morrow JD (2003) The isoprostanes: unique products of arachidonic acid oxidation-a review. Curr Med Chem 10(17):1723–1740
45. Mark RJ, Lovell MA, Markesbery WR, Uchida K, Mattson MP (1997) A role for 4-hydroxynonenal, an aldehydic product of lipid peroxidation, in disruption of ion homeostasis and neuronal death induced by amyloid beta-peptide. J Neurochem 68(1):255–264
46. Gaudet RJ, Alam I, Levine L (1980) Accumulation of cyclooxygenase products of arachidonic acid metabolism in gerbil brain during reperfusion after bilateral common carotid artery occlusion. J Neurochem 35(3):653–658
47. Andreyev AY, Fahy B, Fiskum G (1998) Cytochrome c release from brain mitochondria is independent of the mitochondrial permeability transition. FEBS Lett 439(3):373–376
48. Verhagen AM, Ekert PG, Pakusch M, Silke J, Connolly LM, Reid GE et al (2000) Identification of DIABLO, a mammalian protein that promotes apoptosis by binding to and antagonizing IAP proteins. Cell 102(1):43–53
49. Endres M, Wang ZQ, Namura S, Waeber C, Moskowitz MA (1997) Ischemic brain injury is mediated by the activation of poly(ADP-ribose)polymerase. J Cereb Blood Flow Metab 17(11):1143–1151
50. Saito A, Maier CM, Narasimhan P, Nishi T, Song YS, Yu F et al (2005) Oxidative stress and neuronal death/survival signaling in cerebral ischemia. Mol Neurobiol 31(1-3):105–116
51. Moro MA, Cardenas A, Hurtado O, Leza JC, Lizasoain I (2004) Role of nitric oxide after brain ischaemia. Cell Calcium 36(3-4):265–275
52. Iadecola C, Forster C, Nogawa S, Clark HB, Ross ME (1999) Cyclooxygenase-2 immunoreactivity in the human brain following cerebral ischemia. Acta Neuropathol 98(1):9–14
53. Wang X, Lo EH (2003) Triggers and mediators of hemorrhagic transformation in cerebral ischemia. Mol Neurobiol 28(3):229–244
54. Lees KR, Bluhmki E, von Kummer R, Brott TG, Toni D, Grotta JC et al (2010) Time to treatment with intravenous alteplase and outcome in stroke: an updated pooled analysis of ECASS, ATLANTIS, NINDS, and EPITHET trials. Lancet 375(9727):1695–1703
55. Hacke W, Kaste M, Bluhmki E, Brozman M, Dávalos A, Guidetti D et al (2008) Thrombolysis with alteplase 3 to 4.5 hours after acute ischemic stroke. N Engl J Med 359(13):1317–1329
56. Nicole O, Docagne F, Ali C, Margaill I, Carmeliet P, MacKenzie ET et al (2001) The proteolytic activity of tissue-plasminogen activator enhances NMDA receptor-mediated signaling. Nat Med 7(1):59–64
57. Balami JS, Sutherland BA, Buchan AM (2013) Complications associated with recombinant tissue plasminogen activator therapy for acute ischaemic stroke. CNS Neurol Disord Drug Targets 12(2):155–169
58. Rha JH, Saver JL (2007) The impact of recanalization on ischemic stroke outcome: a meta-analysis. Stroke 38(3):967–973
59. Becker KJ, Brott TG (2005) Approval of the MERCI clot retriever: a critical view. Stroke 36(2):400–403
60. Investigators PPST (2009) The penumbra pivotal stroke trial: safety and effectiveness of a new generation of mechanical devices for clot removal in intracranial large vessel occlusive disease. Stroke 40(8):2761–2768

61. Koh JS, Lee SJ, Ryu CW, Kim HS (2012) Safety and efficacy of mechanical thrombectomy with solitaire stent retrieval for acute ischemic stroke: a systematic review. Neurointervention 7(1):1–9
62. Nogueira RG, Lutsep HL, Gupta R, Jovin TG, Albers GW, Walker GA et al (2012) Trevo versus Merci retrievers for thrombectomy revascularisation of large vessel occlusions in acute ischaemic stroke (TREVO 2): a randomised trial. Lancet 380(9849):1231–1240
63. Ciccone A, del Zoppo GJ (2014) Evolving role of endovascular treatment of acute ischemic stroke. Curr Neurol Neurosci Rep 14(1):416
64. The National Institute of Neurological Disorders and Stroke rt-PA Stroke Study Group (1995) Tissue plasminogen activator for acute ischemic stroke. N Engl J Med 333(24):1581–1588
65. IST-3 Collaborative Group, Sandercock P, Wardlaw JM, Lindley RI, Dennis M, Cohen G, Murray G, Innes K, Venables G, Czlonkowska A, Kobayashi A, Ricci S, Murray V, Berge E, Slot KB, Hankey GJ, Correia M, Peeters A, Matz K, Lyrer P, Gubitz G, Phillips SJ, Arauz A (2012) The benefits and harms of intravenous thrombolysis with recombinant tissue plasminogen activator within 6 h of acute ischaemic stroke (the third international stroke trial [IST-3]): a randomised controlled trial. Lancet 379(9834):2352–2363
66. Jauch EC, Saver JL, Adams HP, Bruno A, Connors JJ, Demaerschalk BM et al (2013) Guidelines for the early management of patients with acute ischemic stroke: a guideline for healthcare professionals from the American Heart Association/American Stroke Association. Stroke 44(3):870–947
67. Logallo N, Kvistad CE, Thomassen L (2015) Therapeutic potential of tenecteplase in the management of acute ischemic stroke. CNS Drugs 29(10):811–818
68. Parsons M, Spratt N, Bivard A, Campbell B, Chung K, Miteff F et al (2012) A randomized trial of tenecteplase versus alteplase for acute ischemic stroke. N Engl J Med 366(12):1099–1107
69. Huang X, Cheripelli BK, Lloyd SM, Kalladka D, Moreton FC, Siddiqui A et al (2015) Alteplase versus tenecteplase for thrombolysis after ischaemic stroke (ATTEST): a phase 2, randomised, open-label, blinded endpoint study. Lancet Neurol 14(4):368–376
70. Coutts SB, Dubuc V, Mandzia J, Kenney C, Demchuk AM, Smith EE et al (2015) Tenecteplase–tissue-type plasminogen activator evaluation for minor ischemic stroke with proven occlusion. Stroke 46(3):769–774
71. Logallo N, Kvistad CE, Nacu A, Thomassen L (2016) Novel thrombolytics for acute ischemic stroke: challenges and opportunities. CNS Drugs 30(2):101–108
72. Hacke W, Albers G, Al-Rawi Y, Bogousslavsky J, Davalos A, Eliasziw M et al (2005) The Desmoteplase in Acute Ischemic Stroke Trial (DIAS): a phase II MRI-based 9-hour window acute stroke thrombolysis trial with intravenous desmoteplase. Stroke 36(1):66–73
73. Furlan AJ, Eyding D, Albers GW, Al-Rawi Y, Lees KR, Rowley HA et al (2006) Dose Escalation of Desmoteplase for Acute Ischemic Stroke (DEDAS): evidence of safety and efficacy 3 to 9 hours after stroke onset. Stroke 37(5):1227–1231
74. Hacke W, Furlan AJ, Al-Rawi Y, Davalos A, Fiebach JB, Gruber F et al (2009) Intravenous desmoteplase in patients with acute ischaemic stroke selected by MRI perfusion-diffusion weighted imaging or perfusion CT (DIAS-2): a prospective, randomised, double-blind, placebo-controlled study. Lancet Neurol 8(2):141–150
75. Albers GW, von Kummer R, Truelsen T, Jensen J-KS, Ravn GM, Grønning BA et al (2015) Safety and efficacy of desmoteplase given 3–9 h after ischaemic stroke in patients with occlusion or high-grade stenosis in major cerebral arteries (DIAS-3): a double-blind, randomised, placebo-controlled phase 3 trial. Lancet Neurol 14(6):575–584
76. Barreto AD, Alexandrov AV, Lyden P, Lee J, Martin-Schild S, Shen L et al (2012) The Argatroban and Tissue-Type Plasminogen Activator Stroke Study: final results of a pilot safety study. Stroke 43(3):770–775
77. Barreto AD, Alexandrov AV, Shen L, Sisson A, Bursaw AW, Sahota P et al (2013) CLOTBUST-Hands Free: pilot safety study of a novel operator-independent ultrasound device in patients with acute ischemic stroke. Stroke 44(12):3376–3381

78. Flint AC, Duckwiler GR, Budzik RF, Liebeskind DS, Smith WS (2007) Mechanical thrombectomy of intracranial internal carotid occlusion: pooled results of the MERCI and Multi MERCI Part I trials. Stroke 38(4):1274–1280
79. Saver JL, Jahan R, Levy EI, Jovin TG, Baxter B, Nogueira RG et al (2012) Solitaire flow restoration device versus the Merci Retriever in patients with acute ischaemic stroke (SWIFT): a randomised, parallel-group, non-inferiority trial. Lancet 380(9849):1241–1249
80. Saver JL, Goyal M, Bonafe A, Diener HC, Levy EI, Pereira VM et al (2015) Stent-retriever thrombectomy after intravenous t-PA vs. t-PA alone in stroke. N Engl J Med 372(24):2285–2295
81. Jovin TG, Chamorro A, Cobo E, de Miquel MA, Molina CA, Rovira A et al (2015) Thrombectomy within 8 hours after symptom onset in ischemic stroke. N Engl J Med 372(24):2296–2306
82. Asadi H, Yan B, Dowling R, Wong S, Mitchell P (2014) Advances in medical revascularisation treatments in acute ischemic stroke. Thrombosis 2014:14
83. Tsivgoulis G, Katsanos AH, Alexandrov AV (2014) Reperfusion therapies of acute ischemic stroke: potentials and failures. Front Neurol 5:215
84. Hussein HM, Georgiadis AL, Vazquez G, Miley JT, Memon MZ, Mohammad YM et al (2010) Occurrence and predictors of futile recanalization following endovascular treatment among patients with acute ischemic stroke: a multicenter study. AJNR Am J Neuroradiol 31(3):454–458
85. Roundtable STAI (1999) Recommendations for standards regarding preclinical neuroprotective and restorative drug development. Stroke 30(12):2752–2758
86. Grupke S, Hall J, Dobbs M, Bix GJ, Fraser JF (2015) Understanding history, and not repeating it. Neuroprotection for acute ischemic stroke: from review to preview. Clin Neurol Neurosurg 129:1–9
87. Schenck JF, Zimmerman EA (2004) High-field magnetic resonance imaging of brain iron: birth of a biomarker? NMR Biomed 17(7):433–445
88. Halliwell B (2006) Oxidative stress and neurodegeneration: where are we now? J Neurochem 97(6):1634–1658
89. Iwamura M, Inamoto N (1967) Novel formation of nitroxide radicals by radical addition to nitrones. Bull Chem Soc Jpn 40(3):703
90. Janzen EG, Blackburn BJ (1969) Detection and identification of short-lived free radicals by electron spin resonance trapping techniques (spin trapping). Photolysis of organolead, -tin, and -mercury compounds. J Am Chem Soc 91(16):4481–4490
91. Poyer JL, Floyd RA, McCay PB, Janzen EG, Davis ER (1978) Spin-trapping of the trichloromethyl radical produced during enzymic NADPH oxidation in the presence of carbon tetrachloride or bromotrichloromethane. Biochim Biophys Acta 539(3):402–409
92. Floyd RA, Soong LM (1977) Spin trapping in biological systems. Oxidation of the spin trap 5,5-dimethyl-1-pyrroline-1-oxide by a hydroperoxide-hematin system. Biochem Biophys Res Commun 74(1):79–84
93. Harbour JR, Bolton JR (1975) Superoxide formation in spinach chloroplasts: electron spin resonance detection by spin trapping. Biochem Biophys Res Commun 64(3):803–807
94. Robert AF, Hema KC, Ting H, Rheal T (2011) Anti-cancer activity of nitrones and observations on mechanism of action. Anticancer Agents Med Chem 11(4):373–379
95. Costa DS, Martino T, Magalhaes FC, Justo G, Coelho MG, Barcellos JC et al (2015) Synthesis of N-methylarylnitrones derived from alkyloxybenzaldehydes and antineoplastic effect on human cancer cell lines. Bioorg Med Chem 23(9):2053–2061
96. Inoue Y, Asanuma T, Smith N, Saunders D, Oblander J, Kotake Y et al (2007) Modulation of Fas-FasL related apoptosis by PBN in the early phases of choline deficient diet-mediated hepatocarcinogenesis in rats. Free Radic Res 41(9):972–980
97. Floyd RA, Towner RA, Wu D, Abbott A, Cranford R, Branch D et al (2010) Anti-cancer activity of nitrones in the Apc(Min/+) model of colorectal cancer. Free Radic Res 44(1):108–117

98. Floyd RA, Kopke RD, Choi CH, Foster SB, Doblas S, Towner RA (2008) Nitrones as therapeutics. Free Radic Biol Med 45(10):1361–1374
99. Floyd RA, Hensley K, Bing G (2000) Evidence for enhanced neuro-inflammatory processes in neurodegenerative diseases and the action of nitrones as potential therapeutics. J Neural Transm Suppl 60:387–414
100. Stiles M, Moiseyev GP, Budda ML, Linens A, Brush RS, Qi H et al (2015) PBN (phenyl-N-tert-butylnitrone)-derivatives are effective in slowing the visual cycle and rhodopsin regeneration and in protecting the retina from light-induced damage. PLoS One 10(12), e0145305
101. Doggrell SA (2006) Nitrone spin on cerebral ischemia. Curr Opin Investig Drugs 7(1):20–24
102. Chioua M, Sucunza D, Soriano E, Hadjipavlou-Litina D, Alcazar A, Ayuso I et al (2012) Alpha-aryl-N-alkyl nitrones, as potential agents for stroke treatment: synthesis, theoretical calculations, antioxidant, anti-inflammatory, neuroprotective, and brain-blood barrier permeability properties. J Med Chem 55(1):153–168
103. Novelli GP, Angiolini P, Tani R, Consales G, Bordi L (1986) Phenyl-t-butyl-nitrone is active against traumatic shock in rats. Free Radic Res Commun 1(5):321–327
104. Edenius C, Strid S, Borgå O, Breitholtz-Emanuelsson A, Vallén KL, Fransson B (2007) Pharmacokinetics of NXY-059, a nitrone-based free radical trapping agent, in healthy young and elderly subjects. J Stroke Cerebrovasc Dis 11(1):34–42
105. Dehouck MP, Cecchelli R, Richard Green A, Renftel M, Lundquist S (2002) In vitro blood-brain barrier permeability and cerebral endothelial cell uptake of the neuroprotective nitrone compound NXY-059 in normoxic, hypoxic and ischemic conditions. Brain Res 955(1-2):229–235
106. Lees KR, Zivin JA, Ashwood T, Davalos A, Davis SM, Diener H-C et al (2006) NXY-059 for acute ischemic stroke. N Engl J Med 354(6):588–600
107. Diener HC, Lees KR, Lyden P, Grotta J, Davalos A, Davis SM et al (2008) NXY-059 for the treatment of acute stroke: pooled analysis of the SAINT I and II Trials. Stroke 39(6):1751–1758
108. Savitz SI, Fisher M (2007) Future of neuroprotection for acute stroke: in the aftermath of the SAINT trials. Ann Neurol 61(5):396–402
109. Savitz SI (2007) A critical appraisal of the NXY-059 neuroprotection studies for acute stroke: a need for more rigorous testing of neuroprotective agents in animal models of stroke. Exp Neurol 205(1):20–25
110. Lapchak PA, Araujo DM, Song D, Wei J, Zivin JA (2002) Neuroprotective effects of the spin trap agent disodium-[(tert-butylimino)methyl]benzene-1,3-disulfonate N-oxide (Generic NXY-059) in a rabbit small clot embolic stroke model: combination studies with the thrombolytic tissue plasminogen activator. Stroke 33(5):1411–1415
111. Lapchak PA, Araujo DM, Song D, Wei J, Purdy R, Zivin JA (2002) Effects of the spin trap agent disodium-[tert-butylimino)methyl]benzene-1,3-disulfonate N-oxide (Generic NXY-059) on intracerebral hemorrhage in a rabbit large clot embolic stroke model: combination studies with tissue plasminogen activator. Stroke 33(6):1665–1670
112. Green AR (2008) Pharmacological approaches to acute ischaemic stroke: reperfusion certainly, neuroprotection possibly. Br J Pharmacol 153(Suppl 1):S325–S338
113. Bath PM, Gray LJ, Bath AJ, Buchan A, Miyata T, Green AR (2009) Effects of NXY-059 in experimental stroke: an individual animal meta-analysis. Br J Pharmacol 157(7):1157–1171
114. Feuerstein GZ, Zaleska MM, Krams M, Wang X, Day M, Rutkowski JL et al (2008) Missing steps in the STAIR case: a translational medicine perspective on the development of NXY-059 for treatment of acute ischemic stroke. J Cereb Blood Flow Metab 28(1):217–219
115. Ning X, Temming RP, Dommerholt J, Guo J, Ania DB, Debets MF et al (2010) Protein modification by strain-promoted alkyne–nitrone cycloaddition. Angew Chem Int Ed 49(17):3065–3068
116. MacKenzie DA, Sherratt AR, Chigrinova M, Kell AJ, Pezacki JP (2015) Bioorthogonal labelling of living bacteria using unnatural amino acids containing nitrones and a nitrone derivative of vancomycin. Chem Commun 51(62):12501–12504

117. Barton DHR, Day MJ, Hesse RH, Pechet MM (1975) A new rearrangement of ketonic nitrones; a convenient alternative to the Beckmann rearrangement. J Chem Soc Perkin Trans 1(18):1764–1767
118. Ayuso MI, Chioua M, Martínez-Alonso E, Soriano E, Montaner J, Masjuán J et al (2015) CholesteroNitrones for stroke. J Med Chem 58(16):6704–6709
119. Pulsinelli WA, Brierley JB (1979) A new model of bilateral hemispheric ischemia in the unanesthetized rat. Stroke 10(3):267–272
120. Garcia-Bonilla L, Cid C, Alcazar A, Burda J, Ayuso I, Salinas M (2007) Regulatory proteins of eukaryotic initiation factor 2-alpha subunit (eIF2 alpha) phosphatase, under ischemic reperfusion and tolerance. J Neurochem 103(4):1368–1380
121. Ayuso MI, Gonzalo-Gobernado R, Marco-Contelles J, Montoya JJ, Alcazar A, Montaner J (2016) Neuroprotection of cholesteronitrones treatment in an experimental stroke model. Euro Stroke J 1(1S):387–388
122. Ayuso MI, Martínez-Alonso E, Chioua M, Escobar-Peso A, Montaner J, Marco-Contelles J, Alcázar A (2016) Quinolinyl nitrone RP19 induces neuroprotection after transient brain ischemia. Manuscript in preparation
123. Lapchak PA (2013) Drug-like property profiling of novel neuroprotective compounds to treat acute ischemic stroke: guidelines to develop pleiotropic molecules. Transl Stroke Res 4(3):328–342
124. Lapchak PA (2010) A critical assessment of edaravone acute ischemic stroke efficacy trials: is edaravone an effective neuroprotective therapy? Expert Opin Pharmacother 11(10):1753–1763
125. Intelligent Pharma. Pythia software. http://intelligentpharma.com/upload/discovery_tools/pythia_stand_alone0714.pdf
126. Suarez I, Bodega G, Fernandez B (2002) Glutamine synthetase in brain: effect of ammonia. Neurochem Int 41(2-3):123–142
127. Greenberg DA, Jin K (2013) Vascular endothelial growth factors (VEGFs) and stroke. Cell Mol Life Sci 70(10):1753–1761
128. Sun Y, Yu P, Zhang G, Wang L, Zhong H, Zhai Z, et al. Therapeutic effects of tetramethylpyrazine nitrone in rat ischemic stroke models. J Neurosci Res 2012;90(8):1662–9
129. Li S, Chen H, Wang X, Wu J, Jiang J, Wang Y. Pharmacokinetic study of a novel stroke therapeutic, 2-[[(1,1-dimethylethyl)oxidoimino]methyl]-3,5,6-trimethylpyrazine, by a simple HPLC-UV method in rats. Eur J Drug Metab Pharmacokinet 2011;36(2):95–101
130. Lapchak PA, Schubert DR, Maher PA. De-risking of stilbazulenyl nitrone (STAZN), a lipophilic nitrone to treat stroke using a unique panel of in vitro assays. Transl Stroke Res 2011;2(2):209–17
131. Ley JJ, Belayev L, Saul I, Becker DA, Ginsberg MD. Neuroprotective effect of STAZN, a novel azulenyl nitrone antioxidant, in focal cerebral ischemia in rats: dose-response and therapeutic window. Brain Res 2007;1180:101–10

Chapter 10
Neuroprotective Strategies via Modulation of Innate Immune Receptors

George Trendelenburg

Abstract Innate immune receptors, such as the Toll-like receptors (TLRs) and the NOD-like receptors (NLRs) represent a class of central sensors not only for infective microorganisms, but also for danger signals released during cerebral ischemia. The activation of these pattern recognition receptors (PRRs) mediates the initiation of the innate immune response and thereby also contributes to the ischemic injury. Accordingly, there is increasing interest in modulating these initial steps of immune activation for the sake of neuroprotection. Whereas TLRs are membranous receptors, NLRs are mainly cytosolic. Members of both classes of PPRs were shown to be activated by various factors released under circumstances of sterile inflammation, e.g., by ischemic tissue damage. Moreover, blocking these receptors or deletion of endogenous danger molecules in transgenic mice has been shown to efficiently reduce the ischemic neuroinflammatory response or the resulting ischemic brain damage. The role of the PRRs in ischemic stroke is discussed as well as the ways to modulate these pathways with the aim to develop successful future stroke therapies.

Keywords Stroke • Cerebral ischemia • Neuroprotection • Inflammasome • Innate immune response • Toll-like receptors • NALP • TLR • Neuroinflammation

Abbreviations

DAMP	Danger-associated molecular patterns
ECM	Extracellular matrix
HSP	Heat-shock protein

G. Trendelenburg (✉)
Department of Neurology, University Medical Center Göttingen,
Robert-Koch Str. 40, 37075 Göttingen, Germany

Kreiskrankenhaus Freiberg gGmbH, Donatsring 20, 09599 Freiberg, Germany
e-mail: george.trendelenburg@med.uni-goettingen.de

P.A. Lapchak, J.H. Zhang (eds.), *Neuroprotective Therapy for Stroke and Ischemic Disease*, Springer Series in Translational Stroke Research,
DOI 10.1007/978-3-319-45345-3_10

NLR	NOD-like receptor
NLRP	NOD-like receptor protein
PAMP	Pathogen-associated molecular pattern
PRR	Pattern recognition receptor
ROS	Reactive oxygen species
TLR	Toll-like receptor

1 Danger Theory and General Concept of Danger Signals (DAMPs)

Receptors of the innate immune system have been shown to be involved in various diseases with an acute inflammatory component and represent an important molecular part of the innate immune system. They are thought to detect invading pathogens by the recognition of pathogen-associated molecular patterns (PAMPs) but are also able to detect signals of tissue stress under pathogenic situations, such as ischemic injury. Accordingly, these signals are named 'help-me' signals, 'danger signals,' or danger-associated molecular patterns (DAMPs). In congruence with Matzingers danger theory [1, 2] of the immune system, their presence is required to activate the immune response, which thereby could discriminate between symbiotic and detrimental microorganisms, or could discriminate between tissue injury and situations of normal development. This recognition of the pathogenic potential of invading microorganisms is of central significance for higher order species, as multicellular organisms are exposed to various pathogen attacks, but often also benefit from symbiotic life constellations. Thus, it could be extrapolated that the resulting response to an 'immune-on'-signal (as assumed to be present under situations of danger) needs to be of high efficiency and must rather follow a digital rapid all-or-nothing concept than a 'wait-and-see' concept. According to that idea one could be easily imagine that this rapid activation could also carry the innate immune response to a vast and overwhelming immune response if a rapid release of huge amounts of these signals occur at once. Thereby, the immune activation could contribute to tissue injury, whereas on the same time it could also protect the healthy tissue of the border zone against damage when this tissue is potentially exposed to necrotic material, cellular debris, or toxic substances released from dying cells (Fig. 10.1).

2 Types of Innate Immune Receptors (PRRs)

The best known group of innate immune receptors comprises the toll-like receptor family. There are 13 toll-like receptor family members which are located at the outer membrane (e.g., TLR2, TLR4) or at the membranes of intracellular organelles (TLR3, TLR7, TLR8, TLR9). The central role of these receptors not only for the activation of

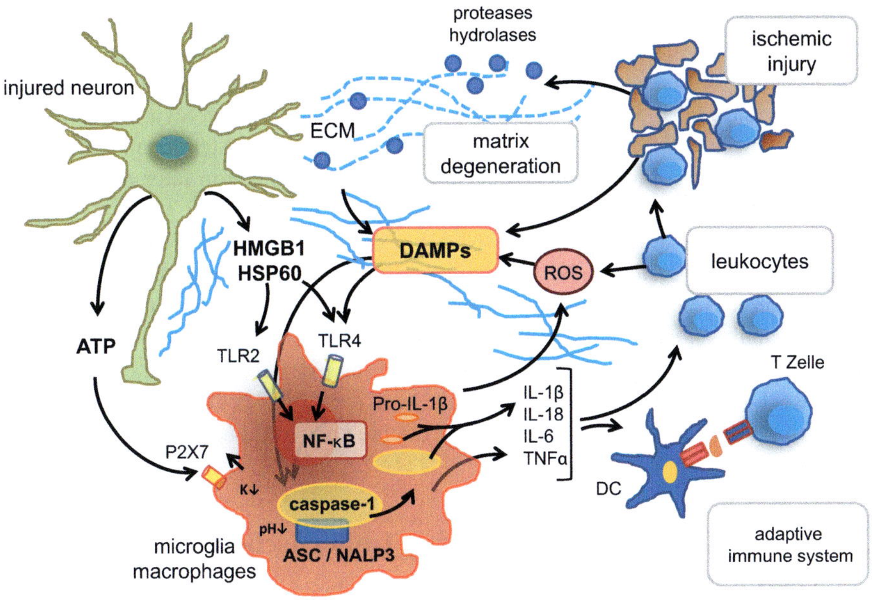

Fig. 10.1 Molecular mechanisms in ischemic brain injury (modified with permission from Iadecola et al. [7]). Danger associated patterns (DAMPs) are released by injured neurons and are thought to activate inflammatory cells such as macrophages or microglia. After activation of innate immune receptors (TLRs or NLRs), these cells release proinflammatory cytokines (e.g., IL-1β or TNFα) which further contribute to ischemic damage

the innate immune response and neuroinflammation (often via activation of TNF-alpha-stimulated induction of proinflammatory cytokines) but also for the regulation of the following adaptive immune response (e.g., by the activation of TLRs of Treg cells) has been demonstrated in various studies [3, 4]. Moreover, these receptors play an essential role already during the chronic exposure of immune cells with bacterial pathogens in the gut: TLR-deficiency results in an altered gut microbiome composition [5] and also is thought to be implicated in autoinflammatory diseases.

Corresponding to the membranous Toll-like receptors, a class of pattern recognition receptors (PRRs) was identified in the cytosol: NOD-like receptors (NLRs) are typically located in the cytosol of the cells, but—similar to TLRs—are also able to detect PAMPs or DAMPS. Depending on the exact type of NLR, direct binding of the activating PAMP or DAMP was experimentally shown to activate some NLR-associated multiprotein complexes, called inflammasomes [6]. Alternatively, the activation of the corresponding NLR-multiprotein complex could also occur by the action of a second signal released under circumstances of stress, danger, or by the pathogen exposure. As it is true for TLRs, different NLRs respond to different signals. Prototypical activation of NLRs results in the aggregation of a multiprotein complex which is called inflammasome [6], which typically comprises a central scaffold protein named ASC (for apoptosis speckle like protein), which by homotypic binding via CARD and

PYD-domains activates autocleavage of the pro-caspase-1 protein into the active proinflammatory caspase-1 (or ICE). Caspase-1 then cleaves pro-interleukin-1β (proIL-1β) into the active IL-1β, which is an important pyrogen and proinflammatory interleukin. IL-1ß, together with other proinflammatory cytokines and in concert with other parts of the innate and adaptive immune system could contribute to ischemic injury, however, possibly could also participate in later mechanisms of repair [7].

Inflammasome subunits are differentially expressed in different cell types of the central nervous system: the NLPR3 inflammasome is mainly expressed in immune cells, whereas the NLRP1 inflammasome is mainly expressed in neurons [8]. Making analysis even more complex, certain human and rodent inflammasome subunits differ substantially in their structure, and data based on the use of the 'caspase-1-knockout' mice in the past have been performed actually with transgenic mice, that were caspase-1 and caspase-11 double knockout mice [9].

3 Mechanisms of PRR Activation Under Ischemic Conditions

There are different ways in which PRRs could influence ischemic events: in the acute phase of a vessel occlusion PRRs contribute to ischemic damage and inhibition. Accordingly, a protective effect in the acute phase of ischemic injury was shown for several TLRs (e.g., TLR2 and TLR4) as well as NLRs (e.g., NALP3) [10]. Moreover, there are arguments for a relevant role in later stages following the acute injury event: PRRs are thought to be involved in rebuilding tissue homeostasis, plasticity, and mechanisms of repair.

Various reports define the role of TLRs in ischemic brain damage and neurodegeneration, whereas the consequences of NLR activation during ischemic brain injury are only partially understood up until now. However, the activation of TLRs and NLRs is functionally connected, as the classical two-step model of inflammasome activation comprises an initial 'priming step,' which is normally derived by TLR stimulation and NFkB-mediated induction of pro-IL-1β mRNA and potentially further transcripts of inflammasome subunits, such as ASC or NLRP3 [10]. The second stimulus then activates the inflammasome either via direct binding of DAMPs to NLRs, or via a postulated interconnected signal (e.g., mitochondrial dysfunction with release of ROS, cathepsin, K^+ efflux, drop of the pH, or a yet unknown signal). There are several DAMPs which potentially activate TLR signaling during ischemic brain injury: heat-shock protein (HSP) 70, Hmgb1, Mrp14, fibrinogen, or hyaluronan and others.

The second step for activation of inflammasome during ischemic injury is possibly mediated by a drop of pH during ischemic tissue stress, by reactive oxygen species (ROS), K^+ release, or DAMPS, which may directly bind to NLRs. Inflammasome activation in model organisms recently revealed a fascinating mechanism of innate immune activation by prion-like protein polymerization of inflammasome subunits such as ASC [11]. Together with the work of Lu et al. [12], a unified model of ASC-dependent inflammasome activation evolved: after the insult reaches a threshold, NLRP3 is released by autoinhibition, allowing the interaction of this PYD domain with the PYD domain of ASC, which results in ASC-prion-like nucleation. However, the consequences of this prion-like polymerization for perpetuation of autoinflammatory diseases, or atherosclerosis, and the significance of inflammasome degradation by autophagy still remain to be elucidated [10].

Moreover, activation of the inflammasome and proinflammatory consequences are modified by another part of the innate immune system: the complement system. Accordingly, the complement subunit C3a was shown to regulate NLRP3-inflammasome activation in monocytes and the complement factor C5a was shown to represent a potent primer for cholesterol crystal-induced Il-1β release [13, 14].

The activation of the inflammasome is also triggered by dysfunction of intracellular organelles, such as the mitochondria or the lysosome [15]. Accordingly, lysosomal membrane rupture after exposure to cell-toxic aggregates such as monosodium urate (MSU), or amyloid-beta in Alzheimer's disease activate the inflammasome [16].

The central contribution of Toll-like receptors for ischemic brain damage is well documented by various studies, but also the NOD-like receptors contribute to the ischemic brain injury via activation of the inflammasome and via interleukin-1β-mediated effects [17].

4 Modulation of PRRs in Stroke or Ischemia

The innate immune system could be modified in ischemic stroke by different ways. Potentially, a decrease of the systemic proinflammatory 'load' could be envisaged by decreasing pathogen-associated molecular patterns on body surfaces or within the systemic vasculature, e.g., by the use of prophylactic antibiotic treatment in patients with a high risk of stroke. Despite several studies using prophylactic antibiotic treatment in patients with acute stroke, there is only insufficient information about the potential benefit of this kind of treatment when applied before any stroke occurs. However, there are hints that antibiotic treatment affects outcome in stroke patients in specific subgroups in which systemic thrombolysis is applied [18].

Another strategy is focused on the endogenous modulating capabilities of the immune system. It is well known that short ischemic periods lead to ischemic tolerance (ischemic preconditioning), which is thought to be caused by reprogramming of the cerebral tissue and the immune system. The contribution of Toll-like receptors for preconditioning has been shown in different studies [19]. Thus, the strength of a stimulus (e.g., the degree of TLR activation or the duration of vessel occlusion) could lead to either protection (preconditioning) or severe brain injury but may involve the same molecular pathways. The significance of TLR signaling for both situations has been demonstrated by various studies [10, 20–22]. Accordingly, NLRP3 deficiency protects against renal ischemia and cerebral ischemia in rodents, as well as in in vitro models of stroke [23, 24].

Interestingly, glyburide, a widely used antidiabetic drug inhibits NLRP3 inflammasome in vitro and its use in stroke patients correlated with a better clinical outcome [25]. Another small molecule (16673-34-0), which is an intermediate substrate in the glyburide synthesis inhibits NLRP3 inflammasome in cardiomyocytes and protects against ischemia–reperfusion injury in the mouse [26].

However, one have to keep in mind that almost all successful strategies of neuroprotection in experimental stroke failed in the clinic [27]. Potentially, this could explained by the use of wrong or artificial models, short observation times, publication bias, use of the wrong outcome parameters, or too small group numbers in animal stroke studies. Thus, before the use of immune-modulating therapies should be tested in the clinic, a more profound knowledge of immune effects in stroke is required, which also takes into account opposing effects in different cell types or different organs, resulting effects for different doses, and time-dependent consequences of immune-modulating strategies.

A ‘crosstalk’ between different organs, such as the central nervous system, the thymus, or the gut has been identified: apoptotic thymocyte cell death causes severe immune depression after stroke [28], and recent studies note that the immunological barrier in body surfaces not only protects against invasion of foreign pathogens, but also modulates the composition of commensal microbiota. Accordingly, modification of the innate immune system by the use of transgenic mice is able to alter the gut microbiota and results in altered severity and frequency of various autoimmune diseases [29] and antibiotic treatment affects mortality after stroke [30].

Thus, it is increasingly recognized that not only local molecular mechanisms of ischemic brain injury should be taken into account when analyzing the effects of the immune activation in stroke, but also the contribution of substantial systemic factors, such as the ‘bacterial burden’ on the various body surfaces, the basic systemic inflammatory level, and the activation state of the immune system.

References

1. Matzinger P (2002) The danger model: a renewed sense of self. Science 296:301–305
2. Matzinger P (2007) Friendly and dangerous signals: is the tissue in control? Nat Immunol 8:11–13
3. Medzhitov R, Preston-Hurlburt P, Janeway CA Jr (1997) A human homologue of the Drosophila Toll protein signals activation of adaptive immunity. Nature 388:394–397
4. Liu D, Rhebergen AM, Eisenbarth SC (2013) Licensing adaptive immunity by NOD-like receptors. Front Immunol 4:486
5. Round JL, Lee SM, Li J, Tran G, Jabri B, Chatila TA, Mazmanian SK (2011) The Toll-like receptor 2 pathway establishes colonization by a commensal of the human microbiota. Science 332:974–977
6. Martinon F, Burns K, Tschopp J (2002) The inflammasome: a molecular platform triggering activation of inflammatory caspases and processing of proIL-beta. Mol Cell 10:417–426

7. Iadecola C, Anrather J (2011) The immunology of stroke: from mechanisms to translation. Nat Med 17:796–808
8. de Rivero Vaccari JP, Lotocki G, Alonso OF, Bramlett HM, Dietrich WD, Keane RW (2009) Therapeutic neutralization of the NLRP1 inflammasome reduces the innate immune response and improves histopathology after traumatic brain injury. J Cereb Blood Flow Metab 29:1251–1261
9. Kayagaki N, Warming S, Lamkanfi M, van de Walle L, Louie S, Dong J, Newton K, Qu Y, Liu J, Heldens S, Zhang J, Lee WP, Roose-Girma M, Dixit VM (2011) Non-canonical inflammasome activation targets caspase-11. Nature 479:117–121
10. Trendelenburg G (2014) Molecular regulation of cell fate in cerebral ischemia: role of the inflammasome and connected pathways. J Cereb Blood Flow Metab 34:1857–1867
11. Cai X, Chen J, Xu H, Liu S, Jiang QX, Halfmann R, Chen ZJ (2014) Prion-like polymerization underlies signal transduction in antiviral immune defense and inflammasome activation. Cell 156:1207–1222
12. Lu A, Magupalli VG, Ruan J, Yin Q, Atianand MK, Vos MR, Schröder GF, Fitzgerald KA, Wu H, Egelman EH (2014) Unified polymerization mechanism for the assembly of ASC-dependent inflammasomes. Cell 156:1193–1206
13. Asgari E, Le Friec G, Yamamotot H, Perucha E, Sacks SS, Köhl J, Cook HT, Kemper C (2013) C3a modulates Il-1β secretion in human monocytes by regulating ATP efflux and subsequent NLRP3 inflammasome activation. Blood 122:3471–3481
14. Samstad EO, Niyonzima N, Nymo S, Aune MH, Ryan L, Bakke SS, Lappegard KT, Brekke OK, Lambris JD, Damas JK, Latz E, Mollnes TE, Espevik T (2014) Cholesterol crystals induce complement-dependent inflammasome activation and cytokine release. J Immunol 192:2837–2845
15. Misawa T, Takahama M, Kozaki T, Lee H, Zou J, Saitoh T, Akira S (2013) Microtubule-driven spatial arrangement of mitochondria promotes activation of the NLRP3 inflammsome. Nat Immunol 14:454–460
16. Heneka MT, Jummer MP, Stutz A, Delekate A, Schwartz S, Vieira-Saecker A, Griep A, Axt D, Remus A, Tzeng TC, Gelpi E, Halle A, Latz E, Golenbock DT (2013) NLRP3 is activated in Alzheimer's disease and contributes to pathology in APP/PS1 mice. Nature 493:674–678
17. Allan SM, Tyrrell PJ, Rothwell NJ (2005) Interleukin-1 and neuronal injury. Nat Rev Immunol 5:629–640
18. Westendorp WF, Vermeij JD, Zock E, Hooijenga IJ, Kruyt ND, Bosboom HJ, Kwa VI, Weisfelt M, Remmers MJ, ten Houten R, Schreuder AH, Vermeer SE, van Dijk EJ, Dippel DW, Dijkgraaf MG, Spanjaard L, Vermeulen M, van der Poll T, Prins JM, Vermeij FH, Roos YB, Kleyweg RP, Kerkhoff H, Brouwer MC, Zwinderman AH, van de Beek D, Nederkoorn PJ, PASS Investigators (2015) The Preventive Antibiotics in Stroke Study (PASS): a pragmatic randomised open-label masked endpoint clinical trial. Lancet 385:1519–26
19. Stenzel-Poore MP, Stevens SL, King JS, Simon RP (2007) Preconditioning reprograms the response to ischemic injury and primes the emergence of unique endogenous neuroprotective phenotypes: a speculative synthesis. Stroke 38:680–685
20. Stevens LS, Stenzel-Poore MP (2006) Toll-like receptors and tolerance to ischemic injury in the brain. Biochem Soc Trans 34:1352–1355
21. Ziegler G, Freyer D, Harhausen D, Khojasteh U, Nietfeld W, Trendelenburg G (2011) Blocking TLR2 in vivo protects against accumulation of inflammatory cells and neuronal injury in experimental stroke. J Cereb Blood Flow Metab 31:757–766
22. Andresen L, Theodorou K, Grünewald S, Czech-Zechmeister B, Könnecke B, Lühder F, Trendelenburg G (2016) Evaluation of the therapeutic potential of anti-TLR4-antibody MTS510 in experimental stroke and significance of different routes of application. PLoS One 11, e0148428
23. Yang F, Wang Z, Wei X, Han H, Meng X, Zhang Y, Shi W, Li F, Xin T, Pang Q, Yi F (2014) NLRP3 deficiency ameliorates neurovascular damage in experimental ischemic stroke. J Cereb Blood Flow Metab 34:660–667
24. Iyer SS, Pulskens WP, Sadler JJ, Butter LM, Teske GJ, Ulland TK, Eisenbarth SC, Florquin S, Flavell RA, Leemans JC, Sutterwala FS (2009) Necrotic cells trigger a sterile inflammatory response through the Nlrp3 inflammasome. Proc Natl Acad Sci U S A 106:20388–20393

25. Kunte H, Schmidt S, Eliasziw M, del Zoppo GJ, Simard JM, Masuhr F, Weih M, Dirnagl U (2007) Sulfonylureas improve outcome in patients with type 2 diabetes and acute ischemic stroke. Stroke 38:2526–2530
26. Marchetti C, Chojnacki J, Toldo S, Mezzaroma E, Tranchida N, Rose SW, Federici M, Van Tassell BW, Zhang S, Abbate A (2014) A novel pharmacologic inhibitor of the NLRP3 inflammasome limits myocardial injury after ischemia-reperfusion in the mouse. J Cardiovasc Pharmacol 63:316–322
27. O'Collins VE, Macleod MR, Donnan GA, Horky LL, van der Worp BH, Howells DW (2006) 1,026 experimental treatments in acute stroke. Ann Neurol 59:467–477
28. Prass K, Meisel C, Höflich C, Braun J, Halle E, Wolf T, Ruscher K, Victorov IV, Priller J, Dirnagl U, Volk HD, Meisel A (2003) Stroke-induced immunodeficiency promotes spontaneous bacterial infections and is mediated by sympathetic activation reversal by poststroke T helper cell type 1-like immunostimulation. J Exp Med 198:725–736
29. Henao-Mejia J, Elinav E, Jin C, Hao L, Mehal WZ, Strowig T, Thaiss CA, Kau AL, Eisenbarth SC, Jurczak MJ, Camporez JP, Shulman GI, Gordon JI, Hoffman HM, Flavell RA (2012) Infammasome-mediated dysbiosis regulated progression of NAFLD and obesity. Nature 482:179–185
30. Winek K, Engel O, Koduah P, Heimesaat MM, Fischer A, Bereswill S, Dames C, Kershaw O, Gruber AD, Curato C, Oyama N, Meisel C, Meisel A, Dirnagl U (2016) Depletion of cultivatable gut microbiota by broad-spectrum antibiotic pretreatment worsens outcome after murine stroke. Stroke 47:1354–1363

Chapter 11
Harnessing the Power of the Human Immune System via Multi-omic Immune Profiling in Stroke Treatment and Recovery

Taura L. Barr, V. Gionis, and R. Giersch

Abstract Precision health is an unrealized opportunity in the practice and care of stroke patients. To achieve truly person-centered care as the Precision Medicine Initiative (PMI) outlines, a new approach to stroke research is necessary. Multi-omic profiling of the immune system response to human stroke provides a significant advantage over single system and/or multibiomarker analyses. By combining powerful 'omic' technologies with machine learning and pattern recognition for interpretation of this multi-omic data, we are better able to understand the complexity of stroke physiology. These measurements allow for new methods in the diagnosis, treatment stratification, and design of personalized approaches to stroke recovery. The foundation by which these novel approaches can be utilized to understand human stroke complexity and translated into programs of research and clinical practice paradigms is being created, and will ultimately change the way we diagnosis and treat stroke patients. This review will briefly outline current approaches and how to leverage these discoveries to achieve person-centered stroke care.

Keywords Precision health • Multi-omic profiling • Immunogenomics • Stroke

1 Introduction

The human response to stroke is complex, and no one animal model can account for the variability seen in patient recovery. Given the tremendous advancements in 'omic'-based technology and big data analytics we can now study the human patient, to some extent, as we did the animal model in years past. This approach is gaining traction through the use of peripheral blood multi-omic profiling and has shown great success in achieving precision diagnostics and treatments for cancer [1–4].

T.L. Barr, Ph.D., R.N., F.A.H.A. (✉) • V. Gionis • R. Giersch
CereDx Inc., Eight Medical Center Drive, Morgantown, WV 26506, USA
e-mail: tbarr@ceredx.com

P.A. Lapchak, J.H. Zhang (eds.), *Neuroprotective Therapy for Stroke and Ischemic Disease*, Springer Series in Translational Stroke Research,
DOI 10.1007/978-3-319-45345-3_11

Unlike cancer patients, stroke patients are difficult to study due to the acuity of their condition, the high cost of clinical trials, and the complexity of carrying out a rapid evaluation in an emergency setting while generating high-quality clinical data. Another challenge is the lack of access to brain tissue. These limitations leave clinical researchers with a paucity of 'omic'-related stroke data to guide the design of future bench and clinical studies. In addition to the difficulty of capturing multi-omic data in a stroke population, there is also the challenge of interpreting the multilevel physiological responses in a timely manner to guide acute clinical decision making. Precision medicine requires that we comprehensively evaluate the patient's physiological responses to disease within the context of their environment and behaviors to make informed decisions about their care. The Precision Medicine Initiative (PMI) will enable scientists and clinicians to collect the necessary data to make sense of the variability in patient responses and identifying targetable multi-omic phenotypes for treatment. This approach has the potential to revolutionize acute stroke therapy, particularly for diagnostics and the extension of the therapeutic time window [5]. Stroke patients are not homogenous; even subgrouping patients based on type and stroke location doesn't account for the many external and internal factors that contribute to patient variability, making it very difficult to not only translate bench studies, but also find ways to improve clinical practice. To overcome this challenge and realize the full potential of precision medicine for stroke, a new approach to the study and treatment of stroke is necessary. Using similar methodology as current cancer trials, stroke can be evaluated using the peripheral blood multi-omic and immune system responses, in conjunction with big data analytics to identify the patterns of stroke response over time. When used in combination with patient symptom and other phenotype data, this information is expected to identify individual patient variables that can be used to guide treatment, similar to the concepts of directed cancer immunotherapies [6, 7]. This approach also allows for the quantification of the patients' internal neuroprotection capabilities and the patient's physiological health to aid in clinical decision making. Subsequent neuroprotection strategies based on these identified innate healing mechanisms and the patient's individual therapeutic milieu can be realized by monitoring these multisystem responses over time. However, as a scientific community, a significant amount of work is required to collect, analyze, and interpret this type of data and to make it work for the assessment of acute neurological disease just as we have seen done in cancer treatments and therapies.

2 Profiling the Multi-omic Stroke Response and the Complexity of the Immune System

Complexity is a marker of a healthy adaptable system. In fact, studies have shown that when variability in physiological response is decreased, recovery is poor [8, 9]. Heart rate variability is probably the most widely used marker of physiological variability; however, there are many physiological systems that can be monitored in the stroke patient that have yet to be explored. The response to stroke is coordinated, with each system doing its part and when one falls short, another steps in to take over. We are only limited in the understanding of this process. Even in the patients that have a poor post stroke recovery, it is evident that complex neuroprotective mechanisms were at work. A reductionist approach to teasing out the nuances in this response is futile, and the long history of failed clinical trials is a reminder that our approach must change. Rather than preformed hypotheses about specific neuroprotective targets and exploring single markers in depth, science as a whole is moving toward unbiased explorations of these complex human systems to generate novel hypotheses from which to design and conduct bench and clinical studies [10]. In addition to a change in the scientific approach, there is also a shift in the dynamics of research teams, by combining disciplines outside of the health sciences and moving away from one-person leadership to team science; this is the only way to make sense of the massive amount of data that is generated from these studies.

To date, most of the science around neuroprotection has focused on the acute phase of stroke injury and neuron-specific responses [11]. However, there are many points of influence over the course of a patient's recovery that can be augmented to enhance recovery, not just in the initial hours and days. The traditional approach to identifying neuroprotective candidates has been via identification of neuroinflammatory mediators that contribute to initial or secondary injury. However, the immune responses that are secondary to the initial inflammatory reactions (adaptive immunity and chronic innate signaling) are targetable in the acute, subacute, and chronic phases of recovery and offer greater promise in terms of contributing to enhancements in patient outcome. Much of what is known about the immune response to stroke is limited to the initial hours and days following injury, with very little work on the subacute to chronic periods of recovery. It is understood that the initial inflammatory cascade is followed by an intense immunosuppression that protects the brain from autoimmune reactions, but increases the patient's risk for infection and a 'stunted' immune response during the weeks and months following the initial injury [12]. The degree of immune suppression is what is most important for the patient and requires a balance between innate and adaptive mechanisms, i.e., a healthy variable system. A nonvariable immune response or one in which the innate and adaptive immune systems are not communicating with one another is maladaptive and a predictor of poor recovery. Very generally, this maladaptation can be assayed poststroke by evaluating the neutrophil/lymphocyte ratio [13, 14]. It is plausible that a more specific marker of a nonvariable immune response can be identified by intense immune investigation and a multi-omic evaluation of the immune system. Multi-omic responses can be assayed in the peripheral blood by genomic, transcriptomic, proteomic, and metabolomic approaches. Immunogenomics is a new area of science that provides a greater characterization of the immune system as a whole and attempts to quantify the ever-changing immune landscape [15, 16]. Evaluating the multi-omic response (genomic, transcriptomic, proteomic, metabolomics, and immungenomic) to human stroke will for the first time provide a comprehensive evaluation of the systems responding to and changing to stroke over time. This information can be used to design bench studies to determine causality and function as well as in clinical trials to test the use of this information in clinical practice to guide diagnosis and repurposing of immune-modifying drugs for stroke treatment [17, 18]. There is also the possibility of targeting specific immune cell populations with nanomaterials to enhance immune modulation [19]. Much work is needed, but there is a foundation for which this new area of stroke research can be expanded [20–23].

Pattern recognition is imperative when interpreting the vast amount of data collected using multi-omic approaches. Cognitive computing, such as that offered by IBM Watson, is a form of artificial intelligence (AI) that uses mathematical algorithms to "learn" patterns and preferences over time. These algorithms typically include machine-learning approaches that are expected to play a critical role in facilitating the use of multi-omic analyses to guide patient care [24, 25]. Our team recently employed a basic machine-learning approach using principal component analysis to discover the pattern of a five-gene profile for stroke diagnosis [26]. Not only was this useful for determining which markers carried the most weight, but also for understanding the relationships among the markers and how they were changing over time in the context of stroke severity. This approach is not without its limitations and requires further optimization of the learning algorithms. The expected outcome of these learning algorithms is that additional biomarkers will be added to this 5-gene profile to increase sensitivity and specificity for guiding precision stroke diagnostics and treatments. A controversial topic in this area is the reliance on statistical significance to guide the development of these learning algorithms. In fact, the American Statistical Society (ASA) recently published an article on the misuse of statistical significance. In that article they urged scientists to critically evaluate their statistical approach and move beyond the traditional $p<0.05$ as their sole criterion for acceptance [27]. Another option for analytical integration is deep learning or the study of neural networks. Deep learning is a branch of machine learning that has yet to be fully utilized for the interpretation of multi-omic data but has been critical to the understanding of high-level nervous system responses [28]. There is much potential in these AI-based approaches; however, the challenge that lies ahead is to move these complex analytical approaches to the bedside by improving interpretability while maintaining data accuracy [29].

3 Leveraging Multi-omic Responses to Aid Stroke Diagnosis, Treatment, and Recovery

3.1 Stroke diagnosis

Improving prehospital stroke diagnosis is a critical need, particularly in rural underserved geographic areas. Thirty percent of patients are misdiagnosed in the prehospital setting [30]; and these estimates are likely higher for younger patients, women, and minorities [31]. The Institute of Medicine (IOM) recently published a report on the importance of accurate diagnosis to decrease hospital readmissions and improve patient quality of life. They provide a list of recommendations which include developing and deploying approaches to reduce diagnostic errors in clinical practice and provide dedicated funding for research on the diagnostic process [32].

Accurate prehospital diagnosis of stroke is a problem that needs immediate attention given the projected increase in strokes expected to occur by 2020. Our laboratory and others have confirmed that specific omic's based blood profiles can differentiate stroke from nonstroke in the acute care setting [33]. Rather than targeting the inflammatory response to stroke, which is often nonspecific, we have chosen to study the systems level response to stroke and have discovered that a stroke-specific pattern can in fact be identified [34, 35]. This approach relies on the use of multi-omic (RNA, protein, and immune cell) analyses that produce a profile with a higher sensitivity and specificity for stroke diagnosis than the current gold standard. The technologies that enable the analysis of these multi-omic multiplexed systems are not yet available at the point of care. However, technologies which allow for simultaneous quantification of RNA, DNA, and protein targets, such as that offered by Nanostring, are being rapidly developed.

3.2 *Stroke Treatment and Recovery*

What if patient outcomes were viewed as varying degrees of adaptation to a stressor rather than good versus poor outcomes? Some patients do a better job of adapting; their physiological systems are healthy enough to survive the initial insult and create a new 'normal' by which their systems respond to future events. The secret to neuroprotection to enhance recovery lies with the patients who have very severe strokes but recover well. Not only can we identify individual multi-omic profiles of recovery from these patients, but we can also use this information to guide treatment decisions throughout the patient journey. Along these same lines, it is interesting to consider whether the brain learns about its environment through interacting with immune cells that are constantly surveying and carrying loads of information around with them. This is not illogical, given that the human brain and the immune systems evolved together and from a very rudimentary perspective it was the immune system that came first and then layers upon layers of complexity followed to bring us to the 'human' as we know it today. By understanding the evolution and development of the immune system, we can gain a deep understanding about neuroprotection and how we can target it for stroke therapeutics. For example, using pathway analysis to identify significant pathways responding to stroke in the peripheral blood, we found Ephrin signaling to be a key player in the first 0–24 h following the insult [35]. Ephrin signaling guides many biological processes during embryo development as well as some neuronal growth processes in adulthood. Ephrins serve as navigational guides for cell migration and can conduct bidirectional signaling. Interestingly, some recent studies in stroke have identified a potential role for Ephrins in stroke neuroprotection, which validates the powerful approach of multi-omic profiling of peripheral blood in patients to identify relevant physiological targets for treatment [36].

A final concept for further evaluation of neuroprotective stroke treatments is the timing of physiological responses and their associated immune implications. This can only be achieved by time-series analyses and following patients throughout the course of their recovery to identify points of inflection by which immune mediators move from being protective to detrimental or vice versa. In immunology and physiology, many factors have dual roles and depending upon internal and external signaling environments their impact can be very different. Identifying these patterns will be critical for creating treatment strategies that fit the patient throughout their journey and may even be dependent upon the patient's inherent immune complexity (i.e., immune repertoire). The immune repertoire refers to the diverse combinations of T and B cell lymphocytes produced by somatic recombination. Greater diversity equates to a healthier immune system. Next-generation sequencing of the immune repertoire is on the horizon and expected to provide a greater classification of immune responses as well as novel immune mechanisms and pathways [37].

Every patient's recovery journey is unique. Not only are there physiological variables that contribute to recovery, but there are also social, emotional, behavioral, environmental, and genetic factors that all interact to either aid or hinder the recovery process. A comprehensive assessment of these domains combined with a multi-omic physiologic approach would enhance and individualize current stroke recovery programs. Lastly, nontraditional and integrative medicine approaches may offer additional neuroprotection for recovering patients, and this needs to be considered when designing stroke clinical trials [38].

4 Future Opportunities

Given the rapid advancements in multi-omics-based technology it will be imperative to remain flexible and open to improvements in methods as they arise, while balancing the need to maintain rigor and continued validation and optimization. Even the design of clinical trials is changing, particularly in oncology where traditional approaches are being replaced by adaptive trial designs that allow for ongoing optimization throughout the course of study, reducing time and cost [39]. The ISPY 1/2 (investigation of serial studies to predict your therapeutic response with imaging and molecular analysis) trial is an example of such a trial that allows for a biomarker-driven therapeutic approaches for the treatment of cancer [40–42]. Other adaptive clinical trials include MIDAS (the multicandidate iterative design with adaptive selection) [43] and BATTLE (Biomarker-integrated Approaches of Targeted Therapy for Lung Cancer Elimination) [44]. These trials have confirmed the feasibility, safety, and potential to improve chemotherapeutic response and optimize individual patient recovery in many cancers. There are many options to modify adaptive trial designs to fit the study and disease of interest and are quickly gaining traction in personalized medicine trials [45]. N of 1 clinical trials are an additional option for rare, or difficult to treat cancers and other chronic and comorbid diseases allowing clinicians to determine the effectiveness of therapies in individual patients [46]. This approach is likely difficult for acute stroke therapies but may be feasible for tailoring subacute to chronic treatment and rehabilitation approaches. Given that many stroke patients struggle with refractory or difficult-to-treat poststroke depression and anxiety, it may be interesting to consider whether an N of 1 approach could allow for drug repurposing, identification of novel neuroprotective strategies, and improve overall outcome for these patients. The reporting of N of 1 has unfortunately been inadequate. Thus, a recommendation was made in 2015 to follow a checklist to decrease variability and improve overall acceptance of reported data (A Consolidated Standards of Reporting Trials (CONSORT) extension for N-of-1 trials [47].

Exciting technologies expected to play a greater role in clinical research and medicine in the upcoming years are sensor and nanotechnology combined with cloud computing. With the digitalization of society, the internet of things (IoT) and wearable devices, these technologies have the potential to offer real-time feedback to patients and providers to enhance personalized approaches to treatment, medication adherence, and overall wellness strategies to improve recovery and quality of life [48]. The application of these technologies to health (the Internet of BioThings) is new with very little data to support the efficacy of this approach for chronic conditions. However, preliminary data in hypertension suggests the approach is easy to use, feasible, and well accepted in the patient population [49]. Data integration, provider bandwidth, and security issues are the largest challenges we face as we continue to implement and test the concept of the IoT in chronic cardiovascular conditions.

5 Conclusions/Summary

In comparison to the tremendous advancements in precision health for cancer-related conditions, there is a shortage of information to guide personalized approaches to stroke diagnosis and treatment. Mutli-omic profiling combined with big data analytics and adaptive clinical trial designs has the potential to revolutionize the approach to acute, subacute, and chronic stroke care. Exciting opportunities lie ahead to allow real-time monitoring via sensor-based technology and cloud computing that could be utilized at the bench and the bedside. As a scientific community, the success of precision stroke care requires a concerted effort to collect, analyze, and interpret relevant physiologic phenotypes that can be used at the point of care while leveraging the existing successes from cancer immunogenomics to accelerate biomarker-driven diagnostics and therapeutics for stroke.

References

1. Ding K, Wu S, Ying W, Pan Q, Li X, Zhao D et al (2015) Leveraging a multi-omics strategy for prioritizing personalized candidate mutation-driver genes: a proof-of-concept study. Sci Rep 5:17564
2. Kim ST, Lee J, Hong M, Park K, Park JO, Ahn T et al (2015) The NEXT-1 (Next generation pErsonalized tX with mulTi-omics and preclinical model) trial: prospective molecular screening trial of metastatic solid cancer patients, a feasibility analysis. Oncotarget 6(32):33358–33368
3. Pineda S, Gomez-Rubio P, Picornell A, Bessonov K, Marquez M, Kogevinas M et al (2015) Framework for the Integration of Genomics, Epigenomics and Transcriptomics in Complex Diseases. Hum Hered 79(3-4):124–136

4. Wang L, Xiao Y, Ping Y, Li J, Zhao H, Li F et al (2014) Integrating multi-omics for uncovering the architecture of cross-talking pathways in breast cancer. PLoS One 9(8), e104282
5. Sandercock PA (2015) Does personalized medicine exist and can you test it in a clinical trial? Int J Stroke 10(7):994–999
6. Yu KH, Snyder M (2016) Omics profiling in precision oncology. Mol Cell Proteomics 15(8):2525–2536
7. Greenplate AR, Johnson DB, Ferrell PB Jr, Irish JM (2016) Systems immune monitoring in cancer therapy. Eur J Cancer 61:77–84
8. Boskovic A, Belada N, Knezevic B (2014) Prognostic value of heart rate variability in post-infarction patients. Vojnosanit Pregl 71(10):925–930
9. Song T, Qu XF, Zhang YT, Cao W, Han BH, Li Y et al (2014) Usefulness of the heart-rate variability complex for predicting cardiac mortality after acute myocardial infarction. BMC Cardiovasc Disord 14:59
10. Benoist C, Germain RN, Mathis D (2006) A plaidoyer for 'systems immunology'. Immunol Rev 210:229–234
11. Lo EH, Ning M (2016) Mechanisms and challenges in translational stroke research. J Investig Med 64(4):827–829
12. Famakin BM (2014) The immune response to acute focal cerebral ischemia and associated post-stroke immunodepression: a focused review. Aging Dis 5(5):307–326
13. Brooks SD, Spears C, Cummings C, VanGilder RL, Stinehart KR, Gutmann L et al (2014) Admission neutrophil-lymphocyte ratio predicts 90 day outcome after endovascular stroke therapy. J Neurointerv Surg 6(8):578–583
14. Celikbilek A, Ismailogullari S, Zararsiz G (2014) Neutrophil to lymphocyte ratio predicts poor prognosis in ischemic cerebrovascular disease. J Clin Lab Anal 28(1):27–31
15. Holt RA (2015) Immunogenomics: a foundation for intelligent immune design. Genome Med 7(1):116
16. Martino D, Allen K (2015) Meeting the challenges of measuring human immune regulation. J Immunol Methods 424:1–6
17. Amantea D, Bagetta G (2016) Drug repurposing for immune modulation in acute ischemic stroke. Curr Opin Pharmacol 26:124–130
18. Fu Y, Liu Q, Anrather J, Shi FD (2015) Immune interventions in stroke. Nat Rev Neurol 11(9):524–535
19. Allen S, Liu YG, Scott E (2016) Engineering nanomaterials to address cell-mediated inflammation in atherosclerosis. Regen Eng Transl Med 2(1):37–50
20. Wu CC, Chen BS (2016) Key immune events of the pathomechanisms of early cardioembolic stroke: multi-database mining and systems biology approach. Int J Mol Sci 17(3):305
21. Wong YH, Wu CC, Lai HY, Jheng BR, Weng HY, Chang TH et al (2015) Identification of network-based biomarkers of cardioembolic stroke using a systems biology approach with time series data. BMC Syst Biol 9(Suppl 6):S4
22. Bielekova B, Vodovotz Y, An G, Hallenbeck J (2014) How implementation of systems biology into clinical trials accelerates understanding of diseases. Front Neurol 5:102
23. Ren C, Guingab-Cagmat J, Kobeissy F, Zoltewicz S, Mondello S, Gao M et al (2014) A neuroproteomic and systems biology analysis of rat brain post intracerebral hemorrhagic stroke. Brain Res Bull 102:46–56
24. Libbrecht MW, Noble WS (2015) Machine learning applications in genetics and genomics. Nat Rev Genet 16(6):321–332
25. Konig IR, Auerbach J, Gola D, Held E, Holzinger ER, Legault MA et al (2016) Machine learning and data mining in complex genomic data-a review on the lessons learned in Genetic Analysis Workshop 19. BMC Genet 17(Suppl 2):1
26. Petrone AB, O'Connell GC, Regier MD, Chantler PD, Simpkins JW, Barr TL (2016) The role of Arginase 1 in post-stroke immunosuppression and ischemic stroke severity. Transl Stroke Res 7(2):103–110

27. Wasserstein RL, Lazar NA (2016) The ASA's statement on p-values: context, process, and purpose. The American Statistician. pp 129-133. http://dx.doi.org/10.1080/00031305.2016.1154108
28. Mamoshina P, Vieira A, Putin E, Zhavoronkov A (2016) Applications of deep learning in biomedicine. Mol Pharm 13(5):1445–1454
29. Yip KY, Cheng C, Gerstein M (2013) Machine learning and genome annotation: a match meant to be? Genome Biol 14(5):205
30. Karlinski M, Gluszkiewicz M, Czlonkowska A (2015) The accuracy of prehospital diagnosis of acute cerebrovascular accidents: an observational study. Arch Med Sci 11(3):530–535
31. Newman-Toker DE, Ernest M, Ernest V, Rosanna C, Hines AL (2014) Missed diagnosis of stroke in the emergency department: a cross-sectional analysis of a large population-based sample. Diagnosis 1(2):155–166. doi:10.1515/dx-2013-0038
32. National Academies of Sciences E and Medicine (2015) Improving diagnosis in health care. The National Academies Press, Washington, DC
33. Gelabert MA (1990) [Infections caused by gram positive bacteria: a new epidemics in patients with positive anti-HIV antibodies and drug addicts]. Arch Esp Urol 43(2):103–105
34. Barr TL, Conley Y, Ding J, Dillman A, Warach S, Singleton A et al (2010) Genomic biomarkers and cellular pathways of ischemic stroke by RNA gene expression profiling. Neurology 75(11):1009–1014
35. Barr TL, VanGilder R, Rellick S, Brooks SD, Doll DN, Lucke-Wold AN et al (2015) A genomic profile of the immune response to stroke with implications for stroke recovery. Biol Res Nurs 17(3):248–256
36. Yang J, Luo X, Huang X, Ning Q, Xie M, Wang W (2014) Ephrin-A3 reverse signaling regulates hippocampal neuronal damage and astrocytic glutamate transport after transient global ischemia. J Neurochem 131(3):383–394
37. Six A, Mariotti-Ferrandiz ME, Chaara W, Magadan S, Pham HP, Lefranc MP et al (2013) The past, present, and future of immune repertoire biology - the rise of next-generation repertoire analysis. Front Immunol 4:413
38. Peterson-Burch FR-R, K; Barr, TL (2017) Rethinking Recovery: Incorporating Holistic Nursing Perspectives in Post-Stroke Care, Holistic Nursing Practice, In Press, 2017
39. Hatfield I, Allison A, Flight L, Julious SA, Dimairo M (2016) Adaptive designs undertaken in clinical research: a review of registered clinical trials. Trials 17(1):150
40. Esserman LJ, Berry DA, DeMichele A, Carey L, Davis SE, Buxton M et al (2012) Pathologic complete response predicts recurrence-free survival more effectively by cancer subset: results from the I-SPY 1 TRIAL—CALGB 150007/150012, ACRIN 6657. J Clin Oncol 30(26): 3242–3249
41. Printz C (2013) I-SPY 2 may change how clinical trials are conducted: researchers aim to accelerate approvals of cancer drugs. Cancer 119(11):1925–1927
42. Barker AD, Sigman CC, Kelloff GJ, Hylton NM, Berry DA, Esserman LJ (2009) I-SPY 2: an adaptive breast cancer trial design in the setting of neoadjuvant chemotherapy. Clin Pharmacol Ther 86(1):97–100
43. Yuan Y, Guo B, Munsell M, Lu K, Jazaeri A (2016) MIDAS: a practical Bayesian design for platform trials with molecularly targeted agents. Stat Med. doi:10.1002/sim.6971
44. Kim ES, Herbst RS, Wistuba II, Lee JJ, Blumenschein GR Jr, Tsao A et al (2011) The BATTLE trial: personalizing therapy for lung cancer. Cancer Discov 1(1):44–53
45. Antoniou M, Jorgensen AL, Kolamunnage-Dona R (2016) Biomarker-guided adaptive trial designs in Phase II and Phase III: a methodological review. PLoS One 11(2), e0149803
46. Collette L, Tombal B (2015) N-of-1 trials in oncology. Lancet Oncol 16(8):885–886
47. Vohra S, Shamseer L, Sampson M, Bukutu C, Schmid CH, Tate R, CONSORT extension for reporting N-of-1 trials (CENT) et al (2015) Statement. J Clin Epidemiol 2015
48. Sieverink F, Siemons L, Braakman-Jansen A, van Gemert-Pijnen L (2016) Internet of things & personalized healthcare. Stud Health Technol Inform 221:129
49. Godbehere P, Wareing P (2014) Hypertension assessment and management: role for digital medicine. J Clin Hypertens (Greenwich) 16(3):235

Chapter 12
Polarization of Microglia/Macrophages in Brain Ischaemia: Relevance for Stroke Therapy

Diana Amantea, Rosaria Greco, Cristina Tassorelli, and Giacinto Bagetta

Abstract The innate immune system plays a pivotal role in ischemic stroke pathobiology, involving soluble and cellular mediators activated locally or recruited from the periphery. Upon injury, subtle modifications of the local environment trigger a rapid activation of microglia that peaks few days after the insult and may persist for several weeks. Initially, the alternatively activated M2 phenotype predominates, whereas, upon priming by ischemic neurons, microglia shift towards the M1 phenotype characterized by reduced phagocytic capacity and release of inflammatory cytokines. Maximally activated microglia can eventually turn into a round amoeboid phenotype morphologically indistinguishable from blood-derived macrophages.

A typical hallmark of cerebral ischaemia is the increased permeability of cerebral microvessels that, together with the upregulation of adhesion molecules on post-capillary venules and the choroid plexus, facilitate brain recruitment of leukocytes. Bone marrow-derived monocytes rapidly extravasate via a chemokine receptor 2 [CCR2]-dependent pathway and, once in the injured tissue, differentiate into non-inflammatory M2 macrophages, mediating neuroprotection and repair of the neurovascular unit. M2 macrophages peak few days after the insult in the core region, whereas the pro-inflammatory M1 phenotype predominates in the peri-infarct areas to gradually increase in number in the core, outnumbering M2 cells over time. The dualistic role exerted by microglia/macrophages suggests that a mere inhibition of

D. Amantea, Ph.D. (✉) • G. Bagetta
Section of Preclinical and Translational Pharmacology, Department of Pharmacy, Health and Nutritional Sciences, University of Calabria, Rende (CS), Calabria, Italy
e-mail: diana.amantea@unical.it

R. Greco
Laboratory of Neurophysiology of Integrative Autonomic Systems, Headache Science Centre, "C. Mondino" National Neurological Institute, Pavia, Italy

C. Tassorelli
Laboratory of Neurophysiology of Integrative Autonomic Systems, Headache Science Centre, "C. Mondino" National Neurological Institute, Pavia, Italy

Department of Brain and Behavioural Sciences, University of Pavia, Pavia, Italy

P.A. Lapchak, J.H. Zhang (eds.), *Neuroprotective Therapy for Stroke and Ischemic Disease*, Springer Series in Translational Stroke Research,
DOI 10.1007/978-3-319-45345-3_12

their activation/recruitment might not represent a promising strategy to rescue ischemic brain injury. By contrast, as thoroughly reviewed here, a rational modulation of their polarization status, aimed at adjusting the M1/M2 ratio coherently with the spatio-temporal progression of injury, has recently been validated in animal models.

Keywords Immune system • Ischemic stroke • Macrophages • Monocytes • Microglia • M1/M2 phenotypes

Abbreviations

ATP	Adenosine 5′-triphosphate
BBB	Blood–brain barrier
BDNF	Brain-derived neurotrophic factor
CB	Cannabinoid receptor
CCL2	Monocyte chemoattractant protein-1
CCR2	CC chemokine receptor 2
CNS	Central nervous system
CXCR1	Fractalkine receptor
DAMPs	Danger-associated molecular patterns
GDNF	Glia-derived neurotrophic factor
GFP	Green fluorescent protein
IGF	Insulin-like growth factor
IL	Interleukin
IL-1R	Interleukin-1 receptor
INF	Interferon
Ly6C	Lymphocyte antigen 6 complex, locus C
MCAo	Middle cerebral artery occlusion
MCP	Monocyte chemoattractant protein
MHC	Major histocompatibility complex
MIP	Macrophage inflammatory protein
miRNAs	MicroRNAs
MMPs	Matrix metalloproteinases
MR	Mineralcorticoid receptor
nAChR	Nicotinic acetylcholine receptor
PACAP	Pituitary adenylate cyclase-activating polypeptide
PBMCs	Peripheral blood mononuclear cells
PPAR	Peroxisome proliferator-activated receptor
PRRs	Pattern recognition receptors
ROS	Reactive oxygen species
RXR	Retinoid X receptor
SR-A	Class A scavenger receptor

TGF	Transforming growth factor
TLR	Toll-like receptor
TNF	Tumour necrosis factor
TREM	Triggering receptor expressed on myeloid cells
USPIO	Ultrasmall superparamagnetic iron oxide particles

1 Introduction

The pathological consequences of ischemic stroke involve a complex cascade of events, including both neuronal and non-neuronal mechanisms that actively participate in the progression of brain damage. In contrast to the old "neuronocentric" view, in which neuronal activity and networking were considered the sole regulators of cerebral activities, during the past decades, a more integrated concept was developed based on the involvement of virtually all the cell types of the neuro-glio-vascular unit in brain function and dysfunction [1–5]. Accordingly, glial cells, such as astrocytes and microglia, as well as vascular endothelium and perivascular elements are affected by the ischemic insult and may participate in cell-specific signalling and execution cascades leading to brain damage [6, 7]. In addition, blood-borne immune cells play a crucial, dualistic role in the evolution of cerebral ischaemia, contributing to both detrimental and reparative processes [8]. Thus, the early evidence of local activation of microglia and cerebral recruitment of circulating leukocytes has stimulated a growing number of studies aimed at exploring the involvement of the immune system in ischemic pathobiology. In fact, although the brain was traditionally thought to be completely devoid of immunologic reactions, this old conviction has been abandoned as neuroscientists and immunologists have begun to unveil that immune mediators are intimately related to neurons, as recently highlighted in a series of neurodegenerative conditions including stroke [1]. It is now clear that after cerebral ischaemia, a complex crosstalk occurs between the cellular elements of the brain, namely neurons, astrocytes, microglia and dendritic cells, and blood-borne immune cells, including neutrophils, macrophages and T cells [9]. This is further underscored by recent blood expression profiling studies reporting that most genes upregulated early after stroke injury in patients are implicated in the regulation of the innate immune system [10–13]. In addition, circulating levels of markers of acute inflammation correlate with the severity of behavioural and histological outcomes [14–18]. Thus, immune mediators activated not only in the brain but also in the periphery participate in the early and late responses to cerebral ischaemia. In this context, an important issue to be considered is the dualistic nature of immune cells that may acquire either detrimental or protective phenotypes, thus prompting neuronal damage or repair depending on specific systemic or microenvironmental stimuli generated during the spatio-temporal evolution of ischemic brain damage. Unravelling the intricate brain-immune interplay will improve our understanding of the involvement of the immune system in ischemic tissue damage, protection and repair, and will open new avenues to the discovery of novel pharmacological targets [19].

2 The Innate Immunological Response of the Brain to Stroke

After blood vessel occlusion, the reduced blood supply to the cerebral tissue triggers multiple inflammatory cascades in the brain and in the peripheral immune system [8]. Activation of platelets and of the complement system represents an early trigger of the inflammatory response [20, 21]. Moreover, swelling of perivascular astrocyte end feet is the first cellular event detected after cerebral ischaemia [22]. In fact, as early as few minutes after the ischemic insult, cytokines and reactive oxygen species (ROS) released by neurons and glial cells alter astrocytes phenotype inducing their hypertrophy and proliferation [23]. Upon the insult, injured neurons and other damaged brain cells release a series of molecules that act as danger-associated molecular patterns (DAMPs) or alarmins by binding and activating specific pattern recognition receptors (PRRs) on various cells [24]. Adenosine 5′-triphosphate (ATP), released or leaked from injured cells, activates purinergic P2Y [1] receptors on astrocytes prompting the release of inflammatory cytokines, such as interleukin (IL)-1β, or matrix metalloproteinases (MMPs) that participate in blood–brain barrier (BBB) rupture and vasogenic edema [25–29]. In line with the dualistic nature of most immune cells, also astrocytes may participate in protective and reparative responses by providing extracellular glutamate uptake [30, 31], BBB reconstitution [32, 33] and release of neurotrophic factors [34, 35]. Therefore, the exact role played by reactive astrogliosis in the progression of ischemic cerebral injury depends on the polarization of astrocytes towards distinct phenotypes [36–38] and on their interaction with the surrounding environment generated by neighbouring neurons and microglia/macrophages [39, 40].

Despite astrocytes play a crucial role in the modifications and remodelling of the neurovascular unit following the ischemic event, immunity in the brain is mainly sustained by cells from the mononuclear phagocyte system [41]. The only haematopoietic cells resident in the central nervous system (CNS) parenchyma are microglia that execute homeostatic and surveillance tasks in the healthy brain. Although in the past there was some debate regarding the precise nature of microglia progenitors, nowadays a generally accepted concept is that microglia are of myeloid origin [42]. In particular, microglia originate from yolk sac-derived macrophages that seed the brain rudiment during early foetal development. Moreover, emerging evidence suggests that other progenitors may supersede the yolk sac contribution, including blood-borne monocytes or other bone marrow-derived cells that enter the CNS, especially during pathological conditions characterized by BBB rupture, either to generate the post-natal microglial compartment or to maintain it into adulthood [42, 43].

During cerebral ischaemia, microglial reactivity is modulated by a series of receptors, including purinergic ATP receptors, such as P2X purinoceptor 7 (P2X7) and platelet adenosine diphosphate receptor P2Y12 receptors, Toll-like receptor (TLR)-4, fractalkine receptor (CX3CR1), peroxisome proliferator-activated receptor (PPAR)-γ, cannabinoid receptor (CB)-2, triggering receptor expressed on myeloid cells (TREM)-2 and/or CD200 receptor [44–52]. Activation of microglial cells occurs very rapidly after ischaemia and is driven by subtle modifications of the local environment [53, 54]. Microglia become activated within minutes after the insult and their reactivity and proliferation peak few days after the insult and may persist for several weeks, as observed both in pre-clinical stroke models [55–57] and in patients [46, 58, 59]. The temporal and spatial activation of microglia in the brain of mice subjected to transient middle cerebral artery occlusion (MCAo) was recently described by an original multimodal imaging approach [60]. The complex functional changes of activated microglial cells are accompanied by a gradual morphological transformation from the highly ramified shape, typical of the surveillance mode, to cells with larger somata, displaying shorter and coarser cytoplasmic processes [61]. Maximally activated microglia can eventually turn into a round amoeboid phenotype associated with increased proliferation, production of inflammatory mediators and upregulation of myeloid markers, rendering these cells morphologically indistinguishable from blood-derived macrophages [55, 62–65]. In the ischemic brain, ramified microglia and "amoeboid" microglia were found surrounding intact and dying neurons, respectively [55, 66].

Once activated, microglia upregulate expression levels of CD11b and ionized calcium-binding adapter molecule 1 (Iba1), and gain expression of molecules associated with antigen presentation, such as major histocompatibility complex (MHC)-II [67]. Functional activation is consistent with increased expression and release of inflammatory cytokines, namely IL-1 and tumour necrosis factor (TNF)-α [26, 57, 68–70], reactive oxygen and nitrogen species [71, 72] and proteases [73]. These harmful mediators underlie the detrimental role of microglia in ischemic inflammation and injury, as highlighted by the evidence that pharmacological- or microRNA-induced suppression of microglial activity limits ischemic cerebral injury [61, 74–76]. Activated microglia also participate in BBB rupture through the release of ROS, MMPs and pro-inflammatory cytokines [46]. Using time-lapse two-photon microscopy in vivo, a recent study dynamically observed that microglia in the penumbra rapidly expanded cellular protrusions towards blood vessels. These perivascular microglia phagocytosed endothelial cells, contributing to the disintegration of blood vessels and, eventually, to the rupture of the BBB [77, 78]. Nevertheless, the causal relationship between microglia activation and BBB disruption has been questioned by some studies. In fact, BBB disruption can, in turn, cause microglia activation, and microglia may also play a protective role through secretion of progranulin, involved in reducing BBB damage and brain oedema after ischemic stroke [79–81].

Accordingly, there is evidence suggesting that microglia can also be potentially beneficial under certain circumstances. Selective ablation of proliferating resident microglia results in elevated brain lesion size associated with an increase in apoptotic neurons in transient focal cerebral ischaemia [82]. Conversely, intrarterial injection of exogenous microglia isolated from brain cultures increases neuronal survival in the ischemic hippocampus of gerbils [83], whereas intracerebroventricular injection of exogenous microglia decreases neuronal loss induced by transient focal ischaemia in rats [84]. The positive effects of microglia on neuronal survival have been attributed to their ability to release anti-inflammatory and neurotrophic molecules, including IL-10, transforming growth factor (TGF)-β, brain-derived neurotrophic factor (BDNF), glia-derived neurotrophic factor (GDNF) and insulin-like growth factor (IGF)-1 that participate in the resolution of inflammation and to the reparative mechanisms implicated in late tissue recovery [41, 82, 83, 85, 86]. Some of these mediators are also involved in the biphasic role of different activation states of microglia on neurogenesis [87].

The dualistic nature of the microglial response is ascribable to the ability of these myeloid cells to polarize towards a number of phenotypes, ranging from the classic M1 to the alternatively activated M2 phenotypes [88]. The pro-inflammatory M1 phenotype is usually activated by interferon (INF)-γ or through TLRs modulation, whereas the "beneficial" M2 phenotype is prompted by regulatory factors, such as IL-4, IL-10, IL-13, and TGF-β [89, 90]. At early stages after ischemic injury, the M2 phenotype predominates, whereas, upon priming by the ischemic neurons, microglia shift towards the pro-inflammatory M1 phenotype characterized by reduced phagocytic capacity and by the release of TNF-α, IL-1β, monocyte chemoattractant protein (MCP)-1, macrophage inflammatory protein (MIP)-1α and IL-6 [53] (Fig. 12.1). Thus, blocking this process by maintaining the M2-polarization status of microglia or promoting the M1-to-M2 polarization shift represents a promising strategy for a rational immunomodulation in the early stages after ischemic stroke. Although its relevance in patients has not been assessed yet, this approach has successfully been validated in animal models (see below).

2.1 Blood-Borne Myeloid Cells

A typical hallmark of cerebral ischaemia is the increased permeability of cerebral microvessels that allows brain transfer of blood-borne cells and molecules that do not normally cross the BBB [91]. Hypoxic/ischemic injury differentially affects microvascular endothelial cells and perivascular cells, i.e. pericytes and astrocytes, and this may have marked impact on barrier stability [92]. Upregulation of adhesion molecules on post-capillary venules and the choroid plexus, together with the secretion of chemokines, facilitates the recruitment of circulating immune cells to the ischemic or haemorrhagic brain [93–96]. Moreover, the activation of proteases (i.e. MMPs) and the local increase of inflammatory cytokines (e.g. IL-1 and TNF-α) further elevate the expression of adhesion molecules and BBB breakdown leading to leukocytes extravasation in the injured brain [97–100]. Thus, upon activation by soluble factors released upon injury, the immune response engages specialized blood-borne cells, such as neutrophils, monocytes/macrophages, dendritic cells and T lymphocytes, that eventually migrate to the lesioned brain [8, 98, 101]. Indeed, growing evidence documents that, regardless of their ability to actually infiltrate the brain, circulating immune cells contribute to the evolution of cerebral ischemic damage.

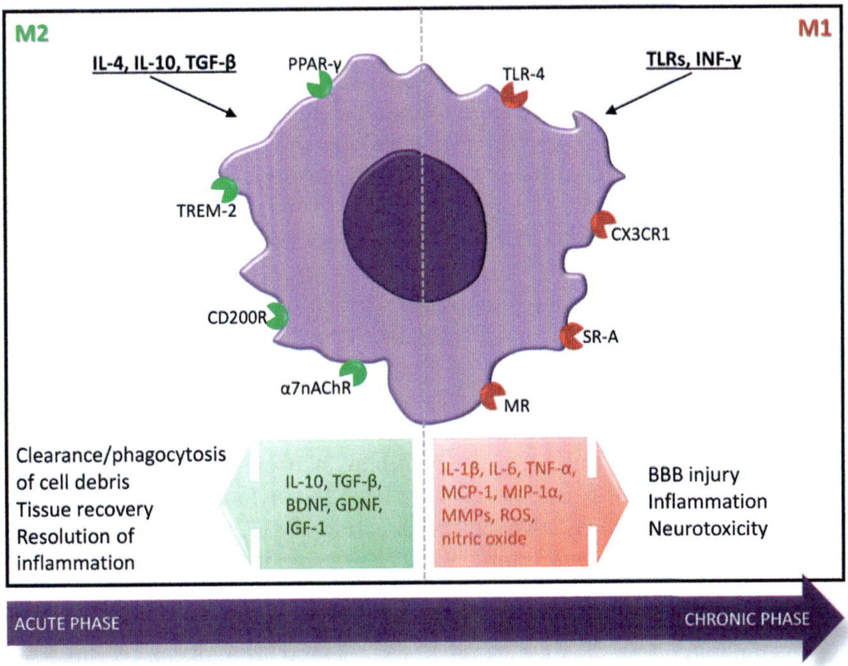

Fig. 12.1 Polarization of microglia/macrophages towards the protective M2 or the pro-inflammatory M1 phenotype after ischemic stroke. At early stages after the ischemic insult, locally activated microglia/macrophages and blood-borne monocytes/macrophages adopt the M2 phenotype characterized by enhanced phagocytic activity, reduced production of inflammatory mediators and pro-survival properties towards hypoxic/ischemic neurons. Therefore, the early recruitment of microglia/macrophages may represent an endogenous attempt to clear debris and to limit brain damage. This M2-mediated response is transient and extinguishes by a week after the insult, being replaced by an abundant wave of M1-polarized cells that display reduced phagocytosis and elevated release of pro-inflammatory and neurotoxic mediators

The relevance of the peripheral immune system in stroke pathobiology is underscored by the evidence that peripheral blood genomic profiles are associated with brain profiles and correlate with cerebral inflammation and injury [13, 102, 103]. More importantly, the majority of the genes identified in all the stroke-specific profiles published to date are related to the regulation of the immune system [10–12, 102]. A recent work originally identified a panel of five genes overlapping among these studies; four of these genes (i.e. arginase 1, lymphocyte antigen 96, MMP-9, S100 calcium binding protein A12) are implicated in the regulation of the innate immune response [13]. Accordingly, the majority of the genes acutely regulated in the blood of stroke patients are expressed in neutrophils and, to a lesser extent, in monocytes [10]. The importance of innate immunity in ischemic stroke is further highlighted by the evidence that monocytes/macrophages and neutrophils are the first cells to infiltrate the injured brain, reaching a peak within 24–72 h after the insult [95, 104, 105]. Moreover, higher peripheral leukocyte and neutrophil counts, but not lymphocyte counts, are associated with larger infarct volumes in patients [106], whereas cerebral accumulation of neutrophils correlates with brain damage severity and poor neurological outcome both in humans [107] and in rodents [108–110].

The majority of leukocytes are recruited into the ischemic brain following a specific temporal profile, with activation of microglia and migration of monocytes/macrophages preceding infiltration of neutrophils, dendritic cells and lymphocytes to the lesioned regions [95]. In particular, brain recruitment of circulating monocytes is coordinated by inflammatory cytokines, chemokines and de novo expressed adhesion molecules.

Activation of CX3CR1 signalling during stroke has been suggested to promote recruitment of peripheral monocyte/macrophages and expansion/activation of CNS microglia and macrophages, associated with a shift towards pro-inflammatory phenotypes and thus the exacerbation of cerebral damage [44, 111, 112]. In addition, activation of CC chemokine receptor 2 (CCR2) by MCP-1 (CCL2) has been implicated in brain recruitment of bone marrow-derived monocytes to the ischemic brain. CCR2+ monocytes have been shown to play either detrimental roles [113–116] or reparative functions acting via TGF-β1 to maintain the integrity of the neurovascular unit [39, 117] or promoting M2 polarization of macrophages following brain ischaemia [118]. Indeed, once in the damaged tissue, blood-borne monocytes can differentiate into macrophages that, in turn, may develop into different polarizations states, ranging from the classically activated M1 subtype to the alternatively activated M2 phenotypes. M2 macrophages peak 3–5 days after the insult in the core region, whereas the pro-inflammatory M1 phenotype predominates in the peri-infarct areas to gradually increase in number in the core, outnumbering M2 cells over time [119]. However, in human brain, accumulation of monocytes seems to be more delayed as compared to rodents, since inflammatory mononuclear cells and macrophages have been detected in the brain of ischemic patients from 2 days up to 53 years [120, 121]. Although immunosuppression occurs after acute stroke, a significant elevation of the number of circulating monocytes has been reported in patients [122–125]. The ability of these cells to differentiate into specific phenotypes characterized by a distinct expression of CD14 and CD16 antigens in human has been correlated with outcome severity [125].

An intriguing hypothesis that has recently emerged refers to the putative splenic origin of recruited monocytes during ischemic injury, based on the identification of a major monocyte reservoir in the spleen of rodents. In fact, ischaemia-induced splenic atrophy has been associated with the release of inflammatory mediators and of spleen-derived inflammatory cells into the circulation, possibly followed by their migration into the brain, thus increasing the inflammatory response and promoting secondary cerebral damage [126–130]. There is evidence that the spleen might contribute to both pro-inflammatory lymphocyte antigen 6 complex, locus C $(Ly6C)^{hi}$ and regulatory $Ly6C^{lo}$ monocytes after brain ischaemia in mice [131, 132]. Nevertheless, further work is needed to provide a direct demonstration supporting the concept that monocytes migrating to the brain are actually mobilized from the spleen and contribute to brain injury.

Although brain recruitment of leukocytes has been widely demonstrated, some studies have argued that not all the immune cells actually infiltrate the cerebral parenchyma, since polymorphonuclear granulocytes recruited to the infarcted area do not cross the vascular endothelium but remain restricted to luminal surfaces or perivascular spaces of cerebral vessels [133]. Neutrophils have been show to exert detrimental effects through the production of ROS, the release of MMPs [134–136] and the induction of microvessel obstruction/thrombosis [137, 138]. Moreover, recruited neutrophils scan for activated platelets and participate in the integration of signals from the endothelium and the circulation before inflammation proceeds [139]. Recent findings have suggested that neutrophil–lymphocyte ratio may reflect post-stroke outcome and represents a useful biomarker to predict short-term mortality in ischemic stroke patients [103]. This further highlights that peripheral blood profiles may be exploited for disease diagnosis and prognosis, for monitoring treatment and intervention responses and, notably, for the identification of pharmacological targets.

2.2 *Differentiating Brain Resident Microglia from Blood-Borne Macrophages*

One of the major challenges in microglia/macrophage research arises from the aptitude of microglia to evolve into a round amoeboid phenotype after injury [140]. In fact, the majority of macrophages in the infarct area are derived from local microglia, demonstrating a significant predominance of local defence mechanisms over blood-borne immune cells [62]. In turn, haematogenous macrophages infiltrating the infarcted brain are able to acquire a ramified morphology indistinguishable from resident microglia [62]. This greatly hampers our comprehension of the contribution of in situ microglia proliferation versus blood-borne monocyte infiltration to the local immune response, since both cell types may differentiate into a macrophage-like phenotype upon cerebral ischaemia [46, 62]. Interestingly, a series of recent innovative studies have demonstrated that infiltrating monocyte-derived macrophages perform indispensable roles that cannot be provided by their resident counterparts [141]. However, to date, only few studies have addressed this issue reaching the critical goal of differentiating the response of the mononuclear phagocyte system of the brain from that elicited in the periphery.

After cerebral ischaemia, brain resident microglia and infiltrating monocyte/macrophages respond to different activation periods. Microglia are the dominating cell type in post-stroke neuroinflammation during the first days after ischemic stroke [62, 65, 142]. In fact, brain resident microglia plays a major role in phagocytosis within 3 days after photothrombosis-induced focal ischaemia in mice, whereas bone marrow-derived macrophages provide removal of necrotic tissue after 6 days [65]. These findings were confirmed by a more sophisticated experimental approach using bone marrow chimeric mice generated by transplanting green fluorescent protein

(GFP) transgenic bone marrow into irradiated wild-type recipients subjected to transient MCAo. In this model, microglia are activated and proliferate as early as 24 h after the ischemic attack, peaking at 4 days and gradually declining to return to baseline levels 28 days after [62]. By contrast, brain infiltration of GFP+ macrophages begins 4 days and peaks 7 days after the insult [143]. At these time points, the number of microglial cells is reduced due to their vulnerability to the detrimental effects of ischaemia, whereas infiltrating monocytes predominate and display higher phagocytic capacity as compared with local microglia [144]. Blood-borne myeloid cell infiltration was also tracked by tagging/labelling these cells with ultrasmall superparamagnetic iron oxide particles (USPIO). Infiltration of USPIO-labelled mononuclear phagocytes was observed by day 4 in the core region and by day 7 in the boundary areas of the brain of rats subjected to permanent focal cerebral ischaemia [145, 146]. This time course was also partially confirmed by flow cytometry (Gelderblom et al. [95]; but see also Stubbe et al. [147]) that allows to differentiate microglia from haematogenous macrophages on the basis of their diverse expression level of CD45 [148, 149]. In fact, CD11b-immunopositive cells identified as $CD45^{lo}$, $CD45^{int}$ or $CD45^{hi}$ represent resting microglia, activated microglia, or monocytes/macrophages, respectively [150]. Murine monocytes can be further classified as $Ly6C^{hi}$ pro-inflammatory monocytes or $Ly6C^{lo}$ anti-inflammatory monocytes [89, 151] and these markers have been used to characterize brain migration of bone marrow-derived monocytes/macrophages [117]. CCR2 expressed on $Ly6C^{hi}$ monocytes plays a pivotal role in their extravasation and transmigration into the brain after cerebral ischaemia-reperfusion [118, 152]. Intriguingly, pharmacological blockade of this receptor results in worsened functional outcomes and larger infarct volumes, associated with reduced M2 polarization and increased peroxynitrite production in macrophages in mice undergone transient MCAo [118]. Thus, during the acute post-stroke period, $Ly6C^{hi}$ monocytes enter the circulation and then the ischemic brain to protect the neurovascular unit and to limit inflammatory and oxidative injury by promoting M2 polarization of macrophages [117, 118]. Similarly, a recent study demonstrated that a subset of immature pro-inflammatory $Ly6C^{hi}CD43^{lo}$ monocytes recruited to the ischemic tissue suffered progressive differentiation to macrophages, downregulating the expression of Ly6C and acquiring the expression of markers of an alternative phenotype, suggesting their role in tissue repair during the subacute phase after stroke [153]. Thus, both inflammatory and non-inflammatory, repair-associated macrophage populations may derive from inflammatory $Ly6C^{hi}/CCR2^{+}$ monocytes. By contrast, $Ly6C^{lo}$ patrolling monocytes have been suggested to be redundant in the progression and recovery of ischemic stroke [154]. Since they do not give rise to significant numbers of non-inflammatory macrophages, $Ly6c^{lo}/CX3CR1^{hi}$ monocytes may instead represent a form of terminally differentiated blood-resident macrophage monitoring the vascular lumen for endothelial integrity [155–157]. The time course of microglia activation and macrophage infiltration strongly depends on the type of ischaemia, being permanent vessel occlusion associated with a more pronounced microglia activation at 1 and 5 days and with an earlier brain infiltration of blood-borne immune cells [93, 158]. Thus, being the first cells activated after brain injury, microglia clear cell debris and injured neurons by phagocytosis and participate in the resolution of inflammation by engulfing invading neutrophils [50, 85]. By contrast, phagocytosis also underlies detrimental effects by microglia through engulfment of viable neurons in the ischemic penumbra [159]. Actually, both microglia and haematogenous macrophages infiltrating the ischemic brain [62, 160] may differently affect the evolution of tissue damage, depending on their aptitude to switch between pro-inflammatory M1 and protective M2 phenotypes [53, 161–163]. Classically activated M1 microglia/macrophages trigger and maintain the inflammatory reaction sustaining neurotoxicity during the ischemic insult. By contrast, M2-polarized cells provide protective functions, including debris clearance, angiogenesis, tissue remodelling and repair [88, 164].

3 Polarization of Monocyte- or Microglia-Derived Macrophages in Stroke

Due to their tremendous plasticity and ability to integrate microenvironment signals, activation of microglia/macrophages can be represented by a multi-dimensional model characterized by a finely regulated cellular programming that is both tissue- and disease-specific, characterized by different activation states and mixed phenotypes [163, 165, 166]. In fact, the M1/M2 dichotomy is an oversimplified conceptual framework that only represents two extreme activation states, since in vivo microglia/macrophages may adopt a spectrum of different, often overlapping functional phenotypes [89, 165]. Accordingly, a growing body of evidence reveals the existence of diverse M2 subpopulations, including M2a, M2b, M2c and M2d, characterized by unique features and distinct biological functions [89, 167, 168]. Since these subpopulations have not been fully characterized in brain disorders, the broad M1/M2 classification is commonly accepted and allows the comprehension of the distinct roles of microglia/macrophages during cerebral injury, including stroke [169].

Although ischaemia-induced upregulation of pro-inflammatory and cytotoxic mediators due to classically activated M1 cells has been extensively reported, growing evidence highlights a concomitant elevation of the expression of M2 markers (CD206, Arginase 1, Ym1, IL-10 and TGF-β) in the ischemic brain [51, 53, 66, 161, 170]. This has been ascribed to locally activated microglia/macrophages that adopt the M2 phenotype early after the ischemic insult, but also to an increased infiltration of alternatively activated (M2) blood-borne monocytes into the brain parenchyma [53, 66]. Interestingly, the deterioration of functional outcome, that occurs in aged mice undergone ischemic stroke, has been linked to reduction of M2 polarization of microglia/macrophages [171]. M2 polarization is induced by factors released by ischemic neurons and is associated with a healthier phenotype characterized by enhanced phagocytic activity, reduced production of inflammatory mediators and pro-survival properties towards hypoxic/ischemic neurons [53, 163]. Therefore, the early recruitment of microglia/macrophages may represent an endogenous attempt to clear debris and to limit brain damage. However, this M2-mediated response is transient and extinguishes by a week after the insult, being replaced by an abundant wave of

M1-polarized cells that display reduced phagocytosis and elevated release of harmful mediators such as TNF-α and nitric oxide [53]. This highlights that the M2-to-M1 shift occurring during chronic inflammation after ischaemia contributes to the expansion of cerebral injury and to the impairment of neuronal recovery (Fig. 12.1).

A crucial issue that needs further research to be clarified is the identity of the molecular switches that control the phenotypic changes of microglia/macrophages. To this aim, a series of receptors have been implicated in the polarization of these myeloid cells after ischemic stroke (Fig. 12.1). The absence of Class A scavenger receptor (SR-A) was associated to lower expression of M1 microglia/macrophages-related genes, i.e. TNF-α, inducible nitric oxide synthase, MCP-1 and IL-1β, and preservation of alternatively activated M2 markers, resulting in reduced infarct size and improved neurological function in mice subjected to focal ischaemia [162]. Similarly, the fractalkine receptor CX3CR1 has been associated with a pro-inflammatory milieu in mice characterized by elevation of M1/M2 ratio [111]. Conversely, deficiency of the mineralocorticoid receptor (MR) reduces the expression of M1 markers while elevating M2-polarized myeloid cells in the ischemic brain, thus resulting in a better stroke outcome in MR−/− mice [161]. Another receptor that has been demonstrated to be relevant for immunoregulation in stroke is the α-7 nicotinic acetylcholine receptor (nAChR), since its activation has been shown to provide neuroprotection in rodent models through reduction of M1/M2 macrophage ratio [172, 173]. Further work has demonstrated that a subpopulation of bone marrow-derived monocytes/macrophages, recruited via CCR2 and acting through TGF-β1, preserves the integrity of the neurovascular unit in murine stroke models [117]. Recent findings have also highlighted that the hydroxycarboxylic acid receptor 2 promotes a neuroprotective phenotype in monocytes/macrophages infiltrating the ischemic brain; thus, activation of this receptor might underlie the neuroprotective effects of the ketone bodies or ketogenic diet [174]. Among soluble factors, IL-10 is critical in modulating microglia polarization towards M2 phenotype as suggested by the evidence that IL-10 knockout mice showed worsened histological and functional outcomes and reduced M2 microglia/macrophage markers expression [175]. Moreover, IL-4, a potent M2-polarizing cytokine, released by sublethally ischemic neurons may offset the ischaemia-induced M1 polarization process. In fact, upon priming by endogenous or exogenously administered IL-4, microglia/macrophages participate in the removal of ischemic debris and produce trophic factors that may provide brain repair, probably through PPAR-γ activation [52, 176, 177]. In fact, PPAR-γ-mediated CD36 upregulation contributes to the modulation of microglia phenotype, triggering phagocytosis of apoptotic neutrophils, and thus promoting resolution of inflammation following ischemic stroke [178].

These studies underscore the crucial concept that a mere inhibition of microglia/macrophages activation and recruitment, disregarding their dualistic role, does not represent an efficient strategy to rescue ischemic brain injury. Conversely, recent work has validated the proof of concept that therapeutic interventions that promote M2 polarization provide neuroprotection in stroke models (Fig. 12.2).

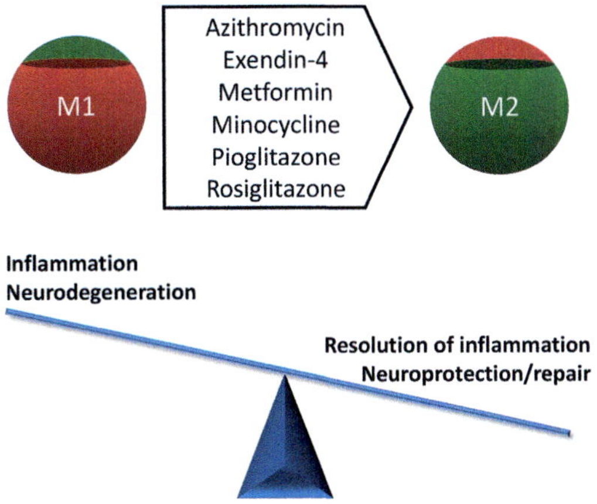

Fig. 12.2 Drugs that induce M1-to-M2 polarization shift promote resolution of inflammation and neurorepair in ischemic stroke models

The majority of drugs that display the peculiar ability to reprogram microglia/macrophages in the setting of stroke were (often unintentionally) identified through the drug repurposing approach [179]. Among these, the neuroprotective properties of certain antibacterial drugs have been demonstrated in ischemic stroke models. By inhibiting MMPs and microglia reactivity, minocycline exerts anti-inflammatory effects during the subacute phase of stroke. In addition, the protective effects of early minocycline treatment last for weeks after the insult due to ability of this drug to trigger alternative activation of microglia/macrophages and, thus, neurovascular remodelling [180, 181]. In fact, minocycline reduces typical markers of M1 activation, such as TNF-α and IL-1β, while it increases levels of M2 markers, namely TGF-β, IL-10 and Ym1, in the brain of adult spontaneously hypertensive rats subjected to transient MCAo [180]. Another antibacterial drug that shows potential to be repurposed for ischemic stroke, based on its immunomodulatory properties, is azithromycin. We have recently shown that a single administration of this macrolide antibiotic prevents BBB rupture, ameliorates neurological outcome and reduces brain infarct damage in mice subjected to transient MCAo [170]. Notably, neuroprotection exerted by azithromycin was associated with the induction of M2 polarization in microglia and in blood-borne macrophages [170].

Another class of drugs that exerts beneficial effects in stroke is represented by certain antidiabetics that, regardless of their classical molecular targets, promote M2-polarization of myeloid cells. These include the thiazolidinediones, metformin and Exendin-4, whose neuroprotective efficacy has been validated in stroke models. By promoting microglial polarization towards the M2 phenotype and oligodendrogenesis, rosiglitazone has been shown to improve long-term white matter integrity and to promote resolution of inflammation following a focal ischemic insult [178, 182]. Conversely, activation of PPAR-γ on haematogenous monocyte/macrophages has been suggested to underlie the ability of pioglitazone to trigger an anti-inflammatory profile in these myeloid cells, thus providing beneficial effects in focal cerebral ischaemia, including stabilization of neovessels in the infarct border and reduction of secondary intracerebral haemorrhage [183].

Interestingly, a recent study demonstrated the efficacy of pioglitazone in reducing the risk of future cardiovascular events in patients with a recent history of ischemic stroke or transient ischemic attack [184]. Activation of PPAR-γ has also been associated with the expression of M2-related genes induced in microglia by IL-4, thus providing an endogenous defence mechanism elicited by this cytokine to afford brain repair after stroke [52]. Interestingly, PPARγ, and more specifically its heterodimer with the retinoid X receptor (RXR), has also been implicated in N2 polarization of neutrophils. These immunomodulatory properties have been involved in the neuroprotection exerted by the PPAR-γ/RXR agonists bexarotene and rosiglitazone in rodent models of ischemic stroke [185, 186]. In mice subjected to MCAo, the antidiabetic drug metformin provides tissue repair and functional recovery by triggering M2 skewing of microglia/macrophages through activation of adenosine 5′-monophosphate-activated protein kinase [187]. Upregulation of M2 markers was also associated with the neuroprotective effects exerted by the glucagon-like receptor 1 agonist, Exendin-4, in young healthy and in aged diabetic/obese mice subjected to focal cerebral ischaemia [188]. Moreover, a recent study using stem cell transplantation to deliver pituitary adenylate cyclase-activating polypeptide (PACAP) into the mouse brain demonstrated that this neuroprotective peptide reduces brain infarct size and improves functional recovery by redirecting the microglial response towards a neuroprotective M2 phenotype [189].

In addition to these M2-inducing strategies, recent findings have also assessed the potential benefit of the administration of M2 cells to stroke animals or patients. Preliminary work in rodents has demonstrated that the efficacy of systemic administration of human umbilical cord blood mononuclear cells in reducing infarct volume and promoting functional recovery is mainly due to the presence of monocytes [190]. Although intravenous injection of M2 macrophages during the subacute stage of ischaemia proved to be ineffective in a rat model of transient MCAo [191], myeloid cell lineage appeared to be an important component within bone marrow-derived mononuclear cells that contribute to improved outcomes after stroke in rodent stroke models [192]. More importantly, intrathecal administration of autologous M2 cells in stroke patients was reported to be safe and resulted in a better neurological recovery, which appeared to be mediated by the immunomodulatory activity of M2 macrophages [193]. Nevertheless, further clinical work is needed to assess the feasibility and efficacy of these procedures in patients.

4 Modulation of Microglia/Macrophages by microRNAs

Increasing evidence demonstrates that microRNAs (miRNAs) play an important role in the pathophysiology of stroke through a series of mechanisms, including the modulation of microglia/macrophages activation and/or polarization [46, 194]. Peripheral blood mononuclear cells (PBMCs) from patients with acute ischemic stroke display decreased levels of miR-122, miR-148a, let-7i, miR-19a and miR-320d, while miR-363 and miR-487b levels were shown to be increased [195]. Importantly, all these miRNAs appear to play a role in inflammation-related pathways, such as TLR and NF-kB signalling. Let-7c-5p and miR-424 are significantly decreased in the blood of patients with ischemic stroke as well as in brain and plasma of experimental animals [75, 196]. The protective effects observed after overexpression of let-7c-5p in experimental stroke have been ascribed to its ability to suppress microglia activation [196]. Moreover, lentiviral overexpression of miR-424 inhibited neuronal apoptosis and microglia activation, through suppression of ionized calcium-binding adaptor molecule-1 expression and reduction of TNF-α production [75]. Also miR-203 suppresses ischaemia-induced microglia activation by targeting MyD88, an adapter protein implicated in most TLRs and interleukin-1 receptor (IL-1R) pathways. Through negative feedback, enforced expression of miR-203 inhibits downstream NF-kB signalling and microglia activation, thereby lessening neuronal injury [197].

This issue has been further extended by recent studies highlighting the association of specific miRNAs with different microglia/macrophages phenotypes [194, 198]. MiR-155 is widely recognized as a pro-inflammatory miRNA and has been associated with M1 phenotype in both microglia and macrophages [198–201]. After cerebral ischaemia, miR-155 is upregulated and underlies pro-inflammatory responses in microglia [201–203]. In fact, disruption of brain homeostatic balance (M1) is associated with elevated expression of miR-155, and reduction of miR-124, whereas its reposition (M2) and maintenance (M0) have been correlated with increased expression of miR-145 and miR-124, respectively [194].

5 Conclusions

The innate immune system plays a pivotal role in the progression of ischemic cerebral injury, with microglia and macrophages exerting either detrimental or reparative functions depending on their polarization towards specific phenotypes. The dualistic role exerted by these myeloid cells suggests that a mere inhibition of their activation/recruitment is not adequate to obtain significant neuroprotection, whereas a rational modulation of their polarization status, aimed at adjusting the M1/M2 ratio coherently with the spatio-temporal progression of injury, may represent a promising strategy [164]. This concept is supported by a number of studies demonstrating that the immunomodulatory properties of various repurposed drugs result in neuroprotection in experimental models of ischemic stroke [1] (Fig. 12.2). Nevertheless, most of our knowledge on polarization of innate immunity cells derives from rodent studies and, although some similarities have been postulated, the applicability to human microglia and monocytes/macrophages of the polarization features of rodent cells needs to be further investigated [89, 204, 205]. More importantly, planning future proof of concept studies in stroke patients is pivotal for validating the clinical efficacy of M1-to-M2 polarization shift in human.

References

1. Amantea D (2016) Editorial overview: neurosciences: brain and immunity: new targets for neuroprotection. Curr Opin Pharmacol 26:1–4
2. Chhabria K, Chakravarthy VS (2016) Low-dimensional models of "neuro-glio-vascular unit" for describing neural dynamics under normal and energy-starved conditions. Front Neurol 7:24
3. Ren K, Dubner R (2016) Activity-triggered tetrapartite neuron-glial interactions following peripheral injury. Curr Opin Pharmacol 26:16–25
4. Perea G, Navarrete M, Araque A (2009) Tripartite synapses: astrocytes process and control synaptic information. Trends Neurosci 32(8):421–431
5. Dityatev A, Rusakov DA (2011) Molecular signals of plasticity at the tetrapartite synapse. Curr Opin Neurobiol 21(2):353–359
6. Dirnagl U (2012) Pathobiology of injury after stroke: the neurovascular unit and beyond. Ann N Y Acad Sci 1268(1):21–25
7. Del Zoppo GJ (2010) The neurovascular unit in the setting of stroke. J Intern Med 267(2):156–171
8. Iadecola C, Anrather J (2011) The immunology of stroke: from mechanism to translation. Nat Med 17(7):796–808
9. Amantea D, Tassorelli C, Petrelli F, Certo M, Bezzi P, Micieli G et al (2014) Understanding the multifaceted role of inflammatory mediators in ischemic stroke. Curr Med Chem 21(18):2098–2117
10. Tang Y, Xu H, Du X, Lit L, Walker W, Lu A et al (2006) Gene expression in blood changes rapidly in neutrophils and monocytes after ischemic stroke in humans : a microarray study. J Cereb Blood Flow Metab 26(8):1089–1102
11. Barr TL, Conley Y, Ding J, Dillman A, Warach S, Singleton A et al (2010) Genomic biomarkers and cellular pathways of ischemic stroke by RNA gene expression profiling. Neurology 75(11):1009–1014
12. Oh SH, Kim OJ, Shin DA, Song J, Yoo H, Kim YK et al (2012) Alteration of immunologic responses on peripheral blood in the acute phase of ischemic stroke: blood genomic profiling study. J Neuroimmunol 249(1-2):60–65
13. Brooks S, Van Gilder R, Frisbee J, Barr T (2014) Genomics for the advancement of clinical translation in stroke. In: Micieli G, Amantea D (eds) Rational basis for clinical translation in stroke therapy. CRC, Boca Raton, pp 123–136
14. Fassbender K, Rossol S, Kammer T, Daffertshofer M, Wirth S, Dollman M et al (1994) Proinflammatory cytokines in serum of patients with acute cerebral ischemia: kinetics of secretion and relation to the extent of brain damage and outcome of disease. J Neurol Sci 122(2):135–139
15. Smith CJ, Emsley HCA, Gavin CM, Georgiou RF, Vail A, Barberan EM et al (2004) Peak plasma interleukin-6 and other peripheral markers of inflammation in the first week of ischaemic stroke correlate with brain infarct volume, stroke severity and long-term outcome. BMC Neurol 4:2
16. Whiteley W, Jackson C, Lewis S, Lowe G, Rumley A, Sandercock P et al (2009) Inflammatory markers and poor outcome after stroke: a prospective cohort study and systematic review of interleukin-6. PLoS Med 6(9), e1000145
17. Basic Kes V, Simundic A-M, Nikolac N, Topic E, Demarin V (2008) Pro-inflammatory and anti-inflammatory cytokines in acute ischemic stroke and their relation to early neurological deficit and stroke outcome. Clin Biochem 41(16-17):1330–1334
18. Chang L-T, Yuen C-M, Liou C-W, Lu C-H, Chang W-N, Youssef AA et al (2010) Link between interleukin-10 level and outcome after ischemic stroke. Neuroimmunomodulation 17(4):223–228
19. Amantea D, Micieli G, Tassorelli C, Cuartero MI, Ballesteros I, Certo M et al (2015) Rational modulation of the innate immune system for neuroprotection in ischemic stroke. Front Neurosci 9:147

20. Atkinson C, Zhu H, Qiao F, Varela JC, Yu J, Song H et al (2006) Complement-dependent P-selectin expression and injury following ischemic stroke. J Immunol 177:7266–7274
21. Nieswandt B, Pleines I, Bender M (2011) Platelet adhesion and activation mechanisms in arterial thrombosis and ischaemic stroke. J Thromb Haemost 1:92–104
22. Dodson RF, Chu LW, Welch KM, Achar VS (1977) Acute tissue response to cerebral ischemia in the gerbil: an ultrastructural study. J Neurol Sci 33:161–170
23. Sofroniew MV (2009) Molecular dissection of reactive astrogliosis and glial scar formation. Trends Neurosci 32:638–647
24. Shichita T, Ito M, Yoshimura A (2014) Post-ischemic inflammation regulates neural damage and protection. Front Cell Neurosci 8:319
25. Viscomi MT, Florenzano F, Latini L, Amantea D, Bernardi G, Molinari M (2008) Methylprednisolone treatment delays remote cell death after focal brain lesion. Neuroscience 154(4):1267–1282
26. Amantea D, Bagetta G, Tassorelli C, Mercuri NB, Corasaniti MT (2010) Identification of distinct cellular pools of interleukin-1beta during the evolution of the neuroinflammatory response induced by transient middle cerebral artery occlusion in the brain of rat. Brain Res 1313:259–269
27. del Zoppo GJ, Hallenbeck JM (2000) Advances in the vascular pathophysiology of ischemic stroke. Thromb Res 98:1276–1283
28. McColl B, Rothwell NJ, Allan SM (2008) Systemic inflammation alters the kinetics of cerebrovascular tight junction disruption after experimental stroke in mice. J Neurosci 28:9451–9462
29. Rosenberg GA, Estrada EY, Dencoff JE (1998) Matrix metalloproteinases and TIMPs are associated with blood–brain barrier opening after reperfusion in rat brain. Stroke 35:2576–2581
30. Rossi DJ, Brady JD, Mohr C (2007) Astrocyte metabolism and signaling during brain ischemia. Nat Neurosci 10:1377–1386
31. Stanimirovic DB, Ball R, Durkin JP (1997) Stimulation of glutamate uptake and Na, K-ATPase activity in rat astrocytes exposed to ischemia-like insults. Glia 19:123–134
32. del Zoppo GJ (2009) Inflammation and the neurovascular unit in the setting of focal cerebral ischemia. Neuroscience 158:972–982
33. Kinoshita A, Yamada K, Kohmura E, Hayakawa T (1990) Effect of astrocyte-derived factors on ischemic brain edema induced by rat MCA occlusion. APMIS 98:851–857
34. Shen LH, Li Y, Chopp M (2010) Astrocytic endogenous glial cell derived neurotrophic factor production is enhanced by bone marrow stromal cell transplantation in the ischemic boundary zone after stroke in adult rats. Glia 58:1074–1081
35. Miyamoto N, Maki T, Shindo A, Liang AC, Maeda M, Egawa N et al (2015) Astrocytes promote oligodendrogenesis after white matter damage via brain-derived neurotrophic factor. J Neurosci 35:14002–14008
36. Amantea D, Tassorelli C, Russo R, Petrelli F, Morrone LA, Bagetta G et al (2011) Neuroprotection by leptin in a rat model of permanent cerebral ischemia: effects on STAT3 phosphorylation in discrete cells of the brain. Cell Death Dis 2, e238
37. Rusnakova V, Honsa P, Dzamba D, Ståhlberg A, Kubista M, Anderova M (2013) Heterogeneity of astrocytes: from development to injury—single cell gene expression. PLoS One 8(8), e69734
38. Zamanian JL, Xu L, Foo LC, Nouri N, Zhou L, Giffard RG et al (2012) Genomic analysis of reactive astrogliosis. J Neurosci 32:6391–6410
39. Gliem M, Krammes K, Liaw L, Van Rooijen N, Hartung H, Jander S (2015) Macrophage-derived osteopontin induces reactive astrocyte polarization and promotes re-establishment of the blood brain barrier after ischemic stroke. Glia 63:2198–2207
40. Kang SS, Keasey MP, Cai J, Hagg T (2012) Loss of neuron-astroglial interaction rapidly induces protective CNTF expression after stroke in mice. J Neurosci 32:9277–9287
41. Ransohoff RM, Brown MA (2012) Innate immunity in the central nervous system. J Clin Invest 122(4):1164–1171

42. Ginhoux F, Lim S, Hoeffel G, Low D, Huber T (2013) Origin and differentiation of microglia. Front Cell Neurosci 7:45
43. Mildner A, Schmidt H, Nitsche M, Merkler D, Hanisch U-K, Mack M et al (2007) Microglia in the adult brain arise from Ly-6ChiCCR2+ monocytes only under defined host conditions. Nat Neurosci 10(12):1544–1553
44. Dénes A, Ferenczi S, Halász J, Környei Z, Kovács KJ (2008) Role of CX3CR1 (fractalkine receptor) in brain damage and inflammation induced by focal cerebral ischemia in mouse. J Cereb Blood Flow Metab 28(10):1707–1721
45. Dentesano G, Serratosa J, Tusell JM, Ramón P, Valente T, Saura J et al (2014) CD200R1 and CD200 expression are regulated by PPAR-γ in activated glial cells. Glia 62(6):982–998
46. Ma Y, Wang J, Wang Y, Yang G (2016) The biphasic function of microglia in ischemic stroke. Prog Neurobiol pii: S0301-0082(15)30070-8. doi:10.1016/j.pneurobio
47. Melani A, Amadio S, Gianfriddo M, Vannucchi MG, Volontè C, Bernardi G et al (2006) P2X7 receptor modulation on microglial cells and reduction of brain infarct caused by middle cerebral artery occlusion in rat. J Cereb Blood Flow Metab 26(7):974–982
48. Matsumoto H, Kumon Y, Watanabe H, Ohnishi T, Takahashi H, Imai Y et al (2007) Expression of CD200 by macrophage-like cells in ischemic core of rat brain after transient middle cerebral artery occlusion. Neurosci Lett 418(1):44–48
49. Lehnardt S, Massillon L, Follett P, Jensen FE, Ratan R, Rosenberg PA et al (2003) Activation of innate immunity in the CNS triggers neurodegeneration through a Toll-like receptor 4-dependent pathway. Proc Natl Acad Sci U S A 100(14):8514–8519
50. Kawabori M, Kacimi R, Kauppinen T, Calosing C, Kim JY, Hsieh CL et al (2015) Triggering receptor expressed on myeloid cells 2 (TREM2) deficiency attenuates phagocytic activities of microglia and exacerbates ischemic damage in experimental stroke. J Neurosci 35(8):3384–3396
51. Zarruk JG, Fernández-López D, García-Yébenes I, García-Gutiérrez MS, Vivancos J, Nombela F et al (2012) Cannabinoid type 2 receptor activation downregulates stroke-induced classic and alternative brain macrophage/microglial activation concomitant to neuroprotection. Stroke 43(1):211–219
52. Zhao X, Wang H, Sun G, Zhang J, Edwards NJ, Aronowski J (2015) Neuronal Interleukin-4 as a modulator of microglial pathways and ischemic brain damage. J Neurosci 35(32):11281–11291
53. Hu X, Li P, Guo Y, Wang H, Leak RK, Chen S (2012) Microglia/macrophage polarization dynamics reveal novel mechanism of injury expansion after focal cerebral ischemia. Stroke 43:3063–3070
54. da Fonseca ACC, Matias D, Garcia C, Amaral R, Geraldo LH, Freitas C et al (2014) The impact of microglial activation on blood–brain barrier in brain diseases. Front Cell Neurosci 8:362
55. Zhang Z, Chopp M, Powers C (1997) Temporal profile of microglial response following transient (2 h) middle cerebral artery occlusion. Brain Res 744(2):189–198
56. Stoll G, Jander S, Schroeter M (1998) Inflammation and glial responses in ischemic brain lesions. Prog Neurobiol 56(2):149–171
57. Denes A, Vidyasagar R, Feng J, Narvainen J, McColl BW, Kauppinen RA et al (2007) Proliferating resident microglia after focal cerebral ischaemia in mice. J Cereb Blood Flow Metab 27(12):1941–1953
58. Gerhard A, Schwarz J, Myers R, Wise R, Banati RB (2005) Evolution of microglial activation in patients after ischemic stroke: a [11C](R)-PK11195 PET study. Neuroimage 24(2):591–595
59. Price CJS, Wang D, Menon DK, Guadagno JV, Cleij M, Fryer T et al (2006) Intrinsic activated microglia map to the peri-infarct zone in the subacute phase of ischemic stroke. Stroke 37(7):1749–1753
60. Zinnhardt B, Viel T, Wachsmuth L, Vrachimis A, Wagner S, Breyholz H et al (2015) Multimodal imaging reveals temporal and spatial microglia and matrix metalloproteinase activity after experimental stroke. J Cereb Blood Flow Metab 35:1711–1721

61. Hailer NP (2008) Immunosuppression after traumatic or ischemic CNS damage: it is neuroprotective and illuminates the role of microglial cells. Prog Neurobiol 84:211–233
62. Schilling M, Besselmann M, Leonhard C, Mueller M, Ringelstein EBB, Kiefer R (2003) Microglial activation precedes and predominates over macrophage infiltration in transient focal cerebral ischemia: a study in green fluorescent protein transgenic bone marrow chimeric mice. Exp Neurol 183(1):25–33
63. Davis EJ, Foster TD, Thomas WE (1994) Cellular forms and functions of brain microglia. Brain Res Bull 34(1):73–78
64. Jung S, Schwartz M (2012) Non-identical twins—microglia and monocyte-derived macrophages in acute injury and autoimmune inflammation. Front Immunol 3:89
65. Schroeter M, Jander S, Huitinga I, Witte OW, Stoll G (1997) Phagocytic response in photochemically induced infarction of rat cerebral cortex. The role of resident microglia. Stroke 28:382–386
66. Perego C, Fumagalli S, De Simoni M-G (2011) Temporal pattern of expression and colocalization of microglia/macrophage phenotype markers following brain ischemic injury in mice. J Neuroinflammation 8:174
67. Kuric E, Ruscher K (2014) Dynamics of major histocompatibility complex class II-positive cells in the postischemic brain--influence of levodopa treatment. J Neuroinflammation 11:145
68. Barone FC, Arvin B, White RF, Miller A, Webb CL, Willette RN et al (1997) Tumor necrosis factor-alpha. A mediator of focal ischemic brain injury. Stroke 28(6):1233–1244
69. Lambertsen KL, Meldgaard M, Ladeby R, Finsen B (2005) A quantitative study of microglial-macrophage synthesis of tumor necrosis factor during acute and late focal cerebral ischemia in mice. J Cereb Blood Flow Metab 25(1):119–135
70. Rothwell N, Allan S, Toulmond S (1997) The role of interleukin 1 in acute neurodegeneration and stroke: pathophysiological and therapeutic implications. J Clin Invest 100(11):2648–2652
71. Chéret C, Gervais A, Lelli A, Colin C, Amar L, Ravassard P et al (2008) Neurotoxic activation of microglia is promoted by a nox1-dependent NADPH oxidase. J Neurosci 28(46):12039–12051
72. Green SP, Cairns B, Rae J, Errett-Baroncini C, Hongo JA, Erickson RW et al (2001) Induction of gp91-phox, a component of the phagocyte NADPH oxidase, in microglial cells during central nervous system inflammation. J Cereb Blood Flow Metab 21(4):374–384
73. del Zoppo GJ, Milner R, Mabuchi T, Hung S, Wang X, Berg GI et al (2007) Microglial activation and matrix protease generation during focal cerebral ischemia. Stroke 38(2 Suppl):646–651
74. Zhang L, Dong L-Y, Li Y-J, Hong Z, Wei W-S (2012) The microRNA miR-181c controls microglia-mediated neuronal apoptosis by suppressing tumor necrosis factor. J Neuroinflammation 9:211
75. Zhao H, Wang J, Gao L, Wang R, Liu X, Gao Z et al (2013) MiRNA-424 protects against permanent focal cerebral ischemia injury in mice involving suppressing microglia activation. Stroke 44(6):1706–1713
76. Fagan SC, Cronic LE, Hess DC (2011) Minocycline development for acute ischemic stroke. Transl Stroke Res 2(2):202–208
77. Jolivel V, Bicker F, Binamé F, Ploen R, Keller S, Gollan R et al (2015) Perivascular microglia promote blood vessel disintegration in the ischemic penumbra. Acta Neuropathol 129(2):279–295
78. Yenari MA, Xu L, Tang XN, Qiao Y, Giffard RG (2006) Microglia potentiate damage to blood–brain barrier constituents: improvement by minocycline in vivo and in vitro. Stroke 37(4):1087–1093
79. Egashira Y, Suzuki Y, Azuma Y, Takagi T, Mishiro K, Sugitani S et al (2013) The growth factor progranulin attenuates neuronal injury induced by cerebral ischemia-reperfusion through the suppression of neutrophil recruitment. J Neuroinflammation 10:105
80. Jackman K, Kahles T, Lane D, Garcia-Bonilla L, Abe T, Capone C et al (2013) Progranulin deficiency promotes post-ischemic blood–brain barrier disruption. J Neurosci 33(50):19579–19589

81. Kanazawa M, Kawamura K, Takahashi T, Miura M, Tanaka Y, Koyama M et al (2015) Multiple therapeutic effects of progranulin on experimental acute ischaemic stroke. Brain 138(Pt 7):1932–1948
82. Lalancette-Hébert M, Gowing G, Simard A, Weng YC, Kriz J (2007) Selective ablation of proliferating microglial cells exacerbates ischemic injury in the brain. J Neurosci 27(10):2596–2605
83. Imai F, Suzuki H, Oda J, Ninomiya T, Ono K, Sano H et al (2007) Neuroprotective effect of exogenous microglia in global brain ischemia. J Cereb Blood Flow Metab 27(3):488–500
84. Kitamura Y, Takata K, Inden M, Tsuchiya D, Yanagisawa D, Nakata J et al (2004) Intracerebroventricular injection of microglia protects against focal brain ischemia. J Pharmacol Sci 94(2):203–206
85. Neumann J, Sauerzweig S, Rönicke R, Gunzer F, Dinkel K, Ullrich O et al (2008) Microglia cells protect neurons by direct engulfment of invading neutrophil granulocytes: a new mechanism of CNS immune privilege. J Neurosci 28(23):5965–5975
86. O'Donnell SL, Frederick TJ, Krady JK, Vannucci SJ, Wood TL (2002) IGF-I and microglia/macrophage proliferation in the ischemic mouse brain. Glia 39(1):85–97
87. Tobin MK, Bonds JA, Minshall RD, Pelligrino DA, Testai FD, Lazarov O (2014) Neurogenesis and inflammation after ischemic stroke: what is known and where we go from here. J Cereb Blood Flow Metab 34:1573–1584
88. Fumagalli S, Perego C, Pischiutta F, Zanier ER, De Simoni M-G (2015) The ischemic environment drives microglia and macrophage function. Front Neurol 6(April):81
89. Italiani P, Boraschi D (2014) From monocytes to M1/M2 macrophages: phenotypical vs. functional differentiation. Front Immunol 5:514
90. Perry VH, Nicoll JAR, Holmes C (2010) Microglia in neurodegenerative disease. Nat Rev Neurol 6(4):193–201
91. Schoknecht K, David Y, Heinemann U (2015) The blood–brain barrier-gatekeeper to neuronal homeostasis: clinical implications in the setting of stroke. Semin Cell Dev Biol 38:35–42
92. Engelhardt S, Huang S-F, Patkar S, Gassmann M, Ogunshola OO (2015) Differential responses of blood–brain barrier associated cells to hypoxia and ischemia: a comparative study. Fluids Barriers CNS 12:4
93. Chu HXHX, Kim HAHA, Lee S, Moore JPJP, Chan CTCT, Vinh A et al (2014) Immune cell infiltration in malignant middle cerebral artery infarction: comparison with transient cerebral ischemia. J Cereb Blood Flow Metab 34(3):450–459
94. Chamorro Á, Meisel A, Planas AM, Urra X, van de Beek D, Veltkamp R (2012) The immunology of acute stroke. Nat Rev Neurol 8(7):401–410
95. Gelderblom M, Leypoldt F, Steinbach K, Behrens D, Choe C-U, Siler DA et al (2008) Temporal and spatial dynamics of cerebral immune cell accumulation in stroke. Stroke 40(5):1849–1857
96. Mracsko E, Javidi E, Na S-Y, Kahn A, Liesz A, Veltkamp R (2014) Leukocyte invasion of the brain after experimental intracerebral hemorrhage in mice. Stroke 45(7):2107–2114
97. del Zoppo GJ (2010) The neurovascular unit, matrix proteases, and innate inflammation. Ann N Y Acad Sci 1207:46–49
98. Rodrigues SF, Granger DN (2014) Leukocyte-mediated tissue injury in ischemic stroke. Curr Med Chem 21(19):2130–2137
99. Amantea D, Russo R, Gliozzi M, Fratto V, Berliocchi L, Bagetta G et al (2007) Early upregulation of matrix metalloproteinases following reperfusion triggers neuroinflammatory mediators in brain ischemia in rat. Int Rev Neurobiol 82:149–169
100. Ishikawa M, Zhang JH, Nanda A, Granger DN (2004) Inflammatory responses to ischemia and reperfusion in the cerebral microcirculation. Front Biosci 9:1339–1347
101. Gill D, Veltkamp R (2016) Dynamics of T cell responses after stroke. Curr Opin Pharmacol 26:26–32
102. Moore DF, Li H, Jeffries N, Wright V, Cooper RA, Elkahloun A et al (2005) Using peripheral blood mononuclear cells to determine a gene expression profile of acute ischemic stroke: a pilot investigation. Circulation 111(2):212–221

103. Asano S, Chantler PD, Barr T (2016) Gene expression profiling in stroke: relevance of blood-brain interaction. Curr Opin Pharmacol 26:80–86
104. Garcia JH, Liu KF, Yoshida Y, Lian J, Chen S, del Zoppo GJ (1994) Influx of leukocytes and platelets in an evolving brain infarct (Wistar rat). Am J Pathol 144(1):188–199
105. Clark RK, Lee EV, White RF, Jonak ZL, Feuerstein GZ, Barone FC (1994) Reperfusion following focal stroke hastens inflammation and resolution of ischemic injured tissue. Brain Res Bull 35(4):387–392
106. Buck BH, Liebeskind DS, Saver JL, Bang OY, Yun SW, Starkman S et al (2008) Early neutrophilia is associated with volume of ischemic tissue in acute stroke. Stroke 39(2):355–360
107. Akopov S, Simonian N, Grigorian G (1996) Dynamics of polymorphonuclear leukocyte accumulation in acute cerebral infarction and their correlation with brain tissue damage. Stroke 27:1739–1743
108. Atochin DN, Fisher D, Demchenko IT, Thom SR (2000) Neutrophil sequestration and the effect of hyperbaric oxygen in a rat model of temporary middle cerebral artery occlusion. Undersea Hyperb Med 27(4):185–190
109. Connolly ES, Winfree CJ, Springer TA, Naka Y, Liao H, Du Yan S et al (1996) Cerebral protection in homozygous null ICAM-1 mice after middle cerebral artery occlusion. Role of neutrophil adhesion in the pathogenesis of stroke. J Clin Invest 97(1):209–216
110. Matsuo Y, Onodera H, Shiga Y, Nakamura M, Ninomiya M, Kihara T et al (1994) Correlation between myeloperoxidase-quantified neutrophil accumulation and ischemic brain injury in the rat. Effects of neutrophil depletion. Stroke 25(7):1469–1475
111. Fumagalli S, Perego C, Ortolano F, De Simoni M-G (2013) CX3CR1 deficiency induces an early protective inflammatory environment in ischemic mice. Glia 61(6):827–842
112. Tang Z, Gan Y, Liu Q, Yin J-X, Liu Q, Shi J et al (2014) CX3CR1 deficiency suppresses activation and neurotoxicity of microglia/macrophage in experimental ischemic stroke. J Neuroinflammation 11:26
113. Schilling M, Strecker J-K, Ringelstein EB, Schäbitz W-R, Kiefer R (2009) The role of CC chemokine receptor 2 on microglia activation and blood-borne cell recruitment after transient focal cerebral ischemia in mice. Brain Res 1289:79–84
114. Schilling M, Strecker J-K, Schäbitz W-R, Ringelstein EB, Kiefer R (2009) Effects of monocyte chemoattractant protein 1 on blood-borne cell recruitment after transient focal cerebral ischemia in mice. Neuroscience 161(3):806–812
115. Hughes PM, Allegrini PR, Rudin M, Perry VH, Mir AK, Wiessner C (2002) Monocyte chemoattractant protein-1 deficiency is protective in a murine stroke model. J Cereb Blood Flow Metab 22(3):308–317
116. Dimitrijevic OB, Stamatovic SM, Keep RF, Andjelkovic AV (2007) Absence of the chemokine receptor CCR2 protects against cerebral ischemia/reperfusion injury in mice. Stroke 38(4):1345–1353
117. Gliem M, Mausberg AK, Lee J-I, Simiantonakis I, van Rooijen N, Hartung H-P et al (2012) Macrophages prevent hemorrhagic infarct transformation in murine stroke models. Ann Neurol 71(6):743–752
118. Chu HX, Broughton BRS, Kim HA, Lee S, Drummond GR, Sobey CG (2015) Evidence that Ly6C(hi) monocytes are protective in acute ischemic stroke by promoting M2 macrophage polarization. Stroke 46(7):1929–1937
119. Benakis C, Garcia-Bonilla L, Iadecola C, Anrather J (2014) The role of microglia and myeloid immune cells in acute cerebral ischemia. Front Cell Neurosci 8:461
120. Beschorner R, Schluesener HJ, Gözalan F, Meyermann R, Schwab JM (2002) Infiltrating CD14+ monocytes and expression of CD14 by activated parenchymal microglia/macrophages contribute to the pool of CD14+ cells in ischemic brain lesions. J Neuroimmunol 126(1-2):107–115
121. Mena H, Cadavid D, Rushing EJ (2004) Human cerebral infarct: a proposed histopathologic classification based on 137 cases. Acta Neuropathol 108(6):524–530
122. Klehmet J, Harms H, Richter M, Prass K, Volk HD, Dirnagl U et al (2009) Stroke-induced immunodepression and post-stroke infections: lessons from the preventive antibacterial therapy in stroke trial. Neuroscience 158(3):1184–1193

123. Chamorro A, Amaro S, Vargas M, Obach V, Cervera A, Torres F et al (2006) Interleukin 10, monocytes and increased risk of early infection in ischaemic stroke. J Neurol Neurosurg Psychiatry 77(11):1279–1281
124. Haeusler KG, Schmidt WUH, Föhring F, Meisel C, Helms T, Jungehulsing GJ et al (2008) Cellular immunodepression preceding infectious complications after acute ischemic stroke in humans. Cerebrovasc Dis 25(1-2):50–58
125. Urra X, Villamor N, Amaro S, Gómez-Choco M, Obach V, Oleaga L et al (2009) Monocyte subtypes predict clinical course and prognosis in human stroke. J Cereb Blood Flow Metab 29(5):994–1002
126. Jin R, Zhu X, Liu L, Nanda A, Granger DN, Li G (2013) Simvastatin attenuates stroke-induced splenic atrophy and lung susceptibility to spontaneous bacterial infection in mice. Stroke 44(4):1135–1143
127. Ajmo CT, Vernon DOL, Collier L, Hall AA, Garbuzova-Davis S, Willing A et al (2008) The spleen contributes to stroke-induced neurodegeneration. J Neurosci Res 86(10):2227–2234
128. Liesz A, Hagmann S, Zschoche C, Adamek J, Zhou W, Sun L et al (2009) The spectrum of systemic immune alterations after murine focal ischemia: immunodepression versus immunomodulation. Stroke 40(8):2849–2858
129. Offner H, Subramanian S, Parker SM, Wang C, Afentoulis ME, Lewis A et al (2006) Splenic atrophy in experimental stroke is accompanied by increased regulatory T cells and circulating macrophages. J Immunol 176(11):6523–6531
130. Seifert HA, Hall AA, Chapman CB, Collier LA, Willing AE, Pennypacker KR (2012) A transient decrease in spleen size following stroke corresponds to splenocyte release into systemic circulation. J Neuroimmune Pharmacol 7(4):1017–1024
131. Kim E, Yang J, Beltran CD, Cho S (2014) Role of spleen-derived monocytes/macrophages in acute ischemic brain injury. J Cereb Blood Flow Metab 34(8):1411–1419
132. Bao Y, Kim E, Bhosle S, Mehta H, Cho S (2010) A role for spleen monocytes in post-ischemic brain inflammation and injury. J Neuroinflammation 7:92
133. Enzmann G, Mysiorek C, Gorina R, Cheng Y-J, Ghavampour S, Hannocks M-J et al (2013) The neurovascular unit as a selective barrier to polymorphonuclear granulocyte (PMN) infiltration into the brain after ischemic injury. Acta Neuropathol 125(3):395–412
134. Justicia C, Panés J, Solé S, Cervera A, Deulofeu R, Chamorro A et al (2003) Neutrophil infiltration increases matrix metalloproteinase-9 in the ischemic brain after occlusion/reperfusion of the middle cerebral artery in rats. J Cereb Blood Flow Metab 23(12):1430–1440
135. Gidday JM, Gasche YG, Copin J-C, Shah AR, Perez RS, Shapiro SD et al (2005) Leukocyte-derived matrix metalloproteinase-9 mediates blood–brain barrier breakdown and is proinflammatory after transient focal cerebral ischemia. Am J Physiol Heart Circ Physiol 289(2):H558–H568
136. Bao Dang Q, Lapergue B, Tran-Dinh A, Diallo D, Moreno J-A, Mazighi M et al (2013) High-density lipoproteins limit neutrophil-induced damage to the blood–brain barrier in vitro. J Cereb Blood Flow Metab 33(4):575–582
137. Granger DN, Benoit JN, Suzuki M, Grisham MB (1989) Leukocyte adherence to venular endothelium during ischemia-reperfusion. Am J Physiol 257(5 Pt 1):G683–G688
138. Mori E, del Zoppo GJ, Chambers JD, Copeland BR, Arfors KE (1992) Inhibition of polymorphonuclear leukocyte adherence suppresses no-reflow after focal cerebral ischemia in baboons. Stroke 23(5):712–718
139. Sreeramkumar V, Adrover JM, Ballesteros I, Cuartero MI, Rossaint J, Bilbao I et al (2014) Neutrophils scan for activated platelets to initiate inflammation. Science 346(6214):1234–1238
140. Graeber MB, Li W, Rodriguez ML (2011) Role of microglia in CNS inflammation. FEBS Lett 585(23):3798–3805
141. London A, Cohen M, Schwartz M (2013) Microglia and monocyte-derived macrophages : functionally distinct populations that act in concert in CNS plasticity and repair. Front Cell Neurosci 7(34):1–10
142. Thiel A, Heiss WD (2011) Imaging of microglia activation in stroke. Stroke 42:507–512

143. Schilling M, Besselmann M, Müller M, Strecker JK, Ringelstein EB, Kiefer R (2005) Predominant phagocytic activity of resident microglia over hematogenous macrophages following transient focal cerebral ischemia: an investigation using green fluorescent protein transgenic bone marrow chimeric mice. Exp Neurol 196(2):290–297
144. Ritzel RM, Patel AR, Grenier JM, Crapser J, Verma R, Jellison ER et al (2015) Functional differences between microglia and monocytes after ischemic stroke. J Neuroinflammation 12:106
145. Kleinschnitz C, Bendszus M, Frank M, Solymosi L, Toyka KV, Stoll G (2003) In vivo monitoring of macrophage infiltration in experimental ischemic brain lesions by magnetic resonance imaging. J Cereb Blood Flow Metab 23(11):1356–1361
146. Rausch M, Sauter A, Fröhlich J, Neubacher U, Radü EW, Rudin M (2001) Dynamic patterns of USPIO enhancement can be observed in macrophages after ischemic brain damage. Magn Reson Med 46(5):1018–1022
147. Stubbe T, Ebner F, Richter D, Engel O, Klehmet J, Royl G et al (2013) Regulatory T cells accumulate and proliferate in the ischemic hemisphere for up to 30 days after MCAO. J Cereb Blood Flow Metab 33:37–47
148. Sedgwick JD, Ford AL, Foulcher E, Airriess R (1998) Central nervous system microglial cell activation and proliferation follows direct interaction with tissue-infiltrating T cell blasts. J Immunol 160:5320–5330
149. Campanella M, Sciorati C, Tarozzo G, Beltramo M (2002) Flow cytometric analysis of inflammatory cells in ischemic rat brain. Stroke 33:586–592
150. Gliem M, Schwaninger M, Jander S (2016) Protective features of peripheral monocytes/macrophages in stroke. Biochim Biophys Acta 1862(3):329–338
151. Nahrendorf M, Pittet MJ, Swirski FK (2010) Monocytes: protagonists of infarct inflammation and repair after myocardial infarction. Circulation 121(22):2437–2445
152. Chu HX, Kim HA, Lee S, Broughton BRS, Drummond GR, Sobey CG (2016) Evidence of CCR2-independent transmigration of Ly6C(hi) monocytes into the brain after permanent cerebral ischemia in mice. Brain Res 1637:118–127
153. Miró-Mur F, Pérez-de-Puig I, Ferrer-Ferrer M, Urra X, Justicia C, Chamorro A et al (2016) Immature monocytes recruited to the ischemic mouse brain differentiate into macrophages with features of alternative activation. Brain Behav Immun 53:18–33
154. Michaud J-P, Pimentel-Coelho PM, Tremblay Y, Rivest S (2014) The impact of Ly6Clow monocytes after cerebral hypoxia-ischemia in adult mice. J Cereb Blood Flow Metab 34(7):e1–e9
155. Ginhoux F, Guilliams M (2016) Tissue-resident macrophage ontogeny and homeostasis. Immunity 44(3):439–449
156. Auffray C, Fogg D, Garfa M, Elain G, Join-Lambert O, Kayal S et al (2007) Monitoring of blood vessels and tissues by a population of monocytes with patrolling behavior. Science 317(5838):666–670
157. Thomas G, Tacke R, Hedrick CC, Hanna RN (2015) Nonclassical patrolling monocyte function in the vasculature. Arterioscler Thromb Vasc Biol 35(6):1306–1316
158. Zhou W, Liesz A, Bauer H, Sommer C, Lahrmann B, Valous N et al (2013) Postischemic brain infiltration of leukocyte subpopulations differs among murine permanent and transient focal cerebral ischemia models. Brain Pathol 23:34–44
159. Neher JJ, Emmrich JV, Fricker M, Mander PK, Théry C, Brown GC (2013) Phagocytosis executes delayed neuronal death after focal brain ischemia. Proc Natl Acad Sci U S A 110(43):E4098–E4107
160. Jander S, Schroeter M, Saleh A (2007) Imaging inflammation in acute brain ischemia. Stroke 38(2 Suppl):642–645
161. Frieler RA, Meng H, Duan SZ, Berger S, Schütz G, He Y, Xi G, Wang MM, Mortensen RM (2011) Myeloid-specific deletion of the mineralocorticoid receptor reduces infarct volume and alters inflammation during cerebral ischemia. Stroke 42(1):179–185
162. Xu Y, Qian L, Zong G, Ma K, Zhu X, Zhang H et al (2012) Class A scavenger receptor promotes cerebral ischemic injury by pivoting microglia/macrophage polarization. Neuroscience 218:35–48

163. Ballesteros I, Cuartero MI, de la Parra J, Perez-Ruiz A, Hurtado O, Lizasoain I et al (2014) Polarization of macrophages/microglia toward an M2 phenotype as a therapeutic strategy for stroke treatment. In: Micieli G, Amantea D (eds) Rational basis for clinical translation in stroke therapy. CRC, Boca Raton, pp 381–392
164. Amantea D (2016) Polarizing the immune system towards neuroprotection in brain ischemia. Neural Regen Res 11(1):81–82
165. Mantovani A, Biswas SK, Galdiero MR, Sica A, Locati M (2013) Macrophage plasticity and polarization in tissue repair and remodelling. J Pathol 229(2):176–185
166. Schultze JL (2016) Reprogramming of macrophages—new opportunities for therapeutic targeting. Curr Opin Pharmacol 26:10–15
167. Colin S, Chinetti-Gbaguidi G, Staels B (2014) Macrophage phenotypes in atherosclerosis. Immunol Rev 262(1):153–166
168. Martinez FO, Gordon S (2014) The M1 and M2 paradigm of macrophage activation: time for reassessment. F1000Prime Rep 6:13
169. Hu X, Leak RK, Shi Y, Suenaga J, Gao Y, Zheng P et al (2015) Microglial and macrophage polarization—new prospects for brain repair. Nat Rev Neurol 11(1):56–64
170. Amantea D, Certo M, Petrelli F, Tassorelli C, Micieli G, Corasaniti MT et al (2016) Azithromycin protects mice against ischemic stroke injury by promoting macrophage transition towards M2 phenotype. Exp Neurol 275:116–125
171. Suenaga J, Hu X, Pu H, Shi Y, Hassan SH, Xu M et al (2015) White matter injury and microglia/macrophage polarization are strongly linked with age-related long-term deficits in neurological function after stroke. Exp Neurol 272:109–119
172. Han Z, Li L, Wang L, Degos V, Maze M, Su H (2014) Alpha-7 nicotinic acetylcholine receptor agonist treatment reduces neuroinflammation, oxidative stress, and brain injury in mice with ischemic stroke and bone fracture. J Neurochem 131(4):498–508
173. Han Z, Shen F, He Y, Degos V, Camus M, Maze M et al (2014) Activation of α-7 nicotinic acetylcholine receptor reduces ischemic stroke injury through reduction of pro-inflammatory macrophages and oxidative stress. PLoS One 9(8), e105711
174. Rahman M, Muhammad S, Khan MA, Chen H, Ridder DA, Müller-Fielitz H et al (2014) The β-hydroxybutyrate receptor HCA2 activates a neuroprotective subset of macrophages. Nat Commun 5:3944
175. Pérez-de Puig I, Miró F, Salas-Perdomo A, Bonfill-Teixidor E, Ferrer-Ferrer M, Márquez-Kisinousky L et al (2013) IL-10 deficiency exacerbates the brain inflammatory response to permanent ischemia without preventing resolution of the lesion. J Cereb Blood Flow Metab 33(12):1955–1966
176. Xiong X, Barreto GE, Xu L, Ouyang YB, Xie X, Giffard RG (2011) Increased brain injury and worsened neurological outcome in interleukin-4 knockout mice after transient focal cerebral ischemia. Stroke 42(7):2026–2032
177. Liu X, Liu J, Zhao S, Zhang H, Cai W, Cai M et al (2016) Interleukin-4 is essential for microglia/macrophage M2 polarization and long-term recovery after cerebral ischemia. Stroke 47(2):498–504
178. Ballesteros I, Cuartero MI, Pradillo JM, de la Parra J, Pérez-Ruiz A, Corbí A et al (2014) Rosiglitazone-induced CD36 up-regulation resolves inflammation by PPARγ and 5-LO-dependent pathways. J Leukoc Biol 95(4):587–598
179. Amantea D, Bagetta G (2016) Drug repurposing for immune modulation in acute ischemic stroke. Curr Opin Pharmacol 124–130
180. Yang Y, Salayandia VM, Thompson JF, Yang LY, Estrada EY, Yang Y (2015) Attenuation of acute stroke injury in rat brain by minocycline promotes blood–brain barrier remodeling and alternative microglia/macrophage activation during recovery. J Neuroinflammation 12:26
181. Liao TV, Forehand CC, Hess DC, Fagan SC (2013) Minocycline repurposing in critical illness: focus on stroke. Curr Top Med Chem 13(18):2283–2290
182. Han L, Cai W, Mao L, Liu J, Li P, Leak RK et al (2015) Rosiglitazone promotes white matter integrity and long-term functional recovery after focal cerebral ischemia. Stroke 46(9): 2628–2636

183. Gliem M, Klotz L, van Rooijen N, Hartung H-P, Jander S (2015) Hyperglycemia and PPARγ antagonistically influence macrophage polarization and infarct healing after ischemic stroke. Stroke 46(10):2935–2942
184. Kernan WN, Viscoli CM, Furie KL, Young LH, Inzucchi SE, Gorman M et al (2016) Pioglitazone after ischemic stroke or transient ischemic attack. N Engl J Med 374(14): 1321–1331
185. Certo M, Endo Y, Ohta K, Sakurada S, Bagetta G, Amantea D (2015) Activation of RXR/PPARγ underlies neuroprotection by bexarotene in ischemic stroke. Pharmacol Res 102:298–307
186. Cuartero MI, Ballesteros I, Moraga A, Nombela F, Vivancos J, Hamilton JA et al (2013) N2 neutrophils, novel players in brain inflammation after stroke: modulation by the PPARγ agonist rosiglitazone. Stroke 44(12):3498–3508
187. Jin Q, Cheng J, Liu Y, Wu J, Wang X, Wei S et al (2014) Improvement of functional recovery by chronic metformin treatment is associated with enhanced alternative activation of microglia/macrophages and increased angiogenesis and neurogenesis following experimental stroke. Brain Behav Immun 40:131–142
188. Darsalia V, Hua S, Larsson M, Mallard C, Nathanson D, Nyström T et al (2014) Exendin-4 reduces ischemic brain injury in normal and aged type 2 diabetic mice and promotes microglial M2 polarization. PLoS One 9(8), e103114
189. Brifault C, Gras M, Liot D, May V, Vaudry D, Wurtz O (2015) Delayed pituitary adenylate cyclase-activating polypeptide delivery after brain stroke improves functional recovery by inducing m2 microglia/macrophage polarization. Stroke 46(2):520–528
190. Womble TA, Green S, Shahaduzzaman M, Grieco J, Sanberg PR, Pennypacker KR et al (2014) Monocytes are essential for the neuroprotective effect of human cord blood cells following middle cerebral artery occlusion in rat. Mol Cell Neurosci 59:76–84
191. Desestret V, Riou A, Chauveau F, Cho T-H, Devillard E, Marinescu M et al (2013) In vitro and in vivo models of cerebral ischemia show discrepancy in therapeutic effects of M2 macrophages. PLoS One 8(6), e67063
192. Yang B, Parsha K, Schaar K, Xi X, Aronowski J, Savitz SI (2016) Various cell populations within the mononuclear fraction of bone marrow contribute to the beneficial effects of autologous bone marrow cell therapy in a rodent stroke model. Transl Stroke Res 7(4):322–330
193. Chernykh ER, Shevela EY, Starostina NM, Morozov SA, Davydova MN, Menyaeva EV et al (2016) Safety and therapeutic potential of M2-macrophages in stroke treatment. Cell Transplant 25(8):1461–1471
194. Cardoso AL, Guedes JR, de Lima MCP (2016) Role of microRNAs in the regulation of innate immune cells under neuroinflammatory conditions. Curr Opin Pharmacol 26:1–9
195. Jickling GC, Ander BP, Zhan X, Noblett D, Stamova B, Liu D (2014) mRNA expression in peripheral blood cells following acute ischemic stroke and their predicted gene targets. PLoS One 9(6):e99283
196. Ni J, Wang X, Chen S, Liu H, Wang Y, Xu X et al (2015) MicroRNA let-7c-5p protects against cerebral ischemia injury via mechanisms involving the inhibition of microglia activation. Brain Behav Immun 49:75–85
197. Yang Z, Zhong L, Zhong S, Xian R, Yuan B (2015) miR-203 protects microglia mediated brain injury by regulating inflammatory responses via feedback to MyD88 in ischemia. Mol Immunol 65(2):293–301
198. Cardoso AL, Guedes JR, Pereira de Almeida L, Pedroso de Lima MC (2012) miR-155 modulates microglia-mediated immune response by down-regulating SOCS-1 and promoting cytokine and nitric oxide production. Immunology 135(1):73–88
199. O'Connell RM, Chaudhuri AA, Rao DS, Baltimore D (2009) Inositol phosphatase SHIP1 is a primary target of miR-155. Proc Natl Acad Sci U S A 106(17):7113–7118
200. Guedes J, Cardoso ALC, Pedroso de Lima MC (2013) Involvement of microRNA in microglia-mediated immune response. Clin Dev Immunol 2013:186872
201. Su W, Hopkins S, Nesser NK, Sopher B, Silvestroni A, Ammanuel S et al (2014) The p53 transcription factor modulates microglia behavior through microRNA-dependent regulation of c-Maf. J Immunol 192(1):358–366

202. Liu D-Z, Tian Y, Ander BP, Xu H, Stamova BS, Zhan X et al (2010) Brain and blood microRNA expression profiling of ischemic stroke, intracerebral hemorrhage, and kainate seizures. J Cereb Blood Flow Metab 30(1):92–101
203. Wen Y, Zhang X, Dong L, Zhao J, Zhang C, Zhu C (2015) Acetylbritannilactone modulates microRNA-155-mediated inflammatory response in ischemic cerebral tissues. Mol Med 21:197–209
204. Smith AM, Dragunow M (2014) The human side of microglia. Trends Neurosci 37(3):125–135
205. Durafourt BA, Moore CS, Zammit DA, Johnson TA, Zaguia F, Guiot M-C et al (2012) Comparison of polarization properties of human adult microglia and blood-derived macrophages. Glia 60(5):717–727

Chapter 13
Endoplasmic Reticulum Stress: An Opportunity for Neuroprotective Strategies After Stroke

Morgane Louessard, Eloise Lemarchand, Carine Ali, Denis Vivien, and Benoit Denis Roussel

Abstract Worldwide, ischaemic stroke is a major cause of death and the leading cause of acquired disability. To date, pharmacological thrombolysis with tissue-type plasminogen activator (alone or with mechanical thrombectomy) remains the gold standard acute treatment for ischaemic stroke. Many strategies of neuroprotection have been tested in the past, but none has ever succeeded in clinical trials. Accumulating evidence suggests that stroke activates the endoplasmic reticulum (ER) stress response, which likely deleteriously contribute to stroke damages. However, this remains to be clearly established. In the present chapter, we provide a snapshot on ER stress in ischaemic stroke, and intend to show that knowledge on ER stress in ischaemic context affecting other organs than the brain can provide interesting therapeutic avenues for neuroprotection.

Keywords ER stress • Ischaemia • Neuroprotection

Abbreviations

ATF	Activating transcription factor
ATP	Adenosine triphosphate
Bip	Binding immunoglobulin protein
CHOP	CAAT/enhancer binding protein homologous transcription factor
CReP	Constitutive repressor of eIF2α phosphorylation
EDEM	Endoplasmic reticulum degradation enhancing alpha mannosidase-like protein

M. Louessard • E. Lemarchand • C. Ali • D. Vivien • B.D. Roussel (✉)
Normandie Universitė, UNICAEN, INSERM UMR-S U919 "Serine Proteases and Pathophysiology of the Neurovascular Unit",
GIP Cyceron, Boulevard Becquerel, Caen 14074, France
e-mail: broussel@cyceron.fr; vivien@cyceron.fr

P.A. Lapchak, J.H. Zhang (eds.), *Neuroprotective Therapy for Stroke and Ischemic Disease*, Springer Series in Translational Stroke Research,
DOI 10.1007/978-3-319-45345-3_13

eIF2	Eukaryotic initiation factor 2
ER	Endoplasmic reticulum
ERAD	Endoplasmic reticulum associated degradation
Ero1	ER oxidoreductin I
GADD34	Growth arrest and DNA damage-inducible protein 34
GLS	Golgi localization signal
Grp	Glucose related protein
Hsp	Heat shock protein
I/R	Ischaemia/reperfusion
IP3R	Inositol-1,4,5-trisphosphate receptor
IRE1	Inositol-requiring protein 1
KEAP1	Kelch-like ECH-associated protein 1
NE	Nuclear envelope
OGD	Oxygen and glucose deprivation
PERK	Protein kinase RNA-like endoplasmic reticulum kinase
PP1	Protein phosphatase 1
RER	Rough endoplasmic reticulum
RIDD	IRE1-dependent decay of mRNA
ROS	Reactive oxygen species
RyR	Ryanodine receptor
S1P	Site-1 protease
S2P	Site-2 protease
SER	Smooth endoplasmic reticulum
SERCA	Sarco endoplasmic reticulum Ca^{2+}-transport ATPase
tPA	Tissue-type plasminogen activator
UPR	Unfolded protein response
UPRE	Unfolded protein response element
XBP1	X-box binding protein 1

1 The ER Structure and Functions

1.1 General Information

The endoplasmic reticulum (ER) is a major organelle in eukaryotic cells, consisting of a single membrane system with a continuous intraluminal space. The ER lumen occupies 10 % of the total cell volume while its membrane represents almost half of the total cell membranes [1]. It is thus a compartment completely separated from the cytosol, allowing the ER to maintain essential conditions (proteins for the molecular machinery, high concentration of calcium ions and oxidizing condition) for its various functions. The ER is involved in synthesis, folding and post-translational modification of proteins and also in protein transport and calcium homeostasis (the ER is one of the main calcium reserve in cells) [2, 3].

The ER is traditionally divided into three parts: the nuclear envelope (NE), the rough ER (RER) and the smooth ER (SER). By using electron microscopy, RER appears granular and tubular due to the attachment of ribosomes on its surface. It is found in all eukaryotic cells that synthesize and fold proteins [4]. Contrary to the RER, the SER does not contain ribosomes and has a more expanded and tortuous shape. The SER is the place of lipid synthesis, cell detoxification and calcium storage [4]. The NE surrounds the entire nucleus. It is an essential part of the ER as it defines the border between the nucleus and the cytoplasm. This physical limit is a fundamental characteristic of eukaryotic cells which can regulate gene expression at different levels [5]. The ER can also be divided into specialized regions depending on its morphology, its protein composition or its proximity to other organelles [1].

Even if all cytosolic proteins do not need to traffic through the ER, it is still an essential compartment for the maturation and the proper folding of the majority of proteins [1]. The oxidative environment of the ER lumen is optimal for the various mechanisms required for protein maturation, including the formation of disulphide bonds and their folding by chaperones proteins. Many cellular damages are linked with chaperone dysfunctions, representing then an essential target for neuroprotection (see part IV). Under physiological conditions, chaperones are involved in the synthesis, folding and degradation of newly synthesized proteins and can be divided into three main groups:

1.1.1 Chaperones of the Hsp70 (Heat Shock Protein 70) Family

They include Grp78 (Glucose Regulated Protein 78, also called Bip for Binding immunoglobulin protein) as its most abundant member, and Grp170 (Glucose Regulated Protein 170) [6]. Grp78 binds to proteins during their folding to prevent their aggregation and participates in the ERAD (Endoplasmic-reticulum-associated degradation) mechanism by directing misfolded proteins to the proteasome by retro-translocation via the translocon [7, 8]. Moreover, Grp78 is involved in calcium homeostasis and is a major player in the induction of the UPR (unfolded protein response; see part II) [2]. The roles of Grp170 are less known, but it seems to be involved in protein maturation during ER stress (such as hypoxia and glucose deprivation).

1.1.2 Grp94

It is the only known representative chaperone of the Hsp90 family. Grp94 roles overlap in many ways with those of Grp78 and both chaperones likely work in pairs [9]. While Grp78 binds to unfolded proteins since their entrance into the ER, Grp94 operates on partially folded proteins released by Grp78. Interestingly, Grp94 primarily promotes the biosynthesis of specific secreted or membrane proteins, such as immunoglobulins [7] or Toll-like receptors [10].

1.1.3 Calnexin/Calreticulin

They are lectin site chaperones that bind to proteins during the whole process of folding [11, 12]. Calnexin and calreticulin bind to all newly synthesized N-glycosylated proteins, allowing misfolded proteins to stay longer in the ER, leaving time to reach a correct folding [13]. Another important feature of these chaperones is the catching of calcium from the ER in very large quantities, hence their name.

The ER has a quality control machinery of newly synthesized proteins that discriminates correctly folded from misfolded proteins and allows them to reach a stable conformation or to be degraded. The main pathway for misfolded proteins destruction is called ERAD. ERAD includes two groups of proteins: chaperone proteins and glycan modifying enzymes (EDEM proteins (ER-Degradation-Enhancing alpha-Mannosidase-like protein)). Successive enzymatic reactions (digestion and addition of sugars) on the hydrocarbon chain allow the control of protein folding. Correctly folded glycoproteins are then transported to the Golgi apparatus. If the right protein structure cannot be achieved and maintained, EDEM proteins interact with these misfolded proteins and translocate them to the cytosol where they are poly-ubiquitinated and degraded by the proteasome [14, 15].

The ER is the main calcium storage site in many cells, playing then an important role in the regulation of calcium homeostasis. While the concentration of free calcium in the cytoplasm is 1 nM, the ER-calcium concentration is between 100 and 800 μM [16]. This difference of concentration is regulated by the presence of channel receptors and pumps on the ER surface. There are two types of channel receptors that allow calcium release into the cytosol:

1.1.4 IP3R (INOSITOL-1,4,5-Trisphosphate Receptor) Channels

They belong to a superfamily of ionic channels with 6 transmembrane domains. There are 3 isoforms in mammals which have different affinities for IP3 (inositol-1,4,5-trisphosphate) [17]. Even if IP3 is the main activator, calcium can also activate IP3R with a biphasic manner: activation at low concentrations of cytosolic calcium versus inhibition at high calcium concentrations.

1.1.5 RyR Channels (Ryanodine Receptor)

Three isoforms of RyR have been identified in mammals (RyR1, RyR2 and RyR3), sharing 65 % homology. Numerous agents including calcium and ATP modulate the activity of RyR. Calcium exerts a biphasic effect on its activity, like for IP3R [18].

To ensure a high concentration of calcium within the ER, the SERCA pump (Sarco endoplasmic reticulum Ca^{2+}-transport ATPase), through an active ATP-dependent process, withdraws calcium from the cytosol to the ER [19].

1.2 The ER in Neurones

Neurones have a complex ER that extends all over the cell and formed by both RER and SER, depending on its localization. The RER is present from the nuclear outer membrane to the soma and is involved in protein synthesis. In dendritic spines and in the axon, the ER becomes SER and is involved in pre- and post-synaptic calcium signalling involved in synaptic plasticity [20].

In neurones, the nuclear envelope has a density of nuclear pores greater than in other cell types [21]. In the soma and at the beginning of dendrites, the ER forms sub-membrane tanks when close to the plasma membrane, whereas in axons, the RE is organized into interconnected tubules that run parallel to the axon axe. This ER axonal tubular network extends into the axon terminal, where it is closely linked to mitochondria. They form an interconnected dynamic network that collaborates with the genesis of the calcium signal, the RE extends into the dendrites down to the spines where it ends as "spine apparatus" [22]. This "spine apparatus" plays an important role in calcium regulation and in the delivery of components of the plasma membrane, as well as those of the axon terminal [23].

All situations leading to the impairment of ER functions prevent the correct protein folding. Unfolded proteins cannot move to the Golgi apparatus, then accumulate within the ER lumen and activate an adaptive programme named UPR (unfolded protein response). This adaptive cell response aims to reduce the amount of unfolded protein and to restore the ER homeostasis [24]. Many diseases are associated with UPR dysregulation or over activation: Alzheimer's disease, Parkinson's disease, Huntington's disease and also stroke [2, 3]. Modulation of the UPR activation appears then as an interesting and innovative therapeutic target for neuroprotection.

2 The ER Stress Response/The UPR

Alzheimer's disease, Parkinson's disease, Huntington's disease and ischaemia are known to activate ER stress and the UPR (Fig. 13.1). The UPR activates a set of cytoplasmic and nuclear signalling pathways characterized by three steps [25]: (1) a translational response to induce a quick reduction of the general protein synthesis by decreasing mRNA translation; (2) a transcriptional response to induce the expression of ER genes, like chaperone proteins, to increase the ER-folding capacity and (3) activation of the ERAD system to degrade unfolded proteins.

To control the accumulation of proteins in the ER, there are three anchored sensors at its membrane: PERK (protein kinase RNA-like endoplasmic reticulum kinase), IRE1 (Inositol-Requiring protein 1) and ATF6 (activating transcription factor 6) [25]. Under physiological conditions, Grp78 binds to the intraluminal domain of the three sensors and keeps them inactive. Grp78 regulates the signal by an exchange system of the balance between misfolded proteins and the sensors [26].

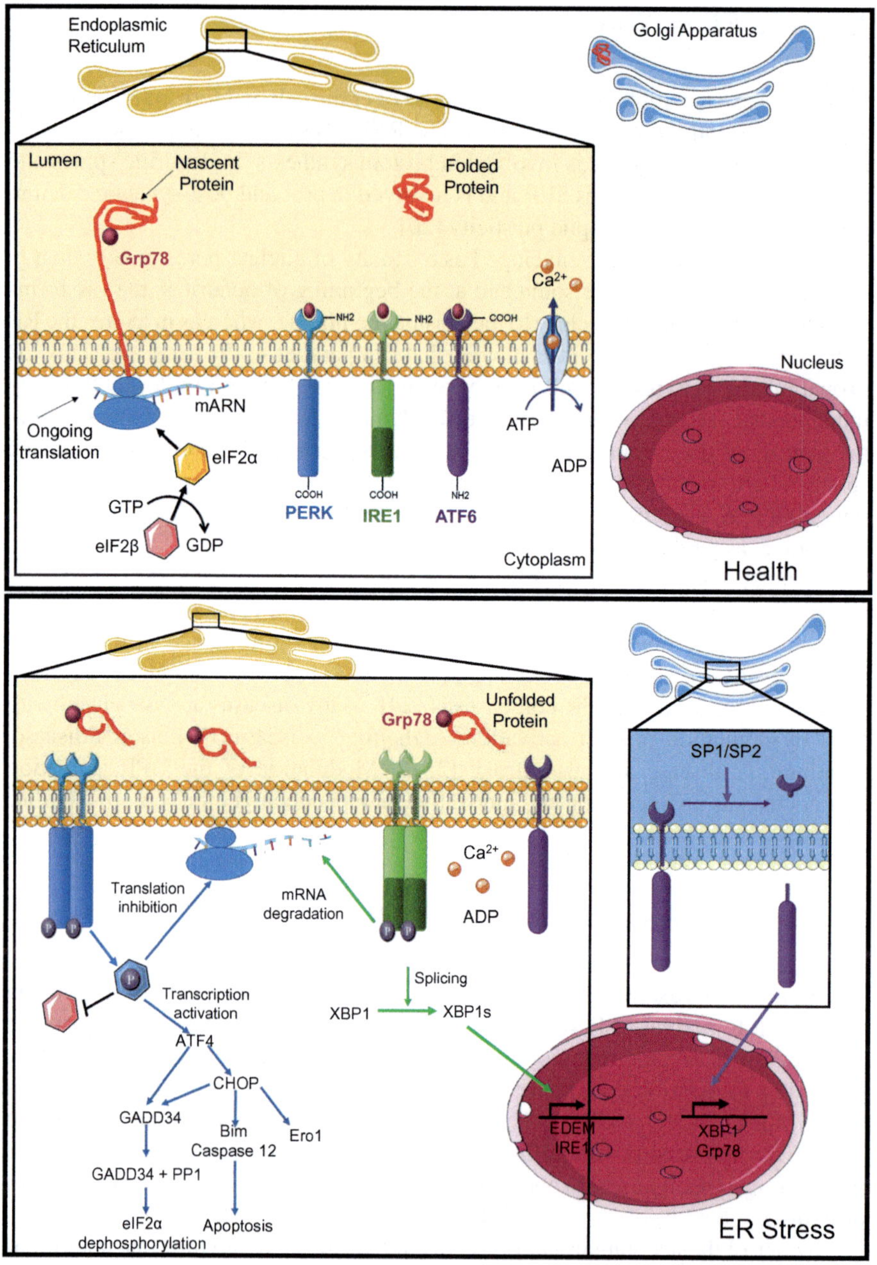

Fig. 13.1 In healthy conditions (*upper panel*), eIF2β binds with eif2α to form the eIF2 complex to initiate protein translation. Newly synthesized proteins are translocated into the ER lumen and fold with the help of chaperones like Grp78. Once correctly folded they traffic along the secretory pathway via the Golgi apparatus. ER stress sensors (PERK, ATF6 and IRE1) are inactivated by their association to Grp78. Under ER stress conditions (*bottom panel*), misfolded proteins accumulate within the ER lumen and sequester Grp78 from the three sensors. PERK phosphorylates

Following ER stress, the amount of misfolded proteins increases and the competitive binding of Grp78 to these proteins induces its dissociation from PERK, ATF6 and IRE1 allowing their respective activation [27].

ATF6, IRE1 and PERK control the regulation of the transcriptional and translational phases of the UPR [28]. The first phase is the attenuation of protein translation by PERK and aims to limit the addition of newly synthesized proteins within an already overloaded ER. During the second phase, IRE1 and ATF6 trigger a transcriptional activation of genes coding for chaperone and resident enzymes of the ER, as well as for components of the exporting and protein degradation pathways. The UPR not only helps cells to respond immediately to the presence of misfolded proteins but also prepare them to support further ER stress by changing its transcriptional programme in the long term [29]. If cells cannot restore a correct folding capacity, ER stress will eventually activate death signals [25, 30] (Fig. 13.1).

2.1 *Origin of ER Stress Activation*

Every situation that alters ER functions modifies the normal protein folding process. Unfolded proteins are no longer able to move towards the Golgi apparatus, then accumulate in the ER lumen and trigger the UPR. ER stress plays a particularly important physiological role in maintaining cellular homeostasis.

Activation of the UPR has many origins like glucose deficiencies, imbalance of redox status or hypoxia [25]. Protein folding within the ER is energetically costly, explaining the sensitivity of the ER to glucose deficient conditions. In addition, the absence of glucose alters post-translational modifications and more specifically glycosylation [31]. Furthermore, a decrease in extracellular glucose concentration results in the induction of ER stress genes through the PERK signalling pathway [32]. Hypoxia causes an alteration in protein conformations within the ER, especially due to the drop of ATP (Adenosine Tri-Phosphate) and leads to the activation of the UPR. Hypoxia affects various components of the protein folding machinery and induces an alteration of disulphide bridges formation [33], an increased expression of chaperones [34] and the inhibition of protein synthesis [35, 36].

The ER transiently produces ROS (reactive oxygen species) during disulphide bridges formation, which makes the ER sensitive to oxidative stress. Oxidative stress

Fig. 13.1 (continued) eIF2α allowing the inhibition of translation but paradoxically it will induce the transcription of ATF4, CHOP and GADD34. CHOP activates pro-apoptotic genes such as Bim and caspase 12 to trigger apoptosis. CHOP also activates Ero1, leading to oxidative stress. GADD34 binds to PP1c to form an active eIF2α phosphatase, allowing a negative feedback of the PERK pathway to restore protein translation. IRE1 also contributes to the inhibition of translation by degraded mRNA associate with the ER. IRE1 splices one intron into the XBP1 mRNA sequence. XBP1s is an active transcription factor which activates the transcription of EDEM and IRE1. ATF6 migrates into the Golgi apparatus where it is cleaved by protease SP1 and SP2. ATF6 cytosolic portion acts as a transcription factors and activates the expression of XBP1 and IRE1 leads to protein aggregation that may be due to uncontrolled disulphide bridges formation or to the inactivation of proteins involved in ERAD [37]. These oxidative stresses induce UPR mainly by the activation of PERK. The PERK signalling pathway also engages the transcription factor Nrf2 to counter oxidative stress [38, 39].

2.2 *PERK Pathway*

2.2.1 Translational Response Mediated by PERK

The first event occurring after the activation of the UPR is the oligomerization and trans-autophosphorylation of PERK (by the detachment of Grp78), followed by the phosphorylation on the serine 51 of the α-subunit of eukaryotic initiation factor 2 (eIF2α), leading to the shutdown of translation [25, 40]. In eukaryotes, eIF2α is associated with eIF2β and eIF2γ to form the eIF2 complex, an essential element for the formation of the translation initiation complex [41]. In PERK-deficient cells, phosphorylation of eIF2α and inhibition of protein synthesis are indeed totally missing during ER stress [32] (Fig. 13.1).

2.2.2 Transcriptional Response Mediated by PERK

The global attenuation of protein synthesis paradoxically induces mRNA translation of UPR elements, including the activating transcription factor 4 (ATF4) [28]. Some studies show the presence of upstream open reading frames (uORFs) in the mRNA of eIF2α-phosphorylated target genes, like ATF4, CHOP (CAAT/enhancer binding protein homologous transcription factor; also known as GADD153 (growth arrest and DNA damage-inducible protein 153)) and GADD34 (growth arrest and DNA damage-inducible protein 34). These targets are more efficiently translated when eIF2α is phosphorylated. Indeed, under low phosphorylation of eIF2α, the presence of uORFs slows the reading of mRNA by the ribosome and is either poorly or not translated. In conditions of high phosphorylation of eIF2α, ribosomes are preferentially recruited at the uORFs and mRNAs containing them are more efficiently translated [42].

CHOP is an important protein activated by the PERK pathway and is crucial for the cell fate. Indeed, in high stress conditions (intensity and/or duration) CHOP activates, among others, the transcription of pro-apoptotic genes such as Bim and caspase-12 to trigger cell death and thus to eliminate damaged cells [43]. CHOP also activates the transcription of Ero1 (ER oxidoreductin I), which catalyses the exchange between proteins and the PDI to facilitate the formation of disulphide bridges [44]: taken together, CHOP will dramatically increase the oxidative stress (Fig. 13.1).

A feedback of the PERK pathway exists. ATF4, alone or in combination with CHOP, activates the transcription of GADD34, a binding factor of the catalytic subunit of the protein phosphatase 1 (PP1c) to form an active eIF2α phosphatase, allowing then a negative feedback of the PERK pathway [44, 45]. But, persistent or prolonged expression of GADD34 increases protein translation and potentially exacerbates protein misfolding. Thus, mammalian cells have developed complex mechanisms, including the GADD34 trafficking from the cytosol to the ER, and the poly-ubiquitination of the α-amino group at the N-terminus of GADD34, to actively degrade proteins via the proteasome [46, 47].

The CReP protein (Constitutive Repressor of eIF2α phosphorylation) can also bind to the PP1c to form an active phosphatase of eIF2α and induces its dephosphorylation. Whereas GADD34 expression is triggered by CHOP, CReP is constitutively expressed. Of note, the dephosphorylation of eIF2α by the CReP-PP1c complex is slower than the one induced by the GADD34-PP1c complex [48].

2.3 *ATF6 Pathway*

Under physiological conditions, Grp78 is associated with ATF6 at its GLS (Golgi localization signal) sequences and is sequestrated within the ER. ATF6 is also restraint in the ER by interaction between its glycosylated residues and calreticulin [49]. Under stress conditions, ATF6 is released from Grp78 and migrates to the Golgi apparatus where it is cleaved by proteases S1P (Site-1 protease) and S2P (Site-1 protease) [50]. S1P cleaves ATF6 in its luminal portion releasing its C-terminal part and leaving its N-terminal portion anchored in the membrane. S2P cleaves ATF6 in its transmembrane portion thus releasing the N-terminal cytoplasmic part. The cleaved cytosolic fragment acts as a transcription factor and also activates the expression of ER stress genes, such as Grp78 and other chaperones [25, 50]. The multiple steps involved in ATF6 trafficking, followed by the proteolytic processing and its entry into the nucleus, make the ATF6 pathway slower than the PERK pathway. Additionally, the half-life of Grp78 is longer than that of ATF4, CHOP or GADD34. Activation of the ATF6 pathway and the delayed increase of Grp78 expression allow cells to anticipate the arrival of other misfolded proteins, ensuring their survival [51] (Fig. 13.1).

Regarding feedbacks of the pathway, there are two isoforms of ATF6: ATF6α and ATF6β. Both are induced by ER stress, but ATF6α is the predominant isoform in many cells. While studies in mice suggest that ATF6α and ATF6β have redundant functions [52], others show that the expression of ATF6β is slightly delayed compared to that of ATF6α and that the β isoform can reduce the α isoform's functions [53]: the β isoform would be at the origin of a negative feedback.

2.4 *IRE1 Pathway*

Following ER stress, Grp78 dissociates from IRE1 allowing its homodimerization and activation by autophosphorylation [54]. IRE1 is an enzyme with both kinase and endonuclease activities. Its main function is the splicing of 26 nucleotides in the intron of the transcription factor XBP1 (X-box binding protein 1) mRNA, thereby generating a shift in the reading frame, and producing an active transcription factor [55, 56]. In mammals, this splicing seems to occur in the nucleus and IRE1 localizes at the inner nuclear membrane [57]. The spliced XBP1 mRNA (XBP1-s) encodes a transcription factor for UPR response genes. There is no difference in the regulation of translation of the XBP1-s mRNA and XBP1-u (unspliced form), but a competition between XBP1-s and XBP1-u for their dimerization with cofactors. Moreover, XBP1-u can inhibit the transcription induced by XBP1-s. Rapid degradation of XBP1-u by the proteasome is, however, observed via the activation of IRE1/XBP1-s pathway [58]. The XBP1 promoter also contains an unfolded protein response element (UPRE) sequence, suggesting a positive feedback loop [55]. Indeed, XBP1 can activate his own transcription, which keeps the IRE1 signalling pathway activated after the repression of the ATF6 and PERK pathways (Fig. 13.1).

IRE1 also degrades mRNA associated to the ER by a mechanism called IRE1-dependent decay of mRNA (RIDD) and can also contribute to the reduction of translation [59, 60].

2.5 UPR: Friend or Foe?

The origin of ER stress encountered by a cell dictates the nature of the UPR. Under physiological conditions, a secretory cell will experience considerable variations in the flow of newly synthesized proteins that will cross the ER according to its needs. When protein synthesis goes from a low flow to a high flow, the cell needs to increase its ability to fold proteins in the ER to avoid the overloading. In this case, the UPR regulates transcription factors which target genes of the ER machinery proteins. But this response includes a time delay, since new chaperones cannot be available instantly. To avoid further accumulation of misfolded proteins, the UPR must adjust downward the transcription of secreted proteins. The modulation of both transcription and translation by the UPR provides a coordinated response adapted to the cell situation. However, if the ER stress is too high and/or too long then this beneficial phenomenon becomes deleterious by inducing different cell death pathways [29].

3 ER Stress and Stroke

ER stress is a conserved mechanism which is thought to be beneficial if cells restore a correct cellular homeostasis; otherwise its overactivation becomes harmful and induces a cell death response. ER stress is activated in many neurodegenerative diseases [2] including stroke, for which restoring the blood flow is mandatory for the recovery of cellular homeostasis.

During cerebral ischaemia, the energetic depletion impairs calcium homeostasis within the ER. In neurones, the ER is the main calcium pool, and disturbance of calcium homeostasis is known to activate ER stress [61, 62]. Moreover, the energetic failure stops SERCA pumps. Therefore, the calcium initially stored in the ER translocates to the cytosol and contributes to the uncontrolled increase of its concentration causing chaperones dysfunction, and finally ER stress [63].

Activation of the UPR has been revealed in both in vitro and in vivo models of stroke [62, 64–67]. In these studies, protein synthesis is rapidly inhibited and accompanied by the phosphorylation of eIF2α [68, 69]. In parallel to the inhibition of protein translation, the UPR induces an increased expression of ER-response genes [25]. Many studies have clearly demonstrated the activation of ER stress in in vitro and in vivo models of stroke. ER stress markers such as PERK, IRE1, GADD34, ATF4, CHOP and Grp78 are shown to be clearly upregulated in various models, including in cultures of cortical neurones subjected to oxygen deprivation [70], cultures of rat astrocytes subjected to oxygen and glucose deprivation (OGD) [71], and primary cultures of mixed rat brain cortical cells under OGD [72]. ER-stress related genes and proteins are also increased in rodent models of middle cerebral artery occlusion [65, 67] and in transient models of common carotid arteries bilateral occlusion [62, 64, 73]. However, some discrepancies remain about the role of ER stress in stroke: some studies demonstrate a deleterious effect of the induction of ER stress after stroke [64, 71, 74], whereas few others show a beneficial effect [65, 70]. The consequences of ER stress likely depend on its level of activation. In the first phase of stroke, ER stress is moderate and promotes cell survival by decreasing general translation and promoting the degradation of misfolded proteins. But if ER stress is prolonged, it becomes deleterious by inducing cytoplasmic calcium overload, increasing toxic products and activating cell death pathways [75, 76]. All these studies use different models of stroke and the subsequent ER stress activation is then different between them, depending on the severity of the model.

4 ER Stress and Other Ischaemic Diseases

4.1 Cardiac Ischaemia

More recently, several studies have demonstrated the activation of ER stress in the myocardium after ischaemia or ischaemia/reperfusion (I/R). A microarray study shows that numerous ER stress response genes are induced 24 h after myocardial infarction in ischaemic mouse hearts in vivo [77]. ER stress markers are also increased in both mouse hearts subjected to I/R ex vivo, and in surviving cardiac myocytes bordering the infarct zone in a mouse model of myocardial infarction [78]. XBP1 splicing is also detected in cultured neonatal rat ventricular myocytes subjected to ischaemia [78, 79]. Dominant-negative XBP1 increases apoptosis in isolated cardiomyocytes in response to I/R, suggesting a cardioprotective effect of ER stress in cardiac ischaemia [77]. Moreover, the overexpression of activated ATF6 in transgenic mouse hearts decreases ischaemic damages and increases ventricular pump functions in an ex vivo I/R model [80].

Even if ER stress can be activated by cardiac ischaemia and I/R, it is not clear whether ER stress is protective or deleterious in this context [81]. Perhaps mild or brief episodes of ischaemia would favour the activation of pro-survival aspects of ER stress, whereas severe or long episodes would lead to an activation of pro-apoptotic factors. Further studies are required to delineate the circumstances under which ER stress is protective or deleterious in the myocardium and what ER stress-inducible genes and pathways are important contributors to these outcomes in the heart.

4.2 *Renal Ischaemia*

To date, only few studies have demonstrated the implication of ER stress in renal ischaemia. CHOP expression is increased after renal ischaemia and leads to apoptotic cell death [82], confirmed by another study showing an overexpression of both Grp78 and CHOP after I/R and confirming the link between CHOP and apoptotic cell death [83]. Moreover, these two studies demonstrate that deletion of CHOP prevents cell death in renal ischaemia. It is also suggested that ischaemic preconditioning attenuates oxidative and ER stresses leading to kidney protection [84]. All these studies demonstrate a deleterious effect of ER stress activation following renal ischaemia, consistent with studies showing a protective effect of ER stress inhibition [85–87].

5 ER Stress and Neuroprotection in Ischaemic Diseases

5.1 *Through UPR Pathways*

Modulation of ER stress holds a promise of therapeutic benefits in ischaemic conditions (Table 13.1). Stroke highly activates the PERK-eIF2α pathway [62, 88], leading to the deleterious activation of CHOP [64, 71]. Indeed, in an in vivo model of stroke, PERK deficient mice do not present eIF2α phosphorylation or protein aggregation [62]. Moreover, PERK activation participates to the loss of motor neurones of the upper airways in a model of obstructive apnoea sleep inducing recurrent cerebral hypoxia in mice [89]. Prevention of the activation of this pathway by salubrinal, an inhibitor of the eIF2α phosphatase GADD34 [90], decreases ER stress and kainate-induced neurotoxicity in vitro [68] and reduces infarct size in an in vivo stroke model [69].

Many studies consider CHOP as a pro-apoptotic factor. One of the first studies was performed by Tajiri and colleagues in 2004 who showed that hippocampal neurones from CHOP deficient mice were more resistant to hypoxia-reoxygenation than those from wild-type littermates [64], and this effect is also confirmed in an in vivo model of stroke [69, 74]. CHOP is also increased in astrocytes under OGD, leading to cell death [71]. So targeting CHOP by an inhibitor of p38MAPK which reduces its activity holds a possible strategy to inhibit UPR-induced cell death.

Other drugs have also demonstrated promising results in experimental models. Dantrolene, a ryanodine receptor antagonist used as muscle relaxant, significantly decreases the infarct volume and provides a neuroprotective effect in a rat stroke model, by reducing ER stress-mediated apoptosis in the ischaemic area [91]. Berberine, an alkaloid, also attenuates ischaemia/reperfusion (I/R) injury in kidney by inhibiting oxidative and ER stresses [85]. The flavonoid glycoside Baicalin, isolated from Scutellaria baicalensis, also protects kidney from I/R injury by reducing oxidative and ER stresses through the activation of the anti-oxidative Nrf2 signalling pathway [86]. Intermedin, a member of the calcitonin/calcitonin gene-related peptide family, down-regulates Grp78, CHOP and caspase 12 protein and prevents apoptosis after renal ischaemia [87] and the non-competitive inhibitor of nicotinic acetylcholine receptor Catestatin decreases ER stress markers after cardiac ischaemia [92]. Regarding cardiac ischaemia, elatoside C inhibits ER stress-associated markers (Grp78, CHOP, Caspase-12 and JNK) and provides a significant protection against ischaemia-induced cardiomyocytes death [93, 94].

5.2 *Through Chaperones*

Chaperone proteins also appear as good targets to counteract the deleterious events associated to ER stress, since they are implicated in both ER stress inhibition (by the binding of Grp78 to the ER stress sensors) and in protein folding (Table 13.1). For example, Grp78 expression increases up to 2 weeks after I/R with a maximum between 24 and 72 h [66, 95, 96] and protects neurones from death [67, 97]. Moreover, overexpression of Grp78 in astrocytes has a protective effect after OGD [72] and by opposition, deletion of Grp78 increases apoptotic mechanism in in vitro models of excitotoxicity or oxidative stress in primary culture of hippocampal neurones [98]. Grp170, another chaperon of the Hsp70 family, might have an anti-apoptotic role in cerebral [99, 100], renal [101], and cardiac [102] ischaemic injuries.

Developing chaperone-inducer compounds is also a strategy to protect cells from ischaemia. The injection of BIX (BiP protein Inducer X) before stroke induces Grp78 and reduces brain damages and the numbers of apoptotic neurones in the penumbra [97]. Sodium 4-phenylbutyrate (PBA), a chemical chaperone is also protective by decreasing the load of unfolded proteins within the ER during cellular stress, and preserves cells against ischaemic spinal cord damages [103].

5.3 *Through Other Cell Death Pathways*

As stated before, UPR activation often leads to cell death by activating several pathways, mainly initiated by CHOP and linked to apoptosis by the PI3K-Akt pathway [74]. The inhibition of CHOP effectors, like BIM_{EI} (Bcl-2 interacting mediator of cell death), a pro-apoptotic BCl-2 family member, also leads to cellular protection and is a therapeutic target [43] (Table 13.1).

Table 13.1 Protective drugs against ER stress-induced cell death in ischaemic diseases

Drugs	Mechanisms/targets	Pathology	Ref
Salubrinal	Inhibition of GADD34	Cerebral ischaemia	[68, 69]
p38MAPK antagonist	Inhibition of CHOP activity		[109]
BIX *(BiP protein Inducer X)*	Inducer of Grp78	Cerebral ischaemia	[97]
Sodium 4-phenylbutyrate *(PBA)*	Chemical chaperone	Spinal cord ischaemia	[103]
Dantrolene	Ryanodine receptor antagonist Inhibition of ER stress-mediated apoptosis	Cerebral ischaemia	[91]
Berberine *(alkaloid)*	Inhibition of oxidative and ER stresses	Renal ischaemia	[85]
Baicalin *(flavonoid glycoside)*	Activation of the anti-oxidative Nrf2 signalling pathway	Renal ischaemia	[86]
Carnosic acid	Activation of the anti-oxidative Nrf2 signalling pathway	Cerebral ischaemia	[105]
Intermedin *(calcitonin/calcitonin gene-related peptide family)*	Down-regulation of ER stress-associated markers (Grp78, CHOP, Caspase-12)	Renal ischaemia	[87]
Catestatin	Non-competitive inhibitor of nicotinic acetylcholine receptor Inhibition of ER stress	Cardiac ischaemia	[92]
Elatoside C	Inhibition of ER stress-associated markers (Grp78, CHOP, Caspase-12 and JNK)	Cardiac ischaemia	[93, 94]
BIM_{EI} *(Bcl-2 interacting mediator of cell death)*	Inhibition of CHOP effectors	Cerebral ischaemia	[43]

Oxidative stress is also induced after stroke, and as described previously, leads to protein aggregation due to uncontrolled disulphide bond formation or inactivation of proteins involved in ERAD [37]. Crosstalk between ER stress and oxidative stress potentiates each other [104] and so, activates endogenous anti-oxidant pathways: this represents another strategy for protecting cells after ischaemia. Compounds inducing the endogenous Nrf2-KEAP1 (Kelch-like ECH-associated protein 1) anti-oxidant pathway show neuroprotection after stroke [105, 106]. Moreover, others anti-oxidant drugs like edaravone and apocynin have shown promising results on the inhibition of ER stress in stroke models [107, 108].

6 Conclusion

The growing interest into the ER stress field opens new horizons for original therapeutic approaches. It is now clear that ER stress leads to UPR activation in cerebral, renal or cardiac ischaemia. Strategies to counteract the deleterious overactivation of

ER stress have been developed, in particular in stroke. CHOP overexpression activates death pathways, but CHOP inhibitors are not yet fully developed or specific. For the moment, the use of chemical chaperones or chaperones inducer like BIX is a promising way for stroke: as chaperone induction is at the "end" of the UPR pathway, inducing their expression earlier during the acute disease seems to be protective. Other agents have demonstrated good results, but their mechanism of action is not fully understood or even not described at all: these agents decrease many ER markers without specificity. More work is necessary to decipher ER stress mechanisms and actions to propose specific inhibitors/strategies to counteract its deleterious actions in ischaemia, without affecting its beneficial effects.

References

1. Voeltz GK, Rolls MM, Rapoport TA (2002) Structural organization of the endoplasmic reticulum. EMBO Rep 3(10):944–950
2. Roussel BD, Kruppa AJ, Miranda E, Crowther DC, Lomas DA, Marciniak SJ (2013) Endoplasmic reticulum dysfunction in neurological disease. Lancet Neurol 12(1):105–118
3. Hetz C, Mollereau B (2014) Disturbance of endoplasmic reticulum proteostasis in neurodegenerative diseases. Nat Rev Neurosci 15(4):233–249
4. Estrada de Martin P, Novick P, Ferro-Novick S (2005) The organization, structure, and inheritance of the ER in higher and lower eukaryotes. Biochem Cell Biol 83(6):752–761
5. Vertel BM, Walters LM, Mills D (1992) Subcompartments of the endoplasmic reticulum. Semin Cell Biol 3(5):325–341
6. Dudek J, Benedix J, Cappel S, Greiner M, Jalal C, Muller L et al (2009) Functions and pathologies of BiP and its interaction partners. Cell Mol Life Sci 66(9):1556–1569
7. Melnick J, Dul JL, Argon Y (1994) Sequential interaction of the chaperones BiP and GRP94 with immunoglobulin chains in the endoplasmic reticulum. Nature 370(6488):373–375
8. Brodsky JL, Werner ED, Dubas ME, Goeckeler JL, Kruse KB, McCracken AA (1999) The requirement for molecular chaperones during endoplasmic reticulum-associated protein degradation demonstrates that protein export and import are mechanistically distinct. J Biol Chem 274(6):3453–3460
9. Yang Y, Li Z (2005) Roles of heat shock protein gp96 in the ER quality control: redundant or unique function? Mol Cells 20(2):173–182
10. Randow F, Seed B (2001) Endoplasmic reticulum chaperone gp96 is required for innate immunity but not cell viability. Nat Cell Biol 3(10):891–896
11. Ellgaard L, Frickel EM (2003) Calnexin, calreticulin, and ERp57: teammates in glycoprotein folding. Cell Biochem Biophys 39(3):223–247
12. Michalak M, Groenendyk J, Szabo E, Gold LI, Opas M (2009) Calreticulin, a multi-process calcium-buffering chaperone of the endoplasmic reticulum. Biochem J 417(3):651–666
13. Helenius A, Aebi M (2004) Roles of N-linked glycans in the endoplasmic reticulum. Annu Rev Biochem 73:1019–1049
14. Molinari M, Calanca V, Galli C, Lucca P, Paganetti P (2003) Role of EDEM in the release of misfolded glycoproteins from the calnexin cycle. Science 299(5611):1397–1400
15. Oda Y, Hosokawa N, Wada I, Nagata K (2003) EDEM as an acceptor of terminally misfolded glycoproteins released from calnexin. Science 299(5611):1394–1397
16. Orrenius S, Zhivotovsky B, Nicotera P (2003) Regulation of cell death: the calcium-apoptosis link. Nat Rev Mol Cell Biol 4(7):552–565
17. Mikoshiba K (2007) IP3 receptor/Ca2+ channel: from discovery to new signaling concepts. J Neurochem 102(5):1426–1446

18. Amador FJ, Stathopulos PB, Enomoto M, Ikura M (2013) Ryanodine receptor calcium release channels: lessons from structure-function studies. FEBS J 280(21):5456–5470
19. Brostrom MA, Brostrom CO (2003) Calcium dynamics and endoplasmic reticular function in the regulation of protein synthesis: implications for cell growth and adaptability. Cell Calcium 34(4-5):345–363
20. Berridge MJ (2002) The endoplasmic reticulum: a multifunctional signaling organelle. Cell Calcium 32(5-6):235–249
21. Talamas JA, Capelson M (2015) Nuclear envelope and genome interactions in cell fate. Front Genet 6:95
22. Spacek J, Harris KM (1997) Three-dimensional organization of smooth endoplasmic reticulum in hippocampal CA1 dendrites and dendritic spines of the immature and mature rat. J Neurosci 17(1):190–203
23. Broadwell RD, Cataldo AM (1984) The neuronal endoplasmic reticulum: its cytochemistry and contribution to the endomembrane system. II. Axons and terminals. J Comp Neurol 230(2):231–248
24. Mollereau B, Rzechorzek NM, Roussel BD, Sedru M, Van den Brink DM, Bailly-Maitre B, et al. (2016) Adaptive preconditioning in neurological diseases - therapeutic insights from proteostatic perturbations. Brain Res. pii: S0006-8993(16)30092-0. doi:10.1016/j.brainres
25. Marciniak SJ, Ron D (2006) Endoplasmic reticulum stress signaling in disease. Physiol Rev 86(4):1133–1149
26. Schroder M, Kaufman RJ (2005) The mammalian unfolded protein response. Annu Rev Biochem 74:739–789
27. Bertolotti A, Zhang Y, Hendershot LM, Harding HP, Ron D (2000) Dynamic interaction of BiP and ER stress transducers in the unfolded-protein response. Nat Cell Biol 2(6):326–332
28. Harding HP, Zhang Y, Bertolotti A, Zeng H, Ron D (2000) Perk is essential for translational regulation and cell survival during the unfolded protein response. Mol Cell 5(5):897–904
29. Mollereau B, Manie S, Napoletano F (2014) Getting the better of ER stress. J Cell Commun Signal 8(4):311–321
30. Ron D, Walter P (2007) Signal integration in the endoplasmic reticulum unfolded protein response. Nat Rev Mol Cell Biol 8(7):519–529
31. Helenius A (1994) How N-linked oligosaccharides affect glycoprotein folding in the endoplasmic reticulum. Mol Biol Cell 5(3):253–265
32. Scheuner D, Song B, McEwen E, Liu C, Laybutt R, Gillespie P et al (2001) Translational control is required for the unfolded protein response and in vivo glucose homeostasis. Mol Cell 7(6):1165–1176
33. Gess B, Hofbauer KH, Wenger RH, Lohaus C, Meyer HE, Kurtz A (2003) The cellular oxygen tension regulates expression of the endoplasmic oxidoreductase ERO1-Lalpha. Eur J Biochem 270(10):2228–2235
34. Paris S, Denis H, Delaive E, Dieu M, Dumont V, Ninane N et al (2005) Up-regulation of 94-kDa glucose-regulated protein by hypoxia-inducible factor-1 in human endothelial cells in response to hypoxia. FEBS Lett 579(1):105–114
35. Blais JD, Filipenko V, Bi M, Harding HP, Ron D, Koumenis C et al (2004) Activating transcription factor 4 is translationally regulated by hypoxic stress. Mol Cell Biol 24(17):7469–7482
36. Koumenis C, Naczki C, Koritzinsky M, Rastani S, Diehl A, Sonenberg N et al (2002) Regulation of protein synthesis by hypoxia via activation of the endoplasmic reticulum kinase PERK and phosphorylation of the translation initiation factor eIF2alpha. Mol Cell Biol 22(21):7405–7416
37. Stadtman ER, Levine RL (2003) Free radical-mediated oxidation of free amino acids and amino acid residues in proteins. Amino Acids 25(3-4):207–218
38. Harding HP, Zhang Y, Zeng H, Novoa I, Lu PD, Calfon M et al (2003) An integrated stress response regulates amino acid metabolism and resistance to oxidative stress. Mol Cell 11(3):619–633
39. Cullinan SB, Diehl JA (2004) PERK-dependent activation of Nrf2 contributes to redox homeostasis and cell survival following endoplasmic reticulum stress. J Biol Chem 279(19):20108–20117

40. Harding HP, Zhang Y, Ron D (1999) Protein translation and folding are coupled by an endoplasmic-reticulum-resident kinase. Nature 397(6716):271–274
41. Roy AL, Chakrabarti D, Gupta NK (1987) Protein synthesis in rabbit reticulocytes: Mg2+-inhibition of ternary complex (Met-tRNA(f).eIF-2.GTP) formation by reticulocyte eIF-2. Biochem Biophys Res Commun 146(1):114–120
42. Vattem KM, Wek RC (2004) Reinitiation involving upstream ORFs regulates ATF4 mRNA translation in mammalian cells. Proc Natl Acad Sci U S A 101(31):11269–11274
43. Puthalakath H, O'Reilly LA, Gunn P, Lee L, Kelly PN, Huntington ND et al (2007) ER stress triggers apoptosis by activating BH3-only protein Bim. Cell 129(7):1337–1349
44. Marciniak SJ, Yun CY, Oyadomari S, Novoa I, Zhang Y, Jungreis R et al (2004) CHOP induces death by promoting protein synthesis and oxidation in the stressed endoplasmic reticulum. Genes Dev 18(24):3066–3077
45. Novoa I, Zeng H, Harding HP, Ron D (2001) Feedback inhibition of the unfolded protein response by GADD34-mediated dephosphorylation of eIF2alpha. J Cell Biol 153(5):1011–1022
46. Brush MH, Shenolikar S (2008) Control of cellular GADD34 levels by the 26S proteasome. Mol Cell Biol 28(23):6989–7000
47. Zhou W, Brush MH, Choy MS, Shenolikar S (2011) Association with endoplasmic reticulum promotes proteasomal degradation of GADD34 protein. J Biol Chem 286(24):21687–21696
48. Dalton LE, Healey E, Irving J, Marciniak SJ (2012) Phosphoproteins in stress-induced disease. Prog Mol Biol Transl Sci 106:189–221
49. Hong M, Li M, Mao C, Lee AS (2004) Endoplasmic reticulum stress triggers an acute proteasome-dependent degradation of ATF6. J Cell Biochem 92(4):723–732
50. Shen J, Chen X, Hendershot L, Prywes R (2002) ER stress regulation of ATF6 localization by dissociation of BiP/GRP78 binding and unmasking of Golgi localization signals. Dev Cell 3(1):99–111
51. Rutkowski DT, Arnold SM, Miller CN, Wu J, Li J, Gunnison KM et al (2006) Adaptation to ER stress is mediated by differential stabilities of pro-survival and pro-apoptotic mRNAs and proteins. PLoS Biol 4(11), e374
52. Yamamoto K, Sato T, Matsui T, Sato M, Okada T, Yoshida H et al (2007) Transcriptional induction of mammalian ER quality control proteins is mediated by single or combined action of ATF6alpha and XBP1. Dev Cell 13(3):365–376
53. Thuerauf DJ, Morrison L, Glembotski CC (2004) Opposing roles for ATF6alpha and ATF6beta in endoplasmic reticulum stress response gene induction. J Biol Chem 279(20):21078–21084
54. Okamura K, Kimata Y, Higashio H, Tsuru A, Kohno K (2000) Dissociation of Kar2p/BiP from an ER sensory molecule, Ire1p, triggers the unfolded protein response in yeast. Biochem Biophys Res Commun 279(2):445–450
55. Yoshida H, Matsui T, Yamamoto A, Okada T, Mori K (2001) XBP1 mRNA is induced by ATF6 and spliced by IRE1 in response to ER stress to produce a highly active transcription factor. Cell 107(7):881–891
56. Calfon M, Zeng H, Urano F, Till JH, Hubbard SR, Harding HP et al (2002) IRE1 couples endoplasmic reticulum load to secretory capacity by processing the XBP-1 mRNA. Nature 415(6867):92–96
57. Lee K, Tirasophon W, Shen X, Michalak M, Prywes R, Okada T et al (2002) IRE1-mediated unconventional mRNA splicing and S2P-mediated ATF6 cleavage merge to regulate XBP1 in signaling the unfolded protein response. Genes Dev 16(4):452–466
58. Lee AH, Iwakoshi NN, Glimcher LH (2003) XBP-1 regulates a subset of endoplasmic reticulum resident chaperone genes in the unfolded protein response. Mol Cell Biol 23(21):7448–7459
59. Hollien J, Weissman JS (2006) Decay of endoplasmic reticulum-localized mRNAs during the unfolded protein response. Science 313(5783):104–107
60. Maurel M, Chevet E, Tavernier J, Gerlo S (2014) Getting RIDD of RNA: IRE1 in cell fate regulation. Trends Biochem Sci 39(5):245–254
61. Hayashi T, Saito A, Okuno S, Ferrand-Drake M, Dodd RL, Chan PH (2005) Damage to the endoplasmic reticulum and activation of apoptotic machinery by oxidative stress in ischemic neurons. J Cereb Blood Flow Metab 25(1):41–53

62. Owen CR, Kumar R, Zhang P, McGrath BC, Cavener DR, Krause GS (2005) PERK is responsible for the increased phosphorylation of eIF2alpha and the severe inhibition of protein synthesis after transient global brain ischemia. J Neurochem 94(5):1235–1242
63. Szydlowska K, Tymianski M (2010) Calcium, ischemia and excitotoxicity. Cell Calcium 47(2):122–129
64. Tajiri S, Oyadomari S, Yano S, Morioka M, Gotoh T, Hamada JI et al (2004) Ischemia-induced neuronal cell death is mediated by the endoplasmic reticulum stress pathway involving CHOP. Cell Death Differ 11(4):403–415
65. McCaig D, Imai H, Gallagher L, Graham DI, Harland J, Moira Brown S et al (2005) Evolution of GADD34 expression after focal cerebral ischaemia. Brain Res 1034(1-2):51–61
66. Rissanen A, Sivenius J, Jolkkonen J (2006) Prolonged bihemispheric alterations in unfolded protein response related gene expression after experimental stroke. Brain Res 1087(1):60–66
67. Morimoto N, Oida Y, Shimazawa M, Miura M, Kudo T, Imaizumi K et al (2007) Involvement of endoplasmic reticulum stress after middle cerebral artery occlusion in mice. Neuroscience 147(4):957–967
68. Sokka AL, Putkonen N, Mudo G, Pryazhnikov E, Reijonen S, Khiroug L et al (2007) Endoplasmic reticulum stress inhibition protects against excitotoxic neuronal injury in the rat brain. J Neurosci 27(4):901–908
69. Nakka VP, Gusain A, Raghubir R (2010) Endoplasmic reticulum stress plays critical role in brain damage after cerebral ischemia/reperfusion in rats. Neurotox Res 17(2):189–202
70. Halterman MW, Gill M, DeJesus C, Ogihara M, Schor NF, Federoff HJ (2010) The endoplasmic reticulum stress response factor CHOP-10 protects against hypoxia-induced neuronal death. J Biol Chem 285(28):21329–21340
71. Benavides A, Pastor D, Santos P, Tranque P, Calvo S (2005) CHOP plays a pivotal role in the astrocyte death induced by oxygen and glucose deprivation. Glia 52(4):261–275
72. Badiola N, Penas C, Minano-Molina A, Barneda-Zahonero B, Fado R, Sanchez-Opazo G et al (2011) Induction of ER stress in response to oxygen-glucose deprivation of cortical cultures involves the activation of the PERK and IRE-1 pathways and of caspase-12. Cell Death Dis 2, e149
73. Doutheil J, Althausen S, Gissel C, Paschen W (1999) Activation of MYD116 (gadd34) expression following transient forebrain ischemia of rat: implications for a role of disturbances of endoplasmic reticulum calcium homeostasis. Brain Res Mol Brain Res 63(2):225–232
74. Yuan Y, Guo Q, Ye Z, Pingping X, Wang N, Song Z (2011) Ischemic postconditioning protects brain from ischemia/reperfusion injury by attenuating endoplasmic reticulum stress-induced apoptosis through PI3K-Akt pathway. Brain Res 1367:85–93
75. Su Y, Li F (2015) Endoplasmic reticulum stress in brain ischemia. Int J Neurosci 126(8):681–91
76. DeGracia DJ, Montie HL (2004) Cerebral ischemia and the unfolded protein response. J Neurochem 91(1):1–8
77. Harpster MH, Bandyopadhyay S, Thomas DP, Ivanov PS, Keele JA, Pineguina N et al (2006) Earliest changes in the left ventricular transcriptome postmyocardial infarction. Mamm Genome 17(7):701–715
78. Thuerauf DJ, Marcinko M, Gude N, Rubio M, Sussman MA, Glembotski CC (2006) Activation of the unfolded protein response in infarcted mouse heart and hypoxic cultured cardiac myocytes. Circ Res 99(3):275–282
79. Szegezdi E, Duffy A, O'Mahoney ME, Logue SE, Mylotte LA, O'Brien T et al (2006) ER stress contributes to ischemia-induced cardiomyocyte apoptosis. Biochem Biophys Res Commun 349(4):1406–1411
80. Martindale JJ, Fernandez R, Thuerauf D, Whittaker R, Gude N, Sussman MA et al (2006) Endoplasmic reticulum stress gene induction and protection from ischemia/reperfusion injury in the hearts of transgenic mice with a tamoxifen-regulated form of ATF6. Circ Res 98(9):1186–1193
81. Glembotski CC (2007) Endoplasmic reticulum stress in the heart. Circ Res 101(10):975–984

82. Dong B, Zhou H, Han C, Yao J, Xu L, Zhang M et al (2014) Ischemia/reperfusion-induced CHOP expression promotes apoptosis and impairs renal function recovery: the role of acidosis and GPR4. PLoS One 9(10), e110944
83. Noh MR, Kim JI, Han SJ, Lee TJ, Park KM (2015) C/EBP homologous protein (CHOP) gene deficiency attenuates renal ischemia/reperfusion injury in mice. Biochim Biophys Acta 1852(9):1895–1901
84. Mahfoudh-Boussaid A, Zaouali MA, Hadj-Ayed K, Miled AH, Saidane-Mosbahi D, Rosello-Catafau J et al (2012) Ischemic preconditioning reduces endoplasmic reticulum stress and upregulates hypoxia inducible factor-1alpha in ischemic kidney: the role of nitric oxide. J Biomed Sci 19:7
85. Yu W, Sheng M, Xu R, Yu J, Cui K, Tong J et al (2013) Berberine protects human renal proximal tubular cells from hypoxia/reoxygenation injury via inhibiting endoplasmic reticulum and mitochondrial stress pathways. J Transl Med 11:24
86. Lin M, Li L, Zhang Y, Zheng L, Xu M, Rong R et al (2014) Baicalin ameliorates H2O2 induced cytotoxicity in HK-2 cells through the inhibition of ER stress and the activation of Nrf2 signaling. Int J Mol Sci 15(7):12507–12522
87. Wang Y, Tian J, Qiao X, Su X, Mi Y, Zhang R et al (2015) Intermedin protects against renal ischemia-reperfusion injury by inhibiting endoplasmic reticulum stress. BMC Nephrol 16:169
88. Zhu Y, Fenik P, Zhan G, Sanfillipo-Cohn B, Naidoo N, Veasey SC (2008) Eif-2a protects brainstem motoneurons in a murine model of sleep apnea. J Neurosci 28(9):2168–2178
89. Zhu H, Fan X, Yu Z, Liu J, Murata Y, Lu J et al (2010) Annexin A2 combined with low-dose tPA improves thrombolytic therapy in a rat model of focal embolic stroke. J Cereb Blood Flow Metab 30(6):1137–1146
90. Boyce M, Bryant KF, Jousse C, Long K, Harding HP, Scheuner D et al (2005) A selective inhibitor of eIF2alpha dephosphorylation protects cells from ER stress. Science 307(5711):935–939
91. Li F, Hayashi T, Jin G, Deguchi K, Nagotani S, Nagano I et al (2005) The protective effect of dantrolene on ischemic neuronal cell death is associated with reduced expression of endoplasmic reticulum stress markers. Brain Res 1048(1-2):59–68
92. Liao F, Zheng Y, Cai J, Fan J, Wang J, Yang J et al (2015) Catestatin attenuates endoplasmic reticulum induced cell apoptosis by activation type 2 muscarinic acetylcholine receptor in cardiac ischemia/reperfusion. Sci Rep 5:16590
93. Wang M, Meng XB, Yu YL, Sun GB, Xu XD, Zhang XP et al (2014) Elatoside C protects against hypoxia/reoxygenation-induced apoptosis in H9c2 cardiomyocytes through the reduction of endoplasmic reticulum stress partially depending on STAT3 activation. Apoptosis 19(12):1727–1735
94. Wang M, Sun GB, Zhang JY, Luo Y, Yu YL, Xu XD et al (2015) Elatoside C protects the heart from ischaemia/reperfusion injury through the modulation of oxidative stress and intracellular Ca(2)(+) homeostasis. Int J Cardiol 185:167–176
95. Oida Y, Shimazawa M, Imaizumi K, Hara H (2008) Involvement of endoplasmic reticulum stress in the neuronal death induced by transient forebrain ischemia in gerbil. Neuroscience 151(1):111–119
96. Osada N, Kosuge Y, Ishige K, Ito Y (2010) Characterization of neuronal and astroglial responses to ER stress in the hippocampal CA1 area in mice following transient forebrain ischemia. Neurochem Int 57(1):1–7
97. Kudo T, Kanemoto S, Hara H, Morimoto N, Morihara T, Kimura R et al (2008) A molecular chaperone inducer protects neurons from ER stress. Cell Death Differ 15(2):364–375
98. Yu Z, Luo H, Fu W, Mattson MP (1999) The endoplasmic reticulum stress-responsive protein GRP78 protects neurons against excitotoxicity and apoptosis: suppression of oxidative stress and stabilization of calcium homeostasis. Exp Neurol 155(2):302–314
99. Kuwabara K, Matsumoto M, Ikeda J, Hori O, Ogawa S, Maeda Y et al (1996) Purification and characterization of a novel stress protein, the 150-kDa oxygen-regulated protein (ORP150), from cultured rat astrocytes and its expression in ischemic mouse brain. J Biol Chem 271(9):5025–5032

100. Tamatani M, Matsuyama T, Yamaguchi A, Mitsuda N, Tsukamoto Y, Taniguchi M et al (2001) ORP150 protects against hypoxia/ischemia-induced neuronal death. Nat Med 7(3):317–323
101. Bando Y, Tsukamoto Y, Katayama T, Ozawa K, Kitao Y, Hori O et al (2004) ORP150/HSP12A protects renal tubular epithelium from ischemia-induced cell death. FASEB J 18(12):1401–1403
102. Aleshin AN, Sawa Y, Kitagawa-Sakakida S, Bando Y, Ono M, Memon IA et al (2005) 150-kDa oxygen-regulated protein attenuates myocardial ischemia-reperfusion injury in rat heart. J Mol Cell Cardiol 38(3):517–525
103. Mizukami T, Orihashi K, Herlambang B, Takahashi S, Hamaishi M, Okada K et al (2010) Sodium 4-phenylbutyrate protects against spinal cord ischemia by inhibition of endoplasmic reticulum stress. J Vasc Surg 52(6):1580–1586
104. Nakka VP, Prakash-Babu P, Vemuganti R (2016) Crosstalk Between Endoplasmic Reticulum Stress, Oxidative Stress, and Autophagy: Potential Therapeutic Targets for Acute CNS Injuries. Mol Neurobiol 53(1):532–544
105. Satoh T, Kosaka K, Itoh K, Kobayashi A, Yamamoto M, Shimojo Y et al (2008) Carnosic acid, a catechol-type electrophilic compound, protects neurons both in vitro and in vivo through activation of the Keap1/Nrf2 pathway via S-alkylation of targeted cysteines on Keap1. J Neurochem 104(4):1116–1131
106. Shih AY, Li P, Murphy TH (2005) A small-molecule-inducible Nrf2-mediated antioxidant response provides effective prophylaxis against cerebral ischemia in vivo. J Neurosci 25(44):10321–10335
107. Qi X, Okuma Y, Hosoi T, Nomura Y (2004) Edaravone protects against hypoxia/ischemia-induced endoplasmic reticulum dysfunction. J Pharmacol Exp Ther 311(1):388–393
108. Chen H, Song YS, Chan PH (2009) Inhibition of NADPH oxidase is neuroprotective after ischemia-reperfusion. J Cereb Blood Flow Metab 29(7):1262–1272
109. Lee MR, Dominguez C (2005) MAP kinase p38 inhibitors: clinical results and an intimate look at their interactions with p38alpha protein. Curr Med Chem 12(25):2979–2994

Chapter 14
Effects of Neuroprotectants Before and After Stroke: Statins and Anti-hypertensives

Roberto Federico Villa, Federica Ferrari, and Antonio Moretti

Abstract Neuroprotection, as an anjunct or alternative therapy to thrombolysis, may be a rational strategy to improve *penumbra* survival. Nevertheless, so far none of the neuroprotectants found active in preclinical studies have been translated into clinical use.

The hypolipidaemic statins and the anti-hypertensive drugs are widely used as preventive treatments before and after stroke and whose neuroprotectant effects were demonstrated in experimental and some clinical studies.

Indeed, pleiotropic effects were reported on both cerebral endothelium (through the NO-induced vasodilation, and the anti-thrombotic, anti-oxidant and anti-inflammatory activities) and brain parenchyma (anti-excitotoxic, neurotrophic, anti-oxidant activities).

In this chapter we first describe the pharmacological activities of statins and various anti-hypertensive classes (drugs active on the RAS, CCBs, sympatholytics and diuretics), giving particular emphasis on their pleiotropic mechanisms. We then examine the evidence for their primary and secondary stroke-preventive activities. Overall, favourable effects were shown in non-thrombolyzed patients, though without clear differences among the four classes of anti-hypertensives. However, insufficient and inconclusive findings were reported in patients treated with thrombolysis, thus making more ad hoc experimental studies and RCTs highly recommended.

Keywords Neuroprotection • Statins • Anti-hypertensives • Primary prevention • Secondary prevention • Thrombolysis

R.F. Villa, D.Sc., M.D., Ph.D., A.C.C.P. (✉) • F. Ferrari, B.Sc., M.Sc., Ph.D., A.C.C.P.
A. Moretti, D.Pharm., Ph.D.
Laboratory of Pharmacology and Molecular Medicine of Central Nervous System, Department of Biology and Biotechnology, University of Pavia, Via Ferrata, 9, Pavia 27100, Italy
e-mail: robertofederico.villa@unipv.it

P.A. Lapchak, J.H. Zhang (eds.), *Neuroprotective Therapy for Stroke and Ischemic Disease*, Springer Series in Translational Stroke Research,
DOI 10.1007/978-3-319-45345-3_14

1 Introduction

The only currently approved therapy for acute ischaemic stroke (AIS) is thrombolysis with intravenous (IVT) recombinant tissue plasminogen activator (rtPA). Endovascular thrombolysis with mechanical recanalization devices (EVT) is also increasingly applied. The aim of thrombolysis is to recanalize an occluded vessel as quickly as possible, in order to salvage the still-viable brain tissue (*penumbra*). Thrombolysis beneficial effects have been definitely demonstrated in terms of vessel reperfusion and functional recovery. However, this therapy has a number of drawbacks and restrictions that limit its extensive use, as recently reviewed in [1, 2].

There is, therefore, a need to identify and develop effective and safe neuroprotective and neurorestoring therapies targeting brain parenchyma (to stabilize *penumbra*) as alternative or adjunct treatments to thrombolysis [3]. The disappointing fact that preclinical studies on neuroprotectants were not translated into clinical use during the past three decades [4] should not prevent further pursuing that goal. A pharmacological strategy based on well-known drugs for primary and secondary stroke prevention may be useful due to their availability in the ischaemic brain, thus hopefully resulting in therapeutic effects. Greater access of drugs to the ischaemic tissue after reperfusion may also be reasonably expected.

The hypolipidaemic statins and the anti-hypertensives are the most frequent among the various pharmacological classes that patients are treated with, both in pre- and post-stroke phases [5]. Both classes showed compelling evidence of primary and secondary stroke-preventing activities and of neuroprotective effects independently from their primary action.

Therefore, the objectives of this chapter are threefold: (1) reviewing the aforementioned evidence regarding statins and anti-hypertensives, (2) examining the mechanisms of their neuroprotective effects, and (3) analyzing the possible interactions of neuroprotectants with the thrombolytic therapy.

2 Statins

2.1 Hypercholesterolaemia as a Risk Factor for AIS

As stated in the 1985 Nobel Lecture by Michael Brown and Joseph Goldstein (who received the Prize for their fundamental work on cholesterol homeostasis and the role of low-density lipoproteins), "cholesterol is a Janus-faced molecule. The very property that makes it useful in cell membranes, namely its absolute insolubility in water, also makes it lethal" [6].

Indeed, besides hypertension, abnormalities in plasma cholesterol and lipoproteins are the most firmly established risk factors for atherosclerosis and are clearly correlated with coronary heart disease (CHD). By contrast, the relationship of dyslipidaemia with acute ischaemic stroke (AIS) is controversial. Some observational studies found a positive association with total cholesterol (T-C), low-density lipoprotein cholesterol (LDL-C), very low-density lipoprotein cholesterol (VLDL-C), intermediate-density lipoprotein (IDL) and an inverse association with protective high-density lipoprotein (HDL-C) [7–9]. These results on AIS were not replicated by other studies even in the presence of a significant association with CHD [10–12]. Moreover, the positive association between LDL-C and large artery stroke of athero-thrombotic origin, but not cardioembolic stroke [12, 13], highlights the significance of the classification of AIS subtypes. It was also suggested that, besides hypertension and other risk factors, *low* plasma cholesterol may be associated with a risk factor for intracerebral haemorrhage (ICH) [14].

In conclusion, much of the discrepancy of some of the above-mentioned reports on the association between dyslipidaemia and AIS might be explained by the heterogeneity of the types of stroke (ischaemic vs. haemorrhagic), the subtypes of ischaemic stroke, and the patients' age.

2.2 *Pharmacology and Pharmacokinetics of Statins*

Despite the conflicting results previously reported, the *consensus* is almost unanimous regarding the benefit of cholesterol-lowering treatment aimed at reducing the progression or inducing regression of atherosclerosis, thus decreasing the risk of carotid atherosclerotic plaque and the incidence and mortality of CHD and AIS.

The most active drugs are statins. They are competitive, reversible inhibitors of 3-hydroxy-3-methylglutaryl coenzyme A (HMG-CoA) reductase, the first and rate-limiting step in the biosynthesis of cholesterol in liver microsomes:

$$3\ \text{hydroxy}\ \ 3\ \ \text{methylglutaryl CoA} + 2\text{NADPH} + 2\text{H}+ \rightarrow \text{mevalonate} + 2\text{NADP}^{+} + \text{CoA}$$

Since the 5′-hydroxypentanoic pharmacophore of the statins resembles HMG-CoA, it can bind competitively to the catalytic domain of HMG-CoA reductase thereby preventing HMG-CoA from the access to the active site of the enzyme (see [15] for an historical account of statins).

In vitro the available statins inhibit the enzyme activity in the nanomolar range with atorvastatin and pitavastatin being more potent (Table 14.1).

LDL receptor is subjected to feedback regulation, i.e. its production is controlled by the need of cholesterol by cells. When the statin decreases cholesterol inside the cells, the LDL-receptor expression is enhanced and more receptor is found on the liver cell surface resulting in decreasing plasma LDL-C. In other words, the ultimate aim of statin is to stimulate the normal gene to produce higher number of LDL receptors in the hepatocytes [16].

While statins share the same mechanism and the effects on cholesterol, LDL and triglycerides, they differ regarding a number of physicochemical and pharmacokinetic characteristics (Table 14.1). Some derive from fungi, but others are synthetic. Their side chain can be either open (acid) or closed ring (lactone). The latter is inactive and must be opened to the β-hydroxy acid form by liver and plasma carboxyesterases.

Statins also vary regarding lipophilicity (they are all lipophilic except pravastatin and rosuvastatin) and various pharmacokinetic parameters: bioavailability, half-life and metabolism.

Statins are readily (though variably) absorbed from the intestine and undergo greater or lesser extensive first-pass metabolism in the liver through the action of cytochrome P450 (CYP) isoenzymes resulting in reduced systemic bioavailability. This is particularly remarkable for the more lipophilic statins (atorvastatin, lovastatin, simvastatin) which are metabolized via CYP3A4. By contrast, the disposition of pitavastatin and rosuvastatin is barely influenced by CYP enzymes and they are mostly excreted unchanged. In addition to metabolism, drug transporters (OATP) also play a significant role in statin disposition, including pitavastatin and rosuvastatin. Finally, with the exception of pravastatin, statins are highly bound to plasma proteins.

These pharmacokinetic characteristics can result in substantial consequences on the effects of statins. Let us consider two examples:

1. Since the access to CNS by passive diffusion depends on lipophilicity, simvastatin and lovastatin have a high potential to enter CNS, followed by fluvastatin. Nevertheless, focal cerebral ischaemia is associated with the dysfunction of neurovascular unit and the disruption of blood–brain barrier (BBB). This may facilitate statin entering and treatment of ischaemic disease [17, 18].
2. The interactions with other drugs, which in turn may influence the efficacy and side effects of statins. The selectivity of statin target (HMG-CoA reductase) prevents the interactions at pharmacodynamic level. Instead, they can take place at pharmacokinetic levels, especially regarding CYP enzymes in liver microsomes (Table 14.1) which catalyze the metabolism of all statins except the hydrophilic pravastatin and rosuvastatin (not extensively metabolized in the microsomes) and pitavastatin (excreted in unchanged form). More common interactions consist in the competitive inhibition of these enzymes leading to increased plasma level and potential side effects of statins. In other cases, their induction will result in a reduced plasma level and clinical activity of statins (reviewed in [19, 20]). These effects may be more serious in elderly patients who are frequently affected by comorbidities and inevitably receive multiple medications.

2.3 *The Pleiotropic Effects of Statins*

Apart from their cholesterol-reducing property, statins may also act through numerous cholesterol-independent, pleiotropic effects that occur before the reduction of serum LDL-C as reviewed in [21–27].

Most of these effects are secondary to the inhibition of the synthesis of isoprenoid, hydrophobic molecules with long acyl chains, which are downstream products of meva-

Table 14.1 Physicochemical and pharmacokinetic characteristics of statins

	Atorvastatin	Fluvastatin	Pitavastatin	Rosuvastatin	Lovastatin	Pravastatin	Simvastatin
Source	Synthetic	Synthetic	Synthetic	Synthetic	Fungal	Fungal	Fungal
Form	Acid	Acid	Acid	Acid	Lactone	Acid	Lactone
Pro-drug	No	No	No	No	Yes	No	Yes
Lipophilicity log d $_{7.4}$ (a)	Lipophilic 1.53	Lipophilic 1.75	Lipophilic 1.50	Hydrophilic −0.25 to −0.50	Lipophilic 3.91	Hydrophilic −0.47	Lipophilic 4.40
Absorption %	30	98	75	50	30	35	60–85
Bioavailability %	12 %	30 % (b)	51 %	20 %	5 %	18 %	<5 %
$t_{½}$, h	14	2.3	12	19	3	1.3–2.7	3
Protein binding %	>98	>99	>99	88	>98	50	>95
Cytochrome P-450 Isoenzymes	CYP 3A4	CYP 2C9	CYP 2C8, 2C9, Glucuronidation (minor) (c)	CIP2C9, 2C19 (minor) (c)	CYP 3A4	mostly sulfonation	CYP 3A4
OATP (d) transporter	Yes	Yes	Yes	Yes	Yes	Yes	Yes
Active metabolites	Yes	No	Yes (minimal)	Yes (minimal)	Yes	Yes (minimal)	Yes
IC50, nM (e)	1.16	3–10	0.1	0.16	2–4	4	1–2 (active metabolite

(a) Logaritm of octanol–water distribution coeffcient at pH 7.4
(b) Extended release can have lower bioavailability
(c) Mostly excreted unchanged
(d) OATP = organic anion-transporting polypeptide
(e) In vitro inhibition of HMG-CoA reductase on cultureed hepatocytes (Sirtori C. The pharmacology of statins. Pharm Rev. 2014;88:3–11)

Other Refs

Bellosta S, Corsini A. Statin drug interactions and related adverse reactions. Expert Opin Drug Saf. 2012;11:933–46

Sierra S, Ramos MC, Molina P, Esteo C, Vàzquez JA, Burgos JS. Statins as neuroprotectants: a comparative in vitro study of lipophilicity, blood–brain-barrier penetration, lowering of brain cholesterol, and decrease of neuron cell death. J Alzheimer Dis. 2011;23:307–18

McFarland AJ, Anoopkumar-Dukie S, Arora DS, Grant GD, McDermott CM, Perkins AV et al. Molecular mechanisms underlying the effects of statins in the central nervous system. Int J Mol Sci. 2014;15:20607–37 lonate in the cholesterol biosynthesis pathway. By inhibiting the synthesis of mevalonate (and consequently that of isoprenoids) statins induce the accumulation of the inactive, cytoplasmatic form of small G proteins, thereby inhibiting these signalling molecules.

A number of experimental investigations suggest that statins possess the following pleiotropic activities:

1. Improve endothelial dysfunction which is an early and crucial state of atherosclerosis. This effect is related to the increased availability of the potent vasodilator nitric oxide (NO) through the upregulation of its synthesizing enzyme (endothelial NO synthase, eNOS). Contributing to this result is the anti-oxidant effect: statins downregulate the activity and expression of the isoforms of NADPH oxidase, thus reducing the production of superoxide anion ($O_2^{\cdot-}$) and the oxidative stress in the vascular wall. In this context, atorvastatin reduced lipoperoxidation, and nitration and increased GSH levels in dog cortex [28].
2. Exert anti-inflammatory effects as shown by the decrease in the expression and circulating levels of inflammatory mediators: C-reactive protein (CRP), IL-6, IL-1β and TNF. Statins also reduced inflammatory cells in atheromatous plaques leading to plaque-stabilizing effects (review by [29]) and slowed down the progression of carotid atherosclerosis (Cochrane meta-analysis by [30]). Statins inhibit atherogenesis by reducing cholesterol accumulation and migration and proliferation of vascular smooth cells [31].
3. Have anti-thrombotic activity [32].
4. Promote angiogenesis, thus improving collateral blood supply [33].
5. Exert neuroprotection as shown by numerous experimental studies in neuronal cell cultures (e.g. showing increasing neurogenesis and synaptogenesis and protecting against NMDA-mediated excitotoxicity) and in various animal stroke models in which they reduced ischaemic lesion and ameliorated functional recovery [34–37] also in combination with tPA [38].
6. Exhibit neurotrophic effect by upregulating BDNF in neurons, microglia and astrocytes via the PPARα-CREB pathway and improve memory in mice [39].
7. Reduce the extracellular level of β-amyloid (Aβ, the main hallmark of Alzheimer's disease) both by cholesterol-dependent and isoprenoid-dependent mechanisms (review by [40, 41]) and regulate microglia-mediated inflammatory response to Aβ [42].

In summary, through their pleiotropic effects at vascular and parenchymal level, statins may affect multiple steps in the complex cascade of events that lead to atherosclerosis, thrombosis and cerebral ischaemia.

There is still some dispute whether the clinical activities of statins in the primary and secondary prevention of stroke are attributable to their pleiotropic effects or to cholesterol lowering.

On the one hand, a beneficial effect of statins in terms of good functional outcome was showed in patients with normal range of LDL [43, 44], without correlation between T-C and LDL-C and stroke outcome. Moreover, in a primary prevention RCT, rosuvastatin treatment was associated with a 48 % reduction in the risk of stroke (mostly AIS) in a large healthy population with normal LDL-C (<130 mg/dL), but high CRP levels [45]. This finding was matched by the observation of the role of CRP (but not of LDL-C) in statin-induced reduction of coronary atheroma and cardiovascular events [46], thus underlying once more their effects on inflammation as relevant mechanism of action. There is also experimental evidence that *acute*, intravascular statins induce parenchymal protection in animal AIS models [17, 47].

Arguing against a cholesterol-independent mechanism of stroke prevention, a thorough meta-analysis and meta-regression of 82 randomized trials on various cholesterol-lowering treatments found that the reduction of total stroke was proportional to the percent of T-C and LDL-cholesterol decline, as shown by odds ratio (OR) = 0.88 (0.83–0.94, $P < 0.001$) with 0.8 % reduction in the relative risk of stroke for each 1 % reduction of cholesterol [48]. Another meta-analysis also showed that the anti-inflammatory effect of statins (in terms of CRP reduction) is actually correlated with the magnitude of LDL decrease [49].

It is therefore reasonable to conclude that, while the pivotal role in statins action is played by LDL-cholesterol lowering, the pleiotropic anti-inflammatory and anti-thrombotic effects on the atherosclerotic plaques may also contribute extensively.

2.4 *Statins and Primary AIS Prevention*

Statins have been definitely proved to reduce cardiovascular risk and improve clinical output in CHD. Likewise, a number of studies demonstrated the beneficial effects of long-term statin treatment in preventing stroke incidence in subjects with or at high risk of atherosclerosis (reviews by [50–52]).

Table 14.2 displays the results of eight meta-analyses of numerous controlled trials on *AIS risk prevention*. The following conclusions can be drawn:

1. Statin long-term pre-treatment substantially reduces the risk of stroke, either total or non-fatal stroke (mainly ischaemic), but not that of fatal stroke.
2. This effect is associated with LDL-C lowering.
3. Intensive treatment is more effective than non-intensive.
4. The beginning of the statin era has lowered stroke incidence in familial hypercholesterolaemia.

Previous treatment with statins was also associated with improved *functional outcome and reduced mortality* in AIS patients (Table 14.3). The beneficial effects were reported up to 1 year post-stroke and in both large-artery and small vessel strokes. Moreover, statin pre-treatment was associated with reduced stroke severity and infarct volume [53, 54].

The aforementioned findings were confirmed by various studies in which a favourable outcome was either reported at presentation [55, 56], at discharge [57–60] or at 12 months (but not 3 months) [61]. The beneficial effect of statins was subtype dependent, since it was evident in atherothrombotic/large artery and lacunar, but not in cardioembolic strokes [62]. This finding was confirmed by Choi et al. [55] who reported lower initial NIHSS score in statin users than in non-users in athero-

Table 14.2 Pre-stroke statins and prevention of AIS risk: meta-analyses

Authors	No. of studies	Types of stroke	Statins (a)	AIS risk, RR/OR	Association with cholesterol or LDL reduction	Adverse effects
Law et al. (2003)	8 RCT 9 cohorts	Thromboembolic	At, Fl, Lov, Pra, Ros, Sim	−28 % (−35 to 20) −15 % (−21 to −6)	(b)	Rhabdomyolysis (rare)
O'Regan et al. (2008)	11	Overall ischaemic	At, Fl, Lov, Pra, Sim	0.81 (0.69–0.94)	Yes	
Amarenco and Labreuche (2009)	23 13	Fatal and non-fatal Fatal	At, Lov, Pra, Sim	0.81 (0.75–0.87) 0.90 (0.76–1.05) NS	Yes	
CCT Collaboration (2010)	21 5	Overall ischaemic	At, Fl, Lov, Pra, Sim same (c)	0.80 (0.73–0.88) 0.69 (0.50–0.95)	(b) (b)	↑ ICH: RR 1.15 NS
CTT Collaboration (2012)	22	Overall ischaemic	At, Fl, Lov, Pra, Sim	0.79 (0.74–0.85)	(b)	
De Caterina et al. (2010)	49	Total Fatal Non-fatal		0.85 (0.78–0.92) 0.98 (0.85–1.13) NS 0.81 (0.74–0.89)	↓1 % of T-C = ↓ 0.8 % RR for stroke	
Chan et al. (2011)	7	Fatal and non-fatal	At, Pra, Ros, Sim (c)	0.80 (0.71–0.89)	Yes	Elevated liver enzymes: OR 1.35
Wang and Zhang (2014)	15 9	Overall stroke Fatal	At, Fl, Pra, Ros, Sim	0.80 (0.74–087) 0–90 (0.67–1.21) NS		
Barkas et al. (2015)	4 (d)	ischaemic		OR 7.658 (e) OR 0.251 (f)		

RR = relative risk; OR = odds ratio; CCT = Cholesterol Treatment Trialist collaboration
(a) At = atorvastatin; Fl = fluvastatyin; Lov = lovastatin; Pit = pitavastatin ; Pra = pravastatin; Ros = rosuvastatin; Sim = simvastatin
(b) Per 1 mmol/L reduction of LDL-C
(c) Intensive vs. less intensive statin treatment
(d) Subjects with heterozygous familial hypercholesterolaemia; (e) 2 trials in pre-statin era; (f) 2 trials with statins
Differences were statistically significant ($P<0.05$) or highly significant ($P\leq 0.001$) except when indicated (*N*)

Refs

Amarenco P, Labreuche J, Touboul PJ. High-density lipoprotein-cholesterol and risk of stroke and carotid atherosclerosis: a systematic review. Atherosclerosis. 2008;196:489–96

Barkas F, Elisaf M, Milionis H. Statins decrease the risk of stroke in individuals with heterozygous familial hypercholesterolemia: a systematic review and meta-analysis. Atherosclerosis. 2015;243:60–4

Chan DKY, O'Rourke F, Shen Q, Mak JCS, Hung WT et al. Meta-analysis of the cardiovascular benefits of intensive lipid lowering with statins. Acta Neurol Scand. 2011;124:188–95

Cholesterol Treatment Trialists' (CTT) Collaboration. Efficacy and safety of more intensive lowering of LDL cholesterol: a meta-analysis of data from 170,000 participants in 26 randomised trials. Lancet. 2010;376:1670–81

Cholesterol Treatment Trialists' (CTT) Collaboration. The effects of lowering LDL cholesterol with statin therapy in people at low risk of vascular disease: meta-analysis of individual data from 27 randomised trials. Lancet. 2012;380:581–90

De Caterina R, Scarano M, Marfisi R, Lucisano G, Palma F. Cholesterol-lowering interventions and stroke: insights from a meta-analysis of randomized controlled trials. J Am Coll Cardiol. 2010;55:198–211

Law MR, Wald NJ, Rudnicka AR. Quantifying effect of statins on low density lipoprotein cholesterol, ischaemic heart disease, and stroke; systematic review and meta-analysis. BMJ. 2003;326:1423

O'Regan C, Wu P, Arora P, Perri D, Mills EJ. (2008). Statin therapy in stroke prevention: a meta-analysis involving 121,000 patients. Am J Med. 2008;121:24–33

Wang W, Zhang B. Statins for the prevention of stroke: a meta-analysis of randomized controlled trials. PLoS One. 2014;9:e92388

Table 14.3 Pre-stroke statins and outcome in AIS patients: meta-analyses

Authors	Studies no.	Types and subtypes of stroke	Functional outcome OR (a)	Mortality OR
Biffi et al. (2011)	12	Ischaemic Large-artery Small vessel	90 d: 1.62 (1.39–1.88) 2.01 (1.14–3.54) 2.11 (1.32–3.39)	
Nì Chròinin et al. (2013)	9 11 9	Ischaemic	at discharge: 1.64 (1.14–2.36) 90 days: 1.41 (1.29–1.56) 1 year: 1.12 (0.90–1.40) NS	at discharge/30 days: 0.63 (0.54–0.74) 90 days: 0.71 (0.62–0.82) 1 year: 0.80 (0.67–0.85)
Hong and Lee (2015)	19 5 8	Ischaemic	90 days: 1.50 (1.29–1.75) effect on initial stroke severity: 1.24 (1.05–1.48)	7–90 days: 0.42 (0.21–0.82)

OR = odds ratio; ICH = intracerebral haemorrhage
(a) mod Rankin scale (mRS 0-2); single statins were not reported
Differences were statistically significant ($P<0.05$) or highly significant ($P\leq0.001$) except when indicated (NS)

Refs

Biffi, A, Devan WJ, Anderson CD, Cortellini L, Furie KL, Rosand J et al. Statin treatment and functional outcome after ischemic stroke. Case-control and meta-analysis. Stroke. 2011;42:1314–9
Hong K-S, Lee JS. Statins in acute ischemic stroke: a systematic review. J Stroke. 2015;12:282–301
Nì Chròinìn D, Asplund K, Asberg S, Callaly E, Cuadrad-Godia E, Díez-Tejedor E et al. Statin therapy and outcome after ischemic stroke. Systematic review and meta-analysis of observational studies and randomized trials. Stroke. 2013;44:448–56

sclerotic/large artery, small-vessel occlusion and other subtypes of stroke, but not in cardioembolic ones. Intensive treatment with statin also prevented MRI-assessed ischaemic brain damage or microvascular alterations (white matter hyperintensities) in familial hypercholesterolaemia patients [63, 64]. Moreover, among different classes of cardiovascular drugs, only the use of statins was significantly associated with improved NIHSS score (OR 0.740, 0.580–0.944) [65].

Adverse effects: statins can be associated with increased risk of muscle-related adverse effects (myalgia, myopathy and, more rarely, rhabdomyolysis) and hepatotoxicity. They are caused by interactions with other drugs at the level of CYP enzymes or drug transporters (OATP) resulting in increased statins plasma concentration, which is an index of their potential toxicity. Drugs potentially interacting with statins belong to various, different pharmacological classes: cardiovascular, immunosuppressants, fibrates, SSRIs, protease inhibitors, macrolides. Should a statin-treated patient be prescribed one of these drugs, the statin dose must be properly reduced (reviewed in 19).

Concern was also raised by some reports of a significant increase in the odds of intracerebral haemorrhagic (ICH) or symptomatic haemorrhagic stroke (SICH) with statin therapy, but these preliminary findings were not confirmed. As shown in

Table 14.2, the CCT trial found a non-significant trend towards ICH increase. Moreover, two meta-analysis of 23 [66] and 31 [67] RCTs, respectively, documented the lack of association of statins with a statistically significant increment in the risk of ICH. Relative risks were, respectively, 1.10 (0.86–1.4) for primary and secondary prevention together [66] and 0.96 (0.75–1.23) for only the primary prevention [67]. Of note, there has also been a report of improved functional outcome in patients with haemorrhagic stroke pre-treated with statins [68].

2.5 *Statins and Primary Prevention in AIS Patients Treated with Thrombolysis*

The aforementioned promising results on statin stroke prevention suggested a possible improvement of pre-stroke statins on functional activities also in tPA-treated patients.

Experimentally, simvastatin administered before a large clot embolic stroke in rabbits subsequently treated with tPA significantly reduced the tPA-induced haemorrhagic incidence and volume [69].

As displayed in Table 14.4, studies in patients with AIS treated with i.v. thrombolysis yielded controversial findings. A large, recent meta-analysis [70] and a study on dose-related statins [71] reported beneficial effects of pre-stroke statin on functional outcome. Investigations on early clinical recovery [72] and on reperfusion were consistent with favourable effects. Conversely, three previous meta-analyses [73–75] failed to find them. Moreover, meta-analyses and the Scheitz's study [71] on high doses showed a detrimental statin effect on the rate of ICH and SICH. This finding was also replicated in EVT-receiving patients [76], although recent trials with more advanced thrombectomy devices are lacking. However, a pooled observational study (not shown on Table 14.4) did not find significant differences between statin users and non-users regarding functional outcome and ICH in stroke patients treated with IVT [77].

2.6 *Statins and Secondary Prevention of Recurrent Stroke*

Survivors of AIS are at high (though variable across studies) risk of recurrence [78]. A systematic review and meta-analysis of 13 studies reported a cumulative risk of 3.1 % at 30 days, 11.1 % at 1 year, 26.4 % at 5 years and 39.2 % at 10 years after the first stroke [79]. Main predictors were age, previous TIA and diabetes mellitus. Association was also reported with atherogenic dyslipidaemia, particularly in patients with strokes of large-artery atherosclerosis subtype [80].

Experimentally, in middle cerebral artery occlusion (MCAO) statin administration (1) reduced infarct area and upregulated eNOS in rat blood vessels [81], (2)

Table 14.4 Pre-stroke statins and outcome in AIS patients treated with thrombolysis

Authors	Type of study	Statins (a)	Good functional outcome (b) OR	Mortality OR	SICH OR
Cordenier et al. (2011)	Meta-analysis (11studies)		1.01 (0.64–1.61) NS	0.56 (0.40–0.78)	2.34 (1.31–4.17)
Meseguer et al. (2012)	Meta-analysis (11studies)	At, Fl, Lov, Pit, Pra, Ros, Sim	0.99 (0.88–1.12) NS		1.31 (0.97–1.76) NS
Martinez-Ramirez et al. (2012)	Meta-analysis (3 studies)		1.09 (0.73–1.61) NS	1.32 (0.84–2.07)	1.99 (1.03–3.84)
McKinney and Kostis (2012)	Meta-analysis (31 RCTs)				0.96 (0.75–1.23)
Hong and Lee (2015)	Meta-analysis	–	1.44 (1.10–1.89) (12 studies)	0.87 (0.58–1.32) (5 studies)	1.63 (1.04–2.56) (10 studies)
Scheitz et al. (2014)	Pooled 2 registries	At, Fl, Pra, Ros, Sim	1.80 (1.29–2.519) 1.88 (1.19–2.98) low dose 1.78 (1.11–2.84) medium dose 1.66 (0.88–3.16) high dose	–	1.93 (0.83–447) NS 1.10 (0.24–5.14) low dose 2.68 (0.89–8.06) med. dose 6.20 (1.72–22.39) highdose
Campos et al. (2013)	Observational	–			0.74 NS
Tsivgoulis et al. (2015)	SITS-EAST	–	mRS: 0.81 (0.52–1.27) NS ECR: 1.91 (1.25–1.92) (c)	0.92 (0.57–1.49) NS	1.13–1.89 NS
Ford et al. (2011)	MRI-assessed reperfusion	–	No IVT: MTT 48 % (statin) vs. 13 % (non-statin) (d) ΔNIHSS 8.8 ± 4.0 points (statin) vs. 4.4 ± 5.7 points (non-statin) (e) IVT: MTT 48 % (statin) vs. 15 % (non-statin) (d) ΔNIHSS 8.9 ± 4.2 (statin) vs. 4.4 ± 4.9 (non-statin) (e)	–	–

IVT = intravenous thrombolysis; OR = odds ratio; SICH = symptomatic intracerebral haemorrhage; MRI = magnetic resonance imaging
(a) At = atorvastatin; Fl = fluvastatin; Lov = Lovastatin; Pit = Pitavastatin; Pra = Pravastatin; Ros = Rosuvastatin; Sim = Simvastatin; (b) mod Rankin scale (mRS 0-2); (c) ECR = early clinical recovery: reduction in the baseline National Institute of Health Stroke Scale (NIHSS) score of ≥10 poitns at 24 h; (d) MTT = mean

transit time; (e) ΔNIHSS = change in functional improvement from admission to 1 month after stroke according to the NIHSS. Differences were statistically significant ($P < 0.05$) or highly significant ($P \leq 0.001$) except when indicated (NS)

Refs

Campos M, Garcìa-Bonilla L, Hernàndez-Guillamon M, Barcelo V, Morancho A, Quintana M et al. Combining statins with tissue plasminogen activator treatment after experimental and human stroke: a safety study on hemorrhagic transformation. CNS Neurosci Ther. 2013;19:863–70

Cordenier A, De Smedt A, Brouns R, Uyttenboogaart M, De Raedt S, Luijckx GJ et al. Pre-use of statins on stroke outcome: a meta-analysis of observational studies. Ann Neurol Belg. 2011;111:261–7

Ford AL, Hongyu A, D'Angelo G, Ponisio R, Bushard P, Vo KD et al. Preexisting statin use is associated with greater reperfusion in hyperacute ischemic stroke. Stroke. 2011;42:1307–13

Hong K-S, Lee JS. Statins in acute ischemic stroke: a systematic review. J Stroke. 2015;12:282–301

Martinez-Ramirez S, Delgado-Mederos R, Marin R, Suàrez-Calvet M, Sàinz MP, Alejaldre A et al. Statin pretreatment may increase the risk of symptomatic intracranial haemorrhage in thrombolysis for ischemic stroke: results from a case-control study and a meta-analysis. J Neurol. 2012;259:111-8

McKinney JS, Kostis WJ. Statin therapy and the risk of intracerebral hemorrhage. A meta-analysis of 31 randomized controlled trials. Stroke. 2012;43:2149–56

Meseguer E, Mazighi M, Lapergue B, Labreuche J, Sirimarco G, Gonzalez-Valcarcel J et al. Outcomes after thrombolysis in AIS according to prior statin use. Neurology. 2012;79:1817–23

Tsivgoulis G, Kadlecovà P, Kobayashi A, Czlonkoska A, Broznam M, Švigelj V et al. Safety of statin pretreatment in intravenous thrombolysis for acute ischemic stroke. Stroke. 2015;46:2681–4 induced both parenchymal and vascular protection and (3) improved functional recovery in mice [82].

Neuroprotection by statin in stroke animal models was also corroborated by (1) the interaction with Rho-kinase signalling pathway [83] and (2) the proteomic analysis of rat brain homogenates showing that the reduction of infarct volume and neurological improvement were associated with oxidative stress attenuation and BBB protection [84].

Clinically, the effects of statins in preventing *the risk of recurrent AIS* are displayed in Table 14.5, while Table 14.6 outlines the outcome in recurrent AIS (review by 85).

In the SPARCL trials, high-dose atorvastatin (80 mg daily) in patients with a recent AIS (70 %) or TIA (30 %) significantly reduced the recurrent risk of all AIS by 16 % and that of fatal AIS (but not of non-fatal) by 43 %. Comparable effects were found in men vs. women, in patients aged >65 years vs. those <65 years, and in those with carotid stenosis vs. no carotid stenosis. The statin preventive effect of recurrent AIS was related to the reduction of LDL-C, thus supporting their role as a surrogate therapeutic target, but not to baseline stroke subtypes (large-vessel vs. small-vessel disease) [86]. The latter finding was not replicated by Hosomi et al. [87] who documented reduced atherothrombotic AIS rate (but not of other subtypes) in patients treated with pravastatin. It is unknown whether this observation can be generalized or is peculiar to Asian patients, whose stroke profile is different from Caucasians due to higher prevalence in the former ones of lacunar/small vessel infarction and cerebral haemorrhage.

Of note in SPARCL, atorvastatin treatment was associated with higher incidence of ICH [88], a finding not supported by the large meta-analysis by McKinney and Kostis [67]. Increased rate of ICH was also documented in a subset of patients in the Heart Protection Study (HPS) [89] with a history of prior ICH, suggesting the need of caution in using statins in these particular subjects [5].

Regarding *the outcome* (Table 14.6), the majority of studies indicate better functional outcome (or less poor [90]) and reduced mortality as a result of statin treatment. Two observations are worth mentioning: (1) the effects of in-hospital or post-stroke statin treatment seem to be higher than the pre-stroke one, and (2) withdrawal of therapy is an independent predictor of worsened outcome and mortality, in agreement with previous reports [91, 92]. This further underlines the importance of prolongation of post-stroke treatment as recommended by the Guidelines [5]. Notably, in-hospital survival was associated with early statin treatment as shown by progressively increased hazard ratios from 0.51 (initiation on Day 1) to 0.57 (Day 2), and 0.68 (Day 3 or later) [93]. The same authors reported a dose–response relationship with high doses more active than low-moderate doses both in the pre-stroke and in-hospital setting. This further strengthens the casual relationship between statin use and post-stroke survival. Statin use was also significantly associated with increased likelihood of discharge at home (or to rehabilitation centre) and decreased death in hospital. Again, the in-hospital statin initiation (OR 2.02) provided significantly better outcome than the pre-stroke treatment (OR 1.21). For patients treated before stroke and in-hospital the OR was 2.08 [57].

Table 14.5 Secondary stroke prevention with statins: effect on recurrent stroke risk

Authors	Type of study	Types and subtypes of strokes	Statins (a)	Follow-up, years	Risk, of recurrent stroke, HR	Risk of mortality, HR	Risk of ICH, HR
Amarenco and Labreuche (2009)	Meta-analysis		At, Pra, Sim	4.9–6.1	0.82 (0.77–0.87)		
McKinney and Kostis (2012)	Meta-analysis		–	–			1.26 (0.91–1.73) NS
Amarenco et al. (2006)	RCT (SPARCL)	All Non fatal Fatal	At (b)	Median 4.9	0.84 (0.71–0.99) 0.87 (0.73–1.03) NS 0.57 (0.35–0.95)	0.78 (0.66–0.94)	
Amarenco et al. (2007)	Post-hoc analysis of SPARCL		At (b)	Median 4.9	0.66 (0.52–0.83 (e) 0.89 (0.72–1.09) (f)		
Goldstein et al. (2008)	Post-hoc analysis of SPARCL		At(b)	Median 4.9	0.79 (0.66–0.95)		1.68 (1.09–2.59)
Milionis et al. (2009)	Retrospective		At, Fl, Pra, Sim (c)	10	0.61 (0.35–0.92)	0.43 (0.29–0.61)	
Makihara et al. (2013)	Multicentre Fukuda Registry	All Non CE	–	Median 2	0.70 (0.53–0.92) 0.66 (0.49–0.89)	0.67 (0.50–0.69)	
Hosomi et al. (2015)	RCT, multicentre J-STARS	All subtypes Atherothrombotic/LAA	Pra (d)	4.9 ± 1.4	0.95 (0.71–1.28) NS 0.33 (0.15–0.74)		

HR = hazard ratio; ICH = intracerebral heamorrhage; SPARCL = Stroke Prevention by Aggressive Reduction in Cholesterol Level; RCT = randomized clinical trial; CE = cardio-embolic; LAA = large artery atherosclerosis

(a) At = atorvastatin; Fl = fluvastatin; Pra = pravastatin; Sim = simvastatin

Time between stroke and the beginning of statin treatment: (b) = 1–6 months , (c) at post-discharge, (d) = 1 month; (e) for ≥50 % decrease of LDL; (f) for <50 % decrease of LDL. Differences were statistically significant ($P < 0.05$) or highly significant ($P \leq 0.001$) except when indicated (NS)

(continued)

Table 14.5 (continued)

Refs

Amarenco P, Labreuche J, Touboul PJ. High-density lipoprotein-cholesterol and risk of stroke and carotid atherosclerosis: a systematic review. Atherosclerosis. 2008;196:489–96

Amarenco P, Bogousslavsky J, Callahan III A, Goldstein LB, Hennerici M, Rudolph AE et al. High-dose atorvastatin after stroke or transient ischemic attack. N Engl J Med. 2006;355:549–59

Amarenco P, Goldstein LB, Szarek M, Sillisen H, Rudolph AE, Callahan III A et al. Effects of intense low-density lipoprotein cholesterol reduction in patients with stroke or transient ischemic attack: the Stroke Prevention by Aggressive Reduction in Cholesterol Levels (SPARCL) trial. Stroke. 2007;38:3198–204

Goldstein LB, Amarenco P, Szarek M, Callahan III A, Hennerici M, Sillesen H et al. Hemorrhage stroke in stroke prevention by aggressive reduction in cholesterol levels study. Neurology. 2008;70:2364–70

Hosomi N, Nagai Y, Kohriyama T, Ohtsuki T, Aoki S, Nezu T et al. The Japan Statin Treatment Against Recurrent Stroke (J-STARS): a multicenter, randomized, open-label, parallel-group study. EBioMedicine. 2015;2:1071–8

Makihara N, Kamouchi M, Hata J, Matsuo R, Ago T, Kuroda J et al. Statins and the risk of stroke recurrence and death afeter ischemic stroke: the Fukuota Stroke Resgistry. Atherosclerosis. 2013;231:211–5

McKinney JS, Kostis WJ. Statin therapy and the risk of intracerebral hemorrhage. A meta-analysis of 31 randomized controlled trials. Stroke. 2012;43:2149–56

Milionis HJ, Giannopoulos S, Kosmidou M, Panoulas V, Manios E, Kyritsis AP et al. Statin therapy after first stroke reduces 10-year stroke recurrence and improves survival. Neurology. 2009;72:1816–22

Table 14.6 Secondary stroke prevention with statins: effects on the outcome in recurrent AIS

Authors	Type of study	Statins (a)	Administration time	Follow-up	Good functional outcome (b) OR	Mortality, OR
Hong and Lee (2015)	Meta-analysis	_	In-hospital	At discharge/90 days	1.31 (1.12–1.53)	0.41 (0.29–0.58)
Nì Chròinìn et al. (2013)	Meta-analyses: Observational studies One RCT	_	In-hospital (0–72 h)	discharge/30 days 90 days 1 year 90 days	1.64 (1.14–2.36) 1.41 (1.29–1.56) 1.12 (0.90–1.40) NS 1.50 (1.0–2.24)	0.63 (0.54–0.74) 0.71 (0.62–0.82) 0.80 (0.67–0.95) 0.98 (0.67–1.42)
Song et al. (2014)	National, prospective registry	_	In-hospital: All strokes Minor Non minor	90 days	0.95 (0.81–1.11) (c) 1.02 (0.78–1.34) NS (c) 0.73 (0.57–0.92) (c)	0.51 (0.38–0.67) 0.68 (0.40–1.14) 0.44 (0.31–0.62)
Flint et al. (2012)	Multicentre database	Lov, Sim	– Pre-stroke – Pre-and in-hospital – In-hospital – Withdrawal	1 year	_	0.85 (0.79–0.93) 0.59 (0.53–0.65) 0.55 (0.50–0.61) 2.5 (2.1–2.9)
Hijalmarsson et al. (2012)	Retrospective (mean age 78 years)	_	Pre-stroke Post-stroke	30 days 1 year	0.76 (0.46–1.25) NS 2.09 (1.25–3.52)	1.80 (0.75–4.29) NS 0.33 (0.20–0.54)
Tziomalos et al. (2015)	Prospective	At, Ros, Sim (d)	At discharge	1 year	mRS :low dose 2.7 ± 2.2 Intermediate 2.2 ± 2.2 High 1.8 ± 2.2 NS	_
Moonis et al. (2014)	Retrospective (national database)	_	Pre-stroke Post-stroke	at discharge	1.67 (1.12–2.49) 2.63 (1.61–4.31)	_
Choi et al. (2015)	Prospective database		Pre-stroke In-hospital	at discharge	1.55 (1.25–1.92) 1.26 (1.14–1.40)	_
Hong and Lee (2015)	Withdrawal meta-analysis	_	In-hospital	90 days	1.83 (1.01–3.30) (c)	2.5 (2.1–2.9) (1 year)

(continued)

Table 14.6 (continued)

OR = odds ratio; RCT = randomized clinical trial
(a) At = atorvastatin; Lov = lovastatin; Ros = rosuvastatin; Sim = simvastatin
(b) Good functional outcome: mod Rankin scale (mRS 0–2)
(c) Poor functional outcome: mRS > 2 (= dependency)
(d) Patients were treated with atorvastatin or equipotent doses of the other statins
Differences were statistically significant ($P<0.05$) or highly significant ($P\leq0.001$) except when indicated (NS)

Refs

Choi JC, Lee JS, Park TH, Cho YJ, Parl JM et al. Effect of pre-stroke statin use on stroke severity and early functional recovery: a retrospective cohort study. BMC Neurol. 2015;15:120

Hjalmarsson C, Bokemark L, Manhem K, Mehlig K, Andersson B. The effect of statins on acute and long-term outcome after ischemic stroke in the elderly. Am J Geriatr Pharmacother. 2012;10:313–22

Hong K-S, Lee JS. Statins in acute ischemic stroke: a systematic review. J Stroke. 2015;12:282–301

Flint AC, Kamel H, Navi BB, Rao VA, Faigeles BS, Conell C et al. Statin use during ischemic stroke hospitalization is strongly associated with improved poststroke survival. Stroke. 2012;43:147–154

Moonis M, Kumar R, Henninger N, Kane K, Fisher M. Pre and post-stroke use of statins improves stroke outcome. Indian J Community Med. 2014;39:214–7

Nì Chròinìn D, Asplund K, Asberg S, Callaly E, Cuadrad-Godia E, Díez-Tejedor E et al. Statin therapy and outcome after ischemic stroke. Systematic review and meta-analysis of observational studies and randomized trials. Stroke. 2013;44:448–56

Song B, Wang Y, Zhao X, Liu L, Wang C et al. Association between statin use and short-term outcome based on severity of ischemic stroke : a cohort study. PLoS ONE. 2014;9:e84389

Tziomalos K, Giampatzis V, Bouziana SD, Spanou M, Papadopoulou M, Kazantzidou P et al. Effects of different classes of antihypertensive agents on the outcome of acute ischemic stroke. J Clin Hypertens. 2015;17:275–80

2.7 Statins and Secondary Prevention in AIS Patients Treated with Thrombolysis

To our knowledge, experimental and clinical studies on the effects of statins combined with or following tPA are scant.

Experimentally, the post-stroke combination of high-dose atorvastatin with rtPA in a rat embolic MCAO reduced the infarct volume at 7 days and the incidence of ICH, enhanced CBF in the ipsilateral hemisphere (without affecting that in the contralateral one), and improved neurological outcome. These results could be attributed to the effects on cerebral microvascular integrity by the upregulation of eNOS expression [38, 94]. Conversely, given alone, simvastatin improved behaviour following embolization through a neuroprotective mechanism, but did not exert additive or synergistic effects in combination with tPA [69].

Among the few available clinical studies (Table 14.7) three [95–97] reported significantly good or excellent functional outcome and neurological improvement by in-hospital, post-thrombolysis statin administration in statin-naive patients, whereas no statin use was associated with poor outcome. However, one study [98] did not find favourable results. Moreover, statin treatment started before stroke and continued in the acute phase, and cardioembolic stroke was the only factor associated with SICH [95].

2.8 Conclusions

Over the past decade a number of experimental and clinical studies (RCTs, observational and meta-analyses) reported the benefit of primary and secondary statin therapies regarding the risk and outcomes of first and recurrent ischaemic strokes. These effects were obtained in patients with a different range of ages and blood pressures, irrespective of sex, previous vascular disease or baseline LDL level [61]. Favourable effects on stroke correlate with the substantial reduction of the risk for CHD in the context of the management of various vascular risk factors (hypertension, diabetes mellitus, atrial fibrillation). At variance with the above-mentioned statement, three meta-analyses found that the outcome in thrombolyzed patients pre-treated with statins did not significantly improve. The evidence (shown in post hoc analysis of SPARCL and HPS) linking long-term statins to ICH (Table 14.5) was based on a relatively low number of events and not confirmed by the large meta-analyses of Hackam et al. [66] and Mc Kinney and Kostis [67].

Two controversial issues have arisen:

1. The mechanism(s) of action of statins: as discussed before, lipid lowering is considered to be the main mechanism of statins (although in some cases it was not correlated to stroke prevention), but clinical activities are also attributable to their pleiotropic effects. An example of this is the atheroprotective effects on the

Table 14.7 Pre- and/or post-stroke statin and outcome in AIS patients treated with thrombolysis (ivt or evt)

Authors	Type of study and thrombolysis	Statins (a)	Statin administration time	Follow-up days	Functional outcome: (b) good, (c) excellent or (d) poor; major neurological improvement (e), OR	Mortality, OR	SICH, OR
Cappellari et al. (2011)	Retrospective (single centre) IVT	At, Pra, Ros, Sim No statin	Pre-stroke +10.6 h post IVT In-hospital 11.5 h post IVT same	90 90 90	(b) NS (b) 6.18 (1.43–26.62) (d) 4.13 (1.49–11.42)	_ _ _	6.65 (1.58–29.12) (f) _ _
Cappellari et al. (2013)	Retrospecitve (multicentre) IVT	At, Sim, others	In-hospital 24–72 h post IVT same	7 90	(e) 1.43 (1.11–1.85) (b) 1.63 (1.18–2.26)	_ 0.48 (0.28–0.82)	0.52 (0.20–1.34) NS _
Nì Chròinin et al. (2013)	Meta-analysis IVT		Pre-stroke In-hospital 0–72 h post-stroke	90 90	(b) 1.01 (0.88–1.15) NS (b) NS	1.25 (1.02–1.52) 1.14 (0.90.1.44) NS	
Restrepo et al. (2009)	Prospective database (single centre) EVT/ PME	At	Pre-and post-stroke In-hospital (48 h to 19 days)	90 90	(c) 0.7 (0.1–3.6) NS (e) 45 % in statin pts vs. 39 % in no-statin NS 1.7 (0.5–6.6) (c) (e) 44 % in statin pts vs. 25 % in non-statin		1.4 (0.4–4.9) NS 1.5 (0.6–4.0) NS

OR = odds ratio; IVT = intravenous thrombolysis; EVT = endovascular thrombolysis; PME = percutaneous mechanical embolectomy
(a) At = atorvastatin; Pra = pravastatin; Ros = rosuvastatin; Sim = simvastatin
(b) good functional outcome (mRS 0–2)
(c) excellent outcome (mRS 0–1)
(d) poor functional outcome (mRS 3–6)
(e) neurological improvement according to the National Institute of Health Stroke Scale, NIHSS (reduction ≥4 points from baseline)
(f) In cardioembolic stroke. Differences were statistically significant ($P<0.05$) or highly significant ($P\leq0.001$) except when indicated (NS)

Refs

Cappellari M, DeLuca C, Tinazzi M, Tomelleri G, Carletti M, Fiaschi A et al. Does statin in the acute phase of ischemic stroke improve outcome after intravenous thrombolysis? A retrospective study. J Neurol Sci. 2011;308:128–34

Cappellari M, Bovi P, Moretto G, Zini A, Nencini P, Sessa M et al. The THRombolysis and STatins (THRaST) study. Neurology. 2013;80:655–61

Nì Chròinìn D, Asplund K, Asberg S, Callaly E, Cuadrad-Godia E, Díez-Tejedor E et al. Statin therapy and outcome after ischemic stroke. Systematic review and meta-analysis of observational studies and randomized trials. Stroke. 2013;44:448–56

Restrepo L, Bang OY, Ovbiagele B, Ali L, Kim D, Liebeskind DS et al. Impact of hyperlipidemia and statins on ischemic stroke outcomes after intra-arterial fibrinolysis and percutaneous mechanical embolectomy. Cerebrovasc Dis. 2009;28:384–90

carotid and major cerebral arteries resulting in the suppression of atherosclerotic process. Haemorrhagic complications could also be attributed to pleiotropic effects of statins through their anti-thrombotic activity.
2. The balance between risks and benefit of statin administration in AIS patients with particular reference to intracerebral haemorrhage: the ICH risk in patients on pre-stroke statin and treated with thrombolysis (Table 14.4) might be related to the upregulation of tPA and reduction of the activity of plasminogen activator inhibitor 1 (PAI), thereby resulting in the stimulation of the fibrinolytic system [99]. ICH could especially occur in patients with small vessel disease who often have silent intracerebral microhaemorrhages progressing into manifest macrohaemorrhages.

The crucial question is, therefore, whether pre-stroke statin treatment should be discontinued after AIS. As noted by Scheitz et al. [100] on the basis of a survey of 12 trials, available data do not indicate that statins long-term treatment confers a substantially higher risk of ICH, except in chronic haemorrhagic patients with risk of recurrence, for whom extreme caution should be exercised when considering continuation of statin therapy. On the other hand, abrupt discontinuation increases infarct volume and worsens clinical outcome. Even a brief withdrawal of statins for >2 days after AIS causes a significantly enhanced mortality [93].

Indeed Guidelines recommend continuation of statin therapy during and after AIS. Regrettably, the patients' long-term compliance to secondary statin treatment after first stroke remains uncertain and probably inadequate [85, 101].

Finally, despite the dearth of available studies, the in-hospital initiation of statin treatment in the acute phase of AIS—whether or not in combination with rtPA—might suggests a favourable short- and long-term outcome, in some cases even more robust than that obtained with pre-stroke administration. In the lack of ad hoc RCTs, this issue is still debated. A cautionary view urges that, because of safety issues (risk of ICH, myopathy, elevation of liver enzymes), statins should not be used indiscriminately, but only in atherosclerotic strokes and weighing the characteristics of the individual [102, 103]. Conversely, others emphasize the favourable effects of statins in terms of reduced stroke recurrence and better functional outcome (probably related to the pleiotropic effects) in the first post-stroke days [104].

In view of the central importance of neuroprotective therapies for post-stroke recovery in thrombolyzed, as well as in non-thrombolyzed patients, there is a need for appropriate, definite RCTs. Indeed, the hard part is yet to come.

3 Anti-hypertensives

3.1 Hypertension as a Risk Factor for AIS

Hypertension has been arbitrarily defined by the World Health Organization as systolic blood pressure (SBP)>140 mmHg and diastolic blood pressure (DBP)>90 mmHg. It is long known that hypertension is one of the most important modifiable risk factors for

AIS, the relationship between BP levels and first stroke risk appearing to be log-linear [105]. In particular, Lewington et al. [106] showed that over the range of 115/75 to 185/115 mmHg, each 20-mmHg elevation in SBP doubles the risk of death from stroke.

The ways by which hypertension may cause stroke are manifold. Several studies have shown that the CBF remains steady to its normal value (approximately 50 mL/100 g/min in humans) within a fixed range of mean BP (60–150 mmHg). Upon neuronal activity, CBF is rapidly increased through the activation of autoregulation mechanisms, based on the dilation and constriction of cerebral vessels to meet the metabolic requirements of the brain parenchyma.

Chronic hypertension induces complex pathological changes in the medias of brain arteries, producing vascular hypertrophy and increasing the vascular resistance to protect the cerebral microcirculation. On the other hand, these structural changes predispose to cerebral ischaemia by impairing vasodilation, promoting atherosclerosis in large cerebral arteries, and lipohyalinosis in small penetrating arterioles [107].

Recently, the Reasons for Geographic and Racial Differences in Stroke (REGARDS) study, a longitudinal cohort study involving 30,239 community-dwelling black and white individuals aged >45 years, showed that hypertensive individuals had over twice a residual risk of stroke compared with those who were normotensive without medication, this risk increasing with the number of employed anti-hypertensive classes: being the SBP target value <120 mmHg, patients treated with 1 class of antihypertensive medication showed 42 % higher stroke risk than those who were at that same BP level without medications. The risk of stroke was, respectively, 1.60-fold and 2.48-fold higher among hypertensive individuals who were taking 2 and ≥3 classes [108].

Therefore, although decades of studies on the impact of hypertension and of anti-hypertensive treatments on AIS risk, many aspects remain unclear, suggesting that the real problem is not hypertension per se but its pathogenetic cause, unfortunately unknown in the majority of cases.

3.2 *The Mechanism of Action of Anti-hypertensives and Their Pleiotropic Effects*

3.2.1 ACE-Is and ARBs

The renin–angiotensin system (RAS) plays an important role in the regulation of volaemia and BP. This system is activated by the release from the kidneys of the aspartyl protease renin, which cleaves the first ten amino acids present on the angiotensinogen, an inactive precursor of hepatic origin. These cleaved residues are referred to as angiotensin I (Ang I), which is the substrate of the angiotensin-converting enzyme (ACE) that transforms angiotensin I into vasoactive angiotensin II (Ang II) through the removal of further two amino acids [109].

However, Ang II exerts several other physiologic effects through the binding to its specific receptors AT_1R and AT_2R: it has a direct effect both on blood vessels, being a potent vasoconstrictor, and on kidneys, where it regulates the renal blood flow and

Na^+ reabsorption; moreover, Ang II is dipsogen and stimulates the adrenal gland to produce aldosterone, an hormone that induces the reabsorption of water and of Na^+ in the distal tubules and collecting ducts of nephrons, therefore increasing water retention and BP. Besides these physiological effects whose imbalance may explain high BP occurrence, by interacting with AT_1R, Ang II seems also to promote thrombosis [110, 111] and atherosclerosis [112, 113], overall increasing the risk of AIS.

The anti-hypertensive action of ACE inhibitors (ACE-Is) relies on the blockade of the conversion of Ang I to Ang II and on the prevention of the deleterious effects of Ang II. The Heart Outcomes Prevention Evaluation (HOPE) study has suggested that these drugs might be beneficial on endothelium, possessing an anti-atherosclerotic effect in addition to their BP-lowering properties [114]. Moreover, in RCTs, ACE-Is short-term treatment significantly reduced concentrations of fibrinogen and of plasminogen activator inhibitor-I [115, 116].

The newer drug class of Angiotensin II Receptor Blockers (ARBs), which are antagonists of the AT_1R, has been developed to overcome the adverse effects of ACE-Is arising from the non-selectivity of the enzyme ACE towards Ang I, being also bradykinin and other tachykinins substrates of this enzyme. Another advantage of ARBs over ACE-Is is that, apart from their anti-hypertensive effect, ARBs exert also a neuroprotective action through the AT_2R stimulation as a consequence of the selective AT_1R inhibition [117]. To this purpose, many experimental studies have confirmed the positive effects of AT_2R agonists on functional deficits and infarct volume in several stroke models, such as focal cerebral ischaemia–reperfusion by MCAO [118, 119] and temporary endothelin-1-induced stroke of MCA in conscious spontaneously hypertensive rats [120, 121].

Moreover, there are pieces of evidence supporting the beneficial effects of ARBs beyond lowering BP and AT_2R stimulation: in rats exposed to MCAO, sub-hypotensive doses of candesartan were shown to promote neuroplasticity by enhancing the expression of vascular endothelial growth factor (VEGF), brain-derived neurotrophic factor (BDNF; [122]), synaptophysin and PSD-95 [123], and to reduce BBB disruption and oedema [124].

Recently, several studies have produced new insights about the RAS system (reviews by [125–127]), and novel molecules have been identified that seem involved in neuroprotection, next to the classical components of the RAS system (ACE/Ang II/AT_1 axis). The most studied molecules in this context have been those of the *ACE2/Ang* (1–7)*/Mas axis* [128–130]. Briefly, the enzyme ACE2 converts Ang II into the peptide Ang (1–7), which in turn binds to its receptor Mas exerting several cerebroprotective effects, such as the stimulation of vasodilation and of anti-thrombotic, anti-inflammatory and anti-oxidative mechanisms (reviews in [131, 132]).

These beneficial effects have been observed also in experimental ischaemic stroke models [127, 133–135] and, interestingly, all the components of the ACE2/Ang (1–7)/Mas axis have been shown to be overexpressed after AIS in rats [136], suggesting that these molecules should be considered as part of endogenous mechanisms of tissue repair and promising molecular targets of new therapeutic strategies.

3.2.2 Calcium Channel Blockers

Calcium channel blockers (CCBs) are antagonists of the voltage-gated L-type Ca^{2+} channels (LTCCs) and were firstly developed for the treatment of *angina pectoris*, having been afterwards proven beneficial also for ischaemic heart disease and hypertension.

The BP-lowering mechanism of action of these drugs is based on the prevention of free Ca^{2+} entry in the smooth muscle cells of the vessels and thus on the prevention of the activation of the Ca^{2+}-calmodulin myosin light-chain kinase, which is responsible for the phosphorylation of myosin light chain, increasing the activity of actin-myosin kinase and muscle contraction. Therefore, CCBs act by relaxing vascular smooth muscles and decreasing the systemic vascular resistance, with negligible effects on venous capacitance vessels.

LTCCs are expressed in many tissues and not only on smooth muscle cells: as reviewed by Striessnig et al. [137], Cav1.1 channels are present almost exclusively in skeletal muscle where they are responsible for depolarization-induced Ca^{2+} release from the sarcoplasmic reticulum; Cav1.2 and Cav1.3 are expressed in many organs, such as heart and brain, while Cav1.4 channel is present in the retina.

Apart from their anti-hypertensive action, many studies proposed that CCBs might exert neuroprotective effects, not only for AIS [138] but also for neurodegenerative diseases like Parkinson's disease [139]. This hypothesis about CCBs as neuroprotectant arose several years ago, starting from a study by Steen and colleagues published in 1983 about the neuroprotection induced by nimodipine in animal models of ischaemia [140].

Neuroprotection by CCBs may be achieved by many molecular mechanisms:

1. By dilating the collateral blood vessels. Actually, this effect may reveal itself a double-edge sword, being autoregulation of CBF preserved only in non-ischaemic areas, where consequent intracerebral flow steal and hypotension may occur.
2. By blocking the excessive increase of intracellular Ca^{2+} concentrations and by this way counteracting glutamate-Ca^{2+} overload neurotoxicity [141]. Nevertheless, it is now acknowledged that glutamate massive release peaks within 30 min from AIS occurrence [142], being glutamate promptly metabolized [143].
3. By exerting anti-oxidant properties, preventing the formation of intracellular reactive oxygen species produced after neuronal Ca^{2+} overload [144] and increasing the activity of anti-oxidant enzymes, such as glutathione peroxidase, glutathione reductase and superoxide dismutase [145]. This hypothesis was confirmed by Yamato et al. [146], which demonstrated that nifedipine reduced ischaemic lesion volume and oxidative stress after MCAO in rats.
4. By preventing the progression of cellular apoptosis and preserving mature neurons, as shown by Maniskas et al. [147] in mice treated with intra-arterial verapamil immediately after MCAO.

3.2.3 Sympatholytics

In the therapy of hypertension, both drugs acting on α and β adrenergic receptors are currently used. Nevertheless, as regards AIS, clinical trials focused mainly on β-blockers. The anti-hypertensive mechanism of this latter drug class is to decrease myocardial contractility, heart rate, cardiac output and renin release from the kidney, thus reducing the production of Ang II.

As for CCBs, also for β-blockers, the first studies about their potential neuroprotective actions are dated back to mid-1980s [148–151]. Apart from the purported hypothesis that β-blocker neuroprotection may be mediated by the attenuation of catecholamines elevation induced by brain injury [152], other experimental observations suggest that the involved mechanisms are manifold.

Song et al. [153] showed that betaxolol abolished oedema formation in a mice MCAO model after 3 h of ischaemia and 8 h of reperfusion, the drug having been administered intraventricularly 15 min before the occlusion. Moreover, β-blockers were shown to suppress the elevation of glutamate concentration in the ipsilateral striatum in a model of rat transient focal ischaemia [154] and esmolol reduced neuronal hippocampal damage in CA_1 [155].

Overall, these studies have been made on β_1 receptor antagonists, being those primarily involved in BP lowering; recently, an increasing interest has been put also on the antagonists of the β_2 receptor because of their anti-inflammatory effects. Mracsko et al. [156] demonstrated that β_2 receptors mediate the catecholaminergic-induced apoptosis of, and decrease interferon-γ release from, leukocytes from spleens of MCAO ischaemic mice, these effects being reverted by a β_2-blocker added in vitro.

Finally, controversial experimental results have been reported for α_2 agonists: α_2 receptors mediate pre-synaptic negative feedback regulation by inhibiting the central release of catecholamines upon stressful stimuli [157]. Early studies [158, 159] showed that α_2 agonists improve the morphologic and functional outcome when administered during experimental ischaemia in the rat likely through (1) the inhibition of voltage-operated Ca^{2+} channels, (2) the activation of G protein-linked inward rectifying K^+ channels and (3) membrane hyperpolarization. Beneficial effect of the α_2 agonist clonidine was observed also in preventing memory deficits, brain damage, brain oxidative stress and the lowering of AChE activity in ischaemia-induced vascular dementia in mice [160]. Moreover, pre-conditioning with the α_2 agonist clonidine 6, 18 and 24 h before forebrain ischaemia in rats improved neurological deficit scores and reduced infarct size [161].

Nevertheless, Brede et al. [162] failed to demonstrate a neuroprotective effect of clonidine in an MCAO model in the rat and further studies are recommended, also because of the known interference of this drug with brain energy metabolism [163, 164].

3.2.4 Diuretics

A brief mention is deserved also by diuretics as regards their potential neuroprotective effects, even if these are more directly linked to their principal mechanism of action: diuretics are prescribed for hypertensive patients because of their hypovolaemic effect, achieved by the increased excretion of water from the body through urine.

Although thiazide diuretics are currently the most used class in primary and secondary AIS prevention, neuroprotection research focused on loop diuretics and in particular on bumetanide, because these drugs inhibit the electroneutral Na^+–K^+–Cl^- co-transporter (NKCC) present in the Henle's nephron loop that was demonstrated to participate through its brain isoform NKCC1 to the ischaemia-induced cerebral oedema formation [165].

Bumetanide was shown to decrease not only brain swelling but also brain infarction [166] and both grey and white matter ischaemic damage [167] in rats after 2 h of MCAO and reperfusion; these results have been confirmed by Wang et al. [168], who also demonstrated that this drug downregulated the expression level of the NKCC1 protein in the rat brain cortex. Moreover, Migliati and colleagues [169] showed that bumetanide decreased the perivascular pool of aquaporin-4 48 h after MCAO in mice.

3.3 Primary AIS Prevention

3.3.1 ACE-Is and ARBs

Both ACE-Is and ARBs have been proven to be effective in AIS primary prevention in several experimental and clinical settings (Table 14.8), being the results obtained in the first studies, conducted on smaller sample sizes, confirmed in broader clinical trials.

The North-East Melbourne Stroke Incidence Study (NEMESIS), Australian, population-based stroke incidence study, recruited 718 patients with first-ever AIS of which 25 % were taking ACE-Is before stroke. Pre-stroke administration of ACE-Is was independently associated with a reduced risk of severe neurologic deficits (OR 0.56, 0.35–0.91) and mortality within 28 days (OR 0.46, 0.24–0.87; [170]).

However, Yu et al. [171] later reported controversial results, observing in a retrospective chart review of 553 consecutive AIS patients that pre-morbid Ang II-suppressing medications were associated with more severe strokes (OR 2.13, 1.00–4.52). Nevertheless, both ACE-Is and β-adrenergic receptor blockers were included within the Ang II-suppressing agents, originating a possible confounding factor in data interpretation. In fact, following studies confirmed that (1) patients who were not taking pre-stroke ACE-Is showed higher in-hospital mortality (OR 2.7, 1.4–5.4; [172]) and (2) in-hospital survival was positively associated with ACE-I pre-treatment (OR 3.42, 1.9–8.1; $p=0.012$) in the GIFA (Italian Group of Pharmacoepidemiology in the Elderly) study [173].

Table 14.8 Pre-stroke ACE-is and ARBS: primary risk prevention and outcome in AIS patients

Authors	Type of study	Total no. of patients	Drug; no. of treated patients	AIS risk	Neurological deficits	Mortality
Chitravas et al. (2007)	Population-based incidence study	718	ACE-Is; 179	–	OR severe neurological deficits (NIHSS score≥8); $n=511$ 0.56 (95 % CI, 0.35–0.91) $p=0.019$	OR early death (within 28 days); $n=696$ 0.46 (95 % CI, 0.24–0.87) $p=0.017$
Hassan et al. (2010)	Retrospective cohort study	327	ACE-Is (perindopril, captopril, enalapril); 119	Patients with vs. without prior ACE-Is use: 65.6 % vs. 70.7 % $p=0.336$	–	OR in-hospital mortality in patients without prior ACE-Is use; 2.7 (95 % CI, 1.4–5.4) $p=0.003$
Tuttolomondo et al. (2011)	Retrospective chart study	1096	ACE-Is; 431	–	OR cognitive function (HAMT score>6); 5.55 (95 % CI, 2.34–12.9); $p<0.0001$	OR no in-hospital mortality; 3.42 (95 % CI, 1.88–8.1) $p=0.012$
Miyamoto et al. (2012)	Retrospective study	151	ARBs; 94	–	OR for mRS=0–2; 3.81 (95 % CI, 1.11–13.07); $p=0.033$ OR for BI score >75; 2.81 (95 % CI, 1.02–8.09); $p=0.045$	–
Desmaele et al. (2016)	Retrospective study	1974	ARBs; 255	–	OR on NIHSS scale: 1.24 (95 % CI, 0.891–1.727); $p=0.203$	–
Sundbøll et al. (2014)	Population-based cohort study	83,736	ACE-Is + ARBs; 21,808 current users, 3939 former users	–	–	30-day mortality RR 0.85 (95 % CI, 0.81–0.89) for current users 0.99 (95 % CI, 0.90–1.09) for former users

The ONTARGET Investigators (2008)	Double-blind, randomized, controlled trial	25,620	ARB (telmisartan); 8542 ACE-I (ramipril); 8576 Combination (telmisartan + ramipril); 8502	RR for death from stroke: 0.91 (95 % CI, 0.79–1.05) for telmisartan vs. ramipril 0.93 (95 % CI, 0.81–1.07) for combination therapy	–	–

ACE-Is: angiotenin-converting enzyme inhibitors; AIS: acute ischaemic stroke; ARBs: angiotensin receptor blockers; BI: Barthel Index; CI: confidence interval; HAMT: Hodkinson Abbreviated Mental Test; mRS: modified Rankin Scale; NIHSS: National Institute of Health Stroke Scale; ONTARGET: Ongoing Telmisartan Alone and in Combination with Ramipril Global Endpoint Trial; OR: odds ratio; RR: risk ratio

Refs

Chitravas N, Dewey HM, Nicol MB, Harding DL, Pearce DC, Thrift AG. Is prestroke use of angiotensin-converting enzyme inhibitors associated with better outcome? Neurology. 2007;68:1687–93

Desmaele S, Cornu P, Barbé K, Brouns R, Steurbaut S, Dupont AG. Relationship between pre-stroke cardiovascular medication use and stroke severity. Eur J Clin Pharmacol. 2015. doi: 10.1007/s00228-015–2001-1

Hassan Y, Aziz NA, Al-Jabi SW, Looi I, Zyoud SH. Impact of angiotensin-converting enzyme inhibitors administration prior to acute ischemic stroke onset on in-hospital mortality. J Cardiovasc Pharmacol Ther. 2010;15:274–81

Miyamoto N, Tanaka Y, Ueno Y, Tanaka R, Hattori N, Urabe T. Benefits of prestroke use of angiotensin type 1 receptor blockers on ischemic stroke severity. J Stroke Cerebrovasc Dis. 2012;21:363–8

Sundbøll J, Schmidt M, Horváth-Puhó E, Christiansen CF, Pedersen L, Bøtker HE et al. Preadmission use of ACE inhibitors or angiotensin receptor blockers and short-term mortality after stroke. J Neurol Neurosurg Psychiatry. 2015;86:748–54

ONTARGET Investigators, Yusuf S, Teo KK, Pogue J, Dyal L, Copland I et al. Telmisartan, ramipril, or both in patients at high risk for vascular events. N Engl J Med. 2008;358:1547–59

Tuttolomondo A, Di Sciacca R, Di Raimondo D, Pedone C, La Placa S, Pinto A et al. Effects of clinical and laboratory variables and of pretreatment with cardiovascular drugs in acute ischaemic stroke: a retrospective chart review from the GIFA study. Int J Cardiol. 2011;151:318–22

More recently, Sundbøll et al. [174] performed a large nationwide population-based cohort study with follow-up of 100,043 patients hospitalized for a first time stroke. Even though they did not distinguish between ACE-I and ARB use at the time of admission, pre-stroke treated patients showed lower 30-day mortality for current users (last prescription within 90 days; OR 0.85 (0.81–0.89)) and for former users (last prescription between 90 and 180 days; OR 0.99 (0.90–1.09)) compared with patients not taking either ACE-Is or ARBs before AIS.

Also ARBs alone have been proven beneficial in primary AIS prevention. Seminal clinical trials have been:

1. The LIFE (Cardiovascular Morbidity and Mortality in the Losartan Intervention for Endpoint Reduction in Hypertension) study, a randomized, double-blinded trial that enrolled 9193 patients aged 55–80 years with essential hypertension (BP 160–200/95–115 mmHg) and left ventricular hypertrophy [175]. This study firstly compared the effects of an ARB, i.e. losartan, with the β-blocker atenolol and showed that fatal and non-fatal stroke was reduced by 25 % in the losartan group ($p=0.001$), although the BP reduction was similar.
2. The Study on Cognition and Prognosis in the Elderly (SCOPE), a randomized, double-blind, placebo-controlled trial that enrolled 4964 hypertensive elderly patients (aged 70–89 years) with SBP = 160–179 mmHg and/or DBP = 90–99 mmHg [176]. Also in this trial, the studied ARB, i.e. candesartan, showed a 27.8 % reduction in non-fatal stroke vs. placebo ($p=0.04$).

The positive effects of pre-treatment with ARBs on AIS outcome has been provided by the more recent study by Miyamoto et al. [177], a regression analysis indicating that in patients with a modified Rankin Scale (mRS) score of 0–2, therefore with a favourable outcome, pre-treatment with ARBs was associated with better outcomes.

In any case, when comparing the effects of ACE-Is and ARBs, the advantage of using one drug class over the other was not clearly highlighted.

In the ONgoing Telmisartan Alone and in combination with Ramipril Global Endpoint Trial (ONTARGET), the efficacy of telmisartan, ramipril and their combination was compared [178]. In this study, death from cardiovascular causes occurred similarly in the three groups, even if there was no additional advantage from the combination of telmisartan *plus* ramipril instead of ramipril alone; on the contrary, the combination of the two drugs induced hyperkalemia, low blood pressure and worsening of kidney function compared with the single drug administration.

Interestingly, Desmaele et al. [65] conducted a retrospective study on 1974 patients with a suspected stroke, reporting that the pre-stroke ARB use did not show a significant relationship with the NIHSS score (OR 1.240, 0.891–1.727; $p<0.203$).

In summary, even though the experimental results suggest that ARBs may exert higher neuroprotective effects over ACE-Is, clinical reports indicate that there are little differences between the two anti-hypertensive drug classes and that ACE-Is should be preferred as a first-line drug, being ARB use restricted to patients intolerant to ACE-Is [179].

3.3.2 Calcium Channel Blockers

The beneficial effects of CCBs in primary stroke prevention has been widely confirmed: nitrendipine was shown to reduce the incidence of fatal and non-fatal stroke by 38 % [180] and nifedipine that of stroke or TIA by 30 % in the A Coronary disease Trial Investigating Outcome with Nifedipine GITS (ACTION) study performed on patients with stable angina and hypertension [181].

Nevertheless, a *consensus* is lacking about the superiority of CCBs over other anti-hypertensive drug classes in primary AIS prevention. Experimentally, both short- and long-term pre-treatments with losartan protected stroke-prone spontaneously hypertensive rats from stroke more than amlodipine that, on the contrary, was more effective in counteracting cell apoptosis through the enhancement of antioxidants defences [182].

On the other hand, in the clinical setting, a meta-analysis of >40 trials by Lawes et al. [183] reported that stroke risk was reduced with respect to placebo with pooled relative risk reductions for β-blockers and/or diuretics, ACE-Is and CCBs of 35 %, 28 % and 39 %, respectively; however, these differences between the considered anti-hypertensive drug classes were not found when the comparison was made between the different anti-hypertensives and not vs. placebo.

A following outcome analysis of 12 trials [184] is partially in accordance with the previous results: amlodipine-treated patients showed reduced stroke and myocardial infarction than those treated with other anti-hypertensive drugs, including ARBs.

However, because of the evident difficulties in highlighting clear superiority between anti-hypertensives in primary AIS prevention, a strategy put forward has been to identify effective combination therapies. In this context, the Combination Therapy of Hypertension to Prevent Cardiovascular Events (COPE) trial has been developed to examine the combination effects of the CCB benidipine with a thiazide diuretic, an ARB and a β-blocker [185]. Results indicated that BP lowering was achieved similarly in the various groups, but the incidence of fatal or non-fatal strokes was reduced primarily in the benidipine-thiazide diuretic treated group ($p=0.0109$). These results have been confirmed by a post hoc analysis of the COPE trial [186] addressing potential differences in relation to the stroke subtype: the hazard ratios both for haemorrhagic and ischaemic stroke were higher in benidipine-β-blocker-treated patients than in those treated with benidipine-thiazide diuretic (HR 4.60, 1.00–21.31 for haemorrhagic stroke; HR 2.56, 1.00–6.60 for ischaemic stroke) while no differences were reported for benidipine-ARB combination therapy with respect to the other treatment regimens.

3.3.3 Sympatholytics

Considering the protective effects towards first-ever stroke by ARBs (see Sect. 3.3.1), it has been already pointed out that in the LIFE study [175] losartan proved to be more effective in reducing fatal and non-fatal stroke with respect to the β-blocker atenolol.

However, experimentally, esmolol and landiolol pre-stroke administration improved both the histological and neurological outcomes after 7 days from transient forebrain ischaemia in rats [187]. In Clinics:

1. Dziedzic et al. [188] showed that β-blocker pre-treatment was associated with reduced risk of glucose level on admission ≥7.8 mmol/L (OR 0.22, 0.07–0.74) and fasting glucose 1 day after admission ≥7.0 mmol/L (OR 0.21, 0.05–0.91), suggesting that these drugs may reduce the risk of hyperglycaemia, a complication that occurs in 60 % of AIS patients and is correlated with a poor clinical outcome [189];
2. Sykora et al. [190] failed to detect in 5212 patients of Virtual International Stroke Trials Archive (VISTA) database either a significant reduction of mortality after 3 months or an improvement of functional outcome in patients treated pre-stroke with β-blockers, even if this treatment was associated with reduced frequency of post-stroke pneumonia (adjusted RR 0.77; 0.6–0.98).

Therefore, β-blockers were demonstrated to exert some degree of neuroprotection in both experimental and clinical settings, although less compared to ACE-Is, ARBs and CCBs; this is confirmed also by the studies evaluating the beneficial effects of anti-hypertensives in secondary prevention of recurrent stroke (see 3.5).

As regards primary AIS prevention exerted by drugs acting on α adrenoreceptors, a few studies addressed this issue. In this context, interesting is the study by Lai et al. [191] who observed in 7502 patients aged ≥50 years and treated with α antagonists an increased incidence risk of ischaemic stroke (adjusted IRR 1.40, 1.22–1.61) within 21 days from the initiation period.

3.3.4 Diuretics

Several pieces of evidence support the hypothesis that diuretic pre-treatment is beneficial in primary stroke prevention:

1. In the Anti-hypertensive and Lipid-Lowering Treatment to Prevent Heart Attack Trial (ALLHAT), a randomized, double-blind, active-controlled clinical trial that involved 33,357 hypertensive patients at cardiovascular risk aged ≥55 years, patients treated with the ACE-I lisinopril had higher rate of stroke with respect to the thiazide-like diuretic chlortalidone (6.3 % vs. 5.6 %; RR 1.15, 1.02–1.30) after 6 years of follow-up [192].
2. In the GIFA study [173], in-hospital survival was positively associated not only with pre-stroke treatment with ARBs (see 3.3.1) but also with diuretics (OR 5.12, 1.9–12.8; $p=0.014$).
3. In a study evaluating the prognosis of 140 AIS patients [193], pre-stroke use of thiazide diuretics demonstrated to improve the functional status as assessed by mRS score after 3 months of follow-up (mRS ≤ 2: 42.4 % vs. 26.9 %; OR 3.34, 0.13–0.86; $p=0.02$).
4. In a study involving 482 AIS patients [194], diuretic treatment before stroke was associated with non-severe stroke (OR 0.22, 0.09–0.51; $p<0.001$).

Negative effects were otherwise observed in the NEMESIS study (see 3.3.1; 170), where pre-stroke diuretics were associated with an increased risk of neurologic deficits (OR 1.81, 1.13–2.90) and at borderline significance with early death after stroke (OR 1.68, 0.95–2.96).

Also the Second Australian National Blood Pressure Study (ANBP2), a practice-based open-label trial that compared the effects of the diuretic hydrochlorothiazide with the ACE-I enalapril [195] in 6083 patients aged ≥65 years showed that enalapril was marginally beneficial over hydrochlorothiazide in preventing cardiovascular events *plus* all-cause mortality, including stroke (11 % reduction; RR 0.89, 0.79–1.00; p=0.05). Having been this study published in the same years of the ALLHAT study, a scientific debate arose highlighting that the two trials are not directly comparable above all because of the difference between the evaluated drugs and the population of hypertensive patients [196, 197].

Nowadays, diuretics are recommended in primary stroke prevention: in the Italian Stroke Organization (ISO)-SPREAD Guidelines [198], diuretics and CCBs should be preferred, above all in older patients with isolated systolic hypertension (ISH), in accordance with a seminal Systolic Hypertension in the Elderly Program (SHEP) study [199].

In any case, as recommended by American Heart Association/American Stroke Association (AHA/ASA) Guidelines [51], at present BP lowering remains the goal to prevent stroke occurrence, and treatments should be individualized in relation to the single patient clinical picture and medication tolerance.

3.4 *Primary Prevention in AIS Patients and Interaction with Thrombolysis*

To our knowledge, there is a scant number of studies in literature directly addressing potential interactions between anti-hypertensive drugs with rt-PA in AIS, even if several were focused on the effects of anti-hypertensives in the acute phase, when these drugs have been investigated for their BP-lowering properties (for ACE-Is and ARBs see ([200–204]; for CCBs see [205–210]); for diuretics see [211, 212]; for β-blockers see the seminal BEST trial [213–215]).

In this regard, the recent Third International Stroke Trial (IST-3) randomized 3035 AIS patients to rtPA within 6 h from the onset of symptoms and showed that high baseline BP and BP variability during the first 24 h were associated with early adverse events and early deaths; otherwise, employing BP-lowering treatment during the first 24 h was associated with a reduced risk of poor outcome at 6 months (OR 0.78, 0.65–0.93; p=0.007), independently from the administration of rtPA [216].

This observation is interesting because the interactions between anti-hypertensives and fibrinolytic system have been observed (reviewed in [217]). For example, the PAI is increased both by Ang II and aldosterone and consequently by ACE-Is, but not by ARBs [218]; also CCBs may impact thrombolysis, because of their direct action on the inhibition of platelet activation [219]. Moreover, Gremmel et al. [220] recently demonstrated that the effect of clopidogrel on ADP-inducible platelet activation is decreased by dihydropyridinic CCB treatment in patients undergoing angioplasty and stenting for cardiovascular disease, likely because of the inhibition of cytochrome 3A4 and the interference with the conversion of clopidogrel to its active form [221]. Additionally, the ACE-I lisinopril, the ARB irbesartan and the CCB amplodipine were shown to decrease differently the mean plasma levels of PAI, thrombomodulin and p-selectin in hypertensive patients after 1 and 6 months of follow-up [222].

A reported correlation supported by several clinical pieces of evidence is between ACE-Is and rt-PA and is orolingual angioedema, i.e. one of the adverse reactions to rt-PA that might be life-threatening because of airway obstruction [223, 224]. In a recent meta-analysis, ACE-Is were associated with a high relative risk of orolingual angioedema after intravenous rt-PA (12.9, 4.5–37.0; [225]). Accordingly, Hurford et al. [226] observed that only ACE-I pre-treatment was a significant independent predictor of angioedema (OR 2.3, 1.1–4.7) in 530 patients receiving tPA for AIS.

Finally, a recent study addressed what anti-hypertensive treatment regimen between labetalol, nicardipine and hydralazine grants a better time-to-target BP before alteplase administration in AIS patients [227]; results showed that median time-to-target BP was 10, 22 and 25 min, respectively, labetalol being the most effective in particular if 20 mg was administered with respect to 10 mg p.o.

3.5 *Secondary Prevention of Recurrent Stroke*

The control of hypertension is one of the major secondary prevention strategies of recurrent AIS stroke (review in [228]): in a meta-analysis of 16 randomized controlled trials including 40,292 patients [229], BP lowering reduced the risk of recurrent stroke by 18 % (9–26 %), being each 10-mmHg SBP reduction associated with 33 % (9–51 %) reduction in the risk of recurrent stroke.

However, although BP lowering is acknowledged as a primary goal in preventing recurrent strokes, conclusive evidence of what class of anti-hypertensive drugs should be recommended is lacking (Table 14.9).

As regards the inhibition of RAS, results are controversial:

1. The Perindopril Protection Against Recurrent Stroke Study (PROGRESS) trial was the first study to provide reliable evidence that BP lowering could be a strategy for preventing recurrent stroke [230]. The PROGRESS was a randomized placebo-controlled trial on 6105 patients with prior stroke or transient ischaemic attack (TIA) and reported that over a mean follow-up of 4 years, the overall relative risk of recurrent stroke was reduced by 28 % (17–38 %) in the perindopril-treated group compared to placebo.
2. The Prevention Regimen For Effectively Avoiding Second Strokes (PRo-FESS) trial investigated the effects of telmisartan in 20,322 AIS patients but treated patients showed reduced risk of recurrent stroke only by 5 % (4–14 %; [231]).

Table 14.9 Anti-hypertensives for the prevention of recurrent ischaemic stroke

Authors	Total no. of patients	Drug; no. of treated patients	Follow-up; years	Risk of recurrent AIS	Mortality
PROGRESS Collaborative Group (2001)	6105	Perindopril alone or in combination with thiazidic diuretic indapamide; 3051	4	RR reduction 28 % (95 % CI, 17–38 %); $p<0.0001$	RR reduction of death from stroke 16 % (95 % CI, −27–44 %)
		– Single therapy with Perindopril; 1281 – Combination Therapy perindopril + indapamide; 1770		RR reduction for perindopril: 43 % (95 % CI, 30–54 %) RR reduction for combination: 5 % (95 % CI, −19–23)	
Yusuf et al. (2008) Pro-FESS Study Group	20,332	Telmisartan; 10,146	2.5	HR 0.95 (95 % CI, 0.86–1.04) $p=0.23$	HR for death from cardio- and cerebrovascular causes 0.94 (95 % CI, 0.87–1.01); $p=0.11$
Schrader et al. (2005) MOSES Study Group	1405	Eprosartan; 681 Nitrendipine; 671	2.5	IDR eprosartan vs. nitrendipine for recurrent fatal and non-fatal strokes 0.75 (95 % CI, 0.58–0.97); $p=0.026$	–
Liu et al. (2005) FEVER Study Group	9711	Felodipine	3.3	Risk reduced by 27 % $p=0.001$	–
Liu et al. (2009) PATS Study Group	5665	Indapamide; 2840	2	HR 0.69 (95 % CI, 0.54–0.89); $p<0.001$	Reduced death from all causes by 13 % (95 % CI, −30–9); p =0.13 reduced death from stroke by 26 % (95 % CI, −47–3); $p=0.071$
De Lima et al. (2014) (meta-analysis)	1° RCT: 1473 2° RCT: 720	Atenolol 1° RCT 732 2° RCT 372	2.6 2.3	RR 0.95 (95 % CI, 0.76–1.18)	RR death from all causes 0.94 (95 % CI, 0.68–1.32)

(continued)

Table 14.9 (continued)

AIS: acute ischaemic stroke; CI: confidence interval; FEVER: Felodipine Event Reduction Study; HR: Hazard Ratio; IDR: incidence density ratio; MOSES: Morbidity and Mortality After Stroke, Eprosartan Compared with Nitrendipine for Secondary Prevention; PATS: Post-stroke Antihypertensive Treatment Study; PROGRESS: Perindopril Protection Against Recurrent Stroke Study; RCT: randomized clinical trial; RR: risk ratio

Refs

De Lima LG, Saconato H, Atallah AN, da Silva EM. Beta-blockers for preventing stroke recurrence. Cochrane Database Syst Rev. 2014 Oct 15;10:CD007890

Liu L, Zhang Y, Liu G, Li W, Zhang X, Zanchetti A; FEVER Study Group. The Felodipine Event Reduction (FEVER) Study: a randomized long-term placebo-controlled trial in Chinese hypertensive patients. J Hypertens. 2005;23:2157–72

Liu L, Wang Z, Gong L, Zhang Y, Thijs L, Staessen JA et al. Blood pressure reduction for the secondary prevention of stroke: a Chinese trial and a systematic review of the literature. Hypertens Res. 2009;32:1032–40

PROGRESS Collaborative Group. Randomised trial of a perindopril-based blood-pressure-lowering regimen among 6,105 individuals with previous stroke or transient ischaemic attack. Lancet. 2001;358:1033–41

Schrader J, Lüders S, Kulschewski A, Hammersen F, Plate K, Berger J et al. Morbidity and Mortality After Stroke, Eprosartan Compared with Nitrendipine for Secondary Prevention: principal results of a prospective randomized controlled study (MOSES). Stroke. 2005;36:1218–26

Yusuf S, Diener HC, Sacco RL, Cotton D, Ounpuu S, Lawton WA et al. Telmisartan to prevent recurrent stroke and cardiovascular events. N Engl J Med. 2008;359:1225–37

Moreover, the Pro-FESS MRI sub-study failed to find a significant benefit in telmisartan-treated patients over placebo when evaluating the progression of both periventricular and sub-cortical white matter lesions [232].

Hopefully, the upcoming CEREBRAL study, a randomized open-label controlled trial with blinded end-point assessment and a prospective cohort study in elderly hypertensive AIS patients, will give new insights about the effectiveness of ACE-Is and ARBs in preventing the occurrence of a fatal or non-fatal secondary cerebrovascular event and the progression of cerebrovascular lesions as evaluated by magnetic resonance imaging [233].

Interestingly, Levijoki et al. [234] demonstrated that valsartan *plus* levosimendan, a vasodilator acting on ATP-sensitive K^+ channels, had a cumulative beneficial effect in extending mean survival times after the first ischaemic event, and clinical investigations on this interaction should be undertaken in the future.

Also as regards CCBs, reduction in recurrent stroke has been demonstrated (see Felodipine Event Reduction (FEVER) study [235]), but results remain so far controversial when beneficial effects have been assessed with respect to other anti-hypertensive drug classes:

1. the large-scale randomized Morbidity and Mortality After Stroke, Eprosartan Compared With Nitrendipine for Secondary Prevention (MOSES) trial [236] showed that the ARB eprosartan was superior to the CCB nitrendipine in reducing recurrent strokes and TIA (IDR 0.75, 0.58–0.97; $p=0.026$) and
2. the risk of primary composite events such as death, cardiovascular or cerebrovascular event (IDR 0.79,0.66–0.96; $p=0.014$), despite the two drugs resulted in similar achieved follow-up BP levels. However, a relevant limit to this study should be reported, i.e. that the observed risk reduction was mainly due to TIA and by single patients with multiple events;
3. the Cilnidipine effect on high blood pressure and cerebral pERfusion in Ischemic Stroke patients with Hypertension (CHERISH) trial [237] assessed the effects of the ARB losartan vs. that of the CCB cilnedipine on CBF of AIS hypertensive patients after 4 weeks of follow-up, but failed to demonstrate a significant difference, being the CBF enhanced by both drugs.

Also diuretics and β-blockers have been studied for recurrent stroke prevention: in the intervention Cochrane review by De Lima et al. [238], β-blockers showed no statistical reduction in risks of fatal and non-fatal stroke with respect to placebo (RR 0.94, 0.76–1.18), while in the early Post-Stroke Anti-hypertensive Study [239], the diuretic indapamide reduced the risk of recurrent stroke by 27 % (11–40 %). These latter results have been consolidated by an updated analysis of PATS [240], where the authors made also a systematic review of 10 trials, reporting that a significant reduction in recurrent stroke was achieved with diuretics (alone or in combination with ACE-Is) but not with ARBs, CCBs and β-blockers alone.

Considering all these heterogeneous results, the ISO-SPREAD guidelines [198] recommend BP control in patients with previous AIS or TIA, preferably with drugs acting on RAS, CCBs and diuretics to prevent secondary stroke. On the other hand, the AHA/ASA Guidelines [5] do not recommend an optimal drug regimen, however specifying that (1) diuretics or diuretics *plus* ACE-Is could be superior over the other anti-hypertensive drug classes for secondary stroke prevention and (2) the choice of drugs should be individualized on the basis of pharmacological properties, mechanism of action and consideration of patient characteristics.

Finally, another controversial point about BP-lowering beneficial effects is emerged starting from an observational analysis of the Pro-FESS trial [241], which reported a J-shaped association, with both the highest and the lowest BP levels being related to the increased risks of recurrent stroke. In this regard, Arima and Chalmers [242] failed to demonstrate an association between recurrent stroke occurrence and larger SBP reductions in a post hoc PROGRESS analysis. Further studies should be conducted to clarify this point.

3.6 Secondary Prevention in AIS Patients Treated with Thrombolysis

As for primary prevention, secondary prevention in patients treated with thrombolysis is scarcely addressed even if, surely, the studies about the anti-hypertensive effects in preventing stroke recurrence include also patients treated with alteplase. Likely, this could occur because:

1. Only a small percentage (≤10 %) of AIS patients receive thrombolysis [2] because of the restricted exclusion criteria for the use of rtPA [243].
2. international guidelines [5, 51] focus on the importance of BP lowering instead of a precise recommendation for a specific anti-hypertensive drug to be employed.

3.7 Conclusions

BP lowering is one of the main strategies to prevent AIS and improve patients' outcome. This beneficial effect of BP lowering has been confirmed in recent systematic review and meta-analysis by the Blood Pressure Lowering Treatment Trialists' Collaboration (BPLTTC; [244]), which showed that, in mild hypertensive patients, the reduction of BP was positively correlated with less stroke events over 5 years (OR 0.72; CI 0.55–0.94).

Several anti-hypertensive drugs are nowadays available and have proven to possess heterogeneous neuroprotective properties alongside the BP-lowering effects. This observation is confirmed not only by the several experimental studies but also by the clinical practice, where anti-hypertensives are first row agents in both primary and secondary AIS prevention. Nevertheless, a conclusive evidence of what class of hypertensive should be recommended is lacking: even if in primary stroke prevention diuretics and CCBs seem to be preferred, the main goal remains BP normalization.

As regards the major interest of this chapter, i.e. *the pharmacological influence on therapy and patients' outcomes with respect to* the interactions between the drugs used before stroke and thrombolysis in patients suffering from both first and recurrent stroke, this has not been directly addressed so far for the anti-hypertensives. Considering that alteplase is the only drug currently approved to be used in AIS, a further clarification of the interactions with other widely employed drugs, such as anti-hypertensives, is strongly recommended through ad hoc experimental and clinical studies.

Acknowledgements The authors thank Ms. Giuliana Di Gregorio and Dr. Filippo Grandini for editing the chapter.

Conflict of Interest Statement The authors declare that there are no conflicts of interest.

Funding Italian Ministry for Education, University and Research (MIUR) and Investigator Fellowship by University College "*Collegio Ghislieri*", Pavia, Italy, for Dr. Ferrari.

References

1. Mikulik R, Wahlgren N (2015) Treatment of acute stroke: an update. J Intern Med 278:145–165
2. Moretti A, Ferrari F, Villa RF (2015) Pharmacological therapy of acute ischaemic stroke: achievement and problems. Pharmacol Ther 153:79–89
3. Lapchak PA (2015) Critical early thrombolytic and endovascular reperfusion therapy for acute ischemic stroke victims: a call for adjunct neuroprotection. Transl Stroke Res 6:345–354
4. Moretti A, Ferrari F, Villa RF (2015) Neuroprotection for ischaemic stroke: current status and challenges. Pharmacol Ther 146:23–34
5. Kernan WN, Ovbiagele B, Black HR, Bravata DM, Chimowitz MI, Ezekowitz MD et al (2014) Guidelines for the prevention of stroke in patients with stroke and transient ischemic attack: a guideline for healthcare professionals from the American Heart Association/American Stroke Association. Stroke 45:2160–2236
6. Stryer L (1995) Biochemistry. WH Freeman, New York, 692 p
7. Amarenco P, Labreuche J, Touboul PJ (2008) High-density lipoprotein-cholesterol and risk of stroke and carotid atherosclerosis: a systematic review. Atherosclerosis 196:489–496
8. Iso H, Jacobs DR Jr, Wentworth D, Neaton JD, Cohen JD (1989) Serum cholesterol levels and six-year mortality from stroke in 350,977 men screened for the multiple risk factor intervention trial. N Engl J Med 320:904–910
9. Zhang Y, Tuomilehto J, Jousilahti P, Wang Y, Antikainen R, Hu G (2012) Total and high-density lipoprotein cholesterol and stroke risk. Stroke 43:1768–1774
10. Canouï-Poitrine F, Luc G, Bard JM, Ferrieres J, Yarnell J, Arveiler D et al (2010) Relative contribution of lipids and apolipoproteins to incident coronary heart disease and ischemic stroke: the PRIME study. Cerebrovasc Dis 30:252–259
11. Di Angelantonio E, Sarwar N, Perry P, Kaptoge S, Ray KK, Thompson A et al (2009) Major lipids, apolipoprotein, and risk of vascular disease. JAMA 302:1993–2000
12. Lewington S, Withlock G, Clarke R, Sherliker P, Emberson J, Halsey J, Prospective Studies Collaboration et al (2007) Blood cholesterol and vascular mortality by age, sex, and blood pressure: a meta-analysis of individual data from 61 prospective studies with 55000 vascular deaths. Lancet 370:1829–1839

13. Hachinski V, Graffagnino C, Beaudry M, Bernier G, Buck C, Donner A et al (1996) Lipids and stroke: a paradox resolved. Arch Neurol 53:303–308
14. Sturgeon JD, Folsom AR, Longstreth WT, Shahar E, Rosamond WD, Cushman M (2007) Risk factors for intracerebral hemorrhage in a pooled prospective study. Stroke 38:2718–2725
15. Endo A (1992) The discovery and development of HMG-CoA reductase inhibitors. J Lipid Res 33:1569–1582
16. Goldstein JL, Brown MS (1984) Progress in understanding the LDL receptor and HMG-CoA reductase, two membrane proteins that regulate the plasma cholesterol. J Lipid Res 25:1450–1461
17. Carone D, Librizzi L, Cattalini A, Sala G, Conti E, Cuccione E et al (1615) Pravastatin acute neuroprotective effects depend on blood brain barrier integrity in experimental cerebral ischemia. Brain Res 2015:31–41
18. Thompson BJ, Ronaldson PT (2014) Drug delivery to the ischemic brain. Adv Pharmacol 71:165–2012
19. Bellosta S, Corsini A (2012) Statin drug interactions and related adverse reactions. Expert Opin Drug Saf 11:933–946
20. Hirota T, Ieiri I (2015) Drug-drug interactions that interfere with statin metabolism. Expert Opin Drug Metab Toxicol 11:1435–1447
21. Cimino M, Gelosa P, Gianella A, Nobili E, Tremoli E, Sironi L (2007) Statins: multiple mechanisms of action in the ischemic brain. Neuroscientist 13:208–213
22. Endres M (2005) Statins and stroke. J Cereb Blood Flow Metab 25:1093–1110
23. Li Q, Zhuang QK, Yang J-N, Zhang Y-Y (2014) Statins exert neuroprotection on cerebral ischemia independent of their lipid-lowering action: the potential molecular mechanisms. Eur Rev Med Pharmacol Sci 18:1113–1126
24. Malfitano AM, Marasco G, Proto MC, Laezza C, Gazzerro P, Bifulco M (2014) Statins in neurological disorders: an overview and update. Pharmacol Res 88:74–83
25. Miida T, Takahashi A, Ikeuchi T (2007) Prevention of stroke and dementia by statin therapy: Experimental and clinical evidence of their pleiotropic effects. Pharmacol Ther 113:378–393
26. van Aelst L, D'Souza-Schorey C (1997) Rho GTPases and signaling networks. Genes Dev 11:2295–2322
27. Zhao J, Zhang X, Dong L, Wen Y, Cui L (2014) The many roles of statins in ischemic strokes. Curr Neuropharmacol 12:564–574
28. Barone E, Cenini G, Di Domenico F, Martin S, Sultana R et al (2011) Long-term high-dose atorvastatin decreases brain oxidative and nitrosative stress in a preclinical model of Alzheimer disease: a novel mechanism of action. Pharmacol Res 63:172–180
29. Puato M, Zambon A, Faggin E, Rattazzi M, Pauletto P (2014) Statin treatment and carotid plaque composition: a review of clinical studies. Curr Vasc Pharmacol 12:518–526
30. Bedi US, Singh M, Singh PP, Bhuriya R, Bahekar A, Molnar J et al (2010) Effects of statin on progression of carotid atherosclerosis as measured by intimal-thickness: a meta-analysis of randomized controlled trials. J Cardiovasc Pharmacol Ther 15:268–273
31. Llorente-Cortès V, Martinez-Gonzàlez J, Badimon L (1998) Esterified cholesterol accumulation induced by aggregated LDL uptake in human vascular muscle cells is reduced by HMG-CoA reductase inhibition. Arterioscl Thromb Vasc Biol 18:738–746
32. Violi F, Calvieri C, Ferro D, Pignatelli P (2013) Statins as antithrombotic drugs. Circulation 127:251–257
33. Ovbiagele B, Saver JL, Starkman S, Kim D, Ali LK, Jahan R et al (2007) Statin enhancement of collateralization in acute stroke. Neurology 68:2129–2131
34. Fisher M, Moonis M (2012) Neuroprotective effects of statins: evidence from preclinical and clinical studies. Curr Treat Options Cardiovasc Med 14:252–259
35. Garcìa-Bonilla L, Campos M, Giralt D, Salat D, Chacòn P, Hernandez-Guillamon M et al (2012) Evidence for the efficacy of statins in animal stroke models: a meta-analysis. J Neurochem 122:233–243
36. Goldstein LB (2009) Statins and ischemic stroke severity: cytoprotection. Curr Atheroscl Rep 11:296–300

37. Montecucco F, Quercioli A, Mirabelli-Badenier M, Viviani GL, Mach F (2012) Statins in the treatment of acute ischemic stroke. Curr Pharmacol Biotechnol 13:68–76
38. Zhang L, Chopp M, Jia L, Cui Y, Lu M, Zhang ZG (2009) Atorvastatin extends the therapeutic window for tPA to 6h after the onset of embolic stroke in rats. J Cerebr Blood Flow Metab 29:1816–1824
39. Roy A, Jana M, Kundu M, Corbett GT, Rangaswamy SB et al (2015) HMG-CoA reductase inhibitors bind to PPARα to upregulate neurotrophin expression in the brain and improve memory in mice. Cell Metab 22:253–265
40. Mendoza-Oliva A, Zepeda A, Arias C (2014) The complex actions of statins in brain and their relevance for Alzheimer's disease treatment: an analytical review. Curr Alzheimer Res 11:817–833
41. Yamamoto N, Fujii Y, Kasahara R, Tanida M, Ohora K et al (2016) Simvastatin and atorvastatin facilitates amyloid β-protein degradation in extracellular spaces by increasing neprilysin secretion from astrocytes through activation of MAPK/Erk1/2 pathways. Glia
42. Cordle A, Landreth G (2005) 3-Hydroxy-3-methylglutaryl-coenzyme A inhibitors attenuate beta-amyloid-induced microglial inflammatory responses. J Neurosci 25:299–307
43. Stead LG, Vaidvanathan L, Kumar G, Bellolio MF, Brown RD, Suravaram S et al (2009) Statins in ischemic stroke: just low-density lipoprotein lowering or more? J Stroke Cerebrovsc Dis 18:124–127
44. Yoon SS, Dambrosia J, Chalela J, Ezzeddine M, Warach S, Haymore J et al (2004) Rising statin use and effect on ischemic stroke outcome. BMC Med 2:4
45. Everett BM, Glynn RJ, MacFayden JG, Ridker PM (2010) Rouvastatin in the prevention of stroke among men and women with elevated levels of C-reactive protein. Justification for the use of statins in prevention: an interventional trial evaluating rosuvastatin (JUPITER). Stroke 121:143–150
46. Puri R, Nissen SE, Libby P, Shao M, Ballantyne CM, Barter PJ et al (2013) C-Reactive protein, but not low-density lipoprotein cholesterol levels, associate with coronary atheroma regression and cardiovascular events after maximally intensive statin therapy. Circulation 128:2395–2403
47. Beretta S, Pastori C, Sala G, Piazza F, Ferrarese C, Cattalini A et al (2011) Acute lipophilicity-dependent effect of intravascular simvastatin in the early phase of focal cerebral ischemia. Neuropharmacology 60:878–885
48. De Caterina R, Scarano M, Marfisi R, Lucisano G, Palma F (2010) Cholesterol-lowering interventions and stroke: insights from a meta-analysis of randomized controlled trials. J Am Coll Cardiol 55:198–211
49. Kinlay S (2007) Low-density lipoprotein-dependent and -independent effects of cholesterol-lowering therapies on C-reactive protein: a meta-analysis. J Am Coll Cardiol 49:2003–2009
50. Amarenco P, Labreuche J (2009) Lipid management in the prevention of stroke: review and updated meta-analysis of statins for stroke prevention. Lancet Neurol 8:453–463
51. Meschia JF, Bushnell C, Boden-Albala B, Braun LT, Bravata DM, Chaturvedi S et al (2014) Guidelines for the primary prevention of stroke. A statement for healthcare professionals from the American Heart Association/American Stoke Association. Stroke 45:3754–3832
52. Paciaroni M, Bogousslavsky J (2009) Statins and stroke prevention. Expert Rev Cardiovasc Ther 97:1231–1243
53. Nicholas JS, Swearingen CJ, Thomas JC, Rumboldt Z, Tumminello P, Patel SJ (2008) The effect of statin pretreatment on infarct volume in ischemic stroke. Neuroepidemiology 31:48–56
54. Schook SJ, Gupta R, Vora NA, Tievsky AL, Katzan I, Krieger DW (2006) Statin use is independently associated with smaller infarct volume in nonlacunar MCA territory stroke. J Neuroimaging 16:341–346
55. Choi JC, Lee JS, Park TH, Cho YJ, Parl JM et al (2015) Effect of pre-stroke statin use on stroke severity and early functional recovery: a retrospective cohort study. BMC Neurol 15:120
56. Martinez-Sànchez P, Fuentes B, Martinez-Martinez M, Ruiz-Ares G, Fernàndez-Travieso J, Sanz-Cuesta BE et al (2013) Treatment with statins and ischemic stroke severity: does the dose matter? Neurology 80:1800–1805

57. Flint AC, Kamel H, Navi BB, Rao VA, Faigeles BS, Conell C et al (2012) Inpatients statin use predicts improved stroke discharge disposition. Neurology 78:1678–1683
58. Fuentes B, Martinez-Sànchez P, Diez-Tejedor E (2009) Lipid-lowering drugs in ischemic stroke prevention and their influence on acute stroke outcome. Cardiovasc Dis 27(Suppl 1):126–133
59. Reeves MJ, Gargano JW, Luo Z, Mullard AJ, Jacobs BS (2008) Effect of pretreatment with statins on ischemic stroke outcome. Stroke 39:1779–1785
60. Sacco S, Tini D, Bignamini AA, Zaninelli A, Gensini GF, Carolei A et al (2011) Effect of prior medical treatment on ischemic stroke severity and outcome. Funct Neurol 26:133–139
61. Hjalmarsson C, Bokemark L, Manhem K, Mehlig K, Andersson B (2012) The effect of statins on acute and long-term outcome after ischemic stroke in the elderly. Am J Geriatr Pharmacother 10:313–322
62. Martinez-Sànchez P, Rivera-Ordonez C, Fuentes B, Ortega-Casarrubios MA, Idrovo L, Díez-Tejedor E (2009) The beneficial effects of statin treatment by stroke subtype. Eur J Neurol 16:127–133
63. Schmitz SA, O'Regan DP, Fitzpatrick J, Neuwirth C, Potter E, Tosi I et al (2007) MRI at 3 Tesla detects no evidence for ischemic brain damage in intensively treated patients with homozygous familial hypercholesterolemia. Neuroradiology 49:927–931
64. Soljanlahtti S, Autti T, Lauerma K, Raininko R, Keto P, Turtola H et al (2005) Familial hypercholesterolemia patients treated with statins at no increased risk for intracranial vascular lesions despite increased cholesterol burden and extracranial atherosclerosis. Stroke 36:1572–1574
65. Desmaele S, Cornu P, Barbé K, Brouns R, Steurbaut S, Dupont AG (2015) Relationship between pre-stroke cardiovascular medication use and stroke severity. Eur J Clin Pharmacol. doi:10.1007/s00228-015-2001-1
66. Hackam DG, Woodward M, Newby LK, Bhatt DL, Shao M, Smith EE et al (2011) Statins and intracerebral hemorrhage: collaborative systematic review and meta-analysis. Circulation 124:2233–2242
67. McKinney JS, Kostis WJ (2012) Statin therapy and the risk of intracerebral hemorrhage. A meta-analysis of 31 randomized controlled trials. Stroke 43:2149–2156
68. Gomis M, Ois A, Rodriguez-Campello A, Cuadrado-Godia E, Jimènez-Conde J, Subirana I et al (2010) Outcome of intracerebral haemorrhage patients pre-treated with statins. Eur J Neurol 17:443–448
69. Lapchak PA, Han MK (2009) The 3-hydroxy-3-methylglutaryl coenzyme A reductase inhibitor simvastatin reduces thrombolytic-induced intracerebral hemorrhage in embolized rabbits. Brain Res 1303:144–150
70. Hong K-S, Lee JS (2015) Statins in acute ischemic stroke: a systematic review. J Stroke 12:282–301
71. Scheitz JF, Seiffge DJ, Tutüncu S, Gensicke H, Audebert HJ, Bonati LH et al (2014) Dose-related effects of statins on symptomatic intracerebral hemorrhage and outcome after thrombolysis for ischemic stroke. Stroke 45:509–514
72. Tsivgoulis G, Kadlecovà P, Kobayashi A, Czlonkoska A, Broznam M, Švigelj V et al (2015) Safety of statin pretreatment in intravenous thrombolysis for acute ischemic stroke. Stroke 46:2681–2684
73. Cordenier A, De Smedt A, Brouns R, Uyttenboogaart M, De Raedt S, Luijckx GJ et al (2011) Pre-use of statins on stroke outcome: a meta-analysis of observational studies. Ann Neurol Belg 111:261–267
74. Martinez-Ramirez S, Delgado-Mederos R, Marin R, Suàrez-Calvet M, Sàinz MP, Alejaldre A et al (2012) Statin pretreatment may increase the risk of symptomatic intracranial haemorrhage in thrombolysis for ischemic stroke: results from a case-control study and a meta-analysis. J Neurol 259:111–118
75. Meseguer E, Mazighi M, Lapergue B, Labreuche J, Sirimarco G, Gonzalez-Valcarcel J et al (2012) Outcomes after thrombolysis in AIS according to prior statin use. Neurology 79:1817–1823

76. Meier N, Nedeltchev K, Brekenfeld C, Galimanis A, Fischer U, Findling O et al (2009) Prior statin use, intracranial hemorrhage, and outcome after intra-arterial thrombolysis for acute ischemic stroke. Stroke 40:1729–1737
77. Engelter ST, Soinne L, Ringleb P, Sarikaya H, Border R, Berrouschot J et al (2011) IV thrombolysis and statins. Neurology 77:888–895
78. Burn J, Dennis M, Bamford J, Sandercock P, Wade D, Warlow C (1994) Long-term risk of recurrent stroke after a first-ever stroke. The Oxfordshire Community Stroke Project. Stroke 25:333–337
79. Mohan KM, Wolfe CD, Rudd AG, Heuschmann PU, Kolominski-Rabat PL, Grieve AP (2011) Risk and cumulative stroke recurrence: a systematic review and meta-analysis. Stroke 42:1489–1494
80. Zhao L, Wang R, Song B, Gao Y, Fang H, Lu J et al (2015) Association between atherogenic dyslipidemia and recurrent stroke risk in patients with different subtypes of ischemic stroke. Int J Stroke 10:752–758
81. Sironi L, Cimino M, Guerrini U, Calvio AM, Lodetti B, Asdente M et al (2003) Treatment with statins after induction of focal ischemia in rats reduces the extent of brain damage. Arteriosc1 Thromb Vasc Biol 23:322–327
82. Potey C, Ouk T, Petrault M, Berezowski V, Salleron J, Bordet R et al (2015) Early treatment with atorvastatin exerts parenchymal and vascular protective effects in experimental cerebral model. Br J Pharmacol 172:5188–5198
83. Lapchak PA, Han MK (2010) Simvastatin improves clinical scores in a rabbit multiple infarct ischemic stroke model: synergism with a ROCK inhibitor but not the thrombolytic tissue plasminogen activator. Brain Res 1344:217–225
84. Campos-Martorell M, Salvador N, Monge M, Canals F, Garcia-Bonilla L, Hernandez-Guillamon M et al (2014) Brain proteomics identifies potential simvastatin targets in acute phase of stroke in a rat embolic model. J Neurochem 130:301–332
85. Laloux P (2013) Risk and benefit of statins in stroke secondary prevention. Curr Vasc Pharmacol 11:812–816
86. Amarenco P, Benavente O, Goldstein LB, Callahan A, Sillesen H et al (2009) Results of the stroke prevention by aggressive reduction in cholesterol levels (SPARCL) trial by stroke subtypes. Stroke 40:1405–1409
87. Hosomi N, Nagai Y, Kohriyama T, Ohtsuki T, Aoki S, Nezu T et al (2015) The Japan Statin Treatment Against Recurrent Stroke (J-STARS): a multicenter, randomized, open-label, parallel-group study. EBioMedicine 2:1071–1078
88. Goldstein LB, Amarenco P, Szarek M, Callahan A III, Hennerici M, Sillesen H et al (2008) Hemorrhage stroke in stroke prevention by aggressive reduction in cholesterol levels study. Neurology 70:2364–2370
89. Collins R, Armitage J, Parish S, Sleight P, Peto R (2004) Effects of cholesterol-lowering with simvastatin on stroke and other major vascular events in 20536 people with cerebrovascular disease or other high-risk conditions. Lancet 363:757–767
90. Song B, Wang Y, Zhao X, Liu L, Wang C et al (2014) Association between statin use and short-term outcome based on severity of ischemic stroke : a cohort study. PLoS One 9, e84389
91. Blanco M, Nombela F, Castellanos M, Rodriguez-Yanez M, Garcìa-Gil M, Leira R et al (2007) Statin treatment withdrawal in ischemic stroke: a controlled randomized study. Neurology 69:904–910
92. Colivicchi F, Bassi A, Santini M, Caltagirone C (2007) Discontinuation of statin therapy and clinical outcome after ischemic stroke. Stroke 38:2652–2657
93. Flint AC, Kamel H, Navi BB, Rao VA, Faigeles BS, Conell C et al (2012) Statin use during ischemic stroke hospitalization is strongly associated with improved poststroke survival. Stroke 43:147–154
94. Zhang L, Zhang ZG, Ding GL, Jiang Q, Liu X, Meng H et al (2005) Multitarget effects of statin-enhanced thrombolytic therapy for stroke with recombinant human tissue-type plasminogen activator in the rat. Circulation 112:3486–3494

95. Cappellari M, DeLuca C, Tinazzi M, Tomelleri G, Carletti M, Fiaschi A et al (2011) Does statin in the acute phase of ischemic stroke improve outcome after intravenous thrombolysis? A retrospective study. J Neurol Sci 308:128–134
96. Cappellari M, Bovi P, Moretto G, Zini A, Nencini P, Sessa M et al (2013) The THRombolysis and STatins (THRaST) study. Neurology 80:655–661
97. Restrepo L, Bang OY, Ovbiagele B, Ali L, Kim D, Liebeskind DS et al (2009) Impact of hyperlipidemia and statins on ischemic stroke outcomes after intra-arterial fibrinolysis and percutaneous mechanical embolectomy. Cerebrovasc Dis 28:384–390
98. Nì Chròinìn D, Asplund K, Asberg S, Callaly E, Cuadrad-Godia E, Díez-Tejedor E et al (2013) Statin therapy and outcome after ischemic stroke. Systematic review and meta-analysis of observational studies and randomized trials. Stroke 44:448–456
99. Orem C, Uydu HA, Ylmaz R, Gokce M, Baykan M, Eminagaoglu S et al (2004) The effects of atorvastatin treatment on the fibrinolytic system in dyslipidemic patients. Jpn Heart J 45:977–987
100. Scheitz JF, Nolte CH, Endres M (2013) Should statins be paused or discontinued after thrombolysis or acute intracerebral hemorrhage? No! Stroke 44:1472–1476
101. Brewer L, Mellon L, Hall P, Dolan E, Horgan F, Shelley E et al (2015) Secondary prevention after ischaemic stroke: the ASPIRE-S study. BMC Neurol 15:216
102. Furie KL (2012) High-dose statins should only be used in atherosclerotic strokes. Stroke 43:1994–1995
103. Selim MH, Molina CA (2012) High-dose statin for every stroke. The good, the bad and the unknown. Stroke 43:1996–1997
104. Moonis M (2012) High-dose statins should be used in all acute ischemic strokes. Stroke 43:1992–1993
105. Collins R, MacMahon S (1994) Blood pressure, antihypertensive drug treatment and the risks of stroke and of coronary heart disease. Br Med Bull 50:272–298
106. Lewington S, Clarke R, Qizilbash N, Peto R, Collins R, Prospective Studies Collaboration (2002) Age-specific relevance of usual blood pressure to vascular mortality: a meta-analysis of individual data for one million adults in 61 prospective studies. Lancet 360:1903–1913
107. Dickinson CJ (2001) Why are strokes related to hypertension? Classic studies and hypotheses revisited. J Hypertens 19:1515–1521
108. Howard VJ, Tanner RM, Anderson A, Irvin MR, Calhoun DA, Lackland DT et al (2015) Apparent treatment-resistant hypertension among individuals with history of stroke or transient ischemic attack. Am J Med 128:707–714
109. Barreras A, Gurk-Turner C (2003) Angiotensin II receptor blockers. Proc (Bayl Univ Med Cent) 16:123–126
110. Gromotowicz-Poplawska A, Stankiewicz A, Kramkowski K, Gradzka A, Wojewodzka-Zelezniakowicz M, Dzieciol J et al (2016) The acute prothrombotic effect of aldosterone in rats is partially mediated via angiotensin II receptor type 1. Thromb Res 138:114–120
111. Kamińska M, Mogielnicki A, Stankiewicz A, Kramkowski K, Domaniewski T, Buczko W et al (2005) Angiotensin II via AT1 receptor accelerates arterial thrombosis in renovascular hypertensive rats. J Physiol Pharmacol 56:571–585
112. Moraes JA, Frony AC, Dias AM, Renovato-Martins M, Rodrigues G, Marcinkiewicz C et al (2015) Alpha1beta1 and integrin-linked kinase interact and modulate angiotensin II effects in vascular smooth muscle cells. Atherosclerosis 243:477–485
113. Wu H, Cheng XW, Hu L, Hao CN, Hayashi M, Takeshita K et al (2014) Renin inhibition reduces atherosclerotic plaque neovessel formation and regresses advanced atherosclerotic plaques. Atherosclerosis 237:739–747
114. Yusuf S, Sleight P, Pogue J, Bosch J, Davies R, Dagenais G (2000) Effects of an angiotensin-converting-enzyme inhibitor, ramipril, on cardiovascular events in high-risk patients. The Heart Outcomes Prevention Evaluation Study Investigators. N Engl J Med 342:145–153
115. Fogari R, Zoppi A, Lazzari P, Preti P, Mugellini A, Corradi L et al (1998) ACE inhibition but not angiotensin II antagonism reduces plasma fibrinogen and insulin resistance in overweight hypertensive patients. J Cardiovasc Pharmacol 32:616–620

116. Pahor M, Franse LV, Deitcher SR, Cushman WC, Johnson KC, Shorr RI et al (2002) Fosinopril versus amlodipine comparative treatments study: a randomized trial to assess effects on plasminogen activator inhibitor-1. Circulation 105:457–461
117. Li J, Culman J, Hörtnagl H, Zhao Y, Gerova N, Timm M et al (2005) Angiotensin AT_2 receptor protects against cerebral ischemia-induced neuronal injury. FASEB J 19:617–619
118. Alhusban A, Fouda AY, Pillai B, Ishrat T, Soliman S, Fagan SC (2015) Compound 21 is pro-angiogenic in the brain and results in sustained recovery after ischemic stroke. J Hypertens 33:170–180
119. Lee S, Brait VH, Arumugam TV, Evans MA, Kim HA, Widdop RE et al (2012) Neuroprotective effect of an angiotensin receptor type 2 agonist following cerebral ischemia in vitro and in vivo. Exp Transl Stroke Med 4:16
120. McCarthy CA, Vinh A, Broughton BR, Sobey CG, Callaway JK, Widdop RE (2012) Angiotensin II type 2 receptor stimulation initiated after stroke causes neuroprotection in conscious rats. Hypertension 60:1531–1537
121. McCarthy CA, Vinh A, Miller AA, Hallberg A, Alterman M, Callaway JK et al (2014) Direct angiotensin AT2 receptor stimulation using a novel AT2 receptor agonist, compound 21, evokes neuroprotection in conscious hypertensive rats. PLoS One 9, e95762
122. Gutiérrez-Fernández M, Fuentes B, Rodríguez-Frutos B, Ramos-Cejudo J, Otero-Ortega L, Díez-Tejedor E (2015) Different protective and reparative effects of olmesartan in stroke according to time of administration and withdrawal. J Neurosci Res 93:806–814
123. Ishrat T, Pillai B, Soliman S, Fouda AY, Kozak A, Johnson MH et al (2015) Low-dose candesartan enhances molecular mediators of neuroplasticity and subsequent functional recovery after ischemic stroke in rats. Mol Neurobiol 51:1542–1553
124. Panahpour H, Nekooeian AA, Dehghani GA (2014) Candesartan attenuates ischemic brain edema and protects the blood-brain barrier integrity from ischemia/reperfusion injury in rats. Iran Biomed J 18:232–238
125. Gironacci MM (2015) Angiotensin-(1-7): beyond its central effects on blood pressure. Ther Adv Cardiovasc Dis 9:209–216
126. Kalra J, Prakash A, Kumar P, Majeed AB (2015) Cerebroprotective effects of RAS inhibitors: Beyond their cardio-renal actions. J Renin Angiotensin Aldosterone Syst 16:459–468
127. Regenhardt RW, Bennion DM, Sumners C (2014) Cerebroprotective action of angiotensin peptides in stroke. Clin Sci (Lond) 126:195–205
128. Donoghue M, Hsieh F, Baronas E, Godbout K, Gosselin M, Stagliano N et al (2000) A novel angiotensin-converting enzyme-related carboxypeptidase (ACE2) converts angiotensin I to angiotensin 1-9. Circ Res 87:E1–E9
129. Nguyen G, Delarue F, Burcklé C, Bouzhir L, Giller T, Sraer JD (2002) Pivotal role of the renin/prorenin receptor in angiotensin II production and cellular responses to renin. J Clin Invest 109:1417–1427
130. Santos RA, Simoes e Silva AC, Maric C, Silva DM, Machado RP, de Buhr I et al (2003) Angiotensin-(1-7) is an endogenous ligand for the G protein-coupled receptor Mas. Proc Natl Acad Sci U S A. 100:8258-63.
131. Bennion DM, Haltigan E, Regenhardt RW, Steckelings UM, Sumners C (2015) Neuroprotective mechanisms of the ACE2-angiotensin-(1-7)-Mas axis in stroke. Curr Hypertens Rep 17:3
132. Jiang T, Gao L, Lu J, Zhang YD (2013) ACE2-Ang-(1-7)-Mas axis in brain: a potential target for prevention and treatment of ischemic stroke. Curr Neuropharmacol 11:209–217
133. Bennion DM, Haltigan EA, Irwin AJ, Donnangelo LL, Regenhardt RW, Pioquinto DJ et al (2015) Activation of the neuroprotective angiotensin-converting enzyme 2 in rat ischemic stroke. Hypertension 66:141–148
134. Jiang T, Yu JT, Zhu XC, Zhang QQ, Tan MS, Cao L et al (2014) Angiotensin-(1-7) induces cerebral ischaemic tolerance by promoting brain angiogenesis in a Mas/eNOS-dependent pathway. Br J Pharmacol 171:4222–4232
135. Mecca AP, Regenhardt RW, O'Connor TE, Joseph JP, Raizada MK, Katovich MJ et al (2011) Cerebroprotection by angiotensin-(1-7) in endothelin-1-induced ischaemic stroke. Exp Physiol 96:1084–1096

136. Lu J, Jiang T, Wu L, Gao L, Wang Y, Zhou F et al (2013) The expression of angiotensin-converting enzyme 2-angiotensin-(1-7)-Mas receptor axis are upregulated after acute cerebral ischemic stroke in rats. Neuropeptides 47:289–295
137. Striessnig J, Ortner NJ, Pinggera A (2015) Pharmacology of L-type calcium channels: novel drugs for old targets? Curr Mol Pharmacol 8:110–122
138. Kobayashi T, Mori Y (1998) Ca^{2+} channel antagonists and neuroprotection from cerebral ischemia. Eur J Pharmacol 363:1–15
139. Pasternak B, Svanström H, Nielsen NM, Fugger L, Melbye M, Hviid A (2012) Use of calcium channel blockers and Parkinson's disease. Am J Epidemiol 175:627–635
140. Steen PA, Newberg LA, Milde JH, Michenfelder JD (1983) Nimodipine improves cerebral blood flow and neurologic recovery after complete cerebral ischemia in the dog. J Cereb Blood Flow Metab 3:38–43
141. Choi DW (1988) Glutamate neurotoxicity and diseases of the nervous system. Neuron 1: 623–634
142. Dirnagl U (2012) Pathobiology of injury after stroke: the neurovascular unit and beyond. Ann N Y Acad Sci 1268:21–25
143. Villa RF, Gorini A, Ferrari F, Hoyer S (2013) Energy metabolism of cerebral mitochondria during aging, ischemia and post-ischemic recovery assessed by functional proteomics of enzymes. Neurochem Int 63:765–781
144. Hall ED, Andrus PK, Althaus JS, VonVoigtlander PF (1993) Hydroxyl radical production and lipid peroxidation parallels selective post-ischemic vulnerability in gerbil brain. J Neurosci Res 34:107–112
145. El-Abhar HS, Shaalan M, Barakat M, El-Denshary ES (2002) Effect of melatonin and nifedipine on some antioxidant enzymes and different energy fuels in the blood and brain of global ischemic rats. J Pineal Res 33:87–94
146. Yamato M, Shiba T, Ide T, Honda Y, Yamada K, Tsutsui H (2011) Nifedipine treatment reduces brain damage after transient focal ischemia, possibly through its antioxidative effects. Hypertens Res 34:840–845
147. Maniskas ME, Roberts JM, Aron I, Fraser JF, Bix GJ (2015) Stroke neuroprotection revisited: Intra-arterial verapamil is profoundly neuroprotective in experimental acute ischemic stroke. J Cereb Blood Flow Metab 36(4):721–730. doi:10.1177/0271678X15608395
148. Capraro JA, Reedy DP, Latchaw JP, Slugg RM, Stowe NT, Lesser RP et al (1984) Treatment of acute focal cerebral ischemia with propranolol. Stroke 15:486–491
149. Latchaw JP, Little JR, Slugg RM, Lesser RP, Stowe N (1985) Treatment of acute focal cerebral ischemia and recirculation with d-propranolol. Neurosurgery 16:18–22
150. Little JR, Latchaw JP Jr, Slugg RM, Lesser RP, Stowe NT (1982) Treatment of acute focal cerebral ischemia with propranolol. Stroke 13:302–307
151. Standefer M, Little JR (1986) Improved neurological outcome in experimental focal cerebral ischemia treated with propranolol. Neurosurgery 18:136–140
152. Ley EJ, Scehnet J, Park R, Schroff S, Dagliyan G, Conti PS et al (2009) The in vivo effect of propranolol on cerebral perfusion and hypoxia after traumatic brain injury. J Trauma 66: 154–161
153. Song D, Xu J, Du T, Yan E, Hertz L, Waltz W et al (2014) Inhibition of brain swelling after ischemia-reperfusion by β-adrenergic antagonists: correlation with increased K^+ and decreased Ca^{2+} concentrations in extracellular fluid. Biomed Res Int 2014:873590
154. Goyagi T, Nishikawa T, Tobe Y (2011) Neuroprotective effects and suppression of ischemia-induced glutamate elevation by β_1-adrenoreceptor antagonists administered before transient focal ischemia in rats. J Neurosurg Anesthesiol 23:131–137
155. Danura T, Takeda Y, Shiraishi K, Naito H, Mizoue R, Sato S et al (2013) Quantitative evaluation of the neuroprotective effects of a short-acting β-adrenoceptor antagonist at a clinical dose on forebrain ischemia in gerbils: effects of esmolol on ischemic depolarization and histologic outcome of hippocampal CA1. J Neurosurg Anesthesiol 25:292–298

156. Mracsko E, Liesz A, Karcher S, Zorn M, Bari F, Veltkamp R (2014) Differential effects of sympathetic nervous system and hypothalamic-pituitary-adrenal axis on systemic immune cells after severe experimental stroke. Brain Behav Immun 41:200–209
157. Gilsbach R, Röser C, Beetz N, Brede M, Hadamek K, Haubold M et al (2009) Genetic dissection of alpha2-adrenoceptor functions in adrenergic versus nonadrenergic cells. Mol Pharmacol 75:1160–1170
158. Berkman MZ, Zirh TA, Berkman K, Pamir MN (1998) Tizanidine is an effective agent in the prevention of focal cerebral ischemia in rats: an experimental study. Surg Neurol 50:264–271
159. Hoffman WE, Cheng MA, Thomas C, Baughman VL, Albrecht RF (1991) Clonidine decreases plasma catecholamines and improves outcome from incomplete ischemia in the rat. Anesth Analg 73:460–464
160. Gupta S, Sharma B (2014) Pharmacological modulation of I_1-imidazoline and α_2-adrenoceptors in sub acute brain ischemia induced vascular dementia. Eur J Pharmacol 723:80–90
161. Zhang Y (2004) Clonidine preconditioning decreases infarct size and improves neurological outcome from transient forebrain ischemia in the rat. Neuroscience 125:625–631
162. Brede M, Braeuninger S, Langhauser F, Hein L, Roewer N, Stoll G et al (2011) α_2-Adrenoceptors do not mediate neuroprotection in acute ischemic stroke in mice. J Cereb Blood Flow Metab 31:e1–e7
163. Benzi G, Gorini A, Ghigini B, Arnaboldi R, Villa RF (1993) Synaptosomal non-mitochondrial ATPase activities and drug treatment. Neurochem Res 18:719–726
164. Villa RF, Arnaboldi R, Gorini A, Geroldi D (1989) Action of piracetam and clonidine on different mitochondrial populations from hippocampus. Farmaco 44:215–226
165. O'Donnell ME, Tran L, Lam TI, Liu XB, Anderson SE (2004) Bumetanide inhibition of the blood–brain barrier Na^+-K^+-Cl^- cotransporter reduces edema formation in the rat middle cerebral artery occlusion model of stroke. J Cereb Blood Flow Metab 24:1046–1056
166. Yan Y, Dempsey RJ, Sun D (2001) Na^+-K^+-Cl^- cotransporter in rat focal cerebral ischemia. J Cereb Blood Flow Metab 21:711–721
167. Chen H, Luo J, Kintner DB, Shull GE, Sun D (2005) Na^+-dependent chloride transporter (NKCC1)-null mice exhibit less gray and white matter damage after focal cerebral ischemia. J Cereb Blood Flow Metab 25:54–66
168. Wang G, Huang H, He Y, Ruan L, Huang J (2014) Bumetanide protects focal cerebral ischemia-reperfusion injury in rat. Int J Clin Exp Pathol 1487–94
169. Migliati ER, Amiry-Moghaddam M, Froehner SC, Adams ME, Ottersen OP, Bhardwaj A (2010) Na^+-K^+-$2Cl^-$ cotransport inhibitor attenuates cerebral edema following experimental stroke via the perivascular pool of aquaporin-4. Neurocrit Care 13:123–131
170. Chitravas N, Dewey HM, Nicol MB, Harding DL, Pearce DC, Thrift AG (2007) Is prestroke use of angiotensin-converting enzyme inhibitors associated with better outcome? Neurology 68:1687–1693
171. Yu AY, Keezer MR, Zhu B, Wolfson C, Côté R (2009) Pre-stroke use of antihypertensives, antiplatelets, or statins and early ischemic stroke outcomes. Cerebrovasc Dis 27:398–402
172. Hassan Y, Aziz NA, Al-Jabi SW, Looi I, Zyoud SH (2010) Impact of angiotensin-converting enzyme inhibitors administration prior to acute ischemic stroke onset on in-hospital mortality. J Cardiovasc Pharmacol Ther 15:274–281
173. Tuttolomondo A, Di Sciacca R, Di Raimondo D, Pedone C, La Placa S, Pinto A et al (2011) Effects of clinical and laboratory variables and of pretreatment with cardiovascular drugs in acute ischaemic stroke: a retrospective chart review from the GIFA study. Int J Cardiol 151:318–322
174. Sundbøll J, Schmidt M, Horváth-Puhó E, Christiansen CF, Pedersen L, Bøtker HE et al (2015) Preadmission use of ACE inhibitors or angiotensin receptor blockers and short-term mortality after stroke. J Neurol Neurosurg Psychiatry 86:748–754
175. Dahlöf B, Devereux RB, Kjeldsen SE, Julius S, Beevers G, de Faire U et al (2002) Cardiovascular morbidity and mortality in the Losartan Intervention For Endpoint reduction in hypertension study (LIFE): a randomised trial against atenolol. Lancet 359:995–1003

176. Lithell H, Hansson L, Skoog I, Elmfeldt D, Hofman A, Olofsson B et al (2003) The Study on Cognition and Prognosis in the Elderly (SCOPE): principal results of a randomized double-blind intervention trial. J Hypertens 21:875–886
177. Miyamoto N, Tanaka Y, Ueno Y, Tanaka R, Hattori N, Urabe T (2012) Benefits of prestroke use of angiotensin type 1 receptor blockers on ischemic stroke severity. J Stroke Cerebrovasc Dis 21:363–368
178. Yusuf S, Teo KK, Pogue J, Dyal L, Copland I et al (2008) Telmisartan, ramipril, or both in patients at high risk for vascular events. N Engl J Med 358:1547–1559
179. Ferrari R, Boersma E (2013) The impact of ACE inhibition on all-cause and cardiovascular mortality in contemporary hypertension trials: a review. Expert Rev Cardiovasc Ther 11:705–717
180. Wang JG, Staessen JA, Gong L, Liu L (2000) Chinese trial on isolated systolic hypertension in the elderly. Systolic Hypertension in China (Syst-China) Collaborative Group. Arch Intern Med 160:211–220
181. Lubsen J, Wagener G, Kirwan BA, de Brouwer S (2005) Poole-Wilson PA; ACTION investigators. Effect of long-acting nifedipine on mortality and cardiovascular morbidity in patients with symptomatic stable angina and hypertension: the ACTION trial. J Hypertens 23:641–648
182. He DH, Zhang LM, Lin LM, Ning RB, Wang HJ, Xu CS et al (2014) Long-term prehypertension treatment with losartan effectively prevents brain damage and stroke in stroke-prone spontaneously hypertensive rats. Int J Mol Med 33:301–309
183. Lawes CM, Bennett DA, Feigin VL, Rodgers A (2004) Blood pressure and stroke: an overview of published reviews. Stroke 35:1024–1033
184. Wang JG, Li Y, Franklin SS, Safar M (2007) Prevention of stroke and myocardial infarction by amlodipine and angiotensin receptor blockers. A quantitative overview. Hypertension 50:181–188
185. Matsuzaki M, Ogihara T, Umemoto S, Rakugi H, Matsuoka H, Shimada K et al (2011) Prevention of cardiovascular events with calcium channel blocker-based combination therapies in patients with hypertension: a randomized controlled trial. J Hypertens 29:1649–1659
186. Umemoto S, Ogihara T, Rakugi H, Matsumoto M, Kitagawa K, Shimada K et al (2013) Effects of a benidipine-based combination therapy on the risk of stroke according to stroke subtype: the COPE trial. Hypertens Res 36:1088–1095
187. Goyagi T, Horiguchi T, Nishikawa T, Tobe Y, Masaki Y (2012) Neuroprotective effects of selective β_1 adrenoceptor antagonists, landiolol and esmolol, on transient forebrain ischemia in rats; a dose-response study. Brain Res 1461:96–101
188. Dziedzic T, Pera J, Zur-Wyrozumska K, Klimkowicz-Mrowiec A, Szczudlik A, Slowik A (2012) Beta-blockers use and risk of hyperglycemia in acute stroke patients. Atherosclerosis 223:209–211
189. Allport L, Baird T, Butcher K, Macgregor L, Prosser J, Colman P et al (2006) Frequency and temporal profile of poststroke hyperglycemia using continuous glucose monitoring. Diabetes Care 29:1839e44
190. Sykora M, Siarnik P, Diedler J, VISTA Acute Collaborators (2015) β-Blockers, pneumonia, and outcome after ischemic stroke: evidence from Virtual International Stroke Trials Archive. Stroke 46:1269–1274
191. Lai CL, Kuo RN, Chen HM, Chen MF, Chan KA, Lai MS (2016) Risk of ischemic stroke during the initiation period of α-blocker therapy among older men. CMAJ 188(4):255–260. doi:10.1503/cmaj.150624
192. Appel LJ (2002) The verdict from ALLHAT-thiazide diuretics are the preferred initial therapy for hypertension. JAMA 288:3039–3042
193. Shih HM, Lin WC, Wang CH, Lin LC (2014) Hypertensive patients using thiazide diuretics as primary stroke prevention make better functional outcome after ischemic stroke. J Stroke Cerebrovasc Dis 23:2414–2418
194. Tziomalos K, Giampatzis V, Bouziana SD, Spanou M, Kostaki S et al (2015) Comparative effects of more versus less aggressive treatment with statins on the long-term outcome of patients with acute ischemic stroke. Atherosclerosis 243:65–70

195. Wing LM, Reid CM, Ryan P, Beilin LJ, Brown MA, Jennings GL et al (2003) A comparison of outcomes with angiotensin-converting enzyme inhibitors and diuretics for hypertension in the elderly. N Engl J Med 348:583–592
196. Cutler JA (2003) The ANBP2 and ALLHAT: conflicting or consistent? J Clin Hypertens 5:192–195
197. Doggrell SA (2003) ACE inhibitors versus diuretics: ALLHAT versus ANBP2. Expert Opin Pharmacother 4:825–828
198. Toni D, Mangiafico S, Agostoni E, Bergui M, Cerrato P, Ciccone A et al (2015) Intravenous thrombolysis and intra-arterial interventions in acute ischemic stroke: Italian Stroke Organisation (ISO)-SPREAD guidelines. Int J Stroke 10:1119–1129
199. Staessen J, Fagard R, Amery A (1991) Isolated systolic hypertension in the elderly: implications of Systolic Hypertension in the Elderly Program (SHEP) for clinical practice and for the ongoing trials. J Hum Hypertens 5:469–474
200. Hornslien AG, Sandset EC, Bath PM, Wyller TB, Berge E, Scandinavian Candesartan Acute Stroke Trial Study Group (2013) Effects of candesartan in acute stroke on cognitive function and quality of life: results from the Scandinavian Candesartan Acute Stroke Trial. Stroke 44:2022–2024
201. Hornslien AG, Sandset EC, Igland J, Terént A, Boysen G, Bath PM et al (2015) Effects of candesartan in acute stroke on vascular events during long-term follow-up: results from the Scandinavian Candesartan Acute Stroke Trial (SCAST). Int J Stroke 10:830–835
202. Nakamura T, Tsutsumi Y, Shimizu Y, Uchiyama S (2010) Renin-angiotensin system blockade safely reduces blood pressure in patients with minor ischemic stroke during the acute phase. J Stroke Cerebrovasc Dis 19:435–440
203. Sandset EC, Bath PM, Boysen G, Jatuzis D, Kõrv J, Lüders S et al (2011) The angiotensin-receptor blocker candesartan for treatment of acute stroke (SCAST): a randomised, placebo-controlled, double-blind trial. Lancet 377:741–750
204. Sandset EC, Jusufovic M, Sandset PM, Bath PM, Berge E, SCAST Study Group (2015) Effects of blood pressure-lowering treatment in different subtypes of acute ischemic stroke. Stroke 46:877–879
205. Ahmed N, Näsman P, Wahlgren NG (2000) Effect of intravenous nimodipine on blood pressure and outcome after acute stroke. Stroke 31:1250–1255
206. Ahmed N, Wahlgren NG (2003) Effects of blood pressure lowering in the acute phase of total anterior circulation infarcts and other stroke subtypes. Cerebrovasc Dis 15:235–243
207. Cao TS, Huynh VM, Tran VH (2015) Effects of lercanidipine versus amlodipine in hypertensive patients with cerebral ischemic stroke. Curr Med Res Opin 31:163–170
208. Horn J, de Haan RJ, Vermeulen M, Limburg M (2001) Very Early Nimodipine Use in Stroke (VENUS): a randomized, double-blind, placebo-controlled trial. Stroke 32:461–465
209. Kimura Y, Kitagawa K, Oku N, Kajimoto K, Kato H, Tanaka M et al (2008) Hemodynamic influences of azelnidipine, a novel calcium channel blocker, on cerebral circulation in hypertensive patients with ischemic white matter lesions. Hypertens Res 31:2147–2154
210. Watanabe M, Hirano T, Okamoto S, Shiraishi S, Tomiguchi S, Uchino M (2010) Azelnidipine, a long-acting calcium channel blocker, could control hypertension without decreasing cerebral blood flow in post-ischemic stroke patients. A 123I-IMP SPECT follow-up study. Hypertens Res 33:43–48
211. Diringer MN, Scalfani MT, Zazulia AR, Videen TO, Dhar R (2011) Cerebral hemodynamic and metabolic effects of equi-osmolar doses mannitol and 23.4% saline in patients with edema following large ischemic stroke. Neurocrit Care 14:11–17
212. Webb AJ, Wilson M, Lovett N, Paul N, Fischer U, Rothwell PM (2014) Response of day-to-day home blood pressure variability by antihypertensive drug class after transient ischemic attack or nondisabling stroke. Stroke 45:2967–2973
213. Barer DH, Cruickshank JM, Ebrahim SB, Mitchell JR (1988) Low dose beta blockade in acute stroke ("BEST" trial): an evaluation. Br Med J (Clin Res Ed) 296:737–741

214. Liu-DeRyke X, Levy PD, Parker D Jr, Coplin W, Rhoney DH (2013) A prospective evaluation of labetalol versus nicardipine for blood pressure management in patients with acute stroke. Neurocrit Care 19:41–47
215. Phelan C, Alaigh V, Fortunato G, Staff I, Sansing L (2015) Effect of β-adrenergic antagonists on in-hospital mortality after ischemic stroke. J Stroke Cerebrovasc Dis 24:1998–2004
216. Berge E, Cohen G, Lindley RI, Sandercock P, Wardlaw JM, Sandset EC et al (2015) Effects of blood pressure and blood pressure-lowering treatment during the first 24 hours among patients in the Third International Stroke Trial of Thrombolytic Treatment for Acute Ischemic Stroke. Stroke 46:3362–3369
217. Fogari R, Zoppi A (2006) Antihypertensive drugs and fibrinolytic function. Am J Hypertens 19:1293–1299
218. Nakamura S, Nakamura I, Ma L, Vaughan DE, Fogo AB (2000) Plasminogen activator inhibitor-1 expression is regulated by the angiotensin type 1 receptor in vivo. Kidney Int 58:251–259
219. Hernández-Hernández R, Armas-Padilla MC, Velasco M, Carvajal AR, Armas de Hernández MJ, Guerrero-Pajuelo J et al (1999) Effects of amlodipine and enalapril on platelet function in patients with mild to moderate hypertension. Int J Clin Pharmacol Ther 37:323–331
220. Gremmel T, Durstberger M, Eichelberger B, Koppensteiner R, Panzer S (2015) Calcium-channel blockers attenuate the antiplatelet effect of clopidogrel. Cardiovasc Ther 33:264–269
221. Katoh M, Nakajima M, Yamazaki H, Yokoi T (2000) Inhibitory potencies of 1,4-dihydropyridine calcium antagonists to P-glycoprotein-mediated transport: comparison with the effects on CYP3A4. Pharm Res 17:1189–1197
222. Tiryaki O, Usalan C, Buyukhatipoglu H, Sayiner ZA, Kilisli H (2012) Effects of lisinopril, irbesartan, and amlodipine on the thrombogenic variables in the early and late stages of the treatment in hypertensive patients. Clin Exp Hypertens 34:145–152
223. Hill MD, Lye T, Moss H, Barber PA, Demchuk AM, Newcommon NJ et al (2003) Hemi-orolingual angioedema and ACE inhibition after alteplase treatment of stroke. Neurology 60:1525–1527
224. Ottomeyer C, Sick C, Hennerici MG, Szabo K (2009) Orolingual angioedema under systemic thrombolysis with rt-PA: an underestimated side effect. Nervenarzt 80:459–463
225. Lin SY, Tang SC, Tsai LK, Yeh SJ, Hsiao YJ, Chen YW et al (2014) Orolingual angioedema after alteplase therapy of acute ischaemic stroke: incidence and risk of prior angiotensin-converting enzyme inhibitor use. Eur J Neurol 21:1285–1291
226. Hurford R, Rezvani S, Kreimei M, Herbert A, Vail A, Parry-Jones AR et al (2015) Incidence, predictors and clinical characteristics of orolingual angio-oedema complicating thrombolysis with tissue plasminogen activator for ischaemic stroke. J Neurol Neurosurg Psychiatry 86:520–523
227. McKay C, Hall AB, Cortes J (2015) Time to blood pressure control before thrombolytic therapy in patients with acute ischemic stroke: comparison of labetalol, nicardipine, and hydralazine. J Neurosci Nurs 47:327–332
228. Carcel C, Anderson CS (2015) Timing of blood pressure lowering in acute ischemic stroke. Curr Atheroscler Rep 17:520
229. Arima H, Anderson C, Omae T, Woodward M, MacMahon S, Mancia G et al (2014) Degree of blood pressure reduction and recurrent stroke: the PROGRESS trial. J Neurol Neurosurg Psychiatry 85:1284–1285
230. PROGRESS Collaborative Group (2001) Randomised trial of a perindopril-based blood-pressure-lowering regimen among 6,105 individuals with previous stroke or transient ischaemic attack. Lancet 358:1033–1041
231. Yusuf S, Diener HC, Sacco RL, Cotton D, Ounpuu S, Lawton WA et al (2008) Telmisartan to prevent recurrent stroke and cardiovascular events. N Engl J Med 359:1225–1237
232. Weber R, Weimar C, Blatchford J, Hermansson K, Wanke I, Möller-Hartmann C et al (2012) Telmisartan on top of antihypertensive treatment does not prevent progression of cerebral white matter lesions in the prevention regimen for effectively avoiding second strokes (PRoFESS) MRI substudy. Stroke 43:2336–2342

233. Yamano S, Horii M, Takami T, Sakuma M, Morimoto T, Okada S et al (2015) Comparison between angiotensin-converting enzyme inhibitors and angiotensin receptor blockers on the risk of stroke recurrence and longitudinal progression of white matter lesions and silent brain infarcts on MRI (CEREBRAL study): rationale, design, and methodology. Int J Stroke 10:452–456
234. Levijoki J, Kivikko M, Pollesello P, Sallinen J, Hyttilä-Hopponen M, Kuoppamäki M et al (2015) Levosimendan alone and in combination with valsartan prevents stroke in Dahl salt-sensitive rats. Eur J Pharmacol 750:132–140
235. Liu L, Zhang Y, Liu G, Li W, Zhang X, Zanchetti A, FEVER Study Group (2005) The Felodipine Event Reduction (FEVER) Study: a randomized long-term placebo-controlled trial in Chinese hypertensive patients. J Hypertens 23:2157–2172
236. Schrader J, Lüders S, Kulschewski A, Hammersen F, Plate K, Berger J et al (2005) Morbidity and Mortality After Stroke, Eprosartan Compared with Nitrendipine for Secondary Prevention: principal results of a prospective randomized controlled study (MOSES). Stroke 36:1218–1226
237. Hong K-S, Kang DW, Bae HJ, Kim YK, Han MK, Park JM et al (2010) Effect of cilnidipine vs losartan on cerebral blood flow in hypertensive patients with a history of ischemic stroke: a randomized controlled trial. Acta Neurol Scand 121:51–57
238. De Lima LG, Saconato H, Atallah AN, da Silva EM (2014) Beta-blockers for preventing stroke recurrence. Cochrane Database Syst Rev 10, CD007890
239. PATS Collaborating Group (1995) Post-stroke antihypertensive treatment study. A preliminary result. Chin Med J (Engl) 108:710–717
240. Liu L, Wang Z, Gong L, Zhang Y, Thijs L, Staessen JA et al (2009) Blood pressure reduction for the secondary prevention of stroke: a Chinese trial and a systematic review of the literature. Hypertens Res 32:1032–1040
241. Ovbiagele B, Diener HC, Yusuf S, Martin RH, Cotton D, Vinisko R et al (2011) Level of systolic blood pressure within the normal range and risk of recurrent stroke. JAMA 306:2137–2144
242. Arima H, Chalmers J (2011) PROGRESS: prevention of recurrent stroke. J Clin Hypertens 13:693–702
243. Fugate JE, Rabinstein AA (2013) Contraindications to intravenous rtPA for acute stroke: a critical reappraisal. Neurol Clin Pract 3:177–185
244. Sundström J, Arima H, Jackson R, Turnbull F, Rahimi K, Chalmers J et al (2015) Effects of blood pressure reduction in mild hypertension: a systematic review and meta-analysis. Ann Intern Med 162:184–191

Chapter 15
Toward Effective Combination Therapy and Pleiotropic Drugs

Sergio Amaro, Xabier Urra, and Ángel Chamorro

Abstract Acute ischaemic stroke is a major cause of morbidity and mortality. The most effective treatment is early reperfusion, although less than half of patients who are finally treated obtain permanent benefits. The combination of rapid reperfusion and neuroprotective therapies would be of added value for maximizing the beneficial effects of rapid reperfusion and to reduce the harmful consequences of reperfusion injury, including the no-reflow phenomenon, after recanalization. Given the relevance of oxidative stress in the physiopathology of brain ischaemia, the use of antioxidant molecules, such as uric acid, could translate into effective neuroprotective effects. Uric acid is the most potent natural antioxidant, and its exogenous administration is neuroprotective and has synergistic effects when administered alongside alteplase, a thrombolytic drug. In humans, uric acid therapy is safe and has measurable neuroprotective effects, especially in patients with pre-treatment hyperglycaemia or in women. Given the encouraging preclinical and clinical data regarding the potential neuroprotective effects of uric acid administration in combination with reperfusion in acute brain ischaemia, the development of adequately powered pivotal confirmatory clinical trials is warranted.

Keywords Stroke • Ischaemia • Neuroprotection • Reperfusion • Oxidative stress • Uric acid

S. Amaro, M.D., Ph.D. • X. Urra, M.D., Ph.D.
Hospital Clinic of Barcelona Neurosciences Institute, Functional Unit of Cerebrovascular Diseases, Hospital Clínic of Barcelona, 170 Villarroel Street, Barcelona 08036, Spain

Institut d'Investigacions Biomediques August Pi i Sunyer, Barcelona, Spain
e-mail: samaro@clinic.ub.es; xurra@clinic.ub.es

Á. Chamorro, M.D., Ph.D. (✉)
Functional Unit of Cerebrovascular Diseases, Clínic Institute of Neurosciences (ICN), Hospital Clínic de Barcelona, Barcelona 08036, Spain

Institut d'Investigacions Biomediques August Pi i Sunyer, Barcelona, Spain

Medical Department, School of Medicine, University of Barcelona, Barcelona, Spain
e-mail: achamorro@clinic.ub.es

P.A. Lapchak, J.H. Zhang (eds.), *Neuroprotective Therapy for Stroke and Ischemic Disease*, Springer Series in Translational Stroke Research,
DOI 10.1007/978-3-319-45345-3_15

Abbreviations

AIS	Acute ischaemic stroke
BBB	Blood–brain barrier
MCA	Middle cerebral artery
ROS	Reactive oxygen species
rtPA	Recombinant tissue plasminogen activator
UA	Uric acid

1 Introduction

Acute ischaemic stroke (AIS) is a devastating disease that represents the second leading cause of death worldwide and a major health problem. The age-standardized rates of AIS mortality have decreased in the past two decades, but the absolute number of people who have a stroke every year, stroke survivors, related deaths, and the overall global burden of stroke is high and increasing [1]. The treatment of AIS is a major medical challenge given the high incidence of the disease and the complexity of its biology. Consequently, there is an imperative need for effective preventive therapy, early critical care, and rehabilitation in patients with stroke. The emergent therapy of AIS has made significant progress in the last years, in particular after the demonstration of the clinical value of early reperfusion strategies. These therapies include the intravenous administration of recombinant tissue plasminogen activator (rtPA) within 4'5 h from stroke onset and/or the use of endovascular mechanical thrombectomy within 6 h from stroke onset [2]. Early reperfusion is associated with a significant increase in the odds of good functional outcome at long term and a significant reduction in the risk of death [2]. However, inter-organizational, intra-organizational, medical, and psychological barriers hamper a broader implementation of early reperfusion therapies for acute ischaemic stroke [3]. In the case of rtPA, these factors determine that only about 20 % of stroke patients arrive in time and/or are eligible to be treated with early reperfusion therapies, and less than 10 % of all stroke patients are actually treated with rtPA [4]. In the best scenario, less than half of those who are finally treated obtain permanent benefits from reperfusion therapy, while up to one-third of the patients continue with long-term substantial disability [2]. This notion was further confirmed in five randomized controlled trials performed in 2015 which demonstrated the superiority of endovascular thrombectomy over best medical treatment to improve outcome in patients with AIS and large proximal artery occlusions [5–9], but these trials also brought into attention that more than half of the randomized patients did not improve the functional outcome following the endovascular procedure. Importantly, these relatively modest clinical results were in sheer contrast with the observation that three out of four patients that received mechanical thrombectomy displayed a full recanalization of the occluded vessel by the end of the procedure. Although the proportion of patients who reperfused following endovascular therapy was not reported in these trials, adequate brain perfusion was presumably a prevalent problem, for previous studies suggested that one out of four patients may not reperfuse despite recanalization [11–14]. These data underlines the need for additional strategies aimed to maximize the benefits of reperfusion therapies [15].

2 Adequate Reperfusion Is a Major Target for Neuroprotection

The aim of systemic or endovascular thrombolysis is to reopen an occluded artery, restore anterograde perfusion, salvage ischaemic brain, and improve clinical outcome. Early vessel recanalization is associated with a four to fivefold increase in the odds of good long-term functional outcome, and a four to fivefold reduction in the odds of death [16]. However, as introduced before, arterial recanalization is not always followed by clinical improvement. In the case of effective vessel recanalization, the lack of favourable response may be related to fast infarct progression before recanalization, inadequate reperfusion (no reflow), downstream embolization, or reperfusion injury. Untimely reperfusion may be injurious and facilitate the development of cerebral oedema, brain haemorrhage, or both in experimental conditions [17]. In human stroke, the relevance of reperfusion injury is less well established, but brain perfusion MRI studies suggest that 40–50 % of the patients show hyperperfusion within hours of stroke onset [18]. The tenfold increase of cerebral bleeding encountered in thrombolysed patients could also be due in part to reperfusion injury. Indeed, experimental and clinical data show that recanalization itself may partially override the beneficial effects of tissue reperfusion when the return of oxygen to the ischaemic brain paradoxically facilitates oxidative stress [19]. Arguably, neuroprotection could be accomplished in patients with AIS using drugs that harness the effects of reperfusion injury. Thus, improving the current treatment of AIS will require not only fixing the plumbing but also impeding the tissue and cellular consequences induced by the ischaemia [15, 20].

Cerebral blood flow interruption results in a rapid failure of cell bioenergetics, followed by an ever-increasing list of reported cellular perturbations in neurons, astrocytes, endothelial cells, oligodendrocytes, and subpopulations of these cells [21]. The immediate consequence of diminished energy reserves is perturbation of membrane electrochemical gradients, depolarization, and activation of transcription factors, phospholipases, endonucleases, and proteases that lead to deranged intracellular signalling, compromised cellular function, and loss of structural integrity. Excessive release of glutamate results in an increased activation of glutamate receptors and an additional influx of calcium, accentuating cellular injury. Many different neuroprotection approaches targeting different aspects of the ischaemic cascade were tested previously in animal stroke models and clinical development programmes [22]. However, despite many successful treatment experiments in animals regarding both infarct size reduction and improved functional outcome, no neuroprotective drug demonstrated unequivocal efficacy in clinical trials that would have satisfied the threshold to apply for regulatory approval [23]. A frequently quoted explanation of previous failures is that most neuroprotective drugs had a narrow specific therapeutic target, such as competitive and non-competitive receptor antagonists, while the current prevailing view is that drugs with broader and pleiotropic effects on the ischaemic cascade are potentially more appealing [24].

It was initially expected that reperfusion would simply be a reversal of the ischaemia-induced disruption of the cellular milieu to re-establish function. However, experimental evidence shows that reperfusion also triggers a set of unique, potentially pathologic events including, for example, increased prostaglandin synthesis, elevated production of second messengers, inflammation, and mitochondrial dysfunction, as indicated by elevated reactive oxygen species (ROS) and the opening of the mitochondrial permeability transition pore. These effects are mainly a consequence of oxidative stress. The ischaemic brain is particularly vulnerable to oxidative stress compared with other organs as the result of its high consumption of oxygen, its rich content of iron and unsaturated lipids, and its relatively low antioxidant capacity [25]. Under physiologic conditions, the mitochondria-generated O_2, H_2O_2, and OH^- play important roles in regulating those pathways integral to signalling and metabolism, and these ROS are normally inactivated by endogenous scavenging systems. Excessive ROS production, however, can overwhelm free radical scavenging systems that may have been compromised by ischaemia. Indeed, during ischaemia, there is a surge in production of superoxide (O_2^-), nitric oxide (NO), and peroxynitrite ($ONOO^-$) for at least 6–12 h that facilitates mitochondrial dysfunction, excitotoxicity, lipid peroxidation, and inflammation [26]. The increased production of superoxide anion, hydrogen peroxide, hydroxyl radical, peroxynitrite, or nitrogen dioxide has been localized in infiltrating phagocytes, vascular cells, and glial cells (astrocytes, oligodendrocytes, and microglia) [27, 28]. The levels of superoxide start to rise during the ischaemic phase, followed by a much larger increase during the early reperfusion period, both in neurons and in endothelial cells [29]. The sources of high concentrations of superoxide include the mitochondrion [30], the activity of cyclooxygenase enzymes [31], NADPH oxidase expressed by neurons [32], endothelial cells [33], and infiltrating neutrophils, and the hypoxic-dependent conversion of xanthine dehydrogenase into xanthine oxidase [21]. In humans, there is an increase in the levels of nitric oxide metabolites (nitrates and nitrites) in blood and cerebrospinal fluid within 24 h of stroke onset which is correlated with infarction volume and the risk of early neurological worsening independent of other important factors. Neuronal nitric oxide synthase plays the predominant role in the formation of nitric oxide in the early phase of ischaemia and reperfusion, but endothelial nitric oxide synthase plays this role in the vascular endothelium during the early reperfusion phase. At later stages of stroke, formation of nitric oxide is totally dependent on the expression of inducible nitric oxide synthase. In addition to direct toxic effects, the enhanced release of free radicals resulting from ischaemia and reperfusion can modify the expression of several genes that up-regulate the redox-sensitive transcription factor nuclear factor-kB. This mechanism may result in an increased expression of pro-inflammatory cytokines and endothelial adhesion molecules, which play a relevant role in the fate of brain tissue under ischaemic conditions. Moreover, infiltrating inflammatory cells may elaborate more ROS that increase oxidative stress, induce blood–brain barrier damage, and facilitate further brain injury.

Common to all these mechanisms of reperfusion injury are concomitant changes in the microvascular structure which may be major determinants for the impairment of brain reperfusion, and this inadequate matching of blood flow to neural activity may produce damage to neurons or glial cells beyond that caused by the initial ischaemia [34]. Models of transient experimental brain ischaemia performed on albino rabbits more than 40 years ago identified the lack of reflow at the microcirculation following the ischaemic insult [35]. These pioneer studies described the presence of narrowed brain capillary lumina filled with entrapped erythrocytes and mottled areas of nonperfused brain tissue interspersed with well-perfused tissue. The no-reflow phenomenon was primarily attributed to spasm or swelling of the cells in or around the vessel wall, while an increase in the viscosity of the blood was considered a secondary contributing factor. Yet, the rich expression of thromboplastins in the brain tissue also lent support to speculate that fibrin clots would form within the capillary bed because the ischaemic brain injury would expose these thromboplastins to the circulation, although a weakness of this hypothesis was that brain perfusion did not improve when animals received large doses of heparin. Later experimental studies found capillary-obstructing polymorphonuclear leukocytes in the microvascular bed and this finding led to postulate their role in early microvascular injury [36]. Other putative mechanisms advocated to explain the no-reflow phenomenon were refinements of the original descriptions, including the clogging of the perivascular space secondary to fibrinogen extravasation and activation of tissue factor due to increased blood. The increased production brain barrier (BBB) permeability, distal microembolism originating from a disintegrated proximal thrombus, or microvascular obstruction that would result from compression by swollen astrocyte end-feet, although later work suggested that this effect had been overestimated. A major step forward to understand the mechanisms of the no-reflow phenomenon was the identification of a rich expression of NADPH oxidases in endothelial, vascular smooth muscle, and adventitial cells of cerebral arteries and in brain capillary pericytes [37], as these enzymes are a major source of oxygen and nitrogen radicals such as superoxide and peroxynitrite [38]. Pericytes are cells apposed to central nervous system capillaries and contain contractile proteins that can control the diameter of capillaries, and thus are essential regulators of the microcirculatory blood flow [39]. Pericytes express a number of receptors for vasoactive mediators such as catecholamines, angiotensin-II, vasoactive intestinal peptide, endothelin-1, and vasopressin, and also modulate the growth and development of endothelial cells, play a part in angiogenesis and differentiation of the BBB, and take part in immune responses. Some pericytes constrict at the start of ischaemia because in the absence of ATP, Ca^{2+} is not able to pump out of the cell. Contrary to this view, a recent study that used optical imaging techniques in living mice claimed that capillary pericytes were not contractile on the vascular tree, while it suggested that the regulation of cerebral blood flow in physiological and pathological conditions was mediated by arteriolar smooth muscle cells [40]. Alternatively, pericytes would not be adequate targets for pharmacological intervention as these cells stay in rigor following the onset of brain ischaemia because no ATP is available to relax their contractile filaments. However, other studies suggest that pericytes are essential regulators of the microcirculatory blood flow and showed that superoxide and peroxynitrite caused pericyte contraction [41]. A schematic outline of the effects of oxidative stress in the physiopathology of brain ischaemia is shown in Fig. 15.1.

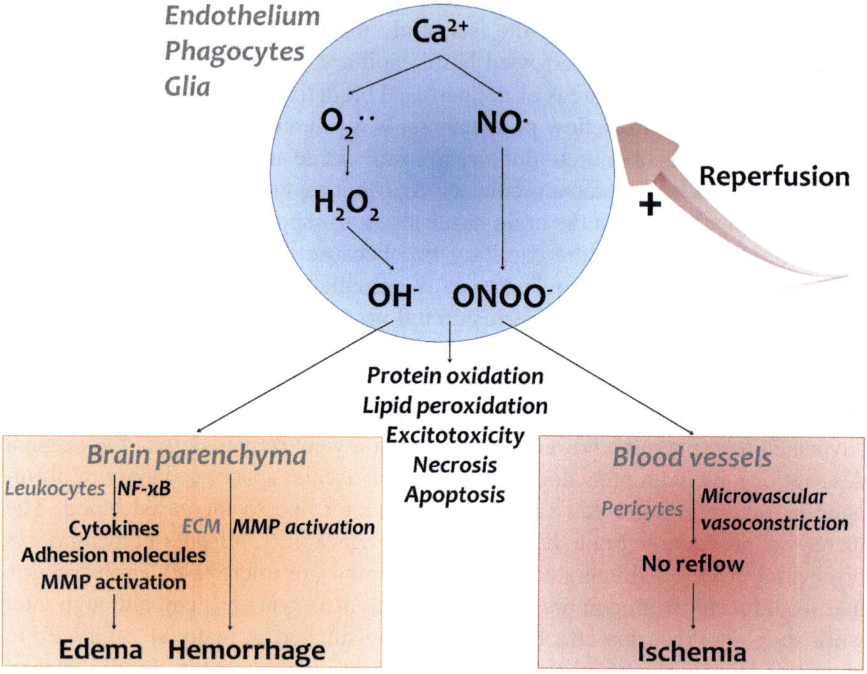

Fig. 15.1 Consequences of oxidative and nitrosative stress in cerebral ischaemia. During ischaemia, there is a surge in production of superoxide (O_2^-), nitric oxide (NO), and peroxynitrite ($ONOO^-$). The increased production of superoxide anion, hydrogen peroxide, hydroxyl radical, peroxynitrite, or nitrogen dioxide has been localized in infiltrating phagocytes, vascular cells, and glial cells. Excessive ROS production during reperfusion can overwhelm free radical scavenging systems that may have been compromised by ischaemia. The enhanced release of free radicals facilitates mitochondrial dysfunction, excitotoxicity, lipid peroxidation, and can modify the expression of several genes that up-regulate the transcription factor nuclear factor-kB (NF-kB). This results in an increased expression of pro-inflammatory cytokines and endothelial adhesion molecules. Moreover, infiltrating leukocytes may elaborate more ROS that increase oxidative stress, induce blood–brain barrier damage, and facilitate further brain injury. Increased activity of matrix metalloproteinases further increases tissue damage, mainly through degradation of the extracellular matrix (ECM). Superoxide and peroxynitrite also favour pericyte contraction that may impair the restoration of blood flow in the microcirulation despite large vessel recanalization

From a clinical standpoint, several observational studies have described decreased antioxidant capacity in patients with acute stroke in parallel with reduced levels of vitamin C, vitamin E, and superoxide dismutase [42]. During reperfusion, as it may occur more frequently after thrombolytic therapy, oxidative stress reaches higher peaks and has a more sustained duration than other pathogenic mechanisms of ischaemic cell death, further supporting the research of antioxidant therapies in these patients [43]. Preclinical and clinical data sustain that the amplification of the antioxidant capacity in humans could translate into effective strategies to minimize the global burden of stroke [44]. Indeed, several antioxidant therapies have shown neuroprotection in experimental models of brain ischaemia [45]. These antioxidants pertain to major classes that include inhibitors of free radical production, free radical scavengers, and boosters of free radical degradation. Regrettably, numerous neuroprotective drugs were unable to demonstrate beneficial effects in Phase II/III clinical trials despite previous encouraging preclinical results [25]. Reasons for the failure of achieving quantifiable neuroprotection in clinical trials include the fact that most of the previous trials allowed an excessive delay to the onset of treatment and excluded or limited the co-administration of thrombolytic therapy. The lack of combination treatment with thrombolytic agents might have hindered the access of the neuroprotective agent to the ischaemic brain in adequate drug concentrations [46]. The delayed treatment onset may have resulted in the exhaustion of the mechanism of damage aimed by the neuroprotectant and/or the lack of salvageable penumbral tissue in the treated patients.

3 Uric Acid, a Powerful Antioxidant in Humans

Uric Acid (UA) is a product of the catabolism of purine nucleotides, the principal constituents of DNA, RNA, and cellular energy stores, such as ATP. In most mammals, UA is degraded by hepatic enzyme uricase (urate oxidase) to allantoin [47]. In humans, the uricase gene is non-functional, due to two parallel but distinct mutations occurring approximately 15 million years ago and resulting in higher serum UA levels. Consequently, humans have higher UA levels than most mammals, in concentration almost tenfold higher than other antioxidants, an observation that has been interpreted as evidence that there must have been an evolutionary advantage for early primates in having elevated UA levels [45]. The main antioxidant properties of UA include scavenging of hydroxyl radicals, hydrogen peroxide, and peroxynitrite; suppression of the Fenton reaction; chelation of transition metals; and prevention of lipid peroxidation [48]. As a result of these wide antioxidant properties, endogenous UA contributes up to 60 % of the total plasma antioxidant activity in healthy subjects. A summary of the pathways of formation, degradation, and elimination of UA is shown in Fig. 15.2.

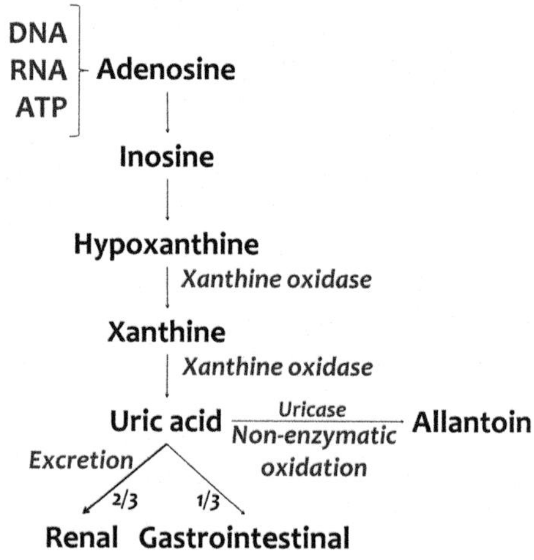

Fig. 15.2 Metabolism of uric acid. Uric Acid is a product of the catabolism of purine nucleotides, the principal constituents of DNA, RNA, and cellular energy stores, such as ATP. Uricase (urate oxidase) is a hepatic enzyme that degrades uric acid in most mammals, but this enzyme has undergone inactivation in humans because of a gene mutation. Uric acid loss is primarily via renal and gastrointestinal excretion

4 Uric Acid Administration as a Neuroprotectant Drug

The potential value of the administration of UA is supported by preclinical experimental and laboratory data and also by phase 1 studies. In vitro, UA protects cultured rat hippocampal neurons against cell death induced by exposure to glutamate and cyanide [49]. In these studies, treated neurons suppressed the accumulation of ROS, stabilized calcium homeostasis, and preserved mitochondrial functions. Moreover, in vivo the administration of UA up to 1 h after middle cerebral artery (MCA) intraluminal occlusion and reperfusion was neuroprotective in rats [49]. Importantly, the neuroprotection afforded by exogenous UA administration was validated experimentally using a model of clot injection in the rat ischaemic brain, as this model is more representative of the human situation than the intraluminal occlusion with a filament of the MCA artery. In the aforementioned thromboembolic model, the exogenous administration of UA shows neuroprotective effects when administered 20 min after the injection of thrombi in the MCA preceding rtPA administration after 3 h of reperfusion [50]. Importantly, the combination of UA and rtPA showed in these studies synergistic effects which included reduced tyrosine nitration, less neutrophil infiltration in the brain, reduced infarct volume, and improved behaviour [51].

The free-radical scavenging properties of UA were also assessed in a randomized double-blind placebo controlled crossover study in healthy volunteers [52]; UA (500 mg) was given to subjects that performed high intensity aerobic exercise for 20 min to incite oxidative stress. A single bout of high-intensity exercise caused a significant increase in circulating levels of the well-validated marker of oxidative stress 8-iso-prostaglandin F2α, an increase that was abolished without safety concerns in UA-treated subjects. Another randomized, placebo-controlled double-blind study showed a significant increase in serum free radical scavenging capacity during UA and vitamin C infusion in healthy volunteers, but the study showed that the antioxidant effect of UA was superior to vitamin C [53]. Serum UA concentration exhibited a two-component decay curve, with an elimination half-life of about 11 h, and the healthy volunteers did not experience adverse reactions after UA administration.

The potential value of the administration of UA is also supported by several observational studies in patients with AIS, and also by randomized phase II studies. UA is rapidly consumed after stroke onset [54], and even more remarkable reductions of serum UA levels in patients with arterial recanalization [55]. In acute stroke patients higher serum UA levels at stroke onset are associated with a statistically significant increase in the odds of good clinical [56] and less infarction growth [57]. Contrarily, lower UA levels were associated with an increased risk of malignant MCA infarction and/or hemorrhagic transformation [57].

The administration of UA was assessed in patients with AIS in a phase II randomized controlled trial, and in a phase IIb-III trial (URICO-ICTUS trial). The pilot phase II placebo-controlled trial showed in 28 patients with AIS that the combination of 1000 mg UA and rtPA was safe and prevented an early fall of circulating UA levels [54]. UA also prevented an increment of stress biomarkers such as metalloproteinase (MMP)-9, which was highly correlated with a better stroke outcome at 3 months, suggesting that the administration of UA could translate into clinically relevant biological effects [58]. The URICO-ICTUS trial recently confirmed the safety of UA therapy in 421 patients treated with rtPA within 4.5 h of stroke onset [59]. In this phase II-b trial, a total of 39.3 % of 211 patients treated with UA and a total of 33.0 % of 200 patients treated with placebo had an excellent outcome at 90 days, defined as a modified Rankin Score (mRS) 0–1, or 2 if the premorbid score was 2 [relative risk (RR) 95 % confidence intervals (CI) 1.22 (0.96–1.56) $p=0.09$] [59]. Hence, adding UA to rtPA resulted in a 6.3 % absolute effect size on the main outcome measure compared with placebo, a treatment effect within absolute effect sizes of 2–8 % that the Stroke Therapy Academic Industry Roundtable considered acceptable for neuroprotective therapies [24]. Sensitivity analyses of the primary outcome measure of the trial also showed that more patients treated with UA shifted into better functional categories of the mRS (odds ratio 1.40 [95 % CI 0.99–1.98]; $p=0.05$), and decreased their residual disability by a median 1 point on the mRS (95 % CI 0.42–1.58; $p=0.04$), an effect similar to the benefit shown by alteplase over placebo in the NINDS trial [59]. Additionally, more patients treated with UA than with placebo avoided the incidence of early ischaemic stroke worsening and were fully independent for the activities of daily living at 90 days [RR (95 % CI) 1.21 (0.99–1.48), $p=0.057$] [59]. Reassuringly, as URICO-ICTUS included an elderly population with severe stroke and high comorbidity, in whom the administration of UA was safe and did not increase mortality, the incidence of bleeding complications, gouty attacks, or any other adverse events compared with placebo.

The results from predefined exploratory analyses showed that women, patients with mild to moderate stroke on admission, or with higher pre-treatment levels of glucose, obtained statistically significant larger clinical benefits compared with placebo-treated patients, with absolute increments of the rate of excellent outcome at 90 days of 12.8 % in women, 14.7 % in patients with hyperglycaemia at stroke onset, and 16.8 % in patients with moderate stroke [59–61]. These absolute increments paralleled significantly lower serum levels of UA before treatment onset in women, and a less pronounced fall of serum UA concentration at follow up in patients with moderate stroke. URICO-ICTUS showed that UA therapy lessened infarct growth more effectively than placebo in selected groups of patients, and the superiority of UA over placebo was increased if arterial recanalization occurred during thrombolytic therapy, or pre-treatment glucose concentration was elevated [59, 60]. In the placebo group, the pro-oxidant effects of re-entry of oxygen after arterial recanalization, and an increased concentration of glucose in the ischaemic tissue, may have boosted the generation of ROS that led to the increment of infarct growth, as shown in clinical [62] and experimental studies [63]. The larger benefits of hyperglycaemic patients were also consistent with preclinical studies showing that glucose enhances the generation of damaging free radicals after brain ischaemia, facilitating cellular acidosis, increasing blood–brain barrier permeability, mitochondrial dysfunction, and cellular oedema [64]. In relation to sex differences, women had lower concentration of UA than men at baseline, and UA levels were persistently lower in women allocated to the placebo group. At the end of UA infusion women disclosed a greater allantoin-to-UA ratio, especially in those with reduced infarct growth, suggesting a greater exposition to the effects of free radicals and a more efficient non-enzymatic oxidative consumption of UA [59, 61]. Altogether, the subgroup analyses of URICO-ICTUS suggested that UA therapy might be particularly advantageous in individuals with constitutionally impaired antioxidant capacity (women), less antioxidant consumption (mild to moderate strokes), and greater oxidative load (hyperglycaemic patients).

Recently, the effects of UA therapy in stroke patients treated with mechanical thrombectomy in the URICOICTUS trial was presented in abstract form [65]. Of 411 patients included in the trial, 45 received mechanical thrombectomy because a CT-angiogram showed persistence of a proximal vessel occlusion following alteplase administration. In this subgroup, the proportion of good functional outcome at 90 days (modified Rankin Score 0–2), full independence at 90 days (Barthel Index score 95–100), ischaemic worsening during the first 72 h, mortality at 90 days, and risk of intracerebral bleeding was compared between the patients allocated to UA ($n=24$) or placebo ($n=21$). Adjusted odds ratios and 95 % confidence intervals were estimated using logistic regression models adjusted by the stratification factors and baseline traits. Good outcome was observed in 16 patients (67 %) receiving UA, and in 10 (48 %) receiving placebo (adjusted odd ratio 6.12 [95 % CI 1.06–34.56). There were also statistically significant differences between treatment groups with respect to full independence (16 [67 %] patients who received UA vs. 9 [43 %] who received placebo) (adjusted odd ratio 9.20 [95 % CI 1.53–55.20]), and ischaemic stroke worsening within 72 h (none [0 %] vs. 4 [19 %]). No clinically relevant or statistically significant differences were reported between groups with respect to death (4 [17 %] patients who received UA vs. 4 [19 %] who received placebo), or symptomatic intracerebral haemorrhage (1 [4 %] vs. none [0 %]). These result led to the conclusion that UA therapy was safe and more effective than placebo to improve functional outcome and daily living activities in stroke patients treated with mechanical thrombectomy.

5 Concluding Remarks: Towards an Effective Combined Approach

The emergence of positive clinical trials demonstrating the efficacy of mechanical thrombectomy offers a new window of opportunity for the demonstration of the elusive benefits of neuroprotective therapies in acute stroke [59–61, 65]. The combination of rapid reperfusion and neuroprotective therapies would be of added value for maximizing the beneficial effects of rapid reperfusion and to reduce the harmful consequences of reperfusion injury, including the no-reflow phenomenon, after recanalization [15]. The benefits of this combined approach may be hypothetically heightened in patients with good collateral circulation, a condition in which the arrival of the neuroprotectant drug to the ischaemic tissue may be facilitated while successful reperfusion is achieved [15, 66]. As commented in this chapter, a number of evidences support the value of the addition of UA to acute reperfusion strategies. Given the encouraging preclinical and clinical data regarding the potential neuroprotective effects of UA administration in brain ischaemia, the development of adequately powered pivotal confirmatory clinical trials is clearly warranted [15, 66].

References

1. Feigin VL, Forouzanfar MH, Krishnamurthi R, Mensah GA, Connor M, Bennett DA et al (2014) Global and regional burden of stroke during 1990-2010: findings from the Global Burden of Disease Study 2010. Lancet 383(9913):245–254
2. Powers WJ, Derdeyn CP, Biller J, Coffey CS, Hoh BL, Jauch EC et al (2015) 2015 American Heart Association/American Stroke Association focused update of the 2013 guidelines for the early management of patients with acute ischemic stroke regarding endovascular treatment: a guideline for healthcare professionals from the American Heart Association/American Stroke Association. Stroke 46(10):3020–3035
3. Choi JC, Kang SY, Kang JH, Ko YJ, Bae JM (2007) Are in-hospital delays important obstacles in thrombolytic therapy following acute ischemic stroke? J Clin Neurol 3(2):71–78
4. Dirks M, Niessen LW, Huijsman R, van Wijngaarden J, Minkman MM, Franke CL et al (2007) Promoting Acute Thrombolysis for Ischaemic Stroke (PRACTISE). Int J Stroke 2(2):151–159
5. Jovin TG, Chamorro A, Cobo E, de Miquel MA, Molina CA, Rovira A et al (2015) Thrombectomy within 8 hours after symptom onset in ischemic stroke. N Engl J Med 372(24):2296–2306
6. Saver JL, Goyal M, Bonafe A, Diener HC, Levy EI, Pereira VM et al (2015) Stent-retriever thrombectomy after intravenous t-PA vs. t-PA alone in stroke. N Engl J Med 372(24):2285–2295
7. Goyal M, Demchuk AM, Menon BK, Eesa M, Rempel JL, Thornton J et al (2015) Randomized assessment of rapid endovascular treatment of ischemic stroke. N Engl J Med 372(11):1019–1030
8. Campbell BC, Mitchell PJ, Kleinig TJ, Dewey HM, Churilov L, Yassi N et al (2015) Endovascular therapy for ischemic stroke with perfusion-imaging selection. N Engl J Med 372(11):1009–1018
9. Berkhemer OA, Fransen PS, Beumer D, van den Berg LA, Lingsma HF, Yoo AJ et al (2015) A randomized trial of intraarterial treatment for acute ischemic stroke. N Engl J Med 372(4):11–20
10. Soares BP, Tong E, Hom J, Cheng SC, Bredno J, Boussel L et al (2010) Reperfusion is a more accurate predictor of follow-up infarct volume than recanalization: a proof of concept using CT in acute ischemic stroke patients. Stroke 41(1):e34–e40

11. Baird AE, Donnan GA, Austin MC, Fitt GJ, Davis SM, McKay WJ (1994) Reperfusion after thrombolytic therapy in ischemic stroke measured by single-photon emission computed tomography. Stroke 25(1):79–85
12. Khatri P, Neff J, Broderick JP, Khoury JC, Carrozzella J, Tomsick T (2005) Revascularization end points in stroke interventional trials: recanalization versus reperfusion in IMS-I. Stroke 36(11):2400–2403
13. Albers GW, Thijs VN, Wechsler L, Kemp S, Schlaug G, Skalabrin E et al (2006) Magnetic resonance imaging profiles predict clinical response to early reperfusion: the diffusion and perfusion imaging evaluation for understanding stroke evolution (DEFUSE) study. Ann Neurol 60(5):508–517
14. De Silva DA, Fink JN, Christensen S, Ebinger M, Bladin C, Levi CR et al (2009) Assessing reperfusion and recanalization as markers of clinical outcomes after intravenous thrombolysis in the echoplanar imaging thrombolytic evaluation trial (EPITHET). Stroke 40(8):2872–2874
15. Chamorro Á, Dirnagl U, Urra X, Planas AM (2016) Neuroprotection in acute stroke: targeting excitotoxicity, oxidative and nitrosative stress, and inflammation. Lancet Neurol 15(8):869–881
16. Rha JH, Saver JL (2007) The impact of recanalization on ischemic stroke outcome: a meta-analysis. Stroke 38(3):967–973
17. Schaller B, Graf R (2004) Cerebral ischemia and reperfusion: the pathophysiologic concept as a basis for clinical therapy. J Cereb Blood Flow Metab 24(4):351–371
18. Kidwell CS, Saver JL, Mattiello J, Starkman S, Vinuela F, Duckwiler G et al (2001) Diffusion-perfusion MRI characterization of post-recanalization hyperperfusion in humans. Neurology 57(11):2015–2021
19. Warach S, Latour LL (2004) Evidence of reperfusion injury, exacerbated by thrombolytic therapy, in human focal brain ischemia using a novel imaging marker of early blood-brain barrier disruption. Stroke 35(11 Suppl 1):S2659–S2661
20. Fisher M, Albers GW, Donnan GA, Furlan AJ, Grotta JC, Kidwell CS et al (2005) Enhancing the development and approval of acute stroke therapies: Stroke Therapy Academic Industry roundtable. Stroke 36(8):1808–1813
21. Lipton P (1999) Ischemic cell death in brain neurons. Physiol Rev 79(4):1431–1568
22. Hossmann KA (2006) Pathophysiology and therapy of experimental stroke. Cell Mol Neurobiol 26(7-8):1057–1083
23. Gladstone DJ, Black SE, Hakim AM, Heart and Stroke Foundation of Ontario Centre of Excellence in Stroke Recovery (2002) Toward wisdom from failure: lessons from neuroprotective stroke trials and new therapeutic directions. Stroke 33(8):2123–2136
24. Fisher M (2011) New approaches to neuroprotective drug development. Stroke 42(1 Suppl):S24–S27
25. O'Collins VE, Macleod MR, Donnan GA, Horky LL, van der Worp BH, Howells DW (2006) 1,026 experimental treatments in acute stroke. Ann Neurol 59(3):467–477
26. Crack PJ, Taylor JM (2005) Reactive oxygen species and the modulation of stroke. Free Radic Biol Med 38(11):1433–1444
27. Guix FX, Uribesalgo I, Coma M, Munoz FJ (2005) The physiology and pathophysiology of nitric oxide in the brain. Prog Neurobiol 76(2):126–152
28. Iadecola C, Zhang F, Casey R, Clark HB, Ross ME (1996) Inducible nitric oxide synthase gene expression in vascular cells after transient focal cerebral ischemia. Stroke 27(8):1373–1380
29. Kim GW, Kondo T, Noshita N, Chan PH (2002) Manganese superoxide dismutase deficiency exacerbates cerebral infarction after focal cerebral ischemia/reperfusion in mice: implications for the production and role of superoxide radicals. Stroke 33(3):809–815
30. Sims NR, Anderson MF (2002) Mitochondrial contributions to tissue damage in stroke. Neurochem Int 40(6):511–526
31. Kawano T, Anrather J, Zhou P, Park L, Wang G, Frys KA et al (2006) Prostaglandin E2 EP1 receptors: downstream effectors of COX-2 neurotoxicity. Nat Med 12(2):225–229

32. Vallet P, Charnay Y, Steger K, Ogier-Denis E, Kovari E, Herrmann F et al (2005) Neuronal expression of the NADPH oxidase NOX4, its regulation in mouse experimental brain ischemia. Neuroscience 132(2):233–238
33. Ago T, Kitazono T, Kuroda J, Kumai Y, Kamouchi M, Ooboshi H et al (2005) NAD(P)H oxidases in rat basilar arterial endothelial cells. Stroke 36(5):1040–1046
34. Dalkara T, Arsava EM (2012) Can restoring incomplete microcirculatory reperfusion improve stroke outcome after thrombolysis? J Cereb Blood Flow Metab 32(12):2091–2099
35. Ames A 3rd, Wright RL, Kowada M, Thurston JM, Majno G (1968) Cerebral ischemia II. The no-reflow phenomenon. Am J Pathol 52(2):437–453
36. del Zoppo GJ, Schmid-Schönbein GW, Mori E, Copeland BR, Chang CM (1991) Polymorphonuclear leukocytes occlude capillaries following middle cerebral artery occlusion and reperfusion in baboons. Stroke 22(10):1276–1283
37. Kuroda J, Ago T, Nishimura A, Nakamura K, Matsuo R, Wakisaka Y et al (2014) Nox4 is a major source of superoxide production in human brain pericytes. J Vasc Res 51(6):429–438
38. Miller AA, Drummond GR, Schmidt HH, Sobey CG (2005) NADPH oxidase activity and function are profoundly greater in cerebral versus systemic arteries. Circ Res 97(10):1055–1062
39. Hall CN, Reynell C, Gesslein B, Hamilton NB, Mishra A, Sutherland BA et al (2014) Capillary pericytes regulate cerebral blood flow in health and disease. Nature 508(7494):55–60
40. Hill RA, Tong L, Yuan P, Murikinati S, Gupta S, Grutzendler J (2015) Regional blood flow in the normal and ischemic brain is controlled by arteriolar smooth muscle cell contractility and not by capillary pericytes. Neuron 87(1):95–110
41. Yemisci M, Gursoy-Ozdemir Y, Vural A, Can A, Topalkara K, Dalkara T (2009) Pericyte contraction induced by oxidative–nitrative stress impairs capillary reflow despite successful opening of an occluded cerebral artery. Nat Med 15(9):1031–1037
42. Cherubini A, Polidori MC, Bregnocchi M, Pezzuto S, Cecchetti R, Ingegni T et al (2000) Antioxidant profile and early outcome in stroke patients. Stroke 31(10):2295–2300
43. Chamorro A, Planas AM, Muner DS, Deulofeu R (2004) Uric acid administration for neuroprotection in patients with acute brain ischemia. Med Hypotheses 62(2):173–176
44. Amaro S, Chamorro A (2011) Translational stroke research of the combination of thrombolysis and antioxidant therapy. Stroke 42(5):1495–1499
45. Ames BN, Cathcart R, Schwiers E, Hochstein P (1981) Uric acid provides an antioxidant defense in humans against oxidant- and radical-caused aging and cancer: a hypothesis. Proc Natl Acad Sci U S A 78(11):6858–6862
46. Ly JV, Zavala JA, Donnan GA (2006) Neuroprotection and thrombolysis: combination therapy in acute ischaemic stroke. Expert Opin Pharmacother 7(12):1571–1581
47. Amaro S, Planas AM, Chamorro A (2008) Uric acid administration in patients with acute stroke: a novel approach to neuroprotection. Expert Rev Neurother 8(2):259–270
48. Davies KJ, Sevanian A, Muakkassah-Kelly SF, Hochstein P (1986) Uric acid-iron ion complexes. A new aspect of the antioxidant functions of uric acid. Biochem J 235(3):747–754
49. Yu ZF, Bruce-Keller AJ, Goodman Y, Mattson MP (1998) Uric acid protects neurons against excitotoxic and metabolic insults in cell culture, and against focal ischemic brain injury in vivo. J Neurosci Res 53(5):613–625
50. Romanos E, Planas AM, Amaro S, Chamorro A (2007) Uric acid reduces brain damage and improves the benefits of rt-PA in a rat model of thromboembolic stroke. J Cereb Blood Flow Metab 27(1):14–20
51. Penumbra Pivotal Stroke Trial Investigators (2009) The penumbra pivotal stroke trial: safety and effectiveness of a new generation of mechanical devices for clot removal in intracranial large vessel occlusive disease. Stroke 40(8):2761–2768
52. Waring WS, Convery A, Mishra V, Shenkin A, Webb DJ, Maxwell SR (2003) Uric acid reduces exercise-induced oxidative stress in healthy adults. Clin Sci (Lond) 105(4):425–430
53. Waring WS, Webb DJ, Maxwell SR (2001) Systemic uric acid administration increases serum antioxidant capacity in healthy volunteers. J Cardiovasc Pharmacol 38(3):365–371

54. Amaro S, Soy D, Obach V, Cervera A, Planas AM, Chamorro A (2007) A pilot study of dual treatment with recombinant tissue plasminogen activator and uric acid in acute ischemic stroke. Stroke 38(7):2173–2175
55. Hong JM, Bang OY, Chung CS, Joo IS, Gwag BJ, Ovbiagele B (2010) Influence of recanalization on uric acid patterns in acute ischemic stroke. Cerebrovasc Dis 29(5):431–439
56. Chamorro A, Obach V, Cervera A, Revilla M, Deulofeu R, Aponte JH (2002) Prognostic significance of uric acid serum concentration in patients with acute ischemic stroke. Stroke 33(4):1048–1052
57. Amaro S, Urra X, Gómez-Choco M, Obach V, Cervera A, Vargas M et al (2011) Uric acid levels are relevant in patients with stroke treated with thrombolysis. Stroke 42(1 Suppl):S28–S32
58. Amaro S, Obach V, Cervera A, Urra X, Gómez-Choco M, Planas AM et al (2009) Course of matrix metalloproteinase-9 isoforms after the administration of uric acid in patients with acute stroke: a proof-of-concept study. J Neurol 256(4):651–656
59. Chamorro A, Amaro S, Castellanos M, Segura T, Arenillas J, Martí-Fábregas J et al (2014) Safety and efficacy of uric acid in patients with acute stroke (URICO-ICTUS): a randomised, double-blind phase 2b/3 trial. Lancet Neurol 13(5):453–460
60. Amaro S, Llull L, Renú A, Laredo C, Perez B, Vila E et al (2015) Uric acid improves glucose-driven oxidative stress in human ischemic stroke. Ann Neurol 77(5):775–783
61. Llull L, Laredo C, Renú A, Pérez B, Vila E, Obach V et al (2015) Uric acid therapy improves clinical outcome in women with acute ischemic stroke. Stroke 46(8):2162–2167
62. Parsons MW, Barber PA, Desmond PM, Baird TA, Darby DG, Byrnes G et al (2002) Acute hyperglycemia adversely affects stroke outcome: a magnetic resonance imaging and spectroscopy study. Ann Neurol 52(1):20–28
63. Suh SW, Shin BS, Ma H, Van Hoecke M, Brennan AM, Yenari MA et al (2008) Glucose and NADPH oxidase drive neuronal superoxide formation in stroke. Ann Neurol 64(6):654–663
64. Dungan KM, Braithwaite SS, Preiser JC (2009) Stress hyperglycaemia. Lancet 373(9677): 1798–1807
65. Chamorro A, Amaro S, Castellanos M, San Román L, Gomis M, Blasco J, et al (2016) Uric acid therapy in stroke patients treated with mechanical thrombectomy: an exploratory analysis of the URICOICTUS trial. ESOC 2006: the second European Stroke Organisation Conference; 2016 May 10-12; Barcelona, Spain
66. Amaro S, Chamorro Á (2016) Should uric acid be administered alongside thrombolysis for stroke patients? Expert Rev Cardiovasc Ther 14(4):407–409

Chapter 16
TRPM7 Channels as Potential Therapeutic Targets for Stroke

Hong-Shuo Sun and Zhong-Ping Feng

Abstract Transient receptor potential melastatin 7 (TRPM7) channels are calcium-permeable nonselective cation channels. TRPM7 channels are widely expressed in different tissues including the brain and are thought to regulate calcium homeostasis based on the metabolic state of the cell. TRPM7 partakes in a wide range of processes in cell biology that affects cell adhesion, cell growth, and proliferation, as well as in skeleton formation and embryonic development. TRPM7 plays a key role in brain damages and neuronal cell death in ischemia and hypoxia. As one of the key components of the nonglutamate mechanism of stroke besides the traditional glutamate mechanism that initiates the excitotoxicity, TRPM7 also triggers the intracellular ionic imbalance and neuronal cell death in anoxia in vitro and ischemia in vivo, respectively. We have shown that TRPM7 channels, expressed in adult hippocampal and cortical neurons, triggered neuronal cell death under ischemic conditions, and gene silencing of TRPM7 with siRNA increased neuronal survival and improved neurobehavioral outcomes to cerebral ischemia in vivo. Recently, we also showed that TRPM7 plays a neuroprotective role in brain damage to hypoxia in vivo. We have demonstrated the contributions of TRPM7 channels in many processes in both physiology and pathophysiology, as well as the current understanding of the role of TRPM7 in stroke.

H.-S. Sun, M.D., M.Sc., Ph.D. (✉)
Department of Surgery, Faculty of Medicine, University of Toronto,
1 King's College Circle, Toronto, Ontario, Canada, M5S 1A8

Department of Physiology, Faculty of Medicine, University of Toronto,
1 King's College Circle, Toronto, Ontario, Canada, M5S 1A8

Department of Pharmacology, Faculty of Medicine, University of Toronto,
1 King's College Circle, Toronto, Ontario, Canada, M5S 1A8

Institute of Medical Science, Faculty of Medicine, University of Toronto,
1 King's College Circle, Toronto, Ontario, Canada, M5S 1A8
e-mail: hss.sun@utoronto.ca

Z.-P. Feng (✉)
Department of Physiology, Faculty of Medicine, University of Toronto,
1 King's College Circle, Toronto, Ontario, Canada, M5S 1A8
e-mail: zp.feng@utoronto.ca

P.A. Lapchak, J.H. Zhang (eds.), *Neuroprotective Therapy for Stroke and Ischemic Disease*, Springer Series in Translational Stroke Research,
DOI 10.1007/978-3-319-45345-3_16

Keywords Ion channels • TRP • TRPM7 • Neuroprotection • Cerebral ischemia • Stroke • In vivo

1 Introduction

Currently, stroke is the second most common cause of death and leading cause of chronic disability in the world according to the World Health Organization (WHO) [1–3], with death rate of about 30 %. The number of stroke incidences is expected to significantly increase multiple folds in 2050. Although our aging population is most at risk, stroke can occur at any age and an effective treatment for the disease is lacking. Stroke has profound social and economic impacts in individuals and our society worldwide [4]. Furthermore, damage to the human infant brain due to lack of oxygen (neonatal hypoxic–ischemic brain injury) can cause neonatal stroke and hypoxic–ischemic encephalopathy (HIE) [5–7]. HIE, an early onset brain and behavioral disorder in children, represents a group of impairments including neurodevelopmental delay, cognitive and motor impairments, and epilepsy. These disorders and disabilities in survivors of neonatal hypoxic–ischemic brain injury also have significant impact on families, society, and the healthcare system worldwide as well.

Stroke or cerebral ischemia [8, 9], also called brain attack in layman's term, is a sudden loss of the brain functions caused by interruption of the cerebral blood flow supplying the brain. Stroke could be divided into either ischemic stroke which is due to interruption of the blood flow to the brain or hemorrhagic stroke which is mostly due to rupture of the cerebral blood vessels in the brain. Most of the stroke are ischemic contributing approximately 80 % of the stroke cases; the remaining ~20 % of the stroke cases are from hemorrhagic stroke and neurotrauma. Among the ischemic stroke, most areas affected are from the middle cerebral artery perfusion regions. In addition, stroke could be further classified as global ischemia and focal ischemia. The global ischemia is seen in cardiac arrest and drowning, with characteristic and hall-marked delayed hippocampal neuronal cell death in CA1 area seen both in human and rodents, which causes memory loss. On the other hand, the focal ischemia is mostly caused by thrombosis where blood clots locally and less by embolisms where the clots are from outside of the brain, flow and trap in the brain as seen in cardiac arrhythmia (atrial fibrillation is the most common one in cardiac arrhythmia and estimated up to 15 % of stroke caused by the atrial fibrillation).

Many events happen during stroke, which trigger calcium overload, ionic imbalance, and eventually neuronal cell death [8, 9]. The traditional glutamate model of excitotoxicity emphasizes events happened at the NMDA and AMPA receptor channels [10, 11]. New data now suggests that the nonglutamate mechanism may also cause neuronal ionic imbalance and hypoxic–ischemic neuronal cell death [10, 11]. These include transient receptor potential (TRP) channels [12–14], ATP-sensitive potassium (K_{ATP}) channels [15–18], hemichannels [19, 20], acid-sensing ion channels

(ASICs) [21], volume-regulated anion channels (VRACs) [22], sodium-calcium exchangers (NCXs) [23, 24], and other nonselective cation channels [25]. Thus, the nonglutamate mechanism may also be the targets for neuroprotection. Currently, the ion channel is the third largest target for drug development.

Stroke treatment strategy could be falling into these two categories, either in improving perfusion using thrombolytic agent, such as tPA (tissue plasminogen activator) or limiting injured areas using so-called neuroprotection agents. Currently, stroke treatment is still very limited to only the tPA, which is the only clot-busting drug approved by the FDA 20 years ago in 1996 [26, 27]. The tPA was initially developed for the treatment of heart attack. However, only limited stroke cases have the benefits from the tPA treatment mainly due to the limited therapeutic window of the tPA [28]. Thus, it is important to develop the so-called neuroprotectants for stroke, which is largely targeted on ion channels and receptors.

During stroke or cerebral ischemic, the major event occurs in the brain is a simultaneous massive release of the excitatory neurotransmitter glutamate from the central neurons, which in turn triggers in intracellular calcium overload and eventual neuronal cell death [8, 9]. The concept of excitotoxicity in stroke has been in the focus of stroke research for several decades. The excessive glutamate released from neurons causes increase of the intracellular calcium (Ca^{2+}) concentration [$(Ca^{2+})_i$] through calcium-permeable NMDA (*N*-methyl-D-aspartic acid), AMPA (DL-a-amino-3-hydroxy-5-methyl-4-isoxazole propionic acid) receptors [29], and L-type voltage-dependent Ca^{2+} channels. These are considered as the major calcium entry into the neurons and the cause of calcium overload in the neurons during stroke. This calcium overload or a broad spectrum of ion imbalance inside the neuronal cell during stroke is thought to set off a wide range of sequential episodes that lead to irreversible impairments to mitochondria, protein synthesis, cytoskeleton, and plasma membrane of the neurons, and eventually trigger the neuronal cell death. Interrupting this calcium overload in stroke is conceptually thought to be able to provide neuroprotection clinically for the stroke victims. In the laboratory setting, pharmacologically blocking these calcium-mediating receptor channels surely prevents this intracellular calcium overload during stroke and provides sufficient neuroprotection in the experimental animal stroke models. These promising findings obtained from the animal stroke models in the laboratory have been translated into numerous clinical trials in potentially stroke treatment. However, the disappointing results from the antiexcitotoxic therapies (AET) clinical trials [30], which mainly include blockers for the NMDA and AMPA receptors, appeared to be unsuccessful and even showing harmful side effects in the stroke patients (as glutamate is the major excitatory neurotransmitter in the brain, blocking it may have the unwanted psychological effects). This will not be the focus of this chapter. Because of the unfavorable outcomes of AET clinical trials [30] and the limitation of the glutamate mechanism, researchers have been searching for alternative mechanism. Tymianski group subsequently described the role of interaction between the NMDA receptor and its downstream signaling pathway PSD-95 in cerebral ischemia and developed the PSD-95 inhibitors for treatment of stroke. The PSD-95 inhibitors successfully reduced the brain damages after cerebral ischemia and hypoxia using multiple animal models and species [31–36].

Recently, the newly described nonglutamate mechanism in stroke, which also causes ionic imbalance and neuronal cell death during stroke, caught the attention as the potential therapeutic targets for drug development in stroke. As mentioned, these include: TRP channels [12–14], K_{ATP} channels [15–18], ASICs [21], hemichannels [19, 20], VRACs [22], NCXs [23, 24], and other nonselective cation channels [25].

Relevant to the drug development for stroke, for decades, much of the attention has been given on developing effective pharmacological treatments for the acute ischemic stroke. Preclinical research has been focused on excitotoxicity and targeted on glutamate mechanism of the stroke. Many of the antiglutamate receptor drugs only worked well on the animal stroke models in the laboratory setting; however, the antiexcitotoxic therapies (AET) failed the clinical trials [30] for potential stroke treatment, thus diminishing the initial excitement and also the hope for the bench-to-bedside translational research in stroke. Even the underlying mechanisms for cerebral ischemia and stroke are commencing to be better understood, there is still no effective clinical or experimental treatment that could improve the outcome for stroke sufferers. Thus, continuing in searching new stroke drug targets toward drug development for stroke is the priority. Stroke researchers have been aiming to find the reason(s) underlying the failure of AET stroke drug clinical trials and to identify new potential therapeutic targets for stroke and hypoxia. As pathophysiological events in stroke can lead to ionic imbalance and neuronal cell death [8, 9], researchers have turned their attention to outside of the glutamate mechanism for stroke drug searching and identification. Ischemic neuronal cell death is now understood to result from glutamate-mediated excitotoxicity via NMDA receptors and also the newly described nonglutamate mechanisms of stroke [10, 11]. The nonglutamate mechanism of stroke may contribute at least partially to the failure of antiexcitotoxicity clinical trials. We may need to focus on both nonglutamate and glutamate mechanisms of stroke and, here, we first need to evaluate the spatial and temporal relationship of several ion channels in nonglutamatergic mechanisms in ischemic cell damage and neuroprotection against stroke using in vivo animal models of human disease based on the STAIR Protocol [37].

Ion channels are the third largest target for drug development and play many essential roles in physiological and pathophysiological brain functions. Among the nonglutamate mechanism, one group of ion channels, Transient Receptor Potential (TRP) channels, was originally described in fruit fly (*Drosophila*) [38, 39]. TRPM7 channel [40], a member of the TRP super family, is located at and conducts calcium through the cell membrane including neurons. TRPM7 plays an important role in the brain both under physiological conditions, such as cell growth (proliferation) [41, 42] and maturation [43], and under pathophysiological conditions, such as ischemia- or hypoxia-induced neuronal cell death [12, 13, 44] and survival of brain tumor cells [45, 46]. However, there is a lack of pharmacological inhibitor for TRPM7 channels until recently the discovery of the specific inhibitor [43, 47]. This chapter further discusses the pharmacology of TRPM7 and the potential for drug development in stroke. Here, we provide a current view on TRPM7, one of the nonglutamate mechanisms of stroke, from our recent in vivo studies.

Based on the recommendations of stroke drug testing protocols from the Stroke Therapy Academic Industry Roundtable (STAIR) committee [37], it is crucial to validate the preclinical stroke drug development in "Proof of Principle" using multiple animal stroke models and also multiple species. Recent in vivo studies using rodent stroke models have identified some important nonglutamate mechanisms in stroke, which include TRPM7 [12, 13], K_{ATP} channels [15–18], ASICs [21], hemichannels [19, 20], TRPM2 [14], and VRACs [22]. In this chapter, we will focus on the nonglutamate mechanism TRPM7 channels and our current understanding of its molecular, biophysical, and pharmacological properties, as well as its physiological and pathophysiological roles and its therapeutic potential for drug development in stroke.

2 Classification, Structures, and Distributions

2.1 Classification

The TRP channel was originally discovered in *Drosophila* where they were found to play an important role in phototransduction. The name transient receptor potential is based on the transient response to light in the fruit flies carrying mutant *trp* gene [38, 39]. The superfamily of TRP channels consists of a group of nonselective cation channels [40, 48–50]. Presently, approximately 30 mammalian TRP channels have been identified and classified based on their homologous sequences [50]. The TRP channels are divided into six subfamilies: (1) TRPC (canonical), (2) TRPV (vanilloid), (3) TRPM (melastatin), (4) TRPA (ankyrin), (5) TRPML (mucolipin), and (6) TRPP (polycystin). Diverse chemical and physical stimuli activate different TRP channels. Through different gating mechanisms, the TRP channels play many important cellular functions under various physiological and pathological conditions [40, 48–50].

TRPM channel subfamily has eight members, such as TRPM1-8; TRPM7 channel is the seventh member from this TRP channel subfamily. TRPM7 channels play a role in a wide range of processes in cell biology; in here we focus on its role in mediating neuronal cell death in cerebral ischemia and hypoxia in vivo [12, 13].

2.2 Gene and Protein Structures

The human *TRPM7* gene is encoded by 39 exons and positioned on chromosome 15 in the q21.2 region, with about 126.5 kb of DNA sequence spanning. The mouse *TRPM7* gene is 95 % identical to the human *TRPM7* gene [41]. The mouse *TRPM7* homolog is encoded by 39 exons and located on chromosome 2 on cytoband F2, with about 84.6 kb of DNA sequence spanning.

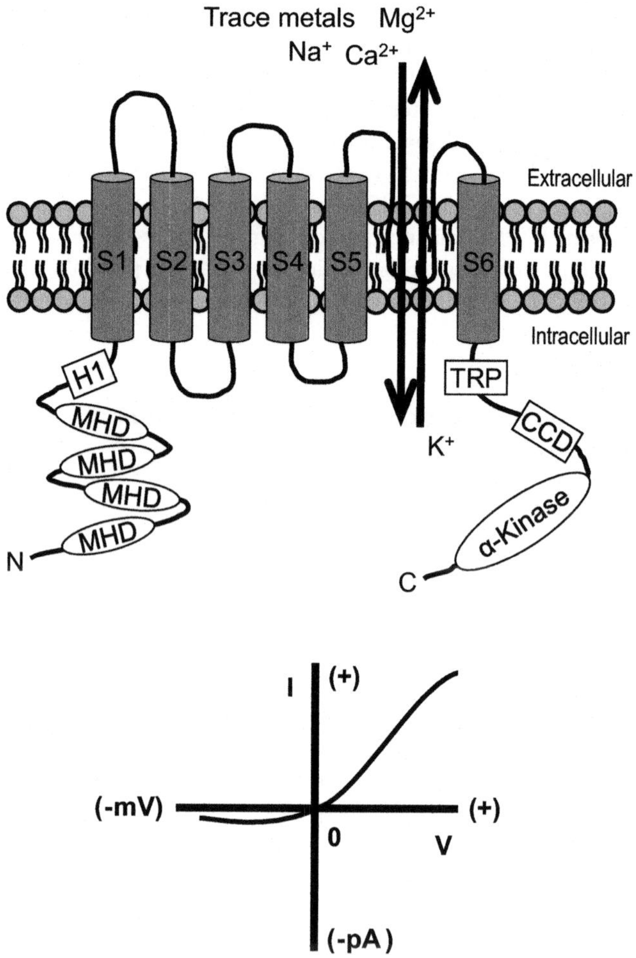

Fig. 16.1 Schematic of TRPM7 channel transmembrane topology. TRPM7 channel has a tetrameric assembly with four subunits. Each subunit has six transmembrane (TM) spanning segments (S1–S6) and a pore-forming reentrant P-loop locates between the S5 and S6 segments. There are N- and C-terminal regions in the TRPM7. The intracellular N-terminus has the additional hydrophobic region termed H1 and four regions of TRPM family homology domain (MHD). The C-terminus has a TRP box, a coiled-coil domain (CCD) and α-kinases domain. Inset: representative TRPM7 I–V curve

With approximately 212.4 kDa in predicted molecular weight, TRPM7 channel is a large protein. Full-length TRPM7 proteins encode 1864 amino acids in human and 1863 amino acids in mouse, respectively. TRPM7 channel has a tetrameric assembly with four subunits [50]. Each TRPM7 subunit has six transmembrane (TM) spanning segments (S1–S6); a pore-forming re-entrant P-loop locates between the S5 and S6 segments [50, 51] (Fig. 16.1). There are N- and C- terminal regions in the TRPM7. The intracellular N-terminus has the additional hydrophobic region termed H1 and four regions of TRPM family homology domain (MHD). The biological significance of the N-terminus is largely unknown. However, the functional importance of the C-terminus is better understood; it has a TRP box, a coiled-coil domain (CCD) and catalytic domain. The TRP box is a positive regulator of some TRP channel and may interact with Phosphatidylinositol 4,5-bisphosphate (PIP_2) [40]. The coiled-coil domain CCD may involve subunit–subunit interactions. The enzymatic domain positioned at the distal of C-terminus has an atypical serine/threonine protein kinase domain and is a distinct structural element of the TRPM7 channel. This kinase domain belongs to a homologous family of α-kinases [52, 53]; it may not directly concern the channel activity [50, 52, 54] but may regulate the channel function by Mg^{2+} nucleotides [53, 55].

2.3 *Tissue and Cellular Distribution*

TRPM7 channels are widely expressed in different tissues [48, 56], including brain, heart, blood vessels, lung, liver, skeletal muscle, stomach, intestine, spleen, peripheral blood mononuclear cells, macrophages, pancreas, prostate, placenta, cartilage, bone, and bone marrow [48, 56, 57]. TRPM7 mRNA is widely spread in the brain and highly expressed in heart, pituitary, bone, and adipose tissue [48, 56, 57].

In the immunohistochemistry, the TRPM7 channels are diffusingly distributed at the cell membrane of the hippocampal and cortical neurons [12, 13, 58]. The TRPM7 protein is also labeled in various cells, such as neuroblastoma cells U87 [45], U251 [46] and N1E-115 [59], and vascular smooth muscle cells [60].

3 Biophysical Properties, Regulatory Mechanisms, Pharmacology

3.1 *Biophysical Properties*

TRPM7 is a calcium-permeable nonselective cation channel and has its own distinguishable biophysical properties that make TRPM7 channel different from other TRP channels. TRPM7 channel has a prominent outward rectification current–voltage (I–V) curve and a reversal potential at 0 mV [50, 61, 62]. TRPM7 conducts a small inward current at negative membrane potentials, by mediating divalent cations (e.g., Ca^{2+} and Mg^{2+}) down their concentration gradients [41]. Current density of TRPM7 is usually small <20 pA/pF [41, 62]. TRPM7 conducts a large outward current at positive membrane potential, by strong driving force from intracellular cations to exit the cell. This property of TRPM7 outwardly rectifying is caused by a voltage-dependent block of monovalent cation influx by extracellular divalent cations. For example, TRPM7 conducts inward monovalent cations in the absence of divalent cations [41], and, as the result, its I–V relation becomes quasi-linear indicating the lack of voltage-dependent gating property and channel selectivity.

Distinct to many other calcium-permeating ion channels, TRPM7 is also permeable to a series of trace metal ions, including $Zn^{2+} \approx Ni^{2+} \gg Ba^{2+} > Co^{2+} > Mg^{2+} \geq Mn^{2+} \geq Sr^{2+} \geq Cd^{2+} \geq Ca^{2+}$ [56]. TRPM7 channel is constitutively active for both sensing the extracellular concentration of divalent cations and maintaining the homeostasis of intracellular magnesium during ischemic episodes and seizure that trigger the intense neuronal activity [58].

Extracellular pH regulates TRPM7 channel activity. A decrease in extracellular pH activates TRPM7 current [63, 64]. Protons also increase the inward current of TRPM7 by competing with the calcium and magnesium binding sites and releasing the divalent cation block [63]. During ischemic stroke, acidic pathophysiological conditions could activate the TRPM7 channels [64–66], even the protons' effect on endogenous TRPM7 channels remains controversial.

Shear stress could also activate the TRPM7 channels in vascular smooth muscle cells [67], even though TRPM7 channel is not a mechano-activated ion channel.

3.2 Regulatory Mechanisms

TRPM7 channel activities can be regulated by various extracellular and intracellular factors [61, 62]. For example, magnesium, regardless extracellular or intracellular, plays an important role in regulating the TRPM7 channel activity. Intracellular Mg^{2+} and Mg^{2+}-complexed nucleotides (such as MgATP and MgGTP) inhibit TRPM7 channel activities [41, 55, 62]. In contrast, Mg^{2+} chelators (e.g., HEDTA or Na-ATP) applied intracellularly or omitting Mg^{2+} and Mg^{2+}-complexed nucleotides in intracellular pipette solutions increase the TRPM7 channel activities [55, 61]. The inhibitory effects of Mg^{2+} or rather Mg^{2+}-nucleotides may be mediated by binding to C-terminal kinase domain of the TRPM7 channels [54, 55, 61]. The Mg^{2+}-nucleotides binding to the kinase domain is responsible for regulating the TRPM7 channel activity, which is favorite theory in TRPM7 channel regulation especially during ischemia [55].

Many other factors, such as phosphatidylinositol-4,5-bisphosphate (PIP_2) (a substrate of phospholipase C (PLC) [68]), carbachol (an agonist for $G_{\alpha q}$-linked muscarinic type 1 (M1) receptors), and phosphorylation can also regulate the TRPM7 channel activities. In related to the channelopathy, TRPM7 mutations (a variant of TRPM7 with a missense mutation (T1482I)) have been identified in patients with Guamanian amyotrophic lateral sclerosis (ALS-G) and Parkinsonism-dementia (PD-G) [69].

3.3 Pharmacological Properties

This is the fastest changing area for the TRPM7 channels. There are now various pharmacological modulators, both activators and blockers, available for TRPM7 channels [47, 70, 71]. Global deletion of TRPM7 gene in knockout mice is embryonically lethal [72, 73]; however, gene silencing using small interfering RNA (siRNA) to knockdown the TRPM7 channels is well tolerated in central neurons [12, 43], peripheral neurons [74], glial cells [45], vascular endothelial cells [42], vascular smooth muscle cells [75, 76], gastrointestinal tract interstitial pacemaker cells [77], and human epithelial cells [78]. TRPM7 antibodies are also available from many commercial sources as well for studying TRPM7 channel expressions in many cells and tissues [12, 13, 45]. Trivalent ions, such as Gd^{3+} (IC_{50} ~ 1.4–2.5 μM) and La^{3+} (IC_{50} ~ 17 μM) [44], as well as 2-Aminoethoxydiphenyl borate (2-APB) (IC_{50} ~ 50 μM), which is a well-known nonspecific blocker of many TRP channels [79], can block the TRPM7 channels. Inhibitors of 5-lipoxygenase (5-LOX), NDGA (Nordihydroguaiaretic acid, IC_{50} ~ 6.3 μM), AA861(IC_{50} ~ 6.0 μM), and MK886 (IC_{50} ~ 8.6 μM), can also inhibit the TRPM7 current in HEK-293 cells [80]. Nonselective TRPM7 blocker carvacrol and activator naltriben are also available commercially. Nonselective TRPM7 blocker carvacrol, a pungent natural compound derivative used as a food additive, blocks TRPM7 current in TRPM7 overexpression HEK293 cells, in hippocampal neurons and glial cells [13, 43, 45, 81, 82]; carvacrol also provides neuroprotection in focal ischemia and hypoxic–ischemic injury [13, 82]. Even carvacrol is a nonselective TRPM7 channel blocker; it is a valuable pharmacological tool for studying the TRPM7 channel functions in vitro and in vivo. In addition, a recent study from Fleig and colleagues reported a potent and relatively specific TRPM7 channel inhibitor waixenicin A [47]. This novel TRPM7 blocker inhibits TRPM7 currents in TRPM7 overexpression HEK293 cells and TRPM7-like currents in mouse hippocampal neurons [43]. Waixenicin A also enhanced neurite outgrowth in hippocampal neurons in vitro [43], which rewrites the chapter for current TRPM7 pharmacology. Thus, waixenicin A as a specific pharmacological agent for TRPM7 channels potentiates the anticipation for studying the physiological and pharmacological roles of the TRPM7 channels in vivo and potential therapeutic applications, especially in brain diseases.

4 Physiological Functions

Currently, TRPM7 channel physiology has been better understood, thank for better molecular and pharmacological reagents. The knowledge of TRPM7 physiology is the second fast expanding area. TRPM7 channels play many important roles under physiological conditions. TRPM7 channels are calcium-permeable nonselective divalent cation channels [40]. TRPM7 channels are widely expressed in different tissues including brain, heart, and blood vessels [48, 56] and are thought to contribute to a ubiquitous homeostatic mechanism that regulates Ca^{2+} fluxes based on the metabolic state of the cell. TRPM7 plays a role in a wide range of processes in cell biology that affects cell adhesion, cell growth, and proliferation [41, 42], as well as in skeleton formation [83] and embryonic development [72]. Global knockdown of TRPM7 is embryonic lethal [72], indicating that TRPM7 channels are essential for development. Interestingly, overexpression of TRPM7 channels reduces cell viability [41, 80, 84]. TRPM7 also plays a key role in neuronal cell death in anoxia in vitro [44], and ischemia in vivo [12]. Recently, we have also shown that: (1) inhibition of TRPM7 in vitro enhances neuronal outgrowth in culture hippocampal cells under normoxic condition [43], (2) TRPM7 plays a role in vivo in brain damage to hypoxia [13], and (3) TRPM7 also plays an important role in cell survival in glioma cells in vitro [45, 46]. However, there is a lack of pharmacological inhibitor for TRPM7 channels until recently the discovery of the specific inhibitor [43, 47], which will help us to understand not only the physiology but also the pathophysiology of the TRPM7 channels.

5 Pathophysiological Relevance in Cerebral Ischemia and Hypoxia

Intracellular ionic imbalance induced by deregulated monovalent or divalent cation influx has been implicated in various cellular and molecular mechanisms underlying neuronal cell death during cerebral ischemia and hypoxia [86]. These cell death mechanisms include excitotoxicity, apoptosis, and oxidative stress. Many ion channels and receptor channels on the cell membrane are the cation influx pathways from extracellular components and are closely related in neuronal cell death under stress. Conventionally, Ca^{2+}-permeable NMDA and AMPA receptor channels are widely accepted as the traditional glutamate mechanism of stroke that induces the excitotoxicity and calcium overload during ischemia. For decades, the glutamate mechanism has been the center of stroke research and the promising therapeutic target for stroke [10, 11, 29]. However, numerous antiexcitotoxic therapies (AET) in clinical trials for stroke patients yielded unsuccessful results and even disappointing outcomes due to the unwanted psychological side effects. The side effects are thought to be from blocking these glutaminergic receptors, as the glutamate is the main neurotransmitter in the human brain. The shortcomings of AET clinical trials in stroke have led the stroke researchers in searching for the new glutamate independent mechanism(s) outside the traditional glutamate mechanism of stroke. New data later suggest that the nonglutamate mechanism may also cause neuronal ionic imbalance and trigger the hypoxic and/or ischemic neuronal cell death [10, 11] (Fig. 16.2). These nonglutamate mechanisms include many newly discovered ion channels and nonselective cation channels, such as transient receptor potential (TRP) channels [12–14], ATP-sensitive potassium channels (K_{ATP}) [15–18], acid-sensing ion channels (ASIC) [21], hemichannels [19, 20], volume-regulated anion

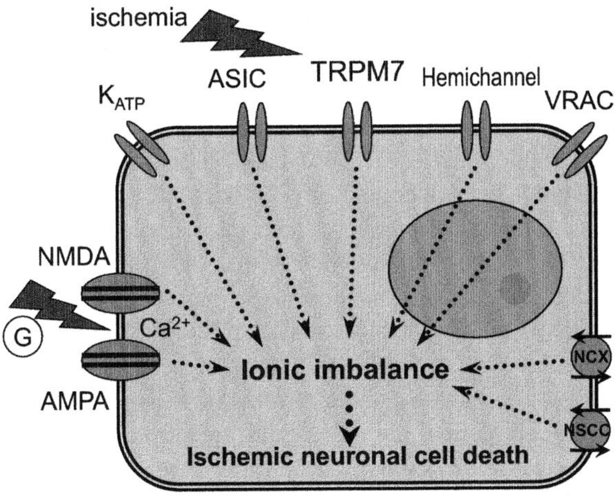

Fig. 16.2 TRPM7 channel as a nonglutamate mechanism in cerebral ischemia. Both traditional glutamate mechanism (NMDA and AMPA) and nonglutamate mechanisms are shown in the molecular pathways during cerebral ischemia. Energy failure in ischemia leads to depolarization of central neurons, activates glutamate mechanism and releases glutamate (G), which triggers the calcium overload and excitotoxicity. Other nonglutamate mechanisms including TRPM7 channels also contribute to the overall ionic imbalance and eventually neuronal cell death

channels (VRAC) [22], sodium-calcium exchangers [23, 24], and other nonselective cation channels [25]. Here, we mainly focus on the TRPM7 channel-mediated neuronal cell death and brain damage in stroke and hypoxia in vivo.

The involvement of TRPM7 in anoxic-induced neuronal cell death was initially from in vitro data. The earliest study described the so-called original I_{OGD} currents; when primary cultured cortical neurons were subjected to oxygen-glucose deprivation (OGD) challenge in vitro for a prolonged period of time, there was an increase in Ca^{2+} influx through the later identified TRPM7 channels and neuronal cell deaths [44]. Using the TRPM7 siRNA directed against TRPM7 channels in the primary mouse cortical neurons, the TRPM7 mRNA expression was decreased, so as the subsequent neuronal cell death under the anoxia/OGD condition. The TRPM7 effects on the anoxic cell death were consistently unmasked with cocktail of blockers for glutamate NMDA and AMPA receptors and L-type calcium channels using MK-801, CNQX, and nimodipine [44]. The study was well organized using the electrophysiological approach and calcium imaging technique, in addition to the molecular biology. This indicated the independent role of TRPM7 in mediating Ca^{2+} influx and subsequent neuronal cell death during the prolonged anoxia challenge. OGD is a commonly used and simplified in vitro model in studying “ischemia,” but merely the anoxia model of cell death in vitro, missing many important factors in the body such as network connectivity and blood, etc. Thus, in vivo studies are needed to further discover the pathophysiological role of the TRPM7 channels in ischemia and hypoxia.

In an unrelated in vivo study, TRPM7 mRNA and protein levels were increased after focal ischemia [87], an effect regulated by the TrkA receptor pathway, suggesting that TRPM7 may be involved in brain ischemia. Later, Sun and colleagues [12] reported that suppression of TRPM7 channels in hippocampal CA1 neurons using virally mediated gene silencing siRNA in vivo reduced delayed neuronal cell death and preserved functional outcomes after global cerebral ischemia in rats. As there was no TRPM7 blocker (antagonist) at that time, the study was used virally mediated gene silencing with shRNA approach to knockdown the TRPM7 in CA1 pyramidal neurons of hippocampus in adult rat brains. The TRPM7 shRNA was packaged with adeno-associated viral vectors (AAV serotype-1) and was delivered to hippocampal CA1 area in vivo using stereotaxic microinjection technique. The authors were able to show first that AAV could infect the adult hippocampal CA1 neurons in vivo. Second, they convincingly showed suppression of TRPM7 channels in: (1) mRNA level in conjunction with Laser Capture Microdissection targeted for infected hippocampal CA1 cells, (2) protein level with Western Blot analysis using micro-dissected CA1 tissues, (3) protein and location with immunohistochemistry in conjunction with Laser Confocal microscope in hippocampal brain slices, and (4) functional level with electrophysiology using acutely prepared hippocampal brain slices. Third, the transient knockdown TRPM7 channels in the adult rat hippocampi had no ill effects on cell survival, neuronal cells and its dendritic morphology, and other cell membrane properties such as neuronal excitability, or synaptic plasticity. Fourth, the group demonstrated that virally mediated TRPM7 suppression significantly reduced hippocampal CA1 neuronal death in vivo and preserved functional outcomes using the well-accepted 4-vessel occlusion (4VO) global ischemia model of stroke. Fifth, they showed that the morphological integrity and ultrafine dendritic structures of the survived hippocampal CA1 neuronal cells were well preserved. Sixth, the authors also showed that, in electrophysiology, the survived hippocampal CA1 cells were able to maintain their electrophysiological properties as exemplified in Long-Term Potentiation (LTP) and well-preserved cell membrane potentials. Last, they were able to show that the hippocampal-associated behaviors, such as fear-associated and spatial-navigation memory tasks, were preserved using the well-established Morris Water Maze [12, 88] and Fear Conditioning [12, 89] tests. This was the first in vivo study showing the important role of TRPM7 channels in mediating ischemic neuronal cell death in cerebral ischemia.

Recently, Sun group also demonstrated the important role of TRPM7 channels in mediating brain damage in vivo using the hypoxic–ischemic brain injury model [13]. In that study, they showed that inhibition of the TRPM7 channels using nonselective inhibitor carvacrol (also see Pharmacological Properties before) significantly and dose dependently reduced the brain damage in hypoxic–ischemic brain injury model and preserved behavioral outcomes, including (1) geotaxis reflex [13, 16, 36, 90–92] in evaluating the vestibular and/or proprioceptive functions, (2) cliff avoidance reaction [13, 16, 36, 90, 91, 93] in testing the maladaptive impulsive behavior, and (3) grip test [13, 16, 36, 90, 94] in assessing the force and fatigability. The effects of TRPM7 inhibition in reducing brain damages in the hypoxic–ischemic injury were partially mediated by inhibiting the pro-apoptotic signaling (such as capase-3 and Bcl/Bax) and promoting pro-survival signaling (such as Akt) in the study [13].

In addition to neuronal roles, another noticeable pathophysiological feature worth mentioning is the role of TRPM7 channels in glial cells. Even it may not be directly related to ischemia or hypoxia here, the TRPM7 also plays a role in cell survival in glioma cell lines in vitro [45, 46]. The same group has also showed that inhibition of TRPM7 channels using nonselective inhibitor carvacrol as well as xyloketal B, a marine compound with TRPM7 inhibitory effect, reduced cell proliferation, migration and invasion in glioma cells in vitro. The effects of TRPM7 inhibition in cell survival in glioma cells were reported partially by regulating the PI3K/Akt, MEK/ERK, MMP-2, cofilin, and caspase-3 signaling pathways [45, 46].

6 Clinical Potentials and Therapeutic Perspectives

Currently, there is no effective treatment for stroke, besides the tPA which with a limited therapeutic window. tPA is currently the only US Food and Drug Administration (FDA) approved drug for treatment of acute ischemic stroke. tPA dissolves the clots and relieves vascular occlusion [26–28, 95]. The effectiveness of tPA, as a potent drug for stroke, is still limited to its brief therapeutic window (up to 4.5 h), and intrinsic adverse effect of potential bleeding in certain stroke cases. Considering the disappointing AET clinical trials targeting glutamate-induced excitotoxicity, we need to search for alternative therapeutic targets for stroke.

Several lines of evidence support our proposal that nonglutamate mechanism TRPM7 may be a potential target for drug development in ischemic stroke [10, 11]. We have to further validate the TRPM7 as a potential drug development target for stroke and develop pharmacological reagents to further study the diverse physiological and pathophysiological functions of TRPM7 channels using in vivo animal models, even many cellular and animal studies are compelling.

The future for TRPM7 research is promising with the development of specific pharmacological modulators for TRPM7 channels, however, extensive testing will be needed in this direction. As we understand more about the molecular and cellular roles of TRPM7 played in stroke and hypoxia, we expect to see more preclinical testing in the near future using both in vitro and in vivo models, and in multiple species as per Stroke STAIR protocol [37].

7 Conclusions

The newly described nonglutamatergic mechanisms involved in ischemic neuronal cell death have been emphasized and will be further rigorously investigated using many experimental models. With the failure of AET clinical trials using NMDA and

AMPA antagonists for stroke treatment and disappointment of clinical outcomes, we expect to see more studies toward the understanding of nonglutamate mechanisms and their molecular aspects in stroke. The TRPM7 channel is a promising drug development target for stroke and hypoxia, same as for many other nonglutamatergic mechanisms. Further extensive preclinical testing will be required to confirm the therapeutic potential for TRPM7 channels in stroke.

References

1. WHO. WHO: the top 10 causes of death. 2014. Ref type: internet communication (needs website link inserted)
2. Carandang R, Seshadri S, Beiser A, Kelly-Hayes M, Kase CS, Kannel WB, Wolf PA (2006) Trends in incidence, lifetime risk, severity, and 30-day mortality of stroke over the past 50 years. JAMA 296(24):2939–2946
3. Mozaffarian D, Benjamin EJ, Go AS, Arnett DK, Blaha MJ, Cushman M, Das SR, de FS, Despres JP, Fullerton HJ, Howard VJ, Huffman MD, Isasi CR, Jimenez MC, Judd SE, Kissela BM, Lichtman JH, Lisabeth LD, Liu S, Mackey RH, Magid DJ, McGuire DK, Mohler ER III, Moy CS, Muntner P, Mussolino ME, Nasir K, Neumar RW, Nichol G, Palaniappan L, Pandey DK, Reeves MJ, Rodriguez CJ, Rosamond W, Sorlie PD, Stein J, Towfighi A, Turan TN, Virani SS, Woo D, Yeh RW, Turner MB (2016) Heart disease and stroke statistics-2016 update: a report from the American Heart Association. Circulation 133(4):e38–e60
4. Stroke (2000) http://www.strokeassociation.org/STROKEORG/AboutStroke/About-Stroke_UCM_308529_SubHomePage.jsp, website
5. Vannucci SJ (2007) Cerebral ischemia and the developing brain: introduction. Stroke 38:723
6. Nelson KB, Lynch JK (2004) Stroke in newborn infants. Lancet Neurol 3(3):150–158
7. Vannucci RC (2000) Hypoxic-ischemic encephalopathy. Am J Perinatol 17(3):113–120
8. Dirnagl U, Iadecola C, Moskowitz MA (1999) Pathobiology of ischaemic stroke: an integrated view. Trends Neurosci 22(9):391–397
9. Lipton P (1999) Ischemic cell death in brain neurons. Physiol Rev 79(4):1431–1568
10. Tymianski M (2011) Emerging mechanisms of disrupted cellular signaling in brain ischemia. Nat Neurosci 14(11):1369–1373
11. Besancon E, Guo S, Lok J, Tymianski M, Lo EH (2008) Beyond NMDA and AMPA glutamate receptors: emerging mechanisms for ionic imbalance and cell death in stroke. Trends Pharmacol Sci 29(5):268–275
12. Sun HS, Jackson MF, Martin LJ, Jansen K, Teves L, Cui H, Kiyonaka S, Mori Y, Jones M, Forder JP, Golde TE, Orser BA, Macdonald JF, Tymianski M (2009) Suppression of hippocampal TRPM7 protein prevents delayed neuronal death in brain ischemia. Nat Neurosci 12(10):1300–1307
13. Chen W, Xu B, Xiao A, Liu L, Fang X, Liu R, Turlova E, Barszczyk A, Zhong X, Sun CL, Britto LR, Feng ZP, Sun HS (2015) TRPM7 inhibitor carvacrol protects brain from neonatal hypoxic-ischemic injury. Mol Brain 8(1):11
14. Alim I, Teves L, Li R, Mori Y, Tymianski M (2013) Modulation of NMDAR subunit expression by TRPM2 channels regulates neuronal vulnerability to ischemic cell death. J Neurosci 33(44):17264–17277
15. Liu R, Wang H, Xu B, Chen W, Turlova E, Dong N, Sun CL, Lu Y, Fu H, Shi R, Barszczyk A, Yang D, Jin T, Mannucci E, Feng ZP, Sun HS (2016) Cerebrovascular safety of sulfonylureas: the role of KATP channels in neuroprotection and the risk of stroke in patients with type 2 diabetes. Diabetes 65(9):2795–2809
16. Sun HS, Xu B, Chen W, Xiao A, Turlova E, Alibraham A, Barszczyk A, Bae CY, Quan Y, Liu B, Pei L, Sun CL, Deurloo M, Feng ZP (2015) Neuronal K(ATP) channels mediate hypoxic preconditioning and reduce subsequent neonatal hypoxic-ischemic brain injury. Exp Neurol 263:161–171

17. Sun HS, Feng ZP, Miki T, Seino S, French RJ (2006) Enhanced neuronal damage after ischemic insults in mice lacking Kir6.2-containing ATP-sensitive K+ channels. J Neurophysiol 95(4):2590–2601
18. Sun HS, Feng ZP, Barber PA, Buchan AM, French RJ (2007) Kir6.2-containing ATP-sensitive potassium channels protect cortical neurons from ischemic/anoxic injury in vitro and in vivo. Neuroscience 144(4):1509–1515
19. Thompson RJ, Zhou N, MacVicar BA (2006) Ischemia opens neuronal gap junction hemichannels. Science 312(5775):924–927
20. Thompson RJ, Jackson MF, Olah ME, Rungta RL, Hines DJ, Beazely MA, Macdonald JF, MacVicar BA (2008) Activation of pannexin-1 hemichannels augments aberrant bursting in the hippocampus. Science 322(5907):1555–1559
21. Xiong ZG, Zhu XM, Chu XP, Minami M, Hey J, Wei WL, Macdonald JF, Wemmie JA, Price MP, Welsh MJ, Simon RP (2004) Neuroprotection in ischemia: blocking calcium-permeable acid-sensing ion channels. Cell 118(6):687–698
22. Alibrahim A, Zhao LY, Bae CY, Barszczyk A, Lf SC, Wang GL, Sun HS (2013) Neuroprotective effects of volume-regulated anion channel blocker DCPIB on neonatal hypoxic-ischemic injury. Acta Pharmacol Sin 34(1):113–118
23. Matsuda T, Arakawa N, Takuma K, Kishida Y, Kawasaki Y, Sakaue M, Takahashi K, Takahashi T, Suzuki T, Ota T, Hamano-Takahashi A, Onishi M, Tanaka Y, Kameo K, Baba A (2001) SEA0400, a novel and selective inhibitor of the Na+-Ca2+ exchanger, attenuates reperfusion injury in the in vitro and in vivo cerebral ischemic models. J Pharmacol Exp Ther 298(1):249–256
24. Pignataro G, Tortiglione A, Scorziello A, Giaccio L, Secondo A, Severino B, Santagada V, Caliendo G, Amoroso S, di Renzo G, Annunziato L (2004) Evidence for a protective role played by the Na+/Ca2+ exchanger in cerebral ischemia induced by middle cerebral artery occlusion in male rats. Neuropharmacology 46(3):439–448
25. Simard JM, Chen M, Tarasov KV, Bhatta S, Ivanova S, Melnitchenko L, Tsymbalyuk N, West GA, Gerzanich V (2006) Newly expressed SUR1-regulated NC(Ca-ATP) channel mediates cerebral edema after ischemic stroke. Nat Med 12(4):433–440
26. Zivin JA (1999) Thrombolytic stroke therapy: past, present, and future. Neurology 53(1):14–19
27. Zivin JA (2009) Acute stroke therapy with tissue plasminogen activator (tPA) since it was approved by the U.S. Food and Drug Administration (FDA). Ann Neurol 66(1):6–10
28. Marler JR, Goldstein LB (2003) Medicine. Stroke—tPA and the clinic. Science 301(5640):1677
29. Choi DW (1988) Glutamate neurotoxicity and diseases of the nervous system. Neuron 1(8):623–634
30. Davis SM, Lees KR, Albers GW, Diener HC, Markabi S, Karlsson G, Norris J (2000) Selfotel in acute ischemic stroke: possible neurotoxic effects of an NMDA antagonist. Stroke 31(2):347–354
31. Sattler R, Xiong Z, Lu WY, Hafner M, Macdonald JF, Tymianski M (1999) Specific coupling of NMDA receptor activation to nitric oxide neurotoxicity by PSD-95 protein. Science 284(5421):1845–1848
32. Aarts M, Liu Y, Liu L, Besshoh S, Arundine M, Gurd JW, Wang YT, Salter MW, Tymianski M (2002) Treatment of ischemic brain damage by perturbing NMDA receptor-PSD-95 protein interactions. Science 298(5594):846–850
33. Sun HS, Doucette TA, Liu Y, Fang Y, Teves L, Aarts M, Ryan CL, Bernard PB, Lau A, Forder JP, Salter MW, Wang YT, Tasker RA, Tymianski M (2008) Effectiveness of PSD95 inhibitors in permanent and transient focal ischemia in the rat. Stroke 39(9):2544–2553
34. Cook DJ, Teves L, Tymianski M (2012) Treatment of stroke with a PSD-95 inhibitor in the gyrencephalic primate brain. Nature 483(7388):213–217
35. Cook DJ, Teves L, Tymianski M (2012) A translational paradigm for the preclinical evaluation of the stroke neuroprotectant Tat-NR2B9c in gyrencephalic nonhuman primates. Sci Transl Med 4(154):154ra133
36. Xu B, Xiao AJ, Chen W, Turlova E, Liu R, Barszczyk A, Sun CL, Liu L, Tymianski M, Feng ZP, Sun HS (2015) Neuroprotective effects of a PSD-95 inhibitor in neonatal hypoxic-ischemic brain injury. Mol Neurobiol, doi:10.1007/s12035-015-9488-4

37. Stroke Therapy Academic Industry Roundtable (STAIR) (1999) Recommendations for standards regarding preclinical neuroprotective and restorative drug development. Stroke 30:2752–2758
38. Montell C, Rubin GM (1989) Molecular characterization of the Drosophila trp locus: a putative integral membrane protein required for phototransduction. Neuron 2(4):1313–1323
39. Montell C, Rubin GM (1988) The Drosophila ninaC locus encodes two photoreceptor cell specific proteins with domains homologous to protein kinases and the myosin heavy chain head. Cell 52(5):757–772
40. Clapham DE (2003) TRP channels as cellular sensors. Nature 426(6966):517–524
41. Nadler MJ, Hermosura MC, Inabe K, Perraud AL, Zhu Q, Stokes AJ, Kurosaki T, Kinet JP, Penner R, Scharenberg AM, Fleig A (2001) LTRPC7 is a Mg.ATP-regulated divalent cation channel required for cell viability. Nature 411(6837):590–595
42. Inoue K, Xiong ZG (2009) Silencing TRPM7 promotes growth/proliferation and nitric oxide production of vascular endothelial cells via the ERK pathway. Cardiovasc Res 83(3):547–557
43. Turlova E, Bae CY, Deurloo M, Chen W, Barszczyk A, Horgen FD, Fleig A, Feng ZP, Sun HS (2016) TRPM7 regulates axonal outgrowth and maturation of primary hippocampal neurons. Mol Neurobiol 53(1):595–610
44. Aarts M, Iihara K, Wei WL, Xiong ZG, Arundine M, Cerwinski W, Macdonald JF, Tymianski M (2003) A key role for TRPM7 channels in anoxic neuronal death. Cell 115(7):863–877
45. Chen WL, Barszczyk A, Turlova E, Deurloo M, Liu B, Yang BB, Rutka JT, Feng ZP, Sun HS (2015) Inhibition of TRPM7 by carvacrol suppresses glioblastoma cell proliferation, migration and invasion. Oncotarget 6(18):16321–16340
46. Chen WL, Turlova E, Sun CL, Kim JS, Huang S, Zhong X, Guan YY, Wang GL, Rutka JT, Feng ZP, Sun HS (2015) Xyloketal B suppresses glioblastoma cell proliferation and migration in vitro through inhibiting TRPM7-regulated PI3K/Akt and MEK/ERK signaling pathways. Mar Drugs 13(4):2505–2525
47. Zierler S, Yao G, Zhang Z, Kuo WC, Porzgen P, Penner R, Horgen FD, Fleig A (2011) Waixenicin A inhibits cell proliferation through magnesium-dependent block of transient receptor potential melastatin 7 (TRPM7) channels. J Biol Chem 286(45):39328–39335
48. Clapham DE, Runnels LW, Strubing C (2001) The TRP ion channel family. Nat Rev Neurosci 2(6):387–396
49. Pedersen SF, Owsianik G, Nilius B (2005) TRP channels: an overview. Cell Calcium 38(3-4):233–252
50. Wu LJ, Sweet TB, Clapham DE (2010) International Union of Basic and Clinical Pharmacology. LXXVI. Current progress in the mammalian TRP ion channel family. Pharmacol Rev 62(3):381–404
51. Ramsey IS, Delling M, Clapham DE (2006) An introduction to TRP channels. Annu Rev Physiol 68:619–647
52. Yamaguchi H, Matsushita M, Nairn AC, Kuriyan J (2001) Crystal structure of the atypical protein kinase domain of a TRP channel with phosphotransferase activity. Mol Cell 7(5):1047–1057
53. Takezawa R, Schmitz C, Demeuse P, Scharenberg AM, Penner R, Fleig A (2004) Receptor-mediated regulation of the TRPM7 channel through its endogenous protein kinase domain. Proc Natl Acad Sci U S A 101(16):6009–6014
54. Schmitz C, Perraud AL, Johnson CO, Inabe K, Smith MK, Penner R, Kurosaki T, Fleig A, Scharenberg AM (2003) Regulation of vertebrate cellular Mg2+ homeostasis by TRPM7. Cell 114(2):191–200
55. Demeuse P, Penner R, Fleig A (2006) TRPM7 channel is regulated by magnesium nucleotides via its kinase domain. J Gen Physiol 127(4):421–434
56. Monteilh-Zoller MK, Hermosura MC, Nadler MJ, Scharenberg AM, Penner R, Fleig A (2003) TRPM7 provides an ion channel mechanism for cellular entry of trace metal ions. J Gen Physiol 121(1):49–60
57. Fonfria E, Murdock PR, Cusdin FS, Benham CD, Kelsell RE, McNulty S (2006) Tissue distribution profiles of the human TRPM cation channel family. J Recept Signal Transduct Res 26(3):159–178

58. Wei WL, Sun HS, Olah ME, Sun X, Czerwinska E, Czerwinski W, Mori Y, Orser BA, Xiong ZG, Jackson MF, Tymianski M, Macdonald JF (2007) TRPM7 channels in hippocampal neurons detect levels of extracellular divalent cations. Proc Natl Acad Sci U S A 104(41):16323–16328
59. Clark K, Langeslag M, van Leeuwen B, Ran L, Ryazanov AG, Figdor CG, Moolenaar WH, Jalink K, van Leeuwen FN (2006) TRPM7, a novel regulator of actomyosin contractility and cell adhesion. EMBO J 25(2):290–301
60. Yogi A, Callera GE, Tostes R, Touyz RM (2009) Bradykinin regulates calpain and proinflammatory signaling through TRPM7-sensitive pathways in vascular smooth muscle cells. Am J Physiol Regul Integr Comp Physiol 296(2):R201–R207
61. Runnels LW, Yue L, Clapham DE (2001) TRP-PLIK, a bifunctional protein with kinase and ion channel activities. Science 291(5506):1043–1047
62. Penner R, Fleig A (2007) The Mg2+ and Mg(2+)-nucleotide-regulated channel-kinase TRPM7. Handb Exp Pharmacol 179:313–328
63. Jiang J, Li M, Yue L (2005) Potentiation of TRPM7 inward currents by protons. J Gen Physiol 126(2):137–150
64. Li M, Du J, Jiang J, Ratzan W, Su LT, Runnels LW, Yue L (2007) Molecular determinants of Mg2+ and Ca2+ permeability and pH sensitivity in TRPM6 and TRPM7. J Biol Chem 282(35):25817–25830
65. Rehncrona S (1985) Brain acidosis. Ann Emerg Med 14(8):770–776
66. Rehncrona S, Kagstrom E (1983) Tissue lactic acidosis and ischemic brain damage. Am J Emerg Med 1(2):168–174
67. Oancea E, Wolfe JT, Clapham DE (2006) Functional TRPM7 channels accumulate at the plasma membrane in response to fluid flow. Circ Res 98(2):245–253
68. Runnels LW, Yue L, Clapham DE (2002) The TRPM7 channel is inactivated by PIP(2) hydrolysis. Nat Cell Biol 4(5):329–336
69. Hermosura MC, Nayakanti H, Dorovkov MV, Calderon FR, Ryazanov AG, Haymer DS, Garruto RM (2005) A TRPM7 variant shows altered sensitivity to magnesium that may contribute to the pathogenesis of two Guamanian neurodegenerative disorders. Proc Natl Acad Sci U S A 102(32):11510–11515
70. Chubanov V, Schafer S, Ferioli S, Gudermann T (2014) Natural and synthetic modulators of the TRPM7 channel. Cells 3(4):1089–1101
71. Hofmann T, Schafer S, Linseisen M, Sytik L, Gudermann T, Chubanov V (2014) Activation of TRPM7 channels by small molecules under physiological conditions. Pflugers Arch 466(12):2177–2189
72. Jin J, Desai BN, Navarro B, Donovan A, Andrews NC, Clapham DE (2008) Deletion of Trpm7 disrupts embryonic development and thymopoiesis without altering Mg2+ homeostasis. Science 322(5902):756–760
73. Ryazanova LV, Rondon LJ, Zierler S, Hu Z, Galli J, Yamaguchi TP, Mazur A, Fleig A, Ryazanov AG (2010) TRPM7 is essential for Mg(2+) homeostasis in mammals. Nat Commun 1:109
74. Krapivinsky G, Mochida S, Krapivinsky L, Cibulsky SM, Clapham DE (2006) The TRPM7 ion channel functions in cholinergic synaptic vesicles and affects transmitter release. Neuron 52(3):485–496
75. He Y, Yao G, Savoia C, Touyz RM (2005) Transient receptor potential melastatin 7 ion channels regulate magnesium homeostasis in vascular smooth muscle cells: role of angiotensin II. Circ Res 96(2):207–215
76. Kim BJ, Park KJ, Kim HW, Choi S, Jun JY, Chang IY, Jeon JH, So I, Kim SJ (2009) Identification of TRPM7 channels in human intestinal interstitial cells of Cajal. World J Gastroenterol 15(46):5799–5804
77. Kim BJ, Lim HH, Yang DK, Jun JY, Chang IY, Park CS, So I, Stanfield PR, Kim KW (2005) Melastatin-type transient receptor potential channel 7 is required for intestinal pacemaking activity. Gastroenterology 129(5):1504–1517

78. Numata T, Shimizu T, Okada Y (2007) TRPM7 is a stretch- and swelling-activated cation channel involved in volume regulation in human epithelial cells. Am J Physiol Cell Physiol 292(1):C460–C467
79. Li M, Jiang J, Yue L (2006) Functional characterization of homo- and heteromeric channel kinases TRPM6 and TRPM7. J Gen Physiol 127(5):525–537
80. Chen HC, Xie J, Zhang Z, Su LT, Yue L, Runnels LW (2010) Blockade of TRPM7 channel activity and cell death by inhibitors of 5-lipoxygenase. PLoS One 5(6):e11161
81. Parnas M, Peters M, Dadon D, Lev S, Vertkin I, Slutsky I, Minke B (2009) Carvacrol is a novel inhibitor of Drosophila TRPL and mammalian TRPM7 channels. Cell Calcium 45(3):300–309
82. Yu H, Zhang ZL, Chen J, Pei A, Hua F, Qian X, He J, Liu CF, Xu X (2012) Carvacrol, a food-additive, provides neuroprotection on focal cerebral ischemia/reperfusion injury in mice. PLoS One 7(3):e33584
83. Elizondo MR, Arduini BL, Paulsen J, MacDonald EL, Sabel JL, Henion PD, Cornell RA, Parichy DM (2005) Defective skeletogenesis with kidney stone formation in dwarf zebrafish mutant for trpm7. Curr Biol 15(7):667–671
84. Su LT, Agapito MA, Li M, Simonson WT, Huttenlocher A, Habas R, Yue L, Runnels LW (2006) TRPM7 regulates cell adhesion by controlling the calcium-dependent protease calpain. J Biol Chem 281(16):11260–11270
85. Olah ME, Jackson MF, Li H, Perez Y, Sun HS, Kiyonaka S, Mori Y, Tymianski M, Macdonald JF (2009) Ca2+-dependent induction of TRPM2 currents in hippocampal neurons. J Physiol 587(Pt 5):965–979
86. Simard JM, Tarasov KV, Gerzanich V (2007) Non-selective cation channels, transient receptor potential channels and ischemic stroke. Biochim Biophys Acta 1772(8):947–957
87. Jiang H, Tian SL, Zeng Y, Li LL, Shi J (2008) TrkA pathway(s) is involved in regulation of TRPM7 expression in hippocampal neurons subjected to ischemic-reperfusion and oxygen-glucose deprivation. Brain Res Bull 76(1-2):124–130
88. Morris RG, Garrud P, Rawlins JN, O'Keefe J (1982) Place navigation impaired in rats with hippocampal lesions. Nature 297(5868):681–683
89. Cheng VY, Martin LJ, Elliott EM, Kim JH, Mount HT, Taverna FA, Roder JC, Macdonald JF, Bhambri A, Collinson N, Wafford KA, Orser BA (2006) Alpha5GABAA receptors mediate the amnestic but not sedative-hypnotic effects of the general anesthetic etomidate. J Neurosci 26(14):3713–3720
90. Xiao AJ, Chen W, Xu B, Liu R, Turlova E, Barszczyk A, Sun CL, Liu L, Deurloo M, Wang GL, Feng ZP, Sun HS (2015) Marine compound xyloketal B reduces neonatal hypoxic-ischemic brain injury. Mar Drugs 13(1):29–47
91. Ten VS, Bradley-Moore M, Gingrich JA, Stark RI, Pinsky DJ (2003) Brain injury and neuro-functional deficit in neonatal mice with hypoxic-ischemic encephalopathy. Behav Brain Res 145(1-2):209–219
92. Sanches EF, Arteni NS, Spindler C, Moyses F, Siqueira IR, Perry ML, Netto CA (2012) Effects of pre- and postnatal protein malnutrition in hypoxic-ischemic rats. Brain Res 1438:85–92
93. Bouslama M, Renaud J, Olivier P, Fontaine RH, Matrot B, Gressens P, Gallego J (2007) Melatonin prevents learning disorders in brain-lesioned newborn mice. Neuroscience 150(3):712–719
94. Liu XH, Yan H, Xu M, Zhao YL, Li LM, Zhou XH, Wang MX, Ma L (2013) Hyperbaric oxygenation reduces long-term brain injury and ameliorates behavioral function by suppression of apoptosis in a rat model of neonatal hypoxia-ischemia. Neurochem Int 62(7):922–930
95. Bambauer KZ, Johnston SC, Bambauer DE, Zivin JA (2006) Reasons why few patients with acute stroke receive tissue plasminogen activator. Arch Neurol 63(5):661–664

Chapter 17
Cholinergic Protection in Ischemic Brain Injury

Victor V. Uteshev

Abstract The cholinergic system is essential for maintenance of cognitive, autonomic, and immune homeostasis in mammals. Existing preclinical studies suggest that endogenous cholinergic tone elevated by ischemic or traumatic brain injury may serve as a combination autotherapy aiming at multiple distinct cellular and molecular pathways with converging neuroprotective and anti-inflammatory efficacies. These hardwired endogenous protective mechanisms can be augmented by cholinergic treatments including nicotinic receptor agonists, positive allosteric modulation, and vagus nerve stimulation. The α7 subtype of nicotinic acetylcholine receptors (nAChRs) is uniquely positioned as a promising therapeutic target in stroke and traumatic brain injury because of the neuroprotective anti-inflammatory efficacy of α7 nAChR activation and the ubiquitous expression of α7 nAChRs in mammalian neuronal, glial, and immune tissues. This article further explores the therapeutic promise of endogenous α7-dependent neuroprotective and anti-inflammatory injury-induced autotherapy which may act as an important physiological function of these ubiquitous receptors and may hold significant translational potential.

Keywords Neuroprotection • Ischemic • Traumatic • Brain injury • Stroke • Immune • Inflammatory • Vagus • Vagus nerve stimulation • Stem cells • Astrocytes • Glia • Neurogenesis • Alpha7 • Choline • Nicotinic • Allosteric • PNU120596 • PNU-120596

Abbreviations

ATP	Adenosine triphosphate
CaMKII	Ca^{2+}/calmodulin-dependent kinase-II
CDP-choline	Cytidine diphosphate-choline
DMV	The dorsal motor nucleus of the vagus

V.V. Uteshev, Ph.D. (✉)
Institute for Healthy Aging, Center for Neuroscience Discovery, University of North Texas Health Science Center, 3500 Camp Bowie Blvd., CBH 501, Fort Worth, TX 76107, USA
e-mail: Victor.Uteshev@unthsc.edu

P.A. Lapchak, J.H. Zhang (eds.), *Neuroprotective Therapy for Stroke and Ischemic Disease*, Springer Series in Translational Stroke Research,
DOI 10.1007/978-3-319-45345-3_17

JAK	Janus kinase
NAb	The nucleus ambiguus
NfkB	Nuclear factor kappa-light-chain-enhancer of activated B cells
PI3K/Akt	Phosphatidylinositol-4,5-bisphosphate 3-kinase/serine/threonine-specific protein kinase
RK	Extracellular signal-regulated kinase
STAT	Signal transducer and activators of transcription

1 Protection of Neuroprotection

The following recent comment by a reviewer from the National Institutes of Health (NIH) highlights neuroprotection-phobia, an unfortunate development in the field of neuropharmacology as the term "neuroprotection" has become derogatory: "*Another neuroprotective agent demonstrated only in rats with a treatment window of only 6 h does not significantly advance the field... There should be some discussion about why the failure of translation of acute neuroprotectants from rats to humans will not apply here... The potential out-licensing value of the compound for treatment of acute ischemic stroke will likely be highly dependent upon demonstrating a uniquely wide treatment window or other property that will differentiate the compound from the numerous other neuroprotectants that failed translation.*" The reviewer's concerns arose from an unfortunate fact that despite tremendous investments of funds and efforts made in the past two decades, neuroprotective treatments of brain injury have so far failed in all clinical trials. This fact, however, does not justify the use of a rejection template to indiscriminately define pending neuroprotective strategies as failures. Such a template would not be supported by the NIH guidelines and would prejudge and jeopardize both the existing neuroprotection-based clinical trials (e.g., NCT02315443, NCT02195791; accessed May 2016) as well as the NIH efforts to protect and improve human health.

The truth is the impact of clinically effective neuroprotectants cannot be overestimated. In contrast to tissue plasminogen activator (tPA), neuroprotectants can be effective in both ischemic and hemorrhagic strokes and thus, can be given prior to brain imaging. As a result, in an event of ischemic stroke, neuroprotectants can serve as an important bridge to intraarterial interventions (e.g., thrombectomy). If given in the first 4 h after the onset of stroke, just like tPA, a potent neuroprotectant could raise multifold the percent of patients successfully treated with thrombolysis and/or thrombectomy providing a tremendous benefit to stroke patients. Thus, a concomitant use of neuroprotectants with thrombolysis and thrombectomy offers a valuable clinical advantage. Furthermore, because neuroprotection is expected to help reduce neuronal injury, degeneration, and death at early phases after stroke, there is not a rational basis to expect that neuroprotection will be incompatible with late-phase rehabilitation treatments (e.g., targeted plasticity [1] and neurogenesis through conversion of stem [2] or glial [3] cells into mature neurons).

The other concern of the same reviewer was that the delay between symptoms and test treatments should not be less than 6 h to be clinically relevant. Although the efficacy of all stroke treatments is time sensitive and the treatments should be applied as promptly as possible as permitted by clinical practice [4], the ambulance practitioners can do intravenous (i.v.) drug administration in much less time than 6 h after the onset of stroke. For example, the 2014 FAST-MAG study [5] has shown that i.v. medications could be successfully delivered in the field by ambulance personnel in 73 % of acute strokes within 1 h and 99 % within 2 h of onset. Thus, setting the minimal treatment window to 1–2 h seems appropriate and achievable in practice [5]. Furthermore, our preliminary results suggest that even though the effect of a single acute treatment may vanish in several hours depending on the time course of drug clearance and metabolism, multiple subchronic treatments may offer additional benefits such that the time of the first treatment may become less critical so long as the first treatment is followed by a prolonged (e.g., days to weeks) subchronic therapy that may naturally extend into the phase of rehabilitation. Although effective treatments of stroke with unknown time of symptom onset (e.g., wake-up stroke) will likely remain to be a considerable challenge (see NCT00167765, NCT01183533; accessed May 2016), a less than 6 h delay between the symptom onset and treatment seems to be quite achievable in the modern clinical practice.

Similar to neurobehavioral deficits that result from acute ischemic or traumatic events, neurological disorders associated with relatively slow but progressive neuronal dysfunction and death, like in Parkinson's disease (PD), can benefit from the use of neuroprotectants. At the early phase of PD, neuroprotection and inhibition of inflammation initiated by drug therapies would aim at disease modification, delaying disease progression, and chronic inflammation [6, 7] thus, facilitating subsequent therapies. The early phase of PD is primarily associated with nonmotor symptoms [6–8] (e.g., constipation, loss of sense of smell, and cardiovascular deficits) and may precede the phase of substantial damage of dopaminergic terminals and/or somatodendritic region [9]. Preclinical animal models suggest that neuroprotection is most effective at the early stage of PD [10, 11] but once the population of dopaminergic neurons is substantially depleted, the efficacy of neuroprotectants is expected to be markedly reduced and other approaches should be preferred: e.g., neurogenesis [2, 9]. Importantly, cholinergic neuroprotectants can act as potent central [12–14] and peripheral [15–17] anti-inflammatory agents and treatments that selectively enhance activation of α7 nicotinic acetylcholine receptors (nAChRs) are especially attractive because of the ubiquitous expression of these receptors in central and peripheral neuronal, glial, and immune tissues. As discussed later, activation of α7 nAChRs are expected to stimulate multiple distinct cholinergic pathways producing significant therapeutic efficacy via direct neuroprotection and central/peripheral anti-inflammatory action [14, 15, 17, 18]. This notion is supported by epidemiological studies which indicate that the incidence of PD is significantly reduced in tobacco smokers. Animal studies have also demonstrated significant neuroprotective and anti-inflammatory efficacies of nicotine. In mammals, α7 nAChRs are an important sensor of nicotine and a promising target for neuroprotective and anti-inflammatory combination therapies in progressive neurological disorders including PD [19].

An important additional advantage of neuroprotection as a strategy is that neuroprotective drug formulations (e.g., intranasal spray) [20] do not require a professional medical assistance and can be administered by patients themselves, their friends, or relatives at the critical first 3–4 h after the onset of stroke. An early response to a stroke event and symptoms can significantly impede stroke maturation and help reduce longer term poststroke sequelae including edema, breakdown of blood–brain barrier, synaptic degeneration, and excessive inflammation. The current failure of neuroprotectants in clinics does not imply the failure of neuroprotection as a therapeutic concept because injury-induced endogenous neuroprotective mechanisms are evident [20–24] (Figs. 17.1 and 17.2). Although naturally efficacious, these protective mechanisms can be augmented by poststroke cholinergic treatments to achieve higher therapeutic efficacy as some preclinical data suggest [20]. Promising approaches include endogenous brain protective mechanisms that can simultaneously recruit multiple cholinergic cellular and molecular pathways thus, initiating a natural combination therapy [17, 18, 25]. Despite a long trail of failures, these recent research findings have revived the hope of creating clinically efficacious solutions for ischemic and traumatic brain injuries to meet high standards of clinical and social needs [26]. In these research efforts, neuroprotection, as a therapeutic concept working alone or in combination with intraarterial interventions (e.g., thrombectomy), needs to be objectively reevaluated and if necessary, protected from cynical attacks. Neuroprotectants can become important allies, not adversaries [27], while clinical relevance should become obligatory in all preclinical trials [28].

2 Converting Nonneuronal Cells to Neurons

At the late phases of progressive neurodegeneration, the affected neuronal population and synaptic networks endure severe irreversible damage and neuroprotective strategies are unlikely to deliver significant therapeutic efficacy. In the case of ischemic stroke, this progression would correspond to expansion of ischemic core, shrinkage of ischemic penumbra, and formation of ischemic cavity despite the ongoing endogenous neurogenesis [2, 29, 30] and neuroprotective/anti-inflammatory mechanisms [18, 20, 25]. By contrast, cell therapies that can replace damaged neurons and revamp functional neuronal networks become quite promising. Ischemic stroke initiates reactive gliosis and creates a layer of glial scar tissue formed from reactive astrocytes, microglia, and chondroitin sulfate proteoglycans (CSPGs) that surround the ischemic infarct core [31–34]. The formation of ischemic glial scar is beneficial to neuronal networks because it restricts the expansion of ischemic core and thus, protects further ischemic damage. However, at later stages, the scar begins to impede poststroke recovery by inhibiting angiogenesis, synaptogenesis, axonal growth, and maturation [31, 35–38]. As a result, the patient's recovery after ischemic stroke becomes obstructed.

Clinically effective measures that could control the size of glial scars and the rate of glial scar formation would provide an important tool to optimization of the poststroke recovery process. In this search, novel strategies utilizing reprogramming

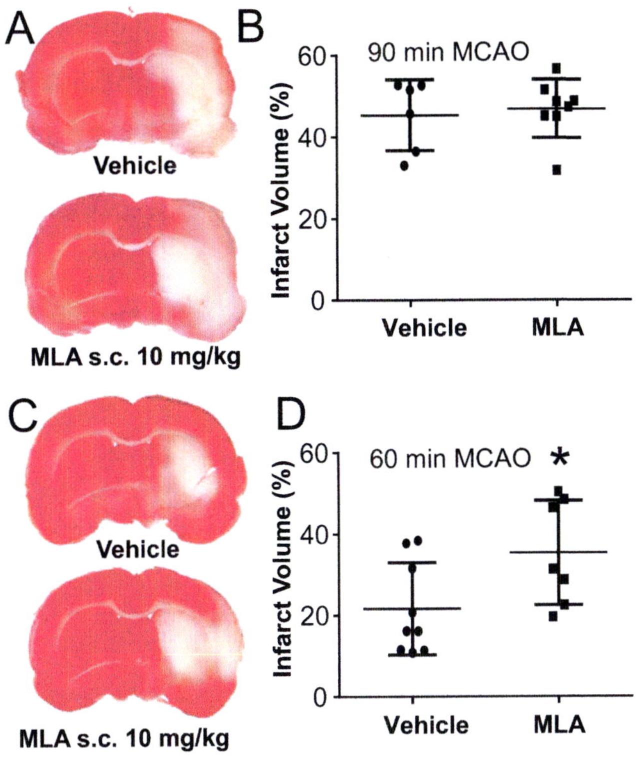

Fig. 17.1 Endogenous α7-dependent protection after transient middle cerebral artery occlusion (tMCAO): ischemic infarct volume. Methyllycaconitine (MLA; s.c., 10 mg/kg), a selective α7 nAChR antagonist, failed to significantly increase infarct volume after a 90-min tMCAO (unpaired, two-tailed Student's t test; $t=0.3572$; df = 12; $p=0.7271$; 95 % CI −10.75, 7.72; $n=6$–8; **a**, **b**) suggesting that α7 activation does not significantly save neurons in the absence of PNU in this experimental paradigm (circles vs. squares; **b**). These results also suggested that ischemic brain injury resulting from a 90-min tMCAO is too extensive (~45 %; **a**, **b**) and the use of PNU-like α7-PAMs is absolutely required to achieve significant α7-dependent protection in these settings. To test whether a less severe injury is more receptive to endogenous protective mechanisms, we used a 60-min tMCAO. Reducing the duration of tMCAO from 90 to 60 min resulted in a ~50 % reduction of infarct volume (~22 vs. ~45 %; circles; **b**, **d**). MLA significantly increased infarct volume after a 60-min tMCAO from 20 to 35 % unpaired, two-tailed Student's t test; $t=2.272$; df = 14; $p=0.0393$; 95 % CI 0.7724, 26.72; $n=7$–9; **c**, **d**), an increase of ~75 % (circles vs. squares; **d**), indicating that endogenous α7-dependent protection is greater after a less severe ischemic injury. Statistical significance was defined by the p value (* <0.05). *(Reprinted from Sun et al., 2016 (20) with permission from Springer)*

of brain tissues and cells have been particularly intriguing and promising [39, 40]. Specifically, recent advances in bioengineering have extended the approach of reprogramming fibroblasts to induce pluripotency [39] and now enable direct conversion of glial cells to functional mature neurons with confirmed electrophysiological properties [41–46]. The rationale for this approach arises from the ability of forced expression of various transcription factors (e.g., Neurogenin 2 (NEUROG2),

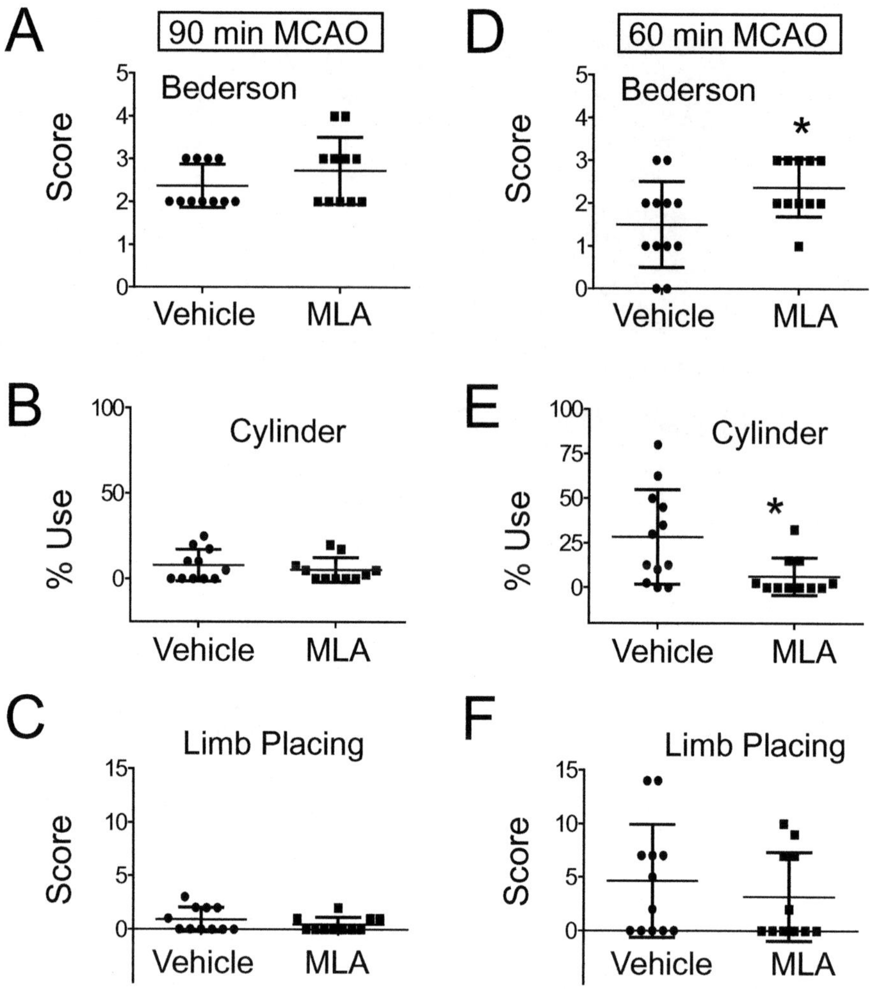

Fig. 17.2 Endogenous α7-dependent protection after tMCAO: behavioral tests. The results of behavioral tests were consistent with infarct volume data: MLA (squares as compared to circles, i.e., vehicle groups) failed to produce significant neurological deficits after a 90-min tMCAO (unpaired, two-tailed Mann–Whitney; $n = 11$–12; **a**–**c**): Bederson ($p = 0.3556$; 95 % CI 0.0, 1.0), cylinder ($p = 0.6127$; 95 % CI −10.0, 2.5), and limb placing ($p = 0.4478$; 95 % CI −1.0, 0.0). By contrast, significant neurological deficits after MLA treatment were detected in two tests after a 60-min tMCAO (unpaired, two-tailed Mann-Whitney; $n = 11$–12; **d**, **e**): Bederson ($p = 0.0427$; 95 % CI 0.0, 2.0) and cylinder ($p = 0.0227$; 95 % CI –45.0, 0.0). The remaining test (limb placing) conducted after a 60-min tMCAO failed to detect significant differences between MLA-treated and control groups: $p = 0.5940$; 95 % CI −7.0, 2.0 (**f**). Statistical significance was defined by the p value (* <0.05). *(Reprinted from Sun et al., 2016 (20) with permission from Springer)*

Distal-less homeobox 2 (DLX2), Achaete-scute homolog 1 (ASCLl1), NeuroD1, and SOX2) to allow this conversion in animal models [41–47]. Furthermore, a recent study has determined that a "chemical barrier" created by CSPGs can be reduced by administration of chondroitinase ABC (chABC) benefiting nerve regrowth "along preexisting axonal pathways" after focal ischemia in a rat model of ischemic stroke [3]. Directing de novo neuronal formation focally, near the site of ischemic injury by transcription factors under the control of a set of promoters could ensure a full control of the fate of newly formed neurons, neurotransmitters, and synapses to achieve an optimal excitatory-inhibitory balance of newly formed neuronal networks [45, 48]. The impact of a technology that could convert glial scar cells into subsets of mature neurons integrated in both proximal and distal preexisting functional neuronal networks cannot be overestimated and could help stroke patients in the late phases of poststroke recovery when neuroprotective mechanisms are largely ineffective [45, 48].

Importantly, reactive astrocytes activated by ischemic injury densely occupy space in the vicinity of ischemic lesion and thus, the challenge of precise delivery of these endogenous neuronal precursors to the site of neurogenesis is naturally eliminated if astrocytes can be recruited by a poststroke neurogenesis therapy. A similar concept of injury-guided delivery of efficacious drug therapy to the site of brain injury has been proposed previously and utilizes positive allosteric modulation of α7 nAChRs [17, 18, 25] (see the subsection 3.3). Prior to clinical application, the efficacy of these approaches needs to be confirmed in animal stroke models with a broad spectrum of biological variables (e.g., sex, age, conditions) as instructed by the Stroke Treatment Academic Industry Roundtable (STAIR) [49, 50] and RIGOR [51] guidelines.

The use of exogenous and/or endogenous stem cells as a source of neurogenesis in regenerative medicine has been viewed as a promising approach to treatment of ischemic stroke and aiding stroke recovery [2, 30, 52–55]. Although the exact mechanisms of stem cell-based therapies remain uncertain, neuroprotection is a likely component [56, 57]. Furthermore, stem cells also express functional α7 nAChRs and selective activation of these receptors modulate stem cell maturation, differentiation, migration, and function [58–61] providing another dimension to potential α7 nAChR-based antistroke therapies. Some of the challenges associated with the use of stem cells arise from its limited capacity for optimization, the dependence on the time and mode of delivery of exogenous stem cells to the ischemic cavity in a clinically appropriate biodegradable scaffolding [62], and the ability to provide a stable reproducible recruitment of endogenous neural stem cells in the injured adult brain [2, 30]. These challenges, although not unresolvable, are substantial and considerably impede advances in stem cell technology in application to stroke.

3 Therapeutics of Choline and Choline Derivatives

3.1 Biosynthesis of Phosphatidylcholine and Tissue Repair

The recent International Citicoline Trial on Acute Stroke (ICTUS) [63] has tested and failed to confirm an otherwise intriguing concept that stimulation of the brain phosphatidylcholine (PCh) synthesis and PCh-dependent tissue repair can generate significant therapeutic efficacy after ischemic stroke. The failure of ICTUS has delivered a blow to the long-standing hope for a translational efficacy of endogenous CDP-choline and its synthetic analog, citicoline (i.e., cytidine-5′-diohosphocholine), a highly bioavailable and safe choline derivative, a potent cell membrane stabilizer and inhibitor of free fatty acid toxicity [64–67]. Multiple reasons may have contributed to the failure of ICTUS [68]. These may include the ceiling effects associated with the high percent of patients older than 70 years old and patients with severe stroke (i.e., smaller penumbra), the high percent of tPA use, and the high quality care in specialized stoke units provided to all patients including control groups, as discussed previously [68]. While discouraging to stroke patients and researchers, the results of the ICTUS study have not destroyed the therapeutic concept of PCh-dependent tissue repairs. In fact, a more recent clinical trial has determined significant pro-cognitive effects of long-term treatments with citicoline [69].

Other preclinical data from models of glaucoma [70, 71] and vascular dementia [72–74] have also supported the concept that endogenous choline, choline-derived endogenous compounds (e.g., CDP-choline), and choline-derived synthetic supplements (e.g., citicoline) may benefit recovery from brain injury. The exact therapeutic mechanisms are not fully understood, but may include biosynthesis of phospholipids (because PCh is an elementary building block of neuronal and glial membranes in the mammalian brain including human brain [75]) and neuroprotective/anti-inflammatory action of α7 nAChRs (see the subsection 3.2) [14, 15, 17, 25]. Thus, despite obstacles, the use of choline derivatives remains a promising therapeutic concept. However, a uniform elevation of choline and its derivatives throughout the mammalian body as a result of citicoline treatment may only have a very limited impact on the focal rate of biosynthesis in the ischemic penumbra and this critical limitation may also have contributed to the failure of ICTUS. Furthermore, choline is not known to readily accumulate in the mammalian brain after i.v. injections [76]. The observation that successful preclinical animal studies required doses of citicoline much higher than those used in clinical trials [4] suggest that inadequate blood/brain levels of choline and choline derivatives may have been, at least in part, responsible for the dichotomy between failed clinical and successful preclinical trials. If this notion holds any value then, theoretically, a focal injection of choline or its derivatives into the ischemic core/penumbra would be expected to have a greater therapeutic potential as compared to systemic injections because a greater presence of PCh can be achieved focally where it is most needed, i.e., near the site of ischemic injury. In fact, systemic elevation of choline derivatives may not even be necessary at all because ischemic and traumatic brain injuries significantly elevate ambient levels of choline via a natural breakdown of cell membrane phospholipids [77–79]. Thus, systemic administration of choline derivatives as an acute treatment of ischemic brain injury after ischemic stroke seems redundant because high levels of choline and its derivatives may be already present focally, near the site of injury.

The existence of endogenous neuroprotection and tissue repair (evident from spontaneous improvements in some stroke victims [30, 80–83]) prevents establishing robust negative controls for drug treatments because these endogenous autotherapies are induced by the original injury, infection, and/or inflammation. Thus, the concept of unprotected (i.e., control) brain may be somewhat illusory and perhaps, most treatments can be viewed as combination treatments administered on top of injury-induced endogenous autotherapies. A treatment that augments endogenous protection and/or tissue repair may be highly efficacious because of the evolutionarily optimized spatiotemporal distribution of the contributing endogenous protective mechanisms. By contrast, treatments that disregard the presence of endogenous protective and restorative mechanisms may not be as efficacious as they may interfere with the endogenous protective pathways and cause adverse reactions.

While stimulation of biosynthesis of cell membrane phospholipids and brain tissue repair provides a solid rational basis for choline-based therapies, choline is much more than a source of phospholipids and derivatives. In addition to being a precursor and metabolite of ACh, a major mammalian neurotransmitter, choline is a selective agonist of α7 nAChRs, a receptor subtype uniquely positioned as a promising target for neuroprotective and anti-inflammatory combination therapies in stroke [18, 20, 25, 84–86], traumatic brain injury (TBI) [12, 16, 17, 87, 88], and other neurological conditions [87, 89–106]. Thus, the ability of acute infection, inflammation, and brain injury to stimulate cholinergic pathways augmenting neuroprotective and anti-inflammatory α7-dependent mechanisms may hold significant therapeutic potential [14, 15, 17, 18, 20, 25, 84, 85, 107]. In fact, at least some therapeutic effects of CDP-choline may arise from α7 nAChR-dependent mechanisms [108–110].

3.2 Neuroprotective and Anti-inflammatory Efficacies of α7 nAChRs

It is beneficial for central neurons to express α7 nAChRs [17, 18, 25, 104, 111–113] because α7 activation by endogenous (i.e., choline and ACh) and/or exogenous (e.g., nicotine) agonists may augment brain resistance to ischemic and traumatic injury by supporting survival and function of individual neurons and neuronal networks [12, 13, 16, 20, 84, 85, 87, 94, 101, 114–124]. In addition, activation of α7 nAChRs may produce strong pro-cognitive [103, 125–128] and analgesic effects [129–131]. Consistent with these conclusions, neuroprotective effects of nicotine are not detected in α7 knock-out mice [116]. Although α7 nAChRs belong to so-called neuronal subtype of nAChRs, these receptors are commonly expressed in brain glial and immune cells where α7 activation produces strong anti-inflammatory efficacy [12–14, 104]. Similarly, in the periphery, activation of α7 nAChRs by ACh released from the vagus nerve significantly inhibits peripheral immune cells, reduces synthesis and release of pro-inflammatory cytokines, and protects the integrity of blood–brain barrier [15, 16, 107, 122, 132–134]. In fact, functional α7 nAChRs are absolutely required for the vagal anti-inflammatory efficacy [135, 136].

The efferent motor branch of the vagus nerve innervates most of the immune-related subdiaphragmatic organs (e.g., spleen, liver, gastrointestinal tract) making vagus nerve stimulation (VNS) a promising treatment of ischemic and traumatic brain injuries in patients and animal models [137–139]. The vagus nerve appears to potently control systemic inflammatory responses to toxins, pathogens, cytokines, and other inflammatory mediators [134] contributing to the endogenous cholinergic anti-inflammatory reflex [132]. Thus, both electrical and/or pharmacological stimulation of the vagus nerve may decrease production of pro-inflammatory cytokines and reduce inflammation [15, 17, 140]. These treatments may be useful for both acute and chronic inflammation. Chronic elevation in levels of pro-inflammatory cytokines is harmful for human health and results from various usually ill-defined sources including genetic susceptibility to autoimmune disorders; ischemic or traumatic injury; chronic infection and prolonged exposure to stress, environmental, and biochemical hazards [104, 141–147].

These concepts and experimental data suggest that therapeutic benefits of the direct α7-dependent neuroprotection can be considerably extended by central [12, 13, 148–153] and peripheral [15–17, 107, 132, 133, 135, 136, 154] α7-dependent anti-inflammatory mechanisms which can be boosted by selective α7-dependent agents [18, 20, 25, 86, 106] as well as electrical and pharmacological VNS [17, 107, 123, 140]. As a result, because of the ubiquitous expression of α7 nAChRs in neuronal, glial, and immune tissues, selective α7 nAChR agents are uniquely positioned as promising combinational neuroprotective, anti-inflammatory, pro-cognitive, and analgesic treatments. Importantly, the observation that subcutaneous injection of methyllycaconitine (MLA; a selective α7 nAChRs antagonist) significantly worsens neurobehavioral outcomes of focal ischemia in a transient suture middle cerebral artery occlusion (tMCAO) model of ischemic stroke in rats [20] (Figs. 17.1 and 17.2) supports the existence of endogenous α7-dependent neuroprotective and anti-inflammatory mechanisms triggered by ischemic injury [17, 18, 25]. These protective mechanisms may be considerably augmented by positive allosteric modulators (PAMs) of α7 nAChRs (see the subsection 3.3) [12, 16–18, 20, 25, 84, 85].

Despite targeting only the α7 subtype of nAChRs, these therapies may be intrinsically combinational as they aim at multiple distinct α7-dependent cellular and molecular pathways with converging therapeutic outcomes including direct neuroprotection [20, 25, 84, 85] and inhibition of central [12–14] and peripheral [15, 107] inflammation (Fig. 17.3). These therapeutic concepts may have already generated novel promising therapeutic targets, i.e., expanding the research focus from neuronal to neuroimmune networks [12–14, 17, 25, 155] and centrally/peripherally

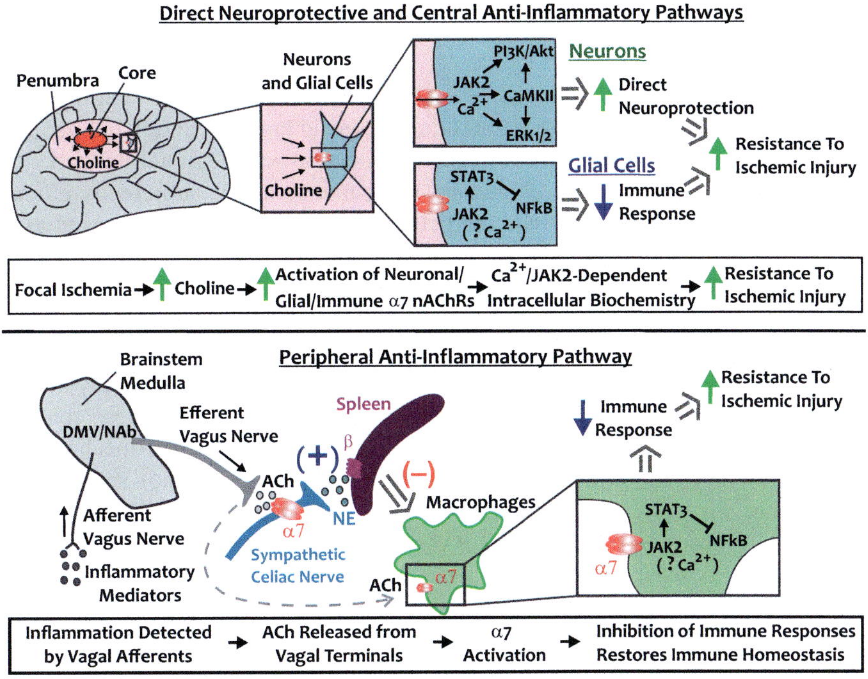

Fig. 17.3 Distinct α7-dependent neuroprotective and anti-inflammatory therapeutic mechanisms induced by injury, infection, and/or inflammation. The endogenous α7-dependent autotherapy may result from injury-, infection-, and/or inflammation-induced activation of multiple distinct cholinergic pathways [13–15, 17, 18, 25, 104, 192–195]. These include direct neuroprotection as well as central and peripheral anti-inflammatory effects. Although the biological functions of neurons, glial, and immune cells are very different, the intracellular signaling initiating neuroprotective and anti-inflammatory mechanisms downstream of α7 nAChR activation involve some of the same second messengers in neuronal and nonneuronal tissues including JAK2, STAT3, PI3K/Akt, CaMKII, and ERK1/2 [14, 104, 106, 114, 154, 192, 195–199] (*DMV* the dorsal motor nucleus of the vagus; *NAb* the nucleus ambiguus)

originating anti-inflammatory reflexes [14, 15, 17, 86, 106, 135, 136, 152, 156, 157] may achieve significant therapeutic benefits. These considerations are directly relevant to cholinergic therapies for stroke and TBI because of the key role the cholinergic mechanisms play in the central and peripheral nervous systems.

The cholinergic system is required for the maintenance of cognitive, autonomic, and immune homeostasis [1, 12, 17, 18, 158]. Recent preclinical studies suggest that cholinergic therapies may be highly efficacious after stroke [18, 20, 25, 84, 85] and TBI [12, 16, 87, 88, 107, 121–123]. Cholinergic homeostasis in the central nervous system is important for healthy neurological and immune functions: excessive cholinergic activity is linked to depression [159–162] and infection [163–166], whereas deficient cholinergic activation is associated with cognitive impairment [167–171] and inflammation [163, 166]. Although many of these effects may result from activation of central muscarinic acetylcholine receptors (mAChRs) [172–174], the contribution of nAChRs has been well documented [104, 175–178].

3.3 Therapeutic Efficacy of Positive Allosteric Modulators

As discussed earlier, activation of α7 nAChRs by endogenous (i.e., choline and acetylcholine) or exogenous (e.g., nicotine) agonists inhibits central and peripheral inflammation and supports resistance of neurons and neuronal networks to injury and death. However, endogenous agonists may not be sufficiently efficacious; whereas nicotine is a toxic, addictive substance that cannot be used as a medicine. Thus, novel safe activators of endogenous α7-dependent cholinergic tone that avoid adverse effects of nicotine may provide effective treatments after brain injury. We have recently reported a proof of concept to demonstrate that positive allosteric modulators (α7 PAMs) of α7 nAChRs augment the therapeutic efficacy of endogenous cholinergic neuroprotective and anti-inflammatory mechanisms after ischemic and traumatic brain injury [12, 17, 18, 20]. These conclusions have been since expanded by other researches [16]. Similar mechanisms (i.e., augmenting the endogenous α7-dependent cholinergic tone) seem to be responsible for pro-cognitive [103, 125–127, 179–181] and anti-nociceptive [129–131] efficacy of α7 PAMs. α7 PAMs do not directly activate α7 nAChRs, but inhibit α7 desensitization and enhance α7 activation by endogenous nicotinic agonists (i.e., choline and ACh) near the time and site of injury, infection, and/or inflammation [17, 18, 25]. Thus, α7 PAMs may introduce a novel therapeutic paradigm that converts endogenous choline and ACh into potent therapeutic agents after ischemic and traumatic brain injury by inhibiting α7 nAChR desensitization [12, 16–18, 20, 25].

The rational basis for the use of α7 PAMs after ischemic and traumatic brain injuries arises from a number of factors: (1) Functional α7 nAChRs are ubiquitously expressed in neuronal [182–186], glial, and immune tissues [13, 135, 136, 149, 155, 187–191]; (2) Ischemic and traumatic injuries, pathogens, and inflammation elevate neuroprotective and anti-inflammatory α7-dependent cholinergic tone which can be augmented by α7 PAMs [12, 16, 20, 84, 85, 155]; and (3) α7-dependent neuroprotective and anti-inflammatory mechanisms appear to be complementary to one another and converge to a significant therapeutic efficacy after ischemic and/or traumatic injury [16–18, 86, 106]. Therefore, significant neuroprotective and anti-inflammatory efficacies may be achieved after brain injury by recruiting only a single player (i.e., α7 nAChRs) that activates multiple distinct endogenous neuroprotective and anti-inflammatory cholinergic pathways (Fig. 17.3) [17]. This unique property of α7 PAMs is unlikely to be random, rather, it may serve as a common evolutionarily shaped mammalian motif (Figs.17.1 and 17.2). Intriguingly, because α7 activation inhibits reactive gliosis after ischemic or traumatic injuries [12, 13], α7 agents may be able to inhibit and/or modulate formation and growth of glial scars after ischemic stroke (see the subsection 2). This potential therapeutic mechanism may thus link these two independent approaches in a promising combination therapy expanding the range of α7-dependent therapeutic benefits. However, this opportunity has not yet been explored and remains speculative.

4 Possible Mechanisms of α7 nAChR Protective Action

The endogenous α7-dependent autotherapy may result from injury-, infection-, and/or inflammation-induced activation of multiple distinct cholinergic pathways [13–15, 17, 18, 25, 104, 192–195]. These include the direct neuroprotection as well as central and peripheral anti-inflammatory mechanisms (Fig. 17.3). Although the biological functions of neurons, glial, and immune cells are very different, the intracellular signaling initiating neuroprotective and anti-inflammatory mechanisms downstream of α7 nAChR activation involves some of the same second messengers in neuronal and nonneuronal tissues including JAK2, STAT3, PI3K/Akt, CaMKII, and ERK1/2 (Fig. 17.3) [14, 104, 106, 114, 154, 192, 195–200].

The direct α7-dependent neuroprotection arises from activation of the JAK2 ($\pm Ca^{2+}$)/PI3K/Akt-dependent [14, 114, 194, 196, 197, 201], ERK1/2-dependent [192, 198, 202–204], and/or CaMKII-dependent [20, 192, 199, 202, 205, 206] intracellular signaling antiapoptotic pathways. Although CaMKII activity is often associated with pro-cognitive benefits arising from activation of CREB [192, 198], recent findings have linked activation of CaMKII-dependent pathways to neuroprotective and/or anti-inflammatory efficacies [20, 192, 206]. However, the involvement of CaMKII in protective mechanisms after ischemic and traumatic injury is not without a controversy as the CaMKII-dependent intracellular signaling can result in both cytoprotection and cytotoxicity [207] highlighting the complexity of intracellular biochemistry in neuronal and nonneuronal tissues. Remarkably, inhibition of constitutive CaMKII activity significantly increased infarct volume after transient suture MCAO supporting the existence of endogenous CaMKII-dependent protection after focal ischemia [20].

The role of ERK1/2 in protecting brain tissues from ischemic damage remains unclear [208]. Both beneficial and detrimental effects of ERK1/2 activation on brain tissues after ischemia have been reported [192, 202–204, 209]. The reason for this dichotomy is unknown but may include the dissimilar pathways downstream of α7 activation in neurons vs. astrocytes/microglia as well as the distinct impacts of endogenous vs. exogenous α7 activation on the outcomes of treatments [208].

In neurons, activation of α7 nAChRs appears to support energy consolidation within mitochondria and thus, may prolong mitochondrial function under ischemic conditions to allow neurons better meet the energy demand of ischemic/hypoglycemic conditions to delay the failure of Na^+/K^+-ATPase pumps. In the absence of functional ATPase pumps, the neuronal transmembrane electrochemical gradient cannot be maintained and the terminal anoxic depolarization quickly sets in [84, 210]. Thus, both anoxic depolarization and the following brain damage may be reduced or even prevented by moderate activation of α7 nAChRs [84, 116, 118, 120]. By contrast, excessive activation of α7 nAChRs can be toxic [28, 211].

The central and peripheral anti-inflammatory signaling downstream of α7 nAChR activation stimulates the JAK2/STAT3 pathway resulting in inhibition of NFkB-induced pro-inflammatory cytokines (Fig. 17.3) [14, 104, 154, 192, 195, 200, 212]. The endogenous α7-dependent anti-inflammatory efficacy can be significantly augmented by electrical and/or pharmacological stimulation of the vagus nerve [15, 17, 107, 140, 154, 200, 212]. The use of α7 PAMs may allow a measured intervention aimed at inhibiting excessive injury/infection-induced immune response both centrally [14] and peripherally [15]. Theoretically, direct neuroprotective and central/peripheral anti-inflammatory pathways can be targeted individually or in cohort. A timely comprehensive treatment of stroke victims with neuroprotective and anti-inflammatory therapies may impede stroke maturation and help reduce long-term poststroke sequelae including edema, synaptic degeneration, and chronic inflammation thereby improving clinical outcomes.

The relevance of α7 nAChR-mediated ionic and Ca^{2+} influx to activation of anti-inflammatory pathways remains controversial because until recently α7-mediated ion currents have not been detected in immune cells obtained from acute preparation [213]. As a result, it has been proposed that immune-related α7 nAChRs are not linked to functional ion channels [214]. While the expression of functional α7 nAChR-gated ion channels in cultured immune cell preparations has been confirmed [149, 191], evidence for a similar expression of functional α7 channels in immune brain cells obtained from acute preparation has been missing until only recently [153, 195]. These latter studies utilized α7 PAMs and electrophysiological patch-clamp recordings to detect immune-related α7 nAChR-mediated ionic currents and demonstrate that those currents are remarkably similar to α7 nAChR-mediated currents observed in central neurons. These findings supported the notion that neuronal and immune α7 nAChRs may not be very different. One important consequence of this conclusion is that immune cells could be directly targeted by α7 PAMs and α7 agonists just like neurons [17, 18], providing a rational basis for α7 PAM-based therapies that could augment endogenous anti-inflammatory mechanisms through pharmacological VNS and direct α7-dependent inhibition of immune cells (Fig. 17.3) [17].

5 Conclusions

It is an intriguing concept that the mammalian brain can protect itself from ischemic and traumatic injury by activating endogenous injury-, infection-, and inflammation-induced protective mechanisms that in turn can be augmented by efficacious cholinergic treatments. An effective combination therapy is usually expected to target multiple cellular and molecular pathways to generate a comprehensive disease-modifying outcome. In this manuscript, I discussed an approach where the therapeutic target is a single subtype of nAChRs (i.e., α7) ubiquitously expressed in neuronal, glial, and immune tissues. Injury, infection, and/or inflammation can activate α7 nAChRs in neuronal, glial, and immune tissues resulting in complementary therapeutic efficacies. Importantly, the endogenous α7-dependent therapeutic cholinergic tone can be significantly augmented by selective α7 nAChR agents and stimulation of the vagus nerve. This ability to modulate the hardwired endogenous brain protective cholinergic pathways may allow researchers to directly team up with a powerful ally—nature. While rehabilitation programs can be highly effective [1], the potential positive impact of a successful early poststroke therapy on the patient progress cannot be overestimated as the majority of brain damage occurs within the first 2 h after the onset of stroke. Thus, strategies that enhance the brain's innate capacity to resist ischemic injury appear quite attractive as these approaches could allow clinicians to recruit and augment endogenous protective mechanisms that are already in place to selectively target ischemic and traumatic injury with a high spatiotemporal precision.

Acknowledgments I thank the William and Ella Owens Foundation for support and Dr. William Culp for insightful discussions.

References

1. Hays SA, Rennaker RL, Kilgard MP (2013) Targeting plasticity with vagus nerve stimulation to treat neurological disease. Prog Brain Res 207:275–299
2. Lindvall O, Kokaia Z (2010) Stem cells in human neurodegenerative disorders—time for clinical translation? J Clin Invest 120(1):29–40
3. Hill JJ, Jin K, Mao XO, Xie L, Greenberg DA (2012) Intracerebral chondroitinase ABC and heparan sulfate proteoglycan glypican improve outcome from chronic stroke in rats. Proc Natl Acad Sci U S A 109(23):9155–9160
4. Wardlaw JM, Murray V, Berge E, del Zoppo G, Sandercock P, Lindley RL et al (2012) Recombinant tissue plasminogen activator for acute ischaemic stroke: an updated systematic review and meta-analysis. Lancet 379(9834):2364–2372
5. Saver JL, Starkman S, Eckstein M, Stratton SJ, Pratt FD, Hamilton S et al (2015) Prehospital use of magnesium sulfate as neuroprotection in acute stroke. N Engl J Med 372(6):528–536
6. Murman DL (2012) Early treatment of Parkinson's disease: opportunities for managed care. Am J Manag Care 18(7 Suppl):S183–S188
7. Haddad BR, Gu L, Mirtti T, Dagvadorj A, Vogiatzi P, Hoang DT et al (2013) STAT5A/B gene locus undergoes amplification during human prostate cancer progression. Am J Pathol 182(6):2264–2275
8. Salvatore MF, Disbrow EA, Emborg ME (2014) Peripheral and cognitive signs: delineating the significance of impaired catecholamine metabolism in Parkinson's disease progression. J Neurochem 131(2):129–133
9. Cheng HC, Ulane CM, Burke RE (2010) Clinical progression in Parkinson disease and the neurobiology of axons. Ann Neurol 67(6):715–725
10. Quik M, Huang LZ, Parameswaran N, Bordia T, Campos C, Perez XA (2009) Multiple roles for nicotine in Parkinson's disease. Biochem Pharmacol 78(7):677–685
11. AlDakheel A, Kalia LV, Lang AE (2014) Pathogenesis-targeted, disease-modifying therapies in Parkinson disease. Neurotherapeutics 11(1):6–23
12. Gatson JW, Simpkins JW, Uteshev VV (2015) High therapeutic potential of positive allosteric modulation of alpha7 nAChRs in a rat model of traumatic brain injury: proof-of-concept. Brain Res Bull 112C:35–41
13. Parada E, Egea J, Buendia I, Negredo P, Cunha AC, Cardoso S et al (2013) The Microglial alpha7-Acetylcholine Nicotinic Receptor Is a Key Element in Promoting Neuroprotection by Inducing Heme Oxygenase-1 via Nuclear Factor Erythroid-2-Related Factor 2. Antioxid Redox Signal 19(11):1135–1148
14. Egea J, Buendia I, Parada E, Navarro E, Leon R, Lopez MG (2015) Anti-inflammatory role of microglial alpha7 nAChRs and its role in neuroprotection. Biochem Pharmacol 97(4):463–472
15. Pavlov VA, Tracey KJ (2015) Neural circuitry and immunity. Immunol Res 63(1-3):38–57

16. Dash PK, Zhao J, Kobori N, Redell JB, Hylin MJ, Hood KN et al (2016) Activation of alpha 7 cholinergic nicotinic receptors reduce blood-brain barrier permeability following experimental traumatic brain injury. J Neurosci 36(9):2809–2818
17. Uteshev VV, Tenovuo O, Gaidhani N (2016) The cholinergic potential, the vagus nerve and challenges in treatment of traumatic brain injury. Curr Pharm Des 22(14):2083–2092
18. Uteshev VV (2016) Allosteric modulation of nicotinic acetylcholine receptors: the concept and therapeutic trends. Curr Pharm Des 22(14):1986–1997
19. Fuenzalida M, Perez MA, Arias HR (2016) Role of nicotinic and muscarinic receptors on synaptic plasticity and neurological diseases. Curr Pharm Des 22(14):2004–2014
20. Sun F, Johnson SR, Jin K, Uteshev VV (2016) Boosting endogenous resistance of brain to ischemia. Mol Neurobiol. doi:10.1007/s12035-016-9796-3
21. Mattson MP, Scheff SW (1994) Endogenous neuroprotection factors and traumatic brain injury: mechanisms of action and implications for therapy. J Neurotrauma 11(1):3–33
22. Dirnagl U, Simon RP, Hallenbeck JM (2003) Ischemic tolerance and endogenous neuroprotection. Trends Neurosci 26(5):248–254
23. Wang Y, Reis C, Applegate R 2nd, Stier G, Martin R, Zhang JH (2015) Ischemic conditioning-induced endogenous brain protection: applications pre-, per- or post-stroke. Exp Neurol 272:26–40
24. Dezfulian C, Garrett M, Gonzalez NR (2013) Clinical application of preconditioning and postconditioning to achieve neuroprotection. Transl Stroke Res 4(1):19–24
25. Uteshev VV (2014) The therapeutic promise of positive allosteric modulation of nicotinic receptors. Eur J Pharmacol 727:181–185
26. Uteshev VV (2016) Editorial: effective and promising treatments for neurological disorders and cancer. Curr Pharm Des 22(14):1979–1980
27. Tymianski M (2010) Can molecular and cellular neuroprotection be translated into therapies for patients?: yes, but not the way we tried it before. Stroke 41(10 Suppl):S87–S90
28. Uteshev V (2016) Are positive allosteric modulators of alpha7 nAChRs clinically safe? J Neurochem 136(2):217–219
29. Gould E, Tanapat P (1997) Lesion-induced proliferation of neuronal progenitors in the dentate gyrus of the adult rat. Neuroscience 80(2):427–436
30. Marlier Q, Verteneuil S, Vandenbosch R, Malgrange B (2015) Mechanisms and functional significance of stroke-induced neurogenesis. Front Neurosci 9:458
31. Silver J, Miller JH (2004) Regeneration beyond the glial scar. Nat Rev Neurosci 5(2):146–156
32. Yiu G, He Z (2006) Glial inhibition of CNS axon regeneration. Nat Rev Neurosci 7(8):617–627
33. Sofroniew MV (2009) Molecular dissection of reactive astrogliosis and glial scar formation. Trends Neurosci 32(12):638–647
34. Benner EJ, Luciano D, Jo R, Abdi K, Paez-Gonzalez P, Sheng H et al (2013) Protective astrogenesis from the SVZ niche after injury is controlled by Notch modulator Thbs4. Nature 497(7449):369–373
35. Pekny M, Nilsson M (2005) Astrocyte activation and reactive gliosis. Glia 50(4):427–434
36. Robel S, Berninger B, Gotz M (2011) The stem cell potential of glia: lessons from reactive gliosis. Nat Rev Neurosci 12(2):88–104
37. Hamill CE, Goldshmidt A, Nicole O, McKeon RJ, Brat DJ, Traynelis SF (2005) Special lecture: glial reactivity after damage: implications for scar formation and neuronal recovery. Clin Neurosurg 52:29–44
38. Fawcett JW, Asher RA (1999) The glial scar and central nervous system repair. Brain Res Bull 49(6):377–391
39. Takahashi K, Yamanaka S (2006) Induction of pluripotent stem cells from mouse embryonic and adult fibroblast cultures by defined factors. Cell 126(4):663–676
40. Yu J, Vodyanik MA, Smuga-Otto K, Antosiewicz-Bourget J, Frane JL, Tian S et al (2007) Induced pluripotent stem cell lines derived from human somatic cells. Science 318(5858):1917–1920
41. Vierbuchen T, Ostermeier A, Pang ZP, Kokubu Y, Sudhof TC, Wernig M (2010) Direct conversion of fibroblasts to functional neurons by defined factors. Nature 463(7284): 1035–1041

42. Costa MR, Gotz M, Berninger B (2010) What determines neurogenic competence in glia? Brain Res Rev 63(1–2):47–59
43. Heinrich C, Blum R, Gascon S, Masserdotti G, Tripathi P, Sanchez R et al (2010) Directing astroglia from the cerebral cortex into subtype specific functional neurons. PLoS Biol 8(5):e1000373
44. Blum R, Heinrich C, Sanchez R, Lepier A, Gundelfinger ED, Berninger B et al (2011) Neuronal network formation from reprogrammed early postnatal rat cortical glial cells. Cereb Cortex 21(2):413–424
45. Guo Z, Zhang L, Wu Z, Chen Y, Wang F, Chen G (2014) In vivo direct reprogramming of reactive glial cells into functional neurons after brain injury and in an Alzheimer's disease model. Cell Stem Cell 14(2):188–202
46. Badan I, Buchhold B, Hamm A, Gratz M, Walker LC, Platt D et al (2003) Accelerated glial reactivity to stroke in aged rats correlates with reduced functional recovery. J Cereb Blood Flow Metab 23(7):845–854
47. Yang H, Sun M, Zhao S, Zhu M, Xie Y, Niu C et al (2013) Construction of chiral tertiary alcohol stereocenters via the [2,3]-Meisenheimer rearrangement: enantioselective synthesis of the side-chain acids of homoharringtonine and harringtonine. J Org Chem 78(2):339–346
48. Lu J, Bradley RA, Zhang SC (2014) Turning reactive glia into functional neurons in the brain. Cell Stem Cell 14(2):133–134
49. Saver JL, Albers GW, Dunn B, Johnston KC, Fisher M (2009) Stroke Therapy Academic Industry Roundtable (STAIR) recommendations for extended window acute stroke therapy trials. Stroke 40(7):2594–2600
50. Fisher M, Feuerstein G, Howells DW, Hurn PD, Kent TA, Savitz SI et al (2009) Update of the stroke therapy academic industry roundtable preclinical recommendations. Stroke 40(6): 2244–2250
51. Lapchak PA, Zhang JH, Noble-Haeusslein LJ (2013) RIGOR guidelines: escalating STAIR and STEPS for effective translational research. Transl Stroke Res 4(3):279–285
52. Mir O, Savitz SI (2013) Stem cell therapy in stroke treatment: is it a viable option? Expert Rev Neurother 13(2):119–121
53. Liu X, Ye R, Yan T, Yu SP, Wei L, Xu G et al (2014) Cell based therapies for ischemic stroke: from basic science to bedside. Prog Neurobiol 115:92–115
54. Burns TC, Verfaillie CM, Low WC (2009) Stem cells for ischemic brain injury: a critical review. J Comp Neurol 515(1):125–144
55. Thored P, Arvidsson A, Cacci E, Ahlenius H, Kallur T, Darsalia V et al (2006) Persistent production of neurons from adult brain stem cells during recovery after stroke. Stem Cells 24(3):739–747
56. Stone LL, Grande A, Low WC (2013) Neural repair and neuroprotection with stem cells in ischemic stroke. Brain Sci 3(2):599–614
57. Hao L, Zou Z, Tian H, Zhang Y, Zhou H, Liu L (2014) Stem cell-based therapies for ischemic stroke. BioMed Res Int 2014:468748
58. Campbell NR, Fernandes CC, Halff AW, Berg DK (2010) Endogenous signaling through alpha7-containing nicotinic receptors promotes maturation and integration of adult-born neurons in the hippocampus. J Neurosci 30(26):8734–8744
59. Liu Z, Neff RA, Berg DK (2006) Sequential interplay of nicotinic and GABAergic signaling guides neuronal development. Science 314(5805):1610–1613
60. Schraufstatter IU, DiScipio RG, Khaldoyanidi SK (2009) Alpha 7 subunit of nAChR regulates migration of human mesenchymal stem cells. J Stem Cells 4(4):203–215
61. Narla S, Klejbor I, Birkaya B, Lee YW, Morys J, Stachowiak EK et al (2013) alpha7 nicotinic receptor agonist reactivates neurogenesis in adult brain. Biochem Pharmacol 86(8):1099–1104
62. Jin K, Mao X, Xie L, Galvan V, Lai B, Wang Y et al (2010) Transplantation of human neural precursor cells in Matrigel scaffolding improves outcome from focal cerebral ischemia after delayed postischemic treatment in rats. J Cereb Blood Flow Metab 30(3):534–544
63. Davalos A, Alvarez-Sabin J, Castillo J, Diez-Tejedor E, Ferro J, Martinez-Vila E et al (2012) Citicoline in the treatment of acute ischaemic stroke: an international, randomised, multicentre, placebo-controlled study (ICTUS trial). Lancet 380(9839):349–357

64. D'Orlando KJ, Sandage BW Jr (1995) Citicoline (CDP-choline): mechanisms of action and effects in ischemic brain injury. Neurol Res 17(4):281–284
65. Adibhatla RM (2013) Citicoline in stroke and TBI clinical trials. Nat Rev Neurol 9(3):173
66. Schabitz WR, Weber J, Takano K, Sandage BW, Locke KW, Fisher M (1996) The effects of prolonged treatment with citicoline in temporary experimental focal ischemia. J Neurol Sci 138(1–2):21–25
67. Clark WM, Warach SJ, Pettigrew LC, Gammans RE, Sabounjian LA (1997) A randomized dose-response trial of citicoline in acute ischemic stroke patients. Citicoline Stroke Study Group. Neurology 49(3):671–678
68. Overgaard K (2014) The effects of citicoline on acute ischemic stroke: a review. J Stroke Cerebrovasc Dis 23(7):1764–1769
69. Alvarez-Sabin J, Santamarina E, Maisterra O, Jacas C, Molina C, Quintana M (2016) Long-term treatment with citicoline prevents cognitive decline and predicts a better quality of life after a first ischemic stroke. Int J Mol Sci 17(3):390
70. Ottobelli L, Manni GL, Centofanti M, Iester M, Allevena F, Rossetti L (2013) Citicoline oral solution in glaucoma: is there a role in slowing disease progression? Ophthalmologica 229(4):219–226
71. Parisi V, Centofanti M, Ziccardi L, Tanga L, Michelessi M, Roberti G et al (2015) Treatment with citicoline eye drops enhances retinal function and neural conduction along the visual pathways in open angle glaucoma. Graefes Arch Clin Exp Ophthalmol 253(8):1327–1340
72. Cotroneo AM, Castagna A, Putignano S, Lacava R, Fanto F, Monteleone F et al (2013) Effectiveness and safety of citicoline in mild vascular cognitive impairment: the IDEALE study. Clin Interv Aging 8:131–137
73. Gareri P, Castagna A, Cotroneo AM, Putignano S, De Sarro G, Bruni AC (2015) The role of citicoline in cognitive impairment: pharmacological characteristics, possible advantages, and doubts for an old drug with new perspectives. Clin Interv Aging 10:1421–1429
74. Alvarez-Sabin J, Ortega G, Jacas C, Santamarina E, Maisterra O, Ribo M et al (2013) Long-term treatment with citicoline may improve poststroke vascular cognitive impairment. Cerebrovasc Dis 35(2):146–154
75. van Meer G, de Kroon AI (2011) Lipid map of the mammalian cell. J Cell Sci 124(Pt 1):5–8
76. Tolvanen T, Yli-Kerttula T, Ujula T, Autio A, Lehikoinen P, Minn H et al (2010) Biodistribution and radiation dosimetry of [(11)C]choline: a comparison between rat and human data. Eur J Nucl Med Mol Imaging 37(5):874–883
77. Friedman SD, Brooks WM, Jung RE, Hart BL, Yeo RA (1998) Proton MR spectroscopic findings correspond to neuropsychological function in traumatic brain injury. AJNR Am J Neuroradiol 19(10):1879–1885
78. Kiewert C, Mdzinarishvili A, Hartmann J, Bickel U, Klein J (2010) Metabolic and transmitter changes in core and penumbra after middle cerebral artery occlusion in mice. Brain Res 1312:101–107
79. Sarter M, Bruno JP, Parikh V, Martinez V, Kozak R, Richards JB (2006) Forebrain dopaminergic-cholinergic interactions, attentional effort, psychostimulant addiction and schizophrenia. EXS 98:65–86
80. Rothrock JF, Clark WM, Lyden PD (1995) Spontaneous early improvement following ischemic stroke. Stroke 26(8):1358–1360
81. Winters C, van Wegen EE, Daffertshofer A, Kwakkel G (2015) Generalizability of the Proportional Recovery Model for the upper extremity after an ischemic stroke. Neurorehabil Neural Repair 29(7):614–622
82. Saver JL, Altman H (2012) Relationship between neurologic deficit severity and final functional outcome shifts and strengthens during first hours after onset. Stroke 43(6): 1537–1541
83. Yu TS, Washington PM, Kernie SG (2016) Injury-induced neurogenesis: mechanisms and relevance. Neuroscientist 22(1):61–71
84. Kalappa BI, Sun F, Johnson SR, Jin K, Uteshev VV (2013) A positive allosteric modulator of alpha7 nAChRs augments neuroprotective effects of endogenous nicotinic agonists in cerebral ischaemia. Br J Pharmacol 169(8):1862–1878

85. Sun F, Jin K, Uteshev VV (2013) A type-II positive allosteric modulator of alpha7 nAChRs reduces brain injury and improves neurological function after focal cerebral ischemia in rats. PLoS One 8(8):e73581
86. Norman GJ, Morris JS, Karelina K, Weil ZM, Zhang N, Al-Abed Y et al (2011) Cardiopulmonary arrest and resuscitation disrupts cholinergic anti-inflammatory processes: a role for cholinergic alpha7 nicotinic receptors. J Neurosci 31(9):3446–3452
87. Guseva MV, Hopkins DM, Scheff SW, Pauly JR (2008) Dietary choline supplementation improves behavioral, histological, and neurochemical outcomes in a rat model of traumatic brain injury. J Neurotrauma 25(8):975–983
88. Huber KB, Uteshev VV, Pauly JR (2016) Targeting the cholinergic system for neuroprotection and/or enhancement of functional recovery following neurotrauma. Curr Pharm Des 22(14):2072–2082
89. Buccafusco JJ, Letchworth SR, Bencherif M, Lippiello PM (2005) Long-lasting cognitive improvement with nicotinic receptor agonists: mechanisms of pharmacokinetic-pharmacodynamic discordance. Trends Pharmacol Sci 26(7):352–360
90. Buccafusco JJ, Terry AV Jr, Decker MW, Gopalakrishnan M (2007) Profile of nicotinic acetylcholine receptor agonists ABT-594 and A-582941, with differential subtype selectivity, on delayed matching accuracy by young monkeys. Biochem Pharmacol 74(8):1202–1211
91. Kitagawa H, Takenouchi T, Azuma R, Wesnes KA, Kramer WG, Clody DE et al (2003) Safety, pharmacokinetics, and effects on cognitive function of multiple doses of GTS-21 in healthy, male volunteers. Neuropsychopharmacology 28(3):542–551
92. Leiser SC, Bowlby MR, Comery TA, Dunlop J (2009) A cog in cognition: how the alpha 7 nicotinic acetylcholine receptor is geared towards improving cognitive deficits. Pharmacol Ther 122(3):302–311
93. Olincy A, Stevens KE (2007) Treating schizophrenia symptoms with an alpha7 nicotinic agonist, from mice to men. Biochem Pharmacol 74(8):1192–1201
94. Arendash GW, Sengstock GJ, Sanberg PR, Kem WR (1995) Improved learning and memory in aged rats with chronic administration of the nicotinic receptor agonist GTS-21. Brain Res 674(2):252–259
95. Meyer EM, Tay ET, Papke RL, Meyers C, Huang G, de Fiebre CM (1997) Effects of 3-[2,4-dimethoxybenzylidene]anabaseine (DMXB) on rat nicotinic receptors and memory-related behaviors. Brain Res 768:49–56
96. Van Kampen M, Selbach K, Schneider R, Schiegel E, Boess F, Schreiber R (2004) AR-R 17779 improves social recognition in rats by activation of nicotinic alpha7 receptors. Psychopharmacology (Berl) 172(4):375–383
97. Woodruff-Pak DS, Li Y, Kem WR (1994) A nicotinic agonist (GTS-21), eyeblink classical conditioning, and nicotinic receptor binding in rabbit brain. Brain Res 645:309–317
98. Bitner RS, Bunnelle WH, Decker MW, Drescher KU, Kohlhaas KL, Markosyan S et al (2010) In vivo pharmacological characterization of a novel selective alpha7 neuronal nicotinic acetylcholine receptor agonist ABT-107: preclinical considerations in Alzheimer's disease. J Pharmacol Exp Ther 334(3):875–886
99. Boess FG, De Vry J, Erb C, Flessner T, Hendrix M, Luithle J et al (2007) The novel alpha7 nicotinic acetylcholine receptor agonist N-[(3R)-1-azabicyclo[2.2.2]oct-3-yl]-7-[2-(methoxy) phenyl]-1-benzofuran-2-carboxamide improves working and recognition memory in rodents. J Pharmacol Exp Ther 321(2):716–725
100. Ren K, Thinschmidt J, Liu J, Ai L, Papke RL, King MA et al (2007) alpha7 Nicotinic receptor gene delivery into mouse hippocampal neurons leads to functional receptor expression, improved spatial memory-related performance, and tau hyperphosphorylation. Neuroscience 145(1):314–322
101. Takeuchi H, Yanagida T, Inden M, Takata K, Kitamura Y, Yamakawa K et al (2009) Nicotinic receptor stimulation protects nigral dopaminergic neurons in rotenone-induced Parkinson's disease models. J Neurosci Res 87(2):576–585
102. Brown KL, Comalli DM, Biasi MD, Woodruff-Pak DS (2010) Trace eyeblink conditioning is impaired in alpha7 but not in beta2 nicotinic acetylcholine receptor knockout mice. Front Behav Neurosci 4:166

103. Hurst RS, Hajos M, Raggenbass M, Wall TM, Higdon NR, Lawson JA et al (2005) A novel positive allosteric modulator of the alpha7 neuronal nicotinic acetylcholine receptor: in vitro and in vivo characterization. J Neurosci 25(17):4396–4405
104. Bencherif M, Lippiello PM, Lucas R, Marrero MB (2011) Alpha7 nicotinic receptors as novel therapeutic targets for inflammation-based diseases. Cell Mol Life Sci 68(6):931–949
105. Targowska-Duda KM, Wnorowski A, Budzynska B, Jozwiak K, Biala G, Arias HR (2016) The positive allosteric modulator of alpha7 nicotinic acetylcholine receptors, 3-furan-2-yl-N--p-tolyl-acrylamide, enhances memory processes and stimulates ERK1/2 phosphorylation in mice. Behav Brain Res 302:142–151
106. Pavlov VA, Ochani M, Yang LH, Gallowitsch-Puerta M, Ochani K, Lin X et al (2007) Selective alpha7-nicotinic acetylcholine receptor agonist GTS-21 improves survival in murine endotoxemia and severe sepsis. Crit Care Med 35(4):1139–1144
107. Lopez NE, Krzyzaniak MJ, Costantini TW, Putnam J, Hageny AM, Eliceiri B et al (2012) Vagal nerve stimulation decreases blood-brain barrier disruption after traumatic brain injury. J Trauma Acute Care Surg 72(6):1562–1566
108. Gurun MS, Parker R, Eisenach JC, Vincler M (2009) The effect of peripherally administered CDP-choline in an acute inflammatory pain model: the role of alpha7 nicotinic acetylcholine receptor. Anesth Analg 108(5):1680–1687
109. Knott V, de la Salle S, Smith D, Choueiry J, Impey D, Smith M et al (2015) Effects of acute CDP-choline treatment on resting state brain oscillations in healthy volunteers. Neurosci Lett 591:121–125
110. Uslu G, Savci V, Buyukuysal LR, Goktalay G (2014) CDP-choline attenuates scopolamine induced disruption of prepulse inhibition in rats: involvement of central nicotinic mechanism. Neurosci Lett 569:153–157
111. Lendvai B, Kassai F, Szajli A, Nemethy Z (2013) alpha7 nicotinic acetylcholine receptors and their role in cognition. Brain Res Bull 93:86–96
112. Wallace TL, Porter RH (2011) Targeting the nicotinic alpha7 acetylcholine receptor to enhance cognition in disease. Biochem Pharmacol 82(8):891–903
113. Umana IC, Daniele CA, McGehee DS (2013) Neuronal nicotinic receptors as analgesic targets: it's a winding road. Biochem Pharmacol 86(8):1208–1214
114. Akaike A, Takada-Takatori Y, Kume T, Izumi Y (2010) Mechanisms of neuroprotective effects of nicotine and acetylcholinesterase inhibitors: role of alpha4 and alpha7 receptors in neuroprotection. J Mol Neurosci 40(1–2):211–216
115. Del Barrio L, Martin-de-Saavedra MD, Romero A, Parada E, Egea J, Avila J et al (2011) Neurotoxicity induced by okadaic acid in the human neuroblastoma SH-SY5Y line can be differentially prevented by alpha7 and beta2* nicotinic stimulation. Toxicol Sci 123(1):193–205
116. Egea J, Rosa AO, Sobrado M, Gandia L, Lopez MG, Garcia AG (2007) Neuroprotection afforded by nicotine against oxygen and glucose deprivation in hippocampal slices is lost in alpha7 nicotinic receptor knockout mice. Neuroscience 145(3):866–872
117. Kaneko S, Maeda T, Kume T, Kochiyama H, Akaike A, Shimohama S et al (1997) Nicotine protects cultured cortical neurons against glutamate-induced cytotoxicity via alpha7-neuronal receptors and neuronal CNS receptors. Brain Res 765(1):135–140
118. Li Y, Papke RL, He YJ, Millard WJ, Meyer EM (1999) Characterization of the neuroprotective and toxic effects of alpha7 nicotinic receptor activation in PC12 cells. Brain Res 830(2): 218–225
119. Roncarati R, Scali C, Comery TA, Grauer SM, Aschmi S, Bothmann H et al (2009) Procognitive and neuroprotective activity of a novel alpha7 nicotinic acetylcholine receptor agonist for treatment of neurodegenerative and cognitive disorders. J Pharmacol Exp Ther 329(2):459–468
120. Shimohama S, Greenwald DL, Shafron DH, Akaike A, Maeda T, Kaneko S et al (1998) Nicotinic à7 receptors protect against glutamate neurotoxicity and neuronal ischemic damage. Brain Res 779:359–363

121. Smith DC, Modglin AA, Roosevelt RW, Neese SL, Jensen RA, Browning RA et al (2005) Electrical stimulation of the vagus nerve enhances cognitive and motor recovery following moderate fluid percussion injury in the rat. J Neurotrauma 22(12):1485–1502
122. Smith DC, Tan AA, Duke A, Neese SL, Clough RW, Browning RA et al (2006) Recovery of function after vagus nerve stimulation initiated 24 hours after fluid percussion brain injury. J Neurotrauma 23(10):1549–1560
123. Clough RW, Neese SL, Sherill LK, Tan AA, Duke A, Roosevelt RW et al (2007) Cortical edema in moderate fluid percussion brain injury is attenuated by vagus nerve stimulation. Neuroscience 147(2):286–293
124. Verbois SL, Scheff SW, Pauly JR (2003) Chronic nicotine treatment attenuates alpha 7 nicotinic receptor deficits following traumatic brain injury. Neuropharmacology 44(2):224–233
125. McLean SL, Idris NF, Grayson B, Gendle DF, Mackie C, Lesage AS et al (2012) PNU-120596, a positive allosteric modulator of alpha7 nicotinic acetylcholine receptors, reverses a sub-chronic phencyclidine-induced cognitive deficit in the attentional set-shifting task in female rats. J Psychopharmacol 26(9):1265–1270
126. Thomsen MS, El-Sayed M, Mikkelsen JD (2011) Differential immediate and sustained memory enhancing effects of alpha7 nicotinic receptor agonists and allosteric modulators in rats. PLoS One 6(11):e27014
127. Callahan PM, Hutchings EJ, Kille NJ, Chapman JM, Terry AV Jr (2013) Positive allosteric modulator of alpha 7 nicotinic-acetylcholine receptors, PNU-120596 augments the effects of donepezil on learning and memory in aged rodents and non-human primates. Neuropharmacology 67:201–212
128. Olincy A, Harris JG, Johnson LL, Pender V, Kongs S, Allensworth D et al (2006) Proof-of-concept trial of an alpha7 nicotinic agonist in schizophrenia. Arch Gen Psychiatry 63(6):630–638
129. Freitas K, Carroll FI, Damaj MI (2012) The antinociceptive effects of nicotinic receptors alpha7-positive allosteric modulators in murine acute and tonic pain models. J Pharmacol Exp Ther 344(1):264–275
130. Freitas K, Ghosh S, Ivy Carroll F, Lichtman AH, Imad DM (2013) Effects of alpha 7 positive allosteric modulators in murine inflammatory and chronic neuropathic pain models. Neuropharmacology 65:156–164
131. Munro G, Storm A, Hansen MK, Dyhr H, Marcher L, Erichsen HK et al (2012) The combined predictive capacity of rat models of algogen-induced and neuropathic hypersensitivity to clinically used analgesics varies with nociceptive endpoint and consideration of locomotor function. Pharmacol Biochem Behav 101(3):465–478
132. Borovikova LV, Ivanova S, Zhang M, Yang H, Botchkina GI, Watkins LR et al (2000) Vagus nerve stimulation attenuates the systemic inflammatory response to endotoxin. Nature 405(6785):458–462
133. Tracey KJ (2009) Reflex control of immunity. Nat Rev Immunol 9(6):418–428
134. Oke SL, Tracey KJ (2009) The inflammatory reflex and the role of complementary and alternative medical therapies. Ann N Y Acad Sci 1172:172–180
135. Vida G, Pena G, Deitch EA, Ulloa L (2011) alpha7-cholinergic receptor mediates vagal induction of splenic norepinephrine. J Immunol 186(7):4340–4346
136. Wang H, Yu M, Ochani M, Amella CA, Tanovic M, Susarla S et al (2003) Nicotinic acetylcholine receptor alpha7 subunit is an essential regulator of inflammation. Nature 421 (6921):384–388
137. Pruitt DT, Schmid AN, Kim LJ, Abe CM, Trieu JL, Choua C et al (2016) Vagus nerve stimulation delivered with motor training enhances recovery of function after traumatic brain injury. J Neurotrauma 33(9):871–879
138. Hays SA, Khodaparast N, Hulsey DR, Ruiz A, Sloan AM, Rennaker RL 2nd et al (2014) Vagus nerve stimulation during rehabilitative training improves functional recovery after intracerebral hemorrhage. Stroke 45(10):3097–3100

139. Khodaparast N, Hays SA, Sloan AM, Fayyaz T, Hulsey DR, Rennaker RL 2nd et al (2014) Vagus nerve stimulation delivered during motor rehabilitation improves recovery in a rat model of stroke. Neurorehabil Neural Repair 28(7):698–706
140. Bernik TR, Friedman SG, Ochani M, DiRaimo R, Ulloa L, Yang H et al (2002) Pharmacological stimulation of the cholinergic antiinflammatory pathway. J Exp Med 195(6):781–788
141. Wanderas MH, Moum BA, Hoivik ML, Hovde O (2016) Predictive factors for a severe clinical course in ulcerative colitis: results from population-based studies. World J Gastrointest Pharmacol Ther 7(2):235–241
142. Herrero MT, Estrada C, Maatouk L, Vyas S (2015) Inflammation in Parkinson's disease: role of glucocorticoids. Front Neuroanat 9:32
143. Kloting N, Bluher M (2014) Adipocyte dysfunction, inflammation and metabolic syndrome. Rev Endocr Metab Disord 15(4):277–287
144. Myhre O, Utkilen H, Duale N, Brunborg G, Hofer T (2013) Metal dyshomeostasis and inflammation in Alzheimer's and Parkinson's diseases: possible impact of environmental exposures. Oxid Med Cell Longev 2013:726954
145. Gomez-Mejiba SE, Zhai Z, Akram H, Pye QN, Hensley K, Kurien BT et al (2009) Inhalation of environmental stressors & chronic inflammation: autoimmunity and neurodegeneration. Mutat Res 674(1–2):62–72
146. Tansey MG, McCoy MK, Frank-Cannon TC (2007) Neuroinflammatory mechanisms in Parkinson's disease: potential environmental triggers, pathways, and targets for early therapeutic intervention. Exp Neurol 208(1):1–25
147. Tansey MG, Goldberg MS (2010) Neuroinflammation in Parkinson's disease: its role in neuronal death and implications for therapeutic intervention. Neurobiol Dis 37(3):510–518
148. Yue Y, Liu R, Cheng W, Hu Y, Li J, Pan X et al (2015) GTS-21 attenuates lipopolysaccharide-induced inflammatory cytokine production in vitro by modulating the Akt and NF-kappaB signaling pathway through the alpha7 nicotinic acetylcholine receptor. Int Immunopharmacol 29(2):504–512
149. Shytle RD, Mori T, Townsend K, Vendrame M, Sun N, Zeng J et al (2004) Cholinergic modulation of microglial activation by alpha 7 nicotinic receptors. J Neurochem 89(2):337–343
150. Teaktong T, Graham A, Court J, Perry R, Jaros E, Johnson M et al (2003) Alzheimer's disease is associated with a selective increase in alpha7 nicotinic acetylcholine receptor immunoreactivity in astrocytes. Glia 41(2):207–211
151. Shi FD, Piao WH, Kuo YP, Campagnolo DI, Vollmer TL, Lukas RJ (2009) Nicotinic attenuation of central nervous system inflammation and autoimmunity. J Immunol 182(3):1730–1739
152. Han Z, Shen F, He Y, Degos V, Camus M, Maze M et al (2014) Activation of alpha-7 nicotinic acetylcholine receptor reduces ischemic stroke injury through reduction of pro-inflammatory macrophages and oxidative stress. PLoS One 9(8):e105711
153. Shen JX, Yakel JL (2012) Functional alpha7 nicotinic ACh receptors on astrocytes in rat hippocampal CA1 slices. J Mol Neurosci 48(1):14–21
154. Wu H, Li L, Su X (2014) Vagus nerve through alpha7 nAChR modulates lung infection and inflammation: models, cells, and signals. BioMed Res Int 2014:283525
155. Munro G, Hansen R, Erichsen H, Timmermann D, Christensen J, Hansen H (2012) The alpha7 nicotinic ACh receptor agonist compound B and positive allosteric modulator PNU-120596 both alleviate inflammatory hyperalgesia and cytokine release in the rat. Br J Pharmacol 167(2):421–435
156. Goehler LE, Gaykema RP, Hansen MK, Anderson K, Maier SF, Watkins LR (2000) Vagal immune-to-brain communication: a visceral chemosensory pathway. Auton Neurosci 85(1–3):49–59
157. Szelenyi J, Vizi ES (2007) The catecholamine cytokine balance: interaction between the brain and the immune system. Ann N Y Acad Sci 1113:311–324
158. Pruitt D, Hays S, Schmid A, Choua C, Kim L, Trieu J et al (2014) Controlled-cortical impact reduces volitional forelimb strength in rats. Brain Res 1582:91–98

159. Mineur YS, Obayemi A, Wigestrand MB, Fote GM, Calarco CA, Li AM et al (2013) Cholinergic signaling in the hippocampus regulates social stress resilience and anxiety- and depression-like behavior. Proc Natl Acad Sci U S A 110(9):3573–3578
160. Shytle RD, Silver AA, Lukas RJ, Newman MB, Sheehan DV, Sanberg PR (2002) Nicotinic acetylcholine receptors as targets for antidepressants. Mol Psychiatry 7(6):525–535
161. Janowsky DS, el-Yousef MK, Davis JM (1974) Acetylcholine and depression. Psychosom Med 36(3):248–257
162. Saricicek A, Esterlis I, Maloney KH, Mineur YS, Ruf BM, Muralidharan A et al (2012) Persistent beta2*-nicotinic acetylcholinergic receptor dysfunction in major depressive disorder. Am J Psychiatry 169(8):851–859
163. Meisel C, Schwab JM, Prass K, Meisel A, Dirnagl U (2005) Central nervous system injury-induced immune deficiency syndrome. Nat Rev Neurosci 6(10):775–786
164. Kox M, Pompe JC, Pickkers P, Hoedemaekers CW, van Vugt AB, van der Hoeven JG (2008) Increased vagal tone accounts for the observed immune paralysis in patients with traumatic brain injury. Neurology 70(6):480–485
165. Engel O, Akyuz L, da Costa Goncalves AC, Winek K, Dames C, Thielke M et al (2015) Cholinergic pathway suppresses pulmonary innate immunity facilitating pneumonia after stroke. Stroke 46(11):3232–3240
166. Tracey KJ (2002) The inflammatory reflex. Nature 420(6917):853–859
167. Cabrera SM, Chavez CM, Corley SR, Kitto MR, Butt AE (2006) Selective lesions of the nucleus basalis magnocellularis impair cognitive flexibility. Behav Neurosci 120(2):298–306
168. Muir JL, Everitt BJ, Robbins TW (1994) AMPA-induced excitotoxic lesions of the basal forebrain: a significant role for the cortical cholinergic system in attentional function. J Neurosci 14(4):2313–2326
169. Muir JL, Page KJ, Sirinathsinghji DJ, Robbins TW, Everitt BJ (1993) Excitotoxic lesions of basal forebrain cholinergic neurons: effects on learning, memory and attention. Behav Brain Res 57(2):123–131
170. Demeter E, Hernandez-Garcia L, Sarter M, Lustig C (2011) Challenges to attention: a continuous arterial spin labeling (ASL) study of the effects of distraction on sustained attention. Neuroimage 54(2):1518–1529
171. Bartus RT, Dean IRL, Beer B, Lippa A (1982) The cholinergic hypothesis of geriatric memory dysfunction. Science 217:408–417
172. Petersen RC (1977) Scopolamine induced learning failures in man. Psychopharmacology (Berl) 52(3):283–289
173. Ghoneim MM, Mewaldt SP (1977) Studies on human memory: the interactions of diazepam, scopolamine, and physostigmine. Psychopharmacology (Berl) 52(1):1–6
174. Sunderland T, Tariot PN, Weingartner H, Murphy DL, Newhouse PA, Mueller EA et al (1986) Pharmacologic modelling of Alzheimer's disease. Prog Neuropsychopharmacol Biol Psychiatry 10(3–5):599–610
175. Newhouse PA, Potter A, Corwin J, Lenox R (1992) Acute nicotinic blockade produces cognitive impairment in normal humans. Psychopharmacology (Berl) 108(4):480–484
176. Zurkovsky L, Taylor WD, Newhouse PA (2013) Cognition as a therapeutic target in late-life depression: potential for nicotinic therapeutics. Biochem Pharmacol 86(8):1133–1144
177. Levin ED, McClernon FJ, Rezvani AH (2006) Nicotinic effects on cognitive function: behavioral characterization, pharmacological specification, and anatomic localization. Psychopharmacology (Berl) 184(3–4):523–539
178. Arias HR, Ravazzini F, Targowska-Duda KM, Kaczor AA, Feuerbach D, Boffi JC et al (2016) Positive allosteric modulators of alpha7 nicotinic acetylcholine receptors affect neither the function of other ligand- and voltage-gated ion channels and acetylcholinesterase, nor beta-amyloid content. Int J Biochem Cell Biol 76:19–30
179. Andersen ND, Nielsen BE, Corradi J, Tolosa MF, Feuerbach D, Arias HR et al (2016) Exploring the positive allosteric modulation of human alpha7 nicotinic receptors from a single-channel perspective. Neuropharmacology 107:189–200

180. Stevens KE, Zheng L, Floyd KL, Stitzel JA (2015) Maximizing the effect of an alpha7 nicotinic receptor PAM in a mouse model of schizophrenia-like sensory inhibition deficits. Brain Res 1611:8–17
181. Nikiforuk A, Kos T, Potasiewicz A, Popik P (2015) Positive allosteric modulation of alpha 7 nicotinic acetylcholine receptors enhances recognition memory and cognitive flexibility in rats. Eur Neuropsychopharmacol 25(8):1300–1313
182. Dani JA, Bertrand D (2007) Nicotinic acetylcholine receptors and nicotinic cholinergic mechanisms of the central nervous system. Annu Rev Pharmacol Toxicol 47:699–729
183. McGehee DS, Role LW (1995) Physiological diversity of nicotinic acetylcholine receptors expressed by vertebrate neurons. Annu Rev Physiol 57:521–546
184. Breese CR, Adams C, Logel J, Drebing C, Rollins Y, Barnhart M et al (1997) Comparison of the regional expression of nicotinic acetylcholine receptor alpha7 mRNA and [125I]-alpha-bungarotoxin binding in human postmortem brain. J Comp Neurol 387(3):385–398
185. Whiteaker P, Davies AR, Marks MJ, Blagbrough IS, Potter BV, Wolstenholme AJ et al (1999) An autoradiographic study of the distribution of binding sites for the novel alpha7-selective nicotinic radioligand [3H]-methyllycaconitine in the mouse brain. Eur J Neurosci 11(8):2689–2696
186. Clarke PB (1993) Nicotinic receptors in mammalian brain: localization and relation to cholinergic innervation. Prog Brain Res 98:77–83
187. Hao J, Simard AR, Turner GH, Wu J, Whiteaker P, Lukas RJ et al (2010) Attenuation of CNS inflammatory responses by nicotine involves alpha7 and non-alpha7 nicotinic receptors. Exp Neurol 227(1):110–119
188. Koval L, Lykhmus O, Zhmak M, Khruschov A, Tsetlin V, Magrini E et al (2011) Differential involvement of alpha4beta2, alpha7 and alpha9alpha10 nicotinic acetylcholine receptors in B lymphocyte activation in vitro. Int J Biochem Cell Biol 43(4):516–524
189. Kawashima K, Fujii T (2004) Expression of non-neuronal acetylcholine in lymphocytes and its contribution to the regulation of immune function. Front Biosci 9:2063–2085
190. De Rosa MJ, Dionisio L, Agriello E, Bouzat C, Esandi MC (2009) Alpha 7 nicotinic acetylcholine receptor modulates lymphocyte activation. Life Sci 85(11–12):444–449
191. Sharma G, Vijayaraghavan S (2001) Nicotinic cholinergic signaling in hippocampal astrocytes involves calcium-induced calcium release from intracellular stores. Proc Natl Acad Sci U S A 98(7):4148–4153
192. Shin SS, Dixon CE (2015) Alterations in cholinergic pathways and therapeutic strategies targeting cholinergic system after traumatic brain injury. J Neurotrauma 32(19):1429–1440
193. Echeverria V, Yarkov A, Aliev G (2016) Positive modulators of the alpha7 nicotinic receptor against neuroinflammation and cognitive impairment in Alzheimer's disease. Prog Neurobiol
194. Kalkman HO, Feuerbach D (2016) Modulatory effects of alpha7 nAChRs on the immune system and its relevance for CNS disorders. Cell Mol Life Sci 73(13):2511–2530
195. Baez-Pagan CA, Delgado-Velez M, Lasalde-Dominicci JA (2015) Activation of the macrophage alpha7 nicotinic acetylcholine receptor and control of inflammation. J Neuroimmune Pharmacol 10(3):468–476
196. Quik M, Zhang D, McGregor M, Bordia T (2015) Alpha7 nicotinic receptors as therapeutic targets for Parkinson's disease. Biochem Pharmacol 97(4):399–407
197. Shaw S, Bencherif M, Marrero MB (2002) Janus kinase 2, an early target of alpha 7 nicotinic acetylcholine receptor-mediated neuroprotection against Abeta-(1-42) amyloid. J Biol Chem 277(47):44920–44924
198. Thomsen MS, Hansen HH, Timmerman DB, Mikkelsen JD (2010) Cognitive improvement by activation of alpha7 nicotinic acetylcholine receptors: from animal models to human pathophysiology. Curr Pharm Des 16(3):323–343
199. Bok J, Wang Q, Huang J, Green SH (2007) CaMKII and CaMKIV mediate distinct prosurvival signaling pathways in response to depolarization in neurons. Mol Cell Neurosci 36(1):13–26
200. de Jonge WJ, van der Zanden EP, The FO, Bijlsma MF, van Westerloo DJ, Bennink RJ et al (2005) Stimulation of the vagus nerve attenuates macrophage activation by activating the Jak2-STAT3 signaling pathway. Nat Immunol 6(8):844–851

201. Kihara T, Shimohama S, Sawada H, Honda K, Nakamizo T, Shibasaki H et al (2001) alpha 7 nicotinic receptor transduces signals to phosphatidylinositol 3-kinase to block A beta-amyloid-induced neurotoxicity. J Biol Chem 276(17):13541–13546
202. Toulorge D, Guerreiro S, Hild A, Maskos U, Hirsch EC, Michel PP (2011) Neuroprotection of midbrain dopamine neurons by nicotine is gated by cytoplasmic Ca2+. FASEB J 25(8): 2563–2573
203. Ren K, Puig V, Papke RL, Itoh Y, Hughes JA, Meyer EM (2005) Multiple calcium channels and kinases mediate alpha7 nicotinic receptor neuroprotection in PC12 cells. J Neurochem 94(4):926–933
204. Toborek M, Son KW, Pudelko A, King-Pospisil K, Wylegala E, Malecki A (2007) ERK 1/2 signaling pathway is involved in nicotine-mediated neuroprotection in spinal cord neurons. J Cell Biochem 100(2):279–292
205. Gubbins EJ, Gopalakrishnan M, Li J (2010) Alpha7 nAChR-mediated activation of MAP kinase pathways in PC12 cells. Brain Res 1328:1–11
206. Ashpole NM, Chawla AR, Martin MP, Brustovetsky T, Brustovetsky N, Hudmon A (2013) Loss of calcium/calmodulin-dependent protein kinase II activity in cortical astrocytes decreases glutamate uptake and induces neurotoxic release of ATP. J Biol Chem 288(20):14599–14611
207. Coultrap SJ, Vest RS, Ashpole NM, Hudmon A, Bayer KU (2011) CaMKII in cerebral ischemia. Acta Pharmacol Sin 32(7):861–872
208. Sawe N, Steinberg G, Zhao H (2008) Dual roles of the MAPK/ERK1/2 cell signaling pathway after stroke. J Neurosci Res 86(8):1659–1669
209. Liu Y, Hu J, Wu J, Zhu C, Hui Y, Han Y, et al. (2012) alpha7 nicotinic acetylcholine receptor-mediated neuroprotection against dopaminergic neuron loss in an MPTP mouse model via inhibition of astrocyte activation. J Neuroinflammation 9:98
210. White SH, Brisson CD, Andrew RD (2012) Examining protection from anoxic depolarization by the drugs dibucaine and carbetapentane using whole cell recording from CA1 neurons. J Neurophysiol 107(8):2083–2095
211. Uteshev VV, Patlak JB, Pennefather PS (2000) Analysis and implications of equivalent uniform approximations of nonuniform unitary synaptic systems. Biophys J 79(6):2825–2839
212. Gallowitsch-Puerta M, Pavlov VA (2007) Neuro-immune interactions via the cholinergic anti-inflammatory pathway. Life Sci 80(24–25):2325–2329
213. Araque A, Martin ED, Perea G, Arellano JI, Buno W (2002) Synaptically released acetylcholine evokes Ca2+ elevations in astrocytes in hippocampal slices. J Neurosci 22(7): 2443–2450
214. Skok MV (2009) Editorial: to channel or not to channel? Functioning of nicotinic acetylcholine receptors in leukocytes. J Leukoc Biol 86(1):1–3

Chapter 18
Remote Ischemic Conditioning: A Highly Translatable Therapy for Acute Stroke

Chizoba J. Ezepue and David C. Hess

Abstract Ischemic preconditioning is the most potent protectant against ischemic injury. Remote ischemic conditioning (RIC), a further development of ischemic preconditioning, is safe, feasible, and highly translational and represents one of the most promising strategies for neuroprotection and vasculoprotection during acute ischemic stroke. RIC grew out of the field of cardioprotection but there is now strong preclinical evidence and emerging clinical evidence of its activity and efficacy in acute ischemic stroke. While the mechanisms of protection are not completely known, RIC triggers endogenous protective mechanism and increases cerebral blood flow. RIC is ideal as a prehospital therapy and has potential as an adjunctive therapy to mechanical thrombectomy. Its ease of use makes RIC an attractive therapy to be used in community hospitals and during helicopter transport.

Keywords Remote ischemic conditioning • Preconditioning • Acute ischemic stroke • Remote ischemic perconditioning

1 Introduction

Every year, approximately 795,000 people in the U.S experience new or recurrent stroke, resulting in 130,000 deaths [1]. Globally, it is the third cause of death worldwide, and a main cause of chronic, severe adult disability [2]. Roughly 87 % of all strokes are ischemic strokes, when blood flow to the brain is blocked [1]. Efforts to develop therapies to counter the damaging effects of cerebral ischemia have been limited, with tissue plasminogen activator (tPA) being the only Food and Drug Administration (FDA)-approved therapy for select patients with acute ischemic stroke as long as 4.5 h after onset [3, 4].

C.J. Ezepue, M.D. • D.C. Hess, M.D. (✉)
Department of Neurology, Medical College of Georgia, Augusta University,
1120 15th Street, Augusta, GA 30912, USA
e-mail: cezepue@augusta.edu; dhess@augusta.edu

P.A. Lapchak, J.H. Zhang (eds.), *Neuroprotective Therapy for Stroke and Ischemic Disease*, Springer Series in Translational Stroke Research,
DOI 10.1007/978-3-319-45345-3_18

Unfortunately, because the timing of stroke is usually unpredictable, less than 5 % of patients are treated with tPA, and most stroke victims receive only supportive care [4]. Endovascular approaches with mechanical thrombectomy (MT) are very effective therapies but are applied to less than 5 % of stroke patients and even with MT more than 50 % of patients are still disabled [5–8]. As more patients undergo reperfusion with tPA and MT, there is growing need to combine reperfusion strategies with neuroprotective agents.

While many attempts have been made to translate neuroprotection from the bench to the bedside, none of these attempts have been successful [9]. One of the reasons so many neuroprotective agents have failed in clinical trial is that they target one pathway or one receptor (e.g., oxygen free radicals, NMDA). Targeting multiple pathways may be a better approach [10]. The brain has a remarkable capacity for self-preservation induced by stress and injurious stimuli, such as ischemia [4, 11]. The brain and vasculature is able to self-protect and turn on endogenous protective pathways producing an "ischemic tolerant" phenotype. A promising strategy is to harness these endogenous self-protective pathways and mechanisms. The most effective way to turn on these pathways is "preconditioning" of the brain.

2 History of Preconditioning

Preconditioning is the phenomenon that small doses of an injurious agent protect against subsequent larger injurious or lethal doses. The term "preconditioning" and the term "tolerance" are often interchanged. Janoff first introduced the terms "preconditioning" and "tolerance" to refer to the stabilization of lysosomes observed in rodents "preconditioned" with minor doses of traumatic injury prior to a lethal rotational traumatic injury in a Noble Collip drum [12]. The concept of tolerance or preconditioning has been known since Ancient times. Mithridates VI, King of Pontus (c 132–63 B.C.), a rival and enemy of the Romans, protected himself against poisoning by daily ingesting small doses of toxins and poisons [13–15]. The preconditioning was so effective that when cornered by the Romans, Mithridates was unable to commit suicide by overdose and had to be killed by the sword. His universal antidote was known as "Mithridatuum" and "Mithridiatism" refers to tolerance to a poison or toxin by gradually taking larger doses of the toxin. In the sixteenth century, the Swiss physician-astrologer, Paracelsus, often called the father of toxicology, is credited with formulating the concept that the "dose makes the poison" [16]. Preconditioning is immortalized in literature and philosophy by Nietzsche—"what doesn't kill you makes you stronger" [17]. Exercise training is also a form of preconditioning. There are many forms of preconditioning agents and stimuli, the most notable being hypoxia and ischemia.

3 History of the Concept of Ischemic Preconditioning and Remote Ischemic Conditioning

One of the most potent, innate protective mechanisms against cerebral ischemic injury is ischemic preconditioning [18–20]. Ischemic preconditioning was first described over 30 years ago by Murry and colleagues [21], who demonstrated that in a dog, multiple brief episodes of ischemia (4 cycles of 5 min occlusion followed by reperfusion) of the left anterior descending artery reduced the size of the infarction induced by subsequent prolonged occlusion of that vessel. Li and colleagues were the first to document the efficacy of ischemic preconditioning in rat hearts in 1992 [22]. They demonstrated that brief episodes of preconditioning ischemia reduced the size of a myocardial infarction in a rat model of low coronary collateral blood flow [22]. However, the beneficial effects of preconditioning appeared to have been very short-lived in this model [22].

Przyklenk and colleagues [23] observed that brief episodes of ischemia–reperfusion in a different region of the same organ provide protection to the nearby ischemic heart, the first demonstration of "regional" preconditioning. Using a canine model, they observed that repetitive brief occlusion of the circumflex artery protected against subsequent left anterior descending artery occlusion during 60 min of occlusion [23]. The benefits of IC were not limited to those myocytes that were subjected to brief ischemia; rather, brief ischemia in one vascular bed protected remote, virgin myocardium from subsequent sustained coronary artery occlusion. This implied a role for a humoral blood-borne substance that is activated, released, and transported within the heart.

In 1997, remote preconditioning was extended to limb ischemia by Birnbaum and colleagues [24] who demonstrated that rabbit hindlimb ischemia, produced by combined brief cycles of blood flow restriction with electrical stimulation of the hindlimb, protected the heart against later ischemia [11]. That same year, a minimally invasive approach of inducing hindlimb ischemia as a remote preconditioning stimulus was introduced by Oxman and colleagues [25] who demonstrated that applying a thin-elastic tourniquet to the hindlimb to induce 10 min of limb ischemia, significantly decreased reperfusion arrhythmias in a rat heart following a sustained ischemic insult. Five years later, Kharbanda and colleagues [26] were the first to demonstrate that brief and clinically practical period of limb ischemia induced remote preconditioning of human arterial vessels in vivo. In 2007, Schmidt and colleagues [27] demonstrated in a pig model that 4 cycles of 5 min of ischemic conditioning applied with a tourniquet to the hindlimb, reduced infarct size, preserved global systolic and diastolic functions, and protected against malignant arrhythmias after 40 min of left anterior descending occlusion. While prior studies had introduced the ischemic stimulus prior to the index ischemic event, Schmidt and colleagues performed the remote conditioning during the ischemia so-called perconditioning. Together, these preclinical studies have paved the way for a number of randomized clinical trials of remote limb preconditioning to protect the heart.

One of the advantages of RIC is that it can be performed in a nonvital organ, avoiding the high risk involved with inducing ischemia in vital organs such as the brain or the heart [28–30]. It can also be applied before the ischemic insult in the target site (remote ischemic preconditioning RIPreC), during the ischemic insult (remote ischemic perconditioning RIPerC) or at the onset of reperfusion (remote ischemic postconditioning RIPostC) [9, 31–33]. All three remote ischemic conditioning (RIC) strategies have similar therapeutic potential and trigger innate mechanisms to protect against brain injury after ischemic stroke [34–36]. However, due to the unpredictability of the timing of the ischemic event, RIPreC has limited clinical application [11, 36]. Recently, a number of studies have demonstrated that RIPerC and RIPostC offer a more feasible approach for the treatment of acute ischemic stroke in the clinical settings as they can be applied during or after the ischemic period [18, 35–38].

4 Mechanisms of RIC

Although the mechanism through which an episode of brief ischemia and reperfusion in an organ or tissue exerts protection against a subsequent sustained insult of ischemia–reperfusion injury in a remote organ or tissue remains to be fully elucidated, several hypothesis have been proposed [39]. Three theories mostly from work in the heart have been advanced [39, 40]: (1) neurogenic transmission with involvement of muscle afferents and the autonomic nervous system, (2) the involvement of humoral or blood-borne factors in the mechanism of RIC, and (3) systemic response which suppresses inflammation and apoptosis. The protective pathways converge on the mitochondria (see Fig. 18.1).

4.1 Neural Mechanism

The neural hypothesis proposes that preconditioning of the organ or tissue remote from the heart generates an endogenous substance such as adenosine, bradykinin, or calcitonin gene-related peptide (CGRP), which then activates a local afferent neural pathway stimulating an efferent neural pathway terminating at the heart and mediating cardioprotection [41, 42]. Gho and colleagues [43] showed the reduction in myocardial infarct size induced by brief ischemia and reperfusion of the anterior mesenteric artery (MAO) could be reversed in the presence of hexamethonium, an antagonist of nicotinic acetylcholine receptors and an autonomic ganglia blocker. The efferent pathway involves the vagal nerve as RI PreC-mediated cardioprotection is dependent upon activation of vagal preganglionic fibers [44]. Moreover, application of capsaicin, a stimulator of type C afferent sensory fibers mimics the cardioprotective effect of RIC [45]. Similarly, direct electrical stimulation of peripheral nerves is cardioprotective, likely via release of humoral factors [46].

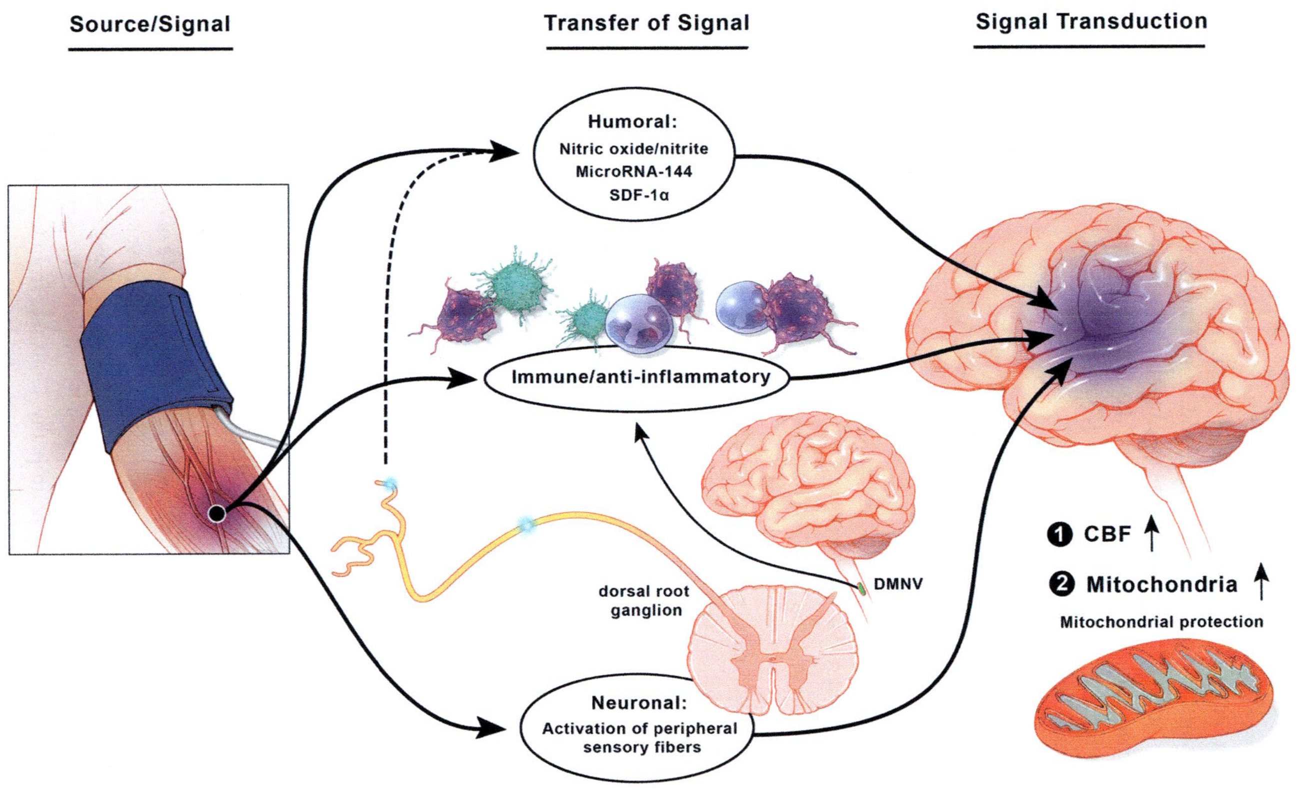

Fig. 18.1 Depiction of mechanisms of RIC from signal generation in the limb to signal transmission to the effects on the target organ—the brain and the cerebrovasculature

4.2 Humoral Mechanism

The transmission of the signal from the peripheral organ (e.g., limb) to the remote target organ (brain) is likely via humoral blood-borne factors. The best evidence for this is in transfer experiments where blood from a human or animal subjected to limb RIC protects the heart in an ex vivo Langendorff cardiac model [47]. Although the exact mediators remain unknown, the best evidence to date exists for stromal derived factor 1 a (SDF-1) [48], micro RNA 144 [49], interleukin 10 [50], and plasma nitrite [51].

4.3 Nitric Oxide

In RIC, the nitric oxide (NO) system appears to play a role in the mechanism of protection [52]. In a mouse liver ischemia–reperfusion model, Abu-Mara and colleagues [53] demonstrated that NO plays an essential role in reducing liver damage during the early phase of IR injury. Mice treated with RIC were protected from liver ischemic damage and have increased microvascular blood flow [53]. This protection was abolished with C-PTIO (2-(4-carboxyphenyl)-4,4,5,5-tetra methyl imidazoline-1-oxyl-3-oxide), a direct NO scavenger [53]. The mice treated with RIC had elevated NO levels in the blood. Moreover, the neuroprotective effect of preconditioning was lost in endothelial NO synthase knockout mice [54]. This protection involved preservation of the sinusoidal structure and maintenance of blood flow through the hepatic microcirculation [53]. In addition, NO and nitrite are involved in a cell signaling pathway mediating RIC-induced cardioprotection [52]. Rassaf et al. [51] reported that in mice subjected to 4 cycles of no-flow ischemia with subsequent reactive hyperemia, RIC protected hearts and increased nitrite levels in the plasma and hearts in mice. Similarly, nitrite levels increased in the plasma of healthy human volunteers during RIC of the arm [51]. RIPerC increases relative cerebral blood flow (CBF) after thromboembolic stroke [56] much in the same manner as highlighted earlier with the hepatic microvasculature.

4.4 Combined Neural and Humoral Mechanism

Both the neural and humoral hypotheses are combined into a revised hypothesis that a humoral factor is induced and released by peripheral nerves [11]. This was demonstrated in a study of patients with diabetes mellitus by Jensen and colleagues [55]. In this study, RIC was administered to patients with diabetic peripheral neuropathy and patients with diabetes but no neuropathy. Dialysate from their blood was subsequently transferred to isolated hearts. Dialysate from patients with neuropathy offered less cardioprotection than did dialysate from patients without neuropathy, suggesting that peripheral nerves are required for the release of a humoral factor.

4.5 *Systemic Anti-inflammatory Response*

Finally, there is evidence to suggest that RIC may be part of a protective systemic response which affects immune cells and inflammatory response [39, 56]. Kostantinov and colleagues using a microarray method demonstrated that a simple model of brief forearm ischemia suppresses proinflammatory gene expression in circulating leukocytes within 15 min of RIC and at 24 h after conditioning. At 24 h, this change was accompanied by a significant reduction in the expression of leukocyte CD11b [56]. In a rat myocardial infarction model, chronic repeated daily RIC reduced the number of neutrophils and macrophages in the heart, and the expression of monocyte chemotactic protein 1, indicating a long-term reduction of inflammation [57].

5 Remote Ischemic Conditioning in Acute Myocardial Infarction

Acute myocardial infarction shares many common mechanisms and treatment approaches with acute ischemic stroke. Treatment of acute ischemic stroke has followed many of the treatment strategies used in acute myocardial infarction (AMI) but has lagged by at least a decade. Thrombolytic agents and mechanical reperfusion strategies are both successfully used in AMI with percutaneous coronary intervention (PCI) analogous to mechanical thrombectomy in acute ischemic stroke.

The search for adjunctive treatments in AMI has been disappointing. Cardioprotection has been as vexing a problem as neuroprotection with many failed clinical trials [58, 59]. However, RIC seems to be one of the most promising cardioprotectants and adjuncts to PCI. In a landmark randomized trial of ST elevation MI (STEMI) in the prehospital setting in Denmark (CONDI 1) RIC (4 cycles of 5 min) in the ambulance before PCI improved myocardial salvage [60]. Similarly, other randomized trials have shown that RIC started in the hospital before PCI reduced infarct size, improved myocardial salvage, and reduced myocardial edema [61] or reduced infarct size [62]. In all these trials, RIC was safe and well tolerated. Interesting in the CONDI trial, one RIC treatment used as an adjunct to PCI was associated with reduction in long-term major adverse cardiovascular and cerebrovascular events (MACCE) [63]. A similar reduction in long-term MACCE was observed in patients receiving RIC as an adjunct to elective coronary stenting [64]. A larger trial of RIC as an adjunct to PCI is ongoing in Europe, the CONDI2 trial, with a sample size of 4300 subjects with the primary endpoint being cardiac mortality [65] (Clinicaltrial.gov NCT01857414). There is also evidence that RIC is effective in combination with thrombolysis in AMI. In the ERIC LYSIS trial on the island of Mauritius where PCI is not widely available, RIC in combination with tPA reduced myocardial infarct size as measured by troponin and CPK [66]. In summary, RIC is safe, feasible, well tolerated, and has "activity" when used as an adjunct to both PCI and thrombolysis in AMI.

6 Review of Preclinical Studies of RIC in Stroke

The finding that brief periods of tissue ischemia can induce protection in other tissues has generated a number of preclinical studies which provide strong evidence that remote ischemic conditioning is an effective neuroprotectant with implications for acute stroke care [40]. We and others have shown that RIC is a powerful "neuroprotectant" and "vasculoprotectant" in a number of preclinical animal models [34, 67–70]. We have tested RIC in an embolic clot ischemic stroke model in the mouse and have demonstrated that RIC is effective alone and in combination with tissue plasminogen activator (tPA) in both young, males, and ovariectomized females and aged animals [34, 69, 71]. RIC reduces infarct size, improves functional outcome, improves cerebral blood flow, and reduces tPA-related ICH. Other investigators have shown RIC to be effective in rat mechanical occlusion models [67] and in Table 18.1.

There are some key questions that will be important to address for clinical translation. Does repeated RIC enhance the neuroprotective effect? Since RIPreC, RiPerC, and RIPostC are all effective in rodent cerebral ischemic models (Table 18.1) combining RIPostC with RIPerC might lead to an enhanced effect. In a rat TMCAo model, RIPostC for 14 days added to RiPerC led to reductions in infarct size and improved outcome compared to RiPerC alone [36]. Can RIC be combined with other interventions? Using a preclinical 2×2 factorial design using a mouse TE model in which minocycline was administered at 1 h after stroke onset and RIPerC administered at 2 h after onset in mice aged 12 months, the combination had an additive effect in animals that were treated or not treated with rtPA at 4 h after onset [69]. RIPerC increased cerebral blood flow in both animals treated with IV tPA and those not treated with tPA [11].

7 RIC and Cerebral Blood Flow

One of the key mechanisms of neuroprotection with ischemic preconditioning is an increase in CBF. In a mouse MCAo model, brief episodes of ischemia (15 min) protected against a later 45 min period of ischemia and increased CBF by laser Doppler and MRI arterial spin labeling (ASL) [72]. Ischemic preconditioning also increased CBF and perfusion when delivered 24 h before permanent MCA occlusion [73] RIC increased relative CBF as measured by laser contrast speckle imaging after thromboembolic stroke in mice [34]. This CBF increase occurred in young healthy males [34], older males [69], and ovariectomized females [68]. Chronic RIC increased CBF by SPECT patients with intracranial atherosclerosis [74]. Moreover, in a bilateral carotid artery stenosis model in mice, daily RIC for 2 weeks increased CBF and this increase persisted [70]. What is the mechanism of this increased CBF? Studies are ongoing in our laboratory to address whether RIC is increasing angiogenesis and/or arteriogenesis (collateral blood flow).

Table 18.1 Preclinical studies of remote ischemic conditioning in animal models of acute ischemic stroke

Reference	Stroke model	Timing of conditioning	Limb ischemia protocol	Infarct size reduction	Behavioral testing	Other	Quality
[30]	Bilateral CCA occlusion (30 min) with distal MCA occlusion Male SD rat	PreC 48, 12 h, or immediately before ischemia	Femoral occlusion: 2–3 cycles of 5 or 15 min	Immediate PreC: infarct size reduced from 48 % of cortex to 9 % with 3 cycles/15 min; to 25 % with 2 cycles of 15 min and no reduction with 2 cycles of 5 min	Not done	PreC at 48 h: infarct reduced to 23 % with 3 cycle/15 min; none with 2 cycles 15 min; PreC at 12 reduced to 25 % and none with 2 cycles/15 min with 3 cycles 15 min	Randomized
[81]	Bilateral CCA occlusion with distal MCA occlusion Male SD rat	PostC at reperfusion, 3, 6 h postreperfusion	Femoral occlusion: 3 cycles of 15 min	Infarct size at 2 days reduced by 67 % at reperfusion, 43 % at 3 h, not at 6 h; no reduction in infarct size at 2 months	Vibrissae-forearm placement improved out to 60 days	Blocked by capsaicin and cycloheximide	Randomized, blinded
[82]	Bilateral CCA occlusion with distal MCA occlusion Male SD rat	PreC before onset	Femoral occlusion: 3 cycles of 15 min	Infarct size reduced at 2 days and 2 months (40 %)	Improvement in three behavioral tests	Blocked by capsaicin and hexamethonium	Randomized, blinded
[83]	MCA suture occlusion (120 min) Male Wistar rat	PreC 24, 48, or 72 h before ischemia	Infrarenal aortic occlusion (bilateral leg ischemia) 3 cycles/10 min	Infarct size (24 h) reduced with PreC 24 h prior; no effect with PreC 48 or 72 h prior	Improvement in graded neurological exam scores with PreC 24 h prior	Blocked by hexamethonium	Randomized, blinded
[84]	MCA suture occlusion (90 min) Male SD rat	PerC at time of MCA occlusion	Bilateral femoral occlusion: 3 cycles of 10 min	Infarct size at 24 h reduced by 66 %	Not done	PerC reduced cerebral edema and blood brain barrier permeability	Randomized, blinded
[18]	MCA suture occlusion (120 min) Male SD rat	PreC (before occlusion) and PerC (prior to reperfusion)	Femoral occlusion with tourniquet: 4 cycles of 5 min	Infarct size reduced; PerC superior to PreC in infarct size reduction	Not done		Randomized

(continued)

Table 18.1 (continued)

Reference	Stroke model	Timing of conditioning	Limb ischemia protocol	Infarct size reduction	Behavioral testing	Other	Quality
[85]	MCA suture occlusion (90 min)	PostC at reperfusion, 3, or 6 h	Femoral occlusion for 15 s, 2/3 cycles of 15 min; 3 cycles of 5 min	Infarct size at 72 h reduced most with PostC at 6 h, 3 cycles of 5 min	Garcia scores improved at 24, 48, and 72 h	Effect abolished by mitochondrial KATP channel blocker	Randomized, blinded
	Male SD rats						
[86]	MCA suture occlusion (120 min)	PreC 1 h before	Femoral tourniquet: 3 cycles of 5 min	Infarct size by TTC and MRI-DWI at 24 h reduced with PreC	Neurological deficit scores improved at 24 h with PreC	PreC effect abolished with selective adenosine A1 receptor antagonist	Randomized, blinded
	Male SD rats						
[34]	Thromboembolic clot with/without tPA at 4 h	PerC at 2 h post occlusion	Femoral occlusion: 5 cycles of 5 min	Infarct size reduced by 25 % with PerC at 2 h; additive effect with tPA at 4 h (50 %)	Neuro scores improved by PerC and with tPA + RI PerC	Increase in relative CBF with PerC	Randomized, blinded, sample size estimation
	Male C57Bl mouse						
[71]	TE clot model with/without tPA at 4 h	PerC at 2 h	BP cuff on hindlimb	Infarct size reduced in both tPA and non tPA treated mice	Neuro scores improved	Increase in CBF by LCSI	Randomized, blinded, sample size estimation
	Female C57/Bl OVX females						
[87]	MCA suture occlusion 2 h	PostC at reperfusion	Bilateral femoral occlusion	Infarct size reduced	Functional outcome improved	Reduced endoplasmic reticulum, stress response and apoptosis	Randomized
	Male rats		3 cycles of 10 min occlusions				
[88]	MCA occlusion 1 h	PreC	Tourniquet on hindlimb	Reduced damage of CA1 in hippocampus	Improved memory on Morris water maze		Randomized, blinded
	Male SD rats		3 cycles of 5 min occlusion				

[69]	TE clot model with/ without tPA with/ without minocycline 12 month old male mice (C57/Bl)	PerC at 2 h	BP cuff on hindlimb	Infarct size and functional outcome improved alone and additive effect on infarct size with minocycline	Neuro scores and adhesive test improved	Increase in CBF by LCSI	Randomized, blinded, sample size estimation
[36]	MCA suture occlusion 90 min in SD rats	PerC at onset of occlusion alone or combined with 14 days daily postC	Cuff on hindlimb 10 min × 3	Reduced infarct size at 7 days with per and per/post; reduced infarct size at 14 days with per/post	Motor function improved	Increase in neuroglobin with postC	Randomized, blinded

PreC preconditioning, *PostC* postconditioning, *PerC* preconditioning, *SD* Sprague Dawley. In "quality" column, studies reviewed for reporting of three measures of study quality: randomization, blinding of endpoints, and whether sample size analysis was performed for a hypothesized effect size

8 Clinical Trials of RIC in Acute Ischemic Stroke

The first translation of RIC in acute ischemic stroke was the Danish prehospital trial. Hougaard and colleagues [9, 75] designed a randomized clinical trial investigating the impact of remote limb conditioning (RIPerC) in acute stroke patients in a prehospital setting in Denmark. The primary endpoint was penumbral salvage defined as perfusion–diffusion mismatch not progressing to infarction after 1 month and secondary endpoints were the final infarct size, infarct growth, and clinical outcomes at 3 months. Patients were randomized in a single-blind fashion to 4 cycles of 5 min of occlusion with a blood pressure cuff on the arm or no intervention. RIPerC was induced during transportation in the ambulance with four inflations of a standard upper limb blood pressure cuff to 200, or 25 mmHg above the patient's systolic blood pressure if their blood pressure was >175 mmHg. If the transportation time was too short for 4 cycles of inflation and deflation, the procedure was discontinued on arrival to the stroke unit. All eligible patients underwent baseline brain magnetic resonance imaging (MRI) and received intravenous tissue plasminogen activator within 4.5 h if eligible with repeat MRI performed at 24 h. Of the 443 randomized patients, 247 received RIPerC and 196 received standard treatment. Baseline diffusion-weighted imaging and perfusion-weighted imaging did not differ between the intervention and control group. Transient ischemic attack was more frequent in patients who received RIPerC than in patients who did not (42 of 247 versus 16 of 196, $P=0.006$), and patients who received RIPerC had lower baseline NIHSS scores than patients who did not (median 4 versus 5, $P=0.016$). MRI data showed no significant effect of RIPerC between the intervention and control group on penumbral salvage, final infarct size, or infarct growth. However, after adjustment for baseline perfusion and diffusion lesion severity, voxel-wise logistical analysis showed that RIPerC reduced tissue risk of infarction ($P=0.0003$). There were no effects on functional outcome between the intervention and control groups. RIPerC was safe and well tolerated.

This first prehospital stroke trial of RIPerC is a landmark trial and there are some important learning points. First, the trial enrolled a mild stroke population. No prehospital stroke severity scale was used so the baseline mean hospital NIHSS [5] was much lower than FAST-MAG a prehospital trial of magnesium where the Los Angeles Motor Scale was used where the baseline hospital NIHSS was much higher [11, 76]. Second, many subjects did not receive the full dose of RIC. Since RIC was discontinued when the patient arrived at the ED, only 41 % of the subjects received the full RIC dose of 4 cycles. Third, due to a misunderstanding, many of the patients randomized to the control group were not followed up long term due to failure to approach them for informed consent. Fourth, there was a high rate of stroke mimics (24.5 % versus 3.9 % in FAST MAGS). Finally, the study only focused and reported on efficacy endpoints on the subset of patients receiving IV tPA (272/443), 61 % of subjects. A number of lessons can be learned from the trial—the need to include a prehospital severity scale and to use an automated RIC device so that the RIC can be continued once the patient arrives to the ED and the full dose can be administered.

9 Other Clinical Trials of RIC in Stroke

The safety and feasibility of RIC has also been tested in a phase 1b trial of remote ischemic preconditioning with 33 patients with recent subarachnoid hemorrhage. Koch and colleagues [77] found repetitive ischemic limb preconditioning using blood pressure cuff on the leg with inflation as long as 10 min, to be feasible, safe, and well tolerated in critically ill patients with subarachnoid hemorrhage. Likewise Gonzalez and colleagues [78] assessed the feasibility and safety of RIC for aneurysmal subarachnoid hemorrhage in 20 patients enrolled within a median of 4–5 days post-SAH (range 1–13). The authors found that the RIC procedure was well tolerated by patients and did not cause any injury. No patient developed delayed ischemic neurological deficit (DIND) during their enrollment in the trial and patients had promising outcomes at follow-up [78].

In a proof-of-concept study of upper limb preconditioning on the rate of stroke recurrence in patients with intracranial atherosclerosis, Meng and colleagues [74] demonstrated that repeated bilateral arm preconditioning (5 cycles of bilateral upper limb ischemia for 5 min followed by reperfusion for another 5 min) performed twice a day for a total of 300 consecutive days reduced recurrent stroke in patients [74]. This has been confirmed in a second trial of bilateral arm conditioning in octagenarians and nonagenarians with a reduction of stroke and TIA in the RIC group [79]. In these trials, there was an improvement in the modified Rankin score of the index strokes by RIC suggesting that RIC in the form of RIPostC was promoting recovery from stroke.

10 Future Directions and Conclusions

The feasibility, safety, and tolerability of RIC make it an attractive and promising strategy in acute ischemic stroke. There is a solid foundation of the safety and "activity" of RIC in the cardiac field and a growing preclinical body of evidence in rodent stroke models. The Danish prehospital trial also shows a hint of activity and sets the foundation for future prehospital stroke trials. RIC increases CBF in animal models suggesting an important mechanism of action and measurements of CBF by MRI arterial spin labeling may be an important imaging biomarker in future studies. Validated plasma biomarkers of the conditioning response are still needed.

There are also other potential opportunities to apply RIC in the acute stroke setting. With the increase in transfers of stroke patients from community hospitals to Comprehensive Stroke Centers some with long helicopter or ground transfer times, there will be a growing need to develop therapies that "freeze the penumbra" and target reperfusion therapy. RIC is an ideal therapy to administer in the helicopter (see Fig. 18.2). Since RIC targets reperfusion injury and increases CBF, it is an ideal intervention to use before, during, or after MT. RIC has already been shown to be safe and feasible during air transport of STEMI patients prior to PCI [80].

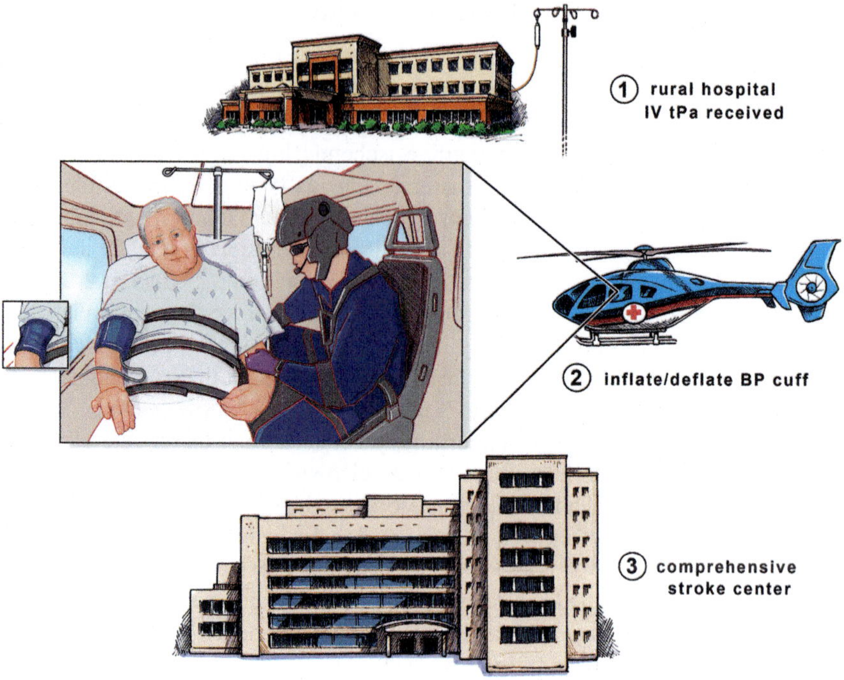

Fig. 18.2 RIC has potential to be used during helicopter transfer of stroke patients from community hospitals where IV tissue plasminogen activator is administered to comprehensive stroke centers for mechanical thrombectomy

References

1. Mozaffarian D, Benjamin EJ, Go AS, Arnett DK, Blaha MJ, Cushman M et al (2015) Heart disease and stroke statistics—2015 update: a report from the American Heart Association. Circulation 131(4):e29–e322
2. Moretti A, Ferrari F, Villa RF (2015) Neuroprotection for ischaemic stroke: current status and challenges. Pharmacol Ther 146:23–34
3. Park JM, Hess DC (2015) Breaking out from the neuroprotective logjam: combined treatment with remote ischemic conditioning and minocycline in the prehospital setting. Neural Regen Res 10(4):537
4. Iadecola C, Anrather J (2011) Stroke research at a crossroad: asking the brain for directions. Nat Neurosci 14(11):1363–1368
5. Saver JL, Goyal M, Bonafe A, Diener H-C, Levy EI, Pereira VM et al (2015) Stent-retriever thrombectomy after intravenous t-PA vs. t-PA alone in stroke. N Engl J Med 372(24): 2285–2295
6. Goyal M, Menon BK, van Zwam WH, Dippel DW, Mitchell PJ, Demchuk AM et al (2016) Endovascular thrombectomy after large-vessel ischaemic stroke: a meta-analysis of individual patient data from five randomised trials. Lancet 387(10029):1723–1731
7. Campbell BC, Mitchell PJ, Kleinig TJ, Dewey HM, Churilov L, Yassi N et al (2015) Endovascular therapy for ischemic stroke with perfusion-imaging selection. N Engl J Med 372(11):1009–1018

8. Jovin TG, Chamorro A, Cobo E, de Miquel MA, Molina CA, Rovira A et al (2015) Thrombectomy within 8 hours after symptom onset in ischemic stroke. N Engl J Med 372(24):2296–2306
9. Hougaard K, Hjort N, Zeidler D, Sørensen L, Nørgaard A, Thomsen R et al (2013) Remote ischemic perconditioning in thrombolysed stroke patients: randomized study of activating endogenous neuroprotection–design and MRI measurements. Int J Stroke 8(2):141–146
10. Albers GW, Goldstein LB, Hess DC, Wechsler LR, Furie KL, Gorelick PB et al (2011) Stroke Treatment Academic Industry Roundtable (STAIR) recommendations for maximizing the use of intravenous thrombolytics and expanding treatment options with intra-arterial and neuroprotective therapies. Stroke 42(9):2645–2650
11. Hess DC, Blauenfeldt RA, Andersen G, Hougaard KD, Hoda MN, Ding Y et al (2015) Remote ischaemic conditioning [mdash] a new paradigm of self-protection in the brain. Nat Rev Neurol 11(12):698–710
12. Janoff A (1964) Alterations in lysosomes (intracellular enzymes) during shock; effects of preconditioning (tolerance) and protective drugs. Int Anesthesiol Clin 2:251–269
13. Ring J, Gutermuth J (2011) 100 years of hyposensitization: history of allergen-specific immunotherapy (ASIT). Allergy 66(6):713–724
14. Valle G, Carmignani M, Stanislao M, Facciorusso A, Volpe AR (2012) Mithridates VI Eupator of Pontus and mithridatism. Allergy 67(1):138–139; author reply 9–40
15. Valle G, Stanislao M, Facciorusso A, Carmignani M, Volpe AR (2009) Mithridates VI Eupator, father of the empirical toxicology. Clin Toxicol 47(5):433
16. Waddell WJ (2010) History of dose response. J Toxicol Sci 35(1):1–8
17. Nietzsche F (2008) [1895] Twilight of the idols: or how to philosophize. Oxford University Press, Oxford
18. Hahn CD, Manlhiot C, Schmidt MR, Nielsen TT, Redington AN (2011) Remote ischemic perconditioning: a novel therapy for acute stroke? Stroke 42(10):2960–2962
19. Zhang Y, Liu X, Yan F, Min L, Ji X, Luo Y (2012) Protective effects of remote ischemic preconditioning in rat hindlimb on ischemia-reperfusion injury. Neural Regen Res 7(8):583–587
20. Failure of extracranial-intracranial arterial bypass to reduce the risk of ischemic stroke. Results of an international randomized trial. The EC/IC Bypass Study Group (1985) N Engl J Med 313(19):1191–1200
21. Murry CE, Jennings RB, Reimer KA (1986) Preconditioning with ischemia: a delay of lethal cell injury in ischemic myocardium. Circulation 74(5):1124–1136
22. Li Y, Whittaker P, Kloner RA (1992) The transient nature of the effect of ischemic preconditioning on myocardial infarct size and ventricular arrhythmia. Am Heart J 123(2):346–353
23. Przyklenk K, Bauer B, Ovize M, Kloner RA, Whittaker P (1993) Regional ischemic 'preconditioning' protects remote virgin myocardium from subsequent sustained coronary occlusion. Circulation 87(3):893–899
24. Birnbaum Y, Hale SL, Kloner RA (1997) Ischemic preconditioning at a distance reduction of myocardial infarct size by partial reduction of blood supply combined with rapid stimulation of the gastrocnemius muscle in the rabbit. Circulation 96(5):1641–1646
25. Oxman T, Arad M, Klein R, Avazov N, Rabinowitz B (1997) Limb ischemia preconditions the heart against reperfusion tachyarrhythmia. Am J Physiol 273(4):H1707–H1712
26. Kharbanda R, Mortensen U, White P, Kristiansen S, Schmidt M, Hoschtitzky J et al (2002) Transient limb ischemia induces remote ischemic preconditioning in vivo. Circulation 106(23):2881–2883
27. Schmidt MR, Smerup M, Konstantinov IE, Shimizu M, Li J, Cheung M et al (2007) Intermittent peripheral tissue ischemia during coronary ischemia reduces myocardial infarction through a KATP-dependent mechanism: first demonstration of remote ischemic perconditioning. Am J Physiol Heart Circ Physiol 292(4):H1883–H1890
28. Candilio L, Malik A, Hausenloy DJ (2013) Protection of organs other than the heart by remote ischemic conditioning. J Cardiovasc Med 14(3):193–205
29. Yuan H-J, Zhu X-H, Luo Q, Wu Y-N, Kang Y, Jiao J-J et al (2012) Noninvasive delayed limb ischemic preconditioning in rats increases antioxidant activities in cerebral tissue during severe ischemia-reperfusion injury. J Surg Res 174(1):176–183

30. Ren C, Gao X, Steinberg GK, Zhao H (2008) Limb remote-preconditioning protects against focal ischemia in rats and contradicts the dogma of therapeutic time windows for preconditioning. Neuroscience 151(4):1099–1103
31. Aimo A, Borrelli C, Giannoni A, Pastormerlo LE, Barison A, Mirizzi G et al (2015) Cardioprotection by remote ischemic conditioning: mechanisms and clinical evidences. World J Cardiol 7(10):621
32. Le Page S, Bejan-Angoulvant T, Angoulvant D, Prunier F (2015) Remote ischemic conditioning and cardioprotection: a systematic review and meta-analysis of randomized clinical trials. Basic Res Cardiol 110(2):1–17
33. Hausenloy DJ, Yellon DM (2011) The therapeutic potential of ischemic conditioning: an update. Nat Rev Cardiol 8(11):619–629
34. Hoda MN, Siddiqui S, Herberg S, Periyasamy-Thandavan S, Bhatia K, Hafez SS et al (2012) Remote ischemic perconditioning is effective alone and in combination with intravenous tissue-type plasminogen activator in murine model of embolic stroke. Stroke 43(10):2794–2799
35. Zhao H (2009) Ischemic postconditioning as a novel avenue to protect against brain injury after stroke. J Cereb Blood Flow Metab 29(5):873–885
36. Ren C, Wang P, Wang B, Li N, Li W, Zhang C et al (2015) Limb remote ischemic perconditioning in combination with post-conditioning reduces brain damage and promotes neuroglobin expression in the rat brain after ischemic stroke. Restor Neurol Neurosci 33(3):369–379
37. Yu Z, Liu N, Liu J, Yang K, Wang X (2012) Neuroglobin, a novel target for endogenous neuroprotection against stroke and neurodegenerative disorders. Int J Mol Sci 13(6):6995–7014
38. Veighey K, MacAllister RJ (2012) Clinical applications of remote ischemic preconditioning. Cardiol Res Pract 2012:620681
39. Hausenloy DJ, Yellon DM (2008) Remote ischemic preconditioning: underlying mechanisms and clinical application. Cardiovasc Res 79(3):377–386
40. Hess DC (2013) Remote limb preconditioning and postconditioning: will it translate into a promising treatment for acute stroke? Stroke 44(4):1191–1197
41. Yetgin T, Manintveld OC, Groen F, Tas B, Kappetein AP, van Geuns R-J et al (2012) The emerging application of remote ischemic conditioning in the clinical arena. Cardiol Rev 20(6):279–287
42. Lim SY, Hausenloy DJ (2012) Remote ischemic conditioning: from bench to bedside. Front Physiol 3:27
43. Gho BC, Schoemaker RG, van den Doel MA, Duncker DJ, Verdouw PD (1996) Myocardial protection by brief ischemia in noncardiac tissue. Circulation 94(9):2193–2200
44. Mastitskaya S, Marina N, Gourine A, Gilbey MP, Spyer KM, Teschemacher AG et al (2012) Cardioprotection evoked by remote ischaemic preconditioning is critically dependent on the activity of vagal pre-ganglionic neurones. Cardiovasc Res 95(4):487–494
45. Jones WK, Fan GC, Liao S, Zhang JM, Wang Y, Weintraub NL et al (2009) Peripheral nociception associated with surgical incision elicits remote nonischemic cardioprotection via neurogenic activation of protein kinase C signaling. Circulation 120(11 Suppl):S1–S9
46. Redington KL, Disenhouse T, Strantzas SC, Gladstone R, Wei C, Tropak MB et al (2012) Remote cardioprotection by direct peripheral nerve stimulation and topical capsaicin is mediated by circulating humoral factors. Basic Res Cardiol 107(2):241
47. Shimizu M, Tropak M, Diaz RJ, Suto F, Surendra H, Kuzmin E et al (2009) Transient limb ischaemia remotely preconditions through a humoral mechanism acting directly on the myocardium: evidence suggesting cross-species protection. Clin Sci (Lond) 117(5):191–200
48. Davidson SM, Selvaraj P, He D, Boi-Doku C, Yellon RL, Vicencio JM et al (2013) Remote ischaemic preconditioning involves signalling through the SDF-1alpha/CXCR4 signalling axis. Basic Res Cardiol 108(5):377
49. Li J, Rohailla S, Gelber N, Rutka J, Sabah N, Gladstone RA et al (2014) MicroRNA-144 is a circulating effector of remote ischemic preconditioning. Basic Res Cardiol 109(5):423
50. Cai ZP, Parajuli N, Zheng X, Becker L (2012) Remote ischemic preconditioning confers late protection against myocardial ischemia-reperfusion injury in mice by upregulating interleukin-10. Basic Res Cardiol 107(4):277

51. Rassaf T, Totzeck M, Hendgen-Cotta UB, Shiva S, Heusch G, Kelm M (2014) Circulating nitrite contributes to cardioprotection by remote ischemic preconditioning. Circ Res 114(10):1601–1610
52. Hess DC, Hoda MN, Khan MB (2016) Humoral mediators of remote ischemic conditioning: important role of eNOS/NO/nitrite. Brain Edema XVI. Springer, Basel, pp 45–48
53. Abu-Amara M, Yang SY, Quaglia A, Rowley P, De Mel A, Tapuria N et al (2011) Nitric oxide is an essential mediator of the protective effects of remote ischaemic preconditioning in a mouse model of liver ischaemia/reperfusion injury. Clin Sci 121(6):257–266
54. Abu-Amara M, Yang SY, Quaglia A, Rowley P, Fuller B, Seifalian A et al (2011) Role of endothelial nitric oxide synthase in remote ischemic preconditioning of the mouse liver. Liver Transpl 17(5):610–619
55. Jensen RV, Støttrup NB, Kristiansen SB, Bøtker HE (2012) Release of a humoral circulating cardioprotective factor by remote ischemic preconditioning is dependent on preserved neural pathways in diabetic patients. Basic Res Cardiol 107(5):1–9
56. Konstantinov IE, Arab S, Kharbanda RK, Li J, Cheung MM, Cherepanov V et al (2004) The remote ischemic preconditioning stimulus modifies inflammatory gene expression in humans. Physiol Genomics 19(1):143–150
57. Wei M, Xin P, Li S, Tao J, Li Y, Li J et al (2011) Repeated remote ischemic postconditioning protects against adverse left ventricular remodeling and improves survival in a rat model of myocardial infarction. Circ Res 108(10):1220–1225
58. Yellon DM, Hausenloy DJ (2007) Myocardial reperfusion injury. N Engl J Med 357(11):1121–1135
59. Kloner RA, Schwartz LL (2011) State of the science of cardioprotection: challenges and opportunities—proceedings of the 2010 NHLBI workshop on cardioprotection. J Cardiovasc Pharmacol Ther 16(3–4):223–232
60. Botker HE, Kharbanda R, Schmidt MR, Bottcher M, Kaltoft AK, Terkelsen CJ et al (2010) Remote ischaemic conditioning before hospital admission, as a complement to angioplasty, and effect on myocardial salvage in patients with acute myocardial infarction: a randomised trial. Lancet 375(9716):727–734
61. White SK, Frohlich GM, Sado DM, Maestrini V, Fontana M, Treibel TA, et al (2015) Remote ischemic conditioning reduces myocardial infarct size and edema in patients with ST-segment elevation myocardial infarction. JACC Cardiovasc Interv 8(1 Pt B):178–188
62. Prunier F, Angoulvant D, Saint Etienne C, Vermes E, Gilard M, Piot C et al (2014) The RIPOST-MI study, assessing remote ischemic perconditioning alone or in combination with local ischemic postconditioning in ST-segment elevation myocardial infarction. Basic Res Cardiol 109(2):400
63. Sloth AD, Schmidt MR, Munk K, Kharbanda RK, Redington AN, Schmidt M et al (2014) Improved long-term clinical outcomes in patients with ST-elevation myocardial infarction undergoing remote ischaemic conditioning as an adjunct to primary percutaneous coronary intervention. Eur Heart J 35(3):168–175
64. Davies WR, Brown AJ, Watson W, McCormick LM, West NE, Dutka DP et al (2013) Remote ischemic preconditioning improves outcome at 6 years after elective percutaneous coronary intervention: the CRISP stent trial long-term follow-up. Circ Cardiovasc Interv 6(3):246–251
65. Hausenloy DJ, Kharbanda R, Rahbek Schmidt M, Moller UK, Ravkilde J, Okkels Jensen L et al (2015) Effect of remote ischaemic conditioning on clinical outcomes in patients presenting with an ST-segment elevation myocardial infarction undergoing primary percutaneous coronary intervention. Eur Heart J 36(29):1846–1848
66. Yellon DM, Ackbarkhan AK, Balgobin V, Bulluck H, Deelchand A, Dhuny MR et al (2015) Remote ischemic conditioning reduces myocardial infarct size in STEMI patients treated by thrombolysis. J Am Coll Cardiol 65(25):2764–2765
67. Hess DC, Hoda MN, Bhatia K (2013) Remote limb preconditioning and postconditioning: will it translate into a promising treatment for acute stroke? Stroke 44(4):1191–1197
68. Hoda MN, Bhatia K, Hafez SS, Johnson MH, Siddiqui S, Ergul A et al (2014) Remote ischemic perconditioning is effective after embolic stroke in ovariectomized female mice. Transl Stroke Res 5(4):484–490

69. Hoda MN, Fagan SC, Khan MB, Vaibhav K, Chaudhary A, Wang P et al (2014) A 2 × 2 factorial design for the combination therapy of minocycline and remote ischemic perconditioning: efficacy in a preclinical trial in murine thromboembolic stroke model. Exp Transl Stroke Med 6:10
70. Khan MB, Hoda MN, Vaibhav K, Giri S, Wang P, Waller JL et al (2015) Remote ischemic postconditioning: harnessing endogenous protection in a murine model of vascular cognitive impairment. Transl Stroke Res 6(1):69–77
71. Hoda MN, Bhatia K, Hafez SS, Johnson MH, Siddiqui S, Ergul A, et al (2014) Remote ischemic perconditioning is effective after embolic stroke in ovariectomized female mice. Transl Stroke Res 5(4):484–490
72. Hoyte LC, Papadakis M, Barber PA, Buchan AM (2006) Improved regional cerebral blood flow is important for the protection seen in a mouse model of late phase ischemic preconditioning. Brain Res 1121(1):231–237
73. Zhao L, Nowak TS (2006) CBF changes associated with focal ischemic preconditioning in the spontaneously hypertensive rat. J Cereb Blood Flow Metab 26(9):1128–1140
74. Meng R, Asmaro K, Meng L, Liu Y, Ma C, Xi C et al (2012) Upper limb ischemic preconditioning prevents recurrent stroke in intracranial arterial stenosis. Neurology 79(18):1853–1861
75. Hougaard KD, Hjort N, Zeidler D, Sørensen L, Nørgaard A, Hansen TM et al (2014) Remote ischemic perconditioning as an adjunct therapy to thrombolysis in patients with acute ischemic stroke a randomized trial. Stroke 45(1):159–167
76. Saver JL, Starkman S, Eckstein M, Stratton SJ, Pratt FD, Hamilton S et al (2015) Prehospital use of magnesium sulfate as neuroprotection in acute stroke. N Engl J Med 372(6):528–536
77. Koch S, Katsnelson M, Dong C, Perez-Pinzon M (2011) Remote ischemic limb preconditioning after subarachnoid hemorrhage a phase Ib study of safety and feasibility. Stroke 42(5):1387–1391
78. Gonzalez NR, Connolly M, Dusick JR, Bhakta H, Vespa P (2014) Phase I clinical trial for the feasibility and safety of remote ischemic conditioning for aneurysmal subarachnoid hemorrhage. Neurosurgery 75(5):590
79. Meng R, Ding Y, Asmaro K, Brogan D, Meng L, Sui M, et al (2015) Ischemic conditioning is safe and effective for octo- and nonagenarians in stroke prevention and treatment. Neurotherapeutics 12(3):667–677
80. Martin-Gill C, Wayne M, Guyette FX, Olafiranye O, Toma C (2016) Feasibility of remote ischemic peri-conditioning during air medical transport of STEMI patients. Prehosp Emerg Care 20(1):82–89
81. Ren C, Yan Z, Wei D, Gao X, Chen X, Zhao H (2009) Limb remote ischemic postconditioning protects against focal ischemia in rats. Brain Res 1288:88–94
82. Wei D, Ren C, Chen X, Zhao H (2012) The chronic protective effects of limb remote preconditioning and the underlying mechanisms involved in inflammatory factors in rat stroke. PLoS One 7(2):e30892
83. Malhotra S, Naggar I, Stewart M, Rosenbaum DM (2011) Neurogenic pathway mediated remote preconditioning protects the brain from transient focal ischemic injury. Brain Res 1386:184–190
84. Ren C, Gao M, Dornbos D 3rd, Ding Y, Zeng X, Luo Y et al (2011) Remote ischemic postconditioning reduced brain damage in experimental ischemia/reperfusion injury. Neurol Res 33(5):514–519
85. Sun J, Tong L, Luan Q, Deng J, Li Y, Li Z et al (2012) Protective effect of delayed remote limb ischemic postconditioning: role of mitochondrial K(ATP) channels in a rat model of focal cerebral ischemic reperfusion injury. J Cereb Blood Flow Metab 32(5):851–859
86. Hu S, Dong H, Zhang H, Wang S, Hou L, Chen S et al (2012) Noninvasive limb remote ischemic preconditioning contributes neuroprotective effects via activation of adenosine A1 receptor and redox status after transient focal cerebral ischemia in rats. Brain Res 1459:81–90
87. Liu X, Zhao S, Liu F, Kang J, Xiao A, Li F et al (2014) Remote ischemic postconditioning alleviates cerebral ischemic injury by attenuating endoplasmic reticulum stress-mediated apoptosis. Transl Stroke Res 5(6):692–700
88. Hu X, Lu Y, Zhang Y, Li Y, Jiang L (2013) Remote ischemic preconditioning improves spatial learning and memory ability after focal cerebral ischemia-reperfusion in rats. Perfusion 28(6):546–551

Chapter 19
Hypothermia for Acute Ischemic Stroke

Roy Poblete and Gene Sung

Abstract The use of hypothermia as a medical therapy has been used and tested for centuries. In the modern era, its use in neurological diseases has attained significant interest and potential benefit has been found in a number of areas, particularly cerebral ischemia. In the setting of global cerebral ischemia, clinical benefit has been observed and so it seems that this may be the case in focal cerebral ischemia as well. This chapter reveals the current state of the evidence.

Keywords Therapeutic hypothermia • Acute ischemic stroke • Neuroprotection • Cerebral ischemia

Abbreviations

CBF	Cerebral blood flow
CLABSIs	Central-line associated bloodstream infections
CMRO2	Cerebral metabolic rate of oxygen
ECG	Electrocardiogram
ICP	Intracranial pressure
MCA	Middle cerebral artery
mRS	Modified Rankin Scale
NSAIDs	Nonsteroidal anti-inflammatory drugs
ROS	Reactive oxygen species
tPA	Tissue plasminogen activator

R. Poblete, M.D.
Department of Neurology, Keck Hospital of the University of Southern California, 1540 Alcazar Street, Suite 215, Los Angeles, CA 90033, USA
e-mail: roy.poblete@med.usc.edu

G. Sung, M.D., M.P.H. (✉)
Department of Neurology, University of Southern California, Clinic Tower A4E111, 1200 North State Street, Los Angeles, CA 90033, USA
e-mail: gsung@usc.edu

P.A. Lapchak, J.H. Zhang (eds.), *Neuroprotective Therapy for Stroke and Ischemic Disease*, Springer Series in Translational Stroke Research,
DOI 10.1007/978-3-319-45345-3_19

1 Therapeutic Hypothermia and Its First Historical Applications

The use of therapeutic hypothermia is not exclusive to modern medicine. Its first descriptions are found in the oldest known medical text—the Edwin Smith papyrus (c. 3000 BCE)—to treat ulcerated lesions and open head wounds [1]. The ancient Greeks and Romans were also known to apply cold remedies to the injured. Hippocrates described packing patients in ice and snow to reduce hemorrhage [2], Homer's *Iliad* depicted cold water packs as a wound dressing [1], and total body cooling was advocated for the treatment of tetanus during the fourth and fifth centuries BCE [3]. Anecdotally and throughout the historical record, victims of accidental hypothermia—most notably in drowning victims—have made unexpected recovery despite mimicking death [4]. In cases of accidental hypothermia, patients can achieve full neurologic recovery despite sustaining core temperatures slightly less than 14 °C [5]. It was not until the late eighteenth century that the first physiologic experiments with therapeutic hypothermia were completed by Dr. James Currie of Liverpool. In his now famous work, Currie examined the effects of various methods of cooling in human subjects and later promoted the use of cold water for the treatment of fever [6].

Hypothermia for neurologic disease was first introduced in the 1930s by the American neurosurgeon, Dr. Temple Fay. His first applications of whole-body refrigeration were in 169 patients with intractable chronic pain associated with metastatic cancers, but later found uses in traumatic brain injury and intracerebral infection [7]. Other early applications of therapeutic hypothermia in neurology patients were found in the field of surgery. Neurosurgeon Irving Cooper developed a cryoablation technique for extrapyramidal disorders and intracranial tumors [8, 9], while whole body hypothermia was pioneered as a neuroprotectant during induced circulatory arrest for intracranial aneurysm surgery [10–12].

2 Extrapolation from Cardiac Arrest Population

In twentieth-century medicine, therapeutic hypothermia has been most firmly established in the context of cardiac disease. In the setting of cardiac failure, whether spontaneous or iatrogenic, cooling was hypothesized to slow neuronal metabolism and protect the brain from ischemic injury. In the 1950s, induction of moderate hypothermia was used successfully for open-heart surgery [13], and its benefit following cardiac arrest had been reported [14, 15]. Its routine use, however, was subsequently abandoned because of insufficient evidence and difficulties with its application [16]. Guidelines went unchanged until two randomized clinical trials published in 2002 demonstrated the benefit of induced hypothermia in reducing mortality and improving neurologic outcome following out-of-hospital cardiac arrest due to ventricular fibrillation [17, 18]. Apart from ventricular arrhythmias, its application to other situations of cardiac arrest is continually being investigated [19]. Questions regarding the optimal timing and duration of hypothermia, methods of cooling, and ideal target temperature are still a matter of debate.

To date, induced hypothermia remains the only treatment modality with proven neuroprotective properties following global ischemic events such as cardiac arrest and newborn hypoxic encephalopathy [20, 21]. Given its effectiveness in treating diffuse neurologic injury, its application to focal brain injuries—including acute ischemic stroke—is of significant clinical interest. Therapeutic whole-body cooling has already been investigated in traumatic brain [22, 23], spinal cord injury [24, 25], and multiple sclerosis [26]; however, with nearly 1,000,000 hospitalizations per year in the United States for stroke [27], its use in acute cerebral infarction is of added importance.

3 The Pathophysiology of Stroke and the Effect of Hypothermia

Relative to other body systems, the brain is metabolically a highly active organ. Although it only accounts or 2 % of total body weight, it receives 15 % of cardiac output and consumes 20 % of total body oxygen [28]. At the time of stroke onset, brain tissue normally supplied by the occluded vessel experiences relative hypoperfusion. The degree of decreased cerebral blood flow (CBF) is variable, but relative CBF reductions of 60 % compared to the contralateral side are felt to predict the ischemic core of nonsalvageable tissue [29]. Through local compensatory mechanisms that include vessel dilation and increased transfer of oxygen from blood to tissue, the brain is initially able to maintain cerebral blood volume and preserve cerebral oxygen metabolism [30]. If blood flow is not restored, metabolic crisis can ensue. This is characterized by a reduced cerebral metabolic rate of oxygen (CMRO2) and glucose in penumbral tissue [31], potentially leading to enlargement of the ischemic core.

Historically, the neuroprotective effect of induced hypothermia was thought to largely relate to reducing cerebral metabolism and avoiding metabolic crisis. Since our first clinical experiences with therapeutic hypothermia, much has been derived about the cellular mechanisms that lead from brain ischemia to apoptosis and necrosis as a final common pathway. These pathophysiologic processes have largely been elucidated through work involving animal models of both global and focal cerebral ischemia.

3.1 Cerebral Metabolism

As discussed earlier, cerebral oxygen metabolism is initially preserved prior to onset of metabolic crisis [30]. In the acute phase of ischemia, hypothermia may reduce CMRO2 approximately 6 % for every −1 °C change in temperature, potentially maintaining viable cells in a low metabolic state [32, 33]. Following the acute injury, depression of the mitochondrial aerobic metabolic rate can last several weeks; however, induced hypothermia results in faster metabolic recovery associated with increases in high-energy phosphates and reduced concentrations of toxic metabolites [34, 35].

3.2 ATP Depletion, Ca^{2+} Influx, and the Release of Excitatory Neurotransmitters

With cessation of blood flow to a neuron, ATP depletion occurs within a short period of time and results in the failure of the Na^+–K^+ ATPase pump [36]. Subsequent loss of the normal ionic gradient leads to increases in extracellular potassium [37], calcium influx, and intracellular acidosis which is exacerbated by phosphate, lactate, and H^+ production through anaerobic glycolysis [36]. Excess concentrations of intracellular Ca^{2+} give rise to a number of pathophysiologic consequences. Mitochondrial injury and dysfunction lead to the release of enzymes into the cytoplasm that causes further intracellular damage. Ca^{2+} overload will additionally promote spontaneous membrane depolarization and release of large amounts of glutamate into the extracellular space [38]. The result is a persistent hyperexcitable state characterized by recurrent Ca^{2+} influx and cyclic neuronal damage. Hypothermia may help maintain physiologic ionic gradients in the setting of ischemia and may interfere with processes that promote calcium influx and the release of excitotoxic neurotransmitters [39–41].

3.3 Free-Radical Production

The generation of reactive oxygen species (ROS) is an important consequence of ischemic injury and has many potential sources. Although they are produced during oxidative stress, large amounts of oxygen radicals are released with mitochondrial damage, membrane phospholipid metabolism, and reperfusion injury [42, 43]. The process is cyclical, with excess free radicals leading to peroxidation of lipids, proteins, and nucleic acids, and further generation of ROS. There is evidence that hypothermia may act to slow free radical production, allowing innate enzymatic systems to clear excess ROS [40, 44].

3.3.1 Inflammation

Proinflammatory mediators such as tumor necrosis factor-alpha and interleukin-1 are increased during acute ischemic injury and even more so during reperfusion. In models of ischemia and reperfusion, levels of inflammatory markers can remain elevated for up to 5 days [38]. In the presence of upregulated inflammatory signaling, adhesion and migration of inflammatory cells across the blood brain barrier is stimulated and may lead to further production of toxic metabolites. Hypothermia suppresses proinflammatory cytokine production [45] and indirectly reduces the inflammatory response by slowing free-radical production and cellular injury.

3.4 Apoptosis

Apoptosis is the final common pathway of a variety of pathophysiologic processes that occur after ischemic injury and is thought to be associated with mitochondrial dysfunction and caspase-mediated DNA fragmentation. The effect of hypothermia is primarily mediated through inhibition of caspase activation as well as through a reduction in mitochondrial damage and dysfunction [46–48]. Because the actions of hypothermia are broad, suppression of apoptosis may be the result of blockade of many different upstream pathways.

4 Definitions

4.1 Levels of Hypothermia

Therapeutic hypothermia is defined by an intentional reduction in a patient's core body temperature below 36.0 °C. The degree of hypothermia is often classified into the following categories: mild (34.0–35.9 °C), moderate (32.0–33.9 °C), moderate–deep (30.0–31.9 °C), and deep (<30.0 °C) [49].

4.2 Phases of Temperature Modulation

Four phases define the implementation of therapeutic hypothermia: induction, maintenance, de-cooling (or rewarming), and normothermia. Successful achievement of all four phases in sequential order is not a passive process and requires continuous monitoring and implementation of temperature modulation strategies.

5 Methods of Temperature Modulation

Patient temperature is modulated through pharmacologic interventions or through the use of cooling devices. When rapid induction of therapeutic hypothermia (<36 °C) is desired, it is achieved through the use of devices that act through two distinct methods: surface cooling or intravascular (core, internal) cooling. A third method, selective cooling, discrepantly lowers brain temperature greater than systemic or core temperature with use of noninvasive devices. At present, its application is largely investigational but may be clinically encountered. The efficacy of using one method over another to improve patient outcomes is unknown.

5.1 *Pharmacologic Interventions*

Medications, given either orally or intravenous (IV), can successfully modulate temperature, but are typically reserved for fever control in the ICU or to maintain normothermia. Antipyretic agents, such has acetaminophen, aspirin, and other nonsteroidal anti-inflammatory drugs (NSAIDs) can lower the hypothalamic temperature set point by inhibiting cyclooxygenase-mediated prostaglandin synthesis in the brain. The physiologic response is to activate mechanisms of heat dissipation, mainly vasodilation and sweating [50]. Pharmacologic interventions alone can maintain normothermia in neurologically injured patients [51–53] but is ineffective in achieving or sustaining hypothermia in small studies [49]. They are often used before and after cooling to prevent fever and can be used in combination with cooling devices in cases refractory to hypothermia induction.

5.2 *Surface Cooling*

Surface cooling is the least invasive, easiest to implement, and an inexpensive method of inducing hypothermia; therefore, it is a first-line strategy at many centers. Heat is conducted away from the patient using air, cold water or ice, or volatile liquids as a transfer media. Traditionally, initiation typically involves the removal of heat-trapping coverings from the patient and application of ice or cold water packs under the patient's axilla and around other superficial vascular sites, such as the groin or neck. Disadvantages to conventional surface cooling including prolonged times to achieving goal temperature, difficulties maintaining that target after induction, and increased patient shivering. A number of commercially available modern surface cooling devices have been developed to induce hypothermia at a rapid rate and regulate maintenance temperature. All are fashioned as surface pads that adhere closely to the skin to promote heat exchange. The Arctic Sun (Medivance) was introduced earliest and is still widely used. CoolBlue (Innercool Therapies), ThermoWrap (MTRE Advanced Technologies), and KoolKit (Cincinnati SubZero Products) are among more recently introduced products, with Cincinnati SubZero products shown to improve duration of normothermia in neurocritical care patients [54]. Thermosuit System (Life Recovery Systems) has shown promise in swine models, achieving a target core temperature of 33 °C in 9.0 ± 5.3–11.9 min in a small study [55], but clinical data on its safety and efficacy in humans are incomplete.

5.3 *Intravascular Cooling*

The primary alternate to surface cooling is inducing hypothermia through intravascular cooling. Other terms often used synonymously include core and internal cooling. Infusion of cold saline to induce hypothermia is by definition an internal cooling method, but it is often used in combination with conventional surface cooling protocols. Cold fluid infusion is the best studied induction method, having been shown to reduce temperature by 2–4 °C in cardiac and neurologically injured patients [56, 57], and can be easily and quickly implemented when induced hypothermia is indicated. Modern internal devices for acute ischemic stroke utilize intravascular cooling catheters that employ metal or cold fluid-filled balloons to conduct heat away from the patient. Other internal induction methods, such as peritoneal lavage and extracorporeal cooling have not been widely studied in stroke. In comparison to surface cooling, the purported advantage to endovascular cooling is a shorter induction time and longer time spent at goal temperature [58–60]. Improvement in precision temperature management during the maintenance phase is largely achieved by electronic feedback temperature control systems that are integrated with the cooling catheter. Disadvantages in using intravascular devices include purchasing costs, the requirement of an invasive procedure, difficulties in use, and concerns over adverse events. Complications can arise during placement, or may be delayed as in the case of catheter-related bloodstream infections, hematoma development, and venous thrombosis [61]. Commercially available products include the CoolGard temperature management system (Alsius), InnerCool (Philips Healthcare), and the Thermogard cooling catheters (Zoll Medical).

5.4 *Selective Cooling*

Selective cooling strategies have emerged more recently in attempts at differentially lowering brain temperature while avoiding the costs and complications associated with whole-body induced hypothermia. In a small, randomized study, Want et al. demonstrated the effectiveness of a cooling helmet to significantly decrease brain temperature and maintain selective hypothermia for 48–72 h [62]. In the intervention group of eight patients, it took a mean duration of 3.4 h to achieve a brain temperature <34 °C and 6.67 h before systemic hypothermia (<36 °C) occurred. In the larger Pre-ROSC IntraNasal Cooling Effectiveness (PRINCE) study, Castrén et al. investigated the use of the *RhinoChill*, a transnasal evaporative cooling device, in cardiac arrest patients [63]. Compared to standard care, patients randomized to device treatment achieved a brain temperature of <34 °C in a significantly shorter time (mean 1.7 h). Significant complications have not been reported with either device. Given the ease of use, safety profile, and effectiveness of such noninvasive devices, their application to the prehospital setting is under investigation [64].

6 Adverse Events Associated with Therapeutic Hypothermia

Any observed benefit from systemic or selective hypothermia must be considered against the potential treatment complications. Adverse events can occur during any of the four phases of therapeutic hypothermia and requires vigilance from all providers. The safety of one cooling method over another, or one temperature goal versus another, in preventing treatment complications is not sufficiently known.

6.1 Shivering

Shivering is the most commonly encountered side effect of whole-body cooling. In milder forms, shivering is intermittent and easily suppressed with nonparalytic pharmacologic interventions; however, more severe cases can be refractory to conventional therapy and have detrimental physiological effects. Systemically, shivering results in significant increases in resting energy expenditure, rate of oxygen consumption, and heat energy production [65–67]. Skin is subject to breakdown when exposed to persistent cold and moisture, and shivering events can be misinterpreted as pain, agitation, posturing, or seizure. Within the brain, decreases in brain tissue oxygenation were observed with onset of shivering in traumatic brain injury and subarachnoid hemorrhage patients [68], while shivering has also been associated with poorer neurologic exams [67]. These adverse effects, if left untreated, can interfere with successful temperature cooling and largely offset any potential benefits intended by therapeutic hypothermia [69].

Active cutaneous warming (air, blanket, or radiant heating) is the most effective nonpharmacological antishivering technique [70]; however, it may interfere with maintenance of hypothermia. Shivering is typically controlled through the use of surface counter-warming combined with pharmacologic interventions that include magnesium, buspirone, meperidine and other opiates, benzodiazepines, propofol, alpha-agonists such as clonidine and dexmedetomidine, and paralytic agents in refractory cases. Magnesium is an appealing choice given its low side effect profile and possible neuroprotective properties. The optimal treatment strategy is unknown. There is significant heterogeneity in the literature in regards to study methods and results [71], with various antishivering protocol subsequently suggested [72, 73].

6.2 Cardiac and Hemodynamic Complications

Cardiac and hemodynamic adverse events are the most feared systemic complication of whole-body cooling. Bradycardia is the most pronounced effect of mild-to-moderate hypothermia and is largely responsible for the 25–40 % reduction in cardiac output observed [74, 75]. At core temperatures near 32 °C, heart rate can decrease to 40–45 beats per minute [60]. The physiologic response consists of peripheral vascoconstriction and increased myocardial contractility; therefore if hemodynamic instability occurs during hypothermia, hypovolemia should be considered and a fluid challenge is warranted [60]. Careful consideration and evaluation of volume status, underlying cardiac function, and evidence of systemic hypoperfusion should be given before administering chronotropic agents for bradycardia, as this may be associated with reductions in cardiac contractility.

Clinically significant arrhythmias are most likely to occur at temperatures below 30 °C as mild hypothermia may actually increase myocardial membrane stability [76]. ECG changes typically precede the development of arrhythmias and are characterized by PR-prolongation, J-waves, QRS-complex widening, and increased QTc interval. These findings do not typically require intervention.

6.3 *Metabolic Panel Abnormalities*

Induction and maintenance of therapeutic hypothermia is associated with a variety of electrolyte derangements that include hypokalemia, hypomagnesemia, and hypophosphatemia. Low electrolyte levels are a result of physiologic diuresis in response to hypothermia in combination with tubular dysfunction and ionic shifts [60, 77]. Close monitoring of magnesium levels during therapeutic hypothermia for acute ischemic stroke may be of special importance. Magnesium is potentially neuroprotective after brain injury [78–80] while hypomagnesemia has been associated with higher mortality rates in the critically ill [81, 82]. Caution should be given when replacing electrolytes, especially during the rewarming phase, as decreased renal losses and reversed intracellular shifts may result in elevated electrolytes.

Hypothermia induces hyperglycemia by simultaneously reducing insulin secretion from the pancreas while increasing insulin resistance [60]. Diabetes mellitus, a vascular risk factor, is a common comorbidity in patients with acute ischemic stroke. Diabetic patients are prone to having poorly controlled blood sugars during acute illness, and this risk is amplified with induced hypothermia. The ideal blood glucose range during hypothermia is unknown.

6.4 *Infectious Risks*

Infectious risks are primarily the result of the anti-inflammatory properties of hypothermia. Hypothermia suppresses proinflammatory cytokine production [45] and indirectly reduces the inflammatory response by slowing free-radical production and cellular injury. Impairment of immunologic function is a result of suppression of leukocyte migration and interference with cellular killing. Hyperglycemia may be an exacerbating factor.

In safety and feasibility trials for induced hypothermia for acute ischemic stroke, pneumonia was the most commonly reported nosocomial infection [83–85]. There has been additional concern for increased risk of central-line associated bloodstream infections (CLABSIs) with use of intravascular cooling devices; however, this has not been confirmed in clinical trials. As with any critically ill patient, implementation of management bundles to reduce the risk of nosocomial infections, such as ventilator-associated pneumonia and CLABSIs, is recommended.

6.5 Coagulopathy and Bleeding Risk

The coagulopathy of hypothermia is characterized by thrombocytopenia, platelet dysfunction, plasminogen activator inhibition, and interference with other coagulation pathways [86–89]. Bleeding diathesis is thought to develop at temperatures less than 33 °C [86, 88, 90] and is most problematic for patients with active bleeding. The risk of significant bleeding from mild hypothermia (>33 °C) is minimal.

Acute ischemic stroke patients subjected to therapeutic hypothermia should be considered high risk for bleeding complications. Patients receiving systemic thrombolysis at the time of admission have the potential to compound the coagulopathies of hypothermia if induction is initiated without adequate intervening time. Additionally, large territory infarcts are highly susceptible to bleeding; 20 % of large vessel infarcts develop hemorrhagic transformation even without reperfusion therapies [91]. In clinical safety and feasibility trials for induced moderate hypothermia in the setting of systemic alteplase use, symptomatic intracerebral hemorrhage has been observed [85, 92, 93]; however, these studies were insufficiently powered to make definite conclusions regarding bleeding risk.

7 Preclinical Data

Over 500 experimental treatments have demonstrated efficacy in animal models of focal cerebral ischemia [94], and that number is continually increasing. Despite robust preclinical data, IV thrombolysis using recombinant tissue plasminogen activator (rt-PA, or alteplase) has remained the only proven therapy for acute ischemic stroke over the last 20 years. In contrast to the majority of hypothesized neuroprotective agents, therapeutic hypothermia has been appealing because of its actions on a wide range of pathophysiologic mechanisms [95] and its success in treating global ischemia postcardiac arrest. Repeated meta-analysis has shown hypothermia to be the most potent neuroprotective agent in experimental models of stroke [95–97]. Overall, in animal models, hypothermia reduces infarct size by 44 %. Greatest efficacy is seen with cooling to temperatures ≤31 °C and when induction occurs early in the disease course; however, substantial benefit is still observed with cooling to 35 °C and with initiation between 90 and 180 min [95].

8 A Review of Clinical Trials for Therapeutic Hypothermia in the Treatment of Acute Ischemic Stroke

8.1 Early Clinical Feasibility and Safety Trials

Early clinical feasibility and safety trials for induced hypothermia were supported by experience in animal models of cerebral ischemia [98, 99] and the demonstration of neurologic benefit in cardiac arrest patients [17, 18]. Observational analysis conducted in the 1990s of acute ischemic stroke patients from Denmark showed that every 1 °C increase in admission temperature was associated with a 2.2 % rise in relative risk of poor outcome [100, 101]. Subsequently published in 2000, the Copenhagen Stroke Study was designed as a safety and feasibility trial for therapeutic hypothermia in awake patients with acute ischemic stroke [102]. In the study, 17 patients admitted within 12 h of symptom onset achieved mild hypothermia to a mean temperature of 35.5 °C for 6 h using cooling blankets. Shivering was reported as the most common side effect but was well controlled with meperidine. Results from the Copenhagen Stroke Study were encouraging but did not examine safety and feasibility of lower target temperatures in patients with large hemispheric infarcts.

Near the time of the Danish feasibility study, Schwab et al. investigated therapeutic hypothermia for sedated patients with severe middle cerebral artery (MCA) infarcts. In a pilot study of 15 patients with large MCA stroke, it was found that recorded cerebral temperatures were at least 1 °C higher than core body temperature [49]. Systemic cooling effectively brought brain temperatures to a goal of 33–34 °C; however, administration of single pharmacologic agents (paracetamol, metamizol) did not adequately induce or sustain hypothermia. In a follow-up feasibility and safety trial, 25 patients with large MCA infarcts were successfully induced and maintained in moderate hypothermia (target 33 °C) using a combination of surface cooling and cold saline infusions [83]. To evaluate the hypothesis that moderate hypothermia would counteract the pathophysiologic effects of malignant edema, intracranial pressure (ICP) and cerebral perfusion pressure were serially recorded. Significant adverse events reported were rebound ICP elevations seen following rewarming, and pneumonia observed in 40 % of patients. Because of the study design, it is unclear whether these complications were the result of hypothermia, or rather, associated with stroke severity. A survival rate of 56 % in study patients was promising, as it compared favorably to previously reported case fatality rates of approximately 80 % after severe MCA infarcts [103, 104].

Schwab et al. conducted a second feasibility and safety trial in 50 patients with massive hemispheric infarction [84]. Target bladder temperatures below 33 °C were successfully reached after an average of 6.5 h using both surface cooling and cold saline infusions. Moderate hypothermia was maintained for an average of 55 h with a passive rewarming phase of approximately 17 h. Similar to their prior study, a survival rate of 62 % compared favorably against historical controls; however, neither study was powered to determine efficacy. The most common systemic adverse events reported were thrombocytopenia (70 %), bradycardia (62 %), and pneumonia (48 %). Again, it was demonstrated that rewarming was associated with rebound ICP elevations, with faster rewarming associated with greater increases in ICP.

8.2 The Introduction of Modern Cooling Devices for Induction

Endovascular cooling methods for induced moderate hypothermia was demonstrated in a 2001 study by Georgiadis et al. [105]. A then novel device (CoolGard system, Alsius Corporation) for induction and temperature maintenance was used. The system consisted of a cooling catheter with fluid-filled balloons at the distal end, inserted in to the femoral vein and integrated with an external temperature management system. In six patients with severe MCA infarct, use of the device resulted in rapid cooling (1.4 ± 0.6 °C per hour) to moderate hypothermia (32.2–33.4 °C). One of the six patients died at 59 h after the initiation of hypothermia because of uncontrollable intracranial hypertension, with no other deaths reported during the acute phase of stroke. No device-related side effects were observed except for a case of diaphragmatic spasm. Importantly, the development of groin hematoma was not seen, including in two patients who had received IV thrombolytics 29 and 32 h before catheter insertion. Other complications reported had been previously observed with systemic hypothermia achieved by surface cooling.

Despite advances in achieving moderate hypothermia in acute stroke, mild hypothermia remained as an alternate target. At the time, the ideal temperature to improve outcomes and minimize systemic complications was not known. Additionally, moderate hypothermia, especially when induced by endovascular cooling methods, was thought to require heavy sedation and mechanical ventilation. Zweifler et al. demonstrated that a novel surface cooling device (Arctic Sun Energy Transfer Pads, Medivance) could be used to rapidly induce mild hypothermia in awake, nonintubated patients [106]. In the study of ten healthy volunteers, a target temperature of 34–35 °C was successfully achieved in 90 ± 53 min with a rate of cooling of approximately 1.4 °C per hour. Induction and maintenance phases occurred without the use of anesthesia, and all subjects received acetaminophen and intravenous meperidine as needed for comfort and shivering. Commonly reported side effects were nausea (30 %) and elevated systolic blood pressure.

In lead-up to a larger multicenter clinical trial (The Intravascular Cooling for the Treatment of Stroke-Longer window trial, discussed later), Guluma et al. showed that an intravascular cooling catheter could induce moderate hypothermia (target core temperature of 33 °C) in awake, nonintubated patients [107]. The study protocol used superficial counter-warming blankets in addition to buspirone and meperidine during induction and a 24-h maintenance phase. Hypothermia protocol was well tolerated, with minimal shivering and no reports of oversedation.

8.3 Induced Hypothermia in Combination with Systemic Thrombolytic Therapy

To date, intravenous rtPA remains the only FDA approved pharmacologic intervention for the treatment of acute ischemic stroke. Benefits of reduced mortality and improved functional outcomes have been confirmed in meta-analysis of available randomized clinical trials [108]. For therapeutic hypothermia to become clinically viable, its safety and efficacy as an adjunct to IV thrombolytics would need to be demonstrated. Although the feasibility and safety of cooling to mild and moderate hypothermia had been shown in small studies, trials prior to 2001 had largely excluded stroke patients who had received alteplase. Safety concerns existed over a theoretical risk of intracerebral hemorrhage in the setting of hypothermia-induced thrombocytopenia and coagulopathy.

Cooling for Acute ischemic Brain Damage (COOL AID) was an open-label pilot study published in 2001 that addressed the combined use of therapeutic hypothermia with IV thrombolytics in acute ischemic stroke [92]. Nineteen patients with severe stroke (NIHSS score >15) who had received IV alteplase were enrolled. Patients were further eligible for induced hypothermia if the NIHSS remained >8 following thrombolytic therapy and informed consent, resulting in ten patients (NIHSS 19.8 ± 3.3) being cooled and nine patients serving as control. In the treatment group, time from symptom onset to thrombolysis was 3.1 ± 1.4 h and time from symptom onset to initiation of hypothermia was 6.2 ± 1.3 h. Using surface cooling methods, hypothermia to a core target temperature of 32 °C was achieved in 3.5 ± 1.5 h for a mean duration of 47.4 ± 20.4 h. Patients receiving hypothermia had a 3-month modified Rankin Scale (mRS) score of 3.1 ± 2.3 compared to 4.2 ± 1.6 in the control group. Noncritical sinus bradycardia was the most common complication in the cooling group with other adverse events reported as ventricular ectopy, hypotension, infections, and atrial fibrillation with rapid ventricular response in two patients with chronic atrial fibrillation. Death occurred in three patients who were cooled, with one death attributed to the development of a large intraparenchymal hematoma.

COOL AID II was subsequently conducted as a multicenter randomized pilot trial to evaluate the feasibility of intravascular cooling to moderate hypothermia in patients with acute ischemic stroke [93]. Forty patients presenting within 12 h of symptom onset were enrolled in the study: 18 were randomized to hypothermia (mean NIHSS score 18.2 ± 4.4) and 22 to receive standard medical management (mean NIHSS score 16.9 ± 5.5). Thirteen patients were treated with thrombolysis in the cooling group (72.2 %) versus 15 patients (71.4 %) in the control group. In comparison to COOL AID, which used surface cooling, subjects in COOL AID II achieved target core temperature (33 °C) more rapidly (77 ± 44 min); however, five patients intended to be cooled did not successfully reach target temperature because of suboptimal catheter placement ($n = 4$) and shivering ($n = 1$). In the hypothermia group, symptomatic hemorrhagic transformation of ischemic stroke occurred in two patients and retroperitoneal hematoma developed in one. No significant bleeding events were observed in those who were not cooled. Clinical outcomes, including mortality, NIHSS score, and modified Rankin score at 3 months were similar in both groups; however, the study was not powered to detect such differences. Considered together, COOL AID and COOL AID II demonstrated the feasibility of induced hypothermia using either surface or endovascular cooling methods in acute stroke patients, including those who received thrombolysis. Piironen et al. have since shown the utility of surface cooling to mild hypothermia following systemic rtPA in a small randomized controlled trial [109].

Published in 2010, The Intravascular Cooling in the Treatment of Stroke-Longer tPA window (ICTuS-L) study was a larger multicenter, randomized controlled trial that aimed to confirm the safety of endovascular hypothermia in patients receiving thrombolytic therapy after stroke [85]. Early experience with treatment protocol was gained during the previously published phase I ICTuS trial [110]. Compared to the ICTuS-L study, ICTuS was an uncontrolled pilot study of 15 patients who were cooled using intravascular methods; however, only 5 patients in the cohort (28 %) received rtPA. ICTuS-L enrolled 59 patients divided into 2 cohorts. Patients who presented within 3 h of symptom onset received IV alteplase, followed by randomization to either 24 h of endovascular cooling to 33 °C or standard medical management. Those presenting between 3 and 6 h from onset were randomized into four groups: alteplase alone, hypothermia alone, combined alteplase and hypothermia, or standard medical management without alteplase or hypothermia. In total, 28 patients were randomized to receive endovascular cooling. Baseline NIHSS score was 14.3 ± 5.0 in the hypothermia groups and 13.7 ± 5.1 in the normothermia groups. Pneumonia occurred more frequently in the hypothermia groups versus those who were not cooled (25 % versus 7 %, $P < 0.05$). The incidence of intracerebral hemorrhage at 48 h was similar in both groups. All four patients who developed symptomatic intracerebral hemorrhage were treated with rtPA within 3 h of symptom onset, while only one of the four patients was cooled. There was no statistical difference in mortality or functional outcomes by mRS at 3 months. Because these studies were not powered to do so, no definite conclusions could be made regarding the safety or efficacy of hypothermia as an adjunct to alteplase.

8.4 Induced Hypothermia in Combination to Decompressive Hemicraniectomy

Large hemispheric MCA infarcts are characterized by the development of early malignant cerebral edema, potential herniation, high mortality, and severe disability in survivors. In response, aggressive, but life-saving therapies have been proposed. In randomized, controlled trials and pooled analysis, decompressive hemicraniectomy has shown a significant mortality benefit when done early (within 48 h) after malignant MCA strokes [111–114]. With feasibility demonstrated in early safety trials for hypothermia in large hemispheric infarcts, a 2006 study by Els et al. evaluated combined hypothermia with surgical decompression [115]. Twenty-five patients with an infarct size greater than two-thirds of the hemisphere were randomized to combination hemicraniectomy and mild hypothermia (target temperature 35 °C), or hemicraniectomy alone. Surgery was performed within 15 ± 6 h of symptom onset, followed by immediate cooling in the hypothermia group. Mortality and clinical outcome as measured by NIHSS score at 6 months was not statistically significant between groups. No case fatalities directly associated with combination occurred, and no significant adverse events were otherwise reported.

8.5 *Induced Hypothermia in Combination with Intra-arterial Treatment*

Intra-arterial treatment (IAT) for stroke due to large-vessel occlusion has emerged as an effective method of acutely restoring cerebral blood flow (CBF) to potentially viable penumbral tissue. IAT is performed using a combination of intra-arterial alteplase infusion and mechanical clot retrieval. In select patients, the effectiveness of mechanical thrombectomy in restoring CBF and improving functional outcomes has been demonstrated in a number of pivotal trials [116–120], with vessel recanalization (TICI 2b or 3) in 58.7–88 % of cases [121]. Combining IAT with immediate postrecanalization cooling has been proposed as a promising strategy to minimize reperfusion injury [122, 123] and is considered against the risk of hemorrhagic transformation of stroke, especially in the presence of a large ischemic core.

The Endovascular Reperfusion and Cooling in Cerebral Acute Ischemia (ReCCLAIM I) study was a phase I single-arm, open label clinical trial to examine the feasibility of endovascular hypothermia after reperfusion with mechanical thrombectomy [123]. In the study, 20 patients with a median NIHHSS score of 19 and a median baseline imaging ASPECTS score of 6 were enrolled and analyzed. Patients who received IV thrombolysis were excluded. The average time from symptom onset to groin puncture was 5.4 ± 1.8 h with 85 % of patients achieving TICI 2B or 3 grade reperfusion. Surface cooling was initiated during IAT, followed by rapid endovascular cooling using a Quattro catheter (Zoll Medical). Moderate hypothermia (target 33 °C) was achieved in 64 ± 50 min from the initiation of cooling. Using HARM MRI protocol [124], evidence of blood–brain barrier breakdown was seen in three patients. Intracranial hemorrhage was found in three patients, with one patient being symptomatic. In comparison to historical controls at the study institution, hypothermia was potentially protective against hemorrhage (OR 0.09, $p<0.01$). Patient outcomes were highly variable, with six reported patient fatalities due to withdrawal of care, and six patients who achieved mRS of 0–2 at 3 months.

A larger two-center (center A and B) cohort study was conducted by Hong et al. to further examine the effects of therapeutic hypothermia after successful recanalization with TICI 2b or 3 reperfusion [125]. Patients enrolled at center A were induced to mild hypothermia for 48 h (goal temperature of 34.5 °C) with intravascular cooling methods used in 95 %, while patients enrolled at center B received standard care without hypothermia. Patients in the hypothermia group ($n=39$) had an initial median NIHSS of 17, a baseline ASPECTS score of 6, and received mechanical thrombectomy in 31 cases (79.5 %), with no statistical baseline differences between groups. Compared to ReCCLAIM I, induction time was longer (378 ± 355 min). Patients who received cooling had less hemorrhagic transformation (61.5 % versus 86.1 %) and less cerebral edema (46.2 % versus 83.3 %) compared to the control group, with no other differences in medical complication rates. No difference in mortality rates were seen, but hypothermia patients had a higher proportion of good outcome (mRS ≤ 2) at 3 months.

Efficacy trials for mechanical thrombectomy in acute large vessel ischemic stroke utilized advanced neuroimaging techniques to select for patients who would most benefit from IAT [116–120]. Application of such imaging-based selection strategies may prove advantageous in future studies investigating combination IA clot retrieval and induced hypothermia or hypothermia alone.

8.6 Phase III Trials for Therapeutic Hypothermia

To date, there have been no published large phase III efficacy trials for induced hypothermia following acute ischemic stroke. Such research is urgently needed, as animal models and small phase I and II trials have proven feasibility, acceptable safety, and substantial potential clinical benefit with therapeutic cooling to mild or moderate hypothermia. Two ongoing phase III trials may help confirm or refute these preliminary findings.

ICTuS 2/3 is a novel phase 2 and 3 trial aimed at confirming the safety and feasibility of study protocol, then determining if combination IV thrombolysis with moderate endovascular hypothermia using study protocol is superior to thrombolysis alone [126]. ICTuS 2 ($n=400$) will target four milestones: (1) target temperature reached within 6 h of symptom onset, (2) no increased risk of pneumonia compared to controls, (3) no increased signs or symptoms of fluid overload due to chilled saline infusions, and (4) sufficient enrollment. If these milestones are met, study protocol will be seamlessly applied in ICTuS 3 ($n=1200$) for a total analyzed sample of 1600 patients. Endovascular cooling methods were chosen based on the rapid induction rates achieved with cooling catheters, while a target temperature of 33 °C is planned based on preclinical data suggesting possible superiority of moderate hypothermia compared to mild hypothermia [126]. Eligible adult patients will have an NIHSS score ≥ 7 and ≤20 (for right brain lesions) or ≤24 (for left brain lesions) and will have received systemic rtPA. Patients with planned use of IAT, including mechanical thrombectomy, will be excluded. The primary outcome measure will be the proportion of patients achieving a mRS of 0 or 1 at 3-months following stroke, while mortality and adverse events will also be reported.

EuroHYP-1 is a multicenter, randomized, phase III clinical trial of therapeutic hypothermia plus best medical treatment versus best medical treatment alone for acute stroke being conducted in Europe [127]. Eligible adult patients ($n=1500$) are allowed an NIHSS score ≥6 and ≤18 and a GCS motor score ≥5, while patients who do not receive IV thrombolysis may still be randomized to therapeutic hypothermia. In contrast to the ICTuS 2/3 study design, the occurrence of mechanical thrombectomy is not part of the delineated exclusion criteria. In patients randomized to hypothermia, cooling to a target body temperature of 34–35 °C will be initiated within 6 h of symptom onset and within 90 min after the start of alteplase, if given. Refrigerated normal saline infusions or surface cooling will be used for induction, with use of either surface or endovascular methods for the maintenance phase of 24 h. The primary outcome is mRS score at 3 months. Secondary outcomes include mortality rate, as well as final infarct size, infarct growth, cerebral swelling, and hemorrhagic transformation on repeat brain imaging at 48 h. An incremental cost-utility analysis is also planned.

Investigators of ICTuS 2/3 and EuroHYP-1 have coordinated data collection, with pooled analysis to follow the conclusion of both trials. Results are eagerly awaited.

9 Conclusion

There is an abundance of information that supports the use of hypothermia for acute ischemic stroke. However, there is no conclusive clinical data as of yet. Similar to the application of mechanical thrombectomy and surgical decompression, level I evidence for the benefit of therapeutic hypothermia in acute ischemic stroke has the potential to significantly alter the scope of practice for this common and devastating condition.

References

1. Tisherman SA, Rodriquez A, Safar P (1999) Therapeutic hypothermia in traumatology. Surg Clin North Am 79(6):1269–1289
2. Polderman KH (2004) Application of therapeutic hypothermia in the ICU: opportunities and pitfalls of a promising treatment modality. Part 1: indications and evidence. Intensive Care Med 30(4):556–575
3. Wang H, Olivero W, Wang D, Lanzino G (2006) Cold as a therapeutic agent. Acta Neurochir 148:565–570
4. Guly H (2011) History of accident hypothermia. Resuscitation 82:122–125
5. Gilbert M, Busund R, Skagseth A et al (2000) Resuscitation from accidental hypothermia of 13.7 degrees C with circulatory arrest. Lancet 355(9201):375–376
6. Forrester JM (2000) The origins and fate of James Currie's cold water treatment for fever. Med Hist 44:57–74
7. Fay T (1959) Early experiences with local and generalized refrigeration of the human brain. J Neurosurg 16:239–259
8. Cooper IS, Lee A (1961) Cryostatic congelation: a system for producing a limited controlled region of cooling or freezing of biologic tissues. J Nerv Ment Dis 133:259–263

9. Gage AA (1998) History of cryosurgery. Semin Surg Oncol 14:99–109
10. Rosomoff HL (1956) Hypothermia and cerebral vascular lesions. I. Experimental interruption of the middle cerebral artery during hypothermia. J Neurosurg 13:332–343
11. Spetzler RF, Hadley MN, Rigamonti D et al (1988) Aneurysms of the basilar artery treated with circulatory arrest, hypothermia and barbiturate cerebral protection. J Neurosurg 68:868–879
12. Solomon RA (1991) Principles of aneurysm surgery: cerebral ischemic protection, hypothermia, and circulatory arrest. Clin Neurosurg 41:351–363
13. Sealy WC, Brown IW, Young WG (1958) A report on the use of both extracorporeal circulation and hypothermia for open heart surgery. Ann Surg 147(5):603–613
14. Benson DW, Williams GR, Spencer F et al (1959) The use of hypothermia after cardiac arrest. Anesth Analg 38:423–428
15. Williams GR, Spencer FC (1958) The clinical use of hypothermia following cardiac arrest. Ann Surg 148:462–468
16. Nolan JP, Morley PT, Vanden Hoek TL et al (2003) Therapeutic hypothermia after cardiac arrest. An advisory statement by the advanced life support task force of the international liason committee on resuscitation. Circulation 108:118–121
17. Bernard SA, Gray TW, Buist MD et al (2002) Treatment of comatose survivors of out-of-hospital cardiac arrest with induced hypothermia. N Engl J Med 346:557–563
18. The Hypothermia after Cardiac Arrest Study Group (2002) Mild therapeutic hypothermia to improve the neurologic outcome after cardiac arrest. N Engl J Med 346:549–556
19. Arrich J, Holzer M, Havel C, et al (2016) Hypothermia for neuroprotection in adults after cardiopulmonary resuscitation. Cochrane Database Syst Rev 2:Art. No: CD004128. doi:10.1002/14651858.CD004128.pub4.
20. Shankaran S, Laptook AR, Ehrenkranz RA et al (2005) Whole-body hypothermia for neonates with hypoxic-ischemic encephalopathy. N Engl J Med 353:1574–1584
21. Azzopardi DV, Strohm BM, Edwards AD et al (2009) Moderate hypothermia to treat perinatal asphyxial encephalopathy. N Engl J Med 361:1349–1358
22. Clifton GL, Valadka A, Zygun D et al (2011) Very early hypothermia induction in patients with severe brain injury (the National Acute Brain Injury Study: hypothermia II): a randomized trial. Lancet Neurol 10(2):131–139
23. Andrews PJ, Sinclair HL, Rodriguez A et al (2015) Hypothermia for intracranial hypertension after traumatic brain injury. N Engl J Med 373:2403–2412
24. Levi AD, Casella G, Green BA et al (2010) Clinical outcomes using modest intravascular hypothermia after acute cervical spinal cord injury. Neurosurgery 66(4):670–677
25. Dididze M, Green BA, Dalton Dietrich W et al (2013) Systemic hypothermia in acute cervical spinal cord injury: a case-controlled study. Spinal Cord 51(5):395–400
26. Schwid SR, Petrie MD, Murray R et al (2003) A randomized controlled study of the acute and chronic effects of cooling therapy for MS. Neurology 60(12):1955–1960
27. Hall MG, Levan S, DeFrances CJ (2012) Hospitalization for stroke in U.S. hospitals, 1989-2009. NCHS Data Brief, no 95. 2012. National Center for Health Statistics, Hyattsville
28. Magistretti PJ, Pellerin L (1996) Cellular mechanisms of brain energy metabolism. Relevance to functional brain imaging and to neurodegenerative disorders. Ann N Y Acad Sci 777:380–387
29. Hatazawa J, Shimosegawa E, Toyoshima H et al (1999) Cerebral blood volume in acute brain infarction. Stroke 30:800–806
30. Powers WJ, Grubb RL, Raichle ME (1984) Physiological responses to focal cerebral ischemia in humans. Ann Neurol 16(5):546–552
31. Heiss WD, Huber M, Fink GR et al (1992) Progressive derangement of periinfarct viable tissue in ischemic stroke. J Cereb Blood Flow Metab 12(2):193–203
32. Busto R, Dietrich WD, Globus MY et al (1987) Small differences in intraischemic brain temperature critically determine the extent of ischemic neuronal injury. J Cereb Blood Flow Metab 7(6):729–738

33. Welsh FA, Sims RE, Harris VA (1990) Mild hypothermia prevents ischemic injury in gerbil hippocampus. J Cereb Blood Flow Metab 10(4):557–563
34. Lanier WL (1995) Cerebral metabolic rate and hypothermia: their relationship with ischemic neurologic injury. J Neurosurg Anesthesiol 7:216–221
35. Kimura T, Sako K, Tanaka K et al (2002) Effect of mild hypothermia on energy state recovery following transient forebrain ischemia in the gerbil. Exp Brain Res 145:83–90
36. Small DL, Morley P, Buchan AM (1999) Biology of ischemic cerebral cell death. Prog Cardiovasc Dis 42:185–207
37. Nedergaard M, Hansen AJ (1993) Characterization of cortical depolarizations evoked in focal cerebral ischemia. J Cereb Blood Flow Metab 13(4):568–574
38. Siesjo BK, Bengtsson F, Grampp W et al (1989) Calcium, excitotoxins, and neuronal death in brain. Ann N Y Acad Sci 568:234–251
39. Winfree CJ, Baker CJ, Connolly ES et al (1996) Mild hypothermia reduces penumbral glutamate levels in the rat permanent focal cerebral ischemia model. Neurosurgery 38:1216–1222
40. Globus MY-T, Alonso O, Dietrich WD et al (1995) Glutamate release and free radical production following brain injury: effects of post-traumatic hypothermia. J Neurochem 65:1704–1711
41. Baker AJ, Zornow MH, Grafe MR et al (1991) Hypothermia prevents ischemia-induced increases in hippocampal glycine concentrations in rabbits. Stroke 22:666–673
42. Dugan LL, Choi DW (1999) Free radicals in hypoxia-ischemia. In: Siegel GJ. Agranoff BW, Albers RW et al (eds) Basic neurochemistry: molecular, cellular and medical aspects, 6th edn. Lippincott-Raven, Philadelphia
43. Lipton P (1999) Ischemic cell death in brain neurons. Physiol Rev 79(4):1431–1568
44. Globus MY-T, Busto R, Lin B et al (1995) Detection of free radical activity during transient global ischemia and recirculation: effects of intraischemic brain temperature modulation. J Neurochem 65:1250–1256
45. Kimura A, Sakurada S, Ohkuni H et al (2002) Moderate hypothermia delays proinflammatory cytokine production of human peripheral blood mononuclear cells. Crit Care Med 30:1499–1502
46. Xu L, Yenari MA, Steinberg GK et al (2002) Mild hypothermia reduces apoptosis of mouse neurons in vitro early in the cascade. J Cereb Blood Flow Metab 22:21–28
47. Adachi M, Sohma O, Tsuneishi S et al (2001) Combination effect of systemic hypothermia and caspase inhibitor administration against hypoxic-ischemic brain damage in neonatal rats. Pediatr Res 50:590–595
48. Ning XH, Chen SH, Xu CS et al (2002) Hypothermic protection of the ischemic heart via alterations in apoptotic pathways as assessed by gene array analysis. J Appl Physiol 92:2200–2207
49. Schwab S, Spranger M, Aschoff A et al (1997) Brain temperature monitoring and modulation in patients with severe MCA infarction. Neurology 48:762–767
50. Mackowiak PA (1998) Concepts of fever. Arch Intern Med 158:1870–1881
51. Koennecke HC, Leistner S (2001) Prophylactic antipyretic treatment with acetaminophen in acute ischemic stroke: a pilot study. Neurology 57:2301–2303
52. Dippel DW, van Breda EJ, van der Worp HB et al (2003) Effect of paracetamol (acetaminophen) and ibuprofen on body temperature in acute ischemic stroke PISA, a phase II double-blind, randomized, placebo-controlled trial. BMC Cardiovasc Disord 3:2
53. Cormio M, Citerio G (2007) Continuous low dose diclofenac sodium infusion to control fever in neurosurgical critical care. Neurocrit Care 6:82–89
54. Mayer SA, Kowalksi RG, Presciutti M et al (2004) Clinical trial of a novel surface cooling system for fever control in neurocritical care patients. Crit Care Med 32:2508–2515
55. Janata A, Weihs W, Bayegan K et al (2008) Therapeutic hypothermia with a novel surface cooling device improves neurologic outcome after prolonged cardiac arrest in swine. Crit Care Med 36(3):895–902
56. Kim F, Olsufka M, Carlbom D et al (2005) Pilot study of rapid infusion of 2L of 4°C normal saline for induction of mild hypothermia in hospitalized, comatose survivors of out-of-hospital cardiac arrest. Circulation 112:715–719

57. Polderman KH, Rijnsburger ER, Peerdeman SM et al (2005) Induction of hypothermia in patients with various types of neurological injury with use of large volumes of ice-cold intravenous fluid. Crit Care Med 33:2744–2751
58. Holzer M (2008) Devices for rapid induction of hypothermia. Eur J Anaesthesiol Suppl 42:31–38
59. Seder DB, van der Kloot TE (2009) Methods of cooling: practical aspects of therapeutic temperature management. Crit Care Med 37(7 Suppl):S211–S222
60. Polderman KH, Herold I (2009) Therapeutic hypothermia and controlled normothermia in the intensive care unit: practical considerations, side effects, and cooling methods. Crit Care Med 37(3):1101–1120
61. Simosa HF, Petersen DJ, Agarwal SK et al (2007) Increased risk for deep venous thrombosis with endovascular cooling in patients with traumatic head injury. Am Surg 73:461–464
62. Want H, Olivero W, Lanzino G et al (2004) Rapid and selective cerebral hypothermia achieved using a cooling helmet. J Neurosurg 100(2):272–277
63. Castrén M, Nordberg P, Svensson L et al (2010) Intra-arrest transnasal evaporative cooling: a randomized, prehospital, multicenter study (PRINCE: Pre-ROSC IntraNasal Cooling Effectiveness). Circulation 122(7):729–736
64. Nordberg P, Taccone FS, Castrén M et al (2013) Design of the PRINCESS trial: pre-hospital resuscitation intra-nasal cooling effectiveness survival study (PRINCESS). BMC Emerg Med 13:21. doi:10.1186/1471-227X-13-21
65. Hardy JD, Stolwijk JA (1966) Partitional calorimetric studies of man during exposures to thermal transients. J Appl Physiol 21:1799–1806
66. Claessens-van Ooijen AM, Westerterp KR, Wouters L et al (2006) Heat production and body temperature during cooling and rewarming in overweight and lean men. Obesity 14:1914–1920
67. Badjatia N, Kowalski R, Schmidt M et al (2007) Predictors and clinical implications of shivering during therapeutic normothermia. Neurocrit Care 6:186–191
68. Oddo M, Frangos S, Maloney-Wilensky E et al (2010) Effect of shivering on brain tissue oxygenation during induced normothermia in patients with severe brain injury. Neurocrit Care 12(1):10–16
69. Badjatia N, Strongilis E, Gordon E et al (2008) Metabolic impact of shivering during therapeutic temperature modulation: the bedside shivering assessment scale. Stroke 39:3242–3247
70. Park B, Lee T, Berger K et al (2015) Efficacy of nonpharmacological antishivering interventions: a systematic analysis. Crit Care Med 43(8):1757–1766
71. Park SM, Mangat HS, Berger K et al (2012) Efficacy spectrum of antishivering medications: meta-analysis of randomized controlled trials. Crit Care Med 40(11):3070–3082
72. Chamorro C, Borrallo JM, Romera MA et al (2010) Anesthesia and analgesia protocol during therapeutic hypothermia after cardiac arrest: a systematic review. Anesth Analg 110:1328–1335
73. Choi HA, Ko SB, Presciutti M et al (2011) Prevention of shivering during therapeutic temperature modulation: the Columbia anti-shivering protocol. Neurocrit Care 14:389–394
74. Thoresen M, Whitelaw A (2000) Cardiovascular changes during mild therapeutic hypothermia and rewarming in infants with hypoxic-ischemic encephalopathy. Pediatrics 106:92–99
75. Reuler JB (1978) Hypthermia: pathophysiology, clinical settings, and management. Ann Intern Med 89:519–527
76. Boddicker KA, Zhang Y, Zimmerman MB et al (2005) Hypothermia improves defibrillation success and resuscitation outcomes from ventricular fibrillation. Circulation 111:3195–3201
77. Polderman KH, Peerdeman SM, Girbes ARJ (2001) Hypophosphatemia and hypomagnesemia induced by cooling in patients with severe head injury. J Neurosurg 94:697–705
78. McIntosh TK, Vink R, Yamakami I et al (1989) Magnesium protects against neurological deficit after brain injury. Brain Res 482:252–260
79. Polderman KH, van Zanten ARH, Girbes ARJ (2003) The importance of magnesium in critically ill patients: a role in mitigating neurological injury and in the prevention of vasospasms. Intensive Care Med 29:1202–1203

80. Van den Bergh WM, Algra A, van Kooten F et al (2005) Magnesium sulfate in aneurysmal subarachnoid hemorrhage: a randomized controlled trial. Stroke 36:1011–1015
81. Soliman HM, Mercan D, Lobo SS et al (2003) Development of ionized hypomagnesemia is associated with higher mortality rates. Crit Care Med 31:1082–1087
82. Rubeiz GJ, Thill-Baharozian M, Hardie D et al (1993) Association of hypomagnesaeia and mortality in acutely ill medical patients. Crit Care Med 21:203–209
83. Schwab S, Schwarz S, Spranger M et al (1998) Moderate hypothermia in the treatment of patients with severe middle cerebral artery infarction. Stroke 29:2461–2466
84. Schwab S, Georgiadis D, Berrouschot J et al (2001) Feasibility and safety of moderate hypothermia after massive hemispheric infarction. Stroke 32(9):2033–2035
85. Hemmen TM, Raman R, Guluma KZ et al (2010) Intravenous thrombolysis plus hypothermia for acute treatment of ischemic stroke (ICTuS-L): final results. Stroke 41(10):2265–2270
86. Michelson AD, MacGregor H, Barnard MR et al (1994) Hypothermia-induced reversible platelet dysfunction. Thromb Haemost 71:633–640
87. Watts DD, Trask A, Soeken K et al (1998) Hypothermic coagulopathy in trauma: effect of varying levels of hypothermia on enzyme speed, platelet function, and fibrinolytic activity. J Trauma 44:846–854
88. Valeri CR, MacGregor H, Cassidy G et al (1995) Effects of temperature on bleeding time and clotting time in normal male and female volunteers. Crit Care Med 23:698–704
89. Ferrara A, MacArthur JD, Wright HK et al (1990) Hypothermia and acidosis worsen coagulopathy in patients requiring massive transfusion. Am J Surg 160:515–518
90. Reed RL, Bracey AW, Hudson JD et al (1990) Hypothermia and blood coagulation: dissociation between enzyme activity and clotting factor levels. Circ Shock 32:141–152
91. Hernandez-Perez M, Perez de la Ossa N, Aleu A et al (2014) Natural history of acute stroke due to occlusion of the middle cerebral artery and intracranial internal carotid artery. J Neuroimaging 24:354–358
92. Krieger DW, De Georgia MA, Abou-Chebl A et al (2001) Cooling for acute ischemic brain damage (cool aid): an open pilot study of induced hypothermia in acute ischemic stroke. Stroke 32(8):1847–1854
93. De Georgia MA, Krieger DW, Abou-Chebl A et al (2004) Cooling for acute ischemic brain damage (COOL AID): a feasibility trial of endovascular cooling. Neurology 63(2):312–317
94. O'Collins VE, Macleod MR, Donnan GA et al (2006) 1,026 experimental treatments in acute stroke. Ann Neurol 59:467–477
95. van der Worp HB, Sena ES, Donnan GA et al (2007) Hypothermia in animal models of acute ischaemic stroke: a systematic review and meta-analysis. Brain 130:3063–3074
96. van der Worp HB, Macleod MR, Kollmar R (2010) Therapeutic hypothermia for acute ischemic stroke: ready to start large randomized trials? J Cereb Blood Flow Metab 30:1079–1093
97. Dumitrascu OM, Lamb J, Lyden PD (2016) Still cooling after all these years: meta-analysis of pre-clinical trials of therapeutic hypothermia for acute ischemic stroke. J Cereb Blood Flow Metab 36(7):1157–1164. pii:0271678X16645112.
98. Baker CJ, Onesti ST, Solomon RA (1992) Reduction by delayed hypothermia of cerebral infarction following middle cerebral artery occlusion in the rat: a time-course study. J Neurosurg 77:438–444
99. Baena RC, Busto R, Dietrich WD et al (1997) Hyperthermia delayed by 24 hours aggravates neuronal damage in rat hippocampus following global ischemia. Neurology 48:768–773
100. Reith J, Jørgensen HS, Pedersen PM et al (1996) Body temperature in acute stroke: relation to stroke severity, infarct size, mortality, and outcome. Lancet 347(8999):422–425
101. Jørgensen HS, Reith J, Pedersen PM et al (1996) Body temperature and outcome in stroke patients. Lancet 348(9021):193
102. Kammersgaard LP, Rasmussen BH, Jørgensen HS et al (2000) Feasibility and safety of inducing modest hypothermia in awake patients with acute stroke through surface cooling: a case-control study: the Copenhagen Stroke Study. Stroke 31(9):2251–2256

103. Hacke W, Schwab S, Horn M et al (1996) 'Malignant' middle cerebral artery territory infarction: clinical course and prognostic signs. Arch Neurol 53(4):309–315
104. Berrouschot J, Sterker M, Bettin S et al (1998) Mortality of space-occupying ('malignant') middle cerebral artery infarction under conservative intensive care. Intensive Care Med 24(6):620–623
105. Georgiadis D, Schwarz S, Kollmar R et al (2001) Endovascular cooling for moderate hypothermia in patients with acute stroke: first results of a novel approach. Stroke 32(11):2550–2553
106. Zweifler RM, Voorhees ME, Mahmood MA et al (2003) Induction and maintenance of mild hypothermia by surface cooling in non-intubated subjects. J Stroke Cerebrovasc Dis 12(5):237–243
107. Guluma KZ, Hemmen TM, Olsen SE et al (2006) A trial of therapeutic hypothermia via endovascular approach in awake patients with acute ischemic stroke: methodology. Acad Emerg Med 13(8):820–827
108. Wardlaw JM, Murray V, Berge E, et al (2014) Thrombolysis for acute ischaemic stroke. Cochrane Database Syst Rev 7:CD000213. doi:10.1002/14651858.CD000213.pub3
109. Piironen K, Tiainen M, Mustanoja S et al (2014) Mild hypothermia after intravenous thrombolysis in patients with acute stroke. Stroke 45:486–491
110. Lyden PD, Allgren RL, Ng K et al (2005) Intravascular cooling in the treatment of stroke (ICTuS): early clinical experience. J Stroke Cerebrovasc Dis 14(3):107–114
111. Vahedi K, Vicaut E, Mateo J et al (2007) Sequential-design, multicenter, randomized, controlled trial of early decompressive craniectomy in malignant middle cerebral artery infarction (DECIMAL Trial). Stroke 38(9):2506–2517
112. Jüttler E, Schwab S, Schmiedek P (2007) Decompressive surgery for the treatment of malignant infarction of the middle cerebral artery (DESTINY): a randomized, controlled trial. Stroke 38(9):2518–2525
113. Hofmeijer J, Kappelle LJ, Algra A et al (2009) Surgical decompression for space-occupying cerebral infarction (the Hemicraniectomy After Middle Cerebral Artery infarction with Life-threatening Edema Trial [HAMLET]): a multicentre, open, randomised trial. Lancet Neurol 8(4):326–333
114. Back L, Nagaraja V, Kapur A et al (2015) Role of decompressive hemicraniectomy in extensive middle cerebral artery strokes: a meta-analysis of randomized trials. Intern Med J 45(7):711–717
115. Els T, Oehm E, Voigt S et al (2006) Safety and therapeutical benefit of hemicraniectomy combined with mild hypothermia in comparison with hemicraniectomy alone in patients with malignant ischemic stroke. Cerebrovasc Dis 21:79–85
116. Berkhemer OA, Fransen PS, Beumer D et al (2015) A randomized trial of intraarterial treatment for acute ischemic stroke. N Engl J Med 372(1):11–20
117. Campbell BC, Mitchel PJ, Kleinig TJ et al (2015) Endovascular therapy for ischemic stroke with perfusion-imaging selection. N Engl J Med 372(11):1009–1018
118. Goyal M, Demchuk AM, Menon BK et al (2015) Randomized assessment of rapid endovascular treatment of ischemic stroke. N Engl J Med 372(11):1019–1030
119. Saver JL, Goyal M, Bonafe A et al (2015) Stent-retriever thrombectomy after intravenous t-PA vs. t-PA alone in stroke. N Engl J Med 372(24):2285–2295
120. Jovin TG, Chamorro A, Cobo E et al (2015) Thrombectomy within 8 hours after symptom onset in ischemic stroke. N Engl J Med 372(24):2296–2306
121. Palaniswami M, Yan B (2015) Mechanical thrombectomy is now the gold standard for acute ischemic stroke: implications for routine clinical practice. Interv Neurol 4:18–29
122. Hwang YA, Jeon JS, Kim YW, et al (2016) Impact of immediate post-reperfusion cooling on outcome in patients with acute stroke and substantial ischemic changes. J Neurointervent Surg. 1–5
123. Horn CM, Sun CHJ, Nogueira RG et al (2014) Endovascular reperfusion and cooling in cerebral acute ischemia (ReCCLAIM I). J Neurointerv Surg 6:91–95
124. Warach S, Latour LL (2004) Evidence of reperfusion injury, exacerbated by thrombolytic therapy, in human focal brain ischemia using a novel imaging marker of early blood-brain barrier disruption. Stroke 35(11 Suppl 1):2659–2661

125. Hong JM, Lee JS, Song HJ et al (2014) Therapeutic hypothermia after recanalization in patients with acute ischemic stroke. Stroke 45:134–140
126. Lyden PD, Hemmen TM, Grotta J (2014) Endovascular therapeutic hypothermia for acute ischemic stroke: ICTuS 2/3 protocol. Int J Stroke 9(1):117–125
127. van der Worp HB, Macleod MR, Bath PMW et al (2014) EuroHYP-1: European multicenter, randomized, phase III clinical trial of therapeutic hypothermia plus best medical treatment vs. best medical treatment alone for acute ischemic stroke. Int J Stroke 9(5):642–645

Chapter 20
Modern Endovascular Treatment of Ischemic Disease

Steven L. Gogela, Todd A. Abruzzo, and Mario Zuccarello

Abstract Acute ischemic strokes still account for substantial morbidity and mortality worldwide despite decades of research aimed at reducing the burden of this disease. Since the establishment of intravenous thrombolysis as the standard of care, advances in device technology have rapidly progressed while randomized clinical trials have struggled to keep pace. Early trials were hampered by inconsistent use of technology and poor patient selection, leading to overall futility. However, a string of recent studies utilizing the most modern technology has unequivocally proven the superiority of endovascular treatment for acute ischemic strokes caused by large vessel occlusion in select patients. We describe the sequential progression of this therapeutic paradigm in parallel with the clinical trials that have proven its efficacy and identify clinical questions that remain yet unanswered.

Keywords Stroke intervention • Endovascular stroke treatment • Acute ischemic stroke • Neurointerventional stroke treatment • Stent retriever

1 Introduction

Stroke is a well-known source of extensive morbidity and mortality in modern society. As recently as 2013, the annual incidence was estimated at 795,000 new strokes each year, with more than three-fourths of these falling in the acute ischemic stroke category [1]. Worldwide, cerebral infarctions represent the second most common

S.L. Gogela, M.D. (✉)
Department of Neurosurgery, College of Medicine, University of Cincinnati (UC), Cincinnati, OH, USA
e-mail: sgogela@gmail.com

T.A. Abruzzo, M.D. • M. Zuccarello, M.D.
Department of Neurosurgery, College of Medicine, University of Cincinnati (UC), Cincinnati, OH, USA

Comprehensive Stroke Center at UC Neuroscience Institute, Cincinnati, OH, USA

Mayfield Clinic, Cincinnati, OH, USA
e-mail: Mario.Zuccarello@uc.edu

P.A. Lapchak, J.H. Zhang (eds.), *Neuroprotective Therapy for Stroke and Ischemic Disease*, Springer Series in Translational Stroke Research,
DOI 10.1007/978-3-319-45345-3_20

cause of mortality and are responsible for the second greatest loss of disability-adjusted life years in high-income countries [2–4].

Despite the profound effect that this disease has had on society, treatment modalities have been limited to medical therapy alone until recent times. The fact that cerebral ischemia sets off a cascade of cellular processes with the potential to cause severe and permanent neurological injury is well known, the particulars of which are out of the scope of this chapter and are addressed elsewhere in this comprehensive text. Overall, the goal in acute ischemic stroke (AIS) treatment is the rapid institution of reperfusion in an effort to limit infarction size and salvage ischemic tissue before permanent injury. In this chapter, we elucidate the progression of device technology and evidence-based management of this common condition that has led to the rise of endovascular intervention as the primary treatment of AIS for select patients.

2 Medical Thrombolysis in Stroke

Ischemic strokes result from a state of cerebral hypoperfusion and are most commonly secondary to arterial occlusion. Initial treatment efforts sought to reverse this condition by recanalizing the arterial occlusion with systemically administered thrombolytic drugs. Early pilot studies that assessed systemic administration of thrombolytic drugs were fraught with high rates of intracerebral hemorrhage (ICH) [5, 6]. Nearly two decades later, additional pilot studies emphasizing the early administration of controlled doses of intravenous (IV) recombinant human tissue plasminogen activator (rt-PA) delivered an encouraging safety profile and opened the door for randomized trials using this treatment modality [7, 8].

Eventually, the National Institute of Neurological Disorders and Stroke (NINDS) rt-PA Stroke Study Group published the results of a landmark randomized placebo-controlled clinical trial in the *New England Journal of Medicine* that established the use of IV thrombolysis in AIS treatment. This trial applied numerous exclusion criteria that included: an undetermined time of onset, ICH seen on a screening computed tomography (CT) scan of the head, prior stroke or ICH, serious head trauma within 3 months, blood pressure exceeding 185 mmHg systolic or 110 mmHg diastolic, recent surgery or arterial puncture, improving or minor symptomatology, suspicion for subarachnoid hemorrhage, anticoagulation, various lab abnormalities, or seizure at stroke onset [9].

Following these rigorous exclusion criteria, 624 patients were randomized: half received the study dose (0.9 mg/kg) of IV t-PA and half IV placebo. Despite a higher rate of symptomatic ICH (sICH), patients who received IV t-PA within 3 h of stroke onset had significantly improved clinical outcomes and reduced disability at all measured time points [9]. These positive findings established IV thrombolysis as the standard of care for a select group of patients. However, a significant proportion of the stroke population presented outside of the 3-h window and were thus ineligible for this treatment. A later study extended the therapeutic window to 4.5 h after stroke onset after finding significantly improved clinical outcomes in patients who received IV t-PA in the 3–4.5-h temporal window [10, 11].

In reality, this extension to a 4.5-h window did not provide substantial societal benefit. One study found that only 0.5 % of 1858 consecutive patients presenting with AIS within 3–4.5 h of stroke onset were eligible for thrombolysis: overall, 22 % of observed stroke patients presented within 3 h and 3.4 % between 3 and 4.5 h [12]. This limited impact on the AIS population highlights the importance of timely patient transportation and improved awareness and recognition of stroke symptoms in the general population. Regardless, IV t-PA therapy was now the proven standard of care and a major step forward in stroke treatment.

3 Initial Approaches to Endovascular Stroke Treatment

Studies that assessed vascular imaging of patients after the administration IV t-PA found a significant shortcoming of the medication: specifically, it was an ineffective treatment modality for recanalization of large vessel occlusions (LVO). Rates of resolution for arterial occlusions with IV t-PA alone were reported as low as 4 % in the distal intracranial internal carotid artery (ICA) and 30 % in the middle cerebral artery (MCA) [13]. Researchers recognized that the failure of intravenously administered thrombolytics to recanalize intracerebral arterial occlusions was at least partially the consequence of inefficient convective transport of drug to the arterial occlusion site caused by the target lesion. Early pioneers in the field of stroke intervention conceptualized site-directed transcatheter administration of thrombolytics directly to the arterial occlusive thrombus as a means of overcoming the circulatory roadblocks imposed by arterial occlusive pathology. This led to the birth of endovascular treatment for AIS patients as an alternative solution for those either not eligible for or unresponsive to IV t-PA.

Intra-arterial (IA) treatment of ischemic stroke naturally began with the application of thrombolytic medication (i.e., t-PA) to the site of the arterial occlusive thrombus (Fig. 20.1). After publication of several small cohort studies in support of IA t-PA, two large-scale trials then established IA thrombolysis as a viable treatment. First, the Prolyse in Acute Cerebral Thromboembolism (PROACT) II trial was a randomized controlled trial (RCT) of IA r-pro-urokinase (rproUK) vs. standard medical therapy for MCA occlusions [14]. Patients meeting inclusion criteria underwent diagnostic cerebral angiography within 6 h of onset and were randomized if a Thrombolysis in Myocardial Infarction (TIMI) score of 0 (occlusion) or 1 (contrast penetration with minimal perfusion) was diagnosed in the symptomatic first (M1) or proximal second (M2) segment of the MCA (Table 20.1). All patients in both treatment arms received a low-dose bolus and low-dose infusion of IV heparin during the interventional procedure once occlusion was verified for a total of 4 h.

The adaptation of low-dose heparin protocols in the IA treatment arm was an important aspect of trial design that reduced the prohibitively high rates of procedure-related cerebral hemorrhage found in earlier trials of IA thrombolytic therapy. The PROACT II study found statistically significant improvement in recanalization and clinical outcome measures in patients who received IA thrombolytic. MCA

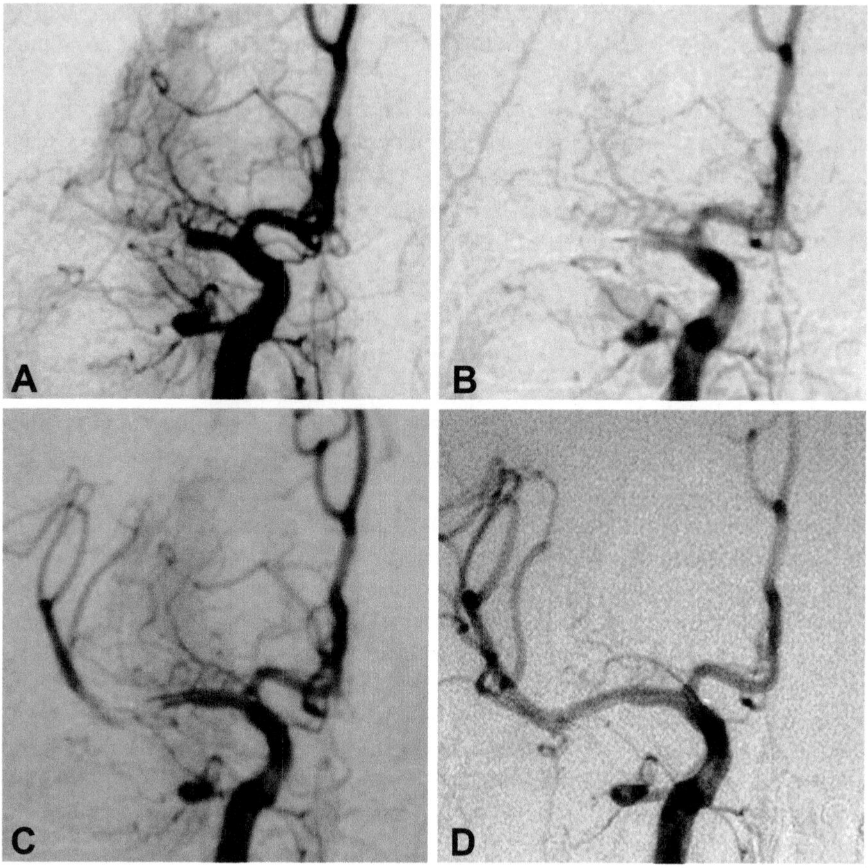

Fig. 20.1 Series of angiograms detailing intra-arterial thrombolysis of the middle cerebral artery. (**a**) Clot visualization. (**b**–**d**) tPA 1 mg/cm^3 infused distal to (**b**), into (**c**), and proximal (**d**) to clot. Total dose 22 mg

recanalization (TIMI 2–3) occurred in 66 % of the 121 patients in the IA rproUK cohort and 18 % of the 59 patients in the control group ($p<0.001$). TIMI 3 (complete recanalization) rates at 2 h were 19 % and 2 %, respectively ($p<0.003$; Table 20.1). This resulted in a good clinical outcome (mRS score of 0–2 at 90 days) in 40 % of rproUK and 25 % of control patients ($p=0.04$) despite sICH rates of 10 % and 2 % ($p=0.06$), respectively. Mortality was statistically similar between the two groups [14].

Years later, the Middle Cerebral Artery Embolism Local Fibrinolytic Intervention Trial (MELT) Japan was organized to compare treatment outcomes for patients with MCA occlusions who received either IA urokinase ($n=57$) or placebo ($n=57$) within 3 h of stroke onset. The trial was stopped early by the Independent Monitoring Committee after approval of IV tPA in Japan and thus did not reach statistical significance. The study identified trends toward higher rates of mRS 0–2 scores at 90 days, improved recanalization, and modest rates of sICH in patients treated with IA thrombolysis [15]. These results were combined with the PROACT II data in a meta-analysis that found statistically significant improvement in all

Table 20.1 Summary comparing reperfusion grading scales used in endovascular stroke trials [59–61]

Reperfusion grade	Thrombolysis in myocardial infarction (TIMI) risk score	Thrombolysis in cerebral infarction (TICI)	Modified TICI (mTICI)
0	Absence of antegrade flow: no perfusion	No perfusion	No perfusion
1	Faint antegrade flow with incomplete filling: penetration without perfusion	Penetration with minimal perfusion	Limited antegrade reperfusion: limited distal branch filling
2	Delayed antegrade flow with filling of distal branches: partial perfusion	Partial perfusion: 1B: <2/3 of vascular territory visualized	2a: reperfusion of <1/2 of occluded territory
		2B: complete filling of vascular territory but filling slower than usual	2b: reperfusion of >1/2 of occluded territory
3	Normal flow to all distal branches: complete perfusion	Complete reperfusion	Complete reperfusion

standard functional outcomes, giving further support for transcatheter IA thrombolytic therapy in AIS caused by MCA occlusion [16].

Early experience with IA thrombolytic therapy led interventionists to realize that mechanical disruption of the arterial occlusive thrombus by microguidewire manipulation and repetitious penetration of the target occlusion with the treating microcatheter could accelerate and increase the likelihood of recanalization by improving convective transport of drug across a greater surface area of thrombus. Additional adjuvant endovascular techniques being popularized simultaneously included the use of balloon angioplasty for LVO resistant to IA thrombolysis. In patients with persistent occlusion despite IA t-PA, Ringer et al. reported that balloon angioplasty led to improved rates of reperfusion both at the site of thrombus and in the more distal vasculature, with more effective transport of thrombolytic medication. Although the authors noted some added risk of associated vessel rupture and reperfusion hemorrhage, this technique was recommended as a salvage procedure [17].

4 Early Endovascular Trials

With its growing popularity among interventionalists, it was clear that a randomized study of endovascular stroke treatment was due. First, the Emergency Management of Stroke Bridging Trial established the safety, feasibility, and potential efficacy of combination IV plus IA t-PA treatment of LVO within 3 h of onset [18]. Based on this retrospective study, the first of three Interventional Management of Stroke (IMS) trials was conducted. IMS-I was a prospective trial with only a treatment arm for 80 patients aged 18–80 with AIS caused by LVO [19]. Inclusion criteria included a National Institute of Health Stroke Scale (NIHSS) score $\geq$10 and presentation within 3 h of stroke onset. Patients received IV t-PA (o.6 mg/kg) within 3 h, followed by IA t-PA (up to 22 mg) at the site of occlusion if it persisted on diagnostic angiography. Following endovascular treatment, Thrombolysis in Cerebral Infarction (TICI) perfusion scores of 2–3 were achieved in 56 % and a Modified Rankin Score (mRS) of 0–2 at 90 days in 43 % (Table 20.1). Comparing IMS-I to the NINDS IV t-PA trial, rates of sICH (6.3 % vs. 6.6 %) and 3-month mortality (16 % vs. 24/21 % placebo/IV t-PA) were similar [19]. These results again confirmed the safety of transcatheter IA thrombolysis in stroke. However, there was widespread dissatisfaction concerning the 51 % rate of TICI 2–3 reperfusion.

The first mechanical devices for endovascular stroke intervention were developed in parallel to this growing clinical experience with transcatheter IA thrombolysis, beginning with the EKOS MicroLYSUS® (Ekos Corporation, Bothell, WA) catheter. This microcatheter, outfitted with an ultrasound transducer at the tip, allowed the delivery of both t-PA and endovascular ultrasound directly into the arterial occlusive thrombus. The addition of ultrasound was theorized to increase permeability of the thrombus, drive t-PA further into the thrombus matrix, and uncoil fibrin chains, thus exposing peptide sequence targets to substrate specific plasminogen active sites. Given the disappointing reperfusion rates of IMS-I, exploration of the benefits of the EKOS catheter was conducted in a prospective setting with an expectation of being able to improve this metric.

The design of IMS-II was identical to IMS-I except for the additional use of the EKOS catheter whenever clinically possible. In this prospective, single-armed study, 81 patients received IV t-PA within 3 h and subsequently interventional treatment if persistent occlusion was found on angiography [20]. Next, IA t-PA was administered at the site of thrombus using either a standard microcatheter or the EKOS MicroLYSUS® infusion catheter; a maximum dose of 22 mg was given over 2 h or until reperfusion was achieved. Similar to IMS-I, this study achieved significantly improved clinical outcomes compared with the NINDS trial as measured by the Barthel Index and Global Outcome Test. At 90 days follow up, the IMS-II results showed a nonsignificant trend toward a mRS 0–1 and 46 % of patients had achieved mRS 0–2. Differences in mortality at 3 months (16 %) and rates of sICH (9.9 %) were statistically similar to the NINDS data. TICI 2–3 reperfusion was achieved in 60 % of patients (Table 20.1) [20]. Of note, because the EKOS catheter failed to access the intracranial occlusion site in a large number of patients, it is no longer commonly used as an endovascular tool in stroke intervention. Overall, the combined results of IMS-1 and IMS-II indicated the need for randomized study of endovascular intervention in the treatment of AIS.

5 Rapid Advancement of Endovascular Technology

As endovascular treatment of stroke was building momentum, so was the development of interventional device technology that aimed to improve the rate and efficiency of cerebrovascular reperfusion. The first device to receive US Food and Drug Administration (FDA) approval for cerebral arterial occlusion was the Mechanical Embolus Removal in Cerebral Ischemia (MERCI®) Retriever (Concentric Medical, Mountainview, CA) in 2004 (Fig. 20.2). This technology consists of a self-expanding

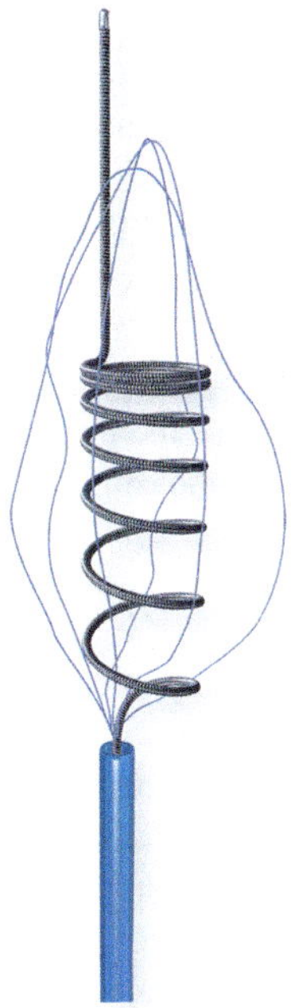

Fig. 20.2 Merci V-series retriever with seven loops of platinum filaments. This device is no longer used in endovascular stroke treatment. Reproduced with permission

array of stacked helical loops of NiTiNOL (Nickel-Titanium Naval Ordinance Laboratory) tethered to a pusher wire. The helical array of stacked loops is advanced linearly across the arterial occlusive thrombus through a microcatheter in a collapsed state by applying forward tension to the in-line pusher wire. While maintaining forward pressure on this wire, the stacked helical loops of self-expanding shape memory alloy are deployed distal to the arterial occlusive thrombus by withdrawal of the constraining microcatheter. The expanded, stacked helical loops are used to engage and extract the arterial occlusive thrombus by retracting the in-line pusher wire and coaxial microcatheter into a recovery catheter, which is stationed in the upstream cervicocerebral access artery (typically the cervical internal carotid artery). Adjunctive balloon occlusion and manual aspiration of the upstream cervicocerebral access artery were recommended to reverse blood flow and reduce risk of distal embolization during the process of thrombus extraction [21].

In a prospective, single-arm trial of 151 patients who underwent treatment with the MERCI device within 8 h of stroke onset, 48 % achieved recanalization. Good clinical outcome (mRS 0–2 at 90 days) was realized in 46 % of patients who achieved recanalization (TICI 2–3) and only in 10 % of those who did not ($p < 0.00001$) [22]. In a later single-arm, prospective study with newer generation MERCI devices, 57 % of 131 patients with AIS and persistent LVO after IV t-PA achieved recanalization with the MERCI device alone; this increased to 70 % when adjunctive IA t-PA or other devices were utilized for persistent occlusion. Good clinical outcomes (mRS 0–2) were attained in 36 % of patients at 90 days and sICH occurred in 10 %. In a subset analysis of patients with intracranial ICA occlusions, good clinical outcomes were observed in 39 % of patients who were recanalized and only 3 % of those who were not [23].

Overall, outcomes were similar to the PROACT II study involving IA t-PA alone [24, 25]. Later, a coaxial Distal Access Catheter (DAC) (Concentric Medical, Mountainview, CA) was introduced as an adjunct to the MERCI Retriever. The DAC improved the efficiency in navigating the tortuous distal ICA and the mechanical factors that favor successful thrombus extraction.

The next device to gain FDA approval for stroke intervention was the Penumbra System® (Penumbra Inc., Alameda, CA) in 2007. This device consists of a large bore thromboaspiration catheter that is advanced over a microguidewire to the proximal face of the arterial occlusive thrombus. Vacuum suction is applied to the hub of the thromboaspiration catheter to remove thrombus through the lumen. Next, a pear-shaped "plunger" mounted on a pusher wire—coined a "separator" device—is repeatedly withdrawn into the shaft of the thromboaspiration catheter to clear thrombus while maintaining a continuous suction force at the catheter tip. The catheter is incrementally advanced to remove additional thrombus [26]. The clinical efficacy of the Penumbra system was tested in a prospective, single-arm, industry-sponsored study of 125 patients with NIHSS score ≥ 8 who were randomized within 3 h of stroke onset, and were either ineligible for IV t-PA or had persistent LVO despite IV t-PA. Although the method for determining recanalization was not completely clear, the trial reported that 82 % of patients had successful partial or complete recanalization after treatment. Other results included sICH in 11 %, an mRS 0–2 in 25 %, and mortality in 33 % [26]. A retrospective, multicenter meta-analysis of 157 patients treated with the Penumbra System, which included those in the same database, later confirmed these promising results. Specifically, TICI 2–3 reperfusion was achieved in 87 %, mRS 0–2 in 41 % at 90 days, sICH in 6.4 %, and mortality in 20 % (Table 20.1) [27].

The Penumbra System® MAX™ was later introduced with a larger lumen and ovalization-resistant shaft (Fig. 20.3). This construction achieved a greater suction

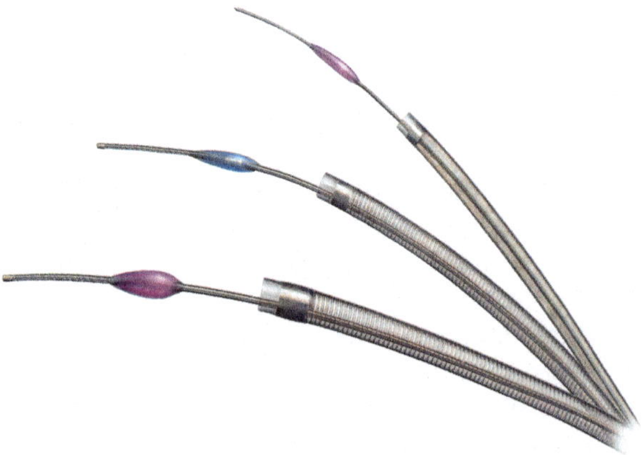

Fig. 20.3 Three sizes of the Penumbra MAX system of aspiration catheters with separator devices. Note that separators previously used to clear the catheter of clot during piecemeal removal are no longer commonly used with the ADAPT or SOLUMBRA techniques. Reproduced with permission from Penumbra Inc.

force and decreased median recanalization time from 45 min in the initial trial to 20 min with the updated system [26, 28, 29].

This latest generation of thromboaspiration catheters fostered the development of a direct-aspiration first-pass technique (ADAPT) for rapid recanalization of cerebral arteries [30]. This technique is a modification of the traditional thromboaspiration method whereby the arterial occlusive thrombus is extracted en bloc, in its entirety. Aspiration is achieved either through a large bore catheter (thrombus ingestion), or by wedging the thrombus into the shaft of a thromboaspiration catheter and removing the thrombus and catheter as a single unit (Figs. 20.4 and 20.5). The greatest advantage of this approach relates to its remarkable speed and simplicity. Compared with other mechanical revascularization approaches, the ADAPT technique is relatively inexpensive and less traumatic. Since the likelihood of a good clinical outcome decreases 12–15 % every 30 min [31, 32], ADAPT has become the default initial approach of many neurointerventionists today. Initial reports of this technique showed mTICI 2b-3 reperfusion in 78 % of patients. More recent experience with the latest iteration of thromboaspiration technologies indicate that reperfusion

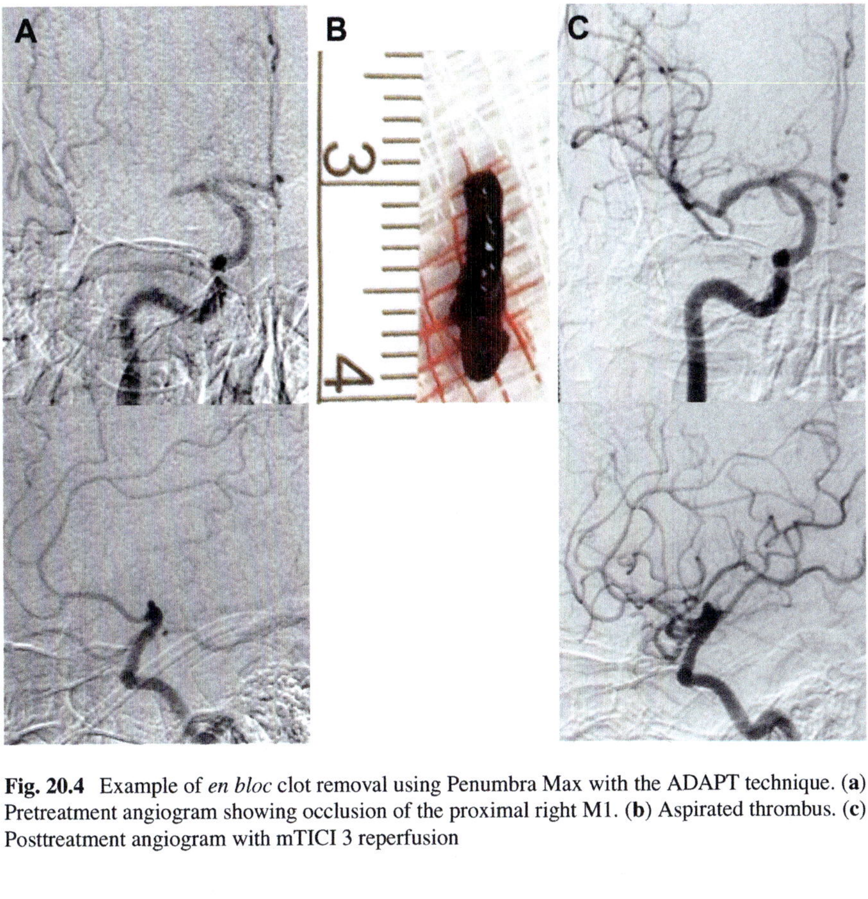

Fig. 20.4 Example of *en bloc* clot removal using Penumbra Max with the ADAPT technique. (**a**) Pretreatment angiogram showing occlusion of the proximal right M1. (**b**) Aspirated thrombus. (**c**) Posttreatment angiogram with mTICI 3 reperfusion

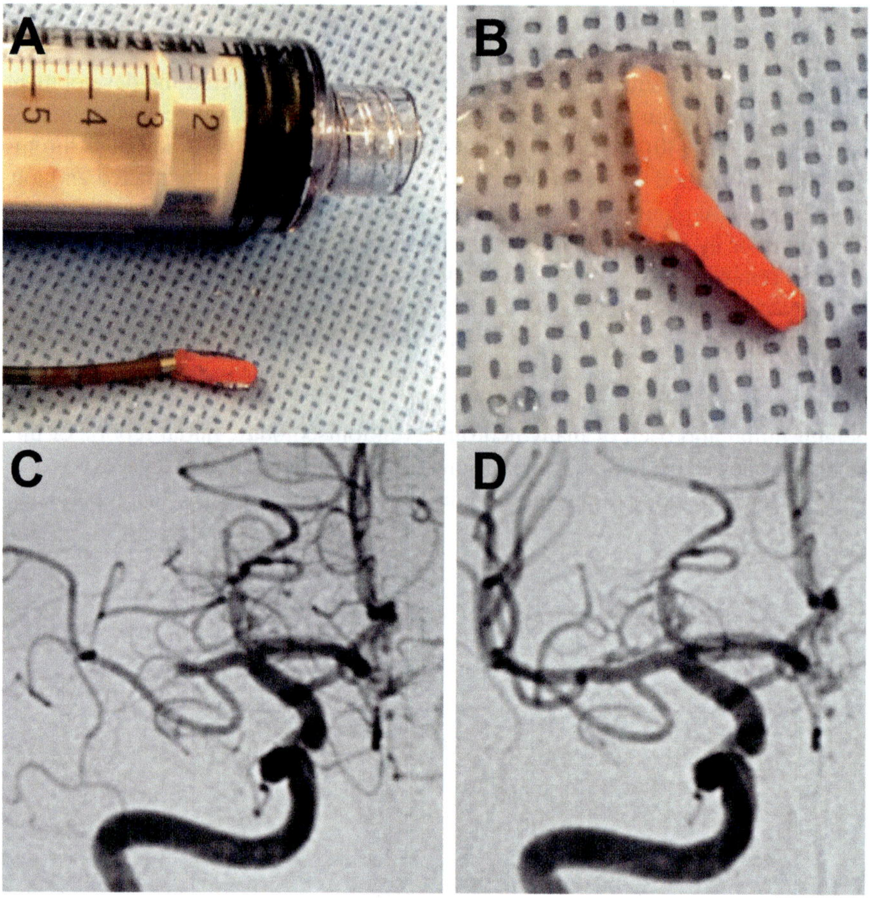

Fig. 20.5 Thromboaspiration using ADAPT technique and Penumbra Max for a proximal right M2 thrombus. (**a**) Thrombus as it was removed with catheter. (**b**) *en bloc* thrombus removal. Pre- (**c**) and posttreatment (**d**) angiogram showing mTICI 3 reperfusion

rates may be superior to those achieved with stentrievers without the inherent risk of clot manipulation (Table 20.1, Fig. 20.6) [30, 33].

A parallel development in endovascular stroke therapy was conceived when interventionists began treating cerebral thromboemboli with implantable self-expanding stents. This was a transient, intermediate stage in the evolution of modern interventional stroke therapy. Interventionists using this technology to treat thromboembolic stroke realized that attempts to reposition such stents often resulted in thrombus retention by the partially deployed stent. Case reports of thrombus extraction by these partially deployed self-expanding stents paved the way for the development of retrievable self-expanding stents. When mounted on a pusher wire, this class of devices became known as stent retrievers or stentrievers.

Stentrievers are deployed across arterial occlusive thrombi through a microcatheter and bridge the occlusion. This enables immediate cerebral reperfusion through the

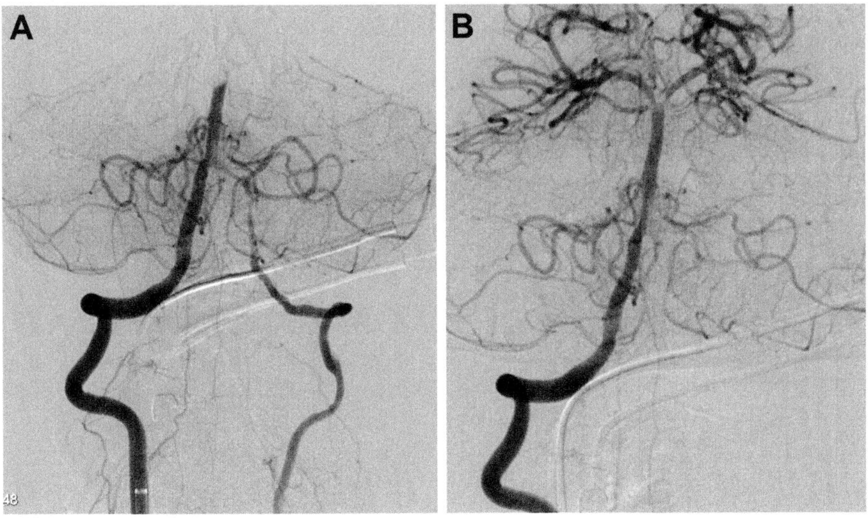

Fig. 20.6 Treatment of a 64-year-old man who presented with an NIHSS score of 32, comatose, with extensor posturing in all extremities; last seen normal 2 h earlier. (**a**) At presentation showing complete occlusion of the basilar artery. (**b**) He underwent a single-pass ADAPT technique using a Penumbra Max thromboaspiration catheter, with mTICI 3 reperfusion being obtained 45 min after groin access. Shortly after procedure, the patient had an NIHSS score of 1, was ambulatory, and was discharged home within 72 h

stented channel and exposes the partially recanalized thrombus to systemically circulating thrombolytic drug in patients who receive intravenous rt-PA. The expanding stent struts compress and intercalate into the matrix of the arterial occlusive thrombus, forming an hourglass or cone-shaped structure that extends between normal arterial sections through the thrombus. By applying traction to the pusher wire, the stentriever construct and associated thrombus are then removed together through a coaxial recovery catheter stationed in the cervicocerebral access artery. This process is contingent upon the retention of thrombus by the stentriever (Fig. 20.7). Adjunctive balloon occlusion and aspiration of the cervicocerebral access artery using a dual-lumen balloon occlusion guide catheter is believed to increase the likelihood of successful thrombus extraction without fragmentation or distal embolization. Opponents of this adjunctive technology argue that it increases the complexity and duration of the thrombus extraction procedure, potentially negating any of its technical benefits.

A modification of this technique combines an intracranial coaxial thromboaspiration catheter (rather than an extracranial balloon occlusion catheter) with a stentriever. This brings the aspiration source to the intracranial occlusion site and, theoretically, further reduces the risk of thrombus fragmentation and distal embolization. This adjunctive method is less complex, less traumatic, and less time consuming than the extracranial balloon guide catheter method. Since early reports of this method were based on the combined use of the Penumbra and Solitaire™ stentriever systems, the method became known as SOLUMBRA. Early experience with the SOLUMBRA method achieved mTICI 2b-3 reperfusion in 95 % of treated patients (Table 20.1) [34]. Notably, because stentrievers are removed with the arterial

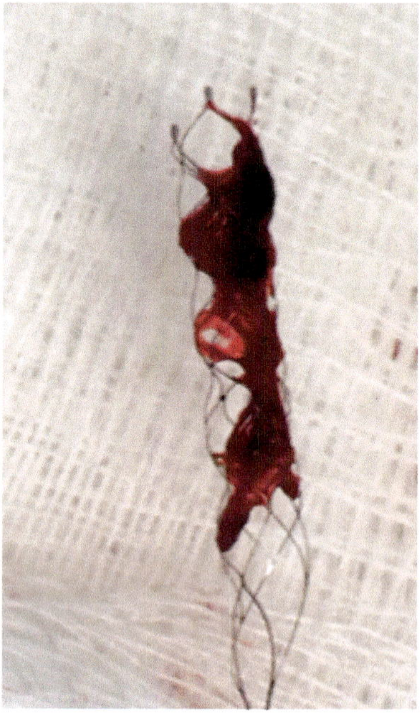

Fig. 20.7 Image representative of a thrombus being incorporated into a stentriever. Note that the clot often is fragmented by stent expansion before its removal

occlusive thrombus and no vascular implant is left behind, the need for dual antiplatelet medication, as required for implanted stent devices, is obviated.

In 2012, The Solitaire™ Flow Restoration (FR) Device (eV3 Endovascular, Irvine, CA) became the first stent retriever to gain FDA clearance (Fig. 20.8). This device was directly compared with the MERCI Retriever in a noninferiority study entitled Solitaire FR with the Intention for Thrombectomy (SWIFT) [35]. This randomized, prospective, industry-sponsored trial included 113 patients who were either ineligible for or who had persistent LVO after IV t-PA and who could receive endovascular treatment within 8 h of stroke onset. Study authors found that primary and secondary analyses strongly favored the Solitaire FR over the MERCI Retriever: TICI 2–3 reperfusion was achieved in 61 % vs. 24 % ($p<0.001$; Table 20.1), good clinical outcomes (mRS 0–2) at 90 days were reached in 58 % vs. 33 % (p noninferiority=0.0001, p superiority=0.02), and mortality at 90 days was decreased to 17 % from 28 % with the MERCI (p noninferiority=0.0001, p superiority=0.02) [35].

Later in 2012, the Trevo® ProVue Retriever (Stryker Neurovascular, Kalamazoo, MI) gained FDA approval and was found to be clinically superior to the MERCI device in a noninferiority study similar to the Solitaire FR (Fig. 20.8) [36]. The Thrombectomy Revascularization of large Vessel Occlusions in acute ischemic stroke (TREVO 2) trial was a randomized, prospective, industry-sponsored study of 178 patients who underwent endovascular treatment with either the Trevo or MERCI device. Inclusion criteria were similar to the SWIFT trial [35, 36]. Primary and secondary outcome measures unequivocally demonstrated the superiority of the Trevo device. Comparing TREVO with MERCI, respectively, TICI 2–3 reperfusion was 86 % vs. 60 % (p superior)<0.0001; Table 20.1), good clinical outcomes (mRS 0–2 at 90 days) were 33 % vs. 24 % ($p=0.01$), and mortality at 90 days was similar at 33 % vs. 24 % ($p=0.18$) [36]. Combined with the

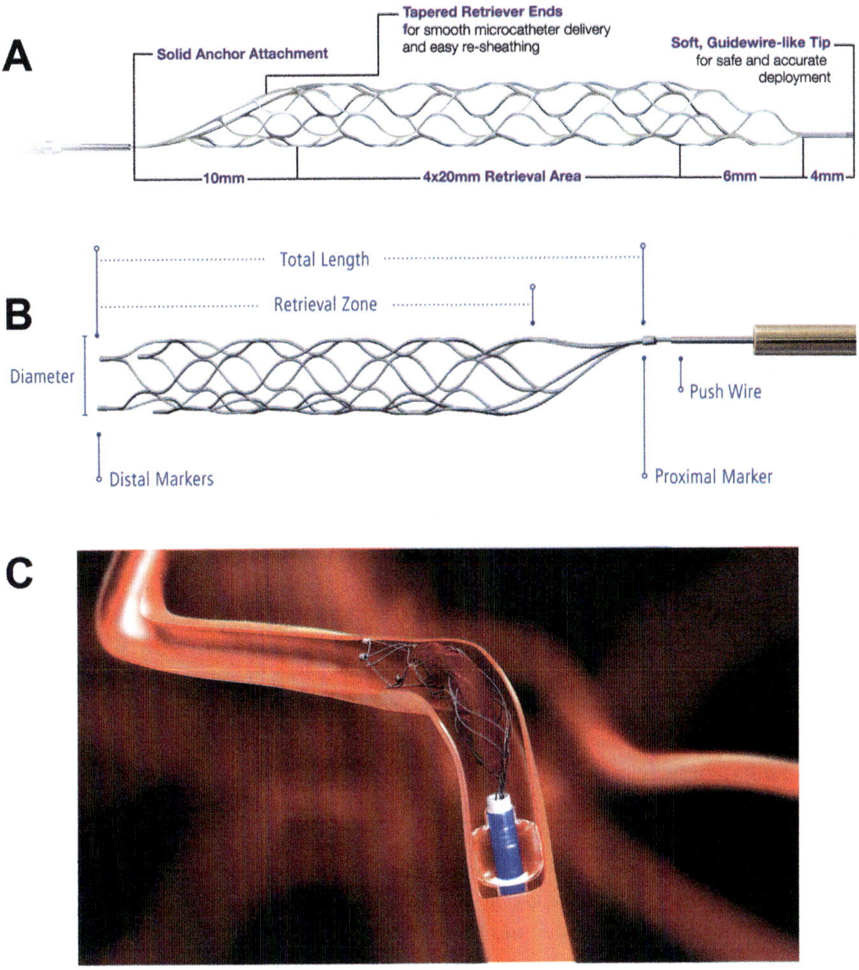

Fig. 20.8 (**a**) TrevoPro Vue self-expanding stentriever device. Reproduced with permission. (**b**) Solitaire self-expanding stentriever device. Reproduce with permission. (**c**) Balloon guide catheter with aspiration during thrombectomy with Solitaire stentriever. Inset, clot integrated into Solitaire stentriever device. Reproduced with permission

SWIFT study, these results showed substantial improvement in recanalization rates with stent retrievers compared with early endovascular device technology.

6 Early Randomized Controlled Trials of Stroke Intervention

The rapid development in stroke device technology during the past decade has found large-scale trials struggling to keep up with this torrential pace. At the 2013 International Stroke Conference in Honolulu, Hawaii, three separate RCTs were presented that assessed the success of endovascular treatment for stroke. The SYNTHESIS EXPANSION was a prospective RCT comparing outcomes of 362 patients with AIS who were randomized (1:1) within 4.5 h of onset to either IV t-PA per standard dosing or to endovascular treatment alone. Patients in the endovascular arm predominantly received IA thrombolysis alone (66 %), although interventionalists had the option to utilize thrombectomy devices (34 %) per their preference. Notably, stent retrievers were used in only 14 %. The primary outcome (mRS 0–1 at 3 months) was reached in 30 % of the intervention arm and 35 % of the IV t-PA patients (OR, 0.71; 95 % CI, 0.44–1.14; p=0.16). No statistical significance was found in outcome measures and recanalization rates were not published [37].

The IMS-III trial is the largest to date that has assessed the efficacy of endovascular management of AIS. This prospective RCT was stopped short of its planned 900-patient enrollment because of futility after an interim analysis of 656 patients determined that there was a <20 % probability of reaching the primary endpoint—a 10 % increase in patients with mRS 0–2 at 90 days. Those with AIS who met enrollment criteria were randomized (1:2) to either IV t-PA (0.9 mg/kg, n=222) or a combination of IV t-PA (0.6 mg/kg) and endovascular therapy (n = 434) [38, 39]. Notably, large vessel occlusion was not a requirement for enrollment because patient selection was based on clinical criteria. The stroke intervention had to begin within 5 h and end by 7 h after stroke onset. Although IMS-III failed to show statistically significant improvement in the primary outcomes, it was powered sufficiently for substantial subset analyses that have since helped guide further clinical study. One important finding was that the proportion of patients with a good clinical outcome (mRS 0–2 at 90 days) significantly increased with successful recanalization. Further, as previously noted, every 30-min delay in reperfusion reduced the likelihood of a good outcome in this population by 12–15 % [32].

The third of these conjointly presented studies was the Magnetic Resonance and Recanalization of Stroke Clots Using Embolectomy (MR RESCUE) trial. This prospective RCT randomized 118 patients with anterior circulation AIS due to LVO within 8 h who were either ineligible for IV t-PA or had persistent vascular occlusion after this treatment [40]. Patients received standard medical care (n=54) or underwent mechanical embolectomy with either the MERCI or Penumbra device (n=64) with the additional option of IA t-PA in the interventional group. Clinical outcomes were statistically similar between patients who received medical care and those who underwent embolectomy. When the study further evaluated clinical outcomes according to those with or without a favorable ischemic penumbra pattern on MR imaging, no statistically significant difference in clinical outcomes were found between cohorts [40].

Overall, these trials yielded discouraging results for study investigators who were hoping to show improved outcomes with endovascular management of

Table 20.2 Summary of three large-scale RCTs assessing the efficacy of endovascular treatment of acute ischemic stroke

Trial	Number enrolled (*n*) (intervention:IV t-PA alone)	Outcome analysis	Rates of endovascular reperfusion	% of stent retrievers in treatment arm
SYNTHESIS EXPANSION	362 (181:181)	mRS 0–1 at 90 d: 30.4 % (intervention) vs. 34.8 % (IV t-PA); $p=0.16$	Not published	14 %
IMS-III	656 (434:222)	mRS 0–2 at 90 d: 40.8 % (intervention) vs. 38.7 % (IV t-PA); 95 % CI −6.1 to 9.1	41 % mTICI 2b-3	1.5 %
MR RESCUE	118 (64:54)	mRS 0–2 at 90 d: 17–21 % (intervention) vs. 10–26 % (IV t-PA); $p=0.48$	67 % TICI 2a-3	0 %

These early trials failed to show statistically significant improvement of IV t-PA outcomes in the NINDS trial [9, 32, 37–40]. d = day

stroke (Table 20.2). The studies were heavily critiqued and several factors have been blamed for their futility. Most importantly, device technology was advancing very rapidly during enrollment so that the most current technology was noticeably underutilized. Stent retrievers, in particular, were seldom used despite evidence of their superiority over earlier devices. Specifically, of those who received endovascular treatment, stent retrievers were used in 1.5 % within IMS-III (41 % received IA t-PA alone), 14 % within SYNTHESIS, and in no patient within MR RESCUE [34–37, 39, 40]. This led to underwhelming rates of recanalization, such as the 41 % rate of TICI 2b-3 achieved in IMS-III patients (Table 20.1) [39].

Another shortcoming of these initial trials was the inclusion of patients with minor ischemic deficits who would likely achieve a good outcome with IV thrombolytic therapy alone. Therefore, it was difficult to identify statistical differences between treatment groups and likely contributed to overall trial futility. Furthermore, imaging confirmation of vascular occlusion was not a requirement in IMS-III or SYNTHESIS before randomization; this then diminished the statistical power of intention-to-treat analysis because some patients randomized to endovascular treatment lacked a LVO and thus did not receive IA therapy [41]. Compared with later trials, initiation of the intervention was also delayed, which likely contributed to dampened outcomes. Given the substantial improvement in reperfusion achieved with newer device technologies, there was clearly an indication for further study.

7 Breakthrough of Modern Mechanical Thrombectomy Devices

With the revelation of futility in these three early endovascular trials, many centers continued to offer endovascular therapy in select patients with AIS despite significant controversy. The next major advance came from the Multicenter Randomized CLinical trial of Endovascular treatment for Acute ischemic stroke in the Netherlands (MR CLEAN). Within this prospective RCT, 500 patients with AIS due to confirmed LVO were randomized (1:1) to either usual care ($n=267$) or combination IV and IA treatment ($n=233$). Occlusion of the intracranial ICA, M1, M2, A1, or A2 was confirmed with CT angiography (CTA), magnetic resonance angiography (MRA), or digital subtraction angiography (DSA). Endovascular treatment consisted of IA t-PA, mechanical thrombectomy, or combination therapy [42]. Stent retrievers were used in 81.5 % of patients randomized to the endovascular arm and in 97 % of patients who actually underwent intervention. Compared with earlier large-scale trials, recanalization improved with 59 % of patients achieving mTICI 2b-3 reperfusion (Table 20.1). Study authors found statistically significant improvement in the primary outcome measure (mRS at 90 days) in favor of the intervention: mRS 0–2 at this time point was 32.6 % in the intervention group and 19.1 % in the control group (adjusted OR, 2.16; 95 % CI, 1.39–3.38) [42]. Mortality and sICH rates were similar between groups. No safety concerns were raised within a 6-h window, but the clinical benefit of endovascular treatment was no longer statistically significant after 6 h and 19 min past stroke onset [42].

MR CLEAN's positive findings in favor of endovascular stroke treatment were presented at the World Stroke Conference in Istanbul, Turkey in October 2014 to much fanfare. Soon after, safety committees for three ongoing endovascular trials halted enrollment to perform an early interim analysis. For all three trials, the data revealed that each had reached predetermined criteria for stopping enrollment due to proven efficacy. This news received a standing ovation when the trials were presented sequentially at the 2015 ISC meeting in Nashville, Tennessee.

The Endovascular Treatment for Small Core and Anterior Circulation Proximal Occlusion with Emphasis on Minimizing CT to Recanalization Times (ESCAPE) trial was a prospective RCT that randomized patients up to 12 h after stroke onset. Patients who received both a noncontrast CT and CTA were included if they had a modest infarct burden; occlusion of the ICA, M1, or at least two M2 branches; and collateral filling of at least 50 % of the MCA pial arterial circulation on CTA (moderate-to-good collaterals) [43]. The control group received IV t-PA within 4.5 h per established guidelines ($n=150$) while the endovascular arm ($n=165$) received standard therapy in addition to mechanical thrombectomy. The use of stent retrievers in 86 % and proximal balloon guide catheter suction was recommended but not required. In ESCAPE's interim analysis of endovascular vs. control groups, clinical outcomes improved (mRS 0–2 at 90 days in 53 % vs. 29.3 %, $p<0.001$), 90-day mortality decreased (10.4 % vs. 19.0 %, $p=0.04$), and rates of sICH were similar (3.6 % vs. 2.7 %, $p=0.75$), respectively. TICI 2b-3 reperfusion was achieved in 72.4 % of the endovascular arm (Table 20.1). Unfortunately, too few of patients underwent randomization in the 6- to 12-h window to allow for statistical analysis of this subset [43].

Simultaneously, the Solitaire with the Intention for Thrombectomy as Primary Endovascular Treatment (SWIFT PRIME) trial was an industry-sponsored, prospective RCT to establish the efficacy of Solitaire devices. CTA or MRA was used to confirm intracranial ICA or M1 occlusion. Although CT perfusion evidence of territorial ischemia was required initially, an addendum later allowed sites without CT perfusion capability to randomize with an Alberta Stroke Program Early CT Score (ASPECTS) greater than 6 [44]. Before halting enrollment, 196 patients had undergone randomization to either IV t-PA within 4.5 h ($n=98$) or to IV t-PA plus Solitaire thrombectomy ($n=98$). Primary outcome analysis favored endovascular intervention: a significantly improved proportion of patients reached mRS 0–2 at 90 days (60.2 % vs. 35.5 %, $p<0.001$) and had improved distribution of mRS score at 90 days ($p<0.001$). Rates of mortality and sICH were similar between the two cohorts. Study interventionalists achieved TICI 2b-3 reperfusion in 88 % of the endovascular group (Table 20.1) [44].

The Extending the Time for Thrombolysis in Emergency Neurological Deficits—Intra-Arterial (EXTEND-IA) trial was another prospective RCT that assessed outcomes with the Solitaire device, which was supplied for the trial by Covidien. Patients were required to have LVO of the ICA, M1, or M2 division of the MCA on CTA, and meet three criteria on either CT or MR perfusion imaging: a mismatch ratio greater than 1.2, an absolute mismatch volume exceeding 10 mL, and an ischemic core lesion volume below 70 mL [45]. Of the 70 enrolled patients, 35 received IV t-PA alone and 35 had a combination of IV t-PA and thrombectomy using the Solitaire FR retrievable stent. Outcomes again significantly favored the intervention group. Intervention achieved better reperfusion at 24 h (median decrease in perfusion-lesion 100 % vs. 37 %, $p<0.001$), greater neurologic improvement by day 3 ($p=0.002$), and higher rates of functional independence (mRS 0–2 at 90 days in 71 % vs. 40 %, $p=0.01$). There were no differences in rates of mortality or sICH. Treatment results were again excellent, as TICI 2b-3 reperfusion was achieved in 86 % of the intervention group (Table 20.1) [45].

Soon after the unveiling of these trials, three more studies were presented at the inaugural European Stroke Organization Conference in Glasgow, United Kingdom in April 2015. First, the Randomized Trial of Revascularization with Solitaire FR Device vs. Best Medical Therapy in the Treatment of Acute Stroke Due to Anterior Circulation Large Vessel Occlusion Presenting within Eight Hours of Symptom Onset (REVASCAT) was a prospective, industry-sponsored RCT carried out at four centers in Catalonia, Spain. Patients with CTA, MRA, or angiographic evidence of occlusion of the intracranial ICA or M1 were randomized to receive either medical therapy alone (IV t-PA, $n=103$) or endovascular treatment with the Solitaire stent retriever in addition to medical treatment ($n=103$) [46]. The trial was stopped early when it was determined that there was no longer clinical equipoise given the growing evidence in favor of endovascular management of stroke [43–45]. With this premature cessation of enrollment, statistical analysis changed from hypothesis testing to estimation because the necessary power for the first was no longer achievable. Results lent further support to endovascular treatment of AIS. Compared with the medical arm, patients in the thrombectomy arm had an improved distribution of mRS (adj. OR 1.7; 95 % CI, 1.05–2.8) and better functional outcomes (mRS 0–2 at 90 days in 43.7 % vs. 28.2 %, adj. OR 2.1; 95 % CI, 1.1–4.0). No difference in rates of mortality or sICH occurred between the groups and the thrombectomy group had a 66 % rate of TICI 2b-3 reperfusion (Table 20.1) [46].

Two other trials presented at this meeting that favored thrombectomy are yet to be published. The THERAPY Trial, a prospective RCT, compared the Penumbra aspiration system with standard IV t-PA. Inclusion criteria stipulated that a long, occlusive clot measuring at least 8 mm must be present on vascular imaging before randomization. Stopped early because of the positive results of other trials, THERAPY's primary outcome measures failed to reach statistical significance; however, based on ordinal analysis, findings statistically favored endovascular therapy (OR 2.28; 95 % CI, 1.05–4.96, $p=0.038$) [47].

Given the growing literature in support of stroke intervention, the French RCT called THRACE was also stopped early; study findings were presented in

Table 20.3 Summary of seven large-scale RCTs that have shown the efficacy of endovascular treatment of acute ischemic stroke

Trial	Enrollment (*n*) (intervention:IV t-PA alone)	Outcome analysis	Rates of endovascular reperfusion	% of stent retrievers in treatment arm
MR CLEAN	500 (233:267)	mRS 0–2 at 90d: 32.6 % (intervention) vs. 19.1 % (IV t-PA); OR, 2.16; 95 % CI, 1.39–3.38	59 % mTICI 2b-3	81.5 % overall (97 % of those treated)
ESCAPE	315 (165:150)	mRS 0–2 at 90d: 53 % (intervention) vs. 29.3 % (IV t-PA); $p<0.001$	72.4 % mTICI 2b-3	86.1 %
SWIFT PRIME	196 (98:98)	mRS 0–2 at 90d: 60.2 % (intervention) vs. 35.5 % (IV t-PA); $p<0.001$	88 % mTICI 2b-3	100 %
EXTEND-IA	70 (35:35)	mRS 0–2 at 90d: 71 % (intervention) vs. 40 % (IV t-PA); $p=0.01$	86 % mTICI 2b-3	100 %
REVASCAT	206 (103:103)	mRS 0–2 at 90d: 43.7 % (intervention) vs. 28.2 % (IV t-PA); OR 2.1; 95 % CI, 1.1–4.0	66 % mTICI 2b-3	100 %
THERAPY	102	(Unpublished)	(Unpublished)	0 % (Penumbra aspiration system)
THRACE	414	(Unpublished)	(Unpublished)	100 %

The THERAPY and THRACE trial results have not been published at this time [42–48]

Glasgow and publication is forthcoming. Inclusion criteria were intentionally broad, including all patients with moderate-to-severe stroke due to LVO who could begin endovascular treatment within 5 h of stroke onset. Preliminary results reported statistically significant improvement in clinical outcome after intervention, with mRS 0–2 at 90 days in 54.2 % of patients in the treatment arm vs. 42.1 % in the medical group ($p = 0.02$) [48]. Overall, these trials add support for the growing data of improved clinical outcomes with mechanical thrombectomy (Table 20.3).

8 Current Standard of Care

Given the glaring need for improved stroke treatment and the groundbreaking data from these trials, the perceived "standard of care" at comprehensive stroke centers in the United States and abroad has undergone a seismic shift in recent times as endovascular treatment proved to be efficacious. Although evidence for endovascular treatment modalities is impressive, one must remember that interventional treatment is only validated for select patients with AIS due to LVO. Many patients will not qualify for intervention and should instead receive maximum medical therapy, including IV t-PA, per published guidelines if appropriate. At this time, prompt endovascular management is recommended for patients who meet the following criteria [49]:

1. Good baseline functional status (prestroke mRS of 0–1).
2. Documented moderate to severe AIS with NIHSS score of at least 8 and ASPECTS score of at least 6.
   (a) Administration of IV t-PA within 4.5-h window if appropriate.
3. Occlusion of intracranial ICA or M1 (American Heart Association Class I, Level of Evidence A), proximal M2, intracranial vertebral artery or basilar artery (Class II, Level of Evidence C).
4. Age ≥ 18 years.
   (a) Mechanical thrombectomy may be reasonable for some patients <18 years of age with acute ischemic stroke who have demonstrated large-vessel occlusion and in whom treatment can be initiated (groin puncture) within 6 h of symptom onset, but the benefits are not established in this age group (*American Heart Association Class IIb; Level of Evidence C*).
5. Ability to begin endovascular intervention within 6 h of stroke onset.

AIS patients who receive their initial care outside of a Comprehensive Stroke Center (CSC) should receive supportive emergency care and undergo a noncontrast head CT followed by IV thrombolytic therapy (if eligible) as rapidly as possible. Patients with moderate to severe AIS who are candidates for interventional therapy must be transported to a CSC or other thrombectomy-capable facility without delay [49]. Future advances in prehospital assessment and triage may favor direct transport of moderate to severe AIS to CSCs; this strategy could bypass facilities unable to offer thrombectomy as long as it would not substantially delay administration of IV thrombolytic therapy to eligible patients.

Confirmation of LVO with vascular imaging has become a widespread standard of care after all five positive RCTs required CTA, MRA, or angiographic evidence of LVO before randomization. Nonetheless, once a noncontrast CT has been obtained and hemorrhage is ruled out, patients who qualify for IV t-PA should be administered as soon as possible while additional workup ensues. Furthermore, patients who undergo IV thrombolysis should not be observed before initiation of endovascular treatment, as temporal delay will substantially decrease the likelihood of a good clinical outcome [32, 49]. Currently, the use of perfusion-based imaging has not been validated, but remains an option for patient selection at centers where this is commonplace.

Although no consensus has been reached regarding treatment for patients under age 18 years, benefits in lower mortality for those over 80 years negate a justification for withholding treatment solely on the basis of advanced age [49–51]. Several guidelines have also been established regarding particulars of stroke intervention. Regardless, time to reperfusion is of vast importance and interventionalists should seek mTICI 2b-3 reperfusion as quickly as possible, certainly within 6 h of onset.

Current evidence overwhelmingly favors the use of modern thromboembolectomy devices over intra-arterial thrombolysis for the management of LVO. The first positive clinical trials that supported endovascular intervention for AIS primarily used stent retrievers because advances in thromboaspiration catheter technology lagged behind. Industry sponsorship and conflicts of interest within the community of neurointerventional physicians contributed to a growing body of literature that was biased in favor of stent retrievers and had promoted these devices over alternative mechanical thrombectomy devices. In our experience, modern thromboaspiration technologies and techniques are at least as efficacious as stent retrievers and thus are preferred at our center. Furthermore, compared with stent retrievers, current thromboaspiration technologies are significantly less costly and less traumatic to intracranial vessels [53]. IA t-PA may be used either in salvage efforts along with thromboembolectomy for the management of LVO or alone for the management of moderate to severe AIS caused by arterial occlusions that are too distal or too small to approach with devices. Unfortunately, no IA dose for thrombolytic drugs has been well established or FDA-approved for this indication [49].

As for the method of procedural anesthesia, some data has indicated that conscious sedation may result in fewer complications with similar rates of recanalization. However, a direct comparison with general anesthesia is currently lacking and observational studies have been confounded by stroke severity [50, 53, 54]. Choice of anesthetic method should thus be based on individual patient characteristics [49].

Fig. 20.9 (continued) into the distal superior M2 branch. (**e**) Shows relatively minor stroke burden given his presenting ischemic pattern and unclear time course. He was discharge home with mild deficits

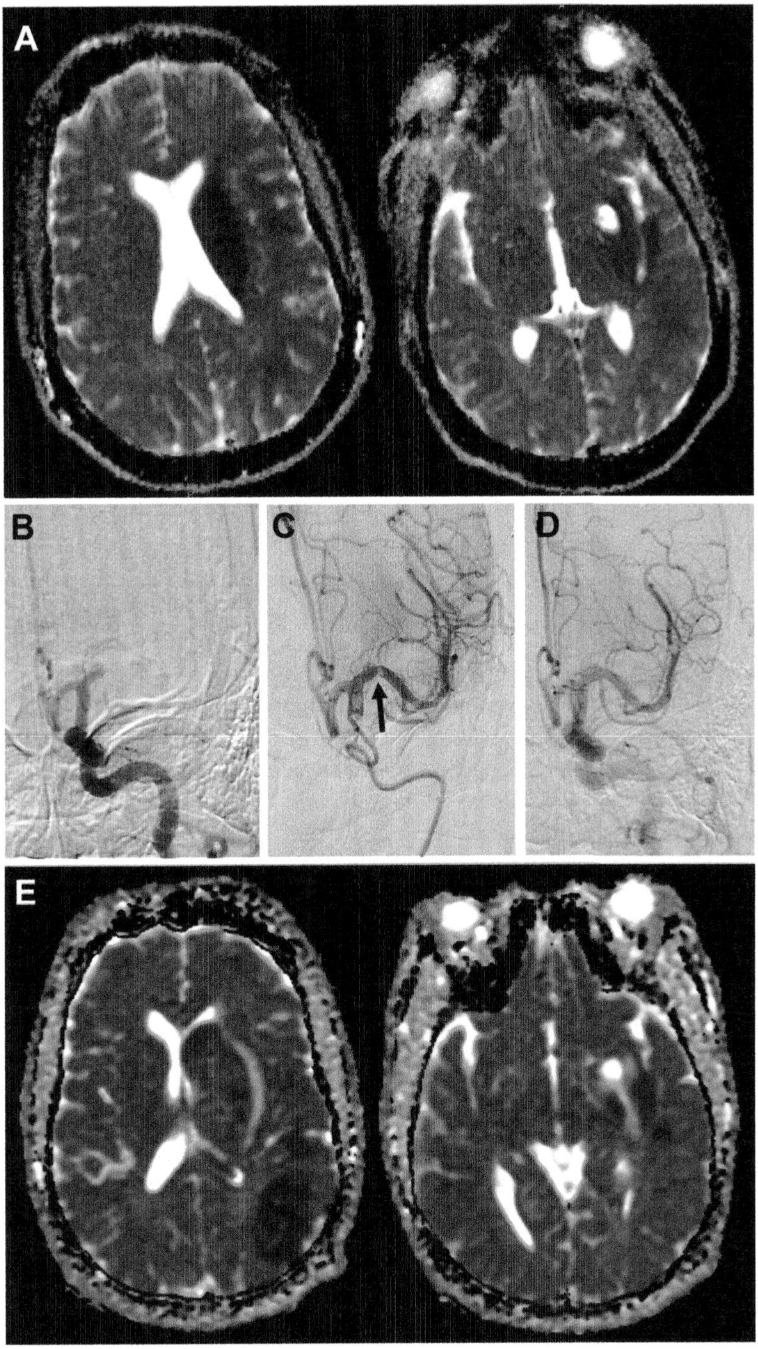

Fig. 20.9 Treatment course of a 37-year-old man who awoke with a complete left MCA syndrome and presented with an NIHSS score of 21. (**a**) Initial MRI showed small core infarct size with large symptomatic penumbra. (**b**) Pretreatment angiogram revealed an occlusive proximal left M1 thrombus. (**c**) Saddle embolus (*arrow*) is noted on angiogram at an early MCA bifurcation after using a Penumbra Max ADAPT technique. (**d**) Next the SOLUMBRA technique with a combination of the Penumbra Max thromboaspiration catheter and Solitaire stentriever resolved the occlusive embolus. mTICI 2b reperfusion was achieved, as a small portion of the clot embolized

9 Directions for Future Research and Clinical Guidelines

Although the role for endovascular treatment of AIS has been firmly established for select patients, many circumstances require further study and clarification. Positive trials have generally excluded patients with NIHSS scores less than 6, yet patients with LVO and minor clinical symptoms may be at significant risk for rapid clinical deterioration and thus could benefit from prompt recanalization to prevent this decline [50]. Further, although improved outcomes have been well established for patients with intracranial ICA or proximal MCA occlusions, the population of patients with M2 occlusions could potentially benefit from embolectomy. Although RCTs have focused on the anterior cerebral circulation, most would agree that properly selected patients with basilar artery occlusion could strongly benefit from endovascular treatment and that a specific clinical trial directed at this question is not warranted given the severity of infarctions in this vascular territory.

The absolute temporal window of therapeutic opportunity for endovascular stroke treatment remains unclear (Fig. 20.9). Although existing evidence has established benefit if reperfusion is achieved within 6 h of stroke onset, additional trials are ongoing to elucidate any potential benefit in patients presenting within 8–24 h of stroke onset. Clinical benefit outside the 6-h window is theoretically possible in some patients with large penumbral regions supported by unusually robust collateral perfusion. Additionally, neuroprotective agents may artificially extend the temporal window of therapeutic opportunity and are currently under active investigation. The use of advanced perfusion-imaging techniques to identify patients who may benefit from reperfusion, particularly those with an unknown time of onset (e.g., "wake up" strokes) and those who present outside the conventional 6-h window, has not been validated and likewise warrants further investigation.

Management of carotid stenosis proximal to an intracranial LVO in AIS has been of particular interest among stroke investigators. Severe ICA stenosis or occlusion has anecdotally been associated with delayed access to intracranial emboli by interventionalists. However, utilization of carotid stenting or angioplasty was inconsistent in these patients in the five published positive RCTs [42–46]. Some centers have reported success with tandem sequential treatment of extracranial carotid stenosis and intracranial LVO. However, carotid stenting requires dual antiplatelet therapy, which can increase the risk of hemorrhagic conversion in these patients [55–57]. Additionally, the safety and efficacy of carotid endarterectomy within 48 h of receiving t-PA for AIS has been proven in patients with severe ipsilateral ICA stenosis [58]. Overall, the ideal procedural sequence for addressing proximal ICA stenosis associated with intracranial LVO has yet to be established and requires further study.

Taken from a broader context, efforts must continue to increase accessibility to this groundbreaking stroke technology. This expansion includes the rapid identification and diagnosis of AIS, expedient initiation of appropriate stroke treatment, and transfer to a CSC if necessary. Put simply, a strategy toward greater availability of acute stroke treatment could have a much greater societal impact than minor sequential improvements in technology [2, 4, 13]. Nonetheless, rapid improvements in interventional device technology have revolutionized the treatment paradigm for AIS and further device enhancements are sure to continue into the future as indications broaden and clinical experience progresses.

10 Conclusions

With the advent of modern mechanical thrombectomy devices, multiple randomized trials have now proven the efficacy of mechanical thrombectomy in properly selected patients for the treatment of acute ischemic stroke caused by large vessel occlusion. By combining enhanced patient selection criteria with highly effective recanalization and low periprocedural morbidity, investigators have revolutionized stroke management and established endovascular intervention as the key component of this treatment paradigm. Further evolution of interventional device technology, neuroimaging selection tools, neuroprotective pharmaceuticals, and clinical experience will hopefully continue to expand patient eligibility, improve technical reperfusion results, and further increase the proportion of patients with good clinical outcomes.

References

1. Go AS, Mozaffarian D, Roger VL, Benjamin EJ, Berry JD, Borden WB et al (2013) Executive summary: heart disease and stroke statistics—2013 update: a report from the American Heart Association. Circulation 127:143–152
2. Lopez AD, Mathers CD, Ezzati M, Jamison DT, Murray CJ (2006) Global and regional burden of disease and risk factors, 2001: systematic analysis of population health data. Lancet 367:1747–1757
3. Lozano R, Naghavi M, Foreman K, Lim S, Shibuya K, Aboyans V et al (2012) Global and regional mortality from 235 causes of death for 20 age groups in 1990 and 2010: a systematic analysis for the global burden of disease study 2010. Lancet 380:2095–2128
4. Feigin VL, Lawes CM, Bennett DA, Anderson CS (2003) Stroke epidemiology: a review of population-based studies of incidence, prevalence, and case-fatality in the late 20th century. Lancet Neurol 2:43–53
5. Fletcher AP, Alkjaersig N, Lewis M et al (1976) A pilot study of urokinase therapy in cerebral infarction. Stroke 7:135–142
6. Meyer JS, Gilroy J, Barnhart J (1965) Therapeutic thrombolysis in cerebral thromboembolism: randomized evaluation of intravenous streptokinase. In: Siekert W, Whisnant JP (eds) Cerebral vascular diseases. Grune & Stratton, New York, pp 200–213
7. Brott TG, Haley EC Jr, Levy DE et al (1992) Urgent therapy for stroke. I. Pilot study of tissue plasminogen activator administered within 90 minutes. Stroke 23:632–640
8. Haley EC Jr, Levy DE, Brott TG et al (1992) Urgent therapy for stroke. II. Pilot study of tissue plasminogen activator administered 91-180 minutes from onset. Stroke 23:641–645
9. Tissue plasminogen activator for acute ischemic stroke. The National Institute of Neurological Disorders and Stroke rt-PA Stroke Study Group (1995). N Engl J Med 333:1581–1587
10. Hacke W, Kaste M, Bluhmki E, Brozman M, Dávalos A, Guidetti D et al (2008) Thrombolysis with alteplase 3 to 4.5 hours after acute ischemic stroke. N Engl J Med 359:1317–1329

11. Wahlgren N, Ahmed N, Dávalos A, Hacke W, Millán M, Muir K et al (2008) Thrombolysis with alteplase 3-4.5 h after acute ischemic stroke (SITS-ISTR): an observational study. Lancet 372:1303–1309
12. de Los Ríos la Rosa F, Khoury J, Kissela BM, Flaherty ML, Alwell K, Moomaw CJ, et al (2012) Eligibility for intravenous recombinant tissue-type plasminogen activator within a population: the effect of the European Cooperative Acute Stroke Study (ECASS) III Trial. Stroke 43:1591–1595
13. Bhatia R, Hill MD, Shobha N, Menon B, Bal S, Kochar P et al (2010) Low rates of acute recanalization with intravenous recombinant tissue plasminogen activator in ischemic stroke: real-world experience and a call for action. Stroke 41:2254–2258
14. Furlan A, Higashida R, Wechsler L, Gent M, Rowley H, Kase C et al (1999) Intra-arterial prourokinase for acute ischemic stroke; The PROACT II study: a randomized controlled trial. JAMA 282(21):2003–2011
15. Ogawa A, Mori E, Minematsu K, Taki W, Takahashi A, Nemoto S et al (2007) Randomized trial of intraarterial infusion of urokinase within 6 hours of middle cerebral artery stroke: the middle cerebral artery embolism local fibrinolytic intervention trial (MELT) Japan. Stroke 38:2633–2639
16. Fields JD, Khatri P, Nesbit GM, Liu KC, Barnwell SL, Lutsep HL et al (2011) Meta-analysis of randomized intra-arterial thrombolytic trials for the treatment of acute stroke due to middle cerebral artery occlusion. J Neurointerv Surg 3(2):151–155
17. Ringer AJ, Qureshi AI, Fessler RD, Guterman LR, Hopkins LN (2001) Angioplasty of intracranial occlusion resistant to thrombolysis in acute ischemic stroke. Neurosurgery 48:1282–1288
18. Ernst R, Pancioli A, Tomsick T, Kissela B, Woo D, Kanter D et al (2000) Combined intravenous and intra-arterial recombinant tissue plasminogen activator in acute ischemic stroke. Stroke 31:2552–2557
19. IMS Study Investigators (2004) Combined intravenous and intra-arterial recanalization for acute ischemic stroke: the Interventional Management of Stroke Study. Stroke 35:904–911
20. IMS II Trial Investigators; (2007) The Interventional Management of Stroke (IMS) II Study. Stroke 38:2127–2135
21. Gobin YP, Starkman S, Duckwiler GR, Grobelny T, Kidwell CS, Jahan R et al (2004) MERCI 1: a phase 1 study of mechanical embolus removal in cerebral ischemia. Stroke 35:2848–2854
22. Smith WS, Sung G, Starkman S, Saver JL, Kidwell CS, Gobin YP et al (2005) Safety and efficacy of mechanical embolectomy in acute ischemic stroke: results of the MERCI trial. Stroke 36:1432–1438
23. Smith WS, Sung G, Saver J, Budzik R, Duckwiler G, Liebeskind DS et al (2008) Mechanical thrombectomy for acute ischemic stroke: final results of the multi MERCI trial. Stroke 39:1205–1212
24. Flint AC, Duckwiler GR, Budzik RF, Liebeskind DS, Smith WS, MERCI and Multi MERCI Writing Committee (2007) Mechanical thrombectomy of intracranial internal carotid occlusion: pooled results of the MERCI and Multi MERCI part I trials. Stroke 38:1274–1280
25. Josephson SA, Saver JL, Smith WS, Merci and Multi Merci Investigators (2009) Comparison of mechanical embolectomy and intraarterial thrombolysis in acute ischemic stroke within the MCA: MERCI and Multi MERCI compared to PROACT II. Neurocrit Care 10:43–49
26. The Penumbra Pivotal Stroke Trial Investigators (2009) The penumbra pivotal stroke trial: safety and effectiveness of a new generation of mechanical devices for clot removal in intracranial large vessel occlusive disease. Stroke 40:2761–2768
27. Tarr R, Hsu D, Kulcsar Z, Bonvin C, Rufenacht D, Alfke K et al (2010) The POST trial: initial post-market experience of the Penumbra system: revascularization of large vessel occlusion in acute ischemic stroke in the United States and Europe. J Neurointerv Surg 2:341–344
28. Simon SD, Grey CP (2014) Hydrodynamic comparison of the Penumbra system and commonly available syringes in forced-suction thrombectomy. J Neurointerv Surg 6:205–211

29. Frei D, Gerber J, Turk A, McPherson M, Heck D, Hui F et al (2013) The SPEED study: initial clinical evaluation of the Penumbra novel 054 reperfusion catheter. J Neurointerv Surg 5(Suppl 1):i74–i76
30. Turk AS, Frei D, Fiorella D, Mocco J, Baxter B, Siddiqui A et al (2014) ADAPT FAST study: a direct aspiration first pass technique for acute stroke thrombectomy. J Neurointerv Surg 6:260–264
31. Khatri P, Abruzzo T, Yeatts SD, Nichols C, Broderick JP, Tomsick TA et al (2009) Good clinical outcome after ischemic stroke with successful revascularization is time-dependent. Neurology 73(13):1066–1072
32. Khatri P, Yeatts SD, Mazighi M, Broderick JP, Liebeskind DS, Demchuck AM et al (2014) Time to angiographic reperfusion and clinical outcome after acute ischemic stroke: an analysis of data from the Interventional Management of Stroke (IMS III) phase 3 trial. Lancet Neurol 13:567–574
33. Navia P, Larrea JA, Pardo E, Arce A, Martinez-Zabaleta M, Diez-Gonzalez N et al (2015) Initial experience using the 3MAX cerebral reperfusion catheter in the endovascular treatment of acute ischemic stroke of distal arteries. J Neurointerv Surg 7:A28–A29
34. Castaño C, Dorado L, Guerrero C, Millán M, Gomis M, Perez de la Ossa N et al (2010) Mechanical thrombectomy with the solitaire AB device in large artery occlusions of the anterior circulation: a pilot study. Stroke 41:1836–1840
35. Saver JL, Jahan R, Levy EI, Jovin TG, Baxter B, Nogueira RG et al (2012) Solitaire flow restoration device versus the Merci Retriever in patients with acute ischaemic stroke (SWIFT): a randomised, parallel-group, non-inferiority trial. Lancet 380:1241–1249
36. Nogueira RG, Lutsep HL, Gupta R, Jovin TG, Albers GW, Walker GA, et al (2012) Trevo versus Merci retrievers for thrombectomy revascularisation of large vessel occlusions in acute ischaemic stroke (TREVO 2): a randomised trial. Lancet 380:1231–1240
37. Ciccone A, Valvassori L, Nichelatti M, Sgoifo A, Ponzio M, Sterzi R et al (2013) Endovascular treatment for acute ischemic stroke. N Engl J Med 368(10):904–913
38. Khatri P, Hill MD, Palesch YY, Spilker J, Jauch EC, Carrozzella JA et al (2008) Methodology of the Interventional Management of Stroke (IMS)III trial. Int J Stroke 3(2):130–137
39. Broderick JP, Palesch YY, Demchuk AM, Yeatts SD, Khatri P, Hill MD et al (2013) Endovascular therapy after intravenous t-PA versus t-PA alone for stroke. N Engl J Med 368(10):893–903
40. Kidwell CS, Jahan R, Gornbein J, Alger JR, Nenov V, Ajani Z et al (2013) A trial of imaging selection and endovascular treatment for ischemic stroke. N Engl J Med 368(10):914–923
41. Qureshi AI, Abd-Allah F, Aleu A, Connors JJ, Hanel R, Hassan AE et al (2014) Endovascular treatment for acute ischemic stroke patients: implications and interpretation of IMS III, MR RESCUE, and SYNTHESIS EXPANSION trials: a report from the Working Group of International Congress of Interventional Neurology. J Vasc Interv Neurol 7(1):56–75
42. Berkhemer OA, Fransen PSS, Beumer D, van den Berg LA, Lingsma HF, Yoo AJ et al (2015) A randomized trial of intraarterial treatment for acute ischemic stroke. N Engl J Med 372:11–20
43. Goyal M, Demchuk AM, Menon BK, Eesa M, Rempel JL, Thornton J et al (2015) Randomized assessment of rapid endovascular treatment of ischemic stroke. N Engl J Med 372(11):1019–1030
44. Saver JL, Goyal M, Bonafe A, Diener HC, Levy EI, Pereira VM et al (2015) Stent-retriever thrombectomy after intravenous t-PA vs. t-PA alone in stroke. N Engl J Med 372(24):2285–2295
45. Campbell BC, Mitchell PJ, Kleinig TJ, Dewey HM, Churilov L, Yassi N et al (2015) Endovascular therapy for ischemic stroke with perfusion-imaging selection. N Engl J Med 372(11):1009–1018
46. Jovin TG, Chamorro A, Cobo E, de Miquel MA, Molina CA, Rovira A et al (2015) Thrombectomy within 8 hours after symptom onset in ischemic stroke. N Engl J Med 372(24):2296–2306
47. Mocco J, Zaidat O, von Kummer R, Yoo A, Gupta R, Lopes D, et al (2015) Results of the THERAPY trial: a prospective, randomized trial to define the role of mechanical thrombectomy as adjunctive treatment to IV rt-PA in acute ischemic stroke. Int J Stroke 10(S2):10 In: The European Stroke Organisation Conference; 2015 Apr 17–19; Glasgow, UK

48. Bracard S, Guillemin F, Ducrocq X. THRACE study: intermediate analysis results. Int J Stroke 2015;10(S2): 31. In: The European Stroke Organisation Conference; 2015 Apr 17–19; Glasgow, UK
49. Powers WJ, Derdeyn CP, Biller J, Coffey CS, Hoh BL, Jauch EC et al (2015) 2015 American Heart Association/American Stroke Association Focused Update of the 2013 Guidelines for the early management of patients with acute ischemic stroke regarding endovascular treatment: a guideline for healthcare professionals From the American Heart Association/American Stroke Association. Stroke 46:3020–3035
50. Campbell BCV, Donnan GA, Lees KR, Hacke W, Khatri P, Hill MD et al (2015) Endovascular stent thrombectomy: the new standard of care for large vessel ischaemic stroke. Lancet Neurol 14:846–854
51. Emberson J, Lees KR, Lyden P, Blackwell L, Albers G, Bluhmki E et al (2014) Effect of treatment delay, age, and stroke severity on the effects of intravenous thrombolysis with alteplase for acute ischaemic stroke: a meta-analysis of individual patient data from randomised trials. Lancet 384:1929–1935
52. Eugene F, Gauvrit JY, Ferre JC, Gentric JC, Besseghir A, Ronziere T et al (2015) One-year MR angiographic and clinical follow-up after intracranial mechanical thrombectomy using a stent retriever device. AJNR Am J Neuroradiol 36(1):126–132
53. Anastasian ZH (2014) Anaesthetic management of the patient with acute ischaemic stroke. Br J Anaesth 113(S2):ii9–ii16.
54. Brinjikji W, Murad MH, Rabinstein AA, Cloft HJ, Lanzino G, Kallmes DF (2015) Conscious sedation versus general anesthesia during endovascular acute ischemic stroke treatment: a systematic review and meta-analysis. Am J Neuroradiol 36:525–529
55. Malik AM, Vora NA, Lin R, Zaidi SF, Aleu A, Jankowitz BT et al (2011) Endovascular treatment of tandem extracranial/intracranial anterior circulation occlusions: preliminary single-center experience. Stroke 42:1653–1657
56. Heck DV, Brown MD (2015) Carotid stenting and intracranial thrombectomy for treatment of acute stroke due to tandem occlusions with aggressive antiplatelet therapy may be associated with a high incidence of intracranial hemorrhage. J Neurointerv Surg 7:170–175
57. Aghaebrahim A, Jovin T, Jadhav AP, Noorian A, Gupta R, Nogueira RG (2014) Endovascular recanalization of complete subacute to chronic atherosclerotic occlusions of intracranial arteries. J Neurointerv Surg 6:645–648
58. McPherson CM, Woo D, Cohen PL, Pancioli AM, Kissela BM, Carrozzella JA et al (2001) Early carotid endarterectomy for critical carotid artery stenosis after thrombolysis therapy in acute ischemic stroke in the middle cerebral artery. Stroke 32(9):2075–2080
59. Antman EM, Cohen M, Bernink PJ, McCabe CH, Horacek T, Papuchis G et al (2000) The TIMI risk score for unstable angina/non-ST elevation MI: a method for prognostication and therapeutic decision making. JAMA 284(7):835–842
60. Hagashida RT, Furlan AJ, Roberts H, Tomsick T, Connors B, Barr J et al (2003) Trial design and reporting standards for intra-arterial cerebral thrombolysis for acute ischemic stroke. Stroke 34(8):e109–e137
61. Zaidat OO, Yoo AJ, Khatri P, Tomsick TA, von Kummer R, Saver JL et al (2013) Recommendations on angiographic revascularization grading standards for acute ischemic stroke: a consensus statement. Stroke 44(9):2650–2663

Chapter 21
Cerebral Microbleeds and Thrombolysis for Acute Ischemic Stroke

JungSeok Lee and Mark Fisher

Abstract Cerebral microbleeds (CMB) are small MRI signal voids due to tiny hemosiderin deposits derived from red blood cell extravasation. The relationship between CMB and postthrombolysis intracerebral hemorrhage (ICH), the most feared complication of thrombolysis for acute ischemic stroke, has received considerable attention. Individual studies relating CMB and postthrombolysis ICH have generally failed to show a significant association. However, recent meta-analyses indicate that CMB presence or burden on pretreatment MRI is associated with postthrombolysis ICH. The exact nature of the CMB–postthrombolysis ICH relationship is complicated by the heterogeneous nature of CMB and their associated risk factors hypertension, cerebral amyloid angiopathy, and chronic kidney disease. CMB may be due to a primary disruption of the vessel wall (*primary CMB*) or may be a consequence of ischemic brain injury (*secondary CMB*). Clinicians making decisions regarding whether or not to offer intravenous thrombolysis to patients with acute ischemic stroke should attempt to distinguish primary from secondary CMB, as it is the former that likely create the greatest risk for postthrombolysis ICH.

Keywords Cerebral microbleeds • Hemosiderin • Ischemia • Capillary • Smooth muscle cell • Intracerebral hemorrhage

J. Lee, M.D., Ph.D.
Department of Neurology, JeJu National University School of Medicine, 1 Ara-1 dong, Jeju, South Korea

Department of Neurology, University of California Irvine, Irvine, CA, USA
e-mail: nrlee71@naver.com

M. Fisher, M.D. (✉)
Departments of Neurology, Anatomy & Neurobiology, and Pathology & Laboratory Medicine, University of California Irvine, Irvine, CA, USA

Department of Neurology, UC Irvine Medical Center, 101 The City Drive South, Shanbrom Hall, Room 121, Orange, CA 92868, USA
e-mail: mfisher@uci.edu

P.A. Lapchak, J.H. Zhang (eds.), *Neuroprotective Therapy for Stroke and Ischemic Disease*, Springer Series in Translational Stroke Research,
DOI 10.1007/978-3-319-45345-3_21

1 Introduction

Intravenous tissue-type plasminogen activator (tPA) is recommended as first-line therapy for patients with acute ischemic stroke, especially large artery occlusion presenting within 4.5 h of symptom onset [1, 2]. However, postthrombolysis intracranial hemorrhage (ICH) remains the most feared complication of thrombolysis for acute ischemic stroke. Approximately 7 % of the patients with acute ischemic stroke receiving intravenous tPA experience postthrombolysis ICH [3]. Clinical markers such as longer stroke onset-to-treatment times, age, high systolic blood pressure, hyperglycemia, and higher clinical stroke severity have emerged as risk factor for postthrombolysis ICH [4]. Recent studies suggest that neuroimaging markers of cerebral small vessel disease (SVD) such as cerebral microbleeds (CMB), multiple lacunes, and white matter hyperintensities (WMH) may also be risk factors for postthrombolysis ICH [4, 5]. CMB, indicating a microhemorrhage-prone microangiopathy, may indicate an increased risk of postthrombolysis ICH. However, the association between CMB (CMB presence and CMB burden) and the risk of postthrombolysis ICH in patients with acute ischemic stroke receiving intravenous thrombolysis remains somewhat controversial [4–6].

2 Radiographic Characteristics of Cerebral Microbleeds

Cerebral microbleeds (CMB) is a term that primarily refers to a radiographic brain lesion showing a small MRI signal void [7]. It appears as a small, homogenous, rounded lesion of low signal intensity on MRI sequences sensitive to paramagnetic susceptibility effect (GRE-T2*WI, SWI). The maximum diameter is 10 mm and, in some studies, a minimum diameter of 2 mm [7, 8].

There are two major scales used for assessing CMB. One is the Microbleed Anatomical Rating Scale (MARS) [9]. In this scale, CMB are scored as "definite" or "possible" CMB. Definite CMB are defined as small, rounded or circular, well-defined hypointense lesions within parenchyma with a clear margin, ranging from 2 to 10 mm on GRE-T2*WI. The other scale is the Brain Observer Microbleed Scale (BOMBS) [10]. In this scale, CMB are defined as homogenous, round foci, <10 mm diameter (no minimum size specified) of low signal intensity on GRE-T2*WI.

Recent studies indicate that CMB location provides important clues for the specific underlying vascular pathologic states, particularly hypertensive vasculopathy (for deep hemispheric or infra-tentorial CMB) and cerebral amyloid angiopathy (CAA) (for strictly lobar CMB) [7, 11]. The Rotterdam scan study showed an association of the APOE ε4 allele with strictly lobar CMB (APOE ε4 versus APOE ε3ε3; OR, 1.87; 95 %CI, 1.5–2.81) but not deep hemispheric or infra-tentorial CMB (APOE ε4 versus APOE ε3ε3; OR, 1.17; 95 %CI, 0.70–1.93) [11], consistent with CAA being the underlying pathology for this group of CMB.

2.1 Cerebral Microbleeds and Cerebral Microhemorrhage

There have been few studies of the radiological–pathological relationship for CMB [12, 13]. Fazekas et al. showed focal hemosiderin deposition in 21 of 34 areas of radiologically defined CMB. No specific pathology was found in 13/34 CMB. Hemosiderin deposits were also noted, without radiologically defined CMB, in two brains. These findings were based on 1.5T gradient-echo T2*-weighted images (GRE-T2*WI) with section thickness of 5 mm and with a gap of 0.5 mm [12].

Schrag et al., using 3T susceptibility-weighted image (SWI) with section thickness of 2 mm, showed 31 radiologically defined CMB from 10 cases. In ten lesions, intact red blood cells (RBC) were found. In 16 lesions, old hematomas were found. In three lesions, hemosiderin granules were detected only by microscopy. In one lesion, dissection in the vessel wall was observed. The remaining lesion appeared to be caused by a microaneurysm [13]. SWI sequence was more sensitive for detection of CMB than conventional GRE-T2*WI sequence [14].

More recent studies have suggested additional processes underlying radiologically defined CMB [15]. CMB can be indicative of a primarily ischemic process, and may result from hemorrhagic infarction [16] or hemorrhagic microinfarction [17]. CMB originating in this manner are referred to as *secondary CMB*, in contrast to those CMB resulting from a primary disruption of the vessel wall (*primary CMB*) [15]. Ischemia-induced release of iron stores in oligodendrocytes may result in radiologically defined CMB, particularly in putamen. Fisher et al. showed putamen microhemorrhages with capillary involvement in the elderly with no clinical history of stroke [18]. Janaway et al. showed putamen hemosiderin deposition in 99 % of the aged 65 and older (198/200 cases) [19].

2.2 Postthrombolysis Intracerebral Hemorrhage

ICH is the most feared complication of treatment with intravenous recombinant tissue-plasminogen activator (IV-tPA) in acute ischemic stroke. ICH in this setting occurs within or at margin of ischemic or infarcted brain tissue [20] and is categorized as hemorrhagic infarct (HI) or parenchymal hemorrhage (PH) based on the radiologic anatomy of the lesions. The European Cooperative Acute Stroke Study 2 (ECASS2) described HI1 (small petechiae without mass effect), HI2 (more confluent petechiae without mass effect), PH1 (<30 % of the infarcted area with mild space-occupying effect), and PH2 (>30 % of the infarcted area with significant space-occupying effect) [21].

Another type of ICH is remote ICH, defined as ICH in a brain region without visible ischemic injury [21, 22]. Two cohort studies showed remote ICH in roughly 2–3 % of postthrombolysis ICH cases [22, 23]. One report suggested that patients who developed new CMB had a significantly increased risk of remote ICH than patients without new CMB [24].

Direct comparison of the symptomatic postthrombolysis ICH rate among different studies may be hampered due to different recruitment methods such as tPA therapeutic window (the time from the onset of stroke to the beginning of the treatment), National Institutes of Health Stroke Scale (NIHSS) score, and definition of symptomatic postthrombolysis ICH [4]. In the NINDS study, symptomatic postthrombolysis ICH rate was 6.4 % in the tPA group compared with 0.6 % in placebo group [25]. In the European Cooperative Acute Stroke Study (ECASS) II, symptomatic postthrombolysis ICH occurred in 36 (8.8 %) of the intravenous tPA group and 13 (3.4 %) of the placebo group [26]. In the Alteplase Thrombolysis for Acute Noninterventional Therapy in Ischaemic Stroke (ATLANTIS) study, 7.0 % of the patients showed symptomatic postthrombolysis ICH [27]. Intravenous tPA was begun within 3 h of the onset of stroke in the NINDS study, while the therapeutic window was within 5 h in ATLANTIS study.

2.3 *Cerebral Small Vessel Disease and Postthrombolysis Intracerebral Hemorrhage*

The term cerebral small vessel disease (SVD) refers to a group of pathological processes affecting small arteries, arterioles, capillaries, and venules of the brain [28]. MRI manifestations of cerebral SVD include white matter hyperintensities (WMH), lacunes, and CMB [29]. In one retrospective study of 449 patients with acute anterior circulation stroke, the rate of postthrombolysis ICH was significantly more frequent in patients with moderate-to-severe deep WMH than in patients without WMH, even after correction for confounders such as age, stroke severity, and types of thrombolysis (intravenous, intra-arterial, and combined) (OR, 2.9; 95 %CI, 1.29–6.59; $p=0.015$) [30]. Palumbo et al. showed a significant association between postthrombolysis ICH and either severe WMH (OR, 2.75; 95 %CI, 1.15–6.53) or multiple lacunes (OR, 3.40; 95 %CI, 1.50–7.68) in 936 acute ischemic stroke patients receiving intravenous tPA. Patients with multiple lacunes had higher mortality at 90 days compared to those with one or no lacunes (OR, 2.9; 95 %CI, 1.3–6.2; $p=0.008$) [31]. A Belgian study, analyzing data from 400 patients receiving intravenous tPA for acute ischemic stroke, showed a nonsignificant trend toward association of WMH and postthrombolysis ICH (OR, 1.9; 95 %CI, 0.78–4.68; $p=0.16$). However, WMH was independently associated with poor outcome (OR, 2.39; 95 %CI, 1.21–4.72; $p=0.01$) [32]. In the study of Shi et al. analyzing data from 105 patients with acute ischemic stroke treated by mechanical thrombectomy, moderate-to-severe WMHs in the deep white matter was an independent predictor of hemorrhagic transformation (OR, 3.43; 95 %CI, 1.23–9.57; $p=0.019$) and PH (OR, 6.26; 95 %CI, 1.74–22.45; $p=0.005$) on multivariate logistic analysis [33]. There is still very limited data supporting a relationship between WMH burden and postthrombolysis ICH. Therefore, detecting moderate-to-severe WMH should not preclude intravenous thrombolysis in patients with acute ischemic stroke.

2.4 Cerebral Microbleeds and Postthrombolysis Intracerebral Hemorrhage

Two meta-analyses have explored the relationship between CMB presence and postthrombolysis ICH [34, 35]. The first by Charidimou et al. included 10 studies comprising 2028 patients with acute ischemic stroke treated with thrombolysis [34]. The second meta-analysis by Tsivgoulis et al. included 9 studies comprising 2479 patients with acute ischemic stroke treated with intravenous thrombolysis only [35]. The first and second meta-analyses shared 6 studies including 1623 patients with acute ischemic stroke treated with intravenous thrombolysis only. These two meta-studies indicated that CMB presence on pretreatment MRI was associated with postthrombolysis ICH [34, 35]. However, this result should be interpreted with caution, because differences in research methods may be a confounding factor that could influence the association. Interestingly, all individual studies except one (Dannenberg et al.) showed no significant relationship between CMB presence on pretreatment MRI and postthrombolysis ICH [36–48] (Table 21.1).

In the meta-analysis by Charidimou et al., the pooled odds ratio (OR) for CMB presence on pretreatment MRI and postthrombolysis ICH was 2.26 (95 %CI, 1.46–3.49; $p<0.0001$) [34, 36–45] (Table 21.1). These results remained in subgroup pooled analyses including 1704 patients treated only with intravenous tPA (pooled OR, 2.87; 95 %CI, 1.76–4.69; $p<0.0001$) [36–41, 44, 45].

In the pairwise meta-analysis by Tsivgoulis et al., the pooled relative risk (RR) for CMB presence on pretreatment MRI and postthrombolysis ICH was 2.36 (95 %CI, 1.21–4.61; $p=0.01$). This latter meta-analysis divided patients into three groups: CMB absence (0 CMB), low/moderate CMB burden (1–10 CMBs), and high CMB burden (>10 CMBs). The risk of postthrombolysis ICH in patients with high CMB burden was increased when compared to both patients with 0–10 CMB (pooled RR, 12.10; 95 %CI, 4.36–33.57; $p<0.001$) and 1–10 CMBs (pooled RR, 7.01; 95 %CI, 3.20–15.38; $p<0.001$) [34–41, 46–48]. Tsivgoulis et al. also performed an individual-patient data meta-analysis (regarded as the gold standard for systemic review) using original data from three available studies ($n=521$) to produce more reliable results. In this individual-patient data meta-analysis, high CMB burden (>10 CMBs) was associated with postthrombolysis ICH when compared to patients with 0–10 CMBs adjusting for potential confounders such as MRI sequence, MRI field strength, baseline stroke severity, onset-to-treatment time, demographics, and vascular risk factors (adjusted OR, 18.17; 95 %CI, 2.39–138.22; $p=0.005$) [35, 36, 41, 47].

In 2014 and 2015, four new studies reported results regarding the association between CMB presence (or burden) and symptomatic postthrombolysis ICH; these studies had conflicting results [36, 41, 46, 48]. Two of these studies showed a significant association between CMB presence (or burden) and symptomatic postthrombolysis ICH [36, 46], whereas the remaining two studies indicated no relationship between CMB and symptomatic postthrombolysis ICH [41, 48]. In study of Dannenberg et al., the OR for CMB presence on pretreatment MRI and postthrombolysis ICH was 7.06 (95 %CI, 1.87–26.66) [35, 36]. However, this study showed

Table 21.1 MRI parameters and CMB prevalence

Study	Study design	Number of patients	Sequence	Field strength (Tesla)	Section thickness (mm)	CMB prevalence (%)	Thrombolysis regimen	Meta-analysis
Kidwell et al. [42]	Retrospective	41	T2*WI-GRE	1.5	7	12	IV/IA	C
			EPI-SWI				tPA or UK	
Derex et al. [37]	Retrospective	44	T2*WI-GRE	1.5	5	18.2	IV tPA	C,T
Kakuda et al. [38]	Prospective	70	T2*WI-GRE	1.5	5	15.7	IV tPA	C,T
Kim et al. [43]	Retrospective	65	T2*WI-GRE		5	38.5	IV tPA, IA UK	C
Fiehler et al. [39]	Prospective	570	T2*WI-GRE		5–7	15.1	IV tPA	C,T
Moriya et al. [44]	Retrospective	71	T2*WI-GRE	1.5		19.7	IV tPA	C
Kimura et al. [40]	Prospective	224	T2*WI-GRE	1.5	6	32.1	IV tPA	C,T
Dannenberg et al. [36]	Prospective	326	T2*WI-GRE	3	5	24.8	IV tPA	C,T
Gratz et al. [41]	Prospective	392	SWI	1.5 or 3		20.2	IV tPA or EVT	C,T
Yan et al. [45]	Retrospective	225	SWI	3	2	36.1	IV tPA	C
Yan et al. [46]	Retrospective	333	SWI	3	2	39.7	IV tPA	T
Turc et al. [48]	Prospective	717	T2*WI-GRE	1.5		20.9	IV tPA	T

CMB cerebral microbleeds, *T2*WI-GRE* gradient echo T2*-weighted images, *EPI-SWI* echo planar susceptibility-weighted imaging, *SWI* susceptibility-weighted imaging, *IV tPA* intra-venous tissue plasminogen activator, *IA UK* intra-arterial urokinase, *EVT* endovascular treatment, *C* meta-analysis (Charidimou et al.), *T* meta-analysis (Tsivgoulis et al.)

the presence of single CMB did not increase the risk of postthrombolysis ICH [36]. Yan et al. study showed CMB burden (≥3 CMBs) was independently associated with PH (OR, 6.70; 95 %CI, 2.05–21.88; $p=0.002$) and poor clinical outcome (OR, 2.28; 95 %CI, 1.02–5.09; $p=0.044$) [46]. In the study of Gratz et al., CMB presence on pretreatment SWI did not increased the risk for symptomatic asymptomatic postthrombolysis ICH or influence the clinical outcome of patients receiving thrombolysis. There was a marginally significant association between high CMB burden and ICH outside the infarct (OR, 1.004; 95 %CI, 1.000–1.008; $p=0.048$) [41]. The study of Turc et al. ($n=717$) showed no significant association between CMB presence or burden and postthrombolysis ICH in multivariate analysis, irrespective of symptomatic postthrombolysis ICH definition [48]. The rates of symptomatic postthrombolysis ICH ranged from 3.8 % (ECASS-3 definition) to 9.1 % (NINDS definition). In this study, the number of CMB and their distribution were rated according to MARS. In addition, there were some differences in population characteristics among studies. In the study of Dannenberg et al. [36], the patients were relatively older and more frequently had presumed CAA compared to the Turc et al. study [48]. In the study of Yan et al. [46], the patients had a higher frequency of CMB presence (39.9 %) and higher CMB burden compared to the study of Turc et al. [48].

There appears to be four major risk factors for CMB: hypertension, CAA, chronic kidney disease (CKD), and cerebral autosomal dominant arteriopathy with subcortical infarcts and leukoencephalopathy (CADASIL). Hypertension, CAA, and CADASIL have both arteriolar smooth muscle cell (SMC) and capillary (i.e., blood–brain barrier) involvement. CKD has been characterized by SMC alterations, while capillary involvement in CKD has not yet been investigated [49]. It is thought that postthrombolysis ICH is more likely to develop in the presence of preexisting arteriolar SMC injury [49].

3 Conclusions

Recent studies have shown that CMB burden, especially presence of more than ten CMB on pretreatment MRI, increases the risk of postthrombolysis ICH during IV thrombolysis. However, a definitive answer cannot yet be provided as to whether excessive CMB burden on pretreatment MRI represents an absolute exclusion criterion for IV thrombolysis in acute ischemic stroke. Therefore, the data on the influence of CMB burden on postthrombolysis ICH should be treated with caution. In that regard, clinicians making decisions regarding whether or not to offer intravenous thrombolysis to the patients with acute ischemic stroke should realize that the number of CMB present is less important than the nature of the process driving CMB development [49]. In other words, clinicians will need to distinguish CMB that have developed as a consequence of ischemia (secondary CMB) versus primary CMB resulting from an underlying process that directly injures the vessel wall, such as hypertension and CAA. It is the patient with primary CMB who would appear to be at greatest risk for postthrombolysis ICH.

Disclosures Supported by NIH NS 20989

Dr. Fisher receives research support from Boehringer-Ingelheim and Otsuka Pharmaceutical Company, and serves on Clinical Events Committees for Covidien.

References

1. del Zoppo GJ, Saver JL, Jauch EC, Adams HP (2009) Expansion of the time window for treatment of acute ischemic stroke with intravenous tissue plasminogen activator: a science advisory from the American Heart Association/American Stroke Association. Stroke 40:2945–2948
2. Lees KR, Bluhmki E, von Kummer R, Brott TG, Toni D, Grotta JC et al (2010) Time to treatment with intravenous alteplase and outcome in stroke: an updated pooled analysis of ECASS, ATLANTIS, NINDS, and EPITHET trials. Lancet 375:1695–1703
3. Mullen MT, Pisapia JM, Tilwa S, Messe SR, Stein SC (2012) Systematic review of outcome after ischemic stroke due to anterior circulation occlusion treated with intravenous, intra-arterial, or combined intravenous intra-arterial thrombolysis. Stroke 43:2350–2355
4. Derex L, Nighoghossian N (2008) Intracerebral haemorrhage after thrombolysis for acute ischaemic stroke: an update. J Neurol Neurosurg Psychiatry 79:1093–1099
5. Charidimou A, Kakar P, Fox Z, Werring DJ (2013) Cerebral microbleeds and the risk of intracerebral haemorrhage after thrombolysis for acute ischaemic stroke: systematic review and meta-analysis. J Neurol Neurosurg Psychiatry 84:277–280
6. Shoamanesh A, Kwok CS, Lim PA, Benavente OR (2013) Postthrombolysis intracranial hemorrhage risk of cerebral microbleeds in acute stroke patients: a systematic review and meta-analysis. Int J Stroke 8(5):348–356
7. Greenberg SM, Vernooij MW, Cordonnier C, Viswanathan A, Al-Shahi Salman R, Warach S et al (2009) Cerebral microbleeds: a field guide to detection and interpretation. Lancet Neurol 8:165–174
8. Cordonnier C, Al-Shahi Salman R, Wardlaw J (2007) Spontaneous brain microbleeds: systematic review, subgroup analyses and standards for study design and reporting. Brain 130: 1988–2003
9. Gregoire SM, Chaudhary UJ, Brown MM, Yousry TA, Kallis C, Jager HR et al (2009) The Microbleed Anatomical Rating Scale (MARS): reliability of a tool to map brain microbleeds. Neurology 73(21):1759–1766
10. Cordonnier C, Potter GM, Jackson CA et al (2009) Improving inter-rater agreement about brain microbleeds: development of the Brain Observer MicroBleed Scale (BOMBS). Stroke 40:94–99
11. Vernooij MW, van der Lugt A, Ikram MA, Wielopolski PA, Niessen WJ, Hofman A et al (2008) Prevalence and risk factors of cerebral microbleeds: the Rotterdam Scan Study. Neurology 70:1208–1214
12. Fazekas F, Kleinert R, Roob G, Kleinert G, Kapeller P, Schmidt R et al (1999) Histopathologic analysis of foci of signal loss on gradient-echo T2*-weighted MR images in patients with spontaneous intracerebral hemorrhage: evidence of microangiopathy related microbleeds. AJNR Am J Neuroradiol 20:637–642
13. Schrag M, McAuley G, Pomakian J, Jiffry A, Tung S, Mueller C et al (2010) Correlation of hypointensities in susceptibility-weighted images to tissue histology in dementia patients with cerebral amyloid angiopathy: a postmortem MRI study. Acta Neuropathol (Berl) 119:291–302
14. Haacke EM, DelProposto ZS, Chaturvedi S, Sehgal V, Tenzer M, Neelavalli J et al (2007) Imaging cerebral amyloid angiopathy with susceptibility-weighted imaging. AJNR Am J Neuroradiol 28:316–317
15. Fisher M (2014) Cerebral microbleeds: where are we now? Neurology 83:1304–1305
16. Jeon SB, Kwon SU, Cho AH, Yun SC, Kim JS, Kang DW (2009) Rapid appearance of new cerebral microbleeds after acute ischemic stroke. Neurology 73:1638–1644

17. Tanskanen M, Mäkelä M, Myllykangas L, Rastas S, Sulkava R, Paetau A (2012) Intracerebral hemorrhage in the oldest-old: a population-based study (Vantaa 851). Front Neurol 3:103
18. Fisher M, French S, Ji P, Kim RC (2010) Cerebral microbleeds in the elderly: a pathological analysis. Stroke 41:2782–2785
19. Janaway BM, Simpson JE, Hoggard N et al (2014) Brain haemosiderin in older people: pathological evidence for an ischaemic origin of magnetic resonance imaging (MRI) microbleeds. Neuropathol Appl Neurobiol 40:258–269
20. Wahlgren N, Ahmed N, Davalos A, Ford GA, Grond M, Hacke W (2007) Thrombolysis with alteplase for acute ischaemic stroke in the Safe Implementation of Thrombolysis in Stroke-Monitoring Study (SITS-MOST): an observational study. Lancet 369:275–282
21. Trouillas P, von Kummer R (2006) Classification and pathogenesis of cerebral hemorrhages after thrombolysis in ischaemic stroke. Stroke 37:556–561
22. Gao Y, Churilov L, Teo S, Yan B (2015) Remote intracerebral haemorrhage post intravenous thrombolysis: experience from an Australian stroke centre. J Clin Neurosci 22(2):352–356
23. Mazya MV, Ahmed N, Ford GA, Hobohm C, Mikulik R, Nunes AP et al (2014) Remote or extraischemic intracerebral hemorrhage–an uncommon complication of stroke thrombolysis: results from the safe implementation of treatments in stroke-international stroke thrombolysis register. Stroke 45:1657–1663
24. Shoamanesh A, Yan S, Charidimou A (2015) New cerebral microbleeds and mechanism of post-thrombolysis remote intracerebral hemorrhage: "red meets white" revisited. Front Neurol 6:203
25. The National Institute of Neurological Disorders and Stroke rt-PA Stroke Study Group (1995) Tissue plasminogen activator for acute ischemic stroke. N Engl J Med 333:1581–1587
26. Hacke W, Kaste M, Fieschi C, von Kummer R, Davalos A, Meier D et al (1998) Randomised double-blind placebo-controlled trial of thrombolytic therapy with intravenous alteplase in acute ischaemic stroke (ECASSII). Lancet 352:1245–1251
27. Clark WM, Wissman S, Albers GW, Jhamandas JH, Madden KP, Hamilton S (1999) Recombinant tissue-type plasminogen activator (alteplase) for ischemic stroke 3 to 5 hours after symptom onset: the ATLANTIS Study, a randomized controlled trial: alteplase thrombolysis for acute noninterventional therapy in ischemic stroke. JAMA 21:2019–2026
28. Pantoni L, Fierini F, Poggesi A (2014) Thrombolysis in acute stroke patients with cerebral small vessel disease. Cerebrovasc Dis 37(1):5–13
29. Werring DJ, Gregoire SM, Cipolotti L (2010) Cerebral microbleeds and vascular cognitive impairment. J Neurol Sci 299(1–2):131–135
30. Neumann-Haefelin T, Hoelig S, Berkefeld J, Fiehler J, Gass A, Humpich M, Kastrup A et al (2006) Leukoaraiosis is a risk factor for symptomatic intracerebral hemorrhage after thrombolysis for acute stroke. Stroke 37:2463–2466
31. Palumbo V, Boulanger JM, Hill MD, Inzitari D, Buchan AM (2007) Leukoaraiosis and intracerebral hemorrhage after thrombolysis in acute stroke. Neurology 68:1020–1024
32. Aries MJH, Uyttenboogaart M, Vroomen PC, de Keyser J, Luijckx GJ (2010) tPA treatment for acute ischaemic stroke in patients with leukoaraiosis. Eur J Neurol 17:866–870
33. Shi ZS, Loh Y, Liebeskind DS, Saver JL, Gonzalez NR, Tateshima S et al (2012) Leukoaraiosis predicts parenchymal hematoma after mechanical thrombectomy in acute ischemic stroke. Stroke 43:1806–1811
34. Charidimou A, Shoamanesh A, Wilson D, Gang Q, Fox Z, Jäger HR et al (2015) Cerebral microbleeds and postthrombolysis intracerebral hemorrhage risk updated meta-analysis. Neurology 85(11):927–934
35. Tsivgoulis G, Zand R, Katsanos AH, Turc G, Nolte CH, Jung S et al (2016) Risk of symptomatic intracerebral hemorrhage after intravenous thrombolysis in patients with acute ischemic stroke and high cerebral microbleed burden: a meta-analysis. JAMA Neurol 73(6):675–683 (published online April 18, 2016)
36. Dannenberg S, Scheitz JF, Rozanski M, Erdur H, Brunecker P, Werring DJ et al (2014) Number of cerebral microbleeds and risk of intracerebral hemorrhage after intravenous thrombolysis. Stroke 45:2900–2905

37. Derex L, Nighoghossian N, Hermier M, Adeleine P, Philippeau F, Honnorat J et al (2004) Thrombolysis for ischemic stroke in patients with old microbleeds on pretreatment MRI. Cerebrovasc Dis 17:238–241
38. Kakuda W, Thijs VN, Lansberg MG, Bammer R, Wechsler L, Kemp S et al (2005) Clinical importance of microbleeds in patients receiving IV thrombolysis. Neurology 65:1175–1178
39. Fiehler J, Albers GW, Boulanger JM, Derex L, Gass A, Hjort N et al (2007) Bleeding Risk Analysis in Stroke Imaging Before Thrombolysis (BRASIL): pooled analysis of T2*-weighted magnetic resonance imaging data from 570 patients. Stroke 38:2738–2744
40. Kimura K, Aoki J, Shibazaki K, Saji N, Uemura J, Sakamoto Y (2013) New Appearance of extraischemic microbleeds on T2*-weighted magnetic resonance imaging 24 hours after tissue-type plasminogen activator administration. Stroke 44:2776–2781
41. Gratz PP, El-Koussy M, Hsieh K, von Arx S, Mono ML, Heldner MR et al (2014) Preexisting cerebral microbleeds on susceptibility-weighted magnetic resonance imaging and post-thrombolysis bleeding risk in 392 patients. Stroke 45:1684–1688
42. Kidwell CS, Saver JL, Villablanca JP, Duckwiler G, Fredieu A, Gough K et al (2002) Magnetic resonance imaging detection of microbleeds before thrombolysis: an emerging application. Stroke 33:95–98
43. Kim HS, Lee DH, Ryu CW, Lee JH, Choi CG, Kim SJ et al (2006) Multiple cerebral microbleeds in hyperacute ischemic stroke: impact on prevalence and severity of early hemorrhagic transformation after thrombolytic treatment. AJR Am J Roentgenol 186:1443–1449
44. Moriya Y, Takahashi W, Kijima C et al (2013) Predictors for hemorrhagic transformation with intravenous tissue plasminogen activator in acute ischemic stroke. Tokai J Exp Clin Med 38:24–27
45. Yan SQ, Wan JP, Guo Y, Lou M (2014) Impact of cerebral microbleeds on outcomes of acute ischemic stroke treated with intravenous thrombolysis (in Chinese). Zhejiang Da Xue Xue Bao Yi Xue Ban 43:20–27
46. Yan S, Jin X, Zhang X, Zhang S, Liebeskind DS, Lou M (2015) Extensive cerebral microbleeds predict parenchymal haemorrhage and poor outcome after intravenous thrombolysis. J Neurol Neurosurg Psychiatry 86(11):1267–1272
47. Goyal N, Tsivgoulis G, Male S, Bhaskaran D, Alexandrov AV, Zand R (2015) Feasibility of rapid short sequence magnetic resonance imaging for screening stroke mimics within the iv-tpa treatment window: a pilot study. Poster presented at international stroke conference 2015; February 11–12, 2015; Nashville
48. Turc G, Sallem A, Moulin S, Tisserand M, Machet A, Edjlali M et al (2015) Microbleed status and 3-month outcome after intravenous thrombolysis in 717 patients with acute ischemic stroke. Stroke 46(9):2458–2463
49. Fisher M (2016) Cerebral microbleeds and thrombolysis: clinical consequences and mechanistic implications. JAMA Neurol 73(6):632–635. doi:10.1001/jamaneurol.2016.0576 (Epub ahead of print)

Chapter 22
Targeting Pericytes and the Microcirculation for Ischemic Stroke Therapy

Ain A. Neuhaus, Brad A. Sutherland, and Alastair M. Buchan

Abstract Ischaemic stroke is a major global cause of disability and death, yet the therapeutic options currently available for stroke are very limited. The only effective acute treatments of ischemic stroke revolve around restoring patency to the occluded artery through degradation (intravenous thrombolysis) or mechanical removal of the clot [Int J Stroke 10:1168–1178, 2015; Brain 136:3528–3553, 2013]. However, there is an increasing body of evidence to suggest that even following recanalization of the large vessel, the post-ischemic microvasculature remains dysfunctional and does not necessarily allow effective reperfusion. Contributors to this phenomenon include astrocyte swelling and compression of microvessels, obstruction of flow due to inflammatory changes, leukocyte adhesion and thrombosis, and the constriction of capillaries by pericytes dying in rigor [Shock 8:95–101, 1997; J Cereb Blood Flow 36:451–455, 2016]. This chapter will provide an overview of microvascular function in health, describe the pathological changes that occur following ischemia and reperfusion, and explore the role of the microvasculature, with a focus on pericytes as a potential therapeutic target in ischemic stroke.

A.A. Neuhaus, M.A.
Acute Stroke Programme, Radcliffe Department of Medicine, University of Oxford, Oxford, UK
e-mail: ain.neuhaus@rdm.ox.ac.uk

B.A. Sutherland, Ph.D.
School of Medicine, University of Tasmania, Hobart, Australia

Acute Stroke Programme, Radcliffe Department of Medicine, University of Oxford, Oxford, UK
e-mail: brad.sutherland@utas.edu.au

A.M. Buchan, DSc (Oxon) LLD (Hon) FRCP FRCPEd (✉)
Acute Stroke Programme, Radcliffe Department of Medicine, University of Oxford, Oxford, UK

Medical Sciences Division, University of Oxford,
Level 3, John Radcliffe Hospital, Headley Way, Oxford, Oxfordshire OX3 9DU, UK

Acute Vascular Imaging Centre, University of Oxford, Oxford University Hospitals, Oxford, UK
e-mail: alastair.buchan@medsci.ox.ac.uk

P.A. Lapchak, J.H. Zhang (eds.), *Neuroprotective Therapy for Stroke and Ischemic Disease*, Springer Series in Translational Stroke Research,
DOI 10.1007/978-3-319-45345-3_22

Keywords Pericyte • Microvasculature • No-reflow • Stroke • Ischemic stroke • Cerebral ischemia • stroke treatment

1 Introduction

Perhaps more so than in any other organ, cerebral blood flow (CBF) is tightly regulated to ensure adequate supply and match the very high energy demands of the brain—despite comprising only 2 % of body weight, the brain receives approximately 20 % of cardiac output [1, 2]. It is critical that rapid increases in cerebral metabolism are matched by a hyperemic response to avoid even transient periods of hypoxia [2]. This phenomenon, known as neurovascular coupling, requires communication between active neurons, glial cells, and the vasculature, which collectively form the neurovascular unit (NVU; Fig. 22.1). The key function of the NVU is to facilitate nutrient

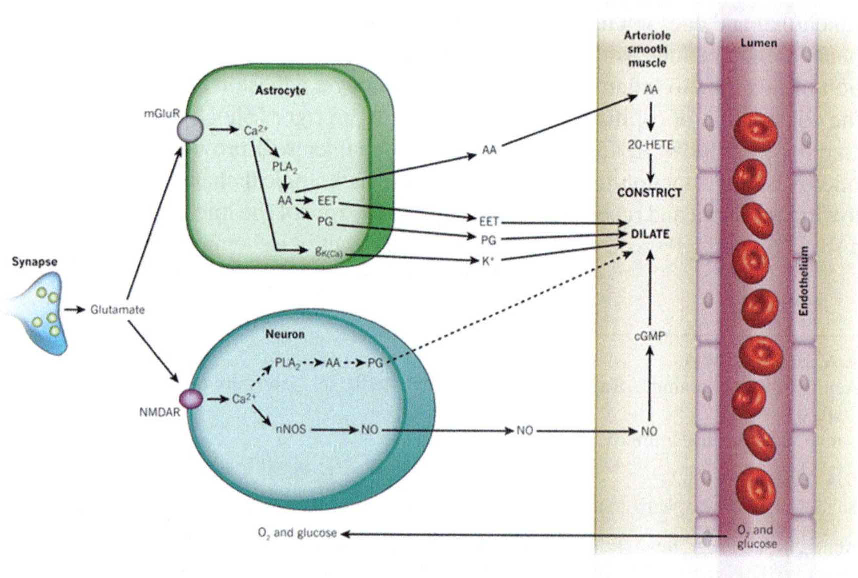

Fig. 22.1 The neurovascular unit is responsible for neurovascular coupling. Neuronal activity triggers Ca^{2+} responses in both other neurons, through NMDA receptors (NMDAR), and in astrocytes, through metabotropic glutamate receptors (mGluR). This leads to the release of numerous vasoactive mediators, including nitric oxide (NO) produced by neuronal nitric oxide synthase (nNOS), potassium ions (K^+), phospholipase A_2-derived arachidonic acid (AA), and AA derivatives including various prostaglandin species (PG) and epoxyeicosatrienoic acids (EET). Through various surface membrane receptors (in the case of AA derivatives), direct hyperpolarization (K^+-mediated inward rectifying K^+ channel activation), or activation of cylic GMP production (NO), these mediators cause relaxation of smooth muscle and pericytes. Under certain conditions, astrocytes may also mediate vasoconstriction through AA-derived 20-hydroxyeicosatetraenoic acid (20-HETE). When vasodilation occurs, O_2 and glucose supply to the parenchyma is increased. Reproduced from [2] with permission from Nature Publishing Group

exchange, maintain stability of the neuronal milieu, and increase or reduce blood supply depending on the needs of the tissue it supplies. The mechanisms by which the latter occurs remain highly controversial, yet are clearly of immense importance to understanding stroke pathology on a microvascular level.

The primary drivers of metabolic demand are neurons, due to the ATP demands of synaptic transmission [3], but it is not certain whether they directly communicate with the vasculature. It has been proposed that astrocytes act as intermediaries, sensing synaptic activity through the "tripartite synapse" and releasing vasoactive substances onto the vasculature [4, 5]. While such an arrangement seems advantageous, due to the much higher incidence of astrocyte endfeet covering the vasculature compared to neuronal processes [6], some recent experiments have failed to demonstrate changes to CBF responses when astrocyte activity is inhibited [7]. Equally uncertain are the molecular mediators of neurovascular coupling: proposed candidates include prostaglandins, K^+ ions, nitric oxide, and numerous others, but their relative contributions and importance remain poorly understood (for an in-depth overview, see [2, 8]).

The role of the NVU in normal physiology is particularly relevant to cerebrovascular disorders. Stroke is a leading cause of morbidity and mortality, with 6.5 million stroke-related deaths and over 25 million stroke survivors in 2013, a global burden that is increasing [9]. Approximately 87 % of strokes are ischemic, with the remainder being hemorrhagic in origin. In either case, the loss of nutrient supply will result in widespread cell death in the affected brain region. Brain ischemia is caused by a thrombus or an embolus occluding an artery within the brain. The most common occlusion site in human ischemic stroke is the middle cerebral artery (MCA), which prevents blood flow to the downstream territory that this artery supplies, including most of the cerebral cortex through its superficial branches and the striatum through its deep lenticulostriate branches [10]. Strokes of the anterior cerebral artery and posterior circulation are less common, but regardless of the occlusion site, the ultimate consequence of this reduction in perfusion is cell dysfunction and eventual death unless blood flow is rapidly restored. This is not restricted to neurons and affects the entire NVU, leading to abnormal regulation of cerebral blood flow [2]. Moreover, some of these effects are not necessarily reversed even when perfusion is restored, making the NVU an important target for novel stroke therapies [11].

2 Hemodynamic Effects of Stroke

Arterial occlusion leads to a cessation of anterograde blood flow through the affected vessel, but cerebral perfusion can be maintained if collateral vessels dilate and provide an alternative blood supply [12]. When collateral blood supply also falls below the level required to sustain neuronal viability, the severe reduction in perfusion of the brain parenchyma results in the development of a stroke [12]. Where blood flow is completely abolished, the parenchyma suffers rapid and irreversible damage, forming the core of the infarct [13]. Surrounding the core is the penumbra—a region where blood flow has been reduced, but not to the critical level where neurons begin to die, estimated to be approximately 20 % of baseline CBF [13, 14]. The penumbra was first described in the late 1970s [15, 16], as part of a detailed description of the various blood flow thresholds leading to changes in multiple aspects of neuronal function, including protein synthesis, synaptic activity, and membrane integrity [17]. Neurons in the penumbra remain electrically silent on limited energy supply, but can resume their activity when the tissue returns to normal oxygenation status. Therefore, the penumbra is the main target of therapeutic interventions, as it can be salvaged (or at least its death delayed) with neuroprotective agents and the restoration of blood flow through the affected artery.

Arteries are strong blood vessels with concentric rings of vascular smooth muscle (VSM) and endothelial cells surrounded by a thick tunica media layer. Arteries can withstand the shear force associated with a large volume of pulsating blood at high pressures. When an occlusion occurs within an artery, downstream blood flow ceases and arteries lose their elasticity and increase in stiffness. One recent clinical study revealed that carotid artery elasticity, as measured by velocity vector imaging, was reduced in carotid arteries ipsilateral to the stroke [18]. In isolated MCA studies, ischemia can diminish vascular tone and disrupt the actin cytoskeleton of the smooth muscle [19]. As a result, the cerebral circulation loses its autoregulatory properties—the ability to maintain constant flow despite variations in systemic blood pressure—leading to a loss of fine control over cerebral blood flow [20] and potential worsening of ischemia when blood pressure is reduced. This has been corroborated by studies where pharmacologically inducing hypertension reduced brain injury following MCA occlusion (MCAO) by enhancing cerebral blood flow [21]. Interestingly, the drop in perfusion pressure in downstream arteries of the occlusion site can lead to myogenic vasodilation, possibly through the decreased clearance of carbon dioxide and lactate [22].

Downstream of arteries are penetrating arterioles which dive deep into the cerebral cortex to supply blood to the capillary bed. After an occlusion in an artery, its arterioles also become devoid of blood flow and, similarly to arteries, lose their elasticity and autoregulation. Parenchymal arterioles under normotensive conditions respond to carotid artery occlusion by active vasodilation, but do not undergo structural remodeling [23]. Using in vivo two-photon imaging, Shih and colleagues showed that arterioles exhibited robust vasodilation throughout MCAO for 90 min [24]. However, under hypertensive conditions, parenchymal arterioles have increased wall thickness and distensibility, leading to a lack of response to ischemia to counteract the perfusion deficit [23]. When an arteriole is directly occluded, a microinfarct forms in its territory, suggesting that there is enough blood flow deficit to produce neuronal damage even when neighboring arterioles and the upstream vasculature remain intact [25]. Arterioles feed into a large capillary network that supplies oxygen and glucose throughout the brain to maintain neurovascular function. Using a vascular angiome model, it was calculated that downstream capillaries had reduced flow up until approximately halfway to the next penetrating arteriole, where full recovery of flow in capillaries occurred [26]. The control of microcirculation through capillaries and its dysfunction during stroke is discussed below.

At the other end of the capillary network are the draining ascending venules taking deoxygenated blood towards the large veins. During an arterial occlusion, there is no blood to drain and so blood flow is shunted through penetrating venules into upstream capillaries to relieve the ischemic region of blood flow depletion [27]. In addition, occlusion of venules can occur, which can in turn reduce blood flow to their upstream capillaries, reverse the direction of flow, and cause an increase in vessel diameter [28]; however, this is restricted to the first few branches off the venule [26]. This reduction in flow can also lead to microinfarcts and localized neuronal damage that is, broadly speaking, similar to the consequences of single arteriole occlusion [25].

The final steps in the cerebrovascular tree are the draining veins taking deoxygenated blood to the heart. Similarly to venules, when an arterial occlusion occurs, there is no blood flow through the veins, which leads to either venous blood pooling and no venous return, or incompletely occluded venous flow [29]. Impaired venous return after arterial occlusion was associated with increased functional deficit and histological damage in a non-human primate model of stroke [29]. However, early venous filling can occur in stroke, reflecting arteriovenous shunting, increased local circulation, and an attempt to restore blood flow to the ischemic territory [30].

Overall, vascular occlusion at any step within the vascular tree can have profound effects leading to loss of vessel function and interaction with neighboring parenchymal cells. While restoration of blood flow can reverse many of these effects, the residual dysfunction may be a problem of considerable clinical relevance.

3 Hemodynamic Effects of Recanalization

The major therapeutic strategy for ischemic stroke is to recanalize the occluded blood vessel and restore blood flow to the ischemic region. Over the last 20 years, recombinant tissue plasminogen activator (rtPA, formulated as alteplase) has been used in acute ischemic stroke therapy for this purpose. It activates plasmin to degrade fibrin, effectively dissolving the occluding clot and restoring blood flow to that artery and the downstream vascular bed. rtPA improves neurological outcome in stroke patients, provided it is administered within the first 4.5 h after stroke onset [31]. However, with rtPA, there is a risk of intracerebral hemorrhage, potentially due to an increase in matrix metalloprotease activity and degradation of the blood–brain barrier [32]. More recently, five trials have shown significant improvement in clinical outcomes by restoring blood flow in ischemic stroke patients using endovascular thrombectomy, where a stent-retriever device is used to retract the clot [33–37]. This treatment has been shown to be superior to rtPA without increasing adverse events such as symptomatic intracerebral hemorrhage [38, 39].

The level of reperfusion after recanalization is a key determinant of neurological outcome following ischemia. There is clinical evidence that greater collateralization is associated with improved recanalization and reperfusion rates, as well as improved efficacy following endovascular thrombectomy after ischemic stroke

[40, 41]. In rodent studies using a transient MCAO model, CBF during reperfusion was more predictive of outcome than CBF during ischemia [42]. In fact, multiple neuroprotective agents have actually provided protection through improvement of blood flow and alteration of other physiological parameters rather than a direct neuropharmacological effect [43]. In addition, the speed at which CBF perfusion is restored is also critical to ischemic stroke pathophysiology. rtPA treatment leads to a gradual increase in CBF as the clot is slowly dissolved while endovascular thrombectomy leads to an abrupt increase in CBF as the clot is suddenly removed [44]. These disparate CBF profiles can alter the pathophysiologic response in the brain: the gradual restoration of CBF with rtPA lends itself to greater ischemic damage for longer, while the abrupt restoration of CBF with endovascular thrombectomy can lead to transient hyperperfusion and reperfusion injury via increased free radical production [45].

An artery that has been occluded and is suddenly recanalized can still feature impaired autoregulation and vascular reactivity for some time. The large amount of free radical production as a result of the restoration of blood flow can damage arterial smooth muscle, leading to loss of basal tone and myogenic reactivity [46, 47].

In parenchymal arterioles during ischemia, in vivo imaging revealed that these vessels dilate; when the arterioles are reperfused, this persistent vasodilation offsets the incomplete recovery of erythrocyte velocity, to allow as much post-ischemic erythrocyte flux as possible [24]. However, in isolated parenchymal arteriole studies, upon reperfusion, parenchymal arterioles constrict, which is not due to an increase in smooth muscle calcium influx, but appears to be due to increased sensitivity to calcium [48] and peroxynitrite generation [49]. This constrictive response in parenchymal arterioles could contribute to post-ischemic hypoperfusion and expansion of the ischemic injury.

Upon reperfusion, stagnant flow, thrombus-like foci, and neutrophil-like cells begin to accumulate within cortical venules [50]. Venules appear to be a major site where leukocytes accumulate during reperfusion following stroke, contributing to an exaggerated inflammatory response at this time. In a rat model, 2 h of MCAO followed by 1 h of reperfusion led to a significant increase in leukocyte rolling and adhesion within cerebral venules [51]. Like venules, cerebral venous pooling can occur after reperfusion, which can lead to poor restoration of flow. Early venous filling while attempting to restore flow to the ischemic region can be an indicator of tissue damage; recanalization can then lead to severe hyperemia in these vessels and associated hemorrhage [30].

4 Capillary Blood Flow

The phenomena described thus far focus on the large blood vessels, typically considered the most relevant to ischemic stroke, and the larger vessels of the microvasculature (the arterioles and venules). Indeed, while even penetrating arteriole occlusions can result in infarct development, occluding single capillaries does not [25]. Nonetheless, capillaries are of immense importance as the main sites of nutrient exchange between tissues and the vasculature, due to three factors. First, their relative proximity to neurons and other parenchymal cells reduces the distance of diffusion and facilitates exchange of gases and metabolites [52]. Second, the very high density of capillaries—in excess of 500 mm of vessel length per mm^3 of tissue in rat cortex—increases the surface area on which exchange can occur [53]. Finally, blood flow through capillaries is slow, at approximately 1 mm per second [54], allowing more time for gas equilibration. Recent work suggests that under resting conditions, arteriolar flow may be a significant contributor to oxygen delivery [55]. However, it is unequivocally clear that during conditions of stress (be it increased demand in the form of neuronal activity, or reduced supply during an ischemic insult), maintaining normal capillary flow is crucial for appropriate supply of oxygen and metabolites to the brain [55].

Due to the small size of capillaries, typically <5 μm in diameter, even very small alterations of capillary diameter would produce a considerable change in flow through the vessel. According to Poiseuille's law, flow through a tube will increase proportionally with radius to the fourth power (r^4); as such, an increase in diameter from 5 to 6 μm would result in a more than twofold increase in flow. Capillary flow, however, does not operate according to the ideal conditions modelled by Poiseuille's law due to the non-Newtonian properties of blood. The size of erythrocytes (approximately 7–8 μm) and leukocytes (up to 10 μm) necessitates deformation to allow the cells to pass through [56, 57]. This problem becomes insurmountable at very small capillary diameters, effectively preventing flow. Moreover, the luminal surfaces of microvessels are coated with a host of glycoproteins and other complex molecules that impede flow, partly due to transient adhesion of leukocytes to the endothelium [58].

The issues described thus far can be evaluated in a single hypothetical vessel. The models become substantially more complicated in a capillary network, where differences in flow between individual vessels affect their overall function even at a constant arterial inflow. This heterogeneity in capillary perfusion was first described by Krogh in 1919, who postulated that not all capillaries were active at rest, and recruitment of additional vessels was a mechanism contributing to hyperemic responses to activity [59]. While this is unlikely to be a simplistic binary model (flow is either maximal or absent), there is considerable evidence for differences in capillary transit time heterogeneity (CTTH), which are reduced during neural activity and the hyperemic response to it [54].

The importance of this has recently been highlighted by modelling studies of the capillary network in silico, demonstrating that under conditions of elevated arteriolar tone and flow into the capillary bed, oxygen extraction fraction can decrease when the capillary bed is not uniformly perfused [60]. Not only can this prevent increases in nutrient supply, but under extreme conditions the effect of elevated CBF becomes inverted and tissue oxygenation is reduced, a situation termed malignant CTTH. Therefore, appropriate coupling between CBF and cerebral metabolism requires regulation at both the arterial and capillary levels, accounting for increased CBF and reduced CTTH, respectively. This has been postulated to have relevance to ischemic [61] and hemorrhagic stroke [62], dementia [63], and traumatic brain injury [64].

5 Pericytes as Regulators of Capillary Blood Flow

Control over CTTH, however, requires that there be some mechanism to directly alter capillary tone. Until recently, the consensus was that the last vessels with tone on the inflow side were terminal arterioles, covered by concentric rings of VSM that establish a basal tone, which can be relaxed to increase CBF [14]. The alternative to this is the capillary pericyte. The French physiologist Rouget first described the histological appearance of pericytes on the capillary wall in the nineteenth century. In the 1920s, they were proposed by Vimtrup to have contractile properties, based on his observations of capillary constriction in larvae (discussed in [65]). Pericytes are defined as spatially isolated vascular mural cells that lie within the basal lamina, with known molecular markers including CD13/aminopeptidase N [66], desmin [67], NG2 [68], and regulator of G-protein signalling 5 [69]. The expression of alpha smooth muscle actin (αSMA) is heterogenous, with higher levels expressed by pericytes towards the arteriolar side; mid-capillary pericytes, in contrast, are negative for αSMA labelling [70], a fact that is highly relevant for interpretation of their function as discussed below.

The physiological role pericytes play was not well understood until recent decades, when there has been an explosion of interest in their contribution to vascular function. They are now known to be the key inducer of the blood–brain barrier (BBB) phenotype in endothelial cells, necessary to maintain the delicate milieu that allows normal neuronal function in the brain [71–73]. Pericytes also regulate angiogenesis [74] and leukocyte recruitment [75] and exhibit pluripotency to differentiate into other cell types [76]. Crucially, there is now evidence to suggest pericytes are also the primary regulators of CBF. It was first shown in 2006 that in acute brain slices, capillary diameter changed in response to electrical stimulation and the application of vasoactive substances [77]. This occurred at pericyte locations, but not at uncovered endothelial segments, implying active constriction of the capillary. It was subsequently shown that pericytes exhibit an outward current (and hyperpolarize) during neural stimulation ex vivo, corresponding to capillary dilations; when neural activity was induced by whisker stimulation in mice, capillaries in the barrel cortex dilated as well [78]. Importantly, this dilation preceded arteriolar dilation, implying that the pericyte is the first cell type to respond during neurovascular coupling [78].

This hypothesis is not without controversy, and it has been suggested that pericytes are either capable of constricting/dilating capillaries but remain unresponsive to most stimuli [79], or that they are fundamentally incapable of altering capillary diameter even when directly stimulated using transgenic and optogenetic methods in vivo [80]. However, these discordant findings are potentially reconciled by differences in terminology. Clearly, there is heterogeneity in the mural cells of the vascular tree, even when we broadly classify them as 'pericytes' or 'VSM cells' [80, 81], including in the expression of contractile proteins such as αSMA [70]. It is plausible and indeed necessary a priori that contractile, CBF-regulating pericytes would be located towards the arterial side of the capillary network and might well exhibit functional and morphological differences from other pericyte populations in smaller capillary branches or venules [80]. As a result, it becomes difficult to define the transition point where VSM stops and pericytes begin, and likewise, drawing distinctions between terminal arterioles and first-order capillaries is not clear-cut. What can be stated, and what both sides of the argument appear to agree on, is that CBF is regulated initially in very small vessels on the arterial side of the capillary network, by spatially isolated contractile mural cells, before the response spreads into larger vessels [78, 80–82].

6 Microvascular Changes in Stroke

The clinical signs and symptoms of stroke primarily reflect neuronal dysfunction; however, it is clear that the other cell types in the ischemic area are also affected. This includes the microvasculature, which can suffer direct ischemic damage. Equally problematic, however, is reperfusion injury, and the oxidative stress and inflammation associated with it. As a consequence of microvascular damage during ischemia, recanalization of the occluded artery does not necessarily result in reperfusion of the microvascular bed, referred to as the "no-reflow" phenomenon (Fig. 22.2). Although no-reflow was first described in global ischemia in rabbits [84], the same effect has been demonstrated in macaque MCAO [85]. It has also been shown in a meta-analysis of several clinical studies that 26 % of stroke patients who are successfully recanalized (based on computed tomography or magnetic resonance angiography) do not achieve reperfusion of the microvasculature, evaluated using perfusion imaging [86]. As described above, even if a relatively small number of capillaries are affected, the effect on tissue oxygenation could be profound due to increased CTTH [61]; indeed, capillary flow heterogeneity has been reported to increase under ischemic conditions [87]. Reduced capillary diameters (<4 μm) can also increase blood viscosity, further exacerbating the issue [88]. There has been a resurgence of interest in no-reflow in recent years, and our understanding of its mechanistic basis has greatly expanded since the early histological studies.

A major step in the so-called "ischemic cascade" (Fig. 22.3) of molecular events occurring following loss of nutrient supply is the dysregulation of Ca^{2+} homeostasis [13]. This has a plethora of downstream effects due to the importance and ubiquity of Ca^{2+} as a second messenger in intracellular signaling. Among these is the regulation of enzyme activity, including proteases and lipases, but also an enormous increase in the production of reactive oxygen and nitrogen species (ROS and RNS, respectively), including superoxide (O_2^-), hydrogen peroxide (H_2O_2), nitric oxide (NO), and others [45]. While some of these mediators are crucially important for normal vascular function—indeed, endothelial-derived NO is a potent vasodilator and protective in stroke [89]—their overproduction and the resulting damage to cell membranes, proteins, and DNA will further impair vascular function [13, 90]. Importantly, the rapid restoration of CBF and tissue oxygen tension seen in transient MCAO in rats can actually exacerbate infarct development through oxidative stress, further emphasizing its importance in considering abnormal reperfusion in the context of recanalization [91].

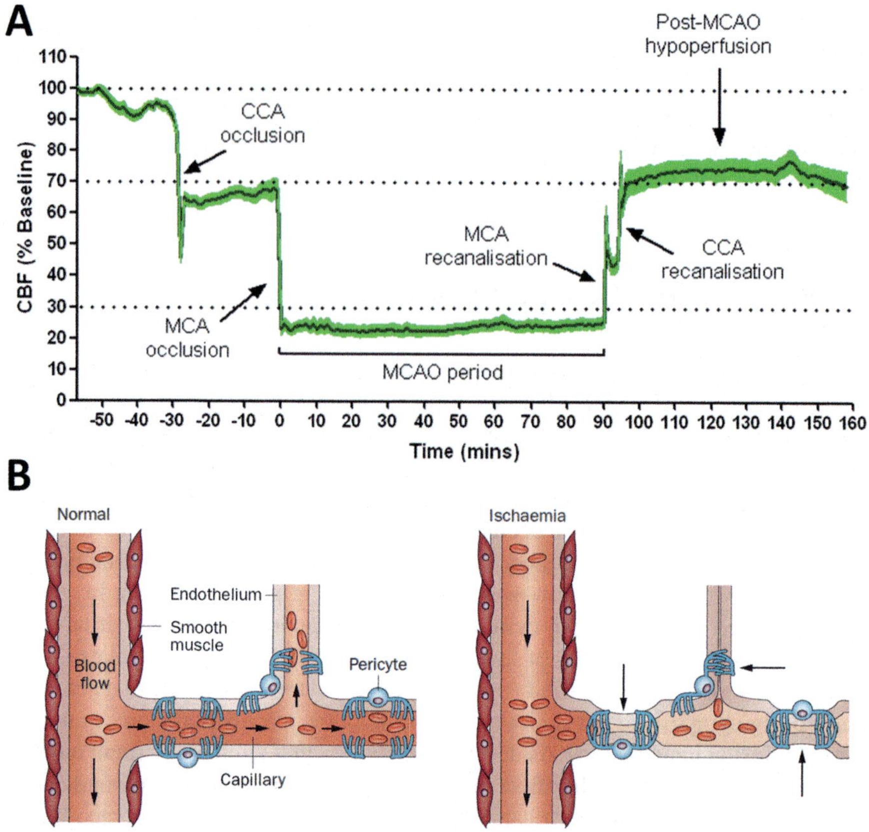

Fig. 22.2 The "no-reflow" phenomenon in the ischemic brain. (**a**) Relative cerebral blood flow (CBF) measurements of the intraluminal filament middle cerebral artery occlusion (MCAO) model as measured by laser Doppler flowmetry show that occlusion of the common carotid artery (CCA) reduces CBF to approximately 65 % of baseline. Subsequent MCAO leads to a drop in CBF to approximately 25 % of baseline which remains throughout the occlusion period. Retraction of the filament leads to MCA recanalization and an increase in CBF which is further enhanced with CCA recanalization. However, post-MCAO CBF in most cases never reaches baseline levels and for the first 60 min hovers around 75 % of baseline, providing direct evidence of the "no-reflow" phenomenon. Data were acquired from 43 male Wistar rats and presented as mean ± SEM. All experiments were conducted in accordance with the Animal (Scientific Procedures) Act, 1986 (UK) under a project license approved by the Home Office (UK) and the University of Oxford (UK). (**b**) Under normal conditions, oxygenated blood flows through a penetrating arteriole (characterized by continuous smooth muscle surrounding the endothelium) into a capillary which possesses distinct pericytes which wrap processes along and around the capillary wall. Upon ischemia, pericytes contract which narrows the capillary lumen, and upon recanalization of the upstream arteriole, pericytes remain contracted in rigor preventing blood flowing through the capillary (*arrows*). This is considered a major mechanism contributing to the "no-reflow" phenomenon. Reproduced from [83] with permission from Nature Publishing Group

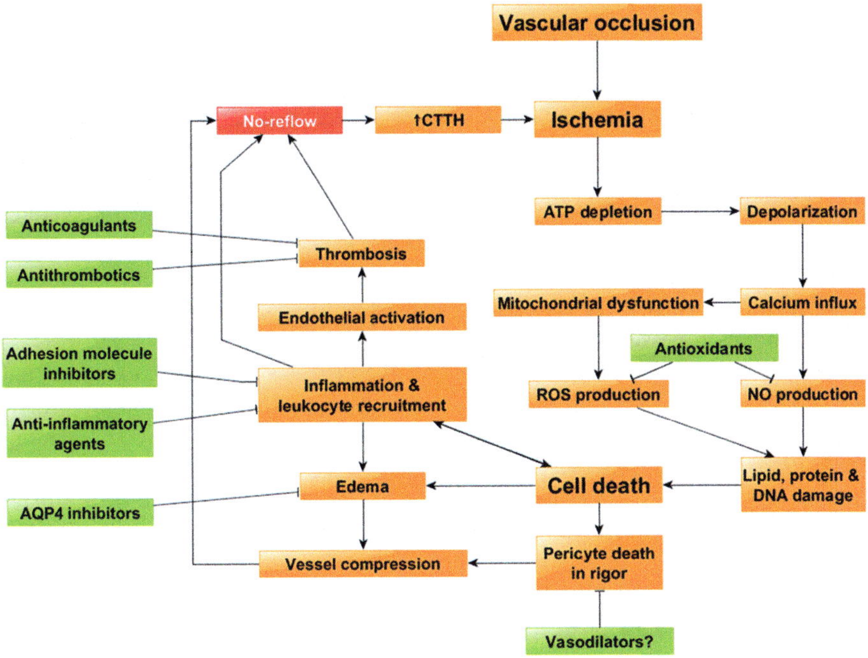

Fig. 22.3 The ischemic cascade. Following arterial occlusion, a number of pathological processes occur in the ischemic region, key examples of which are shown above (in orange). Ca^{2+} dysregulation and oxidative stress damage neurons, astrocytes, endothelial cells, pericytes, and other cell types. During ischemia and following reperfusion, the release of danger signals (such as ATP) by dying cells will cause an inflammatory response and further cell death, as well as vessel obstruction by leukocyte adhesion. Endothelial damage will lead to local thrombotic events, also reducing erythrocyte flux through the capillaries. Edema formation and pericyte constriction can constrict microvessels and thus reduce luminal diameter. These processes are therefore promising targets for preventing no-reflow (*red*), with potential therapeutic approaches shown in *green*

In endothelial cells, one avenue through which this occurs is the recruitment of leukocytes. Endothelial cells upregulate their expression of adhesion molecules and pro-inflammatory molecules as part of the tissue damage response to hypoxia and oxidative stress [92–94]. While the details of innate immune responses in stroke fall outside the scope of this text, the recruitment process creates an additional hemodynamic disturbance in the smallest vessels, as alluded to above—in one study examining baboons after MCAO, approximately 5 % of capillaries were completely occluded by a single neutrophil, and capillary perfusion was restored only in 35 % of vessels [95]. A similar mechanism occurs with platelets, which are activated by a number of mediators known to be released following stroke, including ATP and platelet-activating factor [94]. Platelet activation in combination with an inflammatory endothelial phenotype also results in local intravascular fibrin deposition, further trapping circulating blood cells and exacerbating the flow disturbance [93].

Luminal obstruction is not the only problem, however, as capillary lumen size can also be reduced following ischemia. Stroke is often accompanied by edema formation: first, there is a cytotoxic component, as ion exchange across cell membranes is impaired due to activation of cation influx channels as well as a lack of ATP, preventing Na^+/K^+-ATPase function to counteract the influx. The build-up of intracellular Na^+ leads to water retention and cell swelling. When the blood–brain barrier breaks down and Na^+ enters the extracellular space from the vasculature, there is a net increase in tissue water content and generalized swelling follows [96]. This leads to overall vascular compression, with a key contribution from endothelial cells and aquaporin-4 (AQP4)-expressing astrocyte endfeet, compressing their underlying microvessels [97–99].

However, it was noted in the 1970s that even in instances where extreme glial and endothelial swelling was not observed, there were relatively more small-diameter capillaries in post-ischemic brains compared to controls [100], a finding later validated with intravital microscopy [88]. The authors proposed that post-ischemic reflow abnormalities were better explained by constriction of upstream vessels, but there is intriguing recent evidence that it is, in fact, the capillaries that constrict. Many of the experiments examining pericyte function also looked at their responses to ischemia and observed that simulating stroke in brain slices resulted in pericyte constriction [77, 78]. Importantly, the mechanism for this was “death in rigor”, as capillary diameter was progressively reduced over approximately 20 min, while pericytes were still viable, but did not recover even after their near-complete death by 60 min [78]. Pericytes are also more vulnerable to ischemia than other cell types in vivo [78], and capillary occlusions preferentially occurred at pericyte locations following MCAO in mice [101]. It is important to note that one study found no change in capillary diameter at pericyte locations following MCAO or reperfusion in the mouse [80]; however, the aforementioned caveats concerning pericyte definitions still apply here. Indeed, the study did observe ischemic constrictions in precapillary arterioles and terminal VSM cells, which other groups might call first-order capillaries and pericytes, respectively [80, 81].

7 No-Reflow as a Therapeutic Target

Given that some of the findings described above date back to the 1960s and 1970s, it is unsurprising that there have been many preclinical attempts and clinical trials targeting the pathways that lead to no-reflow. It is important to note that this was not necessarily the underlying rationale of these studies; as there are many distinct signaling pathways operating as part of the ischemic cascade, and many of these occur in multiple cell types in the ischemic territory; interventions designed to target neuronal dysfunction could easily have vascular effects and vice versa.

The most clear-cut argument for a purely vascular effect is with drugs targeting coagulation. Even in a mechanical MCAO model in mice, inhibiting platelet activation with a glycoprotein IIb/IIIa receptor antagonist reduced platelet aggregation and fibrin deposition and increased CBF at 24 h following ischemia [102, 103].

While other studies have disagreed with the exact target, finding no benefit from glycoprotein IIb/IIIa inhibition, they nonetheless showed that inhibiting earlier steps in platelet activation (such as glycoprotein Ib or VI inhibition) reduces infarct volume in mice without increasing risk of hemorrhage [104]. Antiplatelet and anticoagulant drugs are used in stroke and known to be effective in secondary prevention of stroke recurrence [105, 106], and there is evidence that heparin improves reperfusion when combined with streptokinase in acute myocardial infarction [107]. In mice, cilostazol (a phosphodiesterase inhibitor antiplatelet drug with pleiotropic effects) reduced platelet aggregation and leukocyte adhesion following MCAO, resulting in improved neurological function [108].

In the case of inhibiting leukocyte adhesion, it is more complicated to distinguish whether any beneficial effect is derived from reduced inflammation in the tissue rather than improved reperfusion [94]. Nonetheless, there have been experiments demonstrating that inhibition of integrin β2, a molecule crucially important for leukocyte adhesion, significantly improved reperfusion of 7.5–30 μm diameter vessels and showed a trend towards benefit in the smallest vessels [109]. Conversely, there are examples where inhibition of a key leukocyte adhesion molecule, e.g. ICAM-1, results in a neuroprotective effect in the absence of CBF alterations before, during, or after the ischemic period [110].

Similarly, while edema is a major target in ischemic stroke, especially due to its macroscopic effects and the high mortality associated with malignant MCA strokes, the extent to which these interventions exert their effect by improving reflow is very difficult to accurately establish. However, it is certainly known that deletion of AQP4 reduces edema formation following stroke in mice [111], and AQP4-deficient astrocytes are more resistant to swelling under hypotonic conditions favoring water influx [112]. As a result, AQP4 inhibition and other edema-related targets such as sulfonylurea receptor 1 (SUR1) are being actively investigated as potential stroke therapeutics [113].

The effects of oxidative/nitrosative stress on no-reflow are also promising. This seems paradoxical, as clinical trials have failed to show an overall benefit for these agents—most notably in the case of NXY-059, which had very convincing preclinical evidence and improved functional outcome in the SAINT-I trial [114], but became a high-profile failure following its lack of efficacy in the follow-up SAINT-II trial [115]. However, these trials were conducted mainly in the absence of recanalization: for example, 44 % of patients in SAINT-II received rtPA (already an unusually high percentage), but their recanalization/reperfusion status was not known. In the recent URICO-ICTUS trial, the use of uric acid in patients treated with rtPA did not demonstrate a significant beneficial effect, but showed a strong trend that warrants further investigation [116]. It will be particularly interesting to see the effects of antioxidants when combined with thrombectomy, due to the more rapid restoration of blood flow seen with endovascular treatment, which could exacerbate reperfusion injury [44]. Indeed, it is known that in mouse MCAO, treatment with the O_2^- scavenger PBN and NO synthase inhibitor L-NA reduced the number of pericyte constriction sites with trapped erythrocytes in the capillaries and prevented capillary constriction in vivo [101]. Inhibition of RNS (but not ROS) also reduced pericyte death in brain slices exposed to oxygen-glucose deprivation [78].

There are a number of other potential molecular targets that could be used to prevent pericyte constriction. For example, the contractile phenotype of pericytes in vitro is regulated by Rho GTPase expression [117]. The principal downstream effector of Rho is Rho kinase, and the inhibition of Rho kinase is known to substantially improve CBF following MCAO, which is accompanied by reduced lesion volume and neurological impairment [118]. Rho is itself activated by Ca^{2+}, which also activates myosin light chain kinase and the cytoskeletal contractile apparatus, leading to constriction [119]. Incubating brain slices in 0 [Ca^{2+}] also resulted in robust protection of pericytes from oxygen-glucose deprivation [78]. While removal of Ca^{2+} is obviously not feasible in vivo, it does raise the possibility of using Ca^{2+} channel antagonists to improve reperfusion. Much like antioxidant therapies, Ca^{2+} channel antagonists have been tested in clinical trials, where they showed little effect [120]. Again, however, these trials were mainly motivated by putative neuronal effects and warrant re-examination for any vascular benefit, in light of the recent advances with recanalization [44].

8 Conclusion

The advent of endovascular thrombectomy as the new gold standard for acute stroke treatment has been an enormous advance, but there is still great scope for other therapies to further improve patient outcomes, including through improvements in microvascular function. Due to the variety of mechanisms implicated in no-reflow and their complex interactions, it is unlikely that inhibiting a single receptor or activating a single pathway will prove a panacea to the problem. Indeed, even in the broader context of neuroprotective therapies, it is widely agreed that multimodal, combinatorial approaches are more likely to prove successful than single targets [11, 121]. Nonetheless, recent research has emphasized the importance of microvascular changes in stroke, provided us with new targets to investigate in the preclinical setting, and will hopefully lead us to novel stroke therapeutics.

Acknowledgments The authors thank Dr. Hasneen Karbalai for his constructive comments on the manuscript. The authors were supported by funding from the Radcliffe Department of Medicine Scholarship (AAN), Medical Research Council (AMB, BAS), and Oxford Biomedical Research Centre (AMB).

Disclosures AMB is a cofounder of Brainomix Limited, a start-up company from the University of Oxford developing imaging biomarkers for neurological and cerebrovascular disorders.

References

1. Mink JW, Blumenschine RJ, Adams DB (1981) Ratio of central nervous system to body metabolism in vertebrates: its constancy and functional basis. Am J Physiol 241(3):R203–R212
2. Attwell D, Buchan AM, Charpak S, Lauritzen M, Macvicar BA, Newman EA (2010) Glial and neuronal control of brain blood flow. Nature 468(7321):232–243

3. Howarth C, Gleeson P, Attwell D (2012) Updated energy budgets for neural computation in the neocortex and cerebellum. J Cereb Blood Flow Metab 32(7):1222
4. Gordon GRJ, Choi HB, Rungta RL, Ellis-Davies GCR, MacVicar BA (2008) Brain metabolism dictates the polarity of astrocyte control over arterioles. Nature 456(7223):745–749
5. Wang X, Lou N, Xu Q, Tian G-F, Peng WG, Han X et al (2006) Astrocytic Ca2+ signaling evoked by sensory stimulation in vivo. Nat Neurosci 9(6):816–823
6. Mathiisen TM, Lehre KP, Danbolt NC, Ottersen OP (2010) The perivascular astroglial sheath provides a complete covering of the brain microvessels: an electron microscopic 3D reconstruction. Glia 58(9):1094–1103
7. Nizar K, Uhlirova H, Tian P, Saisan PA, Cheng Q, Reznichenko L et al (2013) In vivo stimulus-induced vasodilation occurs without IP3 receptor activation and may precede astrocytic calcium increase. J Neurosci 33(19):8411–8422
8. Hillman EMC (2014) Coupling mechanism and significance of the BOLD signal: a status report. Annu Rev Neurosci 37:161–181
9. Feigin VL, Krishnamurthi RV, Parmar P, Norrving B, Mensah GA, Bennett DA et al (2015) Update on the global burden of ischemic and hemorrhagic stroke in 1990-2013: the GBD 2013 study. Neuroepidemiology 45(3):161–176
10. Bogousslavsky J, Van Melle G, Regli F (1988) The Lausanne Stroke Registry: analysis of 1,000 consecutive patients with first stroke. Stroke 19(9):1083–1092
11. Neuhaus AA, Rabie T, Sutherland BA, Papadakis M, Hadley G, Cai R et al (2014) Importance of preclinical research in the development of neuroprotective strategies for ischemic stroke. JAMA Neurol 71(5):634–639
12. Liebeskind DS (2007) Understanding blood flow: the other side of an acute arterial occlusion. Int J Stroke 2(2):118–120
13. Dirnagl U, Iadecola C, Moskowitz MA (1999) Pathobiology of ischaemic stroke: an integrated view. Trends Neurosci 22(9):391–397
14. Girouard H, Iadecola C (2006) Neurovascular coupling in the normal brain and in hypertension, stroke, and Alzheimer disease. J Appl Physiol 100(1):328–335
15. Astrup J, Siesjö BK, Symon L, Siesjo BK, Symon L (1981) Thresholds in cerebral ischemia—the ischemic penumbra. Stroke 12(6):723–725
16. Astrup J, Symon L, Branston NM, Lassen NA (1977) Cortical evoked potential and extracellular K+ and H+ at critical levels of brain ischemia. Stroke 8(1):51–57
17. Hossmann KA (1994) Viability thresholds and the penumbra of focal ischemia. Ann Neurol 36(4):557–565
18. Hu J-L, Li X, Wang X-M, Cheng Z-P, Chen D-F, Wang S-C et al (2015) Comparison of elastic properties of bilateral carotid arteries in relation to site of acute ischemic stroke using velocity vector imaging. Chin Med J (Engl) 128(21):2960–2963
19. Cipolla MJ, Lessov N, Hammer ES, Curry AB (2001) Threshold duration of ischemia for myogenic tone in middle cerebral arteries: effect on vascular smooth muscle actin. Stroke 32(7):1658–1664
20. Dirnagl U, Pulsinelli W (1990) Autoregulation of cerebral blood flow in experimental focal brain ischemia. J Cereb Blood Flow Metab 10(3):327–336
21. Shin HK, Nishimura M, Jones PB, Ay H, Boas DA, Moskowitz MA et al (2008) Mild induced hypertension improves blood flow and oxygen metabolism in transient focal cerebral ischemia. Stroke 39(5):1548–1555
22. Hossmann K-A (2006) Pathophysiology and therapy of experimental stroke. Cell Mol Neurobiol 26(7–8):1057–1083
23. Sweet JG, Chan S-L, Cipolla MJ (2015) Effect of hypertension and carotid occlusion on brain parenchymal arteriole structure and reactivity. J Appl Physiol 119(7):817–823
24. Shih AY, Friedman B, Drew PJ, Tsai PS, Lyden PD, Kleinfeld D (2009) Active dilation of penetrating arterioles restores red blood cell flux to penumbral neocortex after focal stroke. J Cereb Blood Flow Metab 29(4):738–751
25. Shih AY, Blinder P, Tsai PS, Friedman B, Stanley G, Lyden PD et al (2013) The smallest stroke: occlusion of one penetrating vessel leads to infarction and a cognitive deficit. Nat Neurosci 16(1):55–63

26. Blinder P, Tsai PS, Kaufhold JP, Knutsen PM, Suhl H, Kleinfeld D (2013) The cortical angiome: an interconnected vascular network with noncolumnar patterns of blood flow. Nat Neurosci 16(7):889–897
27. Shih AY, Rühlmann C, Blinder P, Devor A, Drew PJ, Friedman B et al (2015) Robust and fragile aspects of cortical blood flow in relation to the underlying angioarchitecture. Microcirculation 22(3):204–218
28. Nguyen J, Nishimura N, Fetcho RN, Iadecola C, Schaffer CB (2011) Occlusion of cortical ascending venules causes blood flow decreases, reversals in flow direction, and vessel dilation in upstream capillaries. J Cereb Blood Flow Metab 31(11):2243–2254
29. Sasaki M, Honmou O, Radtke C, Kocsis JD (2011) Development of a middle cerebral artery occlusion model in the nonhuman primate and a safety study of i.v. infusion of human mesenchymal stem cells. PLoS One 6(10):e26577
30. Ohta H, Nakano S, Yokogami K, Iseda T, Yoneyama T, Wakisaka S (2004) Appearance of early venous filling during intra-arterial reperfusion therapy for acute middle cerebral artery occlusion: a predictive sign for hemorrhagic complications. Stroke 35(4):893–898
31. Emberson J, Lees KR, Lyden P, Blackwell L, Albers G, Bluhmki E et al (2014) Effect of treatment delay, age, and stroke severity on the effects of intravenous thrombolysis with alteplase for acute ischaemic stroke: a meta-analysis of individual patient data from randomised trials. Lancet 384(9958):1929–1935
32. Balami JS, Hadley G, Sutherland BA, Karbalai H, Buchan AM (2013) The exact science of stroke thrombolysis and the quiet art of patient selection. Brain 136(Pt 12):3528–3553
33. Berkhemer OA, Fransen PSS, Beumer D, van den Berg LA, Lingsma HF, Yoo AJ et al (2015) A randomized trial of intraarterial treatment for acute ischemic stroke. N Engl J Med 372(1):11–20
34. Campbell BCV, Mitchell PJ, Kleinig TJ, Dewey HM, Churilov L, Yassi N et al (2015) Endovascular therapy for ischemic stroke with perfusion-imaging selection. N Engl J Med 372(11):1009–1018
35. Goyal M, Demchuk AM, Menon BK, Eesa M, Rempel JL, Thornton J et al (2015) Randomized assessment of rapid endovascular treatment of ischemic stroke. N Engl J Med 372(11):1–12
36. Jovin TG, Chamorro A, Cobo E, de Miquel MA, Molina CA, Rovira A et al (2015) Thrombectomy within 8 hours after symptom onset in ischemic stroke. N Engl J Med 372(24):2296–2306
37. Saver JL, Goyal M, Bonafe A, Diener H-C, Levy EI, Pereira VM et al (2015) Stent-retriever thrombectomy after intravenous t-PA vs. t-PA alone in stroke. N Engl J Med 372(24):2285–2295
38. Balami JS, Sutherland BA, Edmunds LD, Grunwald IQ, Neuhaus AA, Hadley G et al (2015) A systematic review and meta-analysis of randomized controlled trials of endovascular thrombectomy compared with best medical treatment for acute ischemic stroke. Int J Stroke 10(8):1168–1178
39. Goyal M, Menon BK, van Zwam WH, Dippel DWJ, Mitchell PJ, Demchuk AM, et al (2016) Endovascular thrombectomy after large-vessel ischaemic stroke: a meta-analysis of individual patient data from five randomised trials. Lancet (London, England) 387(10029):1723–1731
40. Leng X, Fang H, Leung TWH, Mao C, Miao Z, Liu L et al (2016) Impact of collaterals on the efficacy and safety of endovascular treatment in acute ischaemic stroke: a systematic review and meta-analysis. J Neurol Neurosurg Psychiatry 87(5):537–544
41. Leng X, Fang H, Leung TWH, Mao C, Xu Y, Miao Z et al (2016) Impact of collateral status on successful revascularization in endovascular treatment: a systematic review and meta-analysis. Cerebrovasc Dis 41(1–2):27–34
42. Sutherland BA, Buchan AM. Alteplase treatment does not increase brain injury after mechanical middle cerebral artery occlusion in the rat. J Cereb Blood Flow Metab. Nature Publishing Group; 2013 Aug 21;(April):1–7.
43. Sutherland BA, Papadakis M, Chen R-L, Buchan AM (2011) Cerebral blood flow alteration in neuroprotection following cerebral ischaemia. J Physiol 589(Pt 17):4105–4114

44. Sutherland BA, Neuhaus AA, Couch Y, Balami JS, Deluca GC, Hadley G et al (2016) The transient intraluminal filament middle cerebral artery occlusion model as a model of endovascular thrombectomy in stroke. J Cereb Blood Flow Metab 36(2):363–369
45. Li C, Jackson RM (2002) Reactive species mechanisms of cellular hypoxia-reoxygenation injury. Am J Physiol Cell Physiol 282(2):C227–C241
46. De Silva TM, Brait VH, Drummond GR, Sobey CG, Miller AA (2011) Nox2 oxidase activity accounts for the oxidative stress and vasomotor dysfunction in mouse cerebral arteries following ischemic stroke. PLoS One 6(12):e28393
47. Maneen MJ, Hannah R, Vitullo L, DeLance N, Cipolla MJ (2006) Peroxynitrite diminishes myogenic activity and is associated with decreased vascular smooth muscle F-actin in rat posterior cerebral arteries. Stroke 37(3):894–899
48. Cipolla MJ, Chan S-LSL, Sweet J, Tavares MJ, Gokina N, Brayden JE (2014) Postischemic reperfusion causes smooth muscle calcium sensitization and vasoconstriction of parenchymal arterioles. Stroke 45(8):2425–2430
49. Cipolla MJ, Sweet JG, Gokina NI, White SL, Nelson MT (2013) Mechanisms of enhanced basal tone of brain parenchymal arterioles during early postischemic reperfusion: role of ET-1-induced peroxynitrite generation. J Cereb Blood Flow Metab 33(10):1486–1492
50. Belayev L, Pinard E, Nallet H, Seylaz J, Liu Y, Riyamongkol P et al (2002) Albumin therapy of transient focal cerebral ischemia: in vivo analysis of dynamic microvascular responses. Stroke 33(4):1077–1084
51. Ritter LS, Orozco JA, Coull BM, McDonagh PF, Rosenblum WI (2000) Leukocyte accumulation and hemodynamic changes in the cerebral microcirculation during early reperfusion after stroke. Stroke 31(5):1153–1161
52. Lovick TA, Brown LA, Key BJ (1999) Neurovascular relationships in hippocampal slices: physiological and anatomical studies of mechanisms underlying flow-metabolism coupling in intraparenchymal microvessels. Neuroscience 92(1):47–60
53. Borowsky IW, Collins RC (1989) Metabolic anatomy of brain: a comparison of regional capillary density, glucose metabolism, and enzyme activities. J Comp Neurol 288(3):401–413
54. Kleinfeld D, Mitra PP, Helmchen F, Denk W (1998) Fluctuations and stimulus-induced changes in blood flow observed in individual capillaries in layers 2 through 4 of rat neocortex. Proc Natl Acad Sci U S A 95(26):15741–15746
55. Sakadžić S, Mandeville ET, Gagnon L, Musacchia JJ, Yaseen MA, Yucel MA et al (2014) Large arteriolar component of oxygen delivery implies a safe margin of oxygen supply to cerebral tissue. Nat Commun 5:5734
56. Chien S (1987) Red cell deformability and its relevance to blood flow. Annu Rev Physiol 49:177–192
57. Schmid-Schönbein GW, Shih YY, Chien S (1980) Morphometry of human leukocytes. Blood 56(5):866–875
58. Pries AR, Secomb TW, Gessner T, Sperandio MB, Gross JF, Gaehtgens P (1994) Resistance to blood flow in microvessels in vivo. Circ Res 75(5):904–915
59. Krogh A (1919) The supply of oxygen to the tissues and the regulation of the capillary circulation. J Physiol 52(6):457–474
60. Jespersen SN, Østergaard L (2012) The roles of cerebral blood flow, capillary transit time heterogeneity, and oxygen tension in brain oxygenation and metabolism. J Cereb Blood Flow Metab 32(2):264–277
61. Østergaard L, Jespersen SN, Mouridsen K, Mikkelsen IK, Jonsdottír KÝ, Tietze A et al (2013) The role of the cerebral capillaries in acute ischemic stroke: the extended penumbra model. J Cereb Blood Flow Metab 33(5):1–14
62. Østergaard L, Aamand R, Karabegovic S, Tietze A, Blicher JU, Mikkelsen IKK et al (2013) The role of the microcirculation in delayed cerebral ischemia and chronic degenerative changes after subarachnoid hemorrhage. J Cereb Blood Flow Metab 33(12):1825–1837
63. Østergaard L, Aamand R, Gutiérrez-Jiménez E, Ho YC, Blicher JU, Madsen SM et al (2013) The capillary dysfunction hypothesis of Alzheimer's disease. Neurobiol Aging 34(4):1018–1031

64. Østergaard L, Engedal TS, Aamand R, Mikkelsen R, Iversen NK, Anzabi M et al (2014) Capillary transit time heterogeneity and flow-metabolism coupling after traumatic brain injury. J Cereb Blood Flow Metab 34(March):1–14
65. Krueger M, Bechmann I (2010) CNS pericytes: concepts, misconceptions, and a way out. Glia 58(1):1–10
66. Kunz J, Krause D, Kremer M, Dermietzel R (1994) The 140-kDa protein of blood-brain barrier-associated pericytes is identical to aminopeptidase N. J Neurochem 62(6):2375–2386
67. Nehls V, Denzer K, Drenckhahn D (1992) Pericyte involvement in capillary sprouting during angiogenesis in situ. Cell Tissue Res 270(3):469–474
68. Ozerdem U, Grako KA, Dahlin-Huppe K, Monosov E, Stallcup WB (2001) NG2 proteoglycan is expressed exclusively by mural cells during vascular morphogenesis. Dev Dyn 222(2):218–227
69. Bondjers C, Kalén M, Hellström M, Scheidl SJ, Abramsson A, Renner O et al (2003) Transcription profiling of platelet-derived growth factor-B-deficient mouse embryos identifies RGS5 as a novel marker for pericytes and vascular smooth muscle cells. Am J Pathol 162(3):721–729
70. Nehls V, Drenckhahn D (1991) Heterogeneity of microvascular pericytes for smooth muscle type alpha-actin. J Cell Biol 113(1):147–154
71. Bell RD, Winkler EA, Sagare AP, Singh I, LaRue B, Deane R et al (2010) Pericytes control key neurovascular functions and neuronal phenotype in the adult brain and during brain aging. Neuron 68(3):409–27
72. Armulik A, Genové G, Mäe M, Nisancioglu MH, Wallgard E, Niaudet C et al (2010) Pericytes regulate the blood–brain barrier. Nature 468(7323):557–561
73. Daneman R, Zhou L, Kebede AA, Barres BA (2010) Pericytes are required for blood-brain barrier integrity during embryogenesis. Nature 468(7323):562–566
74. Armulik A, Abramsson A, Betsholtz C (2005) Endothelial/pericyte interactions. Circ Res 97(6):512–523
75. Stark K, Eckart A, Haidari S, Tirniceriu A, Lorenz M, von Brühl M-L et al (2013) Capillary and arteriolar pericytes attract innate leukocytes exiting through venules and "instruct" them with pattern-recognition and motility programs. Nat Immunol 14(1):41–51
76. Dore-Duffy P, Katychev A, Wang X, Van Buren E (2006) CNS microvascular pericytes exhibit multipotential stem cell activity. J Cereb Blood Flow Metab 26(5):613–624
77. Peppiatt CM, Howarth C, Mobbs P, Attwell D (2006) Bidirectional control of CNS capillary diameter by pericytes. Nature 443(7112):700–704
78. Hall CN, Reynell C, Gesslein B, Hamilton NB, Mishra A, Sutherland BA et al (2014) Capillary pericytes regulate cerebral blood flow in health and disease. Nature 508(7494):55–60
79. Fernández-Klett F, Offenhauser N, Dirnagl U, Priller J, Lindauer U (2010) Pericytes in capillaries are contractile in vivo, but arterioles mediate functional hyperemia in the mouse brain. Proc Natl Acad Sci U S A 107(51):22290–22295
80. Hill RA, Tong L, Yuan P, Murikinati S, Gupta S, Grutzendler J (2015) Regional blood flow in the normal and ischemic brain is controlled by arteriolar smooth muscle cell contractility and not by capillary pericytes. Neuron 87(1):95–110
81. Attwell D, Mishra A, Hall CN, O'Farrell FM, Dalkara T (2016) What is a pericyte? J Cereb Blood Flow Metab 36(2):451–455
82. Chen BR, Kozberg MG, Bouchard MB, Shaik MA, Hillman EMC (2014) A critical role for the vascular endothelium in functional neurovascular coupling in the brain. J Am Heart Assoc 3(3):e000787
83. O'Farrell FM, Attwell D (2014) A role for pericytes in coronary no-reflow. Nat Rev Cardiol 11(7):1–6
84. Ames A, Wright RL, Kowada M, Thurston JM, Majno G (1968) Cerebral ischemia. II. The no-reflow phenomenon. Am J Pathol 52(2):437–453
85. Crowell RM, Olsson Y (1972) Impaired microvascular filling after focal cerebral ischemia in monkeys. J Neurosurg 36(3):303–309
86. Dalkara T, Arsava EM (2012) Can restoring incomplete microcirculatory reperfusion improve stroke outcome after thrombolysis? J Cereb Blood Flow Metab 32(12):2091–2099

87. Tomita Y, Tomita M, Schiszler I, Amano T, Tanahashi N, Kobari M et al (2002) Moment analysis of microflow histogram in focal ischemic lesion to evaluate microvascular derangement after small pial arterial occlusion in rats. J Cereb Blood Flow Metab 22(6):663–669
88. Hauck EF, Apostel S, Hoffmann JF, Heimann A, Kempski O (2004) Capillary flow and diameter changes during reperfusion after global cerebral ischemia studied by intravital video microscopy. J Cereb Blood Flow Metab 24(4):383–391
89. Endres M, Laufs U, Liao JK, Moskowitz MA (2004) Targeting eNOS for stroke protection. Trends Neurosci 27(5):283–289
90. Minnerup J, Sutherland BA, Buchan AM, Kleinschnitz C (2012) Neuroprotection for stroke: current status and future perspectives. Int J Mol Sci 13(9):11753–11772
91. Aronowski J, Strong R, Grotta JC (1997) Reperfusion injury: demonstration of brain damage produced by reperfusion after transient focal ischemia in rats. J Cereb Blood Flow Metab 17(10):1048–1056
92. Michiels C, Arnould T, Remacle J (2000) Endothelial cell responses to hypoxia: initiation of a cascade of cellular interactions. Biochim Biophys Acta 1497(1):1–10
93. Okada Y, Copeland BR, Fitridge R, Koziol JA, del Zoppo GJ (1994) Fibrin contributes to microvascular obstructions and parenchymal changes during early focal cerebral ischemia and reperfusion. Stroke 25(9):1847–1853; discussion 1853–4
94. Iadecola C, Anrather J (2011) The immunology of stroke: from mechanisms to translation. Nat Med 17(7):796–808
95. del Zoppo GJ, Schmid-Schönbein GW, Mori E, Copeland BR, Chang CM (1991) Polymorphonuclear leukocytes occlude capillaries following middle cerebral artery occlusion and reperfusion in baboons. Stroke 22(10):1276–1283
96. Simard JM, Kent TA, Chen M, Tarasov KV, Gerzanich V (2007) Brain oedema in focal ischaemia: molecular pathophysiology and theoretical implications. Lancet Neurol 6(3):258–268
97. Garcia JH, Liu KF, Yoshida Y, Chen S, Lian J (1994) Brain microvessels: factors altering their patency after the occlusion of a middle cerebral artery (Wistar rat). Am J Pathol 145(3): 728–740
98. Kimelberg HK (2005) Astrocytic swelling in cerebral ischemia as a possible cause of injury and target for therapy. Glia 50(4):389–397
99. Ito U, Hakamata Y, Kawakami E, Oyanagi K (2011) Temporary [corrected] cerebral ischemia results in swollen astrocytic end-feet that compress microvessels and lead to delayed [corrected] focal cortical infarction. J Cereb Blood Flow Metab 31(1):328–338
100. Fischer EG, Ames A, Hedley-Whyte ET, O'Gorman S (1977) Reassessment of cerebral capillary changes in acute global ischemia and their relationship to the "no-reflow phenomenon". Stroke 8(1):36–39
101. Yemisci M, Gursoy-Ozdemir Y, Vural A, Can A, Topalkara K, Dalkara T (2009) Pericyte contraction induced by oxidative-nitrative stress impairs capillary reflow despite successful opening of an occluded cerebral artery. Nat Med 15(9):1031–1037
102. Choudhri TF, Hoh BL, Zerwes HG, Prestigiacomo CJ, Kim SC, Connolly ES et al (1998) Reduced microvascular thrombosis and improved outcome in acute murine stroke by inhibiting GP IIb/IIIa receptor-mediated platelet aggregation. J Clin Invest 102(7):1301–1310
103. Abumiya T, Fitridge R, Mazur C, Copeland BR, Koziol JA, Tschopp JF et al (2000) Integrin alpha(IIb)beta(3) inhibitor preserves microvascular patency in experimental acute focal cerebral ischemia. Stroke 31(6):1402–1409; discussion 1409–10
104. Kleinschnitz C, Pozgajova M, Pham M, Bendszus M, Nieswandt B, Stoll G (2007) Targeting platelets in acute experimental stroke: impact of glycoprotein Ib, VI, and IIb/IIIa blockade on infarct size, functional outcome, and intracranial bleeding. Circulation 115(17):2323–2330
105. Chen ZM, Sandercock P, Pan HC, Counsell C, Collins R, Liu LS et al (2000) Indications for early aspirin use in acute ischemic stroke: a combined analysis of 40 000 randomized patients from the Chinese acute stroke trial and the international stroke trial. On behalf of the CAST and IST collaborative groups. Stroke 31(6):1240–1249
106. Kernan WN, Ovbiagele B, Black HR, Bravata DM, Chimowitz MI, Ezekowitz MD et al (2014) Guidelines for the prevention of stroke in patients with stroke and transient ischemic

attack: a guideline for healthcare professionals from the American Heart Association/American Stroke Association. Stroke 45(7):2160–2236
107. Simoons M, Krzemińska-Pakula M, Alonso A, Goodman S, Kali A, Loos U et al (2002) Improved reperfusion and clinical outcome with enoxaparin as an adjunct to streptokinase thrombolysis in acute myocardial infarction. The AMI-SK study. Eur Heart J 23(16):1282–1290
108. Hase Y, Okamoto Y, Fujita Y, Kitamura A, Nakabayashi H, Ito H et al (2012) Cilostazol, a phosphodiesterase inhibitor, prevents no-reflow and hemorrhage in mice with focal cerebral ischemia. Exp Neurol 233(1):523–533
109. Mori E, del Zoppo GJ, Chambers JD, Copeland BR, Arfors KE (1992) Inhibition of polymorphonuclear leukocyte adherence suppresses no-reflow after focal cerebral ischemia in baboons. Stroke 23(5):712–718
110. Vemuganti R, Dempsey RJ, Bowen KK (2004) Inhibition of intercellular adhesion molecule-1 protein expression by antisense oligonucleotides is neuroprotective after transient middle cerebral artery occlusion in rat. Stroke 35(1):179–184
111. Manley GT, Fujimura M, Ma T, Noshita N, Filiz F, Bollen AW et al (2000) Aquaporin-4 deletion in mice reduces brain edema after acute water intoxication and ischemic stroke. Nat Med 6(2):159–163
112. Solenov E, Watanabe H, Manley GT, Verkman AS (2004) Sevenfold-reduced osmotic water permeability in primary astrocyte cultures from AQP-4-deficient mice, measured by a fluorescence quenching method. Am J Physiol Cell Physiol 286(2):C426–C432
113. Simard JM, Woo SK, Bhatta S, Gerzanich V (2008) Drugs acting on SUR1 to treat CNS ischemia and trauma. Curr Opin Pharmacol 8(1):42–49
114. Lees KR, Zivin JA, Ashwood T, Davalos A, Davis SM, Diener H-C et al (2006) NXY-059 for acute ischemic stroke. N Engl J Med 354(6):588–600
115. Shuaib A, Lees KR, Lyden P, Grotta J, Davalos A, Davis SM et al (2007) NXY-059 for the treatment of acute ischemic stroke. N Engl J Med 357(6):562–571
116. Chamorro A, Amaro S, Castellanos M, Segura T, Arenillas J, Martí-Fábregas J et al (2014) Safety and efficacy of uric acid in patients with acute stroke (URICO-ICTUS): a randomised, double-blind phase 2b/3 trial. Lancet Neurol 13(5):453–460
117. Kolyada AY, Riley KN, Herman IM (2003) Rho GTPase signaling modulates cell shape and contractile phenotype in an isoactin-specific manner. Am J Physiol Cell Physiol 285(5): C1116–C1121
118. Rikitake Y, Kim HH, Huang Z, Seto M, Yano K, Asano T et al (2005) Inhibition of Rho kinase (ROCK) leads to increased cerebral blood flow and stroke protection. Stroke 36(10):2251–2257
119. Sakurada S, Takuwa N, Sugimoto N, Wang Y, Seto M, Sasaki Y et al (2003) Ca2+-dependent activation of Rho and Rho kinase in membrane depolarization-induced and receptor stimulation-induced vascular smooth muscle contraction. Circ Res 93(6):548–556
120. Zhang J, Yang J, Zhang C, Jiang X, Zhou H, Liu M (2012) Calcium antagonists for acute ischemic stroke. Cochrane Database Syst Rev 5:CD001928
121. Rogalewski A, Schneider A, Ringelstein EB, Schäbitz W-R (2006) Toward a multimodal neuroprotective treatment of stroke. Stroke 37(4):1129–1136
122. Hossmann KA (1997) Reperfusion of the brain after global ischemia: hemodynamic disturbances. Shock 8(2):95–101; discussion 102–3

Part III
Thrombolysis and Embolectomy

Chapter 23
Thrombolytic and Endovascular Therapies for Acute Ischemic Stroke

Hormozd Bozorgchami and Helmi L. Lutsep

Abstract Although clinical trials have investigated thrombolytic therapies such as streptokinase, urokinase, and desmoteplase in acute ischemic stroke treatment, only trials assessing intravenous (IV) tissue plasminogen activator (tPA) and intraarterial prourokinase were successful in showing efficacy. Since a confirmatory trial with prourokinase was not performed, it has not been approved for treatment and IV tPA remains the standard of care. Advances in fluoroscopic imaging led to additional research in endovascular therapies as performing interventional cerebrovascular procedures became more feasible. After years of device development, enhanced acute stroke imaging, and improvements in systems of care, endovascular therapy has now shown efficacy in acute stroke treatment in randomized trials as well. Trials have shown the efficacy of stent retrievers compared to medical therapy alone in patients with anterior circulation large vessel occlusions. This has resulted in a paradigm shift in the management of acute ischemic stroke.

Keywords Ischemic stroke • Thrombolysis • Thrombolytic • Embolectomy • Thrombectomy • Stent retriever • Stentriever • Reperfusion • tPA • Solitaire • Trevo • Alteplase

1 Thrombolytic Therapy in Acute Stroke

Successful outcomes in patients with cardiac ischemia treated with thrombolytics prompted interest in the treatment of ischemic stroke patients with such agents in the 1990s. However, concerns about hemorrhagic transformation and lack of

H. Bozorgchami (✉) • H.L. Lutsep
Oregon Health & Science University, Portland, OR 97239, USA
e-mail: bozorgch@ohsu.edu; lutseph@ohsu.edu

P.A. Lapchak, J.H. Zhang (eds.), *Neuroprotective Therapy for Stroke and Ischemic Disease*, Springer Series in Translational Stroke Research,
DOI 10.1007/978-3-319-45345-3_23

Table 23.1 Major randomized thrombolytic trials in acute ischemic stroke

Thrombolytic	Time from symptom onset to treatment (h)	Administration route	Name of trial	Trial results positive?
Streptokinase	0–6	IV	MAST-E [1]	No
Streptokinase	0–6	IV	MAST-I [2]	No
Streptokinase	0–4	IV	ASK [3]	No
Prourokinase	0–6	IA	PROACT II [6]	Yes (no confirmatory trial done)
Desmoteplase	3–9	IV	DIAS 3 [7]	No
tPA	0–3	IV	NINDS [8] Parts I and II	Yes
tPA	0–6	IV	ECASS [11]	No
tPA	0–6	IV	ECASS 2 [12]	No
tPA	3–5	IV	ATLANTIS [13]	No
tPA	3–4.5	IV	ECASS3 [16]	Yes

tPA tissue plasminogen activator, *MAST-E* The Multi-center Acute Stroke Trial—Europe, *MAST-I* Multicentre Acute Stroke Trial—Italy, *ASK* Australian Streptokinase, *PROACT* Prolyse in Acute Cerebral Thromboembolism, *DIAS* Desmoteplase in Acute Ischemic Stroke, *NINDS* Neurological Disorders and Stroke, *ECASS* European Cooperative Acute Stroke Study, *ATLANTIS* Alteplase ThromboLysis for Acute Noninterventional Therapy in Ischemic Stroke

systems for promoting the early treatment of acute stroke created challenges for the use of such agents in stroke treatment. Table 23.1 provides an overview of the major randomized thrombolytic trials in acute ischemic stroke.

1.1 Streptokinase

The first thrombolytic to be studied in randomized, controlled stroke trials was streptokinase, a protein produced by several species of streptococci that activates plasminogen. The Multi-center Acute Stroke Trial—Europe (MAST-E) compared intravenous streptokinase to placebo within 6 h of symptom onset [1], while the Multicentre Acute Stroke Trial—Italy (MAST-I) compared streptokinase, with or without aspirin, to aspirin alone and to placebo within 6 h of symptom onset [2]. Both trials showed increased mortality in the streptokinase group. The Australian Streptokinase (ASK) trial compared streptokinase to placebo in patients treated up to 4 h after symptom onset and determined whether outcomes in those treated within 3 h of symptom onset were better than in those receiving the drug after 3 h [3]. Although excess deaths were only seen in the group treated after 3 h, there was no significant benefit over placebo. While the streptokinase dose in the trials, 1.5 million units, may have been too high for the stroke population, hypotension was also an issue [3].

1.2 Urokinase and Prourokinase

1.2.1 Urokinase

Urokinase, a thrombolytic obtained from human kidneys, had been assessed in a small pilot study of 31 ischemic stroke patients reported in 1976 [4]. While biochemical studies revealed plasma thrombolytic activity with only minor disturbances of the coagulation profiles, hemorrhagic complications were seen and the authors reported no clear favorable therapeutic effects.

The Middle Cerebral Artery Embolism Local Fibrinolytic Intervention Trial (MELT) trial in Japan assessed the safety and clinical efficacy of intraarterial urokinase with 6 h of ischemic stroke [5]. While the trial was stopped early due to the approval of tPA in Japan and the primary endpoint did not reach statistical significance, a preplanned secondary outcome measure suggested that the treatment could be associated with favorable outcomes.

1.2.2 Prourokinase

Prourokinase, a proenzyme precursor of urokinase, was investigated as an intraarterially, not intravenously, delivered therapy in part to promote homogeneity of the stroke population being assessed [6]. In the Prolyse in Acute Cerebral Thromboembolism (PROACT) II trial, prourokinase plus heparin was compared to heparin alone for the treatment of middle cerebral artery (MCA) occlusions within 6 h of symptom onset [6]. Patients in the prourokinase treatment group had better vessel recanalization and significantly improved clinical outcomes. Slight or no neurological disability, defined as a modified Rankin Scale (mRS) of 2 or less, was seen in 40 % of the prourokinase group versus 25 % of control individuals ($p=0.04$). However, the drug was not approved by the FDA because a second, confirmatory trial was not performed.

1.2.3 Desmoteplase

The thrombolytic most recently investigated in clinical trials was desmoteplase, a fibrin-dependent plasminogen activator with high fibrin specificity found in the saliva of vampire bats [7]. Desmoteplase or placebo was given to patients between 3 and 9 h after symptom onset in a phase 3 randomized study of those with stroke and occlusion or high-grade stenosis in major cerebral arteries [7]. Median time to stroke treatment was approximately 7 h in the two groups. While no safety concerns emerged, desmoteplase was not associated with improved outcomes as assessed by a mRS of 0–2 at 90 days.

1.3 Tissue Plasminogen Activator

The serine protease tissue plasminogen activator (tPA), which is found on endothelial cells but is currently produced through recombinant DNA technology for clinical use, was initially assessed in myocardial infarction. The drug then entered clinical trials for acute ischemic stroke, where it became the first approved treatment for acute stroke in 1996. tPA continues to hold the distinction of being the only efficacious medical therapy for acute stroke and its use has become part of standard stroke care. Despite this long-standing history, however, controversies continue to surround facets of its use.

1.3.1 Treatment up to 3 h After Symptom Onset

tPA was investigated in ischemic stroke patients treated within 3 h of symptom onset in the two-part, randomized, double blind National Institute of Neurological Disorders and Stroke (NINDS) trial using a dose of 0.9 mg/kg, maximum of 90 mg, versus placebo [8]. Part 1, with 291 patients enrolled, sought to show early improvement with IV tPA. Early improvement was defined as either a 4-point or more improvement in the National Institutes of Health Stroke Scale (NIHSS) at 24 h or complete resolution of symptoms. While the primary endpoint was not met for Part 1, a significant benefit to tPA was seen on all studied outcome scales at 3 months. Part 2, in which 333 patients were enrolled, aimed to show that the proportion of patients that achieved essentially full recovery at 3 months would be different in the tPA and placebo-treated groups. The study met its primary endpoint, revealing that tPA-treated patients were at least 30 % more likely to have no or minimal disability at 3 months on a global test statistic. Symptomatic intracranial hemorrhage occurred in 6.4 % of patients treated with tPA compared to 0.6 % in the placebo group but mortality did not differ in the two groups. Based on these studies, tPA was approved by the Food and Drug Administration (FDA) in June 1996.

After the approval of tPA for stroke treatment, the FDA mandated that a large phase 4 trial be done to assess the outcomes and safety of IV tPA as used in clinical practice. This study, Standard Treatment with Alteplase to Reverse Stroke (STARS), enrolled 389 patients [9]. STARS found favorable outcomes in the tPA-treated patients at 30 days: 35 % had a mRS of 0–1 and 43 % were functionally independent, with a mRS of 0–2. Only 3.3 % had a symptomatic intracranial hemorrhage. The European Union also required a study to assess the safety profile of tPA in clinical practice, the Safe Implementation of Thrombolysis in Stroke-Monitoring Study (SITS-MOST) [10]. This observational study of 6483 patients confirmed the good safety profile and efficacy of tPA when used within 3 h of stroke onset.

1.3.2 Treatment Later than 3 h from Symptom Onset

While two trials, the European Cooperative Acute Stroke Study (ECASS) and ECASS II, had not shown efficacy for IV tPA given up to 6 h after symptom onset [11, 12], investigators still wanted to determine whether the treatment window for IV tPA could be lengthened past 3 h to extend the treatment to more patients. The first of the studies was the Alteplase ThromboLysis for Acute Noninterventional Therapy in Ischemic Stroke (ATLANTIS) [13]. While this trial had originally been designed to assess tPA treatment between 0 and 6 h and then 0–5 h from symptom onset, after the release of the NINDS trial results the time window was modified to include only patients between 3 and 5 h from symptom onset. The trial showed no benefit to treatment within this time window. The primary efficacy outcome, an NIHSS score of 0–1 at 90 days, was not different in the tPA and placebo-treated groups and symptomatic hemorrhage occurred in 7.0 % of the tPA group and 1.1 % of the placebo group.

Despite the discouraging results of ATLANTIS, a pooled analysis of six randomized trials suggested that favorable outcomes could be observed between 3 and 4.5 h [14]. An observational study of 664 patients treated within the 3–4.5 h timeframe and compared to 11865 patients treated within 3 h showed no differences in symptomatic intracerebral hemorrhage rates, mortality, or independence, defined as a mRS 0–2 at 3 months [15]. The randomized ECASS III trial was initiated in Europe to assess the efficacy and safety of tPA given between 3 and 4.5 h after stroke symptom onset [16]. The trial enrolled 821 patients. Inclusion criteria were stricter than those of the NINDS trial, excluding patients over the age of 80, those with a combination of previous stroke and diabetes, any oral anticoagulant treatment, and those with severe strokes defined clinically as an NIHSS of greater than 25. An absolute improvement of 7.2 % points was seen for the primary efficacy outcome, a mRS of 0–1, in the tPA treatment group ($p=0.04$). Symptomatic hemorrhage occurred more often in the tPA group, 2.4 %, than the placebo group, 0.2 % ($p=0.008$) but mortality was not affected. While the authors noted that the effect size in this time frame was clinically meaningful, they emphasized the importance of early treatment to increase the chance of a positive outcome. An analysis allowing derivation of number needed to treat to benefit or to harm over the entire outcome range estimated that the benefits of tPA treatment in the 3–4.5 h window was approximately half that in the under 3-h window with no increase in the conferral of harm [17].

Explanations have been offered for the differing results of ATLANTIS and ECASS III. The ECASS III authors noted that the ATLANTIS cohort treated 3–4.5 h after the onset of symptoms was smaller than in the ECASS III trial and lacked the power to detect an effect size of 7–10 % [16]. ECASS III was published 9 years later than ATLANTIS, after acute stroke care had evolved. Some authors, however, have argued that it was the lower severity of the strokes in the ECASS III treatment group compared to the placebo group that affected the trial outcomes [18] or the lower proportion of patients with a previous stroke in the treatment group [19]. Yet in a post hoc intention-to-treat analysis that adjusted for confounding baseline variables including the baseline NIHSS score, treatment with tPA remained significantly associated with a favorable outcome [16].

One potential concern surrounding the longer time window for tPA treatment, that stroke treatment would be delayed in patients, appears to be unfounded. After guidelines extended the treatment window in October 2008, an updated analysis of patients treated in Europe, Asia, and Australia showed that the implementation of thrombolytic treatment given 3–4.5 h after stroke occurred rapidly but with a simultaneous increase in the number of patients treated within 3 h and no increase in the door to treatment time [20]. While in the US the FDA has not approved the extended time window for IV tPA treatment, guidelines have supported its use [21]. Although one more trial, the open label International Stroke Trial 3, has attempted to extend the treatment window to 6 h, this study did not meet its primary endpoint and the results have not led to changes in practice [22].

1.3.3 Controversies Surrounding tPA Treatment of Ischemic Stroke

Despite its current entrenched status for neurologists treating acute ischemic stroke, the use of IV tPA has engendered controversies over time. Due to questions raised about potential risks of treatment, the generalizability of the trial results and biases stemming from imbalances in the baseline stroke severity between the treatment and control patients, the NINDS commissioned an independent committee to reanalyze the data from the NINDS trial [23]. The committee's findings supported the use of IV tPA within 3 h of stroke onset [23]. Questions have been raised about the necessity of certain eligibility criteria for IV tPA treatment in stroke, such as the exclusion of patients with mild or rapidly improving symptoms [24]. Recent labeling changes have reduced the number of specified contraindications [25] and a phase IIIB trial comparing the efficacy and safety of tPA to aspirin in patients with mild strokes treated within 3 h of symptom onset is ongoing [26]. The possibility of different treatment effects in men and women has also been raised, some studies suggesting greater tPA benefits for women [27, 28].

1.4 Future Directions for IV tPA Treatment of Ischemic Stroke

The advent of tPA treatment has been the catalyst for enormous change in stroke systems of care. The urgency of the treatment has spawned stroke protocols with prescribed time limits for each step of the evaluation and treatment process. Stroke centers able to deliver timely stroke care are now certified for the level of services they are able to provide. The process has created an emphasis on teamwork between disciplines, such as physicians, nurses, and pharmacists, to expedite care. The care of the patient after treatment has developed as well, and neurointensive care units have been built in an effort to provide this specialized expertise. Finally, telemedicine has flourished as a mechanism to bring expertise in stroke treatment to emergency departments that do not have this service readily available.

While IV tPA has been the mainstay of acute stroke treatment, it has limitations. Patients presenting later than 4.5 h are currently not able to receive treatment. It cannot be used in patients in whom bleeding risks are high. In addition, studies have shown that the length of the middle cerebral artery occlusion is an independent predictor of outcome with IV tPA at 3 months [29]. TPA is already being used in combination with endovascular therapy in patients that qualify for both treatments as described in the next section. The addition of transcranial ultrasound, delivered using an operator-independent head frame, to tPA is being investigated in a phase 3 trial [30]. It is possible that other combinations will arise in the future, whether with other devices or medications, to enhance the efficacy or reduce the side effects associated with thrombolysis.

2 Endovascular Treatment in Acute Stroke

Despite extensive data establishing that acute ischemic stroke patients benefit from IV thrombolytic therapy, the majority of patients who present to the emergency room are not candidates for this therapy because of its limited time windows and inclusion/exclusion criteria [31]. Furthermore, the majority of patients who present with large vessel occlusion are often IV tPA failures because the tPA fails to recanalize the occluded large vessel [32, 33]. As new tools for intracerebral catheterization have developed, researchers have developed various methods to recanalize cerebral vessels using endovascular techniques. However, early technologies were plagued with high complication rates and poor quality of revascularization. Recently, more promising methods have demonstrated successful recanalization with improved patient outcomes and minimal hemorrhagic complications.

2.1 Fundamentals of Catheter-Based Angiography

With the advent of Dynamic Fluoroscopy and Digital Subtraction Angiography (DSA), the safety of cerebral angiography dramatically improved. Now it is possible to view catheter movement in real time and to navigate a catheter intracranially by using a "roadmap" image, which reduces the risk for perforating vessels, causing subarachnoid hemorrhage.

For stroke treatment, the patient lays on the angiography table with the right common femoral artery area prepped and draped. Patients are typically sedated with moderate sedation or general anesthesia. Arterial access is achieved, and often a diagnostic catheter is introduced and is used to catheterize the cervical artery suspected to supply the occluded intracranial vessel. Angiography is used to characterize the lesion and the path to the lesion. The diagnostic catheter is then exchanged for a guide catheter system, which is often a large bore catheter that is parked in the cervical vessel. Microcatheters are introduced into the guide catheter, which are then taken to the thrombus to intervene upon it. Once the occlusion is revascularized, final imaging is obtained, then the whole system is withdrawn back to the right common femoral artery. The arteriotomy is then closed.

2.2 *Intraarterial Thrombolysis*

Direct infusion of intraarterial thrombolytics into a thrombus was studied because it was thought this would minimize the systemic effects of thrombolysis while ensuring good penetration of thrombolytics to the thrombus to maximize recanalization. Early on, multiple pilot studies demonstrated that this could be an effective means for stroke therapy. As described earlier, the Japanese MELT trial utilized intraarterial urokinase within 6 h for the treatment of a Middle Cerebral artery M1 segment or M2 segment occlusion [5]. Patients received 5,000 units of IV Heparin in the beginning, and then received infusions of urokinase directly into the thrombus in increments of 120,000 units over 5 min, which could be repeated until the total dose reached 600,000 units. A total of 114 patients were randomized before the study was stopped early because of ethical considerations once IV tPA was approved for stroke therapy. The primary endpoint was not met (mRS $\leq$ 2), but in a secondary analysis, patients who received IA urokinase had an increased chance for excellent functional outcome (mRS $\leq$ 1) over placebo (42.1 % vs. 22.8 %, $P=0.045$) [5]. Intracerebral hemorrhage occurred in 5 patients (9 %) in the urokinase group compared to one patient (2 %) in the control group ($P=0.206$).

Similar studies were performed in The United States with similar measures known as The Prolyse in Acute Cerebral Thromboembolism (PROACT I and PROACT II) trials, which studied the use of intraarterial recombinant prourokinase in patients treated within 6 h of their acute ischemic strokes due to MCA (M1 or M2) occlusion [6]. In PROACT II, patients were randomized between getting a 9 mg IA prourokinase over 2 h followed by an IV infusion of heparin versus heparin only. Despite the study demonstrating benefit of IA prourokinase over placebo (% mRS $\leq$ 2: 40 % vs. 25 %), the FDA did not want to approve the drug on a single study, and the technology was abandoned [6].

Although these studies did demonstrate that intraarterial infusion of thrombolytic agents is an effective treatment for acute ischemic stroke due to a proximal MCA occlusion within 6 h, no further significant research studies were pursued in this field [34]. There was a paradigm shift to pursuing mechanical methods as the medications had some limitations. The first concern was that in the meta-analysis, the hemorrhage rate was 10.5 % in the intraarterial group, while only 2.3 % in the placebo group [34]. Second, in cases performed using the PROACT II methodology, the interventionalist is required to infuse the prourokinase over 2 h, which became tedious, resource consuming, and delays reperfusion. Modern methods can revascularize a vascular territory immediately upon deployment of the device [35].

2.3 *First-Generation Mechanical Thrombectomy Devices*

As early intraarterial thrombolytic trials paved the way for the development of mechanical devices aimed to retrieve thrombi or recanalize vessels, multiple different companies arose and attempted to design the optimal device [36]. Early devices were crude with stiff delivery systems, making it quite challenging to navigate them to the occluded vessels [37]. They utilized different technologies including ultrasound, vigorous irrigation, high-intensity light/heat, baskets, balloons, and snares, which either did not successfully consistently recanalize vessels, or resulted in hemorrhagic complications [38, 39]. Eventually two devices were developed which introduced a viable concept for mechanical intervention.

2.3.1 Merci Device

In 1995, Dr Pierre Gobin and his fellow JP Wensel were treating an acute stroke patient at the University of California, Los Angeles via intraarterial thrombolysis and were not able to recanalize a vessel despite the 2-h infusion [40]. Their frustration led to the development of Concentric Medical in 1999 (Mountain View, CA), which designed the first FDA-approved mechanical thrombectomy device for revascularization of cerebral vessels in patients with acute ischemic strokes called the Merci Retriever, assessed in the Mechanical Embolus Removal in Cerebral Ischemia (MERCI) trials (Fig. 23.1) [40].

The Merci device is a corkscrew-shaped flexible nitinol wire that retrieves thrombi by entangling itself within a clot and removing it with traction. It is the first device to gain international acceptance as a viable tool for mechanical intervention for cerebral large vessel occlusions. The first trial, known as The MERCI Trial, demonstrated that in 151 IV tPA-ineligible patients, 46 % (69 patients) of patients had successful recanalization with the device, which was used within 8 h of stroke onset. Successful recanalization was defined as a Thrombolysis in Myocardial Infarction (TIMI) score of 2 or 3 in all treatable vessels within 6 passes of the device. Approximately 341 devices were used in the study with 11 of them fracturing inside the body. Furthermore, there was a 7.8 % symptomatic intracranial hemorrhage rate. Despite this, good neurologic outcome, defined as an mRS $\leq$ 2, was more likely to occur in patients with successful recanalization compared to those with persistent vessel occlusion (46 % vs. 10 %, $P<0.0001$) [41].

Since the initial MERCI Trial, multiple generations of this device were developed, which added loops of filaments that were attached to a helical nitinol coil as well as varying degrees of softness to the helical loops [42]. The next trial known as the Multi MERCI Trial, included the second-generation L5 device, and allowed patients with persistent occlusion after receiving IV tPA to be enrolled [43]. In this study 57.3 % of the 131 patients enrolled in the study had successful recanalization with the Merci device, but that number improved up to 69 % after additional adjunctive therapy. However, good clinical outcomes (mRS $\leq$ 2) occurred in 36 % of

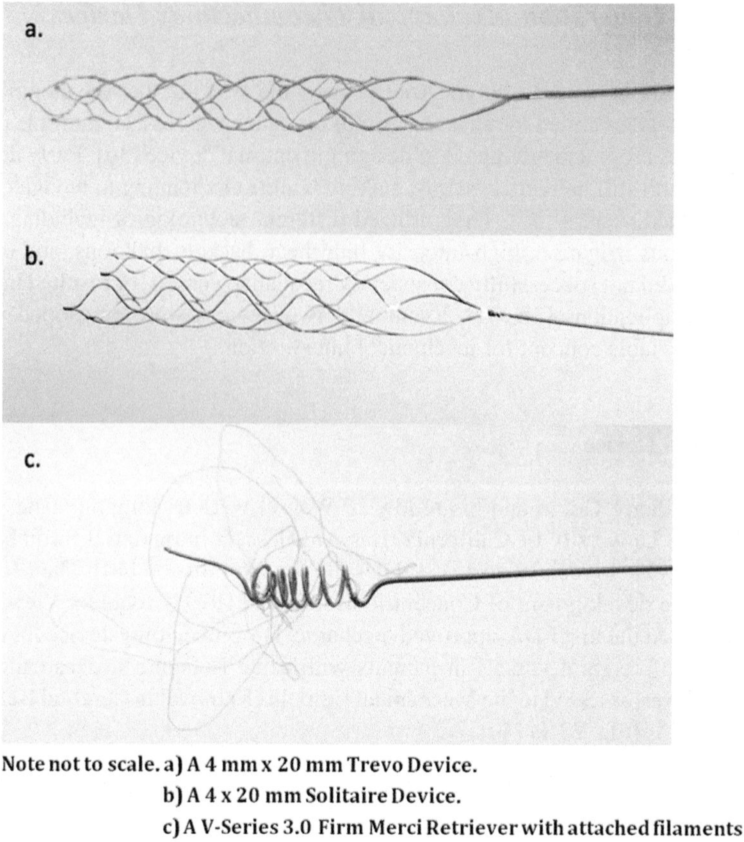

Fig. 23.1 Examples of mechanical thrombectomy devices. Note not to scale. (**a**) A 4 mm × 20 mm Trevo Device. (**b**) A 4 × 20 mm Solitaire Device. (**c**) A V-Series 3.0 Firm Merci Retriever with attached filaments

patients with recanalization and 9.8 % in those without recanalization. There was a symptomatic hemorrhage rate of 9.8 % and an overall mortality rate of 34 %, the majority of which was seen in patients not recanalized. This data supported further that patients had a better chance for a good clinical outcome if they had recanalization of their large vessel occlusion [44].

Since these trials, the Merci Retriever was used in many routine practices as an adjunctive treatment to IV tPA for treatment of acute ischemic stroke due to large vessel occlusions within 8 h of stroke onset. The device is quite operator dependent, as the device's wire can be rotated to tighten or loosen the helical loops of the device as it is being retrieved. As the device's use spread worldwide, The MERCI Registry was initiated to capture "real-world" revascularization rates and outcomes with the use of the Merci device. Unfortunately, despite completing enrollment in 2010, the primary results of the Registry have not been published. Additionally, many critics of endovascular therapy state that the studies endorsing endovascular therapy for acute ischemic stroke via the Merci Retriever were not randomized studies and did not compare patients to a placebo group. They only compared patients with successful revascularization versus unsuccessful revascularization. Further studies were performed to address this, which will be discussed later in the chapter.

2.3.2 Penumbra Device

The second device to obtain FDA approval for endovascular revascularization of cerebral vessels is known as the Penumbra Aspiration System (Penumbra, Inc., Alameda, CA). This system is comprised of 4 different sizes of reperfusion catheters with an associated "Separator Device," a thrombus removal ring and an Aspiration Pump. The Reperfusion Catheter is introduced into a minimum 6 French Guide Catheter and is navigated immediately proximal to the thrombus typically over a microcatheter/microwire system. The Microwire is removed and the Penumbra Separator wire is introduced through the reperfusion catheter. Continuous aspiration through the reperfusion catheter is achieved with the Aspiration pump while the Separator is advanced and withdrawn through the thrombus. If thrombus remained, a second method of direct mechanical retrieval with the thrombus removal ring was used to assist revascularization [45].

In the Penumbra Pivotal Trial, 81.6 % of 125 patients who presented with acute ischemic stroke due to a large vessel occlusion within 8 h had successful revascularization of the treated vessel (TIMI 2–3) [6]. However, there was a 32.8 % mortality rate at 90 days and only 25 % of patients revascularized achieved a good clinical outcome of mRS $\leq$ 2, whereas only 9 % of the persistent occlusion patients had good outcomes. In the PROACT II trial, the placebo group had a 25 % good outcome rate, and both studies enrolled patients with similar stroke severity (median NIHSS 17 in PROACT II while mean NIHSS 17.8 in Penumbra trial) [6, 45]. This discrepancy is likely explained by the fact that the definition of "successful revascularization" differed in the Penumbra study than in other thrombectomy studies. Revascularization in this study was defined as restoration of TIMI 2–3 flow in the target vessel only and did not include occlusions distal to the primary occlusion [46].

2.3.3 EKOS Catheter System

With the potentially successful good outcomes suggested with intraarterial thrombolysis from the PROACT studies, one device company's aim was to enhance the treatment effect of IA thrombolysis with ultrasound [47]. The EKOS MicroLys US Infusion Catheter System, which was made by EKOS Corp (Bothell, WA), is a microcatheter with the ability to transmit sonographic energy directly into the thrombus while tPA is infusing [47]. The company's goal was to demonstrate that the use of high frequency, low-intensity sonography directly into the thrombus from the catheter would help facilitate thrombolysis. An early study by Mahon et al. was performed in 14 patients with acute ischemic stroke due to large vessel occlusion who did not receive IV tPA. Ten patients with anterior circulation strokes and 4 patients with posterior circulation strokes were treated with the device [47]. The protocol used for most of the patients was one in which the catheter would engage the clot, and while active ultrasound was being used, a 2 mg initial bolus of IA alteplase was infused, then a 0.3 mg/min infusion of alteplase was infused over an hour (maximum dose 20 mg) [47]. In other sites in Canada, a maximum dose of 30 mg was used over 120 min. A total of 5 deaths occurred in the study, but none were attributed to the device, nor were there any device-related complications. Since the treatment number was so small, no major conclusions were made from the study other than there may be a trend toward equivalent or better recanalization rates than those seen in PROACT and that use of the device is feasible in stroke treatment [47].

Subsequent to the initial safety and feasibility study, The International Management of Stroke II (IMS II) trial was developed to determine the effectiveness of combined reduced-dose IV tPA followed by additional IA tPA with low-energy sonography via the EKOS Primo Micro-Infusion Catheter within 3 h of onset [48]. IMS II was a 13-center open label single-arm pilot study where patients received reduced-dose (0.6 mg/kg) IV tPA as a 15 % bolus with the remainder infused for 30 min, followed by additional tPA (≤22 mg) and low-energy sonography via the microcatheter at the site of the arterial occlusion [48]. Eighty-one patients entered into the trial with 26 treated with IV tPA only, while 55 were treated with combined IA and IV treatment. Of the 55 patients treated with combined therapies, 36 were treated with the EKOS catheter, while 19 were treated with a standard microcatheter. Of the 36 patients intended to be treated with an EKOS ultrasound catheter, a total of 29 patients received treatment. Twelve patients had complete revascularization in 1 h, and by 2 h 18/29 had revascularization. When compared to IMS I data, there was no significant revascularization difference between the standard microcatheter and the EKOS, but there was a trend toward improved recanalization with the EKOS ultrasound catheter [48]. The device received a general approval for arterial revascularization within the body, but the company did not pursue a specific ischemic stroke or cerebral vessel revascularization indication.

2.4 Development of Reperfusion Scales

Since the adoption of the cardiac TIMI score for cerebral artery recanalization, various scales specialized at grading recanalization of cerebral vessels have been developed (Table 23.2). The TIMI scale has its limitations in the cerebral vasculature, as it does not address the patency of vessels in a distal vascular bed of the treated vessel or address other treatable vessels, which can affect outcomes [49]. In 2003, the original Thrombolysis in Cerebral Infarction (TICI) scale was introduced, which addresses the perfusion issue (Table 23.2) [50]. In this scale, there is a TICI 2a and TICI 2b score, which represent perfusion with incomplete distal branch filling, <2/3rd and >2/3rd distal branch filling, respectively. This original TICI scale

Table 23.2 Reperfusion scales commonly used in acute ischemic stroke trials

TIMI score	
0	No perfusion
1	Perfusion past the initial occlusion but no distal branch filling
2	Perfusion past the initial occlusion with incomplete or slow distal branch filling
3	Full perfusion with filling of all distal branches
TICI score	
0	No perfusion
1	Penetration, but no distal branch filling
2a	Perfusion with incomplete (<2nd/3rd) distal branch filling
2b	Perfusion with incomplete (>2nd/3rd) distal branch filling
3	Full perfusion with filling of all distal branches
m-TICI score	
0	No perfusion
1	Penetration, but no distal branch filling
2	Perfusion with incomplete (<2nd/3rd) distal branch filling
2a	Perfusion with incomplete (>2nd/3rd) distal branch filling
3	Full perfusion with filling of all distal branches

TIMI thrombolysis in myocardial infarction [49], *TICI* thrombolysis in cerebral ischemia [50], *m-TICI* modified thrombolysis in cerebral ischemia [50]

(o-TICI) was then adjusted to a modified TICI (m-TICI) so that TICI 2a and TICI 2b represent partial perfusion with complete distal branch filling, <1/2 and >1/2, respectively [48]. Although the m-TICI score has become more common, there are proposals to adding an additional score to the m-TICI score (TICI 2c), which would represent "near-complete perfusion except for slow flow in a few distal cortical vessels or presence of small distal cortical emboli" [49]. The reasoning behind adding this score is that there is no way to distinguish patients with 51–99 % reperfusion as they would all be included into the TICI 2b group. Although there is no strong evidence to support this, the thought is that patients in the proposed TICI 2c group will likely have a better outcome rate than TICI 2b [49].

2.5 *Second-Generation Mechanical Thrombectomy Devices*

With the first-generation devices demonstrating that vessel recanalization is associated with better outcomes, research efforts focused on developing devices that opened vessels more gently and effectively. Anecdotally, when first-generation devices failed to revascularize vessels, practitioners adapted other devices to be used for thrombectomy. One such method was to fully deploy a stent or to partially deploy an intracranial stent across a thrombus, then withdrawing the entire apparatus through the guide catheter in hopes that the thrombus would adhere to the withdrawn stent [51, 52]. These ideas helped develop a new class of devices known as Stent Retrievers. These devices consist of a closed-cell, self-expanding stent that can be fully deployed and retrieved but not detached. When the device is deployed across a thrombus, it often immediately creates a perfusion channel across the thrombus as the stent applies radial displacement of the thrombus against the vessel wall (Fig. 23.2). This allows for immediate temporary partial revascularization of the affected territory. Practitioners often let the device sit against the thrombus for 3–5 min to provide temporary reperfusion and to give time for the thrombus to incorporate into the stent struts. The stent is then retrieved in a slow and smooth motion while continuous aspiration is applied from the guide catheter and/or intermediate catheter. Two such devices were approved in the United States (Solitaire FR™, Trevo®) after they showed superiority to the Merci Retriever [35, 53].

2.5.1 Solitaire FR Device

The Solitaire FR device (Covidien, ev3, Irvine, CA) is a self-expanding stent that is attached to a guidewire designed to be deployed within an intracranial vessel, which can then be withdrawn out of the body (Fig. 23.1). It was first intended to work as a removable scaffold for coil embolization of aneurysms. The first generation was

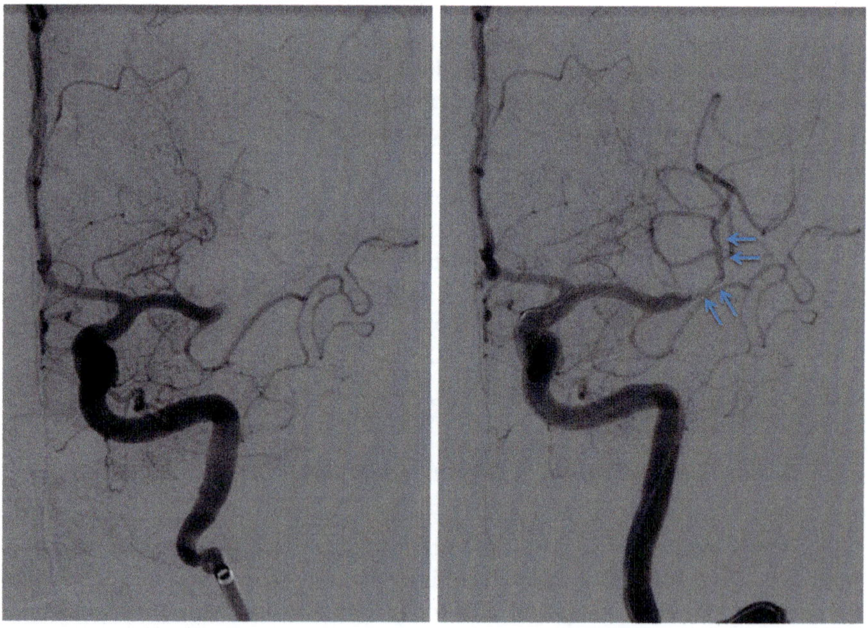

Fig. 23.2 MCA occlusion with a stent retriever forming a perfusion channel. On the left, mid-M1 left middle cerebral artery occlusion prior to intervention. On the right, interval placement of a Trevo ProVue Device across the Left MCA thrombus (*arrows*). The angiogram on the right demonstrates the perfusion channel that is typically created shortly after device placement

detachable, in case the aneurysm embolization required a permanent scaffold. The device was then studied as a mechanical thrombectomy device and when early animal studies demonstrated the effectiveness of this promising new technology, it went on to clinical trials.

In the Solitaire With the Intention For Thrombectomy (SWIFT) trial, patients with significant stroke deficits due to occlusion of proximal cerebral arteries who could get treatment within 8 h were randomized between receiving treatment with the Solitaire FR device versus the Merci device [35]. The primary endpoint of the study was successful recanalization with the assigned device without symptomatic hemorrhage. The definition of successful recanalization in this study was TIMI 2/3 revascularization in all treatable vessels. A total of 144 patients were enrolled in the study including 31 roll-in patients who received Solitaire treatment, while the remaining 113 patients were randomized between the two devices. In randomized patients, 61 % of the patients who received the Solitaire device met the primary endpoint with successful revascularization without symptomatic hemorrhage, while only 24 % of the patients who received Merci met this endpoint. Additionally, patients who received the Solitaire device were more likely to have a good neurologic outcome than patients who received the Merci device (58 % vs. 33 %, $p<0.017$) [35]. This data was compelling enough for FDA approval of the Solitaire device in 2012.

Since the SWIFT trial and FDA approval, there have been studies investigating the 'real-world' outcomes related to the Solitaire device. The North American Solitaire Stent Retriever Acute Stroke Registry (NASA) was a retrospective study looking at angiographic and clinical outcomes of patients treated with the Solitaire FR device [54]. This study reviewed 354 patients who underwent mechanical thrombectomy with the Solitaire FR device for treatment of their acute ischemic strokes in 24 centers. Overall the study found a TIMI 2/3 rate of 83.3 % (operator reported) with a 2 % symptomatic hemorrhage rate. When looking at the operator reported rate of TIMI 2/3 revascularization in SWIFT of 83 %, this data demonstrates that the Solitaire FR device performance in clinical practice is comparable to the randomized trial results [54].

With the advent of stent retriever technology, mechanical thrombectomy became safer and more effective [55]. As it became a more ubiquitous device, many centers were reporting unintentional device detachment during device retrieval [56]. Since the original intent of the device was to have an optional detachment capability in the use for stent-assisted coil embolization, there was a weak point in the device between the stent and the pusher wire. While some practitioners felt that this would act as a fail safe to prevent excessive shear force against a vessel during device retraction, others were concerned that this likely more a hazard than benefit. Therefore, The Solitaire 2 Device was introduced where the area of the previous detachment zone is now reinforced to prevent detachment and the pusher wire was made of nitinol so that it as the wire is pushed, it is much more difficult to form kinks within the wire.

2.5.2 Trevo Device

Shortly after the Solitaire Device entered clinical trials, Concentric Medical, now Stryker Corporation (Mountain View, CA and Kalamazoo, MI) was developing its own stent retriever called the Trevo Device (Fig. 23.1). The two devices are similar in that they are both stents that are attached to a pusher wire and are retrievable. The Solitaire device, however, has a proprietary overlapping stent design called a Parametric design. The Trevo device is constructed from a hollow tube with laser cut struts. The Trevo device initially had a leading wire tip, but on subsequent updates to the device, known as the Trevo CP ProVue, the wire tip was removed and replaced with open struts so that it opened more like a stent. Additionally, the ProVue is now fully radio-opaque, while the Solitaire device utilizes marker bands to identify device positioning.

The Trevo device was studied in an open-label randomized trial comparing it to the Merci Retriever in the TREVO2 Trial [53]. In this study, 178 patients were enrolled in 27 centers with 88 patients allocated to Trevo while 90 were allocated to Merci. The primary endpoint of this study was revascularization success defined as o-TICI 2a or greater in the territory of the occlusion. This endpoint was met in 86 % of patients in the Trevo group while it was met in 60 % of the Merci group. There was no significant difference in symptomatic hemorrhage between the two groups, but vessel perforations happened more in the Merci group than the Trevo group (1 % vs. 10 % $P=0.0182$) [53]. The 90-day good outcome rate (mRS $\leq$ 2) was significantly higher in the Trevo group than the Merci group (40 % vs. 21.8 % $p=0.013$) and there was no significant difference in 90-day mortality. Based on this data, the Trevo device became FDA approved after the Solitaire device in 2012.

Since the SWIFT and TREVO2 studies differ in study design, revascularization definitions/scores, and study masking assignments, a direct comparison of the two devices from these trials is not feasible [57]. Both studies were able to demonstrate that stent retrievers were superior to the Merci device in revascularization and in patient outcomes. However, continued criticisms of these device trials include the issue that a control group of patients receiving best medical therapy was not included.

2.5.3 Direct Aspiration Catheters

At around the time the stent retrievers were developed, The Penumbra Aspiration System evolved to a long intermediate catheter system, where the catheter directly engages the thrombus and aspiration with a machine or manual suction is applied [58]. This method of thrombectomy later became known as the "ADAPT" technique: A direct aspiration first pass technique for stroke thrombectomy [58]. With this model, a soft-tip, larger inner-diameter intermediate catheter is navigated over a microcatheter/microwire system directly adjacent to the thrombus. The microcatheter is removed, and aspiration is applied to the intermediate catheter. Once there is inability to aspirate from the catheter, this confirms that the catheter has engaged the thrombus [58]. The catheter is then advanced a few millimeters to ensure it is firmly engaged to the thrombus. The catheter is slowly withdrawn from the area of occlusion while maintaining continuous aspiration. If this was not successful, then the procedure can be repeated in a similar fashion, or instead a stent retriever can be used in conjunction with the aspiration [58]. Proponents of this method state that it is safe, quick, and low cost since it typically does not require the use of a stent retriever [59]. In the ADAPT FAST study, 98 patients were treated with the ADAPT technique, and TICI 2B/3 was reached in 78 % of cases, and with the addition of a stent retriever, it rose as high as 95 % [59]. Forty percent of patients had a mRS 0–2 and there was a 20 % mortality rate. They reported 2 complications and no symptomatic intracerebral hemorrhages [59]. This was not a randomized trial comparing the technique to another thrombectomy device, nor was it placebo controlled.

2.6 *First-Generation Randomized Controlled Studies*

There was a growing criticism that mechanical thrombectomy devices had never been compared to best medical therapy for the treatment of acute ischemic stroke. Despite the fact that the investigational device studies demonstrated better outcomes than that of historic controls, many in the stroke community still considered mechanical thrombectomy as an experimental procedure [6, 35]. Therefore, several randomized studies were pursued to attempt to demonstrate the efficacy of endovascular stroke treatments. The first three studies, as described later, all failed to demonstrate improved outcomes with mechanical intervention [60–62].

2.6.1 International Management of Stroke 3 Trial

The International Management of Stroke 3 (IMS-3) trial was an international, randomized, open-label trial with a blinded outcome that randomized treatment of acute ischemic stroke between IV tPA versus IV tPA plus endovascular treatment [60]. The study went through multiple amendments throughout its course. All eligible patients received IV tPA within 3 h of stroke onset or last known normal. Randomization into the study must have occurred within 40 min of IV tPA initiation because those who were randomized to receive acute endovascular therapy only received 2/3rd the total IV tPA dose via the systemic approach [60]. Eventually, more data demonstrated that in patients who received the complete systemic dose of IV tPA, additional IA tPA was not dangerous, therefore an amendment changed the dosing of tPA in the endovascular group. Additionally, after 284 patients were already enrolled in the study, an amendment was made to add the use of CT angiography (CTA) prior to randomization in patients with NIHSS 8–9 to determine eligibility with confirmation of large vessel occlusion. The type of endovascular therapy was selected by each site based on practitioner preference, most of which used IA tPA. The primary outcome measure was a mRS 0–2 at 90 days.

After 656 participants were randomized in the study, the study was stopped early by the Data and Safety Monitoring Board (DSMB) because of futility. Based on their analysis, there would be no way there could be statistically different favorable outcomes between the two treatment groups. There was a 40.8 % good outcome rate in the endovascular group while 38.7 % had good outcomes in the IV tPA group (95 % confidence interval –6.1–9.1) [60]. Additionally, when looking at the pre-defined subgroups such as patients with NIHSS ≥ 20, there was a trend toward a better outcome rate with thrombectomy, but no statistical difference was seen (23.8 % vs. 16.8 %, $p=0.06$).

Since the results of the trial were released, there have been many criticisms of the study, which may have explained why the study had failed to demonstrate efficacy of endovascular therapy. Of the 434 patients randomized to endovascular therapy, 100 of those patients (24 %) did not actually receive IA treatment after IV tPA because of various reasons which included early clinical improvement, absence of thrombus, and technical failures [63]. Additionally, many of those who did receive intraarterial treatment did not achieve "good reperfusion," as more data suggests that reperfusion <TICI2B may not be sufficient [64]. For example, in a retrospective analysis, patients who had TICI2b/3 recanalization were discharged home 41 % of the time, while patients with TICI2a recanalization were discharged home 17.4 % of the time ($p=0.008$ [64].

In IMS-3, TICI2b/3 revascularization rates were below 50 %, which is well below the good revascularization rates seen in SWIFT and TREVO2 with the modern stent retrievers [35, 53, 60]. Patients were not all required to have CTA imaging to determine whether vessel occlusion was present. The majority of IMS-3 cases used IA tPA ($n=142$) and a significant number used Merci ($N=95$), while only 5 patients received the Solitaire device [63]. Lastly, the time to revascularization in the endovascular arm was longer than in previous IMS studies (127 min in IMS-3 versus 84 min in IMS-1) [60]. A subanalysis demonstrated that a delay in reperfusion by 30 min was associated with a 10 % reduction in the likelihood of a favorable outcome [65]. Overall, IMS-3, although negative, was insufficient to conclude that endovascular therapy does not improve outcomes in patients with acute ischemic stroke. It merely emphasized the importance of proper patient selection and good quality reperfusion, which was demonstrated eventually in future studies.

2.6.2 MR RESCUE

Published in the same issue of the New England Journal of Medicine as IMS-3, the Mechanical Retrieval and Recanalization of Stroke Clots Using Embolectomy (MR RESCUE) study also failed to demonstrate benefit with endovascular therapy when using imaging criteria as additional selection criteria for stroke intervention [61]. In this study, participants with acute ischemic stroke due to a large vessel occlusion within the anterior circulation were randomized to undergo mechanical thrombectomy (with Merci or the Penumbra System) or to receive standard of care within 8 h from their stroke onset. All patients underwent a pretreatment CT scan or MRI of the brain with perfusion imaging. Patients were randomized according to whether they had a favorable penumbral pattern of small infarct core with substantial salvageable tissue [61]. They enrolled 127 patients into the study to the 2 main study groups, but 9 were excluded from the primary analysis. Of the 118 patients left, 64 underwent thrombectomy (34 with penumbral pattern, 30 with nonpenumbral pattern) while 54 received standard of care (34 with penumbral pattern, 20 with non-penumbral). The percentage of patients with a good outcome at 90 days was not significantly different between the 4 groups, but each group had less than 35 patients in it. Furthermore, the percentage of patients with partial or complete revascularization was no different among the 4 groups, emphasizing the importance of the concept that in order for a mechanical thrombectomy device to be effective, it has to successfully remove blood clots [61, 63]. Yet despite the fact that revascularization success was no different in each group, the authors made the conclusion that a favorable penumbral pattern on neuroimaging did not identify patients who would benefit differentially from endovascular therapy.

2.7 *Development of Imaging Criteria for Patient Selection*

Computed Tomography (CT) has evolved significantly since it was first invented in 1972 [66]. It has had the ability to identify a cerebral hemorrhage in the setting of an acute stroke presentation, and it especially became a necessary modality in stroke treatment when IV tPA was introduced [8]. Shortly after CT was invented, researchers were investigating the use of Magnetic Resonance Imaging (MRI) in the body and were investigating its uses in the medical field [67]. Diffusion weighted imaging (DWI) for stroke was developed in the early 1990s, which is now considered one of the most sensitive modalities in detecting an acute ischemic stroke on brain imaging [68–70]. These advanced imaging modalities are constantly evolving and have significantly influenced stroke care today.

2.7.1 ASPECTS Score

CT technology has progressed in the last 40 years and its resolution has gotten to the point that with today's scanners, we can distinguish basal ganglia structures. However, it is difficult to visualize hyperacute ischemic strokes on CT, as often it is not sensitive enough to identify stroke. Therefore, patients typically are sent for MRI as diffusion-weighted imaging is highly sensitive [69]. However, MRI is time consuming and not always available at many institutions. As a reasonable solution to this problem, researchers at the University of Calgary developed the Alberta Stroke Program Early CT Score (ASPECTS) as a way to assess early ischemic changes on a CT [71]. This is a 10-point quantitative topographic CT scan score that is used to determine the degree of infarct in patients with acute ischemic stroke in the anterior circulation. Essentially, the MCA territory of interest is divided into 10

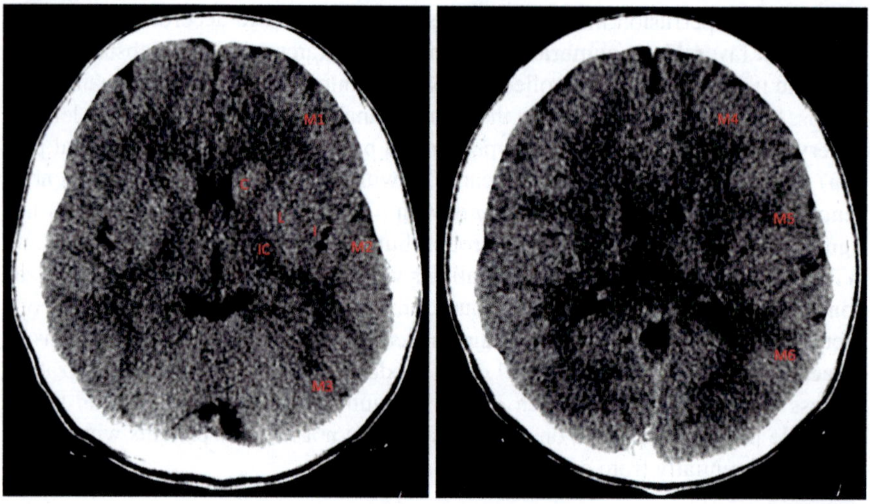

Fig. 23.3 The Alberta Stroke Program Early CT Score (ASPECTS) is a 10-point quantitative CT scan score [71]. Above are two slices from a Head CT (ganglionic and supraganglionic levels) with the regions of the middle cerebral artery (MCA) territory divided into ten parts. *C* caudate, *I* insula, *IC* internal capsule, *L* lentiform nucleus, *M1* anterior MCA cortex, *M2* MCA cortex lateral to insula, *M3* posterior MCA cortex; *M4*, *M5*, *M6* are the anterior, lateral and posterior MCA territories superior to M1, M2, M3

points: 3 subcortical points, and 7 cortical points (Fig. 23.3). A normal head CT has 10 points, but for every one of those ten regions that has evidence of hypodensity from the acute event, 1 point is subtracted. A score of 0 indicates diffuse involvement of infarct throughout the MCA territory. The score was initially used as a more reliable way to determine whether 1/3rd of the MCA territory was already infarcted when determining candidacy for IV thrombolytic therapy [72]. Then it was investigated retrospectively in the PROACT-II study, and the patients who had an ASPECTS>7 were three times more likely to have a favorable outcome with IA thrombolytic treatment compared with controls [73]. Patients with ASPECTS of 7 or less were less likely to benefit from treatment [73]. ASPECTS was later used in multiple clinical trials of acute ischemic stroke treatments, establishing it as a fundamental imaging score used in clinical practices throughout the world [74, 75].

2.7.2 CT Angiogram Collateral Scores

In addition to a noncontrast head CT, often a CT angiogram (CTA) of the head and neck is obtained to confirm that a large vessel occlusion is present. One observation that has been made is that patients can have varying degrees of collateral circulation to a territory affected by a large vessel occlusion [76]. The collaterals for a patient with a proximal MCA occlusion often arise from leptomeningeal vessels that are supplied by the posterior cerebral or anterior cerebral arteries. Occasionally patients can have a collateral pathway from branches of the external carotid artery as well [77]. An observation was made that patients with poor or absent collaterals were more likely to have a poor outcomes after an ischemic stroke due to large vessel occlusion [76]. Likewise, patients with good collaterals in an affected hemisphere have a better chance for a good outcome [76]. Further study was then done at the University of Tennessee, where they performed a retrospective review of 50 patients with acute ischemic stroke due to large vessel occlusion treated with endovascular therapy [78]. In this study, 76 % of those patients received IV tPA as well as endovascular therapy; they had a good recanalization rate of 86.6 % after endovascular therapy. Patients were dichotomized between having "good vs. poor" collateral scores on a standard CTA by two blinded neuroradiologists. Patients who received IV tPA had good collateral scores, and successful recanalization had a 52.7 % good outcome rate [78]. However, patients with recanalization but poor collateral scores without IV tPA only had a good clinical outcome 2 % of the time. This data demonstrates that scoring collateral circulation may be an additional marker to predict outcome with recanalization therapies.

Based on early data suggesting a potential role for collateral scores, researchers mainly out of Canada developed a multiphasic CTA collateral scoring system to grade a patient's collaterals [79]. In a study of 147 patients with acute ischemic stroke, patients received a standard noncontrast CT head, single-phase CTA, multiphase CTA, and a CT perfusion study [79]. The multiphasic CTA would then generate time-resolved images of the pial arteries. The filling of these arteries was scored on a six-point ordinal scale, and the interpretation of the collateral scores was compared to an interpretation of collaterals on CTA . The study found excellent interrater reliability for the multiphase CTA, and discovered that the single-phase CTA often called patients with good collaterals as having poor collaterals [79]. The authors concluded that multiphasic CT angiography is a reliable tool for imaging selection in acute stroke patients with large vessel occlusion [79]. This method of imaging criteria for patient selection in the treatment of acute ischemic stroke was eventually studied prospectively in the Endovascular Treatment for Small Core and Anterior Circulation Proximal Occlusion with Emphasis on Minimizing CT to Recanalization Times (ESCAPE) trial, as described later in this chapter [74].

2.7.3 Perfusion Imaging

Perfusion imaging is an imaging method that determines how well blood flows through tissue. In acute stroke, the most common modalities used are MR perfusion and CT perfusion. CT perfusion was first conceptualized back in the early 1980s but it was not until 1991 when its first use was demonstrated in the brain with the creation of quantifiable color maps [80, 81]. Perfusion imaging was eventually studied as a way to simultaneously estimate the volume of an acute ischemic stroke core of dead brain tissue and the volume of penumbra of brain tissue that is oligemic but not dead yet [82].

MR stroke imaging utilizes diffusion-weighted imaging with an apparent diffusion coefficient (ADC) threshold, which is able to very accurately quantify the area of brain already ischemic from an acute stroke [83]. MR perfusion modalities then utilize a bolus of gadolinium contrast and repetitively scan the brain as the contrast enters the brain. The brain tissue is broken down into 3-dimensional pixels called voxels, and in each voxel the intensity of the contrast is measured with each scan [84]. In the contralateral unaffected large vessel (typically proximal MCA), the software measures the time at which the maximum intensity of contrast is reached, and this is known as the T_{max} [84]. The software then measures the delay of the T_{max} in the abnormal tissue, and the threshold for T_{max} with a 2-s or more delay would then be used to display a perfusion-weighted imaging (PWI) lesion [83, 84]. Automated RAPID software (iSchemiaView, Inc., Menlo Park, CA) has the ability to automatically provide quantitative measurements of the volumes as well as provide the imaging characteristics (Fig. 23.4) [83]. Now that this technology is available, the challenge has been determining the optimal imaging criteria warranting acute ischemic stroke treatment.

A pooled analysis of the Echoplanar Imaging Thrombolytic Evaluation Trial (EPITHET) and Diffusion and Perfusion Imaging Evaluation for Understanding Stroke Evolution (DEFUSE) study was done, which looked at 174 patients (EPITHET $n = 100$, DEFUSE $n = 74$) with acute ischemic stroke who had RAPID MRI imaging prior to reperfusion (IA or IV methods) [82, 84, 85]. The authors categorized the DWI-PWI imaging profiles into three categories: Target Mismatch,

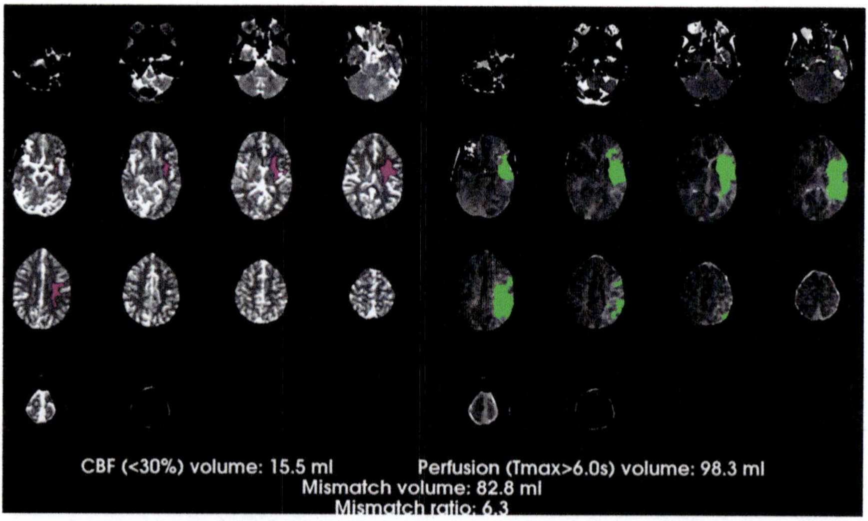

Fig. 23.4 Example of a CT Perfusion study processed with RAPID Software. This is an example of a RAPID software's (Ischema View, Menlo Park, CA) processed perfusion image from a patient with an acute ischemic stroke due to a left middle cerebral artery occlusion. The area in *purple* estimates the volume of ischemic core, while the area in *green* represents an area of hypoperfusion consistent with an ischemic penumbra

Malignant Profile, and No Mismatch. They defined favorable clinical response as a NIHSS of 0–1 or 8 point or greater improvement in the score at 90 days. They saw that in "Target Mismatch" patients, reperfusion was associated with a favorable clinical response, with an odds ratio of 5.6. Reperfusion also was associated with reduced infarct growth of 10±21 mL versus 40±44 mL (p<0.001). On the other hand, reperfusion was not associated with favorable outcomes in patients with a malignant profile as well as in patients with a no mismatch profile [85].

After the EPITHET and DEFUSE studies, a subsequent DEFUSE 2 study was performed to determine if MRI perfusion can be used to predict who is most likely to benefit from endovascular reperfusion therapy within 12 h of stroke onset [83]. In this study, 138 consecutively encountered patients signed informed consent but 28 did not undergo catheter-based angiography. Ninety-eight of the 110 remaining patients who underwent angiography had an endovascular intervention. Fifty-nine patients who received endovascular treatment also had received IV tPA prior to the intervention. In the Target Mismatch group, 46 of 78 patients had successful reperfusion, while 12 of the 21 No Target Mismatch patients had reperfusion. A favorable clinical response was associated with reperfusion (adjusted odds ratio 8.5) in the Target Mismatch Group, while it was not in the No Target Mismatch group [83]. The authors concluded that this data supports a randomized controlled trial of endovascular treatment in patients with Target Mismatch Profiles [83].

Although MRI perfusion does provide accurate estimation of the core lesion in acute stroke, it has disadvantages. MRI availability in most centers is limited, and it does take more time to perform than a CT. Additionally, screening patients for MRI safety, such as confirmation of MRI compatible internal prosthetics, can be challenging in an acute stroke situation. There has been significant research in CT perfusion, as it is likely to be more readily available in most centers and does not require prescreening other than confirming renal health and contrast allergy. The disadvantages to CT are that it requires a large bore IV, uses X-ray radiation, uses contrast that can strain kidneys and can be technologist dependent [86]. Additionally, there had been much debate on how best to estimate the core infarct on CT, as there is no DWI equivalent for CT.

Current CT perfusion protocols utilize an approximately 4-mL/s contrast injection while the scanner scans the region of interest multiple times as the software measures the rate of Hounsfield unit concentration of contrast within the tissue [86]. The system identifies an artery and vein in the Region of Interest (ROI) on an unaffected vessel that is perpendicular to the acquisition plane. This then generates an arterial input function (AIF) and a Venous Output Function (VOF), which are graphs demonstrating essentially the concentration of contrast over time. Color-coded perfusion maps measuring cerebral blood volume (CBV), mean transit time (MTT), and cerebral blood flow (CBF) are generated based on the AIF and VOF comparison between normal and oligemic tissues. MTT is calculated by performing a mathematical technique called deconvolution on the regional time-attenuation curve of each pixel with respect to the arterial curve [86]. In brain ischemia, there are multiple stages which CT perfusion attempts to quantify. In early stages of ischemia, the CBV increases as a compensatory mechanism for ischemia as vasodilation occurs, which is dependent on cerebrovascular reserve [86]. Since CBF is calculated based on CBV/MTT, it too also typically increases. As this reserve is depleted, the CBV becomes decreased, which is suggestive of an infarcted core. The MTT is increased in tissue that is both ischemic and oligemic, so many people look at the decreased CBV as an area of estimated core, while an area of increased MTT is an area of estimated core plus penumbra [86]. However, other research groups feel that the CBF is a more reasonable estimate of a core as a normal or raised CBV may not accurately represent salvageable tissue [87]. Automated software such as RAPID has now been developed to process CT perfusion data. The software defines ischemic core as a CBF reduction to less than 30 % and an area of penumbra defined as a T_{max} greater than 6 s [75]. This software was utilized in several acute ischemic stroke randomized controlled trials.

2.8 *Acute Ischemic Stroke Therapies 2015: The Next Generation of Randomized Data*

After studies such as IMS-3 and MR RESCUE were published, there was increased criticism that endovascular therapy was still not a proven therapy for acute ischemic stroke due to large vessel occlusion. Therefore, as a result, multiple randomized controlled studies were initiated around the world to demonstrate the effectiveness of endovascular therapy as a treatment for acute ischemic stroke (Fig. 23.5) [74, 75, 88–90].

2.8.1 MR CLEAN

The Multicenter Randomized Clinical Trial of Endovascular Treatment for Acute Ischemic Stroke in the Netherlands (MR CLEAN) study was the first of the five randomized controlled clinical trials on Endovascular Stroke treatment to be published [88]. In this study, 500 patients with acute ischemic stroke due to large vessel occlusion that could be potentially be treated with IA therapy within 6 h were randomized between intraarterial stroke therapy with standard medical therapy to standard medical therapy alone. Eligible patients required a stroke due to an occlusion of the distal intracranial carotid artery, middle cerebral artery (M1 or M2), or anterior cerebral artery (A1 or A2) which was confirmed on CT angiography, MR angiography, or conventional angiography. Of the 500 patients enrolled, 445 received intravenous alteplase before randomization, with 87 % of the IA group receiving IV tPA and 91 % receiving it in the control group. Additionally, of the 233 patients randomized to IA treatment, 190 of them received treatment with a stent retriever. Functional independence, as defined as a mRS 0–2, was seen in 32.6 % of the intervention group and in 19.1 % of the control group [88]. There was no significant difference seen in mortality between the two groups [88]. The results of this study

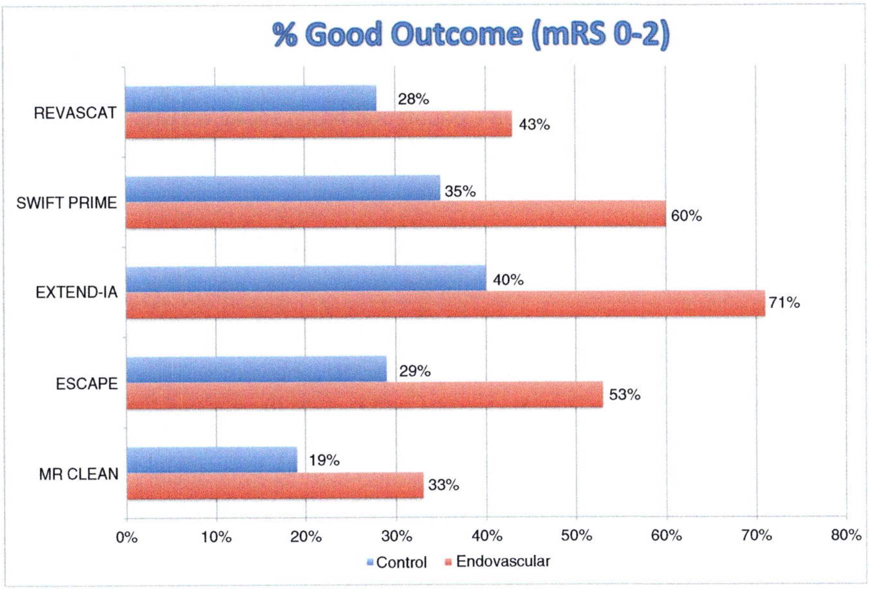

Fig. 23.5 Good outcome rate seen in the 2015 endovascular stroke studies. Summary graph showing the rate of good outcomes (% modified ranking score 0–2) in each of the 2015 endovascular studies that demonstrated efficacy of mechanical embolectomy for the treatment of acute ischemic stroke due to large vessel occlusion within the anterior circulation [74, 75, 88–90]

were so profound that it caused the four other interventional trials to stop enrollment to allow interim analyses to be performed.

2.8.2 REVASCAT

The Randomized Trial of Revascularization with Solitaire FR Device versus Best Medical Therapy in the Treatment of Acute Stroke Due to Anterior Circulation Large Vessel Occlusion Presenting within Eight Hours of Symptom Onset (REVASCAT) study was performed in Spain [89]. This was the last of the five studies on endovascular stroke treatment of 2015 to publish its data. In this study, 206 patients with acute ischemic stroke due to a large anterior circulation vessel occlusion (MCA/ICA) were randomized between treatment with the Solitaire FR device versus best medical therapy. Patients either received IV tPA within 4.5 h with failure to recanalize after 30 min or had a contraindication to IV tPA. They required having an ASPECTS score of 7–10 on CT, or a DWI ASPECTS on MRI of 6–10. After 160 patients were enrolled, the inclusion criteria were modified to include patients age 80–85 with an ASPECTS 9–10. One hundred and three patients were randomized to intervention and 103 were randomized to medical therapy. An interim analysis had been conducted as planned after 25 % of the patients in REVASCAT had completed their 90-day follow up, and enrollment was ended due to loss of equipoise as the other studies were just published. Of those who received intervention, 43.7 % had a good functional outcome (mRS = 0–2) at 90 days while the control group had a 28.2 % good functional outcome rate (adjusted odds ratio 2.1, 95 % CI [1.1, 4.0]). The trial had been embedded within a population-based stroke registry that showed that only eight patients who met eligibility criteria were treated outside of the trial at the participating trials [89].

2.8.3 ESCAPE

The Endovascular Treatment for Small Core and Anterior Circulation Proximal Occlusion with Emphasis on Minimizing CT to Recanalization Times (ESCAPE) trial was a study performed primarily in Canada with the goal to randomize 500 participants to standard care or standard care plus endovascular treatment [74]. However, with the results from MR CLEAN, the study was stopped for efficacy after an unplanned interim analysis was conducted after 316 patients were enrolled. Patients with an acute ischemic stroke up to 12 h after symptom onset were enrolled in the study if they had a small infarct core on CT (ASPECTS 6–10), proximal artery occlusion in the MCA or intracranial ICA seen on CT angiogram, and a good collateral circulation score defined as filling of 50 % or more of the MCA pial arterial circulation. Of the 316 patients enrolled, 165 patients were randomized to intervention, while 150 patients were randomized to control (1 patient information missing). IV tPA was given in 73 % in the control arm while it was given to 79 % of the intervention arm. Functional independence was seen in 53 % of the intervention group, while it was seen only in 29.3 % of the control group ($p < 0.001$) [74]. The median time of reperfusion in the intervention group from onset time was 241 min [74].

2.8.4 SWIFT PRIME

The Solitaire with the Intention for Thrombectomy as Primary Endovascular Treatment (SWIFT PRIME) trial was a study performed primarily in the United States with a few European sites [75]. In this study, all patients received IV tPA within 4.5 h of stroke onset and then were randomized to endovascular treatment with the Solitaire FR device versus medical management once an IV tPA failure was confirmed. Large vessel occlusions of the intracranial ICA or M1 were confirmed by CTA or MRA. Initially, participants were required to meet specific perfusion criteria via perfusion imaging with RAPID software, but after 71 patients were enrolled, centers with limited perfusion capabilities were allowed to use a "small-to-moderate core-infarct strategy" [75]. After 196 participants were enrolled, the study was placed on hold for analysis due to the release of the other positive trials. Of these patients, 80 % had perfusion imaging performed, and the median ASPECTS score on these patients was 9. In the 98 patients that received IV tPA with mechanical thrombectomy, 60.2 % had a good functional outcome (mRS 0–2) at 90 days, while in 98 patients who received IV tPA only in the control group, 35.5 % had a good functional outcome ($p<0.001$) [75]. Successful reperfusion of TIC2b-3 was achieved in 83 % of patients in the intervention group but in only 40 % in the control group ($p<0.001$) [75]. In this study, tandem occlusions with a cervical carotid artery occlusion were excluded, while in other studies they were included and also had the greatest benefit to interventional therapy [74].

2.8.5 EXTEND-IA

In the Extending the Time for Thrombolysis in Emergency Neurological Deficits—Intra-Arterial (EXTEND-IA) trial, the authors planned to randomize 100 patients at 14 centers in Australia and New Zealand to IV tPA plus treatment with Solitaire FR device versus IV tPA alone [90]. However, like the other endovascular studies for 2015, the study was stopped early after enrolling 70 patients because of the MR CLEAN results. Patients were included in the trial if they could receive IV tPA within 4.5 h of onset, and IA therapy within 6 h. Patients were required to have a CTA to confirm an intracranial ICA, or MCA (M1 or M2) occlusion. Furthermore, all patients received a CT perfusion study that was processed with RAPID software and it had to demonstrate salvageable brain tissue. The primary outcome, an 8-point or more reduction in NIHSS or a score of 0–1 at day 3, was seen in 80 % of those in the endovascular group compared to 37 % in the control group ($p=0.002$) [90]. Furthermore, the 90-day good outcome rate (mRS 0–2) was significantly higher in the endovascular group with 71 % having a good outcome while only 40 % of the control group had a good outcome ($p=0.01$) [90]. In this study, the endovascular group had a median stroke onset to revascularization time of 248 min [90].

2.9 *Why Did Thrombectomy Work in 2015 and not 2013?*

With 5 individual randomized controlled studies demonstrating the effectiveness of endovascular therapy for the treatment of acute ischemic strokes due to large vessel occlusion, the standard of care for acute therapy was changed. The American Heart Association Guidelines now state that the treatment of an acute ischemic stroke due to an intracranial ICA or MCA (M1 or M2) occlusion within 6 h should be treated with endovascular therapy (Level 1A evidence) [91].

Compared to 2013 when IMS-3 was published, the difference seen in outcomes today with endovascular therapy is profound, and it has changed the treatment paradigms for acute stroke throughout the world. In a meta-analysis that included patients from the 5 endovascular studies of 2015 who were treated with Solitaire FR, the rate of successful m-TICI 2b/3 revascularization was 77 % [55]. In IMS-3, this rate of TICI2b/3 was 41 % and it was worse in MR RESCUE with 27 % good revascularization in the endovascular group [60, 61]. Furthermore, in the 2015 studies there was more emphasis on patient selection and on minimizing the endovascular treatment of patients outside of the trials at enrolling institutions, which helped avoid selection bias. Patients also had baseline imaging confirming a large vessel occlusion and most of the studies included some form of imaging parameter to estimate volume of salvageable tissue [55].

Now that we have seen improvements in patient selection and in the endovascular therapy itself, the next phase in the treatment paradigm is to focus on systems of care to facilitate rapid access to tertiary centers. This requires educating providers at different phases of care from emergency medical services (EMS), to the emergency room providers as well as transfer coordinators. EMS providers will need to be educated on identifying acute ischemic strokes likely due to large vessel occlusions with prehospital stroke scales such as Rapid Arterial oCclusion Evaluation (RACE) [92]. Additional methods such as EMS notifying a hospital emergency room of an incoming stroke (prehospital notification) have been shown to improve access to acute stroke therapies [93]. In time, the hope will be that more patients will have access to these acute therapies, and the burden of ischemic stroke to society will be dramatically reduced.

References

1. The Multicenter Acute Stroke Trial–Europe Study Group (1996) Thrombolytic therapy with streptokinase in acute ischemic stroke. N Engl J Med 335(3):145–150
2. Multicentre Acute Stroke Trial—Italy (MAST-I) Group (1995) Randomised controlled trial of streptokinase, aspirin, and combination of both in treatment of acute ischaemic stroke. Lancet 346(8989):1509–1514
3. Donnan GA, Davis SM, Chambers BR, Gates PC, Hankey GJ, McNeil JJ et al (1996) Streptokinase for acute ischemic stroke with relationship to time of administration: Australian Streptokinase (ASK) Trial Study Group. JAMA 276(12):961–966
4. Fletcher AP, Alkjaersig N, Lewis M, Tulevski V, Davies A, Brooks JE et al (1976) A pilot study of urokinase therapy in cerebral infarction. Stroke 7(2):135–142
5. Ogawa A, Mori E, Minematsu K, Taki W, Takahashi A, Nemoto S et al (2007) Randomized trial of intraarterial infusion of urokinase within 6 hours of middle cerebral artery stroke: the middle cerebral artery embolism local fibrinolytic intervention trial (MELT) Japan. Stroke 38(10):2633–2639
6. Furlan AHR, Wechsler L, Gent M, Rowley H, Kase C, Pessin M, Ahuja A, Callahan F, Clark WM, Silver F, Rivera F (1999) Intra-arterial prourokinase for acute ischemic stroke. The PROACT II study: a randomized controlled trial. Prolyse in Acute Cerebral Thromboembolism. JAMA 282(21):2003–2011
7. Albers GW, von Kummer R, Truelsen T, Jensen JK, Ravn GM, Gronning BA et al (2015) Safety and efficacy of desmoteplase given 3-9 h after ischaemic stroke in patients with occlusion or high-grade stenosis in major cerebral arteries (DIAS-3): a double-blind, randomised, placebo-controlled phase 3 trial. Lancet Neurol 14(6):575–584
8. The National Institute of Neurological Disorders and Stroke rt-PA Stroke Study Group (1995) Tissue plasminogen activator for acute ischemic stroke. N Engl J Med 333(24):1581–1587
9. Albers GW, Bates VE, Clark WM, Bell R, Verro P, Hamilton SA (2000) Intravenous tissue-type plasminogen activator for treatment of acute stroke: the Standard Treatment with Alteplase to Reverse Stroke (STARS) study. JAMA 283(9):1145–1150
10. Wahlgren N, Ahmed N, Davalos A, Ford GA, Grond M, Hacke W et al (2007) Thrombolysis with alteplase for acute ischaemic stroke in the Safe Implementation of Thrombolysis in Stroke-Monitoring Study (SITS-MOST): an observational study. Lancet 369(9558):275–282

11. Hacke W, Kaste M, Fieschi C, Toni D, Lesaffre E, von Kummer R et al (1995) Intravenous thrombolysis with recombinant tissue plasminogen activator for acute hemispheric stroke. The European Cooperative Acute Stroke Study (ECASS). JAMA 274(13):1017–1025
12. Hacke W, Kaste M, Fieschi C, von Kummer R, Davalos A, Meier D et al (1998) Randomised double-blind placebo-controlled trial of thrombolytic therapy with intravenous alteplase in acute ischaemic stroke (ECASS II). Second European-Australasian Acute Stroke Study Investigators. Lancet 352(9136):1245–1251
13. Clark WM, Wissman S, Albers GW, Jhamandas JH, Madden KP, Hamilton S (1999) Recombinant tissue-type plasminogen activator (Alteplase) for ischemic stroke 3 to 5 hours after symptom onset. The ATLANTIS Study: a randomized controlled trial. Alteplase Thrombolysis for Acute Noninterventional Therapy in Ischemic Stroke. JAMA 282(21):2019–2026
14. Hacke W, Donnan G, Fieschi C, Kaste M, von Kummer R, Broderick JP et al (2004) Association of outcome with early stroke treatment: pooled analysis of ATLANTIS, ECASS, and NINDS rt-PA stroke trials. Lancet 363(9411):768–774
15. Wahlgren N, Ahmed N, Davalos A, Hacke W, Millan M, Muir K et al (2008) Thrombolysis with alteplase 3-4.5 h after acute ischaemic stroke (SITS-ISTR): an observational study. Lancet 372(9646):1303–1309
16. Hacke W, Kaste M, Bluhmki E, Brozman M, Davalos A, Guidetti D et al (2008) Thrombolysis with alteplase 3 to 4.5 hours after acute ischemic stroke. N Engl J Med 359(13):1317–1329
17. Saver JL, Gornbein J, Grotta J, Liebeskind D, Lutsep H, Schwamm L et al (2009) Number needed to treat to benefit and to harm for intravenous tissue plasminogen activator therapy in the 3- to 4.5-hour window: joint outcome table analysis of the ECASS 3 trial. Stroke 40(7):2433–2437
18. Clark WM, Madden KP (2009) Keep the three hour TPA window: the lost study of Atlantis. J Stroke Cerebrovasc Dis 18(1):78–79
19. Alper BS, Malone-Moses M, McLellan JS, Prasad K, Manheimer E (2015) Thrombolysis in acute ischaemic stroke: time for a rethink? BMJ 350:h1075
20. Ahmed N, Wahlgren N, Grond M, Hennerici M, Lees KR, Mikulik R et al (2010) Implementation and outcome of thrombolysis with alteplase 3-4.5 h after an acute stroke: an updated analysis from SITS-ISTR. Lancet Neurol 9(9):866–874
21. Jauch EC, Saver JL, Adams HP Jr, Bruno A, Connors JJ, Demaerschalk BM et al (2013) Guidelines for the early management of patients with acute ischemic stroke: a guideline for healthcare professionals from the American Heart Association/American Stroke Association. Stroke 44(3):870–947
22. Group ISTc, Sandercock P, Wardlaw JM, Lindley RI, Dennis M, Cohen G et al (2012) The benefits and harms of intravenous thrombolysis with recombinant tissue plasminogen activator within 6 h of acute ischaemic stroke (the third international stroke trial [IST-3]): a randomised controlled trial. Lancet 379(9834):2352–2363
23. Ingall TJ, O'Fallon WM, Asplund K, Goldfrank LR, Hertzberg VS, Louis TA et al (2004) Findings from the reanalysis of the NINDS tissue plasminogen activator for acute ischemic stroke treatment trial. Stroke 35(10):2418–2424
24. Tong DC (2012) Avoiding thrombolysis in patients with mild stroke: is it SMART? Stroke 43(3):625–626
25. Tissue Plasminogen Activator prescribing information [cited 2016 March 16, 2016]. http://www.gene.com/download/pdf/activase_prescribing.pdf.
26. A Study of the Efficacy and Safety of Activase (Alteplase) in Patients With Mild Stroke (PRISMS) [updated March 1, 2016; cited 2016 March 16, 2016]. https://clinicaltrials.gov/ct2/show/NCT02072226?term=prisms+and+stroke&rank=5.
27. Kent DM, Price LL, Ringleb P, Hill MD, Selker HP (2005) Sex-based differences in response to recombinant tissue plasminogen activator in acute ischemic stroke: a pooled analysis of randomized clinical trials. Stroke 36(1):62–65
28. Lorenzano S, Ahmed N, Falcou A, Mikulik R, Tatlisumak T, Roffe C et al (2013) Does sex influence the response to intravenous thrombolysis in ischemic stroke?: answers from safe implementation of treatments in Stroke-International Stroke Thrombolysis Register. Stroke 44(12):3401–3406

29. Rohan V, Baxa J, Tupy R, Cerna L, Sevcik P, Friesl M et al (2014) Length of occlusion predicts recanalization and outcome after intravenous thrombolysis in middle cerebral artery stroke. Stroke 45(7):2010–2017
30. Schellinger PD, Alexandrov AV, Barreto AD, Demchuk AM, Tsivgoulis G, Kohrmann M et al (2015) Combined lysis of thrombus with ultrasound and systemic tissue plasminogen activator for emergent revascularization in acute ischemic stroke (CLOTBUST-ER): design and methodology of a multinational phase 3 trial. Int J Stroke 10(7):1141–1148
31. Acute and Preventive Stroke Treatments: American Stroke Association; 2016 [March 3, 2016]. http://www.strokeassociation.org/STROKEORG/AboutStroke/Treatment/Treatment_UCM_310892_Article.jsp.
32. Riedel CH, Zimmermann P, Jensen-Kondering U, Stingele R, Deuschl G, Jansen O (2011) The importance of size: successful recanalization by intravenous thrombolysis in acute anterior stroke depends on thrombus length. Stroke 42(6):1775–1777
33. Hacke W, Zeumer H, Ferbert A, Bruckmann H, del Zoppo GJ (1988) intra-arterial thrombolytic therapy improves outcome in patients with acute vertebrobasilar occlusive disease. Stroke 19(10):1216–1222
34. Fields JD, Khatri P, Nesbit GM, Liu KC, Barnwell SL, Lutsep HL et al (2011) Meta-analysis of randomized intra-arterial thrombolytic trials for the treatment of acute stroke due to middle cerebral artery occlusion. J Neurointerv Surg 3(2):151–155
35. Saver JL, Jahan R, Levy EI, Jovin TG, Baxter B, Nogueira RG et al (2012) Solitaire flow restoration device versus the Merci Retriever in patients with acute ischaemic stroke (SWIFT): a randomised, parallel-group, non-inferiority trial. The Lancet 380(9849):1241–1249
36. Chopko BW, Kerber C, Wong W, Georgy B (2000) Transcatheter snare removal of acute middle cerebral artery thromboembolism: technical case report. Neurosurgery 46(6):1529–1531
37. Sharafuddin MJ, Hicks ME, Jenson ML, Morris JE, Drasler WJ, Wilson GJ (1997) Rheolytic thrombectomy with use of the AngioJet-F105 catheter: preclinical evaluation of safety. J Vasc Interv Radiol 8(6):939–945
38. Grundfest WS, Litvack F, Forrester JS, Goldenberg T, Swan HJ, Morgenstern L et al (1985) Laser ablation of human atherosclerotic plaque without adjacent tissue injury. J Am Coll Cardiol 5(4):929–933
39. Grundfest WS, Litvack IF, Goldenberg T, Sherman T, Morgenstern L, Carroll R et al (1985) Pulsed ultraviolet lasers and the potential for safe laser angioplasty. Am J Surg 150(2):220–226
40. MERCI Story [Webpage]. UC Regents; 2012 [cited 2016 February 28, 2016]. http://www.radnet.ucla.edu/radweb/sections/endovascular/news/MERCI_Story.jsp.
41. Smith WS, Sung G, Starkman S, Saver JL, Kidwell CS, Gobin YP et al (2005) Safety and efficacy of mechanical embolectomy in acute ischemic stroke: results of the MERCI trial. Stroke 36(7):1432–1438
42. Alshekhlee A, Pandya DJ, English J, Zaidat OO, Mueller N, Gupta R et al (2012) Merci mechanical thrombectomy retriever for acute ischemic stroke therapy: literature review. Neurology 79(13 Suppl 1):S126–S134
43. Smith WS, Sung G, Saver J, Budzik R, Duckwiler G, Liebeskind DS et al (2008) Mechanical thrombectomy for acute ischemic stroke: final results of the Multi MERCI trial. Stroke 39(4):1205–1212
44. Nogueira RG, Liebeskind DS, Sung G, Duckwiler G, Smith WS (2009) Merci, et al. Predictors of good clinical outcomes, mortality, and successful revascularization in patients with acute ischemic stroke undergoing thrombectomy: pooled analysis of the Mechanical Embolus Removal in Cerebral Ischemia (MERCI) and Multi MERCI Trials. Stroke 40(12):3777–3783
45. Penumbra Pivotal Stroke Trial I (2009) The penumbra pivotal stroke trial: safety and effectiveness of a new generation of mechanical devices for clot removal in intracranial large vessel occlusive disease. Stroke 40(8):2761–2768
46. Saver JL, Liebeskind DS, Nogueira RG, Jahan R (2010) Need to clarify Thrombolysis In Myocardial Ischemia (TIMI) scale scoring method in the Penumbra Pivotal Stroke Trial. Stroke 41(2):e115–e116

47. Mahon BR, Nesbit GM, Barnwell SL, Clark W, Marotta TR, Weill A et al (2003) North American clinical experience with the EKOS MicroLysUS infusion catheter for the treatment of embolic stroke. AJNR Am J Neuroradiol 24(3):534–538
48. Tomsick T, Broderick J, Carrozella J, Khatri P, Hill M, Palesch Y et al (2008) Revascularization results in the Interventional Management of Stroke II trial. AJNR Am J Neuroradiol 29(3):582–587
49. Goyal M, Fargen KM, Turk AS, Mocco J, Liebeskind DS, Frei D et al (2014) 2C or not 2C: defining an improved revascularization grading scale and the need for standardization of angiography outcomes in stroke trials. J Neurointerv Surg 6(2):83–86
50. Higashida RT, Furlan AJ, Roberts H, Tomsick T, Connors B, Barr J et al (2003) Trial design and reporting standards for intra-arterial cerebral thrombolysis for acute ischemic stroke. Stroke 34(8):e109–e137
51. Kulcsar Z, Bonvin C, Lovblad KO, Gory B, Yilmaz H, Sztajzel R et al (2010) Use of the enterprise intracranial stent for revascularization of large vessel occlusions in acute stroke. Clin Neuroradiol 20(1):54–60
52. Suh SH, Lee KY, Seo KD, Lim SM, Roh HG, Kim BM (2013) Recanalization of acute intracranial artery occlusion using temporary endovascular bypass technique. Neurointervention 8(2):80–86
53. Nogueira RG, Lutsep HL, Gupta R, Jovin TG, Albers GW, Walker GA et al (2012) Trevo versus Merci retrievers for thrombectomy revascularisation of large vessel occlusions in acute ischaemic stroke (TREVO 2): a randomised trial. The Lancet 380(9849):1231–1240
54. Zaidat OO, Castonguay AC, Gupta R, Sun CH, Martin C, Holloway WE et al (2014) North American Solitaire Stent Retriever Acute Stroke registry: post-marketing revascularization and clinical outcome results. J Neurointerv Surg 6(8):584–588
55. Campbell BC, Hill MD, Rubiera M, Menon BK, Demchuk A, Donnan GA et al (2016) Safety and Efficacy of Solitaire Stent Thrombectomy: Individual Patient Data Meta-Analysis of Randomized Trials. Stroke 47(3):798–806
56. Yub Lee S, Won Youn S, Kyun Kim H, Rok DY (2015) Inadvertent detachment of a retrievable intracranial stent: review of manufacturer and user facility device experience. Neuroradiol J 28(2):172–176
57. Broderick JP, Schroth G (2013) What the SWIFT and TREVO II trials tell us about the role of endovascular therapy for acute stroke. Stroke 44(6):1761–1764
58. Turk AS, Spiotta A, Frei D, Mocco J, Baxter B, Fiorella D et al (2014) Initial clinical experience with the ADAPT technique: a direct aspiration first pass technique for stroke thrombectomy. J Neurointerv Surg 6(3):231–237
59. Turk AS, Frei D, Fiorella D, Mocco J, Baxter B, Siddiqui A et al (2014) ADAPT FAST study: a direct aspiration first pass technique for acute stroke thrombectomy. J Neurointerv Surg 6(4):260–264
60. Broderick JP, Palesch YY, Demchuk AM, Yeatts SD, Khatri P, Hill MD et al (2013) Endovascular therapy after intravenous t-PA versus t-PA alone for stroke. N Engl J Med 368(10):893–903
61. Kidwell CS, Jahan R, Gornbein J, Alger JR, Nenov V, Ajani Z et al (2013) A trial of imaging selection and endovascular treatment for ischemic stroke. N Engl J Med 368(10):914–923
62. Ciccone A, Valvassori L, Nichelatti M, Sgoifo A, Ponzio M, Sterzi R et al (2013) Endovascular treatment for acute ischemic stroke. N Engl J Med 368(10):904–913
63. von Kummer R, Gerber J (2013) IMS-3, synthesis, and MR Rescue: no disaster, but down to earth. Clin Neuroradiol 23(1):1–3
64. Jayaraman MV, Grossberg JA, Meisel KM, Shaikhouni A, Silver B (2013) The clinical and radiographic importance of distinguishing partial from near-complete reperfusion following intra-arterial stroke therapy. AJNR Am J Neuroradiol 34(1):135–139
65. Khatri P, Yeatts SD, Mazighi M, Broderick JP, Liebeskind DS, Demchuk AM et al (2014) Time to angiographic reperfusion and clinical outcome after acute ischaemic stroke: an analysis of data from the Interventional Management of Stroke (IMS III) phase 3 trial. Lancet Neurol 13(6):567–574

66. Bates S, British Institute of Radiology (2012) Godfrey Hounsfield: intuitive genius of CT, vol xiv. British Institute of Radiology, London, 261pp
67. Pincock S (2003) US and UK researchers share Nobel prize. Paul C Lauterbur and Peter Mansfield share award for seminal work on MRI. Lancet 362(9391):1203
68. Taylor DG, Bushell MC (1985) The spatial mapping of translational diffusion coefficients by the NMR imaging technique. Phys Med Biol 30(4):345–349
69. Smajlovic D, Sinanovic O (2004) Sensitivity of the neuroimaging techniques in ischemic stroke. Med Arh 58(5):282–284
70. Moseley ME, Cohen Y, Mintorovitch J, Chileuitt L, Shimizu H, Kucharczyk J et al (1990) Early detection of regional cerebral ischemia in cats: comparison of diffusion- and T2-weighted MRI and spectroscopy. Magn Reson Med 14(2):330–346
71. Barber PA, Demchuk AM, Zhang J, Buchan AM (2000) Validity and reliability of a quantitative computed tomography score in predicting outcome of hyperacute stroke before thrombolytic therapy. ASPECTS Study Group. Alberta Stroke Programme Early CT Score. Lancet 355(9216):1670–1674
72. Pexman JH, Barber PA, Hill MD, Sevick RJ, Demchuk AM, Hudon ME et al (2001) Use of the Alberta Stroke Program Early CT Score (ASPECTS) for assessing CT scans in patients with acute stroke. AJNR Am J Neuroradiol 22(8):1534–1542
73. Hill MD, Rowley HA, Adler F, Eliasziw M, Furlan A, Higashida RT et al (2003) Selection of acute ischemic stroke patients for intra-arterial thrombolysis with pro-urokinase by using ASPECTS. Stroke 34(8):1925–1931
74. Goyal M, Demchuk AM, Menon BK, Eesa M, Rempel JL, Thornton J et al (2015) Randomized assessment of rapid endovascular treatment of ischemic stroke. N Engl J Med 372(11):1019–1030
75. Saver JL, Goyal M, Bonafe A, Diener HC, Levy EI, Pereira VM et al (2015) Stent-retriever thrombectomy after intravenous t-PA vs. t-PA alone in stroke. N Engl J Med 372(24):2285–2295
76. Lima FO, Furie KL, Silva GS, Lev MH, Camargo EC, Singhal AB et al (2010) The pattern of leptomeningeal collaterals on CT angiography is a strong predictor of long-term functional outcome in stroke patients with large vessel intracranial occlusion. Stroke 41(10):2316–2322
77. Tolosa E (1968) Collateral circulation in occlusive vascular lesions of the brain. The role of the middle meningeal artery in the collateral circulation in compensating for occlusions of the internal carotid artery or its branches. Prog Brain Res 30:247–254
78. Elijovich L, Goyal N, Mainali S, Hoit D, Arthur AS, Whitehead M et al (2015) CTA collateral score predicts infarct volume and clinical outcome after endovascular therapy for acute ischemic stroke: a retrospective chart review. J Neurointerv Surg
79. Menon BK, d'Esterre CD, Qazi EM, Almekhlafi M, Hahn L, Demchuk AM et al (2015) Multiphase CT Angiography: A New Tool for the Imaging Triage of Patients with Acute Ischemic Stroke. Radiology 275(2):510–520
80. Axel L (1980) Cerebral blood flow determination by rapid-sequence computed tomography: theoretical analysis. Radiology 137(3):679–686
81. Miles KA, Hayball M, Dixon AK (1991) Colour perfusion imaging: a new application of computed tomography. Lancet 337(8742):643–645
82. Davis SM, Donnan GA, Parsons MW, Levi C, Butcher KS, Peeters A et al (2008) Effects of alteplase beyond 3 h after stroke in the Echoplanar Imaging Thrombolytic Evaluation Trial (EPITHET): a placebo-controlled randomised trial. Lancet Neurol 7(4):299–309
83. Lansberg MG, Straka M, Kemp S, Mlynash M, Wechsler LR, Jovin TG et al (2012) MRI profile and response to endovascular reperfusion after stroke (DEFUSE 2): a prospective cohort study. Lancet Neurol 11(10):860–867
84. Albers GW, Thijs VN, Wechsler L, Kemp S, Schlaug G, Skalabrin E et al (2006) Magnetic resonance imaging profiles predict clinical response to early reperfusion: the diffusion and perfusion imaging evaluation for understanding stroke evolution (DEFUSE) study. Ann Neurol 60(5):508–517
85. Lansberg MG, Lee J, Christensen S, Straka M, De Silva DA, Mlynash M et al (2011) RAPID automated patient selection for reperfusion therapy: a pooled analysis of the Echoplanar

Imaging Thrombolytic Evaluation Trial (EPITHET) and the Diffusion and Perfusion Imaging Evaluation for Understanding Stroke Evolution (DEFUSE) Study. Stroke 42(6):1608–1614
86. de Lucas EM, Sanchez E, Gutierrez A, Mandly AG, Ruiz E, Florez AF et al (2008) CT protocol for acute stroke: tips and tricks for general radiologists. Radiographics 28(6):1673–1687
87. d'Esterre CD, Roversi G, Padroni M, Bernardoni A, Tamborino C, De Vito A et al (2015) CT perfusion cerebral blood volume does not always predict infarct core in acute ischemic stroke. Neurol Sci 36(10):1777–1783
88. Berkhemer OA, Fransen PS, Beumer D, van den Berg LA, Lingsma HF, Yoo AJ et al (2015) A randomized trial of intraarterial treatment for acute ischemic stroke. N Engl J Med 372(1):11–20
89. Jovin TG, Chamorro A, Cobo E, de Miquel MA, Molina CA, Rovira A et al (2015) Thrombectomy within 8 hours after symptom onset in ischemic stroke. N Engl J Med 372(24):2296–2306
90. Campbell BC, Mitchell PJ, Kleinig TJ, Dewey HM, Churilov L, Yassi N et al (2015) Endovascular therapy for ischemic stroke with perfusion-imaging selection. N Engl J Med 372(11):1009–1018
91. Powers WJ, Derdeyn CP, Biller J, Coffey CS, Hoh BL, Jauch EC et al (2015) 2015 American Heart Association/American Stroke Association Focused Update of the 2013 Guidelines for the Early Management of Patients with Acute Ischemic Stroke Regarding Endovascular Treatment: A Guideline for Healthcare Professionals From the American Heart Association/American Stroke Association. Stroke 46(10):3020–3035
92. Perez de la Ossa N, Carrera D, Gorchs M, Querol M, Millan M, Gomis M et al (2014) Design and validation of a prehospital stroke scale to predict large arterial occlusion: the rapid arterial occlusion evaluation scale. Stroke 45(1):87–91
93. Prabhakaran S, O'Neill K, Stein-Spencer L, Walter J, Alberts MJ (2013) Prehospital triage to primary stroke centers and rate of stroke thrombolysis. JAMA Neurol 70(9):1126–1132

Chapter 24
Sonothrombolysis for Acute Ischemic Stroke: A Critical Appraisal

Georgios Tsivgoulis, Apostolos Safouris, and Andrei V. Alexandrov

Abstract Sonothrombolysis is the adjunct use of ultrasound during thrombolysis as an augmentation technique that can facilitate clot dissolution. When no thrombolytics are used, we refer to this technique as sonolysis. In the setting of acute ischemic stroke (AIS), various ultrasound protocols have been tested with different devices and parameters, with or without the concomitant use of ultrasound contrast material. After years of clinical research, much is known regarding the safety profile of low-intensity and high-frequency (2 MHz) sonothrombolysis but efficacy remains to be proved. This review provides an overview of clinical trials and meta-analyses of sonothrombolysis. As we are entering a new era of combined intravenous and interventional AIS therapies, it is a challenge for researchers to incorporate sonothrombolysis in the constantly evolving AIS protocols. Proving the efficacy of sonothrombolysis in a well-designed randomized controlled clinical trial of a next-generation noninvasive therapeutic device or a drug–device combination would have substantial impact on future AIS management.

G. Tsivgoulis, M.D.
Department of Neurology, University of Tennessee Health Science Center, 855 Monroe Avenue, Suite 415, Memphis, TN 38163, USA

Second Department of Neurology, National and Kapodistrian University of Athens, School of Medicine, Athens, Greece

Department of Neurology, Attikon Hospital, Rimini 1, Athens 12462, Greece
e-mail: tsivgoulisgiorg@yahoo.gr

A. Safouris, M.D.
Second Department of Neurology, National and Kapodistrian University of Athens, School of Medicine, Athens, Greece

Stroke Unit, Metropolitan Hospital, Ethnarhou Makariou 9, Piraeus 18547, Greece
e-mail: safouris@yahoo.gr

A.V. Alexandrov, M.D. (✉)
Department of Neurology, University of Tennessee Health Science Center, 855 Monroe Avenue, Suite 415, Memphis, TN 38163, USA

Stroke Unit, Metropolitan Hospital, Ethnarhou Makariou 9, Piraeus 18547, Greece
e-mail: avalexandrov@att.net

P.A. Lapchak, J.H. Zhang (eds.), *Neuroprotective Therapy for Stroke and Ischemic Disease*, Springer Series in Translational Stroke Research,
DOI 10.1007/978-3-319-45345-3_24

Keywords Sonothrombolysis • Intravenous thrombolysis • Acute ischemic stroke • Ultrasound • Recanalization • Outcome

Abbreviations

AIS	Acute ischemic stroke
CLOTBUST	The Combined lysis of thrombus in brain ischemia using transcranial ultrasound and systemic t-PA
CLOTBUST-ER	Combined lysis of thrombus with ultrasound and systemic tissue plasminogen activator for emergent revascularization in acute ischemic stroke
ELVO	Emergent large vessel occlusion
IMS	Interventional Management of Stroke Trial
IVT	Intravenous thrombolysis
MCA	Middle cerebral artery
NOR-SASS	The Norwegian randomized controlled sonothrombolysis in acute stroke study
NOR-TEST	Norwegian Tenecteplase Stroke Trial
sICH	Symptomatic intracranial hemorrhage
SONOBUSTER	Sonolysis in Prevention of Brain Infarction During Carotid Endarterectomy and Stenting
TCCD	Transcranial color-coded duplex
TCD	Transcranial Doppler
tPA	Tissue plasminogen activator
TRUMBI	Transcranial Low-Frequency Ultrasound-Mediated Thrombolysis in Brain Ischemia
TUCSON	Transcranial Ultrasound in Clinical Sonothrombolysis trial
μS	Microspheres

1 Introduction

Intravenous thrombolysis (IVT) has revolutionized acute ischemic stroke (AIS) treatment since its approval for clinical practice in 1996 after the NINDS clinical trial [1, 2]. IVT represents the current standard of care for the vast majority of AIS patients during the first 4.5 h following symptom onset [1–4], while mechanical thrombectomy has recently been approved for treatment of a subset of patients with emergent large vessel occlusion (ELVO) [5]. The main limitation of IVT as well as any AIS treatment remains the limited time window in order to salvage the hypoperfused brain tissue or the ischemic penumbra [6]. As a result, only a minority of AIS patients are offered systemic reperfusion therapy within the allowed time window [7, 8]. However, even for this minority of patients, good outcomes are directly related to recanalization [9].

Transcranial Doppler (TCD) provides real-time data on cerebral hemodynamics during AIS and IVT treatment [10]. TCD studies have shown that tissue plasminogen activator (tPA)-induced recanalization depends on occlusion site [11] and clot burden [12]. Furthermore, reocclusion occurs frequently (10–25 %) following initial recanalization [13]. Thus, improving the efficacy of systemic thrombolysis to achieve complete recanalization was the focus of research interest during the past decade [14]. Development of ultrasound-enhanced thrombolysis stems from the clinical observation that patients who received TCD monitoring during IVT recanalized more frequently [15].

Recently, mechanical thrombectomy has been established as an interventional reperfusion therapy for patients with ELVO that is complementary to IVT [16, 17]. Health care systems around the world are reorganizing to offer this new therapeutic modality to AIS patients. However, stroke centers with endovascular treatment capability are scarce and large populations remain underserved even in high-income countries [18]. Consequently, any novel reperfusion therapy that may enhance the lytic effect of tPA will have a substantial impact on the established stroke networks. IVT augmentation with an ultrasound device at lower cost relative to specialized endovascular procedure can be widely available. Sonothrombolysis is still considered investigational [19] but it has established safety [in terms of symptomatic intracranial hemorrhage (sICH)] and potential efficacy (in terms of recanalization rates) in AIS. We will review the currently available data on the application of ultrasound-enhanced thrombolysis as a potential therapeutic modality for AIS.

2 Definitions and Mechanisms

2.1 Technical Issues

Sonothrombolysis is a process of delivering tPA to the binding sites at deeper layers of thrombus, mechanically stretching the thrombus or promoting enzymatic activity (dependently on frequency, kHz or MHz, respectively) that results in augmentation of the residual blood flow and faster thrombus dissolution. In AIS, most clinical research has been done during IVT using diagnostic TCD. TCD uses 1–2 MHz ultrasound that has a trade-off between sufficient penetration (better with lower frequencies) and satisfactory Doppler signal (better with higher frequencies). Transcranial color-coded duplex (TCCD) is progressively replacing TCD in clinical practice and has also been studied in phase II sonothrombolysis trials. TCCD emits a range of frequencies between 1 and 5 MHz and yields both spectral (Doppler imaging) and parenchymal (B-mode imaging) information. All diagnostic ultrasound devices conform to FDA output limits of spatial-peak temporal-average intensity to less than 720 mW/cm^2 [20]. A wide range of frequencies and intensities has been evaluated in vitro and in animal studies. However, after a clinical trial that studied lower frequency (300 kHz) sonothrombolysis and raised significant safety concerns due to notably high rates (36 %) of sICH in the sonothrombolysis group, all subsequent trials used the established diagnostic frequency (2 MHz for TCD; 1–5 MHz for TCCD) [21].

Microsphere-potentiated sonothrombolysis consists of the intravenous administration of ultrasound contrast in parallel with sonothrombolysis. Ultrasound contrast consists of gas-filled microspheres (μS) that enhance the reflection of ultrasound waves due to the great difference in echogenicity between gas and surrounding tissues or fluids. The first generation of commercially available ultrasound contrast media contain room air (Levovist®) and the second-generation contrast media comprise sulfur hexafluoride-filled μS that are more stable and have a longer half-life (SonoVue®) [22]. Newer agents contain lipid instead of gas, have greater concentration of bubbles per unit volume and, theoretically, greater ability to penetrate thrombi (MRX-801) [23]. Microspheres produce stronger returned echoes and this process is helpful first to overcome the barrier of insufficient temporal bone windows, and second to transmit more energy from ultrasound beam to thrombus thus further facilitating actions of sonothrombolysis as described earlier.

Sonolysis consists of ultrasound use without thrombolytic drug and it has been studied in AIS patients with contraindications to IVT. Finally, EKOS MicroLysUS is an infusion catheter carrying a 2.1-MHz ring sonography transducer at its tip, designed to work simultaneously with intraarterial thrombolysis [24].

2.2 *Physiology*

Ultrasound effects on biological tissue have been first reported back in 1930 but sonothrombolysis, the adjunct use of ultrasound in parallel to thrombolytic agents, has been first described in the 1970s [25]. The physiology behind sonothrombolysis is still not fully elucidated. Table 24.1 summarizes the potential mechanisms of ultrasound-enhanced thrombolysis as well as microsphere-potentiated sonothrombolysis. As ultrasound waves propagate through tissues, energy absorption and scattering occur depending on the attenuation coefficient of each tissue. Absorption and scattering of waves lead to conversion of acoustic energy into heat and mechanical energy. Heat and cavitation are unwanted side effects of ultrasound examination and FDA poses a strict limit on maximum intensity. Higher ultrasound frequencies accelerate enzymatic kinetics of thrombolysis and greater risk of increasing temperature of exposed tissues. However, no significant temperature variations have been recorded with low-intensity ultrasound used in clinical practice [26]. On the other hand, it is currently believed that mechanical shearing forces generated by pressure fluctuations may lead to thrombolysis augmentation either through direct fluid motion and radial forces or through acoustic cavitation effects (Table 24.1).

Table 24.1 Potential mechanisms of action of sonothrombolysis and microsphere-potentiated sonothrombolysis

Mechanism of action	Sonothrombolysis	Microsphere-potentiated sonothrombolysis
Cavitation (stable > inertial) [27]	Yes	Yes
Reversible disaggregation of cross-linked fibrin fibers [28]	Yes	Yes
Increased penetration of tPA into the thrombus (microstreaming) [30]	Yes	Yes
Increased binding of tPA to fibrin [31]	Yes	Yes
Improvement of tissue perfusion (NO-dependent mechanism of US-induced vasodilation) [32]	Yes	Yes
Lowers the threshold of thrombolysis (preexisting μS lower the threshold for cavitation) [33]	No	Yes
Beneficial effects on microvasculature (improves impaired microcirculation) [34, 35]	No	Yes

Microbubbles either oscillate (stable cavitation) or collapse (inertial cavitation) at the blood–thrombus interface with stable cavitation being more important for enhancing thrombolysis at lower intensity levels [27].

Braaten et al. have shown via electron microscopy that ultrasound exposure leads to reversible disaggregation of fibrin network, further enhancing penetration of thrombolytics [28]. Attempts to further enhance sonothrombolysis were made with the addition of ultrasound contrast agents with diameter of several microns enabling them to pass through lungs after intravenous injection. These μS lower the threshold of ultrasound energy for attaining cavitation at the blood–thrombus interface, amplifying local delivery of mechanical stress forces that in turn lead to thrombus surface disruption and further tPA penetration [29]. Microspheres offer better targeting of ultrasound beam delivery and transmission of energy momentum to the thrombus thus minimizing exposure of other tissues.

3 Clinical Trials

3.1 *Sonothrombolysis*

Alexandrov et al. first reported in 2000 high IVT recanalization rates (30 % complete, 40 % partial) with concurrent diagnostic TCD monitoring [15]. Treated patients had higher rates of favorable clinical recovery during TCD monitoring and at 24 h. This initial observation was replicated by other centers in the following years. Eggers et al. performed a randomized clinical trial sonothrombolysis in 25 AIS patients using TCCD. They documented a trend toward higher middle cerebral artery (MCA) recanalization rates and improved 3-month outcome in the sonothrombolysis group despite higher rates of sICH (18 %) [36]. Alexandrov et al. have conducted a phase II, multicenter, double-blind, randomized-controlled clinical trial termed CLOTBUST (The Combined Lysis of Thrombus in Brain Ischemia Using Transcranial Ultrasound and Systemic t-PA). A total of 126 AIS patients treated with IVT were randomized to either continuous TCD or placebo monitoring [37]. All patients had MCA occlusion on TCD and sonothrombolysis led to statistically significantly higher rates of complete recanalization within the 2 h of TCD monitoring (46 % vs. 18 %; $p<0.001$) without increased rates of symptomatic intracranial bleeding (4.8 % in both groups). At 3 months, 22 of 53 patients in the sonothrombolysis group who were available for follow-up and 14 of 49 in the control group (29 %) had favorable functional outcomes ($p=0.20$) as indicated by a score of 0–1 on the modified Rankin scale. A subgroup analysis of CLOTBUST trial in patients with moderate-to-severe AIS (baseline NIHSS-score ≥10 points) indicated that patients randomized to sonothrombolysis achieved higher rates of favorable functional outcome at three months in comparison to patients randomized to IVT (37 % vs. 16 %; $p=0.045$) [38].

The Transcranial Low-Frequency Ultrasound-Mediated Thrombolysis in Brain Ischemia (TRUMBI) was a phase II, prospective, nonrandomized clinical trial that evaluated sonothrombolysis with the use of low-frequency (300 kHz) ultrasound [21]. The trial was halted prematurely because of very high rates of any (93 %) or symptomatic (36 %) hemorrhage in patients treated with sonothrombolysis. Hemorrhages were also highly atypical for IVT in the setting of AIS: there were two parenchymal hemorrhages of the contralateral hemisphere, two subarachnoid and one intraventricular hemorrhage. A number of potential explanations of the excessive sICH rate in the TRUMBI trial have been proposed. For example, low frequency ultrasound may cause blood–brain-barrier disruption and increased permeability [39]. In addition, creation of standing waves outside the targeted region (pressure levels in the brain > inertial cavitation threshold) may lead to remote parenchymal bleeding outside the infarcted area [40, 41].

Further development of sonothrombolysis required testing of a therapeutic 2 MHz pulsed wave device [42] in a phase III RCT, the Combined lysis of thrombus with ultrasound and systemic tissue plasminogen activator for emergent revascularization in acute ischemic stroke (CLOTBUST-ER) [43]. In this study, instead of a TCD or TCCD handheld diagnostic device, an operator-independent head frame was used (Fig. 24.1). This head frame carried a set of sixteen 2 MHz transducers that emitted ultrasound through the conventional windows for TCD (foraminal and both temporal). AIS patients with moderate-to-severe stroke (baseline NIHSS-score ≥ 10 points) treated with IVT were either randomized to undergo continuous ultrasound monitoring or sham monitoring for 2 h. As of March 2015 a total of 676 patients were enrolled and preliminary results were presented in the European Stroke Organization Conference 2016 [44] showing a good safety profile (2.1 % vs. 1.5 % sICH rates in the active and sham monitoring group, respectively) but no clinical benefit in terms of improvement of 3-month functional outcome. Publication of the final analyses is imminent.

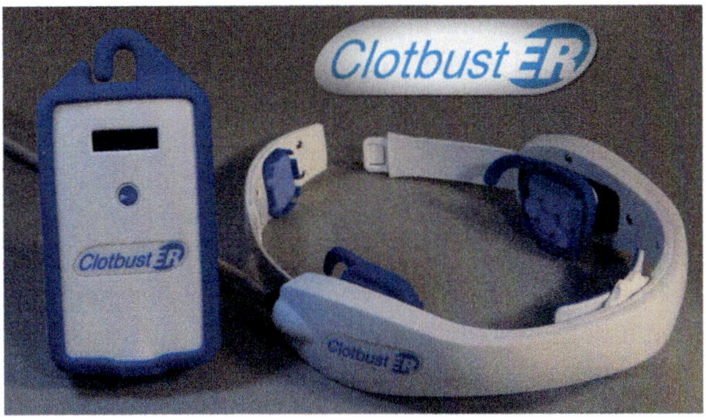

Fig. 24.1 Hands-free sonothrombolysis device tested in CLOTBUSTER trial

3.2 *Microsphere-Potentiated Sonothrombolysis*

Molina et al. first administered three doses of 2.5 g (400 mg/mL) galactose-based μS (Levovist®) that were infused at 2, 20, and 40 min after tPA-bolus on top of 2 h TCD monitoring [45]. A total of 38 patients were enrolled and compared to 73 patients who were previously treated either with ultrasound-enhanced thrombolysis or with intravenous thrombolysis alone. Sustained complete recanalization rate at 2 h after tPA-bolus was significantly higher in the microsphere-potentiated sonothrombolysis subgroup (55 %) compared to sonothrombolysis (41 %) or IVT (24 %) subgroups. No excess in sICH was documented but the study was underpowered to show clinical benefit. The proportion of patients who achieved clinical improvement at 24 h (>3 points in NIHSS score) tended to be higher in the sonothrombolysis group compared with the other two. Larrue et al. used the same ultrasound contrast media and TCCD to deliver sonothrombolysis in a small pilot trial. However, recruitment was halted after including 20 patients because of more frequent detection of hemorrhages on gradient-echo MRI in the treatment group, despite the fact that all these hemorrhages have been asymptomatic [46]. Another pilot study evaluated the safety and efficacy of microsphere-potentiated sonothrombolysis (using TCCD and SonoVue®) in 11 AIS patients and reported no sICH, high recanalization and favorable functional outcome rates [47]. In the only study comparing two different contrast agents, no significant difference has been noted in recanalization rates, clinical outcome, or sICH rates in AIS patients treated with microsphere-potentiated sonothrombolysis [48].

Our research group has performed a pilot randomized study of sonothrombolysis augmentation with perflutren-lipid μS in 12 patients with MCA occlusion and found high rates of complete recanalization at 2 h, significantly higher compared to the IVT control arm of the CLOTBUST trial (50 % versus 18 %) [49]. We have also recorded that μS were found beyond the occlusion in 3 out of 4 patients, in areas with no pretreatment flow. The Transcranial Ultrasound in Clinical Sonothrombolysis (TUCSON) trial was phase IIb RCT that evaluated two different doses of perflutren-lipid μS coupled with TCD-monitoring and IVT in comparison to IVT alone. The trial was stopped due to three cases (27 %) of sICH in the high-dose μS-group, while no sICH was documented in the low-dose or control group [50]. Although patients with sICH had similar screening and pretreatment systolic blood pressure (SBP) levels in comparison to the rest, higher SBP levels were documented in sICH patients at 30, 60, 90 min, and 24–36 h following tPA-bolus, indicating that inadequate blood pressure control may have contributed to the excessive sICH rate that was documented in the high-dose subgroup [50]. Both doses showed a trend toward more efficient recanalization and better outcome [50].

Limited research has been conducted regarding the safety and efficacy of sonothrombolysis in posterior circulation strokes. In a case series of 20 patients with acute (<12 h from symptom onset) basilar artery occlusion, IVT combined with 2-h TCD plus Levovist® led to recanalization in 1 out of 3 patients at 6 h and 1 out of 2 in 24 h [51]. There was a significant difference in NIHSS for early recanalization <6 h (median score 1), recanalization at 6–24 h (median score 11), and no recanalization (median score 30) that corresponded to a significant mortality benefit of recanalizers in 3 months.

Finally, the Norwegian randomized controlled sonothrombolysis in acute stroke study (NOR-SASS) is an arm of the Norwegian Tenecteplase Stroke Trial (NOR-TEST) randomizing patients with AIS to either tenecteplase or alteplase [52]. In NOR-SASS study 30-min continuous IV infusion of SonoVue was used in combination with 60-min diagnostic TCD monitoring in parallel with IVT (either with alteplase drip or tenecteplase bolus). First results are awaited in 2016.

3.3 Sonolysis

Cintas et al. have tested 1-h TCCD monitoring of the occluded MCA in AIS patients that had contraindications for IVT [53]. They recruited 6 patients, 5 of which showed early partial recanalization but only 1 complete recanalization in 24 h. Sonolysis has been evaluated in the prevention of ischemic lesions during carotid interventions in the Sonolysis in Prevention of Brain Infarction During Carotid Endarterectomy and Stenting (SONOBUSTER) [54]. Two-hindered and forty-two patients with internal carotid stenosis >70 % were randomized to either 1-h TCD monitoring or no sonolysis during carotid endarterectomy or stenting. Sonolysis significantly ($p=0.03$) reduced the risk of new ischemic lesions on diffusion-weighted MRI by 55 %. No significant differences on cognitive outcomes or reperfusion hemorrhage rates between the two groups were detected.

3.4 Intraarterial Sonothrombolysis

In a pilot study, 14 AIS patients with anterior or posterior circulation large vessel occlusion received intraarterial alteplase at different doses depending on site, IV heparin and intraarterial sonication with the EKOS catheter. No catheter-related adverse events were noted, two patients developed sICH (14 %) and recanalization rates were over 50 % [24]. The EKOS catheter was also used in the Interventional Management of Stroke II (IMS II) Trial. IMS II trial was an open-label trial of reduced dose IVT (0.6 mg/kg alteplase; 15 % bolus and the rest in 30 min) and intraarterial alteplase administration (<22 mg, bolus and total dose depended on the use of EKOS or not) combined with intravascular sonication with the EKOS catheter during intraarterial alteplase infusion [55]. Complete recanalization was reported in 41 % of AIS patients with anterior circulation large vessel occlusions at 60 min and in 69 % at 2 h [56]. Intraarterial sonothrombolysis with EKOS has been also tested in the IMS III trial, an endovascular AIS treatment trial that was stopped due to futility; 14 patients were treated with EKOS and combined IVT and intraarterial alteplase with 71 % recanalization rates in AIS patients with large vessel occlusion in the arterial circulation [57]. Ribo et al. used an alternative approach by applying rescue intraarterial therapy with alteplase and ultrasound contrast administration with concomitant TCD monitoring after IVT failure in AIS patients. The complete and partial recanalization rate was 78 % at the end of procedure, while sICH occurred in only one patient [58].

The recent success of mechanical thrombectomy trials has marginalized other endovascular techniques such as intraarterial thrombolysis or sonothrombolysis. Two major advantages of thrombectomy over these techniques are the retrieval of thrombus instead of dissolving it, leading to less risk of peripheral embolization and the fact that no thrombolytics are used intraarterially leading to less bleeding risk.

3.5 *Meta-analyses*

Our group has performed a meta-analysis evaluating the safety and efficacy of sonothrombolysis in 2010. A total of nine studies were deemed eligible (5 TCD and 4 TCCD), six of them being randomized and four among them using augmentation with μS [59]. In total, 423 AIS patients with intracranial artery occlusion (documented by ultrasound in all but one study) were included. Sonothrombolysis was found to be associated with nearly threefold increased odds of complete recanalization, mainly driven by the TCD studies. Likewise, sonothrombolysis was associated with higher rates of favorable functional outcome at three months, while no association between sonothrombolysis and risk of sICH was detected. The Cochrane Collaboration published a meta-analysis of randomized controlled trials of sonothrombolysis in 2012 [60]. They included five studies (2 TCD and 3 TCCD; 2 with microsphere-potential sonothrombolysis) including a total of 233 patients. AIS patients randomized to sonothrombolysis were less likely to be dead or disabled at 3 months (although confidence intervals were quite wide) and more likely to recanalize. No safety concerns were raised, with sICH showing no significant difference in the treatment and control arms and no difference in mortality. A third meta-analysis was published in 2013 by Saqqur et al. including ten studies seven of which were randomized. This report confirmed again the safety and efficacy of sonothrombolysis [61]. An overview of the main findings of the three available meta-analyses is presented in Table 24.2.

Table 24.2 Meta-analyses summarizing the safety and efficacy of sonothrombolysis

Meta-analysis (Reference)	Studies (randomized)	Complete recanalization (OR, 95%CI)	sICH (OR, 95%CI)	Good functional outcome (OR, 95%CI)
Tsivgoulis et al., 2010 [59]	9 (6)	Favors ST (2.99; 1, 7–5, 25)	Neutral (1.26; 0.44–3.6)	Favors ST (2.09; 1.17–3.71)[c]
Ricci et al., 2012 [60]	5 (5)	Favors ST (3.57; 2.00–6.25)[a]	Neutral (2.35; 0.95–5.8)[b]	Favors ST (2.00; 1.10–3.70)[c]
Saqqur et al., 2013 [61]	10 (7)	Favors ST (2.95; 1.81–4.81)	Neutral (1.14; 0.56–2.34)	Favors ST (2.2; 1.52–3.19)[e]

sICH symptomatic intracranial hemorrhage, *OR* odds ratio, *CI* Confidence interval, *ST* sonothrombolysis

[a]Outcome was complete or partial recanalization

[b]Symptomatic plus asymptomatic intracranial hemorrhage

[c]Modified Rankin Scale: 0–1

[d]Oxford Handicap Scale ≤2

[e]Modified Rankin Scale: 0–2

4 Pitfalls and Future Directions

It should be acknowledged that despite numerous attempts from different centers around the world, the available sonothrombolysis studies are highly heterogeneous with small sample sizes, using variable treatment protocols and ultrasound devices. It is clear that the vast majority of sonothrombolysis trials required expertise in neurosonology in order to properly identify intracranial occlusion, maintain ultrasound delivery to the site of thrombus and accurately evaluate the final result as complete, partial, or no recanalization [62]. Of note, recanalization has been evaluated using ultrasound criteria in the majority of the aforementioned trials. When compared to digital subtraction angiography in real time, TCD provides an 89 % overall accuracy in detecting complete recanalization and 82 % for partial recanalization, numbers that are far from 100 % [63]. Improved animal models and self-targeting by an ultrasound device at thrombus-residual blood flow interface are needed in order to conduct a future clinical trial with safe and most efficacious ultrasound parameters to potentiate thrombolysis. An ample opportunity for clinical applications of sonothrombolysis exists in the era of mechanical thrombectomy since majority of patients with ELVO are brought to nearest hospitals often lacking endovascular expertize. Therefore, IVT is initiated and patients have to be transferred to a comprehensive stroke center level facility. Even if a patient with ELVO is brought directly to such facility, door-to-groin puncture times often exceed 60 min. Thus, both scenarios leave ample opportunity to apply sonothrombolysis prior to mechanical thrombectomy.

Another promising field is targeted sonothrombolysis: the use of μS as vehicles of antithrombotic molecules to occluded arteries. Sonothrombolysis of the internal carotid occlusion in rats with immunobubbles tagged with the glycoprotein IIb/IIIa receptor antagonist abciximab is feasible [64]. In another animal study, sonothrombolysis with low IV tPA dose and tPA-tagged μS showed equal efficacy with full-dose IVT sonothrombolysis [65]. A recent study has tested μS double tagged with recombinant single-chain urokinase plasminogen activators and thrombus-targeting activated-platelet-specific single-chain antibody [66]. Targeted therapies are a promising therapeutic approach that might combine minimal risk and equal or superior efficacy compared to established treatment modalities.

Ultrasound has provided valuable real-time data on recanalization and reocclusion of intracranial arteries in the acute phase of ischemic stroke but it remains to be answered if it may also improve functional outcomes at 90 days post-tPA treatment. As we are entering a new era of combined intravenous and interventional AIS therapies, it is a challenge for researchers to incorporate sonothrombolysis in the constantly evolving AIS protocols. Proving the efficacy of sonothrombolysis in a well-designed randomized controlled clinical trial of a next-generation noninvasive therapeutic device or a drug–device combination would have substantial impact on future AIS management.

References

1. The National Institute of Neurological Disorders and Stroke rt-PA Stroke Study Group (1995) Tissue plasminogen activator for acute ischemic stroke. N Engl J Med 333(24):1581–1587
2. Hacke W, Kaste M, Bluhmki E, ECASS Investigators. et al (2008) Thrombolysis with alteplase 3 to 4.5 hours after acute ischemic stroke. N Engl J Med 359(13):1317–1329
3. Tsivgoulis G, Safouris A, Alexandrov AV (2015) Safety of intravenous thrombolysis for acute ischemic stroke in specific conditions. Expert Opin Drug Saf 14(6):845–864
4. Demaerschalk BM, Kleindorfer DO, Adeoye OM, Demchuk AM, Fugate JE, Grotta JC, Khalessi AA, Levy EI, Palesch YY, Prabhakaran S, Saposnik G, Saver JL, Smith EE, American Heart Association Stroke Council and Council on Epidemiology and Prevention (2016) Scientific Rationale for the Inclusion and Exclusion Criteria for Intravenous Alteplase in Acute Ischemic Stroke: A Statement for Healthcare Professionals From the American Heart Association/American Stroke Association. Stroke 47(2):581–641
5. Powers WJ, Derdeyn CP, Biller J, American Heart Association Stroke Council (2015) 2015 American Heart Association/American Stroke Association Focused Update of the 2013 Guidelines for the Early Management of Patients With Acute Ischemic Stroke Regarding Endovascular Treatment: A Guideline for Healthcare Professionals From the American Heart Association/American Stroke Association. Stroke 46(10):3020–3035
6. Lisabeth LD, Brown DL, Morgenstern LB (2006) Barriers to intravenous tissue plasminogen activator for acute stroke therapy in women. Gend Med 3(4):270–278
7. Alexandrov AV, Burgin SW, Demchuk AM, El-Mitwalli A, Grotta JC (2001) Speed of intracranial clot lysis with intravenous tissue plasminogen activator therapy: sonographic classification and short-term improvement. Circulation 103(24):2897–2902
8. Hacke W, Donnan G, Fieschi C et al (2004) ATLANTIS Trials Investigators; ECASS Trials Investigators; NINDS rt-PA Study Group Investigators. Association of outcome with early stroke treatment: pooled analysis of ATLANTIS, ECASS, and NINDS rtPA stroke trials. Lancet 363(9411):768–774
9. Rha JH, Saver JL (2007) The impact of recanalization on ischemic stroke outcome: a meta-analysis. Stroke 38(3):967–973

10. Tsivgoulis G, Alexandrov AV, Sloan MA (2009) Advances in transcranial Doppler ultrasonography. Curr Neurol Neurosci Rep 9(1):46–54
11. Saqqur M, Uchino K, Demchuk AM et al (2007) Site of arterial occlusion identified by transcranial Doppler (TCD) predicts the response to intravenous thrombolysis for stroke. Stroke 38(3):948–954
12. Riedel CH, Zimmermann P, Jensen-Kondering U, Stingele R, Deuschl G, Jansen O (2011) The importance of size: successful recanalization by intravenous thrombolysis in acute anterior stroke depends on thrombus length. Stroke 42(6):1775–1777
13. Saqqur M, Molina CA, Salam A, CLOTBUST Investigators et al (2007) Clinical deterioration after intravenous recombinant tissue plasminogen activator treatment: a multicenter transcranial Doppler study. Stroke 38(1):69–74
14. Tsivgoulis G, Katsanos AH, Alexandrov AV (2014) Reperfusion therapies of acute ischemic stroke: potentials and failures. Front Neurol 5:215
15. Alexandrov AV, Demchuk AM, Felberg RA et al (2000) High rate of complete recanalization and dramatic clinical recovery during tPA infusion when continuously monitored with 2-MHz transcranial doppler monitoring. Stroke 31(3):610–614
16. Tsivgoulis G, Safouris A, Katsanos AH, Arthur AS, Alexandrov AV (2016) Mechanical thrombectomy for emergent large vessel occlusion: a critical appraisal of recent randomized controlled clinical trials. Brain Behav 6(2), e00418
17. Tsivgoulis G, Safouris A, Krogias C, Arthur AS, Alexandrov AV (2016) Endovascular reperfusion therapies for acute ischemic stroke: dissecting the evidence. Expert Rev Neurother 16(5):527–534
18. Adeoye O, Albright KC, Carr BG et al (2014) Geographic access to acute stroke care in the United States. Stroke 45(10):3019–3024
19. Jauch EC, Saver JL, Adams HP Jr, American Heart Association Stroke Council, Council on Cardiovascular Nursing, Council on Peripheral Vascular Disease, Council on Clinical Cardiology (2013) Guidelines for the early management of patients with acute ischemic stroke: a guideline for healthcare professionals from the American Heart Association/American Stroke Association. Stroke 44(3):870–947
20. Nelson TR, Fowlkes JB, Abramowicz JS, Church CC (2009) Ultrasound biosafety considerations for the practicing sonographer and sonologist. J Ultrasound Med 28(2):139–150
21. Daffertshofer M, Gass A, Ringleb P et al (2005) Transcranial low-frequency ultrasound-mediated thrombolysis in brain ischemia: increased risk of hemorrhage with combined ultrasound and tissue plasminogen activator: results of a phase II clinical trial. Stroke 36(7): 1441–1446
22. Sonovue summary of product characteristics, European Medicinal Agency. http://www.ema.europa.eu/docs/en_GB/document_library/EPAR_-_Product_Information/human/000303/WC500055380.pdf. Last accessed 18/06/2016
23. Barreto AD, Sharma VK, Lao AY et al (2009) Safety and dose-escalation study design of Transcranial Ultrasound in Clinical SONolysis for acute ischemic stroke: the TUCSON Trial. Int J Stroke 4(1):42–48
24. Mahon BR, Nesbit GM, Barnwell SL et al (2003) North American clinical experience with the EKOS MicroLysUS infusion catheter for the treatment of embolic stroke. Am J Neuroradiol 24(3):534–538
25. Medel R, Crowley RW, McKisic MS, Dumont AS, Kassell NF (2009) Sonothrombolysis: an emerging modality for the management of stroke. Neurosurgery 65(5):979–993
26. Fatar M, Stroick M, Griebe M, Alonso A, Hennerici MG, Daffertshofer M (2006) Brain temperature during 340-kHz pulsed ultrasound insonation: A safety study for sonothrombolysis. Stroke 37(7):1883–1887
27. Datta S, Coussios CC, McAdory LE et al (2006) Correlation of cavitation with ultrasound enhancement of thrombolysis. Ultrasound Med Biol 32(8):1257–1267
28. Braaten JV, Goss RA, Francis CW (1997) Ultrasound reversibly disaggregates fibrin fibers. Thromb Haemost 78(3):1063–1068

29. Dijkmans PA, Juffermans LJ, Musters RJ et al (2004) Microbubbles and ultrasound: from diagnosis to therapy. Eur J Echocardiogr 5(4):245–256
30. Daffertshofer M, Hennerici M (2003) Ultrasound in the treatment of ischaemic stroke. Lancet Neurol 2(5):283–290
31. Sakharov DV, Barrertt-Bergshoeff M, Hekkenberg RT, Rijken DC (1999) Fibrin-specificity of a plasminogen activator affects the efficiency of fibrinolysis and responsiveness to ultrasound: comparison of nine plasminogen activators in vitro. Thromb Haemost 81(4): 605–612
32. Iida K, Luo H, Hagisawa K et al (2006) Noninvasive low-frequency ultrasound energy causes vasodilation in humans. J Am Coll Cardiol 48(3):532–537
33. Tsivgoulis G, Culp WC, Alexandrov AV (2008) Ultrasound enhanced thrombolysis in acute arterial ischemia. Ultrasonics 48(4):303–311
34. Xie F, Tsutsui JM, Lof J et al (2005) Effectiveness of lipid microbubbles and ultrasound in declotting thrombosis. Ultrasound Med Biol 31(7):979–985
35. Alexandrov AV, Barlinn K, Strong R, Alexandrov AW, Aronowski J (2011) Low-power 2-MHz pulsed-wave transcranial ultrasound reduces ischemic brain damage in rats. Transl Stroke Res 2(3):376–381
36. Eggers J, Koch B, Meyer K, König I, Seidel G (2003) Effect of ultrasound on thrombolysis of middle cerebral artery occlusion. Ann Neurol 53(6):797–800
37. Alexandrov AV, Molina CA, Grotta JC, CLOTBUST Investigators et al (2004) Ultrasound-enhanced systemic thrombolysis for acute ischemic stroke. N Engl J Med 351(21):2170–2178
38. Barlinn K, Tsivgoulis G, Barreto AD, Alleman J, Molina CA, Mikulik R, Saqqur M, Demchuk AM, Schellinger PD, Howard G, Alexandrov AV (2014) Outcomes following sonothrombolysis in severe acute ischemic stroke: subgroup analysis of the CLOTBUST trial. Int J Stroke 9(8):1006–1010
39. Reinhard M, Hetzel A, Krüger S et al (2006) Blood-brain barrier disruption by low-frequency ultrasound. Stroke 37(6):1546–1548
40. Baron C, Aubry JF, Tanter M, Meairs S, Fink M (2009) Simulation of intracranial acoustic fields in clinical trials of sonothrombolysis. Ultrasound Med Biol 35(7):1148–1158
41. Alexandrov AV, Barlinn K (2012) Taboos and opportunities in sonothrombolysis for stroke. Int J Hyperthermia 28(4):397–404
42. Saqqur M, Tsivgoulis G, Molina CA, CLOTBUST-PRO Investigators et al (2008) Design of a PROspective multi-national CLOTBUST collaboration on reperfusion therapies for stroke (CLOTBUST-PRO). Int J Stroke 3(1):66–72
43. Schellinger PD, Alexandrov AV, Barreto AD, CLOTBUSTER Investigators et al (2015) Combined lysis of thrombus with ultrasound and systemic tissue plasminogen activator for emergent revascularization in acute ischemic stroke (CLOTBUST-ER): design and methodology of a multinational phase 3 trial. Int J Stroke 10(7):1141–1148
44. European Stroke Organization Conference 2016. Andrei V. Alexandrov, USA. Ultrasound enhanced thrombolysis for ischemic stroke: results of a multi-national Phase III trial (CLOTBUSTER). Webcast http://esoc2016.com/scientific-information-abstracts/webcast#.V2VsdPl97IU. Last accessed 18/06/2016
45. Molina CA, Ribo M, Rubiera M et al (2006) Microbubble administration accelerates clot lysis during continuous 2-MHz ultrasound monitoring in stroke patients treated with intravenous tissue plasminogen activator. Stroke 37(2):425–429
46. Larrue V, Viguier A, Arnaud C et al (2007) Transcranial ultrasound combined with intravenous microbubbles and tissue plasminogen activator for acute ischemic stroke: a randomized controlled study. Stroke 38:472
47. Perren F, Loulidi J, Poglia D, Landis T, Sztajzel R (2008) Microbubble potentiated transcranial duplex ultrasound enhances IV thrombolysis in acute stroke. J Thromb Thrombolysis 25(2):219–223

48. Rubiera M, Ribo M, Delgado-Mederos R et al (2008) Do bubble characteristics affect recanalization in stroke patients treated with microbubble-enhanced sonothrombolysis? Ultrasound Med Biol 34(10):1573–1577
49. Alexandrov AV, Mikulik R, Ribo M et al (2008) A pilot randomized clinical safety study of sonothrombolysis augmentation with ultrasound-activated perflutren-lipid microspheres for acute ischemic stroke. Stroke 39(5):1464–1469
50. Molina CA, Barreto AD, Tsivgoulis G et al (2009) Transcranial ultrasound in clinical sonothrombolysis (TUCSON) trial. Ann Neurol 66(1):28–38
51. Pagola J, Ribo M, Alvarez-Sabín J, Lange M, Rubiera M, Molina CA (2007) Timing of recanalization after microbubble-enhanced intravenous thrombolysis in basilar artery occlusion. Stroke 38(11):2931–2934
52. Nacu A, Kvistad CE, Logallo N et al (2015) A pragmatic approach to sonothrombolysis in acute ischaemic stroke: the Norwegian randomised controlled sonothrombolysis in acute stroke study (NOR-SASS). BMC Neurol 15:110
53. Cintas P, Le Traon AP, Larrue V (2002) High rate of recanalization of middle cerebral artery occlusion during 2-MHz transcranial color-coded Doppler continuous monitoring without thrombolytic drug. Stroke 33(2):626–628
54. Školoudík D, Kuliha M, Hrbáč T, Jonszta T, Herzig R, SONOBUSTER Trial Group (2015) Sonolysis in Prevention of Brain Infarction During Carotid Endarterectomy and Stenting (SONOBUSTER): a randomized, controlled trial. Eur Heart J. pii: ehv492. [Epub ahead of print]
55. IMS, II Trial Investigators (2007) The Interventional Management of Stroke (IMS) II Study. Stroke 38(7):2127–2135
56. Tomsick T, Broderick J, Carrozella J, Interventional Management of Stroke II Investigators (2008) Revascularization results in the Interventional Management of Stroke II trial. AJNR Am J Neuroradiol 29(3):582–587
57. Broderick JP, Palesch YY, Demchuk AM, Interventional Management of Stroke (IMS) III Investigators et al (2013) Endovascular therapy after intravenous t-PA versus t-PA alone for stroke. N Engl J Med 368(10):893–903
58. Ribo M, Molina CA, Alvarez B, Rubiera M, Alvarez-Sabin J, Matas M (2010) Intra-arterial administration of microbubbles and continuous 2-MHz ultrasound insonation to enhance intra-arterial thrombolysis. J Neuroimaging 20(3):224–227
59. Tsivgoulis G, Eggers J, Ribo M et al (2010) Safety and efficacy of ultrasound-enhanced thrombolysis: a comprehensive review and meta-analysis of randomized and nonrandomized studies. Stroke 41(2):280–287
60. Ricci S, Dinia L, Del Sette M et al (2012) Sonothrombolysis for acute ischaemic stroke. Cochrane Database Syst Rev 10, CD008348
61. Saqqur M, Tsivgoulis G, Nicoli F et al (2014) The role of sonolysis and sonothrombolysis in acute ischemic stroke: a systematic review and meta-analysis of randomized controlled trials and case-control studies. J Neuroimaging 24(3):209–220
62. Tsivgoulis G, Alexandrov A (2008) Ultrasound-enhanced thrombolysis: from bedside to bench. Stroke 39(5):1404–1405
63. Tsivgoulis G, Ribo M, Rubiera M et al (2013) Real-time validation of transcranial Doppler criteria in assessing recanalization during intra-arterial procedures for acute ischemic stroke: an international, multicenter study. Stroke 44(2):394–400
64. Alonso A, Dempfle CE, Della Martina A et al (2009) In vivo clot lysis of human thrombus with intravenous abciximab immunobubbles and ultrasound. Thromb Res 124(1):70–74
65. Hua X, Zhou L, Liu P et al (2014) In vivo thrombolysis with targeted microbubbles loading tissue plasminogen activator in a rabbit femoral artery thrombus model. J Thromb Thrombolysis 38(1):57–64
66. Wang X, Gkanatsas Y, Palasubramaniam J et al (2016) Thrombus-targeted theranostic microbubbles: a new technology towards concurrent rapid ultrasound diagnosis and bleeding-free fibrinolytic treatment of thrombosis. Theranostics 6(5):726–738

Chapter 25
Combination Therapy with Thrombolysis

Burak Yulug and Wolf-Rüdiger Schäbitz

Abstract Thrombolysis is the most effective treatment for acute ischemic stroke representing the basic therapeutic standard in the western world. Novel drug treatments are therefore unlikely to be developed as stand-alone approaches and every novel treatment should be considered as combination treatment together with thrombolysis. Such a combination should exert synergistic actions together with t-PA, and probably even more important decrease the potential adverse effects of thrombolysis including reperfusion-related adverse effects such as cytotoxin and free radical release and toxicity as well as BBB degradation and subsequent hemorrhagic complications. Preclinical data show that this approach can make thrombolysis safer and buy time for later treatment initiation. At current, the clinical evidence for this approach is relatively weak, but a systematic translation of preclinical findings into the human situation is ongoing with some recent hopeful examples from clinical trials.

Keywords Neuroprotection • Acute ischemic stroke • Tissue plasminogen activator

Abbreviations

AMPA	α-Amino-3-hydroxy-5-methyl-4-isoxazolepropionic acid
BBB	Blood–brain barrier
CEPO	Carbamylated EPO

B. Yulug
Department of Neurology and Regenerative and Restorative Medical Research Center, Experimental Neurology Laboratory, University of Istanbul-Medipol, Istanbul, Turkey

W.-R. Schäbitz (✉)
Department of Neurology, Bethel-EvKB, University of Münster, Burgsteig 13, Bielefeld 33617, Germany
e-mail: wolf.schaebitz@evkb.de

P.A. Lapchak, J.H. Zhang (eds.), *Neuroprotective Therapy for Stroke and Ischemic Disease*, Springer Series in Translational Stroke Research, DOI 10.1007/978-3-319-45345-3_25

EGCG	Epigallocatechin gallate
eNOS	Endothelial nitric oxide synthase
EPO	Erythropoietin
ER	Endoplasmic reticulum
ERK	Extracellular signal-regulated kinase
FK-506	Tacrolimus
G-CSF	Granulocyte-colony-stimulating factor
ICAM-1	Intercellular adhesion molecule-1
IRAK-1	Interleukin receptor-associated kinase-1
MMPs	Matrix metalloproteinases
NBO	Normobaric hyperoxia
NF-kB	Nuclear factor kappa B
NMDA	N-methyl-D-aspartic acid
NR1-1	NMDA receptor 1
PAR-1	Protease-activated receptor 1
ROCK	Rho kinases
tPA	Tissue plasminogen activator
VEGF	Vascular endothelial growth factor

1 Introduction

Thrombolysis is the most effective treatment for acute ischemic stroke applied within the 4.5 h time window after stroke onset [1]. Due to this short time window and several contraindications (i.e., bleeding, trauma) just a limited number of patients qualify for this type of treatment. Therefore, research focused on alternative opportunities, many of them (i.e., neurotrophic factors, calcium channel blockers, antioxidants, and glutamate receptor blockers) with promising brain protective effects in preclinical studies. Translation into human situation failed due to several reasons including the important factum that therapeutically promising stroke patients normally receive thrombolysis: Every novel therapeutic development will therefore face the situation to be combined with thrombolysis when clinical development is carried to late stages. Ideally, such a combination treatment should exert synergistic actions together with tPA, and probably even more important decrease the potential adverse effects of thrombolysis.

2 The Rationale for a tPA Combination Therapy

Stroke may impact diffusion of drugs to the ischemic target when administered after onset of ischemia due to an impaired body circulation, cerebral metabolism, and critically reduced perfusion of the ischemic penumbra [2–7]. Even candidate drugs with lipophilic property and the ability to cross the blood–brain barrier (BBB) may fail to reach the target region and finally fail to protect the ischemic penumbra [4–7]. Additional factors include a disturbed and increased intracranial pressure as well as induction of secondary brain edema [2–6]. These pathophysiological factors underline the importance of a thrombolytic therapy: Recanalization restores brain perfusion, enabling the candidate drug to circulate to the therapeutic target hereby building the basis for effective brain protection. Previous data show also that tPA has direct detrimental effects such as NMDA (*N*-methyl-D-aspartic acid)-mediated neurotoxicity and direct BBB damage [8, 9]. In addition, minor delays in administration of tPA beyond the therapeutic window enhance and proportionally induce the activation of free radical release with subsequent induction of neuroinflammation and BBB impairment [10–12]. A combination of tissue plasminogen activator (tPA) with drugs counteracting such effects would be a desirable therapeutic target to protect cerebral tissues from reperfusion injury. In the following sections, we summarize prominent preclinical examples of combination approaches with thrombolysis and analyze the translational impact.

3 Drugs with Anti-inflammatory Activity

Minocycline is one of the promising agents already been shown to exert a significant neuroprotective effect in various experimental animal studies [12, 13]. The proposed mechanism of action includes significant inhibition of neuroinflammatory activity, free radical production, and matrix metalloproteinases (i.e., matrix metalloproteinase-9, MMP-9) which finally leads to attenuation of neuronal apoptosis [13, 14]. It has been shown that combined treatment of minocycline at 4 h plus tPA at 6 h after embolic stroke in the rat significantly reduced infarct volumes, brain hemorrhage, and decreased plasma MMP-9 levels in combined treatment groups compared to monotherapy with tPA [14]. Confirming minocycline's previously established multiple neuroprotective mechanisms of action, recent study results also indicated that the inhibitory effect of minocycline on MMP-9 and MMP-2 did not interact with the fibrinolytic activity of tPA [15].

In addition to minocycline, several studies revealed that decreased activation of leukocytes via anti-CD 18 monoclonal antibodies lead to significant benefits in the neurological outcome and prolonged the time window in animals treated with a combined approach compared to tPA alone [16, 85].

FK-506, one of the most widely used immunosupressant agents, has been shown to be neuroprotective in various preclinical studies [17–19]. Its neuroprotective efficacy is attributed to its antiexcitotoxic action involving nitric oxide and calcineurin. A recent study evaluated the combined effect of FK 506 with thrombolysis. In this study, FK-506 proofed to be effective in extending the therapeutic time window for systemic thrombolysis compared to t-PA alone without increasing the risk of hemorrhage [17]. Furthermore, MRI analysis revealed that a combined therapy of FK-506 and t-PA salvaged ischemic lesions directed to infarction compared to animals treated by tPA alone [18].

Highlighting the therapeutic role of a suppression of tPA-induced activation of intercellular adhesion molecule-1 (ICAM-1) and protease-activated receptor 1 (PAR-1) in the pathogenesis of cerebral ischemia, Bowes et al. indicated that combination of alpha-ICAM-1 and tPA synergistically improved neurological outcome. Interestingly, this study revealed that neither alpha-ICAM-1 nor tPA was effective alone [20, 21]. These findings suggest furthermore that a blockage of leukocyte adhesion may extend the therapeutic time window of thrombolytic therapy.

In agreement with this, the combination of tPA with glycoprotein IIb/IIIa (GP IIb/IIIa) receptor inhibitor E revealed significant synergistic effects on neurological outcome improval and decreased tPA-induced intracerebral hemorrhage [22]. Another interesting substance Epigallocatechin gallate (EGCG), the major catechin found in green tea, exerts a neuroprotective efficacy in various animal models, and alleviates potential side effects of delayed tPA treatment such as reperfusion damage including blood–brain barrier disruption and cerebral edema formation. In addition, EGCG exerts its protective functions against delayed tPA activity through upregulation of the plasminogen activator inhibitor-1 and downregulation of matrix metalloproteinases [23]. This indicates that EGCG can be coadministered with tPA to extend the therapeutic window by reducing major side effects.

4 Drugs Targeting Glutamate Toxicity

In addition to the prominent role in stroke pathophysiology, NMDA receptor signaling has been identified to play a key role in tPA-associated neurotoxicity. Specifically, the aminoterminal domain of the NR1–1a, a subunit of the NMDA receptor, mediates tPA-induced neurotoxicity during cerebral ischemia [8, 9, 24]. Based on these data a polyclonal antibody against the interaction site of tPA on the NR1 subunit of NMDA receptor has been shown to prevent in vitro and in vivo proexcitotoxic effects of tPA without altering NMDA-induced neurotransmission associated with improval of long-term neurological outcome in mice [25]. These findings together suggest that immunotherapeutic approaches targeting this subunit may not only protect the brain against stroke but also antagonize tPA-induced toxicity after stroke and finally help to extend the therapeutic window. Recent work demonstrated that blocking glutamate receptors via MK-801 and Dizocilpine in combination with thrombolysis exerted significant attenuation of neurological damage compared to t-PA treatment alone [26, 85, 86]. These results were replicated by additional studies showing that the blockage of excitotoxic aminoacid efflux and NMDA antagonism via elidopril when combined with tPA augmented the benefit of thrombolysis in a rat embolic stroke model [27, 83]. In addition, Suzuki et al. coadministered the AMPA receptor antagonist YM872 in an embolic stroke model in the rat and demonstrated that just the combination of YM872 and tPA exerted significant neuroprotective effects [28, 85].

5 Drugs with Free Radical Scavenging Activity

5.1 *The Role of Uric Acid and tPA Combination in Cerebral Ischemia*

Thus far, several studies confirmed the brain protective effect of uric acid in cerebral ischemia [29]. Uric acid in combination with tPA synergistically reduced infarct volume compared to each single treatment alone. Hemorrhagic transformation rates in monotherapy versus combination therapy groups revealed that intracranial hemorrhages were additively decreased in the combination therapy group compared to uric acid and tPA treatment alone after thromboembolic middle cerebral artery occlusion [30].

5.2 *The Role of 17B-Estradiol and tPA Combination in Cerebral Ischemia*

A clear benefit of the combination therapy was confirmed by another study revealing that combination of 17B-estradiol and tPA reduced infarct volumes via inhibiting neuronal apoptosis through multiple signaling pathways in ovariectomized female rats after ischemia/reperfusion [31] This study also suggested that 17B-estradiol prevented tPA-induced brain injury by decreasing MMP-9 expression, brain edema, and hemorrhage formation [32].

5.3 *The Role of Edaravone and Melatonin in Combination with tPA*

Further studies also indicate that the free radical scavenger edaravone combined with tPA inhibited extravasation of tPA from brain vessels, blocked MMP-9 upregulation and free radical release leading to prevention of BBB leakage and a subsequent prevention of intracranial hemorrhages [33, 34]. In addition to such classic free radical scavengers, melatonin, a well-known natural antioxidant, has been recently shown to inhibit oxidative injury at the neurovascular unit, to decrease MMP-9 activity and to prevent BBB leakage. Similar effects were observed when melatonin and tPA were applied at 6 h after the onset of focal ischemia and reperfusion [35, 36].

5.4 *The Role of tPA Receptor Blockage in Cerebral Ischemia*

Recent data also show that blockage of the tPA receptor via annexin A2 not only improves thrombolytic efficacy of tPA but more importantly reduced the tPA-induced hemorrhagic transformation [37, 38]. Additive or synergistic combination therapies together with thrombolysis may help to reduce the dose-related adverse effects of tPA, resulting in improved drug safety and potentially widen the therapeutic time window.

6 Drugs Targeting RHO Kinases

Rho kinases (ROCK) have a wide variety of action including the regulation of stress fiber formation, smooth muscle contraction, as well as the cell migration [39–45]. Recent studies suggested that ROCK plays a significant role in cerebral vasospasm, vascular inflammation, hypertension, and arteriosclerosis [39–45]. Fasudil, a Rho kinase inhibitor, was shown to exert important functions in ischemic stroke pathophysiology [46, 47]. A recent study demonstrated that the combination therapy of tPA with fasudil prevented the development of hemorrhagic transformation and significantly reduced mortality and increased locomotor activity seven days after reperfusion [46]. The underlying mechanism of action of the combination therapy includes prevention of injury in human brain endothelial cells by decreasing MMP-9 activity. These findings indicate that fasudil prevents not only the hemorrhagic transformation in mice treated with tPA but also helps to extend the time window of thrombolytic therapy by blocking the increased activity of MMP-9 in endothelial cells.

7 Proteasome Inhibitors

Drugs targeting the ubiquitin–proteasome pathways may not only exert significant brain protective effects after cerebral ischemia but rather represent interesting candidates to extend the therapeutic time window for tPA. Proteasome inhibitors such as Bortezomib counteract poststroke neuroinflammation and upregulate endothelial nitric oxide synthase and antioxidative enzyme expression [48, 49]. Moreover, Bortezomib monotherapy was shown to provide significant neuroprotection in experimental stroke and extend the therapeutic time window when combined with tPA [48, 49]. Similarly, Velcade, another selective proteasome inhibitor, proofed to exert synergistic neuroprotective effects when combined with tPA [50]. Treatment with velcade alone or in combination with tPA was demonstrated to maintain BBB integrity by reducing MMP-9 and increasing eNOS activity, finally resulting in significantly reduced lesion volumes and neurological deficits [50, 83].

8 Growth Factors

Several studies have documented the neurorestorative role of hematopoietic growth factors. The most prominent examples Erythropoietin (EPO), carbamylated EPO (CEPO), and granulocyte-colony-stimulating factor (G-CSF) were demonstrated in numerous studies to reduce infarct volumes and improve neurological outcome after experimental stroke [51–54]. Detailed investigation of the mechanism of action revealed that these agents prevent the expression of neuronal and glial proinflammatory cytokines via signaling pathways which may in turn result in antiapoptotic and anti-inflammatory effects [51–54]. Indeed, EPO and CEPO and their combination decreased infarct volumes and improved neurological function in a rat stroke model [52]. Despite these promising data, EPO has been shown to aggravate tPA-induced brain hemorrhage without reduction of ischemic brain damage when administered 6 h after stroke in a rat model of embolic stroke. Deleterious effects of EPO were attributed to upregulation of MMP9, NF-kB, and IRAK-1 [55, 56].

Vascular endothelial growth factor (VEGF) is another interesting growth factor demonstrated to promote angiogenesis, neurogenesis, and to improve outcome after subacute ischemia in the rat suture occlusion model [57, 58, 60]. Moreover, VEGF prevented reperfusion injury and reduced neuronal damage by suppressing the extracellular signal-regulated kinase (ERK) signaling and the endoplasmic reticulum (ER) stress pathway [12, 59–61]. Further studies showed that intracerebroventricularly administered VEGF for short periods reduced not only infarct volumes but also neurological outcome at 24 h after non-tPA-related reperfusion [62] Importantly, a recent study showed that blockage of the VEGF pathway attenuated tPA-related hemorrhage after focal cerebral ischemia [58]. These data suggest that VEGF-induced angiogenesis may play a significant role in hemorrhagic transformation processes during reperfusion injury. VEGF may be considered to have biphasic roles in ischemic stroke pathophysiology and the inhibition of the VEGF signaling pathway may be a promising therapeutic strategy for attenuating hemorrhagic transformation after tPA treatment.

9 Nonpharmacological Approaches

A recent study demonstrated that the combination of normobaric hyperoxia (NBO) and delayed applied tPA significantly reduced reperfusion-related BBB damage and subsequent brain edema resulting in improved neurological function and mortality suggesting a role of NBO as an effective therapy to extend the time window of thrombolysis after ischemic stroke [63]. Another interesting candidate is xenon, a well-tolerable anesthetic gas with neuroprotective activity after stroke. A recent study tested the efficacy of a combination of xenon and tPA in experimental cerebral ischemia. Postreperfusion administration of xenon not only exerted significant neuroprotective effects but also prevented tPA-induced BBB disruption and subsequent hemorrhage [64, 65].

The importance of combining nonpharmacological approaches with tPA has been confirmed by several studies demonstrating that hypothermia in combination with thrombolysis reduced infarct volume, brain edema, and improved functional neurological outcome in animals treated with the combination approach compared to each treatment alone [65–67]. Based on such encouraging preclinical results, the therapeutic hypothermia was tested in humans clinically by Kammersgaard and colleagues who revealed that hypothermia was not associated with poor outcome, death, or an increased incidence of infectious complications [68]. These pilot studies led to larger clinical trials with different cooling methods, timing, and target temperature, confirmed feasibility of combination approach but failed to show any sign for a better clinical outcome in hypothermia-treated patients [69].

10 Translational Aspects

As summarized earlier, preclinical evidence suggests that combination therapies on top of tPA might be an attractive therapeutic target [81]. Drugs combined with thrombolysis counteract reperfusion-related adverse effects such as cytotoxin and free radical release and toxicity as well as BBB degradation and subsequent hemorrhagic complications. Combination treatment consistently resulted in a synergistically improved neurological outcome and protection from further infarction. Preclinical data show that this approach can make thrombolysis safer but also buy time for later treatment initiation. This conclusion is indeed supported by a recent systematic analysis of 63 combination studies suggesting an extension of the therapeutic time window from 4 h up to 8 h poststroke [70]. Not surprisingly, the greatest effect of a combination treatment may be achieved when the add-on therapy is applied before start of i.v. thrombolysis [70]. One would expect now to see this experimental evidence integrated into further clinical development when novel stroke trials are planned—particularly since thrombolysis represents standard of care in every fourth stroke patient in larger centers [71, 80, 82, 84]. However, systematic analysis and modeling of clinical combination trials represent an unexplored terrain with uncertainty about groups sizes, endpoints, and treatment effects.

In the German EPO trial, for example, the neuroprotective effect of EPO was evaluated in 460 patients. The majority (about 60 %) of all patients indeed received more or less unexpected a combined treatment with systemic thrombolysis which was not backed by experimental data [72]. The study failed to show any clinical benefit on the primary endpoint as well as on all other outcome measures independent if patients were combined treated (with tPA) or EPO treated alone. Unexpectedly, EPO cotreatment with tPA caused a significantly increased mortality rate. Deaths were attributed to an increased rate of intracranial hemorrhages in the combination group. These findings raise not just simple safety concerns for EPO in human stroke, but rather point toward the importance of a detailed testing of combination therapies in animal studies before combination approaches are translated from the bench into humans. Interestingly, a preclinical study by Zechariah et al. published after the results of the EPO trial were communicated clearly demonstrated that a combination of EPO and tPA induced disaggregation of the BBB permeability and the extracellular matrix and subsequently lead to an increased rate of hemorrhages in animals treated with the combination approach [56].

Another growth factor (G-CSF) also run due to recruitment issues into a combination design which was not the primary intention. Initial evidence suggested reasonable safety and feasibility for a 72-h intravenous application of G-CSF in acute ischemic stroke patients [73]. The larger AXIS II trial failed to exert significant clinical benefits in patients treated with G-CSF. Although patients cotreated with tPA were not included in the primary analysis to avoid a dilutional effect in the original analysis, separate subgroup analyses of the combination treatment groups fortunately revealed no significant adverse effects of G-CSF on clinical outcome, mortality, and hemorrhage rate [74].

The first truly as combination study planned trial is URICO-ICTUS where uric acid was combined to thrombolysis within 4.5 h of symptom onset. This study reported no significant differences between groups with respect to serious adverse effects and mortality rates while the addition of uric acid to thrombolysis did not exert a significant benefit in the primary outcome [75, 76]. However, it should be noted that the combination of uric acid and tPA showed a trend in the primary outcome (mRS 0-1) and promising results in several secondary outcomes measures.

The newest example of a combination trial is the treatment of Natalizumab, a humanized monoclonal antibody preventing leukocyte migration across the BBB via α4 Integrin, in acute stroke patients [77, 78]. In this study 161 patients received either intravenous natalizumab or placebo within 9 h of stroke onset on top of standard therapy which was represented in this population in a 75 % thrombolysis rate. Although there was no effect on the primary endpoint (infarct growth), the overall analysis including thrombolysed patients revealed positive signals on several secondary endpoints such as the Barthel Index and Stroke Impact Scale at 90 days [79]. Although combination treatment in this approach was not modeled preclinically, this study revealed no serious adverse effects for the combination therapy such as an increased hemorrhage or mortality rate.

11 Conclusion

The rationale of a combination therapy, thrombolysis plus other drug or nonpharmacological approach, is backed by ample preclincal evidence suggesting an enhancement of thrombolysis and a prevention of its potential negative effects. Systematic translation of these findings into the human situation is ongoing with some recent hopeful examples from clinical trials where combination treatments such as uric acid or natalizumab were administered with some favorable signs and importantly no adverse effects. The latter one should be considered before preclinical studies are translated into humans.

References

1. Ahmed N, Wahlgren N, Grond M, Hennerici M, Lees KR, Mikulk R, Parsons M, Roine RO, Toni D, Ringleb P, SITS Investigators (2010) Implementation and outcome of thrombolysis with alteplase 3-4.5 h after an acute stroke: an updated analysis from SITS-ISTR. Lancet Neurol 9:866–874
2. Menzies SA, Betz AL, Hoff JT (1993) Contributions of ions and albumin to the formation and resolution of ischemic brain edema. J Neurosurg 78:257–266
3. Wu SK, Yang MT, Kang KH, Liou HC, Lu DH, Fu WM et al (2014) Targeted delivery of erythropoietin by transcranial focused ultrasound for neuroprotection against ischemia/reperfusion-induced neuronal injury: a long-term and short-term study. PLoS One 9(2):e90107
4. Pardridge WM (2002) Drug and gene delivery to the brain: the vascular route. Neuron 36:555–558
5. Begley DJ (2004) Delivery of therapeutic agents to the central nervous system: the problems and the possibilities. Pharmacol Ther 104:29–45
6. Kilic E, Spudich A, Kilic U, Rentsch KM, Vig R, Matter CM et al (2008) ABCC1: a gateway for pharmacological compounds to the ischaemic brain. Brain 131:2679–2689
7. Jaffer H, Morris VB, Stewart D, Labhasetwar V (2011) Advances in stroke therapy. Drug Deliv Transl Res 1(6):409–419
8. Baron A, Montagne A, Casse F et al (2010) NR2D-containing NMDA receptors mediate tissue plasminogen activator-promoted neuronal excitotoxicity. Cell Death Differ 17:860–871
9. Nicole O, Docagne F, Ali C et al (2001) The proteolytic activity of tissue-plasminogen activator enhances NMDA receptor-mediated signaling. Nat Med 7:59–64
10. Wang X, Tsuji K, Lee SR et al (2004) Mechanisms of hemorrhagic transformation after tissue plasminogen activator reperfusion therapy for ischemic stroke. Stroke 35:2726–2730
11. Jickling GC, Liu D, Stamova B et al (2014) Hemorrhagic transformation after ischemic stroke in animals and humans. J Cereb Blood Flow Metab 34(2):185–199
12. Wang Y, Li Q, Wang J, Zhuang QK, Zhang YY (2015) Combination of thrombolytic therapy and neuroprotective therapy in acute ischemic stroke: is it important? Eur Rev Med Pharmacol Sci 19(3):416–422
13. Pan H, Pan LH, Zhang YY (2011) Neuroprotective effect of minocycline after stroke and its mechanism. J Shenyang Pharm Univ 28:325–329
14. Murata Y, Rosell A, Scannevin RH et al (2008) Extension of the thrombolytic time window with minocycline in experimental stroke. Stroke 39:3372–3377
15. Machado LS, Sazonova IY, Kozak A, Wiley DC, El-Remessy AB, Ergul A, Hess DC, Waller JL, Fagan SC (2009) Minocycline and tissue-type plasminogen activator for stroke: assessment of interaction potential. Stroke 40(9):3028–3033
16. Zhang L, Zhang ZG, Zhang RL, Lu M, Krams M, Chopp M (2003) Effects of a selective CD11b/CD18 antagonist and recombinant human tissue plasminogen activator treatment alone and in combination in a rat embolic model of stroke. Stroke 34(7):1790–1795
17. Maeda M, Furuichi Y, Ueyama N, Moriguchi A, Satoh N, Matsuoka N, Goto T, Yanagihara T (2002) A combined treatment with tacrolimus (FK506) and recombinant tissue plasminogen activator for thrombotic focal cerebral ischemia in rats: increased neuroprotective efficacy and extended therapeutic time window. J Cereb Blood Flow Metab 22(10):1205–1211
18. Okubo S, Igarashi H, Kanamatsu T, Hasegawa D, Orima H, Katayama Y (2007) FK-506 extended the therapeutic time window for thrombolysis without increasing the risk of hemorrhagic transformation in an embolic rat stroke model. Brain Res 1143:221–227
19. Yousuf S, Atif F, Kesherwani V, Agrawal SK (2011) Neuroprotective effects of Tacrolimus (FK-506) and Cyclosporin (CsA) in oxidative injury. Brain Behav 1(2):87–94
20. Bowes MP, Rothlein R, Fagan SC, Zivin JA (1995) Monoclonal antibodies preventing leukocyte activation reduce experimental neurologic injury and enhance efficacy of thrombolytic therapy. Neurology 45:815–819

21. Bowes MP, Zivin JA, Rothlein R (1993) Monoclonal antibody to the ICAM-1 adhesion site reduces neurological damage in a rabbit cerebral embolism stroke model. Exp Neurol 119(2):215–219
22. Zhang L, Zhang ZG, Zhang R, Morris D, Lu M, Coller BS, Chopp M (2003) Adjuvant treatment with a glycoprotein IIb/IIIa receptor inhibitor increases the therapeutic window for low-dose tissue plasminogen activator administration in a rat model of embolic stroke. Circulation 107(22):2837–2843
23. You YP (2016) Epigallocatechin gallate extends the therapeutic window of recombinant tissue plasminogen activator treatment in ischemic rats. J Stroke Cerebrovasc Dis 25(4):990–997
24. Macrez R, Bezin L, Mauff BL et al (2010) Functional occurrence of the interaction of tissue plasminogen activator with the NR1 subunit of N-methyl-D-aspartate receptors during stroke. Stroke 41:2950–2955
25. Macrez R, Obiang P, Gauberti M, Roussel B, Baron A, Parcq J, Cassé F, Hommet Y, Orset C, Agin V, Bezin L, Berrocoso TG, Petersen KU, Montaner J, Maubert E, Vivien D, Ali C (2011) Antibodies preventing the interaction of tissue-type plasminogen activator with N-methyl-D-aspartate receptors reduce stroke damages and extend the therapeutic window of thrombolysis. Stroke 42(8):2315–2322
26. Zivin JA, Mazzarella V (1991) Tissue plasminogen activator plus glutamate antagonist improves outcome after embolic stroke. Arch Neurol 48:1235–1238
27. Lekieffre D, Benavides J, Scatton B, Nowicki JP (1997) Neuroprotection afforded by a combination of eliprodil and a thrombolytic agent, rt-PA, in a rat thromboembolic stroke model. Brain Res 776(1-2):88–95
28. Suzuki M, Sasamata M, Miyata K (2003) Neuroprotective effects of YM872 coadministered with t-PA in a rat embolic stroke model. Brain Res 959:169–172
29. Yu ZF, Bruce-Keller AJ, Goodman Y, Mattson MP (1998) Uric acid protects neurons against excitotoxic and metabolic insults in cell culture, and against focal ischemic brain injury in vivo. J Neurosci Res 53:613–625
30. Romanos E, Planas AM, Amaro S, Chamorro A (2007) Uric acid reduces brain damage and improves the benefits of rt-PA in a rat model of thromboembolic stroke. J Cereb Blood Flow Metab 27(1):14–20
31. Liu R, Liu Q, He S, Simpkins JW, Yang SH (2010) Combination therapy of 17beta-estradiol and recombinant tissue plasminogen activator for experimental ischemic stroke. J Pharmacol Exp Ther 332(3):1006–1012
32. Li M, Zhang Z, Sun W, Koehler RC, Huang J (2011) 17β-estradiol attenuates breakdown of blood-brain barrier and hemorrhagic transformation induced by tissue plasminogen activator in cerebral ischemia. Neurobiol Dis 44(3):277–283
33. Zhang WR, Sato K, Hayashi T, Omori N, Nagano I, Kato S, Horiuchi S, Abe K (2004) Extension of ischemic therapeutic time window by a free radical scavenger, Edaravone, reperfused with tPA in rat brain. Neurol Res 26:342–348
34. Yagi K, Kitazato KT, Uno M, Tada Y, Kinouchi T, Shi-Mada K, Nagahiro S (2009) Edaravone, a free radical scavenger, inhibits MMP-9-related brain hemorrhage in rats treated with tissue plasminogen activator. Stroke 40:626–631
35. Hung YC, Chen TY, Lee EJ, Chen WL, Huang SY, Lee WT, Lee MY, Chen HY, Wu TS (2008) Melatonin decreases matrix metalloproteinase-9 activation and expression and attenuates reperfusion-induced hemorrhage following transient focal cerebral ischemia in rats. J Pineal Res 45(4):459–467
36. Chen HY, Chen TY, Lee MY, Chen ST, Hsu YS, Kuo YL, Chang GL, Wu TS, Lee EJ (2006) Melatonin decreases neurovascular oxidative/nitrosative damage and protects against early increases in the blood-brain barrier permeability after transient focal cerebral ischemia in mice. J Pineal Res 41(2):175–182
37. Jiang Y, Fan X, Yu Z, Cheng C, Wang XS, Lo EH, Sun X, Wang X (2015) Low dose tPA plus annexin A2 combination attenuates tPA delayed treatment-associated hemorrhage and improves recovery in rat embolic focal stroke. Neurosci Lett 602:73–78

38. Jiang Y, Fan X, Yu Z, Liao Z, Wang XS, van Leyen K, Sun X, Lo EH, Wang X (2015) Combination Low-dose tissue-type plasminogen activator plus annexin A2 for improving thrombolytic stroke therapy. Front Cell Neurosci 9:397
39. Rikitake Y, Kim H-H, Huang Z et al (2005) Inhibition of Rho kinase (ROCK) leads to increased cerebral blood flow and stroke protection. Stroke 36(10):2251–2257
40. Amano M, Fukata Y, Kaibuchi K (2000) Regulation and functions of Rho-associated kinase. Exp Cell Res 261:44–51
41. Sato M, Tani E, Fujikawa H, Kaibuchi K (2000) Involvement of ROCK-mediated phosphorylation of myosin light chain in enhancement of cerebral vasospasm. Circ Res 87:195–200
42. Masumoto A, Mohri M, Shimokawa H, Urakami L, Usui M, Takeshita A (2002) Suppression of coronary artery spasm by the ROCK inhibitor fasudil in patients with vasospastic angina. Circulation 105:1545–1547
43. Masumoto A, Hirooka Y, Shimokawa H, Hironaga K, Setoguchi S, Takeshita A (2001) Possible involvement of ROCK in the pathogenesis of hypertension in humans. Hypertension 38:1307–1310
44. Kataoka C, Egashira K, Inoue S, Takemoto M, Ni W, Koyanagi M, Kitamoto S, Usui M, Kaibuchi K, Shimokawa H, Takeshita A (2002) Important role of ROCK in the pathogenesis of cardiovascular inflammation and remodeling induced by long-term blockade of nitric oxide synthesis in rats. Hypertension 39:245–250
45. Miyata K, Shimokawa H, Kandabashi T, Higo T, Morishige K, Eto Y, Egashira K, Kaibuchi K, Takeshita A (2000) ROCK is involved in macrophage-mediated formation of coronary vascular lesions in pigs in vivo. Arterioscler Thromb Vasc Biol 20:2351–2358
46. Ishiguro M, Kawasaki K, Suzuki Y, Ishizuka F, Mishiro K, Egashira Y, Ikegaki I, Tsuruma K, Shimazawa M, Yoshimura S, Iwama T, Hara H (2012) A Rho kinase (ROCK) inhibitor, fasudil, prevents matrix metalloproteinase-9-related hemorrhagic transformation in mice treated with tissue plasminogen activator. Neuroscience 220:302–312
47. Fukuta T, Asai T, Sato A, Namba M, Yanagida Y, Kikuchi T, Koide H, Shimizu K, Oku N (2016) Neuroprotection against cerebral ischemia/reperfusion injury by intravenous administration of liposomal fasudil. Int J Pharm 506(1-2):129–137
48. Zhang L, Zhang ZG, Liı X et al (2006) Treatment of embolic stroke in rats with bortezomib and recombinant human tissue plasminogen activator. Thromb Haemost 95:166–173
49. Zhang L, Zhang ZG, Zhang RL et al (2001) Postischemic (6-hour) treatment with recombinant human tissue plasminogen activator and proteasome inhibitor PS-519 reduces infarction in a rat model of embolic focal cerebral ischemia. Stroke 32:2926–2931
50. Zhang L, Zhang ZG, Buller B, Jiang J, Jiang Y, Zhao D, Liu X, Morris D, Chopp M (2010) Combination treatment with VELCADE and low-dose tissue plasminogen activator provides potent neuroprotection in aged rats after embolic focal ischemia. Stroke 41(5):1001–1007
51. Aluclu MU, Acar A, Guzel A et al (2007) Evaluation of erythropoietin effects on cerebral ischemia in rats. Neuroendocrinol Lett 28:170–174
52. Wang Y, Zhang ZG, Rhodes K et al (2007) Postischemic treatment with erythropoietin or carbamylated erythropoietin reduces infarction and improves neurological outcome in a rat model of focal cerebral ischemia. Br J Pharmacol 151:1377–1384
53. Minnerup J, Heidrich J, Wellmann J, Rogalewski A, Schneider A, Schäbitz WR (2008) Meta-analysis of the efficacy of granulocyte-colony stimulating factor in animal models of focal cerebral ischemia. Stroke 39(6):1855–1861
54. England TJ, Gibson CL, Bath PM (2009) Granulocyte-colony stimulating factor in experimental stroke and its effects on infarct size and functional outcome: A systematic review. Brain Res Rev 62(1):71–82
55. Jia L, Chopp M, Zhang L, Lu M, Zhang Z (2010) Erythropoietin in combination of tissue plasminogen activator exacerbates brain hemorrhage when treatment is initiated 6 hours after stroke. Stroke 41(9):2071–2076
56. Zechariah A, ElAli A, Hermann DM (2010) Combination of tissue-plasminogen activator with erythropoietin induces blood-brain barrier permeability, extracellular matrix disaggregation, and DNA fragmentation after focal cerebral ischemia in mice. Stroke 41(5):1008–1012

57. Sun Y, Jin K, Xie L, Childs J, Mao XO, Logvinova A, Greenberg DA (2003) VEGF-induced neuroprotection, neurogenesis, and angiogenesis after focal cerebral ischemia. J Clin Invest 111:1843–1851
58. Kanazawa M, Igarashi H, Kawamura K et al (2011) Inhibition of VEGF signaling pathway attenuates hemorrhage after tPA treatment. J Cereb Blood Flow Metab 31(6):1461–1474
59. Yang JP, Liu XF, Liu HJ, Xu GL, Ma YP (2008) Extracellular signal-regulated kinase involved in NGF/VEGF-induced neuroprotective effect. Neurosci Lett 434(2):212–217
60. Yang J, Guo L, Liu R, Liu H (2012) Neuroprotective effects of VEGF administration after focal cerebral ischemia/reperfusion: dose response and time window. Neurochem Int 60(6):592–596
61. Narasimhan P, Liu J, Song YS, Massengale JL, Chan PH (2009) VEGF Stimulates the ERK 1/2 signaling pathway and apoptosis in cerebral endothelial cells after ischemic conditions. Stroke 40(4):1467–1473
62. Kaya D, Gürsoy-Ozdemir Y, Yemisci M, Tuncer N, Aktan S, Dalkara T (2005) VEGF protects brain against focal ischemia without increasing blood--brain permeability when administered intracerebroventricularly. J Cereb Blood Flow Metab 25(9):1111–8, Erratum in: J Cereb Blood Flow Metab. 2006;26(3):447
63. Liang J, Qi Z, Liu W, Wang P, Shi W, Dong W, Ji X, Luo Y, Liu KJ (2015) Normobaric hyperoxia slows blood-brain barrier damage and expands the therapeutic time window for tissue-type plasminogen activator treatment in cerebral ischemia. Stroke 46(5):1344–1351
64. David HN, Haelewyn B, Risso JJ, Colloc'h N, Abraini JH (2010) Xenon is an inhibitor of tissue-plasminogen activator: adverse and beneficial effects in a rat model of thromboembolic stroke. J Cereb Blood Flow Metab 30(4):718–728
65. Goossens J, Hachimi-Idrissi S (2014) Combination of therapeutic hypothermia and other neuroprotective strategies after an ischemic cerebral insult. Curr Neuropharmacol 12(5):399–412
66. Kallmunzer B, Schwab S, Kollmar R (2012) Mild hypothermia of 34 degrees C reduces side effects of rt-PA treatment after thromboembolic stroke in rats. Exp Transl Stroke Med 4(1):3
67. Liu LP, Tang XN, Yenari MA (2006) Mild hypothermia reduces hemorrhage and blood brain barrier (BBB) disruption following tissue plasminogen activator treatment in experimental stroke. Neurology 66(5):A242–A243
68. Kammersgaard LP, Rasmussen BH, Jørgensen HS, Reith J, Weber U, Olsen TS (2000) Feasibility and safety of inducing modest hypothermia in awake patients with acute stroke through surface cooling: A case-control study: the Copenhagen Stroke Study. Stroke 31(9):2251–2256
69. Groysman LI, Emanuel BA, Kim-Tenser MA, Sung GY, Mack WJ (2011) Therapeutic hypothermia in acute ischemic stroke. Neurosurg Focus 30(6), E17
70. O'Collins VE, Macleod MR, Donnan GA, Howells DW (2012) Evaluation of combination therapy in animal models of cerebral ischemia. J Cereb Blood Flow Metab 32(4):585–597
71. Asadi H, Williams D, Thornton J (2016) Changing management of acute ischaemic stroke: the new treatments and emerging role of endovascular therapy. Curr Treat Options Neurol 18(5):20
72. Ehrenreich H, Weissenborn K, Prange H, Schneider D, Weimar C, Wartenberg K, Schellinger PD, Bohn M, Becker H, Wegrzyn M, Jähnig P, Herrmann M, Knauth M, Bähr M, Heide W, Wagner A, Schwab S, Reichmann H, Schwendemann G, Dengler R, Kastrup A, Bartels C, EPO Stroke Trial Group (2009) Recombinant human erythropoietin in the treatment of acute ischemic stroke. Stroke 40(12):e647–e655
73. Schäbitz WR, Laage R, Vogt G, Koch W, Kollmar R, Schwab S et al (2010) AXIS: a trial of intravenous granulocyte colony-stimulating factor in acute ischemic stroke. Stroke 41:2545–2551
74. Ringelstein EB, Thijs V, Norrving B, Chamorro A, Aichner F, Grond M, Saver J, Laage R, Schneider A, Rathgeb F, Vogt G, Charissé G, Fiebach JB, Schwab S, Schäbitz WR, Kollmar R, Fisher M, Brozman M, Skoloudik D, Gruber F, Serena Leal J, Veltkamp R, Köhrmann M, Berrouschot J, AXIS 2 Investigators (2013) Granulocyte colony-stimulating factor in patients with acute ischemic stroke: results of the AX200 for Ischemic Stroke trial. Stroke 44(10):2681–2687

75. Chamorro A, Amaro S, Castellanos M, Segura T, Arenillas J, Martí-Fábregas J, Gállego J, Krupinski J, Gomis M, Cánovas D, Carné X, Deulofeu R, Román LS, Oleaga L, Torres F, Planas AM, URICO-ICTUS Investigators (2014) Safety and efficacy of uric acid in patients with acute stroke (URICO-ICTUS): a randomised, double-blind phase 2b/3 trial. Lancet Neurol 13(5):453–460
76. Amaro S, Cánovas D, Castellanos M, Gállego J, Martí-Fèbregas J, Segura T, Chamorro A (2010) The URICO-ICTUS study, a phase 3 study of combined treatment with uric acid and rtPA administered intravenously in acute ischaemic stroke patients within the first 4.5 h of onset of symptoms. Int J Stroke 5(4):325–328
77. Llovera G, Hofmann K, Roth S, Salas-Pérdomo A, Ferrer-Ferrer M, Perego C, Zanier ER, Mamrak U, Rex A, Party H, Agin V, Fauchon C, Orset C, Haelewyn B, De Simoni MG, Dirnagl U, Grittner U, Planas AM, Plesnila N, Vivien D, Liesz A (2015) Results of a preclinical randomized controlled multicenter trial (pRCT): Anti-CD49d treatment for acute brain ischemia. Sci Transl Med 7(299):299ra121
78. Polman CH, O'Connor PW, Havrdova E, Hutchinson M, Kappos L, Miller DH, Phillips JT, Lublin FD, Giovannoni G, Wajgt A, Toal M, Lynn F, Panzara MA, Sandrock AW, AFFIRM Investigators (2006) A randomized, placebo-controlled trial of natalizumab for relapsing multiple sclerosis. N Engl J Med 354(9):899–910
79. Elkins J, Elkind M, Veltkamp R, Montaner J, Singhal A, Becker K, Lansberg M, Chang LH, Tang W, Gheuens S, Mehta L (2016) Natalizumab versus placebo in patients with Acute Ischemic Stroke (AIS): RESULTS from ACTION, a multicenter, double-blind, placebo-controlled, randomized Phase 2 clinical trial. Neurology 86(16 Supplement):S7.005
80. Hacke W, Kaste M, Bluhmki E, Brozman M, Dávalos A, Guidetti D, Larrue V, Lees KR, Medeghri Z, Machnig T, Schneider D, von Kummer R, Wahlgren N, Toni D, ECASS Investigators (2008) Thrombolysis with alteplase 3 to 4.5 hours after acute ischemic stroke. N Engl J Med 359(13):1317–1329
81. Ly JV, Zavalla JA, Donnan GA (2006) Neuroprotection and thrombolysis: combination therapy in acute ischaemic stroke. Expert Opin Pharmacother 7:1571–1581
82. Dorado L, Millán M, Dávalos A (2014) Reperfusion therapies for acute ischemic stroke: an update. Curr Cardiol Rev 10(4):327–335
83. Pandya RS, Mao L, Zhou H, Zhou S, Zeng J, Popp AJ, Wang X (2011) Central nervous system agents for ischemic stroke: neuroprotection mechanisms. Cent Nerv Syst Agents Med Chem 11(2):81–97
84. Jauch EC, Saver JL, Adams HP Jr, American Heart Association Stroke Council, Council on Cardiovascular Nursing, Council on Peripheral Vascular Disease, Council on Clinical Cardiology et al (2013) Guidelines for the early management of patients with acute ischemic stroke: a guideline for healthcare professionals from the American Heart Association/American Stroke Association. Stroke 44:870–947
85. Rogalewski A, Schneider A, Ringelstein EB, Schäbitz WR (2006) Toward a multimodal neuroprotective treatment of stroke. Stroke 37(4):1129–1136
86. Sereghy T, Overgaard K, Boysen G (1993) Neuroprotection by excitatory amino acid antagonist augments the benefit of thrombolysis in embolic stroke in rats. Stroke 24:1702–1708

Chapter 26
Oxygen Carriers: Are They Enough for Cellular Support?

Jennifer L.H. Johnson

Abstract What started as a quest for a blood substitute has become a search for a cellular support model (CSM) that does not introduce toxicities. Since cells need more than an oxygen supply to thrive, a good CSM should also transport carbon dioxide, control ROS exposure, and permit vasodilation without causing additional damage. Current CSMs are either hemoglobin based (HB), perfluorocarbon based (PFCB), or heme-nitric oxide protein based (H-NOXB). HB CSMs have typically exhibited undesirable vasoactivity. PFCB CSMs have tended to cause accumulation-related toxicities due to the large size of the PFCs and the large doses necessary to accomplish effective oxygen delivery. H-NOXB CSMs appear to be promising but are still at a relatively early stage of development. Combining the most efficacious and least toxic aspects of all approaches may turn out to be most successful.

Keywords Oxygen delivery • Carbon dioxide • Nitric oxide • Carbon monoxide • Heme • Hemoglobin • Perfluorocarbon • Cellular support • Vasoactivity

1 Introduction

The human body requires a constant supply of oxygen as well as an efficient carbon dioxide removal mechanism in order for the cells to maintain healthy metabolic processes. Furthermore, there must also be a way to limit damage to healthy cells caused by reactive oxygen species (ROS) which are naturally generated in the presence of oxygen. Blood, due to all of its components, is capable of meeting these needs (as well as many others) under normal biological conditions. However, when blood flow is restricted or completely cutoff, the surrounding cells experience stress and consequently increase their sensitivity to oxygen (O_2) in an attempt to make lower levels of oxygen "go further." In cases where the blood deprivation continues, the cells, having accumulated too much CO_2 and ROS, die. There are many

J.L.H. Johnson, Ph.D. (✉)
Thayer Medical Corporation, 4575 S. Palo Verde Rd., Tucson, AZ 85714, USA
e-mail: jjohnson@nuvoxphama.com

P.A. Lapchak, J.H. Zhang (eds.), *Neuroprotective Therapy for Stroke and Ischemic Disease*, Springer Series in Translational Stroke Research,
DOI 10.1007/978-3-319-45345-3_26

abnormal conditions and disease states under which this happens. In stroke, severe and permanent neural damage can occur in the brain resulting in reduction in patient functionality, varying degrees of paralysis and, in the worst cases, death. Such neural damage is known to be caused in part by a reduction in the opportunity for local brain cells to exchange gases. This results in cellular accumulation of CO_2 and ROS without enough incoming O_2 to allow for healthy aerobic processes. Current treatments for stroke are particularly focused on thrombolysis: helping to soften/disintegrate/dissolve the existing clot, reducing the likelihood of its growth, preventing formation of new clots, and facilitating blood flow around the clot/s (mostly through colateral vasculature). Unfortunately, these treatments are not only limited in their efficacy but also by their limited time window for useful application due to a hemorrhagic side effect, such as with tissue plasminogen activator (tPA).

In order to survey and responsibly assess the various ways researchers have attempted to develop alternatives to blood, a good understanding of blood and the way it supplies O_2 and other respiratory gases to the body is necessary. Red blood cells (RBCs) are the innate vehicles for O_2 in the body. The passenger inside RBCs solely responsible for selective binding and releasing the O_2 (as well as other respiratory gases) is hemoglobin (Hb). Hb is a large tetrameric protein with 4 binding sites for O_2 that has been reported to carry 1.39 ± 0.006 mL O_2/g Hb [20]. Table 26.1 shows calculated values for various RBC and Hb parameters. The values are useful in facilitating comprehension of the magnitude with which RBCs can carry oxygen.

Using Table 26.1 we can establish that there are ~5×10^9 RBCs in 1 mL of blood and each RBC contains ~2.7×10^8 Hb units. Multiplying these numbers, we can calculate that there are approximately 1.35×10^{18} Hb units in 1 mL of blood. If 1 Hb unit weighs 1.1×10^{-19} g then $1.35 \times$ Hb units weigh 0.1485 g. Since Hb can carry 1.39 mL O_2/g Hb [20], it follows that 1 mL of blood can be estimated to carry 0.2064 mL of O_2; or 21 % O_2 (v/v). If instead of focusing on the volume of oxygen carried within a volume of blood, we focus on the volume of oxygen carried within a volume of RBCs, we get a value of 0.2064 mL O_2/0.4 mL RBCs = 51.6 % O_2 (v/v).

Table 26.1 Red Blood Cell (RBC) and Hemoglobin (Hb) Calculations

Percent of Blood Volume that is RBCs [89]	Volume of RBCs in 1 mL of blood[a]	Number of RBCs in 1 mL of blood[b]	Volume of 1 RBC [89][c]	MW of Hb [90][d]	Mass of 1 Hb unit[e]	Number of Hb units in 1 g of Hb	Number of Hb units in 1 RBC
40 %	400 μL	5×10^9 cells	8×10^{-8} μL	67,522 g/mol	1.1×10^{-19} g	9.1×10^{18} Hb units	2.7×10^8 Hb units

[a]Calculated by 40 % of 1 mL
[b]Calculated as (400 μL)/(8×10^{-8}μL)
[c]Assuming a diameter of 8 um and the shape of a torus for an RBC. $V = (\pi r^2)(2\pi R)$
[d]Authors state a standard deviation of ±5 % (±3378 g/mol)
[e]Applying Avogadro's number = 6.022×10^{23} units in 1 mol

It should be emphasized that if there were a treatment able to replace the duties of blood in preventing cellular damage during ischemia, it would have to be able to do more than just carry oxygen. The ideal treatment would need to be capable of at least 6 tasks:

1. Picking up and carrying O_2 and CO_2
2. Releasing O_2 and CO_2 at a safe rate
3. Allowing intended/useful NO and CO (nonvasoconstrictive)
4. Managing ROS where present in excess
5. Maintaining a long circulation time in the vasculature
6. Introducing no additional toxicities

The potential applications of a product that can conveniently meet all of these needs are widespread. Not only would cellular support/protection during stroke be possible, but such a treatment would likely enable tumor radiation sensitization, chemotherapy sensitization, treatment of sickle cell crisis, treatment of carbon monoxide poisoning, and cellular support/protection during hemorrhagic shock, hypovolemic anemia, myocardial infarction, and pulmonary embolism [30].

Historically there were two approaches to creating an alternative to blood as a support system for cells. Initially these were referred to as "Hemoglobin Based" (HB) and "Perfluorocarbon-Based" (PFCB) Blood Substitutes but because no substitute is likely to replace all the functionality of blood itself, it became more acceptable to use terms that described the main intended functionality; hence, "HB" and "PFCB" Oxygen Carriers. Soon after, it was found that carrying oxygen was not the only task a successful treatment needed to do. As is true with any drug or device, a useful treatment needs to do its intended job (in this case support all necessary healthy cellular processes) without introducing unacceptable toxic concerns.

In the mid-2000s a third approach began to emerge [6, 27, 65]. Initially, scientists found a group of nonglobin proteins that also contained heme groups. These proteins were abundant in various forms in many different species and contained the same histidyl-heme groups of hemoglobin (that would normally bind O_2 with high selectivity), but as an apparent protective mechanism from the potent toxicity of the free radical, nitric oxide (NO) had somehow evolved a high selectivity to bind NO instead [6]. At that point the proteins were referred to as Heme-Nitric Oxide Binding (H-NOB) proteins [6, 27]. The name referring to these proteins was later changed when a smaller subgroup was identified. Thus, when it was found that a minority (~24 %) of these proteins, such as Tt H-NOX [8], actually do bind O_2 preferentially, the name became Heme-Nitric Oxide/Oxygen (H-NOX) proteins to encapsulate both sets of binding preferences [65]. Currently, there is a small company in northern California using this technology as their platform for designing oxygen carriers [41].

A last cellular support model worthy of mention is the use of microbubbles or liposomes to directly carry O_2 to needy tissue. Microbubbles have been on the market for years to be used in ultrasound contrast enhancement and some researchers have shown promise in preferentially delivering therapeutic gases to tissue by applying ultrasound energy at the intended site. Both microbubbles and liposomes can be formulated to carry O_2, Nitric Oxide, and even Xenon [8] for cellular support, vascular dilation, and neuroprotection [24]. This approach may indeed work for small targets of delivery but has its largest weakness in that the microbubbles do not endure in circulation. They are crushed in the lungs due to the pressures exerted. Liposomes, on the other hand, can be engineered to last longer in circulation but they do not carry very large amounts of gases. Liposomes carrying PFCs (which are capable of holding larger amounts of respiratory gases) fall more directly under the "PFC" category previously mentioned [9].

Since we have established that O_2 carrying capacity is not the only consideration in healthy cellular support, rather than "oxygen carriers" the approaches previously described are herein instead referred to as cellular support models (CSMs). The three CSMs with the most promise are Hemoglobin Based (HB-CSMs), Perfluorocarbon Based (PFCB-CSMs), and Heme-Nitric Oxide/Oxygen Based (H-NOXB-CSMs).

2 Hemoglobin-Based CSMs (HB-CSMs)

Originally it was thought that the easiest and safest way to create a CSM would be simply to use Hb, the same protein (or some variation) already known to carry oxygen and remove carbon dioxide in the body. HB-CSMs are made by extracting the Hb from outdated human or bovine blood by way of hemolysis [3, 33]. Free Hb is soluble in the blood. Unfortunately, early attempts to directly use free Hb in the 1930s through the 1960s resulted in nephrotoxicity and adverse cardiovascular effects [12] and as a result we learned that free Hb is highly toxic to the body [7, 10]. Without protection from the red blood cell wall, major tissue damage occurs by way of a cascade of events. In short, with no RBC wall, the Hb is maintained in the ferrous redox state which reduces nitric oxide (NO) available for vascular signaling. Vascular dilation is hindered. Eventually the free Heme and iron induce the immune system to create injury due to inflammation and oxidized Hb in the kidney results in peroxidase oxidative nephritic cellular injury [26].

Secondary attempts in the early 1970s were better, in that they developed artificial RBCs (aRBCs) with artificial membranes. These aRBCs functioned by encapsulating Hb and enzymes such as 2-3-diphosphoglycerate (DPG), carbonic anhydrase (CA), and catalase (CAT) inside membranes [12, 13]. The enzymes aid in O_2 binding and release, CO_2 binding and release, and in reducing the production of ROS. Once O_2 is bound to heme, the 2-3-DPG enables its release when the RBC is near respiring tissue. CA helps transport CO_2 out of respiring tissue partially by controlling the local pH. Where CO_2 concentration is high, CA catalyzes the conversion of CO_2 and H_2O to carbonic acid (H_2CO_3). This lowers the pH in the microenvironment and the heme is compelled to grab local CO_2 for removal. Catalase instigates the breakdown of hydrogen peroxide (H_2O_2) into oxygen (O_2) and water (H_2O).

$$2H_2O_2 + \text{Catalase} \rightarrow 2H_2O + O_2$$

In doing so, it reduces the generation of 2 hydroxyl radicals; an alternative breakdown reaction is shown as follows.

$$H_2O_2 \rightarrow OH + OH$$

According to Chang et al. [12, 13], this approach was somewhat successful. Its main drawback was that these aRBCs were relatively large with abnormal membranes; therefore, they were recognized by the immune system as foreign entities and were quickly removed from circulation by the mononuclear phagocyte system (MPS) mostly present in the spleen [12, 13].

In the 1980s, when it was discovered that HIV could be passed between patients during blood transfusions, there was a pronounced push to speed up development of a CSM [12]. HB-CSMs have long been theorized to offer advantages over actual blood [3]: they are easier to obtain, they can be stored for a much longer period of time, and they do not pose any risk of infection. Due to the rush and their perceived advantages, HB-CSMs were preferentially chosen for clinical trials. Table 26.2 shows a number of HB-CSM products that have gone through clinical trials. Some of these trials are currently in process.

In the 1990s Baxter and Northfield Labs were neck in neck in the race to gain FDA approval. In September of 1998, it was reported that Baxter halted its clinical trial for HemAssist™ citing excessive adverse effects (AEs) and the projected expense to address them as the reason [28]. The article went on to say that Baxter stepping aside left Northfield Labs (PolyHeme®) in the lead to market a "blood substitute." Baxter had apparently already purchased Somatogen, cancelled their clinical trial for Optro™, and decided to develop a new HB-CSM using the recombinant Hb technology platform from Somatogen. Thus, by the year 2000, both HemAssist™ and Optro™ were no longer potential candidates.

Partially as a result of the problems seen in the Baxter clinical trial, the FDA required more studies from Northfield Labs [50, 51]. In clinical studies both Baxter and Northfield Labs used RBCs as their control treatment. Northfield Labs actually showed that PolyHeme® performed equally well to RBCs in cellular support. However, just like Baxter, Northfield Labs encountered serious AEs. Northfield Labs attempted to frame their treatment as acceptable in situations where blood replacement is simply not an option. Regardless of Northfield Labs' contention that the adverse events they were encountering were not indicative of a risk outweighing a benefit where transfusion is impossible, the FDA did not approve PolyHeme® [50, 51]. Likely due to the excessive cost to continue, sometime after 2009, Northfield Labs dropped the project. The clinical AEs of Hemopure® (Biopure), Hemospan™ (Sangart), and Hemolink™ (Hemosol) have also resulted in failure for FDA approval [53]. However, HBO_2 Therapeutics (the current owner of Hemopure®) seems to have decided to try again as they are currently enrolling patients in a phase II study for the treatment of severe anemia [57].

A more recent subcategory of HB-CSM products seeks to use HB binding properties for carbon monoxide (CO) in a therapeutic way. Both Sangart and Prolong Phamaceuticals are developing Hb derivatives that reduce the vasoconstriction normally associated with HB-CSMs by releasing CO in small quantities[47, 86, 87]. Prolong's product is called SANGUINATE™ which is currently in phase II clinical studies for multiple indications [47, 60, 62, 63]. Sangart's product is a reengineered

Table 26.2 Clinical issues with HB-CSMs

HB-CSM Treatment/Product				
Product name (Hb name)	Manufacturer	Approach in reducing nephritic toxicity	Hb conc. in Product (% w/v)	Associated clinical safety issues [75][b] (% patients with symptoms in treated group > % patients with symptoms in control group)
HemAssist™ (DCLHb)	Baxter	Intramolecularly crosslinked	10 %	Death (15.4 % > 12.1 %), hypertension (15.1 % > 7.5 %), cardiac arrhythmia (4.6 % > 3.4 %), myocardial infarction (1.2 % > 0.2 %), GI distress (10.1 % > 6.1 %), liver inflammation (5.4 % > 1.6 %), pancreas inflammation (2.2 % > 0 %), hemorrhage (6.5 % > 4.4 %), lipase increase (5.8 % > 1.7 %)
Hemopure® (PolyBvHb)	Biopure/HBO_2 therapeutics	Intramolecularly and Intermolecularly crosslinked	13 %	Death (3.5 % > 2.3 %), hypoxia (10.7 % > 5.7 %), hypertension (23.4 % > 9.5 %), congestive heart failure (7.6 % > 3.4 %), cardiac arrest (2.4 % > 1 %), myocardial infarction (2 % > 0.6 %), cerebrovascular insult/ischemia (2.3 % > 0.5 %), respiratory distress/failure (3.1 % > 1.9 %), acute renal failure (1.4 % > 0.6 %), GI distress (48.7 % > 31.6 %), liver inflammation (2.8 % > 0.8 %), pancreas inflammation (0.7 % > 0.5 %), thrombosis (6.4 % > 2.9 %), hemorrhage (15.3 % > 8.9 %), sepsis (2.1 % > 1 %), lipase increase (6.8 % > 1.9 %)
Hemospan™ (MP4)	Sangart	Activated PEGylation[a]	4.4 %	Death (2.4 % > 0 %), hypertension (8.2 % > 2.2 %), cardiac arrhythmia (17.6 % > 11.1 %), myocardial infarction (2.4 % > 0 %), GI distress (67 % > 44.4 %), amylase increase (8.2 % > 4.4 %)
Optro™ (rHb1.1)	Somatogen/Baxter	Intramolecularly crosslinked	8 %	Hypoxia (4.7 % > 3.8 %), hypertension (12.5 % > 0 %), chest pain (18.8 % > 0 %), GI distress (56.3 % > 23.1 %), liver inflammation (9.4 % > 11.5 %), lipase increase (10.9 % > 3.8 %), amylase increase (6.3 % > 3.8 %)
PolyHeme™ [50, 51] (PolyHb)	Northfield Labs	Intramolecularly and Intermolecularly crosslinked	10 %	Death (11.7 % > 8.5 %), myocardial infarction (4.7 % > 0.4 %), cerebrovascular insult/ischemia (0.5 % > 0.2 %), thrombosis (2 % > 0.9 %)

Hemolink™ (NCT00038454 n.d.) (O-R-PolyHbA$_0$)	Hemosol	Intramolecularly and Intermolecularly crosslinked	10 %	Hypertension (54 %>39 %), Myocardial Infarction (6.7 %>3.6 %), cerebrovascular insult/ischemia (1 %>0.5 %), GI distress (11 %>0.5 %), liver inflammation (3.8 %>0 %), lipase increase (9 %>1 %), amylase increase (16.7 %>10.4 %)
SAUNGUINATE™ [1] (PEG-bHb-CO)	Prolong Pharmaceuticals	PEGylation of CO-bovineHb[b]	4 % [47]	Hypertension (16.6 %>0 %) still in trials
Not Named (MP4CO)	Sangart	PEGylation of CO-humanHb[b]	4.2 % [86]	Unknown

The table shows a number of HB products used in clinical trials and provides indications of why they are not currently marketed [75]

[a]"PEG" is an acronym for polyethylene glycol which can shield a particle from being scavenged by the MPS system

[b]As determined in clinical trials. (Note that % values for prevalence were calculated for reference only and statistical significance is not drawn)

version of Hespan™ referred to as "MP4CO" [5, 56] which has also completed phase I clinical trials. According to ClinicalTrials.gov, a phase II study was approved in 2013 [58], but withdrawn prior to patient enrollment. Unfortunately, no further information could be found.

OxyVita C™ (Oxyvita Inc.) is a relatively new HB-CSM currently in preclinical studies. It is made of large molecular weight spherical polymerized Hb which has been found to prevent extravasation through endothelial tissue and to limit NO scavenging which, in turn, was found to prevent vasoconstriction of cerebral pial arterioles in healthy rats [2]. In this study, both 0.9 % NaCl solution and HEXTEND® (plasma expander) were used as control solutions and they both caused vasoconstriction. However, OxyVita C increased blood pressure while neither of the controls did.

HEMOXYCarrier® (Hemarina) is another early stage HB-CSM product. The active is called Hemarina-M101 and is made from extracellular marine Hb of a very large molecular weight (~3400 kDa). The Hb source is an oceanic invertebrate called *polychaete annelid*. In vitro experiments with M101 have suggested the binding rates of NO and CO are quite different from those of HbA [81]. One preclinical study in healthy rats demonstrated no effect on heart rate or mean arterial blood pressure and a study in healthy hamsters showed no sign of vasoconstriction [81]. Another preclinical study in a rat model of traumatic brain injury did show an increase in mean arterial blood pressure (increase of 27 mmHg) but also restored brain oxygen levels to normal [48]. The Navy found this promising and the authors suggest this agent should be evaluated in larger animal models.

Ratascopa et al. [66] are researching the use of fetal Hb (HbF) as an alternative to human (HbA) or bovine (HbB). HbF is said to bind O_2 with a much higher affinity and to be less toxic than HbA. One in vitro study showed that recombinant HbF is more stable and can be produced in larger quantities than recombinant HbA [66]. This technology is still in a very early stage of development. Silkstone et al. [73] showed that the addition of tyrosine residues into recombinant Hb can provide a reduction in toxicity from the ferryl heme species, but that the particular candidate under investigation, βLys66Tyr, was not a good specimen because it also offered too many additional undesirable effects.

3 Perfluorocarbon-Based CSMs (PFCB-CSMs)

By comparison to Hb, fluorocarbons (FCs) are known to dissolve respiratory gases as opposed to binding them [29, 67, 68]; thus, the uptake and release process is simply a diffusion process from higher to lower concentration areas. Fluorocarbons are unique among molecules in that they are incredibly chemically stable and even more nonpolar than hydrocarbons. True perfluorocarbons (molecules containing only carbon and fluorine) are so energetically balanced that they do not experience even the weakest of intermolecular attractions known as van der Waals forces. However, their external fluorine atoms that envelope the carbon atoms are naturally electrophilic which means they are attractive environments to very small electron-rich molecular configurations. Without Vander walls forces, it is easier for FC molecules to separate from each other in order to make room for other molecules. Due to their fluorine shells, FCs offer a stabilizing and inviting environment for electron dense entities. Some of the smallest electron dense entities which also lack intermolecular attractive forces are respiratory gases such as O_2, CO_2, CO, and NO. Reiss [67] was one of the first to explain that PFCs dissolve such gases proficiently due to the relative positions of their fluorine atoms. Because the fluorine atoms are larger than hydrogen, the "tetrahedral" arrangement normally taken by hydrogens becomes slightly twisted around the carbon chain. Unlike hydrocarbon chains that are flexible, the twist renders the PFC molecule relatively rigid and incapable of intramolecular flex. Since there are also negligible Van der Waals attractions it becomes easier for cavities to form in the liquid allowing the entrance of gases [19]. The most stabilizing portions of PFCs for electronegative gases are the CF_3 groups. They offer ideal pockets to balance electron density which thermodynamically results in a lower energy system. These qualities (the rigidity, the lack of intermolecular attractive forces, and the shell of electrophilic fluorine atoms in PFCs) are responsible for the remarkable solubility of oxygen in PFCs. However, the oxygen is not molecularly bound, and so, does not require any stimulus other than an encounter with relatively low oxygen tension to begin gradual transfer from carrier (the FC) to hypoxic tissue. For these reasons FC emulsions have been and still are under investigation for respiratory gas transport. Table 26.3 shows a list of such FC-based products clinically investigated.

The amount of oxygen a liquid FC can dissolve and carry per unit volume is mostly dictated by three things: CF_3 group density, FC size (MW), and FC boiling point. The solubility is inversely related to the size (MW) and boiling point of the FC but directly related to the CF_3 group density [18, 19, 30, 31]. Since CF_3 groups bring the most stability to gases, the larger the number of CF_3 groups present in a given volume of PFC the more oxygen will dissolve. Linear PFCs offer two CF_3 groups/molecule (one at each end). So, the smaller the chain of the linear PFC, the higher the concentration of CF_3 groups available in a given volume of PFC. As is true with size comparisons of tennis balls and ping pong balls, larger FCs take up more room than smaller FCs. Just as a larger number of ping pong balls than tennis balls can fit into a bucket, a larger number of small FCs than large FCs can fit into a droplet. Boiling point relates inversely to the size of the intermolecular pockets a FC can create to facilitate the entry of gases. The lower the boiling point of the FC, the easier it is for gases to dissolve so long as the ambient temperature does not exceed the boiling point [10, 30, 42, 45, 82, 83, 84, 85].

For example, perfluoropentane (aka dodecafluoropentane, DDFP, MW=288 g/mol) has a 5 carbon chain with 2 CF_3 groups (one on each end) while perfluorooctylbromide (PFOB, MW=499) has an 8 carbon chain with a CF_3 group on only one end and perfluorodecalin (PFD, MW=462) has 10 cyclic carbons with no CF_3 groups present. The number of DDFP molecules in 1 L (5.7 mol) is roughly 46 % larger than the number of PFOB molecules in 1 L (3.9 mol) and 36 % larger than the number of PFDec molecules in 1 L (4.2 mol). Furthermore, the number of CF_3 groups in 1 L of DDFP (11.4 mol) is 3 times greater than the number of CF_3 groups

Table 26.3 Perfluorocarbon-based CSM products that have been clinically tested

Product name	Manufacturer	PFC (% w/v in product)	PFC MW (g/mol)	PFC density at 25 °C (g/mL)	PFC boiling point (°C)	O_2 Solubility at 25 °C (% v/v of PFC) [67, 68]	O_2 Carrying Capacity (% v/v of product)	Typical dose (mL/kg)	Clinical status
Fluosol™	Green Cross Corp.	PFDecalin (14 %)	462	1.92	142	42 %	0.0008 %	NA	Was approved, but no longer manufactured [11] (storage problems, pulmonary edema, and CHF)
		PFTripropylamine (6 %)	521	1.82	130	23 %			
Oxypherol™	Green Cross Corp.	PFTributylamine (aka FC43) (20 %)	671	1.88	178	30 %	0.006 %	NA	
Perftoran™	SPC-Perftoran	PFDecalin (12 %)	462	1.92	142	42 %	0.0008 %	4–30 mL/kg [35, 46]	Approved in Russia; not approved in the USA [35, 46] (many adverse effects)
		PFMethylcyclohexyl piperidine (3 %)	595	1.96	184	No reference found			
Oxygent™	Alliance Pharmaceutical Corp./SolAeroMed	PFOctylbromide (aka perflubron) (58 %)	499	1.93	142	50 %	0.031 %	1.5 – 6 mL/kg [32]	Phase I/II in Asthma[b] [61, 64] (Recruiting)
		PFDecylbromide (2 %)	599	1.86	185	40 %			
Oxyfluor™	HemaGen [69]	PFDichlorooctane (78 %)	471	1.76	156	43 %	0.0018 %	0.5–1 mL/kg [71]	Phase I/II in CPB [71] (terminated for lack of enrollment and funding)
Oxycyte™	Oxygen Biotherapeutics/Tenax therapeutics	t-Butyl PFCyclohexane (60 %)	500	1.97 [78]	147	43 % [17]	0.017 %	3–6 mL/kg [17]	Phase I in TBI[a] [54, 55] (terminated for futility)
NVX-108™	NuvOx Pharma	PFPentane (aka DDFP) (2 %)	288	1.63	29	80 %	1.4 %	0.1 mL/kg	Phase Ib in GBM [59] (recruiting)

[a]Traumatic brain injury
[b]Note that the treatment for asthma does not appear to involve oxygen delivery as a mechanism

in 1 L of PFOB (3.9 mol) and even though there are more moles of PFD (4.2 mol) than PFOB (3.9 mol) in 1 L, there are no CF_3 groups in PFD. The solubility of O_2 follows this trend and is known to be 80 %, 50 %, and 42 % (v/v) for DDFP, PFOB, and PFD, respectively [18, 36].

Most of the PFCB-CSMs contain relatively large molecular weight perfluorocarbons (PFCs) such as in Perftoran™ and Fluosol™ (containing PFDec, MW = 462), Oxycyte™ (containing t-butylPFCycHex, MW = 500), and Oxygent™ (containing PFOB, MW = 499) [42, 43] having boiling points of 142–144 °C. They are also formulated at relatively high concentration (20–60 % w/v) in order to provide adequate oxygen [34, 40, 88]. As previously stated, the solubility of oxygen is reported to be 42 %, 46 %, and 50 % v/v in PFDec, t-butylPFCycHex, and PFOB [18, 36], respectively. These PFCs also are known to experience a large volume of distribution due to their extreme hydrophobicity. Therefore, they distribute to the tissues extensively, are notably taken up by the RES macrophages, and due to the high degree of accumulation, they exhibit relatively long zero-order half-lives [25, 80]. Because DDFP is smaller, dissolves a higher concentration of oxygen (80 % v/v), has a boiling point of 29 °C and thus, volatilizes at biological temperature, a much smaller amount of it is needed to supply sufficient oxygen in vivo [10, 45, 77]. In addition, studies show that it has a 2 min half-life in the blood and is 99 % cleared through the lungs in 2 h after intravenous administration [14]. Thus, on a volume basis, DDFP can dissolve more gases than other liquid PFCs. Once inside the vasculature at 37 °C, DDFP is prevented from expanding to bubble form by the surrounding pressure. Expansion is only possible if the ambient pressure is rapidly reduced, such as occurs in rapid ascent from deep water diving [72].

The three PFC products that are still under investigation at various stages of clinical testing are Oxygent™ (Alliance/SolAeroMed), Oxycyte™ (Oxygen Biotherapeutics/Tenax Therapeutics), and NVX-108™ (NuvOx Pharma). The NuvOx product is also commonly referred to as "DDFPe" which stands for dodecafluoropentane emulsion. The larger PFCs in Oxygent™ and Oxycyte™ tend to accumulate mostly in the spleen and slowly partition back out over longer periods of time to be exhaled. The elimination time for PFCs is dose dependent. In other words, PFCs exhibit zero-order elimination kinetics. Due to the toxicities associated with the higher PFC accumulation, Oxygent™ and Oxycyte™ are limited by their dose and whether an effective dose for oxygen delivery also results in bioaccumulation of toxic levels of the PFC. The smaller PFC in NVX-108 also partitions into the tissues but carries much more oxygen in the process and because it is small, it has a shorter residence time in the lipid tissue.

A variant of Oxygent, S-1226, is currently in the clinic for treatment of Asthma [61]. A preclinical study by El Mays et al. [21] showed that S-1226 (nebulized perfluorooctylbromide + 12 % CO_2) had a superior effect compared to the standard of care, albuterol, in bronchodilation of asthmatic sheep. The significance of effect was immediate following administration and continued out to 20 min but does not appear to focus on oxygen delivery as a mechanism. Previously, Oxygent endured extensive clinical testing for blood supplementation during surgery [76] where reports suggest it was successful in delivering the intended oxygen. While it is true that PFCs accumulate in the tissues which can cause transient flu-like symptoms, this seems preferable to dying of hypoxic conditions. It is somewhat unclear why Oxygent did not gain FDA approval for at least one of the indications tested clinically. Perhaps Alliance spread its resources too thin in testing as many indications as possible which ultimately led to multiple small successes that never materialized into commercialization.

Oxycyte (1 mL/100 g body weight) was tested along with hyperbaric O_2 for neuroprotection in a rat model with complete middle cerebral artery occlusion [70]. While researchers reported no hemodynamic or metabolic adverse effects, the administration of Oxycyte did not reduce necrotic brain volume as compared to untreated rats. In another study, Oxycyte (5 mL/kg body weight) was given to rats with spinal cord injury. Results showed notable neuroprotective effect out to 42 days postinsult. However, clinical studies with Oxycyte for traumatic brain injury (TBI) were terminated due to futility [54, 55]. There appear to be some conflicting results from these various studies.

The NuvOx product, NVX-108, does not largely accumulate in the body and can carry more O_2 than the others. The Navy has shown that NVX-108 raises brain oxygen levels without any vasoactivity in rats experiencing traumatic brain injury [49, 52]. A group at the University of Arkansas has shown that NVX-108 has dramatic neuroprotective abilities in both rabbits and rats experiencing stroke for up to 24 h poststroke onset [9, 16, 23]. Culp et al. [15] have most recently published that administration of NVX-108 also increases the safe time window for administration of tPA out to 9 h. The Arkansas group states that they are sequentially addressing all the preclinical requirements from the Stroke Therapy Academic Industry Roundtable (STAIR) in an attempt to show worthiness of DDFPe (NVX-108) to enter clinical trials [23]. The NuvOx product is currently in its first clinical trial in Australia for treating glioblastoma. It is being used in conjunction with the standard of care (radiation and chemotherapy) in an attempt to oxygenate tumors and thus, sensitize them to the standard treatments. This is a phase 1b trial with a primary focus on identifying the safe dose and a secondary focus on tumor reversal. Data are not yet available.

Figure 26.1 shows the results of an in vitro oxygen uptake experiment for 2 % (w/v) emulsions of 3 different PFCs (DDFP, PFOB, and PFD) [30]. The emulsified PFD (the active in Fluosol™ and Perftoran™) and PFOB (the active in Oxygent™) formulations were determined to absorb no more oxygen than the blank emulsion formulation at both test temperatures of 21 and 37 °C. By contrast, at 60 min, emulsified DDFP (the active in NVX-108™) absorbed approximately three times more oxygen at 21 °C ($p=0.03$) and 7 times more oxygen at 37 °C ($p=0.001$).

It should be noted that the oxygen uptake concentrations measured in such in vitro experiments cannot be scaled quantitatively to an in vivo situation. The additional pressures exerted in the semiclosed circulatory system will undoubtedly result in a different amount of oxygen uptake by DDFP, PFD, and PFOB. Nevertheless, the trend would be expected to remain the same in that DDFP should be able to deliver more oxygen than the same volume of PFD or PFOB. Figure 26.2 shows the accompanying differences in volume expansion of all the samples and controls when introduced into a 37 °C semisealed flask. Although there were expansions observed with

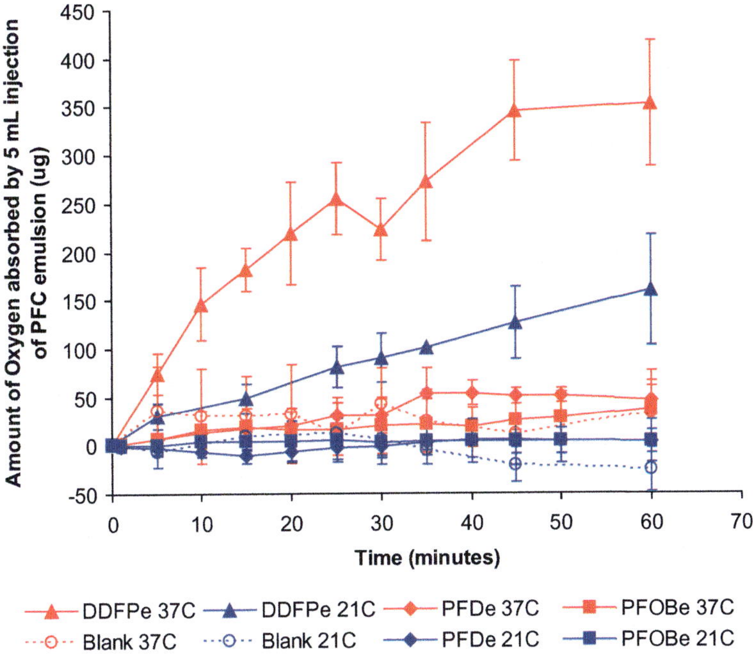

Fig. 26.1 The amount of oxygen absorbed by 5 mL injections of emulsified DDFP (*triangles*), emulsified PFD (*diamonds*), emulsified PFOB (*squares*), and the emulsion formulation blank (*open circles*) at 21 °C (*blue*) and 37 °C (*red*) over the course of 60 min [30]

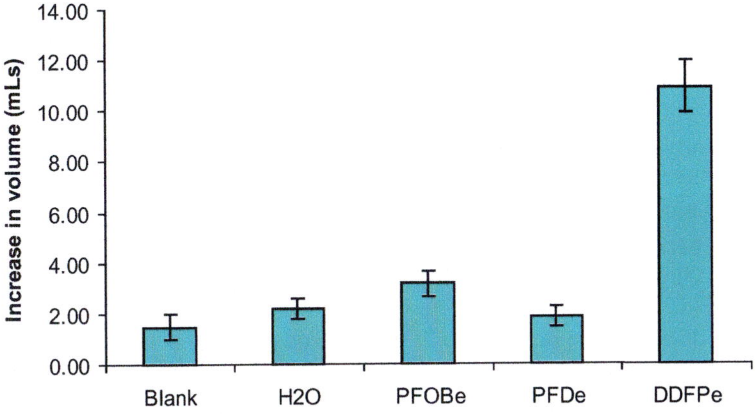

Fig. 26.2 Volume increase upon heating 5 mL injections of 2 % (w/v) PFC emulsions. The study was done with emulsified DDFP, emulsified PFDec, emulsified PFOB, the blank emulsion formulation and water to 37 °C [30]

emulsified PFD and PFOB, neither is significantly larger than the expansion of an equal injection of water ($p=0.35$ and $p=0.06$ for emulsified PFD and PFOB, respectively). There does appear to be a difference between the volume increase of emulsified PFOB and the blank formulation ($p=0.01$) but not between water and the blank formulation ($p=0.12$). The expansion of emulsified DDFP is significantly greater ($p<0.00001$) than all of the other test injections by at least 5 times.

It should be noted that smaller doses of PFCs are not only less invasive for the patient but also for the environment, as it has been clearly documented [14, 41, 79] that PFCs exit the body through the lungs.

The data herein support the contentions of Burkard and Van Liew [10] as well as Lundgren et al. [44, 45] in that DDFP should be able to provide enhanced oxygen delivery over other PFCs due to its expansion from a liquid to a gas at physiological temperature. These in vitro studies coupled with the in vivo results obtained by Lundgren et al. [44] in rats suggest that small doses of the DDFP emulsion may also be useful as a neuroprotectant during stroke.

4 Heme-Nitric Oxide/Oxygen Based CSMs (H-NOXB-CSMs)

The most prominent H-NOXB-CSM, called OMX-4.80P (Omniox, Inc., San Carlos, CA), is in preclinical development. It has been tested in rats, mice, and dogs for radiation sensitization of hypoxic tumors [37–39] and in rats for ischemic stroke [41]. Thus far, studies suggest that OMX-4.80P increases oxygen levels, downregulates HIF-1, and does not cause vasoconstriction. Further testing is warranted.

5 The FDA

Until recently there existed an FDA Draft Guidance for Industry [22, 74] which detailed all of the considerations that should be visited in the development of a CSM. It was originally issued for response in 2004 and contained all of the gathered problems previous CSM products had encountered. It was withdrawn in early 2015 in accordance with the FDA's mission to clear away all Draft Guidances that were outdated and had been lingering in draft form since before 2013 [22]. The Guidance discussed the importance of safety testing for vasoactivity, cardiac toxicity, GI toxicity, Oxidative Stress, pancreatic and liver enzyme elevation, the toxic synergy of endotoxin with hemoglobin, and neurotoxicity when developing HB-CSMs. For PFCB-CSMs the safety list consisted of testing for thrombocytopenia, complement activation, cytokine release, reticuloendothelial system blockade, transient "Flu-like" symptoms, and cerebrovascular accident [74]. The FDA recognizes the problematic environment surrounding the development of a useful CSM

Table 26.4 Current CSMs in the clinic and the mechanisms/approaches for addressing each requirement for useful CSMs

Products	Pickup and Carry O_2 and CO_2	Release O_2 and CO_2 at safe rate	Vasoactivity	Manage ROS levels	Half-life ($T_{1/2}$)	Adverse clinical effects?
SAUNGUINATE™ (PEG-bHb-CO) (Prolong Pharmaceuticals)	Bovine Hb	Same as blood	Uses bound CO to dilate vasculature	None intended	PEGylation increases circulation time	Mild clinical hypertension reported
MP4CO (Sangart)	Human Hb	Same as blood	Uses bound CO to dilate vasculature	None intended	PEGylation increases circulation time	Unknown
Hemopure™ (HBO_2 Therapeutics)	Bovine Hb	Same as blood	Unknown	None intended	Unknown	Hypertension, hemorrhage, and lipase increase
Oxygent™	PFOB	Diffusion	Not vasoactive	None intended	>2 days (due to spleen/tissue saturation)[a]	Accumulation
Oxycyte™	tbPFCH	Diffusion	Unknown	None intended	>5 days (due to spleen/tissue saturation)[a]	Accumulation
NVX-108™	DDFP	Diffusion	Not vasoactive	None intended	~2 h (due to spleen/tissue saturation)[a]	None

[a]Half-lives reported are based on common doses given in Table 26.3 (Note PFCs undergo saturation elimination kinetics; thus, their half-lives are dose dependent.)

and in an attempt to contribute to the cause has now formed a formalized FDA Section specific to developing an Hb-based oxygen therapeutic [4]. Dr. Alayash heads this FDA research and development group which is called the Biochemistry of Hemoglobin (Hb)-based Substitutes Section (BHSS).

6 Addressing More than Carrying Oxygen

Table 26.4 reviews of the CSMs currently in development and attempts to assess how each addresses the issues important to keeping cells alive and healthy.

1. Picking up and carrying O_2 and CO_2
2. Releasing O_2 and CO_2 at a safe rate
3. Allowing intended/useful NO and CO (nonvasoconstrictive)
4. Managing ROS where present in excess
5. Maintaining a long circulation time in the vasculature
6. Introducing no additional toxicities

In conclusion, it is possible that any of the products in Table **26.4** could become commercialized upon succeeding in a Phase III trial as a CSM. None of them appear to be intentionally addressing the control of damage that could result from ROS species which could turn out to be a weakness of them all.

References

1. Abuchowski A (2016) PEGylated bovine carboxyhemoglobin (SANGUNIATE): results of clinical safety testing and use in patients, Part IX: Blood substitutes. In: Elwell CE, Leung TS, Harrison DK (eds) Oxygen transport to tissue, vol XXXVII. Springer, New York, pp 721–730
2. Abutarboush R et al (2014) Effects of the oxygen-carrying solution OxyVita C on the cerebral microcirculation and systemic blood pressures in healthy rats. J Funct Biomater 5:246–258
3. Alayash AI (2014) Blood substitutes: why haven't we been more successful? Trends Biotechnol 32(4):177–185
4. Alayash, A.I. FDA guidance: evaluating the safety and efficacy of hemoglobin-based blood substitutes. In: Biochemistry of hemoglobin (HB)-based substitutes section. The FDA. 2016. www.fda.gov. Accessed 19 March 2016
5. Belcher JD et al (2013) MP4CO, a pegylated hemoglobin saturated with carbon monoxide, is a modulator of HO-1, inflammation, and vaso-occlusion in transgenic sickle mice. Blood 122(15):2757–2764
6. Boon EM, Marletta MA (2005) Ligand Specificity of H-NOX domains: from sGC to Bacterial NO Sensors. J Inorg Biochem 99:892–902
7. Boretti FS et al (2009) Seuqestration of extracellular hemoglobin within a haptoglobin complex decreases its hypersensitive and oxidative effects on dogs and guinea pigs. J Clin Invest 119(8):2271–2280
8. Britton GL et al (2010) In vivo therapeutic gas delivery for neuroprotection with echogenic liposomes. Circulation 122(16):1578–1587
9. Brown AT et al (2014) Dodecafluoropentane emulsion (DDFPe) decreases stroke size and improves neurological scores in a permanent occlusion rat stroke model. Open Neurol J 8:27–33

10. Burkard ME, Van Liew HD (1994) Oxygen transport to tissues by persistent bubbles: theory and simulations. J Appl Physiol 77:2874–2878
11. Castro CI, Briceno JC (2010) Perfluorocarbon-based oxygen carriers: review of products and trials. Artif Organs 38(8):622–634
12. Chang TMS (2013) From hemoglobin based oxygen carrier to oxygen therapeutics, blood substitutes, nanomedicine and artifical cells (Chap. 1). In: Chang TMS (ed) Hemoglobin-based oxygen carriers as red cell substitutes and oxygen therapeutics, vol 1. Springer, Berlin
13. Chang TMS, D'Angillo F, Yu WP, Razack S (2000) Two future generations of blood substitutes based on polyhemoglobin-SOD-catalase and nanoencapsulation. Adv Drug Deliv Rev 40:213–218
14. Correas JM, Meuter AR, Singlas E, Kessler DR, Worah D, Quay SC (2001) Human pharmacokinetics of a perfluorocarbon ultrasound contrast agent evaluated with gas chromatography. Ultrasound Med Biol 27(4):565–570
15. Culp WC, Brown AT, Lowery JD, Arthur MC, Roberson PK, Skinner RD (2015) Dodecafluoropentane emulsion extends window for tPA therapy in a rabbit stroke model. Mol Neurobiol 52:979–984
16. Culp WC et al (2012) Dodecafluoropentane emulsion decreases infarct volume in rabbit ischemic stroke model. J Vasc Interv Radiol 23(1):116–121
17. Denchenko IT, Mahon RT, Allen BW, Piantadosi CA (2012) Brain oxygenation and CNS oxygen toxicity after infusion of perfluorcarbon emulsion. J Appl Physiol 113:224–231
18. Dias AMA, Goncalves CMB, Legido JL, Coutinho JAP, Marrucho IM (2005) Solubility of oxygen in substituted perfluorocarbons. Fluid Phase Equilibria 238:7–12
19. Dias AMA, Freire M, Coutinho JAP, Marrucho IM (2004) Solubility of oxygen in liquid perfluorocarbons. Fluid Phase Equilibria 222–223:325–330
20. Dominguez de Villota E, Garcia Carmona MT, Rubio JJ, Ruiz de Andres S (1981) Equality of the in vivo and in vitro oxygen-binding capacity of haemoglobin in patients with severe respiratory disease. Br J Anaesth 53:1325–1328
21. El Mays TY et al (2014) Nebulized perflubron and carbon dioxide rapidly dilate constricted airways in an ovine model of allergic asthma. Respir Res 15(98):1–9
22. FDA, United States. Withdrawal of draft guidance documents published before december 31, 2013. Federal Register via Government Publishing Office. 2015. www.gpo.gov. Accessed March 19, 2016
23. Fitzgerald RT et al (2015) Dodecafluoropentane emulsion delays and reduces MRI markers of infarction in a rat stroke model: a preliminary report. Magn Reson Imaging 33:236–239
24. Fix SM, Borden MA, Dayton PA (2015) Therapeutic gas delivery via microbubbles and liposomes. J Control Release 209:139–149
25. Flaim SF (1994) Pharmacokinetics and side effects of perfluorocarbon-based blood substitutes. Art Cells, Blood Subst, Immob Biotech 22:1043–1054
26. Gladwin MT, Kanias T, Kim-Shapiro DB (2012) Hemolysis and cell-free hemoglobin drive an intrinsic mechanism for human disease. J Clin Invest 122(4):1205–1208
27. Iyer LM, Anantharaman V, Aravind L (2003) Ancient conserved domains shared by animal soluble guanylyl cyclases and bacterial signaling proteins. BMC Genomics 4:5–12
28. Jaspen B (1998) Baxter drops its Hemassist Program. Chicago Tribune, September 17
29. Johnson EC et al (1995) Effects of a perfluorocarbon emulsion for enhanced O_2 solubility on hemodynamics and O_2 transport in dogs. J Appl Physiol 79(5):1777–1786
30. Johnson JLH, Dolezal MC, Kerschen A, Matsunaga TO, Unger EC (2009) In vitro comparison of dodecafluoropentane (DDFP), perfluorodecalin (PFD), and perfluorooctylbromide (PFOB) in the facilitation of oxygen exchange. Artif Cells Blood Substit Immobil Biotechnol 37:156–162
31. Johnson JLH, Leos RA, Baker AF, Unger EC (2014) Radiosensitization of Hs-766T pancreatic tumor xenografts in mice dosed with dodecafluoropentane nano-emulsion: preliminary findings. J Biomed Nanotechnol 10:1–8
32. Keipert, P.E. (2006) Oxygent, a perfluorochemical-based oxygen therapeutic for surgical patients (Chap. 28) In: R.M. Winslow (ed) Blood substitutes. pp. 312-323

33. Kim HW, Greenburg AG (2013) Hemoglobin-based oxygen carriers as red cell substitutes and oxygen therapeutics. Springer, Berlin
34. Kiral HW, Nicora R, Evitts DP (2002) Comparison of oxygen carrying capacity of a new perfluorocarbon (pfc) blood substitute in rats breathing room air or 100% oxygen. APS Intersociety Meeting, San Diego
35. Kozhura VL, Basarab DA, Timkina MI, Golubev AM, Reshetnyak VI, Monroz VV (2005) Reperfusion injury afer critical intestinal ischemia and its correction with perfluorochemical emulsion "perftoran". World J Gastroenterol 11(45):7084–7090
36. Krafft MP (2001) Fluorocarbons and fluorinated amphiphiles in drug delivery and biomedical research. Adv Drug Deliv Rev 47:209–228
37. Krtolica A et al (2015) AACR 106 Annual meeting. In: OMX-4.80P, a novel H-NOX oxygen carrier that oxygenates hypoxic tumors in multiple tumor models and canine cancer patients, downregulates HIF-1 pathway and increases response to radiation therapy leading to cures. Cancer Research
38. Krtolica A et al (2014) The 19th annual scientific meeting of the society for neuro-oncology. Radiosensitization by OMX-4.80P, a PEGylated H-NOX oxygen carrier that penetrates and oxygenates hypoxic tumors, in preclinical modls of glioblastoma and other hypoxic cancers. Neuro-Oncology
39. Le Moan N et al (2014) Hypoxia reduction in intracranial glioblastoma models by OMX-4.80P, a PEGylated engineered H-NOX oxygen carrier that is long-lasting in circulation and safe. Neuro Oncol 16:v79–v95
40. Leese PT, Noveck RJ, Shorr JS, Woods CM, Flaim KE, Keipert PE (2000) Effects on coagulation function in healthy volunteers. Anesth Analg 91:804–811
41. Leung P et al (2016) Novel and safe oxygen carrier prevents brain damage after focal ischemia and improves functional outcomes. In: The International Stroke Conference 2016
42. Lowe KC (2001) Fluorinated blood substitutes and oxygen carriers. J Fluorine Chem 109:59–65
43. Lowe KC (1991) Pefluorochemicals in medicine. Chem Ind 3(83):1–6
44. Lundgren C, Bergoe G, Olszowka A, Tyssebotn I (2006) Intravascular fluorocarbon-stabilized microbubbles protect against fatal anemia in rats. Artificial Cells, Blood Substitutes, and Biotechnology 34(5):473–468
45. Lundgren C, Bergoe G, Olszowka A, Tyssebotn I (2005) Tissue nitrogen elimination in oxygen-breathing pigs is enhanced by fluorocarbon-derived intravascular micro-bubbles. UHM 32(4):215–226
46. Maevsky EI (2006) Perftoran (Chapter 26). In: Winslow RM (ed) Blood substitutes. Elsevier, Amsterdam, pp 288–297
47. Misra H, Lickliter J, Kazo F, Abuchowski A (2014) PEGylated Carboxyhemoglobin Bovine (SANGUINATE): results of a Phase I clinical trial. Artif Organs 38(8):702–707
48. Moon-Massat P et al (2015) Cerebral vasoactivity and oxygenation with oxygen carrier M101 in rats. J Neurotrauma (Epub), July 10. doi:10.1089/neu.2015.3908
49. Moon-Massat P et al (2014) Effects of perfluorocarbon dodecfluoropentane (NVX-108) on cerebral microvasculature in the healthy rat. Curr Drug Discov Technol 11(3):220–226
50. Moore EE et al (2009) Human polymerized hemoglobin for the treatment of hemorrhagic shock when blood is unavailable: the USA Multicenter Trial. J Am Coll Surg 208(1):1–13
51. Moore EE, Johnson JL, Moore FA, Moore HB (2009) The USA multicenter prehospital hemoglobin-based oxygen carrier resuscitation trial: scientific rationale. Crit Clin Care 25(2):325–356
52. Mullah SA et al (2016) Perfluorocarbon NVX-108 increased cerebral oxygen tension after traumatic brain injury in rats. Brain Res 1634:132–139
53. NCT00038454, Hemosol. www.ClinicalTrials.gov. n.d. www.clinicaltrials.gov. Accessed March 19, 2016
54. NCT00174980, Tenax. www.ClinicalTrials.gov. n.d. Accessed March 22, 2016
55. NCT00908063, Tenax. www.ClinicalTrials.gov. n.d. Accessed March 22, 2016

56. NCT01356485, Sangart. www.ClinicalTrials.gov. n.d. www.clinicaltrials.gov. Accessed March 19, 2016
57. NCT01881503, Biopure/HBO_2 therapeutics. www.ClinicalTrials.gov. n.d. www.clinicaltrials.gov. Accessed March 19, 2016
58. NCT01925001, Sangart. www.ClinicalTrials.gov. n.d. www.clinicaltrials.gov. Accessed March 19, 2016
59. NCT02189109, NuvOx. www.ClnicalTrials.gov. n.d. www.clincialtrials.gov. Accessed March 22, 2016
60. NCT02323685, Prolong. www.Clinical Trials.gov. n.d. www.clinicaltrials.gov. Accessed March 19, 2016
61. NCT02334553, SolAeroMed. www.ClinicalTrials.gov. n.d. www.clinicaltrials.gov. Accessed March 22, 2016
62. NCT02411708, Prolong. www.ClinicalTrials.gov. n.d. www.clinicaltrials.gov. Accessed March 19, 2016
63. NCT02490202, Prolong. www.ClinicalTrials.gov. n.d. www.clinicaltrials.gov. Accessed March 19, 2016
64. NCT02616770, SolAeroMed. www.Clinical Trials.gov. n.d. www.clinicaltrials.gov. Accessed March 22, 2016
65. Plate L, Marletta MA (2013) Nitric Oxide-Sensing H-NOX Proteins Govern Bacterial Communal Behavior. Trends Biochem Sci 38(11):566–575
66. Ratascopa K, Cedervall T, Bulow L (2016) Possibilities of using fetal hemoglobin as a platform for producing hemoglobin-based oxygen carriers (HBOCs). In: Elwell CE, Leung TS, Harrison DK (eds) Oxygen transport to tissue, vol XXXVII. Springer, New York, pp 704–718
67. Reiss JG (2005) Understanding the fundamentals of perfluorocarbons and perfluorocarbon emulsions relevant to in vivo oxygen delivery. Artif Cells Blood Substit Immobil Biotechnol 33:47–63
68. Riess JG (2006) Perfluorocarbon-based oxygen delivery. Artif Cells Blood Substit Immobil Biotechnol 34:567–580
69. Rosoff JD, Soltow LO, Vocelka CR, Schmer G, Chandler WL, Cochran RP (1998) A second generation blood substitute (perfluorodichlorooctane emulsion) does not activate complement during an ex vivo circulation model of bypass. J Cardiothorac Vasc Anesth 12(4):397–401
70. Schneider UC, Karutz T, Schilling L, Woitzik J (2014) Administration of a second generation perfluorochemical in combination with hyperbaric oxygenation does not provide additional benefit in a model of permanent middle cerebral artery occlusion in rats. Springerplus 3(32):1–7
71. Shaw, R.F., T.J. Richard. Rational development of oxyfluor (Chap. 27) In: R.M. Winslow (ed) Blood substitutes. London: Elsevier; 298–311. 2006.
72. Sheppard RL, Regis DP, Mahon RT (2015) Dodecafluoropentane (DDFPe) and decompression sickness-related mortality in rats. Aerosp Med Hum Perform 86(1):21–26
73. Silkstone R.S., et al (n.d.) The βLys6Try variant of human hemoglobin as a component of a blood substitute, Part IX: Blood Substitutes. In: C.E. ELwell, T.S. Leung, D.K. Harrison (eds). Oxygen transport to tissue, XXXVII. Springer, New York, pp. 720–730
74. Silverman T (2004) FDA draft guidance for industry: criteria for safety and efficacy evaluation of oxygen therapeutics as red blood cell substitutes. The United States Food & Drug Administration, www.fda.gov. Accessed March 19, 2016
75. Silverman TA (2009) Hemoglobin-based oxygen carriers: current status and future directions. Transfusion 49(11):2495–2515
76. Spahn DR (1999) Mini-review: blood substitutes, artificial oxygen carriers: perfluorocarbon emulsions. Crit Care 3(5):R93–R97
77. Takanori Y, Schmid-Schonbein GW, Cotter B, DeMaria AN (1999) Flow dynamics of QW7437, a new dodecafluoropentane ultrasound contrast agent, in the microcirculation. J Am Coll Cardiology 34(2):578–586
78. Thompson PP, Clausen G, Kerry R (2009) Emulsions of perfluorocarbons. China Patent CN102395548 A. April 15

79. Toft KG et al (2006) Disposition of perfluorobutane in rats after intravenous injection of Sonazoid. Ultrasound Med Biol 32(1):107–114
80. Tremper KK (1999) Perfluorochemical “blood substitutes”. Anesthesiology 91:1185–1187
81. Tsi AG et al (2012) Microcirculation and No-CO Studies of a Natural Extracellular Hemoglobin Developed for an Oxygen Therapeutic Carrier. Curr Drug Discov Technol 9(3):166–172
82. Van Liew HD, Burkard ME (1995) Behavior of bubbles of slowly permeating gas used for ultrasonic imaging contrast. Invest Radiol 30(5):315–321
83. Van Liew HD, Burkard ME (1995) Bubbles in circulating blood: stabilization and simulations of cyclic changes of size and content. J Appl Physiol 79(4):1379–1385
84. Van Liew HD, Burkard ME (1996) Relationship of oxygen content to PO_2 for stabilized bubbles in the circulation: theory. J Appl Physiol 81(1):500–508
85. Van Liew HD, Raychaudhuri S (1997) Stabilized bubbles in the body: pressure-radius relationships and the limits to stabilization. J Appl Physiol 82:2045–2053
86. Vandegriff KD, Malavalli A, Woolridge J, Lohman J, Winslow RM (2003) A new nonvasoactive PEG-Hb conjugate. Transfusion 43:509–516
87. Vandegriff KD et al (2008) CO-MP4, a polyethylene glycol-conjugated haemoglobin derivative and carbon monoxide carrier that reduces myocardial infarct size in rats. Br J Pharmacol 154(8):1649–1661
88. Wahr JA et al (1996) A pilot study of the effects of a perflubron emulsion, AF 0104, on mixed venous oxygen tension in anesthetized surgical patients. Anesth Analg 82:103–107
89. WebMD; Complete Blood Count (CBC). 2016. www.webmd.com. Accessed March 13, 2016
90. Worley JD, Londo T (1983) Molecular weights of hemoglobin and myoglobin by atomic weight absorption spectrophotometry. J Chem Educ 60(8):650

Chapter 27
A New Paradigm in Protecting Ischemic Brain: Preserving the Neurovascular Unit Before Reperfusion

Natacha Le Moan, Philberta Y. Leung, Natalia Rost, Jonathan A. Winger, Ana Krtolica, and Stephen P. Cary

Abstract Of the ~795,000 strokes that occur each year in the USA, ~695,000 are ischemic strokes (IS) where a clot occludes a major cerebral artery. About half of these IS patients present with so-called penumbra, defined as a hypoperfused tissue immediately surrounding the ischemic core that is severely deprived of oxygen and at risk for deterioration. Collateral vessels can provide sufficient oxygen and nutrients to temporarily maintain neuronal structure in the penumbra but not enough to support function. Thus, the at-risk tissue has the potential for functional recovery if blood flow is restored, but will irreversibly infarct if recanalization is not achieved, resulting in neurological deterioration. Additionally, though collateral circulation can transiently maintain penumbra viability, injury mechanisms such as excitotoxicity and ATP depletion will have already been initiated. Thus, it is imperative to administer therapies that can alleviate ischemia-induced cell death, restore energy metabolism, and halt pathogenic cascades as soon as possible after occlusion in order to protect the at-risk tissue until reperfusion therapies can be employed. Excitingly, the recent breakthroughs in acute IS reperfusion therapy have opened new opportunities for such adjunct neuroprotective treatments. This chapter provides a description of the penumbra tissue, followed by a brief overview of the emerging standard of care for acute IS based on the recent positive clinical trials using IV tPA and mechanical thrombectomy devices. We will then describe the promising use of adjunctive therapies to enhance the benefits of recanalization therapies. In particular, we will discuss the concept of oxygen therapy and oxygen carriers as a valid approach for "combination therapy" to protect the penumbra until

N. Le Moan, Ph.D. (✉) • P.Y. Leung • J.A. Winger • A. Krtolica • S.P. Cary
Omniox Inc., 75 Shoreway Road Suite B, San Carlos, CA 94070, USA
e-mail: nlemoan@omnioxinc.com

N. Rost
J. Philip Kistler Stroke Research Center, Department of Neurology, Massachusetts General Hospital, Boston, MA, USA

P.A. Lapchak, J.H. Zhang (eds.), *Neuroprotective Therapy for Stroke and Ischemic Disease*, Springer Series in Translational Stroke Research,
DOI 10.1007/978-3-319-45345-3_27

reperfusion. Finally, we will discuss the future challenges of clinical trials in acute IS patients and highlight the need for new trial designs to test the potential benefit of combination therapies.

Keywords Penumbra • Oxygen • Cerebral ischemia • Cytoprotection • Neurovascular unit • Oxygen carriers

Abbreviations

ATA	Atmosphere absolute
BBB	Blood–brain barrier
BUN	Blood urea nitrogen
CBF	Cerebral blood flow
DCLHb	Diaspirin cross-linked tetrameric hemoglobin
Hb	Hemoglobin
HBO	Hyperbaric oxygen
HBOC	Hemoglobin-based oxygen carrier
H-NOX	Heme-nitric oxide/oxygen
IS	Ischemic stroke
IV	Intravenous
Mb	Myoglobin
MCAO	Middle cerebral artery occlusion
MSTU	Mobile stroke treatment unit
NBO	Normobaric oxygen
NO	Nitric oxide
NVU	Neurovascular unit
PEG	Polyethylene glycol
PFC	Perfluorocarbon
tPA	Tissue plasminogen activator

1 Therapeutic Target: Salvaging At-Risk Cerebral Tissue

1.1 Evolution of the Penumbra

Within minutes of vascular occlusion, a severe reduction of blood flow and oxygen supply to the brain causes necrosis in a limited area of tissue: the “infarct core.” Immediately surrounding the infarct core is the “penumbra,” a tissue severely deprived of oxygen and at risk for infarction, but where cellular injury is still reversible [1]. The prediction that the existence of a viable penumbra tissue, originally characterized by “misery perfusion,” could lead to the development of therapeutics was first made by JC Baron in a landmark series of papers showing how various cerebral blood flow (CBF) and metabolic rates of oxygen consumption were associated with different outcomes [2–4]. For example, when blood flow drops below 20 mL/100 g/min, reversible functional failure typically occurs; if reduced to a lower threshold (10–12 mL/100 g/min), irreversible morphological damage can occur [5, 6]. Prior efforts therefore focused on the identification of a flow threshold predictive of ultimately infarcted or non-infarcted tissue. However, irreversible brain injury is not only determined by the level of residual flow in the ischemic phase, but also by the duration of flow disturbance and corresponding oxygen availability [7]. Furthermore, these flow thresholds were mainly derived from animal experiments, and their accuracy in predicting tissue fate is still controversial in stroke patients [8, 9]. Thus, quantifying oxygen availability in addition to CBF might better delineate salvageable tissue in stroke patients.

Since stroke treatment can only be successful if hemodynamically compromised brain tissue is detectable and still viable, the capacity to accurately distinguish between infarct and penumbra and a better understanding of the kinetics of penumbra evolution are crucial for identifying patients who would benefit from treatment. Since penumbra infarcts at different rates across patients [10, 11], a more personalized set of selection criteria for intervention than the general time from onset might be warranted. Indeed, recent clinical trials (MR CLEAN, SWIFT-PRIME, ESCAPE, EXTEND-IA, and REVASCAT) [12–16] validated the benefit of using multimodal imaging criteria to stratify patients. However, more work is required on neuroimaging to define the penumbra and core volume thresholds at which therapeutic benefits disappear and investigate whether penumbral and core thresholds may vary between brain structures (e.g., gray matter and white matter). To do so, observational studies incorporating longitudinal and multimodal neuroimaging must be developed. These studies would need to enroll all patients, and then a subsequent analysis of the imaging profile would determine the futility thresholds for each patient. In summary, neuroimaging modalities will require further refinement to identify the most appropriate patients likely to respond to neuroprotective therapy.

1.2 Recanalized Patients: Is There Any Brain Tissue Left to Save?

While clear improvements were observed in the recent trials with recanalization therapy, the penumbra is still lost to infarction from prolonged oxygen deprivation while the patient awaits recanalization. In addition, despite blood flow restoration and the remarkable clinical benefit of endovascular treatments, brain tissue continues to die following recanalization via unclear mechanisms [17–20]. Indeed, rescued penumbra that has undergone severe ischemia may be affected by delayed injury such as edema and inflammation that can lead to infarct growth and selective neuronal loss, regardless of successful or partial reperfusion after recanalization [17, 18]. This progressive infarction of brain tissue may account for suboptimal clinical recovery. As a result, 29–67 % of recanalized stroke patients present a poor outcome (mRS > 3) [12–16], indicating that endovascular treatment can still be considerably improved. To do so, faster workflows to reduce delays in reperfusion along with new endovascular device technology to increase the rates of complete reperfusion are being developed. Importantly, complementary protective agents given before and/or after reperfusion may also help to further improve functional outcome. There is, therefore, renewed interest in the stroke community to develop adjunctive agents to recanalization therapies to preserve the penumbra, and stroke clinicians and scientists agree that the future therapy of IS will encompass a successful combination of protective agents and thrombectomy/thrombolysis.

2 Adjunctive Therapeutic Approaches: Tissue Protection and Recanalization

2.1 From Neuroprotection to Cytoprotection

Despite encouraging data in preclinical stroke models, no clinical trial has definitively demonstrated a significant benefit of any neuroprotectant in stroke patients [21]. Failures to successfully translate neuroprotective results from the laboratory to the clinical setting may have been related to both poor clinical trial design and shortcomings of preclinical studies [22, 23]. In particular, a majority of neuroprotective trials included patient populations that were unlikely to be treated within an optimal time window for efficacy and without neuroimaging evidence of viable tissue. Moreover, a large number of agents were administered as a monotherapy and not combined with recanalization therapies to maximize tissue salvaging. A majority of these agents also targeted a single event in the ischemic cascade typically initiated within minutes following stroke onset and often only targeted the neural element of a complex tissue composed of multiple cell types, both of which are characteristics that likely limited their therapeutic effects. These shortcomings led to discouraging clinical results [24] and dampened enthusiasm for the successful translation of neuroprotectants.

In the past decade, however, the concept of a "neurovascular unit" (NVU) has emerged as a more comprehensive target for acute stroke treatment [25, 26]. The NVU integrates neurons, glia, vascular cells, and matrix components that actively participate in mechanisms of tissue injury and repair. Targeting the NVU is becoming mandatory for a stroke agent, and the field is now turning to multimodal approaches aimed at pairing NVU-protecting agents with recanalization therapies [27, 28]. Such agents would target multiple mechanisms and cell types to extend the time window of recanalization, reduce neurological impairment, potentiate the effect of thrombolysis, reduce reperfusion injury and blood–brain barrier (BBB) breakdown, and decrease intracranial hemorrhages. These therapies with pleiotropic effects are therefore more appropriately called "cytoprotective" and may have the greatest chance of clinical success.

2.2 Promising Cytoprotective Approaches

Recently, the combination of thrombolysis/thrombectomy with cytoprotective agents is garnering more attention and is widely considered a promising approach to treat acute IS patients. In fact, agents previously investigated in failed trials may be reevaluated in combination with recanalization therapies and with improved patient stratification. Although testing combination therapies increases trial complexity and involves more regulatory hurdles, several novel combinatorial clinical trials were recently completed or are currently underway (e.g., statin (THRaST) [29], minocycline (MINOS) [30], uric acid (Urico-Ictus trial) [31] and NA-1 (ENACT trial) [32], hypothermia (ReCCLAIM-1) [33], and 3K3A-APC (RHAPSODY) [34]). Importantly, these clinical studies demonstrate the feasibility of conducting adjunctive trials of cytoprotective agents with intravenous tissue plasminogen activator (IV tPA) or endovascular repair and of selecting a patient subpopulation expected to respond positively to cytoprotective therapy based on their large penumbral and small core volumes.

Oxygen therapy is an additional attractive pleiotropic approach for cytoprotection after IS since viability of the penumbra is greatly influenced by the severity and duration of oxygen deprivation. Indeed, restoring oxygen bioavailability within the penumbra has been shown to promote aerobic metabolism, decrease infarct volume and neuronal death, reduce edema and blood–brain barrier disruption, decrease mortality rate, and prevent stroke-related neurological deficits [35, 36]. Since oxygen availability is a critical parameter for tissue fate, it is logical that oxygen could serve as a promising adjunctive therapy to maximize the benefit of recanalization therapy and widen the treatment time window in stroke patients [37].

3 Oxygen Delivery in Ischemic Disorders: Preclinical Success and Clinical Failures

If delivered in the acute phase of stroke, oxygen therapy could attenuate or decelerate the evolution of the penumbra into infarcted tissue and therefore "buy time" for recanalization therapies. Multiple oxygen delivery methods, including supplemental oxygen, as 100 % oxygen inhaled at either a pressure that is greater than the pressure at sea level (hyperbaric oxygen, HBO) or at ambient pressure (normobaric oxygen, NBO), and several types of oxygen carriers, such as hemoglobin-based oxygen carriers (HBOCs) and perfluorocarbon (PFCs), have been employed in an attempt to restore oxygen to pre-ischemia levels after stroke.

3.1 *HBO-NBO: Preclinical Promise and Clinical Failure of Inhaled Oxygen Therapies*

3.1.1 Preclinical Studies with HBO and NBO

Many preclinical studies have shown the neurological benefit of HBO in both transient [38–45] and permanent [46–48] models of focal ischemia when the treatment was initiated within 2–3 h of occlusion. However, no benefit was observed when HBO was initiated after a critical time window of 6 h after onset, probably due to the conversion of almost the entire penumbra into an infarct by 3–8 h after middle cerebral artery occlusion (MCAO). Compared to HBO, NBO therapy may be more attractive for clinical use due to lower technical requirements, wide availability, and easy administration to a broad range of patients, even in emergency settings such as an ambulance. Multiple preclinical studies have shown the efficacy of NBO in reducing infarct volume and improving neurological scores if administered during occlusion in transient ischemic models [49–54]. However, in permanent occlusion models, conflicting data of NBO efficacy have been reported. The discrepancies are likely due to the variation of penumbra volume and/or duration of penumbra viability between transient and permanent stroke models and the inability of NBO to fully rescue the penumbra in the absence of blood flow [50, 55, 56], reinforcing the rationale for combining oxygen with recanalization therapies. In support of this approach, NBO was shown to be safe with IV tPA and in fact may increase the safety of IV tPA since the combination treatment reduced the mortality, edema, and hemorrhage induced by IV tPA alone [57–60]. Importantly, NBO-IV tPA combination reduced BBB damage and improved neurological functions, even when IV tPA was given 7 h after occlusion, suggesting that NBO can increase the therapeutic window of IV tPA. The benefits of NBO treatment in these studies were largely attributed to the restoration of oxygen, improved metabolism, and decreased oxidative stress in the penumbra [61]. Interestingly, NBO also prevented the delayed selective neuronal loss that can occur in the penumbra after recanalization therapies [62]. This indicates that oxygen could be an attractive complementary therapy to recanalization to decrease the ischemic severity in the penumbra during the vascular occlusion and ultimately reduce long-term neurological deficits that can be observed in successfully reperfused stroke patients.

3.1.2 Failure to Translate HBO and NBO into Clinical Success

Even though preclinical studies demonstrated the efficacy and safety of oxygen therapy in animal models of stroke, NBO and HBO have not been successfully translated to the clinic (Table 27.1). Either HBO treatment was poorly tolerated while control patients showed a trend toward achieving better neurological outcomes [63, 64], or there was no significant long-term improvement in the HBO group [65]. One possibility is that the pressure at which oxygen was administered led to adverse

Table 27.1 Clinical trials with HBO and NBO

Treatment	Schedule	Time from onset (h)	Outcomes	Literature
HBO	100 % O_2 (1.5 ATA 1 h) every 8 h × 15 vs. air	10–148	Trend toward better outcomes in patients treated with air	[63]
	100 % O_2 (1.5 ATA 40 min) every 8 h × 10 vs. air	<24	No significant difference in functional outcomes between the two groups	[65]
	100 % O_2 (2.5 ATA 1 h vs. sham 1.14 ATA)	<24	Better functional outcomes in sham-treated patients at 90 days	[64]
NBO	NBO at 45 L/min (8 h) vs. air	<12	Short-term improvement in NBO-treated patients	[68]
	NBO at 2–3 L/min (72 h) vs. air	<24	Short-term improvement in NBO-treated patients	[69]

effects masking the benefits of HBO treatment. Indeed, in preclinical stroke models, higher pressure levels of HBO resulted in adverse effects, such as increased free radicals and seizures [66, 67], compared to lower levels of pressure [43, 46].

In contrast to HBO, NBO-treated patients exhibited reduced brain lesion volumes 4 h after treatment, and stroke scale scores were significantly better at 24 h when compared to controls [68]. Similarly, a mild improvement in neurological function 1 week after treatment was also observed in the more recent Stroke Oxygen Study (SO_2S) multicenter trial in patients exposed to 72 h of continuous oxygen treatment [69]. However, in both studies, the clinical benefit of oxygen therapy was not sustained. This lack of sustained long-term neurological improvement with NBO may have stemmed from three causes. First, though supplemental oxygen increases the amount of oxygen dissolved in blood plasma and is available for diffusion across the BBB, the limited diffusion distance of oxygen into the brain parenchyma restricts it from reaching all oxygen-deprived tissue efficiently and uniformly. The oxygen diffusion distance from the capillary to the cell in physiological conditions is about 70 μm [70], but can be decreased with a low level of CBF in the penumbra. Therefore, cells far away from blood vessels receive less oxygen, suggesting that the penumbra is not optimally protected with NBO therapy. Second, in all of these clinical studies, supplemental oxygen was administered as a monotherapy. Therefore, since recanalization was not performed to restore blood flow to the ischemic tissue, the benefit of oxygen treatment could not be sustained over time. Finally, while longer periods of oxygen therapy may preserve penumbra viability, oxygen can also induce toxicity in normoxemic tissues by acting as an oxidizing agent and damaging biological molecules [71] or by its inherent vasoactive properties [37]. In summary, despite initial disappointing results in the clinic when supplemental oxygen was used as a monotherapy, restoration of oxygen levels in ischemic brain tissue may prove to be a promising cytoprotective approach when combined with recanalization therapies. Larger controlled and randomized trials in combination with recanalization therapies are required to demonstrate the benefits of supplemental oxygen in extending the therapeutic window for recanalization treatment and in safely expanding the pool of patients that could benefit from IV tPA and endovascular therapies.

3.2 HBOCs/PFCs: Preclinical Promise and Clinical Safety Issues of Artificial Oxygen Carriers

While inhalation of supplemental oxygen enhances oxygen transport by increasing the amount of oxygen dissolved within the plasma and the oxygen-carrying capacity of erythrocytes, it suffers from a significant shortcoming: dependence on the erythrocyte-encapsulated hemoglobin (Hb). Indeed, erythrocyte-encapsulated Hb has a limited ability to deliver an adequate amount of oxygen to oxygen-deprived tissues due to a multitude of factors including size and low oxygen affinity. In an attempt to overcome these limitations, artificial oxygen carriers have been developed with improved oxygen transport and oxygen unloading as an alternate modality [72]. These carriers overcome the limitations exhibited by erythrocyte-encapsulated Hb by featuring characteristics, such as (a) smaller molecular weight for greater tissue penetrance (b), higher oxygen-carrying capacity for delivery of more oxygen (c), higher oxygen affinity for better targeting of oxygen to hypoxic tissues, and/or (d) lower vasoactivity to alleviate toxicity concerns. Two major classes of artificial oxygen carriers have been used to restore oxygen levels in ischemic brain and are discussed below: hemoglobin-based oxygen carriers (HBOCs) made from chemically altered cell-free Hb and perfluorocarbon (PFC)-based oxygen carriers consisting of perfluorocarbon-oxygen emulsions.

Preclinical studies have shown that HBOCs successfully oxygenate ischemic brain tissue and reduce stroke infarct volume up to 70 %. For example, diaspirin cross-linked tetrameric hemoglobin (DCLHb) reduced brain injury in several preclinical stroke models [73, 74]. However, in the clinic, the administration of DCLHb within 18 h of onset and subsequently over 3 days adversely affected outcome in IS patients [75], primarily because of toxicity linked to the agent's nitric oxide (NO) scavenging activity when outside of erythrocytes [76]. The ineffectiveness of DCLHb in clinical trials and its toxicity led many to abandon the approach of delivering oxygen to ischemic tissues using protein carriers.

Another therapeutic approach is to use emulsions based on PFCs to prevent ischemic brain damage. For example, the dodecafluoropentane emulsion (DDFPe) NVX-108, developed by NuVox, delivers oxygen into hypoxic tissues [77], decreases infarct volume and ameliorates neurological scores in a permanent MCAO model, and extends the window of IV tPA therapeutic administration in a rabbit embolic stroke model [78–80]. Since NVX-108 has progressed into a phase I clinical trial to investigate its effect as a radiosensitizer in glioblastoma patients (clinical trial NCT02189109), it may very soon be tested as a promising agent that could bridge recanalization therapies in humans. Though NVX-108 has shown potential preclinically, PFCs generally have very low oxygen-carrying capacities and require coadministration of high levels of oxygen (e.g., 100 % oxygen or carbogen) in order to deliver oxygen efficiently [81]. In addition, oxygen delivery with PFCs is short lived, and repeated dosing would be required to maintain oxygen levels in ischemic tissues until combinatorial therapies could be provided. Therefore, an oxygen carrier able to sustain oxygenation of hypoxic and ischemic tissues over several hours after a single dose might be more useful in a clinical setting.

3.3 Applying Lessons Learned to New Oxygen Carriers

As a consequence of past failures in developing safe and effective oxygen carriers, clinical trials of HBOC products were halted for some time in the USA, although they continued to be evaluated in other countries [82]. Given the unmitigated need of oxygen-bridging agents in the clinic, a new generation of HBOCs with improved physiological and biochemical properties have been developed recently or are currently in development. Structural modifications to the Hb component of the HBOCs overcome the hurdle of extracellular Hb toxicity and increase the safety of HBOCs (reviewed in [82]). Furthermore, the oxygen affinity of HBOCs has been modified to facilitate oxygen transport specifically into ischemic tissues. As a result, HBOCs are returning to the clinic. For example, PEGylated bovine carboxyhemoglobin (SANGUINATE™), developed by Prolong Pharmaceuticals, a biopharmaceutical company developing products for hematology and oncology indications, has progressed from preclinical toxicology to phase II trials in sickle cell and renal disease patients (NCT02411708, NCT02437422) [83]. Furthermore, OXYVITA Inc., a biotech company developing oxygen therapeutics, is seeking to advance Zero-Link™, a solution of large cross-linked bovine Hb polymers, into clinical trials upon FDA approval. The exchange transfusion of Zero-Link™ has shown promise preclinically by decreasing brain damage after transient focal cerebral ischemia [84, 85].

Also in response to the clinical failures of early HBOCs, alternative protein-based oxygen carriers are being developed as therapeutics. For example, Hemarina-M101, developed by Hemarina for wound healing and organ preservation, is a respiratory globin with a high oxygen affinity isolated from a marine invertebrate *Arenicola marina*. Hemarina-M101 can deliver oxygen into hypoxic tissues and is currently under evaluation in the OxyOp clinical trial for hypothermic kidney graft preservation [86, 87]. In addition, OMX, a novel protein-based oxygen carrier [88], has been developed by Omniox Inc. with key properties that are critical for the successful translation of oxygen carriers in the clinic: (a) negligible reactivity toward the natural vasodilator NO to prevent the toxicity associated with NO scavenging by HBOCs, (b) high oxygen affinity to enable release of oxygen at therapeutically relevant levels of hypoxia, (c) low molecular weight (from 20 to 120 kDa) to facilitate tissue penetration, (d) a long circulation half-life that maximizes tissue targeting, and (e) minimal antigenicity. Thus, OMX is a promising oxygen-carrying agent exhibiting biochemical properties suitable for salvaging oxygen-deprived tissues after stroke.

4 OMX Oxygen Carrier: Features and Preclinical Promise

4.1 H-NOX Technology

The oxygen-binding domain of the protein-based OMX oxygen carrier is derived from a member of the heme-nitric oxide/oxygen (H-NOX)-binding protein family. H-NOX proteins contain an iron-bearing heme cofactor identical to that found in Hb and myoglobin (Mb) that serves as the site of oxygen attachment. Like in Hb and Mb, a histidine residue supplied by the protein scaffold binds to one side of the heme, while oxygen binding and release take place on the opposite side (Fig. 27.1). However, the H-NOX protein scaffold is unique; members of the H-NOX family share no genetic or structural relationships with any other oxygen-binding heme proteins [89, 90]. With regard to ligand-binding characteristics, like Hb and Mb, the H-NOX domain binds to oxygen in the reduced, ferrous oxidation state. However, unlike Hb and Mb, oxygen-bound H-NOX exhibits physiologically negligible reactivity toward NO [88, 91], an essential mediator of vascular tone and nitrergic neurotransmission. Consequently, the use of H-NOX to deliver oxygen minimizes concerns about NO-related toxicities like those observed when using HBOC.

Extensive work has shown that the precise spatial arrangement of the amino acids in the oxygen-binding pocket exerts a profound influence on the affinity of the heme for oxygen as well as the reactivity of the heme-bound oxygen toward NO. As a result, H-NOX oxygen affinity can be "tuned" through mutagenesis to produce protein variants capable of oxygen release in specific tissue environments, i.e., the hypoxic environment found in ischemic tissue. One OMX oxygen carrier candidate contains an H-NOX protein whose oxygen affinity has been lowered from ~0.1 to ~2.4 μM by mutation of a heme pocket leucine to phenylalanine, moving it into

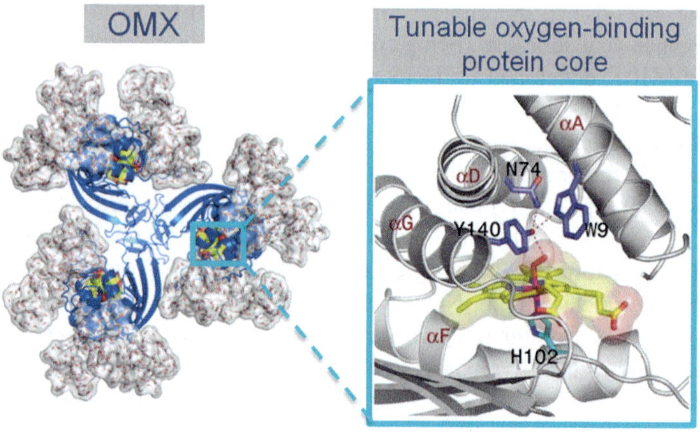

Fig. 27.1 Schematic of OMX. OMX is a PEGylated protein of ~125 kDa with a circulation half-life of ~18 h in rodents and ~35 h in canines

the range of oxygen affinities exhibited by variants of Mb. However, the NO reactivity of this H-NOX protein has been shown to be 10-fold lower than that of an Mb variant with a similar oxygen affinity and 50-fold lower than that of Hb [88, 91].

The stable, compact, and modular nature of the H-NOX oxygen-binding protein also renders it amenable to multimerization and chemical modification. To modulate the size and antigenicity of OMX oxygen carriers, monomeric H-NOX proteins have been multimerized through the addition of a thermostable C-terminal trimerization domain, as well as chemically modified via addition of polyethylene glycol (PEG) moieties of various sizes. Combination of mutagenesis of the H-NOX oxygen-binding pocket with multimerization and PEGylation has enabled the construction of a panel of OMX oxygen carriers with pharmacokinetic and pharmacodynamic profiles that are adaptable to a wide range of specific hypoxic/ischemic indications such as cancer and stroke.

4.2 OMX Mechanism

Release of oxygen by a heme protein is determined primarily by its oxygen-binding affinity (as described by its dissociation constant or K_D) and by the oxygen concentration in the surrounding microenvironment. Once an oxygen-bound protein enters a tissue microenvironment where the oxygen concentration is below a threshold dependent on the K_D, oxygen will dissociate faster than it can rebind, leading to delivery of oxygen into the tissue. For example, if a protein has a high K_D (weak oxygen affinity), it will deliver oxygen in environments with relatively high oxygen concentrations and be depleted of oxygen by the time it reaches hypoxic areas with low oxygen concentrations. Conversely, if a protein has a low K_D (high oxygen affinity), it will remain loaded with oxygen as it transits through environments with relatively high oxygen concentrations and will only begin to deliver oxygen once it reaches a sufficiently hypoxic environment. Thus, erythrocyte-encapsulated hemoglobin, some HBOCs, and other oxygen carriers with relatively weak oxygen affinities release oxygen under physiologic oxygen concentrations found in normal tissues. In contrast, OMX oxygen carriers, which have been engineered to have a higher oxygen affinity, efficiently release oxygen *only* in hypoxic, but not normoxic, tissue microenvironments (Fig. 27.2). Therefore, OMX oxygen carriers will not hyperoxygenate normoxic tissue and consequently do not induce generation of free radicals as tested in a transient MCAO stroke model (not shown).

Another aspect of oxygen delivery by OMX oxygen carriers is their ability to continuously provide oxygen to tissues after delivery of the initial payload. The small size of the OMX oxygen carriers allows them to extravasate through the disrupted vasculature and penetrate deep into hypoxic tissues, thereby creating a gradient of OMX protein that provides a conduit for subsequent facilitated diffusion of additional oxygen to the hypoxic tissue. Overall, OMX overcomes the unfavorable clinical features of prior oxygen carrier therapeutics that failed at increasing oxygen content specifically in hypoxic and ischemic tissues.

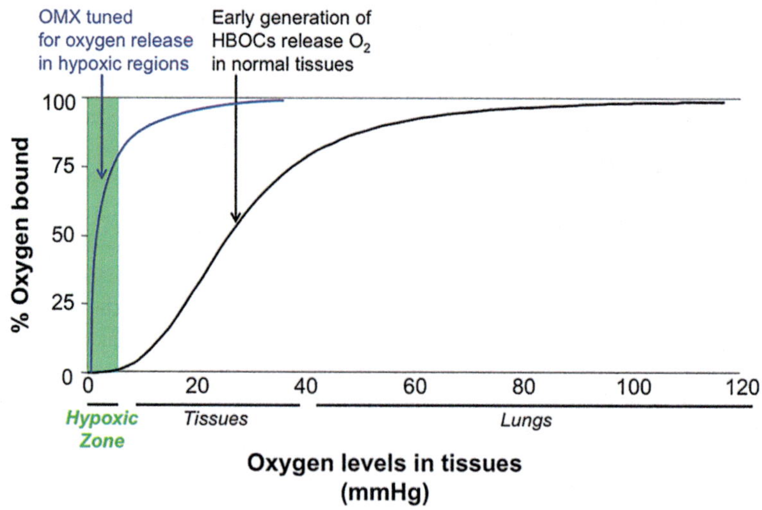

Fig. 27.2 A schematic representation of the oxygen release profiles of oxygen carriers. Due to its high oxygen affinity, OMX efficiently releases oxygen only under hypoxic conditions. In comparison, early generation HBOCs release oxygen under physiologic oxygen concentrations

4.3 *OMX Safety Profile in Preclinical Testing*

Preclinical safety assessments of previous oxygen therapeutics did not translate into safety in clinical trial subjects [63, 75]. The serious safety issues observed in patients can be attributed to the biochemical properties of specific agents, such as vasoconstriction with subsequent hypertension and free radical production [92], and are therefore not common to all oxygen carriers. This clear disconnect between preclinical safety data and clinical safety outcomes might stem from the lack of comprehensive preclinical safety evaluation and the absence of toxicological assessment in the acute treatment phase, which are not routinely performed [93]. Inclusion of biomarkers for heme toxicity exposure, inflammation, oxidative stress (iron overload and inflammation), and organ toxicity as well as assessments of the drug candidate in the disease context in preclinical studies should provide a more comprehensive evaluation of the safety of oxygen carriers. Therefore, nontraditional and non-GLP safety and toxicity studies were conducted to decrease the translational risks of OMX by assessing its effect on oxidative stress, inflammation, arterial pressure, blood flow, hematology, and clinical chemistry during the acute or long-term exposure phase following treatment in rodents, canines, and lambs.

Consecutive daily dosing with supratherapeutic doses of OMX (500 mg/kg, IV bolus) was performed over 27 days in female CD-1 mice. Heme toxicity assessment and tissue histopathology were conducted both at the end of the dosing period and following a 28-day recovery period. After evaluation of brain, heart, kidney, liver, and spleen tissue sections, only a negligible increase in iron positive cells was detected in the kidney (Fig. 27.3) and heart (not shown) of OMX-treated mice.

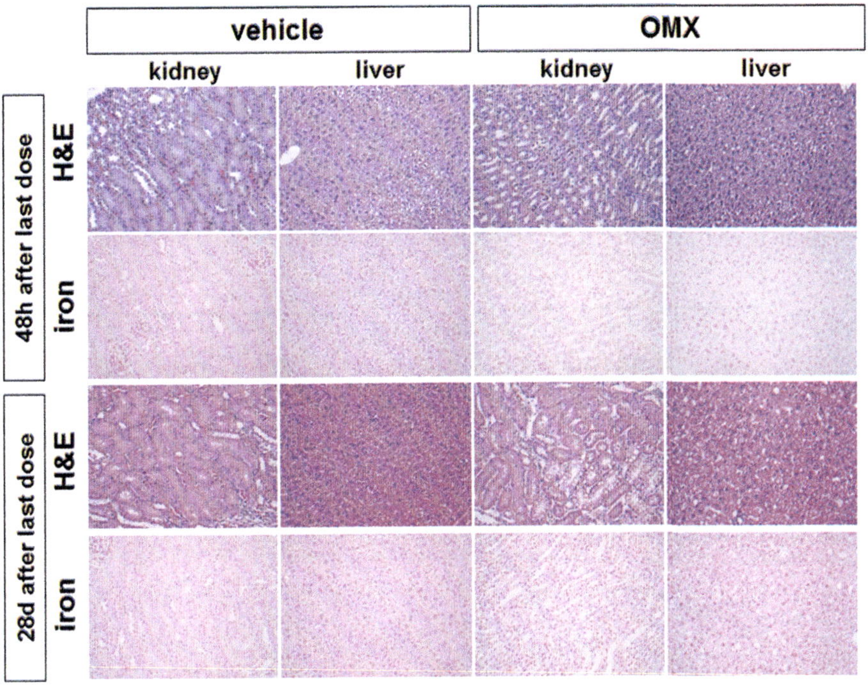

Fig. 27.3 OMX does not increase iron overload in mouse organs. OMX was administered intravenously at 500 mg/kg for 27 consecutive days. Mice were sacrificed and the brain, heart, kidney, liver, and spleen were harvested for histology 48 h after the last dose or following a 28-day recovery period. Heme toxicity was assessed by Perl's Prussian blue staining for iron overload. Histopathology analysis was performed on H&E-stained sections. H&E and Perl's Prussian blue stainings of the kidney and liver are shown

Furthermore, heart rate, pulmonary blood flow (Fig. 27.4), and other hemodynamic variables (not shown) were followed in juvenile lambs breathing 10 % O_2 after treatment with vehicle or OMX. The measured variables did not show a substantial difference between vehicle control and OMX-treated lambs throughout the experiment, indicating that OMX is not vasoactive.

Finally, a preclinical trial was performed to assess the safety and pharmacokinetics of OMX in canines with spontaneous brain cancer. This canine patient population exhibited comorbidities that are shared with human cancer and stroke patients such as frailty condition, advanced age, poor liver and kidney function, and seizures. Blood samples were taken pre-dose, as well as at multiple time points post-dose for hematology, clinical chemistry, and coagulation analyses. There were no significant changes in hematological, clinical chemistry, or coagulation parameters (only blood urea nitrogen [BUN] and hematocrit shown in Table 27.2), demonstrating OMX' highly favorable safety profile.

During this trial, a canine presented to the veterinary hospital with a spontaneous IS. Under a compassionate use waiver and with the owner's consent, this canine

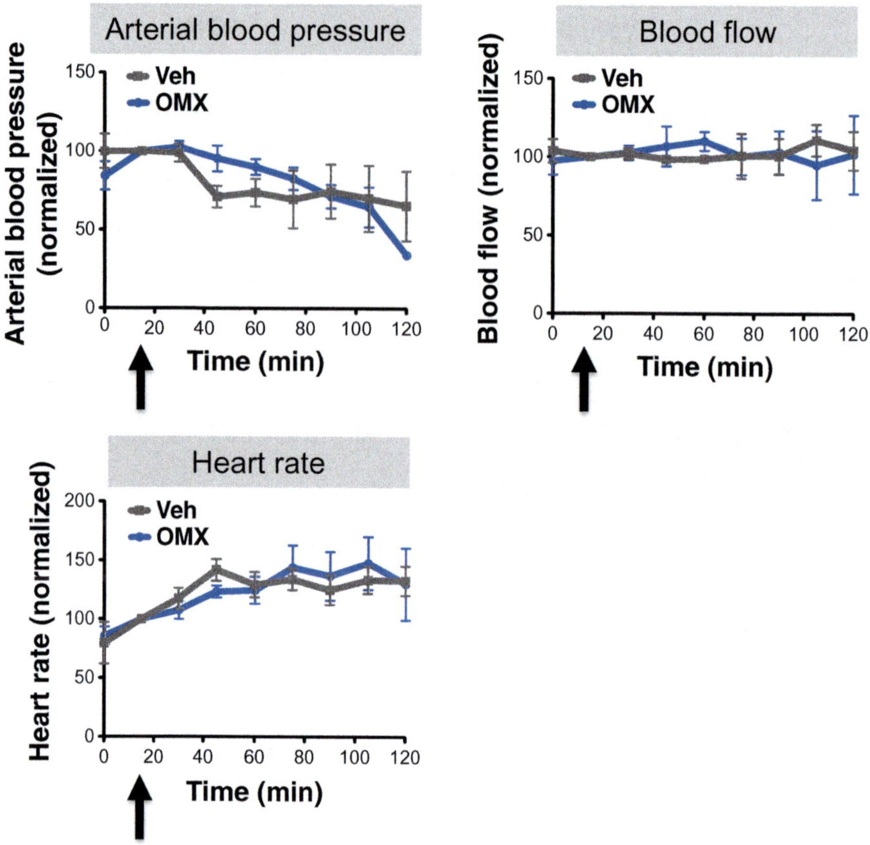

Fig. 27.4 OMX is non-vasoactive. Four-week-old lambs respiring 10 % oxygen were administered a bolus of OMX (200 mg/kg) followed by a slow infusion (70 mg/kg/h), and hemodynamic parameters were measured (blood pressure, blood flow, and heart rates are shown). *Arrows* denote time of Veh or OMX administration during hypoxia

stroke patient was administered OMX to assess its distribution within the brain territory of vascular occlusion. The canine patient did not survive due to the malignant cerebral infarction leading to brain edema and herniation. Immunohistochemistry analysis revealed that, OMX had diffused deep into the stroke tissue via leaky collateral circulation despite permanent vessel occlusion, and accumulated in the peri-infarct area within 6 h after administration, but was retained in the healthy vasculature in normal brain tissue (Fig. 27.5).

Overall, OMX is safe and well tolerated in rodents, juvenile lambs, healthy canines, and canine cancer patients when administered at various doses and with multiple dosing schedules as demonstrated by the lack of immune reaction, clinical symptoms, or dose-limiting toxicities (Table 27.3). In addition to its favorable safety profile, OMX improves radiation-induced anti-tumor activity in rodent can-

Table 27.2 Hematocrit and BUN parameters pre- and post-OMX dosing in rats and dogs

	Species	Dose (mg/kg)	Toxicological assessments (18–24 h)	Toxicological assessments (3–7 days)
Hematocrit	Dog	75	Baseline: 41.4 ± 6.2	Baseline: 41.4 ± 6.2
			OMX: 36.3 ± 6.8	OMX: 35.5 ± 6.1
	Rat	50–500	Veh: 35.5 ± 3.9	Veh: 39.2 ± 1.7
			OMX 50 mg/kg: 37.7 ± 3.7	OMX 50 mg/kg: 41.2 ± 1.4
			OMX 500 mg/kg: 33.6 ± 10.4	OMX 500 mg/kg: 42.2 ± 1.3
BUN	Dog	75	Baseline: 23.1 ± 9.2	Baseline: 23.1 ± 9.2
			OMX: 16 ± 9.5	OMX: 14.7 ± 6.4
	Rat	50–500	Veh: 17 ± 2.7	Veh: 21.3 ± 0.6
			OMX 50 mg/kg: 15.3 ± 5	OMX 50 mg/kg: 18.7 ± 2.2
			OMX 500 mg/kg: 17.3 ± 3.8	OMX 500 mg/kg: 16.7 ± 4.2

A single dose of OMX was administered intravenously at 75 mg/kg in dogs and 50–500 mg/kg in rats. Blood was collected in the acute (18–24 h) and recovery (3–7 days) phases for hematology and clinical chemistry. Only hematocrit and BUN parameters are shown (values are ± SEM, n=3-5 animals per group)

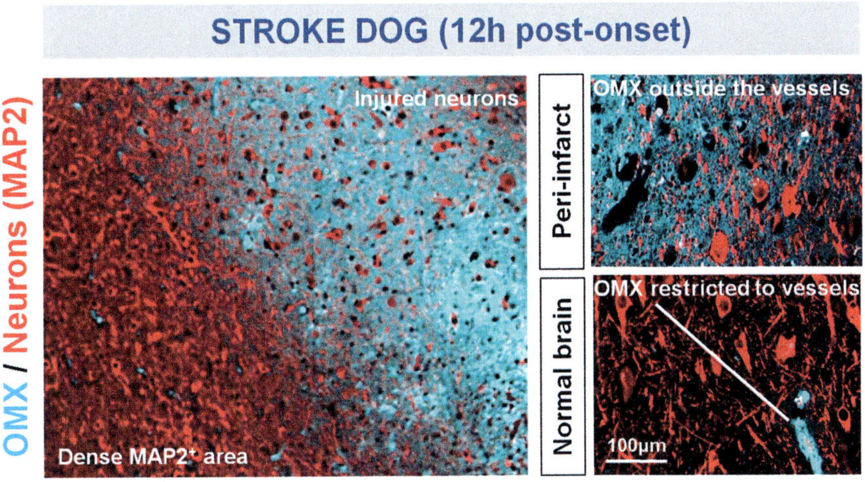

Fig. 27.5 OMX is localized in the brain parenchyma after stroke in a canine patient. A single dose of OMX at 75 mg/kg was administered 6 h after stroke onset; the dog was euthanized due to brain herniation 12 h post-occlusion. The brain was processed for immunohistochemistry, and sections were stained with anti-OMX and MAP2 (neuronal marker) antibodies. OMX is localized to the area of ischemic insult

cer models when used in combination with radiotherapy (manuscript in preparation) and is currently being studied in canine veterinary patients with brain tumors in conjunction with standard of care radiotherapy and as a modulator of the immunosuppressive tumor microenvironment. Moreover, OMX reduces myocardial ischemia and improves cardiac function in a neonatal sheep model of hypoxic

Table 27.3 Summary of preclinical safety and toxicity data for OMX in several species

Species	Dose range (mg/kg)	Dosing frequency	Route	$t_{1/2}$ (h)	Timing of toxicological assessment	Toxicological assessment	Models
Mouse	65–650	Daily, 4 weeks	IV bolus	18 h	24 h–30 days after last dose	H&E and iron staining (Perl's iron) on organs, CBC, serum chemistry	Naïve and tumor bearing
Rat	50–500	Daily, 1 week	IV bolus and IV infusion	18 h	24 h–30 days after last dose	Neurological inflammation (GFAP, Iba1), neurological impairments, oxidative stress (lipid and protein oxidation)	Focal brain ischemia model
Lamb	200	Continuous infusion, 6 h	IV infusion	Ongoing	6 h after last dose	Mean arterial pressure, pH, lactate, pO_2, pCO_2	Hypoxia-induced myocardial dysfunction
Dog	75–100	2–3 times/week, 4 weeks	IV infusion	35 h	24 h–30 days after last dose	CBC and serum chemistry	Canine patients with brain tumor

stress (not shown). Similarly, in rodent stroke models in which salvageable penumbra exists, OMX improves brain oxygenation, prolongs neuronal survival, attenuates infarct growth, and ameliorates sensory–motor function (manuscript in preparation). In conclusion, OMX is a promising agent for acute IS patients to prevent the progressive death of oxygen-deprived brain tissue before recanalization therapies can be administered.

5 Challenges and Opportunities in Combined Trials of Thrombolytic or Endovascular Therapies

5.1 Challenges in Adjunctive Therapy Trials

An important, and paradoxical, evolution of acute IS therapy is that mechanical thrombectomy is expected to become the standard of care, which will lead to improved outcomes in the control arm resulting in a need for larger sample sizes in clinical trials to prove treatment efficacy of additional agents. Therefore, testing future adjunctive therapies will require innovative clinical trial designs to demonstrate a significant benefit for stroke patients by: (a) adopting appropriate multimodal neuroimaging methods to separate responders vs. non-responders (e.g. presence of salvageable brain tissue), (b) developing quantitative assessment of functional outcomes to increase accuracy of the combinatory trial's conclusions, (c) selecting a subgroup of stroke patients with a biological substrate relevant to the mechanism of action of the therapeutic agent while recanalization is taking place, and (d) using an adaptive clinical trial design that allows modification of the key clinical trial characteristics during the trial implementation phase based upon new information acquired within the trial (e.g., number of treatment arms, dosing, randomization ratio, and inclusion/exclusion criteria) to decrease the risk of failure.

Independent from in-hospital patient population selection, another approach to maximize the potential benefit and positive outcome of a stroke agent in an adjunctive trial would be to administer the agent before hospital arrival, either in the field or in the ambulance by paramedics. However, due to the heterogeneity of stroke and the need to rapidly initiate treatment, it is difficult to properly randomize stroke patients according to the multiple factors that can affect the outcome of recanalization therapies, such as time from onset, time to reperfusion, thrombus type, location and size, extent of collateral vessels, comorbid conditions, and age. As a result, baseline criteria would be unbalanced. Therefore, prehospital combinatorial studies will need to include useful endpoints such as the differences in the development of deficits from the field to the hospital, evidence of more brain tissue being preserved, and a slower progression of the infarct core on hospital arrival as demonstrated by multimodal imaging.

5.2 *Opportunities in Prehospital Treatment*

Increasing the number of stroke patients qualifying for recanalization will require extensive education in addition to management of acute stroke treatment with fast transport and efficient teamwork at the hospital. For example, the mobile stroke treatment unit (MSTU) with built-in CT scanners and aided by telemedicine consultation can allow sophisticated triage decisions before hospital arrival, decreased door-to-needle time, and ultimately increased probability of good functional outcomes. However, populations outside of major metropolitan regions and developing countries lacking nearby comprehensive stroke centers are unlikely to benefit from the implementation of MSTUs. Therefore, early initiation of adjunctive cytoprotective therapies may extend the time window for safe recanalization by slowing down the conversion of the penumbra into an infarct such that the proportion of patients who arrive at the hospital with a favorable imaging profile would rise dramatically, thereby increasing the number of patients who could benefit from recanalization. Although the FAST-MAG placebo-controlled trial failed to show differences in disability outcome, it demonstrated for the first time the feasibility of prehospital trials, with a median onset-to-treatment time of 45 min (NCT0005933) [94]. Despite prehospital approaches being resource intensive and logistically challenging, overcoming FDA hurdles is possible when extremely good preclinical and clinical safety profiles are demonstrated prior to the trial. For example, hypothermia (PreCOOL 1 NCT01669408) and cytoprotective drugs, such as glyceryl trinitrate (RIGHT ISRCTN 66434824) [95] and the postsynaptic density-95 protein inhibitor (FRONTIER NCT02315443), were and are currently being evaluated in prehospital trials, confirming the feasibility of ambulance treatment. With the need to maximize the benefit from recanalization treatment with a pharmacological agent that can prevent infarct growth, it is reasonable to assume that the number of stroke agents routinely tested in prehospital trials will dramatically increase in the future.

6 Conclusions

The deep chill of pessimism against developing cytoprotective agents that preserve the penumbra and extend the time window for recanalization treatment has begun to thaw. Acute IS therapy development is reawakening, and we are entering an exciting era of using a new paradigm to protect the ischemic brain. Preclinical stroke research is progressively losing the "neurocentric approach" and shifting toward a more integrated strategy targeting the interactions between neurons and glial and vascular components. In addition, given the complexity of the ischemic cascade pathways, support is quickly growing for cytoprotective agents with pleiotropic mechanisms of action rather than a single target. Finally, these agents will need to act safely in combination with the recanalization therapies that are becoming the new standard of care.

A strong candidate for such a cytoprotective agent is oxygen therapy. Availability of oxygen is a crucial parameter that determines the fate of brain tissue, and oxygen levels are affected by the duration and threshold of collateral blood flow. Therefore, restoring oxygen levels with oxygen carriers independent of blood flow threshold in the penumbra could prevent energy failure and cell death, which is crucial for the successful preservation of penumbra viability and brain function. Oxygen carrier approaches, such as OMX, are therefore a simple, attractive, and promising adjunctive approach to maximize the therapeutic benefit of recanalization therapies to increase treatment efficacy for millions of patients.

Acknowledgments We thank Michael Kent, Pete Dickinson, and Beverly Sturges for canine studies performed at UC Davis, Emin Maltepe and Jeff Fineman for the lamb studies performed at UCSF, and Alexandra Loucks for editorial support.

Conflict of Interest N.L.M., P.Y.L., J.A.W., A.K., and S.P.C. are employees of Omniox, and N.R. is a clinical advisor for Omniox.

N.R. also serves as site principal investigator for the trial of 3K3A-APC (NN104-RHAPSODY).

References

1. Heiss WD (2012) The ischemic penumbra: how does tissue injury evolve? Ann N Y Acad Sci 1268:26–34
2. Baron JC, Bousser MG, Rey A, Guillard A, Comar D, Castaigne P (1981) Reversal of focal “misery-perfusion syndrome” by extra-intracranial arterial bypass in hemodynamic cerebral ischemia. A case study with 15O positron emission tomography. Stroke 12(4):454–459
3. Marchal G, Beaudouin V, Rioux P, de la Sayette V, Le Doze F, Viader F et al (1996) Prolonged persistence of substantial volumes of potentially viable brain tissue after stroke: a correlative PET-CT study with voxel-based data analysis. Stroke 27(4):599–606
4. Marchal G, Furlan M, Beaudouin V, Rioux P, Hauttement JL, Serrati C et al (1996) Early spontaneous hyperperfusion after stroke. A marker of favourable tissue outcome? Brain 119(Pt 2):409–419
5. Hossmann KA (1994) Viability thresholds and the penumbra of focal ischemia. Ann Neurol 36(4):557–565
6. Heiss WD (1992) Experimental evidence of ischemic thresholds and functional recovery. Stroke 23(11):1668–1672
7. Astrup J, Siesjo BK, Symon L (1981) Thresholds in cerebral ischemia—the ischemic penumbra. Stroke 12(6):723–725
8. Baron JC (2001) Perfusion thresholds in human cerebral ischemia: historical perspective and therapeutic implications. Cerebrovasc Dis 11(Suppl 1):2–8
9. Latchaw RE, Yonas H, Hunter GJ, Yuh WT, Ueda T, Sorensen AG et al (2003) Guidelines and recommendations for perfusion imaging in cerebral ischemia: A scientific statement for healthcare professionals by the writing group on perfusion imaging, from the Council on Cardiovascular Radiology of the American Heart Association. Stroke 34(4):1084–1104
10. Markus R, Reutens DC, Kazui S, Read S, Wright P, Pearce DC et al (2004) Hypoxic tissue in ischaemic stroke: persistence and clinical consequences of spontaneous survival. Brain 127(Pt 6):1427–1436
11. Markus R, Reutens DC, Kazui S, Read S, Wright P, Chambers BR et al (2003) Topography and temporal evolution of hypoxic viable tissue identified by 18F-fluoromisonidazole positron emission tomography in humans after ischemic stroke. Stroke 34(11):2646–2652

12. Saver JL, Goyal M, Bonafe A, Diener HC, Levy EI, Pereira VM et al (2015) Stent-retriever thrombectomy after intravenous t-PA vs. t-PA alone in stroke. N Engl J Med 372(24):2285–2295
13. Campbell BC, Mitchell PJ, Kleinig TJ, Dewey HM, Churilov L, Yassi N et al (2015) Endovascular therapy for ischemic stroke with perfusion-imaging selection. N Engl J Med 372(11):1009–18
14. Goyal M, Demchuk AM, Menon BK, Eesa M, Rempel JL, Thornton J et al (2015) Randomized assessment of rapid endovascular treatment of ischemic stroke. N Engl J Med 372(11):1019–30
15. Jovin TG, Chamorro A, Cobo E, de Miquel MA, Molina CA, Rovira A et al (2015) Thrombectomy within 8 hours after symptom onset in ischemic stroke. N Engl J Med 372(24):2296–2306
16. Berkhemer OA, Fransen PS, Beumer D, van den Berg LA, Lingsma HF, Yoo AJ et al (2015) A randomized trial of intraarterial treatment for acute ischemic stroke. N Engl J Med 372(1):11–20
17. Guadagno JV, Jones PS, Aigbirhio FI, Wang D, Fryer TD, Day DJ et al (2008) Selective neuronal loss in rescued penumbra relates to initial hypoperfusion. Brain 131(Pt 10):2666–2678
18. Carrera E, Jones PS, Morris RS, Alawneh J, Hong YT, Aigbirhio FI et al (2013) Is neural activation within the rescued penumbra impeded by selective neuronal loss? Brain 136(Pt 6):1816–1829
19. Haussen DC, Nogueira RG, Elhammady MS, Yavagal DR, Aziz-Sultan MA, Johnson JN et al (2016) Infarct growth despite full reperfusion in endovascular therapy for acute ischemic stroke. J Neurointerv Surg 8(2):117–121
20. Federau C, Mlynash M, Christensen S, Zaharchuk G, Cha B, Lansberg MG et al (2016) Evolution of volume and signal intensity on fluid-attenuated inversion recovery MR images after endovascular stroke therapy. Radiology 151586
21. Minnerup J, Wersching H, Schilling M, Schabitz WR (2014) Analysis of early phase and subsequent phase III stroke studies of neuroprotectants: outcomes and predictors for success. Exp Transl Stroke Med 6(1):2
22. Fisher M, Tatlisumak T (2005) Use of animal models has not contributed to development of acute stroke therapies: con. Stroke 36(10):2324–2325
23. Moskowitz MA, Grotta JC, Koroshetz WJ (2013) The NINDS Stroke Progress Review Group final analysis and recommendations. Stroke 44(8):2343–2350
24. Cheng YD, Al-Khoury L, Zivin JA (2004) Neuroprotection for ischemic stroke: two decades of success and failure. NeuroRx 1(1):36–45
25. del Zoppo GJ (2010) The neurovascular unit in the setting of stroke. J Intern Med 267(2):156–171
26. Zhang L, Zhang ZG, Chopp M (2012) The neurovascular unit and combination treatment strategies for stroke. Trends Pharmacol Sci 33(8):415–422
27. Fisher M, Saver JL (2015) Future directions of acute ischaemic stroke therapy. Lancet Neurol 14(7):758–767
28. Andrew D, Barreto JCG (2015) Combination of thrombolytic therapy with antithrombotics and neuroprotectants. In: Lyden PD (ed) Thrombolytic therapy for acute stroke. Springer, Berlin, pp 65–80
29. Cappellari M, Bovi P, Moretto G, Zini A, Nencini P, Sessa M et al (2013) The THRombolysis and STatins (THRaST) study. Neurology 80(7):655–661
30. Fagan SC, Waller JL, Nichols FT, Edwards DJ, Pettigrew LC, Clark WM et al (2010) Minocycline to improve neurologic outcome in stroke (MINOS): a dose-finding study. Stroke 41(10):2283–2287
31. Chamorro A, Amaro S, Castellanos M, Segura T, Arenillas J, Marti-Fabregas J et al (2014) Safety and efficacy of uric acid in patients with acute stroke (URICO-ICTUS): a randomised, double-blind phase 2b/3 trial. Lancet Neurol 13(5):453–460

32. Hill MD, Martin RH, Mikulis D, Wong JH, Silver FL, Terbrugge KG et al (2012) Safety and efficacy of NA-1 in patients with iatrogenic stroke after endovascular aneurysm repair (ENACT): a phase 2, randomised, double-blind, placebo-controlled trial. Lancet Neurol 11(11):942–950
33. Horn CM, Sun CH, Nogueira RG, Patel VN, Krishnan A, Glenn BA et al (2014) Endovascular Reperfusion and Cooling in Cerebral Acute Ischemia (ReCCLAIM I). J Neurointerv Surg 6(2):91–95
34. Lyden P, Levy H, Weymer S, Pryor K, Kramer W, Griffin JH et al (2013) Phase 1 safety, tolerability and pharmacokinetics of 3K3A-APC in healthy adult volunteers. Curr Pharm Des 19(42):7479–7485
35. Singhal AB (2006) Oxygen therapy in stroke: past, present, and future. Int J Stroke 1(4):191–200
36. Singhal AB (2007) A review of oxygen therapy in ischemic stroke. Neurol Res 29(2):173–183
37. Michalski D, Hartig W, Schneider D, Hobohm C (2011) Use of normobaric and hyperbaric oxygen in acute focal cerebral ischemia—a preclinical and clinical review. Acta Neurol Scand 123(2):85–97
38. Veltkamp R, Warner DS, Domoki F, Brinkhous AD, Toole JF, Busija DW (2000) Hyperbaric oxygen decreases infarct size and behavioral deficit after transient focal cerebral ischemia in rats. Brain Res 853(1):68–73
39. Chang CF, Niu KC, Hoffer BJ, Wang Y, Borlongan CV (2000) Hyperbaric oxygen therapy for treatment of postischemic stroke in adult rats. Exp Neurol 166(2):298–306
40. Hou H, Grinberg O, Williams B, Grinberg S, Yu H, Alvarenga DL et al (2007) The effect of oxygen therapy on brain damage and cerebral pO(2) in transient focal cerebral ischemia in the rat. Physiol Meas 28(8):963–976
41. Henninger N, Fisher M (2006) Normobaric hyperoxia—a promising approach to expand the time window for acute stroke treatment. Cerebrovasc Dis 21(1-2):134–136
42. Henninger N, Kuppers-Tiedt L, Sicard KM, Gunther A, Schneider D, Schwab S (2006) Neuroprotective effect of hyperbaric oxygen therapy monitored by MR-imaging after embolic stroke in rats. Exp Neurol 201(2):316–323
43. Chen LF, Tian YF, Lin CH, Huang LY, Niu KC, Lin MT (2014) Repetitive hyperbaric oxygen therapy provides better effects on brain inflammation and oxidative damage in rats with focal cerebral ischemia. J Formos Med Assoc 113(9):620–628
44. Badr AE, Yin W, Mychaskiw G, Zhang JH (2001) Effect of hyperbaric oxygen on striatal metabolites: a microdialysis study in awake freely moving rats after MCA occlusion. Brain Res 916(1-2):85–90
45. Lou M, Eschenfelder CC, Herdegen T, Brecht S, Deuschl G (2004) Therapeutic window for use of hyperbaric oxygenation in focal transient ischemia in rats. Stroke 35(2):578–583
46. Schabitz WR, Schade H, Heiland S, Kollmar R, Bardutzky J, Henninger N et al (2004) Neuroprotection by hyperbaric oxygenation after experimental focal cerebral ischemia monitored by MRI. Stroke 35(5):1175–1179
47. Sunami K, Takeda Y, Hashimoto M, Hirakawa M (2000) Hyperbaric oxygen reduces infarct volume in rats by increasing oxygen supply to the ischemic periphery. Crit Care Med 28(8):2831–2836
48. Yu M, Xue Y, Liang W, Zhang Y, Zhang Z (2015) Protection mechanism of early hyperbaric oxygen therapy in rats with permanent cerebral ischemia. J Phys Ther Sci 27(10):3271–3274
49. Liu S, Liu W, Ding W, Miyake M, Rosenberg GA, Liu KJ (2006) Electron paramagnetic resonance-guided normobaric hyperoxia treatment protects the brain by maintaining penumbral oxygenation in a rat model of transient focal cerebral ischemia. J Cereb Blood Flow Metab 26(10):1274–1284
50. Kim HY, Singhal AB, Lo EH (2005) Normobaric hyperoxia extends the reperfusion window in focal cerebral ischemia. Ann Neurol 57(4):571–575

51. Liu C, Weaver J, Liu KJ (2012) Rapid conditioning with oxygen oscillation: neuroprotection by intermittent normobaric hyperoxia after transient focal cerebral ischemia in rats. Stroke 43(1):220–226
52. Singhal AB, Dijkhuizen RM, Rosen BR, Lo EH (2002) Normobaric hyperoxia reduces MRI diffusion abnormalities and infarct size in experimental stroke. Neurology 58(6):945–952
53. Singhal AB, Wang X, Sumii T, Mori T, Lo EH (2002) Effects of normobaric hyperoxia in a rat model of focal cerebral ischemia-reperfusion. J Cereb Blood Flow Metab 22(7):861–868
54. Yuan Z, Pan R, Liu W, Liu KJ (2014) Extended normobaric hyperoxia therapy yields greater neuroprotection for focal transient ischemia-reperfusion in rats. Med Gas Res 4:14
55. Veltkamp R, Sun L, Herrmann O, Wolferts G, Hagmann S, Siebing DA et al (2006) Oxygen therapy in permanent brain ischemia: potential and limitations. Brain Res 1107(1):185–191
56. Henninger N, Bouley J, Nelligan JM, Sicard KM, Fisher M (2007) Normobaric hyperoxia delays perfusion/diffusion mismatch evolution, reduces infarct volume, and differentially affects neuronal cell death pathways after suture middle cerebral artery occlusion in rats. J Cereb Blood Flow Metab 27(9):1632–1642
57. Fujiwara N, Murata Y, Arai K, Egi Y, Lu J, Wu O et al (2009) Combination therapy with normobaric oxygen (NBO) plus thrombolysis in experimental ischemic stroke. BMC Neurosci 10:79
58. Liang J, Qi Z, Liu W, Wang P, Shi W, Dong W et al (2015) Normobaric hyperoxia slows blood-brain barrier damage and expands the therapeutic time window for tissue-type plasminogen activator treatment in cerebral ischemia. Stroke 46(5):1344–1351
59. Henninger N, Bratane BT, Bastan B, Bouley J, Fisher M (2009) Normobaric hyperoxia and delayed tPA treatment in a rat embolic stroke model. J Cereb Blood Flow Metab 29(1):119–129
60. Liu W, Hendren J, Qin XJ, Liu KJ (2009) Normobaric hyperoxia reduces the neurovascular complications associated with delayed tissue plasminogen activator treatment in a rat model of focal cerebral ischemia. Stroke 40(7):2526–2531
61. Poli S, Veltkamp R (2009) Oxygen therapy in acute ischemic stroke—experimental efficacy and molecular mechanisms. Curr Mol Med 9(2):227–241
62. Ejaz S, Emmrich JV, Sitnikov SL, Hong YT, Sawiak SJ, Fryer TD et al (2016) Normobaric hyperoxia markedly reduces brain damage and sensorimotor deficits following brief focal ischaemia. Brain 139(Pt 3):751–64
63. Anderson DC, Bottini AG, Jagiella WM, Westphal B, Ford S, Rockswold GL et al (1991) A pilot study of hyperbaric oxygen in the treatment of human stroke. Stroke 22(9):1137–1142
64. Rusyniak DE, Kirk MA, May JD, Kao LW, Brizendine EJ, Welch JL et al (2003) Hyperbaric oxygen therapy in acute ischemic stroke: results of the Hyperbaric Oxygen in Acute Ischemic Stroke Trial Pilot Study. Stroke 34(2):571–574
65. Nighoghossian N, Trouillas P, Adeleine P, Salord F (1995) Hyperbaric oxygen in the treatment of acute ischemic stroke. A double-blind pilot study. Stroke 26(8):1369–1372
66. Chavko M, Xing G, Keyser DO (2001) Increased sensitivity to seizures in repeated exposures to hyperbaric oxygen: role of NOS activation. Brain Res 900(2):227–233
67. Pablos MI, Reiter RJ, Chuang JI, Ortiz GG, Guerrero JM, Sewerynek E et al (1997) Acutely administered melatonin reduces oxidative damage in lung and brain induced by hyperbaric oxygen. J Appl Physiol (1985) 83(2):354–358
68. Singhal AB, Benner T, Roccatagliata L, Koroshetz WJ, Schaefer PW, Lo EH et al (2005) A pilot study of normobaric oxygen therapy in acute ischemic stroke. Stroke 36(4):797–802
69. Roffe C, Ali K, Warusevitane A, Sills S, Pountain S, Allen M et al (2011) The SOS pilot study: a RCT of routine oxygen supplementation early after acute stroke--effect on recovery of neurological function at one week. PLoS One 6(5), e19113
70. Franko AJ, Sutherland RM (1979) Oxygen diffusion distance and development of necrosis in multicell spheroids. Radiat Res 79(3):439–453
71. Weaver J, Liu KJ (2015) Does normobaric hyperoxia increase oxidative stress in acute ischemic stroke? A critical review of the literature. Med Gas Res 5:11

72. Alayash AI (2004) Oxygen therapeutics: can we tame haemoglobin? Nat Rev Drug Discov 3(2):152–159
73. Bowes MP, Burhop KE, Zivin JA (1994) Diaspirin cross-linked hemoglobin improves neurological outcome following reversible but not irreversible CNS ischemia in rabbits. Stroke 25(11):2253–2257
74. Cole DJ, Schell RM, Przybelski RJ, Drummond JC, Bradley K (1992) Focal cerebral ischemia in rats: effect of hemodilution with alpha-alpha cross-linked hemoglobin on CBF. J Cereb Blood Flow Metab 12(6):971–976
75. Saxena R, Wijnhoud AD, Carton H, Hacke W, Kaste M, Przybelski RJ et al (1999) Controlled safety study of a hemoglobin-based oxygen carrier, DCLHb, in acute ischemic stroke. Stroke 30(5):993–996
76. Buehler PW, D'Agnillo F (2010) Toxicological consequences of extracellular hemoglobin: biochemical and physiological perspectives. Antioxid Redox Signal 12(2):275–291
77. Johnson JL, Leos RA, Baker AF, Unger EC (2015) Radiosensitization of Hs-766T pancreatic tumor xenografts in mice dosed with dodecafluoropentane nano-emulsion-preliminary findings. J Biomed Nanotechnol 11(2):274–281
78. Culp WC, Brown AT, Lowery JD, Arthur MC, Roberson PK, Skinner RD (2015) Dodecafluoropentane emulsion extends window for tPA therapy in a rabbit stroke model. Mol Neurobiol 52(2):979–984
79. Culp WC, Woods SD, Skinner RD, Brown AT, Lowery JD, Johnson JL et al (2012) Dodecafluoropentane emulsion decreases infarct volume in a rabbit ischemic stroke model. J Vasc Interv Radiol 23(1):116–121
80. Brown AT, Arthur MC, Nix JS, Montgomery JA, Skinner RD, Roberson PK et al (2014) Dodecafluoropentane emulsion (DDFPe) decreases stroke size and improves neurological scores in a permanent occlusion rat stroke model. Open Neurol J 8:27–33
81. Barbosa FT, Juca MJ, Castro AA, Duarte JL, Barbosa LT (2009) Artificial oxygen carriers as a possible alternative to red cells in clinical practice. Sao Paulo Med J 127(2):97–100
82. Buehler PW, D'Agnillo F, Schaer DJ (2010) Hemoglobin-based oxygen carriers: From mechanisms of toxicity and clearance to rational drug design. Trends Mol Med 16(10):447–457
83. Misra H, Lickliter J, Kazo F, Abuchowski A (2014) PEGylated carboxyhemoglobin bovine (SANGUINATE): results of a phase I clinical trial. Artif Organs 38(8):702–707
84. Mito T, Nemoto M, Kwansa H, Sampei K, Habeeb M, Murphy SJ et al (2009) Decreased damage from transient focal cerebral ischemia by transfusion of zero-link hemoglobin polymers in mouse. Stroke 40(1):278–284
85. Nemoto M, Mito T, Brinigar WS, Fronticelli C, Koehler RC (2006) Salvage of focal cerebral ischemic damage by transfusion of high O_2-affinity recombinant hemoglobin polymers in mouse. J Appl Physiol (1985) 100(5):1688–1691
86. Thuillier R, Dutheil D, Trieu MT, Mallet V, Allain G, Rousselot M et al (2011) Supplementation with a new therapeutic oxygen carrier reduces chronic fibrosis and organ dysfunction in kidney static preservation. Am J Transplant 11(9):1845–1860
87. Le Gall T, Polard V, Rousselot M, Lotte A, Raouane M, Lehn P et al (2014) In vivo biodistribution and oxygenation potential of a new generation of oxygen carrier. J Biotechnol 187:1–9
88. Weinert EE, Plate L, Whited CA, Olea C Jr, Marletta MA (2010) Determinants of ligand affinity and heme reactivity in H-NOX domains. Angew Chem Int Ed Engl 49(4):720–723
89. Pellicena P, Karow DS, Boon EM, Marletta MA, Kuriyan J (2004) Crystal structure of an oxygen-binding heme domain related to soluble guanylate cyclases. Proc Natl Acad Sci U S A 101(35):12854–12859
90. Karow DS, Pan D, Tran R, Pellicena P, Presley A, Mathies RA et al (2004) Spectroscopic characterization of the soluble guanylate cyclase-like heme domains from Vibrio cholerae and Thermoanaerobacter tengcongensis. Biochemistry 43(31):10203–10211
91. Eich RF, Li T, Lemon DD, Doherty DH, Curry SR, Aitken JF et al (1996) Mechanism of NO-induced oxidation of myoglobin and hemoglobin. Biochemistry 35(22):6976–6983

92. Henkel-Honke T, Oleck M (2007) Artificial oxygen carriers: a current review. AANA J 75(3):205–211
93. Paul W, Buehler FDA (2013) Pre-clinical evaluation of hemoglobin based oxygen carriers: animal models and biomarkers. In: Kim HW, Greenburg AG (eds) Hemoglobin-based oxygen carriers as red cell substitutes and oxygen therapeutics. Springer, New York, pp 457–474
94. Saver JL, Starkman S, Eckstein M, Stratton SJ, Pratt FD, Hamilton S et al (2015) Prehospital use of magnesium sulfate as neuroprotection in acute stroke. N Engl J Med 372(6):528–536
95. Ankolekar S, Fuller M, Cross I, Renton C, Cox P, Sprigg N et al (2013) Feasibility of an ambulance-based stroke trial, and safety of glyceryl trinitrate in ultra-acute stroke: the rapid intervention with glyceryl trinitrate in Hypertensive Stroke Trial (RIGHT, ISRCTN66434824). Stroke 44(11):3120–3128

Part IV
Stroke Models and Clinical Trial Considerations

Chapter 28
The Right Rodent for the Job: Infarct Variability Between Strains and Its Impact on Logistics of Experimental Animal Studies

Sarah Rewell and David W. Howells

Abstract This chapter will discuss the variability in infarct size after ischaemic stroke in rat models of stroke, drawing example from our experience with the thread occlusion model. We will describe how the neuroprotective effect of a novel treatment diminished over the course of our testing, with post hoc analysis revealing wide variability in infarct volume in the experiments where the treatment was not shown to be protective. Application of various inclusion criteria failed to reduce variability, only reducing the number of animals. We then compared infarct variability in the Sprague-Dawley strain to other strains of rat used in our laboratory. The spontaneously hypertensive rat proved to be the most consistent strain of rat, having the least variable infarct volume, and stroke being successfully induced in all animals. The ability to include more animals in experimental groups is advantageous in terms of the absolute number of animals used, the time an experiment will take to complete and the cost of preclinical research.

Keywords Rat • Stroke • Variability • Exclusion criteria

1 Introduction

Occlusion of the Middle Cerebral Artery (MCA) is the most common form of ischaemic stroke in patients. Therefore, most animal models have been developed to mimic this type of stroke. A variety of techniques have been developed to induce ischaemia in the MCA territory; however, the most widely used model employed to

S. Rewell
Florey Institute of Neuroscience and Mental Health, Heidelberg, VIC, Australia

D.W. Howells (✉)
Florey Institute of Neuroscience and Mental Health, Heidelberg, VIC, Australia

Faculty of Health, School of Medicine, University of Tasmania,
Medical Science Precinct, 17 Liverpool Street, Hobart, TAS, Australia
e-mail: david.howells@utas.edu.au

P.A. Lapchak, J.H. Zhang (eds.), *Neuroprotective Therapy for Stroke and Ischemic Disease*, Springer Series in Translational Stroke Research,
DOI 10.1007/978-3-319-45345-3_28

study stroke is the intraluminal thread occlusion model [1]. This model involves careful manipulation of the vasculature of the neck to access the carotid arteries, permitting a filament (often with a rounded or cylindrical shaped tip) to be passed via the internal carotid artery (ICA) to occlude the MCA at its origin. Despite almost 30 years [2, 3], and numerous modifications [4–8], the thread occlusion model still has a wide degree of variability in the size of infarct produced, both between laboratories and within studies [9–14]. This has implications for the reliability and reproducibility of preclinical stroke research and the potential for translation to a clinical treatment.

Key factors that may influence infarct size and variability include the site and success of MCA occlusion (directly related to the experimental model) [6, 11, 15–18], occlusion of arteries other than the MCA (e.g. common carotid artery (CCA), anterior choroidal artery [19–21]), the surgical experience of the investigators [22–24], factors related to the surgery (e.g. anaesthetic choice [25–29], control of body temperature [30]) and the time at which outcome is assessed [31]. Additionally, the choice of animal and strain plays a role in variability [7, 13]. Strains of rodent differ markedly in the pattern of their vascular anatomy both at the level of the circle of Willis and more distal MCA branching [32, 33]. This impacts on susceptibility to ischaemia and evolution of ischaemic-related damage [34, 35]. Differences within strains may also be due to vascular variability and may be exacerbated when animals are sourced from different suppliers or when wide age or weight ranges are used [36–39].

Understanding and limiting the amount of variability in animal models is important not only for the testing of potential therapeutics but also for expanding our understanding of ischaemic stroke and the damage that ensues. An animal model where the amount and location of damage are consistent (and ideally can be predicted by cerebral blood flow (CBF) changes and/or behavioural deficits) is imperative if animal models are to be relied upon for preclinical testing. Additionally, as large infarcts that encompass the entire MCA territory are associated with high mortality, controlling the size of the infarct is important ethically and practically. Exploring consistency in infarct size will provide the cornerstone for the work presented in this chapter.

This manuscript examines the variability in infarct size within the Sprague-Dawley strain and the impact of applying inclusion and exclusion criteria aimed to limit variability. We then extended the analysis to examine whether other common strains of rat exhibit the same variability. From the conclusions drawn, we discuss the impact of infarct variability on sample size and subsequent costs of experiments.

2 Influence of Inclusion Criteria on Infarct Volume Variability in Sprague-Dawley Rats

The Sprague-Dawley strain of rat has been widely used in experimental models of stroke, particularly in the evaluation of potential neuroprotective agents [1, 40, 41]. However, within our laboratory, the reproducibility of stroke induction and consistency in infarct size have been found to be highly variable in this strain, even in the hands of multiple experienced stroke surgeons. Whilst improvements were seen when the occluding thread was changed from a poly-L-lysine-coated 3-0 nylon heat-blunted thread to a silicone-coated 4-0 nylon thread, a wide degree of variability of infarct size and behaviour was still noted [7].

Animals presented in Sect. 2 are from two large neuroprotection dose-response experiments examining transient (2 h) and permanent (24 h) middle cerebral artery occlusion (MCAo). These were the final experiments in a series performed by our laboratory to examine the neuroprotective potential of this treatment. Early experiments showed promising reductions in infarct volume at 24 h and 7 days post stroke, albeit with differing levels of protection. The experiments presented here share common experimental design, with MCAo induced by the thread occlusion method using silicone-coated threads and tissue collected for infarct analysis by 2,3,5-triphenyltetrazolium chloride (TTC) staining at 24 h post MCAo. All were randomly allocated to the treatment group (or did not receive treatment depending on the inclusion criteria of the neuroprotection study), and an investigator blinded to treatment assessed infarct volume. Animals were included based upon CBF drop >60 to <85 % relative to baseline and additionally excluded if reperfusion was not noted on laser Doppler at withdrawal of the occluding thread (in the transient MCAo experiment) or if they failed to show signs of neurological deficit at 2 h post MCAo. Animals that failed to meet the inclusion criteria did not receive any treatment and were collected for analysis of infarct volume at 24 h.

In the first dose-response experiment, in which the treatment was infused via the right femoral vein over 4 h starting 6 h after induction of stroke (2 h transient (t) MCAo), we found that the treatment did not affect infarct volume (Fig. 28.1a). In all experimental cohorts, including the saline-treated controls, there was considerable variation in infarct volume. A follow-up experiment designed to determine whether extending the bioavailability of the compound (through a bolus and maintenance dose) might reveal a protective effect after permanent (p)MCAo also found no effect and showed marked infarct volume variability across all cohorts, including controls (Fig. 28.1b). This data was highly variable despite the exclusion of 31 % (tMCAo) and 43 % (pMCAo) animals for failing to meet our CBF (CBF drop at MCAo >60 to <85 % of baseline) or behavioural (>2.5 total neuroscore, with forelimb flexion = 1) inclusion criteria.

Given that many studies use criteria based on CBF and/or behavioural deficit to include or exclude animals from a study [1, 6, 8, 42, 43], we assessed different combinations of these criteria in a post hoc analysis of our data set to examine whether their application provided a practical route to less variable experimental cohorts.

Considering all animals to undergo tMCAo, infarct volume in 122 rats ranged from 0 to 432 mm^3, with a mean of 128.9 ± 111.1 mm^3 and coefficient of variation of 87 % (Table 28.1). Application of the exclusion criteria based on a CBF and behavioural criteria had no impact on average infarct volume or variability in infarct size, with a mean of 129.4 ± 101.0 mm^3 and coefficient of variation of 78 %, but reduced the number of animals included for analysis to 84. Of the 38 animals who did not meet the prespecified inclusion criteria, average infarct volume was not different to those that were treated and only slightly more variable (Table 28.1).

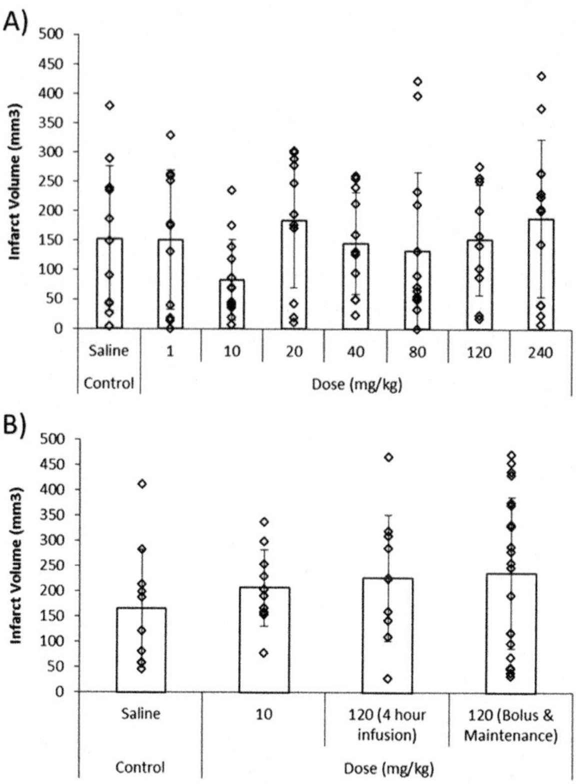

Fig. 28.1 Variability in the ability to induce stroke and the resulting infarct volume masked any protective effect of the treatment after transient (**a**) or permanent (**b**) MCAo. Mean ± standard deviation; each point represents an individual animal

A similar pattern was seen in the pMCAo experiment. When all animals were considered ($n = 88$), mean infarct volume was 194.1 ± 122.5 mm^3, coefficient of variation 63 %. Inclusion of animals based on CBF drop and acute behavioural symptoms resulted in an average infarct volume of 202.1 ± 124.7 mm^3, coefficient of variation of 62 %, including only 50 animals (Table 28.1). Within the saline control group for each experiment, the variability in infarct volume was just as wide, 149.8 ± 129.3 mm^3 (coefficient of variation 86 %) for transient MCAo and 151.9 ± 120.0 mm^3 (coefficient of variation 79 %) for permanent MCAo.

When all experimental data was considered regardless of exclusion criteria, stroke induction was very successful with infarction evident in 81 % and 86 % of tMCAo and pMCAo animals, respectively (Table 28.2). Of those animals to have an infarct, 52 % and 67 % (tMCAo and pMCAo, respectively) involved cortical damage. The most common cause of death in the tMCAo experiment was subarachnoid haemorrhage (SAH) at thread insertion ($n = 14$, 61 % of deaths). After pMCAo, common causes of death were SAH and large hemispheric infarct causing massive oedema (for each: $n = 4$, 36 % of deaths). Depending on the degree of perforation of

Table 28.1 Range of infarct volumes resulting from tMCAo and pMCAo in the Sprague-Dawley rat. Data is broken into those that were included and excluded from the study, together with a focus on the saline control group

Occlusion duration	Criteria	Infarct volume (mm^3)	Coefficient of variation (%)	Minimum infarct volume (mm^3)	Maximum infarct volume (mm^3)	*n* that meet criteria
tMCAo (2 h)	All animals to reach 24 h endpoint (145 underwent surgery)	128.9 ± 111.1	87	0	432.2	122
	Died	–	–	–	–	23
	Treated animals (met inclusion criteria of CBF drop >60 to <85 %; behavioural deficit at 2 h)	129.4 ± 101.0	78	0	421.8	84
	Saline controls	149.8 ± 129.3	86	4.19	378.8	10
	Excluded (did not meet CBF or behavioural criteria)	128.0 ± 132.1	103	0	432.2	38
pMCAo (24 h)	All animals to reach 24 h endpoint (99 underwent MCAo)	194.1 ± 122.5	63	0	469.8	88
	Died	–	–	–	–	11
	Treated animals (met inclusion criteria of CBF drop >60 to <85 %; behavioural deficit at 2 h)	202.1 ± 124.7	62	16.07	464.9	50
	Saline controls	151.9 ± 120.0	79	16.07	410.2	11
	Excluded (did not meet CBF or behavioural criteria)	180.0 ± 119.8	67	0	469.8	38

Table 28.2 Success of stroke induction after transient or permanent MCAo in the Sprague-Dawley rat

	tMCAo	pMCAo
Any infarct	81 % (n=117)	86 % (n=85)
Cortical infarct	52 % (n=76)	67 % (n=66)
No damage	3 % (n=5)	1 % (n=1)
Died	16 % (n=23)	11 % (n=11)
Subarachnoid haemorrhage	61 % (n=14)	36 % (n=4)

the vessel wall, death from SAH can occur quickly (within minutes) or over a more prolonged period. In some animals, death from SAH did not occur until after the thread was withdrawn at reperfusion. Other animals had a slower bleed into the base of the skull, displaying lethargic behaviour and not surviving overnight.

Given the variability in infarct volume across all animals, we sought to determine if there were any criteria that could be applied to the data set in order to limit variability, whilst still including a large number of animals in who stroke had been successfully induced. The first approach was to examine whether CBF drop at MCAo was a good predictor of infarct size after transient or permanent MCAo.

Within the transient MCAo experiment, 61 % of animals had a CBF drop within the range 60–80 % (Fig. 28.2a, b). Infarct volume varied widely, ranging from 0-421 mm^3. Correlation between infarct volume and CBF drop was low: Pearson's correlation $r=0.46$ ($n=122$, $P=0.000$, $r^2=0.212$) (Fig. 28.2b). Animals with a CBF drop greater than 85 % clustered into two groups, those that died and those that had very large infarcts (Fig. 28.2b). It seems likely that those that died might also have had very large infarcts had they survived and had we been able to measure this. In both transient and permanent MCAo, CBF drop was less than 50 % of baseline in only 14 and 9 % of animals (Fig. 28.2a, c). Similarly, in the pMCAo experiment, 50 % of animals had a CBF drop within the range 60–80 %. Again, infarct volume varied, ranging from 31 to 469 mm^3. CBF drop did not correlate with infarct volume. Pearson's correlation $r=0.037$ ($n=85$, $P=0.734$, $r^2=0.001$). Given the range of resulting infarct volumes, CBF drop assessed at the co-ordinates used does not alone appear to be a reliable reason for inclusion.

Behavioural deficit was also explored as a criterion for predicting infarct success (Fig. 28.3). Behavioural deficit (of any degree) was observed in 89 and 91 % of tMCAo and pMCAo animals at 2 h post MCAo, immediately prior to reperfusion for the tMCAo group. The most common score in both experiments was 3 (out of 5), with greater than 60 % of animals scoring 3 at 2 h post stroke (Fig. 28.3a, d). The range in infarct volume for those animals scoring 3 was 0–432 mm^3 for tMCAo and 28–469 mm^3 for pMCAo (Fig. 28.3b, d). By 24 h, many animals' behavioural deficit scores had improved, with 33 % of tMCAo and 14 % of pMCAo animals having no observable deficit (Fig. 28.3c, f). A smaller proportion of animals worsened over 24 h, scoring higher on the basic behavioural scale. These animals tended to have larger infarcts. Thus, behavioural deficit early after the onset of MCAo also does not appear to be a good tool alone for basing inclusion criteria as a wide range of infarct volumes were seen even among animals with a strong behavioural deficit. A very weak correlation between infarct volume and behavioural deficit at 2 h was found for both transient (Spearman's correlation $r=0.263$ ($n=118$, $p=0.004$, $r^2=0.069$)) and

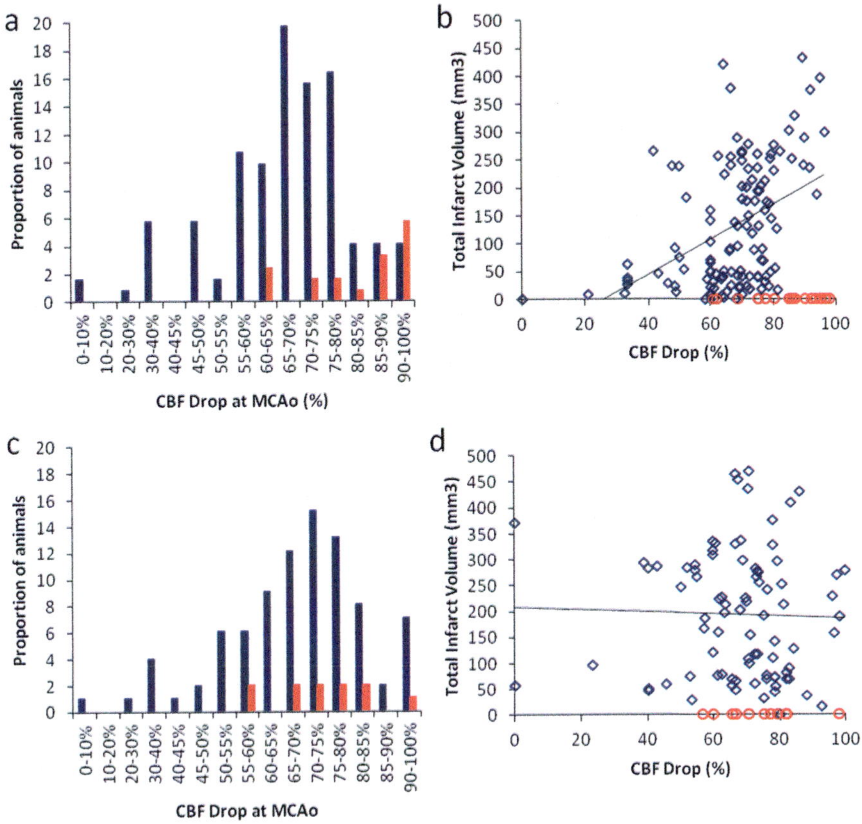

Fig. 28.2 Effect of CBF drop on infarct volume after 2 h transient (**a**, **b**) or permanent (**c**, **d**) MCAo. Animals that died are represented in *red* and given an arbitrary infarct volume of zero (as infarct could not be assessed). Animals that survived 24 h are represented in *blue*

permanent MCAo (Spearman's correlation $r=0.282$ ($n=86$, $p=0.008$, $r^2=0.008$)) (Fig. 28.3b, e). Behavioural deficit assessed at 24 h post MCAo gave a stronger correlation with infarct volume. For transient MCAo Spearman's correlation $r=0.686$ ($n=118$, $p=0.000$, $r^2=0.47$). For permanent MCAo, there was a weaker relationship, Spearman's correlation $r=0.485$ ($n=85$, $p=0.000$, $r^2=0.24$) (Fig. 28.3c, f).

3 Comparison of Variability in Infarct Volume Across Different Strains of Rat

Given the high degree of variability within the Sprague-Dawley (SD) strain, we then looked to see if the same were true for other common strains of rat. Section 3 presents a retrospective analysis of infarct data across multiple strains of rat

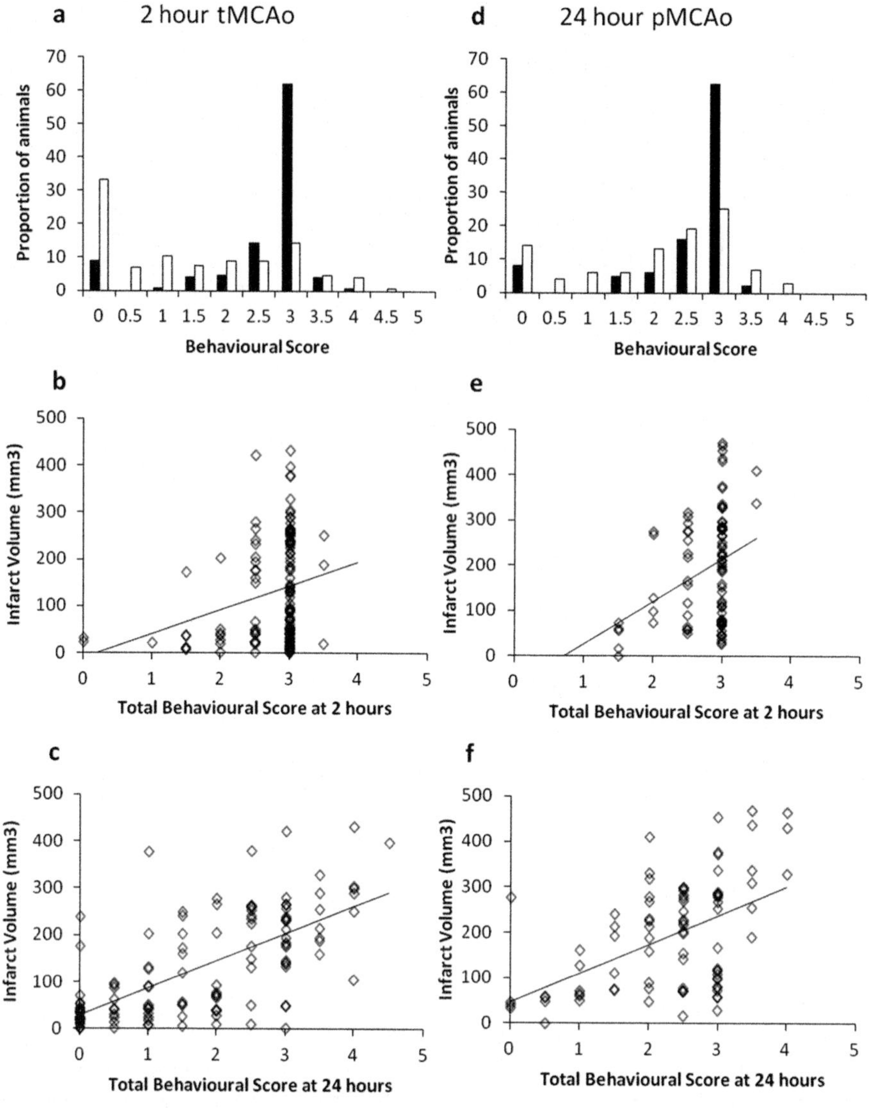

Fig. 28.3 Relationship between infarct volume and behavioural deficit assessed at 2 and 24 h post transient (**a**–**c**) and permanent (**d**–**f**) MCAo. (**a**, **d**) The proportion of animals with each score at each time point (*filled bars* 2 h assessment, *open bars* 24 h assessment). (**b**, **e**) Correlation between infarct volume and behavioural deficit assessed at 2 h post MCAo. (**c**, **f**) Correlation between infarct volume and behavioural deficit assessed at 24 h

(Fig. 28.3, Table **28.3**). SD rats were part of a large neuroprotection trial (unpublished). WKY and SHR rats were parts of larger studies investigating the effect of either co-morbidities, ACE inhibitor treatment or combination therapy on stroke induction and outcome [41, 44, 45]. Only the control animals from each experiment

Table 28.3 Summary of animals used in comparison of infarct volume variability across different rat strains

Strain	Occlusion duration (minutes)	Age of animal at MCAo (weeks)	Histology for infarct analysis	Source experiment	*n*
Sprague-Dawley (SD)	120	12–15	TTC	Unpublished neuroprotection studies	31
Wistar Kyoto (WKY)	120	56	H&E	Porritt et al., Rewell et al. [44, 45]	16
Spontaneously Hypertensive Rat (SHR)	120	12–15	H&E	O'Collins et al. [41]	13
	120	56	H&E	Porritt et al., Rewell et al. [44, 45]	24

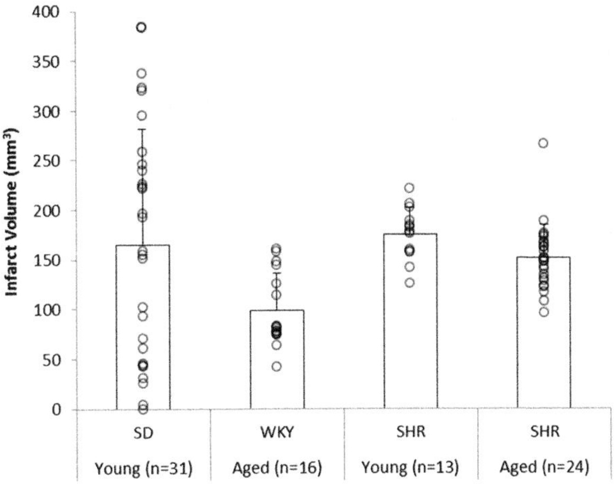

Fig. 28.4 Comparison of infarct volume across different strains of rat after MCAo. All measurements were made 24 h after MCAo. *Bars* represent average ± standard deviation. *Circles* represent individual animals

are included in this analysis. These animals received intravenous saline [41] (unpublished neuroprotection study), strawberry topping delivered orally [44] or 0.1 mol/L sodium citrate delivered via the tail vein [45]. Experimental groups differed in the strain of rat and histological stain used to delineate ischaemic damage. Threads of 0.38 mm (experiments using the SD strain) or 0.35 mm diameter (experiments using the WKY and SHR strains) with a 5 mm silicone length were made and used within 10 days of manufacture [7]. All cohorts had an endpoint of 24 h post MCAo.

Comparing infarct volume across different strains and experiments showed much greater variation in infarct volume within the SD strain compared to both WKY and SHR (Fig. 28.4). This is reflected by the larger coefficient of variation of 71 % for infarct volume within the SD strain, compared to 38, 15 and 22 % for

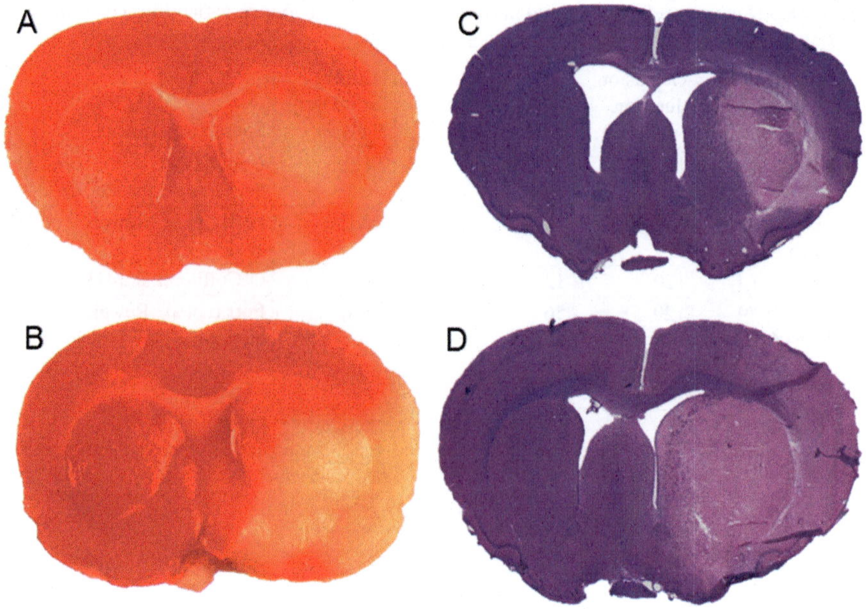

		Stain for infarct analysis	Infarct Volume (mm^3)	Infarct as a proportion of contralateral hemisphere (%)
A	SD	TTC	151mm^3	17.4%
B	SD	TTC	191mm^3	23.5%
C	WKY	H&E	76mm^3	16.4%
D	SHR	H&E	139mm^3	33.5%

Fig. 28.5 Representative images demonstrating TTC (**a**, **b**) and H&E (**c**, **d**) staining after MCAo. Infarct volumes (table) do not reflect the true amount of damage, highlighting the need to express infarct as a proportion of contralateral hemisphere

Wistar Kyoto (WKY) and spontaneously hypertensive rat (young and aged) strains, respectively (Fig. 28.5, Table 28.4). Comparison of infarct volume across different strains of rat also revealed greater success in stroke induction in WKY and SHR strains compared to SD. All WKY and SHR rats to undergo MCAo surgery resulted in a measureable infarct at 24 h post stroke.

Comparison of infarct volume across different strains and experiments also highlighted the need to normalise data by converting infarct volume to a proportion of contralateral hemisphere. This is demonstrated in Figs. 28.4 and 28.5 and Table 28.4 where reporting infarct volume alone does not reflect the true amount of damage. For example, in Fig. 28.5, panels a and c have infarcts that are qualitatively similar (involving the striatum and a small portion of the cortex) and have similar amounts

Table 28.4 Comparison of infarct volume and infarct expressed as a proportion of contralateral hemisphere across different strains of rat after 2 h tMCAo

Strain	Age at MCAo (weeks)	Histology for infarct analysis	*n*	Infarct volume (mm^3)	Infarct as a proportion of contralateral hemisphere	Proportion of animals with cortical involvement (%)
Sprague-Dawley (SD)	12–15	TTC	31	165.1 ± 116.8 (71 %)	21.6 ± 14.9 (69 %)	71
Wistar Kyoto (WKY)	56	H&E	16	98.7 ± 37.2 (37 %)	24.1 ± 10.3 (43 %)	88
Spontaneously hypertensive rat—young	15	H&E	13	175.1 ± 26.6 (15 %)	45.8 ± 7.4 (16 %)	100
Spontaneously Hypertensive Rat—Aged	56	H&E	24	151.3 ± 33.7 (22 %)	37.8 ± 8.5 (22 %)	100

Average ± standard deviation (coefficient of variation)

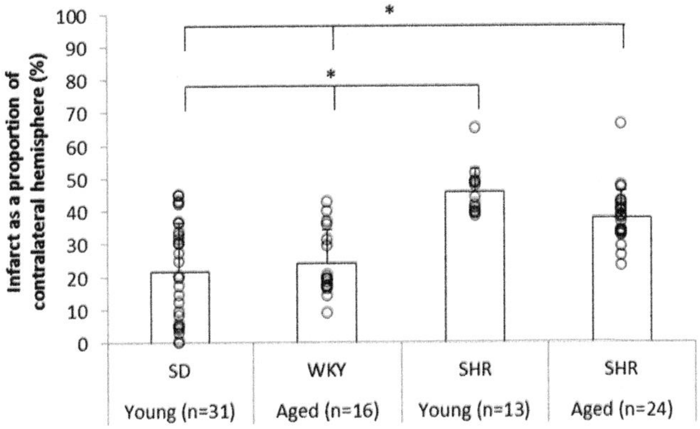

Fig. 28.6 Comparison of different strains of rat after tMCAo where infarct volume is expressed as a proportion of contralateral hemisphere to overcome differences in tissue preparation. *Bars* represent average ± standard deviation. *Circles* represent individual animals. *$p < 0.05$

of damage when expressed as a proportion (17.4 and 16.4 %) yet vastly different infarct volumes (151 and 76 mm^3). Such disparities are likely due to differences in fixation and staining of the brain tissue. Perfusion fixation causes an overall shrinkage of the brain, which must be taken into account when comparing fresh unfixed tissue with fixed tissue.

Converting infarct volume to a proportion of contralateral hemisphere showed the relative differences in infarct size across different strains regardless of staining and fixation method (Fig. 28.6). One-way ANOVA found differences in infarct volume between different strains of rat ($F(3, 177) = 16.974$; $p = 0.000$). Bonferroni post hoc comparison found both young and aged SHR rats to have significantly larger infarcts than the SD and WKY strains ($p < 0.05$). Variability in infarct volume was largest in the Sprague-Dawley strain (21.6 ± 14.9 % of contralateral hemisphere).

Table 28.5 Sample size calculations for a range of effect sizes across different strains of rat based on infarct expressed as a proportion of contralateral hemisphere

Strain	Occlusion duration (minutes)	μ1	σ	50 % Effect	40 % Effect	30 % Effect	20 % Effect	10 % Effect
Sprague-Dawley	120	21.6	14.9	30	47	83	187	747
Wistar Kyoto	120	24.1	10.3	12	18	32	72	287
Spontaneously hypertensive rat—young	120	45.8	7.4	2	3	5	9	41
Spontaneously hypertensive rat—aged	120	37.8	8.5	4	5	9	20	80

Calculations are based on infarct expressed as a proportion of contralateral hemisphere
Two-sided test; power = 0.8; $\alpha = 0.05$

Table 28.6 Sample size calculations for a range of effect sizes across different strains of rat based on infarct volume

Strain	Occlusion duration (minutes)	μ1	σ	50 % Effect	40 % Effect	30 % Effect	20 % Effect	10 % Effect
Sprague-Dawley	120	165.1	116.8	32	50	88	197	786
Wistar Kyoto	120	98.7	37.2	8	14	25	56	223
Spontaneously hypertensive rat—young	120	175.1	26.6	2	2	5	10	37
Spontaneously hypertensive rat—aged	120	151.3	33.7	4	5	9	20	78

Calculations are based on infarct volume
Two-sided test; power = 0.8; $\alpha = 0.05$

Infarcts were of a similar size yet of smaller variability in the WKY (24.1 ± 10.3 % of contralateral hemisphere). Damage was largest and most consistent in the SHR strain of both young and aged groups (45.8 ± 7.4 % and 37.8 ± 8.6 % of contralateral hemisphere, young and aged, respectively). The same pattern was reflected in the proportion of animals with cortical infarction. All SHR animals showed damage to the cortex, whilst 88 % and 71 % of WKY and SD rats had cortical involvement (Fig. 28.5, Table 28.4).

Variability in the success of stroke induction and the resulting infarct volume ultimately impact on the statistical power of an experiment and the number of animals required. Sample size calculations for a range of effect sizes (with 80 % power, α 0.05) demonstrate the large number of animals required in each experimental group when the SD is the chosen rat strain (Tables 28.5 and 28.6). When the WKY is selected, sample sizes are more modest however still relatively large (25–32 animals per cohort required to show an effect of 30 %). The SHR provided manageable group sizes across the range of effect sizes calculated. To detect an effect size of 30 %, 5–9 animals are required per cohort based on infarct volume from young and aged animals.

4 Discussion

Acute markers of successful stroke induction are required for studies in which a treatment is administered early after the onset of ischaemia. It is necessary to distinguish between animals in whom stroke has been successfully induced and those in which the procedure does not result in an infarct before any treatment is initiated. In this analysis, despite animals showing signs of stroke, with decrement in CBF at MCAo and behavioural deficit at 2 h, selection of animals based on these criteria was not able to distinguish between animals with successful and unsuccessful MCAo induction, a wide variability in infarct volume persisting. This is in agreement with a recent analysis by Morris et al., who demonstrated that CBF drop at MCAo did not correlate with infarct volume and that there was no clear rationale for excluding animals with a CBF drop <70 % [20].

For the transient MCAo animals, it could be argued that the 2 h occlusion time applied is close to the threshold for commitment to infarction and thus some animals with appropriate behavioural and laser Doppler signals develop infarcts, whilst others do not. However, the absence of emergence of a strong correlation after permanent MCAo does not support this hypothesis.

A number of methodological and biological factors could account for this. The position of the laser Doppler probe on the skull could influence the degree of CBF decrement observed. Assessing CBF at different (or multiple) sites, or using alternative measurement probes, may improve the predictive value of laser Doppler flowmetry [6, 46–48]. The effectiveness of collateral arteries may also influence the interpretation of laser Doppler recordings, particularly if the probe is placed within the ACA-MCA watershed area. Despite a low correlation between CBF drop and infarct volume, laser Doppler flowmetry is still a useful tool for identifying inadvertent subarachnoid haemorrhage at thread insertion, excluding these animals from treatment [6, 20, 24].

The behavioural assessment used is also likely to play a role in identifying animals in which stroke has been successfully induced. Early behavioural assessments may be influenced and heightened by the effects of anaesthesia. In this study, many animals showed strong behavioural deficit at 2 h despite having no or very small infarcts when assessed at 24 h. The individual components of the test may also be contributing differently to the behavioural score reported.

An animal model of stroke that gives consistent induction of infarction with limited variation is a desirable characteristic when testing potential neuroprotectants. Our experience with the Sprague-Dawley strain of rat indicates that the size of infarct produced after both transient and permanent thread occlusions of the MCA is extremely broad, ranging from very tiny (or no) infarcts to those that encompass the entire MCA-supplied territory. Efforts to limit variability in infarct size by applying criteria of CBF drop at MCAo or degree of behavioural deficit did not affect the range of infarct sizes seen after either transient or permanent occlusion.

Detecting differences in infarct volume between strains of rat is not new [23, 35, 49–53]. However, an appreciation of the differences between strains is important as strains have been reported to respond differently to the effects of potential neuroprotectants [54]. This may be due to differences in the evolution of ischaemic damage between strains. For example, in the SD, lesions appear to evolve more rapidly, developing into larger infarcts than in WKY rats [35]. Additionally, the timing of assessment of infarct may play a role in the degree of variability. Aspey reported that when infarct was assessed at 72 h, the volume of damage was larger and slightly less variable than when measured at 24 h, most likely because of the effects of oedema [50]. Whatever the cause of these differences, they imply that researchers must carefully balance the costs of larger numbers that might be needed at 24 h against the cost of extended animal care and housing.

SD rats had the widest range of infarcts of the three strains examined. They also had the smallest average proportion of damage and more animals with little or no evidence of infarct. Infarct success was greater in the WKY strain, with all animals showing evidence of infarct and 88 % including cortical damage (Table 28.4). The SHR strain was found to have the most consistent and reproducible infarcts. Cortical damage was always noted, and as reported previously, there was no stroke-related mortality at 24 h [45]. These properties suggest this strain might provide an ideal tool for early screening for in vivo drug activity.

Others have found similar patterns in infarct size. Whilst infarcts may appear qualitatively similar across strains, encompassing the striatum and cortex, Sauter et al. showed that when the volume of damage was measured, quantitative differences between strains emerged [54]. Differing portions of the cortex become infarcted, with WKY rats having the smallest volume of cortical damage (30 μL) and SHR having the largest portion of cortical involvement (127 μL). This resulted in a threefold difference in total infarct volume between the smallest (WKY) and largest (SHR) infarct volumes [54]. In addition, the distribution of damage through the length of the brain differs between strains [53]. Aspey et al. reported SD rats to have the smallest and most variable infarct at 24 h compared to Wistar or Fisher rats [50]. However, Duverger, Barone and Bardutzky all reported SD rats to have larger infarcts than WKY rats [13, 35, 53]. The SHR has been reported to have the largest and most consistent infarct volume of all strains of rat [11, 13, 23, 53]. Such differences may be due to variations in the pattern of MCA branching and degree of anastomoses with adjacent arterial territories [32, 36, 55–57]. However, these parameters were not measured in our study.

The results from this analysis, together with that of others, suggest the SD strain to be an unreliable strain in which to induce ischaemic stroke using the thread occlusion model as the infarct size and location are highly variable, even in the hands of multiple experienced surgeons [23]. In practical terms, this suggests that the "noise" generated by data from SDs is too great to effectively power any but the most expensive experiments. The variability in infarct size seen in our study was much tighter than those of Aspey for 2 h tMCAo in both SD and WKY rat strains, yet still large enough to make the sample size unreasonably large even for detecting large effect sizes [50]. Our results support Aspey's contention that WKY rats are the best choice of normotensive strain for stroke studies as infarction is more reliably induced and is reasonably consistent in size.

Whilst the WKY and SHR strains had similar variability, the SHR was favoured as the strain most useful for future studies as all animals showed damage to the cortex, the location of the ischaemic penumbra and most common target for novel therapies. Additionally, the hypertension that develops in the SHR makes it relevant to the stroke population. The larger and more consistent infarct of the SHR may in part be due to the smaller diameter of anastomoses between the MCA, posterior cerebral artery (PCA) and anterior cerebral artery (ACA) in hypertensive rats compared to WKY. This may affect the collateral circulation and influence the size of the infarct resulting from MCAo [58–62]. However, it might also limit the size of the ischaemic penumbra and reduce the chance of detecting a drug effect [63, 64]. Similar effects in humans with long-standing hypertension will have important implications for therapeutic success in different portions of the stroke population.

Sample size calculations for a range of effect sizes highlighted the impact variability of infarct size can have on the quality of a study. Based on these sample size estimates alone, the SD rat is not an appropriate choice of rat strain for thread occlusion models of MCAo (Tables 28.5 and 28.6). The time and expense required to induce MCAo in this large number of animals is not appropriate ethically, financially or practically. The WKY offers a more manageable sample size whilst still maintaining a normotensive background. Based on our analysis, the SHR presents the most attractive choice of strain. The small animal number is required, aided by the fact that all animals have an infarct (including cortical involvement), with limited variability; and the addition of hypertension as a co-morbidity makes the SHR an appropriate strain in which to experimentally model stroke. Moreover, since the SHR brain is protected by early reperfusion, these animals do have tissue that can be targeted for therapy [41, 65, 66].

Variability in outcome and its effect on sample size are important considerations when planning and assessing experiments. Sample size calculations based on variability in our model suggest that the average cohort size used in experimental stroke studies of fewer than 10 animals is too small to detect even large effect sizes [67, 68]. Taking the efficacy of tPA (16 %, correcting for publication bias, randomisation and blinding) [67, 69] as a guide for the potential protective effect of a new therapy, cohort sizes of at least 30 animals would be required for even the most consistent strain used here (young SHR). Such group sizes are much larger than that used traditionally and make individual experiments more costly, both in time to complete and financially [70] However, if larger, well-conducted studies using a model that is consistent and well understood lead to translation of therapies to the clinic, fewer animals will be used overall and not used in vain [67, 70].

5 Methods

All procedures involving animals were approved by the Animal Ethics Committee of Austin Health (Heidelberg, Victoria, Australia) and performed in accordance with institutional and national guidelines (Australian code of practice for the care and use of animals for scientific purposes, 7th edition, 2004). The experiments reported here are in accordance with the Animal Research: Reporting In Vivo Experiments (ARRIVE) guidelines [71]. All animals were sourced from the Animal Resource Centre (Canning Vale, Western Australia). All experiments involved male rats. The age of rat at MCAo differed between strains: Sprague-Dawley 12 weeks, Wistar Kyoto 56 weeks and spontaneously hypertensive rat 15 weeks and 56 weeks.

Anaesthesia was induced by inhalation of 5 % isoflurane (mixed with 50:50 air and oxygen) in an enclosed box and then maintained using a nose cone where 2 % isoflurane (mixed with air and oxygen) was delivered to the spontaneously breathing animal for the duration of surgery. Throughout surgery, body temperature was monitored and controlled to 37.4 °C using a rectal probe coupled to a heat mat (manufactured in house). Postoperative analgesia was provided in the form of a single rectal suppository of paracetamol (5 mg of a 20 mg/kg solution) and provision of paracetamol (120 mg/kg body weight) in the drinking water after surgery.

We used the thread occlusion model of MCAo to induce ischaemia using the methods of Longa with modifications by Spratt [3, 7]. The right MCA was occluded for 120 min before withdrawal of the thread to allow reperfusion (or left in place for permanent occlusion). Briefly, a silicone-coated suture with 5 mm coating length (0.35 mm diameter for WKY and SHR experiments, 0.38 mm diameter for Sprague-Dawley experiments) was inserted approximately 18 mm through a stump created from the external carotid artery to occlude the MCA until mild resistance was felt. This coincided with a drop in blood flow in the area 5 mm lateral and 1 mm posterior to bregma, measured by laser Doppler.

Neurobehavioural deficit was assessed using the methods described by Petullo et al. to score forelimb flexion, torso twisting, lateral push resistance and general

Table 28.7 Scoring criteria for behavioural assessment, based on Petullo et al.

Test	Score	Description
Forelimb flexion	0	*No flexion.* Forelimbs extend equally outstretched towards the bench
	0.5	*Mild.* Left forelimb at approximately 45° angle consistently; or angle is closer to 90° but is not consistent
	1	*Moderate to severe.* Left forelimb is at 90° or greater. Forelimb flexion is consistent
Torso twisting	0	*No signs of body rotation.* Body elongated and extended towards the bench
	0.5	*Mild.* Half twist of the body
	1	*Moderate to severe.* Consistently strong twisting to contralateral side. Head and forelimbs brought towards hind limbs
Lateral push	0	*Equal resistance.* Animal resists being pushed
	0.5	*Weakened resistance.* Animal shows weakened resistance whilst trying to correct
	1	*No resistance.* Animal has severely weakened resistance. Tends to "roll" with left legs collapsing after being pushed to the left
Mobility	0	*Normal mobility.* Animal is able to freely walk, move around cage and rear on hindlimbs
	0.5	*Spontaneous movement reduced*
	1	*Needs stimulus to move*
	2	*Unable to walk*

mobility (Table 28.7) [72]. Assessments were made at 2 h (immediately prior to anaesthesia for reperfusion in animals undergoing transient MCAo) and at 24 h post MCAo.

Tissue was collected for assessment of infarct volume at 24 h post MCAo. Two methods were used to delineate infarction. Sprague-Dawley rats were killed by decapitation after isoflurane overdose. The brain was cut into 2 mm coronal sections and rapidly stained with a 1 % solution of 2,3,5-triphenyltetrazolium chloride (TTC) (Sigma–Aldrich, USA) in normal saline. After 10 min staining per side, sections were fixed in 10 % formalin. WKY and SHR groups were transcardially perfused with 0.9 % saline, followed by 4 % paraformaldehyde. The brain was collected, postfixed in the same fixative and snap frozen, before 40 μm coronal sections were collected for haematoxylin and eosin staining. Digital photographs were taken under fixed lighting conditions and infarct area quantified by researchers blinded to treatment allocation using Image J (NIH).

Infarct volumes were derived by calculating the average infarct area between slices and multiplying by the distance between slices. The impact of oedema was accounted for by adjusting the infarct area by the degree of swelling of the ipsilateral hemisphere relative to the contralateral hemisphere. Corrected infarct volume was then expressed as a proportion of the contralateral hemisphere volume.

All data are presented as mean ± standard deviation. Statistical analyses were conducted using IBM SPSS Statistics v 20. Correlation of normally distributed data was made by Pearson's test and Spearman's test for non-normally distributed data. Sample size calculations were performed using a web-based calculator (http://www.stat.ubc.ca/~rollin/stats/ssize/nt.html). A power of 0.8 and α of 0.05 was used.

References

1. Strom JO, Ingberg E, Theodorsson A, Theodorsson E (2013) Method parameters' impact on mortality and variability in rat stroke experiments: a meta-analysis. BMC Neurosci 14:41, Epub 2013/04/04
2. Koizumi J, Yoshida Y, Nakazawa T, Ooneda G (1986) Experimental studies of ischemic brain edema 1. A new experimental model of cerebral embolism in rats in which recirculation can be introduced in the ischemic area. Jpn J Stroke 8:1–8
3. Longa EZ, Weinstein PR, Carlson S, Cummins R (1989) Reversible middle cerebral artery occlusion without craniectomy in rats. Stroke 20:84–91
4. Laing RJ, Jakubowski J, Laing RW (1993) Middle cerebral artery occlusion without craniectomy in rats. Which method works best? Stroke 24(2):294–297, discussion 7-8
5. Ma J, Zhao L, Nowak TS Jr (2006) Selective, reversible occlusion of the middle cerebral artery in rats by an intraluminal approach Optimized filament design and methodology. J Neurosci Methods 156(1-2):76–83
6. Schmid-Elsaesser R, Zausinger S, Hungerhuber E, Baethmann A, Reulen H-J, Garcia JH (1998) A critical reevaluation of the intraluminal thread model of focal cerebral ischemia: Evidence of inadvertent premature reperfusion and subarachnoid hemorrhage in rats by laser-doppler flowmetry—Editorial Comment: Evidence of inadvertent premature reperfusion and subarachnoid hemorrhage in rats by laser-doppler flowmetry. Stroke 29(10):2162–2170
7. Spratt NJ, Fernandez J, Chen M, Rewell S, Cox S, van Raay L et al (2006) Modification of the method of thread manufacture improves stroke induction rate and reduces mortality after

thread-occlusion of the middle cerebral artery in young or aged rats. J Neurosci Methods 155(2):285–290
8. Takano K, Tatlisumak T, Bergmann AG, Gibson DG III, Fisher M (1997) Reproducibility and reliability of middle cerebral artery occlusion using a silicone-coated suture (Koizumi) in rats. J Neurol Sci 153(1):8–11
9. Dirnagl U (2006) Bench to bedside: the quest for quality in experimental stroke research. J Cereb Blood Flow Metab 26:1465–1478
10. Liu S, Zhen G, Meloni BP, Campbell K, Winn HR (2009) Rodent stroke model guidelines for preclinical stroke trials (1st edition). J Exp Stroke Transl Med 2(2):2–27, Epub 2009/01/01
11. Brint S, Jacewicz M, Kiessling M, Tanabe J, Pulsinelli W (1988) Focal brain ischemia in the rat: methods for reproducible neocortical infarction using tandem occlusion of the distal middle cerebral and ipsilateral common carotid arteries. J Cereb Blood Flow Metab 8(4):474–485
12. Coert BA, Anderson RE, Meyer FB (1999) Reproducibility of cerebral cortical infarction in the wistar rat after middle cerebral artery occlusion. J Stroke Cerebrovasc Dis 8(6):380–387, Epub 2007/09/27
13. Duverger D, MacKenzie ET (1988) The quantification of cerebral infarction following focal ischemia in the rat: influence of strain, arterial pressure, blood glucose concentration, and age. J Cereb Blood Flow Metab 8(4):449–461
14. Hiramatsu K, Kassell NF, Goto Y, Soleau S, Lee KS (1993) A reproducible model of reversible, focal, neocortical ischemia in Sprague-Dawley rat. Acta Neurochir (Wien) 120(1-2):66–71
15. Bederson JB, Pitts LH, Germano SM, Nishimura MC, Davis RL, Bartkowski HM (1986) Evaluation of 2,3,5-triphenyltetrazolium chloride as a stain for detection and quantification of experimental cerebral infarction in rats. Stroke 17(6):1304–1308
16. He Z, Yamawaki T, Yang S, Day AL, Simpkins JW, Naritomi H (1999) Experimental model of small deep infarcts involving the hypothalamus in rats: changes in body temperature and postural reflex. Stroke 30(12):2743–2751, discussion 51. Epub 1999/12/03
17. van der Staay FJ, Augstein KH, Horvath E (1996) Sensorimotor impairments in rats with cerebral infarction, induced by unilateral occlusion of the left middle cerebral artery: strain differences and effects of the occlusion site. Brain Res 735(2):271–284, Epub 1996/10/07
18. Zarow GJ, Karibe H, States BA, Graham SH, Weinstein PR (1997) Endovascular suture occlusion of the middle cerebral artery in rats: effect of suture insertion distance on cerebral blood flow, infarct distribution and infarct volume. Neurol Res 19(4):409–416, Epub 1997/08/01
19. Chen Y, Ito A, Takai K, Saito N (2008) Blocking pterygopalatine arterial blood flow decreases infarct volume variability in a mouse model of intraluminal suture middle cerebral artery occlusion. J Neurosci Methods 174(1):18–24, Epub 2008/07/22
20. Morris GP, Wright AL, Tan RP, Gladbach A, Ittner LM, Vissel B (2016) A Comparative Study of Variables Influencing Ischemic Injury in the Longa and Koizumi Methods of Intraluminal Filament Middle Cerebral Artery Occlusion in Mice. PLoS One 11(2), e0148503, Epub 2016/02/13
21. McLeod DD, Beard DJ, Parsons MW, Levi CR, Calford MB, Spratt NJ (2013) Inadvertent occlusion of the anterior choroidal artery explains infarct variability in the middle cerebral artery thread occlusion stroke model. PLoS One 8(9), e75779, Epub 2013/09/27
22. Durukan A, Tatlisumak T (2007) Acute ischemic stroke: Overview of major experimental rodent models, pathophysiology, and therapy of focal cerebral ischemia. Pharmacol Biochem Behav 87(1):179–197
23. Ren Y, Hashimoto M, Pulsinelli WA, Nowak TS Jr (2004) Hypothermic protection in rat focal ischemia models: strain differences and relevance to "reperfusion injury". J Cereb Blood Flow Metab 24(1):42–53
24. Taninishi H, Jung JY, Izutsu M, Wang Z, Sheng H, Warner DS (2015) A blinded randomized assessment of laser Doppler flowmetry efficacy in standardizing outcome from intraluminal filament MCAO in the rat. J Neurosci Methods 241:111–120, Epub 2014/12/21
25. Head BP, Patel P (2007) Anesthetics and brain protection. Curr Opin Anaesthesiol 20(5):395–399, Epub 2007/09/18

26. Kirsch JR, Traystman RJ, Hurn PD (1996) Anesthetics and cerebroprotection: experimental aspects. Int Anesthesiol Clin 34(4):73–93, Epub 1996/01/01
27. Macleod MR, Fisher M, O'Collins V, Sena ES, Dirnagl U, Bath PM et al (2009) Good laboratory practice: preventing introduction of bias at the bench. Stroke 40(3):e50–e52
28. Macleod MR, O'Collins T, Horky LL, Howells DW, Donnan GA (2005) Systematic review and metaanalysis of the efficacy of FK506 in experimental stroke. J Cereb Blood Flow Metab 25(6):713–721
29. Zausinger S, Baethmann A, Schmid-Elsaesser R (2002) Anesthetic methods in rats determine outcome after experimental focal cerebral ischemia: mechanical ventilation is required to obtain controlled experimental conditions. Brain Res Brain Res Protoc 9(2):112–121, Epub 2002/05/30
30. Reglodi D, Somogyvari-Vigh A, Maderdrut JL, Vigh S, Arimura A (2000) Postischemic spontaneous hyperthermia and its effects in middle cerebral artery occlusion in the rat. Exp Neurol 163(2):399–407, Epub 2000/06/02
31. Liu F, Schafer DP, McCullough LD (2009) TTC, fluoro-Jade B and NeuN staining confirm evolving phases of infarction induced by middle cerebral artery occlusion. J Neurosci Methods 179(1):1–8, Epub 2009/01/27
32. Fox G, Gallacher D, Shevde S, Loftus J, Swayne G (1993) Anatomic variation of the middle cerebral artery in the Sprague-Dawley rat. Stroke 24(12):2087–2092, discussion 92-3. Epub 1993/12/01
33. Woitzik J, Schneider UC, Thome C, Schroeck H, Schilling L (2006) Comparison of different intravascular thread occlusion models for experimental stroke in rats. J Neurosci Methods 151(2):224–231
34. Prieto R, Carceller F, Roda JM, Avendano C (2005) The intraluminal thread model revisited: rat strain differences in local cerebral blood flow. Neurol Res 27(1):47–52
35. Bardutzky J, Shen Q, Henninger N, Bouley J, Duong TQ, Fisher M (2005) Differences in ischemic lesion evolution in different rat strains using diffusion and perfusion imaging. Stroke 36(9):2000–2005
36. Oliff HS, Coyle P, Weber E (1997) Rat strain and vendor differences in collateral anastomoses. J Cereb Blood Flow Metab 17(5):571–576
37. Oliff HS, Weber E, Eilon G, Marek P (1995) The role of strain/vendor differences on the outcome of focal ischemia induced by intraluminal middle cerebral artery occlusion in the rat. Brain Res 675(1-2):20–26
38. Oliff HS, Weber E, Miyazaki B, Marek P (1995) Infarct volume varies with rat strain and vendor in focal cerebral ischemia induced by transcranial middle cerebral artery occlusion. Brain Res 699(2):329–331, Epub 1995/11/20
39. Lourbopoulos A, Karacostas D, Artemis N, Milonas I, Grigoriadis N (2008) Effectiveness of a new modified intraluminal suture for temporary middle cerebral artery occlusion in rats of various weight. J Neurosci Methods 173(2):225–234, Epub 2008/07/19
40. Howells DW, Porritt MJ, Rewell SS, O'Collins V, Sena ES, van der Worp HB et al (2010) Different strokes for different folks: the rich diversity of animal models of focal cerebral ischemia. J Cereb Blood Flow Metab 30(8):1412–1431
41. O'Collins VE, Macleod MR, Cox SF, Van Raay L, Aleksoska E, Donnan GA et al (2011) Preclinical drug evaluation for combination therapy in acute stroke using systematic review, meta-analysis, and subsequent experimental testing. J Cereb Blood Flow Metab 31(3):962–975, Epub 2010/10/28
42. Callaway JK, Knight MJ, Watkins DJ, Beart PM, Jarrott B, Delaney PM (2000) A novel, rapid, computerised method for quantitation of neuronal damage in a rat model of stroke. J Neurosci Methods 102(1):53–60
43. Modo M, Stroemer RP, Tang E, Veizovic T, Sowniski P, Hodges H (2000) Neurological sequelae and long-term behavioural assessment of rats with transient middle cerebral artery occlusion. J Neurosci Methods 104(1):99–109
44. Porritt MJ, Chen M, Rewell SS, Dean RG, Burrell LM, Howells DW (2010) ACE inhibition reduces infarction in normotensive but not hypertensive rats: correlation with cortical ACE activity. J Cereb Blood Flow Metab 30(8):1520–1526

45. Rewell SS, Fernandez JA, Cox SF, Spratt NJ, Hogan L, Aleksoska E et al (2010) Inducing stroke in aged, hypertensive, diabetic rats. J Cereb Blood Flow Metab 30(4):729–733, Epub 2010/01/14
46. Li Y, Zhu S, Yuan L, Lu H, Li H, Tong S (2013) Predicting the ischemic infarct volume at the first minute after occlusion in rodent stroke model by laser speckle imaging of cerebral blood flow. J Biomed Opt 18(7):76024, Epub 2013/07/28
47. Dirnagl U, Kaplan B, Jacewicz M, Pulsinelli W (1989) Continuous measurement of cerebral cortical blood flow by laser-Doppler flowmetry in a rat stroke model. J Cereb Blood Flow Metab 9(5):589–596, Epub 1989/10/01
48. Harada H, Wang Y, Mishima Y, Uehara N, Makaya T, Kano T (2005) A novel method of detecting rCBF with laser-Doppler flowmetry without cranial window through the skull for a MCAO rat model. Brain Res Brain Res Protoc 14(3):165–170
49. Aspey C (1998) Patel, Terruli, Harrison. Middle cerebral artery occlusion in the rat: consistent protocol for a model of stroke. Neuropathol Appl Neurobiol 24(6):487–497
50. Aspey BS, Taylor FL, Terruli M, Harrison MJG (2000) Temporary middle cerebral artery occlusion in the rat: consistent protocol for a model of stroke and reperfusion. Neuropathol Appl Neurobiol 26(3):232–242
51. Herz RC, Jonker M, Verheul HB, Hillen B, Versteeg DH, de Wildt DJ (1996) Middle cerebral artery occlusion in Wistar and Fischer-344 rats: functional and morphological assessment of the model. J Cereb Blood Flow Metab 16(2):296–302
52. Walberer M, Stolz E, Muller C, Friedrich C, Rottger C, Blaes F et al (2006) Experimental stroke: ischaemic lesion volume and oedema formation differ among rat strains (a comparison between Wistar and Sprague-Dawley rats using MRI). Lab Anim 40(1):1–8
53. Barone FC, Price WJ, White RF, Willette RN, Feuerstein GZ (1992) Genetic hypertension and increased susceptibility to cerebral ischemia. Neurosci Biobehav Rev 16(2):219–233, Epub 1992/01/01
54. Sauter A, Rudin M (1995) Strain-dependent drug effects in rat middle cerebral artery occlusion model of stroke. J Pharmacol Exp Ther 274(2):1008–1013, Epub 1995/08/01
55. Kahveci N, Alkan T, Korfali E, Ozluk K (2000) An anatomical and pathological evaluation of middle cerebral artery occlusion in rats. Neurol Res 22(6):609–614
56. Menzies SA, Hoff JT, Betz AL (1992) Middle cerebral artery occlusion in rats: a neurological and pathological evaluation of a reproducible model. Neurosurgery 31(1):100–106, discussion 6-7
57. Rubino GJ, Young W (1988) Ischemic cortical lesions after permanent occlusion of individual middle cerebral artery branches in rats. Stroke 19(7):870–877
58. Coyle P (1984) Diameter and length changes in cerebral collaterals after middle cerebral artery occlusion in the young rat. Anat Rec 210(2):357–364
59. Coyle P (1985) Interruption of the middle cerebral artery in 10-day-old rat alters normal development of distal collaterals. Anat Rec 212(2):179–182
60. Coyle P (1986) Different susceptibilities to cerebral infarction in spontaneously hypertensive (SHR) and normotensive Sprague-Dawley rats. Stroke 17(3):520–525
61. Coyle P (1987) Spatial relations of dorsal anastomoses and lesion border after middle cerebral artery occlusion. Stroke 18(6):1133–1140
62. Coyle P, Feng X (1993) Risk area and infarct area relations in the hypertensive stroke-prone rat. Stroke 24(5):705–709, discussion 10
63. Legos JJ, Lenhard SC, Haimbach RE, Schaeffer TR, Bentley RG, McVey MJ et al (2008) SB 234551 selective ET(A) receptor antagonism: perfusion/diffusion MRI used to define treatable stroke model, time to treatment and mechanism of protection. Exp Neurol 212(1):53–62, Epub 2008/05/09
64. McCabe C, Gallagher L, Gsell W, Graham D, Dominiczak AF, Macrae IM (2009) Differences in the evolution of the ischemic penumbra in stroke-prone spontaneously hypertensive and Wistar-Kyoto rats. Stroke 40(12):3864–3868, Epub 2009/10/03
65. Yao H, Nabika T (2012) Standards and pitfalls of focal ischemia models in spontaneously hypertensive rats: with a systematic review of recent articles. J Transl Med 10:139, Epub 2012/07/10

66. Buchan AM, Xue D, Slivka A (1992) A new model of temporary focal neocortical ischemia in the rat. Stroke 23(2):273–279, Epub 1992/02/01
67. Howells DW, Sena ES, Macleod MR (2014) Bringing rigour to translational medicine. Nat Rev 10(1):37–43, Epub 2013/11/20
68. O'Collins VE, Macleod MR, Donnan GA, Horky LL, van der Worp BH, Howells DW (2006) 1,026 experimental treatments in acute stroke. Ann Neurol 59(3):467–477
69. Sena ES, Briscoe CL, Howells DW, Donnan GA, Sandercock PA, Macleod MR (2010) Factors affecting the apparent efficacy and safety of tissue plasminogen activator in thrombotic occlusion models of stroke: systematic review and meta-analysis. J Cereb Blood Flow Metab 30(12):1905–1913, Epub 2010/07/22
70. Howells DW, Sena ES, O'Collins V, Macleod MR (2012) Improving the efficiency of the development of drugs for stroke. Int J Stroke 7(5):371–377, Epub 2012/06/21
71. Kilkenny C, Browne WJ, Cuthill IC, Emerson M, Altman DG (2010) Improving bioscience research reporting: the ARRIVE guidelines for reporting animal research. PLoS Biol 8(6), e1000412, Epub 2010/07/09
72. Petullo D, Masonic K, Lincoln C, Wibberley L, Teliska M, Yao DL (1999) Model development and behavioral assessment of focal cerebral ischemia in rats. Life Sci 64(13):1099–1108, Epub 1999/04/21

Chapter 29
Rabbit Spinal Cord Ischemia Model for the Development of Neuroprotective Treatments

Daisy Chou, Anja Muehle, Paul A. Lapchak, and Ali Khoynezhad

Abstract Ischemic spinal cord injury (SCI) is one of the most morbid complications of aortic operations and traumatic injury. Numerous animal models have been created to try to understand the pathogenesis of spinal cord injury and to find treatment options. In this chapter, different animal spinal cord injury models are discussed. An overview of potential spinal cord injury therapeutic agents that are studied in the literature is presented as well. Finally, a new minimally invasive rabbit spinal cord ischemia model that uses a transfemoral intra-aortic balloon occlusion and neuromonitoring with motor-evoked potentials (MEPs) is presented. This rabbit model has the potential to aid in novel therapeutic drug development for early ischemic SCI.

Keywords Ischemic spinal cord injury • Intra-aortic balloon occlusion • Neuromonitoring

1 Introduction

Spinal cord injury is one of the most catastrophic complications after aortic operation. Aortic surgery is needed to treat a variety of aortic pathologies, from aneurysms and dissections to traumatic ruptures. Aortic pathologies are the sixth leading cause of death in elderly men in the Western countries. Over 43,000 patients annually die from pathologies of the aorta and its branches. Risk of SCI after aortic procedures is reported up to 14 % in the literature [1–5]. Once an ischemic SCI has occurred, the possibility of a recovery remains low [6, 7]. In addition to the

D. Chou, M.D. • A. Muehle, M.D. •
A. Khoynezhad, M.D., Ph.D., F.A.C.S. (✉)
Division of Cardiothoracic Surgery, Cedars-Sinai Medical Center,
127 S. San Vicente Blvd, Suite A3306, Los Angeles, CA 90048, USA
e-mail: Akhoy@cshs.org

P.A. Lapchak, Ph.D., F.A.H.A.
Department of Neurology and Neurosurgery, Cedars-Sinai Medical Center,
Los Angeles, CA, USA

P.A. Lapchak, J.H. Zhang (eds.), *Neuroprotective Therapy for Stroke and Ischemic Disease*, Springer Series in Translational Stroke Research,
DOI 10.1007/978-3-319-45345-3_29

tremendous social constraints for paraplegic patients, the medical treatment costs can add up to US$ 3 million during one patient's lifetime [8].

SCI can be classified as early or delayed based on onset of symptoms and has different pathogenesis of neuronal damage. Early SCI occurs after blood supply to the spinal cord is interrupted. This lack of nutrition leads to accumulation of toxins and production of free radicals, which causes damage and death of neuronal cells. These toxins and free radicals can cause ongoing cell damage, even after restoration of bloody supply. Delayed SCI occurs secondary to increased production of spinal fluid, which increases pressure in the spinal fluid channel and leads to spinal cord compression and malperfusion, a second "hit" to the neuronal tissue. This causes a delayed paraplegia in patients: the so-called delayed SCI [5].

To prevent or reduce this disabling complication there are several clinically proven adjuncts that can be applied to the patient perioperatively: (1) increasing spinal cord perfusion with anesthetics (fentanyl, ketamine, propofol), methylprednisolone, or papaverine, (2) lowering cerebrospinal fluid pressure with a spinal fluid catheter, (3) reducing body core temperature, (4) establishment of distal aortic perfusion, and (5) segmental artery reimplantation [9–17]. Even with these adjuncts, the risk of SCI remains unacceptably high and research for better solutions and drugs is needed.

2 Potential Therapeutic Agents

Spinal cord injury occurs through a variety of mechanisms including hypoxia and ischemia, excitotoxicity, inflammation, and apoptosis. Extensive research into not only understanding but also finding ways to interfere with each pathway has been conducted. One area of interest that has been studied is the restoration of oxygen and nutrients to ischemic and hypoxemic spinal cord neurons. Oxycyte, which is a new-generation perfluorocarbon, increases oxygen availability. It is administered intravenously and should be given as soon as possible following an acute injury. It has been shown to be neuroprotective in rat and swine models [18–20]. Hopefully, it will prove to be effective in clinical trials as well.

Another potential spinal cord injury mechanism that can be targeted is the excitatory neurotransmitters that lead to toxicity and death of neurons. Levels of the excitatory neurotransmitter glutamate can rise to toxic levels following spinal cord injury. Glutamate binds to one of three receptors, kainate, *N*-methyl D-aspartate (NMDA), and alpha-amino-3-hydroxy-5-methylisoxazoleproprionate (AMPA), which mediate the influx of calcium into the cell and thus control downstream signaling cascades. Antagonists of kainate, NMDA, and AMPA have the potential to mitigate the neuron damage and death. Riluzole, which is currently approved for treatment of amyotrophic lateral sclerosis, is a sodium channel blocker and glutamate receptor modulator that has been shown to modulate excitotoxicity [21–23]. Grossman and colleagues have shown that riluzole improves the motor score of acute cervical spinal cord injury patients and that its complication rate is similar to that of matched patients [24]. Future trials are still needed before riluzole is approved for treatment of acute spinal cord injury.

The downstream effectors of excitotoxicity are potential treatment targets as well. Calpain is a calcium-activated cysteine protease that has been shown to play a role in the synthesis of proteins that leads to apoptosis of injured spinal cord neurons [25, 26]. A number of different calpain inhibitors have been studied over the years, but problems with drug safety and solubility have prevented their use in the clinical setting. One calpain activation modulator with potential use in SCI treatment is melatonin. Samantaray and colleagues have shown that melatonin, a naturally occurring hormone with a variety of effects including antioxidant, anti-nitrosative, and immunomodulation, has the potential to be used in the treatment of acute SCI [27].

The inflammatory response that follows the initial injury phase in acute spinal cord injury is also a therapeutic target. The first pharmaceutical agent investigated for use in spinal cord injury was methylprednisolone, a synthetic glucocorticoid steroid [28]. The anti-inflammatory and immunosuppressant qualities of glucocorticoid steroids were thought to improve the functional recovery of SCI in dogs [29]. Subsequent trials have not proven the efficacy of methylprednisolone in treating SCI, but widespread off-label use of the drug in acute SCI still occurs. Currently, the synthetic agonists of the ligand-activated transcription factor peroxisome proliferator-activated receptor-gamma (PPARγ), thiazolidinediones (TZDs), have been tested as potential neuroprotective agents. In the rat SCI model, Park and colleagues administered the TZD pioglitazone within 2 h of spinal cord injury and found significantly lower expression of inflammatory genes and also improved motor function recovery [30]. These results suggest that pharmacologic agents targeting inflammation may become useful in spinal cord injury treatment in the future.

Multiple agents that target apoptosis have also been investigated. Apoptosis can be triggered by external or internal stimuli and is controlled by cell signaling. One potential therapeutic agent is flavopiridol, a cell cycle inhibitor. Byrnes and colleagues have shown that flavopiridol decreases apoptosis of neurons and oligodendrocytes in rats [31]. Minocycline, an antibiotic, is another agent with anti-apoptotic properties that has been studied in animal models as well as in humans. It has been demonstrated to be safe and improves motor function in acute SCI patients when tested in a phase II clinical trial [32]. Numerous studies involving estrogen, a steroid hormone with anti-apoptotic, anti-inflammatory, and antioxidant properties, have also been conducted. Estrogen treatment has been associated with decreased apoptosis and locomotor function recovery in the rats [33–36]. Lee and colleagues also showed that estrogen prevents RhoA-JNK3 pathway-mediated oligodendrocyte cell death in rats [37]. Additional studies are needed to determine the ideal estrogen dose that provides neuroprotection but avoids its prothrombotic and carcinogenic properties.

Modulation of axon and myelin regeneration is another potential therapeutic target in the treatment of spinal cord injury. Studies have shown that molecules including NoGo, myelin-associated glycoprotein (MAG), and oligodendrocyte myelin glycoprotein (OMgP) inhibit axon regeneration after injury [38]. Thus, inhibiting these molecules may promote axon regeneration and recovery. Currently, ATI-355 (humanized anti-Nogo antibody) and Cethrin (recombinant protein RHO GTPase antagonist) are being studied. Progesterone, another naturally occurring steroid hormone, has potential to be a promyelinating agent. Studies in the rat SCI model using progesterone have shown that it spares white matter tissue, improves motor function, restores myelin levels, and increases the number of oligodendrocyte progenitor cells [39, 40]. Monosialotetrahexosylganglioside sodium (GM-1), or Sygen (trade name), has also been studied in the spinal cord injury model. Sygen is a neuroprotective ganglioside that appears to promote neuron growth, repair, and regeneration through unknown mechanisms [41]. However, GM-1 has not been shown to produce statistically significant improvement in spinal cord injury in clinical trials, so it currently is not recommended for treatment for spinal cord injury [42].

3 Animal Models of Spinal Cord Injury

Multiple animal models for spinal cord injury have been described in the literature. In the past, spinal cord injuries were induced by trauma such as dropping weights onto an animal's spinal cord. However, most spinal cord injury models now involve inducing spinal cord ischemia through a more controlled manner. In smaller animals, such as rats and mice, cross-clamping the aorta is used to induce spinal cord ischemia. Exposure and cross-clamping of the aortic arch through a cervicosternal incision or the descending aorta through a lateral thoracotomy can be used in rats and mice [43, 44]. To induce ischemia, cross-clamp times of 8 and 7.5 min have been reported for rats and mice [43, 44]. This leads to reproducible lower extremity paralysis.

Spinal cord ischemia models in larger animals, such as sheep, pigs, and dogs, exist as well. Larger animals are more expensive, but can be easier to perform operations on. With the introduction and now increasing use of endovascular techniques in vascular disease in humans, this technology is now used in spinal cord ischemia research as well. Spinal cord ischemia models utilizing endovascular techniques such as endograft implantation have been described in the sheep, pig, and rabbit models [45]. Ischemia can be achieved with deployment of aortic endografts that cover the thoracic aorta and thus intercostal and lumbar artery origins or intra-aortic balloon occlusion.

Ischemic SCI has been studied in rabbits due to the unique segmental arterial blood supply from the infrarenal aorta to the spinal cord [46]. Zivin and colleagues first described a highly reproducible model that induced infarction of the lower lumbar and sacral regions of the rabbit spinal cord in 1980 [47]. Most rabbit ischemic SCI studies in the literature induce ischemia by performing laparotomy and cross-clamping the aorta. Although effective in causing easily recognizable and reproducible neurological defects, this method is extremely invasive and is associated with complications such as longer operative time (and thus increased anesthetic requirement), increased postoperative pain, and respiratory complications [48, 49].

4 Minimally Invasive Rabbit Spinal Cord Ischemia Model

The use of intra-aortic balloon to cause infrarenal aortic occlusion is a less invasive alternative to laparotomy and cross-clamping of the aorta. There are studies that describe use of intra-aortic balloon occlusion, but most do not incorporate a neuromonitoring component. Previously, our group described establishment of neuromonitoring with motor-evoked potentials (MEPs) in a leporine SCI model using intra-aortic balloon occlusion [50]. This methodology paper provided information on appropriate stimulatory voltage and placement of electrodes and also recommended the avoidance of inhaled anesthetics, propofol, and prolonged hypotension in order to maintain MEPs [50].

Building upon our previous work, multiple adjustments were made to create a reliable and reproducible rabbit ischemic SCI model. Memantine, a noncompetitive *N*-methyl-D-aspartate receptor antagonist currently used in the treatment of Alzheimer's disease, was used to validate the model. Studies have shown that memantine, given intravenously or orally, significantly reduces neurologic injury secondary to ischemic spinal cord injury after aortic occlusion [51, 52]. Our objective was to establish a minimally invasive rabbit ischemic SCI model utilizing both transfemoral intra-aortic balloon for aortic occlusion and neuromonitoring with MEPs that can be used to study clinical adjuncts to prevent ischemic SCI and subsequent paraplegia.

4.1 Methods

The procedure was performed in female New Zealand white rabbits (3–3.5 kg), obtained from a commercial supplier in California (Fig. 29.1). All treatment drugs were administered intravenously via 24 F lateral ear vein catheter 45 min prior to aortic occlusion and fractionally over 5 min.

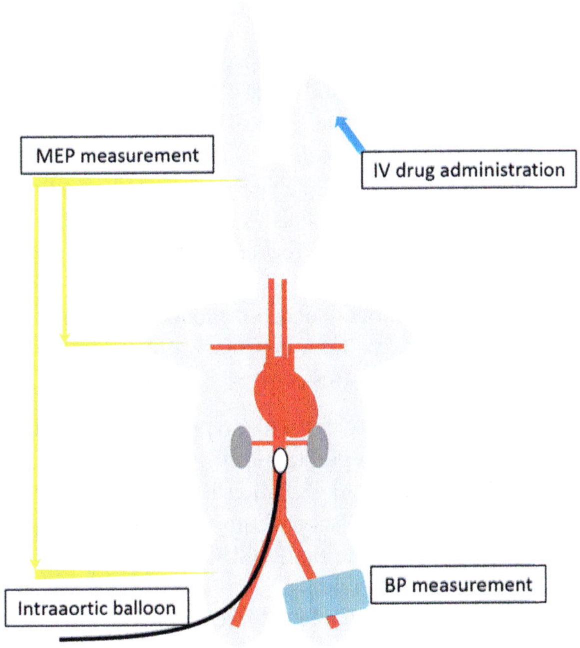

Fig. 29.1 Operative setup demonstrating location of electrodes, intravenous catheter, intra-aortic balloon, and blood pressure cuff

Two treatment groups were used, a memantine control group and a memantine-treated group. The memantine control group (group A) received saline (1 mL/kg, normal sterile injectable saline [Hospira Inc.]). The memantine-treated group (group B) received memantine (20 mg/kg [Sigma-Aldrich], reconstituted in sterile saline [Hospira Inc.]).

Surgical Procedures: Anesthesia was induced using intramuscular injection of 30 mg/kg ketamine and 25 mg/kg xylazine and then maintained by continuous intravenous infusion of 10 mg/mL ketamine and 2 mg/mL xylazine diluted in sterile saline at a rate of 7 mL/h. After induction, the animal was then intubated using a 3.0 mm endotracheal tube and maintained on continuous oxygen to keep oxygen saturation above 95 %. Pulse oximeter monitor was attached to the front foot for continuous monitoring of oxygen saturation and pulse, and a pediatric blood pressure cuff was placed on the left hind limb for measurement of blood pressure. Blood pressure was monitored every 5 min and kept above 95 mmHg with saline boluses (4 cm^3/kg) as required. Heparin 100 IU/kg was infused intravenously prior to incision.

For measurement of motor-evoked potentials (MEPs), an intraoperative neuromonitoring machine (16 channels; Cascade, Cadwell, Kennewick, WA, USA) and commercially available electrodes (Rochester Electro-Medical, Lutz, FL, USA) were used. Transcranial stimulation electrodes were placed onto the scalp in the midline above the eyes and sensor electrodes in each muscle group: bilateral biceps muscles as well as bilateral quadriceps muscles. The transcranial stimuli were triggered prior to incision to confirm accurate placement of electrodes and 1 min after aortic occlusion to detect disappearance of lower extremity motor potentials and thus confirm proper inflation of intra-aortic balloon. Voltage used ranged from 60 to 140 V, and starting voltage of 100 V was used.

Once the right groin was prepped and draped, groin cutdown was performed for femoral artery access. After proximal and distal control of the right femoral artery was gained, the arteriotomy was then made. A 3 F Fogarty balloon was then advanced 18 cm (established on necropsy) through the femoral arteriotomy site into the infrarenal aorta and inflated with 1 mL air. Undetectable left hind leg blood pressure and disappearance of lower extremity motor potentials were used to confirm aortic occlusion.

After the set ischemia time had been reached, the balloon was deflated and then removed. The arteriotomy site was then closed using 7-0 prolene suture. The incision was closed with running subcuticular 4-0 monocryl stitch. The animals were monitored closely during recovery. Intramuscular valium (5 mg) was administered if the animals displayed signs of distress.

Behavioral Function: A trained observer with no previous knowledge of the experimental protocol assessed the neurologic status using the TARLOV score (Table 29.1). Ratings were performed at 4, 24, and 48 h after the procedure. Animals were considered paraplegic with a TARLOV score 0 or 1 at 24 h. Paraplegic animals were euthanized on postoperative day one per IACUC recommendation. All other animals were euthanized after the 48 h rating.

All animals were euthanized with a lethal injection of pentobarbital.

Statistical Analysis: To construct a quantal analysis curve, a wide range of ischemia times (5–55 min) was used to produce a spectrum of paraplegic and non-paraplegic

Table 29.1 TARLOV score

Score	Description
0	Paraplegic with no lower extremity function
1	Poor lower extremity function, weak antigravity movement only
2	Some lower extremity function with good antigravity strength but inability to draw legs under body or hop
3	Ability to draw legs under body and hop but not normally
4	Normal motor function

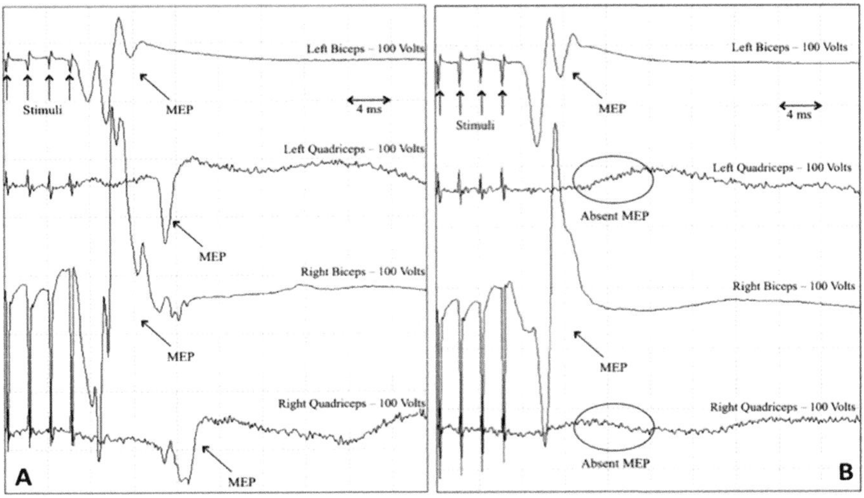

Fig. 29.2 Lower extremity MEPs after stimulation by one cranial electrode at 100 V. (**a**) Before aortic occlusion (baseline) and (**b**) after aortic occlusion

animals, which include death on the continuum of ischemia-induced effects. Ischemia time was plotted against behavior to define the P50 value (ischemia time where 50 % of the rabbits are paraplegic). Each compound tested generated a separate quantal curve, and the P50s of the compound and control were compared. Compounds with statistically significant increase in the P50 value over the control were indicative of neuroprotection.

Statistical analysis was performed using Graph Pad Software. Behavioral results were analyzed using the *t*-test (Graph Pad).

4.2 Results

Baseline motor-evoked potentials (MEPs) were successfully recorded on all animals. Lower extremity MEPs disappeared upon aortic occlusion in 45 of 45 (100 %) of rabbits (Fig. 29.2). All animals that survived the procedure were included for analysis. Twenty-two rabbits were in the control group, and 23 rabbits were in the memantine group. The P50 for the control group was 13.07 ± 2.12 min. The P50 for

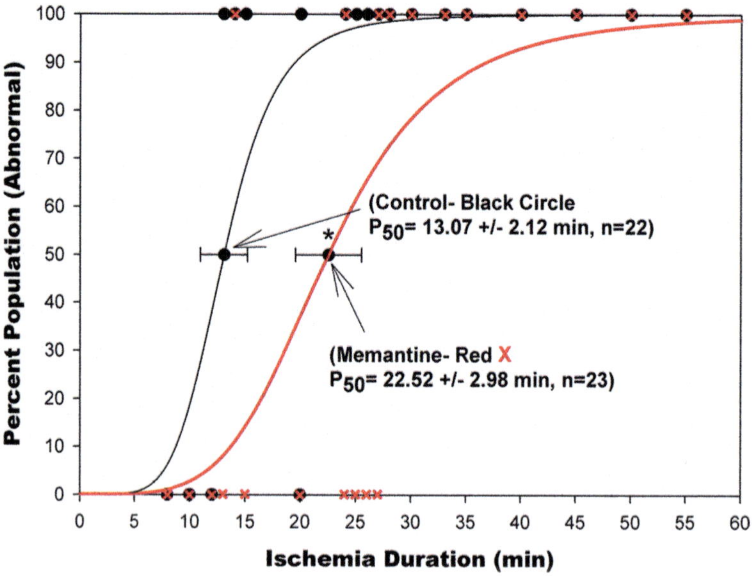

Fig. 29.3 Memantine versus control quantal analysis curves

the memantine group was 22.52±2.98 min (Fig. 29.3). The difference in the P50s was statistically significant, with $p=0.01$. The P50 was calculated based off of the 24 h postoperative rating.

No cases of delayed onset paraplegia were noted.

5 Discussion

Although clinical adjuncts such as distal aortic perfusion, decreasing spinal canal pressure with draining cerebrospinal fluid, and segmental artery reimplantation have decreased the incidence of SCI after aortic surgery, it remains a devastating complication and still occurs in 2–14 % of aortic cases [5, 15, 16]. Reproducible and reliable ischemic SCI animal models are essential to the development of novel neuroprotective techniques. The rabbit model is ideal for studying ischemic SCI, as the consistent segmental nature of the infrarenal aortic blood supply to the spinal cord allows researchers to mimic both open and endovascular aortic repair [46]. There are rabbit ischemic SCI models described in the literature that involve intra-aortic balloon inflation to induce aortic occlusion [49] and also models that utilize neuromonitoring [51], but few incorporate both techniques.

Our group has previously published work on the establishment of a model that uses neuromonitoring for the detection of ischemic SCI in rabbits and an overview of troubleshooting common problems associated with neuromonitoring [50]. We built upon that model and made multiple adjustments. Changes to the anesthesia protocol included the use of intramuscular and intravenous ketamine and xylazine instead of inhaled anesthetics and propofol. Inhaled anesthetics and propofol were associated with more complications, such as hypotension and respiratory arrest. Drug administration changes included intravenous administration of control or study drug at least 45 min prior to aortic occlusion. Changes to intraoperative and postoperative monitoring included recording vital signs every 5 min and administering normal saline boluses to maintain systolic blood pressure above 95 mmHg.

Administration of preoperative intravenous and oral memantine have both been shown to significantly improve the clinical outcomes in rabbits undergoing aortic cross-clamping [51, 52]. Thus, memantine was used to validate our study. The behavioral ratings in this study show that the P50 for the memantine group was significantly higher than that of the control group (22.52 ± 2.98 min versus 13.07 ± 2.12 min, $p = 0.01$). The memantine-treated rabbits in our study were able to tolerate a longer ischemia time without suffering paraplegia, which is similar to the outcomes of memantine-treated rabbits in prior studies. Thus, these results support that the model described in this chapter is reliable and reproducible and achieves aortic occlusion using minimally invasive methods while utilizing neuromonitoring technology to confirm complete aortic occlusion.

6 Conclusion

Pretreatment with intravenous memantine significantly increased the ischemia time necessary to produce neurological deficits and also reduced neurologic injury in our rabbit model. This result validates our minimally invasive rabbit ischemic SCI model. Going forward, this model can be used by researchers to perform preclinical testing to develop potential therapeutics for spinal cord injury.

References

1. Greenberg RK, Lu Q, Roselli EE, Svensson LG, Moon MC, Hernandez AV et al (2008) Contemporary analysis of descending thoracic and thoracoabdominal aneurysm repair: a comparison of endovascular and open techniques. Circulation 118(8):808–817
2. Makaroun MS, Dillavou ED, Kee ST, Sicard G, Chaikof E, Bavaria J et al (2005) Endovascular treatment of thoracic aortic aneurysms: results of the phase II multicenter trial of the GORE TAG thoracic endoprosthesis. J Vasc Surg 41(1):1–9
3. Svensson LG (2005) Paralysis after aortic surgery: in search of lost cord function. Surgeon 3(6):396–405
4. Dillavou ED, Makaroun MS (2008) Predictors of morbidity and mortality with endovascular and open thoracic aneurysm repair. J Vasc Surg 48(5):1114–1119; discussion 1119–1120
5. Khoynezhad A, Kruse M (2008) Spinal cord injury and stroke following thoracic endovascular aortic repair: a risk analysis and review of the literature. Ital J Vasc Endovasc Surg 14(3):221–229
6. Ullery BW, Cheung AT, Fairman RM, Jackson BM, Woo EY, Bavaria J et al (2011) Risk factors, outcomes, and clinical manifestations of spinal cord ischemia following thoracic endovascular aortic repair. J Vasc Surg 54(3):677–684

7. Wong DR, Coselli JS, Amerman K, Bozinovski J, Carter SA, Vaughn WK et al (2007) Delayed spinal cord deficits after thoracoabdominal aortic aneurysm repair. Ann Thorac Surg 83(4):1345–1355; discussion 1355
8. Lim PAC, Tow AM (2007) Recovery and regeneration after spinal cord injury: a review and summary of recent literature. Ann Acad Med Singapore 36(1):49–57
9. Aslan A, Cemek M, Eser O, Altunbaş K, Buyukokuroglu ME, Cosar M et al (2009) Does dexmedetomidine reduce secondary damage after spinal cord injury? An experimental study. Eur Spine J 18(3):336–344
10. Lima B, Nowicki ER, Blackstone EH, Williams SJ, Roselli EE, Sabik JF et al (2012) Spinal cord protective strategies during descending and thoracoabdominal aortic aneurysm repair in the modern era: the role of intrathecal papaverine. J Thorac Cardiovasc Surg 143(4):945–952.e1
11. Onose G, Anghelescu A, Muresanu DF, Padure L, Haras MA, Chendreanu CO et al (2009) A review of published reports on neuroprotection in spinal cord injury. Spinal Cord 47(10):716–726
12. Tran TP, Khoynezhad A (2009) Current management of type B aortic dissection. Vasc Health Risk Manag 5(1):53–63
13. Safi HJ, Miller CC, Huynh TTT, Estrera AL, Porat EE, Winnerkvist AN et al (2003) Distal aortic perfusion and cerebrospinal fluid drainage for thoracoabdominal and descending thoracic aortic repair: ten years of organ protection. Ann Surg 238(3):372–380; discussion 380–381
14. Hsu CC-T, Kwan GNC, van Driel ML, Rophael JA (2012) Distal aortic perfusion during thoracoabdominal aneurysm repair for prevention of paraplegia. Cochrane Database Syst Rev 3, CD008197
15. Khoynezhad A, Donayre CE, Bui H, Kopchok GE, Walot I, White RA (2007) Risk factors of neurologic deficit after thoracic aortic endografting. Ann Thorac Surg 83(2):S882–S889; discussion S890–892
16. Khoynezhad A, Donayre CE, Smith J, Kopchok GE, Walot I, White RA (2008) Risk factors for early and late mortality after thoracic endovascular aortic repair. J Thorac Cardiovasc Surg 135(5):1103–1109, 1109.e1–4
17. Khoynezhad A, Bello R, Smego DR, Nwakanma L, Plestis KA (2005) Improved outcome after repair of descending and thoracoabdominal aortic aneurysms using modern adjuncts. Interact Cardiovasc Thorac Surg 4(6):574–576
18. Yacoub A, Hajec MC, Stanger R, Wan W, Young H, Mathern BE (2014) Neuroprotective effects of perfluorocarbon (oxycyte) after contusive spinal cord injury. J Neurotrauma 31(3):256–267
19. Schroeder JL, Highsmith JM, Young HF, Mathern BE (2008) Reduction of hypoxia by perfluorocarbon emulsion in a traumatic spinal cord injury model. J Neurosurg Spine 9(2):213–220
20. Mahon RT, Auker CR, Bradley SG, Mendelson A, Hall AA (2013) The emulsified perfluorocarbon oxycyte improves spinal cord injury in a swine model of decompression sickness. Spinal Cord 51(3):188–192
21. Wu Y, Satkunendrarajah K, Teng Y, Chow DS-L, Buttigieg J, Fehlings MG (2013) Delayed post-injury administration of riluzole is neuroprotective in a preclinical rodent model of cervical spinal cord injury. J Neurotrauma 30(6):441–452
22. Schwartz G, Fehlings MG (2001) Evaluation of the neuroprotective effects of sodium channel blockers after spinal cord injury: improved behavioral and neuroanatomical recovery with riluzole. J Neurosurg 94(2 Suppl):245–256
23. Springer JE, Azbill RD, Kennedy SE, George J, Geddes JW (1997) Rapid calpain I activation and cytoskeletal protein degradation following traumatic spinal cord injury: attenuation with riluzole pretreatment. J Neurochem 69(4):1592–1600
24. Grossman RG, Fehlings MG, Frankowski RF, Burau KD, Chow DSL, Tator C et al (2014) A prospective, multicenter, phase I matched-comparison group trial of safety, pharmacokinetics, and preliminary efficacy of riluzole in patients with traumatic spinal cord injury. J Neurotrauma 31(3):239–255
25. Banik NL, Powers JM, Hogan EL (1980) The effects of spinal cord trauma on myelin. J Neuropathol Exp Neurol 39(3):232–244

26. Banik NL, Hogan EL, Powers JM, Whetstine LJ (1982) Degradation of cytoskeletal proteins in experimental spinal cord injury. Neurochem Res 7(12):1465–1475
27. Samantaray S, Sribnick EA, Das A, Knaryan VH, Matzelle DD, Yallapragada AV et al (2008) Melatonin attenuates calpain upregulation, axonal damage and neuronal death in spinal cord injury in rats. J Pineal Res 44(4):348–357
28. Bracken MB (1992) Pharmacological treatment of acute spinal cord injury: current status and future prospects. Paraplegia 30(2):102–107
29. Ducker TB, Hamit HF (1969) Experimental treatments of acute spinal cord injury. J Neurosurg 30(6):693–697
30. Park S-W, Yi J-H, Miranpuri G, Satriotomo I, Bowen K, Resnick DK et al (2006) Thiazolidinedione Class of peroxisome proliferator-activated receptor agonists prevents neuronal damage, motor dysfunction, myelin loss, neuropathic pain, and inflammation after spinal cord injury in adult rats. J Pharmacol Exp Ther 320(3):1002–1012
31. Byrnes KR, Stoica BA, Fricke S, Di Giovanni S, Faden AI (2007) Cell cycle activation contributes to post-mitotic cell death and secondary damage after spinal cord injury. Brain 130(11):2977–2992
32. Casha S, Zygun D, McGowan MD, Bains I, Yong VW, John HR (2012) Results of a phase II placebo-controlled randomized trial of minocycline in acute spinal cord injury. Brain 135(4):1224–1236
33. Siriphorn A, Dunham KA, Chompoopong S, Floyd CL (2012) Postinjury administration of 17β-estradiol induces protection in the gray and white matter with associated functional recovery after cervical spinal cord injury in male rats. J Comp Neurol 520(12): 2630–2646
34. Samantaray S, Smith JA, Das A, Matzelle DD, Varma AK, Ray SK et al (2011) Low dose estrogen prevents neuronal degeneration and microglial reactivity in an acute model of spinal cord injury: effect of dosing, route of administration, and therapy delay. Neurochem Res 36(10):1809–1816
35. Sribnick EA, Samantaray S, Das A, Smith J, Matzelle DD, Ray SK et al (2010) Postinjury estrogen treatment of chronic spinal cord injury improves locomotor function in rats. J Neurosci Res 88(8):1738–1750
36. Chen S-H, Yeh C-H, Lin MY-S, Kang C-Y, Chu C-C, Chang F-M et al (2010) Premarin improves outcomes of spinal cord injury in male rats through stimulating both angiogenesis and neurogenesis. Crit Care Med 38(10):2043–2051
37. Lee JY, Choi SY, Oh TH, Yune TY (2012) 17β-Estradiol inhibits apoptotic cell death of oligodendrocytes by inhibiting RhoA-JNK3 activation after spinal cord injury. Endocrinology 153(8):3815–3827
38. Hunt D, Coffin RS, Anderson PN (2002) The Nogo receptor, its ligands and axonal regeneration in the spinal cord; a review. J Neurocytol 31(2):93–120
39. Thomas AJ, Nockels RP, Pan HQ, Shaffrey CI, Chopp M (1999) Progesterone is neuroprotective after acute experimental spinal cord trauma in rats. Spine 24(20):2134–2138
40. De Nicola AF, Gonzalez SL, Labombarda F, González Deniselle MC, Garay L, Guennoun R et al (2006) Progesterone treatment of spinal cord injury: effects on receptors, neurotrophins, and myelination. J Mol Neurosci MN 28(1):3–15
41. Geisler FH, Dorsey FC, Coleman WP (1991) Recovery of motor function after spinal-cord injury—a randomized, placebo-controlled trial with GM-1 ganglioside. N Engl J Med 324(26):1829–1838
42. Geisler FH, Coleman WP, Grieco G, Poonian D, Sygen Study Group (2001) The Sygen multicenter acute spinal cord injury study. Spine 26(24 Suppl):S87–S98
43. Bell MT, Reece TB, Smith PD, Mares J, Weyant MJ, Cleveland JC et al (2014) Reproducable paraplegia by thoracic aortic occlusion in a murine model of spinal cord ischemia-reperfusion. J Vis Exp 85
44. Awad H, Ankeny DP, Guan Z, Wei P, McTigue DM, Popovich PG (2010) A mouse model of ischemic spinal cord injury with delayed paralysis caused by aortic cross-clamping. Anesthesiology 113(4):880–891

45. Böckler D, Kotelis D, Kohlhof P, von Tengg-Kobligk H, Mansmann U, Zink W et al (2007) Spinal cord ischemia after endovascular repair of the descending thoracic aorta in a sheep model. Eur J Vasc Endovasc Surg 34(4):461–469
46. Moore WM, Hollier LH (1991) The influence of severity of spinal cord ischemia in the etiology of delayed-onset paraplegia. Ann Surg 213(5):427–431; discussion 431–432
47. Zivin JA, DeGirolami U (1980) Spinal cord infarction: a highly reproducible stroke model. Stroke 11(2):200–202
48. Cheng MK, Robertson C, Grossman RG, Foltz R, Williams V (1984) Neurological outcome correlated with spinal evoked potentials in a spinal cord ischemia model. J Neurosurg 60(4):786–795
49. Saba T, Manduz S, Sapmaz I, Tunel A, Aker H, Dogan K (2007) Neuroprotective effects of diltiazem in rabbits with occluded aorta. Rev Bras Cir Cardiovasc 22(4):416–424
50. Waterford SD, Rastegar M, Goodwin E, Lapchak PA, Juan V, Haji F et al (2015) Methodology of motor evoked potentials in a rabbit model. Transl Stroke Res 6(5):399–406
51. Ehrlich M, Knolle E, Ciovica R, Böck P, Turkof E, Grabenwöger M et al (1999) Memantine for prevention of spinal cord injury in a rabbit model. J Thorac Cardiovasc Surg 117(2):285–291
52. Panthee N, Ono M, Morota T, Tanaka T, Itoda Y, Ikemura M et al (2014) Paraplegia prevention by oral pretreatment with memantine in a rabbit model. J Thorac Cardiovasc Surg 5

Chapter 30
Stroke Sex Differences: From Basic Research to Clinical Trials

Claire L. Gibson, Philip M.W. Bath, and Raeed Altaee

Abstract Sex is shown to affect various aspects of the clinical spectrum of ischaemic stroke, from how an individual patient may present with ischaemic stroke through to how they respond to treatment strategies. In this chapter, we review the evidence for the effect of sex on ischaemic stroke and focus on how the pathological mechanisms of ischaemic stroke, in particular apoptosis, may differ between the sexes. Understanding sex-specific consequences of ischaemic stroke may enable the development of relevant and effective treatment strategies for an individual suffering an ischaemic stroke.

Keywords Ischaemia • Sex • Steroid hormones • Apoptosis

Abbreviations

AIF	Apoptosis-inducing factor
ENOS	Efficacy of Nitric Oxide in Stroke
NIH	National Institutes of Health
NOS	Nitric oxide synthase
PARP	Poly-ADP ribose polymerase
SITS-ISTR	Stroke-International Stroke Thrombolysis Register
STAIR	Stroke Treatment Academic Industry Roundtable
tPA	Tissue plasminogen activation

C.L. Gibson (✉) • R. Altaee
Department of Neuroscience, Psychology and Behaviour, University of Leicester, Leicester, UK
e-mail: cg95@le.ac.uk

P.M.W. Bath
Stroke, Division of Clinical Neuroscience, University of Nottingham, Nottingham, UK

P.A. Lapchak, J.H. Zhang (eds.), *Neuroprotective Therapy for Stroke and Ischemic Disease*, Springer Series in Translational Stroke Research,
DOI 10.1007/978-3-319-45345-3_30

1 Introduction

There is increasing evidence of differences between men and women in relation to cerebral stroke. Such differences include the risk of stroke but also aetiology, symptoms and outcomes. Women are reported to have a higher overall lifetime risk of stroke in addition to higher rates of post-stroke mortality, disability, depression and dementia, compared to men [1]. This increased risk and worsened outcome post-stroke for women has largely been attributed to the longer life expectancy of women given that age is the strongest independent risk factor for stroke [2] and also a negative predictor for clinical outcome [3]. The average age of onset for stroke is typically 4 years later in women compared to men [4], and elderly women seem to carry the majority of burden, in terms of disability, associated with ischaemic stroke [5]. Thus, over the lifespan, stroke affects a greater number of women due to their increased longevity compared to men. The majority of economically developed countries face an ageing population and consequently increasing stroke incidence rates. Improving our understanding of the mechanisms which may underlie sex differences both in incidence and outcome following stroke may have important ramifications for the development of appropriate treatments and reducing the burden of stroke care to society.

2 Stroke Incidence, Symptoms and Outcome

In terms of stroke incidence rates, differences according to sex do seem to depend on the population studied and their age. For example, the Framingham study reported that up until the age of 45, men experience more strokes and poorer rates of recovery compared to age-matched women [5, 6]. Then during the ages of 45–54 years, women seem to experience an increase in the incidence of ischaemic stroke which coincides with the onset of the female menopause and increased obesity [7]. In a population-based study in Sweden, stroke incidence was found to be 60% lower for women than men at ages 55–64 years but 50% higher in women by the age of 75 years [8]. This trend was also reported in the Oxford Vascular Study in which stroke incidence rates were lower in women aged younger than 75 years but higher in women, compared to men, for those aged over 75 [9]. Although differences do exist between studies and populations, there is general consensus that older women represent the substantial burden in terms of stroke care and treatment.

Sex may also affect the clinical presentation of a stroke patient, and this impacts upon access to appropriate diagnostic and treatment strategies. From the few population-based studies that have been conducted, possible sex differences are suggested in that women tend to present clinically with nonspecific stroke symptoms, such as pain or altered mental status, whereas men tend to present with classic focal neurological symptoms such as sensory abnormalities, ataxia and diplopia [10]. Differences in symptom presentation may be due to the nature of the ischaemic stroke in terms of whether it is caused by the presence of a thrombus or embolism.

Data from a large population-based European study report that the incidence of thrombotic strokes is more frequent in men compared to women with a greater number of embolic strokes occurring in women [11]. In terms of gross pathology, sex differences may occur in the location of the stroke, in terms of hemisphere and causative blood vessel, this then affecting symptom presentation and stroke outcome. However, it is also reported that women delay up to three times longer than men in seeking care for stroke symptoms [12] which is likely to be exacerbated in elderly women who are more likely to be living alone and socially isolated.

Following an ischaemic stroke, there is a high prevalence of mortality and morbidity. As women tend to be older than men when experiencing a stroke, they do comprise the majority of stroke-related mortality and morbidity. However, data is conflicting with regard to whether women experience a greater rate of mortality following stroke than men [13, 14]. Any differences, within studies, in terms of mortality rates between the sexes, appear to be significantly reduced when adjusted for baseline differences in age, stroke severity, subtype and risk factors [15] suggesting these differences, rather than sex per se, account for any observed sex differences in mortality. Sex differences in functional outcome, however, do seem to be significant even after controlling for age and other factors. Women are more likely to have poorer functional outcomes at 3 months, 1 year and 5 years post-stroke [13, 16] compared to men and experience more mental fatigue, depression and dementia; they also have a worse quality of life [17]. Ischaemic stroke triggers a complex set of pathological events leading to cell death which include excitotoxicity, cell necrosis, apoptosis, inflammation, increased oxidative stress and breakdown of the blood-brain barrier along with possible reperfusion injury. Sex may directly influence the mechanisms of injury activated following ischaemic stroke either as a consequence of intrinsic, i.e. chromosomal, or hormonal differences between the sexes.

3 Sex and Stroke: Hormones and Intrinsic Factors

Experimental studies have provided evidence for a female protection against ischaemic stroke even in the presence of specific risk factors such as diabetes and hypertension [18, 19]. This protective effect seen in females is lost following ovariectomy or reproductive senescence, which both result in a decline in circulating sex steroid hormone levels [18]. Thus, it has been postulated that this protective effect of female sex is mediated via the sex steroid hormones. Although experimental studies have provided a breadth of supporting evidence, including mechanistic studies, for a neuroprotective role of estrogens and progesterone following experimental stroke [20, 21], these protective effects have not yet been replicated in intervention design clinical trials [22].

Sex may influence outcome following ischaemic stroke via sex steroid hormones, which vary between the sexes and fluctuate at various stages across the lifespan, or via intrinsic factors, i.e. the sex chromosomes, which are distinct for each sex. Such differences in the genetic complement of cells may, at least in part, explain the sexually dimorphic response seen to ischaemic stroke. Sex-specific cultures of neonatal cells, cultured in the absence of hormonal influences, demonstrate that male-derived cells are more sensitive to ischaemic stroke than female-derived cells, and several molecular mechanisms of injury have been shown to act dimorphically under ischaemic conditions [23, 24].

4 Mechanisms of Injury

Sex differences have been observed in female (i.e. XX)- and male (i.e. XY)-derived cell cultures in the absence of steroid supplementation; XY cultures were more susceptible to excitotoxic death and XX cultures more susceptible to apoptosis [24]. Emerging evidence suggests that there are significant differences between male and female brains regarding the molecular signalling pathways leading to apoptosis activated under ischaemic conditions [25, 26]. Apoptosis occurs via a caspase-dependent or caspase-independent pathway. Caspase-dependent apoptosis is initiated via intrinsic, mitochondria-mediated mechanisms involving cytochrome C release, apoptosome assembly and caspase cleavage, whereas caspase-independent apoptosis is triggered by DNA damage resulting in activation of poly-ADP ribose polymerase (PARP), release of apoptosis-inducing factor (AIF) from the mitochondria and translocation of AIF to the nucleus causing chromatin condensation and DNA fragmentation. Apoptotic pathways have multiple molecular signals, and a number of these have been shown to be sexually dimorphic in terms of their response following injury including AIF, caspase 3, PARP, nitric oxide synthase (NOS), glutathione, Akt, astrocytic aromatase, glial fibrillary acidic protein, angiotensin II type 2 receptor and the P450 enzyme, soluble epoxide hydrolase [27–30]. It seems likely that in males, caspase-independent apoptosis is the main mechanism by which cell death occurs following ischaemia and is largely due to PARP activation triggered initially by activation of neuronal NOS. In females, the caspase-dependent pathways seem the major contributor to cell death as a consequence of cytochrome C and subsequent caspase activation. Thus, therapies aimed at either caspase activation or other elements of the apoptotic pathway are likely to be more effective in one sex as compared to the other.

Although the majority of focus has been on the role of necrosis and apoptosis following ischaemia autophagy, which is considered a novel cell death pathway, has also been recently shown to differ between the sexes. During autophagy, targeted segments of the cell's cytoplasm are engulfed into an autophagosome, which fuses with a lysosome and the contents degraded [31]. In male animals, there seems to be an increase in autophagosome formation and lysosome activity following neonatal hypoxic injury. Importantly, it appears that in neonatal hypoxic injury models, inhibitors of autophagy exhibit a tendency towards a protective effect in male-derived neurons compared to female-derived ones [32]. Understanding if, and by what mechanisms, the cell death pathways activated following cerebral ischaemia differ between the sexes has important consequences for the design and evaluation of potential treatments.

5 Sex Differences in Treatment: Clinical and Preclinical Studies

Currently, thrombolysis with tissue plasminogen activation (tPA) is the only licensed drug approved for the treatment of hyperacute ischaemic stroke. However, due to its narrow therapeutic window (<4.5h), risk of symptomatic intracerebral haemorrhage and the fact that many patients wake up with stroke and are ineligible for treatment, 20% or less of stroke patients receive tPA. However, if sex differences occur in tPA treatment, this may contribute to differences observed in outcome following stroke. A large study reported that, whilst men were more likely than women to receive tPA 10 years ago, sex differences in utility of tPA were eliminated by the end [33]. This may be indicative of an increase overall in tPA use; recent studies have not reported any sex differences in the use of tPA [34]. There is evidence that thrombolytic therapy, i.e. tPA, may be more effective in females than males. Data from multiple clinical tPA clinical studies shows that women are more likely to demonstrate improved outcome 90 days post-stroke with tPA treatment than men [35, 36]. This positive effect of female sex was also seen when data was analysed from the Safe Implementation of Treatments in Stroke-International Stroke Thrombolysis Register (SITS-ISTR [37]).

Only a few clinical trials have specifically investigated whether sex-specific responses to treatment occur, and the current American Heart Association/American Stroke Association guidelines for acute stroke treatment make no distinction in respect to symptom presentation and possible therapeutic strategies between the sexes [38]. However, the Physicians Health Study reported that aspirin reduced the risk of cardiovascular disease, but not stroke, in men [39], and it has been shown that oral anticoagulation with warfarin may differentially affect the risk of ischaemic stroke between the sexes [40]. Whether all male-female differences in efficacy are real remains to be confirmed, and some apparent differential benefit may be artefactual, for example, women rather than men appeared to benefit from transdermal glyceryl trinitrate in the neutral large Efficacy of Nitric Oxide in Stroke (ENOS) trial [41]. However, in the prespecified subgroup of patients treated within 6 h, where GTN appeared to be effective, men were more likely to benefit [42]; these diametrically opposed findings are best put down to chance findings. An alternative explanation for sex differences may reflect pharmacokinetics, as seen for tirilazad, a putative neuroprotectant that is highly lipid soluble. Plasma tirilazad levels were lower in women than men, reflecting that the drug was sequestered in fat. Subsequent trials used higher doses of tirilazad in women to reach target levels although ultimately the drug appeared ineffective, even hazardous [43].

Preclinical studies have provided strong evidence for sex differences in the amount of damage produced following ischaemic stroke. The examination of specific sex effects of a potential treatment has not traditionally been factored into the design of preclinical studies. However, the need for relevant animal groups, such as comorbid animals and different sexes, was incorporated into the STAIR recommendations [44], and a number of recent studies have reported sex-specific effects in treatment efficacy. Minocycline, a tetracycline antibiotic, has been shown in experimental stroke models to be neuroprotective and improve behavioural outcomes but only in males [45]. Additionally, minocycline was not effective in ovariectomized female mice suggesting an intrinsic, rather than hormonal, difference mediating the sex difference in response to minocycline. A subsequent clinical trial showed that oral minocycline only reduced impairment (lower NIH stroke scale score) in males [46].

In relation to apoptosis, PARP activation is one of the key components in the pathway leading to cell death. Experimental studies have shown that pharmacological inhibitors of PARP-1 act differentially in males and females following ischaemic stroke [47]. In males, classical PARP inhibitors are effective post-stroke in terms of reducing infarct volume and improving long-term outcome; however such inhibitors seem to be ineffective in females [48, 49]. Although novel PARP inhibitors, including a water-soluble form, MP-124, do seem to be neuroprotective in both sexes [50] and may warrant clinical investigation. In addition, preclinical studies have shown that reducing nitric oxide production, through pharmacological inhibition or genetic deletion, following ischaemic stroke is beneficial in males but deleterious in females [47].

Numerous studies have investigated the sex steroid hormones, estrogens and progesterone, as potential neuroprotective factors following ischaemic stroke. In recent meta-analyses and systematic reviews, sex differences were identified in the effectiveness of both steroid hormones. Estrogens are reported to be neuroprotective, in terms of reducing lesion volume in ovariectomized females and young, adult males but ineffective in young, intact males [20]. For progesterone, the largest beneficial effect was seen when administered to males [51]. Thus, it is likely that some treatments may only be of real benefit to one sex such as hypothermia which has been shown to be more effective in males following ischaemic stroke than in females [52].

6 Future Directions

There is strong evidence for an effect of sex on stroke risk/incidence, diagnosis, symptom presentation, outcome and effectiveness of treatments. Yet current guidelines for acute stroke treatment do not make any distinction in terms of symptom presentation or therapeutic approach according to sex. Experimental studies have provided a wealth of supporting evidence for sex differences in the amount of damage produced following stroke, and some recent studies have begun to identify sex-specific effects of treatments under investigation. However, evidence is only just beginning to accumulate regarding the sexual dimorphism of the molecular signals activated following ischaemic stroke which ultimately result in cell death. Enhancing our understanding of these pathways and the sexual dimorphism can only be beneficial in terms of designing appropriate treatment strategies.

References

1. Gibson CL (2013) Cerebral ischemic stroke: is gender important? J Cereb Blood Flow Metab 33:1355–1361
2. Rojas JI, Zurru MC, Romano M, Patrucco L, Cristiano E (2007) Acute ischemic stroke and transient ischemic attack in the very old—risk factor profile and stroke subtype between patients older than 80 years and patients aged less than 80 years. Eur J Neurol 14:895–899
3. Seshadri S, Beiser A, Kelly-Hayes M, Kase CS, Au R, Kannel WB et al (2006) The lifetime risk of stroke: estimates from the Framingham study. Stroke 37:345–350
4. Gargano JW, Wehner S, Reeves MJ (2009) Delays among patients with acute stroke. Do presenting symptoms explain sex differences in emergency department? Stroke 40:1114–1120
5. Rosamond W, Flegal K, Furie K, Go A, Greenlund K, Haase N et al (2008) Heart disease and stroke statistics—2008 update: a report from the American Heart Association Statistics Committee and Stroke Statistics Subcommittee. Circulation 117:e25–e146
6. Thorvaldsen P, Asplund KM, Kuulasmaa K, Rajakangas AM, Scholl M (1995) Stroke incidence, case fatality, and mortality in the WHO MONICA project. World Health organization monitoring trends and determinants in cardiovascular disease. Stroke 26:361–367
7. Towfighi A, Saver JL, Engelhardt R, Ovbiagele B (2007) A midlife stroke surge among women in the United States. Neurology 69:1898–1904
8. Lofmark U, Hammartstrom A (2007) Evidence for age-dependent education-related differences in men and women with first-ever stroke. Neuroepidemiology 28:135–141
9. Rothwell PM, Coull AJ, Silver LE et al (2005) Population-based study of event-rate, incidence, case fatality, and mortality for all acute vascular events in all arterial territories (Oxford Vascular Study). Lancet 366:1970–1975
10. Jerath NU, Reddy C, Freeman D, Jerath AU, Brown RD (2011) Gender differences in presenting signs and symptoms of acute ischemic stroke: a population-based study. Gend Med 8:312–319
11. Kolominsky-Rabas PL, Weber M, Gefeller O, Neundoerfer B, Heuschmann PU (2001) Epidemiology of ischemic stroke subtypes according to TOAST criteria: incidence, recurrence, and long-term survival in ischemic stroke subtypes: a population-based study. Stroke 32:2735–2740
12. Mandelzweig L, Goldbourt U, Bokyo V, Tanne D (2006) Perceptual, social, and behavioural factors associated with delays in seeking medical care in patient with symptoms of acute stroke. Stroke 37:1248–1253
13. Di Carlo A, Lamassa M, Baldereschi M et al (2003) Sex differences in the clinical presentation, resource use, and 3-month outcome of acute stroke in Europe: data from a multicenter multinational hospital-based registry. Stroke 34:1114–1119
14. Benatru I, Rouaud O, Durier J et al (2006) Stable stroke incidence rates but improved case-fatality 06 in Dijon, France, from 1985 to 2004. Stroke 37:1674–1679
15. Olsen TS, Dehlendorff C, Andersen KK (2007) Sex-related time-dependent variations in post-stroke survival: evidence of a female stroke survival advantage. Neuroepidemiology 29:218–225
16. Fukuda M, Kanda T, Kamide N, Akutsu T, Sakai F (2009) Gender differences in long-term functional outcome after first-ever ischemic stroke. Intern Med 48:967–973
17. Gray LJ, Sprigg N, Bath PMW, Boysen G, De Deyn PP, Leys D et al (2007) Sex differences in quality of life in stroke survivors. Data from the Tinzaparin in Acute Ischaemic Stroke Trial (TAIST). Stroke 8(11):2960–2964
18. Alkayed NJ, Harukuni I, Kimes AS, London ED, Traystman RJ, Hurn PD (1998) Gender-linked brain injury in experimental stroke. Stroke 29:159–166
19. Toung TJ, Hurn PD, Traystman RJ, Sieber FE (2000) Estrogen decreases infarct size after temporary focal ischemia in a genetic model of type I diabetes mellitus. Stroke 31:2701–2706
20. Gibson CL, Gray LJ, Murphy SP, Bath PMW (2006) Estrogens and experimental ischemic stroke: a systematic review. J Cereb Blood Flow Metab 26:1103–1113

21. Gibson CL, Bath PM (2015) Feasibility of progesterone treatment for ischemic stroke. J Cereb Blood Flow Metab 25:431–439
22. Viscoli CM, Brass LM, Kernan WN, Sarrel PM, Makuch R, Horwitz RJ (2001) A clinical trial of estrogen-replacement therapy after ischemic stroke. N Engl J Med 345:1243–1249
23. Heyer A, Hasselblatt M, von Ahsen N, Hafner H, Siren AL, Ehrenreich H (2005) In vitro gender differences in neuronal survival on hypoxia in 17B-estradiol-mediated neuroprotection. J Cereb Blood Flow Metab 25:317–321
24. Li H, Pin S, Zeng Z, Wang MM, Andreasson KA, McCullough LD (2005) Sex differences in cell death. Ann Neurol 58:317–321
25. Carruth LL, Reisert I, Arnold AP (2002) Sex chromosome genes directly affect brain sexual differentiation. Nat Neurosci 5:933–934
26. Renolleau S, Fau S, Charriau-Marlangue C (2008) Gender-related differences in apoptotic pathways after neonatal cerebral ischemia. Neuroscientist 14:46–52
27. Park EM, Cho S, Frys KA, Glickstein SB, Zhou P, Anrather J et al (2006) Inducible nitric oxide synthase contributes to gender differences in ischemic brain injury. J Cereb Blood Flow Metab 26:392–401
28. Zhu C, Xu F, Wang X, Shibata M, Uchiyama Y, Blomgren K et al (2006) Different apoptotic mechanisms are activated in male and female brains after neonatal hypoxia-ischemia. J Neurochem 96:1016–1027
29. Cordeau P Jr, Lalancette-Herbert M, Weng YC, Krizz J (2008) Live imaging of neuroinflammation reveals sex and estrogen effects on astrocytic response to ischemic injury. Stroke 39:935–942
30. Sakata A, Mogi M, Iwanami J, Tsukuda K, Min L-J, Fujita T et al (2009) Sex-different effect of angiotensin II type 2 receptor on ischemic brain injury and cognitive function. Brain Res 1300:14–23
31. Shintani T, Klionsky DJ (2004) Autophagy in health and disease: a double-edged sword. Science 306:990–995
32. Puyal J, Vaslin A, Mottier V, Clarke PG (2009) Postischemic treatment of neonatal cerebral ischemia should target autophagy. Ann Neurol 66:378–389
33. Towfighi A, Markovic D, Ovbiagele B (2013) Sex differences in revascularization interventions after acute ischemic stroke. J Stroke Cerebrovasc Dis 22:e347–e353
34. Boehme AK, Siegler JE, Mullen MT, Albright KC, Lyerly MJ (2014) Monlezun D, J et al. Racial and gender differences in stroke severity, outcomes, and treatment in patients with acute ischemic stroke. J Stroke Cerebrovasc Dis 23:e255–e261
35. Kent DM, Price LL, Ringleb P, Hill MD, Selker HP (2005) Sex-based differences in response to recombinant tissue plasminogen activator in acute ischemic stroke a pooled analysis of randomized clinical trials. Stroke 36:62–65
36. Kent DM, Buchan AM, Hill MD (2008) The gender effect in stroke thrombolysis: Of CASES, controls and treatment-effect modification. Neurology 71:1080–1083
37. Lorenzano S, Ahmed N, Falcou A, Mikulik R, Tatlisumak T, Roffe C et al (2013) Does sex influence the response to intravenous thrombolysis in ischemic stroke? Answers from safe implementation of treatments in stroke-international stroke thrombolysis stroke register. Stroke 44:3401–3406
38. Jauch EC, Saver JL, Adams HP, Bruno A, Connors JJ, Demaerschalk BM et al (2013) Guidelines for the early management of patients with acute ischemic stroke: a guideline for healthcare professionals from the American Heart Association/American Stroke Association. Stroke 44:870–947
39. Group SCotPHSR (1989) Final report on the aspirin component of the ongoing physicians' health study. Steering committee of the Physicians' Health Study Research Group. N Engl J Med 321:129–135
40. Poli D, Antonucci E, Grifoni E, Abbate R, Gensini GF, Prisco D (2009) Gender differences in stroke risk of atrial fibrillation patients on oral anticoagulant treatment. Thromb Haemostat 101:938–942

41. Bath PMW, Woodhouse L, Scutt P, Krishnan K, Wardlaw JM, Bereczki D et al (2015) Efficacy of nitric oxide, with or without continuing antihypertensive treatment, for management of high blood pressure in acute stroke (ENOS): a partial-factorial randomised controlled trial. Lancet 385(9968):617–628
42. Woodhouse L, Scutt P, Krishnan K, Berge E, Gommans J, Ntaios G et al (2015) Effect of hyperacute administration (within 6 hours) of transdermal glyceryl trinitrate, a nitric oxide donor, on outcome after stroke: subgroup analysis of the Efficacy of Nitric Oxide in Stroke (ENOS) Trial. Stroke 46(11):3194–3201
43. Bath PMW, Blecic S, Bogousslavsky J, Boysen G, Davis S, Diez-Tejedor E et al (2000) Tirilazad mesylate in acute ischemic stroke - a systematic review. Stroke 31(9):2257–2265
44. Fisher M, Feuerstein G, Howells DW, Hurn PD, Kent TA, Savitz SA et al (2009) Update of the stroke therapy academic industry roundtable preclinical recommendations. Stroke 40:2244–2250
45. Li J, McCullough LD (2009) Sex differences in minocycline-induced neuroprotection after experimental stroke. J Cereb Blood Flow Metab 29:670–674
46. Amiri-Nikpour MR, Nazarbaghi S, Hamdi-Holasou M, Rezaei Y (2015) An open-label evaluator-blinded clinical study of minocycline neuroprotection in ischemic stroke: gender-dependent effect. Acta Neurol Scand 131:45–50
47. McCullough LD, Zeng Z, Blizzard KK, Debchoudhury I, Hurn PD (2005) Ischemic nitric oxide and poly (ADP-ribose) polymerase-1 in cerebral ischemia: male toxicity, female protection. J Cereb Blood Flow Metab 25:502–512
48. Couturier JY, Ding-Zhou L, Croci N, Plotkine M, Margaill I (2003) 3-Aminobenzamide reduces brain infarction and neutrophil infiltration after transient focal cerebral ischemia in mice. Exp Neurol 184:973–980
49. Hamby AM, Suh SW, Kauppinen TM, Swanson RA (2007) Use of a poly(ADP-ribose) polymerase inhibitor to suppress inflammation and neuronal death after cerebral ischemia-reperfusion. Stroke 38:632–636
50. Matsuura S, Egi Y, Yuki S, Horikawa T, Satoh H, Akira T (2011) MP-124, a novel poly (ADP-ribose) polymerase-1 (PARP-1) inhibitor, ameliorates ischemic brain damage in a non-human primate model. Brain Res 1410:122–131
51. Gibson CL, Gray LJ, Bath PMW, Murphy SM (2008) Progesterone for the treatment of experimental brain injury; a systematic review. Brain 131:318–328
52. Wagner AK, Bayir H, Ren D, Puccio A, Zafonte RD, Kochanek PM (2004) Relationships between cerebrospinal fluid markers of excitotoxicity, ischemia, and oxidative damage after severe TBI: the impact of gender, age and hypothermia. J Neurotrauma 21:125–136

Chapter 31
Unpuzzling the Comorbid Type 2 Diabetes and Hypertension-Related Cognitive Dysfunction and Stroke

I. Sebastião, E. Candeias, M.S. Santos, C.R. Oliveira, Paula I. Moreira, and Ana I. Duarte

Abstract Type 2 diabetes (T2D) is a highly disabling, major socioeconomic burden, whose long-term complications (particularly those affecting the central nervous system (CNS), as Alzheimer disease (AD)) can be further exacerbated by the frequent development of comorbid hypertension. Although the precise mechanisms involved herein remain elusive, it is conceivable that chronic T2D-related

I. Sebastião
CNC—Center for Neuroscience and Cell Biology, Rua Larga, Faculty of Medicine (Pólo 1, 1st Floor), University of Coimbra, 3004-517 Coimbra, Portugal

E. Candeias (✉) • A.I. Duarte (✉)
Institute for Interdisciplinary Research (IIIUC), University of Coimbra, Casa Costa Alemão—Pólo II, Rua D. Francisco de Lemos, 3030-789 Coimbra, Portugal

CNC—Center for Neuroscience and Cell Biology, Rua Larga, Faculty of Medicine (Pólo 1, 1st Floor), University of Coimbra, 3004-517 Coimbra, Portugal
e-mail: anaimduarte@gmail.com; ana.duarte@cnc.uc.pt

M.S. Santos
CNC—Center for Neuroscience and Cell Biology, Rua Larga, Faculty of Medicine (Pólo 1, 1st Floor), University of Coimbra, 3004-517 Coimbra, Portugal

Life Sciences Department, University of Coimbra, Largo Marquês de Pombal, 3004-517 Coimbra, Portugal

C.R. Oliveira
CNC—Center for Neuroscience and Cell Biology, Rua Larga, Faculty of Medicine (Pólo 1, 1st Floor), University of Coimbra, 3004-517 Coimbra, Portugal

Institute of Biochemistry, Faculty of Medicine, University of Coimbra, 3004-504 Coimbra, Portugal

P.I. Moreira (✉)
CNC—Center for Neuroscience and Cell Biology, Rua Larga, Faculty of Medicine (Pólo 1, 1st Floor), University of Coimbra, 3004-517 Coimbra, Portugal

Institute of Physiology, Faculty of Medicine, University of Coimbra, 3004-504 Coimbra, Portugal
e-mail: pimoreira@fmed.uc.pt; venta@ci.uc.pt

P.A. Lapchak, J.H. Zhang (eds.), *Neuroprotective Therapy for Stroke and Ischemic Disease*, Springer Series in Translational Stroke Research, DOI 10.1007/978-3-319-45345-3_31 brain insulin resistance (IRES) and hyperglycemia may crosstalk with an overactivated brain renin-angiotensin-II-aldosterone system (RAAS) further potentiating the hypertension-related injury and culminating in cognitive dysfunction and AD. Indeed, several studies showed the contribution of abnormal RAAS activation upon hypertension per se to the pathophysiology of CNS disorders, such as stroke and AD. However, most of this available knowledge relies on the indirect effects of pharmacological inhibition of RAAS by drugs belonging either to the angiotensin II receptor blockers (ARBs) or angiotensin-converting enzyme (ACE) inhibitors (ACEi) groups. For instance, antihypertensive drugs have also shown anti-neuroinflammatory properties, widely known to play a pivotal role in brain and cognitive dysfunction.

In sum, albeit during the last two decades a bulk of scientific knowledge has progressively unravelled the molecular mechanisms that lead to chronic T2D-related CNS injury, the information available regarding its exacerbation upon the quite frequent comorbid hypertension is still scarce and somehow controversial. In this perspective, we aim to briefly review some of the subcellular mechanisms (e.g., oxidative stress) that may underlie the hypertension-induced (per se or as a comorbidity to T2D) brain and cognitive dysfunction, vascular dementia (particularly AD), and stroke. We will also briefly discuss the pharmacological evidence on the neuroprotection afforded by antihypertensive drugs.

Keywords Hypertension • Type 2 Diabetes (T2D) • Multifactorial disease • Nephropathy • Angiotensin

1 Introduction

Diabetes mellitus (DM) is one of the most common metabolic diseases, being considered as an epidemic of the twenty-first century [1]. In fact, whereas 12 years ago Wild et al. [2] estimated that 171 million people worldwide (~3 % of the population) suffered from DM, more recent statistics estimated that, in 2011, there were already 366 million diabetics, a number predicted to rise to 552 million patients by 2030 [3]. In this scenario, since 90–95 % of DM cases correspond to type 2 diabetes (T2D), there is a high probability that the number of T2D patients will rise massively over the next 20 years, rendering it an enormous public health concern [4, 5].

T2D is a multifactorial disease, whose most common triggering factors are related with modern lifestyle (as sedentarism, hyperphagia, and obesity) that induce a massive insulin dysregulation and hyperglycemia (the main hallmarks of T2D), ultimately leading to devastating effects on multiple tissues—the so-called long-term complications (such as retinopathy, nephropathy, and peripheral neuropathy) [6]. Albeit less frequently, T2D can be also caused by alterations in the individual's genetic information. A positive family history may also increase significantly the risk for T2D [7], with estimates pointing towards 15–25 % of first-degree relatives of T2D patients developing impaired glucose tolerance or diabetes [8].

Traditionally, the main focus in the treatment of T2D (and, hence, in the reduction of the risk of its long-term vascular and cardiovascular complications) has been the maintenance of a glycemic control within the normal range [9]. Accordingly, several studies showed that a tight control of blood glucose levels, blood pressure, and dyslipidemia greatly reduce the incidence of microvascular complications associated with coronary artery disease and stroke [10, 11]. And to achieve such blood glucose regulation in T2D, one of the most important keys rely on the insulin hormone and, more specifically, on the interplay between insulin secretion and action. Indeed, whereas, physiologically, pancreatic β-cells can adapt to increased insulin action by decreasing its secretion and vice versa [3], as T2D progresses, its successful management may also require the control of blood pressure and lipid levels [12]. Therefore, it is not surprising that, albeit the achievement of an optimal glycemic control could be the main therapeutic focus in the early stages of T2D, this may not be enough to reduce the cardiovascular risk later on [13]. In line with this and with the outcomes of several trials (e.g., Action in Diabetes and Vascular disease: preterAx and iamicro Nmr Controlled Evaluation (ADVANCE), Action to Control Cardiovascular Risk in Diabetes (ACCORD), and Veterans Affairs Diabetes Trial (VADT)), the most recent guidelines establish that pro-euglycemia therapies should imply a patient-adjusted prescription considering specific patient/disease factors [14]. For example, among the several risk factors for T2D, obesity has been one of the closest related with disease progression and development of later complications. Thus, the prescription of an anti-T2D treatment should also consider patients' elevated body weight [15]. This is even more relevant as body weight gain is a frequent secondary effect of some anti-T2D therapies and, for this reason, the development of modern anti-T2D drugs is also taking into account their effects on adipogenesis and fat mass regulation [16].

Importantly, as the blood levels of glycated hemoglobin A_{1C} (HbA_{1C}) constitute an indirect estimation of the average plasma glucose levels in the previous 2–3 months, this gives a more accurate idea on the T2D patients "long-term" (rather than "acute") control of their blood glucose levels [17]. Although there is still some controversy on the ideal levels of HbA_{1C} in blood, another goal in the treatment of T2D should aim at lowering its levels [18]. Noteworthy, according to the United Kingdom Prospective Diabetes Study (UKPDS) and Diabetes Control and Complications Trial (DCCT), the lowering of blood HbA_{1C} levels by 1.0 % was enough to reduce microvascular complications by ~30 %, as well as the risk for myocardial infarction and death after 10 years of intensive glucose control in patients newly diagnosed with T2D [19]. However, and albeit most oral antihyperglycemic agents can also reduce HbA_{1C} values by 1.5–2.0 % from baseline levels of 8.5–9.5 %, it is highly probable that patients presenting baseline HbA_{1C} levels above 9.0 % will not easily achieve the therapeutic goal of <7.0 % and, therefore, may require a combination therapy in a near future [13]. Thus, although the starting blood HbA_{1C} values are important when choosing the best therapeutic approach for each T2D patient, we must bear in mind that this could not be enough as the disease progresses [20].

As referred previously, besides the chronic hyperglycemia (and its related HbA_{1C} levels), T2D also involves insulin resistance [21]. This renders the reduction in insulin resistance and/or the stimulation of insulin secretion other important therapeutic targets in T2D patients [8]. Whereas in the first stages of the disease, the rescue in insulin resistance can be achieved by, e.g., diet, exercise, and/or drug therapy [8], in the later stages these may become inefficient and the patients may also need injections of exogenous insulin [22] (with the inconvenience of, for instance, repeated insulin-induced hypoglycemic episodes).

Besides the relative uncertainty on the precise subcellular impact of badly controlled T2D in central nervous system (CNS), the research conducted in the last decades increasingly highlights chronic T2D as a major risk factor for stroke (also worsening its post-ischemic effects) and its consequent brain damage (more specifically, in terms of progressive cognitive disturbances) [23], but also to the cognitive decline associated with vascular dementia and, most strikingly, to Alzheimer disease (AD) [22]. Indeed, several studies showed that T2D patients have an increased incidence of cognitive impairment (accompanied or not by vascular dementia) and AD [24, 25]. Thus, given the increasingly type 2 diabetic population and the ever-increasing population aging worldwide, it is not surprising that T2D-induced cognitive impairment became a global concern [8]. And although the molecular mechanisms underlying such brain damaging effects of chronic T2D are slowly became unravelled, little is known on the possible involvement of its comorbid pathologies. This is an especially worrying situation since, as further detailed below, chronic T2D is quite often accompanied by dyslipidemia and hypertension [21]. Accordingly, we will provide a brief overview on some of the possible subcellular mechanisms underlying the deleterious effects of hypertension (per se or as a comorbidity to T2D) on brain and cognitive functions, particularly on the role of oxidative stress (and the ultimate impact on vascular dementia (e.g., AD) and stroke). This will be followed by a brief discussion on the few available pharmacological evidence regarding the neuroprotective potential of antihypertensive drugs against such neurodegenerative processes.

1.1 T2D, Hypertension, and the CNS

The previously described scenario can be further aggravated by hypertension, an age-related pathology (like T2D) that affects ~65 % of people aged 65 years and over, being also a major risk factor for T2D. Indeed, more than three decades ago, Mogensen and Christensen [26] reported that T2D diagnosis was often preceded by hypertension, which further aggravated T2D complications. But, most strikingly, hypertension is a common comorbidity of T2D [27]. Accordingly, hypertension not only occurs twice as frequently in diabetics than in nondiabetic people [26], but very recent estimates even point towards ~2/3 of chronic T2D patients suffering from comorbid hypertension [28] that, in the long run, further exacerbate their micro- and macrovascular complications (particularly at the vascular and brain/cognitive levels) [28–31].

Arterial hypertension per se, clinically defined by a systolic blood pressure (BP) ≥140 mmHg and/or a diastolic BP≥90 mmHg [32], has been recently suggested to constitute one of the most important risk factors for cerebrovascular diseases. This may cause cerebral hypoperfusion and ischemia, with the decrease in oxygen delivery to the brain rendering it more vulnerable to stroke [33]. Indeed, hypertension per se constitutes one of the most important, and avoidable, risk factors for stroke [34], a clinical condition defined by ischemic stroke or hemorrhagic stroke [35] and characterized by poor blood flow to the brain that may result in cell death and subsequent cognitive problems (e.g., inability to move or feel, problems in understanding or speaking, vision loss) [36]. Besides stroke, hypertension may be a risk factor also for dementia (including vascular dementia and AD), probably due to its association with amyloid pathology (one of the pathological hallmarks of AD) [37–40]. In line with this, several authors reported that, besides chronic hypertension, the spontaneously hypertensive (SHR) rat model also develops progressive neurobehavioral impairment, several AD risk factors (including dyslipidemia, abdominal obesity, and insulin resistance) [41, 42] and, most strikingly, several AD-like pathologic features (e.g., hyperphosphorylated Tau protein, neurofibrillary tangles, and neuroinflammatory markers) [43]. Notably, hypertension has been also increasingly suggested to contribute to cognitive dysfunction in T2D patients [44], besides exacerbating their risk for cardiovascular disease, stroke, and end stage renal disease [30].

1.1.1 Hypertension and the Renin-Angiotensin-Aldosterone System

The renin-angiotensin-aldosterone system (RAAS) is well known for its critical role in the regulation of arterial blood pressure and in the induction and control of hypertension [45]. RAAS consists of an enzymatic cascade catalyzed by renin and angiotensin-converting enzyme (ACE) that generates the key peptide angiotensin II [46, 47], which may, in turn, be cleaved together with angiotensin I by ACE2, generating angiotensins-(1–7) and -(1–9), respectively [48]. This may constitute a compensatory mechanism, as angiotensin-(1–7) has been recently described to display opposite properties to angiotensin II [49]. Angiotensin II-mediated effects are known to occur via angiotensin II receptors (ATRs), whose two isoforms identified to date are the type 1 receptor (AT1R) and type 2 receptor (AT2R) [50]. Importantly, both angiotensin II and ATRs appear to be ubiquitously expressed throughout the organism, being angiotensin II mainly produced by extra-cardiac tissues [51]. Thus, angiotensin II may exert multiple systemic and local actions in different tissues (particularly in terms of pathological mechanisms underlying target organs damage, e.g., in blood vessels, kidneys, adrenal glands, pancreas, heart and, strikingly, also in CNS) [45, 52–54]. For instance, it has been increasingly suggested that angiotensin II produced by the pancreatic RAAS could negatively influence insulin secretion in vivo by restraining islet blood perfusion and function, whereas such RAAS blocking may exert positive effects herein [55]. This may be conspicuously important in diabetic subjects since they are more prone to comorbid hypertension, and several pancreatic islet RAAS components appear to be upregulated by hypertension and hyperglycemia [56–58].

Traditionally, AT1Rs stimulation has been associated with the progression of cardiovascular diseases (including atherosclerosis, cardiac hypertrophy, and heart failure), most likely through inflammatory, oxidative stress, and tissue remodelling mechanisms [59–61]. However, other studies also reported a high density of AT1Rs in specific brain regions, such as cerebrovascular endothelial cells [62, 63], which may control cerebrovascular flow and blood–brain barrier functioning, ultimately aiding in the regulation of brain's overall function. Additionally, AT1Rs appear to be widely, but selectively, expressed in some neuronal circuits [62] and astrocytes at the forebrain and rostral ventrolateral medulla (RVLM) [64], which may explain how AT1Rs activity regulates multiple brain functions (e.g., stress reactivity, cognition, and control of innate immune response) [65]. Interestingly, angiotensin II microinjection into the RVLM resulted in an AT1R-mediated increase in blood pressure [66, 67], while its blockade reduced blood pressure [68, 69].

Concerning AT2Rs, several authors observed that their expression increases under pathological situations (as cancer and cardiovascular disorders) [70, 71], suggesting that they may have a counterbalancing effect [72, 73]. Indeed, Herrera and Garvin [74] found that the activation of AT2R triggers nitric oxide ($NO^{\bullet}$) release and would therefore potentially antagonize the effects of AT1R activation, leading to cardiovascular protection. However, the precise mechanisms of AT2R-mediated actions remain controversial [60, 75].

1.1.2 Hypertension and Oxidative Stress

Besides SHR rats, the stroke-prone SHR rat strain and many other rodent models are genetic animal models that spontaneously develop hypertension and present increased reactive oxygen species (ROS) levels in vessels [76, 77]. Similarly, increased vascular oxidative stress was also demonstrated in rats with experimentally induced hypertension [78, 79] and in hypertensive patients [80]. However, such massive ROS may be also generated after brain ischemia–reperfusion, further oxidizing proteins and lipids in brain cells, ultimately activating cell death mechanisms, either by necrosis or apoptosis [81]. Indeed, several studies have implicated such transient ischemia-induced massive ROS formation and their detrimental effects in brain in the pathogenesis of ischemic injury [82, 83].

More than two decades ago, Griendling et al. [84] discovered that the activation of the vascular smooth muscle nicotinamide adenine dinucleotide phosphate (NAD(P)H) oxidase—a multi-subunit enzyme and one of the enzymatic sources of the superoxide anion—by angiotensin II was an important intracellular source of ROS [84]). Afterwards, Rajagopalan et al. [85] reported that hypertension caused by in vivo angiotensin II infusion (but not norepinephrine infusion) increased vascular superoxide production [85]. Conversely, several authors found that adenovirus-mediated superoxide dismutase (SOD) overexpression prevented this form of hypertension [86–88].

In addition to ROS, the abnormal formation of reactive nitrogen species (RNS) may also play an important role in the pathogenesis of hypertension. Indeed, accumulating evidence suggests that alterations in NO• synthesis and in NO-sGC-cGMP signalling, or a reduction in the bioavailability of endothelium-derived NO• due to increased oxidative stress may be pivotal contributors to the pathogenesis of hypertension [89, 90] (Fig. 31.1).

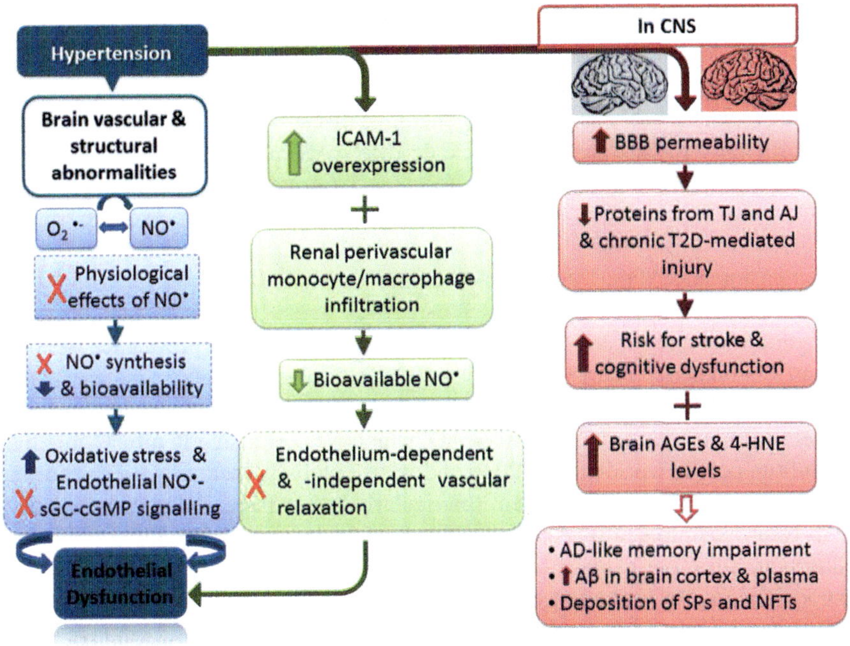

Fig. 31.1 Chronic T2D and its comorbid hypertension-related damage to brain and cognitive function. The excessive ROS formation upon hypertension per se may constitute one of the main mechanisms for brain vascular and structural abnormalities. Changes in NO• synthesis and/or its reduced bioavailability have been correlated with oxidative stress and changes in endothelial NO•-sGC-cGMP signalling, suggesting that the blockade of NO• physiological effects upon its reaction with the superoxide anion may underlie such brain endothelial dysfunction and damage. Additionally, infiltrated renal perivascular monocytes/macrophages and overexpression of ICAM-1 may lower the amount of bioavailable endothelium-derived NO•, inhibiting both endothelium-dependent and -independent vascular relaxation and further promoting endothelial lesion. Hypertension may also render the BBB more permeable and downregulate proteins from TJ and AJ. This, together with the brain damage associated with long-term T2D, may also aggravate the risk for stroke and the prevalence of mild and subtle cognitive dysfunction, in a process further exacerbated by the excessive formation of AGEs and 4-HNE that may culminate in AD-like changes in memory, and in brain cortex and plasma Aβ levels, SPs and NFTs. Abbreviations: *4-HNE* 4-hydroxynonenal, *Aβ* amyloid-β peptide, *AGEs* advanced glycation end-products, *AJ* adherens junction, *BBB* blood–brain barrier, *cGMP* cyclic guanosine monophosphate, *CNS* central nervous system, *ICAM-1* intercellular adhesion molecule-1, *NFTs* neurofibrillary tangles, *NO•* nitric oxide, *SPs* senile plaques, *ROS* reactive oxygen species, *sGC* soluble guanylate cyclase, *T2D* type 2 diabetes, *TJ* tight junctions

1.1.3 Hypertension in CNS Disorders

From the above and given the known expression of all RAAS components in the CNS (namely of AT2Rs in vascular wall and brain areas involved in learning and control of motor activity) [91, 92], it is not surprising that recent studies highlighted the contribution of RAAS to the pathophysiology of CNS disorders (e.g., AD and stroke) [45], as further detailed below. In line with this, several authors showed that the stimulation of AT2Rs protects brain against stroke damage and dementia, at least partially by promoting cerebral blood flow and neuronal differentiation [72, 93, 94]. Moreover, angiotensin II has been increasingly described to participate not only in the regulation of blood pressure, but also in the pathophysiology of CNS illnesses (such as stroke and neurodegenerative diseases) [94]. Accordingly, recent clinical trials demonstrated that blockade of RAAS may delay the triggering of stroke and may also exert beneficial effects on cognitive function [45].

Hypertension and AD

Dementia is an age-related neurodegenerative disorder, whose prevalence has been recently estimated to nearly double every 20 years (reaching approximately 42 million individuals in 2020 and 81 million in 2040) [95]. Among the several types of dementia, the most prevalent form is AD, which was estimated to affect more than 35 million people worldwide in 2010 and whose globally increased prevalence has rendered it a world health problem (particularly in our increased aged population) [96]. Neuropathologically, AD is characterized by intraneuronal clusters of hyperphosphorylated Tau protein (composing the neurofibrillary tangles (NFTs)) and extracellular aggregates of amyloid-beta peptides (Aβ) (that form the senile plaques (SPs)), most likely due to faulty protein processing mechanisms [27] and the subsequent loss of intracellular quality control mechanisms.

Despite all the research efforts and the knowledge accumulated in the past two decades regarding AD pathophysiology, the precise links between its several known risk factors (as aging and T2D) and mechanisms remain incompletely understood. And this may not only impact its precise diagnosis during patients' lifetime, but may also explain the fact that the currently available treatments for AD are only addressed to symptoms, with no successful preventive or delaying strategies available [97].

Risk factors for AD have been divided into genetic and non-genetic, with the last ones constituting ~95 % of all cases. And, besides the previously discussed relevance of aging and T2D, it has been also increasingly described that, among the vascular and vascular-related factors associated with dementia and cognitive decline, high blood pressure and hypertension, total cholesterol, insulin resistance, increased body mass index and obesity may play a role herein [98, 99]. In particular, hypertension is an increasingly studied risk factor for vascular dementia, its role in the pathogenesis of AD being increasingly highlighted [100–102]. More specifically, several observational and experimental studies suggested that RAAS overactivation may be involved in the pathogenesis and progression of AD [103]. Although the precise molecular mechanisms for hypertension-associated AD remain poorly understood, they may possibly involve ischemia, oxidative stress, and inflammation [101, 104]. Moreover, we must bear in mind that the effects of hypertension on brain structure and function in AD patients are still debatable [105] (Fig. 31.1).

Hypertension, T2D, and Stroke

In a very recent study aiming to determine the prevalence and risk factors of ischemic stroke among diabetic patients, it was found that hypertension was a significant independent risk factor for the development of ischemic stroke [106]. Moreover, studies involving experimentally induced stroke (upon middle cerebral artery occlusion) in genetically modified mice demonstrated that AT1R-mediated signalling enhances brain damage [107]. This could be partly due to an increment in oxidative stress and a decrement in cerebral blood flow under such ischemic conditions [107]. Notably, pretreatment with angiotensin receptor blockers (ARBs) (e.g., telmisartan, valsartan, olmesartan) [108] was shown to prevent ischemic brain damage [45], whereas stimulation of AT2R by using an ARB counteracted the effects of AT1R, attenuating brain injury [91] (Fig. 31.2). In line with this, others found that AT2R stimulation promotes neuronal differentiation and regeneration upon similar conditions [109, 110], and enhanced neuronal survival and neurite outgrowth in response to ischemia-induced neuronal injury [111]. Such AT2R-mediated neural differentiation and neurite outgrowth could occur via mitogen-activated protein kinase (MAPK) or NO•, being possibly involved also in brain development [72, 112–114]. Strikingly, Iwanami et al. [115] reported that the expression of AT2R in hematopoietic cells was accompanied by a protective effect against cerebral ischemic damage. This could occur via AT2R stimulation by the ARBs (e.g., telmisartan) in hematopoietic cells, whereas angiotensin II unbound to AT1R under the presence of these drugs would block its downstream signalling [115]. Hence, this reinforces the notion that the stimulation of AT2R-mediated signalling may act as a crucial neuroprotective factor after stroke, being therapeutically promising for the prevention of chronic hypertension-related ischemic brain damage.

This deleterious scenario caused by hypertension can be further aggravated upon diabetes (especially T2D). Indeed, numerous studies reported that T2D exacerbates and/or is a main cause for stroke and myocardial infarction [116–118]. More specifically, individuals suffering from metabolic syndrome or insulin resistance are already under an increased stroke risk, as well as poorer outcome and increased mortality [119]. However, the most clear evidence came from T2D patients, to whom ischemic stroke not only constitutes one of the major potential macrovascular complications, but also consistently suffer from poorer outcomes and prognosis than the nondiabetic individuals after a stroke [116, 120]. This high susceptibility for stroke might be even further potentiated in patients suffering from

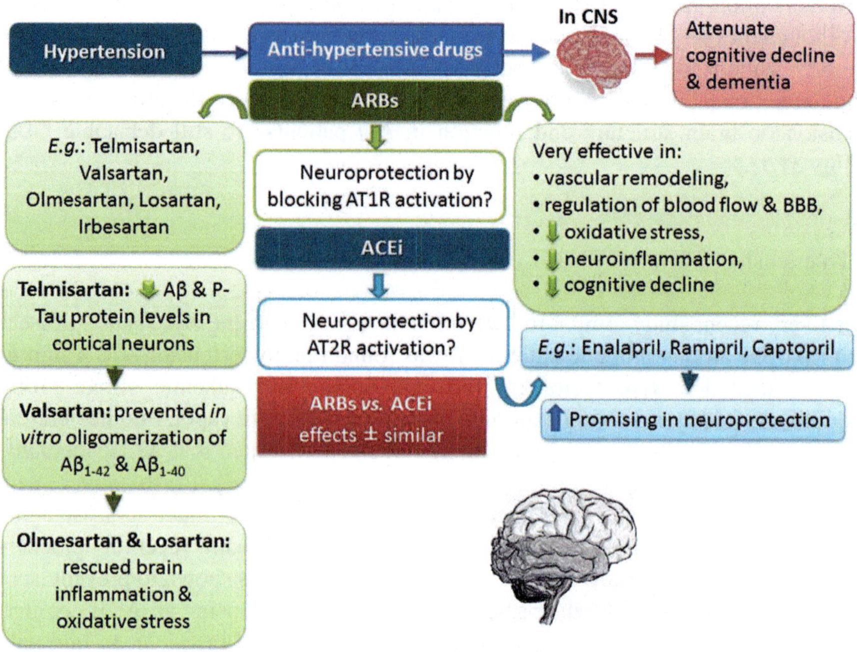

Fig. 31.2 The role of antihypertensive drugs in unravelling the subcellular mechanisms for hypertension-associated stroke or vascular dementia. Antihypertensive drugs have been increasingly shown to ameliorate cognitive outcome and decrease the incidence of dementia. More specifically, the neuroprotective role of ARBs (namely Telmisartan, Valsartan, Olmesartan, Losartan, Irbesartan) may involve the inhibition of AT1Rs, being Telmisartan associated with lower cortical neuronal Aβ and P-Tau protein levels, Valsartan attenuated $A\beta_{1\text{-}42}$ and $A\beta_{1\text{-}40}$ oligomerization of into high-molecular weight species, and Olmesartan and Losartan had anti-inflammatory and anti-oxidative stress effects in brains from diabetic hypertensive rats. In general, ARBs were very effective in vascular remodelling, regulation of cerebral blood flow and BBB, and in decreasing oxidative stress, neuroinflammation and cognitive decline. Conversely, the ACEi class of antihypertensive drugs promotes AT2R activation and their promising neuroprotective potential may occur through this pathway. Importantly, the ultimate effects of both ARBs and ACEi are very similar. Abbreviations: *Aβ* amyloid-β peptide, *ACE* angiotensin-converting enzyme inhibitor, *ACEi* angiotensin-converting enzyme inhibitor, *AR* angiotensin receptor, *ARBs* angiotensin receptor blockers, *AT1R* angiotensin receptor-1, *AT2R* angiotensin receptor-2, *BBB* blood–brain barrier, *CNS* central nervous system, *P-Tau* hyperphosphorylated Tau protein

both T2D and hypertension, in a process that was recently suggested to rely on their cerebral vascular sensitivity to the increment in blood pressure [11]. Unfortunately, and despite the well-documented correlation between hypertension and/or T2D and stroke, their precise underlying mechanisms are still unknown [121]. But, similarly to ischemic brain damage after stroke per se, the massive activation of brain RAAS and its interaction with the previously referred CNS damage associated with hyperglycemia and insulin resistance may be involved herein.

1.2 *The Promising Potential of Antihypertensive Drugs Against CNS Damage Caused by Hypertension Comorbid to T2D*

Many national guidelines recommend the preferable use of RAAS inhibitors (rather than other antihypertensive agents) for the treatment of high-risk hypertensive patients [122–124]. However, the simultaneous use of multiple antihypertensive medications is often necessary to achieve the recommended blood pressure (or at least to maintain its control over a narrow range), in addition to the administration of other antihypertensive drugs (including calcium channel blockers (CCBs) and diuretics) or to the use of high-dose RAAS inhibitors [125].

As previously referred, a multifactorial pathogenesis and progression has been widely accepted for AD and stroke, with the main emphasis put towards three main risk factors: aging process, abnormal brain Aβ or hyperphosphorylated Tau protein accumulation (in the case of AD) and high blood pressure (in stroke) [126], and metabolic/cardiovascular-related risk factors (such as hypertension and T2D) [127]. Accordingly, it seems plausible that medications used to treat such risk factors may also reduce the incidence of AD [128] and stroke [126]. Although this has become increasingly more consensual regarding the use of anti-T2D drugs for the treatment of AD and stroke, concerning the possible benefits of antihypertensive agents against both neurodegenerative conditions (in addition to their blood pressure lowering effects) this has been more debatable. This is mostly due to the poor knowledge on the precise involvement of RAAS in the pathogenesis and progression of neuronal diseases. Nevertheless, several authors hypothesized that RAAS inhibition by ARBs or ACE inhibitors (ACEi) may protect against the neuroinflammatory processes known to play a pivotal role in brain and cognitive dysfunction [129–131] (Fig. 31.2).

1.2.1 ARBs

ARBs are imidazole derivatives whose common class effect relies on the blockade of AT1R [132, 133]. Several clinically available ARBs exist and, according to their chemical structures, each ARB may bind to different AT1R domains, initiating distinct mechanisms for receptor antagonism [134]. Moreover, different ARBs present different lipophilicity degrees, which may in turn underlie their major differences in biological half-life and bioavailability [135, 136]. Currently, ARBs are widely used for the treatment of cardiovascular and renal disorders, as well as in metabolic alterations (including diabetes) [132, 137, 138].

Besides their role in rescuing peripheral inflammation, ARBs also exhibited major beneficial effects on peripheral metabolism and, regardless their administration pathways (they can be administered intraperitoneally, subcutaneously, or orally), also appear to be powerful neuroprotective agents [139]. Indeed, several studies showed that ARBs protect the peripheral and brain vasculature, and reduce hypoxia and brain inflammatory injury [139, 140]. Hence, such attenuation of the abnormal AT1R activity may potentially rescue several brain disorders [141]. In line with this, studies in both animal models and human patients demonstrated a positive association between ARBs treatment and preservation of cognitive function [142], as well as a reduction in the infiltration of inflammatory cells and in paralysis [143]. Strikingly, RAAS inhibition by ARBs was shown to prevent the onset of AD [132, 144, 145], with valsartan (one of the most widely used ARBs) showing an attenuation in the oligomerization of Aβ peptides into high-molecular weight oligomeric ones [146] (Fig. 31.2). Moreover, both valsartan and telmisartan (another ARB) were shown to prevent Aβ-mediated cognitive impairment in AD mouse models [147, 148]. Besides this, neuropathological evidence also suggested that the selective inhibition of AT1R by ARBs may preserve ACE production, which may in turn degrade the Aβ abnormally accumulated in brain [130]. Altogether, these studies suggest that ARBs may prevent the abnormal Aβ deposition in the brain and, as a consequence, they may render a better cognitive outcome than other medications for blood pressure control [140, 149–151] (Fig. 31.2). In line with this, Oshima et al. [45] described that, from the medical records of more than five million patients, those undergoing ARBs therapy had a 35–40 % lower risk for AD and other neurodegenerative disorders. This evidence further emphasizes that ARBs benefits may possibly extend beyond their "traditional" blood pressure-lowering effect [142].

The ARBs known role in attenuating the overactivation of AT1Rs (which is associated, in peripheral tissues, with increased blood pressure and vascular inflammation) rendered these well-tolerated compounds widely prescribed also for the treatment of stroke [133, 145]. Indeed, ARBs have been shown to protect cerebral blood flow and cognition upon stroke [139], to reverse cerebrovascular remodelling and arterial compliance [152–154], reducing brain inflammation and being neuroprotective in brain ischemic conditions, thus attenuating neurological impairment following stroke [155] (Fig. 31.2).

Although the precise mechanisms underlying such ARBs-mediated neuroprotection are still poorly understood and most findings indicate their direct action in brain, we cannot exclude that ARBs-related indirect mechanisms may also contribute to different degrees of preclinical efficacy [156]. More specifically, although neuroprotection by these drugs may arise from the direct blockade of brain AT1Rs, it may be also a consequence of peripheral AT1R inhibition and/or a response to other processes that arise from AT1R antagonism [65].

1.2.2 ACEi

Despite the scarce information on the role for ACEi in hypertension-related neurodegenerative diseases, Vargas et al. [157] and Estato et al. [11] observed recently that the ACEi enalapril attenuated brain inflammation and oxidative stress in diabetic hypertensive rats (Fig. 31.2). Importantly, as these drugs may have similar effects to those arising from AT2R stimulation, it is possible that ACEi share some intermediary processes with ARBs (namely vasodilation, $NO^{\bullet}$ synthesis, apoptosis, and inhibition of cell growth) [71].

Interestingly, only a few studies analyzed the impact of ACEi in delaying AD onset [158], most likely by blunting the abnormal Aβ deposition in CNS [159]. However, this has been challenged by the observation that ACE-deficient mice had no changes in brain Aβ levels [130].

Despite the fact that ACE inhibitors have yet to be further researched, recent clinical studies have shown a certain degree of protection from stroke although seem to be inferior to those of ARBs [160, 161]. Nevertheless, this question on which antihypertensive drugs, ARBs or ACEi, are more effective in the prevention of ischemic brain damage still remains a matter of debate [45].

2 Conclusions

Most of the studies on the role of T2D and hypertension as risk factors for stroke and vascular dementia (including AD) have focused on each "peripheral pathology" per se. This has rendered the knowledge on their precise molecular links (particularly when comorbid) very limited. Nevertheless, the scarce evidence available to date seems to consistently demonstrate that comorbid hypertension further exacerbates the brain and cognitive dysfunction induced by T2D. As such, we may hypothesize that an intricate, highly deleterious network of damaging processes (e.g., oxidative stress and/or inflammation) may result from the crosslink between insulin resistance, hyperglycemia, and RAAS overactivation, leading to neurovascular, brain, and cognitive dysfunction and, ultimately, to stroke and AD.

Under this perspective, unpuzzling these molecular cross-links will not only foster the knowledge on chronic T2D-related stroke and vascular dementia, but will also aid in the establishment of chronically efficient and safer drugs, and of effective preventive strategies against such chronic damage to CNS and cardiovascular system.

Acknowledgements This work was supported by European funds from FEDER, through the Programa Operacional Factores de Competitividade—COMPETE 2020; by Portuguese funds from FCT—Fundação para a Ciência e a Tecnologia (Strategic Project: UID/NEU/04539/2013); and by European Social Fund: Fellowships *SFRH/BPD/84473/2012* to A. I. Duarte and *SFRH/BD/90036/2012* to E. Candeias.

References

1. Skyler JS (2004) Diabetes mellitus: pathogenesis and treatment strategies. J Med Chem 47:4113–4117
2. Wild S, Roglic G, Green A, Sicree R, King H (2004) Global prevalence of diabetes estimates for the year 2000 and projections for 2030. Diabetes Care 27:1047–1053
3. Duarte AI, Candeias E, Correia SC, Santos RX, Carvalho C, Cardoso S et al (1832) Crosstalk between diabetes and brain: glucagon-like peptide-1 mimetics as a promising therapy against neurodegeneration. Biochim Biophys Acta 2013:527–541

4. Vallon V (2015) The mechanisms and therapeutic potential of SGLT2 inhibitors in diabetes mellitus. Annu Rev Med 66:15.1–15.16
5. Diabetes fact sheet 312, http://www.who.int/mediacentre/factsheets/fs312/en/
6. Campbell RK (2011) Clarifying the role of incretin-based therapies in the treatment of type 2 diabetes mellitus. Clin Ther 33:511–527
7. Pierce M, Keen H, Bradley C (1995) Risk of diabetes in offspring of parents with Non-insulin-dependent diabetes. Diabet Med 12:6–13
8. Stumvoll M, Goldstein BJ, van Haeften TW (2005) Type 2 diabetes: principles of pathogenesis and therapy. Lancet 365:1333–1346
9. Reichard P, Nilsson BY, Rosenqvist U (1993) The effect of long-term intensified insulin treatment on the development of microsvacular complications of diabetes mellitus. N Engl J Med 329:304–309
10. Terry T, Raravikar K, Chokrungvaranon N, Reaven PD (2012) Does aggressive glycemic control benefit macrovascular and microvascular disease in Type 2 Diabetes?: Insights from ACCORD, ADVANCE, and VADT. Curr Cardiol Rep 14:79–88
11. Estato V, Obadia N, Carvalho-Tavares J, Freitas FS, Reis P, Neto HCF (2013) Blockade of the renin–angiotensin system improves cerebral microcirculatory perfusion in diabetic hypertensive rats. Microvasc Res 87:41–49
12. Campbell RK, White JR Jr (2008) More choices than ever before—emerging therapies for type 2 diabetes. Diabetes Educ 34:518–534
13. Gavin JR, Stolar MW, Freeman JS, Spellman CW (2010) Improving outcomes in patients with type 2 diabetes mellitus: practical solutions for clinical challenges. J Am Osteopath Assoc 110:S2–S14
14. Morsink LM, Smits MM, Diamant M (2013) Advances in pharmacologic therapies for type 2 diabetes. Curr Atheroscler Rep 15:302
15. Buse J (2000) Combining insulin and oral agents. Am J Med 108:23S–32S
16. Cignarelli A, Giorgino F, Vettor R (2013) Pharmacologic agents for type 2 diabetes therapy and regulation of adipogenesis. Arch Physiol Biochem 119:139–150
17. Woerle HJ, Neumann C, Zschau S, Tenner S, Irsigler A, Schirra J (2007) Impact of fasting and postprandial glycemia on overall glycemic control in type 2 diabetes importance of postprandial glycemia to achieve target HbA1c levels. Diabetes Res Clin Pract 77:280–285
18. Nathan DM, Buse JB, Davidson MB, Heine RJ, Holman RR, Sherwin R et al (2006) Management of hyperglycemia in type 2 diabetes: a consensus algorithm for the initiation and adjustment of therapy. Diabetes Care 29:1963–1972
19. Holman RR, Paul SK, Bethel MA, Matthews DR, Neil HAW (2008) 10-Year follow-up of intensive glucose control in type 2 diabetes. N Engl J Med 359:1577–1589
20. Riddle MC (2005) Glycemic management of type 2 diabetes: an emerging strategy with oral agents, insulins, and combinations. Endocrinol Metab Clin North Am 34:77–98
21. Olverling A, Huang Z, NystrÖm T, Sjoholm A (2013) Acute regulation of pancreatic islet microcirculation and glycaemia by telmisartan and ramipril: discordant effects between normal and Type 2 diabetic rats. Clin Sci 125:433–438
22. Biessels GJ, Deary IJ, Ryan CM (2008) Cognition and diabetes: a lifespan perspective. Lancet Neurol 7:184–190
23. Mooradian AD (1988) Diabetic complications of the central nervous system. Endocr Rev 9:346–356
24. Strachan MWJ, Reynolds RM, Frier BM, Mitchell RJ, Price JF (2008) The relationship between type 2 diabetes and dementia. Br Med Bull 88:131–146
25. Xu WL, Qiu CX, Wahlin Å, Winblad B, Fratiglioni L (2004) Diabetes mellitus and risk of dementia in the Kungsholmen project A 6-year follow-up study. Neurology 63:1181–1186
26. Mogensen CE, Christensen CK (1984) Predicting diabetic nephropathy in insulin-dependent patients. N Engl J Med 311:89–93
27. Frisardi V, Solfrizzi V, Seripa D, Capurso C, Santamato A, Sancarlo D et al (2010) Metabolic-cognitive syndrome: a cross-talk between metabolic syndrome and Alzheimer's disease. Ageing Res Rev 9:399–417

28. Tikkanen I, Narko K, Zeller C, Green A, Salsali A, Broedl UC et al (2015) Empagliflozin reduces blood pressure in patients with type 2 diabetes and hypertension. Diabetes Care 38:420–428
29. Tchistiakova E, Anderson ND, Greenwood CE, MacIntosh BJ (2014) Combined effects of type 2 diabetes and hypertension associated with cortical thinning and impaired cerebrovascular reactivity relative to hypertension alone in older adults. Neuroimage Clin 5:36–41
30. Rosenthal T, Younis F, Alter A (2010) Combating combination of hypertension and diabetes in different Rat models. Pharmaceuticals 3:916–939
31. Olearczyk JJ, Quigley JE, Mitchell BC, Yamamoto T, Kim IH, Newman JW et al (2009) Administration of a substituted adamantly-urea inhibitor of soluble epoxide hydrolase protects the kidney from damage in hypertensive Goto-Kakizaki rats. Clin Sci 116:61–70
32. Chen Y, Chen X, Dang G, Zhao Y, Ouyang F, Su Z et al (2015) Hypertension criterion for stroke prevention—to strengthen the principle of individualization in guidelines. J Clin Hypertens 2015:1–7
33. Flack JM, Peters R, Shafi T, Alrefai H, Nasser SA, Crook E (2003) Prevention of hypertension and its complications: theoretical basis and guidelines for treatment. J Am Soc Nephrol 14:S92–S98
34. Lindekleiv H, Sandvei MS, Romundstad PR, Wilsgaard T, Njolstad I, Ingebrigtsen T et al (2012) Joint effect of modifiable risk factors on the risk of aneurysmal subarachnoid hemorrhage—a cohort study. Stroke 43:1885–1889
35. Chen S, Zeng L, Hu Z (2014) Progressing haemorrhagic stroke: categories, causes, mechanisms and managements. J Neurol 261:2061–2078
36. Donnan GA, Fisher M, Macleod M, Davis SM (2008) Stroke. Lancet 371:1612–1623
37. Skoog I, Gustafson D (2006) Update on hypertension and Alzheimer's disease. Neurol Res 28:605–611
38. Kuller LH, Lopez OL, Jagust WJ, Becker JT, Dekosky ST, Lyketsos C et al (2005) Determinants of vascular dementia in the Cardiovascular Health Cognition Study. Neurology 64:1548–1552
39. Alexander RW (2006) Leukocyte and endothelial angiotensin II type 1 receptors and microvascular thrombotic and inflammatory responses to hypercholesterolemia. Arterioscler Thromb Vasc Biol 26:240–241
40. Carnevale D, Mascio G, D'Andrea I, Fardella V, Bell RD, Branchi I et al (2012) Hypertension induces brain β-amyloid accumulation, cognitive impairment, and memory deterioration through activation of receptor for advanced glycation End products in brain vasculature. Hypertension 60:188–197
41. Korenovaa M, Zilka N, Stozickaa Z, Bugos O, Vanicky I, Novak M (2009) NeuroScale, the battery of behavioral tests with novel scoring system for phenotyping of transgenic rat model of tauopathy. J Neurosci Methods 177:108–114
42. Sabbatini M, Catalani A, Consoli C, Marletta N, Tomassoni D, Avola R (2002) The hippocampus in spontaneously hypertensive rats: an animal model of vascular dementia? Mech Ageing Dev 123:547–559
43. Kurata T, Lukic V, Kozuki M, Wada D, Miyazaki K, Morimoto N et al (2014) Telmisartan reduces progressive accumulation of cellular amyloid beta and phosphorylated Tau with inflammatory responses in aged spontaneously hypertensive stroke resistant Rat. J Stroke Cerebrovasc Dis 23:2580–2590
44. Petrova M, Prokopenko S, Pronina E, Mozheyko E (2010) Diabetes type 2, hypertension and cognitive dysfunction in middle age women. J Neurol Sci 299:39–41
45. Ohshima K, Mogi M, Horiuchi M (2013) Therapeutic approach for neuronal disease by regulating renin-angiotensin system. Curr Hypertens Rev 9:99–107
46. Peach MJ (1977) Renin-angiotensin system: biochemistry and mechanisms of action. Physiol Rev 57:313–370
47. Griendling KK, Murphy TJ, Alexander RW (1993) Molecular biology of the renin-angiotensin system. Circulation 87:1816–1828

48. Chang SY, Chen YW, Chenier I, Le Minh TS, Angiotensin ZSL, Type II II (2011) Receptor deficiency accelerates the development of nephropathy in type I diabetes via oxidative stress and ACE2. Exp Diabetes Res. doi:10.1155/2011/521076
49. de Queiroz TM, Monteiro MMO, Braga VA (2013) Angiotensin-II-derived reactive oxygen species on baroreflex sensitivity during hypertension: new perspectives. Front Physiol 4:105
50. Regenhardt RW, Bennion DM, Sumners C (2014) Cerebroprotective action of angiotensin peptides in stroke. Clin Sci 126:195–205
51. Bader M, Ganten D (2008) Update on tissue renin–angiotensin systems. J Mol Med 86:615–621
52. Sadjadi J, Puttaparthi K, Li L, Welborn MB, Rogers TE, Moe O et al (2002) Upregulation of autocrine-paracrine reninangiotensin systems in chronic renovascular hypertension. J Vasc Surg 36:386–392
53. Campos RR (2009) Oxidative stress in the brain and arterial hypertension. Hypertens Res 32:1047–1048
54. Campos RR, Oliveira-Sales EB, Nishi ÉE, Boim MA, Dolnikoff MS, Bergamaschi CT (2011) The role of oxidative stress in renovascular hypertension. Clin Exp Pharmacol Physiol 38:144–152
55. McGuire DK, Winterfield JR, Rytlewski JA, Ferrannini E (2008) Blocking the renin-angiotensin-aldosterone system to prevent diabetes mellitus. Diabetes Vasc Dis Res 5:59–66
56. Carlsson PO (2001) The renin-angiotensin system in the endocrine pancreas. J Pancreas 2:26–32
57. Skipworth JRA, Szabadkai G, Olde Damink SWM, Leung PS, Humphries SE, Montgomery HE (2011) Review article: pancreatic renin–angiotensin systems in health and disease. Aliment Pharmacol Ther 34:840–852
58. Cheng Q, Leung PS (2011) An update on the islet renin–angiotensin system. Peptides 32:1087–1095
59. Dzau V, Braunwald E (1991) Resolved and unresolved issues in the prevention and treatment of coronary artery disease: a workshop consensus statement. Am Heart J 121:1244–1263
60. Stegbauer J, Coffman TM (2011) New insights into angiotensin receptor actions: from blood pressure to aging. Curr Opin Nephrol Hypertens 20:84–88
61. Ichiki T, Miyazaki R, Kamiharaguchi A, Hashimoto T, Matsuura H, Kitamoto S et al (2012) Resveratrol attenuates angiotensin II-induced senescence of vascular smooth muscle cells. Regul Pept 177:35–39
62. Saavedra JM (1992) Brain and pituitary angiotensin. Endocr Rev 13:329–380
63. Zhou J, Pavel J, Macova M, Yu ZX, Imboden H, Ge L et al (2006) AT1 receptor blockade regulates the local angiotensin II system in cerebral microvessels from spontaneously hypertensive rats. Stroke 37:1271–1276
64. Allen AM, Moeller I, Jenkins TA, Zhuo J, Aldred GP, Chai SY et al (1998) Angiotensin receptors in the nervous system. Brain Res Bull 47:17–28
65. Villapol S, Saavedra JM (2014) Neuroprotective effects of angiotensin receptor blockers. Am J Hypertens 28:289–299
66. Dampney RAL, Tan PSP, Sheriff MJ, Fontes MAP, Horiuchi J (2007) Cardiovascular effects of angiotensin II in the rostral ventrolateral medulla: the push-pull hypothesis. Curr Hypertens Rep 9:222–227
67. Allen AM, Dosanjh JK, Erac M, Dassanayake S, Hannan RD, Thomas WG (2006) Expression of constitutively active angiotensin receptors in the rostral ventrolateral medulla increases blood pressure. Hypertension 47:1054–1061
68. Ito S, Komatsu K, Tsukamoto K, Kanmatsuse K, Sved AF (2002) Ventrolateral medulla AT1 receptors support blood pressure in hypertensive rats. Hypertension 40:552–559
69. Braga VA (2010) Dietary salt enhances angiotensin-II-induced superoxide formation in the rostral ventrolateral medulla. Auton Neurosci 155:14–18
70. Deshayes F, Nahmias C (2005) Angiotensin receptors: a new role in cancer? Trends Endocrinol Metab 16:293–299

71. Jones ES, Vinh A, McCarthy CA, Gaspari TA, Widdop RE (2008) AT2 receptors: functional relevance in cardiovascular disease. Pharmacol Ther 120:292–316
72. Li J, Culman J, Hörtnagl H, Zhao Y, Gerova N, Timm M et al (2005) Angiotensin AT2 receptor protects against cerebral ischemia-induced neuronal injury. FASEB J 19:617–619
73. Padia SH, Carey RM (2013) AT2 receptors: beneficial counter-regulatory role in cardiovascular and renal function. Pflugers Arch 465:99–110
74. Herrera M, Garvin JL (2010) Angiotensin II stimulates thick ascending limb NO production via AT2 receptors and Akt1-dependent nitric-oxide synthase 3 (NOS3) activation. J Biol Chem 285:14932–14940
75. Horiuchi M, Hayashida W, Akishita M, Tamura K, Daviet L, Lehtonen JYA et al (1999) Stimulation of different subtypes of angiotensin II receptors, AT1 and AT2 receptors, regulates STAT activation by negative crosstalk. Circ Res 84:876–882
76. Suematsu M, Suzuki H, Delano FA, Schimd-Schonbein GW (2002) The inflammatory aspect of the microcirculation in hypertension: oxidative stress, leukocytes/endothelial interaction, apoptosis. Microcirculation 9:259–276
77. Tanito M, Nakamura H, Kwon YW, Teratani A, Masutani H, Shioji K et al (2004) Enhanced oxidative stress and impaired thioredoxin expression in spontaneously hypertensive rats. Antioxid Redox Signal 6:89–97
78. Nishiyama A, Yao L, Nagai Y, Miyata K, Yoshizumi M, Kagami S et al (2004) Possible contributions of reactive oxygen species and mitogen-activated protein kinase to renal injury in aldosterone/salt-induced hypertensive rats. Hypertension 24:841–848
79. Virdis A, Neves MF, Amiri F, Viel E, Touyz RM, Schiffrin EL (2002) Spironolactone improves angiotensin-induced vascular changes and oxidative stress. Hypertension 40:504–510
80. Larsen MK, Matchkov VV (2016) Hypertension and physical exercise: the role of oxidative stress. Medicina 52:19–27
81. Broughton BRS, Reutens DC, Sobey CG (2009) Apoptotic mechanisms after cerebral ischemia. Stroke 40:e331–e339
82. Abe K, Yuki S, Kogure K (1988) Strong attenuation of ischemic and postischemic brain edema in rats by a novel free radical scavenger. Stroke 194:480–485
83. Cui J, Holmes EH, Greene TG, Liu PK (2000) Oxidative DNA damage precedes DNA fragmentation after experimental stroke in rat brain. FASEB J 14:955–967
84. Griendling KK, Minieri CA, Ollerenshaw JD, Alexander RW (1994) Angiotensin II stimulates NADH and NADPH oxidase activity in cultured vascular smooth muscle cells. Circ Res 74:1141–1148
85. Rajagopalan S, Kurz S, Münzel T, Tarpey M, Freeman BA, Griendling KK (1996) Angiotensin II–mediated hypertension in the Rat increases vascular superoxide production via membrane NADH/NADPH oxidase activation. J Clin Invest 97:1916–1923
86. Zimmerman MC, Lazartigues E, Sharma RV, Davisson RL (2004) Hypertension caused by angiotensin II infusion involves increased superoxide production in the central nervous system. Circ Res 95:210–216
87. Davisson RL, Zimmerman MC (2010) Angiotensin-II, oxidant signalling, and hypertension: down to a T? Hypertension 55:228–230
88. Lob HE, Marvar PJ, Guzik TJ, Sharma S, McCann LA, Weyand C et al (2010) Induction of hypertension and peripheral inflammation by reduction of extracellular superoxide dismutase in the central nervous system. Hypertension 55:277–283
89. Wolin MS (2005) Loss of vascular regulation by soluble guanylate cyclase is emerging as a key target of the hypertensive disease process. Hypertension 45:1068–1069
90. Paravicini TM, Touyz RM (2006) Redox signaling in hypertension. Cardiovasc Res 71:247–258
91. Iwai M, Liu HW, Chen R, Ide A, Okamoto S, Hata R et al (2004) Possible inhibition of focal cerebral ischemia by angiotensin II type 2 receptor stimulation. Circulation 110:843–848
92. Mogi M, Li JM, Iwanami J, Min LJ, Tsukuda K, Iwai M et al (2006) Angiotensin II type-2 receptor stimulation prevents neural damage by transcriptional activation of methyl methanesulfonate sensitive 2. Hypertension 48:141–148

93. Li G, Rhew IC, Shofer JB, Kukull WA, Breitner JCS, Peskind E et al (2007) Age-varying association between blood pressure and risk of dementia in those aged 65 and older: a community-based prospective cohort study. J Am Geriatr Soc 55:1161–1167
94. Mogi M, Horiuchi M (2013) Effect of angiotensin II type 2 receptor on stroke, cognitive impairment and neurodegenerative diseases. Geriatr Gerontol Int 13:13–18
95. Rizzi L, Rosset I, Roriz-Cruz M (2014) Global epidemiology of dementia: Alzheimer's and vascular types. Biomed Res Int. doi:10.1155/2014/908915
96. Ferri CP, Prince M, Brayne C, Brodaty H, Fratiglioni L, Ganguli M et al (2005) Global prevalence of dementia: a Delphi consensus study. Lancet 366:2112–2117
97. Panza F, Solfrizzi V, Frisardi V, Capurso C, D'Introno A, Colacicco AM et al (2009) Disease-modifying approach to the treatment of Alzheimer's disease. Drugs Aging 26:537–555
98. Panza F, D'Introno A, Colacicco AM, Basile AM, Capurso C, Kehoe PG et al (2004) Vascular risk and genetics of sporadic late-onset Alzheimer's disease. J Neural Transm 111:69–89
99. Solfrizzi V, Capurso C, D'Introno A, Colacicco AM, Frisardi V, Santamato A et al (2008) Dietary fatty acids, age-related cognitive decline and mild cognitive impairment. J Nutr Health Aging 12:382–386
100. Kivipelto M, Helkala EL, Laakso MP, Hanninen T, Hallikainen M, Alhainen K et al (2001) Midlife vascular risk factors and Alzheimer's disease in later life: longitudinal, population based study. BMJ 322:1447–1451
101. Nagai M, Hoshide S, Kario K (2010) Hypertension and dementia. Am J Hypertens 23:116–124
102. Wolf-Maier K, Cooper RS, Kramer H, Banegas JR, Giampaoli S, Joffres MR et al (2004) Hypertension treatment and control in five European countries, Canada, and the United States. Hypertension 43:10–17
103. Yagi S, Akaike M, Ise T, Ueda Y, Iwase T, Sata M (2013) Renin–angiotensin–aldosterone system has a pivotal role in cognitive impairment. Hypertens Res 36:753–758
104. Dai W, Lopez OL, Carmichael OT, Becker JT, Kuller LH, Gach HM (2008) Abnormal regional cerebral blood flow in cognitively normal elderly subjects with hypertension. Stroke 39:349–354
105. Son SJ, Kim J, Lee E, Park JY, Namkoong K, Hong CH et al (2015) Effect of hypertension on the resting-state functional connectivity in patients with Alzheimer's disease (AD). Arch Gerontol Geriatr 60:210–216
106. Al-Rubeaan K, Al-Hussain F, Youssef AM, Subhani SN, Al-Sharqawi AH, Ibrahim HM (2016) Ischemic stroke and its risk factors in a registry-based large cross-sectional diabetic cohort in a country facing a diabetes epidemic. J Diabetes Res 2016:1–9
107. Inaba S, Iwai M, Tomono Y, Senba I, Furuno M, Kanno H et al (2009) Exaggeration of focal cerebral ischemia in transgenic mice carrying human renin and human angiotensinogen genes. Stroke 40:597–603
108. Saavedra JM (2016) Evidence to consider angiotensin II receptor blockers for the treatment of early Alzheimer's disease. Cell Mol Neurobiol. doi:10.1007/s10571-015-0327-y
109. De Gasparo M, Catt KJ, Inagami T, Wright JW, Unger T (2000) International union of pharmacology. XXIII. The angiotensin II receptors. Pharmacol Rev 52:415–472
110. Reinecke K, Lucius R, Reinecke A, Rickert U, Herdegen T, Unger T (2003) Angiotensin II accelerates functional recovery in the rat sciatic nerve in vivo: role of the AT2 receptor and the transcription factor NF-kB. FASEB J 17:2094–2096
111. Laflamme L, de Gasparo M, Gallo JM, Payet MD, Gallo-Payet N (1996) Angiotensin II induction of neurite outgrowth by AT2 receptors in NG108-15 cells—effect counteracted by the AT1 receptors. J Biol Chem 271:22729–22735
112. Gendron L, Laflamme L, Rivard N, Asselin C, Payet MD, Gallo-Payet N (1999) Signals from the AT2 (angiotensin type 2) receptor of angiotensin II inhibit p21ras and activate MAPK (mitogen-activated protein kinase) to induce morphological neuronal differentiation in NG108–15 cells. Mol Endocrinol 13:1615–1626
113. Côté F, Laflamme L, Payet MD, Gallo-Payet N (1998) Nitric oxide, a new second messenger involved in the action of angiotensin II on neuronal differentiation of NG108-15 cells. Endocr Res 24:403–407

114. Côté F, Do TH, Laflamme L, Gallo JM, Gallo-Payet N (1999) Activation of the AT2 receptor of angiotensin II induces neurite outgrowth and cell migration in microexplant cultures of the cerebellum. J Biol Chem 274:31686–31692
115. Iwanami J, Mogi M, Tsukuda K, Min LJ, Sakata A, Jing F et al (2011) Effect of angiotensin II type 2 receptor deletion in hematopoietic cells on brain ischemia-reperfusion injury. Hypertension 58:404–409
116. Kernan WN, Inzucchi SE, Viscoli CM, Brass LM, Bravata DM, Horwitz RI (2002) Insulin resistance and risk for stroke. Neurology 59:809–815
117. McNeill AM, Rosamond WD, Girman CJ, Golden SH, Schmidt MI, East HE et al (2005) The metabolic syndrome and 11-year risk of incident cardiovascular disease in the atherosclerosis risk in communities study. Diabetes Care 28:385–390
118. Ninomiya JK, L'Italien G, Criqui MH, Whyte JL, Gamst A, Chen RS (2004) Association of the metabolic syndrome with history of myocardial infarction and stroke in the third national health and nutrition examination survey. Circulation 109:42–46
119. Hishinuma A, Majima M, Kurabayashi H (2009) Is insulin resistance related to recurrence of stroke or incident of ischemic heart disease in patients with stroke? A preliminary report. J Stroke Cerebrovasc Dis 18:294–297
120. Hu G, Jousilahti P, Sarti C, Antikainen R, Tuomilehto J (2006) The effect of diabetes and stroke at baseline and during follow-up on stroke mortality. Diabetologia 49:2309–2316
121. Kim B, Sullivan KA, Backus C, Feldman EL (2011) Cortical neurons develop insulin resistance and blunted Akt signaling: a potential mechanism contributing to enhanced ischemic injury in diabetes. Antioxid Redox Signal 14:1829–1839
122. Chobanian AV, Bakris GL, Black HR, Cushman WC, Green LA, Izzo JL (2003) Seventh report of the joint national committee on prevention, detection, evaluation, and treatment of high blood pressure. Hypertension 42:1206–1252
123. Mancia G, De Backer G, Dominiczak A, Cifkova R, Fagard R, Germano G et al (2007) Guidelines for the management of arterial hypertension. Eur Heart J 28:1462–1536
124. Mancia G, Fagard R, Narkiewicz K, Redon J, Zanchetti A, Bohm M et al (2013) ESH/ESC Guidelines for the management of arterial hypertension. Eur Heart J 34:2159–2219
125. Ravandi A, Teo KK (2009) Blocking the renin-angiotensin system: dual versus monotherapy. Expert Rev Cardiovasc Ther 7:667–674
126. Parsons C, Murad MH, Andersen S, Mookadam F, Labonte H (2016) The effect of antihypertensive treatment on the incidence of stroke and cognitive decline in the elderly: a meta-analysis. Future Cardiol 12:237–248
127. Brunström M, Carlberg B (2016) Effect of antihypertensive treatment at different blood pressure levels in patients with diabetes mellitus: systematic review and meta-analyses. 2016. BMJ 352:i717. doi:10.1136/bmj.i717
128. Hsu CY, Huang CC, Chan WL, Huang PH, Chiang CH, Chen TJ (2013) Angiotensin-receptor blockers and risk of Alzheimer's disease in hypertension population—a nationwide cohort study. Circ J 77:405–410
129. Kehoe PG, Miners S, Love S (2009) Angiotensins in Alzheimer's disease—friend or foe? Trends Neurosci 32:619–628
130. Kehoe PG, Wilcock GK (2007) Is inhibition of the renin–angiotensin system a new treatment option for Alzheimer's disease? Lancet Neurol 6:373–378
131. Igase M, Kohara K, Miki T (2012) The association between hypertension and dementia in the elderly. Int J Hypertens. doi:10.1155/2012/320648
132. McFarlane SI (2009) Role of angiotensin receptor blockers in diabete: implications of recent clinical trials. Expert Rev Cardiovasc Ther 7:1363–1371
133. Ismail H, Mitchell R, McFarlane SI, Makaryus AN (2010) Pleiotropic effects of inhibitors of the RAAS in the diabetic population: above and beyond blood pressure lowering. Curr Diab Rep 10:32–36
134. Huang Z, Ohno N, Terada N, Saitoh Y, Chen J, Ohno S (2013) Immunohistochemical detection of angiotensin II receptors in mouse cerebellum and adrenal gland using "in vivo cryotechnique". Histochem Cell Biol 140:477–490

135. Kakuta H, Kurosaki E, Niimi T, Gato K, Kawasaki Y, Suwa A et al (2014) Distinct properties of telmisartan on agonistic activities for peroxisome proliferator-activated receptor g among clinically used angiotensin II receptor blockers: drug-target interaction analyses. J Pharmacol Exp Ther 349:10–20
136. Takai S, Kirimura K, Jin D, Muramatsu M, Yoshikawa K, Mino Y et al (2005) Significance of angiotensin II receptor blocker lipophilicities and their protective effect against vascular remodeling. Hypertens Res 28:593–600
137. Burgess E, Muirhead N, de Cotret PR, Chiu A, Pichette V, Tobe S et al (2008) Supramaximal dose of candesartan in proteinuric renal disease. J Am Soc Nephrol 20:893–900
138. Konstam MA, Neaton JD, Dickstein K, Drexler H, Komajda M, Martinez FA et al (2009) Effects of high-dose versus low-dose losartan on clinical outcomes in patients with heart failure (HEAAL study): a randomised, double-blind trial. Lancet 374:1840–1848
139. Saavedra JM, Angiotensin II (2012) AT1 receptor blockers as treatments for inflammatory brain disorders. Clin Sci 123:567–590
140. Saavedra JM, Sánchez-Lemus E, Benicky J (2011) Blockade of brain angiotensin II AT1 receptors ameliorates stress, anxiety, brain inflammation and ischemia: therapeutic implications. Psychoneuroendocrinology 36:1–18
141. Siragy HM (2009) The potential role of the angiotensin subtype 2 receptor in cardiovascular protection. Curr Hypertens Rep 11:260–262
142. Poon IO (2008) Effects of antihypertensive drug treatment on the risk of dementia and cognitive impairment. Pharmacotherapy 28:366–375
143. Lanz TV, Ding Z, Ho PP, Luo J, Agrawal AN, Srinagesh H et al (2010) Angiotensin II sustains brain inflammation in mice via TGF-β. J Clin Invest 120:2782–2794
144. Zanchetti A, Elmfeldt D (2006) Findings and implications of the Study on Cognition and Prognosis in the Elderly (SCOPE)—a review. Blood Press 15:71–79
145. Savoia C, Schiffrin EL (2007) Vascular inflammation in hypertension and diabetes: molecular mechanisms and therapeutic interventions. Clin Sci 112:375–384
146. Wang J, Ho L, Chen L, Zhao Z, Zhao W, Qian X et al (2007) Valsartan lowers brain β-amyloid protein levels and improves spatial learning in a mouse model of Alzheimer disease. J Clin Invest 117:3393–3402
147. Mogi M, Li JM, Tsukuda K, Iwanami J, Min LJ, Sakata A et al (2008) Telmisartan prevented cognitive decline partly due to PPAR-gamma activation. Biochem Biophys Res Commun 375:446–449
148. Tsukuda K, Mogi M, Iwanami J, Min LJ, Sakata A, Jing F et al (2009) Cognitive deficit in amyloid-β–injected mice was improved by pretreatment with a low dose of telmisartan partly because of peroxisome proliferator-activated receptor-γ activation. Hypertension 54:782–787
149. Benicky J, Sánchez-Lemus E, Honda M, Pang T, Orecna M, Wang J et al (2011) Angiotensin II AT1 receptor blockade ameliorates brain inflammation. Neuropsychopharmacology 36:857–870
150. Zorad S, Dou JT, Benicky J, Hutanu D, Tybitanclova K, Zhou J et al (2006) Long-term Angiotensin II AT1 receptor inhibition produces adipose tissue hypotrophy accompanied by increased expression of adiponectin and PPARγ. Eur J Pharmacol 552:112–122
151. Pang T, Benicky J, Wang J, Orecna M, Sanchez-Lemus E, Saavedra JM (2012) Telmisartan ameliorates lipopolysaccharide-induced innate immune response through peroxisome proliferator-activated receptor-γ activation in human monocytes. J Hypertens 30:87–96
152. Zhou J, Ando H, Macova M, Dou J, Saavedra JM (2005) Angiotensin II AT1 receptor blockade abolishes brain microvascular inflammation and heat shock protein responses in hypertensive rats. J Cereb Blood Flow Metab 25:878–886
153. Sironi L, Gelosa P, Guerrini U, Banfi C, Crippa V, Brioschi M et al (2004) Anti-inflammatory effects of AT1 receptor blockade provide end-organ protection in stroke-prone rats independently from blood pressure fall. J Pharmacol Exp Ther 311:989–995
154. Hallevia H, Hazan-Halevyb I, Paranc E (2007) Modification of neutrophil adhesion to human endothelial cell line in acute ischemic stroke by dipyridamole and candesartan. Eur J Neurol 14:1002–1007

155. Grammas P, Martinez J, Miller B (2011) Cerebral microvascular endothelium and the pathogenesis of neurodegenerative diseases. Expert Rev Mol Med. doi:10.1017/S1462399411001918
156. Kasahara Y, Taguchi A, Uno H, Nakano A, Nakagomi T, Hirose H et al (2010) Telmisartan suppresses cerebral injury in a murine model of transient focal ischemia. Brain Res 1340:70–80
157. Vargas R, Rincon J, Pedreanez A et al (2012) Role of angiotensin II in the brain inflammatory events during experimental diabetes in rats. Brain Res 1453:64–76
158. Ohrui T, Tomita N, Sato-Nakagawa T, Matsui T, Maruyama M, Niwa K et al (2004) Effects of brain-penetrating ACE inhibitors on Alzheimer disease progression. Neurology 63:1324–1325
159. Hu J, Igarashi A, Kamata M, Nakagawa H (2001) Angiotensin-converting enzyme degrades Alzheimer amyloid ß-peptide (Aß), retards Aß aggregation, deposition, fibril formation and inhibits cytotoxicity. J Biol Chem 276:47863–47868
160. Bader M (2010) Tissue renin-angiotensin-aldosterone systems: targets for pharmacological therapy. Annu Rev Pharmacol Toxicol 50:439–465
161. Arima H, Chalmers J (2011) PROGRESS: prevention of recurrent stroke. J Clin Hypertens 13:693–702

Chapter 32
Sex-Specific Factors in Stroke

Anjali Chauhan, Hope Moser, and Louise D. McCullough

Abstract Stroke is a sexually dimorphic disease, identified through evidence derived from preclinical and clinical trials. Efficacy of prevention strategies, recognition of symptoms, acute stroke treatment options, and stroke recovery are influenced by sex-specific factors. The terms, "sex" and "gender" are often used interchangeably. By definition, sex is a biological variable, typically characterized as male, female, or intersex and includes indicators of biological sex, including sex chromosomes (XX and XY) and reproductive organs. Gender refers to attitudes, feelings, and behaviors that a culture associates with a person's biological sex. This chapter will focus on sex differences in ischemic stroke and will provide an update of emerging pre-clinical and clinical studies that confirm that sex is an important biological variable in ischemic brain injury.

Keywords Sex differences • Ischemic stroke • Primary prevention • Stroke management • Stroke outcomes

Abbreviations

ADE	Adverse drug event
AF	Atrial fibrillation
AIF	Apoptosis-inducing factor
BASIC	Brain Attack Surveillance in Corpus Christi

A. Chauhan, Ph.D.
McGovern Medical School—The University of Texas Health Science Center at Houston, Houston, TX, USA

H. Moser, R.N., D.N.P.
Memorial Hermann Hospital—Texas Medical Center, Houston, TX, USA

L.D. McCullough, M.D., Ph.D. (✉)
McGovern Medical School—The University of Texas Health Science Center at Houston, Houston, TX, USA

Memorial Hermann Hospital—Texas Medical Center, Houston, TX, USA
e-mail: Louise.D.McCullough@uth.tmc.edu

P.A. Lapchak, J.H. Zhang (eds.), *Neuroprotective Therapy for Stroke and Ischemic Disease*, Springer Series in Translational Stroke Research,
DOI 10.1007/978-3-319-45345-3_32

CAC	Coronary artery calcification
CBF	Cerebral blood flow
CEE	Conjugated equine estrogens
CIMT	Carotid intimal medial thickness
COC	Combined oral contraceptive
CTA	Computed tomography angiogram
CTP	Computed tomography perfusion
DNA	Deoxyribonucleic acid
DNP	Doctor of Nursing Practice
ELITE	Early vs. Late Intervention Trial with Estradiol
HLD	High-density lipoprotein
HRPR	High on treatment residual palatelet activity
HRQoL	Health related quality of life
HT	Hormone therapy
IL-6	Interleukin 6
IMS III	Interventional Management of Stroke III Trial
KEEPS	Kronos Early Estrogen Prevention Study
LDL	Low-density lipoprotein
MD	Medical Doctor
MI	Myocardial infarction
mRNA	Messenger RNA
NAD	Nicotinamide adenine dinucleotide
NIHSS	National Institute of Health Stroke Scale
nNOS	Neuronal nitric oxide synthase
NOAC	Novel oral anticoagulant
PAF	Paroxysmal atrial fibrillation
PAR	Poly ADP-ribose
PARP	Poly ADP-ribose polymerase
PhD	Doctor of Philosophy
PRES	Posterior reversible encephalopathy syndrome
PSD	Poststroke depression
RCVS	Reversible cerebral vasoconstriction syndrome
RN	Registered Nurse
r-tPA	ALTEPLASE
WHI	The Women's Health Initiative
XIAP	X-linked inhibitor of apoptosis
XX	Female sex
XY	Male sex

1 Sex-Specific Risk Factors

Stroke is a sexually dimorphic disease, identified through evidence derived from preclinical and clinical trials. Efficacy of prevention strategies, recognition of symptoms, acute stroke treatment options, and stroke recovery are influenced by

Table 32.1 Risk factors for stroke

Risk factor	Sex-specific risk factor	Risk factor that are stronger or more prevalent in women	Risk factor that are similar in men and women
Pregnancy, preeclampsia, or gestational diabetes	X		
Oral contraceptive or postmenopausal hormone use	X		
Migraine headache with aura		X	
Atrial fibrillation		X	
Diabetes		X	
Hypertension			X
Physical inactivity obesity or unhealthy			X
Age			X
Prior cardiovascular disease			X
Smoking			X
The metabolic syndrome			X
Depression		X	
Psychological stress		X	

sex-specific factors [1]. Recognition of these differences led to the first "sex-specific" guidelines that were developed by the American Heart Association "*Prevention and Management of Stroke in Women*," which discusses the evidence for sex-specific differences in clinical stroke [2] (see Table 32.1).

The terms, "sex" and "gender" are often used interchangeably. By definition, sex is a biological variable, typically characterized as male, female, or intersex and includes indicators of biological sex, including sex chromosomes (XX and XY) and reproductive organs. Gender refers to attitudes, feelings, and behaviors that a culture associates with a person's biological sex [3]. For purposes of this chapter, we will use the term "sex" as we provide an overview of specific topics pertaining to sex-related risk factors in women.

1.1 Pregnancy

The risk of stroke increases during pregnancy and is maintained for at least 12 weeks postpartum [4]. Pregnancy-related conditions such as preeclampsia, hypercoagulability, vasculopathies such as reversible cerebral vasoconstriction syndrome (RCVS), and posterior reversible encephalopathy syndrome (PRES) dramatically increase the risk of a venous thrombotic event or hemorrhagic stroke [2, 5]. The risk for pregnancy-associated complications later in life is uncertain, but there is some evidence that suggests that there is an increased risk to not only the mother but also the child several years or decades after delivery [6]. Tranquilli et al. [7] has documented a correlation between children born to women with preeclampsia have a predisposition to hypertension, insulin resistance, and diabetes mellitus.

Applying a murine model of stroke, the role of parity related to stroke outcome revealed associations with sedentary behavior, weight gain, elevated total cholesterol, and increased triglyceride levels. Compared to the nulliparous mouse brain, the multiparous brain exhibited features of immune suppression with inhibited microglial activity at baseline as well as less glial activation and smaller areas of infarction poststroke (Ritzel unpublished results). Continued preclinical studies on female mice with reproductive experience may better reflect the lifelong patterning of ischemic stroke especially as women are delaying pregnancy, and rates of hypertension are higher in older women, further increasing the risk [8].

1.2 Hormones and Stroke Risk

In women of reproductive age, there is a direct correlation between high dose estrogen and ischemic stroke [9]. Over the past decade, the risk of stroke has decreased largely due to the lower estrogen and use of first- or second-generation progesterone combinations in combined oral contraceptives (COCs) [10]. The risk is proportionately decreased with the lower estrogen dosage; however, the evidence is not clearly consistent with regards to progesterone dose.

Overall, hormonal contraception use is considered relatively safe when the lowest effective dose of estrogen is prescribed with contraindications and risk factors being considered, such as age ≥35 years and concomitant smoking or complex migraine [11]. Migraine headaches are more common among premenopausal women using combination oral contraceptives [12]. Criteria regarding absolute and relative contraindications for the use of COCs have been debated; especially whether COC use in women with migraine is safe. For a summary of the US Medical Eligibility Criteria for Contraceptive Use [13] see Table 32.2. The absolute increase in the risk of stroke is small. Data from a recent prospective cohort study following

Table 32.2 US Medical Eligibility Criteria for Contraceptive Use [13]

Headache	Initiation	Continuation
Nonmigraine (mild or severe).	No restriction	Advantage generally outweighs risk
Migraine		
Without aura		
Age <35 years	Advantage generally outweighs risk	Risk usually outweighs advantage
Age ≥35 years	Risk usually outweighs advantage	Unacceptable health risk
With aura (any age)	Unacceptable health risk	Unacceptable health risk

1. No restriction for the use of combination oral contraceptive
2. Advantages of using COCs generally outweighs the theoretical or proven risks
3. Theoretical or proven risks usually outweighs the advantages of using COCs
4. Unacceptable health risk if the COC used

over 115,000 healthy women ages 25–42 from 1989 to 2011 reported that 15 % of women reported a documented diagnosis of migraine. After adjustment for confounders, migraine was associated with an increased risk for stroke (1.62, 1.37–1.92) (hazard ratio 1.50, 95 % confidence interval 1.33–1.69) [14] compared with women without migraine, suggesting that risk may be independent of COCs. Interestingly, risk factors such as age <50 and ≥50 years, smoking status, hypertension, postmenopausal hormone therapy (HT), and COC use revealed similar risk, suggesting the link was between migraine and stroke. It was recommended that all women with migraine should undergo aggressive screening to assess and aggressively treat their vascular risk factors [14].

2 Sex Differences in Primary Prevention

Gender disparity for enrollment in clinical trials continues to be a significant challenge. Approximately 30 % of patients enrolled in cardiovascular clinical trials are women, making it difficult to develop appropriate guidelines for primary and secondary stroke prevention [15]. The reasons for poor enrollment in women are unknown, but social factors (i.e., difficulty getting to appointments and isolation), increased age, contraindications to therapy due to comorbidities, cognitive deficits, and poor premorbid function have all been implicated.

Sex differences in the efficacy of primary preventive strategies have been seen in large epidemiological trials. As an example, evidence supports that aspirin reduces the risk of ischemic stroke in women by approximately 25 % yet has no effect on cardiac risk [2, 16]. Berger et al. [17] conducted a meta-analysis comparing the sex-related risks and benefits of aspirin used for the primary prevention of cardiovascular disease. Aspirin therapy was associated with a significant 12 % reduction in cardiovascular events and a 17 % reduction in stroke in women, which led to reduced rates of ischemic stroke compared to a significant 14 % reduction in cardiovascular events in men that was secondary to a 32 % reduction in myocardial infarction (MI) with little effect on stroke risk. Adverse drug events (ADEs) such as bleeding, directly related to aspirin use, were not different between the sexes [18].

Sex differences in platelet aggregation and the response to antiplatelet agents have received recent attention. Specifically, biological differences in the pathogenesis of atherosclerosis in the aged female, with >1 risk factor, hormonal influences, and lower body mass have been identified as factors that may influence a diminished effect of commonly used antiplatelet agents such as aspirin, clopidogrel, or ticagrelor which is increasingly used [19]. In patients receiving dual antiplatelet therapy, sex did not impact on the prevalence of high-on treatment residual platelet reactivity (HRPR) in patients treated with aspirin, clopidogrel, or ticagrelor [19]. U.S. Preventive Services Task Force [20] has suggested that the use of aspirin for primary prevention of cardiovascular disease has the greatest potential for positive net benefit in patients if initiated in individuals aged 40–69 years with high cardiovascular risk [21].

3 Sex Differences in Embolic Stroke

Women with atrial fibrillation (AF) not on anticoagulation have shown a higher risk of developing embolic stroke [22] and have higher mortality compared to age-matched men [23]. The higher prevalence of AF in older women is due in part to age-related variations in the pharmacokinetic effect of anticoagulation [22, 24]. CHA2DS2-VASc is a stroke risk stratification score derived from an evidence-based algorithm. The score is recommended in many guidelines and is used to identify low-risk patients. In comparison to CHA2DS2, CHA2DS2-VASc assigns an additional point for female sex, recognizing that women are at increased risk [2].

In a recent cohort study of 600 stroke patients where women were older than men at stroke onset but more men had diabetes and tobacco use. Of the 600 patient, 111 of them had a premorbid diagnosis of AF. After 5 years of follow up there was a statistically nonsignificant mortality rate of 76 % in women and 73 % in men. Additionally, one-third of women and one-half of men were functionally independent. In the patients with known AF, women were treated with warfarin less frequently prior to the stroke (11 % vs. 28 %) but warfarin and novel oral anticoagulant (NOAC) treatment did increase among both women and men at hospital discharge [25] and further reveals the considerable increase in morbidity and mortality in women. Several large epidemiological trials have revealed that women with AF and a moderate to high risk of stroke in an outpatient cardiac population were about 30 % more likely to be treated with aspirin. One of the specific patient characteristics associated with oral anticoagulant use was male sex [26]. Paroxysmal atrial fibrillation must remain within a high level of suspicion, especially in elderly women. Aggressive monitoring with devices such as Holter/loop monitors or implantable cardiac event monitors is advisable.

Sex differences in stroke, systemic embolism, and major bleeding events after treatment with a NOAC have only recently been evaluated. A recent systematic review and meta-analysis evaluated sex differences focusing on residual risk in patients with nonvalvular AF treated with either warfarin or a NOAC [27]. Women with AF taking warfarin were at a significantly greater residual risk of stroke and systemic embolism compared with men. Interestingly, no sex differences in risk of stroke and embolism was seen in patients with AF receiving a NOAC. Major bleeding was less frequent in women with AF treated with a NOAC, suggesting an increased net clinical benefit of NOAC use compared with warfarin in treating women with AF [27].

4 Sex Differences in Acute Stroke Management

Alteplase (r-tPA) is the only FDA-approved pharmacologic treatment for acute ischemic stroke. Most studies have shown that female stroke patients are just as likely as male stroke patients to receive thrombolytic treatment. One observational cohort study did report that women may benefit more from r-tPA treatment, suggesting that while men and women receive weight appropriate doses of r-tPA [28], women are more likely to receive the maximum dose of r-tPA (0.9 mg/kg) due to women holding higher weight [29]. Race has been an element of interest when studying differences in acute stroke management. A recent comparison of Caucasian and African American, male and female stroke patients revealed no differences in the use of IV r-tPA in men but found that African American (AA) women were relatively less likely to receive IV r-tPA than Caucasian women primarily attributed to higher rates of uncontrolled hypertension in AA women [30]. Analysis of thrombolytic trials found that African American women may have greater benefit from r-tPA treatment as compared to men due to the prevalence of cardioembolic stroke and the negative natural history in untreated women [31]. No sex differences in outcome have been reported in studies of mechanical embolectomy [32]; however, the number of treated patients was small and current studies are underpowered to detect sex differences. Endovascular therapy will be widely used with the emergence of the recent, positive, endovascular trials [33]. Recent trials lacked "a priori" design for analysis sex specificity therefore it is difficult to claim whether women and men benefit from these interventions equally until this is specifically examined [34].

5 Aging and Stroke

The most important independent risk factor for stroke is age, as stroke rates double every decade after the age of 55 years in both men and women [35, 36]. Additionally, age is a significant predictor of outcome, independent of sex, vascular risk factors, stroke etiology, timeliness and effectiveness of revascularization, and stroke severity [37]. Stroke recovery is influenced by age, even in studies that attempt to control for this confounder with multivariable modeling. The mean age of first stroke in men is 66.7 year vs. that of woman being 70 years old. Recent epidemiological studies have shown an increase in sex-specific risk beginning in middle age, with women 55–75 years of age have a 20–21 % increase in stroke risk compared to 14–17 % increased risk in men [38, 39]. The changes that occur with reproductive senescence, including the loss of estrogen, likely influence a women's vascular risk [40, 41].

The influence of estrogen and hormone therapy (HT) on a women's stroke risk remains controversial. The absolute risk associated with hormone therapy is stratified based on current age and the length of time since menopause. The Women's Health Initiative (WHI) [42] trial found that in relatively healthy postmenopausal women, estrogen, with or without progestogen, significantly increased the risk of stroke, which was a somewhat surprising result as many prior observational trials suggested that estrogen was beneficial and vasculoprotective. In the WHI, younger women between the ages of 50 and 59 years, taking only conjugated equine estrogens (CEE), showed an improvement in all-cause mortality, decreased incidence of myocardial infarction (MI), but showed no protective effect on stroke or venous thrombosis. Significant controversy ensued which incited review and change in the recommendations and guidelines. Current recommendations state that HT should not be initiated in postmenopausal women for either primary or secondary stroke prevention [2]. Many have criticized the WHI trial on the premise that estrogen therapy may have been initiated too late, showing poor beneficial effect due to diseased endothelium secondary to estrogen's proinflammatory effects [43]. Recent clinical trials suggest that this story is far from over. Findings from two recent clinical trials, the Kronos Early Estrogen Prevention Study (KEEPS) [44] and the Early vs. Late Intervention Trial with Estradiol (ELITE) [45, 46] appear to be consistent with the idea that initiating HT in the perimenopausal period may be safe. These trials were specifically designed to examine the effect of HT on younger, recently postmenopausal women. KEEPS, a multicenter clinical trial of 727 women aged 42–58 years examined estrogen with progesterone transdermal HT 50 mcg/day vs. oral HT 0.45 mg/day vs. placebo over 4 years. Starting HT in healthy, menopausal women between 6 months and 36 months from last menses could reduce the progression of subclinical atherosclerosis as measured by carotid intimal medial thickness (CIMT) and coronary artery calcification (CAC) [44, 47]. The study found no differences in CIMT or CAC progression in estrogen-treated women compared to women treated with placebo. Importantly, estrogen treatment did not increase blood pressure but lowered high-density lipoprotein (HDL) and low-density lipoprotein (LDL) levels. Transdermal therapy improved insulin resistance but an increase in C reactive protein was seen with oral estrogen treatment. Data from the recently published ELITE study, which included 643 women who were followed for 6–7 years after treatment initiation, showed a reduction in CIMT in estrogen treated women when treatment was initiated within 6 years of menopause (average age of 55.4 years), but no effect when treatment was initiated 10 years after menopause (average age of 65.4 years) [46]. Continued follow up of the KEEPS and ELITE cohorts will be needed as enrolled women were young and not yet at high risk for stroke, were only followed for 4 years, and a surrogate marker for stroke, CIMT was used, before the safety of hormone therapy is established.

6 Sex Differences in Cerebral Blood Flow

Common to all stroke etiologies is compromised oxygen delivery distal to the area of infarct causing devastating injury to cerebral tissue. Damage commences immediately and can continue for several hours, with the time of onset to cellular death being directly related to the degree of decreased cerebral blood flow (CBF) [48]. Normal CBF represents 15–20 % of cardiac output or approximately 50 mL/100 g/min. CBF of 10-20 mL/100 g/min will cause functional impairment of neurons but if restored, CBF will save the ischemic penumbra [49]. Conversely, if CBF decreases to less than 10 mL/100 g/min, the ischemic penumbra is rapidly and irreversibly damaged. Cerebral autoregulation maintains constant CBF across the continuum of perfusion pressures [50] to protect the brain from variations in blood flow and pressure [40]. In women, sex hormones, predominantly estrogen is essential to the regulation of vascular reactivity during and after cerebral artery occlusion thus promoting improved tissue perfusion while reducing infarct size [40].

Collateral circulation has been suggested as a critical component in the determination of ischemic core vs. ischemic penumbra during an acute stroke [51–53]. Data from the Interventional Management of Stroke III Trial (IMS III) [54] was reanalyzed to determine the association between collateral status and perfusion parameters [55]. The relationship between computed tomography angiogram (CTA) and CT perfusion (CTP) was examined, supporting the hypothesis that better collaterals are associated with smaller ischemic cores and larger mismatch, reflecting compensation from the collaterals during an acute ischemic event [55]. Studies are needed to examine the extent of collateral formation in men vs. women and if this is a potential source of sex differences seen in stroke recovery.

7 Cell Death Mechanisms

Clinical findings have shown considerable disparity in stroke severity with sex and age. These differences have their roots in biology as experimental studies have shown that sex differences exist in the cellular pathways activated by ischemia, such as autophagy and caspase signaling and activation [56, 57]. Sex difference in autophagy has been demonstrated in both in vitro and in vitro models of cardiac ischemia [58], cerebral neonatal hypoxia ischemia [59], and after brain hemorrhage [60]. These differences are even found in in vitro studies examining cultured neurons derived from male (XY) or female (XX) embryonic mice in cultures that are "hormone free" suggesting they are intrinsic to the sex of the cell. Male neurons are more sensitive to nutrient starvation and died faster than neurons derived from females, highlighting that the sexes utilize different bioenergic pathways, at least under stress conditions [61, 62]. In a neonatal model of ischemic injury, females showed greater activation of autophagic pathways, increased cell turn over, and were protected against ischemic injury [59], consistent with findings in a coronary artery ischemia–reperfusion injury in rats, suggesting autophagy may be beneficial in females [58].

Sex differences in autophagy may be related to more global changes in metabolism between the sexes [63–65]. Studies show that males utilize carbohydrates and amino acids while females mainly rely on fats for fuel [64, 66], a pattern that is determined by the mitochondria. Moreover, we found that pharmacologically inhibiting autophagy only benefits males (unpublished results); however, further studies are needed to establish this effect. One may ask "do the sex differences seen in cell death pathway activation have any relevance to clinical stroke?" Recent data suggest they do. Early work in our laboratory and by others have shown that females seem to be uniquely sensitive to caspase-induced cell death, whereas males cells preferentially die after the overactivation of the DNA repair enzyme PARP with subsequent formation of PAR polymers, NAD loss, and nuclear translocation of AIF [41, 43, 55]. Pharmacological or genetic deletion of nNOS is protective in males but leads to larger infarcts in female mice [67], which remains unexplained but may be due to regulation by microRNAs that enhance endogenous inhibitors of apoptosis such as XIAP [68]. One example that suggests that these do matter in translational drug design is Minocycline. Minocycline is neuroprotective in clinical and experimental stroke studies, due in part to its ability to inhibit poly (ADP-ribose) polymerase. Previous preclinical data have shown that interference with poly (ADP-ribose) polymerase signaling leads to sex-specific neuroprotection, reducing stroke injury only in males [69]. We found that minocycline is ineffective at reducing ischemic damage in females after middle cerebral artery occlusion, likely due to effects on poly (ADP-ribose) polymerase signaling [55]. Interestingly, in a recent small clinical trial, 200 mg of oral minocycline daily for 5 days 6–24 h after stroke onset was associated with an improvement in NIHSS over the 3 months after stroke in men, but treatment had no effect in women. This was a very small trial, with only 53 patients and was not designed or adequately powered to convincingly show sex-specific neuroprotection, however it should serve as a cautionary note to clinicians [70]. Clinical trials must consider possible sex differences in the response to neuroprotective agents, if we hope to translate promising therapies to stroke patients of both sexes.

8 Outcomes After Stroke

Women have poorer outcomes after stroke than men, in part due to the older age of first stroke in women, although results are conflicting [71]. In a large cohort of over 2800 patients followed over 4 years, sex differences were not identified; however, patients were less likely to be discharged home if they were older, separated, or divorced compared to patients that were married, had dysphagia, or cognitive deficits [72]. Data from ischemic stroke patients enrolled in the Brain Attack Surveillance in Corpus Christi (BASIC) project found that women were older than men (71 years vs. 64 years) and women had significantly worse functional outcomes than men even after age adjustment [73]. The most important factor contributing to outcome was prestroke function. Prevention efforts aimed at maintaining functional status in aging women could improve stroke outcomes. Stroke severity modified sex differences in outcomes. Differences were apparent for mild to moderate stroke but not as apparent for severe strokes. Even after adjustment, women still had significantly worse functional outcome than men. Other factors such as social isolation and depression may disproportionally affect women [74]. The detrimental effects of social isolation could be amenable to intervention; however, interventions focused on increasing social support in stroke patients have largely been unsuccessful [75, 76], perhaps due to poor patient selection [77].

8.1 Social Isolation, Loneliness, and Stroke

Substantial evidence from preclinical and clinical studies has demonstrated that social isolation can increase stroke incidence and impair recovery. Social isolation leads to higher rates of recurrent stroke but is often not reported as a risk factor. Prospectively collected stroke center database variables, which included prestroke living situation, were examined to determine if social isolation could be determined from existing data using living arrangement as a proxy. Patients were categorized into four groups, hypothesized to represent increasing levels of social isolation: living with spouse, living with family, living alone with visiting services, or living alone. Patients living alone presented with less severe strokes on admission and had better poststroke functional recovery at 3 months compared with the other cohorts. Upon further detailed examination, patients who lived alone also reported significantly higher prestroke functionality. Preexisting depression was significantly higher in women and not surprising, depressed patients had poorer outcomes at 90 days. Information regarding isolation is notably absent from most large stroke databases. A more comprehensive evaluation of social interaction should be obtained to more accurately measure social isolation [78].

Stroke is associated with worse functional outcomes and negatively affects health-related quality of life (HRQoL) [79]. Stroke-related factors, demographic factors, and physical factors are consistent determinants of HRQoL in stroke patients [80]. Social support is an important factor in adults who have limited social interactions as unhappy or lonely patients are known to be at a higher risk of mortality [81, 82]. Association between immune dysfunction and hypertension has been correlated with loneliness and social isolation [83–85]. Proinflammatory markers such as elevated C-reactive protein and serum IL-6 levels are correlated with social isolation [86, 87] and have been determined to increase overall stroke risk and poststroke mortality [88–90]. Social isolation increased the risk of stroke in both men and women [91]. A recent meta-analysis from 35,925 records showed that poor social interaction was associated with a 32 % increase in stroke risk (pooled RR = 1.32; 95 % CI = 1.04–1.68), with no sex difference in subgroup analysis [82]. In preclinical studies, infarct size is greater in socially isolated male and female mice as compared with mice housed with partners [92, 93]. Isolation was particularly detrimental to female mice [94]. Expression of Mac1 mRNA and GFAP mRNA is increased in socially isolated mice [92]. In a global ischemia model, socially isolated mice demonstrated increased neuronal damage, microglia activation, proinflammatory cytokine expression, and behavior defects as compared to paired house animals [95]. A recent study on aged mice demonstrated that socially isolated mice have larger infarcts, worse neurological deficits, and higher IL-6 levels at day 3 as compared to pair housed animals [96]. Several inflammatory genes associated with poststroke recovery were differentially expressed in socially isolated mice thus highlighting the role of the immune system in mediating these effects on recovery [96].

8.2 Depression and Stroke

Depression is common in general population, with females being at higher risk than men [97, 98]. A number of studies have shown a close association between depression and increased stroke risk [99, 100]. A prospective study with 3852 subjects <55 years old showed that the stroke risk increased in patients between 55 and 64 years with depressive symptoms but not in patients who were older than 65 years. With the multivariate analysis, women showed higher risk of stroke than men [101]. A recent follow-up study in 137, 305 men and 188, 924 women aged ≥30 years showed that depression was associated with increased risk of stroke but there was a somewhat surprising interaction that found that men with depression had higher risk of stroke than women, and more studies are clearly needed [102]. Additionally, recent studies have shown that the use of antidepressant medication increased risk of stroke and all-cause mortality [103, 104] the reasons for this are unknown but may be related to other life-style factors. A previous study by Smoller et al. [105] showed that antidepressant use was associated with increased stroke risk in post-menopausal women and all-cause mortality [105]. Similar results were observed by Pan et al. [99], in 80,574 women between the ages of 54 and 79 years in the Nurses' Health Study without a prior history of stroke. Antidepressant use lead to an increased risk of stroke [99]. However, recent studies show that treatment after stroke may lead to different conclusions regarding antidepressant use. Administration of antidepressants after stroke lowered the 30-day mortality in severe stroke, and no sex differences were found [106]. The results need to be replicated in randomized clinical trials to conclusively determine the risks and benefits of antidepressant use after stroke in both sexes.

Depression is a serious and common psychiatric condition affecting 11–63 % of stroke patients [107–111]. Patients with poststroke depression (PSD), though grossly under diagnosed, have poor functional outcomes and increased morbidity and mortality [112, 113]. Recent studies have shown that female sex is independently associated with a higher risk of suffering PSD [114–117]. Social factors such as depression and isolation and their contribution to both stroke risk and recovery remain understudied.

References

1. Howe MD, McCullough LD (2015) Prevention and management of stroke in women. Expert Rev Cardiovasc Ther 13(4):403–415
2. Bushnell C, McCullough LD, Awad IA, Chireau MV, Fedder WN, Furie KL et al (2014) Guidelines for the prevention of stroke in women: a statement for healthcare professionals from the American heart association/American stroke association. Stroke 45(5):1545–1588
3. Australian Psychological Society (2010) Guidelines for psychological practice with lesbian, gay, and bisexual clients. Comm Prof Conduct 67(1):67–75
4. Kamel H, Navi BB, Sriram N, Hovsepian DA, Devereux RB, Elkind MSV (2014) Risk of a thrombotic event after the 6-week postpartum period. N Engl J Med 370(14):1307–1315
5. Murphy MS, Smith GN (2016) Pre-eclampsia and cardiovascular disease risk assessment in women. Am J Perinatol 33(8):723–731. doi:10.1055/s-0036-1572536, Epub 2016 Mar 31
6. Lin YS, Tang CH, Yang CYC, Wu LS, Hung ST, Hwa HL et al (2011) Effect of pre-eclampsiaeclampsia on major cardiovascular events among peripartum women in Taiwan. Am J Cardiol 107(2):325–330, http://linkinghub.elsevier.com/retrieve/pii/S0002914910018722
7. Tranquilli AL, Landi B, Giannubilo SR, Sibai BM (2012) Preeclampsia: no longer solely a pregnancy disease. Pregnancy Hypertens 2(4):350–357

8. Raio L, Bolla D, Baumann M (2015) Hypertension in pregnancy. Curr Opin Cardiol 30(4):411–415. doi:10.1097/HCO.0000000000000190, Review
9. Roach REJ, Helmerhorst FM, Lijfering WM, Stijnen T, Algra A, Dekkers OM (2015) Combined oral contraceptives: the risk of myocardial infarction and ischemic stroke. In: Roach REJ (ed) The Cochrane database of systematic reviews. Wiley, Chichester, p. CD011054
10. Lidegaard Ø (2016) The risk of arterial thrombosis increases with the use of combined oral contraceptives. Evid Based Med 21(1):2257
11. Tepper NK, Whiteman MK, Zapata LB, Marchbanks PA, Curtis KM (2016) Safety of hormonal contraceptives among women with migraine: a systematic review. Contraception. doi:10.1016/j.contraception.2016.04.016. http://www.ncbi.nlm.nih.gov/pubmed/27153744
12. Aegidius K, Zwart JA, Hagen K, Schei B, Stovner LJ (2006) Oral contraceptives and increased headache prevalence: the Head-HUNT study. Neurology 66(3):349–353
13. Centers for Disease Control and Prevention (CDC) (2012) CDC's U.S. Medical Eligibility Criteria for Contraceptive Use (2010) Revised recommendations for the use of hormonal contraception among women at high risk for HIV infection or infected with HIV. MMWR Morb Mortal Wkly Rep 61:449
14. Kurth T, Winter AC, Eliassen AH, Dushkes R, Mukamal KJ, Rimm EB et al (2016) Migraine and risk of cardiovascular disease in women: prospective cohort study. BMJ 353:i2610. http://www.bmj.com/lookup/doi/10.1136/bmj.i2610
15. Tsang W, Alter DA, Wijeysundera HC, Zhang T, Ko DT (2012) The impact of cardiovascular disease prevalence on women's enrollment in landmark randomized cardiovascular trials: a systematic review. J Gen Intern Med 27(1):93–98
16. Ridker PM, Cook NR, Lee IM, Gordon D, Gaziano JM, Manson JE et al (2005) A randomized trial of low-dose aspirin in the primary prevention of cardiovascular disease in women. N Engl J Med 352(13):1293–1304
17. Berger JS, Roncaglioni MC, Avanzini F, Pangrazzi I, Tognoni G, Brown DL (2006) Aspirin for the primary prevention of cardiovascular events in women and men: a sex-specific meta-analysis of randomized controlled trials. JAMA 295(3):306–313
18. Caso V, Santalucia P, Acciarresi M, Pezzella FR, Paciaroni M (2012) Antiplatelet treatment in primary and secondary stroke prevention in women. Eur J Intern Med 23(7):580–585. http://www.ncbi.nlm.nih.gov/pubmed/22939800
19. Verdoia M, Pergolini P, Rolla R, Nardin M, Barbieri L, Daffara V et al (2016) Gender differences in platelet reactivity in patients receiving dual antiplatelet therapy. Cardiovasc Drugs Ther 30(2):143–150. http://www.ncbi.nlm.nih.gov/pubmed/26868495
20. Bibbins-Domingo K, U.S. Preventive Services Task Force (2016) Aspirin use for the primary prevention of cardiovascular disease and colorectal cancer: Recommendations from the U.S. Preventive Services Task Force. Ann Intern Med 164(12):836–845
21. Dehmer SP, Maciosek MV, Flottemesch TJ, LaFrance AB, Whitlock EP (2016) Aspirin for the primary prevention of cardiovascular disease and colorectal cancer: a decision analysis for the U.S. Preventive Services Task Force. Ann Intern Med 164(12):777–786
22. Björck S, Palaszewski B, Friberg L, Bergfeldt L (2013) Atrial fibrillation, stroke risk, and warfarin therapy revisited: a population-based study. Stroke 44(11):3103–3108. http://www.ncbi.nlm.nih.gov/pubmed/23982711
23. Hart RG, Eikelboom JW, Pearce LA (2012) Sex, stroke, and atrial fibrillation. Arch Neurol 69:1641–1643
24. Volgman AS, Manankil MF, Mookherjee DTR (2009) Women with atrial fibrillation: greater risk, less attention. Gend Med 6(3):419–432
25. Jönsson A-C, Ek J, Kremer C (2015) Outcome of men and women after atrial fibrillation and stroke. Acta Neurol Scand 132(2):125–131. http://www.ncbi.nlm.nih.gov/pubmed/25649996
26. Hsu JC, Maddox TM, Kennedy K, Katz DF, Marzec LN, Lubitz SA et al (2016) Aspirin instead of oral anticoagulant prescription in atrial fibrillation patients at risk for stroke. J Am Coll Cardiol 67(25):2913–2923, http://linkinghub.elsevier.com/retrieve/pii/S0735109716327693

27. Pancholy SB, Sharma PS, Pancholy DS, Patel TM, Callans DJ, Marchlinski FE (2014) Meta-analysis of gender differences in residual stroke risk and major bleeding in patients with nonvalvular atrial fibrillation treated with oral anticoagulants. Am J Cardiol 113(3):485–490. http://www.ncbi.nlm.nih.gov/pubmed/24315113
28. Lasek-Bal A, Puz P, Kazibutowska Z (2014) Efficacy and safety assessment of alteplase in the treatment of stroke—gender differences. Neurol Res 36(9):851–856
29. Diedler J, Ahmed N, Glahn J, Grond M, Lorenzano S, Brozman M et al (2011) Is the maximum dose of 90 mg alteplase sufficient for patients with ischemic stroke weighing >100 kg? Stroke 42(6):1615–1620
30. Boehme AK, Siegler JE, Mullen MT, Albright KC, Lyerly MJ, Monlezun DJ et al (2014) Racial and gender differences in stroke severity, outcomes, and treatment in patients with acute ischemic stroke. J Stroke Cerebrovasc Dis 23(4):e255–e261
31. Kent DM, Price LL, Ringleb P, Hill MD, Selker HP (2005) Sex-based differences in response to recombinant tissue plasminogen activator in acute ischemic stroke: a pooled analysis of randomized clinical trials. Stroke 36(1):62–65, Epub 2004 Nov 29
32. Lutsep HL, Hill MD (2012) Effects of sex on mechanical embolectomy outcome. J Stroke Cerebrovasc Dis 21(3):240–242. http://www.ncbi.nlm.nih.gov/pubmed/20851620
33. Berkhemer OA, Fransen PSS, Beumer D, van den Berg LA, Lingsma HF, Yoo AJ et al (2014) A randomized trial of intraarterial treatment for acute ischemic stroke. N Engl J Med 372(1):11–20
34. Goyal M, Menon BK, Van Zwam WH, Dippel DWJ, Mitchell PJ, Demchuk AM et al (2016) Endovascular thrombectomy after large-vessel ischaemic stroke: a meta-analysis of individual patient data from five randomised trials. Lancet 387(10029):1723–1731
35. Reeves MJ, Bushnell CD, Howard G, Gargano JW, Duncan PW, Lynch G et al (2008) Sex differences in stroke: epidemiology, clinical presentation, medical care, and outcomes. Lancet Neurol 7(10):915–926. http://www.ncbi.nlm.nih.gov/pubmed/18722812
36. Rojas JI, Zurrú MC, Romano M, Patrucco L, Cristiano E (2007) Acute ischemic stroke and transient ischemic attack in the very old--risk factor profile and stroke subtype between patients older than 80 years and patients aged less than 80 years. Eur J Neurol 14(8):895–899
37. Gibson CL (2013) Cerebral ischemic stroke: is gender important? J Cereb Blood Flow Metab 33(9):1355–1361. http://www.ncbi.nlm.nih.gov/pubmed/23756694
38. Mozaffarian D, Benjamin EJ, Go AS, Arnett DK, Blaha MJ, Cushman M et al (2015) Executive summary: heart disease and stroke statistics-2015 update: a report from the American Heart Association. Circulation 131(4):434–441
39. Seshadri S, Beiser A, Kelly-Hayes M, Kase CS, Au R, Kannel WB et al (2006) The lifetime risk of stroke: estimates from the Framingham study. Stroke 37(2):345–350
40. Roy-O'Reilly M, McCullough LD (2014) Sex differences in stroke: the contribution of coagulation. Exp Neurol 259:16–27. http://dx.doi.org/10.1016/j.expneurol.2014.02.011
41. Koellhoffer EC, McCullough LD (2013) The effects of estrogen in ischemic stroke. Transl Stroke Res 4(4):390–401
42. Wassertheil-Smoller S, Hendrix SL, Limacher M, Heiss G, Kooperberg C, Baird A et al (2003) Effect of estrogen plus progestin on stroke in postmenopausal women: the Women's Health Initiative: a randomized trial. JAMA 289(20):2673–2684
43. Liu F, McCullough LD (2012) Interactions between age, sex, and hormones in experimental ischemic stroke. Neurochem Int 61(8):1255–1265
44. Wolff EF, He Y, Black DM, Brinton EA, Budoff MJ, Cedars MI et al (2013) Self-reported menopausal symptoms, coronary artery calcification, and carotid intima-media thickness in recently menopausal women screened for the Kronos early estrogen prevention study (KEEPS). Fertil Steril 99(5):1385–1391. http://www.ncbi.nlm.nih.gov/pubmed/23312232
45. Hodis HN, Mack WJ, Shoupe D, Azen SP, Stanczyk FZ, Hwang-Levine J et al (2015) Methods and baseline cardiovascular data from the Early versus Late Intervention Trial with Estradiol testing the menopausal hormone timing hypothesis. Menopause 22(4):391–401, http://content.wkhealth.com/linkback/openurl?sid=WKPTLP:landingpage&an=00042192-201504000-00007

46. Hodis HN, Mack WJ, Henderson VW, Shoupe D, Budoff MJ, Hwang-Levine J et al (2016) Vascular effects of early versus late postmenopausal treatment with estradiol. N Engl J Med 374(13):1221–1231. http://www.ncbi.nlm.nih.gov/pubmed/27028912
47. Wharton W, Gleason CE, Miller VM, Asthana S (2013) Rationale and design of the Kronos Early Estrogen Prevention Study (KEEPS) and the KEEPS cognitive and affective sub study (KEEPS Cog). Brain Res 1514:12–17, http://linkinghub.elsevier.com/retrieve/pii/S0006899313005246
48. Haast RAM, Gustafson DR, Kiliann AJ (2012) Sex differences in stroke. J Cereb Blood Flow Metab 32(12):2100–2107
49. Baron JC (2001) Perfusion thresholds in human cerebral ischemia: historical perspective and therapeutic implications. Cerebrovasc Dis 11(Suppl 1):2–8. http://www.ncbi.nlm.nih.gov/pubmed/11244194
50. Henriksen OM, Jensen LT, Krabbe K, Guldberg P, Teerlink T, Rostrup E (2014) Resting brain perfusion and selected vascular risk factors in healthy elderly subjects. PLoS One 9(5):e97363. http://www.ncbi.nlm.nih.gov/pubmed/24840730
51. Bang OY, Saver JL, Kim SJ, Kim G-M, Chung C-S, Ovbiagele B et al (2011) Collateral flow predicts response to endovascular therapy for acute ischemic stroke. Stroke 42(3):693–699, http://www.pubmedcentral.nih.gov/articlerender.fcgi?artid=3051344&tool=pmcentrez&rendertype=abstract
52. Saver JL, Jahan R, Levy EI, Jovin TG, Baxter B, Nogueira RG et al (2012) Solitaire flow restoration device versus the Merci Retriever in patients with acute ischaemic stroke (SWIFT): a randomised, parallel-group, non-inferiority trial. Lancet 380(9849):1241–1249. http://www.ncbi.nlm.nih.gov/pubmed/22932715
53. Kim SJ, Son JP, Ryoo S, Lee M-J, Cha J, Kim KH et al (2014) A novel magnetic resonance imaging approach to collateral flow imaging in ischemic stroke. Ann Neurol 76(3):356–369. http://www.ncbi.nlm.nih.gov/pubmed/24985162
54. Broderick JP, Palesch YY, Demchuk AM, Yeatts SD, Khatri P, Hill MD et al (2013) Endovascular therapy after intravenous t-PA versus t-PA alone for stroke. N Engl J Med 368(10):893–903, http://www.pubmedcentral.nih.gov/articlerender.fcgi?artid=3651875&tool=pmcentrez&rendertype=abstract
55. Vagal A, Menon BK, Foster LD, Livorine A, Yeatts SD, Qazi E et al (2015) Association between CT angiogram collaterals and ct perfusion in the Interventional Management of Stroke III Trial. Stroke 47(2):535–538. http://www.ncbi.nlm.nih.gov/pubmed/26658448
56. Hara T, Nakamura K, Matsui M, Yamamoto A, Nakahara Y, Suzuki-Migishima R, Yokoyama M, Mishima K, Saito I, Okano H, Mizushima N (2006) Suppression of basal autophagy in neural cells causes neurodegenerative disease in mice. Nature 441(7095):885–889
57. Komatsu M, Ueno T, Waguri S, Uchiyama Y, Kominami E, Tanaka K (2007) Constitutive autophagy: vital role in clearance of unfavorable proteins in neurons. Cell Death Differ 14(5):887–894
58. Chen C, Hu LX, Dong T, Wang GQ, Wang LH, Zhou XP et al (2013) Apoptosis and autophagy contribute to gender difference in cardiac ischemia-reperfusion induced injury in rats. Life Sci 93(7):265–270
59. Weis SN, Toniazzo AP, Ander BP, Zhan X, Careaga M, Ashwood P et al (2014) Autophagy in the brain of neonates following hypoxia-ischemia shows sex- and region-specific effects. Neuroscience 256:201–209
60. Chen CW, Chen TY, Tsai KL, Lin CL, Yokoyama KK, Lee WS, Chiueh CC, Hsu C (2012) Inhibition of autophagy as a therapeutic strategy of iron-induced brain injury after hemorrhage. Autophagy 8(10):1510–1520
61. Du L, Hickey RW, Bayir H, Watkins SC, Tyurin VA, Guo F et al (2009) Starving neurons show sex difference in autophagy. J Biol Chem 284(4):2383–2396, http://www.jbc.org/cgi/doi/10.1074/jbc.M804396200
62. Straface E, Gambardella L, Brandani M, Malorni W (2012) Sex differences at cellular level: cells have a sex. Handb Exp Pharmacol 214:49–65
63. Widdowson E (1976) The response of the sexes to nutritional stress. Proc Nutr Soc 35:175–180

64. Demarest TG, McCarthy MM (2015) Sex differences in mitochondrial (dys)function: implications for neuroprotection. J Bioenerg Biomembr 47(1-2):173–188
65. Gaignard P, Savouroux S, Liere P, Pianos A, Thérond P, Schumacher M et al (2015) Effect of sex differences on brain mitochondrial function and its suppression by ovariectomy and in aged mice. Endocrinology 156(8):2893–2904
66. Lamont LS (2005) gender differences in amino acid use during endurance exercise. Nutr Rev 63(12 Pt 1):419–422
67. Mccullough LD, Zeng Z, Blizzard KK, Debchoudhury I, Hurn PD (2005) Ischemic nitric oxide and poly (ADP-ribose) polymerase-1 in cerebral ischemia: male toxicity, female protection. J Cereb Blood Flow Metab 25:502–512
68. Siegel C, Yuan M, Zeng Z, McCullough LD (2009) Sex differences in the response to stroke-induced activation of the poly (ADP-ribose) polymerase pathway. Stroke 40(4), e211
69. Yuan M, Siegel C, Zeng Z, McCullough LD (2009) Sex differences in the response to stroke-induced activation of the poly (ADP-ribose) polymerase pathway. Stroke 40(4), e211
70. Amiri-Nikpour MR, Nazarbaghi S, Hamdi-Holasou M, Rezaei Y (2015) An open-label evaluator-blinded clinical study of minocycline neuroprotection in ischemic stroke: gender-dependent effect. Acta Neurol Scand 131(1):45–50. doi: 10.1111/ane.12296. Epub 2014 Aug 23. PubMed PMID: 25155474
71. Lisabeth LD, Reeves MJ, Baek J, Skolarus LE, Brown DL, Zahuranec DB et al (2015) Factors influencing sex differences in poststroke functional outcome. Stroke 46(3):860–863. http://www.ncbi.nlm.nih.gov/pubmed/25633999
72. Nguyen VQC, Prvubettger J, Guerrier T, Hirsch MA, Thomas JG, Pugh TM et al (2015) Factors associated with discharge to home versus discharge to institutional care after inpatient stroke rehabilitation. Arch Phys Med Rehabil. 96(7):1297–1303. http://www.ncbi.nlm.nih.gov/pubmed/25823940
73. Smith MA, Risser JMH, Moy?? LA, Garcia N, Akiwumi O, Uchino K et al (2004) Designing multi-ethnic stroke studies: the brain attack surveillance in Corpus Christi (BASIC) project. Ethn Dis 14(4):520–526. http://www.ncbi.nlm.nih.gov/pubmed/15724771
74. Boden-Albala B, Litwak E, Elkind MS V, Rundek T, Sacco RL (2005) Social isolation and outcomes post stroke. Neurology 64(11):1888–1892. http://www.ncbi.nlm.nih.gov/pubmed/15955939
75. Glass TA, Berkman LF, Hiltunen EF, Furie K, Glymour MM, Fay ME, et al (2004). The Families In Recovery From Stroke Trial (FIRST): primary study results. Psychosom Med 66(6):889–897. http://www.ncbi.nlm.nih.gov/pubmed/15564354
76. Mant J, Carter J, Wade DT, Winner S (2000) Family support for stroke: a randomised controlled trial. Lancet 356(9232):808–813. http://www.ncbi.nlm.nih.gov/pubmed/11022928
77. Friedler B, Crapser J, McCullough L (2015) One is the deadliest number: the detrimental effects of social isolation on cerebrovascular diseases and cognition. Acta Neuropathol 129(4):493–509. http://www.ncbi.nlm.nih.gov/pubmed/25537401
78. Aron AW, Staff I, Fortunato G, McCullough LD (2015) Prestroke living situation and depression contribute to initial stroke severity and stroke recovery. J Stroke Cerebrovasc Dis 24(2):492–499. http://www.ncbi.nlm.nih.gov/pubmed/25524014
79. Carod-Artal FJ, Egido JA (2009) Quality of life after stroke: the importance of a good recovery. Cerebrovasc Dis. 27(Suppl 1):204–214. http://www.ncbi.nlm.nih.gov/pubmed/19342853
80. Kruithof WJ, van Mierlo ML, Visser-Meily JMA, van Heugten CM, Post MWM (2013) Associations between social support and stroke survivors' health-related quality of life—a systematic review. Patient Educ Couns 93(2):169–176. http://www.ncbi.nlm.nih.gov/pubmed/23870177
81. Holt-Lunstad J, Smith TB, Baker M, Harris T, Stephenson D (2015) Loneliness and social isolation as risk factors for mortality a meta-analytic review. Perspect Psychol Sci 10(2):227–237, http://pps.sagepub.com/content/10/2/227.full.pdf
82. Valtorta NK, Kanaan M, Gilbody S, Ronzi S, Hanratty B (2016) Loneliness and social isolation as risk factors for coronary heart disease and stroke: systematic review and meta-analysis of longitudinal observational studies. Heart 102(13):1009–1016, http://heart.bmj.com/content/early/2016/03/15/heartjnl-2015-308790.abstract

83. Grant N, Hamer M, Steptoe A (2009) Social isolation and stress-related cardiovascular, lipid, and cortisol responses. Ann Behav Med 37(1):29–37. http://www.ncbi.nlm.nih.gov/pubmed/19194770
84. Hawkley LC, Cacioppo JT (2010) Loneliness matters: a theoretical and empirical review of consequences and mechanisms. Ann Behav Med 40(2):218–227. http://www.ncbi.nlm.nih.gov/pubmed/20652462
85. Hawkley LC, Thisted RA, Masi CM, Cacioppo JT (2010) Loneliness predicts increased blood pressure: 5-year cross-lagged analyses in middle-aged and older adults. Psychol Aging 25(1):132–141, http://www.pubmedcentral.nih.gov/articlerender.fcgi?artid=2841310&tool=pmcentrez&rendertype=abstract
86. Loucks EB, Berkman LF, Gruenewald TL, Seeman TE (2006) Relation of social integration to inflammatory marker concentrations in men and women 70 to 79 years. Am J Cardiol 97(7):1010–1016. http://www.ncbi.nlm.nih.gov/pubmed/16563907
87. Loucks EB, Sullivan LM, D'Agostino Sr RB, Larson MG, Berkman LF, Benjamin EJ (2006) Social networks and inflammatory markers in the Framingham Heart Study. J Biosoc Sci 38(6):835. http://www.ncbi.nlm.nih.gov/pubmed/16441967
88. Castillo J, Alvarez-Sabín J, Martínez-Vila E, Montaner J, Sobrino T, Vivancos J . (2009) Inflammation markers and prediction of post-stroke vascular disease recurrence: the MITICO study. J Neurol 256(2):217–224. http://www.ncbi.nlm.nih.gov/pubmed/19252763
89. Kaplan RC, McGinn AP, Baird AE, Hendrix SL, Kooperberg C, Lynch J et al (2008) Inflammation and hemostasis biomarkers for predicting stroke in postmenopausal women: The Women's Health Initiative Observational Study. J Stroke Cerebrovasc Dis 17(6):344–355. http://www.ncbi.nlm.nih.gov/pubmed/18984425
90. Stuller KA, Jarrett B, DeVries AC (2012) Stress and social isolation increase vulnerability to stroke. Exp Neurol 233(1):33–39. http://www.ncbi.nlm.nih.gov/pubmed/21281636
91. Nagayoshi M, Everson-Rose SA, Iso H, Mosley TH, Rose KM, Lutsey PL (2014) Social network, social support, and risk of incident stroke: the atherosclerosis risk in communities study. Stroke 45(10):2868–2873. http://www.ncbi.nlm.nih.gov/pubmed/25139878
92. Karelina K, Norman GJ, Zhang N, Morris JS, Peng H, DeVries AC (2009) Social isolation alters neuroinflammatory response to stroke. Proc Natl Acad Sci U S A 106(14):5895–5900
93. Craft TKS, Glasper ER, McCullough L, Zhang N, Sugo N, Otsuka T et al (2005) Social interaction improves experimental stroke outcome. Stroke 36(9):2006–2011. http://www.ncbi.nlm.nih.gov/pubmed/16100018
94. Venna VR, Verma R, O'Keefe LM, Xu Y, Crapser J, Friedler B, McCullough LD (2014) Inhibition of mitochondrial p53 abolishes the detrimental effects of social isolation on ischemic brain injury. Stroke 45(10):3101–3104
95. Weil ZM, Norman GJ, Barker JM, Su AJ, Nelson RJ, Devries AC (2008) Social isolation potentiates cell death and inflammatory responses after global ischemia. Mol Psychiatry. 13(10):913–915. http://www.ncbi.nlm.nih.gov/pubmed/18800053
96. Verma R, Harris NM, Friedler BD, Crapser J, Patel AR, Venna V et al (2016) reversal of the detrimental effects of post-stroke social isolation by pair-housing is mediated by activation of BDNF-MAPK/ERK in Aged Mice. Sci Rep 6:25176. http://www.nature.com/articles/srep25176
97. Van de Velde S, Bracke P, Levecque K (2010) Gender differences in depression in 23 European countries. Cross-national variation in the gender gap in depression. Soc Sci Med 71(2):305–313. http://www.ncbi.nlm.nih.gov/pubmed/20483518
98. Altemus M, Sarvaiya N, Neill Epperson C (2014) Sex differences in anxiety and depression clinical perspectives. Front Neuroendocrinol. 35(3):320–330. http://www.ncbi.nlm.nih.gov/pubmed/24887405
99. Pan A, Okereke OI, Sun Q, Logroscino G, Manson JE, Willett WC et al (2011) Depression and incident stroke in women. Stroke 42(10):2770–2775. http://www.ncbi.nlm.nih.gov/pubmed/21836097
100. O'Donnell MJ, Denis X, Liu L, Zhang H, Chin SL, Rao-Melacini P, et al (2010) Risk factors for ischaemic and intracerebral haemorrhagic stroke in 22 countries (the INTERSTROKE study): a case-control study. Lancet 10;376(9735):112–123. http://www.ncbi.nlm.nih.gov/pubmed/20561675

101. Seifert CL, Poppert H, Sander D, Feurer R, Etgen T, Ander KH, et al (2012) Depressive symptoms and the risk of ischemic stroke in the elderly-influence of age and sex. PLoS One 7(11):e50803. http://www.ncbi.nlm.nih.gov/pubmed/23226388
102. Hamano T, Li X, Lönn SL, Nabika T, Shiwaku K, Sundquist J, Sundquist K (2015) Depression, stroke and gender: evidence of a stronger association in men. J Neurol Neurosurg Psychiatry 86(3):319–323. doi:10.1136/jnnp-2014-307616, Epub 2014 Jun 26
103. Jennum P, Baandrup L, Iversen HK, Ibsen R, Kjellberg J (2016) Mortality and use of psychotropic medication in patients with stroke: a population-wide, register-based study. BMJ Open 6(3):e010662. http://www.pubmedcentral.nih.gov/articlerender.fcgi?artid=4785303&tool=pmcentrez&rendertype=abstract
104. Hansen RA, Khodneva Y, Glasser SP, Qian J, Redmond N, Safford MM (2016) Antidepressant medication use and its association with cardiovascular disease and all-cause mortality in the Reasons for Geographic and Racial Differences in Stroke (REGARDS) study. Ann Pharmacother 50(4):253–261. http://www.ncbi.nlm.nih.gov/pubmed/26783360
105. Smoller JW, Allison M, Cochrane BB, Curb JD, Perlis RH, Robinson JG, et al (2009) Antidepressant use and risk of incident cardiovascular morbidity and mortality among postmenopausal women in the Women's Health Initiative study. Arch Intern Med 169(22):2128–2139. http://www.ncbi.nlm.nih.gov/pubmed/20008698
106. Mortensen JK, Paaske Johnsen S, Larsson H, Andersen G, Johnsen M, Larsson et al (2015) Early antidepressant treatment and all-cause 30-day mortality in patients with ischemic stroke. Cerebrovasc Dis 4040(1-2):81–90. www.karger.com/ced
107. House A, Dennis M, Mogridge L, Warlow C, Hawton K, Jones L (1991) Mood disorders in the year after first stroke. Br J Psychiatry 158:83–92. http://www.ncbi.nlm.nih.gov/pubmed/2015456
108. Gottlieb D, Salagnik I, Kipnis M, Brill S (2002) Post stroke depression, first year post stroke, in middle band patients. Int J Geriatr Psychiatry 17:486–487
109. Robinson RG, Spalletta G (2010) Poststroke depression: a review. Can J Psychiatry 55(6):341–349. http://www.ncbi.nlm.nih.gov/pubmed/20540828
110. Hackett ML, Pickles K (2014) Part I: frequency of depression after stroke: an updated systematic review and metaanalysis of observational studies. Int J Stroke 9(8):1017–25. doi: 10.1111/ijs.12357. Epub 2014 Aug 12. Review. PubMed PMID: 25117911
111. Ayerbe L, Ayis S, Crichton S, Wolfe CDA, Rudd AG (2014) The long-term outcomes of depression up to 10 years after stroke; the South London Stroke Register. J Neurol Neurosurg Psychiatry 85(5):514–521. http://www.ncbi.nlm.nih.gov/pubmed/24163430
112. Ayerbe L, Ayis S, Wolfe CD, Rudd AG (2013) Natural history, predictors and outcomes of depression after stroke: systematic review and meta-analysis. Br J Psychiatry 202(1):14–21. doi: 10.1192/bjp.bp.111.107664. Review. PubMed PMID:23284148
113. Herrmann N, Seitz D, Fischer H, Saposnik G, Calzavara A, Anderson G et al (2011) Detection and treatment of post stroke depression: results from the registry of the Canadian stroke network. Int J Geriatr Psychiatry 26(11):1195–200. http://www.ncbi.nlm.nih.gov/pubmed/21360753
114. Shi Y, Xiang Y, Yang Y, Zhang N, Wang S, Ungvari GS et al (2015) Depression after minor stroke: prevalence and predictors. J Psychosom Res 79(2):143–147. http://www.ncbi.nlm.nih.gov/pubmed/25868687
115. Mazure CM, Weinberger AH, Pittman B, Sibon I, Swendsen J (2014) Gender and stress in predicting depressive symptoms following stroke. Cerebrovasc Dis 38(4):240–246. http://www.ncbi.nlm.nih.gov/pubmed/25401293
116. Alajbegovic A, Djelilovic-Vranic J, Nakicevic A, Todorovic L, Tiric-Campara M (2014) Post stroke depression. Med Arch 68(1):47–50. http://www.ncbi.nlm.nih.gov/pubmed/24838087
117. Zhang WN, Pan YH, Wang XY, Zhao Y (2013) A prospective study of the incidence and correlated factors of post-stroke depression in china. PLoS One 8(11):e78981. http://www.ncbi.nlm.nih.gov/pubmed/24260141

Chapter 33
Current Imaging Strategies for Patient Selection in Acute Ischemic Stroke Trials

Jenny P. Tsai and Gregory W. Albers

Abstract Neuroimaging is an important part of acute ischemic stroke trials. Its key components include vascular, ischemic core, and penumbral imaging. These can be acquired by two modalities: computed tomography (CT) and magnetic resonance imaging (MRI). This chapter provides an overview of common CT- and MRI-based angiographic, parenchymal, and perfusion imaging techniques, and their roles in patient selection in acute ischemic stroke trials.

Keywords Computed tomography • Computed tomography angiography • Computed tomography perfusion • Magnetic resonance angiography • Diffusion-weighted imaging • Perfusion-weighted imaging • Arterial spin labeling • ASPECT score

Abbreviations

ADC	Apparent diffusion coefficient
AHA/ASA	American Heart Association/American Stroke Association
ASL	Arterial spin labeling
ASPECTS	Alberta Stroke Program Early Computed Tomography Score
CBF	Cerebral blood flow
CBV	Cerebral blood volume
CT	Computed tomography
CTA	Computed tomography angiography
CTP	Computed tomography perfusion imaging
DEDAS	Dose Escalation of Desmoteplase for Acute Ischemic Stroke trial
DEFUSE	Diffusion and Perfusion Imaging Evaluation for Understanding Stroke Evolution study

J.P. Tsai • G.W. Albers (✉)
Department of Neurology, Stanford University,
300 Pasteur Dr., MC: 5778, Stanford, CA 94305-5778, USA
e-mail: jptsai@stanford.edu; GAlbers@stanfordhealthcare.org

P.A. Lapchak, J.H. Zhang (eds.), *Neuroprotective Therapy for Stroke and Ischemic Disease*, Springer Series in Translational Stroke Research,
DOI 10.1007/978-3-319-45345-3_33

DIAS	Desmoteplase In Acute Stroke trial
DWI	Diffusion-weighted imaging
EPITHET	Echoplanar Imaging Thrombolytic Evaluation trial
EXTEND	EXtending the time for Thrombolysis in Emergency Neurological Deficits trial
FLAIR	Fluid-attenuated inversion recovery sequence
IMS-III	Interventional Management of Stroke-III trial
kVp	Peak kilovoltage
mAs	Milliamps per second
MCA	Middle cerebral artery
MIP	Maximal intensity projection
MP-CTA	Multiphase computed tomography angiography
MRA	Magnetic resonance angiography
MRI	Magnetic resonance imaging
mSv	Millisievert
MTT	Mean transit time
NCCT	Noncontrast computed tomography
PWI	Perfusion-weighted imaging
SNR	Signal-to-noise ratio
T_{max}	Time-to-maximum of the residue function
TOF	Time-of-flight
tPA	Tissue plasminogen activator
TTP	Time-to-peak
VERiTAS	Vertebrobasilar Flow Evaluation and Risk in Transient Ischemic Attack and Stroke study

A neuroimaging modality best selects patients for an acute ischemic stroke trial by optimizing accuracy, reliability, efficiency, and safety, and should be tailored to the investigational intervention. Neuroimaging strategies have evolved in conjunction with clinical trials and the growing sophistication of neurovascular interventions. In the early 1990s, when the first landmark clinical trials evaluated the safety of intravenous thrombolytic therapy, noncontrast computed tomography (CT) was the optimal imaging strategy. It fulfilled its two roles efficiently: To exclude intracranial hemorrhage, and to identify any evidence of large infarcts occupying ≥1/3 of the middle cerebral artery (MCA) territory [1, 2]. Over the following decades, the stroke community understood the benefits of imaging the arterial occlusive lesion, ischemic core, penumbra, and collaterals in patient selection for endovascular therapy. Multimodal CT- and MRI-based techniques that provided angiographic imaging and demonstration of salvageable ischemic brain tissue have become the preferred strategies for many clinical endovascular therapy trials, especially for ongoing trials that are attempting to expand the treatment window beyond 6 h from symptom onset. Despite continuing advances in both CT and MRI in stroke imaging, the superiority of one over another remains a subject of debate. This chapter addresses the current neuroimaging modalities used for patient selection in acute ischemic stroke and new developments for optimizing their use in clinical research.

1 CT-Based Imaging

Computed tomography (CT) employs collimated X-ray beams to differentiate adjacent structures based on their densities. As its image acquisition is based on differential attenuation properties, adjacent tissues with similar densities can be difficult to discern on the basis of noncontrast CT (NCCT) alone. A frequent example is the progression of acute infarction, where decreasing tissue density becomes increasingly visible over hours as a reflection of irreversible tissue injury.

1.1 Noncontrast CT and ASPECT Score

Despite a well-recognized need to identify evidence of early infarction, there is significant variability in interreader agreement on the extent of ischemic injury. The Alberta Stroke Program Early CT score (ASPECTS) was developed to standardize grading of ischemic injury in the MCA territory. It systematically grades integrity of gray-white differentiation and early ischemic hypodensity in ten cortical and subcortical areas in the symptomatic hemisphere [3]. (Fig. 33.1) ASPECTS based on noncontrast CT is practical, easy to learn, accessible, and shows good interobserver agreement in scoring, with kappa ranging between 0.71 and 0.89 [4, 5]. These advantages allowed it to become a central tool in assessing and communicating patient eligibility for acute stroke treatment.

An ASPECT score ≤7 vs. >7 has an estimated sensitivity of 78 % and specificity of 96 % in predicting functional outcome, and a sensitivity of 90 % and specificity of 62 % in predicting symptomatic hemorrhage [3, 5, 6]. A similar analysis based on a larger cohort suggests a linear association of ASPECTS 6-10 with functional outcome, while scores of 0-5 have limited prognostic value in absence of other clinical markers [7]. ASPECTS also provides an estimate of probability of good functional outcome based on time to reperfusion, ranging from over 50 % within 1 h to 25–30 % by 8 h [8, 9]. Several recent acute ischemic stroke trials adopted a minimum ASPECTS of 6 or 7 on baseline noncontrast CT [10–14]. Few studies include ASPECTS of ≤5, and limited information is available on the optimal acute treatment for these patients.

ASPECTS based on noncontrast CT has several other limitations. It has differential weighting in favor of infarction in the insular and striatocapsular region and is only designed to assess middle cerebral artery strokes [15]. In addition, it provides no information about the location of vascular occlusion, collateral status, or presence of ischemic penumbra. ASPECTS on NCCT provides little information that helps to identify

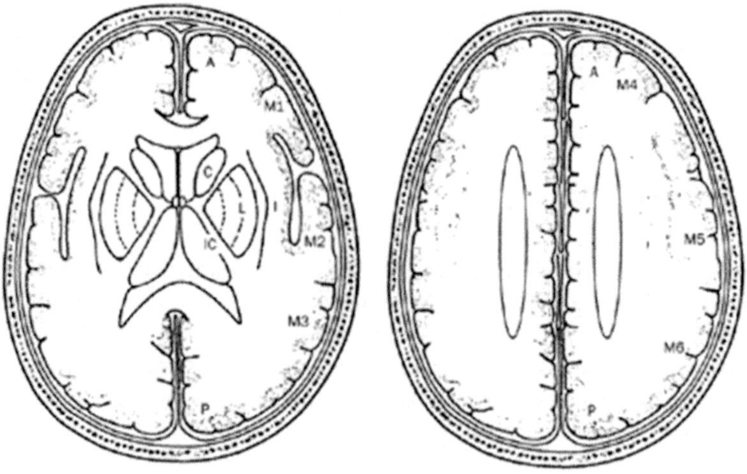

Fig. 33.1 ASPECTS scoring scheme. The Alberta Stroke Program Early Computed Tomography Score (ASPECTS) is a standardized system to evaluate presence of early infarct signs in the middle cerebral artery territory. Two axial slices on noncontrast CT are scored: one at the thalamic and basal ganglia level, and another just rostral to the first. The maximum ASPECTS is 10, and a point is deducted for every cortical or subcortical region with loss of gray-white differentiation in the symptomatic hemisphere. *C*, caudate nucleus; *IC*, internal capsule; *L*, lentiform nucleus; *I*, insular cortex; M1–M3, anterior, lateral insular, and posterior cortical regions; M4–M6, cortical regions immediately superior to M1–M3; *A*, anterior; *P*, posterior; *CT*, computed tomography; *DWI*, diffusion-weighted imaging. Reprinted with permission from "Validity and reliability of a quantitative computed tomography score in predicting outcome of hyperacute stroke before thrombolytic therapy" by Barber PA, Demchuk AM, Zhang J, Buchan AM, et al.. Lancet. 2000 May 13;355(9216):1670–4

which patients have salvageable brain tissue that may benefit from reperfusion therapy beyond 6–8 h [16]. Therefore, in an era of acute and hyperacute stroke treatment, many clinical trials have focused on imaging and rescuing salvageable ischemic tissue using contrast-enhanced CT or MRI.

However, the approach taken in ASPECTS quantification remains useful and applicable in these neuroimaging modalities. ASPECTS has been evaluated on DWI and CT perfusion, and demonstrates greater test performances than on noncontrast CT [16–19]. This methodology has not yet been widely accepted, but may allow greater uniformity in reporting the results of acute stroke imaging.

1.2 CT Angiogram

The Interventional Management of Stroke III (IMS-III) trial was completed over nearly 7 years. The initial IMS-III protocol did not mandate a CT angiogram (CTA) prior to study enrollment. During the study period, clinical practice trended toward increasing use of preintervention CTA, and this change in practice was reflected in sequential protocol amendments [20]. By the fifth and final protocol, all patients were required to have evidence of arterial occlusion on CTA, under the presumption that certain subgroups may show greater benefit from endovascular treatment [20]. Indeed, a post hoc analysis of the IMS-III data demonstrated a trend toward better treatment response in patients with proximal occlusions seen on preintervention CTA [20, 21]. These results paved the way for the subsequent positive endovascular trials, all of which required evidence of an arterial occlusion on CTA or MRA for enrollment [10–14]. The 2015 AHA/ASA guidelines for the Early Management of Patients with Acute Ischemic Stroke Regarding Endovascular Therapy also strongly recommends the use of noninvasive intracranial vascular imaging prior to endovascular therapy (Class 1; Level of Evidence A) [22].

Conventional CTA is acquired with a single scanning phase (Fig. 33.2). Following a contrast bolus injection, rise in Hounsfield units over a preset region of interest in the aorta signals contrast arrival and triggers a timed scan from the aortic arch to vertex [23]. Single-phase CTA allows for visualization of the arterial occlusive site, including a rapid assessment of cervical vessels from their origins. It provides good signal-to-noise ratio (SNR) with limited venous contamination when properly timed [23]. However, it provides limited and at times inaccurate information on collateral status [24].

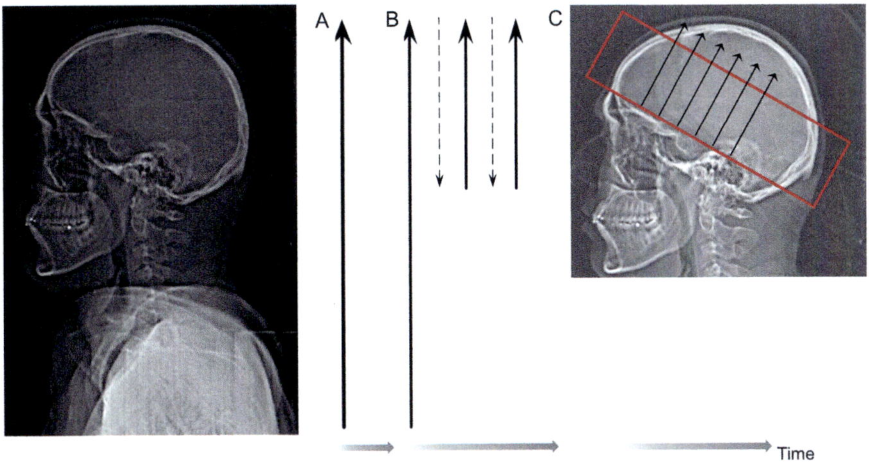

Fig. 33.2 20Comparison of CTA and CTP acquisitions. (**a**) Conventional single-phase CTA is acquired from the aortic arch to the vertex. (**b**) Multiphase CTA is composed of a 3-phase shuttle-mode angiography. The first phase is similar to the conventional single-phase CTA, followed by two additional phases only from skull base to vertex, each separated by 4 s of repositioning and 3.4 s of scanning time. (**c**) This is an example of CTP using a single-slab acquisition, with approximately 8 cm of coverage. The inferior border of the slab should be positioned parallel to and just above the orbital roof. Images are acquired parallel to the longitudinal plane of the slab. Optimal time of acquisition may be dependent on the venous output function (VOF), and typically ranges between 45 and 90 s

1.2.1 Multiphase CT Angiography

Multiphase CTA is a technique developed to address this limitation. It is a time-resolved CTA acquired in three phases, following a single bolus. The first angiogram covers the aortic arch to the vertex in the peak arterial phase, then the multidetector scanner "toggles" between the skull base and the vertex for the additional mid- and late venous phases (Fig. 33.2b). The acquired section thickness is 0.625 mm, and images are reconstructed into axial maximum intensity projections (MIP) [23, 24]. Image postprocessing is automated and requires approximately 2–3 min. The axial MIP images allow for easier visualization of vascular opacification of the Circle-of-Willis, its major branches, and any existing asymmetry in timing and degree of collateral filling. Using collateral status as a surrogate measure, MP-CTA aims to demonstrate evidence of salvageable ischemic tissue. Menon et al. suggested superiority of MP-CTA over SP-CTA and CTP, although this data has yet to be validated in a larger, multicenter study [24]. Meanwhile, Yang et al. demonstrated that MP-CTA outputs can be also generated from CTP source images [23]. MP-CTA is a novel technique and its role as neuroimaging modality in clinical trials and routine practice require further validation.

1.3 CT Perfusion Imaging

CT perfusion imaging (CTP), first proposed in 1980, is a sensitive marker of tissue ischemia, and an attractive option for acute stroke imaging [25]. Despite these advantages, the lack of standardization and traditionally longer postprocessing time has limited its acceptance and routine clinical use in acute stroke imaging for many years. With improvement in both CT scanner technology and fast, user-independent postprocessing software (RAPID; iSchemaView, Menlo Park, CA, USA), CTP has become a reliable imaging modality for acute stroke studies. It has been validated and used across several observational studies and randomized controlled trial [10, 11, 26]. While the technical details of perfusion imaging data acquisition and processing are beyond the scope of this discussion, understanding the basics of perfusion imaging is helpful in understanding the interpretation and applicability of CTP in acute ischemic stroke trials.

1.3.1 Technical Considerations in Using CT Perfusion

The principle behind both CTP and MR perfusion-weighted imaging (PWI) is the tracking of a contrast bolus's passage through its arterial phase, brain parenchyma, and exit into the venous circulation. In CTP, the source data is simply characterized by change in tissue density over time. This variation can be plotted for every voxel of tissue imaged as a time-density function that is directly proportional to contrast concentration [27]. Therefore, CT perfusion requires injection of a radiopaque, iodinated contrast agent followed by repeated image acquisition at fixed time intervals over approximately 70–90 s. There is a little consensus in the protocols currently used, and acquisition parameters vary between institutions. The images are generally acquired with a voltage of 80 kVp and a current between 100 and 200 mAs [28–30]. This setting allows for lower radiation exposure than the conventional 120 kVp used in unenhanced CT, to counter-balance longer duration of radiation exposure with CTP. Average effective radiation dose from a single CTP is approximately 2.5 mSv [28, 29]. (In reference, a New Yorker is exposed to approximately 3.6 mSv of environmental radiation per year.) The radiation dose can be further reduced by approximately 50–1.3 mSv, using a lower current of 50 mAs, without compromising data accuracy [28, 31]. Despite concerns about radiation exposure related to CTP, the dose is relatively limited.

Scanner specification is an important consideration in ensuring that the CTP will capture the entire ischemic core and penumbra and accurately represent the lesion volumes [32]. The craniocaudal lesion coverage of each scanner depends on the width of the scanner's detector row. Common scanners currently used for perfusion imaging range from 64 to 320 detector rows, allowing, respectively, 4–16 cm of coverage in a single slab. An 8-cm coverage in the craniocaudal axis should cover 90 % of the ischemic core in large artery occlusions [33, 34]. (Fig. 33.3) To achieve an 8-cm coverage, a 64-detector row scanner would need to acquire two contiguous and nonoverlapping 4-cm slabs. They can be acquired in two modes. In cine

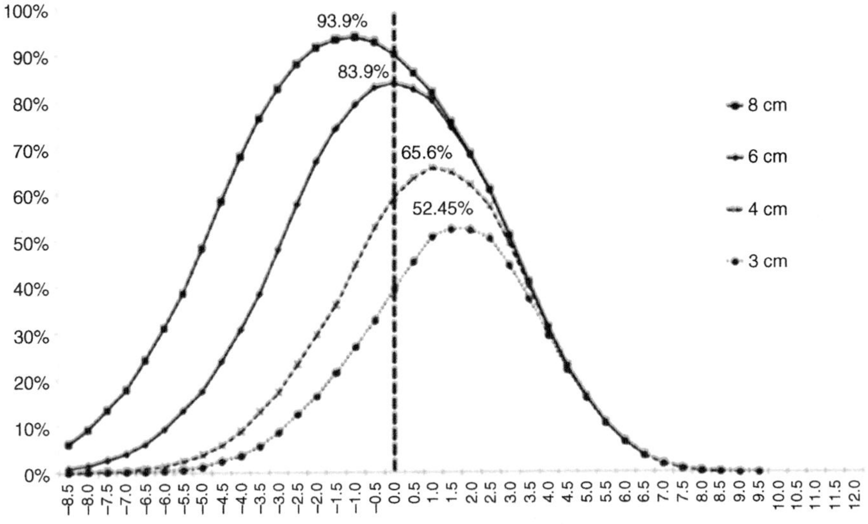

Fig. 33.3 Percent ischemic core coverage per coverage thickness. In comparison to whole brain coverage (100 %), estimates of the ischemic core coverage are illustrated by the *dotted lines*. The *dotted vertical line* indicates the axial central position corresponding to the caudal section used in ASPECTS scoring. Reprinted with permission from "Whole brain CT perfusion in acute anterior circulation ischemia: coverage size matters" by Emmer BJ, Rijkee M, Niesten JM, Wermer MJH, Velthuis BK, van Walderveen MAA. Neuroradiology. Springer Berlin Heidelberg; 2014 Dec;56(12):1121–6

(or toggle) mode, the scanner sequentially acquires the two slabs, each requiring a separate contrast bolus injection [35]. In shuttle mode, only one bolus injection is given, and the scanner alternates between the two slab positions. Though shuttle mode reduces both radiation and contrast exposure, the detector movements result in a lower temporal resolution and slightly lower accuracy [29, 35].

1.3.2 Optimal Parameters for Patient Selection

Perfusion parameters can be either measured or mathematically derived from sources images sequentially acquired over time. If the changes in tissue density and contrast concentration over time are known, the perfusion parameters can be either measured or mathematically derived. Figure 33.4 shows examples of CTP maps, using these parameters:

- Cerebral blood volume (CBV): Total blood volume flowing through a given brain volume, typically measured in mL of blood/100 g of brain tissue. At a tissue level, the remaining volume is accounted for by cells and interstitium.
- Cerebral blood flow (CBF): Total blood volume flowing through a given brain volume per unit time, typically measured in mL of blood/100 g of brain tissue per minute.
- Mean transit time (MTT): Average blood transit time through a given region of interest, typically measured in seconds.
- Time to peak (TTP): Measurement of delay time in contrast arrival in the region of interest, calculated as the derivative of the contrast agent's time-attenuation function without use of a deconvolution function.
- Time to maximum of the residue function (T_{max}): Measurement of delay time in the contrast signal's first arrival in the region of interest, in reference to an arterial input function (AIF) that defines an initial time ($t=0$), and mathematically derived using a deconvolution function.

1.3.3 Thresholds and Performance of CTP Parameters

An ideal candidate for acute stroke intervention is a patient with a perfusion mismatch, which refers to the presence of a small ischemic core and a large penumbra [36, 37]. Their imaging correlates are, respectively, a low-CBV or CBF lesion as ischemic core, and the volumetric difference between the core and a time-based parameter (increased MTT, TTP, or T_{max}) estimates the penumbra [38–42].

There is currently no consensus on which parameters are the optimal measures for either core or penumbra, and at what threshold. In fact, when optimally thresholded, both CBV and CBF offer good approximations of the final infarct core [38, 40, 42–45]. However, when CTP and MRI are acquired within 1 h, CBF

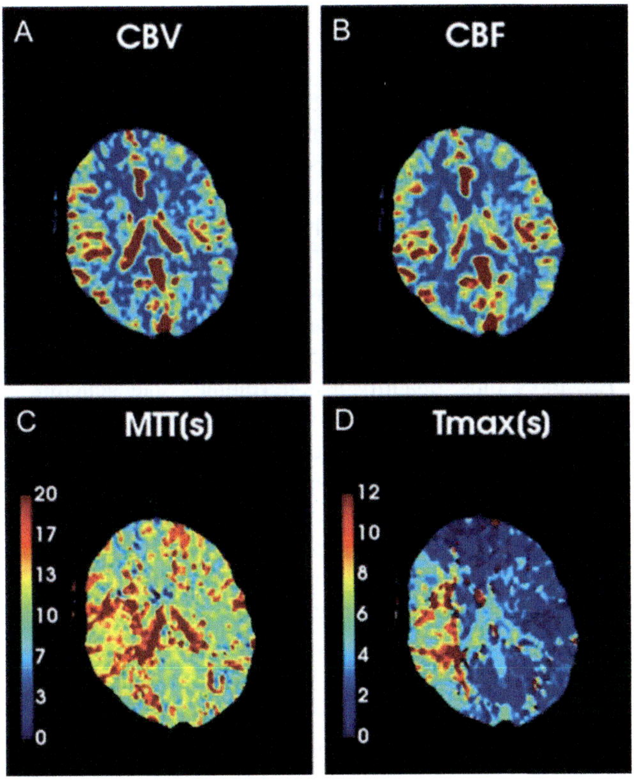

Fig. 33.4 Example of a CT perfusion study demonstrating a right hemispheric area of hypoperfusion related to a right middle cerebral artery occlusion in the M1 segment. Cerebral blood volume (CBV, **a**) and flow (CBV, **b**) are preserved, and indicate absence of a significant ischemic core. Mean transit time (MTT, **c**) and time to maximum of the residue function (T_{max}, **d**) are increased in the area of hypoperfusion due to the arterial occlusion

correlated better than CBV with the DWI lesion, both by lesion location and volume, and provides a translatable ischemic core measure between CTP and DWI [38]. The lesions defined by CBF<30 % of normal showed the highest correlation with the concomitant DWI lesion [38, 46].

Meanwhile, MTT has also been compared to the measures of delay such as T_{max} and TTP, and has shown lower sensitivity and specificity for approximating the penumbra [43, 47]. A T_{max} prolongation to over six seconds (T_{max} >6 s) has been validated as an accurate representation of the ischemic penumbra [47–49]. A recent publication by d'Esterre et al. suggests that CTP may also be able to predict infarct growth using high T_{max} thresholds [50]. Further validation and observational data are needed to apply this approach in clinical trials.

1.4 RAPID

For many years, perfusion imaging postprocessing required a technologist to select a reference arterial input function and perform manual processing steps. The need for manual input resulted in longer processing times and operator dependence of the results [51–53]. In 2011, using data from the Diffusion and Perfusion Imaging Evaluation For Understanding Stroke Evolution (DEFUSE) and Echoplanar Imaging Thrombolytic Evaluation Trial (EPITHET), an automated MRI analysis software was developed at Stanford University. It was named "RAPID" and became the first fully automated software to identify both arterial input and venous output functions, segment ischemic lesions based on prespecified apparent diffusion coefficient (ADC) and T_{max} thresholds, quantify lesion volumes, and generate lesion maps. RAPID also incorporates a motion correction algorithm to increase data accuracy, and total postprocessing time with modern versions of the software is less than 3 min. RAPID outputs are reliable and consistent with manually processed data, for both DWI ($r^2=0.99$) and perfusion ($r^2=0.96$) lesion volumes [52]. In the EXTEND trial, only 3.4 % of RAPID-processed maps were uninterpretable [37].

Though the earliest version was developed based on MRI data, RAPID was subsequently expanded to process CTP images with equal efficiency (Fig. 33.5). The presence of a perfusion lesion was defined by the CTP target mismatch, validated from its MRI predecessor:

- Ischemic core (defined by rCBF < 30 %) volume <70 mL
- Mismatch ratio (between T_{max} >6 s and rCBF < 30 % lesions) ≥1.8
- Mismatch difference (between T_{max} >6 s and DWI lesions) > 15 mL

While the MRI-based RAPID uses decreased apparent diffusion coefficient (ADC < 620) to identify the ischemic core, its CTP equivalent uses a relative CBF less than 30 %, based on its close correlation with the DWI volume estimate. Both versions of RAPID have been used in several observational studies and randomized controlled trials [10, 37, 54, 55].

2 MRI-Based Imaging

MRI plays a key role in acute stroke assessments. It allows for reliable imaging of the infarct core, rapid and contrast-free imaging of the vasculature, and determination of the patient's mismatch status. A number of approaches have been proposed, including the perfusion-and-diffusion (PWI–DWI) mismatch, the clinical–diffusion mismatch, the MR angiography (MRA)–clinical mismatch, and MRA–DWI mismatch. All four use the discrepancy between either clinical or radiological severity and the expected deficits from the DWI lesion. MRI is also the preferred modality for postintervention follow-up imaging. While early infarct size is usually assessed on DWI, late lesions are typically better seen on FLAIR or T2-weighted imaging.

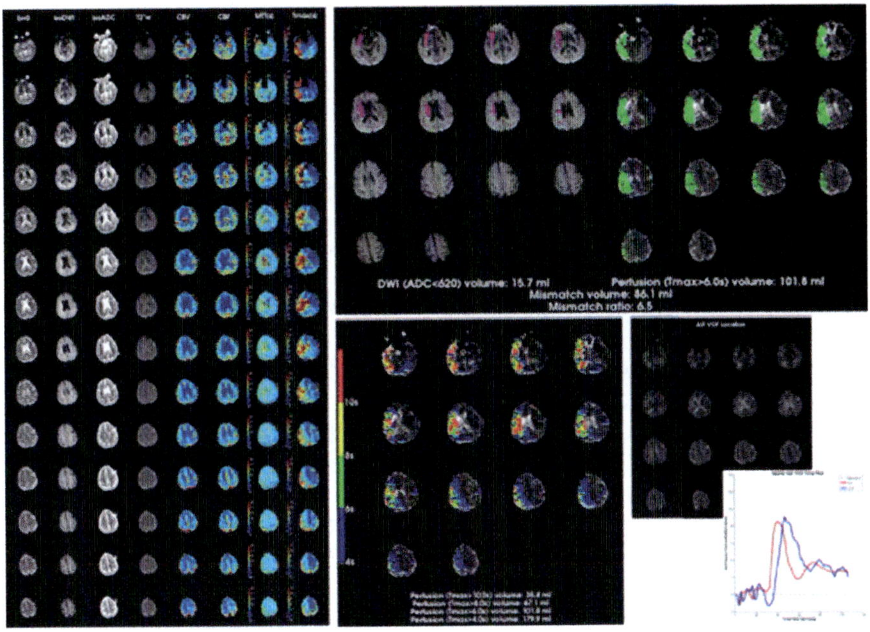

Fig. 33.5 Example of RAPID output images. The RAPID software generates a series of three maps. The column map (*left*) places side-by-side diffusion-weighted images ($b = 0$, isoDWI or $b = 1000$, and isoADC), T2* or gradient echo, and perfusion-weighted images (CBV, CBF, MTT, and T_{max}). In the CT perfusion output series, the diffusion-weighted and T2* images are replaced by the baseline CT, acquired at the onset of the perfusion study. The summary map (*top right*) highlights the ischemic core in *pink*, using an apparent diffusion coefficient (ADC) threshold of <620 mm^2/s. The critically hypoperfused lesion is marked in *green*, using a threshold of T_{max} >6 s. The severity of hypoperfusion within the ischemic territory is shown in a separate map (*middle*), segmented using T_{max} thresholds, from >4 to >10 s. Two reference images (*bottom right*) show the location and time plot of the arterial input and venous output functions

2.1 *Diffusion-Weighted Imaging*

Early data on the use of MRI in imaging hyperacute and acute infarct came from feline and rodent models of middle cerebral artery occlusion [56–59]. Diffusion-weighted spin-echo MRI showed the infarction earlier than T2-weighted imaging. The finding represented decreased diffusivity of water molecules, and corresponded to an area of elevated lactate:N-acetylaspartate ratio and other metabolic changes associated with cerebral ischemia.

Diffusion-weighted imaging (DWI) in patients with acute ischemic stroke showed the superiority in demonstrating early infarcts when compared with both T2-weighted imaging and noncontrast CT [60–62]. DWI offered greater sensitivity, specificity, and interrater reliability for identifying acute infarcts, required no exposure to ionizing radiation, and showed better correlation with the presenting NIHSS and final infarct volume [61–63]. Only an estimated 5 % of infarcts were DWI negative, and most of them are small, in the brainstem, and associated with good clinical outcome [64, 65]. By the early 2000s, DWI was the gold standard for imaging hyperacute and acute stroke [61, 64].

2.1.1 Diffusion Lesion Evolution

Since DWI reflects the underlying cellular changes in acute ischemia, the lesion signal evolves over the first hours to days after infarction. In particular, early DWI lesion changes can affect the accuracy of infarct volume measures and the interpretation of apparent infarct growth. Therefore, timing of follow-up MRI scan and the choice of sequences should take into account the effect of these early changes.

Diffusion reversal is defined as the normalization of diffusion-restricting signal between an early follow-up DWI and final infarct assessment. Diffusion reversal can involve part or all of the lesion. True diffusion reversal is uncommon, with an estimated incidence of 5–7 % [38, 66, 67]. It usually represents <10 % of the lesion volume, almost exclusively occurs in patients with successful reperfusion [66–68]. Transient diffusion reversal can be observed when reimaged within the first hours to days after the stroke, with subsequent return of diffusion restriction or evolution into early FLAIR hyperintensity [68, 69]. The putative mechanism behind this transient reversal is onset then resolution of vasogenic edema [70]. Another possible explanation is that diffusion reversal truly represents tissue normalization after reversed ischemic insult, and that reestablishment of diffusion restriction reflects secondary injury that may potentially be prevented by novel neuroprotective agents [68].

While part of a DWI lesion can reverse, another part of the same lesion can grow into adjacent tissue, even with reperfusion [70, 71]. Therefore, the approach to image analysis can significantly impact interpretation of whether infarct growth has occurred. A simple volumetric assessment of the final infarct burden can lead to misclassification of infarct growth when opposite changes occur in different regions of the lesion. Coregistration of the baseline and follow-up lesions is a more accurate method in demonstrating discrepancies in these regional changes [70]. In addition, timing of the follow-up scan is also important. A follow-up FLAIR lesion within 5 days of the infarct tends to overestimate the final infarct size due to associated edema and can give the apparent impression of lesion growth. Meanwhile, a late follow-up (>30 days) may underestimate the final infarct size due to gliosis or atrophy, and appear stable or decreased in infarct volume [72].

Therefore, the timing and sequences used for follow-up imaging should be carefully determined based on the aim of the neuroprotective trial. Coregistration of the baseline and follow-up lesions should be used as the preferred method for final infarct assessment.

2.2 MR Angiography

As we previously discussed, angiographic evidence of arterial occlusion is important in patient selection for endovascular interventions. Three-dimensional time-of-flight (TOF) MR angiography (MRA) is able to serve a similar purpose. It is routinely performed prior to contrast administration and has the advantage of rapidly acquiring vascular imaging without using a contrast agent [73, 74]. Though TOF MRA can accurately identify a focal occlusion, it may also overestimate the degree of intracranial stenotic lesions, and limit the distinction between a high-grade stenosis and occlusion [73]. TOF MRA also has lower sensitivity in detecting slow or in-plane blood flow distal to the points of stenoses, although the delay in ipsilateral distal vessel filling can be interpreted as a marker of hemodynamically significant stenosis [73–75].

Another important application of MRA is in combination with DWI. The use of an MRA–DWI mismatch has been validated for arterial occlusions in the internal carotid artery (ICA) or the first segment of the proximal middle cerebral artery (M1-MCA), from the DEFUSE and DEFUSE 2 studies [76, 77]. When a proximal arterial occlusion is associated with a small DWI lesion (<50 mL), the MRA-DWI criterion can identify approximately 75 % of patients meeting an MRI perfusion mismatch profile using contrast-based perfusion-weighted imaging [76–78]. MRA-DWI has been shown to select patients with more favorable outcomes following either IV tPA or endovascular therapy [77, 79–81].

2.3 MR Perfusion-Weighted Imaging

Early evidence for the use of MR perfusion-weighted imaging comes from the observation that brain regions with perfusion deficits were at risk for growth of the DWI lesion, to a size predictable by the larger PWI lesion [82–84]. In the DIAS, DEFUSE, and EPITHET studies, patient outcome following intravenous thrombolytic therapy was better in patients with the DWI-PWI mismatch profile, providing some of the earliest clinical evidence to support perfusion imaging in patient selection [36, 85, 86].

Over the past two decades, DWI–PWI mismatch has been the basis of perfusion-imaging-based patient selection. The key perfusion parameters are similar to those used in CTP, although the data acquisition and postprocessing differs between CT and MRI [87–89]. Unlike CT's density-based source data, PWI uses changes in the magnetic signal related to blood flow to derive perfusion data. This change can be detected using an injectable contrast agent with positive magnetic susceptibility. A paramagnetic, gadolinium-based contrast agent is commonly used, but super-paramagnetic iron-oxide compounds are possible alternatives [89]. The perfusion signal acquired can be measured based on the change in the tissue's magnetic susceptibility (or less frequently, relaxivity) due to the presence and passage of the paramagnetic agent. Also, as a consequence of image acquisition based on MRI rather than ionizing radiation, PWI offers whole-brain coverage without added concerns for radiation exposure [90].

As in the case for CTP, the source magnetic signals do not translate simply into the parameters of clinical interest. A number of postprocessing approaches exist. Summary delay parameters such as time-to-peak (TTP) and First Moment (FM) can be obtained from the function of tissue contrast concentration over time [87]. However, their dependence on hemodynamic variables, such as dispersion, delay, and transit time, limit their reliability [87, 91]. The alternative includes a number of deconvolution approaches, all of which aim to improve the reliability of the perfusion estimates based on selection of an arterial input function [27, 87, 88, 91, 92]. The final output parameters from these various postprocessing algorithms are similar to those previously defined from CTP, notably rCBV, rCBF, MTT, and a delay parameter, most commonly TTP or T_{max}.

Independently of ongoing technical refinements in postprocessing approaches, DWI–PWI mismatch has played an important role in patient selection in clinical trials. When RAPID for MRI first allowed automated quantification of lesion volumes, the concept for "target mismatch" was defined [54]:

- Ischemic core (defined by DWI lesion) volume <70 mL
- Mismatch ratio (between T_{max} >6 s and DWI lesions) ≥1.8
- Mismatch difference (between T_{max} >6 s and DWI lesions) >15 mL
- Severely hypoperfused (T_{max} >10 s) lesion volume <100 mL

DEFUSE 2 was the first prospective clinical study to identify patients' target mismatch in real time. Subsequently, the same (or similar) definitions have been applied across other randomized control trials [10, 54, 93]. These studies support DWI–PWI target mismatch, as currently defined, as a reliable predictor of clinical outcome after interventions in acute ischemic stroke. In DEFUSE 2, DWI–PWI mismatch also reliably identified patients with high probability of good functional outcome following endovascular treatment in the extended time window [94]. This evidence suggests that DWI–PWI mismatch may allow for an imaging-based approach to patient selection rather than time-based selection [94, 95].

2.3.1 The Malignant Profile

Another concept originated from the DEFUSE–EPITHET and DEFUSE 2 data is the "malignant profile" [36, 54, 86, 96, 97]. A large volume of severely hypoperfused tissue, marked by T_{max} >10s or a DWI lesion >70–100 mL, appeared destined for infarction and poor outcome irrespective of acute stroke intervention. Patients with malignant profiles were approximately six times more likely to have parenchymal hemorrhage and had only about a 10 % likelihood of good functional outcome [97]. Therefore, based on its natural history, patients with the malignant profile on PWI have be excluded from some clinical trials to maximize both efficacy and safety. However, with the advent of newer techniques for rapid and complete reperfusion (such as endovascular stent retrievers), favorable clinical outcomes have recently been described in some patients with the malignant profile. Other definitions of the malignant profile, using DWI or a combination of CTA and DWI, have been proposed [96–98]. However, they have not yet been applied in large clinical trials.

2.3.2 Arterial Spin Labeling (ASL)

As an alternative to exogenous contrast agents, constituents of normal blood, including deoxyhemoglobin or oxygen, can serve as paramagnetic material allowing for MR-based perfusion imaging [89, 99]. Arterial spin labeling (ASL)'s main advantage is that it does not require administration of an intravenous contrast agent to measure cerebral perfusion [100, 101]. Water molecules in blood are magnetically labeled proximal to the intracranial circulation, and perfusion imaging is acquired at a delay [88, 101, 102]. As an inherent consequence of ASL's use of an endogenous contrast agent, it has lower signal-to-noise ratio (SNR) than contrast perfusion images, which particularly affects the white matter [88, 103]. Its dependence on CBF also biases towards underestimation of perfusion in patients with primarily collateral inflow. These disadvantages, in addition to a long processing time and variabilities in ASL's methodology, have precluded its routine use in current clinical practice and trials [100, 103].

Despite these limitations, the technique behind ASL continues to improve. A relatively recent pseudocontinuous ASL acquisition scheme has nearly doubled the SNR with background suppression, and improved efficiency to decrease acquisition time to as little as <1 min [100]. With better techniques and validation data, ASL is a promising perfusion imaging modality that may gain a greater role for patient screening and selection in the years to come [88, 99–102].

2.4 *The Role of MRI in Future Clinical Trials*

Despite its greater accuracy and reliability, MRI has several limitations in the clinical trial setting. Notably, its traditionally confined space and longer acquisition time decrease patient tolerability and often prolongs time to treatment [104]. Most centers have limited or no access to hyperacute MR imaging, although some are able to provide immediate MR access for patients who arrive with some advanced notice (such as transfer patients in hub and spoke or telemedicine models). The use of MRI also carries the risk of rare but serious complications associated with gadolinium, including nephrogenic systemic fibrosis in patients with renal insufficiency and accumulation in cerebral tissue [105–110]. Despite these challenges, MRI is still the gold standard for assessment of the early ischemic core as well as optimal patient selection of the final infarct volume, and the reference for improvement in other techniques in acute stroke imaging, including multimodal CT. Shorter acute stroke MRI protocols (6–7 min protocols are now in use at some sites) and improvements in ASL techniques will likely increase their practicality and use in clinical trials [111].

3 Novel Neuroimaging Approaches for Patient Selection

Increasing focus in current interventional trials and quality improvement studies is placed on shortening time between stroke recognition and reperfusion therapy. As a result of this effort, novel approaches to image acquisition have been developed in attempt to shorten time in transit. Newer imaging approaches are also looking to fill the gap in patient selection for clinical trials involving the vertebrobasilar system.

3.1 Mobile Stroke Treatment Unit

An increasing number of stroke centers in Europe and the United States are acquiring mobile stroke units. The earlier generation of these units included a mobile CT scanner with capability for a noncontrast CT head to screen for eligibility for intravenous thrombolysis. While the system-wide benefit of the mobile stroke units is still a subject of clinical study, newer models of the mobile stroke unit continue to show improvement in equipment [112–114]. Several now incorporate CT scanners with ability to acquire CTA, and with a built-in hydraulic leveling system to counter the effect of street inclines [112, 115]. While there is still limited experience on the quality of data acquired, these new units may significantly shorten time to thrombolysis, and redefine prehospital phase interventions in both practice and clinical studies [113, 115–117].

3.2 Multimodal in-Suite Imaging

Recent proposals in angiographic suite design include a multimodal neuroimaging system that can allow for CTA and CTP to be performed using the angiographic C-arm [118, 119]. The co-localization of multiple imaging modalities has a potential impact on shortening time to intra-arterial intervention and reperfusion in eligible patients. However, more studies are needed to compare data quality, actual impact on clinical outcome, and cost-effectiveness of the technology.

3.3 Quantitative Magnetic Resonance Angiography

Quantitative magnetic resonance angiography (QMRA) is a cine phase-contrast MRA technique that uses three-dimensional reconstruction of the vasculature to first identify the diseased vessel and then determine its flow velocity and volumetric flow rate. It has been used to study a variety of physiological and pathological changes in blood flow, including vertebrobasilar diseases. Despite accounting for approximately 30 % of all ischemic strokes, the vertebrobasilar system has largely been excluded in the majority of clinical trials [120, 121]. Correlations of perfusion imaging and reperfusion data with clinical outcomes have only been validated for the anterior circulation [122]. Hemodynamic insufficiency detected by QMRA has been associated with higher risk for vertebrobasilar strokes [120, 121]. A commercially available software, Noninvasive Optimal Vessel Analysis (NOVA, VasSol Inc., Chicago, IL, USA) was used in the recently completed VERiTAS study with good reliability [123]. No clinical data is yet available on treatment efficacy using QMRA for patient selection, but it is a promising approach for future clinical trials targeting posterior circulation diseases.

4 Conclusion

Neuroimaging continues to play a central role in patient selection for acute ischemic stroke trials. Selection of the optimal imaging strategy for each clinical trial is a critical part of the study design and requires careful comparison of advantages and potential limitations of each modality considered. With continuing advances in imaging techniques, the definition of an "optimal" neuroimaging strategy for acute stroke imaging will continue to evolve.

References

1. Hacke W, Kaste M, Fieschi C, Toni D, Lesaffre E, Kummer von R et al (1995) Intravenous thrombolysis with recombinant tissue plasminogen activator for acute hemispheric stroke. The European Cooperative Acute Stroke Study (ECASS). JAMA 274(13):1017–1025
2. The National Institute of Neurological Disorders and Stroke rt-PA Stroke Study Group (1995) Tissue plasminogen activator for acute ischemic stroke. N Engl J Med 333(24):1581–1587
3. Barber PA, Demchuk AM, Zhang J, Buchan AM (2000) Validity and reliability of a quantitative computed tomography score in predicting outcome of hyperacute stroke before thrombolytic therapy. ASPECTS Study Group. Alberta Stroke Programme Early CT Score. Lancet 355(9216):1670–1674
4. Pexman JH, Barber PA, Hill MD, Sevick RJ, Demchuk AM, Hudon ME et al (2001) Use of the Alberta Stroke Program Early CT Score (ASPECTS) for assessing CT scans in patients with acute stroke. AJNR Am J Neuroradiol 22(8):1534–1542
5. Barber PA, Hill MD, Eliasziw M, Demchuk AM, Pexman JHW, Hudon ME et al (2005) Imaging of the brain in acute ischaemic stroke: comparison of computed tomography and magnetic resonance diffusion-weighted imaging. J Neurol Neurosurg Psychiatry 76(11):1528–1533
6. Dzialowski I, Hill MD, Coutts SB, Demchuk AM, Kent DM, Wunderlich O et al (2006) Extent of early ischemic changes on computed tomography (CT) before thrombolysis: prognostic value of the Alberta Stroke Program Early CT Score in ECASS II. Stroke 37(4):973–978
7. Weir NU, Pexman JHW, Hill MD, Buchan AM, CASES Investigators (2006) How well does ASPECTS predict the outcome of acute stroke treated with IV tPA? Neurology 67(3):516–518
8. Khatri P, Yeatts SD, Mazighi M, Broderick JP, Liebeskind DS, Demchuk AM, Amarenco P, Carrozzella J, Spilker J, Foster LD, Goyal M, Hill MD, Palesch YY, Jauch EC, Haley EC, Vagal A, Tomsick TA, IMS III Trialists (2014) Time to angiographic reperfusion and clinical outcome after acute ischaemic stroke: an analysis of data from the Interventional Management of Stroke (IMS III) phase 3 trial. Lancet Neurol 13(6):567–574

9. Fransen PSS, Berkhemer OA, Lingsma HF, Beumer D, van den Berg LA, Yoo AJ (2016) Time to reperfusion and treatment effect for acute ischemic stroke: a randomized clinical trial. Jama Neurol 73(2):190–196
10. Saver JL, Goyal M, Bonafe A, Diener HC, Levy EI, Pereira VM et al (2015) Stent-retriever thrombectomy after intravenous t-PA vs. t-PA alone in stroke. N Engl J Med 372(24):2285–2295
11. Campbell BC, Mitchell PJ, Investigators E-I (2015) Endovascular therapy for ischemic stroke. N Engl J Med 372(24):2365–2366
12. Goyal M, Demchuk AM, Menon BK, Eesa M, Rempel JL, Thornton J et al (2015) Randomized assessment of rapid endovascular treatment of ischemic stroke. N Engl J Med 372(11):1019–1030
13. Jovin TG, Chamorro A, Cobo E, de Miquel MA, Molina CA, Rovira A et al (2015) Thrombectomy within 8 hours after symptom onset in ischemic stroke. N Engl J Med 372(24):2296–2306
14. Berkhemer OA, Fransen PS, Beumer D, van den Berg LA, Lingsma HF, Yoo AJ et al (2015) A randomized trial of intraarterial treatment for acute ischemic stroke. N Engl J Med 372(1):11–20
15. Phan TG, Donnan GA, Koga M, Mitchell LA, Molan M, Fitt G et al (2006) The ASPECTS template is weighted in favor of the striatocapsular region. Neuroimage 31(2):477–481
16. Aviv RI, Mandelcorn J, Chakraborty S, Gladstone D, Malham S, Tomlinson G et al (2007) Alberta Stroke Program Early CT Scoring of CT perfusion in early stroke visualization and assessment. AJNR Am J Neuroradiol 28(10):1975–1980
17. Finlayson O, John V, Yeung R, Dowlatshahi D, Howard P, Zhang L et al (2013) Interobserver agreement of ASPECT score distribution for noncontrast CT, CT angiography, and CT perfusion in acute stroke. Stroke 44(1):234–236
18. McTaggart RA, Jovin TG, Lansberg MG, Mlynash M, Jayaraman MV, Choudhri OA et al (2015) Alberta stroke program early computed tomographic scoring performance in a series of patients undergoing computed tomography and MRI: reader agreement, modality agreement, and outcome prediction. Stroke 46(2):407–412
19. Rangaraju S, Streib C, Aghaebrahim A, Jadhav A, Frankel M, Jovin TG (2015) Relationship between lesion topology and clinical outcome in anterior circulation large vessel occlusions. Stroke 46(7):1787–1792
20. Broderick JP, Palesch YY, Demchuk AM, Yeatts SD, Khatri P, Hill MD et al (2014) Evolution of practice during the Interventional Management of Stroke III Trial and implications for ongoing trials. Stroke 45(12):3606–3611
21. Demchuk AM, Goyal M, Yeatts SD, Carrozzella J, Foster LD, Qazi E et al (2014) Recanalization and clinical outcome of occlusion sites at baseline CT angiography in the Interventional Management of Stroke III trial. Radiology 273(1):202–210
22. Powers WJ, Derdeyn CP, Biller J, Coffey CS, Hoh BL, Jauch EC et al (2015) 2015 American Heart Association/American Stroke Association Focused Update of the 2013 Guidelines for the Early Management of Patients With Acute Ischemic Stroke Regarding Endovascular Treatment: A Guideline for Healthcare Professionals From the American Heart Association/American Stroke Association. Stroke 46:3020–3035
23. Yang C-Y, Chen Y-F, Lee C-W, Huang A, Shen Y, Wei C et al (2008) Multiphase CT angiography versus single-phase CT angiography: comparison of image quality and radiation dose. AJNR Am J Neuroradiol 29(7):1288–1295
24. Menon BK, d'Esterre CD, Qazi EM, Almekhlafi M, Hahn L, Demchuk AM et al (2015) Multiphase CT angiography: a new tool for the imaging triage of patients with acute ischemic stroke. Radiology 275(2):510–520
25. Axel L (1980) Cerebral blood flow determination by rapid-sequence computed tomography: theoretical analysis. Radiology 137(3):679–686
26. Campbell BCV, Mitchell PJ, Yan B, Parsons MW, Christensen S, Churilov L et al (2014) A multicenter, randomized, controlled study to investigate EXtending the time for Thrombolysis in Emergency Neurological Deficits with Intra-Arterial therapy (EXTEND-IA). Int J Stroke 9(1):126–132

27. Konstas AA, Goldmakher GV, Lee T-Y, Lev MH (2009) Theoretic basis and technical implementations of CT perfusion in acute ischemic stroke, part 1: Theoretic basis. AJNR Am J Neuroradiol 30(4):662–668
28. Murphy A, So A, Lee T-Y, Symons S, Jakubovic R, Zhang L et al (2014) Low dose CT perfusion in acute ischemic stroke. Neuroradiology 56(12):1055–1062
29. Konstas AA, Goldmakher GV, Lee T-Y, Lev MH (2009) Theoretic basis and technical implementations of CT perfusion in acute ischemic stroke, part 2: technical implementations. AJNR Am J Neuroradiol 30(5):885–892
30. Niesten JM, van der Schaaf IC, Riordan AJ, de Jong HWAM, Horsch AD, Eijspaart D et al (2014) Radiation dose reduction in cerebral CT perfusion imaging using iterative reconstruction. Eur Radiol 24(2):484–493
31. Aviv RI, Parsons M, Bivard A, Jahromi B, Wintermark M (2015) Multiphase CT angiography: a poor man's perfusion CT? Radiology 277(3):922–924
32. Schaefer PW, Barak ER, Kamalian S, Gharai LR, Schwamm L, Gonzalez RG et al (2008) Quantitative assessment of core/penumbra mismatch in acute stroke: CT and MR perfusion imaging are strongly correlated when sufficient brain volume is imaged. Stroke 39(11):2986–2992
33. Emmer BJ, Rijkee M, Niesten JM, Wermer MJH, Velthuis BK, van Walderveen MAA (2014) Whole brain CT perfusion in acute anterior circulation ischemia: coverage size matters. Neuroradiology 56(12):1121–1126
34. Furtado AD, Lau BC, Vittinghoff E, Dillon WP, Smith WS, Rigby T et al (2010) Optimal brain perfusion CT coverage in patients with acute middle cerebral artery stroke. AJNR Am J Neuroradiol 31(4):691–695
35. Wintermark M, Smith WS, Ko NU, Quist M, Schnyder P, Dillon WP (2004) Dynamic perfusion CT: optimizing the temporal resolution and contrast volume for calculation of perfusion CT parameters in stroke patients. AJNR Am J Neuroradiol 25(5):720–729
36. Albers GW, Thijs VN, Wechsler L, Kemp S, Schlaug G, Skalabrin E et al (2006) Magnetic resonance imaging profiles predict clinical response to early reperfusion: the diffusion and perfusion imaging evaluation for understanding stroke evolution (DEFUSE) study. Ann Neurol 60(5):508–517
37. Campbell BCV, Yassi N, Ma H, Sharma G, Salinas S, Churilov L et al (2015) Imaging selection in ischemic stroke: feasibility of automated CT-perfusion analysis. Int J Stroke 10(1):51–54
38. Campbell BCV, Christensen S, Levi CR, Desmond PM, Donnan GA, Davis SM et al (2011) Cerebral blood flow is the optimal CT perfusion parameter for assessing infarct core. Stroke 42(12):3435–3440
39. Campbell BCV, Christensen S, Levi CR, Desmond PM, Donnan GA, Davis SM et al (2012) Comparison of computed tomography perfusion and magnetic resonance imaging perfusion-diffusion mismatch in ischemic stroke. Stroke 43(10):2648–2653
40. Mokin M, Morr S, Fanous AA, Shallwani H, Natarajan SK, Levy EI et al (2015) Correlation between cerebral blood volume values and outcomes in endovascular therapy for acute ischemic stroke. J Neurointerv Surg 7(10):705–708
41. Mui K, Yoo AJ, Verduzco L, Copen WA, Hirsch JA, Gonzalez RG et al (2011) Cerebral blood flow thresholds for tissue infarction in patients with acute ischemic stroke treated with intra-arterial revascularization therapy depend on timing of reperfusion. AJNR Am J Neuroradiol 32(5):846–851
42. Turk AS, Magarick JA, Frei D, Fargen KM, Chaudry I, Holmstedt CA et al (2013) CT perfusion-guided patient selection for endovascular recanalization in acute ischemic stroke: a multicenter study. J Neurointerv Surg 5(6):523–527
43. Wintermark M, Flanders AE, Velthuis B, Meuli R, van Leeuwen M, Goldsher D et al (2006) Perfusion-CT assessment of infarct core and penumbra: receiver operating characteristic curve analysis in 130 patients suspected of acute hemispheric stroke. Stroke 37(4):979–985

44. Murphy BD, Fox AJ, Lee DH, Sahlas DJ, Black SE, Hogan MJ et al (2006) Identification of penumbra and infarct in acute ischemic stroke using computed tomography perfusion-derived blood flow and blood volume measurements. Stroke 37(7):1771–1777
45. Wintermark M, Reichhart M, Thiran J-P, Maeder P, Chalaron M, Schnyder P et al (2002) Prognostic accuracy of cerebral blood flow measurement by perfusion computed tomography, at the time of emergency room admission, in acute stroke patients. Ann Neurol 51(4):417–432
46. Albers GW, Goyal M, Jahan R, Bonafe A, Diener H-C, Levy EI et al (2015) Ischemic core and hypoperfusion volumes predict infarct size in SWIFT PRIME. Ann Neurol 79(1):76–89
47. Olivot JM, Mlynash M, Zaharchuk G, Straka M, Bammer R, Schwartz N et al (2009) Perfusion MRI (Tmax and MTT) correlation with xenon CT cerebral blood flow in stroke patients. Neurology 72(13):1140–1145
48. Olivot JM, Mlynash M, Thijs VN, Kemp S, Lansberg MG, Wechsler L et al (2009) Optimal Tmax threshold for predicting penumbral tissue in acute stroke. Stroke 40(2):469–475
49. Ma H, Parsons MW, Christensen S, Campbell BCV, Churilov L, Connelly A et al (2012) A multicentre, randomized, double-blinded, placebo-controlled Phase III study to investigate EXtending the time for Thrombolysis in Emergency Neurological Deficits (EXTEND). Int J Stroke 7(1):74–80
50. d'Esterre CD, Boesen ME, Ahn SH, Pordeli P, Najm M, Minhas P et al (2015) Time-dependent computed tomographic perfusion thresholds for patients with acute ischemic stroke. Stroke 46(12):3390–3397
51. Lansberg MG, Lee J, Christensen S, Straka M, De Silva DA, Mlynash M et al (2011) RAPID automated patient selection for reperfusion therapy: a pooled analysis of the Echoplanar Imaging Thrombolytic Evaluation Trial (EPITHET) and the Diffusion and Perfusion Imaging Evaluation for Understanding Stroke Evolution (DEFUSE) Study. Stroke 42(6):1608–1614
52. Straka M, Albers GW, Bammer R (2010) Real-time diffusion-perfusion mismatch analysis in acute stroke. J Magn Reson Imaging 32(5):1024–1037
53. Soares BP, Dankbaar JW, Bredno J, Cheng S, Bhogal S, Dillon WP et al (2009) Automated versus manual post-processing of perfusion-CT data in patients with acute cerebral ischemia: influence on interobserver variability. Neuroradiology 51(7):445–451
54. Lansberg MG, Straka M, Kemp S, Mlynash M, Wechsler LR, Jovin TG et al (2012) MRI profile and response to endovascular reperfusion after stroke (DEFUSE 2): a prospective cohort study. Lancet Neurol 11(10):860–867
55. Campbell BC, Christensen S, Parsons MW, Churilov L, Desmond PM, Barber PA et al (2013) Advanced imaging improves prediction of hemorrhage after stroke thrombolysis. Ann Neurol 73(4):510–519
56. Moseley ME, Cohen Y, Mintorovitch J, Chileuitt L, Shimizu H, Kucharczyk J et al (1990) Early detection of regional cerebral ischemia in cats: Comparison of diffusion- and T2-weighted MRI and spectroscopy. Magn Reson Med 14(2):330–346
57. Mintorovitch J, Moseley ME, Chileuitt L, Shimizu H, Cohen Y, Weinstein PR (1991) Comparison of diffusion- and T2-weighted MRI for the early detection of cerebral ischemia and reperfusion in rats. Magn Reson Med 18(1):39–50
58. Minematsu K, Li L, Fisher M, Sotak CH, Davis MA, Fiandaca MS (1992) Diffusion-weighted magnetic resonance imaging: rapid and quantitative detection of focal brain ischemia. Neurology 42(1):235–240
59. Minematsu K, Li L, Sotak CH, Davis MA, Fisher M (1992) Reversible focal ischemic injury demonstrated by diffusion-weighted magnetic resonance imaging in rats. Stroke 23(9):1304–1310; discussion 1310–1
60. Warach S, Schellinger PD (2003) The role of stroke MRI in clinical trials. In: Stroke MRI. Steinkopff, Heidelberg, pp 91–98
61. Lansberg MG, Albers GW, Beaulieu C, Marks MP (2000) Comparison of diffusion-weighted MRI and CT in acute stroke. Neurology 54(8):1557–1561

62. Mohr JP, Biller J, Hilal SK, Yuh WT, Tatemichi TK, Hedges S et al (1995) Magnetic resonance versus computed tomographic imaging in acute stroke. Stroke 26(5):807–812
63. von Baumgarten L, Thierfelder KM, Beyer SE, Baumann AB, Bollwein C, Janssen H et al (2016) Early CT perfusion mismatch in acute stroke is not time-dependent but relies on collateralization grade. Neuroradiology 58(4):357–365
64. Oppenheim C, Stanescu R, Dormont D, Crozier S, Marro B, Samson Y et al (2000) False-negative diffusion-weighted MR findings in acute ischemic stroke. AJNR Am J Neuroradiol 21(8):1434–1440
65. Engelter ST, Wetzel SG, Radue EW, Rausch M, Steck AJ, Lyrer PA (2004) The clinical significance of diffusion-weighted MR imaging in infratentorial strokes. Neurology 62(4):574–580
66. Kate MP, Riaz P, Gioia L, Sivakumar L, Jeerakathil T, Buck B et al (2015) Dynamic evolution of diffusion-weighted imaging lesions in patients with minor ischemic stroke. Stroke 46(8):2318–2321
67. Chemmanam T, Campbell BCV, Christensen S, Nagakane Y, Desmond PM, Bladin CF et al (2010) Ischemic diffusion lesion reversal is uncommon and rarely alters perfusion-diffusion mismatch. Neurology 75(12):1040–1047
68. Campbell BCV, Purushotham A, Christensen S, Desmond PM, Nagakane Y, Parsons MW et al (2012) The infarct core is well represented by the acute diffusion lesion: sustained reversal is infrequent. J Cereb Blood Flow Metab 32(1):50–56
69. Kidwell CS, Saver JL, Starkman S, Duckwiler G, Jahan R, Vespa P et al (2002) Late secondary ischemic injury in patients receiving intraarterial thrombolysis. Ann Neurol 52(6):698–703
70. Inoue M, Mlynash M, Christensen S, Wheeler HM, Straka M, Tipirneni A et al (2014) Early diffusion-weighted imaging reversal after endovascular reperfusion is typically transient in patients imaged 3 to 6 hours after onset. Stroke 45(4):1024–1028
71. Hassen WB, Tisserand M, Turc G, Charron S, Seners P, Edjlali M et al (2016) Comparison between voxel-based and subtraction methods for measuring diffusion-weighted imaging lesion growth after thrombolysis. Int J Stroke 11(2):221–228
72. Olivot JM, Mlynash M, Thijs VN, Kemp S, Lansberg MG, Wechsler L et al (2008) Relationships between infarct growth, clinical outcome, and early recanalization in diffusion and perfusion imaging for understanding stroke evolution (DEFUSE). Stroke 39(8):2257–2263
73. Yang JJ, Hill MD, Morrish WF, Hudon ME, Barber PA, Demchuk AM et al (2002) Comparison of pre- and postcontrast 3D time-of-flight MR angiography for the evaluation of distal intracranial branch occlusions in acute ischemic stroke. AJNR Am J Neuroradiol 23(4):557–567
74. Axel L (1986) Blood flow effects in magnetic resonance imaging. Magn Reson Annu 237–244
75. Lan L, Leng X, Abrigo J, Fang H, Ip VHL, Soo YOY et al (2016) Diminished signal intensities distal to intracranial arterial stenosis on time-of-flight MR angiography might indicate delayed cerebral perfusion. Cerebrovasc Dis 42(3-4):232–239
76. Lansberg MG, Thijs VN, Bammer R, Olivot JM, Marks MP, Wechsler LR et al (2008) The MRA-DWI mismatch identifies patients with stroke who are likely to benefit from reperfusion. Stroke 39(9):2491–2496
77. Mishra NK, Albers GW, Christensen S, Marks M, Hamilton S, Straka M et al (2014) Comparison of magnetic resonance imaging mismatch criteria to select patients for endovascular stroke therapy. Stroke 45(5):1369–1374
78. Deguchi I, Dembo T, Yoshimura S, Sakai N, Okada Y, Kitagawa K et al (2014) Relationship Between Magnetic Resonance Angiography–Diffusion-weighted Imaging Mismatch and Clinical Outcome in Endovascular Treatment for Acute Ischemic Stroke: Subgroup Analysis of the Recovery by Endovascular Salvage for Cerebral Ultra-acute Embolism–Japan Registry. J Stroke Cerebrovasc Dis 23(6):1471–1476

79. Deguchi I, Takeda H, Furuya D, Dembo T, Nagoya H, Kato Y et al (2012) Significance of magnetic resonance angiography-diffusion weighted imaging mismatch in hyperacute cerebral infarction. J Stroke Cerebrovasc Dis 21(2):108–113
80. Inoue T, Tamura A, Tsutsumi K, Saito I, Saito N (2013) Surgical embolectomy for large vessel occlusion of anterior circulation. Br J Neurosurg 27(6):783–790
81. Deguchi I, Dembo T, Fukuoka T, Nagoya H, Maruyama H, Kato Y et al (2012) Usefulness of MRA-DWI mismatch in neuroendovascular therapy for acute cerebral infarction. Eur J Neurol 19(1):114–120
82. Thijs VN, Adami A, Neumann-Haefelin T, Moseley ME, Marks MP, Albers GW (2001) Relationship between severity of MR perfusion deficit and DWI lesion evolution. Neurology 57(7):1205–1211
83. Neumann-Haefelin T, Wittsack HJ, Wenserski F, Siebler M, Seitz RJ, Modder U et al (1999) Diffusion- and perfusion-weighted MRI. The DWI/PWI mismatch region in acute stroke. Stroke 30(8):1591–1597
84. Karonen JO, Vanninen RL, Liu Y, Ostergaard L, Kuikka JT, Nuutinen J et al (1999) Combined diffusion and perfusion MRI with correlation to single-photon emission CT in acute ischemic stroke. Ischemic penumbra predicts infarct growth. Stroke 30(8):1583–1590
85. Hacke W, Furlan AJ, Al-Rawi Y, Dávalos A, Fiebach JB, Gruber F et al (2009) Intravenous desmoteplase in patients with acute ischaemic stroke selected by MRI perfusion-diffusion weighted imaging or perfusion CT (DIAS-2): a prospective, randomised, double-blind, placebo-controlled study. Lancet Neurol 8(2):141–150
86. Davis SM, Donnan GA, Parsons MW, Levi C, Butcher KS, Peeters A et al (2008) Effects of alteplase beyond 3 h after stroke in the Echoplanar Imaging Thrombolytic Evaluation Trial (EPITHET): a placebo-controlled randomised trial. Lancet Neurol 7(4):299–309
87. Meijs M, Christensen S, Lansberg MG, Albers GW, Calamante F (2015) Analysis of perfusion MRI in stroke: to deconvolve, or not to deconvolve. Magn Reson Med. doi:10.1002/mrm.26024
88. Calamante F, Thomas DL, Pell GS, Wiersma J, Turner R (1999) Measuring cerebral blood flow using magnetic resonance imaging techniques. J Cereb Blood Flow Metab 19(7):701–735
89. Barbier EL, Lamalle L, Décorps M (2001) Methodology of brain perfusion imaging. J Magn Reson Imaging 13(4):496–520
90. Lin L, Bivard A, Krishnamurthy V, Levi CR, Parsons MW (2016) Whole-brain CT perfusion to quantify acute ischemic penumbra and core. Radiology 279(3):876–887
91. Wu O, Østergaard L, Weisskoff RM, Benner T, Rosen BR, Sorensen AG (2003) Tracer arrival timing-insensitive technique for estimating flow in MR perfusion-weighted imaging using singular value decomposition with a block-circulant deconvolution matrix. Magn Reson Med 50(1):164–174
92. Yamada K, Wu O, Gonzalez RG, Bakker D, Ostergaard L, Copen WA et al (2002) Magnetic resonance perfusion-weighted imaging of acute cerebral infarction: effect of the calculation methods and underlying vasculopathy. Stroke 33(1):87–94
93. Kidwell CS, Jahan R, Gornbein J, Alger JR, Nenov V, Ajani Z et al (2013) A trial of imaging selection and endovascular treatment for ischemic stroke. N Engl J Med 368(10):914–923
94. Lansberg MG, Cereda CW, Mlynash M, Mishra NK, Inoue M, Kemp S et al (2015) Response to endovascular reperfusion is not time-dependent in patients with salvageable tissue. Neurology 85(8):708–714
95. Fisher M, Albers GW (2013) Advanced imaging to extend the therapeutic time window of acute ischemic stroke. Ann Neurol 73(1):4–9
96. Inoue M, Mlynash M, Straka M, Lansberg MG, Zaharchuk G, Bammer R et al (2012) Patients with the malignant profile within 3 hours of symptom onset have very poor outcomes after intravenous tissue-type plasminogen activator therapy. Stroke 43(9):2494–2496
97. Mlynash M, Lansberg MG, De Silva DA, Lee J, Christensen S, Straka M et al (2011) Refining the definition of the malignant profile: insights from the DEFUSE-EPITHET pooled data set. Stroke 42(5):1270–1275

98. Puetz V, Dzialowski I, Hill MD, Steffenhagen N, Coutts SB, O'Reilly C et al (2010) Malignant profile detected by CT angiographic information predicts poor prognosis despite thrombolysis within three hours from symptom onset. Cerebrovasc Dis 29(6):584–591
99. Yoo R-E, Yun TJ, Rhim JH, Yoon B-W, Kang KM, Choi SH et al (2015) Bright vessel appearance on arterial spin labeling MRI for localizing arterial occlusion in acute ischemic stroke. Stroke 46(2):564–567
100. Bokkers RPH, Hernandez DA, Merino JG, Mirasol RV, van Osch MJ, Hendrikse J et al (2012) Whole-brain arterial spin labeling perfusion MRI in patients with acute stroke. Stroke 43(5):1290–1294
101. Zaharchuk G, El Mogy IS, Fischbein NJ, Albers GW (2012) Comparison of arterial spin labeling and bolus perfusion-weighted imaging for detecting mismatch in acute stroke. Stroke 43(7):1843–1848
102. Amukotuwa SA, Yu C, Zaharchuk G (2015) 3D Pseudocontinuous arterial spin labeling in routine clinical practice: A review of clinically significant artifacts. J Magn Reson Imaging 43(1):11–27
103. Amukotuwa SA, Yu C, Zaharchuk G (2016) 3D Pseudocontinuous arterial spin labeling in routine clinical practice: A review of clinically significant artifacts. J Magn Reson Imaging 43(1):11–27
104. Kim BJ, Kang HG, Kim H-J, Ahn S-H, Kim NY, Warach S et al (2014) Magnetic resonance imaging in acute ischemic stroke treatment. J Stroke 16(3):131–145
105. Murata N, Gonzalez-Cuyar LF, Murata K, Fligner C, Dills R, Hippe D et al (2016) Macrocyclic and other non-group 1 gadolinium contrast agents deposit Low levels of gadolinium in brain and bone tissue: preliminary results from 9 patients with normal renal function. Invest Radiol 1
106. Stojanov D, Aracki-Trenkic A, Benedeto-Stojanov D (2016) Gadolinium deposition within the dentate nucleus and globus pallidus after repeated administrations of gadolinium-based contrast agents-current status. Neuroradiology 58(5):433–441
107. Thomsen HS (2007) Nephrogenic systemic fibrosis (NSF): a late adverse reaction to some of the gadolinium based contrast agents. Cancer Imaging 7(1):130–137
108. Thomsen HS (2016) Nephrogenic systemic fibrosis: a serious adverse reaction to gadolinium - 1997-2006-2016. Part 1. Acta Radiol 57(5):515–520
109. Thomsen HS (2016) Nephrogenic systemic fibrosis: a serious adverse reaction to gadolinium - 1997-2006-2016. Part 2. Acta Radiol 57(6):643–648
110. Thomsen HS (2006) Nephrogenic systemic fibrosis: a serious late adverse reaction to gadodiamide. Eur Radiol 16(12):2619–2621
111. Nael K, Khan R, Choudhary G, Meshksar A, Villablanca P, Tay J et al (2014) Six-minute magnetic resonance imaging protocol for evaluation of acute ischemic stroke: pushing the boundaries. Stroke 45(7):1985–1991
112. John S, Stock S, Masaryk T, Bauer A, Cerejo R, Uchino K et al (2016) Performance of CT angiography on a mobile stroke treatment unit: implications for triage. J Neuroimaging
113. Rajan SS, Baraniuk S, Parker S, Wu T-C, Bowry R, Grotta JC (2015) Implementing a Mobile Stroke Unit Program in the United States. JAMA Neurol 72(2):229–234
114. Walter S, Kostpopoulos P, Haass A, Helwig S, Keller I, Licina T et al (2010) Bringing the hospital to the patient: first treatment of stroke patients at the emergency site. PLoS One 5(10):e13758
115. John S, Stock S, Cerejo R, Uchino K, Winners S, Russman A et al (2016) Brain imaging using mobile CT: current status and future prospects. J Neuroimaging 26(1):5–15
116. Fassbender K, Walter S, Liu Y, Muehlhauser F, Ragoschke A, Kuehl S et al (2003) "Mobile stroke unit" for hyperacute stroke treatment. Stroke 34(6):e44
117. Walter S, Kostopoulos P, Haass A, Keller I, Lesmeister M, Schlechtriemen T et al (2012) Diagnosis and treatment of patients with stroke in a mobile stroke unit versus in hospital: a randomised controlled trial. Lancet Neurol 11(5):397–404

118. Struffert T, Deuerling-Zheng Y, Kloska S, Engelhorn T, Lang S, Mennecke A et al (2015) Dynamic angiography and perfusion imaging using flat detector CT in the angiography suite: a pilot study in patients with acute middle cerebral artery occlusions. AJNR Am J Neuroradiol 36(10):1964–1970
119. Yang P, Niu K, Wu Y, Struffert T, Dorfler A, Schafer S et al (2015) Time-resolved C-arm computed tomographic angiography derived from computed tomographic perfusion acquisition: new capability for One-stop-shop acute ischemic stroke treatment in the angiosuite. Stroke 46(12):3383–3389
120. Amin-Hanjani S, Pandey DK, Rose-Finnell L, Du X, Richardson D, Thulborn KR et al (2016) Effect of hemodynamics on stroke risk in symptomatic atherosclerotic vertebrobasilar occlusive disease. JAMA Neurol 73(2):178–185
121. Amin-Hanjani S, Du X, Zhao M, Walsh K, Malisch TW, Charbel FT (2005) Use of quantitative magnetic resonance angiography to stratify stroke risk in symptomatic vertebrobasilar disease. Stroke 36(6):1140–1145
122. Haase J, Magnussen IB, Ogilvy CS, Ojemann RG, Meyer FB, Quest DO et al (1999) Evaluating patients with vertebrobasilar transient ischemic attacks. Surg Neurol 52(4):386–392
123. Amin-Hanjani S, Rose-Finnell L, Richardson D, Ruland S, Pandey D, Thulborn KR et al (2010) Vertebrobasilar Flow Evaluation and Risk of Transient Ischaemic Attack and Stroke study (VERiTAS): rationale and design. Int J Stroke 5(6):499–505

Index

P.A. Lapchak, J.H. Zhang (eds.), *Neuroprotective Therapy for Stroke and Ischemic Disease*, Springer Series in Translational Stroke Research,
DOI 10.1007/978-3-319-45345-3

F

G

K

L

M

P

S

Printed by Printforce, the Netherlands